6th International Symposium on Measurement Techniques for Multiphase Flows 2008

Journal of Physics: Conference Series Volume 147

Naha, Okinawa, Japan
15-17 December 2008

Volume 1 of 2

ISBN: 978-1-61567-331-5

Printed from e-media with permission by:

Curran Associates, Inc.
57 Morehouse Lane
Red Hook, NY 12571

CURRAN ASSOCIATES INC. proceedings.com

Some format issues inherent in the e-media version may also appear in this print version.

Printed by Curran Associates, Inc. (2009)

For permission requests, please contact the Institute of Physics
at the address below.

Institute of Physics
Dirac House, Temple Back
Bristol BS1 6BE UK

Phone: 44 1 17 929 7481
Fax: 44 1 17 920 0979

techtracking@iop.org

Additional copies of this publication are available from:

Curran Associates, Inc.
57 Morehouse Lane
Red Hook, NY 12571 USA
Phone: 845-758-0400
Fax: 845-758-2634
Email: curran@proceedings.com
Web: www.proceedings.com

TABLE OF CONTENTS

VOLUME 1

VOLUME 2

The 6th International Symposium on Measurement Techniques for Multiphase Flows

Koji Okamoto *et al* 2009 *J. Phys.: Conf. Ser.* **147** 011001 (1p) doi: 10.1088/1742-6596/147/1/011001 Help

Koji Okamoto and Yuichi Murai

Research on multi-phase flows is very important for industrial applications, including power stations, vehicles, engines, food processing, and so on. Also, from the environmental viewpoint, multi-phase flows need to be investigated to overcome global warming. Multi-phase flows originally have non-linear features because they are multi-phased. The interaction between the phases plays a very interesting role in the flows. The non-linear interaction causes the multi-phase flows to be very difficult to understand phenomena. The International Symposium on Measurement Techniques for Multi-phase Flows (ISMTMF) is a unique symposium. The target of the symposium is to exchange the state-of-the-art knowledge on the measurement techniques for non-linear multi-phase flows. Measurement technique is the key technology to understanding non-linear phenomena. The ISMTMF began in 1995 in Nanjing, China. The symposium has continuously been held every two or three years. The ISMTMF-2008 was held in Okinawa, Japan as the 6th symposium of ISMTMF on 15–17 December 2008. Okinawa has a long history as the Ryukyus Kingdom. China and Japan have had cultural and economic exchanges through Okinawa for more than 1000 years. Please enjoy Okinawa and experience its history to enhance our international communication. The present symposium was attended by 124 participants, the program included 107 contributions with 5 plenary lectures, 2 keynote lectures, and 100 oral regular paper presentations. The topics include, besides the ordinary measurement techniques for multiphase flows, acoustic and electric sensors, bubbles and microbubbles, computed tomography, gas–liquid interface, laser-imaging and PIV, oil/coal/drop and spray, solid and powder, spectral and multi-physics. This volume includes the presented papers at ISMTMF-2008. In addition to this volume, ten selected papers will be published in a special issue of *Measurement Science and Technology*.

We would like to express special thanks to all the participants and the contributors to the symposium, and also to the supporting organizations; The Japanese Society for Multiphase Flow, The Chinese Society for Measurement, National Natural Science Foundation of China, The Chinese Academy of Science, and University of the Ryukyus, Okinawa, Japan.

Koji Okamoto
Chair of 6th ISMTMF and proceedings editor
The University of Tokyo, Japan
Yuichi Murai
Proceedings co-editor
Hokkaido University, Japan

Journal of Physics: Conference Series **147** (2009) 012001 doi:10.1088/1742-6596/147/1/012001

Modeling and calibration of a Ring-shaped Electrostatic Meter

Jianyong Zhang [1], Bin Zhou [2], Chuanlong Xu [2], Shimin Wang [2]

[1] University of Teesside, Middlesbrough TS1 3BA, UK

[2] Southeast University, Sipailou 2, Nanjing 210096, CHINA

zhoubinde1980@gmail.com

Abstract. Ring-shaped electrostatic flow meters can provide very useful information on pneumatically transported air-solids mixture. This type of meters are popular in measuring and controlling the pulverized coal flow distribution among conveyors leading to burners in coal-fired power stations, and they have also been used for research purposes, e.g. for the investigation of electrification mechanism of air-solids two-phase flow. In this paper, finite element method (FEM) is employed to analyze the characteristics of ring-shaped electrostatic meters, and a mathematic model has been developed to express the relationship between the meter's voltage output and the motion of charged particles in the sensing volume. The theoretical analysis and the test results using a belt rig demonstrate that the output of the meter depends upon many parameters including the characteristics of conditioning circuitry, the particle velocity vector, the amount and the rate of change of the charge carried by particles, the locations of particles and etc. This paper also introduces a method to optimize the theoretical model via calibration.

1. INTRODUCTION

In coal-fired power station, coal has to be pulverized and pneumatically conveyed to burners. The solid to air ratio in term of mass is usually less than 1.5:1 (Coulthard, J et al., 1998), the equivalent volumetric concentration is less than 0.05% (at 100 ° C). This characterizes the typically dilute (lean) flow. The concentration is so low that the sensors based on the acoustic, microwave and capacitive principles may not be sensitive enough for practical applications, whilst the electrostatic technique appears to have advantages in this aspect (J Y Zhang., 2005; Zhang Bo et al., 2007). Electrification occurs due to the collision, friction and the separation between particles, and between particles and the pipe wall (Bailey, A, G et al., 1993; Armour-Chelu D I et al., 2002; Nifuku M,Sasaki T et al., 1989) in the milling or pulverizing and the pneumatic conveying processes. The electrostatic ring-shaped air solids flow meters are non-intrusive, fast responding, installation-easy and maintenance free. With its high sensitivity, it seems that this type of meters is more suitable for applications under the dilute conditions (Kleber W et al., 1998; Gajewski J B et al., 2006; Chuanlong Xu et al., 2008).

Electrostatic meter can be used to measure the flow distributions and velocities among pipes leading to a boiler and to indicate the average size of particles. It can also be used to for on-line and real-time monitoring the conditions of the milling and transportation systems by indicating the instability and malfunctioning to predict and prevent severe problems such as blockage. Such meter can provide the on-line continuous feedback information for closed-loop control so that the air to solids ratio can be optimized to reduce emissions and improve efficiency (Zhang J Q et al., 2003).

This paper analyses the spatial sensitivity field of the ring-shaped electrostatic meter and establishes a mathematic model expressing the relationship between the meter's output voltage, particle velocities and the amount of charges carried by particles using Finite Element Method (FEM). This paper also demonstrates how a mathematic model can be optimized via the calibration which was carried out at the laboratory of the University of Teesside, U.K. using a 40mm diameter ABB electrostatic meter, trade marked Pfmaster as a part of work for an EU funded project (J Y Zhang et al., 2008).

2. ELECTROSTATIC METER AND EQUIVALENT CIRCUIT

Fig. 1 shows the sectional view of a ring shaped electrostatic sensor, which comprises of a short spool piece and a ring electrode (probe) with the surrounding insulator. The ring electrode is mounted flush with the inner pipe wall offering non-intrusive measurement. Usually the metal pipe is earthed so that it acts as the electrostatic screen and also as the signal reference (signal ground) of the conditioning circuit input.

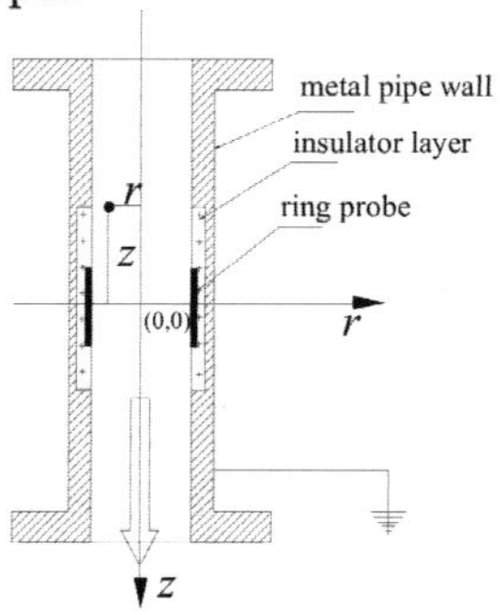

Fig. 1 Sectional View of ring-shaped electrostatic Sensor

Based on electrostatic induction theory, a certain amount of charge Q is induced on the electrode when charged particles are distributed among the sensing zone. The signal is amplified by the conditioning circuit which is usually a charge amplifier. The simplified equivalent circuit is shown in Fig. 2. According to *Kirchhoff's* current law, the relationship between the induced charge and the potential on the electrode is given by,

$$\frac{dQ(t)}{dt} = C\frac{dU_i(t)}{dt} + \frac{U_i(t)}{R} \qquad (1)$$

$U_i(t)$ is the input voltage of the conditioning circuit $R=(R_a \cdot R_i)/(R_a + R_i)$; $C=C_a+C_i+C_c$; where C_a and R_a are the capacitance of the electrode and insulation leakage resistance respectively. C_c is the stray capacitance of the cables connecting the electrode to the charge amplifier. C_i and R_i are the input capacitance and resistance of the conditioning circuit, including the equivalent capacitance and resistance in the feedback path of the charge amplifier due to the Miller effect. The Laplace transform and the Fourier transforms of equation (1) are given by,

$$U_i(s)\Big/Q(s) = \frac{sR}{1+sRC} \qquad U_i(j\omega)\Big/Q(j\omega) = \frac{j\omega R}{1+j\omega RC} \qquad (2)$$

Here $U_i(s)$ and $Q(s)$ are the Laplace transform of $U_i(t)$ *and* $Q(t)$, s the Laplace operator. $U_i(j\omega)$

and $Q(j\omega)$ are the Fourier transforms of $U_i(t)$ and $Q(t)$), ω the angular frequency.

The sensor will behave differently under the following three different conditions.

(1) $|j\omega RC| \ll 1$, then equation (2) can be simplified as,

$$U_i(j\omega) = j\omega R Q(j\omega) \qquad (3)$$

Assuming the zero initial condition, the time-domain analysis of equation (3) is,

$$U_i(t) = R\frac{dQ(t)}{dt} \qquad (4)$$

Which indicates the output voltage of the charge amplifier is proportional to the induced currents.

(2) $|j\omega RC| \gg 1$,

$$U_i(j\omega) = Q(j\omega)/C \qquad (5)$$

Time-domain analysis to equation (5) can be expressed as,

$$U_i(t) = Q(t)/C + C^* \qquad (6)$$

Ignoring C^*, the integral constant, it can be seen that the output voltage of the charge amplifier is proportional to the charge induced on the electrode and inversely proportional to the equivalent capacitance.

(3) If $|j\omega RC|$ is comparable with 1, based on equation (1), we have,

$$U_i(t) = e^{-\frac{t}{RC}} \cdot [\frac{1}{RC^2}(tQ(t) - \int Q(t)dt) + C^{**}] \qquad (7)$$

Which indicates the input voltage of the operational amplifier is determined by both the induced charge and its integration value. C^{**} stands for the integration constant.

In practice, the equivalent capacitance C is high primarily due to the Miller effect, the condition $|j\omega RC| \gg 1$ can usually be met. However in a dynamic electrostatic sensor, only AC signal components are selected by using a high pass filter with a very low lower cut-off frequency (a practical differentiator with low inertia).

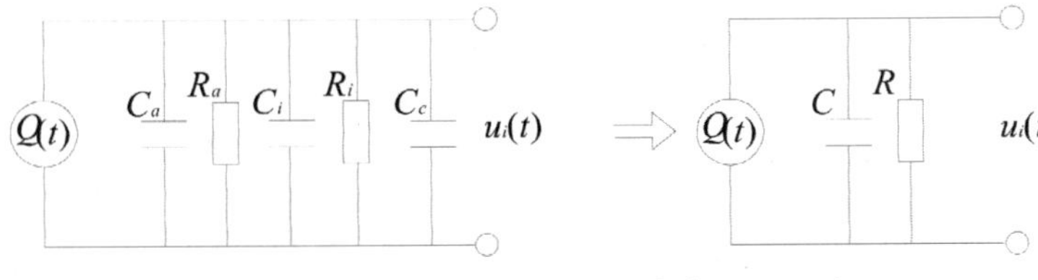

Fig. 2 The equivalent circuit of electrostatic sensor

Therefore the output of the meter can be expressed as

$$U_o(t) = k\frac{du_i(t)}{dt} = k\frac{1}{C}\frac{dQ(t)}{dt} \approx k\frac{1}{C_F}\frac{dQ(t)}{dt} \qquad (8)$$

Where k is a proportional factor determined by the gain of the entire conditioning circuit, C_F is the feedback capacitance of the charge amplifier, $U_o(t)$ the output of the meter.

3. MATHETICAL MODELING

Assuming the magnetic field due to the moving charged particles can be ignored, the electrostatic equilibrium system mentioned above is governed by

$$\begin{cases} \nabla \cdot \left[\varepsilon_0 \varepsilon(x,y,z) \nabla \varphi(x,y,z)\right] = -\rho(x,y,z) \\ \varphi(x,y,z)|_{(x,y,z)\in\Gamma_p} = 0 \\ \varphi(x,y,z,t)|_{(x,y,z)\in\Gamma_e} = \dfrac{1}{C}Q(t) \\ E_\infty = 0 \end{cases} \tag{9}$$

Where $\varphi(x, y, z)$ is the potential distribution inside the region; the boundary Γ_p represents the grounded pipeline wall with zero potential value, Γ_e is the ring electrode which is connected to the conditioning circuitry. $\rho(x, y, z)$ is the charge density distribution; ε_0 is dielectric constant in vacuum, $\varepsilon(x, y, z)$ denotes the relative permittivity distribution of the powder material, E_∞ is the electric field strength at the infinite distance.

Considering the complexity of gas-solids flow, the flow pattern or particulate material distribution is capricious, the analytical solution to equation (9) can hardly be found. FEM can be used to obtain a numerical solution to equation (10), which is the equivalent form to equation (9) with the same boundary conditions.

$$F(\varphi) = \int \{\frac{\varepsilon}{2}[(\frac{\partial\varphi}{\partial x})^2 + (\frac{\partial\varphi}{\partial y})^2 + (\frac{\partial\varphi}{\partial z})^2] - \rho\varphi\} \mathrm{d}x\mathrm{d}y\mathrm{d}z \tag{10}$$

The charge induced on the ring electrode varies with the fluctuations of quasi-static electric field which is produced by the moving charged particles. In the column coordinate system, it can be expressed as,

$$Q(t) = \iiint_{\Omega_r} q(r,\theta,z,t) \cdot s(r,\theta,z) d\Omega \tag{11}$$

where Ω_r denotes the sensing zone of the electrode. Due to the uncertainty of the charge distribution of particles, the electrostatic charge noise $q(r,\theta,z,t)$ is a stochastic variable function of time. $s(r,\theta,z)$ is the space sensitivity distribution function. From equation (11), the induced charge $Q(t)$ on the electrode is the weighted average of $q(r,\theta,z,t)$ by $s(r,\theta,z)$ which can be found using FEM mentioned above.

Due to the symmetrical axial geometry of the sensor, the induced charge on the ring electrode by the point charge at different angular position θ in the same circumference is equal. So that the spatial sensitivity does not vary with θ, the angular coordinate. Hence $s(r,\theta,z)= s(r,z)$. The simulation calculation can be implemented in the rectangular region as depicted in figure 3. The origin (z=0, r=0) is at the geometrical centre of the sensor. The diameter and width of the sensor for the simulation is 200mm and 2mm respectively to match the dimensions of the sensor used for the experiments.

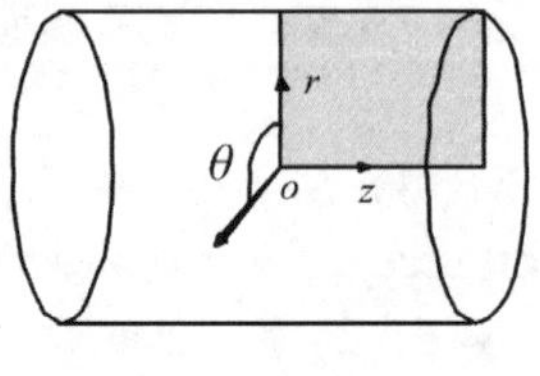

Fig. 3 Simulation region

For the given radial positions (r= 0,30mm,and r=80 mm), $s(r,z)$ decreases with the increase of axial location z which can be seen in Fig.4 (a). It can also be seen that the variation of $s(r,z)$ with the radial coordinate r depends on the axial location z, e.g. at z =50mm, $s(r,z)$ first slowly increases with r and then decreases with r when r approaching the value of pipe radium shown in Fig.4(b). This is due to the fact that the spatial angle included between the charged particle source to the ring electrode varies with spatial position.

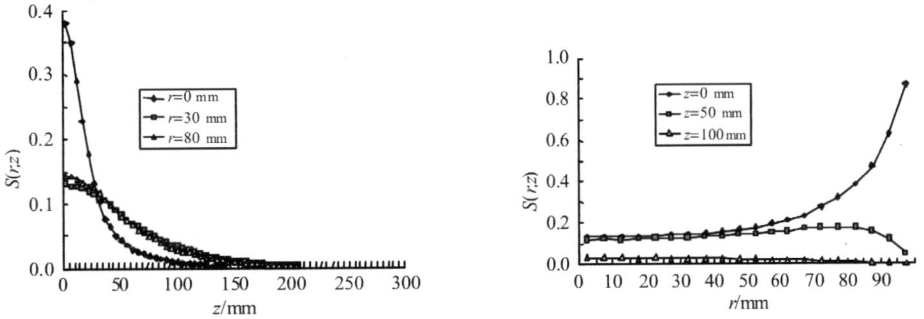

(a) variation of $s(r,z)$ with z (b) variation of $s(r,z)$ with r

Fig. 4 Spatial sensitivity Variation with the charged particle location

The following analytical expression via curve-fitting is derived based on the data contained in Fig. 4. A double Gaussian equation offers high fitting accuracy,

$$S(r,z) = A(r)e^{-B(r)z^2} + C(r)e^{-D(r)z^2} \qquad (12)$$

where A,B,C,D are determined by the geometric shape of the electrode and radial position r of a charged particle inside the pipe .Combining equations (5) (10) and (12), we have,

$$U_i(t) = \frac{1}{C}Q(t) = \iiint_{\Omega_r} q(r,\theta,z,t)[A(r)e^{-B(r)Z^2} + C(r)e^{-D(r)z^2}]d\Omega \qquad (13)$$

$$U_o(t) = \frac{k}{C}\frac{dQ(t)}{dt} = \frac{k}{C}\frac{d\iiint_{\Omega_r} q(r,\theta,z,t)[A(r)e^{-B(r)Z^2} + C(r)e^{-D(r)z^2}]d\Omega}{dt} \qquad (14)$$

If the partial derivatives of $q(r,\theta,z,t)$ and $s(r,z)$ are existed and continuous, the sequence of integration and differentiation can be exchanged, so equation (14) can be rewritten as,

$$U_o(t) = \frac{k}{C}\frac{dQ(t)}{dt} = \frac{k}{C}\iiint_{\Omega_r}\left[[\frac{\partial q}{\partial r}\cdot\frac{dr}{dt} + \frac{\partial q}{\partial \theta}\cdot\frac{d\theta}{dt} + \frac{\partial q}{\partial z}\cdot\frac{dz}{dt} + \frac{\partial q}{\partial t}]\cdot S(r,z) + q\cdot[\frac{\partial S}{\partial r}\cdot\frac{dr}{dt} + \frac{\partial S}{\partial z}\cdot\frac{dz}{dt}]\right]d\Omega \qquad (15)$$

Where $\partial q/\partial r$, $\partial q/\partial \theta$,and $\partial q/\partial z$ are the distribution gradient along the radial, circumferential and axial coordinates of the charges carried by particles. $dr/dt= Vr$, $d\theta/dt= V\theta$,and $dz/dt= Vz$ are the particle velocity projections at the radial, tangential and axial directions. In real powder pneumatic conveyance, the charges carried by the particle q is one of principal sources contributed to the detected signal, which is determined primarily by the solids concentration, $\partial s/\partial r$, $\partial s/\partial z$ the partial derivatives of the spatial sensitivity $s(r,z)$ are dependent upon the geometry of electrode and the location of particles. Besides, it can be seen that the output of the meter is affected by the velocities and the rate of change in charges carried by particles, because the output of the meter is proportional to the derivative of the charges induced on the electrode.

4. EXPERIMENTS RESULTS AND DISCUSSIONS

The results were obtained using a belt rig and a 40mm diameter pneumatic conveying system at the laboratory of the University of Teesside to verify the theoretical analysis presented above. As it is known that the electrostatic charges carried by pneumatically conveyed particles depend on many factors such as humidity, particle size, and chemical compositions of the solids materials, in laboratory, it is difficulty to control all of these variables. However if the results can be collected under the similar conditions, e.g. the similar humidity and temperature, and the solids are discharged completely in a recycling experimental system for each experiment, the results can be used to verify the effects of other factors such as velocity, air to solids ratio and solids location.

4.1 Experiments On Belt Rig

Fig. 5 shows a photograph of the belt rig in the laboratory of the University Teesside, which was used to validate the theoretical analysis discussed in Section 3. A rubber V-belt is wound around two pulleys, one of which is driven by a DC motor. The speed of DC motor is adjustable to set the belt velocity from 5m/s to 25m/s. A brush is attached to the rig frame, so that the electrical charges are generate when the belt passes through it. The charged belt is used to simulate a stream of charged particles travelling along axial direction. The diameter and the width of the electrode were 200mm and 2mm.

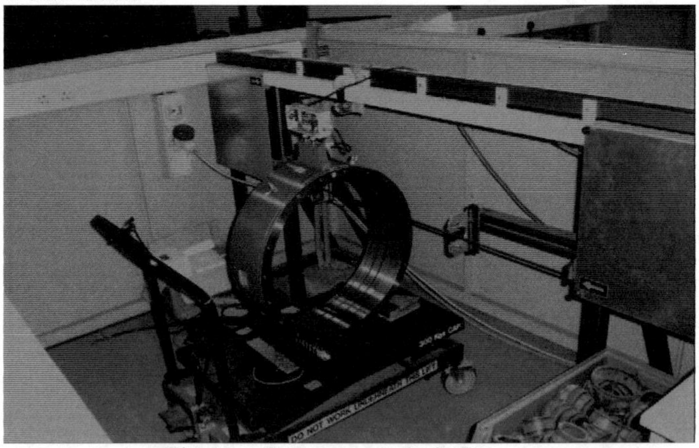

Fig. 5 Conveyance belt fig

The meter was placed on a hydraulic trolley so that the relative position of the belt to the sensor can be adjusted in the horizontal and vertical directions. The data were obtained using the specially designed data acquisition system, and the software program was developed using the National Instrument Labview package.

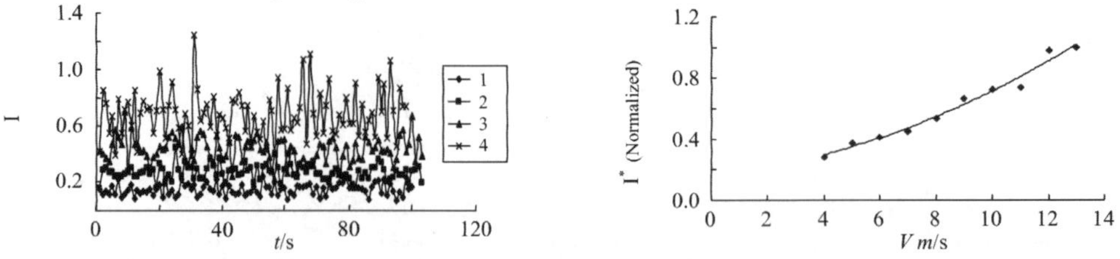

(a) (b)

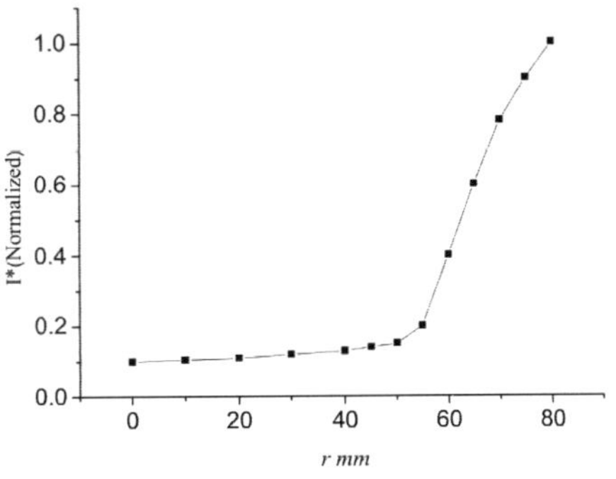

(c)

Fig. 6 Instantaneous value of signal *I*

When the belt moving along the axial direction, restricted by the metal edge of the pulleys, the belt tangential velocity *dθ/dt (Vθ)* is almost zero; When the charge on the belt saturated at a given motor speed, under the relatively dry air conditions, the charge on the belt within the sensing zone is essentially unchanged, so the variation rate of charge with time *∂q/∂t* may be negligible. If we regard the property of rubber material is uniform along the belt, and the even distribution of charge along the belt can be resulted so that *∂q/∂z* will be in theory equal to zero. *∂s/∂z* is an odd function about z, so the integration of this item is also zero. Consequently, equation (15) can be simplified as,

$$U_o(t) = \frac{K}{C}\frac{dQ(t)}{dt} = \frac{K}{C}\iiint_{\Omega_r}\left[\frac{\partial q}{\partial r}\cdot\frac{dr}{dt}\cdot S(r,z) + q\cdot\frac{\partial S}{\partial r}\cdot\frac{dr}{dt}\right]d\Omega_r \qquad (16)$$

Let I be the root mean square value of the sensor output voltage U_o in one data acquisition period, and I^* be the average value of I in much longer data acquisition periods, we have

$$I = \sqrt{\frac{1}{n-1}\sum_{i=1}^{n} U_o^{\,2}} \qquad (17)$$

In figure 6 (a), curves 1,2,3,4 show the instantaneous values of I when the belt ran at a fixed radial position at velocities of 7,9,11 and 13m/s respectively. Fig.6 (b) depicts that I^* increased with velocity over the range of 4 ~ 13 m / s. Fig.6 (c) gives the relationship between I^* *and* r at a belt travelling velocity of 17m/s. All these experimental results support the theoretical analysis in Section 3.

However what on earth caused the signal changes with the location and the velocity vector of the belt remains unclear. The following discussions may be useful for further investigations in this aspect.

There may be three reasons for the signal variation. At first, higher velocity means more intensive friction between the belt and the brush and between the belt and air, which produces more charges; secondly, it was observed that the belt vibration was significantly noticeable in the radial direction with the increased speed of the motor. The ratio of standard deviation to mean value of the signal was significantly higher than that at lower velocity. It can be seen from Equation (16) that the contribution of the radial movement of the belt represented by *dr/dt* (*Vr*)

Journal of Physics: Conference Series **147** (2009) 012001 doi:10.1088/1742-6596/147/1/012001

may be very important and should not be neglected. In other words, even the charge carried by the belt unchanged, the signal could increase with V_r. In addition, due to the nonlinearity of function $s(r,z)$and $\partial s/\partial r$, I^* varies nonlinearly with r.

4.2 Experiments On Pneumatic Conveying System

From the above discussion it can be seen that the output signal of an electrostatic flow meter is affected by many factors. The theoretical analysis can provide essential guidance, however the model can be optimised via calibration procedure. Here demonstrated is an example in which an electrostatic meter was calibrated under different air to solids ratios and solids mass flow rates, and a overall equation was found to govern the relationship between the meter output, solids mass flow rate and air to solids ratio.

A diagram of the test rig is shown in Fig .7 and is based on the suction principle. The solids discharges into the test rig via the screw feeder shown in Fig.8, and the solids mass flow rate into the rig is determined from the 'rate-of-loss' of weight depicted in Fig.9. The air mass flow rate is obtained using an orifice plate located downstream of the cyclone i.e. in its exhaust.

The material used for the experiments was 'Fillite', a commercial product made of fly ash. The average particle size was about 100μm. The maximum solids mass flow rate was over 50kg/hr.

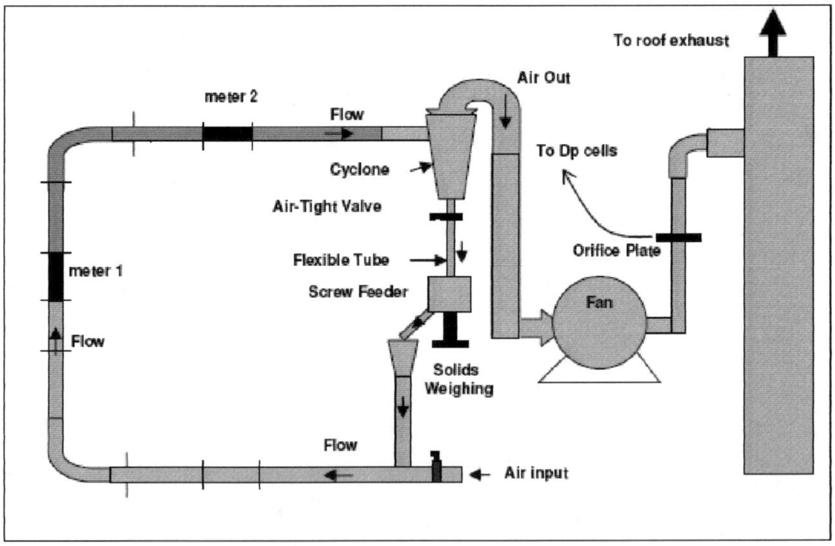

Fig. 7 Block diagram of the University of Teesside pneumatic conveyor

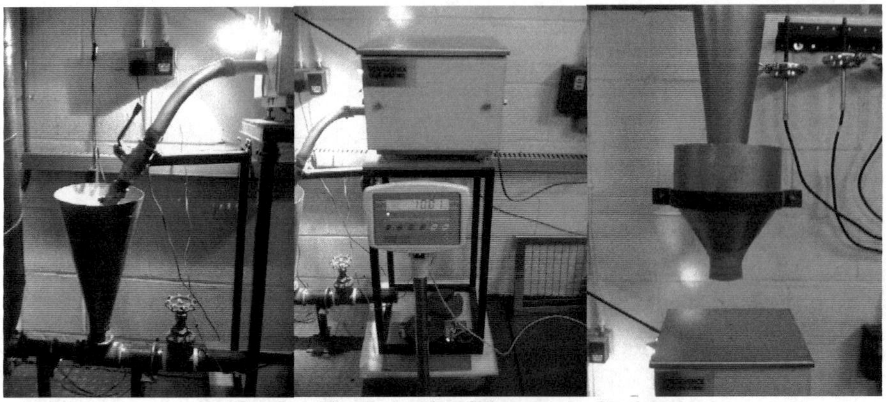

Fig. 8 Photograph of the screw feeder

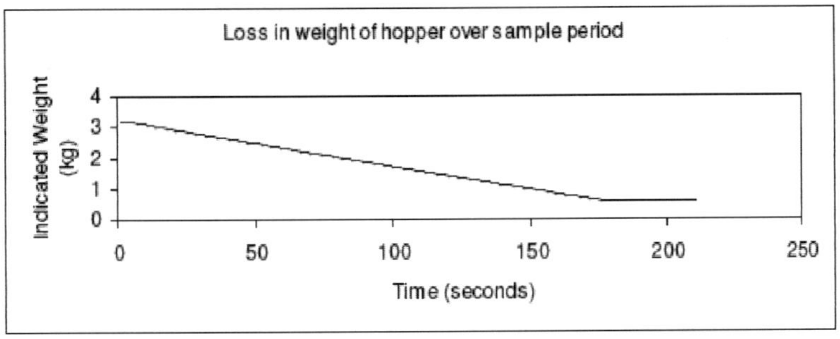

Fig. 9 Loss in weight characteristic

The air to solids ratio was kept approximately constant as the solids mass flow rate was varied for each sequence of tests and the signals were sampled over the selected measuring period, see Figure 10. The tests extended over the full operating range of the pneumatic conveyor and the process was repeated for air : solids ratios of 1.9:1 to 3.9:1.

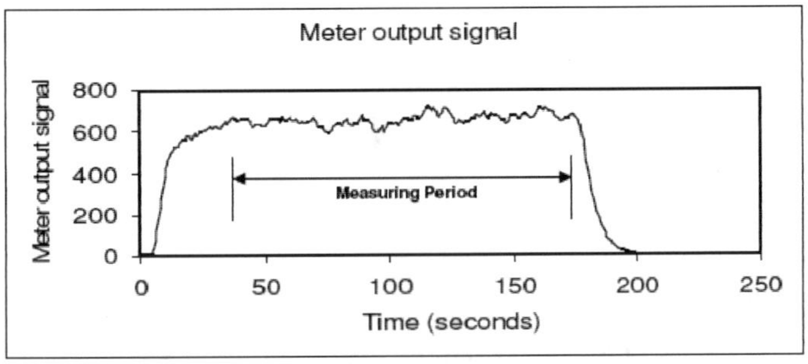

Fig. 10 Meter output-signals over a measurement period

Based on the test results in Fig. 11 and the theoretical analysis in Section 3, the relationship between air-solids ratio R_{as}, solids mass flow rate and the output signal I^* for the in the vertical section of the rig can be expressed as ,

$$I^* = F(R_{as}, \dot{M}) = (A_1 R_{as} + A_2)\dot{M}^2 + (B_1 R_{as} + B_2)\dot{M} + (C_1 R_{as} + C_2) \tag{18}$$

Factors A_1, A_2, B_1, B_2, C_1, and C_2 are constants and were found via calibration using the data provided in Table 1. A_1=-0.1581, A_2=0.7908, B_1=-14.826, B_2=61.511, C_1=-290.05, C_2=1210.5.

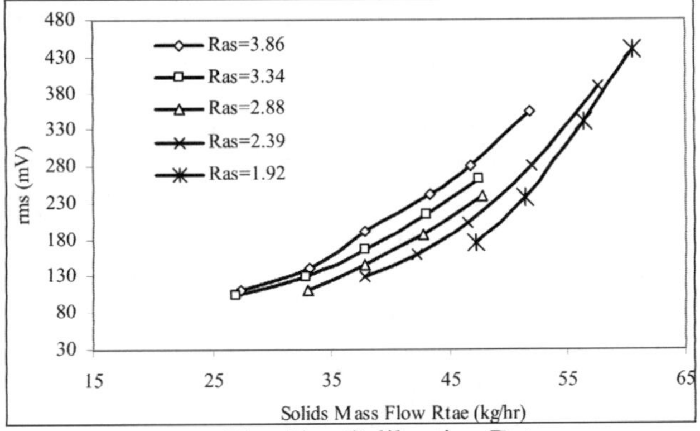

Fig. 11 Calibration Data

Table 1 Calibration data and calibration error

Air/solid (kg/hr/kg/hr)	solids mass flow rate (kg/hr)	Meter Output(mv)	calculated Output(mV)	absolute error (mv)	relative error %
3.78	27.38	111.30	109.19	-2.11	-1.93
3.88	33.12	140.26	147.72	7.46	5.05
3.79	37.80	190.03	183.68	-6.34	-3.45
3.83	43.28	241.60	242.20	0.61	0.25
3.95	46.69	281.31	289.51	8.20	2.83
3.91	51.85	353.58	356.68	3.10	0.87
3.3	32.85	127.70	130.31	2.61	2.01
3.32	37.76	166.32	162.64	-3.69	-2.27
3.32	43.03	213.00	211.32	-1.68	-0.79
3.38	47.44	261.43	266.62	5.19	1.95
3.38	52.38	335.48	336.40	0.92	0.27
2.85	37.79	145.71	141.87	-3.84	-2.71
2.86	42.82	186.82	183.38	-3.44	-1.87
2.88	47.71	237.91	241.28	3.37	1.40
2.38	37.75	128.92	120.98	-7.95	-6.57
2.38	42.21	158.64	151.82	-6.82	-4.49
2.4	46.44	202.87	197.75	-5.12	-2.59
2.39	52.00	279.62	277.70	-1.92	-0.69
2.39	57.60	387.75	385.33	-2.41	-0.63
1.91	47.10	174.90	177.38	2.48	1.40
1.9	51.38	237.63	241.29	3.66	1.52
1.91	56.43	340.91	339.90	-1.01	-0.30
1.96	438.59	440.45	1.86	0.42	0.10

It can be seen that the maximum relative error is less than 7%. It may be still too high for a single phase flow meter, but it is actually acceptable for a multi-phase flow meter.

5. CONCLUSIONS

The characteristics of the ring-shaped electrostatic meter has been analysed using the Finite Element Method and the electrostatic field theory. It concluded that the meter's output depends on the location of the solids, the projections of the solids velocity on the axial and radial coordinates and solids concentration. These conclusions have been verified by experiments carried on the belt rig and the pneumatic rig at the laboratory of the University of Teesside, U.K. The paper also introduced a method to optimise the mathematic model via calibration. Although here provided are only results from meter 1, the equation also applies to other meters in vertical and horizontal sections. In industrial applications, if the air to solids ratio is known, and the output signal in the equation can be obtained from the meter, the equation can give a measured solids mass flow rate with an acceptable error.

ACKNOWLEDGEMENTS

The research work discussed in this paper was part of research project RSC-CR 03005 funded by European Commission: Research funding for Coal and Steel. It was also supported by a grant from the key program of National Natural Science Foundation of China (No. 50836003)

NOMENCLATURE

Q	charge induced on the electrode	[C]
$U_i(t)$	input voltage of the conditioning circuit	[V]
C_a	capacitance of the electrode	[F]
R_a	insulation leakage resistance	[Ω]
C_c	stray capacitance of the cables	[F]
C_i	input capacitance of the conditioning circuit	[F]
R_i	input resistance of the conditioning circuit	[Ω]
$U_i(s)$	Laplace transform of $U_i(t)$	
$Q(s)$	Laplace transform $Q(t)$	
$U_i(j\omega)$	Fourier transforms of $U_i(t)$	
$Q(j\omega)$	Fourier transforms $Q(t)$	
ω	angular frequency	[rad/s]
C^* , C^{**}	integration constants	
C_F	feedback capacitance of amplifier	[F]
$U_o(t)$	output of the meter	[V]
$\varphi(x, y, z)$	potential distribution	[V]
Γ_p	grounded pipeline wall	[m^2]
Γ_e	ring electrode	[m^2]
$\rho(x, y, z)$	charge density distribution	[C /m^2]
ε_0	dielectric constant in vacuum	[8.85×10^{-12}F/m]
$\varepsilon(x, y, z)$	relative permittivity distribution of the powder material	
E_∞	electric field strength at the infinite distance	[V/m]
Ω_r	sensing zone of the electrode	[m^3]
$q(r,\theta,z,t)$	electrostatic charge noise	[C]
$s(r,\theta,z)$	space sensitivity distribution function	
$\partial q/\partial r$	charges distribution radial gradient	[C/m]

$\partial q/\partial \theta$	charges distribution gradient along the circumferential coordinate	[C/rad]
$\partial q/\partial z$	charges distribution gradient along the axial coordinate	[C/m]
Vr	velocity projection at the radial direction	[m/s]
$V\theta$	velocity projection at the tangential direction	[m/s]
Vz	velocity projection at the axial direction	[m/s]
I	root mean square value of the sensor output voltage U_o	[V]
I^*	the average value of I	[V]

REFERENCES

Armour-Chelu D I. (2002). "Comparison of the electric charging properties of particulate materials in gas-solids flows in pipelines," Journal of Electrostatics., 56(1), pp.87-101.

Bailey, A, G. (1993). "Charging of solids and powders," Journal of Electrostatics., 30, pp167-180.

Coulthard, J. (1998). "On-line pulverised fuel monitoring at Methil Power Station using electrostatic sensors," Proceedings 9th International Conference on Flow Measurement. FLOMEKO'98, pp. 265-268

Gajewski J B. (2006). "Non-contact electrostatic flow probes for measuring the flow rate and charge in the two-phase gas-solid flows," J.Chemical Engineering Science., 61(7), pp.2262-2270.

Kleber W. (1998). "Makin B.Triboelectric powder coating:a practical approach for industrial use," J.Part. Sci. Technol., 16(1), pp.43-53.

Nifuku M,Sasaki T. (1989). "Static electrification phenomena in pneumatic transportation of coal," Journal of Electrostatics., 23, pp.45-54.

Xu Chuanlong. (2008). "Velocity measurement of pneumatically conveyed solid particles using an electrostatic sensor," J.Meas. Sci. Technol., 2, pp.1-9.

Zhang Bo. (2007). "On-line monitoring of pneumatic conveyance of particulate solids" J.Chinese Journal of Sci Instrument., 28(11), pp1999-2002.

Zhang J. (2005). "Theoretical and experimental studies of the spatial sensitivity of an electrostatic pulverized fuel meter," Journal of Electrostatics, 63(12), pp.1133-1149.

Zhang J. (2008). "Characteristics of ABB Pulversied Fuel Meters," Journal of Measurement and Control., 41(1), pp.24-27.

Zhang J Q. (2003) "On-line continuous measurement of particle size using electrostatic sensors," J.Powder Technology., 135-136, pp.164-168.

Journal of Physics: Conference Series **147** (2009) 012002 doi:10.1088/1742-6596/147/1/012002

A New Cylindrical Capacitance Sensor for Measurement of Water Cut in a Low-production Horizontal Well

Xingbin Liu[1], Jinhai Hu[1], Wenfeng Xu[2], Lijun Xu[2], Zhonglin Xie[1], Yiwei Li[1]

[1] Daqing Logging & Testing Services Company, Daqing, Heilongjiang 163453, CHINA

[2] School of Instrument Science and Opto-electronic Engineering, Beihang University, Beijing 100083, CHINA

dlts_liuxb@petrochina.com.cn

hujinhai66@126.com

Abstract. In a horizontal well with low flow rate, oil-water two-phase flow is stratified due to gravity. For measuring water cut accurately in a low-production horizontal well, a novel cylindrical capacitance sensor is proposed in this paper. The structure of the sensor is cylindrical and hollow with multi-layer structure which is consisted of inside insulation layer, electrode layer, outside insulation layer and metal casing from inside to outside. And the measurement principle is analyzed in this paper.

The mathematical model is established, which shows that theoretically, there is a good relationship between the sensor response and water holdup. The response curve is monotone and the sensor has a good resolution and a high sensitivity in the whole range of water holdup.

The electric field of cylindrical capacitance sensor was simulated respectively by using ANSYS software when the sensor is filled with pure water, pure oil and oil-water mixture. The results of the simulation are consistent with the mathematical model.

Static experiments with the sensor filled with oil-water mixture were conducted finally. The results have verified the theoretical analysis and show that the proposed sensor is a viable solution to measuring water cut in a low-production horizontal well.

Cylindrical capacitance sensor provides a good reference for the water cut in low-production horizontal well and has a good application prospect.

1. INTRODUCTION

In the development of oilfield, production logging (PL) can not be stressed enough. The purpose of production profile logging, which is a branch of PL, is mainly to find out the production rate and water cut of each production layer in an oil well. But how to measure water cut accurately in horizontal wells is still a challenge in oilfield development. In PL area the term "water cut" is defined as water flow rate divided by total flow rate through the cross section of the casing of the production well, while the water holdup is defined as the water content ratio in a measurement volume filled up with fluid. Owing to the slippage between oil and water, water cut is not equal to water holdup which can be usually detected by a hold up sensor directly, and the water cut can be determined from the measured water holdup and flow rate by using the calibration chart or theoretical models. Also, the

Journal of Physics: Conference Series **147** (2009) 012002 doi:10.1088/1742-6596/147/1/012002

cylindrical capacitance sensor is only sensitive to water holdup, here called "sensor for measuring water cut" is a traditional name.

Some new tools widely used for measuring water cut, like CAT (Gary Frisch et al., 2002), are generally based on the full-bore measurement method in which the array sensors sense the water hold up of entire wellbore. This method is suitable for measuring the oil wells of high flow rate. But in Daqing oilfield, due to the long time development, the production rates of most wells are usually under 50m3/day which is too slow to be measured accurately by using this way. Instead of the full-bore measurement method, the packing method is usually used to measure the flow rate and water cut of a low-production well. The traditional capacitance sensor which is composed of a cylindrical metal electrode coated with insulating layer and a outside shield housing is effectively used to measure water cut in vertical wells, but in low-production horizontal wells, it fails to work due to the stratified flow and high water cut (over 50%). By contrast, the impedance sensor (Liu Xingbin et al., 2006) which depends on the continuous water phase is more suitable for high water cut and stratified flow, but so far, no successful data has been reported by using the impedance sensor.

Recently, with the increasing of the horizontal wells in China oilfields, the logging tools with high-precision and high-reliability are being expected. Much of research work on horizontal wells is being carried out. This research has provided a viable solution to measuring water cut in low-production horizontal wells, aiming at satisfying the requirements for practical measurements.

2. The STRUCTURE AND MEASUREMENT PRICIPLE of the SENSOR

In a horizontal well with low flow rate, oil-water two-phase flow is stratified (G. Oddie et al., 2003, M.Vielma et al., 2002, P. Abduvayt et al., 2004) due to gravity. For stratified flow, the level of the interface between oil and Water crossing the sensor reflects the water holdup. The present sensor can provide a real-time detection of the water level, and then accomplish the continuous measurement of water cut.

As can be seen form Fig.1, the capacitance sensor is cylindrical and hollow with multi-layer structure which is consisted of inside insulation layer, electrode layer, outside insulation layer and metal casing from inside to outside. The inside insulation layer and electrode layer are very thin, just about 1 millimeter. The excitation voltage is applied to the electrode layer, and the metal casing is grounded.

When the fluid flows through the sensor, the heavier phase water at the bottom of the sensor will contact with the casing at the ends of the sensor. The water in the bottom can be considered as grounded electrode, because the tap water and the formation water are good electric conductor in low

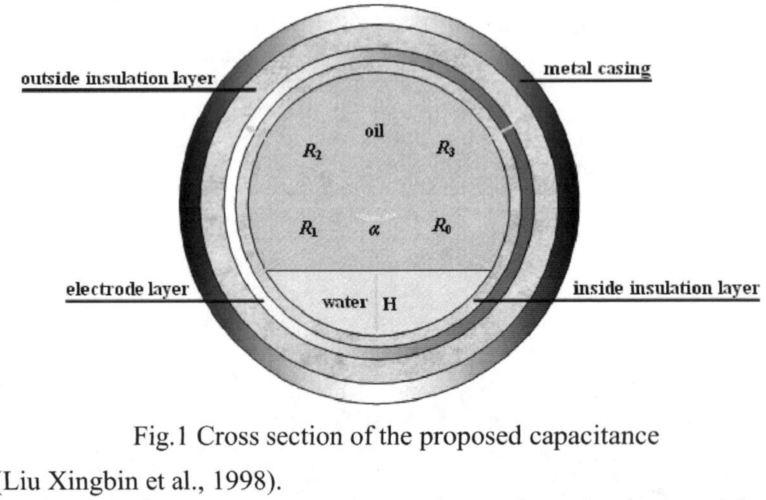

Fig.1 Cross section of the proposed capacitance

frequency range (Liu Xingbin et al., 1998).

Usually formation water and tap water contain numbers of conductive particles. According to electromagnetic theory, if the material satisfies:

$$\sigma / \omega\varepsilon\varepsilon_0 = G >> 1 \tag{1}$$

It can be considered as a conductor. At the low frequency 10 KHZ calculating G value according to typical data can get table 1, which shows that both formation water and tap water can be regarded as good conductors at a low frequency.

Table.1 Electrical characteristics of tap water and formation water at low frequency

testing water	temperature (℃)	salinity(mg/kg)	σ (S/m)	ε	G
tap water	20	400	0.08	78	1.75×103
formation water	60	1000~20000	0.38~5.88	60	1.13×104~1.76×106

As the formation water is conductive at low frequency, the outer surface of water in the sensor is thought to be equipotential. The surface can be divided into two parts: S_1 is contacted with the inside wall of the sensor and S_2 is the oil-water interface.
The value of the capacitance formed by electrode layer, outside insulation layer and metal casing is constant and can be computed by:

$$C_0 = \frac{2\pi\varepsilon_1\varepsilon_0 L}{\ln(R_3/R_2)} \tag{2}$$

The value of the capacitance formed by electrode layer, inside insulation layer and S_1 is variable with water hold up within the sensor, it will become lager with the increase of angle α, and namely, with the increase of water holdup H_w, the value can be computed by:

$$C_1 = \frac{\alpha\varepsilon_2\varepsilon_0 L}{\ln(R_1/R_0)} \tag{3}$$

The value C_2 of the capacitance formed by electrode layer, oil, inside insulation layer and S_2 is relatively small and changes slightly and contributes little to the response of the sensor, to simplify the analysis, it can be neglected.

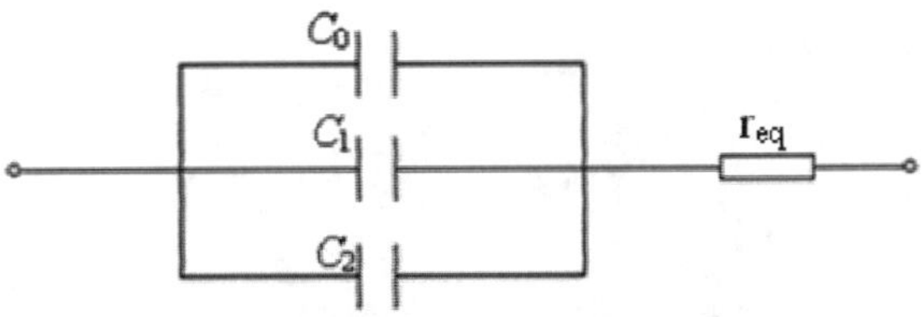

Fig.2 Equivalent circuit of cylindrical capacitance sensor

So the whole response C of a cylindrical capacitance sensor is the sum of C_0 and C_1.

$$C = \frac{2\pi\varepsilon_1\varepsilon_0 L}{\ln(R_2/R_1)} + \frac{\alpha\varepsilon_2\varepsilon_0 L}{\ln(R_1/R_0)} \tag{4}$$

By geometric method, we have

$$H_w = \frac{\alpha - \sin\alpha}{2\pi} \tag{5}$$

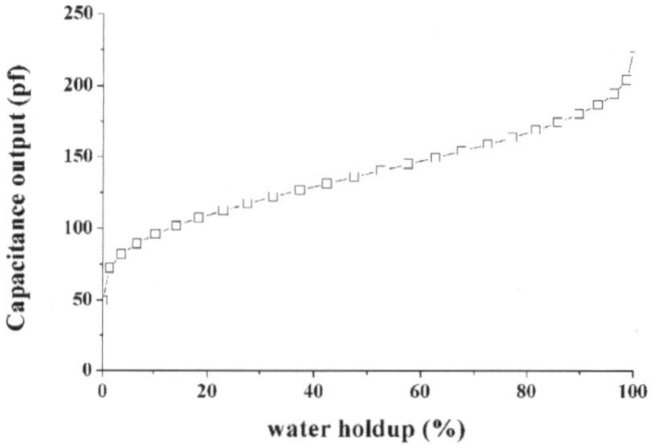

Fig.3 Theoretical dependence of the sensor response on the water holdup

H_w is a function of α, so is C. The approximate-linear curve is obtained from (4) and (5), where R_0 12.5mm, R_1 13.5mm, R_2 14.5mm, R_3 20.5mm, L 80mm, relative permittivity for oil and inside/outer insulation layer 3.5 (parameters of the actual sensor applied in static experiments). Fig.3 shows that the response curve is monotone and the sensor has a good resolution and a high sensitivity in the whole range of water holdup.

3. ANSYS SIMULATIONS

In order to understand the electric field distribution of the sensor with different water holdups and verify the theoretical analysis, the electric field of the sensor is simulated by using ANSYS software. Set the equivalent relative dielectric constants (permittivity) 1000 for water, metal casing and electrode layer, +10V for excitation voltage. Remaining parameters agree with that of the actual sensor applied in static experiments.

When the sensor is filled with oil-water mixture (Fig.4), the complete electric field of the sensor is divided into three parts: the one between electrode layer and metal casing is invariable, the one between electrode layer and S_1 can be regarded as a field of an incomplete cylindrical capacitor, and the one between metal layer and S_2, with very small intensity, could hardly impact on anything. Therefore, the three parts can be simplified for two parts: the constant part (outer) and the variable part (inside), which is consistent with the mathematical model, and the variable part is more important.

When the sensor is full of oil, the variable part disappears, there only exists the outer electric field (Fig.5). The output of the sensor is the minimum at this moment.

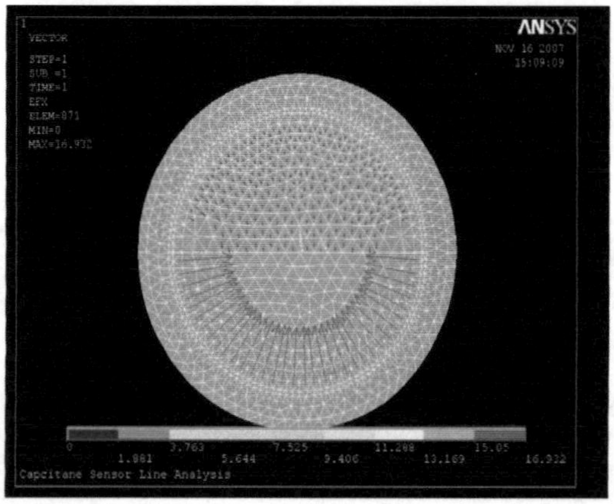

Fig.4 Cross section of the electric field at stratified flow

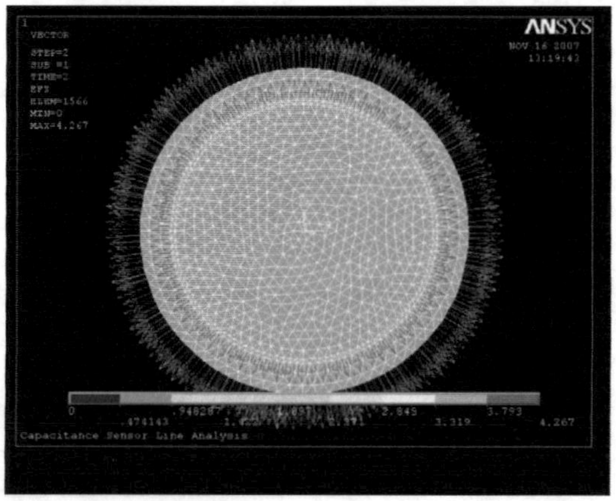

Fig.5 Cross section of the electric field when the sensor is full of oil

When the sensor is full of water, because of the axial symmetry, the profile simulation only shows a part the of electric field distribution. Two evenly radial parts with opposite directions can be seen from Fig.6 and Fig.7. At this moment, the area of the variable part is the maximum, and so is the capacitance of the sensor.

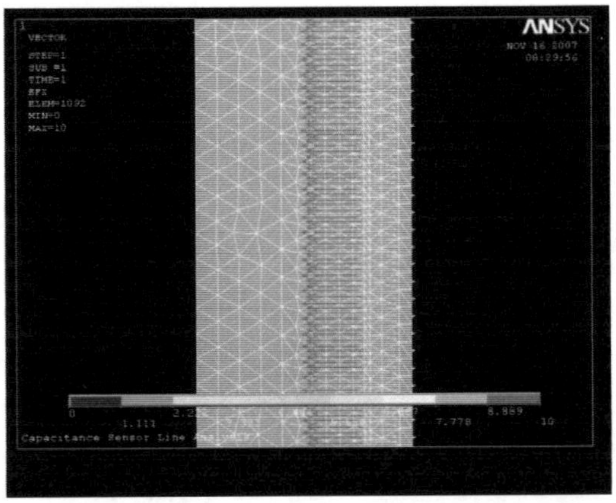

Fig.6 Profile section of the electric field when the sensor is full of water

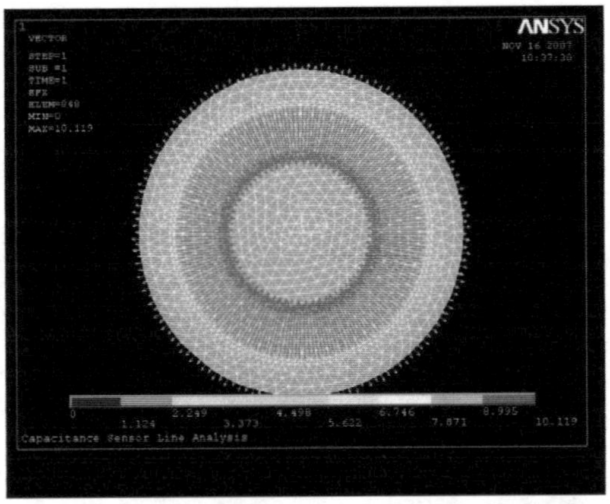

Fig.7 Cross section of the electric field when the sensor is full of water

4. STATIC EXPERIMENT

The static experiment was carried out with oil and water to confirm the theory. Water (tap water) and oil (kerosene) were put into a transparent container to simulate the stratified flow, and then, the cylindrical capacitance sensor was set in the container. The water was injected into the container to increase the level of the interface between the water and the oil gradually, and the capacitance was recorded by a programmable RCL, the relative level of oil-water interface was recorded by a ruler which was fixed on the sensor, which was repeated three times. The experiment result was shown in Fig.8. Similarly, the water was extracted from the container to decrease the interface level, and the capacitance of the capacitor was measured, which was repeated three times. The experiment result was shown in Fig.9. The two figures indicate that the sensor has good water holdup sensitivity and reasonable repeatability.

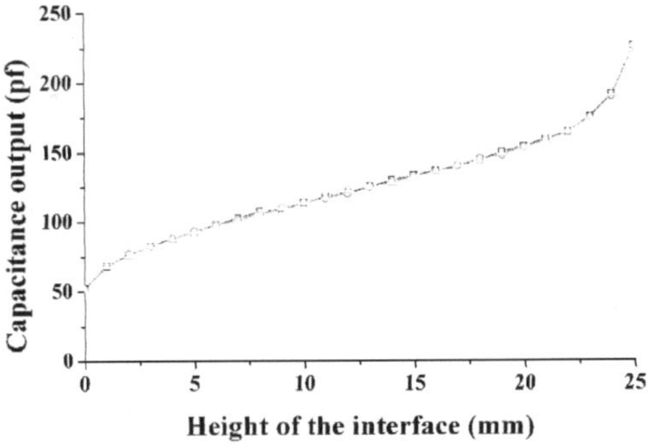

Fig.8 Sensor response of up-travel

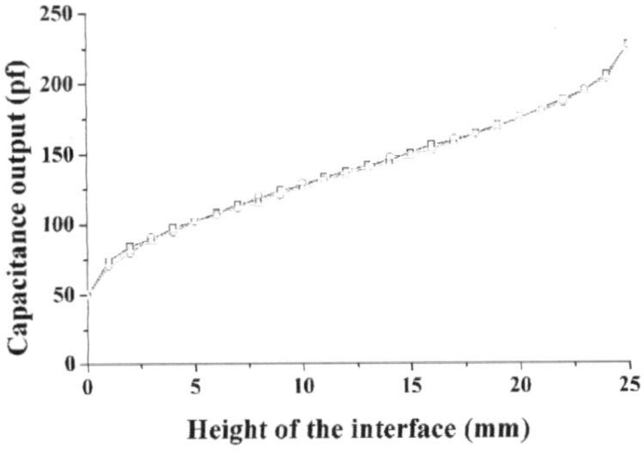

Fig.9 Sensor response of down-travel

Convert the interface level into the holdup, average the three curves of up-travel and the three of down-travel respectively, then put the two curves and the theoretical curve into Fig.10 to comprise together. As can be seen, actual responses of the sensor are well consistent with the theoretical response, but there is a slight difference between the response of up-travel and down-travel, which belongs to "lost motion error ".

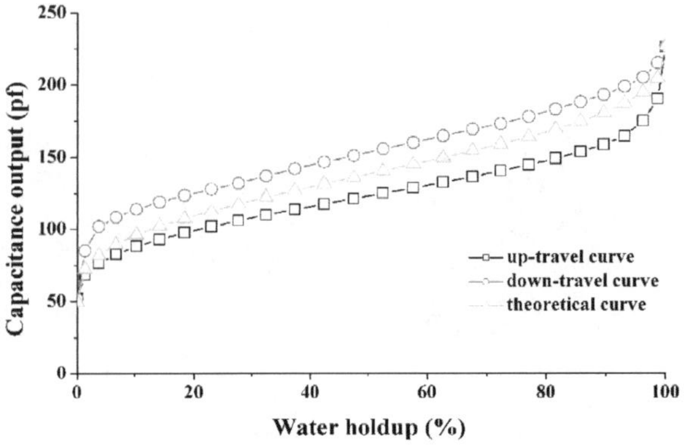

Fig.10 Comparison between the actual response and the theoretical response

In fact, there exists a slight friction between the liquid in the sensor and the inside wall of the sensor, it makes the state of oil-water interface in up-travel (convex) different from that of down-travel (concave), which leads to the lost motion. But in dynamic case, this error will disappear due to the flow of fluid.

5. FURTHER WORKS

Dynamic experiments will be conducted in DLTS (Daqing Logging & Testing Services Company) multiphase flow loop facility. The proposed prototype for dynamic experiments is composed of 4 parts: circuits, cylindrical capacitance sensor, basket packer and motor driver as shown in Fig.11. A packer can trap the fluid into a narrow passage to increase the velocity for measuring the flow rate accurately.

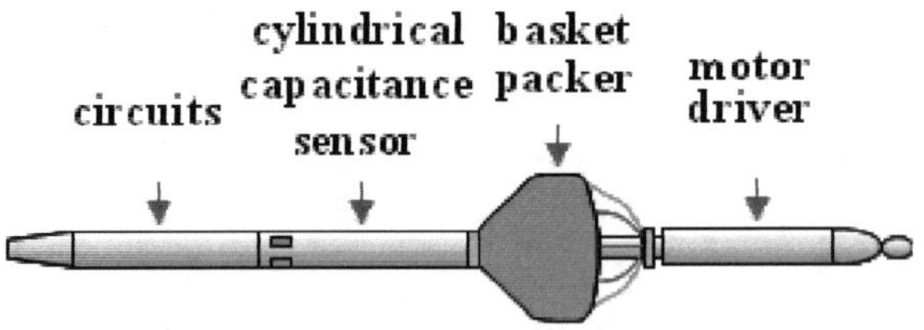

Fig.11 Structure of the experimental prototype

But there is no measurement devices of flow rate installed on the prototype, the purpose to fix the packer is to get the characteristics of the sensor response in packing conditions. The circuits can convert the capacitance value into a frequency signal and supply power to the prototype. The motor controls the mode of the packer.

6. CONCLUSIONS

Theoretically, the cylindrical capacitance sensor with multi-layer structure has a good response to water holdup in the whole range of measurement. The electric simulations have verified the mathematical model and the static experiment results with good repeatability are also well consistent with the mathematical model, which shows that the cylindrical capacitance sensor could better answer the question of water cut measurement in a low-production horizontal well and has a good application prospect. Next, dynamic experiments will be conducted to study the response characteristics of the sensor in dynamic case and the chart of sensor response plotted against water cut will be calibrated.

ACKNOWLEDGEMENTS
The Authors gratefully thank the National High Technology Development 863 Program of China (2007AA06Z231).

NOMENCLATURE

R_0	inside radius of inside insulation layer	[mm]
R_1	outside radius of inside insulation layer (inside radius of electrode layer)	[mm]
R_2	outside radius of electrode layer (inside radius of outside insulation layer)	[mm]
R_3	outside radius of outside insulation layer (inside radius of metal casing)	[mm]
L	length of the electrode layer of the sensor	[mm]
H_w	water holdup	

Greek letters

σ	conductivity of material	[S/m]
ω	work frequency	[rad/s]
α	central angle of semi-cylindrical surface	[rad]
ε	permittivity of vacuum, 8.85×10^{-12}	[F/m]
ε_0	relative permittivity of material	
ε_1	relative permittivity of the outside-insulation-layer material	
ε_2	relative permittivity of the inside-insulation-layer material	

REFERENCES
G. Oddie et al. (2003). "Experimental Study of Two and Three Phase Flows in Large Diameter Inclined Pipes," International Journal of Multiphase Flow,29, pp. 527-558.

Gary Frisch et al. (2002). "Integrating Wellbore Flow Images with a Conventional Production Log Interpretation Method," SPE Annual Technical Conference and Exhibition, San Antonio, Texas, 29 September-2 October 2002.

Liu Xingbin, Qiang Xifu et al. (1998). "The Theoretical Model for Capacitance Water Holdup Meter and Its Application to Production Logging," Flow Measurement and Instrumentation, 9(4), pp 249-257.

Liu Xingbin, Hu Jinhai, Huang Chunhui, et al. (2006). "Conductance Sensor For Measurement of The Fluid Watercut and Flowrate in Production Wells," The 5th International Symposium On Measurement Techniques for Multiphase Flows, Aomen, China, 440-443.

M.Vielma et al. (2002). "Characterization of Oil/water Flows in Horizontal Pipes," SPE Annual Technical Conference and Exhibition, Anaheim, California, U.S.A, 11-14 November 2007.

P. Abduvayt et al. (2004). "Analisis of Oil-water Flow Tests in Horizontal, Hilly-terrain, and Vertical Pipes," SPE Annual Technical Conference and Exhibition, Houston, Texas, September 26-29.

Journal of Physics: Conference Series **147** (2009) 012003 doi:10.1088/1742-6596/147/1/012003

Effect of particle size distribution on formation of linear configuration of dielectric fine particles under the electric field

A Otsuki [1], G Dodbiba [2], T Fujita[2]

[1] Waseda University, Okubo, Shinjuku-Ku, Tokyo, 169-8555, JAPAN
[2] The University of Tokyo, Hongo, Bunkyo-Ku, Tokyo, JAPAN

otsukiakira@aoni.waseda.jp

Abstract. This paper describes the effects of particle size distribution on formation of linear configuration of dielectric fine particles under the electric field. The objective of this study was to verify the mechanism of the measurement by interactive force apparatus and determine the appropriate particle size distribution. The apparatus was proposed for force measurement, which can be applied to particle size measurement. The advantage of size measurement using the apparatus is a direct measurement, not depending on the concentration and optical transparency of the solution. Moreover, the measurement can be conducted in both the aqueous solution and organic solvent. Other common apparatuses for size measurements using laser source (i.e., the dynamic light scattering and laser diffraction) are difficult to measure the size of particles in the solution of high particle concentration and/or low transparency accurately, due to multi-scattering of light in the sample solution or solvent. However, the mechanism of the measurement and appropriate experimental conditions for particle size measurement using the interactive force apparatus have not been fully understood. In the present study, the potential energies of a cluster under the electric field were calculated in order to determine a particle size distribution, which gives linear configuration of particle cluster. The results verify the mechanism of the measurement by interactive force apparatus and determine the appropriate particle size distribution.

1. INTRODUCTION

It is important to manipulate the degree of dispersion and coagulation of nano-particles in various medium (e.g. aqueous solution and organic solvent) for the many process industries. Therefore, there is a need for understanding the dispersion and coagulation of nano-particles in the medium. There are many techniques for evaluating the dispersion and coagulation of nano-particles, i.e. size measurement turbidity measurement, contact angle measurement, zeta potential measurement, force measurement as well as combination of these techniques. However, these methods usually are not suitable for evaluating the dispersion and coagulation of particles in sample solutions of high concentration or the solution with no optical transparency. On the other hand, these kinds of solutions are commonly used in the many industrial procedures, such as separation of fine particles, deposition of fine particles on fibers.

There are many techniques for particle size measurement. The techniques can be divided in two categories, i.e. measurement in dry condition and measurement in wet condition. In dry condition, the microscopic studies based on optical microscopy, SEM and TEM are the most common techniques for size measurement of particles. In wet condition, the techniques using laser source (i.e. the dynamic light scattering and laser diffraction) are the most common techniques for size measurement of particles. Although these techniques have several advantages, they have some drawbacks. The techniques in dry condition are not applicable for measuring the size of particles in solutions. On the other hand, the techniques in wet condition are not suitable for estimating the size of particles in sample solutions of high concentration or the solution with no optical transparency.

The interactive force apparatus (Otsuki et al., 2007) was proposed as an apparatus for particle size measurement. The apparatus is a direct measurement technique, not depending on the concentration and optical transparency of the solution. Moreover, the measurement can be conducted in both the aqueous solution and organic solvent. However, the mechanism of the measurement and the appropriate experimental conditions for particle size measurement were not fully understood. The authors assumed that particles form liner configuration during the measurement. The assumption should be verified in order to elucidate the mechanism of the measurement. In the present study, potential energies of a cluster under the electric field were calculated in order to elucidate the mechanims and determine the appropriate particle size distribution for the measurement.

2. POTENTIAL ENERGIES OF A CLUSTER UNDER THE ELECTRIC FIELD

Potential energies of a cluster composed of fine particles were calculated in order to determine the dominant configuration of particles under the electric field. In the present study, investigation of the formation of linear and triangular configurations composed of magnetic particles under the magnetic field (Aoshima and Satoh, 2005) was modified. In order to investigate the formation of these structures composed of dielectric particles under the electric field, a case where three particles in contact with each other at their surfaces was considered. The dielectric moments of the three particles is assumed to be aligned with the electric field direction. One of the three particles is denoted by *i* and the others are denoted by *j*. Two particles *j* have the same diameter d_j. The influence of the diameter d_i of particle *i* on potential energies of the linear ($u^{(p)}_{lin}$) and triangular ($u^{(p)}_{tri}$) configurations was evaluated by using Eqs. 1 and 2 derived from the Eqs. developed by Aoshima and Satoh (2005) with the assumption that there is no steric layer around the surface of particles:

$$\frac{u^{(p)}_{lin}}{kT\lambda_0} = -2\left(\frac{d_i d_j}{d_0^2}\right)^3\left(\frac{d_0^3}{\left(\frac{d_i+d_j}{2}\right)^3}+\frac{d_0^3}{\left(\frac{d_i+d_j}{2}+d_j\right)^3}\right)-2\left(\frac{d_j^{\,2}}{d_0^2}\right)^3\frac{d_0^3}{d_j^{\,3}} \tag{1}$$

$$\frac{u^{(p)}_{tri}}{kT\lambda_0} = 2\left(\frac{d_i d_j}{d_0^2}\right)^3\frac{d_0^3}{\left(\frac{d_i+d_j}{2}\right)^3}\left(1-3\frac{4\left(\frac{d_i+d_j}{2}\right)^2-d_j^{\,2}}{4\left(\frac{d_i+d_j}{2}\right)^2}\right)+\left(\frac{d_j^{\,2}}{d_0^2}\right)^3\frac{d_0^3}{d_j^{\,3}} \tag{2}$$

where *k* is the Boltzmann's constant, *T* is the absolute temperature, λ_0 is the dimensionless parameters for the particles with the mean diameter d_0.

3. PARTICLE SIZE DISTRIBUTION FUNCTIONS

In order to determine the probabilities of formation of linear and triangular configuration of particles, type of particle size distribution was determined from three types of common distribution functions. The distribution functions were Normal distribution, Rosin-Rammler distribution and Gaudin-Schuhmann distribution as discribed in Eqs. (3)-(5), respectively.

$$Q(d) = \frac{1}{\sqrt{2\pi}\sigma}\exp\left(-\frac{(d-d_0)^2}{2\sigma^2}\right) \tag{3}$$

$$Q(d) = 1 - \exp\left\{-(\frac{d}{d_e})^n\right\} \quad (4)$$

$$Q(d) = (\frac{d}{d_{\max}})^m \quad (5)$$

where d is diameter of particle, d_0 is mean diameter of particle, σ^2 is variance of distribution, d_e is absolute size constant, n is distribution constant, d_{max} is size of the largest particle in the system and m is particle size parameter.

4. RESULTS AND DISCUSSION

Potential energies of linear and triangular formation as a function of particle diameter are shown in Fig. 1. The intersections of potential energies of linear and triangular configurations were summarized in Table 1. It should be noted that there is no intersection where the ratio of d_j / d_0 are 1.60 and 3.20. At these d_j / d_0, linera configuration has always priority over the triangular configuration.

In this study, we used the results of particle size measurement of silica nanoparticles by means of DLS where pH of sample solution was 9.54 as shown in Fig. 2. The three types of particle size distribution functions were fitted with the experiemental data (Fig. 2). Normal distribution, which was the most fit with the experimental data, was selected at d_0 = 19.8 and σ = 5.87.

Frequency of paritcle size distribution where potential energy of linear configuration is less than the one of triangular configuration (Fig. 1) equals to be the probabilities of formation of linear configuration of particles. In order to use the results of calculation of potential energies of linear and triangular formation for determining the probabilities of their formation, the particle size distribution was calculated by using determined parameters (i.e. d_0 and σ) and normalized by mean diameter d_0 (Fig. 3).

Assuming that the particle size distribution was normal distribution as calculated in Fig. 3, the probabilities of formation of linear configuration over triangular under the electric field were also summarized in Table 1. It can be seen that liner configuration is dominant at all conditions over triangular configuration. This implies that the assumption of formation of liner particle configuration during the measurement is valid and particle size measurement by interactive force apparatus can be conducted properly where normal distribution as calculated in Fig. 3. On the other hand, triangular configuration may form up to 4.5 % at ratio of particle size d_j / d_0 = 0.40.

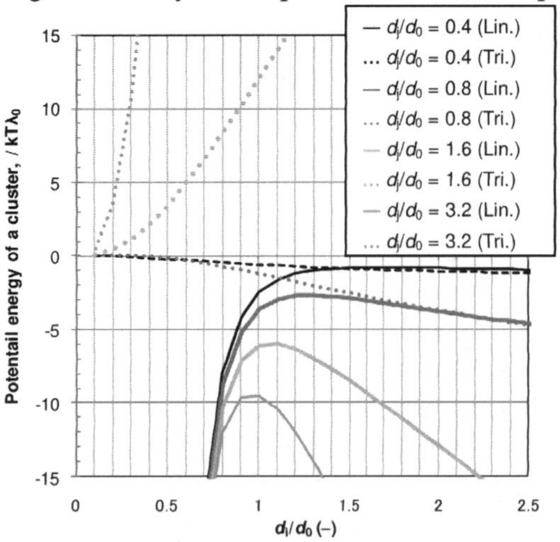

Fig. 1 Potential energies of linear and triangular formation as a function of particle diameter d_i / d_0 at different ratio of d_j / d_0.

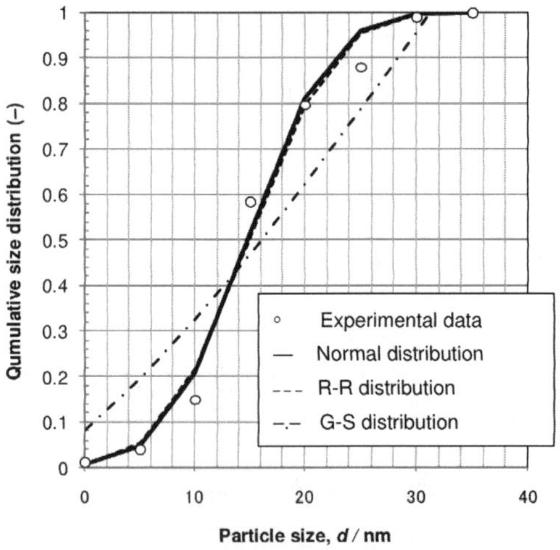

Fig. 2 Particle size distribution of silica nanoparticle where pH of sample solution is 9.45.

Table 1 Probabilities of formation of linear configuration of dielectric fine particles under the electric field.

d_j/d_0	d_i/d_0 at intersection of $u_{lin.}$ and $u_{tri.}$	Probabilities of formation of linear configuration of particles (%)
0.20	2.10	99.9
0.40	1.50	95.5
0.80	2.50	100
1.60	-	100
3.20	-	100

5. CONCLUSIONS

In the present study, the potential energies of a cluster under the electric field were calculated in order to verify the mechanism of the measurement and determine the appropriate particle size distribution. According to the calculations, the mechanism of the measurement and appropriate conditions for measuring the size of fine particles by using interactive force apparatus were summarized as follows:

(1) Liner configuration is dominant at all conditions over triangular configuration. This implies that the assumption of formation of liner particle configuration during the measurement is valid and particle size measurement by interactive force apparatus can be conducted properly where normal distribution as calculated.

(2) Triangular configuration may form up to 4.5 % where particle size distribution was normal distribution as calculated at ratio of particle size d_j / d_0 = 0.40.

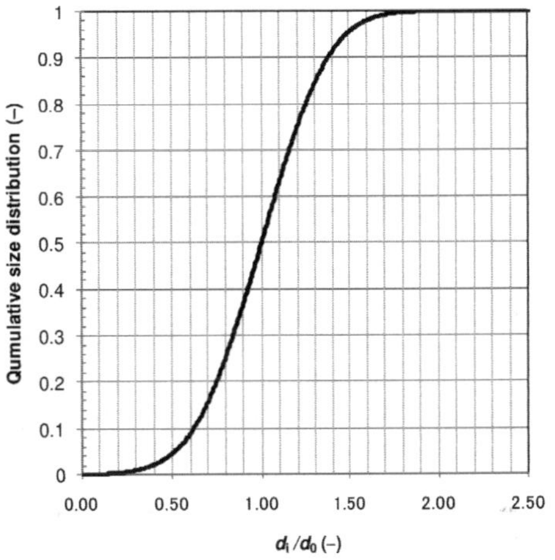

Fig. 3 Calculated particle size distribution of silica nanoparticles based on Normal distribution where d_0 = 1.0, σ = 5.87.

ACKNOWLEDGEMENTS

This work was supported in part by Grants-in Aid for Young Scientists (B) from the Ministry of Education, Culture, Sports, Science and Technology, Japan.

REFERENCES

Aoshima, M. and Satoh A., (2005). "Two-dimensional Monte Carlo simulations of a colloidal dispersion composed of polydisperse ferromagnetic particles in an applied magnetic field," J. Colloid Interface Sci., 288, pp. 475-488.

Otsuki, A. et al., (2007). "Two-Liquid Flotation: Heterocoagulation of Fine Particles in Polar Organic Solvent," Mater. Trans., 48, pp. 1095-1104.

Dense-phase pneumatically conveyed coal particle velocity measurement using electrostatic probes and cross correlation algorithm

C Xu [1], J Zhang [2], D Yang [1], B Zhou [1], G Tang [1] and S Wang [1]
[1] Southeast University, Nanjing, 210096, P. R.China
[2] University of Teesside, Middlesbrough TS1 3BA, UK

smwang@seu.edu.cn, chuanlongxu@seu.edu.cn

Abstract. The fundamental measurement theory of an electrostatic probe and cross-correlation velocity measurement method are introduced in the paper. The effects of the probe's geometric parameters including the length of the probe, the thickness, the length and the relative permittivity of the dielectric pipe, the radius of the screen on the dimensionless calibration coefficient (k) and the statistical error of the transit time (τ_m) of the correlation velocity measurement system using electrostatic probes were investigated theoretically. Finally the measurement system was applied in a 10 mm bore horizontal section of a dense phase pneumatic conveying system under high pressure circulating pulverized coal over a superficial air velocity range of 0.5-7 m/s for a particle concentration 0.052-0.141 m^3/m^3. Experimental results obtained demonstrate that the system is capable of providing solid particle velocity measurements with repeatability better than 10 % under the given experimental conditions.

1. INTRODUCTION

The measurement of gas-solid flow parameters is one of important research fields of multiphase flow, and is also an urgent problem needed to be resolved in fundamental theory, experimental investigation and industrial application of gas-solid flow. However due to the randomness of gas-solid interface in time and space the gas-solid flow characteristics and its measurement methods are far more complicated than single phase flow system. Although lots of research work has been done in the past decades, very few commercialized multiphase flowmeters are employed in industry today, and most of them are still under laboratory research and development. Among these flowmeters, the most promising have been the cross-correlation velocity measurement system which was developed in the 1960s. The non-obstructive measurement method has the advantages of broad measurement range and of good adaptability to hostile industrial environment. Wide ranges of sensing techniques including capacitive, ultrasonic and radiometric probes have been adopted. Useful results have been obtained from the measurement systems based on these probes, and successful applications for the non-contact, on-line continuous measurement of solid particle velocity in pneumatic transport system have been achieved (Xu L A. 1988, Thorn R. etal. 1982). In recent years, electrostatic sensing technique which is based on the particle charging of gas-solid two phase flows has created a major interest in the solid particle flow measurement because of its advantages over the other available correlation measurement techniques in good repeatability, high sensitivity, reliability and low cost (Boeck T. 1989, Yan Y. et al. 1995, Woodhead S. R and Amadi-Echendu J. E.1995, Ma J. and Yan Y. 2000, Gajewski J.B. 1996).

Boeck firstly applied ring shape electrostatic probes together with cross correlation techniques to measuring the particle velocity in a coal injection system, but few details on the measurement system and test results have been given (Boeck T. 1989). Yan and Coulthard studied the sensing mechanism, the spatial sensitivity and spatial filtering effects of the built-in electrostatic probes to optimize design and interpret signals (Yan Y. et al. 1995, Ma J. and Yan Y. 2000). And an electrostatic flow measurement system based on the probes was evaluated by theoretical analysis and by off-line experimental tests and pilot-plant trails. The results obtained showed that the measurement system achieves repeatability better than 2% and linearity within 12% over a velocity range 20-40 m/s for solid particle concentrations of in a range 0.01-0.44%. The correlation measurement systems using ring-shape electrostatic probes were also investigated by Woodhead (Woodhead S. R and Amadi-Echendu J. E. 1995) and Gajewski (Gajewski J.B. 1996), and the reasons for the measurement error of the systems were analyzed in detail.

Although some progress has previously been made on the use of the ring-shape electrostatic probes for metering gas-solid flow parameters, there are still some issues to be discussed, for example the effect of the length of the probe on its sensing characteristics was only considered in establishing its physical and mathematical model. Actually the geometric parameters of the probe including the thickness and relative permittivity of the dielectric pipe, the radius and length of the screen have important effect on the sensing characteristics of the probe, and thus the accuracy of the correlation measurement system based on the electrostatic probes. The objective of this paper is to analyze the effect of the length of the probe, the thickness, the length and relative permittivity of the dielectric pipe, the radius of the screen on the non-dimensional calibration coefficient and transit time of the correlation velocity measurement formula, and to provide the theoretical basis for the optimization design of the electrostatic probe. Presently, the reports on the application of the electrostatic probe is limited to dilute phase gas-solid flow such as the pulverized coal transport system with the particle concentration generally less than 0.5% in coal-fired power station. In recent years, dense phase pneumatic conveying technique, which has the advantages of lower energy consumption, lower wear and abrasion of pneumatic conveyor and higher solid to gas ratio by comparison with dilute phase conveying, has had widespread applications such as coal powder injection systems of blast furnaces in steel plants (Yang W. S et al. 2000) and pulverized coal transport of entrained flow gasifiers in chemical plants (Gong X and Guo X L. 2006). Similarly, it is essential for metering particle velocity of these dense phase pneumatic conveyors to minimize energy consumption and to control the transport of solid particles. Therefore a second objective of the paper is to evaluate a correlation velocity measurement system based on electrostatic probes on a bench-scale dense phase pneumatic conveying system under high pressure.

2. MEASUREMENT PRINCIPLE OF THE ELECTROSTATIC PROBE

During the pneumatic transport process of powder particles, the particles carry charges due to frictional contact charging between particles and the pipe wall and between particles and the airflow. The process of particle charging is complicated, the sign and amounts of charge carried by moving particles are dependent on many factors including particle size and shape, work functions, impurities, surface roughness, relative humidity, volume resistivity, permittivity, the contact energy between particles and pipe wall, the material and configuration of the pneumatic conveyor, and the transport conditions of particles in the pipeline, and hence the charging mechanism of pneumatically transported particles is not completely understood now. An electrostatic probe can be used to detect the charges carried by the moving particles. A number of different geometrical shapes of probes have been used as sensing elements for the flow measurement of solid particles in pipelines. Figure 1 shows some examples of the non-intrusive sensing arrangement. Among them, the ring shape probe has advantages of high sensitivity and can obtain the overall information of charged particles flow in its sensitive zone. Considering the precise layout and installation of the probe, and the operation safety of the dense phase pneumatic conveying system under high pressure, an outer electrostatic probe was adopted. The electrostatic probe is enclosed in a measuring head illustrated in Figure 2. The probe is the main and most important part of the measuring head, whose construction is also described in details in the former works by Gajewski J.B. (Gajewski J.B. 1996). As shown in Figure 2, the length of the probe is denoted by W_e, the radius of the metal screen by R_3, the inner radius of the pipeline by R1 and the outer radius by R_2, the length of the dielectric pipe by l and its relative permittivity by ε. When the charged particles are distributed within the pipelines, the charge and potential are induced on the probe because of electrostatic induction. The total amount of the induced charge q on it can be calculated by the following equation due to the axis-symmetry of the measuring head (Xu C L 2006, Xu C. L. and Wang S. M. 2008)

$$q(t) = \iint \sigma(z+vt, r) \cdot s(z,r) dz dr \qquad (1)$$

where $\sigma(z+vt, r)$, a random function due to the spatial and temporal distribution of the particles in the sensing zone, is the space charge distribution of the particles at (z, r) in the sensing zone, and accordingly is also termed 'electrostatic flow noise'; $s(z, r)$ is the spatial sensitivity distribution function, which is defined as the induced charge on the probe when a unity point charge is differently positioned at (z, r) in the sensing zone of the probe.

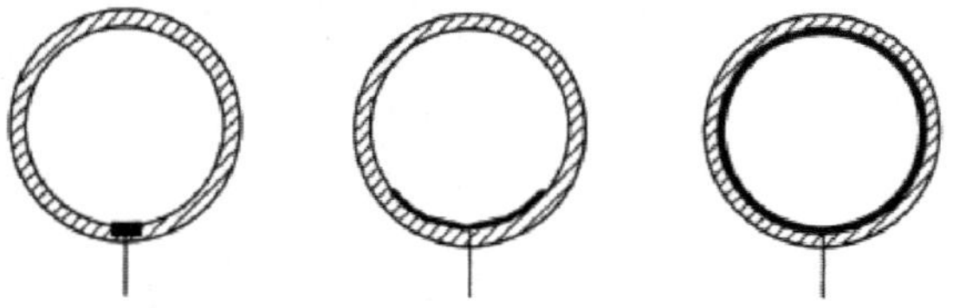

Fig. 1 Simplified diagram of non-intrusive electrostatic probes

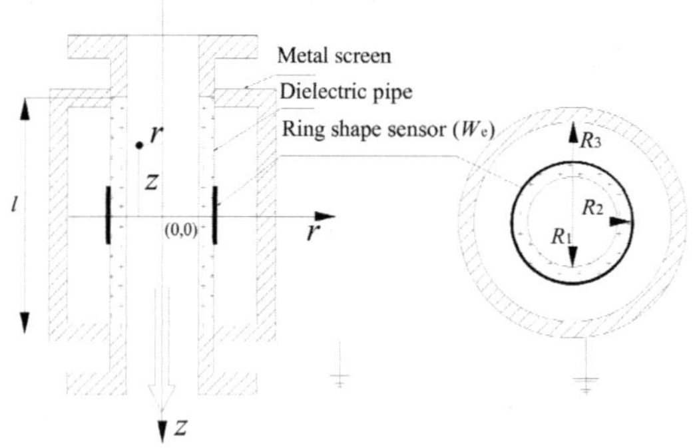

Fig. 2 Measuring head cross-sections of the ring-shape electrostatic probe

The electrostatic probe is connected to a preamplifier to amplify the induced signal, and the probe together with the preamplifier can be simplified as an equivalent circuit, schematically shown in Figure 3. According to Kirchhoff's current law, the following expression applies

$$\frac{dq(t)}{dt} = C\frac{du_i(t)}{dt} + \frac{u_i(t)}{R} \qquad (2)$$

where $u_i(t)$ is the input voltage of the preamplifier, $R=(R_e \cdot R_i)/(R_e+R_i)$, $C=C_e+C_i$, C_e and R_e are the equivalent capacitance and insulation resistance of the probe, and C_i and R_i are the equivalent input capacitance and input resistance of the preamplifier, respectively. Applying the Laplace transform to Equation (2) gives the following equation

$$U_i(s) \Big/ Q(s) = \frac{sR}{1+sRC} \qquad (3)$$

where $U_i(s)$ is the Laplace transform of $u_i(t)$ and $Q(s)$ is the Laplace transform of $q(t)$.

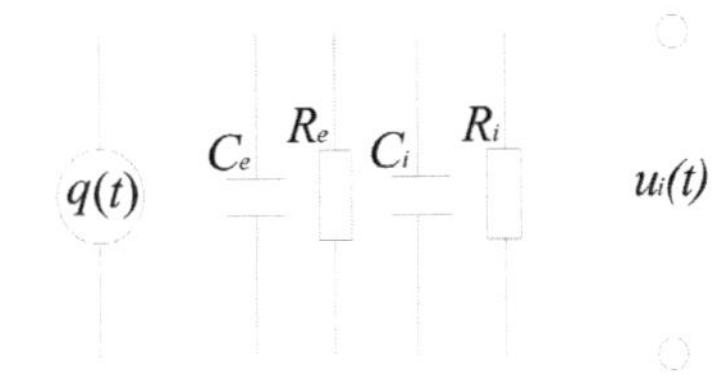

Fig. 3 Equivalent circuit of the electrostatic probe

If the condition $|sRC| << 1$ is fulfilled, and the initial induced charge on the electrostatic probe equals zero, the output signal of the probe can be expressed as

$$u(t) = R\frac{dq(t)}{dt} = R\iint \frac{d\sigma(z+vt,r)}{dt}\cdot s(z,r)dzdr \quad (4)$$

From equation (4), it can be concluded that the output signal of the probe is the change of the induced charge/ the space charge distribution $\sigma(z+vt, r)$ with time, and is also a complicated random signal due to the random properties of electrostatic flow noise. In fact, the signal can not be predicted theoretically due to the unknown particle distribution, the non-uniform particle velocity profile and flow inhomogeneities in pneumatic transport pipelines. Therefore the above theoretical model of the probe only described its operation theory, and can not be directly compared with the experimental data obtained by the probe. Figure 4 shows the output signals of the probe when the particle with a unit charge moves along the axial direction at the different radial coordinates r. It can be seen that the signal nearby the pipe wall has larger amplitude and shorter duration time than the signal on the central axis of the probe.

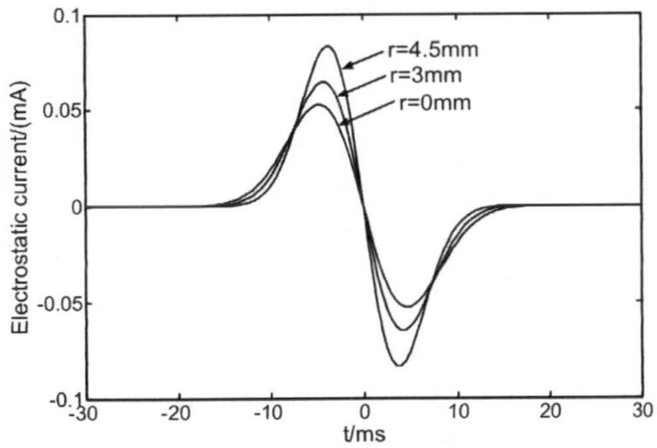

Fig. 4 Theoretical output current of the electrostatic probe from a unit charge traveling through the probe with v=1 m/s

3. CORRELATION VELOCITY MEASUREMENT PRINCIPLE AND MEASUREMENT ERROR ANALYSIS

3.1 Correlation velocity measurement principle

A block diagram of the flowmeter using electrostatic probes is shown in Fig. 5. In the velocity measurement, the actually measured parameter is the flow transit time τ_m determined by cross correlating the two signals derived from the upstream and downstream probes. The correlation velocity v_c, is then calculated from τ_m and the known probe separation L

$$v_c = L/\tau_m \quad (5)$$

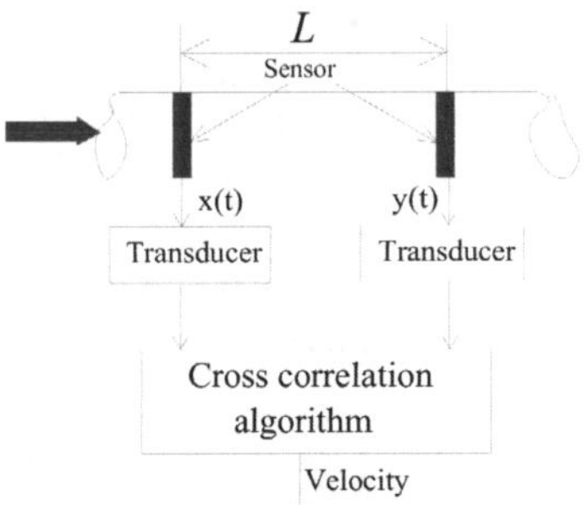

Fig. 5 Block diagram of a cross correlation flowmeter using electrostatic probes

The mean velocity of particles, $v_{m,}$ in pneumatic pipelines can be calculated by

$$v_m = \frac{\iint v(x,y)c(x,y)dxdy}{\iint c(x,y)dxdy} \qquad (6)$$

where $v(x, y)$ and $c(x, y)$ are the particle velocity and concentration profile, respectively.

If the unknown particle distribution, the non-uniform particle velocity profile, non-linear sensing characteristics of the electrostatic probe and flow inhomogeneities in pneumatic transport pipeline are taken into account, a discrepancy exists between the correlation velocity v_c, and the mean particle velocity $v_{m,}$. The relationship between v_c, and v_m can be expressed as

$$v_m = k\,L/\tau_m \qquad (7)$$

where k denotes dimensionless calibration coefficient, whose instability, to a large extent, influences the measurement accuracy of v_m. The correlation properties between the upstream and downstream probes and the systematic error ($\Delta L/L$) should be considered when determining the physical probe spacing L. The reasons are that the larger probe space results in the decrease of cross correlation coefficient of the output signal of the two probes, and contrarily the larger systematic error ($\Delta L/L$) results from the electrical field interaction between the probes. Beck suggested the range of the probe spacing L (Beck M S and Plaskowski A.1987):

$$L = (1/2{\sim}2)D \qquad (8)$$

where D is the pipeline diameter. The systematic error ($\Delta L/L$) could be corrected by calibrating the sensing head using a small-scale insulated hell conveyor with a known and uniform speed (Keech R P. 1982). The geometric parameters of the electrostatic probe affect k and τ_m in complex ways, and hence the measurement error of the mean particle velocity. These will be discussed in the following sections.

3.2 The effect of the geometric parameters of the electrostatic probe on *k*

The effects of the length of the built-in electrostatic probe and particle velocity profile on k were analyzed in (Yan Y. et al. 1995). Actually the geometric parameters of the outer electrostatic probe including the length of the probe, the thickness, and the length and relative permittivity of the dielectric pipe, the radius of the screen have influence on its sensitivity distribution, and hence the accuracy of correlation velocity measurement. The correlation velocity depends upon the total contributions of all the stream lines through the sensing system. Each stream line velocity $v(x,y)$ will be weighted by the spatial sensitivity, correlation coefficient, and concentration of solid particle at that stream line (x, y) (Yan Y. et al. 1995), that is

$$v_c = \frac{\iint v(x,y)s(x,y)\rho(x,y)c(x,y)dxdy}{\iint s(x,y)\rho(x,y)c(x,y)dxdy} \qquad (9)$$

where $s(x,y)$ is the spatial sensitivity distribution related to the geometric parameters of the electrostatic probe; $\rho(x,y)$ is the stream line correlation coefficient, which relates not only to the geometry and dimensions of the sensing head (spatial filtering properties and probe spacing), but also to the flow characteristics and to the operating principles of the signal processing electronics and cross correlation algorithm. For simplification, on the assumption that $\rho(x,y)$ is constant, it can be concluded from equation (9) that v_c can truly represent v_m if sensitivity distribution $s(x, y)$ is uniform, that is, $k = 1$. Figure 6 shows the sensitivity distribution of the electrostatic probe with the lengths being 3 mm, 5 mm and 7 mm, respectively. It can be seen that the length of the probe is one of important factors influencing the spatial sensitivity distribution. The more homogeneous sensitivity distribution attributes to the longer probe. The higher sensitivity exists nearby the pipe wall, which illustrates the probe is more sensitive to the particle adjacent to the pipe wall. Additionally, other geometric parameters of the probe also have important effect on the sensitivity distribution (Xu C. L. and Wang S. M. 2007, Xu C. L. and Wang S. M. 2007). Therefore the non-uniformity of the spatial sensitivity distribution $s(x,y)$ can be reduced by optimizing the geometry of the electrostatic probe.

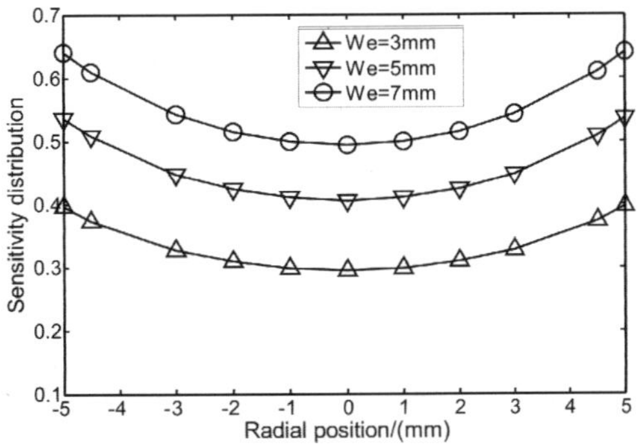

Fig. 6 Sensitivity distribution of the electrostatic probe with different lengths (R_2=10 mm, R_3=18 mm, l=15 mm, ε =3.75)

As far as the gas–solid two-phase flows are concerned, the interaction at the gas–solid phase interface leads to larger variations in particle velocity distribution at different locations of transport pipes. Even in the steady flow condition, the particle concentration and velocity distribution are non-uniform. As a consequence, the gas–solid two-phase flow process is very complicated and, presently, it is impossible to obtain a general solution of the distribution profiles. To simplify the analysis of the effect of the geometric parameters of the probe and particle distribution profile on k, it is assumed that the particle velocity profile is axisymmetrical and obeys a power-law model (Yan Y. et al. 1995)

$$v(x,y) = v(r) = v_{max}(1-\frac{r}{R_1})^{1/n} \qquad (10)$$

where $v_{\max}$ is the particle velocity at the axis of the transport pipe, n is the exponent of the power model, indicating the profile shape of particle velocity distribution over the cross-section of a pneumatic transport pipeline, and this is related to the roughness of the pipe and the Reynolds number.

On the assumption that particle concentration $C(x, y)$ is constant over the cross-section of the pipe, the mean velocity of particles v_m can then be calculated by

$$v_m = \frac{\iint C(x,y)v(x,y)dxdy}{\iint C(x,y)dxdy} = \frac{2n^2}{(n+1)(2n+1)}v_{\max} \qquad (11)$$

Similarly, correlation velocity can be expressed as

$$v_c = \frac{\iint C(x,y)v(x,y)\rho(x,y)S(x,y)dxdy}{\iint C(x,y)\rho(x,y)S(x,y)dxdy} = \frac{\int_0^{R_1} rv(r)S(r)dr}{\int_0^{R_1} rS(r)dr} \qquad (12)$$

Combining equations (10), (11) and (12), the dimensionless calibration coefficient k can be expressed as

$$k = \frac{v_m}{v_c} = \frac{\dfrac{2n^2}{(n+1)(2n+1)} v_{\max} \cdot \int_0^{R_1} rS(r)dr}{\int_0^{R_1} r v_{\max} \left(1 - \dfrac{r}{R_1}\right)^{1/n} S(r)dr} \qquad (13)$$

From equation (13), k can be calculated using numerical methods because the spatial sensitivity $S(r)$ is known for a given electrostatic probe. The charge induced on the ring-shaped probe with different geometric sizes from a single particle having a unity charge was modeled mathematically in (Keech R P. 1982, Xu C. L.and Wang S. M. 2007, Xu C. L. and Wang S. M. 2007), and the spatial sensitivity distribution of the probe was analyzed using finite element method. On the above basis, the effect of the geometric parameters of the sensing head on the dimensionless calibration coefficient k will be further investigated. In the numerical calculation, the radius of the dielectric pipe R_1 equals to 5 mm.

3.2.1 The effect of the length of the probe on *k*

Fig. 7 shows the variation of k with particle velocity profiles when the probe lengths W_e are 3 mm, 5 mm, 7 mm and 20 mm, respectively. And the other parameters of the measuring head are kept at R_2=10 mm, l= 15 mm, ε =3.75 and R_3=18 mm. If the velocity profile is unknown, a constant calibration coefficient may have to be used. In this case, the average value of k over the practical range of velocity profiles should be taken. When the value of n varies in the range 4-12, it is estimated that the relative change in k ($\Delta k/k$) is 1.09% for W_e =3 mm, and 0.29% for W_e =20 mm. It can be concluded that the dimensionless calibration coefficient is not constant for a probe with a given length, and the longer probe contributes to reduce the effect of velocity profiles on k.

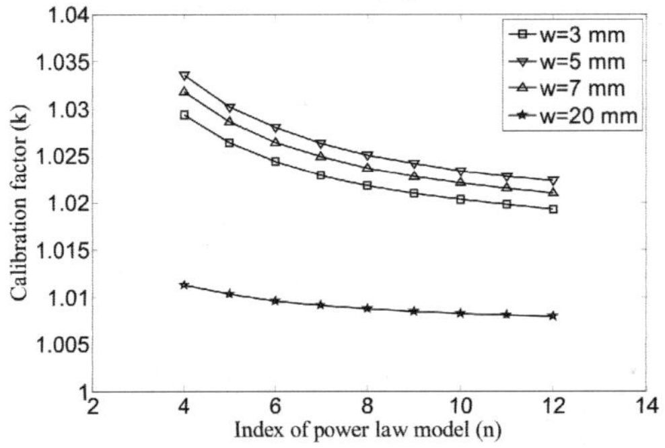

Fig. 7 Variation of k with particle velocity distribution for different probe lengths

3.2.2 The effect of the permittivity of the dielectric pipe on *k*

Fig. 8 shows the variation of k with particle velocity profiles when the relative permittivity of the dielectric pipe ε is 1, 4, 6, 8 and 12, respectively. And the other parameters of the measuring head are kept at W_e =5 mm, R_2=10 mm, R_3=18 mm and l=15 mm. It is estimated that the relative change in k with the variable velocity profile n being in the range 4-12 is not greater than 1.07% for the relative permittivity of the dielectric pipe ε=1~12.

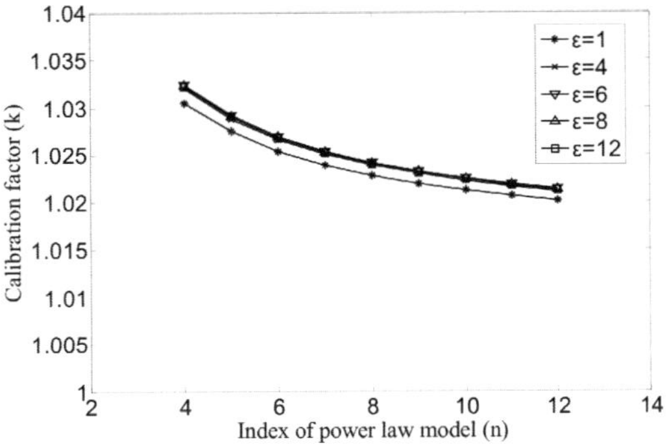

Fig. 8 Variation of *k* with particle velocity distribution for different permittivity of the dielectric pipe

3.2.3 The effect of the thickness of the dielectric pipe on *k*

Fig. 9 shows the variation of *k* with particle velocity profiles when the thicknesses of the dielectric pipe (R_2-R_1) are 3 mm, 5 mm, 8 mm and 10 mm, respectively. And the other parameters of the measuring head are kept at W_e=5 mm, ε =3.75, R_3 =18 mm and l=15 mm. When the value of *n* varies in the range 4-12, it is estimated that the relative change in *k* ($\Delta k/k$) is 1.29% for R_2-R_1=3 mm, and 0.89% for R_2-R_1=10 mm. Therefore it can be concluded that the thicker dielectric pipe contributes to reduce the effect of velocity profiles on *k*.

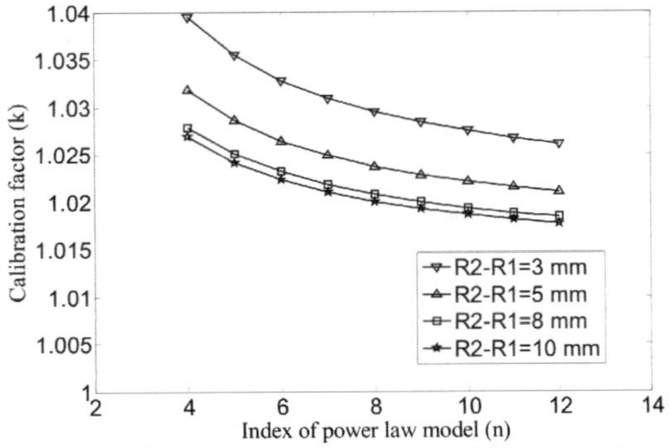

Fig. 9 Variation of *k* with particle velocity distribution for different thicknesses of the dielectric pipe

3.2.4 The effect of the length of the dielectric pipe on *k*

Fig. 10 shows the variation of *k* with particle velocity profiles when the lengths of the dielectric pipe *l* are 10 mm, 15 mm, 20 mm and 25 mm, respectively. And the other parameters of the measuring head are kept at W_e=5 mm, R_2=10 mm, ε =3.75 and R_3=18 mm. When the value of *n* varies in the range 4-12, it is estimated that the relative change in *k* ($\Delta k/k$) is 1.67% for l=10 mm, and 0.67% for l=25 mm. This implies that the longer dielectric pipe contributes to reduce the effect of particle velocity profiles on *k*.

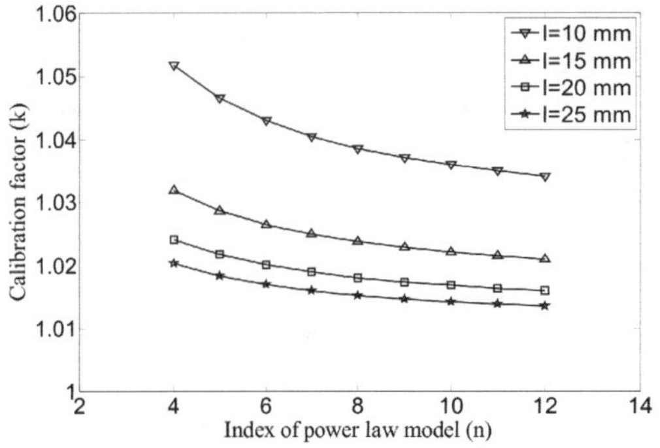

Fig.10 Variation of *k* with particle velocity distribution for different length of the dielectric pipe

3.2.5 The effect of the radius of the screen on *k*

Fig.11 shows the variation of *k* with particle velocity profiles when the radiuses of the screen are 14 mm, 18 mm and 22 mm, respectively. And the other parameters of the measuring head are kept at W_e=5 mm ,R_2=10 mm, ε =3.75 and *l*=15 mm. It is estimated that the relative change in *k* with the variable velocity profile *n* being in the range 4-12 is almost constant (1.07%), which illustrates that the radius of the screen has no effect on the dimensionless calibration coefficient *k*. But it should be noted that the variation of particle velocity profiles causes the relative change in *k*.

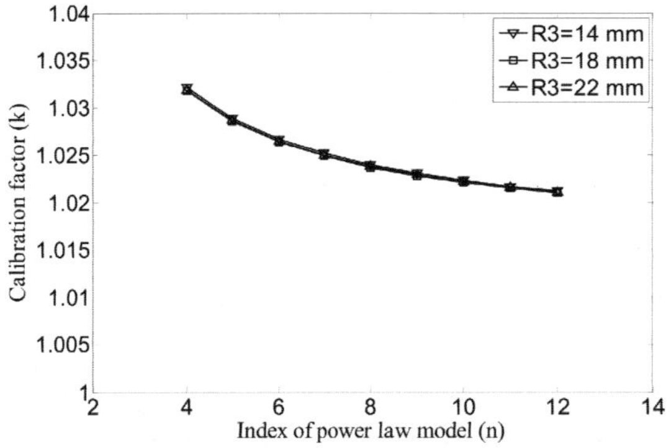

Fig. 11 Variation of *k* with particle velocity distribution for different radiuses of the screen

3.3 The effect of the geometric parameters of the electrostatic probe on τ_m

Statistical error exists in estimating the transit time τ_m from equation (7) due to the randomness of the electrostatic flow noise. The standard deviation $\sigma(\tau_m)$ of the transit time measurement is dependent on the output signal bandwidth of the electrostatic probe (*B*), correlation integration time (*T*), and the correlation coefficient (ρ_m) (Yan Y. et al 1995, Beck M S and Plaskowski A.1987)

$$\sigma(\tau_m) = \left[\frac{0.038}{TB^3}\left(\frac{1}{\rho_m^2} - 1\right)\right]^{1/2} \qquad (14)$$

From equation (14), the standard deviation is proportional to $B^{-1.5}$, which means that a little change in the signal bandwidth *B* may cause the larger standard deviation. Therefore the electrostatic probe should be carefully designed to extract the information of the electrostatic flow noise in a larger frequency bandwidth.

The frequency response characteristics of the measurement system are determined by both the spatial filtering effect of the probe and the preamplifier connected to the probe. The frequency response characteristics can be theoretically expressed as (Xu C L 2006, Xu C. L. and Wang S. M. 2008):

$$\left|U_i(f)\right| = 2\pi f R\left(\frac{a\sqrt{\pi}}{v\sqrt{b}}\exp\left(-\frac{(\pi f)^2}{bv^2}\right)+\frac{c\sqrt{\pi}}{v\sqrt{d}}\exp\left(-\frac{(\pi f)^2}{dv^2}\right)\right) \quad (15)$$

where a, b, c, d are the constants related to the geometric parameters of the measuring head and the radial position r of the point charge in the sensing zone, v is particle velocity, and f is frequency. From equation (15), the frequency response characteristics of the measurement system are dependent on the geometry of the measuring head and particle velocity. From the literatures (Xu C. L. and Wang S. M. 2007, Xu C. L. and Wang S. M. 2007, Xu C. L., Zhao Y. and Wang S. 2006), compared with the relative permittivity of the dielectric pipe ε and the inner radius of the screen R_3, the length and thickness of the dielectric pipe and the length of the probe have important effect on the sensitivity distribution of the electrostatic probe, and hence the constants a, b, c, d and the bandwidth of the measurement system. In the paper, on the basis of the probe's mathematical and finite element model established in (Xu C. L. and Wang S. M. 2007, Xu C. L. and Wang S. M. 2007), the effects of these factors (l, R_2-R_1, W_e) on the bandwidth of the probe are investigated numerically using finite element method. In the numerical simulations, other parameters of the measuring head are kept at ε=3.75, R_1=5 mm and R_3=18 mm, and particle velocity v is 10 m/s.

3.3.1 The effect of the length of the electrostatic probe on the bandwidth

Fig.12 shows the effect of the length of the electrostatic probe on the bandwidth (B) of the measurement system. The lengths W_e are 3 mm, 5 mm and 10 mm, respectively. And the other parameters of the measuring head are kept at R_2=10 mm and l=15 mm. From Figure 12, it can be seen that the bandwidth decreases with the increase of the length of the probe nearby pipe wall (r=4.5 mm), whereas there is almost no change on the central axis of the probe (r=0 mm). This means that the shorter probe attributes to the reduction of the standard deviation $\sigma(\tau_m)$ of the transit time measurement.

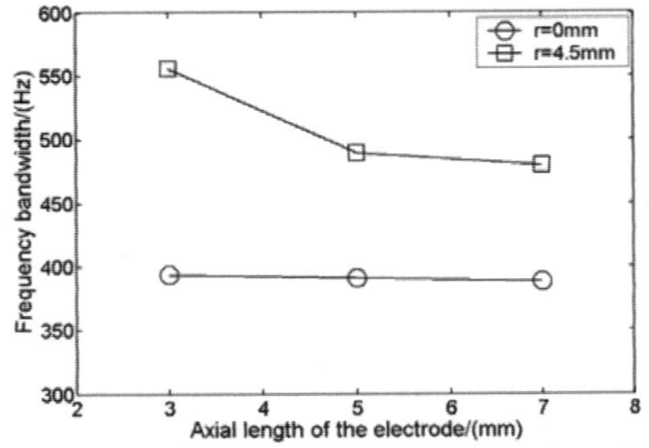

Fig.12 Variation of the bandwidth with the length of the probe

3.3.2 The effect of the thickness of dielectric pipe on the bandwidth

Fig.13 shows the effect of the thickness of the dielectric pipe on the bandwidth (B) of the measurement system. The thicknesses (R_2-R_1) are 3 mm, 5 mm, 8 mm and 10 mm, respectively. And the other parameters are kept at W_e=5 mm and l=15 mm. It can be seen that the bandwidths decrease with the increase of the thickness and then approach constants nearby pipe wall, and on the central axis of the probe.

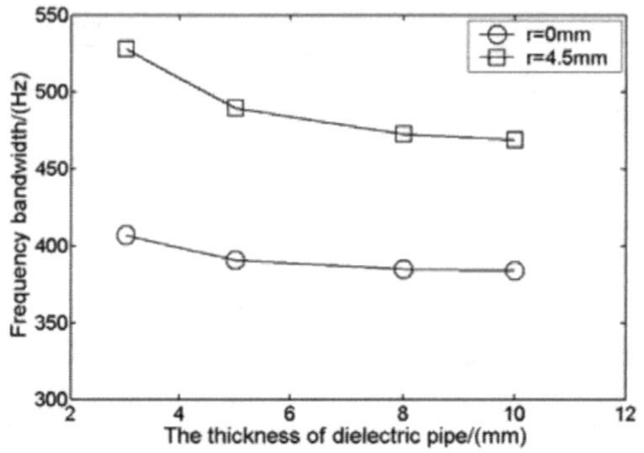

Fig. 13 Variation of the bandwidth with the thickness of the dielectric pipe

3.3.3 The effect of the length of the dielectric pipe on the bandwidth

Fig.14 shows the effect of the length of the dielectric pipe on the bandwidth (B) of the measurement system. The lengths (l) are 10 mm, 15 mm and 20 mm, respectively. And the other parameters of the measuring head are kept at W_e=5 mm and R_2=10 mm. From Figure 14, it can be seen that the bandwidths rapidly decreases with the increase of the length of the dielectric pipe nearby pipe wall (r=4.5 mm) and on the central axis of the probe (r=0 mm), which means that the longer dielectric pipe results in the increase of the standard deviation $\sigma(\tau_m)$ of the transit time measurement.

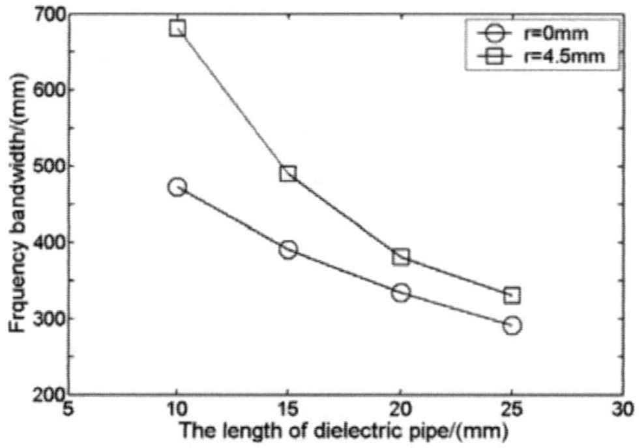

Fig. 14 Variation of the bandwidth with the length of the dielectric pipe

From the results of the sections 3.2 and 3.3, we know that the shorter electrostatic probe not only increases the signal bandwidth (B) of the probe, and improves the statistical error ($\sigma(\tau_m)$) of the transit time, but also reduces the uncertainty of the probe space. But it should be noted that the shorter electrostatic probe leads to the non-uniformity of the sensitivity distribution, which largely influences the stability of the dimensionless calibration coefficient k. The thicker dielectric pipe is useful for the uniformity of the sensitivity distribution, and reduces the effect of particle velocity profiles on the dimensionless calibration coefficient k, but the thicker pipe also reduces the signal bandwidth of the probe, and hence increases the statistical error of the transit time. The longer dielectric pipe results in the similar change in the measurement system to the thicker pipe. Therefore the non-uniformity of the sensitivity distribution and the signal bandwidth of the probe should be synthetically taken into consideration when optimizing the geometry of the electrostatic probe used in a correlation velocity measurement system. Additionally, particle velocity is also an important factor influencing the signal bandwidth of the probe, and the higher velocity produces the wider bandwidth (Xu C. L., Zhao Y. and Wang S. 2006). Hence the statistical error of the transit time will be reduced at the high velocity.

4. EXPERIMENTAL RESULTS AND DISCUSSION

4.1 Experimental rig

A planar view of the main apparatus layout is shown in Fig. 15. The constituent components of the apparatus and their functions are as follows:

- The volume capacity of the hoppers (3) is 0.648 m^3, and by switching the valves in pipelines, the hoppers two can be used to feed and receive powder particles respectively.
- Transport pipeline section consists of a 53.4 m of Φ16x3 mm bore stainless steel pipe.
- Nitrogen cylinder group (8) supplies the transport gas for the dense phase conveyor, and nitrogen exhausted from the pipeline section passes first through the receiving hopper, where bug dust collector is used to disengage the solid particles from the nitrogen flow. The clean nitrogen then passes through a rotor meter (13) to facilitate the measurement of gas flowrate before passing through the electrical control valve (1) to atmosphere. The operating pressure of the receiving hopper can be adjusted by the control valve.
- A computerized data acquisition system (not shown) allowed the signals from all of the probes to be logged simultaneously and saved for further processing. The probes included a weighing cell (2) for the measurement of particle mass in a time interval, 3 rotor meters (10-12) for the measurement of transport nitrogen flowrate, 3 pressure transducers and two electrostatic probes (9).

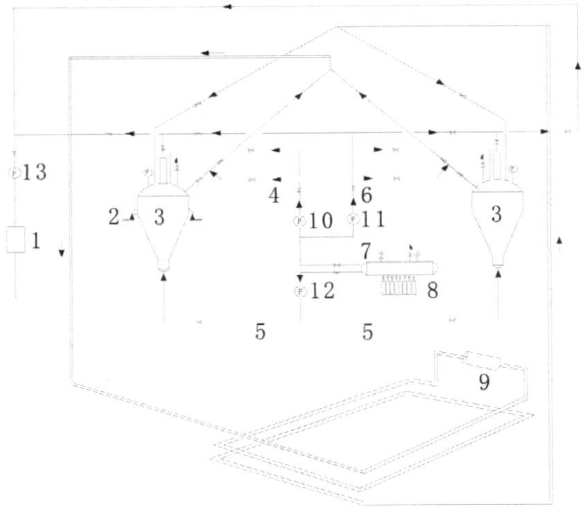

Fig.15 Schematic diagram of Dense-Phase Pneumatic Conveying System of Pulverized Coal
1-electrical control valve 2-weighing cell 3-hopper 4-supplement gas 5-fluidizing gas 6-keeping pressure gas 7- gas buffer tank 8-nitrogen cylinder group 9-electrostatic test pipeline 10~13-rotor meters

The high-pressure nitrogen first passes through the gas buffer tank (7) and then is divided into two ways. The one forms fluidizing gas (5), which first fluidizes the powder particles in the feeding hopper, then carries the fluidized particles passing through the transport pipeline, and finally enters the receiving hopper. The other is used to keep the pressure of the feeding hopper (termed keeping pressure gas (6)) and to change the solid gas mass ratio in the transport pipeline (termed supplement gas (4)). The supplement gas is also called secondary gas.

The correlation velocity measurement system consists of the measuring head, preamplifiers and a computer-based data acquisition system. The output signals of the probes are amplified by the preamplifiers, then are sampled, saved and analyzed by the computer. The sampling rate was 1330 Hz. The geometric parameters of the measuring head are kept at R_1=5 mm, R_2=10 mm, l=100 mm, ε_i=3.75

and R_3=18 mm. The length of the electrostatic probe is 4 mm, and the probe space is 40 mm. The measuring head was installed at the outlet of the horizontal pipeline section with the distance being 15 m from the feeding hopper. The conveyed material is pulverized bituminous coal, whose characteristic parameters are listed in table 1. Fig. 16 shows the coal particle size distribution histogram measured by Centrifugal Particle Size Analyzer (SA-CP3, SHIMADZU). During the experiments, the conveying gas is N_2 with a maximum pressure of up to 4.8 Mpa, and particle concentration is in a range of 5.2-14.1% m^3/m^3. The particle concentration β_s in pipelines can be estimated using the following Equation

$$\beta_s \approx \frac{Q_s}{Q_a + Q_s} \times 100\% \qquad (16)$$

where Q_s is the volumetric flowrate of the pulverized coal with the relation $Q_s = M_s/\rho_s$. ρ_s denotes the particle density and M_s is the coal mass flowrate, which is obtained by weighing the amount of the solid particle collected from the receiving feeder in a suitable time interval.

Table 1 Characteristic parameters of bituminous coal particles

Material	Mean diameter d_s/(μm)	Density ρ_s/(Kg/m^3)	Shape factor Φ_s
Coal	35.6	1350	0.63

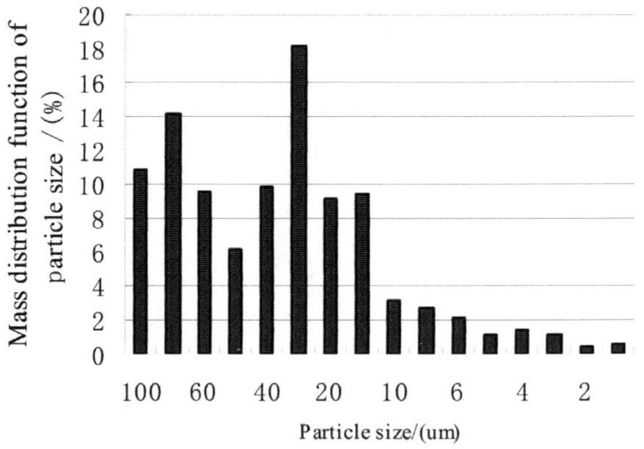

Fig.16 Coal particle size distribution histogram

4.2 Results and Discussion

Due to the lack of the calibration instrumentation for gas-solid flow measurement, the superficial gas velocity is only a traceable reference with which the measured particles mean velocity is compared. The superficial gas velocity v_a can be calculated by

$$v_a = \frac{Q_a}{A} \qquad (17)$$

where Q_a is the volumetric flowrate of transport nitrogen. A is the cross section area of transport pipeline.

The paper adopted the generalized cross correlation algorithm to estimate the transit time τ_m, which can effectively restrain the effect of the noise on the estimation, and improve the accuracy of the transit time measurement. Throughout the entire velocity and concentration ranges of the experimental rig the correlation algorithm gave clearly defined correlogram peaks. A typical correlogram together with its waveforms of the raw signals from the electrostatic probes were recorded during the tests and are shown in Fig. 17. Under all flow conditions, the correlation coefficient representing the similarity between upstream and downstream signals is 0.20- 0.40. This suggests that the dense phase gas-solid flow exhibits unsteady flow regime, particle velocity and concentration profiles over the pipe's cross sectional area.

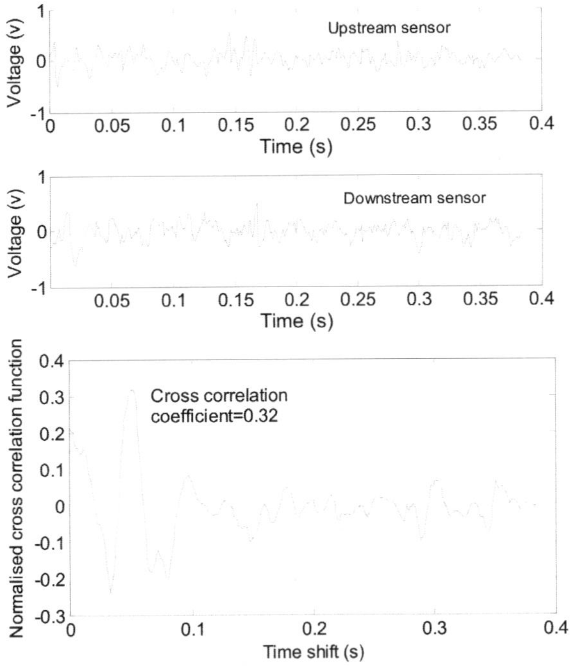

Fig. 17 A typical correlogram and its raw signals at correlation velocity 0.8 m/s

Fig. 18 shows the comparison between the correlation velocity v_c and superficial gas velocity v_a. The correlation velocity is an average of consecutive ten measurements. Over a velocity range of 0.5 - 7 m/s for a particle concentration range 0.052-0.141 m^3/m^3, the measured velocities are slightly higher than the superficial gas velocities. Several reasons for the phenomenon may be found. Firstly, the higher particle concentration over the cross section of the pneumatic pipeline results in the reduction of the area occupied by the transport gas, and the particles are prone to accumulate within the horizontal pipeline due to their intrinsically adhesive properties and own gravities. As a consequence, if the total particle concentration (β) is considered, the true gas velocities should be estimated using the following Equation

$$v_a = \frac{Q_a}{A(1-\beta)} \qquad (18)$$

From Equation (18), it can be concluded that the true gas velocities are higher than the calculated superficial ones because of the reduced effective cross sectional area of the gas flow. Particle size is an important factor influencing the slip between the particles and the gas flow, and Klinzing's studies showed that the slip velocity is almost proportional to the particle size (Klinzing, 1987). In our experiments, the mean size of pneumatically conveyed particles is 35.6 μm, and thus they have better following characteristics. Therefore it is possible that particle velocities are higher than the superficial ones. Secondly, the velocity distribution in the lower section of the horizontal pneumatic pipeline is relatively flat due to the higher particle concentration, but the particle concentration and velocity distribution are exactly reverse in the upper section of the pipeline. The electrostatic probe has non-linear sensing mechanism that is more sensitive to the particle closest to the pipe wall. This means that the particles nearer to the probe are giving a higher weighting factor on the final average velocity reading. Hence the upper particle velocity attributes to the higher measured correlation velocity. Finally the particle concentration profile of the dense phase pneumatic pipeline fluctuates continuously due to the instability of the dense phase gas-solid flow (Liang C., Zhao C. S. and Chen X. P. 2007), which leads to a large change in particle velocity and hence the higher data scattering of the correlation velocity measurement. Although the accuracy of the particle measurement velocity using the spatial filtering method can not be determined due to the lack of the calibration instrumentation and equipment, its repeatability error within ±10% is verified by the experiments under the given experimental conditions. It must be stressed that, since the data in Fig.18 combine both the repeatability of the correlation velocity measurement system and the stability of the solids velocity in the pipeline, the actual repeatability of the system is better than 10%.

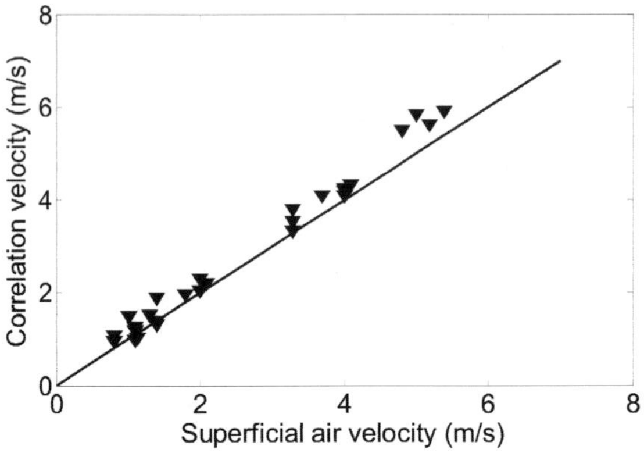

Fig.18 Comparison between the correlation velocity and v_a

5. CONCLUSIONS

The effects of the probe's geometric parameters including the length of the probe, the thickness, the length and the relative permittivity of the dielectric pipe, the radius of the screen on the dimensionless calibration coefficient (k) and the statistical error of the transit time (τ_m) of the correlation velocity measurement system using electrostatic probes were investigated, which provides theoretical basis for the optimization design of the electrostatic probe. Due to the lack of the calibration instrumentation for gas-solid flow measurement, the reference in the experiments is the superficial air velocity with which the measured correlation velocity is compared. Experimental results on a pilot dense phase pneumatic conveying rig of pulverized coal under high pressure indicate that the correlation method is capable of making particles mean velocity measurement with repeatability better than ±10% over a velocity range of 0.5 - 7 m/s for a particle concentration range of 0.052-0.141 m^3/m^3

ACKNOWLEDGEMENTS

The authors wish to express their gratitude to the Special Funds for Major State Basic Research Projects (2004CB17702-04) (China), to Major Program of National Natural Science Foundation of China (50836003) and to Science Foundation of Southeast University (XJ2008304) for funding this research.

REFERENCES

Xu L A. (1988). Cross-correlation Flow Measurement Technique, Tianjin University Press, Tianjin, China, pp. 120-125.

Thorn R, Beck M S and Green R G. (1982). "Non-intrusive methods of velocity measurement in pneumatic conveying," J. Phys. E: Sci. Instrum., 15, pp.1131-1139.

Boeck T. (1989). "Measurement of velocity and mass flow rate of pneumatically conveyed solids by the use of a correlation measuring technique," Proc. Int. Conf. on the Flow of Particulate Solids, Bergen, Germany, April 91-95.

Yan Y., Byrne B., Woodhead S.R., Coulthard J.(1995). "Velocity measurement of pneumatically conveyed solids using electrodynamic probes," Measurement Science and Technology, 6(5), pp. 515-537.

Woodhead S. R, Amadi-Echendu J. E.(1995). "Solid phase velocity measurements utilizing electrostatic probes and cross correlation signal processing," Proceedings of IEEE Instrumentation and Measurement Technology Conference, Waltham, Mass., USA, May 774-777.

Ma J., Yan Y. (2000). "Design and evaluation of electrostatic probes for the measurement of velocity of pneumatically conveyed solids," Flow Measurement and Instrumentation, 11, pp.195-204.

Gajewski J.B. (1996). "Monitoring electrostatic flow noise for mass flow and mean velocity measurement in pneumatic transport," Journal of Electrostatics, 37, pp.261-276.

Yang W. S,Yand Z. C, Sun H. W. (2000). "Practice of PCI system at Lingyuan iron and steel (group) Co.," Iron and Steel, 35(4), pp.10-12. (In Chinese)

Gong X and Guo X L. (2006). "High solids loading pneumatic conveying of pulverized coal," Chinese journal of Chemical Industry and Engineering, 57(3), pp.640-644.

Xu C L. (2006). "Gas-solid two-phase flow parameters and electrostatic measurement," PhD thesis, Southeast University, China, pp.35-57.

Xu C. L. and Wang S. M. (2008). "Velocity measurement of pneumatically conveyed solid particles using an electrostatic probe," Measurement science and technology, 19(2), pp.167-175.

Beck M S and Plaskowski A.(1987). "Cross Correlation Flowmeters-Their Design and Application," AdamHilger, Bristol, pp.123-134.

Keech R P. (1982). "Multi-channel correlation signal processor for velocity measurement ,"Trans. Inst. Meas. Control, 14, pp. 43-52.

Xu C. L., Wang S. M. and Tang G. H.(2007). "Sensing characteristics of electrostatic inductive probe for flow parameters measurement of pneumatically conveyed particles," J. Electrostat, 65, pp.582-592.

Xu C. L. and Wang S. M. (2007). "Finite element analysis of the sensing field of electrodynamic probe," Chinese journal of scientific instrument, 28(7), pp.1311-1318.

Xu C. L., Zhao Y. and Wang S. (2006). "Spatial filtering effect and frequency response characteristics of electrodynamic probe," Journal of southeast university (Natural Science Edition), 36, pp. 556-561. (in Chinese)

Yan Y. and Byrne B.(1997). "Measurement of solids deposition in pneumatic conveying," Powder Technology, 91, pp.131-139.

Klinzing G E. (1987). "Particle size effect on slip velocities with electrostatic forces present," Powder Technol., 51, pp.181-182.

Liang C., Zhao C. S., Chen X. P. (2007). "Flow Characteristics and Shannon Entropy Analysis of Dense-Phase Pneumatic Conveying of Variable Moisture Content under High Pressure," Chemical Engineering and Technology, 30 (7),pp. 926-931.

Journal of Physics: Conference Series **147** (2009) 012005 doi:10.1088/1742-6596/147/1/012005

Application of an Ultrasonic Array Sensor to Air-water Bubbly Flow Measurement

D Ito [1], H Kikura [1], M Aritomi [1], M Mori [2]

[1] Tokyo Institute of Technology, 2-12-1 Ohokayama, Meguro-ku, Tokyo, 152-8550 JAPAN
[2] Tokyo Electric Power Company, Yokohama, JAPAN

ito@2phase.nr.titech.ac.jp

Abstract. The ultrasonic array sensor, which has been successfully applied in the field of medical diagnostics and non-destructive testing, was applied to single-phase and two-phase flow measurement. The applied array sensor has 128 elements on the surface and each of them has a basic frequency of 8 MHz. Flow measurement using the array sensor is very effective, because it can obtain the cross-sectional flow visualization images from outside the pipe without disturbing flows. In this study, the ultrasonic sound pressure distributions were measured by the sound field measuring system, and the ultrasonic beam characteristics and the grouping effects of elements were investigated from the sound pressure field. The velocity profiles of single-phase flows in a rectangular channel were measured with Ultrasonic Velocity Profiler (UVP) by scanning the parallel ultrasonic beam emitted from the array sensor. Furthermore, an ultrasonic array measuring system, which consists of a pulser/receiver, a multiplexer, an A/D converter and a signal processing device, was developed for the improvement of temporal resolution in bubbly flow measurement using the ultrasonic array sensor. Then the bubble distributions in the channel cross-section were confirmed by using the echo signals reflected from bubbles.

1. INTRODUCTION

Bubbly flows, which appear in many industrial plants, are very interesting phenomena in engineering fields. In the previous investigations of bubbly flows, numerous two-phase measuring techniques have been developed and applied to the measurement of flow structure. In bubbly flows, bubble shapes change with time and bubble coalescence and fragmentation occur. Therefore it is neccesary to obtain the instantaneous information of the position, size and velocity of each bubble. Recently, fast X-ray tomography (Hampel et al., 2005; Prasser et al., 2005) and wire-mesh tomography (Prasser et al., 1998; Richter et al., 2002; Prasser et al., 2005) have received considerable attention as useful techniques for bubbly flow measurement. This is because they have high temporal resolution and they can measure cross-sectional distribution. However, these methods have some disadvantages. The fast X-ray method is suitable exclusively for professional users, and the wire-mesh tomography is an intrusive method which has strong influence on flow field. It is difficult to apply these techniques to real industrial systems. So this study focuses on the measurement using pulse ultrasound. Ultrasonic measurement has such advantages as being non-intrusive, applicable to opaque fluids and pipes, low cost, simplicity and ease of use, and it can obtain several major parameters on the measuring line. Murakawa et al. (2007) measured the velocity profiles of air-water bubbly flows in a circular pipe by the ultrasonic Doppler method. They measured velocity profiles of both gas and liquid phases with a multi-wave transducer. Furthermore, Wada et al. (2006) investigated the flow patterns of gas-liquid two-phase flows, which are bubbly flow, slug flow and annular flow, by using the reflected ultrasonic echo signals from the flows.

Cross-sectional measurement using ultrasonic pulses requires a lot of ultrasonic transducers. Thus, installation errors may be larger, and the system becomes complex. An ultrasonic array sensor, which has been applied in many fields, can resolve these problems. In the field of medical diagnostics, the array sensor is used for imaging cross sections within a living body. And the sensor can detect flaws in a sample and recognize their location.

In this study, the ultrasonic array sensor, which can emit pulse ultrasound over the cross-section, is applied to the measurement of single-phase flows and bubbly flows in a rectangular channel. The ultrasonic sound pressure distributions are measured to understand the ultrasonic beam characteristics when the elements are or aren't grouped. The cross-sectional averaged velocity profiles in single-phase flows are measured by the ultrasonic velocity profiling method (Takeda, 1986) in order to confirm the validity of ultrasonic array measurement. Then air-water bubbly flows are measured by the ultrasonic echo technique using the ultrasonic array sensor, pulser/receiver and multiplexer, and the temporal resolution of the ultrasonic array measurement is improved.

2. ULTRASONIC LINEAR ARRAY SENSOR

2.1 Applied Ultrasonic Array Sensor

An ultrasonic linear array sensor (Fig. 1) was developed for ultrasonic array measurement of the flow field. The array sensor has 128 piezoelectric elements which can work individually. The arrangement of these elements is illustrated in Fig. 2. Each element has a size of 1mm × 3mm, so the effective size of this sensor is 128mm × 3mm. The frequency bandwidth is about 5 to 8 MHz. In this study, the frequency of 8 MHz is applied to the measurement because of its higher spatial resolution. In the future, this sensor will be applied to the phased array system for wide area scanning and inclined beam formation.

Fig. 1 Photograph of applied array sensor

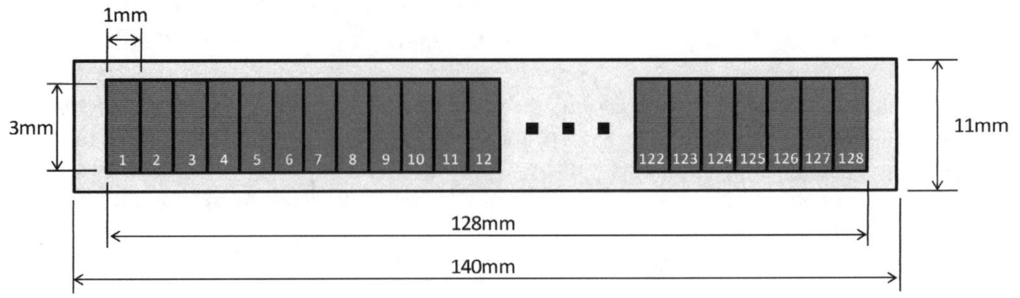

Fig. 2 Arrangement of piezoelectric elements

2.2 Ultrasonic Beam Characteristics

2.2.1 Sound Field Measurement System

The sound pressure field of pulse ultrasound transmitted into water is important in investigating the ultrasonic beam characteristics (e.g. Measurement volume, measuring area and reflected intensity). In this study, sound pressure measurement is performed by the hydrophone technique. Fig. 3 shows the schematic diagram of sound pressure field measuring system. This system consists of pulse transmitting devices (a pulser and an ultrasonic array sensor), receiving devices (an ultrasonic hydrophone and an A/D converter), a three-dimensional stage and a water bath. The pulse signal sent from the pulser is transmitted into water by the ultrasonic sensor. The pulse ultrasound with a frequency of 8 MHz was used for sound pressure field measurement. The transmitted pulse signal has 8 cycles per pulse. The water temperature is maintained at 30 degrees Celsius by the thermostatic water bath. At this temperature, the sound speed in water is 1510 m/s. The ultrasonic hydrophone fixed on the 3-D stage is traversed in the test region and it receives the electrical signals with the piezoelectric element of the hydrophone. This element has a detecting area of 1 × 1 mm^2. This technique can measure the time series data of received signals and is applicable to the transmitting path analysis of ultrasonic pulse in three-dimensional space.

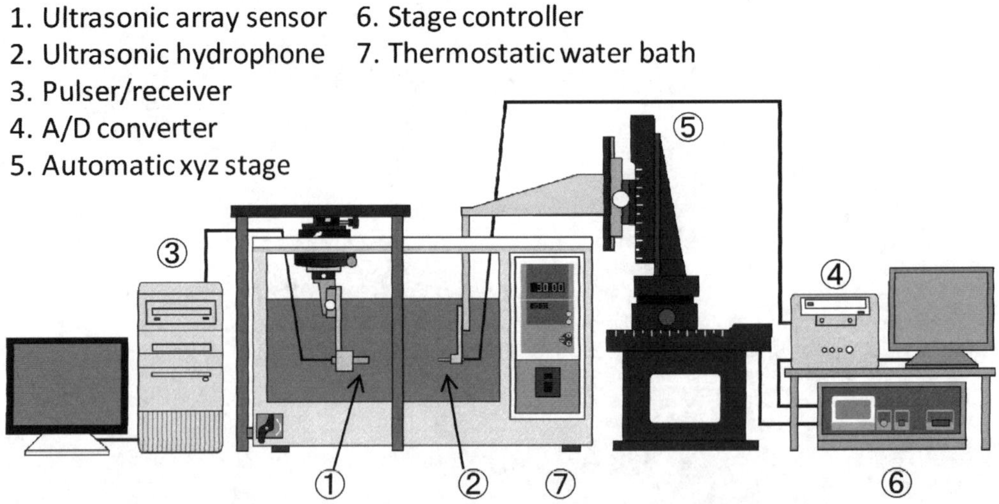

Fig. 3 Schematic diagram of ultrasonic sound pressure field measurement system

Journal of Physics: Conference Series **147** (2009) 012005 doi:10.1088/1742-6596/147/1/012005

2.2.2 Ultrasonic Sound Pressure Distributions

Fig. 4 and Fig. 5 show distributions of ultrasonic sound pressure, as measured with the ultrasonic array sensor. Fig. 4 shows the result of measurement taken from every piezoelectric element on the sensor. Fig. 5 shows the result when every 3 contiguous elements are taken as a unit of measurement. These results revealed a grouping effect of piezoelectric elements taken as a unit of measurement. In these figures, the *x*-axis represents the direction of ultrasonic emission, and the *y*- and *z*-axes show the directions perpendicular to the x-axis. The figure (a) is the sound pressure field on x-y plane at $z = 0$. The figures (b) and (c) are the cross-sectional distributions of ultrasonic beams at $x = 20$ and 40 mm,

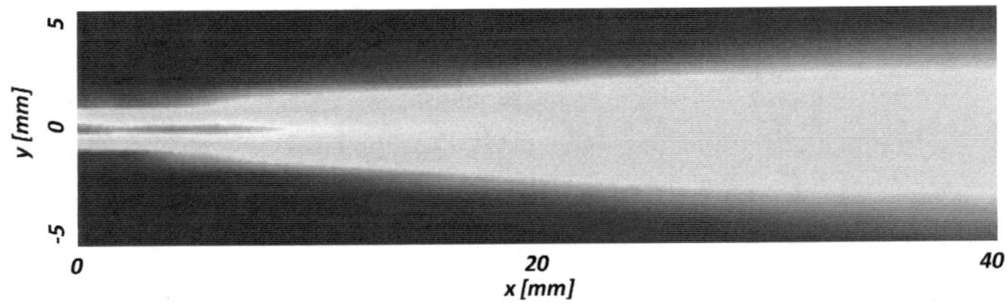

(a) *x-y* plane

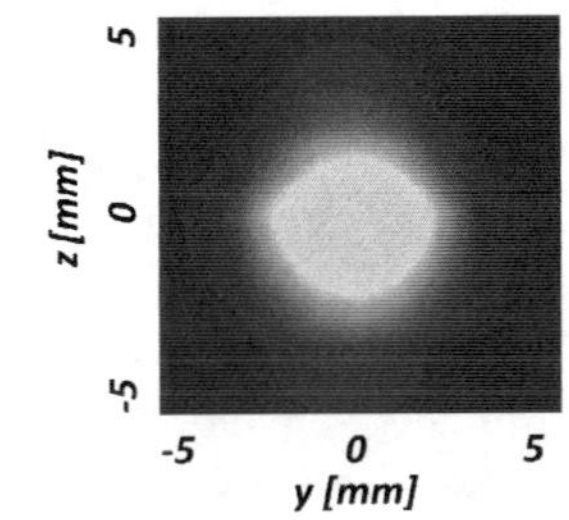

(b) *y-z* plane at *x*=20 mm

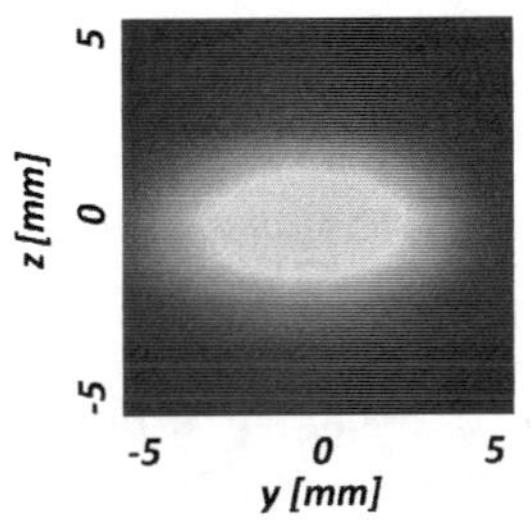

(c) *y-z* plane at *x*=40 mm

Fig. 4 Sound pressure distribution measured with one piezoelectric element

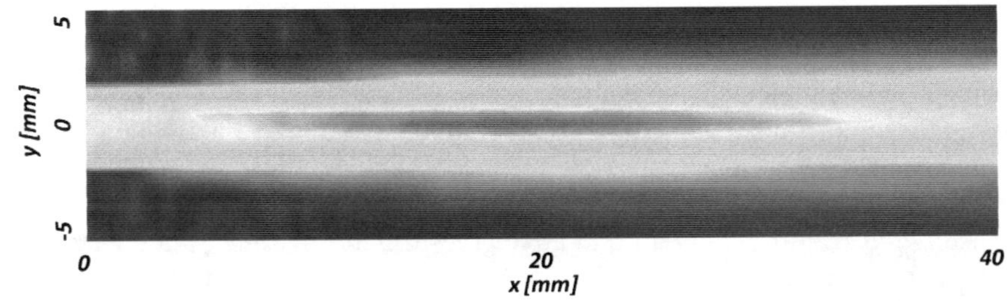

(a) *x-y* plane

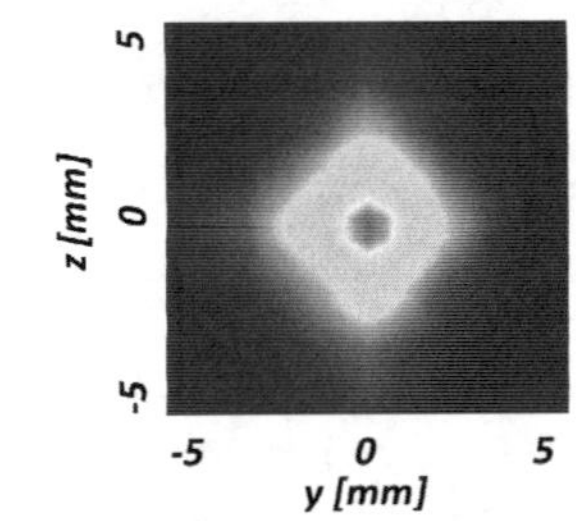

(b) *y-z* plane at *x*=20 mm

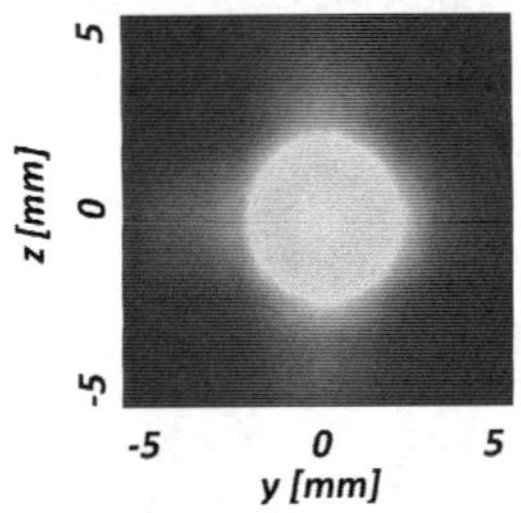

(c) *y-z* plane at *x*=40 mm

Fig. 5 Sound pressure distribution measured with a group of 3 piezoelectric elements

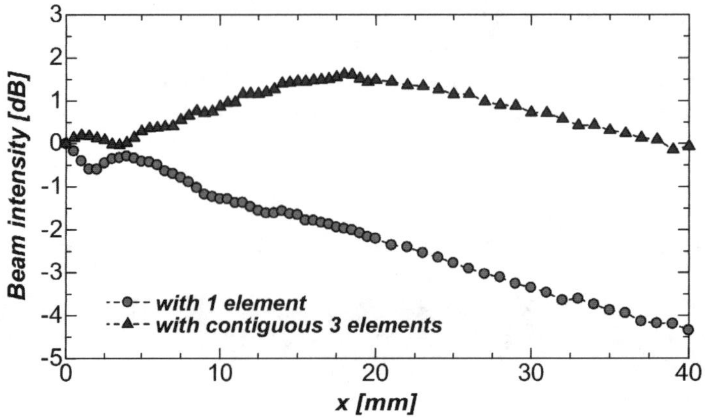

Fig. 6 Beam intensity on the beam center line

respectively. As shown in Fig. 4, when x is less than 10 mm from the front surface of the array sensor, the sound pressure is higher, and the measuring volume is small. However the decrease in the sound pressure and the spreading of the ultrasonic beam are observed when x is more than 10 mm from the sensor. Thus it is better to measure sound pressure in the near field. In Fig. 5, the sound pressure is larger when x is 10 mm to 30 mm. And in this result, the spreading of the beam is less than the use of every element, so the directionality is good. From the results of sound pressure field measurement, the intensity of transmitted beam was estimated. Fig. 6 shows the transition of beam intensity on the beam center line. In the case of every 3 grouped channels, the variation of beam intensity is within 2 dB. However, when a piezoelectric element is individually used, the beam intensity decreases with increasing distance from the sensor, and it is less than -3 dB when x is more than 25 mm. The beam characteristics of ultrasonic pulse transmitted into water can be clarified by measuring the sound pressure field. Therefore the use of every 3 elements is found to be suitable for flow measurement.

3. FLOW MEASUREMENT USING AN ULTRASONIC ARRAY SENSOR

3.1 Experimental Set-up

The experimental apparatus consists of a water-air circulation system, a test section and a measurement system, as shown in Fig. 7. Working fluids are water and air. Water flows into the test channel through a control valve, an orifice flow meter and a flow straightener (0.5 mm mesh stainless plates) by the pump. After water flows through the test section, it overflows and goes back to the storage tank. Water temperature is kept around 20 degrees Celsius by the cooling water. Air is supplied by a compressor. A pressure of air is controlled by a control valve. Air flow quantity is measured by a laminar flow flowmeter. Air inlet is made of five metal needles set at 100 mm downstream from the flow straightener. The outer diameter of each needle is 2 mm and the diameter of holes drilled in each needle is 1 mm. The tip of the needle is round in order to reduce influence on the flow. All the needles are set in parallel to give the same performances.

The measuring section of the ultrasonic array sensor is made of acrylic glass because it has almost the same acoustic impedance as water. In consequence, it is also useful for image measurement. That is the results measured with the ultrasonic array sensor can be compared with those obtained from other optical measurements. (e.g. Particle image velocimetry, Laser Doppler velocimetry). The total length of the test channel is 1,800 mm and the cross section is rectangular, 20 × 100 mm^2, so the hydraulic equivalent diameter is about 33 mm. To improve the ultrasonic transmission, the thickness of the wall is 1 mm at the measuring section, and the incident angle of the ultrasonic array sensor is 45 degrees. A water box is installed in the measuring section. Water is filled between the array sensor and the outer

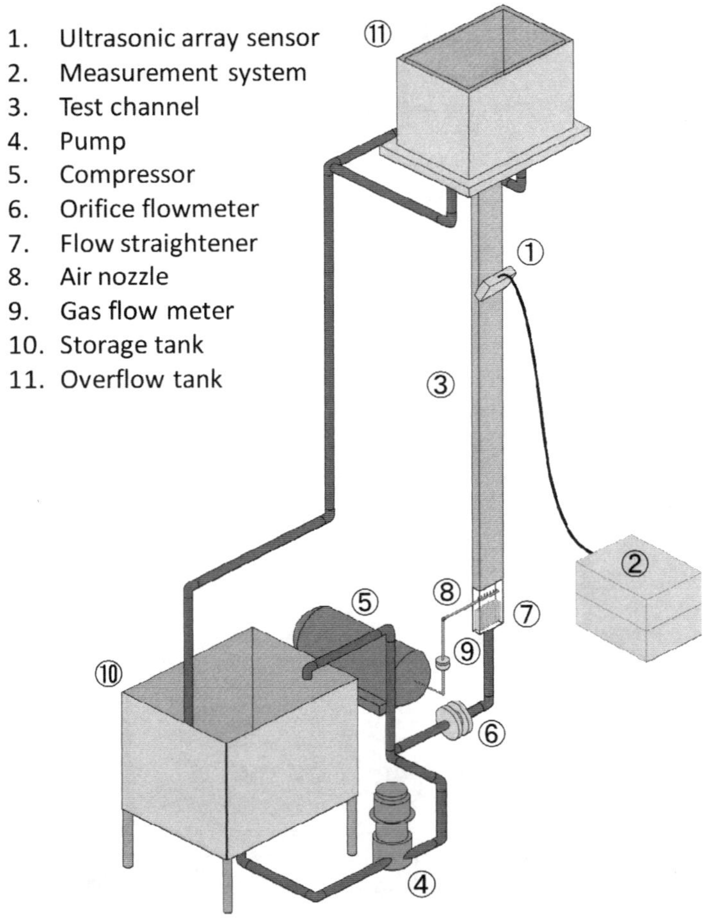

Fig. 7 Schematic diagram of experimental apparatus

wall. For the single phase flow measurement, nylon powders with the diameter of 80 μm and specific gravity of 1.02 are used as ultrasonic reflectors.

3.2 Averaged Velocity Profile Measurement of Single-Phase Flow

Fig. 8 shows the measurement system for velocity profiling using an ultrasonic array sensor. The velocity profiles are measured by the UVP monitor (Met-Flow, UVP-Duo). Two conversion boxes were used to connect the ultrasonic array sensor and UVP. In conversion box1, the ITT-CANON connector is converted to BNC terminal with 128 channels. Here, the 128 elements of the array sensor can operate separately. The conversion box2 was used to put some elements together (Max. 6 channels). Thus, the array sensor is connected with the UVP monitor.

In this study, the cross-sectional average velocity profiles are measured by the array sensor using every piezoelectric element or 3 contiguous elements in order to confirm the grouping effects observed from the sound pressure field measurement. Fig. 9 shows the cross-sectional average velocity profiles measured by the UVP. Fig. 9(a) shows the results when each element worked individually. The velocities on some measuring lines have small irregularity, due to the differences in the sensitivity of each element and the presence of noise. In Fig. 9(b), the 3 contiguous elements are grouped (element No. 1-2-3 is CH.1, element No. 4-5-6 is CH.2...), and the averaged velocity profile can reveal the flow field in the cross-section.

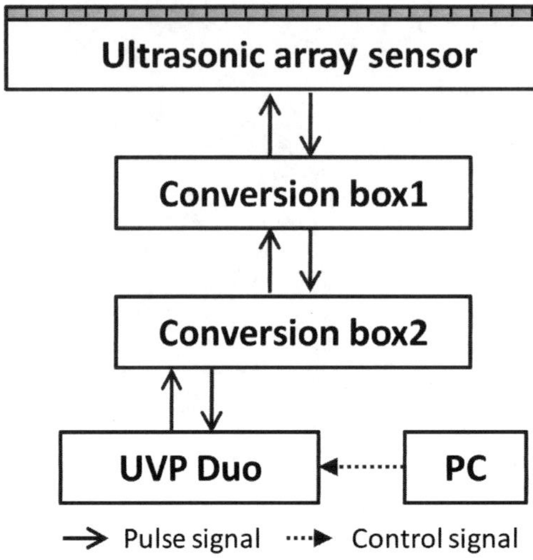

Fig. 8 Ultrasonic array velocity profiling system for single-phase flows

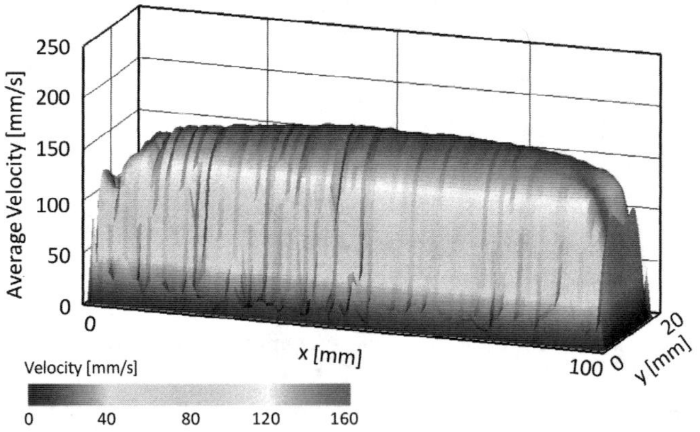

(a) Using each element

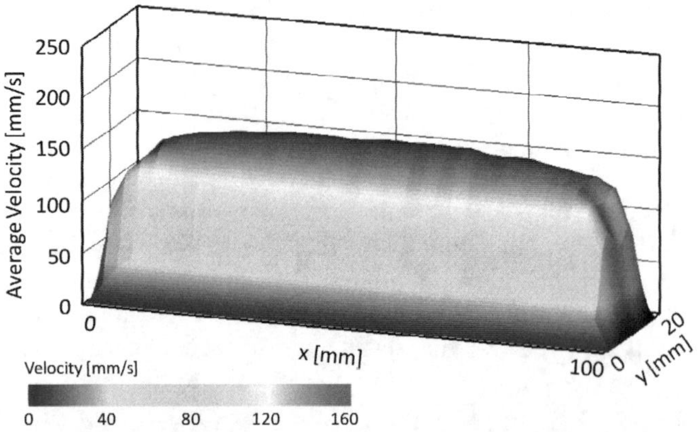

(b) Using every 3 grouped elements

Fig. 9 Cross-sectional average velocity profile (Re=4,000)

3.3 Application to Bubbly Flow Measurement

In the measurement of transient two-phase flows, appropriate temporal resolution is required to clarify the flow phenomena. However, when the UVP array measurement is performed, the temporal resolution of the cross-sectional instantaneous profile depends on the specifications of the UVP monitor. Since the measuring time of cross-sectional velocity distribution using the UVP is no less than about 1 second, their velocity distributions cannot reveal the instantaneous flow field. Therefore, the ultrasonic array measuring system was developed for bubbly flow measurement, as shown in Fig. 10. This system has a multiplexer for switching the pulse signal between the array sensor and the pulser/receiver. The switching time is 6 ms at least. The pulser/receiver has the maximum repetition rate of 8 kHz. The multiplexer and pulser/receiver control the pulse emitting time with a PC, and the received signals were stored in the PC. In this study, the cross-sectional measurement time was improved up to 5 Hz using this system.

The echo signals reflected from bubbly flows are shown in Fig. 11. The reflected signals were obtained by switching the channels of pulse transmission using every 3 contiguous piezoelectric elements of the array sensor, and the positions of bubbles can be detected on the ultrasonic beam from this figure. The cross-sectional distributions are obtained by aligning the echo signals, which were measured by switching each channel in the multiplexer. The cross-sectional echo distributions are shown in Fig. 12. The instantaneous bubble distribution can be measured over the cross-section, and the occurrence and position of bubbles are visualized. Accordingly it was found that the application of the ultrasonic array sensor to bubbly flows leads to the high accuracy bubbly flow measurement. However the temporal resolution of this measurement system is lower than the resolutions of other high-speed measurement system, so further improvements of the system are required for speeding up of channel switching.

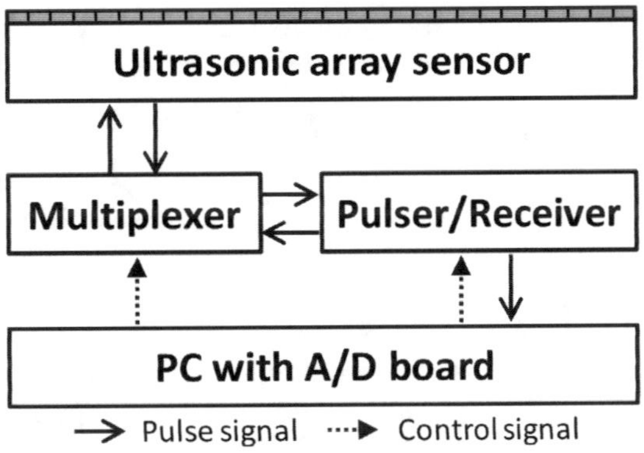

Fig. 10 Ultrasonic array measurement system for bubbly flows

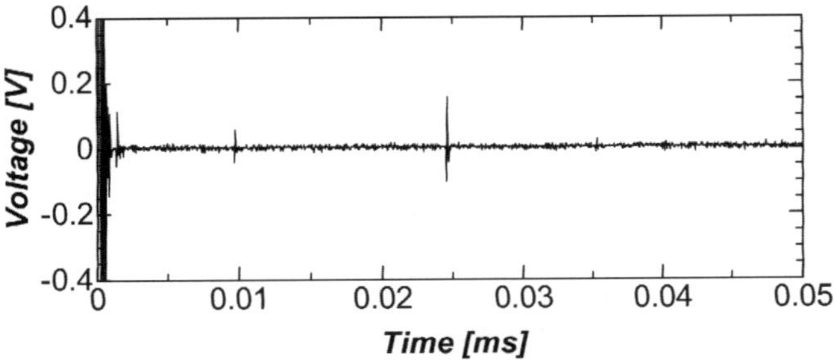

Fig. 11 Typical echo signals (CH.4, element No. 13-14-15, J_L=0.1 m/s, J_G=20 mm/s)

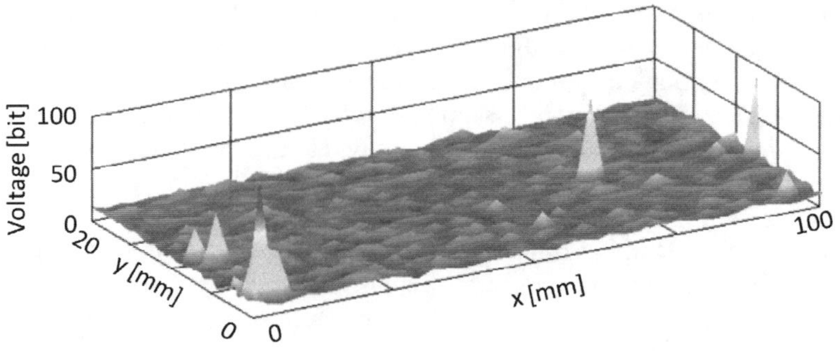

(a) Profile No. 1

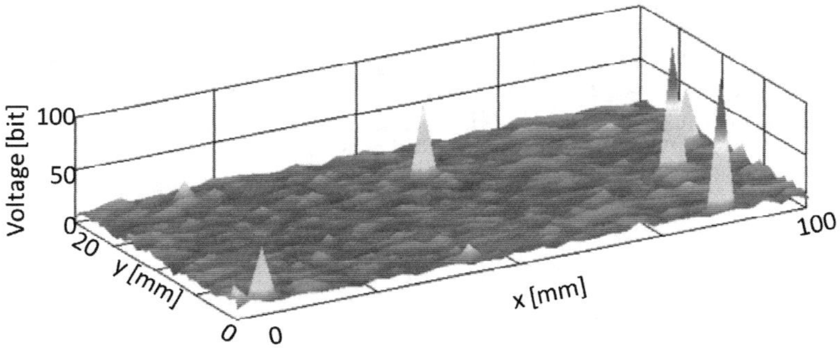

(b) Profile No. 2

Fig. 12 Cross-sectional distributions of reflected echo signals (J_L=0.1 m/s, J_G=20 mm/s)

4. CONCLUSIONS

The ultrasonic linear array sensor was applied to single phase and bubbly flow measurement in a rectangular channel. First, the ultrasonic sound pressure distributions were measured to understand the ultrasonic beam characteristics when the elements are or aren't grouped. Then the cross-sectional averaged velocity profiles in single-phase flows were measured by the ultrasonic velocity profiling method in order to confirm the grouping effect of the elements. Finally, air-water bubbly flows were measured with the ultrasonic echo technique using an array sensor, a pulser/receiver and a multiplexer, and the instantaneous cross-sectional bubble distributions were obtained. However the temporal resolution of the cross-sectional measurement should be improved for much faster measurement.

NOMENCLATURE

J	superficial velocity	[m/s]
Re	Reynolds number	[-]
x	coordinate	[m]
y	coordinate	[m]
z	coordinate	[m]

Subscripts

G	gas phase
L	liquid phase

REFERENCES

Hampel, U. et al. (2005). "Experimental ultra fast X-ray computed tomography with a linearly scanned electron beam source," Flow Meas. Instrum., 16, pp. 65-72.

Murakawa, H. et al. (2008). "Application of ultrasonic multi-wave method for two-phase bubbly and slug flows," Flow Meas. Instrum., 19, pp. 205-213.

Prasser, H.-M. et al. (1998). "A new electrode-mesh tomograph for gas-liquid flows," Flow Meas. Instrum., 9, pp. 111-119.

Prasser, H.-M. et al. (2005). "Comparison between wire-mesh sensor and ultra-fast X-ray tomograph for an air-water flow in a vertical pipe," Flow Meas. Instrum., 16, pp. 73-83.

Richter, S. et al. (2002). "Approach towards spatial phase reconstruction in transient bubbly flow using a wire-mesh sensor," Int. J. Heat Mass Trans., 45, pp. 1063-1075.

Takeda, Y. (1986). "Velocity profile measurement by ultrasonic Doppler shift method," Int. J. Heat Fluid Flow, 7, pp. 313-318.

Wada, S. et al. (2006). "Pattern recognition and signal processing of ultrasonic echo signal on two-phase flow," Flow Meas. Instrum., 17(4), pp. 207-224.

Journal of Physics: Conference Series **147** (2009) 012006 doi:10.1088/1742-6596/147/1/012006

Velocity Profile Measurements in Two-Phase Flow using Multi-wave Sensors

MK Biddinika, D Ito, H Takahashi, H Kikura, M Aritomi

Research Laboratory for Nuclear Reactors, Tokyo Institute of Technology, N1-13, 2-12-1 Ookayama, Meguro-ku, Tokyo 152-8550 JAPAN

biddinika.m.aa@m.titech.ac.jp

Abstract. Two-phase flow has been recognized as one of the most important phenomena in fluid dynamics. In addition, gas-liquid two-phase flow appears in various industrial fields such as chemical industries and power generations. In order to clarify the flow structure, some flow parameters have been measured by using many effective measurement techniques. The velocity profile as one of the important flow parameter, has been measured by using ultrasonic velocity profile (UVP) technique. This technique can measure velocity distributions along a measuring line, which is a beam formed by pulse ultrasounds. Furthermore, a multi-wave sensor can measure the velocity profiles of both gas and liquid phase using UVP method. In this study, two types of multi-wave sensors are used. A sensor has cylindrical shape, and another one has square shape. The piezoelectric elements of each sensor have basic frequencies of 8 MHz for liquid phase and 2 MHz for gas phase, separately. The velocity profiles of air-water bubbly flow in a vertical rectangular channel were measured by using these multi-wave sensors, and the validation of the measuring accuracy was performed by the comparison between the velocity profiles measured by two multi-wave sensors.

1. INTRODUCTION

Two-phase flow is an important phenomena appearing in many industrial applications such as chemical plants and nuclear power generation. However, to understand the phenomena is still difficult because of complex phenomena underlying its behavior.

A number of experimental studies have been carried out to understand the fundamental mechanism of two-phase flow. Some measuring techniques have been developed to study the flow. Two-phase flow measuring methods generally can be divided into two groups including of non-intrusive and intrusive methods. Intrusive methods, such as electrical resistivity method and hot-film anemometry, have been applied to measure the flow at the early stage of two-phase flow measurements while non-intrusive methods have been developed later in order to avoid disturbing the original measured flow.

The widely used non-intrusive method is ultrasonic velocity profile (UVP) method which developed by Takeda (1995). The method employs ultrasonic beam emitted from transducer along the measuring line and the same transducer (sensor) receives the echo of ultrasonic beam which is reflected from the surface of the micro-particles suspended in the liquid. For the two-phase flow measurements, the echo is not reflected only by the micro-particles but also the bubbles' surface.

The UVP method was applied by Aritomi et al. (1996) to measure the velocity profiles in two-phase (bubbly) flow. The UVP was also applied by Suzuki et al. (2002) to measure the liquid structure around bubbles in the bubbly flow. The velocity profile measurements of two-phase flow was also been carried out in rectangular channel using ultrasonic time domain correlation method by Hayashida et al. (2007).

2. EXPERIMENTAL SETUP

The velocity profiles are measured on the vertical rectangular channel as an experimental apparatus. The apparatus is schematically presented in Figure 1.

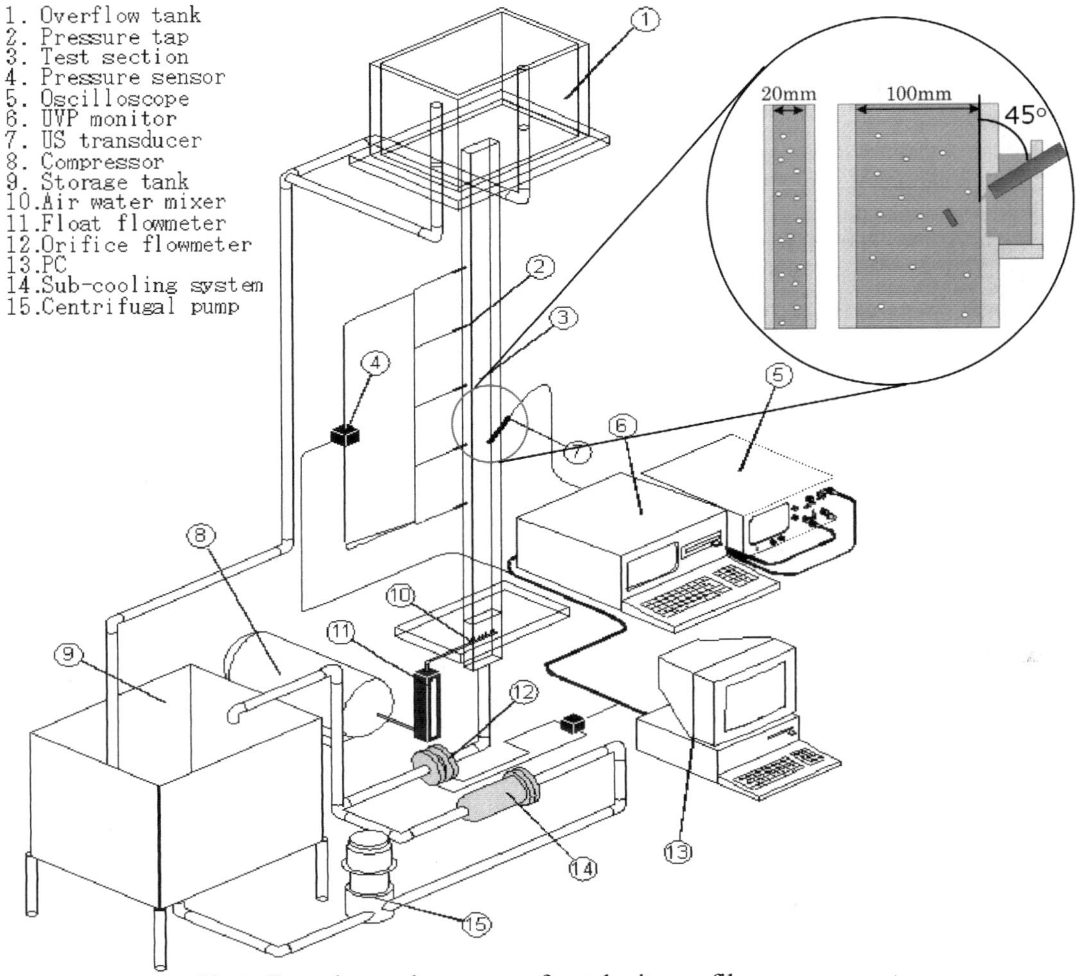

Fig 1. Experimental apparatus for velocity profile measurements

The apparatus consists of the water circulation system, the air supply system, the test section, the temperature control system, and the measurement system. The water circulation system circulates water from storage tank using a centrifugal pump. The water flow is controlled by a valve and measured by orifice flow meter. To construct the two-phase flow, gas phase is provided by an oil-free air compressor in the air supply system. Both water and gas are mixed in the test section. The air is injected into the test section through five 0.19 mm (inner diameter) needles so that the two-phase flow occurs in the section where the measurement is carried out. The needles are also as the air flow rate regulator.

The test section is a vertical rectangular channel made of acrylic glass (Plexiglas). This makes the section transparent so that allows visual observation of the flow. The size of the section is 20 mm x 100 mm x 1700 mm. In order to improve flow uniformity, 122 mm of inlet length is installed which consists of 60 mm-length of 5 mm mesh size grids along the wide side and 60 mm-length of 4 mm mesh size grids along the narrow side. There is a 2 mm gap between the narrow-side and the wide-side grids. An ultrasonic transducer is installed 1020 mm height from the five air injector needles. It is set on the outer surface of the test section with a contact angle of 45° from the vertical side in the downward direction. Between the transducer and the test section, water is filled in order to equalize the sound impedance.

During the measurements, the ambient room temperature and the circulating pump may increase the water temperature. Therefore, the temperature control section is inserted into the storage tank in order to maintain the water temperature.

3. RESULTS AND DISCUSSIONS

3.1 The multi-wave ultrasonic transducers

In this study, two types of multi-wave sensors are used. A cylindrical shape sensor and a square shape sensor. Those unique ultrasonic sensors are shown in Figure 2. The cylindrical shape sensor consists of two concentrically piezoelectric elements with different resonant frequencies. The central element of 3 mm of diameter has a basic frequency of 8 MHz and the outer element has a basic frequency of 2 MHz. The square shape one consists of two of 2 MHz sensors and nine of 8 MHz sensors which compose sandwich-like structure.

The multi-wave transducers are able to emit those two different frequencies independently. The 8 MHz element is for liquid phase measurements and the 2 MHz element is for gas (bubble) phase measurements in two-phase bubbly flows.

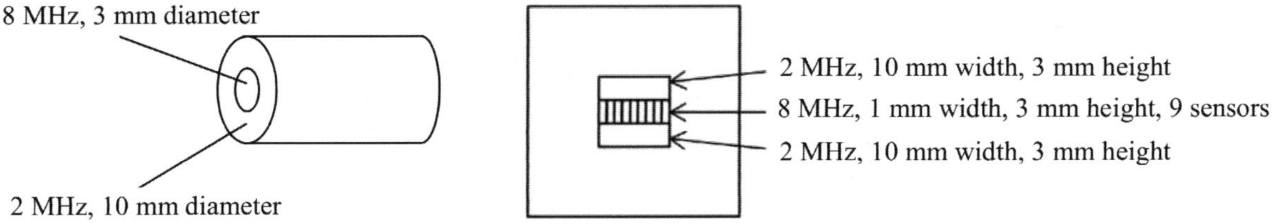

Fig 2. Schematic diagram of the unique multi-wave sensors, cylindrical and square shapes

3.2 The sound pressure distributions

Because of the unique shape of the multi-wave transducers, it is important to observe the characteristic of the transducers. One of the important characteristic is the sound pressure field. Using a hydrophone method, the ultrasound pressure distributions of the multi-wave transducer were carried out.

The ultrasound pressure distributions were measured in two conditions, in water without any disturbance and in water at the back of an acrylic pipe as the same material for channel's wall. The acrylic pipe weakens the ultrasonic intensity. However, the weakening of ultrasonic intensity is larger in the case of 8 MHz than in the case of 2 MHz as shown in Figure 3 and Figure 4 of cylindrical shape sensor.

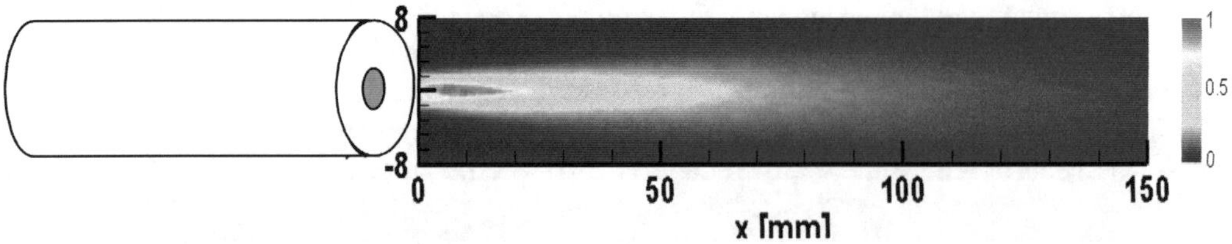

Fig 3. Ultrasound pressure distributions of 8 MHz cylindrical shape sensor

Journal of Physics: Conference Series **147** (2009) 012006 doi:10.1088/1742-6596/147/1/012006

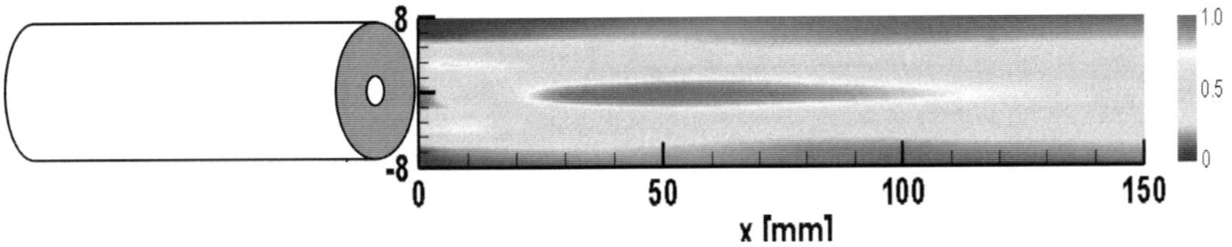

Fig 4. Ultrasound pressure distributions of 2 MHz cylindrical shape sensor

The distributions were also measured for 2 MHz square shape sensors as shown in Figure 5. The distributions measurement was carried out at XZ direction and YZ direction. The XZ distribution shows the ultrasound emissions from the parallel sensors are not parallel to each other. The emissions tend to spread each other.

This phenomena is more pronounced by YZ distribution as shown in Figure 5 as well. In the near-sensor position, those emissions were still parallel as the sensors position. However, in the far-sensor position, the emissions were spread.

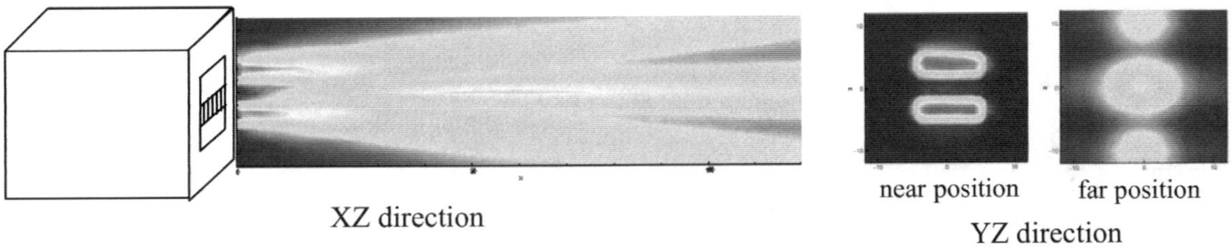

Fig 5. Ultrasound pressure distributions of 2 MHz square shape sensors

3.3 Velocity profile measurement results

The measured velocity profiles are shown in Figure 6 for both two phase and single phase flow measurement using 8 MHz frequency of cylindrical shape transducer. The velocity distributions are averaged from 50,000 instantaneous velocity profiles. The vertical axis is the velocities and the horizontal one is the distance from the wall. The velocity profiles of two phase flow are compared with those of single liquid phase flow which were measured separately with the same superficial liquid velocity (J_L) instead of liquid and gas phase velocity separation from a two phase velocity profile.

In the low J_L, the bubbles play an important role in increasing mean velocity of the two phase flow. By increasing J_L gradually,the gap between velocity profiles of the two phase flow and velocity profiles of its single liquid phase measured separately, are decreased. However, in the relatively high J_L, the bubbles have a tendency to make the mean velocity of the flow flat especially in the near wall region.

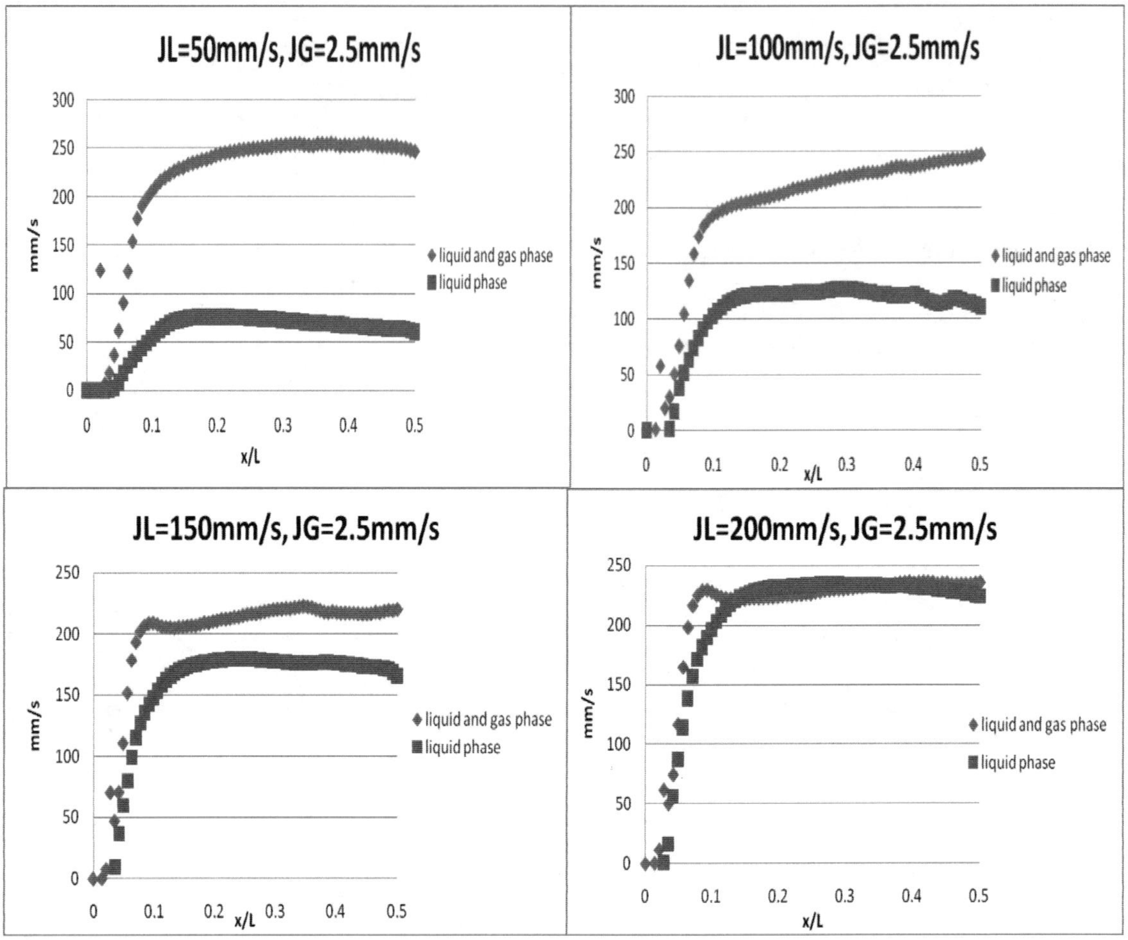

Fig 6. The measured velocity profiles by 8 MHz of cylindrical shape sensor

4. CONCLUSIONS

In this study, two types of multi-wave, 2 MHz and 8 MHz, sensors are used. Those are cylindrical shape and square shape. The 8 MHz sensor has a good agreement in measuring both suspended particle for liquid phase measurement and bubbles for gas phase measurement. By ultrasound pressure distributions measurement, it was observed that the simultaneous ultrasound emissions of 2 MHz square shape sensors tend to spread each other. It makes two-phase flow measurement using this sensor should use only one of 2 MHz sensor instead of simultaneously using two of 2 MHz sensors. Further measurement of ultrasound pressure distributions will be carried out for single use of a 2 MHz square shape sensor.

REFERENCES

Aritomi, M. et al. (1996). "Measurement system of bubbly flow using ultrasonic velocity profile monitor and video data processing unit," J. Nucl. Sci. Technol., 33, pp. 915

Hayashida, T. et al. (2007). "Velocity profile measurements of two-phase flow in rectangular channel using ultrasonic time-domain correlation method," 5th Joint ASME/JSME Fluid Engineering Conference, San Diego, CA, USA, July 30-August 2

Suzuki, Y. et al. (2002). "Microstructure of the flow field around a bubble in counter-current bubbly flow," Exp. Therm. Fluid Sci., 26, pp. 221.

Takeda, Y. (1995). "Velocity profile measurement by ultrasonic Doppler method," Exp. Therm. Fluid Sci., 10, pp. 444-453

Journal of Physics: Conference Series **147** (2009) 012007 doi:10.1088/1742-6596/147/1/012007

An Investigation on Characterizing Two-phase Flow with Broad-Band Ultrasonic Impedance Spectra

XUE Minghua, SU Mingxu, CAI Xiaoshu

University of Shanghai for Science and Technology, 516 Jungong Rd, Shanghai 200093, CHINA

xmh2000@126.com

Abstract. This paper focus on how to extract sufficient impedance information from ultrasonic signals and utilize them effectively in a two-phase characterization method based on broad-band ultrasonic impedance spectra. After an introduction of ultrasonic velocity and attenuation spectra, a classical model and a broad-band ultrasonic signal technique are introduced simultaneously. The relationship between the ultrasonic impedance and spectrum information is investigated both theoretically and experimentally. With an experimental set-up developed in IPTFM, impedance spectra data are measured by reflection using magnitude and the phase spectrum of signals.

1. INTRODUCTION

In recent years, with the dramatic development in particle concerned fields, dense particulate two-phase flow is very common in different industrial areas and production processes. Among particle sizing methods, ultrasonic spectrum analysis, including attenuation and phase velocity spectra, have gained much attention for characterization of concentrated dispersions, due to its penetration capability in highly concentrated media and advantages of fast and non-invasion test, low cost and endurable set-up. With development of physical model and experimental technique, ultrasonic method permits accurate determination of particle size distribution (PSD) and concentration (Su, et al., 2002)

The characterization of two-phase flow by ultrasound requires a formal theoretical basis, a reasonable experimental data, a stable and rapid inversion algorithms. The formal theoretical basis relates the properties of the mixture, particularly the dispersed phase particle size distribution, to the complex wavenumber governing propagation. These propagation problems can be identified two approaches: scattering and coupled phase models (Challis, et al., 2005). A significant scattering theoretical model on acoustical wave propagation in particulate mixtures can be traced back to the Epstein and Carhart (Epstein, et al., 1953). Combined with the work of Allegra and Hawlay (Allegra, et al., 1972), the classical ECAH model was established. This model combined viscous and thermal absorption into scattering, a fully thinking about sound propagation and interaction with particles was made. Riebel (Riebel, et al., 1989) focused on the ultrasonic scattering by particles in short wave limit. He has developed a scattering model by analogism with Lambert-Beer model of light extinction. McClements (McClements, et al., 1991) provided in-depth discussions about droplets size measurement theories in emulsions. He contributed an explicit solution of ECAH in long wave limit. Coupled phase models, as a substitute of scattering models for small hard core particles were studied by Harker (Harker, et al., 1988) and Dukhin (Dukhin, et al., 2002). Such work have provided a possibility for ultrasonic method

to be used in characterizations of colloidal suspensions and emulsions with volume fraction 1% to 50%, size range from 10 nanometer to 1 millimeter.
Most papers focus on the ultrasonic attenuation and/or phase velocity spectrum, which means we should obtain the measurements of frequency dependent ultrasonic attenuation coefficient or acoustic velocity in particulate two-phase flow. In this paper, we use another acoustic parameter, acoustic impedance, describing the ultrasonic spectrum information, establishing the relationship with the classical ultrasonic theoretical models, obtaining the experimental data of acoustic impedance spectrum.

2. PRINCIPLES

There are three major steps in the determination of particle size using ultrasonic spectroscopy : (i) measurement of the ultrasonic parameters spectra of the two-phase flow and (ii) interpretation of the resulting spectra using a suitable theory and (iii) a stable and rapid inversion algorithms. To obtain accurate measurements it is necessary to carry out three steps carefully (Su, et al., 2004).

2.1 Ultrasonic spectroscopy methods

Ultrasonic spectroscopy utilizes measurements of the frequency dependence of the ultrasonic velocity and/or attenuation coefficient of colloidal dispersions to obtain information about the concentration and particle size distribution of colloidal particles. The velocity at which an ultrasonic wave propagates through a particulate suspension, and the amount by which it is attenuated, are governed by interactions between the ultrasonic wave and the particles, e.g. transmission, reflection, absorption, and scattering. Ultrasonic spectroscopy has advantages over many existing particle-sizing technologies because it is non-destructive and non-invasive, is capable of rapid measurements and can be used to characterize systems which are concentrated and optically opaque (McClements, et al. 1997).
Characterizing the two-phase flow relies on being able to measure the frequency dependence of the ultrasonic parameters, including velocity, attenuation, and impedance. The ultrasonic velocity is the distance the ultrasonic wave moves through the sample per unit time, whereas the attenuation coefficient is a measure of the decrease in the amplitude of the ultrasonic wave per unit distance travelled. The ultrasonic velocity and attenuation spectra can be determined by the following:

$$X(\omega)=\exp\left(\frac{-i\omega x}{c(\omega)}\right)\exp\left(-\alpha(\omega)x\right) \tag{1}$$

Where the $X(\omega)$ is the response of the time domain original signal by the fast fourier transform (FFT). X is the position of the ultrasonic wave in the sample at the certain time. The first exponential term represents the phase shift applied to the acoustic signal while propagating distance x through the sample at the phase velocity $c(\omega)$. The second term represents the attenuation of the signal travelling over the same distance. The individual velocity and attenuation spectra can be written as following by the real and imaginary part separation of Eq.(1).

$$c(\omega)=\frac{\omega(x_2-x_1)}{\theta_2(\omega)-\theta_1(\omega)} \tag{2}$$

$$\alpha(\omega)=-\frac{1}{x_2-x_1}\ln\frac{|X(\omega)_2|}{|X(\omega)_1|} \tag{3}$$

Where the θ and symbol $|\ |$ mean the phase and magnitude of the $X(\omega)$. The subscripts 1and 2 refer to the different positions, respectively.
The acoustic impedance of a material is important because it determines the fraction of the ultrasonic wave that is reflected from its surface. The specific acoustic impedance is defined as the ratio of the

Journal of Physics: Conference Series **147** (2009) 012007 doi:10.1088/1742-6596/147/1/012007

acoustic excess pressure (*P*) and the particle velocity (*U*). Generally, the impedance is complex and can be divided into the real and imaginary part.

$$Z = \frac{P}{U} = R_z + i \cdot X_z \tag{4}$$

Where *Rz* is called the resistive component and *Xz* is called the reactive component. Like the ultrasonic velocity and attenuation coefficient, the acoustic impedance is a fundamental physical characteristic. Measurements of acoustic impedance can therefore be used to provide valuable information about the properties of two-phase flow.

2.2 Broad-band ultrasonic spectra

In principle, a spectrum at three frequencies could afford basic data for inversion computation to extract particle size information. But such a spectrum has hardly been used in inverse computations just because it is inadequate to overcome multiple solution problems and random error disturbance in ill condition equations. A broad band spectrum at a series of frequencies, in contrast, shows advantages to ensure stable solutions.

There are at least three techniques to obtain broad band ultrasonic spectrum: transducers array, scanning frequency and the fast fourier transform (FFT). Compared with other two methods, FFT technique has advantages of rapid test and easily handling set-up, which is critical for on-line measurement.

In the FFT method, a signal generator is used which is capable of producing a broad-band electrical pulse which contains a wide range of frequencies. This pulse is applied to the transmitting transducer where it is converted into a broad-band ultrasonic pulse. After the pulse has travelled through the sample it is detected by the receiving transducer. The frequency dependence of magnitude and phase of the pulse are determined using a FFT technique. Generally, the ultrasonic velocity is determined from the phase, the ultrasonic attenuation coefficient is determined from the magnitude, and the ultrasonic impedance is determined from both phase and magnitude.

Fig.1 shows the original signal. Fig.2 and Fig.3 show the magnitude and the phase spectrum using FFT technique.

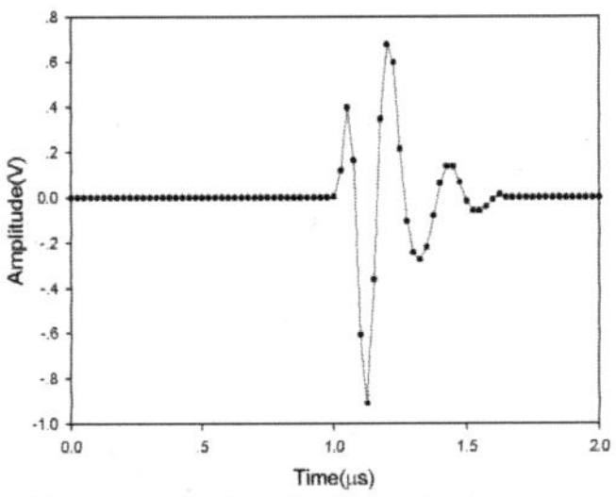

Fig.1 Original signal

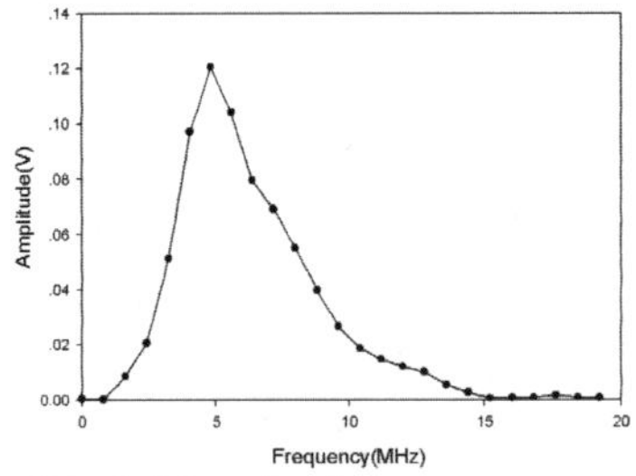

Fig.2 Magnitude spectrum

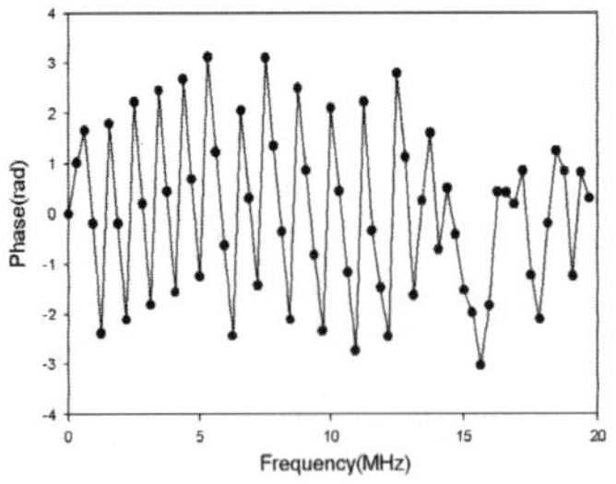

Fig.3 Phase spectrum

2.3 Ultrasonic parameters spectra from theoretical model

Once the ultrasonic parameters spectra of two-phase flow have been measured, it is necessary to convert them into concentration and particle size distribution using an appropriate theory. Theories are based on a mathematical treatment of the physical processes that occur when an ultrasonic wave propagates through an ensemble of particles suspended in two-phase flow. The comprehensive and classical model is based on single scattering theory: Epstein, Carhart, Allegra and Hawlay (ECAH) model.

$$\left(\frac{K}{k_c}\right)^2 = 1 + \frac{3\phi}{ik_c^3 R^3} \sum_{n=0}^{\infty} (2n+1) A_n \tag{5}$$

For the monodisperse particle system, the velocity and attenuation spectra can be represented as:

$$\alpha(\omega)_s = -\frac{3}{2k_c^2}\sum_{j=1}^{N}\frac{\phi_j}{R_j^3}\sum_{n=0}^{\infty}(2n+1)\,\mathrm{Re}(A_n(R_j,\omega)) \tag{6}$$

$$C(\omega)_s = \left(\frac{3}{2\omega k_c^2}\sum_{j=1}^{N}\frac{\phi_j}{R_j^3}\sum_{n=0}^{\infty}(2n+1)\,\mathrm{Im}(A_n(R_j,\omega))+\frac{1}{c_w}\right)^{-1} \tag{7}$$

Where $K = \omega/c_s + i\alpha_s$ is the complex propagation constant, *Cs* is the ultrasonic velocity and α_s is the attenuation coefficient of the two-phase flow. *Kc* is the complex propagation constant of the continuous phase, ϕ is the disperse phase volume fraction, $\omega = 2\pi f$ is the angular frequency, *f* is the frequency, $i = \sqrt{(-1)}$, and *R* is the particle radius. The *An* terms are the scattering coefficients of the various types of waves scattered from the individual particles. Re and Im is real part and imaginary part of the complex number.

The ultrasonic impedance of the two-phase flow can be related to its physical and acoustic complex propagation constant by the following expression:

$$Z = (\omega\rho)/K \tag{8}$$

Where ω is the angular frequency, ρ is two-phase flow density and K is the propagation constant. In the model analysis, the mathematical result is usually given by the style of the propagation constant K. Eq.(8) shows the relationship between the ultrasonic impedance and the theoretical model. The numerical calculation of frequency dependence of the ultrasonic impedance can be obtained by the suitable ultrasonic theoretical models.

3 EXPERIMENTAL METHODS AND DATA ANALYSIS

An ultrasonic measurement set-up showed in Fig.4 was developed to obtain testing data of frequency dependent ultrasonic attenuation. Ultrasonic waves, generated by broad-band ultrasonic transducer(V317-SU) worked with an ultrasonic pulser/receiver (PR5800,Parametric, Inc), transmit through two-phase flow in a cell, and reflected, in which signals could acquired by high-speed data acquisition card, and recording in a personal computer. The ultrasonic transducer could be designed as a style of probe, being convenient for on-line measurement in particulate two-phase flow.

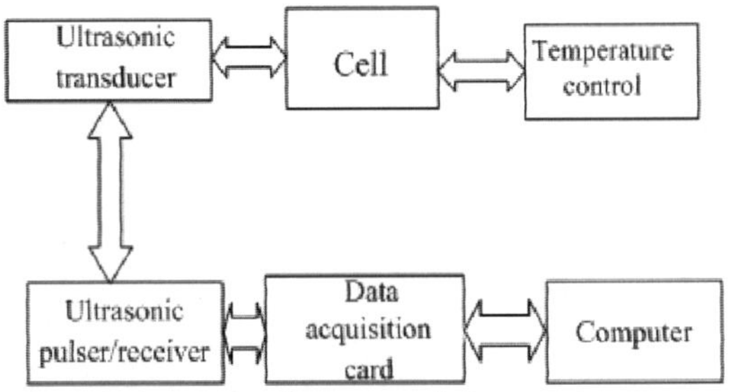

Fig.4 Schematic diagram of ultrasonic system

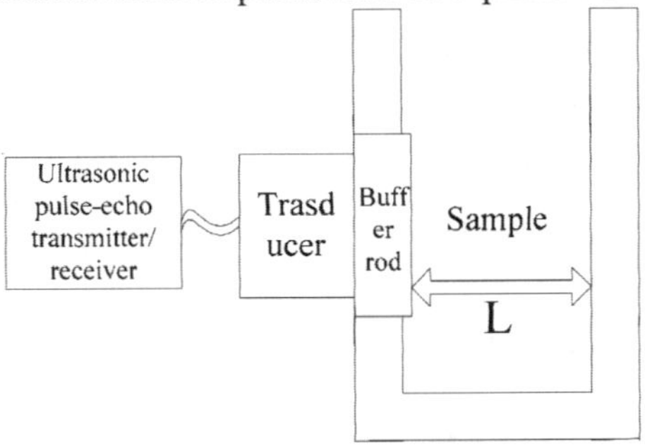

Fig.5 Schematic diagram of cell

Fig.5 shows the schematic diagram of the measurement cell (McClements, et al,. 1991). The transducer generates a pulse of ultrasound which travels along the buffer rod until it reaches the interface where it is partly reflected and partly transmitted. The impedance of a sample is determined by measuring the amplitude of the pluse reflected from the buffer rod/sample interface. The velocity is determined by measuring the time different between the first echo and the second echo, whilst the attenuation coefficient is determined by measuring the amplitude of these two echoes. The magnitude (M) and phase (θ) of the pulse were measured as a function of frequency using Fourier transform analysis. This procedure was carried out for the sample being analyzed, as well as for a calibration material with known acoustic properties, i.e., distilled water (Asylbek, et al,. 2000). Fig.6 shows the measured ultrasonic signal when the buffer rod material is the plexiglass, and Fig.7 shows the measured signal when the buffer rod material is the steel plate. In these experiments sample is the distilled water. The impedance of water is similar to the impedance of plexiglass, so large part of ultrasonic wave can transmit into the sample. On the contrary, the impedance of steel is large compared to the impedance of the plexiglass, so the multiple echoes can be obtained as shown in Fig.7.

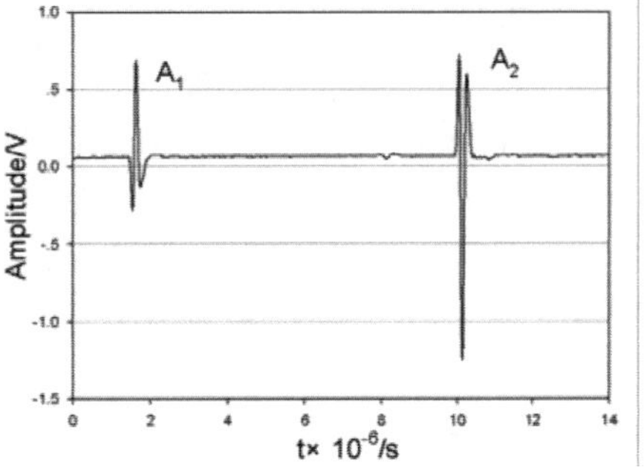

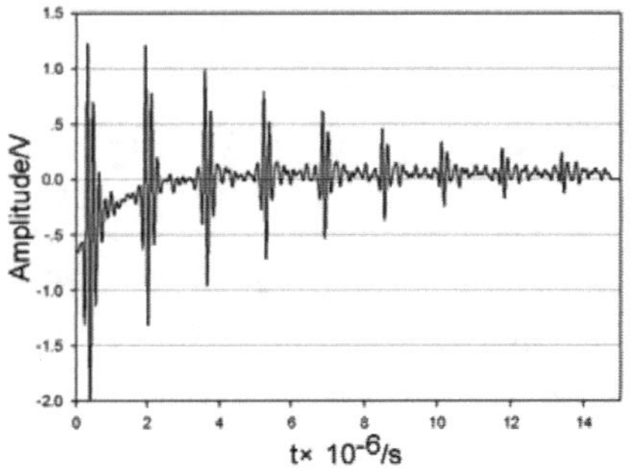

Fig.6 Measured signal (Buffer rod: plexiglass) Fig.7 Measured signal (Buffer rod: steel)

The reflectance was then calculated at each frequency using the following formula:

$$R_s = R_c \frac{M_s}{M_c} \exp i(\theta_s - \theta_c) \tag{9}$$

Where R is the reflectance, and the subscripts S and C refer to the sample and calibration material, respectively. The physical properties of the calibration material (i.e., distilled water) and the buffer rod are known, so the value of Rc can be easily calculated. From the reflectance R, the acoustic impedance can be represented as the following equation:

$$Z_s = Z_{BR} \frac{(1-R_s)}{(1+R_s)} \tag{10}$$

Where Z_{BR} is the acoustic impedance of the buffer rod, and the Z_s is the acoustic impedance of the sample. From Eq.(9), the reflectance Rs is a complex number, and so the acoustic impedance Zs is also complex in Eq.(10). That means it is possible to get the impedance spectra information of two-phase flow by these experimental and computational methods.

For the purpose of obtaining the density and concentration simultaneously, we chose the buffer rod of steel to obtain the impedance spectra. Using this device, the sample density and concentration can be determined by multiple reflections within the steel. More details can be seen the references(Xue, et al,. 2008; Margaret, et al, 2002).

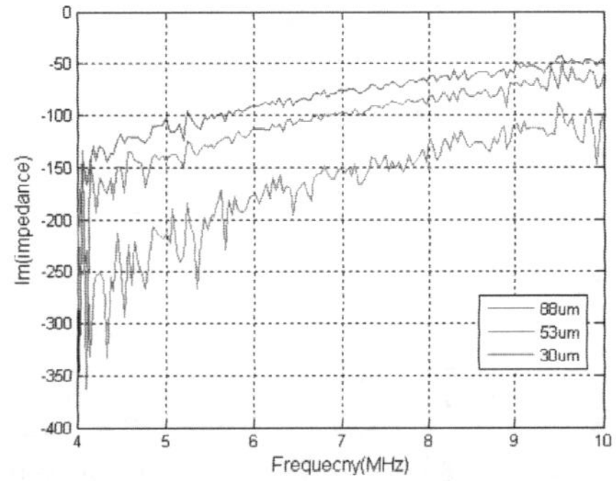

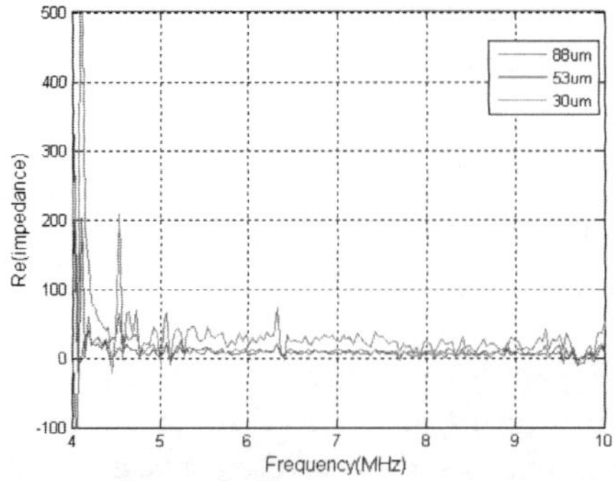

Fig.8 Imaginary part of the acoustic impedance Fig.9 Real part of the acoustic impedance

Fig.8 and Fig.9 shows the real and imaginary part of the acoustic impedance versus frequency for the different particle size of glass beads, which average diameters are 88 μ m, 53 μ m and 30 μ m, but the concentrations are same (1.5vol%). The experimental data shows it is feasible to obtain the ultrasonic impedance spectra by reflection using magnitude and the phase spectrum of signals.

4 CONCLUSIONS

Three ultrasonic parameters spectra methods have been described which are based on measuring the frequency dependent ultrasonic parameters. Broad-band ultrasonic signal technique is introduced and qualified as magnitude and phase spectrum according the fast fourier transform. The relationship between the ultrasonic impedance and spectrum information is investigated both theoretically and experimentally. For the purpose of characterization of impedance spectra, the special experimental set-up is developed. Our theoretical results and experimental data have showed ultrasonic impedance spectra could provide the new method besides these ultrasonic velocity and attenuation spectra methods.

ACKNOWLEDGEMENTS

The Authors gratefully thank the National High Technology Development 863 Program (2006AA03Z349) and the National Natural Science Foundation of China (No.50706029).

REFERENCES

Su M.X. et al. (2002). "The numerical study of acoustical attenuation and velocity in the suspensions of superfine particles. " ACTC ACUSTICA., 27(3), pp 218-222.

R E Challis. et al. (2005). "Ultrasound technique for characterizing colloidal dispersions." Rep Prog Phys. 68, pp 1541-1637.

Epstein P.S. et al. (1953). "The Absorption of Sound in Suspensions and Emulsions: I. Water Fog in Air. " J. Acoust. Soc. Am. 25, pp 553-565.

Allegra J.R. et al. (1972). "Hawley S.A. Attenuation of Sound in Suspensions and Emulsions: Theory and Experiments." J. Acoust. Soc. Am. 51, pp 1545-1564.

Riebel U. et al. (1989). "Loffler F. The fundamentals of particle size analysis by means of ultrasonic spectrometry." Part.Part.Syst.Charact. 6, pp 135-43.

McClements D J. et al.(1991). "Ultrasonic Characterisation of Emulsions and Suspensions. Advances in Colloid and Interface Science." 37, pp 33-72.

Harker A.H. et al.(1988). "Temple. J.A.G. Velocity and Attenuation of Ultrasound in Suspensions of Particles in Fluids." J.Phys. D: Appl.Phys. 21, pp 1576-1588.

Dukhin A.S. et al.(2002). "Ultrasound for Characterizing Colloids Particle Sizing, Zeta Potential, Rheology." Elsevier, pp 117-135.

Su M.X. et al.(2004). "The measurement of particle size and concentration in suspensions by ultrasonic attenuation." ACTC ACUSTICA.29(5), pp 440-444.

McClements D J. et al.(1997). "Ultrasonic Measurements in Particle Size Analysis. Encyclopedia of Analytical Chemistry." ISBN 0471 976709.

McClements D J. et al.(1991). "Ultrasonic pulse echo reflectometer." Ultrasonic. 29, pp 58-62.

Asylbek K. et al.(2000). "Characterization of aerated foods using ultrasonic reflectance spectroscopy." Journal of Food Engineering, 46, pp 235-241.

Xue M H. et al. (2008). "The experimental study on measurement of density for two phase flow with ultrasonic multiple echo reflection method." Journal of Engineering Thermophysics. 29(8):1343-1346.

Margaret S. et al.(2002). "Ultrasonic Sensor to Measure the Density of a Liquid or Slurry during Pipline Transport." Ultrasonics, 40, pp 413-417.

Journal of Physics: Conference Series **147** (2009) 012008 doi:10.1088/1742-6596/147/1/012008

Resolution of Structure Characteristics of Passive Acoustic Emission Signals in Multiphase Flow System

Yi-Jun He, Jing-Dai Wang, Yong-Rong Yang

State Key Laboratory of Chemical Engineering, Department of Chemical and Biochemical Engineering, Zhejiang University, Hangzhou 310027, People's Republic of China

yangyr@zju.edu.cn

Abstract. The underlying structure characteristics of acoustic emissions (AE) measured in a gas-solid fluidized bed was investigated detailedly by resorting to wavelet transform and rescaled range analysis. A general criterion was proposed to resolve AE signals into three characteristic scales, i.e. micro-, meso- and macro-scale, and a so-called "structure diagram" was introduced. Compared with the structure diagram of pressure signals, it was found that AE signals in micro-scale reflect mainly the particles motion while pressure signals in meso-scale reflect mainly the bubbles motion. Energy distribution analysis further revealed that the most energies in AE and pressure signals were distributed mainly by the micro-scale and meso-scale signals respectively. Moreover, the structure characteristics of AE signals collected from gas-solid fluidized bed and liquid-solid stirred tank were compared based on structure diagram and energy distribution analysis. The results indicated that although the same measurement technique was adopted, the structure characteristics of AE signals measured in gas-solid fluidized bed and liquid-solid stirred tank still exhibited larger difference. As an illustrative application of AE technique in process monitoring, a prediction model for particle size distribution was proposed and the satisfactory results were obtained both for laboratory scale and plant scale fluidized beds.

1. INTRODUCTION

Multiphase flow system exists widely in many industrial processes, such as chemical and petrochemical, metallurgy, piping transportation, pharmaceutical, and power engineering. Despite the wide usage of multiphase systems, the thorough understanding of the hydrodynamics of multiphase systems is also a challenging problem, because multiphase flows are almost always unstable and complex spatio-temporal patterns are observed ubiquitously. Advance in computer hardware technology as well as in numerical computation methods, study of the multiphase flow patterns by computational fluid dynamics (CFD) has thus received more and more attention in recent years. However, establishing the exact mathematical model of multiphase systems is usually still impossible, and the experimental validation of the results of CFD simulation is also difficult; on the other hand, CFD simulation is very time-consuming, and its application to large industrial multiphase systems is not yet feasible. Therefore, various measurement techniques, whether invasive or non-invasive, are needed both in academia and in industry, for understanding and analysis of the complex flow patterns, and for design, operation, control and scale-up purposes, respectively. Nevertheless, due to the non-linear, non-equilibrium and multi-scale characteristics of multiphase systems, as well as each

Journal of Physics: Conference Series **147** (2009) 012008 doi:10.1088/1742-6596/147/1/012008

measurement technique has its own particular spatial and time resolution, none of the measurement techniques is capable of providing equally valid information over the complete spectrum of scales relevant to the complete characterization of multiphase flow systems. Hence, it is recommended that several techniques may be simultaneously derived to get complementary information (Boyer et al., 2002).

Fluctuation signals collected from various measurement techniques, such as pressure, optic fiber, capacitance, and radioactive, can usually reflect a wide spectrum of complex fluctuation phenomena in multiphase flow systems. However, due to the lack of more concrete knowledge of the underlying structure characteristics of fluctuation signals, how to mining useful information from these raw fluctuation signals is still a challenging problem. Statistical Characteristics (such as standard deviation, skewness, kurtosis and entropy), obtained by simple statistical analysis of raw signals, are in general directly used for predicting the properties of multiphase systems. Because fluctuation signals are commonly hybrid signals with multi-scale components, the most useful information is usually carried by some specific components at different scales with respect to a specific measurement task, such as particle size distribution, bubble size, and voidage. Therefore, prediction model, established based on the above statistical characteristics, not only has little physical sense, but also probably has poor generalization performance.

In recent years, multi-resolution methodology has been applied to reveal the underlying multi-scale characteristics of fluctuation signals. By resorting to advanced statistical tools, such as wavelet analysis and chaos theory, some efforts have been made to explore the relationship between the decomposed components at different scales and the corresponding physical phenomena, especially for the pressure fluctuation signals in gas-solid fluidized bed system. Lu et al. (1999) used wavelet analysis to pressure fluctuation signals in a bubbling fluidized bed and indicated that the scale 4 detail signal reflects the bubble behaviour. Zhao et al. (2003) adopted wavelet transform to decompose pressure fluctuation signals in a fluidized bed. Hurst analysis of the decomposed fluctuation signals showed that the measured pressure fluctuations can be resolved to three characteristic scales: meso-scale signals with two distinct Hurst exponents; micro- and macro-scale signals with only one characteristic Hurst exponent. Briongos et al. (2006) applied the Hilbert-Huang transform method to perform a multi-resolution analysis of pressure signals collected from a gas-solid fluidized bed, then the concept of averaged instantaneous frequency was used to identify three important dynamic components, such as local bubble, "bulk" and particle dynamics. Furthermore, the bulk dynamic component was used to estimate the bed expansion ratio and bed height. Yang et al. (2008) used wavelet transform to decompose the absolute pressure fluctuation signals into three scales corresponding to macro-scale, meso-scale and micro-scale in three circulating fluidized beds. A redefined variable, homogeneous index HI, obtained from the energy ratio of the micro-scale and meso-scale signals, was used to determine the transition velocities from bubbling to turbulent fluidization.

The passive acoustic emission (AE) technique has received more and more attention in recent years as a potential non-intrusive and real-time process monitoring technique to be used in multiphase flow systems. Due to AE signals are made up of emission from many acoustic sources at different scales, the interpretation of AE signals are often very complicated. Previous studies have demonstrated that AE signals contain rich information with respect to the motion of particles (Boyd et al., 2001; Ren et al., 2008). Multi-scale resolution of AE signals can be thus useful for separating the key feature information of particles motion from the original AE signals, which may help to establish physical meaningful prediction model with respect to specific measurement task. However, the underlying structure characteristics of AE signals is relatively poorly investigated.

In this study, wavelet transform and rescaled range analysis were applied to explore and understand the structure characteristics of AE signals in different multiphase systems. The different structure characteristics of AE signals and pressure fluctuation signals in gas-solid fluidized bed were firstly investigated. Secondly, the structure characteristics of AE signals collected from gas-solid fluidized bed and liquid-solid stirred tank were compared. Finally, as an illustrative application example of AE technique in process monitoring, a prediction model for particle size distribution was proposed.

2. METHODS

2.1 Wavelet transform analysis

The wavelet transform (WT) analysis has become a very powerful time-frequency tool for analysis of non-stationary and transitory signals and has been widely applied in various fields, such as signal processing, image processing, data compression and financial time series. Contrary to the Fourier-related transform methods, WT provides a more flexible way of time-frequency representation of a signal by allowing the use of variable sized windows. In WT, at high frequencies (corresponding to small scales), narrow windows are used to get precise time resolution, whereas at low frequencies (corresponding to large scales), wide windows are used to get finer frequency resolution. Localization in both frequency and time domains is thus the greatest advantage of WT over Fourier-related transform methods. Moreover, WT is often regarded as a mathematical "microscope" that is able to examine different parts of the signal by automatically adjusting the focus.

WT uses the wavelet function and scaling function to perform simultaneously the multi-resolution analysis (MRA) decomposition and reconstruction of the measured signal. The wavelet function serving as a high-pass filter can generate the detailed version of the given signal, while the scaling function serving as a low-pass filter can generate the approximated version of the given signal. The discrete wavelet transform (DWT) can be regarded as an MRA technique, where the original signal can be decomposed into several signals with different scales or resolutions and can reconstruct the signals using inverse discrete wavelet transform. The detailed mathematical description of DWT should be referred to the literature (Zhao et al., 2003).

2.2 Rescaled range analysis

Rescaled range (R/S) analysis, originally developed by an Egyptian hydrologist Hurst (1951) to analyze Nile River's overflows, can identify long-range dependence in highly non-gaussian time series and detect non-periodic cycles. Moreover, it also provides an effective way for studying the fractal characteristics of a time series. Recent years, R/S analysis has been applied by a number of groups to characterize the complex hydrodynamics of multiphase flow systems (Fan et al., 1991; Fan et al., 1993; Briens et al., 1997; Zhao et al., 2003; Ren et al., 2008). In this study, this type of analysis is adopted to analyze the decomposed AE signals on different levels, and a criterion is further established to resolve the AE signals into three characteristic scales in terms of different fractal characteristics. The detailed implementation of R/S analysis should be referred to the literature (Zhao et al., 2003).

By performing R/S analysis, the Hurst exponent, H, can be obtained. If the Hurst exponent equals to 0.5, the time series is random. If the Hurst exponent is greater than 0.5, the time series is persistent. Persistent implies that if the trend of the time series has increased or decreased, then the chances are that it will continue to increase or decrease in the future, respectively. Moreover, the strength of trend-reinforcing behaviour, or persistence, increases as the Hurst exponent approaches 1.0. Conversely, if the Hurst exponent is less than 0.5, the time series is anti-persistence. This means that, whenever the trend of the time series has increased, it is more likely that it will decrease in the close future. The strength of anti-persistent increase as the Hurst exponent approaches 0. It is important to note that persistent time series have little noise whereas anti-persistent time series show presence of high-frequency noise.

3. EXPERIMENTAL

3.1 Gas-solid fluidized bed

The gas-solid multiphase flow experiments were carried out in gas-solid fluidized beds both for cold mode in laboratory scale and hot model in plant scale. A schematic diagram of the laboratory scale cold mode experimental apparatus used in the present study is shown in Fig. 1. It consists of two parts:

a fluidized bed and an AE signal collection system. The fluidized bed assembly includes a plexiglass bed (150mm in ID and 1000mm in height) and a perforated-plate distributor (with a pore diameter of 2.0 mm and a fractional open area ratio of 2.6%). Three types of polyethylene (PE) particles: linear low-density polyethylene (LLDPE), high-density polyethylene (HDPE) and bi-mode polyethylene (BPE), were used as the fluidized particle. Seven different particle sizes of these three types of PE, 0.14, 0.18, 0.36, 0.50, 0.71, 1.19 and 2mm, were selected by sieving. The fluidized media is air at a velocity range of 0~1.1m/s. The AE signal collection and analysis system, developed by the UNILAB Research Center of Chemical Engineering at Zhejiang University, consists of an AE sensor, a preamplifier, a main amplifier and a control computer with an A/D conversion module. The AE sensor, which is made of a piezoelectric accelerometer, was attached close to the wall of the fluidized bed at a location 150mm above the gas distributor. Based on the Shannon sampling theory, a sampling frequency of 500kHz is determined.

The plant scale hot mode experiments were conducted in three industrial fluidized beds for production of LLDPE, HDPE and BPE. The ID and height of LLDPE, HDPE and BPE industrial fluidized beds are 3500mm and 12m, 3500mm and 15m, and 5000mm and 18m, respectively. The AE sensors were installed at a location 1000mm above the gas distributor for these three industrial fluidized beds, with a sampling frequency of 500kHz. The operating gas velocities for LLDPE, HDPE and BPE industrial fluidized beds are 0.6m/s, 0.6m/s and 0.4m/s, respectively.

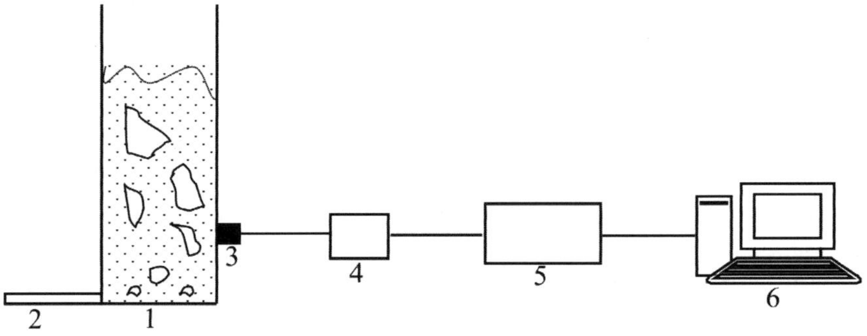

Fig. 1 Schematic diagram of experimental apparatus of gas-solid fluidized bed.
1 – Fluidized bed; 2 – Distributor; 3 – AE sensor; 4 – Preamplifier; 5 – main amplifier; 6 – Computer.

3.2 Liquid-solid stirred tank

The liquid-solid multiphase flow experiments were carried out in a stirred tank. Figure 2 is a schematic diagram of the experimental apparatus used in the present study. It consists of two parts: a stirred tank and an AE signal collection system. The stirred tank assembly includes a tank (with a diameter of 105mm and a height of 150mm) and an impeller of 6 bladed disk turbine (with a diameter of 55mm). The impeller was installed at a height of 15mm above the bottom of the tank. The impeller speed was measured using an electronic constant agitator and is accurate to ± 10 $r \cdot min^{-1}$. The stirred tank is charged with water with as a density of 1000 $kg \cdot m^{-3}$ as the liquid phase, and glass beads with a density of 2900 kgm^{-3} as the solid phase. The glass beads with three different diameters, 0.5 and 0.7mm, were investigated. A sampling frequency of 100kHz is determined in terms of the Shannon sampling theory.

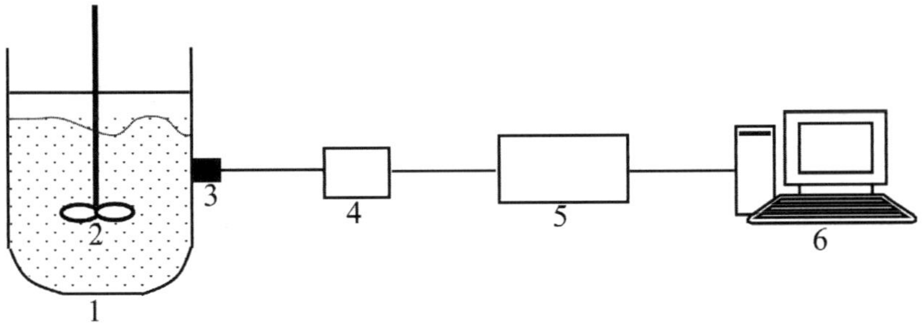

Fig. 2 Schematic diagram of experimental apparatus of liquid-solid stirred tank.
1 – Stirred tank; 2 – Impeller; 3 – AE sensor; 4 – Preamplifier; 5 – main amplifier 6 – Computer.

4. RESULTS AND DISCUSSION

4.1 Structure characteristics comparison between AE signals and pressure signals

One type of PE particles, i.e. BPE, with four different particle sizes of 0.159, 0.365, 0.565 and 0.942mm, was investigated to study the structure characteristics of AE signals. Fluidization was performed in the laboratory scale cold mode fluidized bed at a superficial gas velocity of $0.103 m \cdot s^{-1}$. The original AE signals from four different measurements, corresponding to four different particle sizes, were decomposed to 1-9 level detailed signals ($D_1 - D_9$) and level-9 approximated signal (A_9) using Daubechies 2nd order wavelet transform (Daubechies, 1988). The R/S analysis is further applied to the detailed signals $D_1 - D_9$, as well as the approximated signal A_9. Table 1 shows the Hurst exponents of decomposed signals for different particle sizes.

Table 1. Hurst Exponents of Decomposed Signals for Different Particle Sizes

Particle Size (mm)		Decomposed Signal									
		D_1	D_2	D_3	D_4	D_5	D_6	D_7	D_8	D_9	A_9
0.159	H_1	[–]	[–]	[–]	[–]	[–]	0.984	0.989	0.998	0.992	0.995
	H_2	0.275	0.244	0.288	0.283	0.289	0.496	0.512	0.639	0.671	0.673
	H_3	0.105	0.133	0.195	0.178	0.184	0.168	0.195	[–]	[–]	[–]
0.365	H_1	[–]	[–]	[–]	[–]	[–]	0.996	0.996	0.998	0.996	0.996
	H_2	0.217	0.238	0.275	0.249	0.289	0.494	0.529	0.616	0.666	0.661
	H_3	0.095	0.126	0.18	0.108	0.13	0.168	0.177	[–]	[–]	[–]
0.565	H_1	[–]	[–]	[–]	[–]	[–]	0.995	0.995	0.995	0.996	0.995
	H_2	0.25	0.242	0.218	0.273	0.291	0.489	0.518	0.653	0.662	0.67
	H_3	0.145	0.215	0.165	0.171	0.187	0.191	0.197	[–]	[–]	[–]
0.942	H_1	[–]	[–]	[–]	[–]	[–]	0.997	0.997	0.998	0.998	0.997
	H_2	0.27	0.252	0.316	0.281	0.295	0.488	0.518	0.553	0.662	0.67
	H_3	0.135	0.211	0.168	0.183	0.189	0.167	0.205	[–]	[–]	[–]

It can be seen from Table 1 that two distinct Hurst exponents for the detailed signals $D_1 - D_5$ of all different particle sizes, H_2 from the slope at smaller τ and H_3 from the slope at larger τ, are found to be much less than 0.5. Both Hurst exponents less than 0.5 indicate that the detailed signals $D_1 - D_5$ are much irregular and represent an anti-persistence behaviour in the gas-solid fluidized bed. Because the mechanism of acoustic emissions generated in gas-solid fluidized bed mainly owe to the collisions between particles and wall, which reflects the micro-scale interaction behaviour among particles and between particles and fluid, the irregular and high-frequency detailed signals $D_1 - D_5$ can be considered to imply the complex micro-scale motion of solids phase. Moreover, according to the wavelet analysis, the frequency band of the detailed signals $D_1 - D_5$ lies within [15.63kHz, 500kHz] being fair consistent with the main frequency band in a normal fluidized bed (Jiang et al., 2007), which further supports the above implication.

Three distinct Hurst exponents, H_1, H_2 and H_3 being the slopes at smaller, medium and larger τ respectively, are shown in Table 1 for the detailed signals $D_6 - D_7$ of all different particle sizes. It can be seen from Table 1 that (1) the Hurst exponent H_1 at smaller τ is much greater than 0.5 and almost approaches to 1.0, which indicates a highly persistent hydrodynamics feature of the gas-solid fluidized system, (2) the Hurst exponent H_2 at medium τ roughly equals to 0.5, which indicates a random hydrodynamics feature of the gas-solid fluidized system, and (3) the Hurst exponent H_3 at larger τ is much less than 0.5, which indicates a highly anti-persistent hydrodynamics feature of the gas-solid fluidized system. Generally, there are mainly two types of particles motions in gas-solid fluidized bed, particles motion along with solid phase and particles motion along with bubble phase respectively. Meanwhile, it is in general considered that the motion of bubble phase is more regular than that of solid phase. Therefore, the detailed signals $D_6 - D_7$ imply complex meso-scale interaction between solid phase and bubble phase, where the Hurst exponent H_1 at smaller τ represents a hydrodynamic feature of bubble phase and the Hurst exponent H_3 at larger τ represents a hydrodynamic feature of solid phase. It can also be seen from Table 1 that the Hurst exponent H_2 at medium τ is a little less than 0.5 for the detailed signal D_6 and that is a little greater than 0.5 for the detailed signal D_7. This implies that the detailed signals D_6 and D_7 can be seen as the meso-scale interaction in a fluidized bed mainly dominated by the motions of solid phase and bubble phase respectively.

Two distinct Hurst exponents, H_1 from the slope at smaller τ and H_2 from the slope at larger τ, are shown in Table 1 for the detailed signals $D_8 - D_9$ and the approximated signal A_9 of all different particle sizes. Both Hurst exponents are found to be much greater 0.5, which indicates that the detailed signals $D_8 - D_9$ and the approximated signal A_9 are regular and represent a persistence behavior in the gas-solid fluidized bed. Therefore, the detailed signals $D_8 - D_9$ and the approximated signal A_9 can be considered to represent the whole macro-scale interaction of the fluidized bed.

According to the above R/S analysis of the decomposed AE signals, a complete characterization of hydrodynamics in gas-solid fluidized bed can be described. In the previous study of Zhao et al. (2003), a criterion was established to resolve the pressure fluctuation signals into three characteristic scales in terms of different numbers of Hurst exponents: meso-scale signals with two distinct Hurst exponents; micro- and macro-scale signals with only one Hurst exponent. However, the above criterion is no longer suitable for multi-scale resolution of AE signals. A more general criterion is thus established to resolve AE signals into three characteristic scales: micro-scale signals with all Hurst exponents less than 0.5; meso-scale signals with some Hurst exponents less than 0.5 and some Hurst exponents greater than 0.5; macro-scale signals with all Hurst exponents greater than 0.5.

In order to perform the structure characteristics comparison between AE signals and pressure fluctuation signals, a plot of Hurst exponent H against level, called "structure diagram" in the present study, is adopted. Figs. 3 and 4 show the structure diagrams of AE signals and pressure signals respectively (for ease of explanation, level 10 in both Figs. 3 and 4 is actually a level 9 approximated signal), where the result of Figure 4 is derived and modified from the previous study of Zhao et al. (2003). It can be seen from Fig. 3 that (1) micro-scale signals consist of 1-5 level detailed signals, (2) meso-scale signals consist of 6-7 level detailed signals, and (3) macro-scale signals consist of 8-9 level detailed signals and level 10 approximated signals. It can be seen from Fig. 4 that (1) micro-scale signals consist of 1-2 level detailed signals, (2) meso-scale signals consist of 3-9 level detailed signals, and (3) macro-scale signals consist of level 10 approximated signals.

The energy percentages, R_s, for micro-, meso- and macro-scale AE signals are shown in Table 2 for different particle sizes. It can be seen from Table 2 that the most energy is distributed mainly by micro-scale signals consisting of the level 1-5 detailed signals, and is over 95% of the total energy. It can also be seen that the energy percentage of meso-scale signals increases as the particle size increases, which may be caused by the fact that the effect of the motion of solid phase on the meso-scale interaction in a fluidized bed increases as the particle size increases. However, as Zhao et al.

(2003) stated, the most energy of pressure signals is distributed mainly by meso-scale signals consisting of the level 3-9 detailed signals, and is over 90% of the total energy.

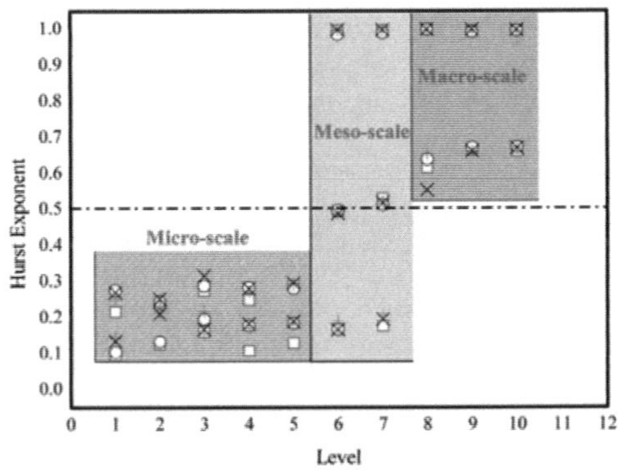

Fig. 3 Structure diagram of AE signals.
Particle size: 0.159mm (○); 0.365mm(□); 0.565mm (△); 0.942mm(×).

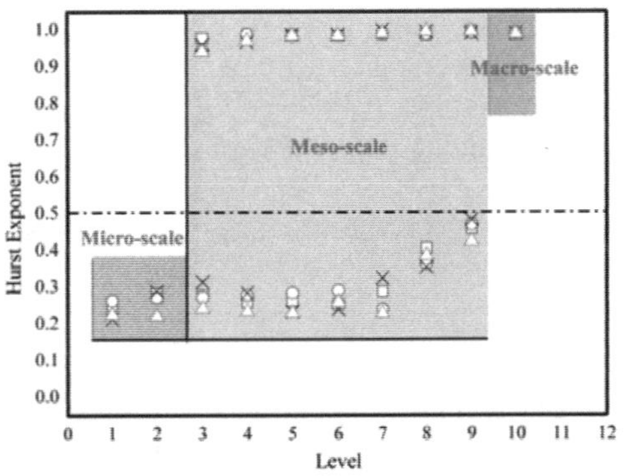

Fig. 4 Structure diagram of pressure signals.
Height: 0.09m(□); 0.20m (○); 0.40m(×); -0.12m (△).

Table 2. Energy Distribution for Multi-scale AE Signals

Particle Size (mm)	0.159			0.365			0.565			0.942		
Scale	Micro	Meso	Macro	Micro	Meso	Macro	Micro	Meso	Macro	Micro	Meso	Macro
R_S(%)	98.96	1.02	0.02	97.95	2.03	0.02	97.65	2.33	0.02	97.36	2.62	0.02

Therefore, the structure characteristics difference between AE signals and pressure signals can thus be summarized that (1) the number of levels of meso-scale in AE signals is much less than that in pressure signals, and (2) the most energies of AE signals and pressure signals are distributed mainly by micro-scale signals and meso-scale signals respectively. These observations are consistent with the measuring principles of AE sensor and pressure probe. AE sensor, i.e. piezoelectric accelerometer, measures vibrations of the wall, originated mainly by collision between particles and wall. This also means that it is not a good way for applying directly AE sensor to measure the information about the bubbles motion. Therefore, signals collected from AE measurement technique represent the micro-scale dynamics of solid phase, which corresponds to the observation that the most energy of AE signals is distributed mainly by micro-scale signals. In contrast, pressure probe measures the fluctuations of pressure, which are generated mainly by the meso-scale motion of bubble phase. The most energy of pressure signals is, therefore, distributed mainly by meso-scale signals.

Based on the above thorough investigations of underlying structure characteristic of AE and pressure signals, it is, therefore, helpful not only to choose a more suitable measurement technique with respect to measurement requirement, but also to establish physical meaningful prediction model. AE measurement technique is more suitable for measuring granular properties (e.g. particle size, particle size distribution, and particle density etc.), whereas pressure measurement technique is more suitable for measuring bubble properties (e.g. bubble size, and wake vortex etc.).

4.2 Structure characteristics comparison of AE signals in gas-solid and liquid-solid systems

In view of the fact that the physical characteristics of different multiphase systems are in general different, the structure characteristics of AE signals collected from different multiphase systems could exhibit larger difference. The structure characteristics of AE signals collected from gas-solid fluidized bed and liquid-solid stirred tank were compared in this section. Two experiments in liquid-solid stirred tank under complete suspension condition were conducted with the solid concentration of $0.13 g \cdot mL^{-1}$ and impeller speed of $8.33 r \cdot s^{-1}$ for two different glass bead diameters of 0.5mm and 0.7mm. The original AE signals from two different measurements with respect to two different glass bead sizes were decomposed into 1-9 level detailed signals and level-9 approximated signal using Daubechies 2nd order wavelet transform. Fig. 5 shows the structure diagram of AE signals by resorting to R/S analysis of the decomposed signals.

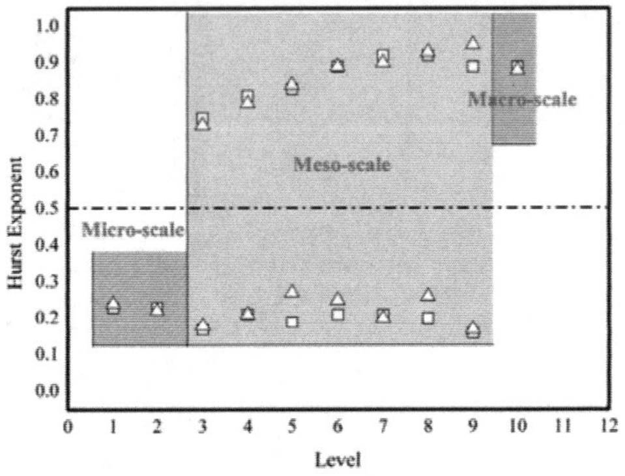

Fig. 5 Structure diagram of AE signals in liquid-solid stirred tank.
Glass bead size: 0.5mm(□); 0.7mm(△).

Compared the structure diagram of AE signals in gas-solid fluidized bed shown in Fig. 3 with the structure diagram of AE signals in liquid-solid stirred tank shown in Fig. 5, it is indicated that (1) the micro-scale signals consisting of 1-2 level detailed signals in liquid-solid stirred tank is much narrower than the micro-scale signals consisting of 1-5 level detailed signals in gas-solid fluidized bed, and (2) the meso-scale signals consisting of 3-9 level detailed signals in liquid-solid stirred tank is much wider than the meso-scale signals consisting of 6-7 level detailed signals in gas-solid fluidized bed. Moreover, the energy percentages of AE signals in liquid-solid stirred tank are 45.5%, 53.2% and 1.3% for micro-, meso- and macro-scale signals respectively. This means that the energy of AE signals in liquid-solid stirred tank is roughly distributed equally by micro- and meso-scale signals, which is different from the AE signals in gas-solid fluidized bed whose most energy is distributed mainly by micro-scale signals. These observed structure characteristics differences between liquid-solid and gas-solid systems could be supported by the fact that (1) in gas-solid fluidized bed, acoustic emissions are generated mainly by the collisions between particles and wall, and (2) in liquid-solid stirred tank, acoustic emissions are generated not only by the collisions between particles and tank, but also by the collisions between liquids and tank. On the other hand, under the complete suspension condition, the most particles are suspended in liquid phase, and the frequency band of collisions between liquids and tank could be comparable to that of collisions between those particles and tank, which may cause the wide meso-scale signals in liquid-solid tank.

In a word, structure characteristics of AE signals measured in different multiphase systems could be distinguished by resorting to structure diagram and energy distribution analysis. Investigating the underlying structure characteristics of signals could provide primary information not only to judge whether this measurement technique is suitable for a particular application, but also to reveal possible difficulties encountered. Moreover, it could be helpful to extract the most characteristic features for a particular application based on structure characteristics of signals.

4.3 Measurement of particle size distribution

The average particle size and particle size distribution (PSD) have significant effects not only on the properties of final products, but also on the performance of gas-solid fluidized beds, especially with respect to the polymerization of olefins in fluidized beds. Therefore, the developments of effective on-line monitoring techniques for both average particle size and PSD measurements are very essential. One of the most widely used methods to determine average particle size, as well as PSD, is manual sieve analysis. However, sieve analysis can not satisfy the real-time requirement because of its long sampling interval. Optical measurement technique can not be suitable for the stringent industrial environment, and its applications in industrial fluidized beds are rarely reported in open literature. Moreover, in view of the fact that the rays (e.g. χ-ray, and γ-ray etc.) are harmful to human health, it hinders their applications as measurement techniques. Therefore, the invention of a novel on-line measurement technique, not only health-friendly but also real-time, has become an urgently task. Recent years, AE measurement technique, considered as a non-intrusive method, has attracted considerable attentions and has been applied to measure average particle size (Halstensen et al., 2000; Boyd et al., 2001; Jiang et al., 2007). However, its application to on-line measurement of PSD is rarely reported.

The earlier analyses of structure characteristics of AE signals indicated that (1) micro-scale signals consisting of 1-5 level detailed signals totally reflect the information about particles motion, and (2) meso-scale signals consisting of 6-7 level detailed signals partially reflect the information about particles motion. In contrast, there are only two levels detailed micro-scale signals of pressure signals that totally represent the information about particles motion, while other seven levels detailed meso-scale signals just partially or even not at all represent the information about particles motion. It could thus be implied that AE technique is more suitable than pressure technique for measure average particle size and PSD. Therefore, as an illustrative application example of AE technique, a prediction model was established to measure PSD, as well as average particle size.

For n identical particles (with a diameter of d_p and a mass of m) impact on an area ΔA of the wall of a bed with the normal velocity of v, the resultant average force $<F(t)>$ in a time interval T is given by

$$< F(t) >= \frac{\int_0^T F(t)dt}{T} = \frac{2mv\int_0^T \sum_{i=1}^{n} \delta(t - t_i)dt}{T} \tag{1}$$

where $\delta(t)$ is Dirac delta function, and t_i is the arrival time of the ith particle on the wall. Let f_p denote the average arrival frequency of the particles on the wall, the number of collisions between particles and wall in a time interval T is $f_p \cdot T$. Thus, the average force $<F(t)>$ in an unit time interval can be expressed as,

$$< F(t) >= \frac{2mv\int_0^T \sum_{i=1}^{n} \delta(t - t_i)dt}{T} = \frac{2mvf_pT}{T} = 2mvf_p \tag{2}$$

Consequently, the acoustic pressure generated on an area ΔA of the wall can be given by

$$P_{AE} = \frac{\eta < F(t) >}{\Delta A} = \frac{2\eta mvf_p}{\Delta A} \tag{3}$$

where η is the transformation efficiency from the collision pressure to acoustic pressure detected by AE sensor. There are several factors that can take significant effects on the transformation efficiency η, such as the distance between collision position and sampling position, and the operation conditions of experimental apparatus. Let ρ_b and w denote the particle density near the wall and the mass fraction of particles with a size of d_p respectively, then the number concentration C (number·m^{-3}) of particles with a size of d_p can be expressed as $C=\rho_b \cdot w/m$. Consequently, the average arrival frequency of the particles on the wall, f_p, can be given by

$$f_p = C \cdot \frac{\Delta A v}{\Delta A} = \frac{\rho_b w v}{m} \tag{4}$$

Therefore, for particles (with d_p in diameter, m in mass, and w in mass fraction) impact on an area ΔA of the wall, the energy of signals received by AE sensor, E, during a time interval T, can be calculated by

$$E(d_p) = \int_0^T P_{AE} \Delta A v dt = \int_0^T 2\eta \rho_b v^3 w dt \tag{5}$$

It can be seen from Eq. (5) that the signal energy E is a function of transformation efficiency η, particle density near the wall ρ_b, particle normal velocity v and particle mass fraction m. By maintaining the operation conditions constant, such as superficial gas velocity and level of material, both the transformation efficiency η and the particle density near the wall ρ_b could be considered roughly to be constant. Meanwhile, under constant operation conditions, the normal velocity v is mainly related to the particle size distribution. It is, therefore, implied that the signal energy can be consider to be a function of PSD under specific operation conditions.

For particles with K different sizes jointly impact on the wall, the total mixed signal energy, E^{mix}, can be calculated based on the linear superposition principle of acoustic energy as follows:

$$E^{\text{mix}} = \sum_{k=1}^{K} E^k \cdot w_k \tag{6}$$

where E^k and w_k are the signal energy and mass fraction of kth type of particles with a size of $d_{p,k}$ respectively. In view of the independent feature of acoustic propagation, different acoustic waves generated from particles with different sizes impact on the wall will maintain their own characteristics (e.g. frequency, and amplitude etc.). Hence, it can be further supposed that the energy of acoustic on different levels can be maintained constant under acoustic superposition. According to wavelet analysis, the conservation relationships of signals for mixed particle sizes and signals for single particle size at different detailed levels, as well as approximated level, can be given as follows:

$$E_{j,D}^{\text{mix}} = \sum_{k=1}^{K} E_{j,D}^k \cdot w_k \,, \ j = 1, \cdots, L \tag{7}$$

$$E_{L,A}^{\text{mix}} = \sum_{k=1}^{K} E_{L,A}^k \cdot w_k \tag{8}$$

where L is the number of wavelet decomposed levels; $E_{j,D}^k$ and $E_{L,D}^k$ are the energies of detailed signal on the level of j and approximated signal on the level of L for the kth type of particles with a size of $d_{p,k}$ respectively; and $E_{j,D}^{\text{mix}}$ and $E_{L,D}^{\text{mix}}$ are the energies of detailed signal on the level of j and approximated signal on the level of L for mixed particle sizes respectively. Let λ_k denote a ratio of signal energy for kth type of particles and signal energy for mixed particle sizes, which is defined as

$$\lambda_k = \frac{E^k}{E^{\text{mix}}} \,, \ k = 1, \cdots, K \tag{9}$$

Therefore, Eqs. (7) and (8) can be further transformed as follows:

$$EP_{j,D}^{\text{mix}} = \sum_{k=1}^{K} \lambda_k \cdot EP_{j,D}^k \cdot w_k \,, \ j = 1, \cdots, L \tag{10}$$

$$EP_{L,A}^{\text{mix}} = \sum_{k=1}^{K} \lambda_k \cdot EP_{L,A}^k \cdot w_k \tag{11}$$

where $EP_{j,D}^k$ and $EP_{L,D}^k$ are the energy percentages of detailed signal on the level of j and approximated signal on the level of L for the kth type of particles respectively; and $EP_{j,D}^{\text{mix}}$ and $EP_{L,D}^{\text{mix}}$ are the energy percentages of detailed signal on the level of j and approximated signal on the level of L for mixed particle sizes respectively. Hence, after calibrations of the values of λ_k (k=1,…,K), as well as the values of $EP_{j,D}^k$ and $EP_{L,D}^k$, where k=1,…,K and j=1, …,L, the mass fractions of different types of particles, $w_k(k=1,\ldots,K)$, can be obtained by solving the linear equations shown in Eqs. (10) and (11). Because the earlier study on structure characteristics of AE signals has implied that the macro-signals consisting of the detailed signals on the levels of 8 and 9 and the approximated signals on the level of 9 do not reflect the information about particles motion at all, it is supposed that decomposing the AE signals into larger than 7 levels will not provide extra information for prediction of PSD, which has been demonstrated based on our experimental results. The number of decomposed levels, L, is thus set equal to 7 in the present study.

To illustrate the effectiveness of the PSD prediction model, the on-line measurement experiments were carried out both in laboratory scale and plant scale gas-solid fluidized beds. Three types of PE particles collected from plant scale plant scale fluidized beds, i.e. LLDPE, HDPE and BPE, were used as the fluidized particle in laboratory scale fluidized beds. The model parameters in Eqs. (10) and (11) were calibrated based on the AE signals collected from laboratory scale apparatus. The results are shown in Table 3. It can be seen from Table 3 that the mass fractions of different particle sizes measured using AE method is consistent with that measured using sieve method. Define the average absolute deviation, AAD, of PSD prediction between AE method and sieve method as

$$AAD = \frac{1}{K}\sum_{k=1}^{K}\left|w_k^{\text{sieve}} - w_k^{\text{AE}}\right| \tag{12}$$

where w_k^{sieve} and w_k^{AE} are the mass fractions of the kth type of particles measured by sieve method and AE method respectively; and K=7 is the number of types of particles. It can be calculated from Table 3 that (1) in the laboratory scale experiments, the values of AAD are 0.62%, 1.60% and 1.57% for LLDPE, HDPE and BPE respectively, and (2) in the plant scale experiments, the values of AAD are 1.26%, 1.86% and 2.14% for LLDPE, HDPE and BPE respectively.

Table 3. Comparisons of Average Particle Size and Particle Size Distribution Measured Using AE Method and Sieve Method in Laboratory Scale and Plant Scale Apparatuses

Type	Method		Mass fractions for different particle sizes (%)							Average particle size (mm)
			2mm	1.19mm	0.71mm	0.5mm	0.36mm	0.18mm	0.14mm	
LLDPE	Sieve method		0.50	6.60	4.00	27.70	42.60	9.90	9.30	0.4396
	AE method	laboratory scale	1.00	6.33	3.90	26.20	43.37	9.30	9.90	0.4407
		plant scale	0.39	8.40	4.40	25.70	40.10	11.00	10.10	0.4440
HDPE	Sieve method		15.80	31.80	15.90	29.80	4.10	2.60	0.00	0.9758
	AE method	laboratory scale	14.80	33.60	15.00	27.60	2.60	4.40	2.00	0.9604
		plant scale	14.10	36.00	12.60	30.00	2.90	2.30	2.10	0.9674
BPE	Sieve method		4.53	9.50	19.74	20.47	32.89	12.87	0.00	0.5877
	AE method	laboratory scale	4.80	11.60	16.80	19.90	30.90	13.00	3.00	0.5917
		plant scale	5.10	13.80	17.90	16.60	31.10	13.30	2.20	0.6153

Moreover, because the average particle size can be calculated based on the measured PSD, the average particle size measured by AE method and sieve method are also compared and shown in Table 3. It can be calculated from Table 3 that (1) in the laboratory scale experiments, the relative deviation of average particle size between AE method and sieve method are 0.25%, 1.57% and 0.67% for LLDPE, HDPE and BPE respectively, and (2) in the plant scale experiments, the relative deviation of average particle size between AE method and sieve method are 0.99%, 0.86% and 4.69% for LLDPE, HDPE and BPE respectively. The results of plant scale experiments for the prediction of average particle size by the PSD prediction model proposed in this study show superior performance to the frequency model proposed by the Jiang et al.(2007). The superiority of PSD prediction model could be supported by the fact that wavelet transform analysis is more suitable to deal with non-stationary signals than Fourier transform analysis. In a word, the results illustrate that the AE measurement in combination with the PSD model provide an effective tool to on-line measurement of both average particle size and PSD.

5. CONCLUSIONS

This work investigated thoroughly the structure characteristics of AE signals measured in gas-solid fluidized bed and liquid-solid stirred tank by resorting to wavelet transform and R/S analysis. A general criterion was established to resolve AE signals to three characteristic scales based on the Hurst exponent characteristics: micro-scale signals with all Hurst exponents less than 0.5; meso-scale signals with some Hurst exponents less than 0.5 and some Hurst exponents greater than 0.5; macro-scale signals with all Hurst exponents greater than 0.5. Meanwhile, structure diagram, a plot of Hurst exponent against level, was introduced.

By the comparison of structure characteristics between AE and pressure signals in gas-solid fluidized bed, it was found that (1) AE signals in micro-scale consisting of 1-5 level detailed signals is much wider than pressure signals in micro-scale consisting of 1-2 level detailed signals, and (2) AE signals in meso-scale consisting of 6-7 level detailed signals is much narrower than pressure signals in meso-scale consisting of 3-9 level detailed signals. Further, energy distribution analysis revealed that the most energies in AE and pressure signals were distributed mainly by the micro-scale and meso-scale signals respectively. These observations imply that AE signals represent mainly micro-scale particles motion while pressure signals represent mainly meso-scale interaction dynamics between particles motion and bubbles motion. Therefore, AE technique could be considered as an effective tool to measure particle-related properties in gas-solid fluidized bed.

The structure characteristics of AE signals collected from gas-solid fluidized bed and liquid-solid stirred tank were also compared based on structure diagram and energy distribution analysis. The results indicated that although the same measurement technique was adopted, the structure characteristics of AE signals measured in gas-solid fluidized bed and liquid-solid stirred tank still exhibited larger difference. Compared with the structure diagram of AE signals in fluidized bed, it was found micro- and meso-scale of AE signals in liquid-solid stirred tank becomes narrow and broadens respectively. On the other hand, the energy of AE signals in liquid-solid stirred tank was roughly distributed equally by micro- and meso-scale signals, which was different from the AE signals in gas-solid fluidized bed whose most energy is distributed mainly by micro-scale signals.

Finally, as an illustrative application of AE technique in process monitoring, a prediction model was proposed to measure particle size distribution (PSD) and average particle size for three types of PE, i.e. LLDPE, HDPE and BPE. The on-line measurements of PSD and average particle size were performed both in laboratory scale and plant scale fluidized beds. The results showed that (1) average absolute deviation between AE method and manual sieve method were no more than 2.14% for PSD prediction, and (2) relative deviation between AE method and sieve method were no more than 4.69% for average particle size prediction.

However, the present study is preliminary and needs to be further investigated. On the one hand, analyses of structure characteristics based on structure diagram and energy distribution need to be further extended to other multiphase systems (e.g. gas-liquid, gas-solid-liquid, and gas-liquid-liquid etc.). On the other hand, it needs to further develop more effective on-line models for process monitoring with respect to AE technique based on the studies of structure characteristics, as well as advanced statistical and intelligent methods.

ACKNOWLEDGEMENTS

The authors are thankful for the support of the National Natural Science Foundation of China (20490205) and National High Technology Research and Development Program of China (2007AA04Z182).

REFERENCES

Boyer, C. et al. (2002). "Measuring techniques in gas-liquid and gas–liquid–solid reactors," Chem. Eng. Sci., 57, pp. 3185-3215.

Boyd, J. W. R. et al. (2001). "The uses of passive measurement of acoustic emissions from chemical engineering processes," Chem. Eng. Sci., 56, pp. 1749-1767.

Briens, C. L. et al. (1997). "Hurst's analysis to detect minimum fluidization and gas maldistribution in fluidized beds," AIChE J., 1997, 43, pp. 1904-1908.

Briongos, J. V. et al. (2006). "Phase space structure and multi-resolution analysis of gas–solid fluidized bed hydrodynamics: Part I—The EMD approach," Chem. Eng. Sci., 61, pp. 6963-6980.

Daubechies, I. (1988). "Orthonormal bases of compactly supported wavelets," Commun. Pure. Appl. Math., 41, pp. 909-996.

Fan, L. T. et al. (1991). "Stochastic analysis of a three-phase fluidized bed : Fractal approach," AIChE J. 1990, 36, pp. 1529-1535.

Fan, L. T. et al. (1993). "Fractal analysis of fluidized particle behavior in liquid-solid fluidized beds," AIChE J., 39, pp. 513-517.

Halstensen, M. et al. (2000). "New developments in acoustic chemometric prediction of particle size distributions "the problem is the solution"," J. Chemom., 14, pp. 463-481.

Hurst, H. E. (1951). "Long-term storage capacity of reservoirs," Trans. Amer. Soc. Civil Engs., 116, pp. 770-808.

Jiang, X. J. et al., (2007). "Study of the Power Spectrum of Acoustic Emission (AE) by Accelerometers in Fluidized Beds," Ind. Eng. Chem. Res., 46, pp. 6904-6909.

Lu, X. S. et al. (1999). "Wavelet analysis of pressure fluctuation signals in a bubbling fluidized bed," Chem. Eng. J., 75, pp. 113–119.

Ren, C. J. et al. (2008). "Determination of critical speed for complete solid suspension using acoustic emission method based on multiscale analysis in stirred tank," Ind. Eng. Chem. Res., 47, pp. 5323-5327.

Yang, T. Y. et al. (2008). "Study of transition velocities from bubbling to turbulent fluidization by statistic and wavelet multi-resolution analysis on absolute pressure fluctuations," Chem. Eng. Sci., 63, pp. 1950-1970.

Zhao, G. B. et al. (2003). "Multiscale resolution of fluidized-bed pressure fluctuations," AIChE J., 49, pp. 869-882.

Study on the Response of the Full-bore Conductance Sensor for Water Cut Measurement

Ronghua Xie[1], Xingbin Liu[2], Jinhai Hu[2], Xuefei Dai[2],Fujun Shan[2],Wenfeng Xu[3]

[1] Daqing Oil Field Limited Company, Daqing, Heilongjiang 163453, CHINA

[2] Daqing Logging & Testing Services Company, Daqing, Heilongjiang 163453, CHINA

[3] School of Instrument Science and Opto-electronic Engineering, Beihang University, Beijing 100083, CHINA

xieronghua@petrochina.com.cn

dlts_liuxb@petrochina.com.cn

hujinhai66@126.com

Abstract. This paper has proposed a new structure of full-bore conductance sensor, which is designed for measuring water cut of the oil-water two-phase flow. The structure of the full-bore conductance sensor and the measurement principle are introduced in the paper. The mental ring-shaped electrode is mounted on the outside wall of the cylindrical insulation body. When the electrode is provided with constant current, according to the electrical theory, the electrode generates a voltage, the value of which is inversely proportional to the conductivity of fluid flowing between the sensor and the casing.

The electrostatic field simulations of the sensor are accomplished by using ANSYS software. The results of the potential distribution simulation show that the potential decays quickly from the electrode along r direction (radial) and z direction (axial)to both sides, and the potential only distributes in a very narrow area near the electrode.

A series of static experiments on the sensor are carried out in laboratory. The experiment results agree with the simulation results. In radial direction, the closer the rod is to the sensor, the more sensitive the sensor becomes and the greater the relative response becomes. In axial direction, the electrode only responds in a certain region on both sides of the electrode and decays rapidly from the electrode to both sides.

And the salinity experiment is conducted in salt solution (3000 ppm), which shows that within the allowable range of experiment error, there is no effect of salinity on the sensor response. response.

1. INTRODUCTION

As an important part of production logging, production profile logging mainly monitors production rate and measures water content of each production layer in an oil well to estimate the exploitation effects of various types of oil layers, so as to provide a reliable basis for water plugging and readjustment of injection-production strategy. In China, most oilfields are developed with multilayered exploitation technique and zonal injection method. Therefore, it is important to learn about the flow rate and water cut of each layer with logging tools (Wang Peilie et al., 2001, Liu Xingpu et al., 1999). In China, two-parameter logging tools have been developed for production profile logging, in which a spinner sensor is used for measuring flow rate and a capacitance sensor for water cut. To overcome the shortcoming of the capacitance sensor which will lose resolution with high water cut, Daqing Logging & Testing Services Company has produced a kind of conductance water cut meter, based on packing method, which appears more preferable with high water cut and has widely used in low-production wells. A packer can mix the two phases more sufficiently, so as to decrease the effects of two-phase slippage and the uneven spatial distribution of dispersed phase. But, this method may change the flow regime and is not adaptive for high-production wells, can not complete the continuous measurement. The full-bore measurement method can overcome above shortcomings and fulfill the continuous measurement of water cut in high-production wells with less impact on the flow state.

2. MEASUREMENT PRINCIPLE OF FULL-BORE CONDUCTANCE SENSOR

The conductivity of the two-phase fluid composed of continuous conductive phase and dispersed non-conductive phase mainly depends on the conductivity of the continuous phase and the volume content (holdup) of the continuous phase, also it is influenced by the spatial distribution of the dispersed phase. Maxwell proposed an analytical model in 1881: When small solid non-conductive particles, small gas-bubbles or small oil-bubbles are distributed uniformly in continuous conductive phase of which the conductivity is σ_c and the holdup is β, the relationship between the mixture phase conductivity σ_m, β and σ_c is:

$$\sigma_m/\sigma_c = 2\beta/(3-\beta) \tag{1}$$

Maxwell model is applicable to finely bubbly flow. In ideal annular flow, the relationship between the mixture phase conductivity and the holdup of continuous phase was put forward by Begovich & Watson in 1978 (Begovich J.M. et al., 1978):

$$\sigma_m/\sigma_c = \beta \tag{2}$$

When water is continuous, for oil-water two-phase upward flowing fluid, the flow regime usually is bubbly flow, so Maxwell model can be applied. But even in bubbly flow, it is possible that the mixed state of oil and water should be very different. With higher velocity, the oil-water two-phase flow which is mixed uniformly can be regarded as finely bubbly flow, while with very low velocity, the uneven distribution of water and oil make the relationship between the conductivity of the mixture and water holdup different from that of uniformly mixed state. Therefore in practice, the analytical model can not be used directly, and it is necessary to calibrate the water cut meter at different flow rates to get a calibration chart.

The structure of the full-bore conductance sensor is shown in Fig.1. The mental ring-shaped electrode is mounted on the outside wall of the cylindrical insulation body. When the electrode is provided with constant current, according to the electrical theory, the electrode generates a voltage, the value of which is inversely proportional to the conductivity of fluid crossing the sensor:

$$u_w/u_m = \sigma_m/\sigma_w \tag{3}$$

Where u_w is the voltage when the casing is full of water, u_m is the voltage when the casing is filled with oil-water two-phase fluid, water holdup β can be obtained by u_w/u_m.

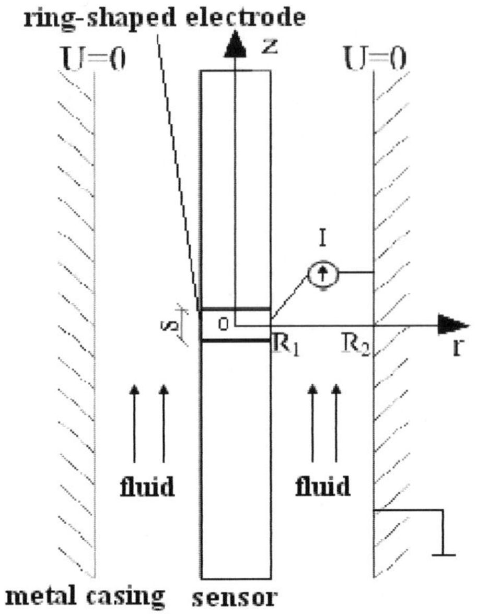

Fig.1 Structure of the full-bore conductance sensor

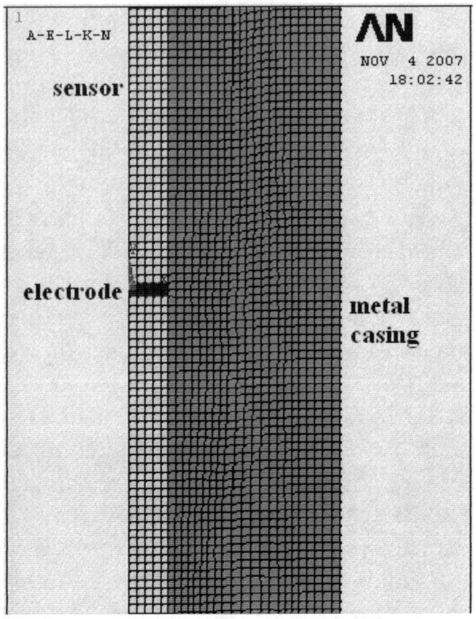

Fig.2 ANSYS model of the full-bore conductance sensor

3. ELECTRIC FIELD SIMULATIONS OF FULL-BORE CONDUCTANCE SENSOR

As shown in Fig.1, suppose that the fluid flowing between the sensor and the casing is uniform, linear and isotropic with conductivity σ. When the excitation electrode is provided with constant current I, a surface potential is generated on the interface where the electrode contacts with the fluid, that is, a sensitive electric field is generated in the area through which the fluid flows. According to the electromagnetic theory, the potential u of the field meets the Laplace equation. In cylindrical coordinate system, taking the axial symmetry of the model into account, the Laplace equation becomes:

$$\nabla^2(r,z) = \frac{\partial^2 u(r,z)}{\partial r^2} + \frac{1}{r}\cdot\frac{\partial u(r,z)}{\partial r} + \frac{\partial^2 u(r,z)}{\partial z^2} = 0 \tag{4}$$

Supposing that when the electrode is enough narrow, the current density on the electrode surface is uniformly distributed. The surface potential u of the excitation electrode meets:

$$\left.\frac{\partial u(r,z)}{\partial r}\right|_{r=R1} = -\frac{I}{2\pi\sigma R_1 S} \quad (-\frac{S}{2} \le z < \frac{S}{2}) \tag{5}$$

Where S is the width of the excitation electrode, R_1 is the radius of the electrode and R_2 is the radius of the well bore. The structure of the sensor is symmetrical to the roz plane and the excitation electrode is just in the center of the plane, there is:

$$\left.\frac{\partial u(r,z)}{\partial z}\right|_{z=0} = 0 \tag{6}$$

Metal casing is grounded, there is:

$$u(r,z)\big|_{r=R_2} = 0 \tag{7}$$

Using ANSYS software, in accordance with above boundary conditions, the electrostatic field of the sensor is simulated.

As the axial symmetry of the experiment model, three-dimensional model can be simplified for two-dimensional axial symmetric plane to complete ANSYS simulation. Set inside radius of the casing 75mm, height 800mm. Taking the contamination of the electrode, the divergence of current field and the grounded casing into account, set radius of the full-bore conductance sensor 14mm, height 725mm, and width of the electrode 5mm. The excitation electrode is also the measurement electrode, which connects to a constant current resource. Fig.2 is the ANSYS model of the full-bore conductance sensor. Fig.3 (a) is the potential distribution of the sensor showing that the potential decays quickly from the electrode along r direction (radial) to both sides. Fig.3 (b) shows the potential decays quickly from the electrode along z direction (axial) to both sides, and the potential only distributes in a very narrow area near the electrode.

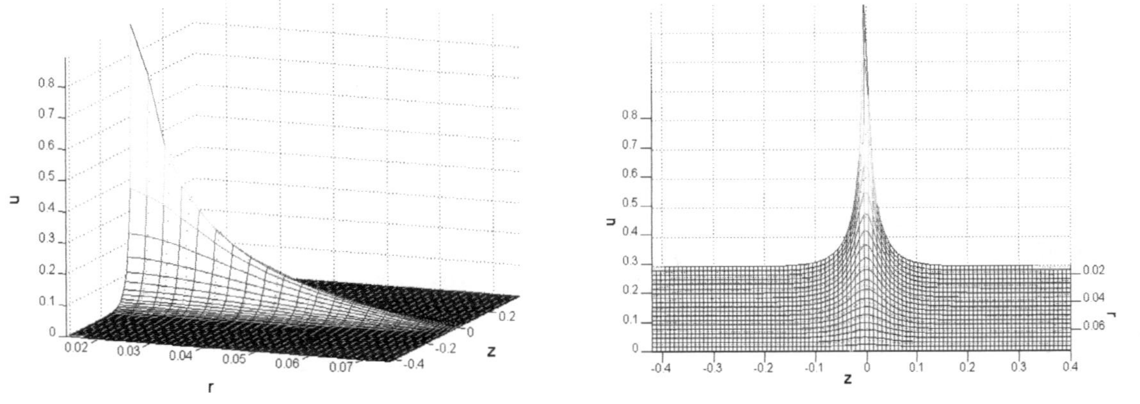

(a) Radial potential distribution (b) Axial potential distribution

Fig.3 Potential distribution of the full-bore conductance sensor

4. STATIC EXPERIMENTS

The purpose of static experiments is to find out the response characteristics of full-bore conductance sensor and verify the results of simulation qualitatively.

Locate the sensor just at the center of a cylindrical metal container vertically, which is full of tap water (the salinity is about several hundreds ppm), ground the cylinder, and supply a constant alternating current to the excitation electrode. Use long cylindrical dielectric rods (slightly higher than the cylinder with diameter 20mm) to simulate non-conductive phase oil. Different number of rods inserted into the cylinder will result in different water holdup which can be calculated easily. When there is no rod in, the voltage of the electrode is named full-water value, and when several rods were inserted, the voltage of the electrode is named mixture value. The mixture value dividing by the full-water value is the relative response.

(1)Insert one rod into the container vertically on a circumference with radius 65mm and the same center as the sensor, record the voltage of the electrode and calculate the volume fraction of water. Then, insert the rods one by one at different position of the circumference, until 5, record the voltage of the electrode and calculate the volume fraction of water at each increment of rod. Change the radius of the circumference (55mm, 45mm, 35mm, and 25mm) and repeat above steps. Fig.4 is the dependence of relative response on water fraction with different radial distances and shows that the relative response is proportional to water holdup and, the sensitivities and the intensities of electric fields are the same with the same distance between the rods and the sensor.

(2) Remove one rod from 65mm to 25mm from the sensor gradually, record the voltage of the electrode at different positions. Then, increase the number of the rods and repeat the above steps. Fig.5 is the dependence of relative response on radial distance from the sensor with different water fractions and shows that the closer the rod is removed to the sensor, the more sensitive the sensor becomes and the greater the relative response changes, that is, the electric field is much stronger near the sensor, which is identical to the simulation in Fig.3 (a).

(3)Put a piece of rod downwards into the container at 65mm from the sensor, record the electrode voltage at each decline of 1cm until the voltage is constant. Then change the point and repeat above steps. Fig.6 is the dependence of relative response on axial distance from the sensor with different radial distance from the sensor. It is obvious that the electric field is stronger only in a certain area near the electrode, and the electrode is sensitive to the change of impedance in this area. On the contrary, there is little response outside this area, and the potential decays rapidly from the electrode to both sides, which is identical to the simulation in Fig.3 (b).

(4) In order to know the effect of salinity on the relative response, the 4th experiment is conducted in salt solution (3000 ppm), in a manner similar to the 1st experiment. Fig.7 shows the relative response

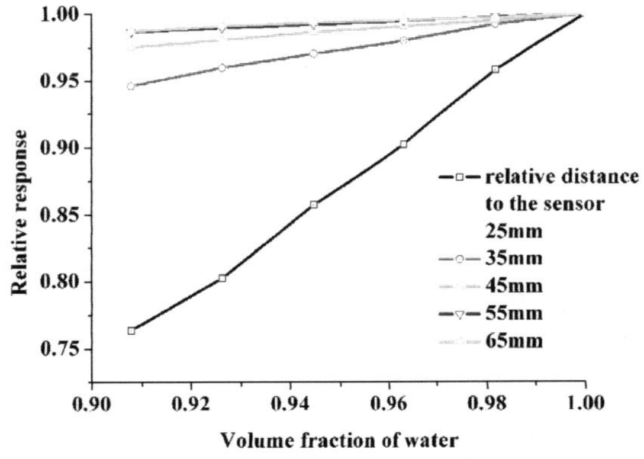

Fig.4 Dependence of relative response on water fraction with different radial distances

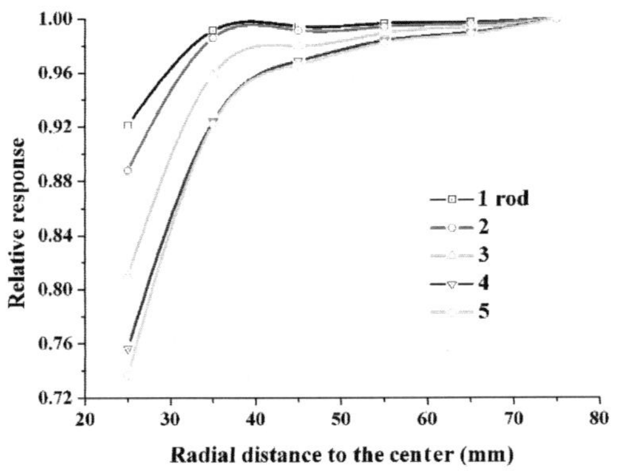

Fig.5 Dependence of relative response on radial distance with different water fractions

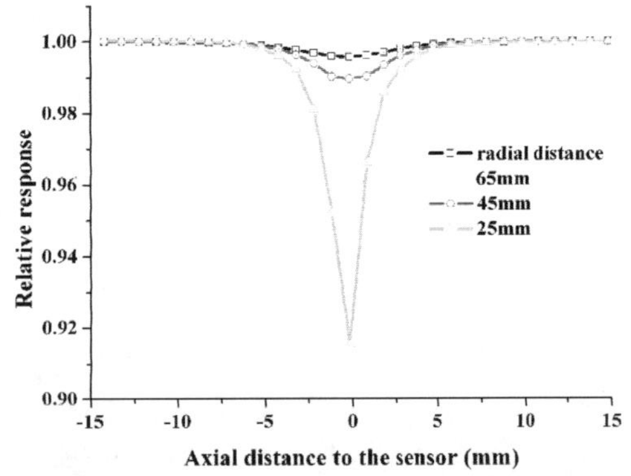

Fig6. Dependence of relative response on axial distance with different radial distances from the sensor

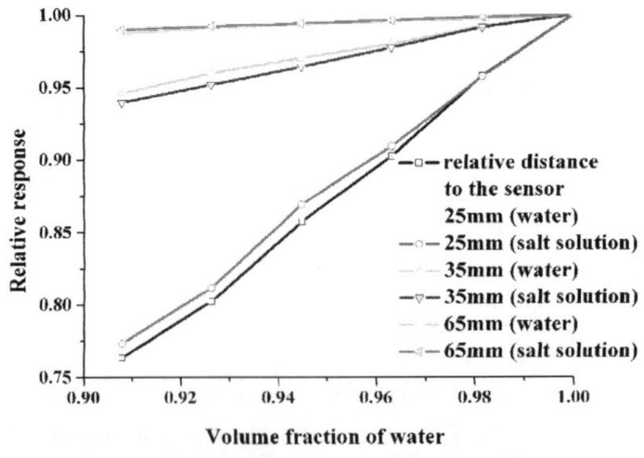

Fig7. Dependence of relative response on water fraction with different radial distance from the sensor

in tap water and that of salt solution is very small. Within the allowable range of experiment error, there is no effect of salinity on relative response (Hu Jinhai et al., 1999).

5. CONCLUSIONS

This paper has proposed a full-bore conductance sensor and carried out simulation and experimental researches on it. Simulation results and experiments results are presented by which the response characteristics of the full-bore conductance sensor are known, the conclusions are listed as follows:
(1)It is learned from the simulation results that the potential decays rapidly from the electrode along r direction (radial direction) to the casing wall, along z direction (axial direction) to both sides, and the potential only distributes in a very narrow area near the electrode.
(2)It is learned from the static experiments that the relative response is proportional to water holdup and, the sensitivities and the intensities of electric fields are the same with the same distance between the rods and the sensor. In radial direction, the closer the rod is removed to the sensor, the more sensitive the sensor becomes and the greater the relative response changes. In axial direction, the electrode only responds in a certain region on both sides of the electrode and decays rapidly from the electrode to both sides. Within the allowable range of experiment error, there is no effect of salinity on relative response.
(3)The experiment results agree with the simulation results.
The electrostatic field simulations and the static experiments show that full-bore conductance sensor proposed in the paper is affected by inhomogeneous distribution of fluid. When the water cut is same, with different spatial distribution of the oil phase, the response of the sensor is different. Under this circumstance, this sensor can not be applied directly to measure water cut in boring casing. Whereas when the measured fluid is distributed uniformly, the sensor can be used to measure water cut. Based on the characteristics of the sensor, it can be applied to study the flow regime. In actual flow fields, when the oil-water fluid is distributed uniformly with high flowrate, the sensor can be used to measure water cut directly.

ACKNOWLEDGEMENTS

The Authors gratefully thank the National High Technology Development 863 Program of China (2007AA06Z231).

REFERENCES

Begovich J.M., Watson J S. (1978). "An Electroconductivity Technique for the Measurement of Axial Variation of Holdups in Three-phase Fluidized Beds," AIChE J， 1978， 24(2), pp 351-354.

Hu Jinhai, Liu Xingbin et al. (1999). "Impedance water cut meter and its application," Well Logging Technology, 23, pp 511-514.

Liu Xingpu, Zhou Yongxin, Li Xiqing. (1999). "Production Profile Logging Technique and Its Application Effect Analysis," Fault-Block Oil & Gas Field, 7(2), pp 47-49.

Wang Peilie, Zhao Zhongjian, Yao Qiang. (2001). "A Review of Production Profile Logging in China," Well Logging Technology, 25(3), pp 163-167.

Journal of Physics: Conference Series **147** (2009) 012010 doi:10.1088/1742-6596/147/1/012010

PIV Measurement of a Contraction Flow using Micro-Bubble Tracer

Masaaki ISHIKAWA[1], Kunio IRABU[1], Isao TERUYA[1], Munehiro NITTA[1]

[1] University of the Ryukyus, 1 Senbaru, Nishihara-cho, Okinawa, 903-0213, Japan

ishi8614@tec.u-ryukyu.ac.jp

Abstract. Recently, a technique using the micro-bubbles is focused. It was applied to many fields such as purification of rivers and lakes, washing the industrial parts, growth of plants and marine products. The characteristics of micro-bubbles are small size, wide surface area, low terminal velocity, and so on. If this micro-bubble is available as tracer of PIV (Particle Image Velocimetry), environment load would become lower because it doesn't need to discard particle. In this paper, we make a micro-bubble generator with Venturi type mechanism. The generated micro-bubbles are applied to a vertical channel flow with contraction. We validate about traceability of the micro-bubble tracer in comparison with the particle tracer.

1. INTRODUCTION

Recently, a technique using the micro-bubbles is focused. It was applied to many fields such as purification of rivers and lakes, washing the industrial parts, growth of plants and marine products (Ohnari, 2006). The characteristics of micro-bubbles are small size, wide surface area, low terminal velocity, and so on. In previous PIV the particle tracer is used. If this micro-bubble is available as tracer of PIV (Particle Image Velocimetry), environment load would become lower because it doesn't need to discard particle.

In this study, we make a micro-bubble generator with Venturi type mechanism. The generated micro-bubbles are applied to a vertical channel flow with contraction, and we validate about traceability of the micro-bubble tracer in comparison with the particle tracer.

2. EXPERIMENTAL APPARATUS

2.1 Micro-bubble generator

Figure 1 shows schematic diagram of micro-bubble generator with Venturi structure (Ishikawa M., 2008). The generation mechanism of micro-bubbles is made from turbulence and strong shear stress at throat part. The inlet and outlet size is 8mm in diameter and throat part is 3mm in diameter. Although Diffuser part should have the gentle taper, we gave expansion structure which has 45 degree angle for easy fabrication. Air is supplied by compressor, it is stored in buffer area, and it is introduced from gap in front of the throat part. Air pressure is about 75kPa, air flow rate is 5L/min. Working fluid is tap water, and the tap water passes the 10μm filter in order to restrain the contaminant. Water pressure is 100kPa, water flow rate is about 5L/min.

Figure 2 shows the relationship between bubble equivalent diameter and number of bubble. Macro zoom lens is used to measure the bubble size. The magnification is about 3 times. Tolerance range of bubble size from image size is 10μm to 1.7mm. From this figure, the peak of a number of

Journal of Physics: Conference Series **147** (2009) 012010 doi:10.1088/1742-6596/147/1/012010

bubbles is obtained at 20-30μm in bubble diameter, and also an average diameter of bubbles is about 50μm.

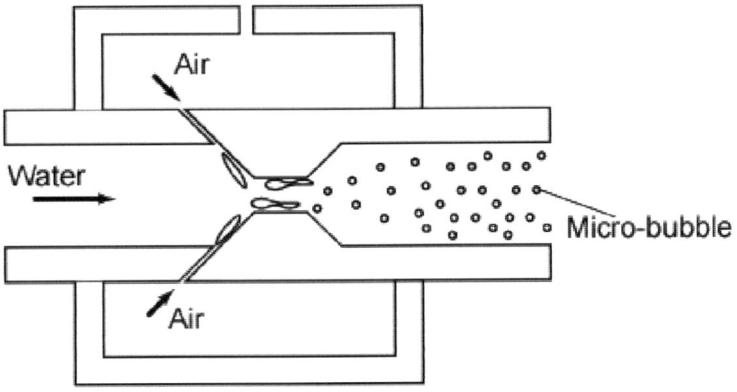

Fig.1 Micro-bubble generator

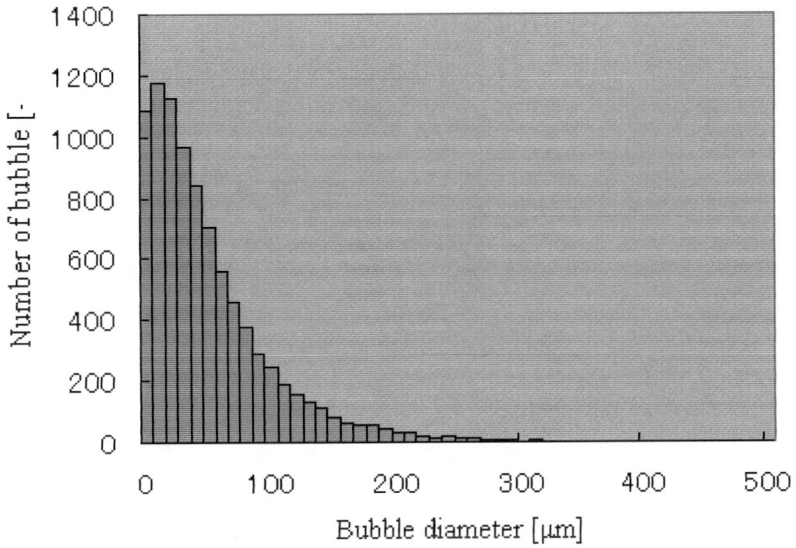

Fig.2 Histogram of bubble equivalent diameter

2.2 Vertical Contraction Channel

Figure 3 shows the experimental apparatus of the vertical contraction channel. Vertical channel is made of transparency acrylic resin. The height is 1700mm and the depth is 20mm. Contraction is set to 1000mm from the bottom. At contraction the width of channel varies from 50mm to 20mm linearly. Working fluid is tap water. Water flow rate is 50L/min (superficial velocity is 2.1m/s at downstream of the contraction). Micro-bubble generator is set inside lower tank. Micro-bubble is generated by the setting in Sec.2.1.

For visualization, we uses a high-speed camera (Photron inc.: FastCam-PCI (256×240 pixels, frame rate: 500fps)). The camera is set in front of contraction part. For illumination, we use a laser (Viasho: YAG laser (250mW, wave length: 532nm). The Laser light is used as laser light sheet with about 1mm in width by using a cylindrical lens.

And also we uses the white high-porous polymer particles (MCI: HP60SS, diameter: 60μm , specific gravity: 1.02) in order to compare with micro-bubble tracer .

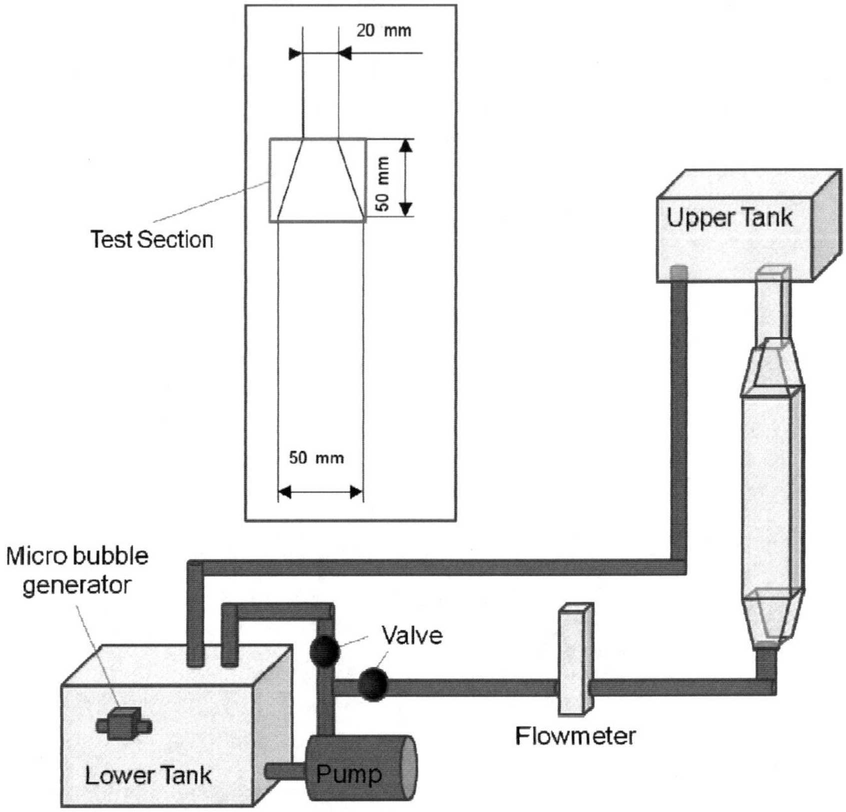

Fig.3 Experimental apparatus of vertical contraction channel

3. RESULTS AND DISCUSSIONS

Figure 4 (a) and (b) show the visualized image using the micro-bubble tracer and particle tracer, respectively. A number of particles in fig.4 (b) are about 100. Number of particles is almost the same in each image in order to the circuit water channel. But from fig.4 (a) in case using micro-bubble tracer, number density of micro-bubble tracer in fig. 4(a) is less than that of the particle tracer in fig.4 (b). This time, we have some problems which generation of micro-bubble is not enough and bubble rise and disappear in the lower tank. From this figures, we find that we have to get higher number density of tracer.

Figure 5 (a) and (b) show instantaneous velocity vector maps obtained by PIV for micro-bubble tracer and particle tracer, respectively. From both figures, we find that flow move to a center of the contraction smoothly and increase the velocity and obtain accelerating distribution in order to reduce the cross-section.

Figure 6 (a) and (b) show the mean velocity vector maps for micro-bubble tracer and particle tracer, respectively. This mean velocity vector is obtained from about 4000 instantaneous velocity vector maps (in about 8sec). From these figures, mean vector maps for micro-bubble tracer is as same as the one of the particle tracer. Although number density of the micro-bubble tracer is lower, we could confirm that the micro-bubbles are available as tracer for such acceleration flow fields.

(a) micro-bubble tracer

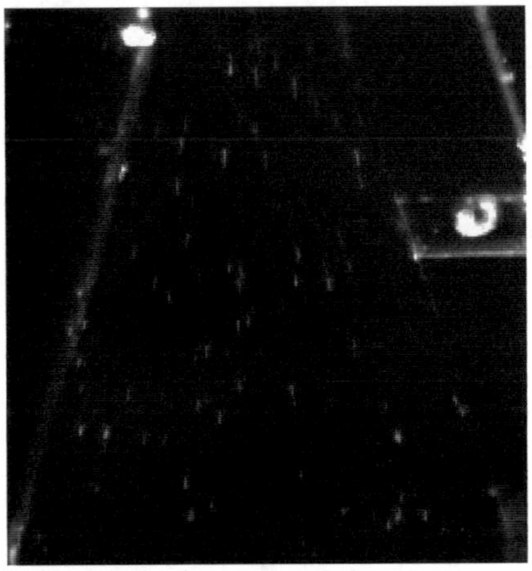
(b) particle tracer

Fig.4 Visualized images

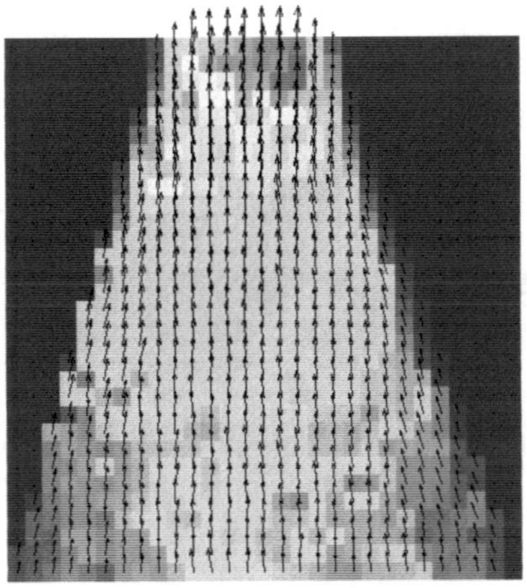
(a) micro-bubble tracer

(b) particle tracer

Fig.5 Instantaneous velocity vector maps

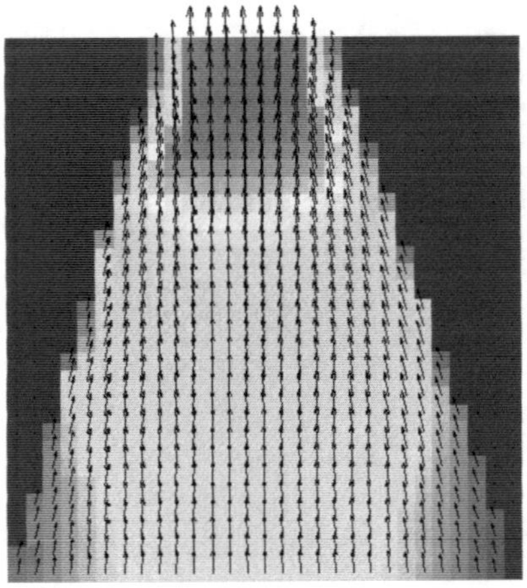
(a) micro-bubble tracer

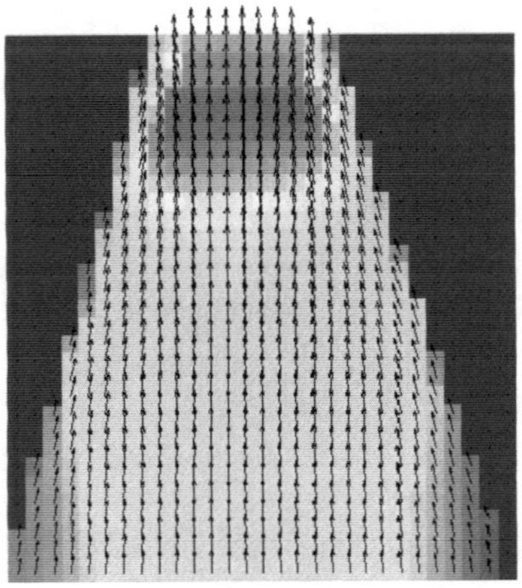
(b) particle tracer

Fig.6 Mean velocity vector maps

4. CONCLUSIONS

Micro-bubbles were used as tracer of PIV, and they were applied to visualization and PIV measurement in vertical channel with contraction. In comparison with micro-bubble tracer and particle tracer, micro-bubble tracer cannot obtain enough number density, but instantaneous velocity vector and mean velocity vector of micro-bubble tracer was almost same as ones of the particle tracer. As our future works, while we will consider the feature of micro-bubbles, we will investigate applicability of the other flow fields except the acceleration flow field.

REFERENCES

Ishikawa, M. et al. (2008). “PIV Measurement of a contraction flow including micro-bubble”, Proc. of JSMF Annual Meeting 2008, Aizu, Japan, pp.288-289 (in Japanese).

Ohnari, H., (2006), “All the Micro-Bubble”, Nippon Jitsugyo Publishing (in Japanese).

Prediction Method for Cavitation Erosion Based on Measurement of Bubble Collapse Impact Loads

S Hattori [1], T Hirose [1], K Sugiyama [2]

[1] Graduate School of Engineering, University of Fukui, 3-9-1 Bunkyo, Fukui-shi, Fukui 910-8507, JAPAN
[2] Ebara Research Co., Ltd., 4-2-1 Honfujisawa, Fujisawa-shi, Kanagawa 251-8502, JAPAN

hattori@mech.fukui-u.ac.jp

Abstract. The prediction of cavitation erosion rates is important in order to evaluate the exact life of components. The measurement of impact loads in bubble collapses helps to predict the life under cavitation erosion. In this study, we carried out erosion tests and the measurements of impact loads in bubble collapses with a vibratory apparatus. We evaluated the incubation period based on a cumulative damage rule by measuring the impact loads of cavitation acting on the specimen surface and by using the "constant impact load - number of impact loads curve" similar to the modified Miner's rule which is employed for fatigue life prediction. We found that the parameter $\sum(F_i^{\alpha} \times n_i)$ (F_i: impact load, n_i: number of impacts and α: constant) is suitable for the evaluation of the erosion life. Moreover, we propose a new method that can predict the incubation period under various cavitation conditions.

1. INTRODUCTION

Cavitation often occurs in the contact area between solids and liquids, in fluid machineries, pipes, ship propellers, valves and so on. The erosion is a phenomenon that erodes the component surface sponge-like. Cavitation erosion is a serious problem that brings a performance reduction of an apparatus or a life reduction by component failure. The measurement of impact loads in bubble collapses (bubble collapse impact loads) helps us to predict the cavitation erosion.

Hattori et al. (1998) measured bubble collapse impact loads using a venturi test facility and a vibratory apparatus, and clarified the relation between the cumulative impact energy $\sum F_i^2$ (F_i: impact load) obtained from the impact load distribution and the erosion volume loss rate. For low $\sum F_i^2$ (near the cavitation damage threshold), however, the relation between $\sum F_i^2$ and the volume loss has not yet been clarified. On the other hand, Iwai et al. (1988) reported a good proportional relation of the incubation period or the volume loss rate in the steady period in a vibratory apparatus with the fatigue damage $\sum(n_i/N_i)$ calculated at stress amplitudes more than a certain threshold value. However, there is an issue that a basic S-N curve against the erosion rate has to be assumed and the threshold value of the impact loads without the influence of erosion has to be found experimentally. Soyama et al. (2007) carried out a cavitation test using a sheet of pure aluminum glued to a pump impeller, and obtained the bubble collapse impact loads from the deformation of the aluminum sheet. They proposed a threshold for the erosion energy so that erosion does not occur. However, the physical meaning of the erosion energy threshold was not clarified.

In this study, we carried out erosion tests and measured the bubble collapse impact loads under

Journal of Physics: Conference Series **147** (2009) 012011 doi:10.1088/1742-6596/147/1/012011

various cavitation conditions. We discuss a prediction method for the incubation period of the cavitation erosion.

2. MATERIALS AND TEST METHOD

2.1 Materials

The test materials are pure aluminum of A1050 and A1070BD-F as accelerated test material, S55C carbon steel for machine structural use and S15C low carbon steel, and SUS304 austenite stainless steel with high corrosion resistance. The chemical compositions and the physical and mechanical properties of these materials are listed Tables 1 and 2, respectively. Erosion test results (except for A1070BD-F) are taken from the date base constructed in our laboratory (Hattori et al., 2004). The size of each test specimen is 25 mm in diameter and 5 mm in thickness. The test specimen surface was mirror-finished by buffing after being polished with emery paper up to grade #1200.

Table 1 Chemical compositions of test materials (mass%)

Material	Si	Fe	Cu	Mn	Mg	Zn	V	Ti	Al
A1050	0.09	0.14	0.01	0.01	0.01	0.01	0	0.01	99.5
A1070BD-F	0.07	0.14	0	0	0.01	0	0.01	0.01	99.75
	C	Si	Mn	P	S	Cr	Ni		
S15C	0.15	0.25	0.47	0.015	0.015	-	-		
S55C	0.54	0.27	0.82	0.016	0.017	-	-		
SUS304	0.071	0.48	1.21	0.025	0.022	18.47	8.21		

Table 2 Physical and mechanical properties of test materials

Material	ρ g/cm^3	E GPa	σ_B MPa	HV
A1050	2.71	71	178	33
A1070BD-F	2.67	71	106	41
S15C	7.87	206	388	151
S55C	7.87	204	701	238
SUS304	7.98	200	618	209

2.2 Test Method

The cavitation erosion tests were carried out in a stationary specimen method by using a vibratory apparatus as specified in the ASTM standard G32-03 (ASTM Designation, G32-03, 2005). A vibrating disk of 16 mm in diameter made of erosion resistant SUS304 steel was screwed into the amplifying horn of an oscillator, and the test specimen was placed in close proximity to the vibrating disk as shown in Fig. 1. To produce various cavitation conditions, the distance between the vibrating disk and the test specimen (standoff distance) was adjusted to be 1, 2 and 4 mm. The resonance frequency of oscillator was 19.5 kHz, and the double (peak to peak) amplitude of the vibrating disk was 50 μm. After using the vibrating disk for 10 hours, the disk was replaced by a new one. The test liquid was deionized water, and kept at 25±2 degrees C with a temperature control device. The test specimen was removed periodically after predetermined time intervals, and weighed with a precision balance

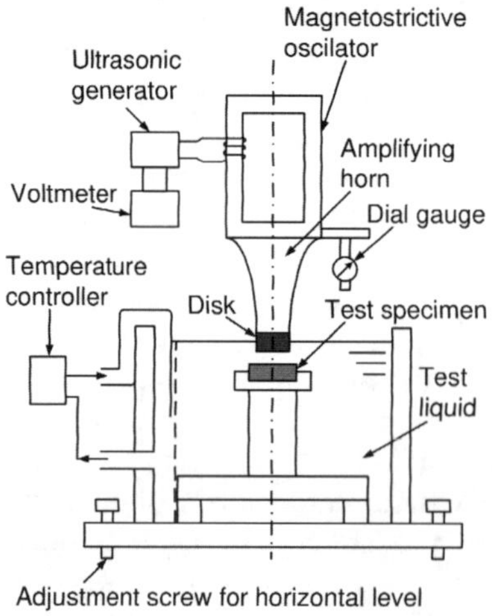

Fig. 1 Vibratory apparatus

(sensitivity of 0.01 mg) after cleaning with acetone in an ultrasonic bath. The test result was expressed by using the MDE (Mean Depth of Erosion), i.e. the mass loss divided by the density of material and the eroded area.

The bubble collapse impact loads were measured under the same conditions as in the cavitation erosion test by using a piezo ceramic transducer (sensor) instead of the test specimen. The bubble collapse impact loads were measured for 1 minute. Fig. 2 shows an illustration of the sensor structure. The piezo ceramic for detecting impact loads is a disk of 5 mm in diameter and 0.2 mm in thickness. The piezo ceramic was sandwiched between the 3 mm detection rod made of Ti and the reflection rod of Cu, and fixed with a conductive adhesive. An epoxy resin agent was used to fill in the space between the part of the pressure detector and the acrylic resin pipe for making it vibration-proof, water-proof and for protection from breakage of the ceramic disk. The sensor performance depends on the surface profiles of the detection rod and the reflection rod and on the adhesive condition of the piezo ceramic. To eliminate the differences in sensor performance, we carried out a steel ball drop test to obtain a calibration coefficient for the sensor before the measurement of bubble collapse impact loads. In the steel ball drop test, a steel ball (0.134g) made of SUS304 was dropped on the detection surface of the sensor, and the output waveform in Fig. 3 was read using an oscilloscope. The maximum voltage ΔV[V] and the hold time of the impact load ΔT[μs] were obtained. The appropriate calibration coefficient of the sensor was obtained from the relation between the voltage and the impact load. The relation between the voltage V and the impact load F of the sensor used in this study was $F=1.83V$ as shown in Fig. 4.

Fig. 5 shows the diagram for the measurement of the bubble collapse impact loads. The output signal obtained from the sensor was passed through the highpass filter to cut the low frequencies, and then fed into the computer with A/D conversion after being processed in the peak-hold circuit. To consider the influence of the voltage reduction in the circuit, we determined the calibration coefficient of the circuit from the relation between the input voltage and the output voltage of a the pulse wave (frequency is 1 kHz and height of pulse wave is adjusted to be 0 to 2.5 V). The pulse wave was generated using a function generator and the output voltage was obtained from the mode value of the height of the pulse wave after it passed through the circuit. The relation between the input voltage V_{in}

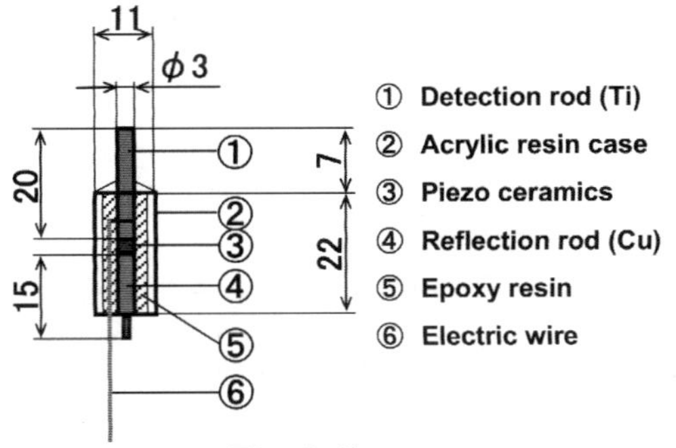

Fig. 2 Sensor

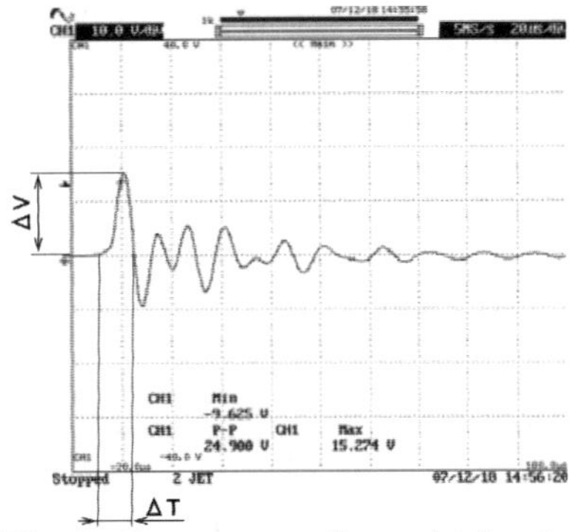

Fig. 3 Output waveform obtained from steel ball drop test

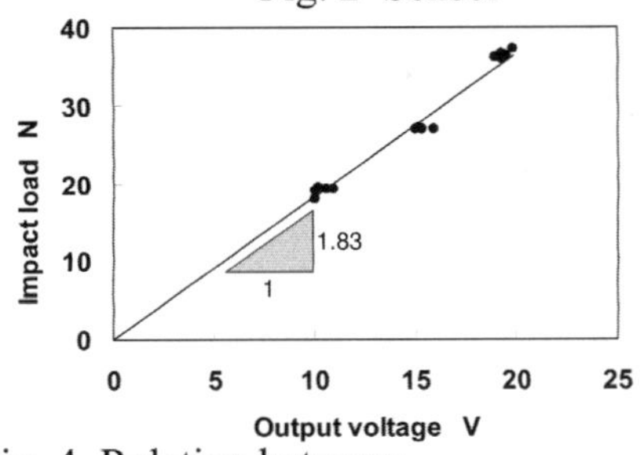

Fig. 4 Relation between output voltage and impact load

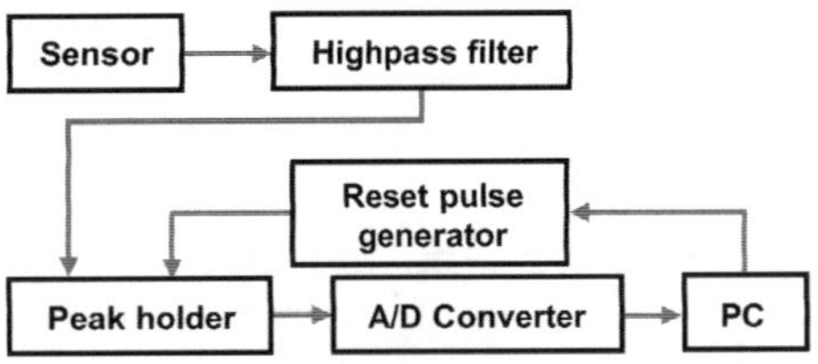

Fig. 5 Diagram for measurement of bubble collapse impact loads

and the output voltage V_{out} of the measuring system used in this study was $V_{in}=3.33V_{out}$. Therefore, we were able to obtain bubble collapse impact loads and their counts using the calibration factor of the sensor and the circuit.

3. TEST RESULTS AND PROPOSAL OF PREDICTION EQUATION

3.1 Test Result of Cavitation Erosion

Fig. 6 shows the MDE curves at standoff distances of 1, 2 and 4 mm between the vibrating disk and the specimen for pure aluminum A1050 and A1070BD-F. Fig. 7 shows the curves for the ferrous steels SUS304, S15C and S55C. Every MDE curve passes through an incubation period with low erosion rate and then increases linearly to reach a maximum rate period for each material. When each material is compared with the same exposure time, the MDE is high for the soft material A1050 and it is low for the hard material SUS304. This corresponds to our previous finding that the erosion rate has a good correlation with the hardness (Hattori et al., 2004).

Table 3 shows the incubation period for each material. The periods were obtained from Figs. 6 and 7 as the point of intersection of the extended straight line of the slope of the maximum rate period with the axis of the exposure time. The incubation period of A1050 with low hardness is short and that of S55C with high hardness is long. This is because cracks initiate easily for low hardness materials, because plastic deformation can occur even at low impact loads, when the various impact loads act on the material surface.

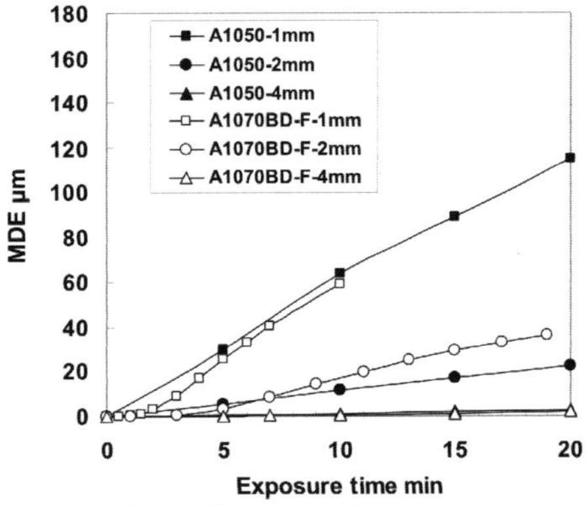

Fig. 6 MDE curves for A1050 and A1070BD-F

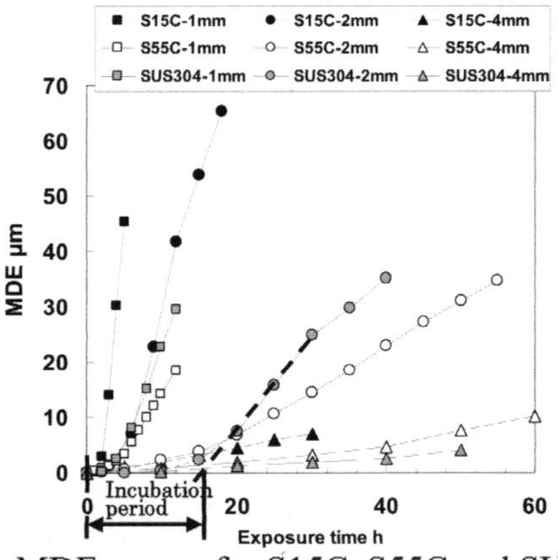

Fig. 7 MDE curves for S15C, S55C and SUS304

Table 3 Incubation period of test materials

Material	Incubation period [min]		
	Stand off distance		
	1mm	2mm	4mm
A1050	0.7	1.3	2.4
A1070BD-F	2.2	3.8	7.5
S15C	128	324	600
S55C	198	660	1410
SUS304	240	840	1320

3.2 Measurement of Bubble Collapse Impact Loads

Fig. 8 shows the result of the bubble collapse impact loads measured with the sensor which was located at standoff distances of 1, 2 and 4 mm. Several thousand counts were detected for small impact loads and several counts with large impact loads. The distributions at the various distances are very similar, and the frequency of the large impact loads is higher at lower standoff distances.

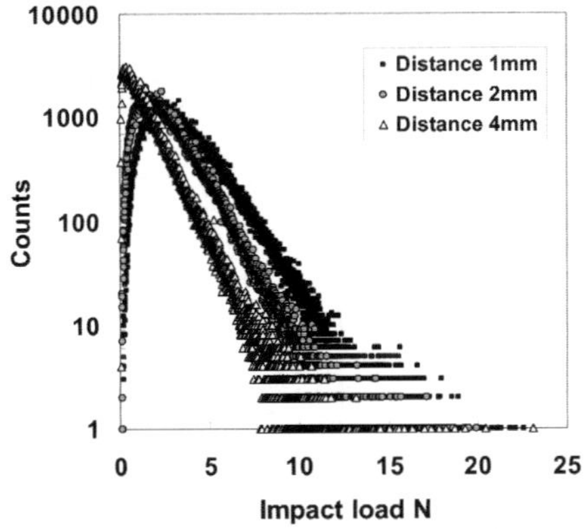

Fig. 8 Distribution of bubble collapse impact loads

3.3 Method for Prediction of Incubation Period

When the impact load F_i acts on the specimen surface during the holding time ΔT, the impact energy e is given (De and Hammitt, 1982; Hattori et al., 1985) by

$$e = \Delta T / \rho c \times F_i^2 \tag{1}$$

where ρ is the density of the test liquid and c is the sound velocity in the liquid. If the holding time ΔT is assumed to be constant irrespective of the magnitude of the impact load acting on the specimen, the cumulative impact energy E for various impact loads F_i on the specimen surface can be assumed to obey

$$E \propto \sum F_i^2 \tag{2}$$

That means the cumulative impact energy E is in proportion to the cumulative square value of impact loads ($\sum F_i^2$). $\sum F_i^2$ can be calculated with $\sum(F_i^2 \times n_i)$ using the impact load F_i and its counts n_i obtained from the measurement of the bubble collapse impact loads.

Fig. 9 shows the relation between $\sum F_i^2$ and the maximum erosion rate $MDER_{max}$ at various standoff distances obtained in this study. $MDER_{max}$ is the maximum slope in the maximum rate period of the MDE curve. $MDER_{max}$ increases linearly with $\sum F_i^2$ for all materials. The linear relation between

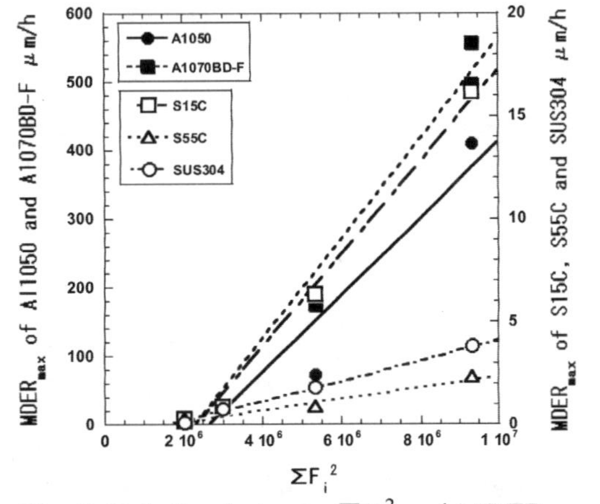

Fig. 9 Relation between $\sum F_i^2$ and $MDER_{max}$

$\sum F_i^2$ and the erosion rate was reported (Hattori et al., 1998), but regarding the result obtained in this study, the erosion rate becomes extremely low at values $\sum F_i^2$ of $1\sim2\times10^6$ and the existence of a threshold value for $\sum F_i^2$ without erosion is observed. To obtain a clear threshold value, however, further long-term tests at low cavitation intensities are required. Therefore, we discuss the prediction method from another viewpoint.

Hammitt et al. (1980) reported that the erosion proceeds by fatigue failure. Vaidya and Preece et al. (1978) reported that striation was observed on the eroded surface of Al-4%Cu alloy. The damage mechanism of components subjected to cavitation bubble collapse impact loads is therefore regarded as fatigue failure under the variable amplitude stress. We thus discussed the cavitation erosion from the viewpoint of fatigue life.

The linear cumulative damage rule is one of the prediction methods for the fatigue life under variable amplitude loading. Fig. 10 shows a schematic S-N curve on a double logarithmic scale. The slope part is given by

$$\sigma_i^{\alpha} \times N_i = C \tag{3}$$

Nakamura et al. (1981) proposed that the S-N curve on a double logarithmic scale should be used to predict the fatigue life. The fatigue damage is given by the cycle ratio n_i/N_i when σ_i is repeated n_i times under variable amplitude stress. It is assumed that the damage at each stress level is independent and is accumulated linearly. It is further assumed that the material ruptures when the sum of the cycle ratios n_i/N_i reaches unity. This sum is given by

$$D = \sum n_i / N_i = 1 \tag{4}$$

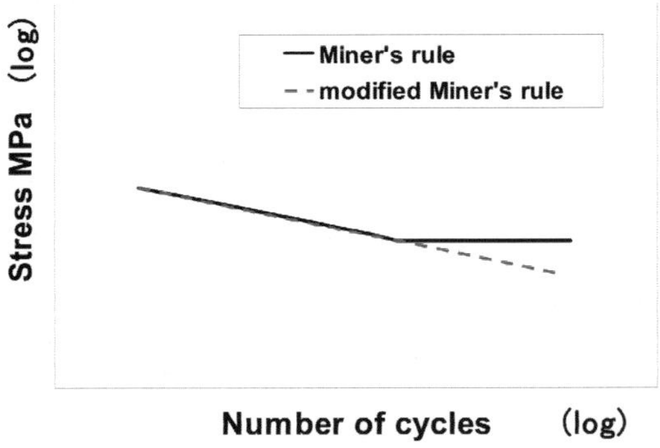

Fig. 10 S-N curves

The solid line in Fig. 10 shows the S-N curve in order to predict the fatigue life using Miner's rule, assuming that fatigue damage does not occur below the fatigue limit. The broken line in Fig. 10 is based on the modified Miner's rule which assumes that all stresses contribute equally to the fatigue damage. The result was that fatigue damage is accumulated due to the stress even below the fatigue limit under variable amplitude stress, when the stress is combined with stresses above the fatigue limit (de Jonge and Nederveen, 1980). The modified Miner's rule is nowadays generally used (de Jonge and Nederveen, 1980) to evaluate fatigue damage.

Since many impact loads at various intensities are measured in cavitation bubble collapses, the modified Miner's rule under the variable amplitude loading is applied to the prediction of cavitation erosion. The incubation period is defined as the point of intersection of the extended straight line of slope of the maximum rate period with the axis of exposure time (ASTM Designation, G32-03, 2005), and the termination of incubation period is well assumed to coincide with macroscopic fatigue failure. The incubation period has been discussed based on the accumulation of fatigue damage.

The prediction method of cavitation erosion was constructed on the basis of the modified Miner's rule by using the relation between the impact load F_i and the number of cycles per unit time n_i obtained from the measurement of bubble collapse impact loads at the various standoff distances. The incubation periods were obtained from the erosion test at the various standoff distances. Since it is impossible to measure the impacted area, we cannot convert the force to a stress. Therefore, the F-N curve was used. N is the impact number at the termination of the incubation period with constant impact load F. Since a test with a constant impact load F_i cannot be carried out for cavitation erosion, the impact number at the termination of incubation period N_i is basically unknown. But, the F-N curve is derived using constant parameters of α and C in

$$F_i^{\alpha} \times N_i = C \qquad (5)$$

The parameter $\sum n_i/N_i$ is the cumulative damage per unit time. The incubation period t_d (the subscript d indicates the standoff distance) finishes when $\sum n_i/N_i$ becomes unity. t_d is therefore given as

$$t_d = \frac{1}{\sum n_i / N_i} \qquad (6)$$

To determine the constants α and C, the unknown number N_i in Eq. (5) is expressed by F_i and the constants α and C, and then substituted into Eq. (6). The following equation results.

$$\frac{1}{C}\sum\left(F_i^{\alpha} \times n_i\right) = \frac{1}{t_d} \qquad (7)$$

For a given material, the constants α and C are independent of the impact load F_i, the number of cycles n_i and the incubation period t_d at standoff distances of 1 and 2 mm. Therefore Eq. (7) with the data at the standoff distance of 1 mm was divided by Eq. (7) with the data at the standoff distance of 2 mm, which gives the following relation.

$$\frac{\sum F_{i,1}^{\alpha} \times n_{i,1}}{\sum F_{i,2}^{\alpha} \times n_{i,2}} = \frac{t_2}{t_1} \qquad (8)$$

Since t_d, F_i and n_i are already given, the constant α can now be determined with a trial & error method using Eq. (8). After α is determined, the constant C can be obtained from Eq. (7). Since the constants α and C are different depending on the material, it is necessary to find them for each material. Table 4 shows the values of α and C for the various materials.

Table 4 α and C of each material

Material	α	C
A1050	2.6	4.5E+05
A1070BD-F	2.0	5.3E+05
S15C	3.7	2.0E+05
S55C	5.1	6.1E+06
SUS304	5.4	1.4E+07

3.4 Predicted Results for Incubation Period

The parameter $\sum F_i^2$ has been used previously to evaluate the erosion (Hattori et al., 1998). We have already discussed the erosion using the parameter $\sum(F_i^{\alpha}\times n_i)$ in Eq. (7). Fig. 11 shows the relation between the parameter $\sum(F_i^{\alpha}\times n_i)$ and the reciprocal of the incubation period obtained from the erosion tests of the various metals at the three standoff distances. Straight lines can be drawn passing through the origin. Therefore, the parameter $\sum(F_i^{\alpha} \times n_i)$ is suitable for the evaluation of cavitation erosion. By using the values of α and C obtained under two different cavitation conditions, we predict the incubation period t_d by Eq. (7) using the values of F_i and n_i obtained at a standoff distance of 4 mm. Fig. 12 shows the relation between the predicted incubation periods and the measured incubation period for the standoff distance of 4 mm, plotted on a double logarithmic scale. A straight line with a slope of 45° on the double logarithmic scale was obtained. This shows that this prediction has a very high accuracy.

In this study, we clarified that the incubation period can be predicted if the constants α and C are obtained by erosion tests and if the bubble collapse impact loads are measured under two different cavitation conditions. Since this result has only been verified at the laboratory level, however, it will be necessary to reseach the applicability of the new erosion prediction method to prototype machine components by measuring the bubble collapse impact loads.

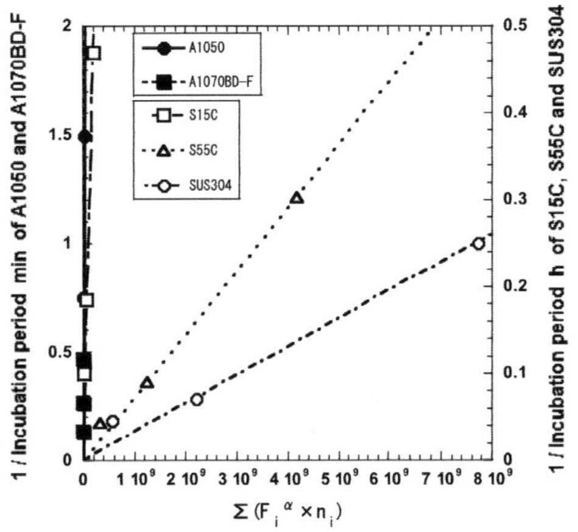

Fig. 11 Relation between $\sum(F_i^{\alpha} \times n_i)$ and the reciprocal incubation period

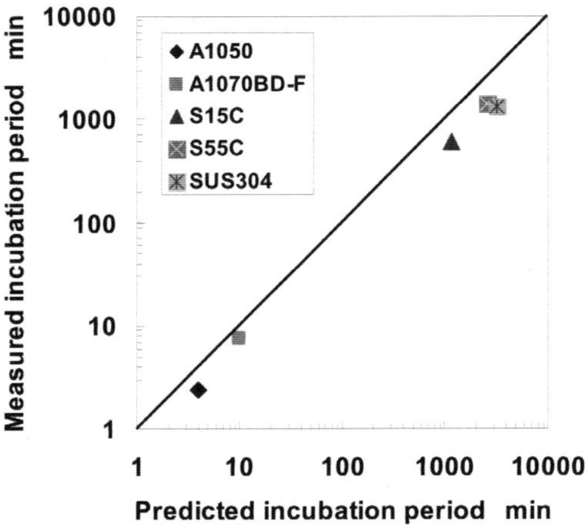

Fig. 12 Prediction accuracy for standoff distance 4 mm

4. CONCLUSIONS

In this study, we propose a new prediction method for the incubation period which is based on two sets of erosion tests and on measurements of bubble collapse impact loads. We clarified the following points.

(1) Since the relation between the parameter $\sum(F_i^{\alpha} \times n_i)$ and the reciprocal of the incubation period obtained for various metals shows a proportional relation, the parameter $\sum(F_i^{\alpha} \times n_i)$ is suitable for the evaluation of cavitation erosion.

(2) After the constants α and C of the F-N curve have been obtained under two different cavitation conditions, we can predict the incubation period under yet another condition, provided that the bubble collapse impact loads are measured again.

REFERENCES

ASTM Designation (2005), G32-03, Annual Book of ASTM Standard, pp. 106-119.

de Jonge, J. B. and Nederveen, A. (1980), ASTM STP 714, pp. 170-184.

De, M. K. and Hammitt, F. G. (1982), Trans. ASTM, Journal of Fluids Engineering, 104, pp. 434-442.

Nakamura, H. (1981), JSMS, Committee on Fatigue of Materials, Research Meeting, No. 158.

Soyama, H. et al. (2007), The 9th AICFM, Jeju, Korea, No. AICFM 9-037.

Hammitt, F. G. (1980) "Cavitation and multiphase flow phenomena", McGraw-Hill Inc., p. 229.

Hattori, S. et al. (1998), Trans. ASME, Journal of Fluids Engineering, Vol. 120, pp. 179-185.

Hattori, S. et al. (1985), Wear, 103, pp. 119-131.

Hattori, S. et al. (2004), Wear, 257, pp. 1022-1029.

Vaidya, S. and Preece, C. M. (1978), Metal Trans. Series A, 9A, 229.

Iwai, Y. et al. (1988), Transactions of the Japan Society of Mechanical Engineers, Series A, Vol. 54, No. 500, pp. 861-867.

The 6th International Symposium on Measurement Techniques for Multiphase Flows IOP Publishing
Journal of Physics: Conference Series **147** (2009) 012012 doi:10.1088/1742-6596/147/1/012012

Wake Structure of Circular Cylinder in Microbubble Mixture

T Suzuki , Y Oishi , Y Murai , Y Tasaka , Y Takeda

Hokkaido University, N13W8 Sapporo, 060-8628, JAPAN

szktkms@ring-me.eng.hokudai.ac.jp

Abstract. Injecting microbubbles near wall can reduce the frictional drag. While it has been confirmed by a variety of experiments to date, the fluid dynamics mechanism of drag reduction has not been comprehensively understood so far. We measure the wake structure of a circular cylinder in microbubble mixture to figure out the characteristic of the interaction between microbubbles and liquid flow accompanying high turbulence. The flow field is restricted to be two-dimensional by confining the bubbly two-phase flow into a thin horizontal channel of 2 mm in height. Microbubbles around 100 μm in diameter are generated with water electrolysis at far upstream the measurement section. Experiment is conducted at Reynolds number higher than 10^4 to clarify the role of microbubbles in highly turbulent situation. We use solid particles as the PTV-tracer in the single-phase flow and assume that microbubbles trace the liquid flow in the microbubble mixture. We measure the turbulent intensity and Reynolds shear stress distribution from the PTV data and demonstrate that microbubbles significantly suppress the turbulence. In addition, the Karman vortex shedding frequency, which is measured from the unsteady stream function, increases when microbubbles are mixed.

1. INTRODUCTION

Microbubble has attracted attention as the way to reduce the skin friction in recent years. It has been confirmed that injecting microbubbles near wall can reduce a skin friction since the first report published by McCormic and Bhattacharyya (1973). This method is expected to apply to the ship and the pipeline. A lot of laboratory experiments have been carried out to investigate the drag reduction effect. Madavan et al. (1984) achieved as much as 80% decrease in the frictional drag using a water channel. Guin et al. (1996) reported that the drag reduction ratio correlates with the void fraction near wall. However the mechanism of microbubble drag reduction has not been fully understood so far. At least the mechanism is not explained universally for a variety of flow configuration, but it should be classified into several regimes. Major parameters to classify it are the Reynolds number, Weber number, and void fraction. Increasing the effective viscosity which associates with the reduction of Reynolds stress is one of the principal candidates to explain the drag reduction mechanisms. However, the drag reduction effect of the previous experiments cannot be explained only by the increment in the effective viscosity. Nowadays the suppression of the turbulent flow has become a focus of the microbubble drag reduction. It means that the microbubbles relax the flow fluctuation only in small scale but do not affect the global property of flow. This is why the microbubbles sensitively alter the turbulent flow.

The turbulent flow in the boundary layer generates a large frictional drag. Kato et al. (1999) investigated the turbulent intensity in the buffer and inner turbulent regions using laser Doppler velocimeter (LDV) and indicated that microbubbles decrease the turbulent intensity when the skin friction decreases. Ferrante and Elghobashi (2004) performed a direct numerical simulation (DNS) of a spatially developing turbulent boundary layer laden with microbubbles and explained that the displacement of streamwise vortices away from the wall induces a drag reduction. In the case of turbulent boundary layer, the role of microbubbles changes gradually in the direction of the flow since the void profile in the wall-perpendicular direction changes as well. Furthermore, a wavy fluctuation in void fraction occurs in the flow direction because of the bubble-bubble interaction. Such a system is inconvenient for the fundamental study in which the role of microbubbles is investigated.

Matsumoto et al. (1988) numerically calculated the bubbly two-phase flow around a blade. They reported the characteristic of bubbly blade about the pressure, Mach number and the void fraction distribution. Also, Hulin et al. (1984) reported the local void fraction and the vortex shedding frequency behind obstacles with a trapezoidal cross-section in gas-liquid flows and confirmed that Strouhal number has a local maximum value at a certain void fraction. In this paper, the characteristic of microbubble in highly turbulent condition is investigated by measuring the wake structure of a circular cylinder. The cylinder wake structure of single-phase flow has been investigated in great detail over the past century. Roshko (1954) indicated the cylinder wake structure between Re=50 and Re=4000 in the single-phase flow. The cylinder wake has the various vortices shedding depending on Reynolds number, and it strongly depends on the spatial development of boundary layer on the cylinder surface from the front stagnation point to the separation point. Therefore, the detection of the role of the microbubbles can be more widely realized with the use of the multiple phenomena happened to the flow around the cylinder. Sugiyama et al. (2001) performed a numerical simulation of two-phase wake flows around a cylinder and confirmed a same relationship between a void fraction and Strouhal number as results obtained by Hulin (1984). In past studies millimetre-sized bubbles have been used for investigation of two-phase wake flow around a cylinder. In the present study through the visualization and the image processing, including particle tracking velocimetry (PTV), the wake dynamic that associates with microbubbles is deduced from the results. We carry out the experiment at high Reynolds numbers to see the interaction between microbubbles and the turbulent structure. In particular, we are going to discuss the reason why a small amount of microbubbles could drastically alter the flow field that is subjected to hard turbulence in the order of 10^4 in Reynolds number.

2. EXPERIMENTAL METHOD

Figure 1 shows the schematic diagram of experimental apparatus which we used in this experiment. The horizontal channel is made of transparent acrylic resin. We used a thin channel of which the height was restricted to be 2 mm for keeping two-dimensional bubbly flow. The width of channel is 80 mm and the entire length is 1000 mm. The working fluid is the water. The test section is 400 mm away from the hydrogen bubble generator. The mean liquid velocity is 2.1 m/s. Figure 2 shows the schematic diagram of the test section. The circular cylinder is made of transparent acrylic resin. The cylinder diameter is 20 mm and the height is 2 mm. The metal halide light is placed under the channel. The backlight image was obtained by the high speed video camera from above. Figure 3 shows the hydrogen bubble generator. Microbubbles are generated by the water electrolysis. The nickel is used both the anode and the cathode of electrolysis. The size of the nickel is 45 x 5 mm. The interval distance between two nickel electrodes is 2 mm. Figure 4 shows the probability density function of the bubble diameter in this experiment. Bubble diameter is between 90 μm and 430 μm. The mean microbubble diameter is 110 μm. A lot of microbubbles are distributed between 90 μm and 120 μm, provided that the microbubbles less than 90 μm could not be recognized by the image analysis because of the spatial resolution as well as the noises involved. The void fraction, i.e. the volume fraction of microbubbles in mixture fluid is controlled to be 0.03 %. This is exactly controlled by the electric current which is consumed for the electrolysis. The experiment was conducted under the condition with microbubbles and with solid particles, respectively. The solid particles are used for measuring the single-phase flow as the tracer particles. The solid particle diameter is 90 μm and the density is 1.01 g/cm^3. Details of experimental condition are as summarized in Table 1.

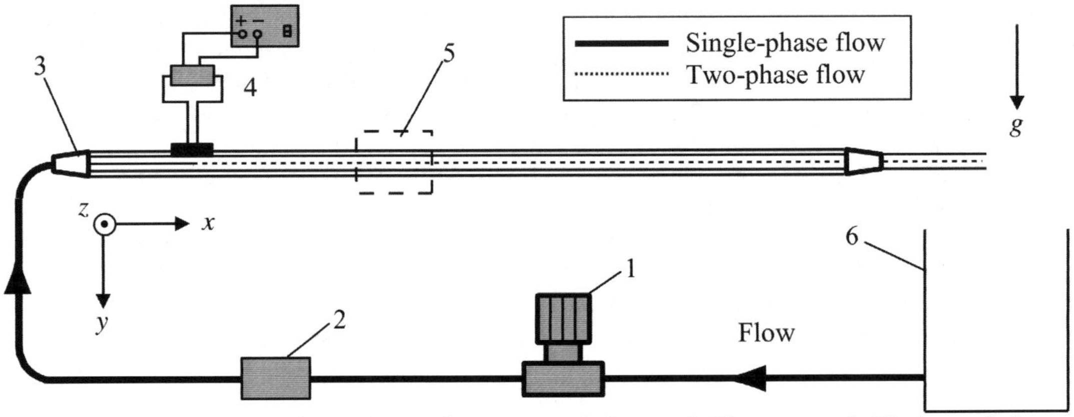

Fig. 1 Schematic diagram of experimental apparatus: 1. Pump, 2. Flowmeter, 3. Horizontal channel, 4. Hydrogen bubble generator, 5. Test section, 6. Removal tank

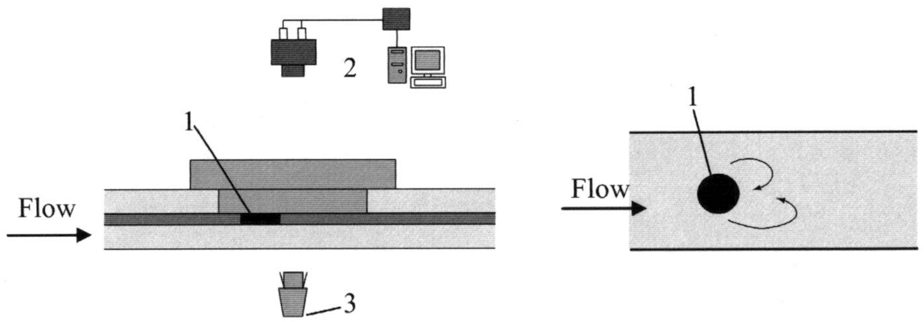

(a) Side view (b) Top view

Fig. 2 Schematic diagram of test section: 1. Circular cylinder, 2. High speed video camera system, 3. Metal halide light

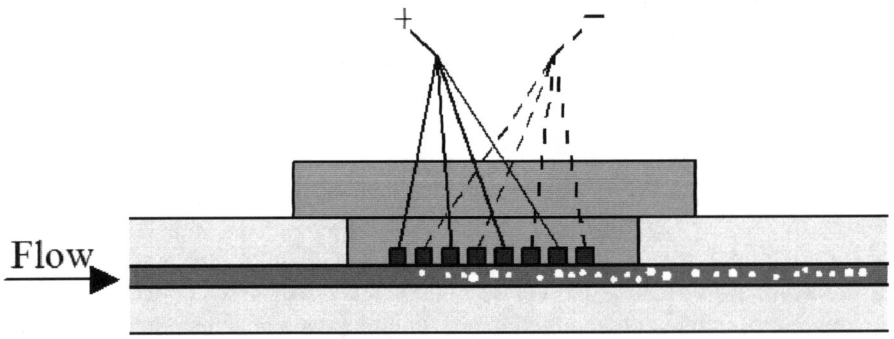

Fig. 3 Hydrogen bubble generator using water electrolysis

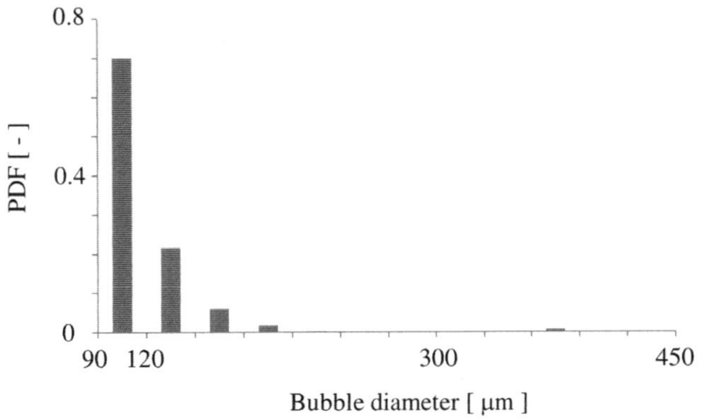

Fig.4 Probability density function of bubble diameter

Table 1 Experimental conditions

Bulk mean velocity	U	2.1	[m/s]
Size of channel		2x80x1000	[mm]
Circular cylinder diameter	D	0.02	[m]
$Re_D = UD/\nu$		46000	[-]
Temperature of water		24	[degrees C]
Density of water		997	[kg/m^3]
Kinematic viscosity	ν	0.91×10^{-6}	[m^2/s]
Void fraction		0.03	[%]
Solid particle diameter		90	[μm]
Density of solid particle		1010	[kg/m^3]
Frame rate		3000	[fps]
Spatial resolution		0.065	[mm/pixel]

3. RESULTS AND DISCUSSION

3.1 Turbulent structure behind circular cylinder

Inoue et al. (1986) investigated the characteristics of two-phase flow around a cylinder using bubbles of which diameter are about 4 mm. They reported that the bubble number density fluctuates because of the Karman vortex shedding from a cylinder. Figure 5 shows trajectories of solid particles and microbubbles around the circular cylinder. The main flow direction is from left to right. In present study the fluctuation of the bubble number density can't be seen from the results. Both solid particles and microbubbles have a recirculation region behind the circular cylinder. We can see the small fluctuation of bubbles' behaviour in the recirculation region. The fluid inertia is dominant there because of the change in local flow direction. The difference in density between the microbubble and the solid particle is significantly affected. Therefore, microbubbles have a fluctuated behaviour there. However, in large structure of which order is about same as a cylinder diameter, microbubbles have trajectories such as following the liquid. The difference of the flow between two types of the dispersion is picked up from the results as bellow.

Figure 6 shows the time-averaged velocity field of solid particles and microbubbles around the circular cylinder. We calculated the velocity of microbubbles and solid particles by PTV and rearranged the velocity field using a spatio-temporal Laplace equation (Ido and Murai, 2006). The period of the analysis is about 0.5 second (1498 frames). The streamwise velocity increases near x/D=0, and this includes extra acceleration due to the blockage effect, i.e. the circular cylinder occupies one-quarter of the channel width. According to the results, it is clear that the velocity of microbubbles is smaller than solid particles. Murai et al. (2006) reported that a bubble in shear layer close to a wall has a streamwise velocity slightly slower than the liquid velocity. They measured this backward slip velocity to be about 10 to 20 % of the local mean liquid velocity. In the present experiment, the velocity of microbubbles decreases by about 14% compared to the velocity of solid particles, i.e. liquid velocity. The reason of the decrease is explained by three factors. One is that the microbubbles are biased near the wall due to lift force, resulting in slower velocity near the wall. The negative slip in the shear flow induces lift force toward the wall, and this migration of bubbles provides negative slip again. Second is that the motion of microbubbles are strongly affected by the turbulence in liquid to have slow-downed velocity in mean value. For instance, the mean streamwise velocity of microbubble advection is significantly reduced when the microbubbles are accumulated into cores of streamwise vortex (Oishi et al, 2007, Suzuki et al, 2008). Third, we need to address that microbubbles in water-H2 gas combination have negative electric charge on individual bubble surface. This may provide a repulsion force to make bubbles scatter around and resists against flow deformation. Although the electric charge is not measured explicitly in this study, it is known to be around 10 mV. Spinning of microbubbles around particular axis generates Lorentz force as well, which has a potential to actively modify the turbulent eddies in the flow. However, we conclude that such an electro-magnetic effect of the bubble interface is not significant in the present flow condition. We have tried to put a strong magnet of 2 Tesla on the channel to see the difference of the microbubble behaviour, but nothing has been recognized about the magnetic effect. In fact the order estimation of Coulomb and Lorentz forces tells us that these forces are less than 10^{-6} times as the buoyant force of bubbles. The difference of the bubble velocity caused by these electro-magnetic forces cannot be sensed and almost ignorable. Consequently, the major reason of the slow velocity for microbubbles should be explained by the fluid dynamics interaction between two phases.

(a) Solid particle

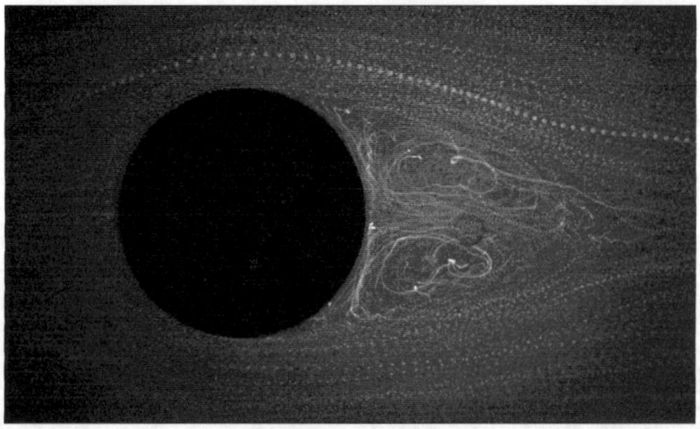

(b) Microbubble

Fig. 5 Trajectories of particles and microbubbles around circular cylinder

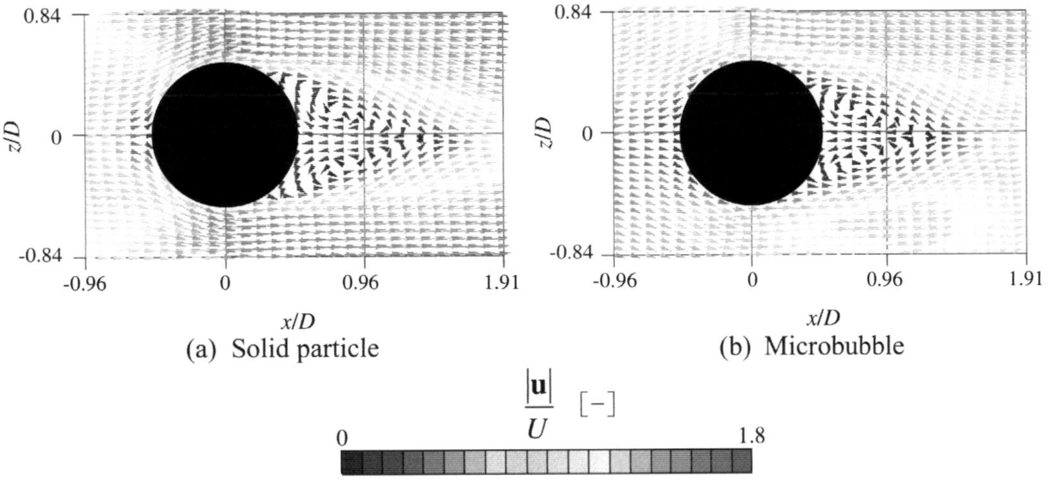

Fig. 6 Time-averaged velocity vector field

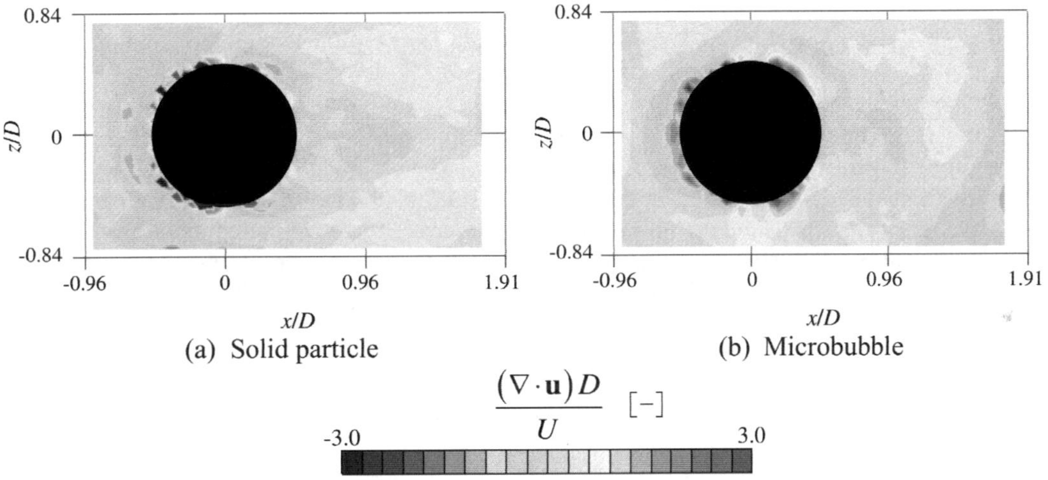

Fig. 7 Divergence map of time-averaged velocity field

Figure 7 shows the time-averaged distribution of two-dimensional divergence of the velocity vector field for the two types of dispersions. The divergence indicates the discrepancy of the dispersion motion from the liquid flow field that satisfies the two-dimensional continuity equation in single-phase flow. Both solid particles and microbubbles have a large magnitude in divergence value near the surface of the cylinder while these two have the near zero value in the wake region. Hence, the microbubbles track the liquid flow uniformly as same as solid particles even if the microbubbles has the backward slip velocity. On the other hand, microbubbles have negative divergence on the front side of the cylinder as shown by the blue arch in the figure. This means that the microbubbles are accumulated into the upstream stagnation point, and also into the boundary layer of the cylinder. As the result, two-phase interaction gets active on the way to progress of the boundary layer around the former half of cylinder. The opposite side of the cylinder has a positive value in divergence, implying that the microbubbles are repulsed from the boundary layer. The position of the change in sign of the divergence is located around 90 degree from the front stagnation point. Since this position and the separation point of the flow are close to each other, it is presumed that the wake structure is strongly affected by the presence of bubbles. It is worth noting that large bubbles would have the similar distribution of the divergence, however, the affected area gets too wide to concentrate the two-phase interaction into the boundary layer. Namely, only microbubbles can alter the flow via their focused action onto the boundary layer.

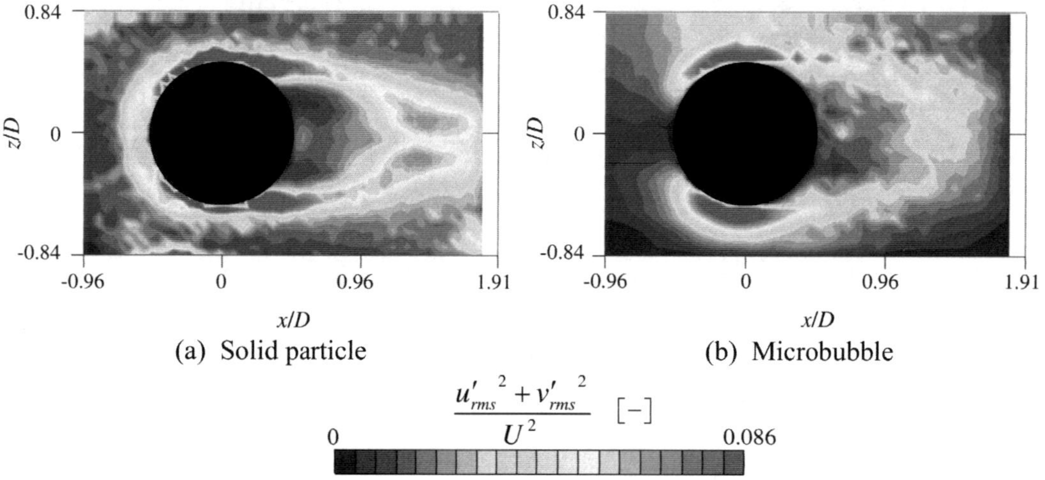

(a) Solid particle (b) Microbubble

Fig. 8 Turbulent Intensity distribution

Figure 8 shows the turbulent intensity distribution, which is normalized by the kinetic energy of inflow. Comparing the two cases, two differences can be pointed out as follows. One is seen near the boundary layer of the cylinder before the separation point. For microbubble flow, the turbulent intensity ceases totally in the front stagnation region. This means that the transition from laminar to turbulent boundary layer is delayed with presence of bubbles, and hence the laminar boundary layer is maintained long there. This effect of microbubbles is also known in pipe flows. The turbulent flow transition of a pipe flow is significantly delayed by inclusion of microbubbles. Namely, the present experiment also supports the idea that the microbubbles contribute to the laminarization of flow. Injection of large bubbles must induce pseudo-turbulence where slip velocity gets large. Hence, the quite turbulence in the region near the front stagnation point in this experiment proves the feature of microbubbles action to fluid. Another difference recognized between the two results is the turbulent intensity near the reattachment region; x/D~1.5. In the case of solid particles, the turbulent intensity gets large there because of turbulent Karman vortex shedding. Injection of microbubbles suppresses the intensity there to be around half of the original condition. We think that this suppression is caused by the alternation of the boundary layer on the cylinder surface, but not by the local interaction between two phases around the reattachment region.

The relationship between the downstream turbulent intensity and the boundary layer structure on the cylinder surface can be found by analyzing the Reynolds shear stress distribution. Figure 9 shows the stress distribution for the two cases. Here the stress is defined simply by the correlation of u' and v', which are the velocity component in the streamwise and the perpendicular directions (but not the velocity component relative to the curved surface of the cylinder). The comparison of these two immediately finds out that the Reynolds shear stress is drastically reduced in the wake region by mixing microbubbles. This indicates that microbubbles relax the motion of turbulent eddies that convey the momentum. This is consistent to the turbulent intensity measured. Moreover, it is very important to mention that the Reynolds shear stress changes its sign near the upstream surface of the cylinder when microbubbles are injected. The negative Reynolds shear stress means inverse transfer of the momentum from slower to faster fluid. This cannot happen in ordinary turbulent shear flow, but it happens to the microbubble mixture as measured here. The mechanism of the inverse transfer of the momentum may be associated with divergence of the flow velocity in the region close to the upstream cylinder surface (see Fig. 7 (b)). Since the local flow diverges there, the thickness of the boundary layer expands more rapidly than the ordinary development rate of the thickness.

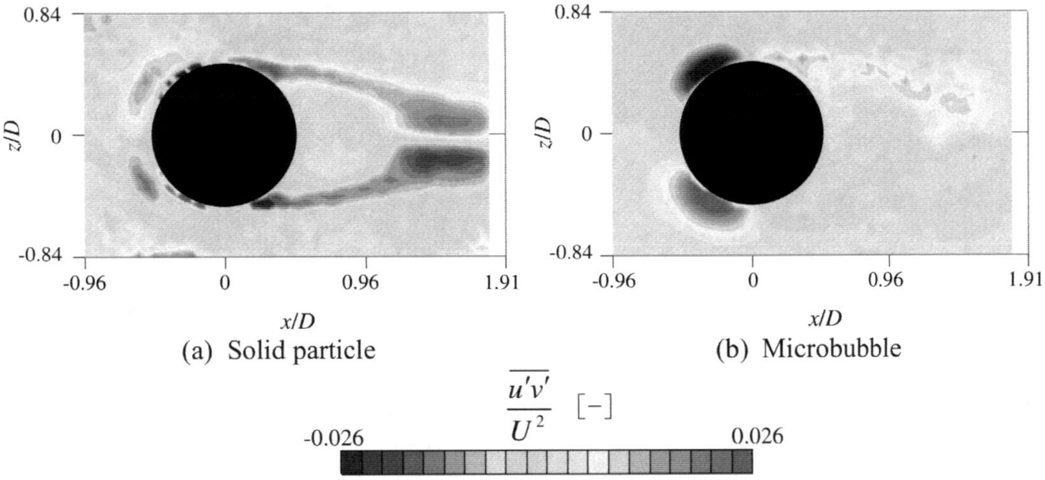

Fig. 9 Reynolds shear stress distribution

3.2. Frequency of Karman vortex shedding

Figure 10 shows the time series distributions of the instantaneous stream function. The stream function is calculated by integrating the velocity distribution. With this expression, it is confirmed that Karman vortex shedding occurs both for the particles and the microbubbles. The flow patterns are similar to each other, however, we can find that the thickness of the boundary layer around the cylinder is obviously large in the case of microbubble flow (see the thickness of green region around the cylinder). In addition, the thickness in the case of microbubble flow expands at around 45 degree from the front stagnation point. This is explained by the divergence effect of the microbubble motion.

Figure 11 shows the power spectra of the flow field obtained by fast Fourier transform. Here target quantity of the spectrum analysis is chosen to be stream function in the wake region because the stream function is insensitive to measurement noise. Also, the employment of the stream function leads to easier detection of organized structure in large scale such as oscillatory wake or Karman vortex. Four points of investigation are set behind the cylinder as shown in the figure 11. From the results, it is again confirmed that the amplitude of the flow fluctuation decreases when microbubbles are injected. There are two peaks in the power spectra both in the solid particle and the microbubble. The flow with solid particles has peak values at 32 Hz and at 38 Hz. The flow with microbubbles has peaks at 35 Hz and at 41 Hz. It is generally well known that Strouhal number defined by Eq. (1) takes the value around 0.21 at $Re_D > 10^3$ in the case of circular cylinder.

$$St = \frac{fD}{U} \qquad (1)$$

Under the value of the Strouhal number at 0.21, Karman vortex shedding frequency should be about 33 Hz in the case of single-phase flow (here, we use the averaged streamwise velocity at x/D=0 as U in the equation). This frequency is close to the first peak frequency detected in the present experiment for the solid particles. Therefore, we can assume that the first peak in the power spectra indicates the Karman vortex shedding frequency. The second peak in each spectrum takes place as the result of modulation which is originated from slow oscillation at 6 Hz (thus, the second frequency is given by adding 6 Hz to the first frequency). The slow oscillation comes up due to the blockage effect, i.e, due to the difference from the cylinder in infinite space. Therefore we focus only on the first peak to discuss the effect of microbubbles.

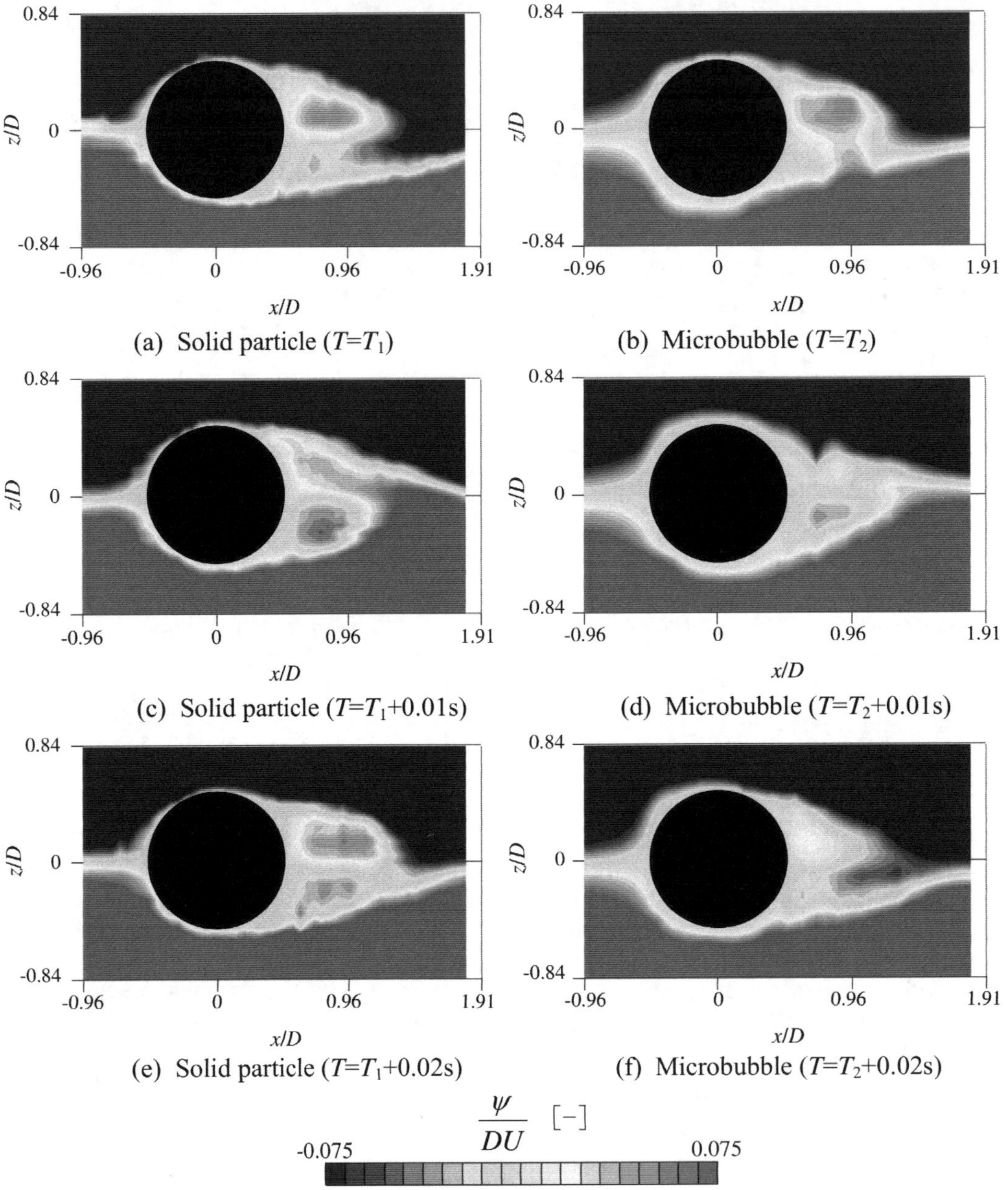

Fig. 10 Instantaneous stream function

In the case of microbubble flow, this first peak frequency increases by 9 % compared to the single-phase flow. Reynolds number of the present measurement is 46000. It is known that Strouhal number decreases with increase in Reynolds number in turbulent flow at $2000<Re<2x10^5$. The effective Reynolds number decreases due to the increment in effective viscosity and the decrement in fluid density in two-phase flow. That can explain the change in frequency of microbubble flow. However, the present void fraction is 0.03% and the effective Reynolds number little changed. The change in frequency can't simply associates with change in Reynolds number in present measurement. The change in Strouhal number by mixing millimetre-sized bubbles can be seen in past studies (Sugiyama et al. 2001, Yokosawa et al. 1986). The present void fraction is quite different from the past one. The change in frequency in the present experiment may be the unique phenomenon of microbubble flow.

On the condition of large concentration of microbubbles, the surface tension works like an elasticity in the liquid phase between microbubbles. This effect may influence with the change in frequency between the solid particle and the microbubble. Yokosawa et al. (1986) reported that the vortex shedding frequency of bubble crowd under the low void fraction was equal to that of the single-phase flow within a few percent errors. We need to collect more data including the effect of fraction of microbubble in the liquid to discuss this matter. The turbulent flow assumption of the relationship between Strouhal number and Reynolds number yields to the following aspect. In single-phase flow, the vortex shedding frequency in turbulent condition decreases with the increase in Reynolds number due to delayed separation for the turbulent momentum transfer. Hence, the increment in the frequency implies the separation occurring early. Microbubbles, therefore, suppress the turbulent momentum transfer to result in early separation of the boundary layer.

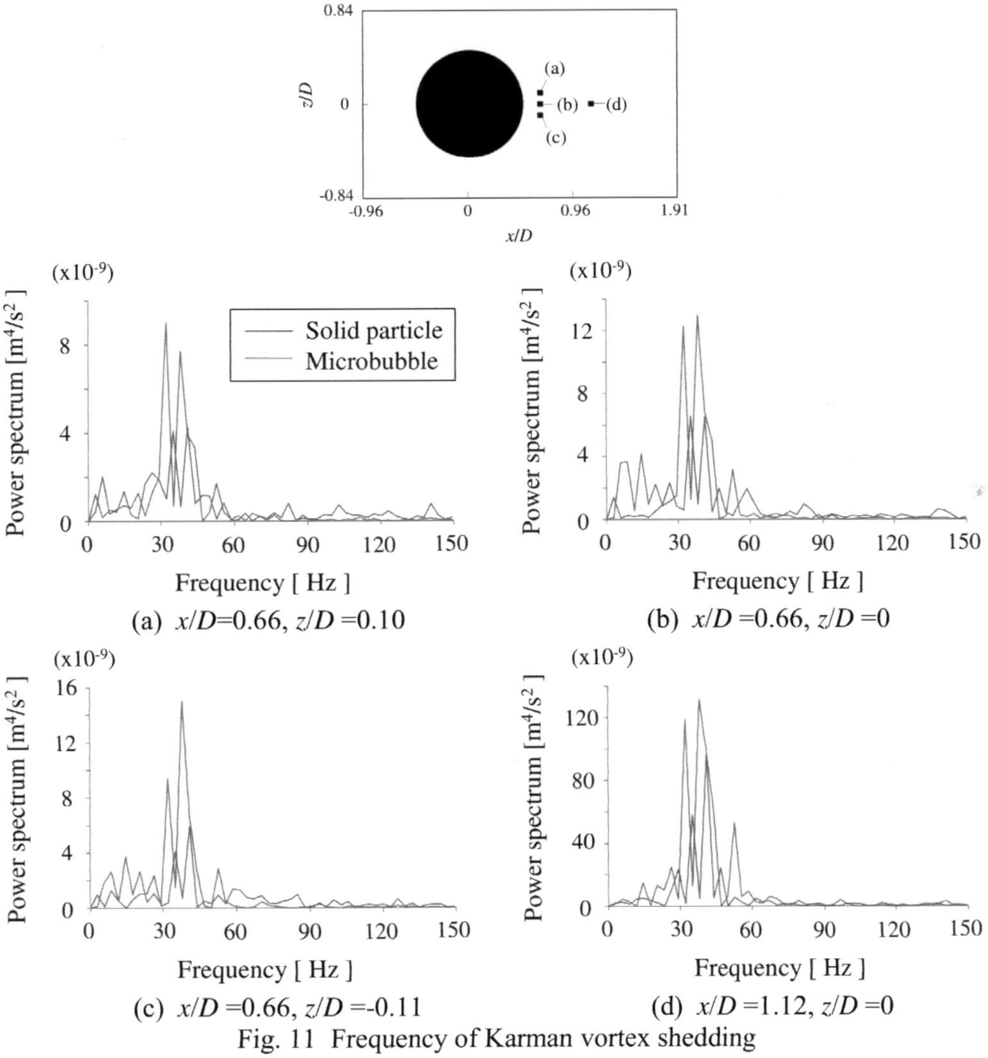

Fig. 11 Frequency of Karman vortex shedding

4. CONCLUDING REMARKS

PTV is applied to the measurement of a wake flow field caused by a circular cylinder. By comparing the velocity vector fields in two cases; with solid particle and with microbubble, we recognize several significant changes of the turbulent flow structure influenced by microbubble. From the turbulent intensity and the Reynolds shear stress, it is confirmed that the microbubbles suppress the flow fluctuation. The Karman vortex shedding frequency, which is obtained from the stream function of the wake flow field, increases with mixing of microbubbles. This is explained by two factors caused by microbubbles. One is divergence effect near the front stagnation region, which makes boundary layer expand along the surface of the cylinder. This effect provides negative Reynolds shear stress to keep the laminar boundary layer long. Another is the microbubble effect for hastening the flow separation. This is originated from suppression of the turbulent momentum transfer before the separation point.

ACKNOWLEDGEMENTS
The authors are grateful for the financial support by Japanese New Energy Development Organization (NEDO).

NOMENCLATURE

$\mathbf{u}$	velocity vector	[m/s]
u	streamwise velocity	[m/s]
v	spanwise velocity	[m/s]
U	bulk mean velocity	[m/s]
D	cylinder diameter	[m]
	Greek Letters	
ψ	stream function	[m^2/s]
ν	kinematic viscosity	[m^2/s]

REFERENCES

Ferrante, A., Elghobashi, S. (2004). " On the physical mechanisms of drag reduction in a spatially developing turbulent boundary layer laden with microbubbles," J. Fluid Mech., Vol.503, pp.345-355.

Guin, M.M., Kato, H., Yamaguchi, H., Maeda, M.,, Miyanaga, M. (1996). " Reduction of skin friction by microbubbles and its relation with near-wall bubble concentration in a channel," Marine Science and Technology, Vol.1, pp241-254.

Hulin, J. P., Foussat, A.J.M., Gaudin, D., Strumolo, G.S. (1984). "Vortex emission behind obstacles in two-phase flows," Encyclopedia of Fluid mechanics, Vol. 3, Chap. 27, pp.771-810.

Ido, T., Murai, Y. (2006). “A Recursive Interpolation Algorithm for Particle Tracking Velocimetry,” Flow Measurement and Instrumentation, Vol.17, pp.267-275.

Inoue, A., Kozawa, Y., Yokosawa, M., Aoki, S. (1986). "Studies on two phase-flow cross flow. Part 1: Flow characteristics around a cylinder," Int. J. Multiphase Flow, Vol. 12, No. 2, pp.149-167.

Kato, H., Iwashima, T., Miyanaga, M., Yamaguchi, H. (1999). " Effect of microbubbles on the sturcuture of turbulence in a turbulent boudary layer," J. Marine Science and Technology, Vol.4, pp155-162.

Madavan, N, K., Deutsch, S., Merkle, C, L. (1984). " Reduction of turbulent skin friction by microbubbles," Physics of Fluid, Vol.27, issue 2, pp.356-363.

Matsumoto, Y., Nishikawa, H., Ohashi, H. (1988). " Performance of cascade of blades in two phase flow," JSME Int. J, Ser 2,Vol. 31, No. 4, pp.652-659.

McCormick, M, E., Bhattachayya, R. (1973). " Drag reduction of a submersible hull by electrolysis," Navel Engineers Journal, Vol.85, No.2, pp.11-16.

Murai, Y., Oishi, Y., Takeda, Y., Yamamoto, F. (2006). " Turbulent shear stress profiles in a bubbly channel flow assessed by particle tracking velocimetry," Exp. Fluids, Vol.41, pp.343-352.

Oishi, Y., Murai, Y., Tasaka, Y., Takeda, Y. (2007). "Motion of microbubbles relative to streamwise vortices in the wall turbulence," Proc. 6th International Conference on Multiphase Flow, abstract pp.55-55, CD-ROM: S1_Wed_C_37.

Roshko, A. (1954). " On the development of turbulent wake from vortex streets," NACA Rep. 1191, pp.1-25.

Sugiyama, K., Takagi, S. Matsumoto, Y. (2001) " Three-dimensional numerical analysis of bubbly flow around a circular cylinder," JSME Int. J, Ser B,Vol. 44, No. 3, pp.319-327.

Suzuki, T., Oishi, Y., Murai, Y., Tasaka, Y., Takeda, Y. (2008). "Interaction between Streamwise Vortex and Bubbles in Horizontal Channel Flow," Proc. 13th Intnl. Symposium on Flow Visualization - Nice, France, CD-ROM paper No.134.

Yokosawa, M., Kozawa, M., Inoue, A., Aoki, S. (1986). "Studies on two phase-flow cross flow. Part 3: Characteristic of unsteady flow behavior," Int. J. Multiphase Flow, Vol. 12, No. 2, pp.185-202.

Journal of Physics: Conference Series **147** (2009) 012013 doi:10.1088/1742-6596/147/1/012013

Mode transition in Bubbly Taylor-Couette flow measured by PTV

K Yoshida [1], Y Tasaka [1], Y Murai [1], T Takeda [1]

[1] Graduate school of engineering, Hokkaido University, Sapporo, JAPAN

koji-y@ring-me.eng.hokudai.ac.jp, tasaka@eng.hokudai.ac.jp, murai@eng.hokudai.ac.jp, yft@eng.hokudai.ac.jp

Abstract. The drag acting to the inner cylinder in Taylor-Couette flow system can be reduced by bubble injection. In this research, relationship between drag reduction and change of vortical structure in a Taylor-Couette flow is investigated by Particle Tracking Velocimetry (PTV). The velocity vector field in the *r-z* cross section and the bubble concentration in the front view (z-θ plane) are measured. This paper describes the change of vortical structures with bubbles, and the mode transition that is sensitively affected by the bubbles is discussed. The bubbles accumulate in the three parts relative to vortex position by the interaction between bubbles and vortices. The status of bubble's distribution is different depending on position. This difference affects mode transition as its trigger significantly. The presence of bubbles affects the transition from toroidal mode to spiral mode but does not induce the transition from spiral mode to toroidal mode. Further we found that Taylor vortex bifurcates and a pair of vortices coalesces when the flow switches between spiral mode and toroidal mode.

1. INTRODUCTION

The use of bubbles as a way for reducing skin frictional drag reduction in shear flows has become a focus for engineering. This method is expected that large energy-saving effect is obtained by applying to large ships and pipelines whose drags are occupied almost with the frictional drag. So far, several mechanisms such as a variation of effective viscosity, decrease of mean density and modulation of turbulent vortices caused by bubbles motion have been reported to date as major factors reducing frictional drag. The actual physical mechanism, however, has been less understood owing to a complicated dynamic process existing in bubbly two-phase turbulent shear flows. To clarify this mechanism, several studies have been performed via experiment. Moriguchi et al. (2002) reported that the drag reduction effect becomes large in high void fraction regions and is generally independent of bubble's diameter. Gabillet et al. (2002) showed that the Reynolds stress rather increases under some conditions with injection bubbles. These studies dealt with channel flow, i.e. spatially developed boundary layer.

In this study, we chose Taylor-Couette flow driven by rotating concentric cylinders. The adopting of this configuration of flow takes following advantages in investigating mechanism of bubble-originated drag reduction. i) It is possible to evaluate the drag reduction mechanism in quasi-steady state because the flow is periodic and closed so that the global energy balance is also closed inside the system. ii) The flow near the bubble generation part doesn't affect the main section of interest so significantly, thus a high reproducibility of experiment is guaranteed.

iii) We can discuss carefully the flow field altered by bubble injection because a number of researches have been published for the structure of single-phase Taylor-Couette flow to be compared with. Upon these advantages, several studies about two-phase Taylor-Couette flow have been performed. Djeridi et al. (2004) observed that the period of circumferential wave is shortened by bubbles in wavy vortex flow region and the axial wave length becomes around 1.5 times as that without bubbles. Further, they also reported that the bubbles accumulate near the center of Taylor vortex in low *Re* number and out-flow region in the vicinity of the inner cylinder wall in high *Re* number. Berg et al. (2003) measured the frictional drag in bubbly Taylor-Couette turbulent flow, and found that bubbles reduce the friction while solid particles did not do so effectively. Murai et al. (2008) presented that the power gain of the drag reduction gets largest when the toroidal mode and the spiral mode coexist (we call this region the composite region). This implies that we may get more understanding of the drag reduction mechanism by investigating the composite region, and this can be best target to physically elucidate the drag reduction mechanism. In addition, the interaction between two phases significantly modifies the original flow structure in this region because the effect of bubbles, which is mainly buoyancy, and inertia force of liquid become comparable, or Froude number is in order of unity. This corresponds with the fact that the flow switches between toroidal mode and spiral mode in time. To understand this interaction in detail, some studies by numerical simulation also have been reported recently. Climent et al. (2007) numerically analyzed the periodic bubble accumulation under one-way coupling approach in wavy vortex mode. Sugiyama et al. (2007) has also carried out nearly direct numerical simulation in this mode and has shown more accurate dynamics in two-phase interaction by means of two-way coupling approach.

In this study, we experimentally investigate the interaction between two phases in the composite region. As far as we surveyed, few studies have succeeded in measuring the velocity and vorticity in liquid phase in this composite region under two-phase situation. Ordinary PIV, which is based on cross correlation interrogation, cannot acquire the liquid behavior in unsteady bubbly Taylor-Couette system because of difficulties in optical separation of visualized image into two phases. Our approach is succeeded by utilizing PTV, i.e. particle-tracking type of PIV, which is combined with spatio-temporal post-processing of discrete velocity vector information. From the parametric study enabled by the present measurement, we are going to discuss how the original vortex structure in Taylor-Couette flow is altered by presence of bubbles, and to conclude which factor has the most important responsibility to promote the drag reduction by bubbles in this paper.

2. EXPERIMENTAL METHOD

2.1 Experimental setup

Figure 1 shows the schematic diagram of experimental setup. The rotating inner cylinder has 120 mm outer diameter and is made of resin painted black to aid flow visualization. The immobilized outer cylinder has 144 mm inner diameter and a thickness 6 mm, and is made of transparent acrylic resin. The gap distance between cylinders is 12 mm. The effective height filled with working fluid is 390 mm; therefore aspect ratio is 32.5, thus the number of Taylor vortex pairs is estimated more than 10. This helps to relax the dependency of initial condition to measuring the two-phase flow in quasi-steady state, or fully developed state. The upper surface is open to atmosphere to release bubbles on the surface. Silicone oil of kinematic viscosity 10 cSt and density 935 kg/m^3 was used as working fluid. Use of silicone oil removes uncertainties to guarantee experimental reproducibility. It restricts bubble coalescence and has few effect of contamination. Bubbles were generated by porous body set up under the cylinder, and flowed up continuously in the gap because of its buoyancy. The gas flow rate was varied from 0 to 1.17×10^{-6} m^3/s, which corresponds to 5% in the bulk mean void fraction at the maximum. The bubble shape is almost spherical because of low capillary number less than 0.05, and of low

Weber number less than unity. The spherical bubble's mean diameter was 0.6 mm in Sauter mean value. A stepper motor was mounted on the shaft connected to the inner cylinder. The rotational speed was varied from 8.3 to 62.5 rad/s, which provides centrifugal acceleration in bubbly mixture stronger than the acceleration of gravity in the fastest case. The apparatus was submerged in rectangular oil jacket to reduce the influence of refraction and temperature change. As tracer particles for PTV, a glass-made fine particle named "FLO-BEADS (CL-2507)" was mixed in the working fluid. The particle properties are 180 μm in average diameter and 0.918 kg/cm^3 in density. The test section of *r-z* cross section was illuminated by metal halide light sheet. The motion of bubbles and tracer particles are recorded by high-speed digital video camera at 1000 frames per second. The detailed photographing condition and experimental condition are shown in Table 1 and 2, respectively.

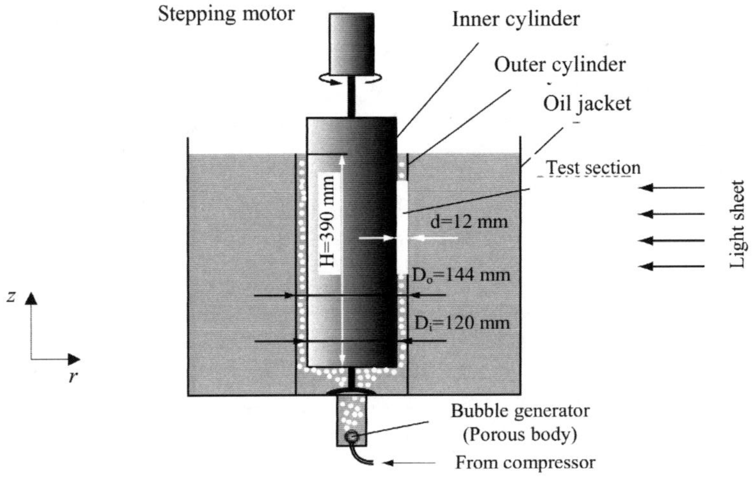

Fig. 1 Schematic of experimental setup

Table 1 Photographing conditions

Frame rate	1000 fps
Shutter speed	1/2000 s
F number	5.6
Frame size	200x1024 pix

Table 2 Experimental conditions

Inner cylinder radius	R_i	60 mm
Outer cylinder radius	R_o	72 mm
Radius ratio	R_i/R_o	0.833
Aspect ratio	L/d	32.5
Density of liquid	ρ_l	935 kg/m^3 (at 298 K)
Kinematic viscosity	ν	10×10^{-6} m^2/s (at 298 K)
Surface tension	σ	19.7×10^{-3} N/m
Mean bubble diameter	d_G	0.6 mm
Gas flow rate	Q_g	0 to 1.17×10^{-6} m^3/s
Re number	*Re*	600 to 4500
Ta number	*Ta*	7.92×10^4 to 3.88×10^6
Fr number	*Fr*	0.65 to 4.89

2.2 Experimental range targeted

It was reported that various flow patterns take place in single phase Taylor-Couette flow depending on inner- and outer-cylinder Reynolds numbers (Andereck et al., 1986). Figure 2 shows the mode transitions along Reynolds number of rotating inner cylinder at fixed outer cylinder, calculated for the present radius ratio of the coaxial cylinders. In this case, Reynolds number is defined by

$$Re = \frac{R_i \omega (R_o - R_i)}{\nu} \quad \cdots\cdots\cdots\cdots \quad (1),$$

where ν is the kinematic viscosity of pure liquid without bubbles. The uncertainty of Re is 5%, which is estimated by the temperature-dependent kinematic viscosity of the silicone oil. The effective viscosity changes slightly when bubbles are mixed. Such effect is quantitatively known in simple shear flows as that of Stokes flow regime (Rust et al., 2002). The two-phase flow inside the present bubbly Couette-Taylor system is considerably inhomogeneous, and thus the original kinematic viscosity of liquid is used for the definition of Re number. In this experiment, Re number was varied from 600 to 4500. The single phase Taylor-Couette flow in this range has five flow modes; i.e., Circular Couette Flow ($Re < 92$, CCF), Taylor Vortex Flow ($92 < Re < 138$, TVF), Wavy Vortex Flow ($138 < Re < 1020$, WVF), Modulated Wavy Vortex Flow ($1020 < Re < 1380$, MWVF), Turbulent Taylor Vortex Flow ($1380 < Re$, TTV). These critical Re numbers are adopted from the result of Djeridi et al. (2004) due to the radius ratio is almost same.

When the interaction between bubbles and vortices is investigated, the ratio of centrifugal acceleration to gravity become important parameter and it is defined by the Froud number

$$Fr = \omega \left(\frac{R_i}{g} \right)^{\frac{1}{2}} \quad \cdots\cdots\cdots\cdots \quad (2).$$

Fr number increases with Re number because both of those are functions of ω.

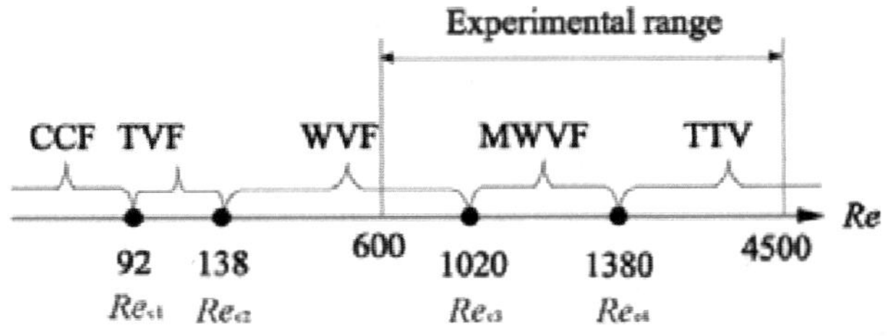

Fig. 2 Flow regime in single phase Taylor-Couette flow

2.3 Measurement of liquid flow field by PTV

The method to obtain velocity vectors of liquid phase is explained in this section. Figure 3 (a) is the raw image photographed, and (b) is the path line of the tracer particle and bubbles. The bubbles exist in the parts of red circle, which are recorded as higher brightness and larger than that of tracer particles. We obtained the velocity vector from photographed images by PTV. The binary cross correlation method was used as PTV algorithm. In this method, the determination of the same particle between two images is enabled by using geometrically resemblance of particle pattern. The binary image is used for detecting the particle information. We distinguished between bubbles and tracer particles by the size. In this experiment, the threshold value of the area between bubbles and tracer particles was determined 15 pixels. This value was determined by using Otsu's method on diameter distribution of bubbles and tracer particle obtained from images. The accuracy of particle detection was also enhanced by doing similar process on brightness. Then we got the vectors from the motion of tracer particles. Further, the obtained vector was rearranged by spatio-temporal Laplace equation (Ido and Murai, 2007) to get regularly arranged vector data as shown in Figure 3(c). Figure 3 (d) shows the contour map of the stream function calculated from the velocity vector obtained. It is confirmed with these figure that the present combination of binarization, PTV, and its post-processing derive fine structure of vortical cells. In the section 3, we discuss about some topics using these quantities measured by the present method.

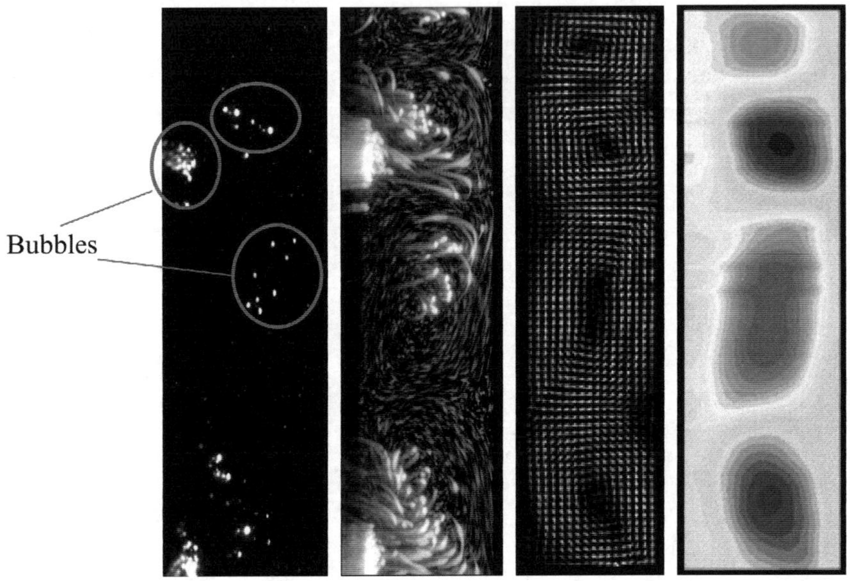

Fig. 3 Instantaneous two-phase flow structure measured in *r-z* section of bubbly Taylor-Couette flow: (a)Original image, (b)Path line of the tracer particle and bubbles, (c)Velocity vector measured, (d)Color contour map of stream function.

3. EXPERIMENTAL RESULTS AND DISCUSSION

3.1 Vortical structure changed by mixing of bubbles

Figure 4 shows the spatio-temporal map of stream function along the center line in the gap (Re=1800, Fr=1.96, Q_g=0 and 0.17 m^3/s). The horizontal axis and vertical axis respectively represent time and vertical position z. Since vortex centers are located near the center line in the gap, the magnitude and the sign of the stream function indicates the vortex intensity and the direction of the rotation. With this representation, spatio-temporal flow structure of vortical cells is easier identified. From these maps, we can see the difference of vortical structure between with and without bubbles. Red and blue bands represent the clockwise vortex and counter clockwise vortex, respectively. We can find that the presence of bubbles provides elongation of vortical wavelength to be around 1.4 times as that without bubbles. This elongation implies the occurrence of the drag reduction, because momentum exchange is restricted by the elongation of vortical wavelength or the decrease in radial momentum transfer of the azimuthal flow. Another change by the mixing of bubbles is that the time fluctuation of vortical structure enlarges. Especially, it occurs at the out flow region, namely around the border between the downside of the blue band and the upside of the red band. This region of the change corresponds with the region in which the bubbles frequently exist. Hence it is simply concluded that the local change by the mixing of bubbles takes place at the region of bubble presence. In one-way numerical simulation, it was found that bubbles are accumulated periodically on the wavy vortex in the azimuthal direction. In experiment, the bubbles accumulated there drive the change of the stripe pattern of the array of cells as shown here. Namely, this alternation of the cell structure is caused by two-way interaction between bubbles and vortex motion, and it would significantly affect the mode transition of Taylor-Couette flow as following discussion.

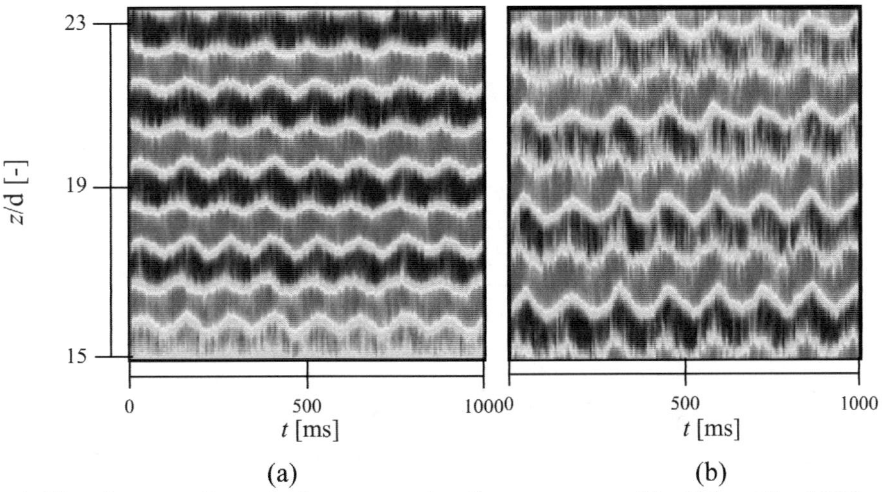

Fig. 4 The change of vortical structure by the presence of bubbles (*Re*=1800, *Fr*=1.96). (a) without bubble, (b)with bubble (Q_g=0.17x10^{-6} m^3/s).

3.2 Mode transition

On focusing on the apparent pattern of bubble distribution, there are three modes found in Taylor-Couette flow; i.e., uniform, toroidal, and spiral modes. Murai et al (2008) investigated the transition diagram of the bubble distribution as functions of gas flow rate and Reynolds number. The first transition from the uniform to the toroidal mode is caused due to liquid-to-bubble interaction while strong Taylor vortex is formed. The second transition from the toroidal to the spiral mode happens due to bubble-to-liquid interaction since local high void fraction in toroidal distribution begins to drive the cell structure resulting in destroying the toroidal mode. In order to find out more detailed process of it, we have captured the switching process between these two modes.

Figure 5 shows temporal change of bubble distribution in this switching region, i.e. coexisting region of toroidal and spiral modes. This image is obtained by rearranging the line image extracted from the front visualized image along the time direction. The Reynolds number is 1800, the Froud number is 1.96 and the gas flow rate is 0.50 x 10^{-6} m^3/s. Horizontal axis represents time and vertical axis represents vertical position, respectively. In Figure 5(a), the toroidal mode occurs at t<4. In this region, torus-shape bubble cluster moves upward gradually by the buoyant effect. Owing to this displacement, then the flow transits to the spiral mode in the lower half of the gap. The bubble cluster connects spirally from the bottom to the top. After the spiral mode is formed, bubble distribution shows a downward phase velocity while individual bubbles rise up continuously in the spiral path. The reason of the downward phase velocity is explained by the azimuthal phase velocity of the spiral structure being slower than the azimuthal speed of the inner cylinder. Namely, the spiral path of the bubbles is also rotated at a certain characteristic speed. The mechanism of the spiral accumulation of bubbles is discussed in Murai et al (2008). Their explanation is based on spatial matching between original Taylor vortex wavelength and the axial pitch of the bubble trajectory that turns around the cylinder during the rotation.

In Figure 5(a), as time elapses at t>8, the toroidal mode begins to be restored from the downside of the gap. That is, the mode transition from spiral to toroidal mode occurs. We found that the ordinary transition from toroidal to spiral mode (T-S transition) can occur from anywhere in the gap. To the contrary, the most opposite transition (S-T transition) occurs from the downside of the gap. Once the ordinary T-S transition occurs, flow pattern keeps the spiral mode whether bubbles exist or not (Fig. 5 (b)). The opposite S-T transition arises when the toroidal mode, which is generated from the lower part of the gap, propagates upward. It is therefore considered that the presence of bubbles significantly affects the vortical structure in the ordinary T-S transition but doesn't affect it in the opposite S-T transition so much. This fact is consistent to the discussion mentioned above, i.e. the ordinary T-S transition is caused by bubble-to-liquid interaction. In other words, the opposite S-T transition is a process in which two-phase interaction ceases for decrease of bubbles in the flow. It is worth noting that there is another independent transition from spiral to toroidal modes at high Re number. This is called third transition and it happens because kinetic energy of turbulent Taylor vortex exceeds the buoyant energy of bubbles. Since bubbles are diffused widely in the gap due to turbulent mixing at this status, the body force is dissipated resulting in the null effect. The S-T transition in the coexisting region is different from it, and it is mainly realized by disappearance of bubbles due to high-speed exhausting of bubbles through the spiral path.

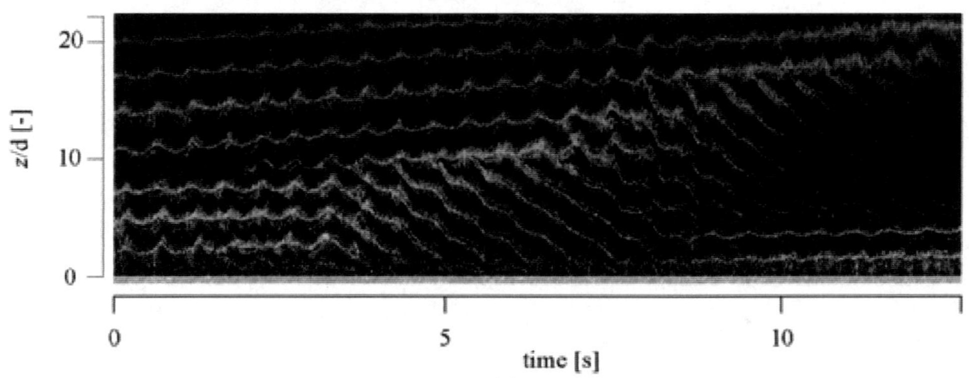

(a)

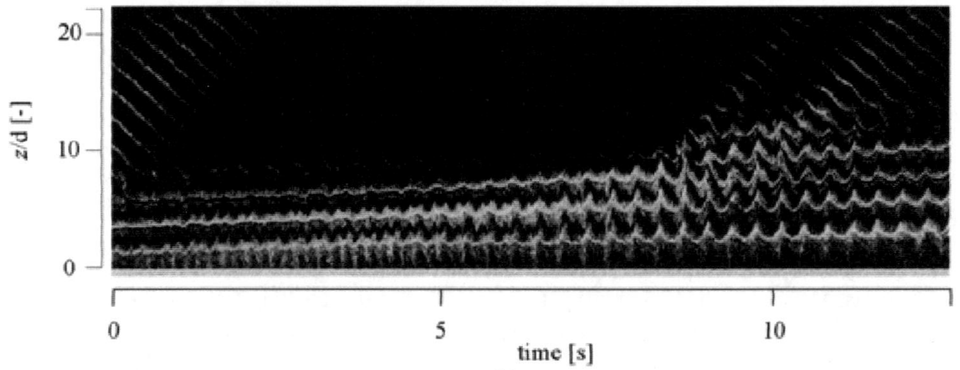

(b)

Fig. 5 Temporal change of the bubble distribution. (Re=1800, Fr=1.96, Q_g=0.50 x 10^{-6} m^3/s)

3.3 Temporal change of vortical structure and bubble distribution

The relationship between bubble distribution and vortical structure during the mode transition is investigated to elucidate how the bubbles affect the transition. Figure 6 (a) shows bubble

Journal of Physics: Conference Series **147** (2009) 012013 doi:10.1088/1742-6596/147/1/012013

distribution in the vicinity of the inner cylinder, obtained by cutting the images of *r-z* cross section. The photographed region of this image is narrower than that of Figure 4 to observe the detailed process of the two-phase interaction. Figure 6 (b) shows spatio-temporal distribution of stream function along the center line in the gap. The stream function is obtained by the measured velocity distribution in r-z cross section, combined with post-processing to interpolate the scattered velocity vector information (see section 2.3). This result corresponds to the composite region of the two modes in bubble distribution, measured at *Re* number of 1800. The *Fr* number is 1.96, and gas flow rate, Q_g, is 0.67x10^{-6} m/s. Red and blue bands represent the clockwise and counter-clockwise vortices, respectively. From figure 6 (a) and (b), we can find that the position of the bubble accumulation and that of vortex have a high correlation. The bubble accumulation takes place in the out-flow region of Taylor vortices. This result agrees with Djeridi et al. (2004), and can be explained by the balance between the centrifugal force of fluid and the drag force of bubbles in the local descending flow there. The most remarkable point, which is newly confirmed here, is that Taylor vortex bifurcates and a pair of vortices coalesces when the flow switches between spiral mode and toroidal mode. Such an oscillatory mechanism does not exist in ordinary Taylor vortex without bubbles.

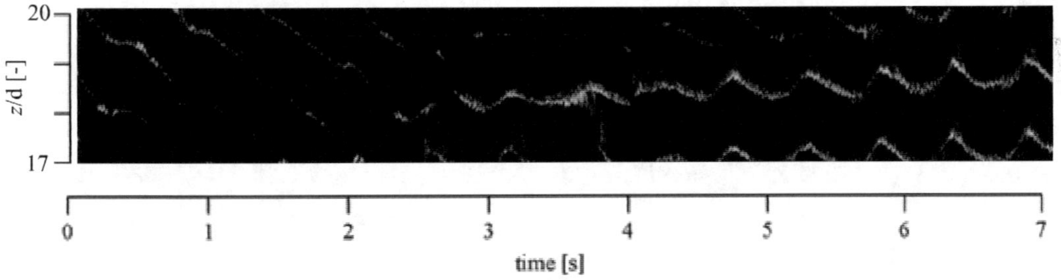

(a)

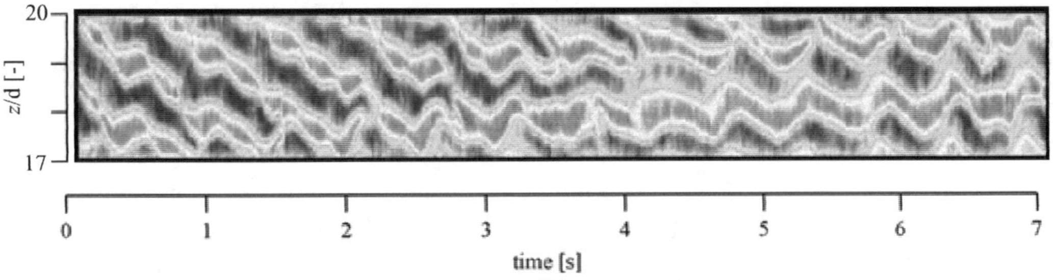

(b)

Fig. 6 (a) Variation of bubble distribution in composite region of spiral mode and toroidal mode, *Re*=1800, *Fr*=1.96, Q_g=0.67x10^{-6} m^3/s. (b) Spatio-temporal distribution of stream function on the center line of the gap. Red band represents clockwise vortex and blue band represents counter clockwise vortex.

Subsequently we measured the average bubble distribution, or probability distribution of bubbles in *r-z* cross section in order to know how bubbles interact with Taylor vortices. Figure 7 (a) is a sample of velocity vector distribution in a single pair of vortices, which is also measured by the present PTV. The bubble position relative to the vortex position is shown in Fig. 7 (b) to (e) for different Reynolds numbers from 800 to 2600. Since the length of the pair of vortices depends on *Re* number, the vertical coordinate in each figure is normalized by the individual length of it at each instant of time. With this expression, where bubbles tend to accumulate can be identified relatively to the vortex structure. At *Re*=800, the bubbles hardly accumulate in the vortices but distribute uniformly inside a thin layer in the vicinity of the inner cylinder. At

Re>1200, this layer of the accumulation is broken, and it separates into three principal regions. One is the vicinity of the inner cylinder of out-flow part. The bubbles around this region clearly concentrate to the root of the outflow region as *Re* number increases. Second region is seen inside the counter clockwise vortex. These bubbles are distributed in the vortex overall, and circulate with the vortex. In this situation, the vortex itself is unmodified by the bubbles while it can be lifted upward due to the buoyancy of bubbles. Third region is inside the clockwise vortex. In this part, the bubbles are distributed disproportionately in the clockwise vortex, i.e. the center of the accumulation is located near the outer cylinder. With this type of accumulation, we can deduce that the vortex is easily deformed due to the localized buoyancy. Therefore, this third region can be concluded to be a key factor that effectively changes the original vortical structure.

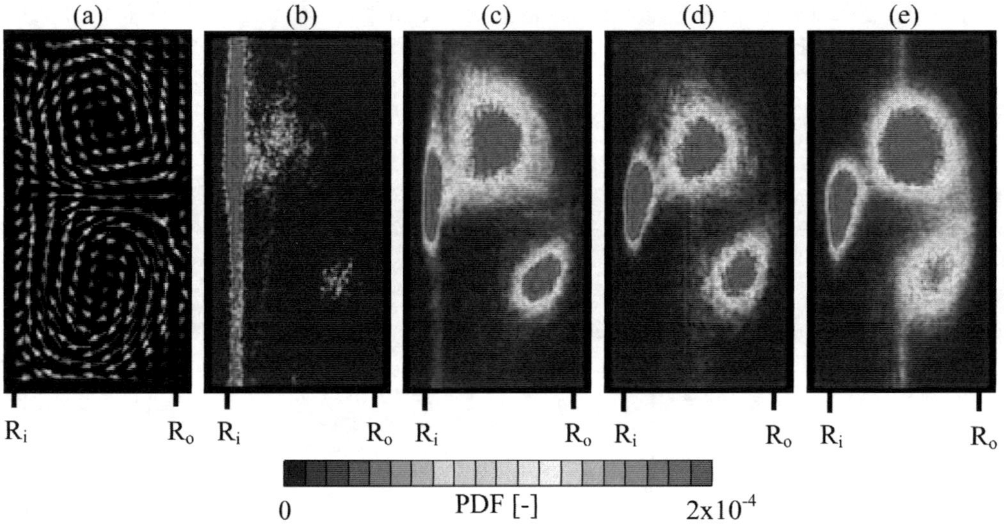

Fig. 7 Bubble position relative to the vortex position: (a) Sample velocity vector of liquid phase, (b),(c),(d),(e) Accumulation position of bubbles in *r-z* cross section: (b) *Re*=800,(c) *Re*=1200,(d) *Re*=1800,(e) *Re*=2600

Figure 8 shows the change of void fraction in the composite region. This graph is extracted from the long sampling data and just represents the moment when the bubble distribution changes from the spiral mode to the toroidal mode. Sum of the brightness in the visualized image over the mode transition, the ordinate of the graph, represents magnitude of the void fraction relatively. While we need to convert it to the real void fraction for quantitative discussion, the brightness value can be a good estimate to evaluate the temporal fluctuation in void fraction since it varies monotonically to the void fraction. As seen in the upper figure showing the stream function, the liquid flow is identified to belong to the spiral mode at $t < 2.2$s, and to the toroidal mode at $t > 2.2$s. As we proved before, the bubble distribution corresponds to the mode of the liquid flow. Thus, the bubbles are distributed as the spiral mode in the first part, and as the toroidal mode in the second part. The result leads to the following discussion. In the spiral mode, the void fraction is kept small. It increases rapidly when the flow transits from the spiral to the toroidal mode. This occurs due to the difference of the vortex structure. In the spiral mode, a single spiral vortex tube exists in the flow, which starts at the bottom and reaches the top of the fluid layer. The bubbles in this mode can rise continuously in a single path that corresponds to the spiral tube. To the contrary, the toroidal mode has an array of torus-shaped vortex tubes that are separated to each other. Under this condition, bubbles are trapped in each tube to have a higher void fraction than the spiral mode. Namely, the rise of bubbles is restricted by the vortex tubes in the toroidal mode. As far as this restriction is stronger than the buoyancy effect, bubbles accumulate in any position as time elapses. When the local void fraction takes higher value than a critical value for the accumulation, the buoyancy exceeds the restriction effect of the toroidal tubes to generate upward jumping of bubbles from a tube to another. This jumping of bubbles, or the axial traverse of bubbles, reduces the local void fraction quickly. Therefore, we can recognize a periodical fluctuation in void fraction in the graph. Furthermore, the data clearly shows that the time of high void fraction corresponds to the time when each vortex displaces to the highest part during the oscillation. Therefore, it is consistent to the previous discussion that the bubbles make the individual vortex centers lifting upward. After the bubbles jump, or the void fraction decreases quickly, the liquid flow field returns to be the toroidal mode temporally (see the oblique patterns of the stream function in the graph). The bubbles then again accumulate in the toroidal mode. Consequently, the iterative process of this periodical interaction generates the highest performance of the drag reduction. This is concluded to be the reason why the compound region, i.e. co-existing region of the spiral and the toroidal modes, provides the highest gain factor of the drag reduction.

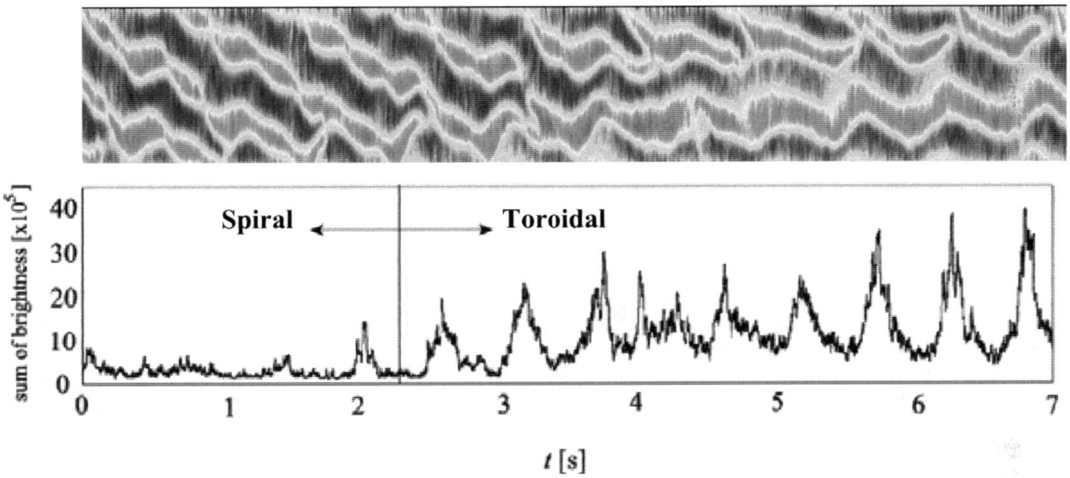

Fig. 8 Temporal change of void fraction during the transition from spiral to toroidal mode.

4. CONCLUSIONS

As a fundamental study on drag reduction enabled by bubbles, the change of flow pattern in bubbly Taylor-Couette flow is investigated by using Particle Tracking Velocimetry. We focus on the flow mode subjected to the competition between the inertia of rotating liquid and the buoyancy of bubbles, which has a Froude number at the order of unity. The two-phase flow in this mode has a periodical switching between toroidal bubble distributions and spiral one, and the drag reduction gets the highest sensitivity to the void fraction. Our flow field measurement targeting this composite mode reveals following remarks. The two patterns of the bubble distribution; the toroidal and the spiral modes, correspond to those of the vortical structure in liquid phase. The transition from the toroidal to the spiral mode occurs due to accumulation of bubbles in torus bands. There are three regions of the bubble localization in r-z cross section; a) the root of the outflow, b) the area inside the counter-clockwise vortex, and c) the spot near the outer cylinder inside the clockwise vortex. Among these three, the effect of b) lifts up the array of vortices, and the effect of c) makes vortex deform to expand the length of the vortex pair. After the mode transits from the toroidal to the spiral one, the local void fraction decreases since the spiral vortex tube conveys bubbles continuously from the bottom to the top. Since the decrease in void fraction allows liquid to return to the toroidal mode, periodical switching between these two modes happens to the flow. Furthermore it is observed that Taylor vortex bifurcates and a pair of vortices coalesces during the mode switching. With these findings, we conclude that the periodical interaction across the two modes provides the best performance of the drag reduction, and this is why the composite region has the highest gain factor to the void fraction.

NOMENCLATURE

R	radius	[mm]
D	diameter	[mm]
d	gap distance	[mm]
g	gravity acceleration	[m/s^2]

Greek Letters

ν	kinematic viscosity	[m^2/s]
α	void fraction	[-]
ω	angular velocity	[rad/s]

Subscripts

i/o	inner/outer cylinder

REFERENCES

Moriguchi, Y. et al. (2002). "Influence of Microbubble Diameter and Distribution on Frictional Resistance Reduction", *J. Marine Sci. Technol*, **7**, 79-85.

Gabillet, C. et al. (2002). "Experimental Study of Bubble Injection in a Turbulent Boundary Layer", *Int. J. Multiphase Flow*, **28**, 553-578.

Andereck, C. D. et al. (1986), "Flow Regimes in a Circular Couette System with Independently Rotating cylinders", *J. Fluid Mech.*, **164**, 155-183.

Djeridi, H. et al. (2004). "Two-Phase Couette-Taylor Flow: Arrangement of the Dispersed Phase and Effects on the Flow Structure", *Phys. Fluids*, **16**, 128-139.

Berg, T. H. et al. (2003). "Smooth and Rough Boundaries in Turbulent Taylor-Couette Flow", *Phys. Rev. E*, **68**, 35307.

Murai, Y. et al. (2008). "Frictional drag reduction in bubbly Couette-Taylor flow", *Phys. Fluids*, **20**, 034101(in press).

Climent, E. et al. (2007). "Preferential accumulation of bubbles in Couette-Taylor flow patterns", *Phys. Fluids*, **19**, 083301.

Sugiyama, K. et al. (2007), "Microbubbly drag reduction in Taylor-Couette flow in the wavy vortex regime", *Phys. Fluid Dyn.* arXiv:physics/0703100.

Rust, A. C. et al. (2002). "Effects of bubble deformation on the viscosity of dilute suspensions", *J. Non-Newtonian Fluid Mech.*, **104**, 53-63.

Ido, T., Murai, Y. (2006). “A Recursive Interpolation Algorithm for Particle Tracking Velocimetry,” *Flow Measurement and Instrumentation*, **17**, 267-275.

Journal of Physics: Conference Series **147** (2009) 012014 doi:10.1088/1742-6596/147/1/012014

PIV Measurement of Turbulent Bubbly Mixing Layer Flow with Polymer Additives

T. NING, F. GUO, B. CHEN*, X. ZHANG

State Key Laboratory of Multiphase Flow in Power Engineering, Xi'an Jiaotong University, Xi'an 710049, China

chenbin@mail.xjtu.edu.cn

Abstract. Based on experimental investigation of single-phase turbulent mixing layer flow with polymer additives, bubbly mixing layer was experimentally investigated by PIV. The velocity ratio between high and low speed is 4:1 and the Reynolds number based on the velocity difference of two steams and hydraulic diameter of the channel ranges is 73333. Gas bubbles with about 0.5% gas fraction were injected into pure water mixing layer with/without polymer additives from three different parts at the end of the splitter plate. The comparison between single phase and bubbly mixing layer shows clearly that the dynamic development of mixing layer is great influenced by the bubble injection. Similar with single phase, the Reynolds stress and vorticity still concentrate in a coniform area of central mixing flow field part and the width will increase with increasing the Reynolds number. Mean Reynolds stress will decrease with bubble injection in high Reynolds numbers and the decreasing of Reynolds stress with polymer additives is much more than pure water case.

1. INTRODUCTION

It is well known that pressure drop from frictional losses can be reduced with gas injection when a shear-thinning fluid is flowing through a pipeline. Ward and Dallavalle (1954) found this phenomenon earliest when they injected air into clay suspensions flowing in the laminar regime. Following experiments about the press drop indicated that hydraulic characteristics of two-phase non-Newtonian flow was quite different with Newton fluid, and the pressure in pipeline dropped a lot (Chhabra et al., 1983; Dziubinski, 1995; Kaminsky, 1998). This phenomenon is broadly applied in various fields such as long distance transportation, drag reduction of large-scle ships, tertiary oil recovery, and reducing the viscous friction of blood flowing to cure the coronary heart disease, etc. Hence, it is very important to study the flow characteristics and drag reduction mechanism through the injection of gas bubbles in conjounction with pplymer additives..

Mixing layer is an classical free shear flow with large coherent vortex structure and it is very helpful to reveal mechanism of drag reduction by studying the influence of turbulent fluctuations and coherent structure with addition of bubble and polymer. Based on the research of pure water turbulent mixing layer(Guo et al., 2007), PIV will be used to investigate the influence of polymer addition and bubbles injection on mixing layer coherent structures and turbulent quantities. It will be the basis to study the interaction between non-newtonian fluids with bubbles, and the drag reduction mechanism with polymer and bubbles in turbulent flow.

Journal of Physics: Conference Series **147** (2009) 012014 doi:10.1088/1742-6596/147/1/012014

2. EXPERIMENTAL SYSTEM

As shown in Fig. 1a, a vertical inflow was seperated into two streams by a specially designed splitter (Fig. 1b) with designed high and low velocity to form the mixing layer. In present paper the velocity ratio is 4:1, velocity difference is 1.25m/s and the Reynolds number for pure water case is 73333 defined as Re= $(\Delta ud)/\nu$, where Δu is velocity difference of two steams and d is hydraulic diameter of the channel. Gas bubbles were injected from high speed side, low speed side and end of splitter with 0.2mm stainless steel tube from a nitrogen bottle. The distance between bubble tube and the end of splitter is 10mm so the disturbance of formation of mixing flow by the bubble injection can be avoided. Odium salicylate (NaSal) was added into the water tank and the concentration is 200ppm. Mixing layer flow with/without polymer additives and bubble injection are measured by a PIV system supplied by La Vision GmbH and compared with pure water cases. The PIV system integrates a double pulsed, 15 Hz, 532 nm, 200 mJ Nd-Yag lasers including sheet optics to generate a light sheet from the laser beam. The focus and the thickness of the laser sheet can be adjusted via the laser-sheet optics and a thickness of approximately 1mm is used throughout the experiment. Silver-coated hollow glass balls with the diameter size distribution of 8-12μm and a density of 1.1g/cm3 are used as tracer particles. Each illumination of the measurement area is recorded using two (2D) ImagerPro4M CCD cameras (8 bit, 15 Hz) with a resolution of 2048 × 2048 pixels.

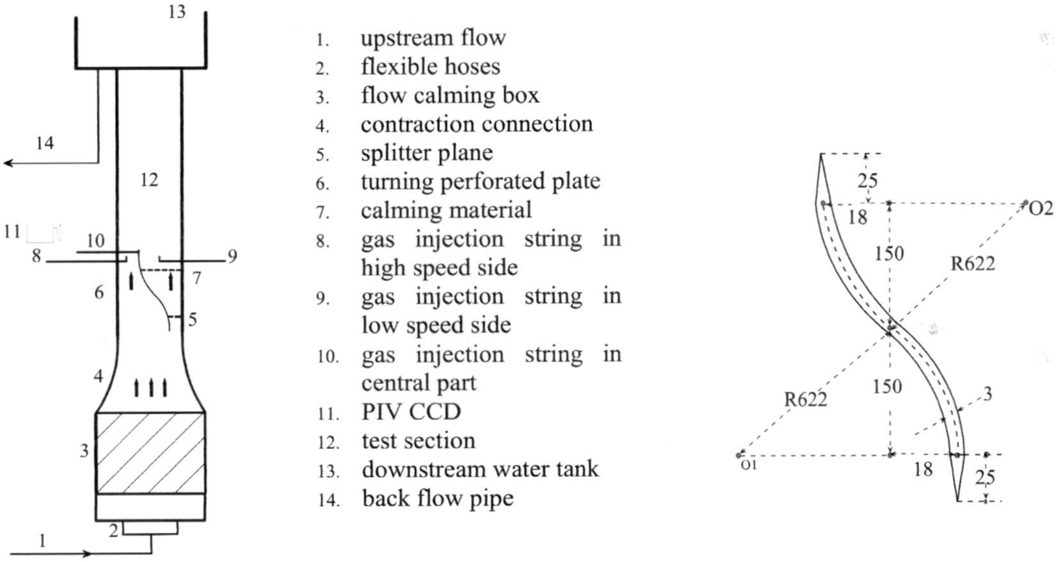

(a) PIV experimental section (b) Splitter design

Fig.1 Experimental system

3. RESULTS AND DISCUSSION

In this section, we analyzed the turbulent flow field of mixing layer with bubbles and polymer additives with velocity difference between high speed and low speed 1.25m/s and Reynolds number is 73333. The fluctuation of turbulent coherent structure and statistics were also discussed, when bubbles and polymer were added into the flow field. The diameter of the bubbles was from 0.5mm to 2mm, and the void fraction defined by $\alpha = A_g / A$ is 0.5%, where A_g is flow rate of gas and A is total flow rate of gas and liquid. In the following figures, the origin is the starting point of the mixing layer (centre of the splitting plate end). The x axis is along the down stream direction, and y axis is from the low speed side to the high speed side. x^* and y^* are non-dimensional length normalized by half width of the channel. Fig. 2a and 2b show the original image of polymer solution mixing layer with/without bubble injection taken by PIV, respectively. The cyan region at the left and central of Fig. 2 is splitter. Because the intensity of PIV incident light is different, the colour of background in Fig. 2a is various.

Fig. 2a showed that the distribution of tracer particles in flow field is homogeneous. In Fig. 2b, it can be seen that bubbles are injected from low speed side, high speed side and the end of splitter simultaneously. The bubbles which injected from high speed side are concentrating in high speed side, and the bubbles from low speed side are diffusion. However, the bubbles from the splitter end aggregative in the central part of mixing layer and migrate to low speed side. The background is painted to black to avoid the reflection. Finally, useful flow information can be contained by processing these original images.

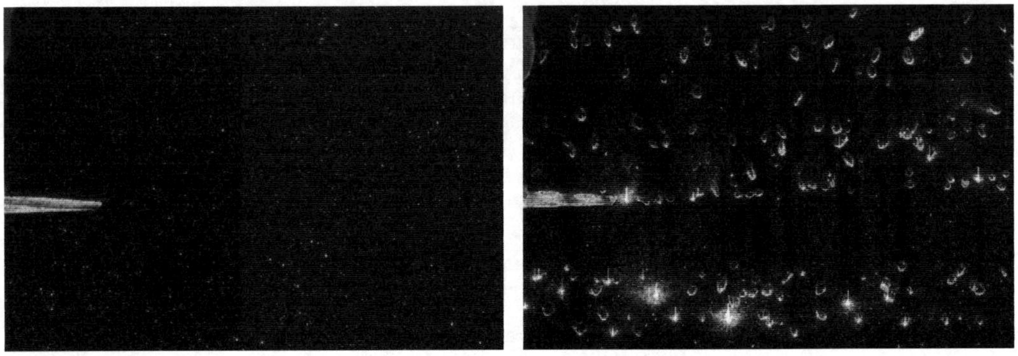

a) Mixing layer with polymer additives b) Mixing layer with polymer additives and bubble injection
Fig. 2 Orginal image of PIV

3.1 Velocity

The profile of average velocity along the down stream direction on same cross-section was shown in Fig. 3. It can be seen that the distribution of velocity was influenced by bubbles injected into both pure water and polymer case. The magnitude of its variation is based on the difference between the initial speed of injected bubbles and the local fluid velocity. It is clearly shown that the modification of pure water case is more obvious than polymer case.

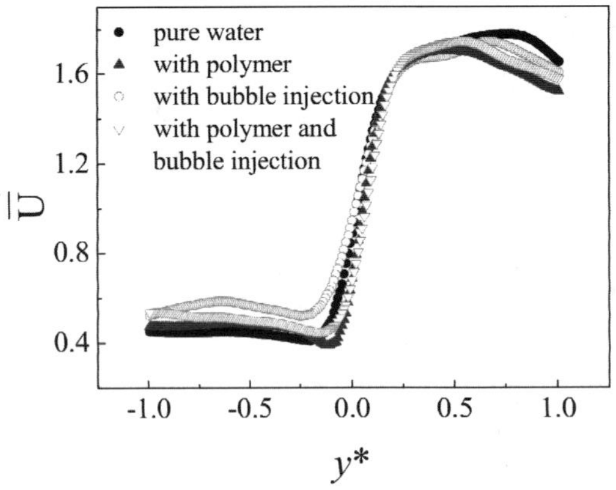

Fig. 3 Profile of the stream wise velocity ($x^*=1$)

Fig. 4 gave the distributions of instantaneous velocity vector $U = u - (\overline{u}_h + \overline{u}_l)/2$ of pure water case, pure water with polymer additives case, pure water with bubbles injection case and the case with both bubbles and polymer additives. The formation, rolling up and evolution of coherent structure can be seen clearly in pure water case. When polymer is added into the water, the coherent structures are clearer, and the small scale vortices are eliminated. When bubbles are injected, the formation of vortex structures is hindered because of the bubbles distribution, and the coherent structures are weakened. However, when bubbles are injected into polymer mixing layer, the coherent structures are re-enhanced, and the large eddy is clear and large again.

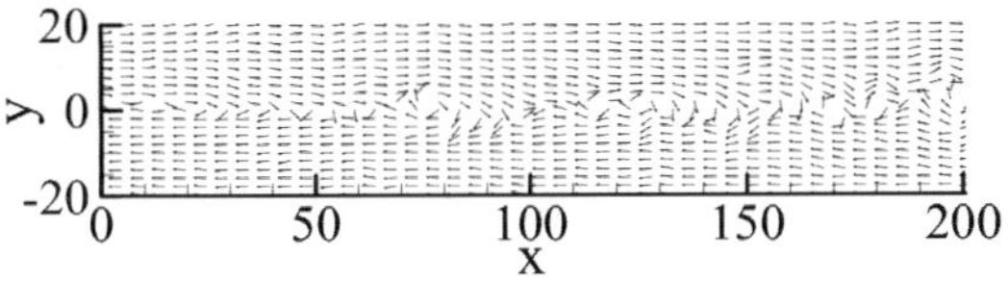

(a) Pure water

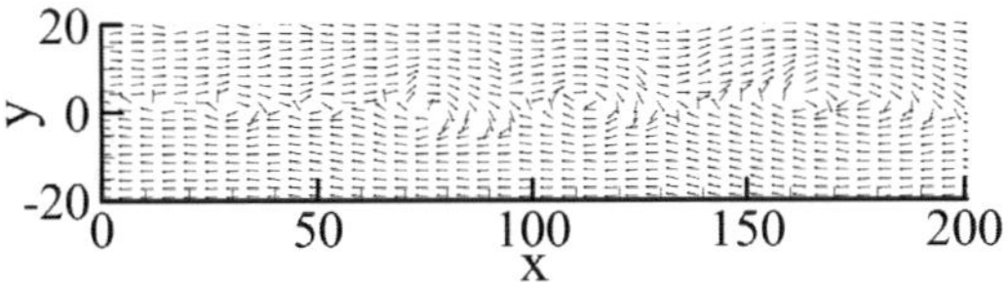

(b) With polymer additives

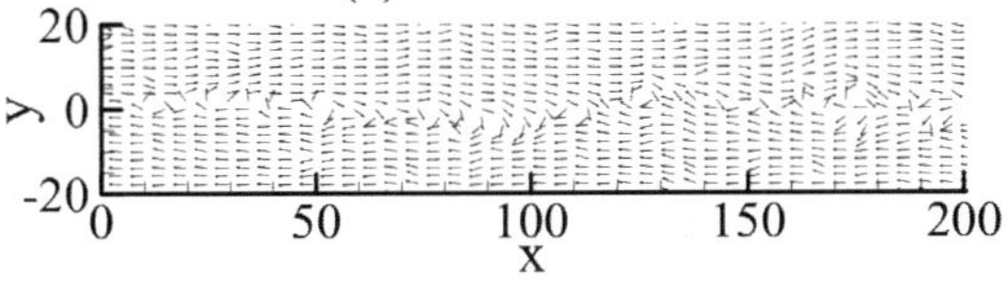

(c) Pure water with bubble injection

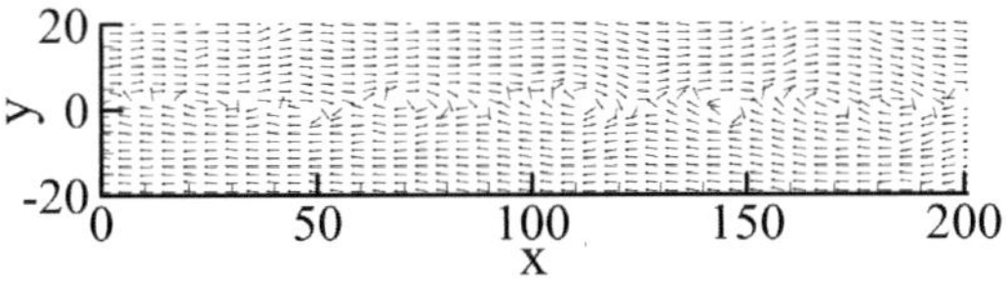

(d) With polymer additives and bubble injection

Fig. 4 Instantaneous velocity vector (Δ u=1.25 m/s)

3.2 Reynolds Shear Stress

Distributions of Reynolds shear stress $\overline{(-u'v')}/(u_1-u_2)^2$ were given in Fig. 5. Similar with pure water case, Reynolds shear stress for all cases are still concentrating in a coniform area of central part of mixing layer and the width of peak value area will increase along the flow direction. When bubbles are injected into the flow field, it is obviously that Reynolds shear stress decreases in both pure water case and polymer case.

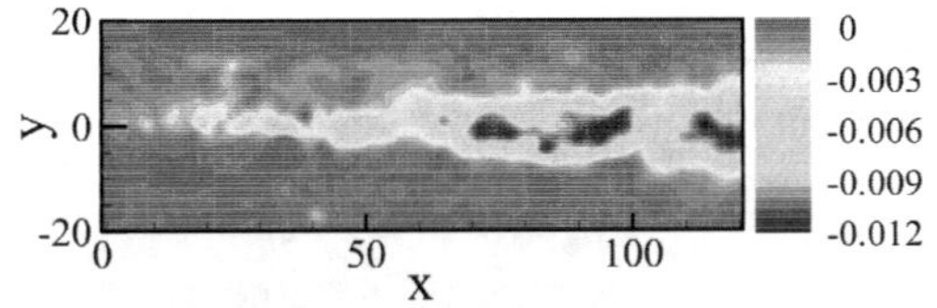

(a) Pure water

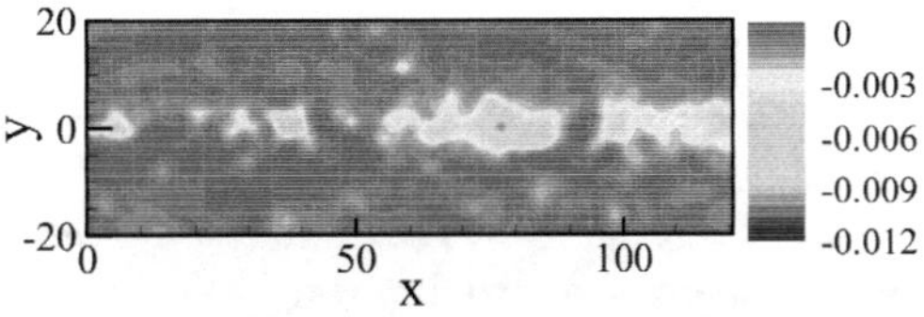

(b) With polymer additives

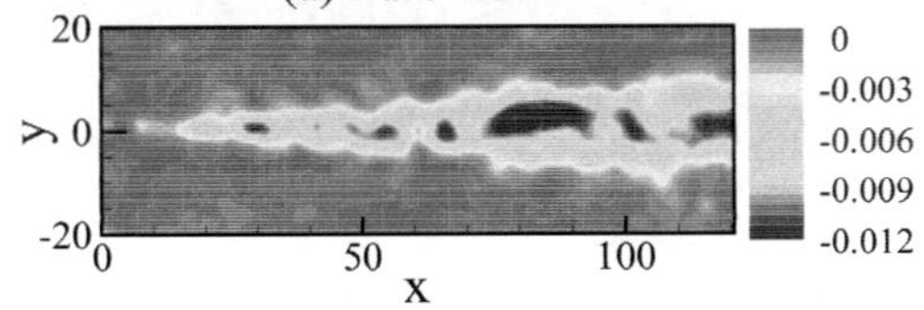

(c) Pure water with bubble injection

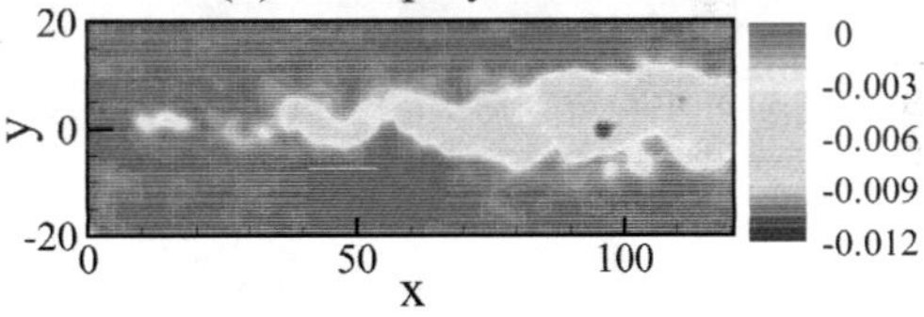

(d) With polymer additives and bubble injection

Fig. 5 Distribution of Reynolds shear stress

Reynolds shear stress of pure water and polymer with and without bubble injection on the same cross-section was compared in Fig. 6. Reynolds stress decreases a lot with bubble injection and the reduction of Reynolds stress when bubbles are injected into polymer solution is much more than pure water case. For polymer case, the increasement of virtual viscocity may be greater to pure water. Therefore, the decreasement of Reynolds stress will be larger. With the injection of bubbles, the effective viscocity is increasing. That may cause the lessening of velocity fluctuation, so the Reynolds stress decreased.

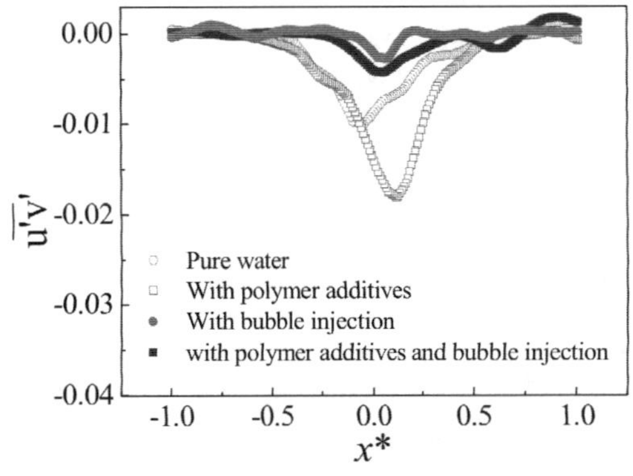

Fig. 6 Profile of Reynolds shear stress (x*=1)

3.3 Vorticity

Fig. 7 gave the average vorticity distribution for all cases. Similar with Reynolds shear stress, the average vorticity is also concentrating in a coniform area of central part of mixing layer. Further more, it decreases with the development of mixing layer along the down-stream direction. With injection of bubbles, the average vorticity decayes, and the coniform area becomes smaller. The vorticity decreases faster because the turbulence is trending to isotropic with the addition of bubbles. It means that the velocity fluctuation in the vertical main stream direction are decreased to balance the disturbance in this aspect.

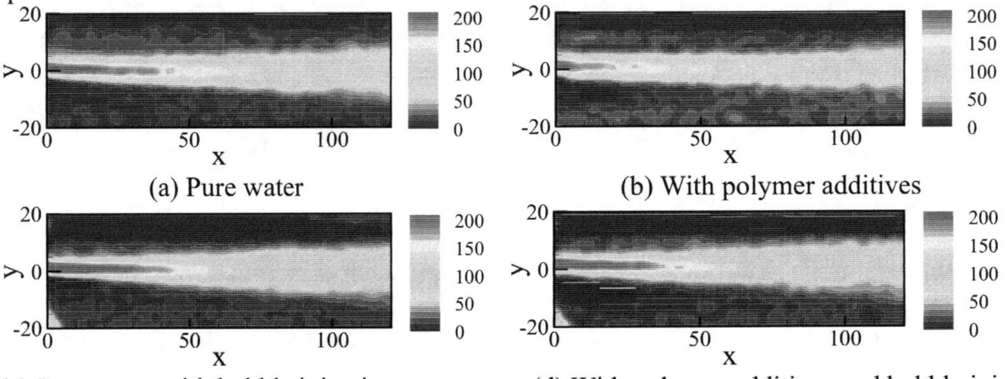

(a) Pure water (b) With polymer additives

(c) Pure water with bubble injection (d) With polymer additives and bubble injection

Fig. 7 Average vorticity field

The peak value of vorticity along down stream direction is shown in Fig. 8 for all cases to reveal the variations of average vorticity. For single phase, the average vorticity decreases directly for the purewater case. But for polymer solution case, it increases a little first and decreases later. For two-phase, cases, peak value will reduce directly in both polymer and water case with the addition of bubbles. The decreasing ratio of pure water case is higher than polymer soution at the initial stage. With the development of mixing layer, there's no significant influence caused by bubble disturbance.

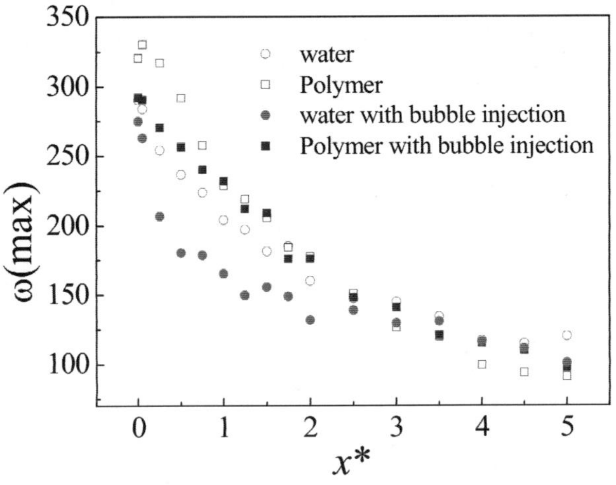

Fig.8 Peak value of average vorticity on each cross-section

4. CONCLUSIONS

An experimental study of gas-liquid two-phase flows of mixing layer with/without polymer additives through a vertical channel had been conducted and compared with pure water mixing layer. It shows clearly that the dynamic development of mixing layer is great influenced by the injection of bubble. When bubbles are injected into pure water or polymer solution, the formation of vortex structures are hindered because of the bubbles distribution. Similar with single phase, the Reynolds stress and vorticity still concentrate in a coniform area of central part of mixing layer. There are some differences that mean Reynolds stress value will decrease with bubble injection and the reduction of Reynolds stress with polymer additives is much more than water. Peak value of average vorticity on each crosssecton increases first and decreases later for single-phase polymer solution, but decreases directly in both polymer solution and pure water with the injection of bubbles.

ACKNOWLEDGEMENTS
The authors greatly acknowledge the financial support by National Natural Science Foundation of China (Grant No. 50676079), Program for New Century Excellent Talents in University (NCET-07-0661) and Key Project of Chinese Ministry of Education (No. 107101).

REFERENCES

Ward, H.C. et al., (1954). "Co-current turbulent–turbulent flow of air and water-clay suspensions in horizontal pipes, " Chem. Eng. Prog. Symp. Series., (10). 1-14

Oliver, D.R. et al., (1968). "Young-Hoon, A., Two-phase non-Newtonian flow. Part I: Pressure drop and holdup, " Trans. IChemE. (46). 106-115

Mahalingam, R. et al., (1972) "Momentum transfer in two-phase flow of gas-pseudo-plastic liquid mixtures, " Ind. Eng. Chem. Fundam., (11). 470-477

Farooqi, S.I. et al., (1980). "Drag reduction by air injection for highly shear-thinning suspensions in horizontal pipeflow, " Trans. IChemE., (58). 16-27

Chhabra, R.P. et al., (1983). "Co-current flow of air and shear-thinning suspensions in pipes of large diameter, " Chem. Eng. Res. Des., (61). 56-61

Dziubinski, M., (1995). "A general correlation for the two-phase pressure drop in intermittent flow of gas and non-Newtonian liquid mixtures in a pipe, (1983). " Trans. Chem. Eng. Res. Des., (73). 528-533

Kaminsky, R.D., (1998). "Predicting single-phase and two-phase non-Newtonian flow behavior in pipes, " Energy Resource Technology. AICHE J., (120). 2-7

Ruiz-Viera, M.J. et al., (2006). "On the drag reduction for the two-phase horizontal pipe flow of highly viscous non-Newtonian liquid/air mixtures: case of lubricating grease," Int. J. Multiphase Flow., (32). 232-247

Guo Fude et al. (2007). "PIV Experimental Investigation of a Turbulent Mixing Layer (I): Single Phase Flow, " 13th China Symposium for Multiphase Flow. 164-171

Guo Fude et al. (2007). "PIV Experimental Investigation of a Turbulent Mixing Layer (II): Multi Phase Flow, " 13th China Symposium for Multiphase Flow. 172-179

Journal of Physics: Conference Series **147** (2009) 012015 doi:10.1088/1742-6596/147/1/012015

Oscillating bubbles in ultrasonic acoustic field

K Matsumoto [1], I Ueno [2]

[1&2]2641 Yamazaki, Noda, Chiba 278-8510, JAPAN

uenolab@gmail.com

Abstract. Behavior of oscillating bubbles is of fundamental study for acoustic cavitations, and is of great importance for medical field. For example, angiographic and diagnosis of cancer of liver. We focused on behavior of multiple air bubbles exposed to ultrasonic wave. The bubbles were injected into the static water from a vertical capillary tube, and then the ultrasonic wave of 20 kHz was applied from above toward the bubbles. Vibrating motion of the bubbles was captured by a high-speed camera at frame rates up to 45000 fps. Excitations of surface wave and shape oscillation with distinct mode number were realized. Correlation between the accelerated bubble behavior and the bubble-bubble distance.

1. INTRODUCTION

To understand a dynamic behavior of the bubble, especially its shape and volume oscillation, under periodic external force is of one of great importance in order to understand and control multiphase flows. Theoretical and experimental works on oscillation of bubble have been widely conducted as fundamental study for such as acoustic cavitations, cavitation noise, wave propagation in bubbly liquid, sonoluminescence production, and boiling heat transfer. A large number of researches have indicated the excitation of volume/shape oscillations and oscillation-mode-transition problems in a single bubble system under small-amplitude external force as reviewed by Plesset & Prosperetti and Feng & Leal. Little knowledge has been accumulated, however, on highly-nonlinear behavior of a single bubble nor interaction on multiple bubbles under oscillations with different mode. The present paper discusses dynamics of a single or multiple bubbles exposed to ultrasonic vibration of wide range of amplitude in water under isothermal condition.

2. EXPERIMENTS

Experimental appratus is shown schematically in Fig. 1. In a series of experiments, a single or successive multiple bubbles were injected by use of micro-syringe into a test fluid in a rectangular tank. Various types of stainless micro-syringe tip of 0.15~2.50 mm in inner diameter were employed to form bubbles of different diameter from 0.3 to 4.0 mm. The test fluid was pure water or water with tiny amount of surfactant.

Ultrasonic oscillator of 20 kHz which intensity of imposed ultrasonic field is 56.65(W/cm^2) was immersed in the fluid from the top, and was triggered after the bubbles detached from the tip of the syringe; the bubbles were exposed to the acoustic pressure field from the top as the bubble rose in liquid. The acoustic pressure was measured with hydrophone. Figures2 and 3 are time series of signal detected by the hydrophone in acoustic field without bubbles and its spectrum respectively. The acousitic pressure field consists of a single component of 20kHz. The bubble behaviors were captured by high-speed camera at frame rates up to 45,000 fps. The bubbles were visualized with continuous light source illuminated from backside.

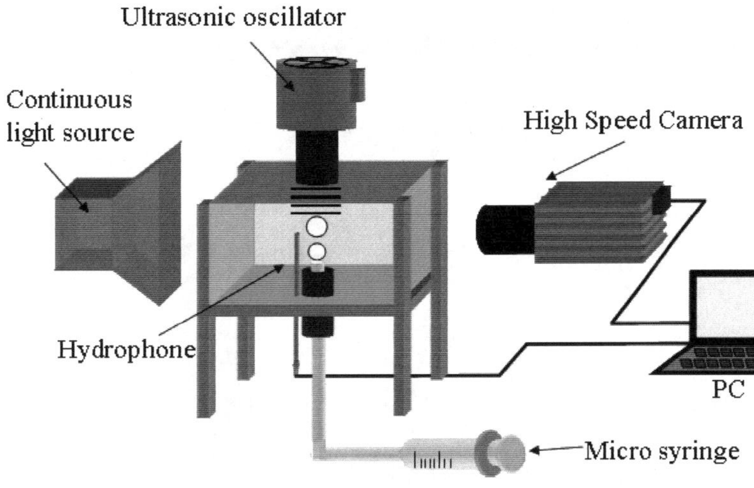

Fig.1 Experimental appratus

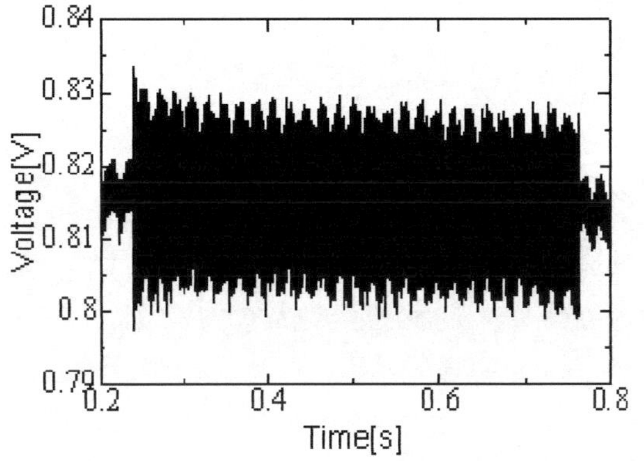

Fig.2 Time series of voltages of hydorophone

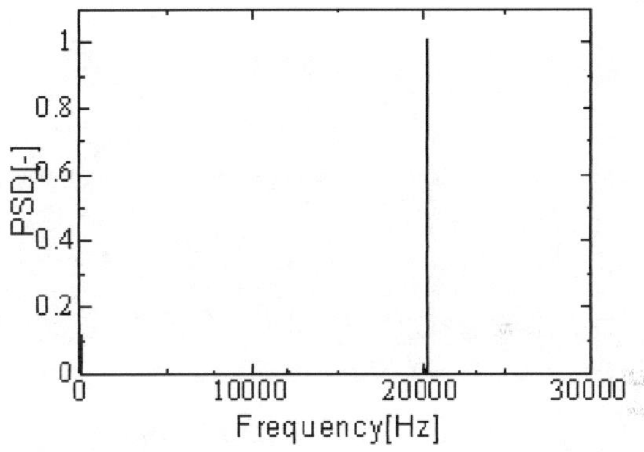

Fig.3 Frequency of voltages

3. RESULTS&DISCUSSONS

In the case of a single bubble under a small-amplitude oscillation, surface wave with a small wavelength emerges over the free surface as shown in Fig. 4. Through experimental observations, the wavelength of the surface wave is about 1.1 ~ 1.5×10^2 μm, and almost independent of the bubble diameter under the present experimental conditions. This surface wave can be explained by considering Kelvin wavelength (Feng & Leal, 1997.& Eisenmenger, 1959.) described as follows;

$$\lambda = \sqrt[3]{\frac{2\pi\sigma}{\rho f_k^2}} = \sqrt[3]{\frac{8\pi\sigma}{\rho f_e^2}} \quad (1)$$

where σ and ρ are the surface tension and the density of solvent, respectively, f_k the frequency of capillary wave, f_e the frequency of ultrasonic vibration. The observed value is in good agreement with the prediction by this equation of 1.2×10^2 μm. In case of large amplitude of the ultrasonic vibration, the bubble exhibits a non-linear shape oscillation with longer wavelength accompanying with the surface wave at a later stage of the oscillation as shown in the last frame in Fig. 4.

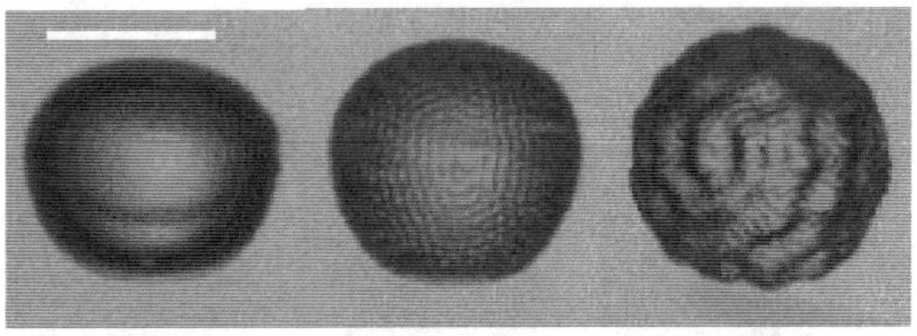

Fig. 4 Time series of surface wave development on a single bubble under ultrasonic vibration of 20 kHz. Scale bar in the left frame corresponds to 1.0 mm.

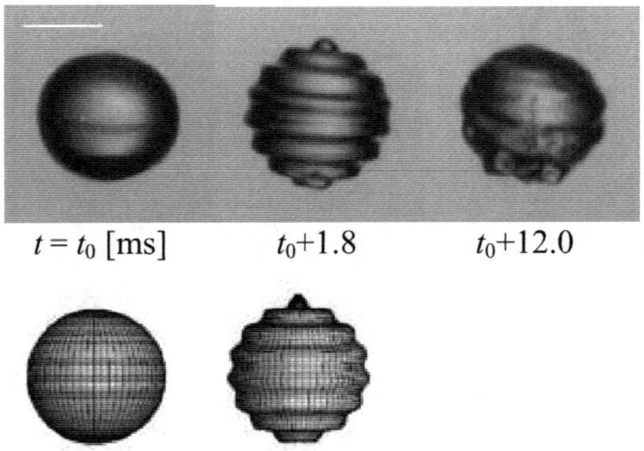

Fig. 5 Shape oscillation of a single bubble with an oscillation mode $(n, m) = (15, 0)$ under large-amplitude ultrasonic vibration of 20 kHz (top), and predicted bubble shape with spherical harmonics of $(n, m) = (15, 0)$ (bottom). Scale bar in the top left frame corresponds to 1.0 mm.

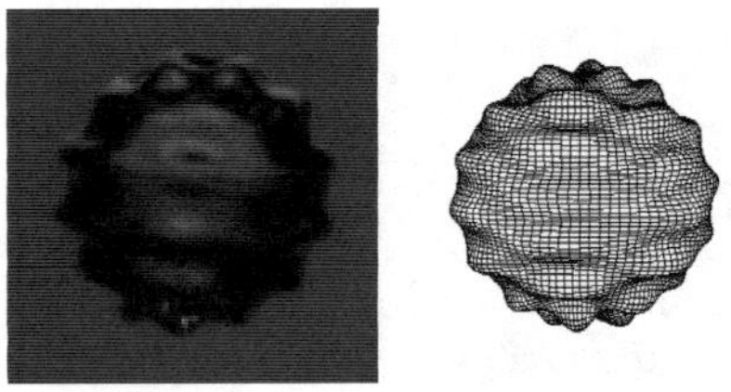

Fig. 6 Non-axisymmetric shape oscillation with a mode $(n, m) = (14, 3)$; experimental result (left) and predicted shape by spherical harmonics (right).

With larger amplitude, axisymmetric shape oscillation with a distinct mode number appears at an early stage. Typical example of shape oscillation of $n = 15$ and $m = 0$ on 1.6-mm-diameter bubble is shown in Fig. 5. It is noted that radial oscillation emerges through the excitation of the oscillation first; irradiation of ultrasonic vibration directly leads the bubble toward shape oscillation of (0, 0) mode, then shape oscillation with distinct mode number. This transition is discussed later in this section. This bubble behavior after the radial oscillation is perfectly described with spherical harmonics (bottom frames in Fig. 5) as follows;

$$r = R(t) + a_n(t) Y_n^m(\theta, \phi) \tag{2}$$

Where R is the instantaneous bubble mean radius, a_n is the amplitude of the surface distortion, Y_n^m is a spherical harmonics, θ and ϕ the polar and azimuthal angles, respectively.

$$\omega_{nm}^2 = \frac{(n-1)(n+1)(n+2)\sigma}{\rho R^3} \qquad (3)$$

Where σ and ρ are the surface tension and the density, respectively. The evaluated value of the oscillating frequency for $(n, m) = (15, 0)$ from the experiments, which corresponds to the case as shown in Fig. 5 is 2.0×10^3 Hz, which is in quite good agreement with the predicted value by Eq. (3) of 2.14×10^3 Hz. The observed values under the present experimental conditions agree well with the predicted one.

This equation(4) is obtained from Rayleigh-Plesset equation and polytropic change equation.

$$\omega_v = \frac{1}{4\pi^2 R}\sqrt{\frac{1}{\rho}\left(3\kappa P_g - \frac{2\sigma}{R}\right)} \qquad (4)$$

Where P_g and κ are the inner pressure of bubble before exposed to the ultrasonic vibration and the polytropic exponent. This equation ignores vapor pressure in the bubble; because that the vapor presuure in the bubble is ignored in this equation ; its pressure is treated to have a small effect in the present scale of the bubble.

Correlation between the frequency of the shape oscillation ω_{nm} and the natural frequency of volume oscillation ω_v

Equation 5 is predicted as follows (Feng & Leal, 1997.)

$$\omega_v = 2\omega_{nm} \qquad (5)$$

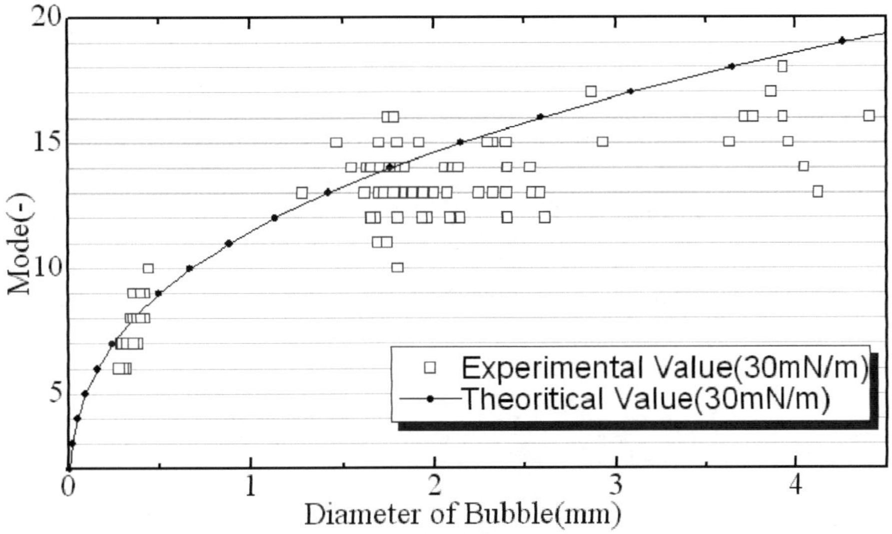

Fig.7: Preferable mode number n of shape oscillation as a function of bubble diameter.Prediction by Rayleigh (1917)-Plesset (1949) & Feng. Z. C. & Leal. L. G. (1997) also indicated. Intensity of imposed ultrasonic field is about 57(W/cm^2) for all data.

One can observe non-axisymmetric shape oscillation as well. Typical example of this oscillation is shown in Fig. 6 in the case of $(n, m) = (14, 3)$. Through the experimental observation, the natural frequency of the non-axisymmetric shape oscillation is almost the same as the one in the case of axisymmetric oscillation if the oscillating bubble evolves the same mode number n. This tendency is explained from Eq. (2), which is independent of the azimuthal mode number m. Figure 7 presents mode number n of the shape oscillation emerged under the present experimental condition. The mode number ranges from 6 to 18. Theoritical value is made from Eq. (3), (4) and (5). In the case that the diameter of bubble is 3.5~4.5, the shape of the bubble near ellipse even under the effect of the surfactant. That's why effective diameters change and the resultant mode number distribute in more scattered manner. On the other hand, in the case that a surface wave instead of the shape oscillation emerges over the free surface of the preceding bubble, surface wave appears on the following bubbles (see the second frame from the left in Fig. 9), and then the bubbles exhibit the shape oscillation with distinct mode (the last frame in Fig. 9). That is, if the energy of ultrasonic vibration is not fully absorbed by the preceding bubble, the acoustic field leads the preceding one to be covered with the surface wave, and still maintains enough energy to trigger a shape oscillation in the following bubbles. One can also observe that contacting bubbles under shape oscillation seldom coalesce.

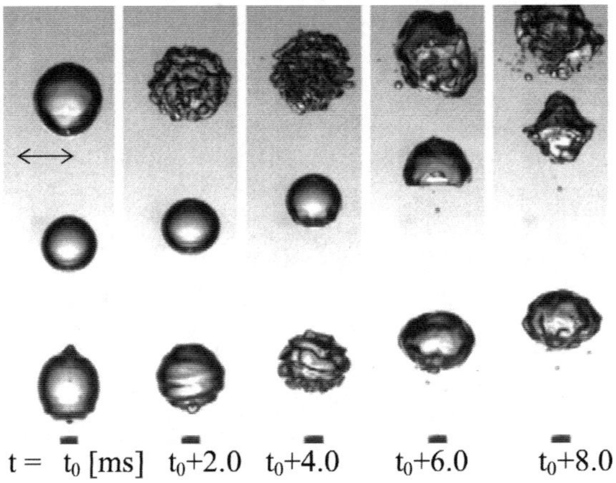

Fig. 8 Effect of the preceding bubble; in case that the preceding bubble exhibits shape oscillation
The black object at the bottom of each frame is the tip of the syringe needle. Scale bar : 2mm

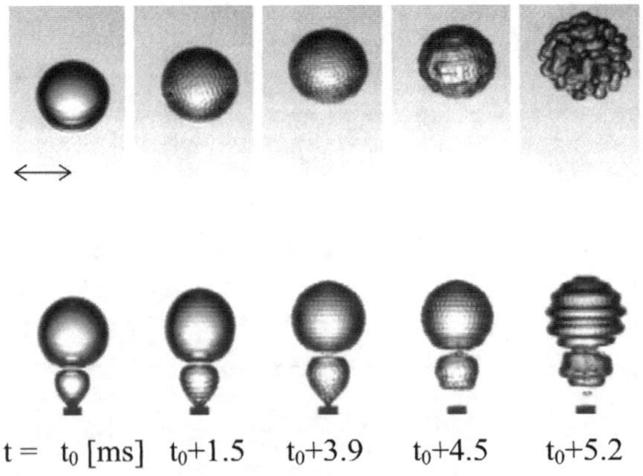

Fig. 9 Effect of the preceding bubble; in case that surface wave emerges on the precedingbubble.

Journal of Physics: Conference Series **147** (2009) 012015 doi:10.1088/1742-6596/147/1/012015

Scale bar in the left frame corresponds to 1.0 mm. The black object at the bottom of each frame is the tip of the syringe needle. Scale bar : 2mm

It is of great importance to discuss the effect of the preceding bubble upon the following one in terms of the distance between them. The authors can qualitatively indicate that the following bubble exhibits its oscillation in almost the same manner as the preceding one if the distance L is larger than about $10R_0$, where R_0 is the mean radius of the bubbles. That is, the following bubble shows a shape oscillation with the same distinct mode as the preceding one. In the case of moderate distance as $3R_0 < L < 10R_0$, it is quite seldom for the following bubble to oscillate in the same manner; almost no distinct shape oscillation is observed and highly non-linear oscillation accompanying with vigorous deformation. If the bubbles are located very closely before irradiation of the acoustic pressure, $L < 3R_0$, the bubbles oscillate in almost the same phase, and the following one is strongly sucked toward the preceding one.

In the case of successive multiple bubbles, the effect of preceding bubble upon the oscillation behavior of following bubble (and vise versa) becomes significant as shown in Fig.8 and .9. In case that the following bubble is rising at a distance between the preceding one in a certain range, the following bubble maintains almost smooth surface despite the preceding bubble vigorously oscillates as shown in the first two frames in Fig. 8. As the following bubble comes closer to the preceding one, the following bubble accelerates towards the preceding one, and deformation and shape oscillation emerge in the following bubble. After those bubbles almost contact, not coalesce, the oscillation in the preceding bubble propagates to the following one, and the both oscillate in quite similar manner. Any distinct shape oscillation does not emerge.

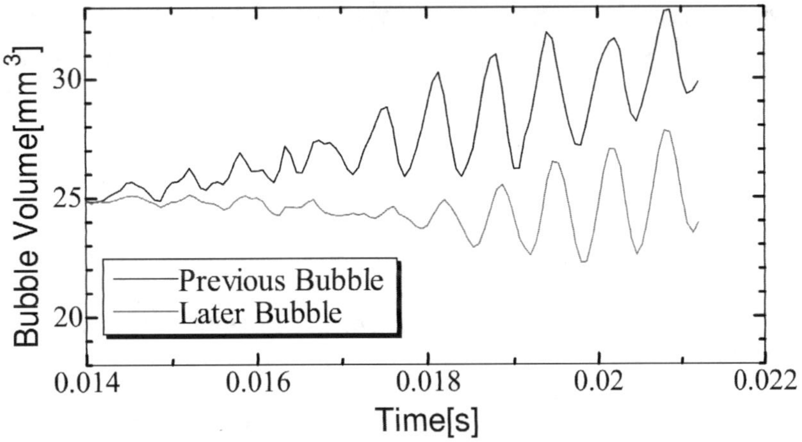

Fig.10 Time series of volume variation of oscillating bubbles

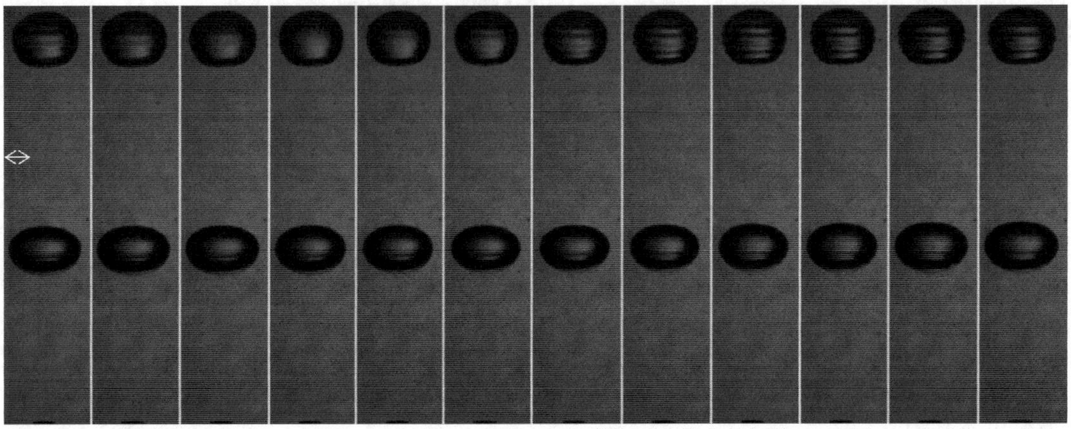

t0(μs) t_0+0.67 t_0+1.34 t_0+2.01 t_0+2.68 t_0+3.35 t_0+4.02 t_0+4.69 t_0+5.36 t_0+6.03 t_0+6.76 t_0+7.37

(a)

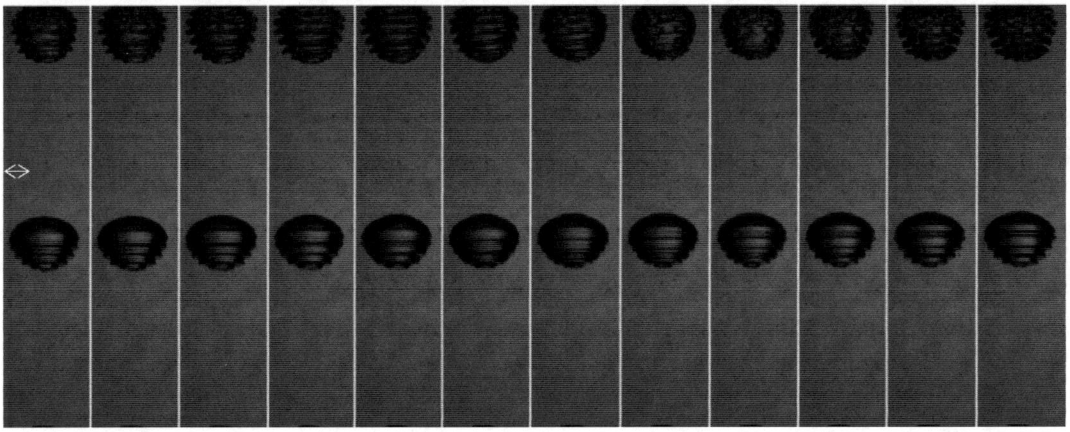

t_0+26.23 t_0+26.9 t_0+27.57 t_0+28.24 t_0+28.91 t_0+29.58 t_0+30.25 t_0+30.92 t_0+31.59 t_0+32.26 t_0+32.93 t_0+33.6

(b)

Fig.11 (a)&(b)Time series of volume variation of snapshots corresponding bubble behaviour Scale bar corresponds to 2mm. Intensity of imposed ultrasonic field is about 57(W/cm^2).

Here the authors focus upon variation of the bubble volume during the oscillation. The volume of the oscillating bubbles are evaluated by extracting the bubble edge (or perimeter) along a horizontal pixel line in the obtained images, and integrating the dish along the vertical axis under the assumption that the bubble oscillation are axisymmetric; this volume evaluation is conducted until the bubble seems to be beyond axisymmetric from the image. In addition, the depression part of the bubble is neglected in the evaluation process. The evaluated volume thus involves some error; the volume must be overestimated. The authors, however, estimate this error be small enough comparing to the major volume variation. Fig.10 indicates a time series of volume variation of the preceding & following bubbles in the case of $L < 10R_0$. The shape of later bubble As an early stage of the oscillation, both bubbles exhibit $(n, m) = (0, 0)$ oscillation perfectly in phase as shown in Fig. 11(a). After a while, the preceding bubble changes into oscillating manner into a combined mode of the surface wave and shape oscillation as shown in Fig.11(b). The following bubble, on the other hand, a shape oscillation with (17, 0) mode dominates the bubble. It should be noted that the preceding & following bubbles exhibit completely different behaviors at last, although they do oscillate almost perfectly in phase with almost the same amplitude.

4. CONCLUDING REMARKS

Behavior of bubble(s) and interaction among successively rising bubbles in water under ultrasonic vibration were discussed. In a single bubble case, excitation of the surface wave over the free surface or shape oscillation with high-order mode was indicated through the observation by use of the high-speed camera. The behavior depends upon the amplitude of the vibration. In the case of multiple bubbles successively rising in a row, the effect of the preceding bubble upon the behavior of the following bubbles was presented.

5. ACKNOWLEDGMENTS

The authors acknowledge Prof. Koichi Suzuki at Tokyo University of Science for their invaluable supports to carry out a series of experiments.

6. REFERENCES

Japan Society of Mechanical Engineers(ed.), *Handbook of Gas-Liquid Two-Phase Flow Technology Second Edition*, Corona Corp., Tokyo, 1989.

Lamb. H., *Hydrodynamics*, Cambridge Univ. Press., Cambridge, 1932.

Plesset. M. S. & Prosperetti. A., Bubble Dynamics and Cavitation, *Ann. Rev. Fluid Mech.*, vol. 9, pp. 145-185, 1977.

Eisenmenger. W., Dynamics Properties of the Surface Tension of Water and Aqueous Solutions of Surface Active Agents with Standing Capillary Waves in the Frequency Range from 10 kc/s to 1.5 Mc/s, *Acoustica*, vol. 9, pp. 327-340, 1959.

Feng. Z. C. & Leal. L. G., Nonlinear Bubble Dynamics, *Ann. Rev. Fluid Mech.*, vol. 29, pp. 201-243, 1997.

7. NOMENCLATURE

a_n	Amplitude of the surface distortion	[mm]
f_e	Frequency of ultrasonic vibration	[Hz]
f_k	Frequency of capillary wave	[Hz]
L	Distance	[mm]
r	Radius of the oscillating bubble	[mm]
R	Radius of the spherical bubble	[mm]
R_0	Mean radius of the bubbles	[mm]
t	Time	[s]
Y_n^m	Spherical harmonics	[-]
θ	Polar angle	[rad]
λ	Kelvin wavelength	[mm]
ρ	Density	[kg/m^3]
σ	Surface tension	[N/m]
ϕ	Azimuthal angle	[rad]
ω_{nm}	Natural angular frequency of (n, m) mode oscillation	[Hz]
ω_v	Angular Frequency of volume oscillation	[Hz]
κ	polytropic exponent	[-]

Journal of Physics: Conference Series **147** (2009) 012016 doi:10.1088/1742-6596/147/1/012016

Measurement of Micro Bubbles Generated by a Pressurized Dissolution Method

S. Hosokawa [1], K. Tanaka [1], A. Tomiyama [1], Y. Maeda [2], S. Yamaguchi [2], Y. Ito [2]

[1] Kobe University, 1-1 Rokkodai, Nada, Kobe 657-8501, JAPAN
[2] Panasonic Electric Works Co., Ltd., 1048 Kadoma, Osaka 571-8686,JAPAN

hosokawa@mech.kobe-u.ac.jp

Abstract. Diameters of micro-bubbles are apt to range from about one mm to several-hundred mm, and therefore, it is difficult to measure a correct diameter distribution using a single measurement method. In this study, diameters of bubbles generated by a pressurized dissolution method are measured by using phase Doppler anemometry (PDA) and an image processing method, which is based on the Sobel filter and Hough transform. The diameter distribution and the Sauter mean diameter of micro bubbles are evaluated based on the diameters measured by both methods. Experiments are conducted for several mass flow rates of dissolved gas and of air bubbles entrained in the upstream of the decompression nozzle to examine effects of the entrained bubbles on bubble diameter. As a result, the following conclusions are obtained: (1) Diameter distribution of micro bubbles can be accurately measured for a wide range of diameter by using PDA and the image processing method. (2) The mean diameter of micro-bubbles generated by gasification of dissolved gas is smaller than that generated by breakup of air bubbles entrained in the upstream of the decompression nozzle. (3) The mean bubble diameter increases with the entrainment of air bubbles in the upstream of the decompression nozzle at a constant mass flow rate of dissolved gas.

1. INTRODUCTION

Micro-bubbles are the bubbles with diameter ranging from one to several-hundred μm, and they are characterized by large interfacial area concentration per unit gas volume and low relative velocity between gas and liquid phases. They have possibilities of improving the efficiencies of chemical reactors, water treatments and washing processes. Generation of fine micro-bubbles of high number density is required in these applications of micro-bubbles (Fujikawa et al., 2003, Sadatomi et al., 2005). Measurements of bubble diameter and bubble number density are, therefore, indispensable for development and improvement of micro-bubble generators. Several measurement methods are available to measure bubble diameter. Intrusive measurement methods such as electrical or optical probe cannot be applied to micro-bubbles due to the small inertia and the high surface tension forces caused by their small diameters. Although phase Doppler anemometry (Dominick and Durst, 1995) is one of powerful tools to measure diameters of spherical particles, it is applicable only to bubbles smaller than a couple of hundreds μm. Image processing measurements are frequently used to measure diameters of micro-bubbles. It is not easy to accurately measure the diameters of bubbles smaller than 10 μm. Hence, it is difficult to measure a correct diameter distribution of micro-bubbles using a single measurement method.

Generation methods of micro-bubbles can be classified into two groups, i.e, methods based on bubble break-up and methods based on gasification of dissolved gas (pressurized dissolution method). The latter method is superior to the former from the point of view of generating fine bubbles. Since the void fraction of micro-bubbles generated by the latter method is determined by the concentration of dissolved gas which depends on the pressure at the dissolution section, dissolution under a high pressure condition is necessary to increase the bubble number density and the void fraction. One of the possible methods to increase the void fraction under a low pressure condition is utilization of breakup of bubbles injected in the upstream of the decompression nozzle. However, the effects of the injected bubbles on the diameter and number density of generated bubbles are not clarified yet.

In this study, diameters of bubbles generated by a pressurized dissolution method are measured by using phase Doppler anemometry (PDA) and an image processing method, which is based on the Sobel filter and Hough transform. The diameter distribution and the Sauter mean diameter of micro-bubbles are evaluated based on the diameters measured by the both methods. Experiments are conducted for several mass flow rates of dissolved gas and of air bubbles entrained in the upstream of the decompression nozzle to examine effects of the entrained bubbles on bubble diameter.

2. EXPERIMENTAL APPARATUS

Schematic of the experimental apparatus is shown in Fig. 1. Air supplied from a compressor (Hitach, SRL-2.2DA6) flows into the dissolution tank and mixes with water supplied from the pump (Iwatani, 25AJT0751B). A part of air dissolves in water through the air-water interface in the dissolution tank, and the other part of air is exhausted from the tank. The schematic of the dissolution tank is shown in Fig. 2. Air and water flow as a bubbly flow in the inner pipe and overflows at the top of the pipe. The falling film is formed on the outer surface of the inner pipe and water flows to decompression nozzle through the tank exit located at the bottom of the outer pipe. The pressure in the dissolution tank is measured by a pressure gauge. Figure 3 shows the concentration of dissolved oxygen measured at the exit of the dissolution tank. The concentration of dissolved oxygen increases with the water level H. That is, the flow rate of dissolved gas can be controlled by changing the water level H, i.e., the interfacial area of the falling film in the dissolved tank. The maximum concentration of dissolved oxygen is achieved for H = 50 mm, and it depends on the liquid volumetric flux J_L at the decompression nozzle. Water with dissolved gas is decompressed at the nozzle shown in Fig. 4, and

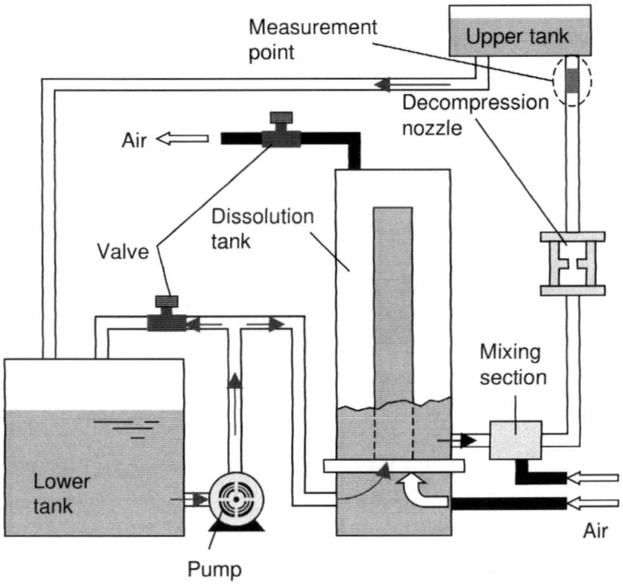

Fig. 1 Experimental apparatus

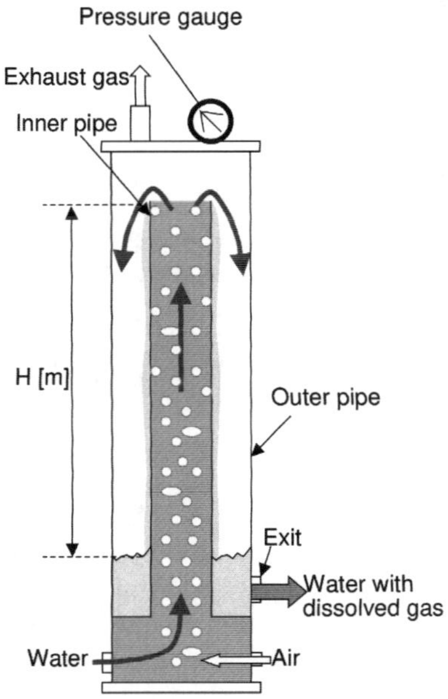

Fig. 2 Dissolution tank

micro-bubbles are generated in the nozzle and in the downstream of the nozzle. Air can be added to the water flow at the mixing section which locates in the upstream of the nozzle to examine the effects of air bubbles on the generation of micro-bubbles. An example of bubble size distribution in the upstream of the nozzle is shown in Fig. 5. The bubble injected from the mixing section mainly ranges from 100 μm to 1 mm and its diameter takes a peak at around 300 μm. The water temperature is kept at 25 ± 0.2 °C through the experiments.

Experiments are conducted for several mass flow rates of dissolved gas and of air bubbles entrained in the upstream of the decompression nozzle to examine effects of the entrained bubbles on the diameter and number density of micro bubbles generated by the pressurized dissolution method. Experimental conditions are summarized in Table 1. The mass flow rate of gasifiable dissolved gas W_{GD} is the mass flow rate of dissolved gas in the upstream of the decompression nozzle subtracting the mass flow rate of dissolved gas under the saturated condition at the atmospheric pressure. The dissolved oxygen in the upstream of the decompression nozzle is measured by a DO meter (HACH, LDO101-01 HQ30d), and W_{GD} is evaluated from the measured oxygen concentration based on the Henry's low. The mass flow rate of bubbles W_{GB} is the mass flow rate of air injected in the mixing section, which is measured by a flow meter (Japan Flow Cell, SPO-4). The W_{GDm} is the maximum value of the mass flow rate of dissolved gas which is achieved by using the present dissolution tank. In cases 1 – 3, the total mass flow rate W_G (= W_{GD} + W_{GB}) is the same whereas the ratio of W_{GD} to W_G is different. On the other hand, the mass flow rate of bubbles W_{GB} is changed from 0 to W_{GDm} for cases 1, 4 and 5 under the constant W_{GD} condition. The liquid volumetric flux at the nozzle is ranged from 9 to 14 m/s. Note that W_{GDm} depends on the liquid volumetric flux J_L since the pressure in the dissolution tank depends on J_L.

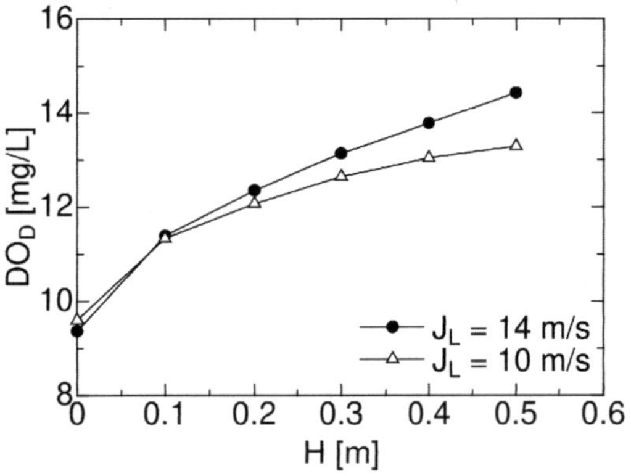

Fig. 3 Performance of the dissolution tank

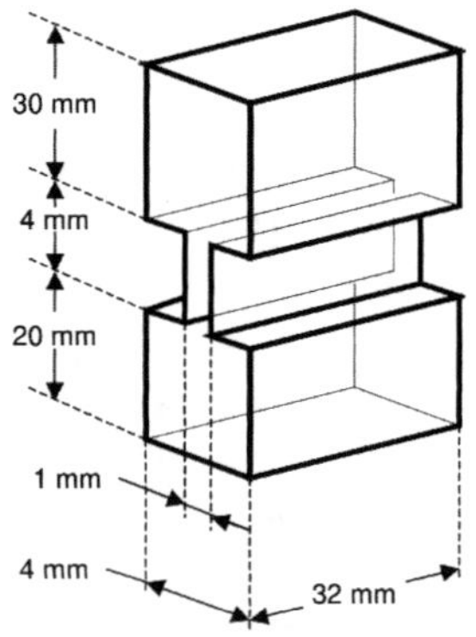

Fig. 4 Decompression nozzle

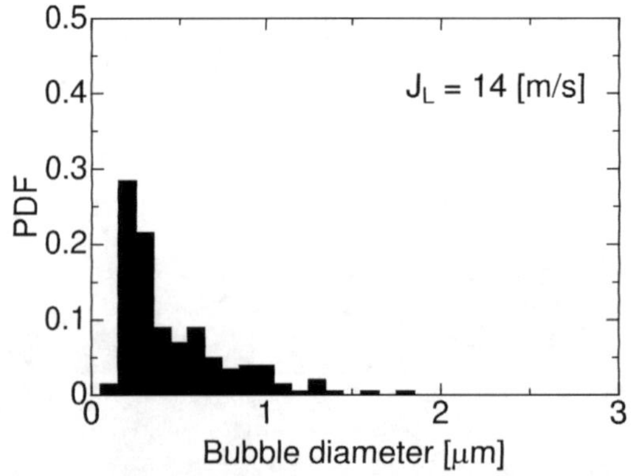

Fig. 5 Bubble size distribution in the upstream of the decompression nozzle (Case 4, J_L = 14 m/s)

Table 1 Experimental condition

	W_G	W_{GD}	W_{GB}	W_{GD}/W_G
Case 1	W_{GDm}	W_{GDm}	0	1
Case 2	W_{GDm}	0.5 W_{GDm}	0.5 W_{GDm}	0.5
Case 3	W_{GDm}	0	W_{GDm}	0
Case 4	1.3 W_{GDm}	W_{GDm}	0.3 W_{GDm}	0.77
Case 5	2 W_{GDm}	W_{GDm}	W_{GDm}	0.5

3. MEASUREMENT OF BUBBLE DIAMETER

The diameter of micro-bubbles generated by using the pressurized dissolution method ranges from about one to several-hundred μm, and therefore, it is difficult to measure the bubble size distribution by using single measurement method. Hence, the bubble diameter is measured by using a combination of a phase Doppler anemometry (PDA) and an image processing method. The measurable diameter ranges of these methods are 0 - 150 μm for PDA and 80 - 1000 μm for the image processing method. The measurement position locates 250 mm downstream of the nozzle exit.

The PDA system (DANTEC, 60X) consists of a transmitter, a receiver and a processor. The scattering angle, i.e., the angle between the transmitter and the receiver, is 90 degree. The sample number of bubbles is 20 000.

An image processing method (Zhang et al., 2008) is used to measure bubble diameter distributions in the range of 80 - 1000 μm. Since bubbles are to be distributed throughout the channel, we have to remove out-of-focus bubbles from recorded images. We therefore adopt the Sobel filter to reject out-of-focus bubbles. Its applicability to bubbly flow was confirmed by Bröder and Sommerfeld, 2007. In the research of Bröder and Sommerfeld, 2007, bubble sizes were comparable to the depth of focus, and therefore, they did not need additional image processing for the detection and separation of overlapped bubble-images. Since bubble sizes in this study are smaller than the depth of focus (1.0 mm), it is necessary to detect and split overlapped bubble-images. Hence, we adopt a normal-line Hough transform to detect and separate the overlapped bubble-images (Yu et al., 2006).

Figure 6 is an example of an image at each processing step. First, the Sobel filter is applied to the original 8-bit grayscale image (Fig. 6 (a)) recorded by the high-speed video camera to obtain the gradient of the brightness level (Fig. 6 (b)). Then, the edges of bubbles located within the depth of focus are detected by binarizing the gradient distribution based on a predefined threshold level (Fig. 6 (c)). Applying the normal-line Hough transform to the binary image yields a vote function, which is used in the Hough transform to express the degree of certainty of the presence of the center at each position. The bubble center is detected by searching the maximum point in the vote function using an interpolation (Fig. 6 (d)). Finally, bubble diameters are evaluated for all the detected bubble centers as shown in Fig. 7. That is, pixels on a bubble edge are detected by scanning the binary image shown in Fig. 6 (c) in a predetermined range of the radius (r_{min} to r_{max}) for each detected bubble center. The bubble radius is evaluated by calculating the mean distance between the center and the pixels on the edge. Since bubble shapes in this experiment are spherical as shown in Fig. 6 (a), the error in the image processing using the Hough transform due to non-sphericity is negligible. The sample number of bubbles is 20 000.

The probability density function $P_{PDA}(d_B)$ of bubble diameter measured by using PDA is merged with the pdf $P_{IP}(d_B)$ measured by using the image processing method by matching the probabilities, α and β, in the diameter range of 80 - 150 μm to obtain the probability density function $P(d_B)$ of bubble diameter in the wide diameter range (Fig. 8):

$$P(d_B) = \begin{cases} \dfrac{\beta}{\alpha+\beta-\alpha\beta} P_{PDA}(d_B) & for \quad d_B < 150\mu m \\ \dfrac{\alpha}{\alpha+\beta-\alpha\beta} P_{IP}(d_B) & for \quad d_B \geq 150\mu m \end{cases} \tag{1}$$

$$\alpha = \int_{d_B=80\mu m}^{150\mu m} P_{PDA}(d_B) dd_B \tag{2}$$

$$\beta = \int_{d_B=80\mu m}^{150\mu m} P_{IP}(d_B) dd_B \tag{3}$$

The Sauter mean diameter d_{32} is calculated from the resultant bubble size distribution. The bubble number density $n_{B,PDA}$ based on the PDA measurement is evaluated from the data acquisition rate N_{PDA} in the PDA measurement:

$$n_{B,PDA} = \frac{N_{PDA}}{S_{PDA} V_L} \tag{4}$$

where S_{PDA} is the cross-section of the PDA measurement volume and V_L the liquid velocity at the measurement point which is evaluated from the measured velocities for bubbles smaller than 30μm. Then the bubble number density is corrected based on $P(d_B)$:

$$n_B = \frac{\alpha+\beta-\alpha\beta}{\beta} n_{B,PDA} \tag{5}$$

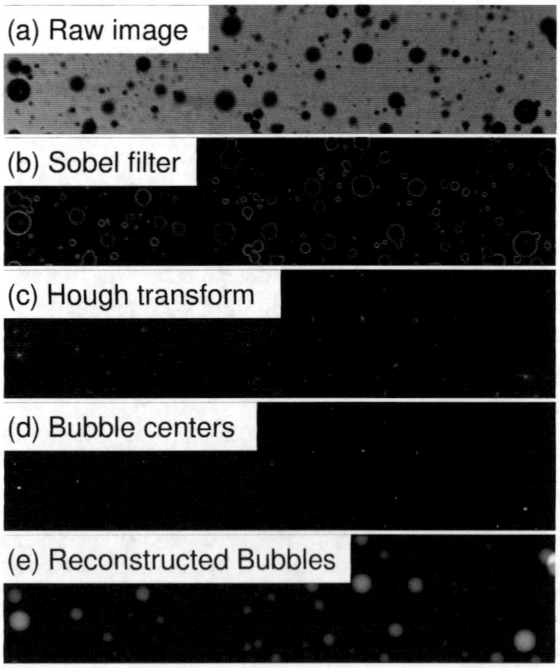

Fig. 6 Examples of raw and processed images

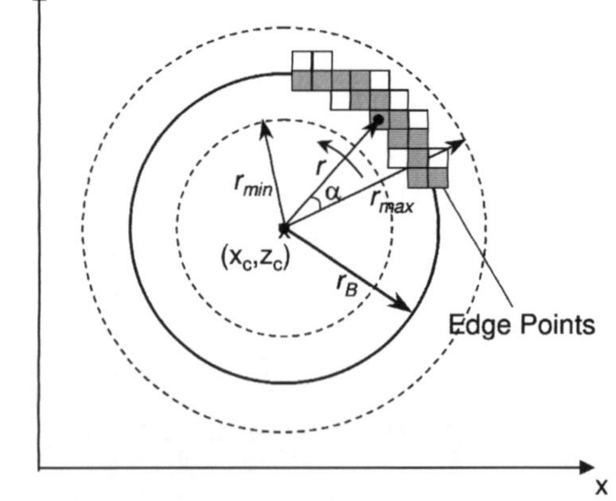

Fig. 7 Calculation of bubble radius

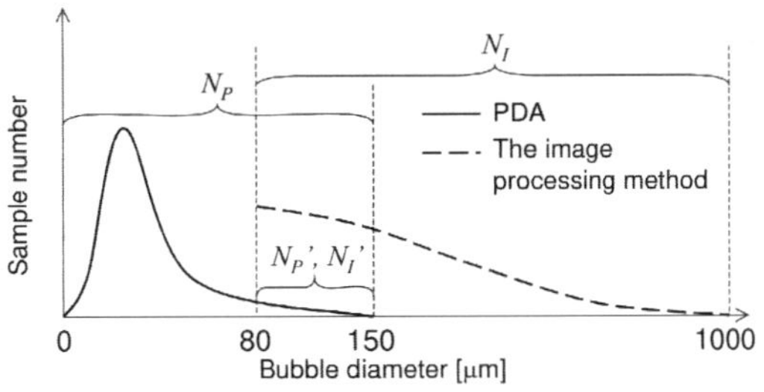

Fig. 8 Schematic of bubble size distributions measured by using PDA and the image processing method

4. RESULTS AND DISCUSSIONS

Figure 9 shows turbidity in the upper tank measured by a turbidity meter (OPTEX, TC-3000). The turbidity tends to increase with the number density of micro-bubbles. The solid line denotes the maximum mass flow rate W_{GDm} of dissolved gas which is attained by using the present dissolution tank. Since the pressure in the dissolution tank increases with J_L, W_{GDm} also increases with J_L. The turbidity increases with J_L except case 3 in which all the gas is supplied as air bubbles (W_{GD}= 0). When W_{GD} / W_G increases at a constant W_G (cases 1 – 3), the turbidity decreases for low J_L (J_L < 12 m/s) and it increases for high J_L ($J_L \geq 12$ m/s). To the contrary, the turbidity increases for low J_L (J_L < 12 m/s) and it decreases for high J_L ($J_L \geq 12$ m/s) when W_{GB} increases at a constant W_{GD} (cases 1, 4, 5). That is, the turbidity decreases with increasing W_{GB} for high J_L whereas the total mass flow rate of gas W_G increases with W_{GB}.

Figure 10 shows the measured Sauter mean bubble diameter d_{32}. For the constant W_G cases (Fig. 10 (a)), the mean bubble diameters for W_{GB}/W_G = 0 (dissolved gas only) are smaller than those for W_{GB}/W_G = 1 (air bubble only). This indicates that the mean diameter of micro-bubbles generated by gasification of dissolved gas is smaller than that generated by breakup of air bubbles entrained in the upstream of the decompression nozzle. The mean bubble diameter increases with W_{GB} not only for the constant W_G cases (Fig. 10 (a)) but also for the constant W_{GD} cases (Fig. 10 (b)). That is, entrainment of air bubbles in the upstream of the decompression nozzle decreases the mean bubble diameter

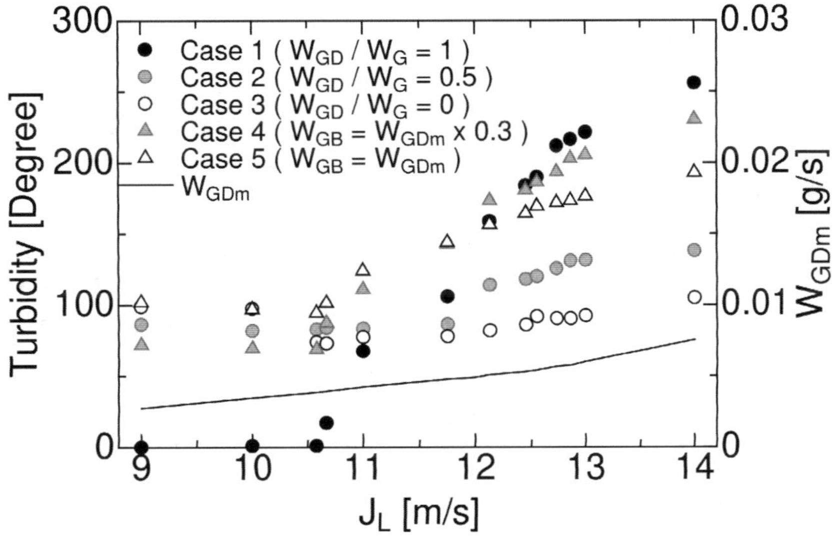

Fig. 9 Turbidity in the upper tank

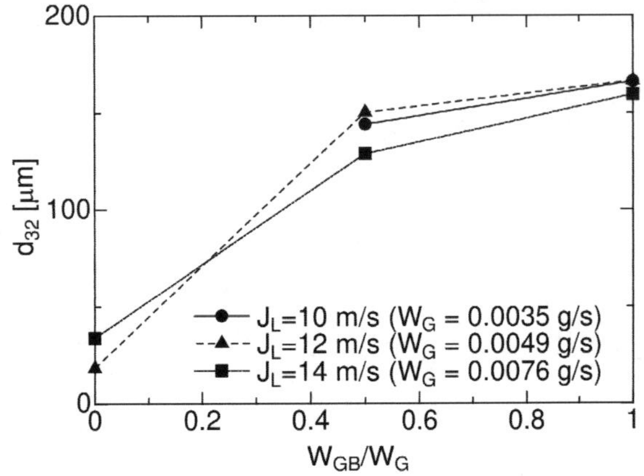

(a) Cases 1, 2 and 3

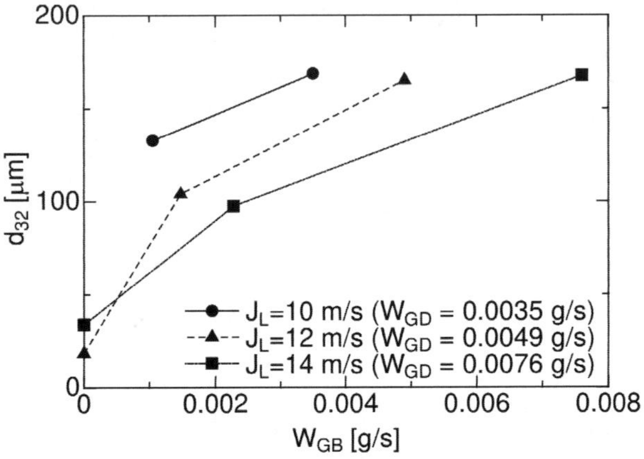

(b) Cases 1, 4 and 5

Fig. 10 The Sauter mean bubble diameter

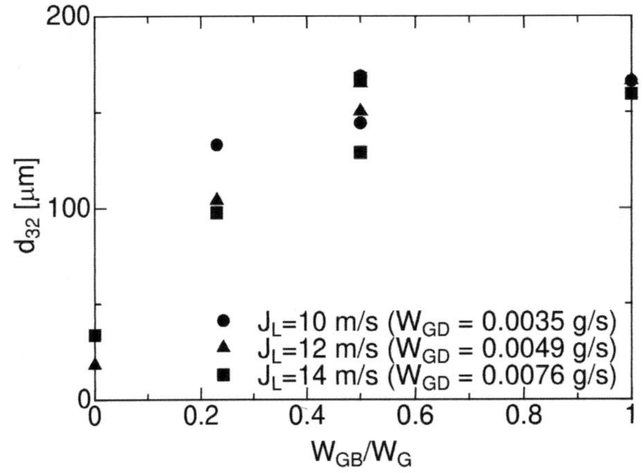

Fig. 11 Relation between d_{32} and W_{GB}/W_G

generated by using the pressurized dissolution method. The Sautr mean diameter of generated bubbles d_{32} is roughly correlated with the ratio of the mass flow rate of entrained air bubbles to the total mass flow rate, W_{GB}/W_G, for all the cases as shown in Fig. 12.

Figure 12 shows the measured bubble number density n_B. For the constant W_G cases (Fig. 12 (a)), the bubble number density decreases with increasing W_{GB}/W_G. This is due to the decrease of bubble diameter shown in Fig. 11 (a) since the total mass flow rate of gas is constant for each J_L condition. For $W_{GB}/W_G = 1$, the bubble number density increases with J_L. This is because the strong shear flow of the liquid phase in the nozzle due to high J_L enhances bubble breakup in the nozzle and increases the bubble number density. Note that the bubble diameter also decreases with increasing J_L and it is clear in the ensemble averaged bubble diameter (61 μm for J_L =10 m/s, 45 μm for J_L = 12 m/s and 32 μm for J_L = 14 m/s) whereas it is not clear in the Sauter mean diameter shown in Fig. 10(a). For the constant W_{GD} cases (Fig. 12 (b)), the bubble number density increases with W_{GB} for low J_L due to the increase of the total mass flow rate of gas W_G. To the contrary, it decreases with increasing W_{GB} for high J_L. This indicates that a part of dissolved gas desorbs from water into air bubbles and the number of micro-bubbles generated by gasification of dissolved gas decreases. The present experimental results clearly indicate that the entrainment of air bubbles in the upstream of the decompression nozzle deteriorates the performance of the micro-bubble generator. Note that the magnitude of the bubble number density is comparable to the number density of bubble nucleus used in numerical simulations of phase change (Hirt et al., 1976).

Figure 13 shows the product of the bubble number density and probability density function, which indicates the bubble number density for each bubble diameter. The number of small bubbles (d_B < 50 μm) decreases with increasing W_{GB}. On the other hand, the number of large bubbles (d_B > 100 μm) increases with W_{GB}. Figure 14 is the ratio of the total volume of each bubble class to the total bubble volume. Addition of air bubbles in the upstream of the decompression nozzle clearly decreases the volume fraction of bubbles smaller than 100 μm and increases that lager than 100 μm. These results indicate that the addition of air bubbles in the upstream of the decompression nozzle strongly affects the generation of micro-bubbles due to gasification of dissolved gas. It decreases the number of the micro-bubbles generated by the gasification due to the absorption of dissolved gas into the entrained air bubbles, whereas the number of bubbles generated by breakup of entrained air bubble increases.

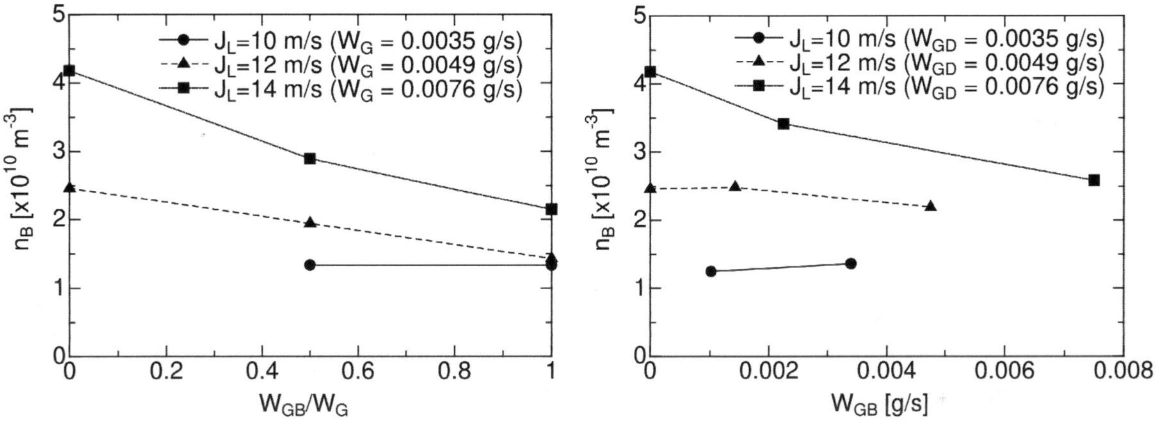

(a) Cases 1, 2 and 3 (b) Cases 1, 4 and 5

Fig. 12 Bubble number density

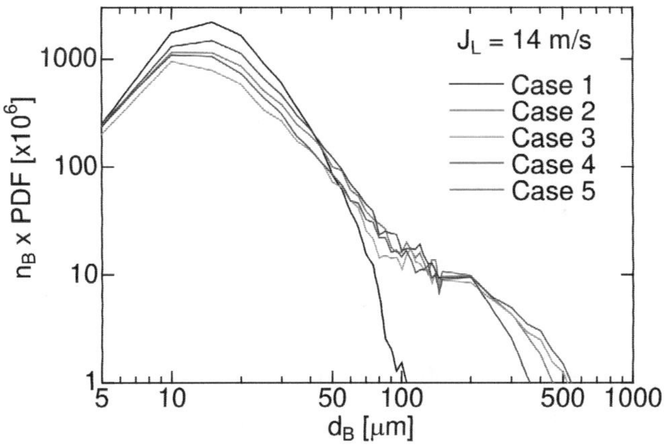

Fig. 13 Number density distribution

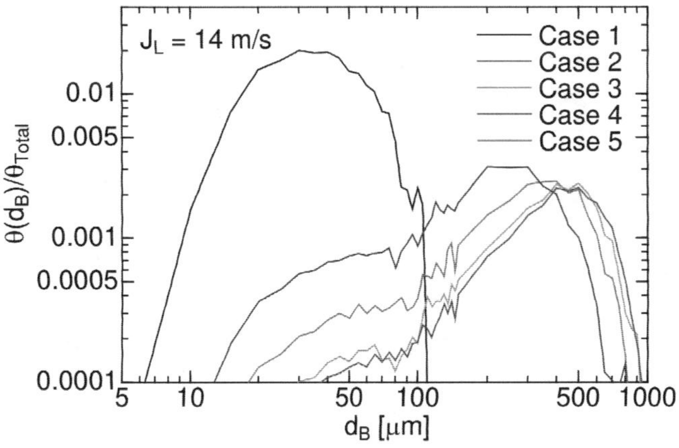

Fig. 14 Volume fraction distribution

Journal of Physics: Conference Series **147** (2009) 012016 doi:10.1088/1742-6596/147/1/012016

5. CONCLUSIONS

Diameters of bubbles generated by a pressurized dissolution method are measured by using phase Doppler anemometry (PDA) and an image processing method, which is based on the Sobel filter and Hough transform. The diameter distribution and the Sauter mean diameter of micro bubbles are evaluated based on the diameters measured by both methods. Experiments are conducted for several mass flow rates of dissolved gas and of air bubbles entrained in the upstream of the decompression nozzle to examine effects of the entrained bubbles on bubble diameter. As a result, the following conclusions are obtained:

(1) Diameter distribution of micro bubbles can be accurately measured for a wide range of diameter by using PDA and the image processing method.

(2) The mean diameter of micro-bubbles generated by gasification of dissolved gas is smaller than that generated by breakup of air bubbles entrained in the upstream of the decompression nozzle.

(3) The mean bubble diameter increases with the entrainment of air bubbles in the upstream of the decompression nozzle at a constant mass flow rate of dissolved gas.

NOMENCLATURE

H	water level in a dissolved tank	[m]
J	volumetric flux	[m/s]
W	mass flow rate	[kg/s]
W_{GDm}	the maximum mass flow rate	[kg/s]
r	bubble radius	[m]
d_B	bubble diameter	[m]
d_{32}	Sauter mean diameter of bubbles	[m]
P	probability density function	
n_B	bubble number density	[1/m^3]

Greek Letters

θ	volume of bubbles	[m^3]

Subscripts

L	liquid phase
G	gas phase
GD	dissoved gas
GB	bubble
PDA	phase Doppler anemometry
IP	image processing

REFERENCES

Fujikawa, S., Zhang, R., Hayama, S. and Peng, G. (2003), "The control of micro-air bubble generation by a rotational porous plate", International Journal of Multiphase Flow, Vol. 29, pp. 1221-1236.

Sadatomi, M., Kawahara, A., Kano, K. and Ohtomo, A. (2005), "Performance of a new micro-bubble generator with a spherical body in a flowing water tube", Experimental Thermal and Fluid Science, Vol. 29, pp. 615-623.

Dominick, J. and Durst, F. (1995), "Measurement of Bubble Size, Velocity and Concentration in Flashing Flow behid a Sudden Constriction", International Journal of Multiphase Flow, Vol. 21, No. 6, pp. 1047 - 1062.

Zhang, Z., Suzuki, K., Hosokawa, S., Tomiyama, A. (2008), "Motion of Small Bubbles near a Grid Spacer in a Two by Three Rod Bundle", Journal of Fluid Science and Technology, Vol. 3, No. 1, pp. 172-182.

Bröder, D. and Sommerfeld, M. (2007), " Planar Shadow Image Velocimetry for the Analysis of the Hydrodynamics in Bubbly Flows ", Measurement Science and Technology, Vol. 18, pp. 2513-2528.

Yu, X., Abe, K., Hirose, Y., Hazuku, T., Takamasa, T. and Oshima, M. (2006), "Measurement of Solid-Liquid Two-Phase Flow Using a New Robust Normal-Line Hough Transform", Proceedings of 5th ISMTMF, Macau, China, Vol. I, pp. 335-340.

Hirt, C.W., Romero, N.C., Torrey, M.D. and Travis, J.R. (1979), "SOLA-DF: A Solution Algorithm for Nonequilibrium Two-Phase Flow", NUREG/CR-0690 (LA-7725-MS).

Journal of Physics: Conference Series **147** (2009) 012017 doi:10.1088/1742-6596/147/1/012017

Flow and heat transfer characteristics of laminar mixed convection of water with sub-millimeter bubbles in a vertical channel

A Kitagawa, K Kimura, H Endo, Y Hagiwara

Kyoto Institute of Technology, Goshokaido-cho, Matsugasaki, Sakyo-ku, Kyoto 606-8585 JAPAN

kitagawa@kit.ac.jp

Abstract. Laminar mixed-convection heat transfer is widely seen in compact heat exchangers. Injection of sub-millimeter bubbles is considered as one of the efficient techniques for enhancing laminar mixed-convection heat transfer for liquids. However, the effects of sub-millimeter-bubble injection on the laminar mixed-convection heat transfer are poorly understood. In this study, we experimentally investigate flow and heat transfer characteristics of the laminar mixed-convection of water with sub-millimeter bubbles in a vertical channel. The thermocouples and a PTV (Particle Tracking Velocimetry) technique are used for the temperature and velocity measurements, respectively. Tap water is used for working fluid and hydrogen bubbles generated by electrolysis of water are used as the sub-millimeter bubbles. The Reynolds number of the main flow ranges from 100 to 150. Our results show that the ratio of the heat transfer coefficient with sub-millimeter-bubble injection to that without injection decreases as the Reynolds number increases. It is found from the liquid velocity measurements that this decrease is mainly due to a decrease in the "bubble advection effect".

1. INTRODUCTION

The development of compact heat exchangers has been strongly desired with miniaturization of industrial equipment. In most cases of compact heat exchangers, the Reynolds number of the main flow is relatively low, so that the flow pattern and the classification of the convective heat transfer are respectively laminar flow and mixed convection. Therefore, detailed information on heat transfer characteristics and heat transfer enhancement techniques for laminar mixed-convection flows is required to develop high-performance compact heat exchangers.

So far, upward mixed-convection flows have been studied by a number of investigators (e.g., Oosthuizen and Hart, 1973; Gryzagoridis, 1975; Abdel-Wahed et al., 1976; Kitamura and Inagaki, 1987). In particular, Kitamura and Inagaki (1987) carried out experiments of the upward mixed convection for water along a vertical flat plate using thermocouples and a hot-film anemometer, and revealed transport mechanisms of heat and momentum for the upward mixed convection. In terms of the mixed-convection heat transfer enhancement for water, Celata et al. (1999) experimentally investigated effects of air injection on the mixed convection in a vertical pipe. In their study, a-few-millimeter bubbles were injected into the water flow. They showed that the heat transfer coefficient with air injection is up to 10 times higher than that without air injection under specific conditions. They also indicated that the heat transfer coefficient with air injection is higher than that using twisted-tape inserts at Re=1000–3000. These results clearly demonstrate that air injection is a highly efficient technique for enhancing mixed-convection heat transfer. However, when air injection is applied to compact heat exchangers, it may happen that a-few-millimeter bubbles attach to the inside wall of a heated flow channel. This bubble attachment leads to a decrease in the heat transfer and an increase in the pressure drop, and consequently the performance of compact heat exchangers is reduced. One of the feasible solutions for this problem is to use bubbles whose diameters are smaller than a channel height (e.g., sub-millimeter bubbles), instead of a-few-millimeter bubbles. However, the flow and heat transfer characteristics of laminar mixed-convection flows with sub-millimeter bubbles are poorly understood.

The purpose of this study is to clarify flow and heat transfer characteristics of the laminar mixed convection of water with sub-millimeter bubbles in a vertical channel. The thermocouples and a PTV (Particle Tracking Velocimetry) technique are used for the temperature and velocity measurements, respectively.

2. EXPERIMENTAL SETUP

2.1 Apparatus

Figure 1 shows a schematic diagram of the apparatus. This apparatus consists of an upstream chamber, a vertical channel, a sub-millimeter-bubble generator, a vertical plate with uniform heat flux (vertical heated plate), a downstream chamber, a double-pipe heat exchanger, a low-temperature thermostatic bath (Asone: LTB-250) and a pump. The vertical channel made of transparent acrylic resin has dimensions of 2000 mm total length, 90 mm width and 8.6 mm height. Tap water of 20 °C is used as the working fluid. The air temperature around the apparatus is kept at 20 °C using air conditioning equipment. The *x*, *y* and *z* axes are, respectively, set along the streamwise, wall-normal and spanwise directions. Here, x=0, y=0 and z=0 are set at the starting point of the heating section, at the surface of the heated plate and at the center of the heated plate in the spanwise direction, respectively.

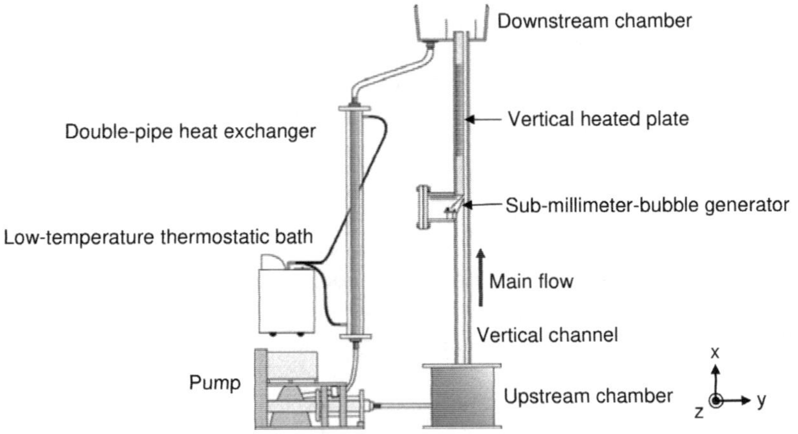

Fig. 1 Schematic diagram of the apparatus

2.2 Sub-Millimeter-Bubble Generator

In this study, we use hydrogen bubbles generated by electrolysis of water as the sub-millimeter bubbles. This bubble generation method using the electrolysis of water has two advantages. One is that the bubbles are generated in the vicinity of the vertical heated plate. The other is that the bubble volumetric flow rate is controlled simply but precisely. The bubble flow rate Q is defined as

$$Q = \frac{IR_0T_\infty}{2Fp}, \quad (1)$$

Journal of Physics: Conference Series **147** (2009) 012017 doi:10.1088/1742-6596/147/1/012017

where I is the electrical current through the electrodes, R_0 is the universal gas constant (R_0=8.31×10^9 Pa·mm^3/(K·mol)), T_∞ is the temperature of the ambient liquid (T_∞=293 K), F is the Faraday constant (F=9.65×10^4 A·s/mol) and p is the pressure (p=1.11×10^5 Pa).

A schematic diagram of the sub-millimeter-bubble generator is shown in Fig. 2. This bubble generator consists of 5-mm thick transparent acrylic resin plates, a 1-mm thick stainless-steel plate, a gold wire (0.1 mm in diameter and 55 mm long) and a carbon rod (15 mm in diameter and 60 mm long) and a charcoal absorber. The gold wire and the carbon rod are used as the cathode and the anode, respectively. The chlorine bubbles generated at the anode are completely removed using the charcoal absorber. The surface of the stainless-steel plate is sandpapered to enhance its hydrophilicity. As a result, the hydrogen bubbles generated at the cathode rise smoothly along this plate. The hydrogen bubbles are injected through a slot of this generator into the vertical channel. The slot is located at 996 mm downstream from the inlet of the vertical channel and the width of the slot is approximately 4 mm. The generation of the bubbles is performed from 390 to 610 s after the heating of the vertical plate starts.

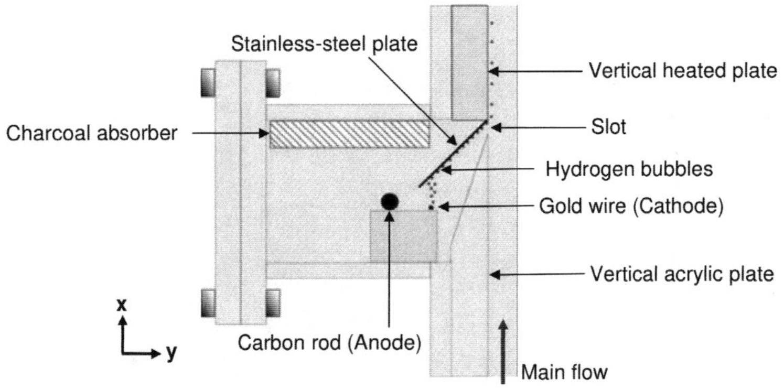

Fig. 2 Schematic diagram of the sub-millimeter-bubble generator

2.3 Vertical Heated Plate

Figure 3 shows a schematic diagram of the vertical heated plate. The stainless-steel foils (110 mm long, 25 mm wide, 0.02 mm thick) are attached to acrylic resin plate 1 (1000 mm high, 90 mm wide, 10 mm thick) at 0.5 mm intervals to electrically heat the plate under uniform heat flux conditions. The foils are connected in series with copper wires and are heated using the DC power supply. The connecting parts of the foils are on the back of the plate to minimize the disturbance to liquid flow. Acrylic resin plate 1 is entirely covered with 25-μm thick Kapton tape for electrical insulation. The air between acrylic resin plate 1 and 8-mm thick acrylic resin plate 3 reduces heat leakage from behind the heated plate. The heat loss through the back of the heated plate is estimated to be less than 2.5 % of the wall heat flux. To measure the surface temperature on the heated plate T_W, thermocouples are

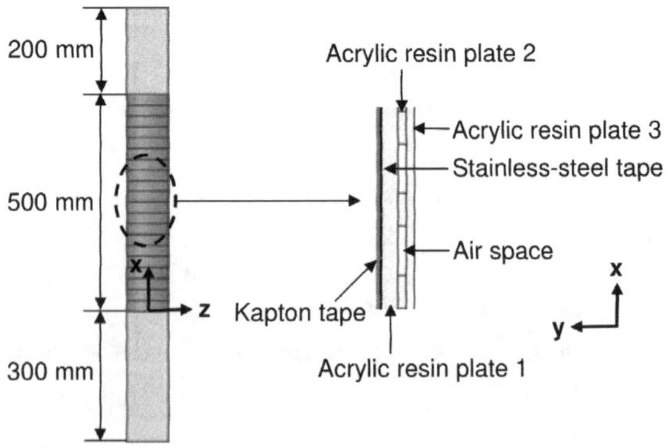

Fig. 3 Schematic diagram of the vertical heated plate

inserted into 3.0 mm diameter holes on the back of the heated plate and the tips, which are covered with Kapton tape, are contact with the back surface of the foils. All the holes are completely filled with silicone sealant to prevent water from flowing into them. The leading edge of the heated plate is located 1000 mm downstream from the inlet of the vertical channel. The starting point of the heating section, namely, x=0 is located 50 mm downstream from the leading edge of the plate. The heating of the plate is performed in the region of $0<x<500$ mm. It is noted that the heated plate is flush with the vertical acrylic plate which is set upstream of the heated plate (see Fig. 2).

2.4 Double-Pipe Heat Exchanger

In our experiments, the liquid temperature in the upstream chamber is kept constant (20 °C) using the double-pipe heat exchanger. A schematic diagram of the double-pipe heat exchanger is shown in Fig. 4. This heat exchanger consists of two pipes. The inner pipe made of aluminum is 40 mm i.d., 1500 mm total length and 1 mm thick, while the outer pipe made of transparent acrylic resin is 70 mm i.d., 1500 mm total length and 3 mm thick. Cool water, whose temperature is controlled by a sensor, is introduced into the outer pipe from the low-temperature thermostatic bath. The heat exchange is made between the warm water flowing in the inner pipe and the cool water flowing in the annulus.

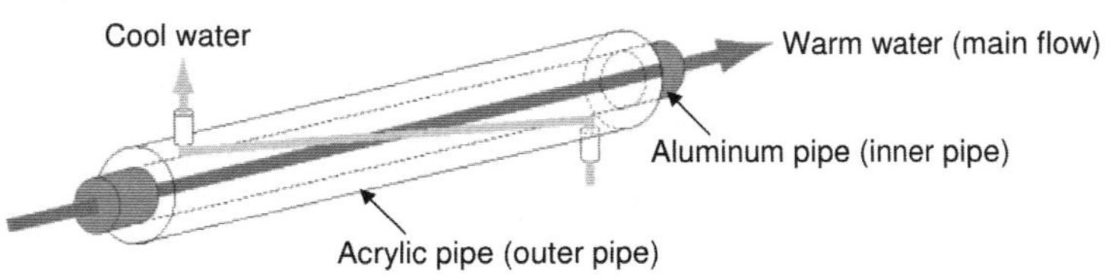

Fig. 4 Schematic diagram of the double-pipe heat exchanger

2.5 Experimental Conditions

Table 1 shows the experimental conditions for the temperature and velocity measurements. In this table, Re is the Reynolds number of the working fluid and is given by

$$Re = \frac{UH}{\nu}, \quad (2)$$

where U is the streamwise mean velocity of the liquid in the cross-section at z=0 mm, H is the height of the vertical channel and ν is the kinematic viscosity of the liquid. In this table, q_w and Q are the wall heat flux and the bubble flow rate, respectively.

Table 1 Experimental conditions for the temperature and velocity measurements

Temperature measurement		
Reynolds number	Re	= 100, 125, 150
Bubble flow rate	Q	= 40 mm^3/s
Wall heat flux	q_w	= 1480 W/m^2
Measurement position	x	= 70, 170, 270 mm, z=0 mm
Velocity measurement		
Reynolds number	Re	= 100, 125, 150
Bubble flow rate	Q	= 40 mm^3/s
Wall heat flux	q_w	= 0, 1480 W/m^2
Measurement position	x	= 270 mm, z =0 mm

3. TEMPERATURE MEASUREMENT TECHNIQUE

3.1 Temperature Measurement System

The temperature measurement system consists of K-type 100 μm thermocouples, a reference junction and a high-speed data acquisition unit (Yokogawa, MX100). The thermocouples, which are accurate within ±0.12 °C, are used to simultaneously measure the surface temperature of the heated plate T_w and the liquid temperature in the upstream chamber T_∞. The thermocouples for measuring T_w are set at x=70, 170, 270 mm and z=0 mm. The high-speed data acquisition unit is directly connected to a personal computer in order to record the thermoelectric force of the thermocouples in real time. A thermostatic bath, which is accurate within ±0.10 °C, is used for calibrating the thermocouple temperature readings. The sampling frequency for the temperature measurement is 5 Hz and the measurement period is 610 s.

3.2 Estimation of Local Heat Transfer Coefficient

The local heat transfer coefficient h_x at each of the measurement positions is obtained from Eq. (3).

$$h_x = \frac{q_w}{T_w - T_\infty}, \quad (3)$$

where q_w is estimated from Eq. (4).

$$q_w = \frac{\rho' l I^2}{S A_{st}}. \quad (4)$$

In Eq. (4), S, l, ρ' and A_{st} are the cross-section area, total length, specific electrical resistance and overall surface area of the foil, respectively. The specific electrical resistance of each foil is 7.4×10^{-7} $\Omega\cdot$m. The total uncertainly in the heat transfer coefficient with respect to both the thermocouple accuracy and the heat loss through the back of the heated plate is approximately 6 %.

4. VELOCITY MEASUREMENT TECHNIQUE

4.1 Velocity Measurement System

In this study, we use the PTV technique to conduct the liquid velocity measurements (e.g., Kitagawa et al., 2005; Murai et al., 2006). We also use a shadow image technique to obtain the bubble interface clearly (e.g., Tokuhiro et al., 1998).

Figure 5 shows a schematic diagram of the velocity measurement system. A digital color CCD camera (Imperx, VGA210-LC), which is set in front of the vertical channel, is used to capture the tracer particle and bubble images. Particles of 60 μm in mean diameter and 1.02 in specific gravity are used as the tracer particles and the particle concentration is approximately 100 ppm. A halogen-light source (Moritex, MHAB-150W) and blue LED arrays (Imac: λ=470 nm) are used for illuminating the tracer particle and bubble, respectively. The former is set at the side of the vertical channel, while the latter is set behind the vertical channel. A red light sheet with a thickness of 3 mm is produced with a line light guide, a colored glass filter, a red-pass filter and a slit. The measurement area is $264<x<276$ mm, $0<y<16$ mm and $-1.5<z<1.5$ mm. Images of 600×480 pixels, which are taken with the color CCD camera, are directly recorded by a personal computer. The spatial resolution is 0.027 mm/pixel. The

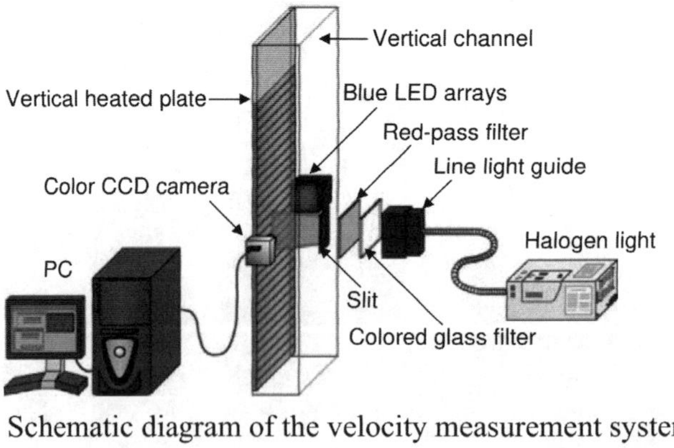

Fig. 5 Schematic diagram of the velocity measurement system

exposure time of the color CCD camera is 0.0005 s and the frame rate is 65 to 110 fps depending on the different experimental conditions.

4.2 Estimation of Bubble Diameter

The diameter of bubbles is calculated from the equivalent area of each bubble image taken by the CCD camera. Each bubble image is binarized by a binary labeling method (e.g., Yamamoto et al., 1996). When the bubble image overlaps with other bubble images, its image is eliminated considering the aspect ratio of the bubble. The uncertainty in the bubble equivalent diameter associated with the binary labeling method is estimated to be approximately 0.4 % of the mean bubble diameter.

4.3 Estimation of Liquid Velocity Vector

The estimation procedure for the liquid velocity vector is described below.

(1) The location of the tracer particle centroid is estimated from the red image using a particle mask correlation method, which was proposed by Takehara and Etoh (1999).

(2) Some of the particle locations obtained in step (1) are detected from light scattered at the bubble interface and the surfaces of the particles that overlap with bubble images. The former depends strongly on the bubble interface motion while the latter is affected by reflection or refraction at the bubble interface. Therefore, both cause errors in the liquid velocity estimation. These particle locations are completely eliminated using binarized bubble images.

(3) The liquid velocity vector is estimated using the velocity gradient tensor method (Ishikawa et al., 2000) in which the velocity gradient tensor is considered in order to obtain the velocity vectors with high accuracy.

When estimating the liquid velocity without bubbles (i.e., single-phase flow), steps (2) is omitted. The uncertainty in the liquid velocity associated with the particle centroid detection is estimated to be 0.83 mm/s. Hence, the error in this value relative to the maximum liquid rise velocity (25.8 mm/s) is 3.2 %.

5. RESULTS AND DISCUSSION

5.1 Flow and Heat Transfer Characteristics without Sub-Millimeter-Bubble Injection

5.1.1 Liquid Mean Velocity

Figure 6 (a)–(c) shows the profiles of the liquid mean velocity at different Reynolds numbers. In this figure, u is the liquid mean velocity in the streamwise direction. The vertical axis is normalized by $u_{0,max}$ which is the maximum value of the streamwise liquid mean velocity without bubble injection and heating. For comparison, the profile of the Poiseuille flow is added to the figure. At each Reynolds number, the experimental result without heating is in agreement with the profile of the Poiseuille flow. The maximum difference between the two is approximately 4 %. Moreover, at each Reynolds number, the liquid mean velocity in the vicinity of the heated wall is increased by heating. It is clear from this

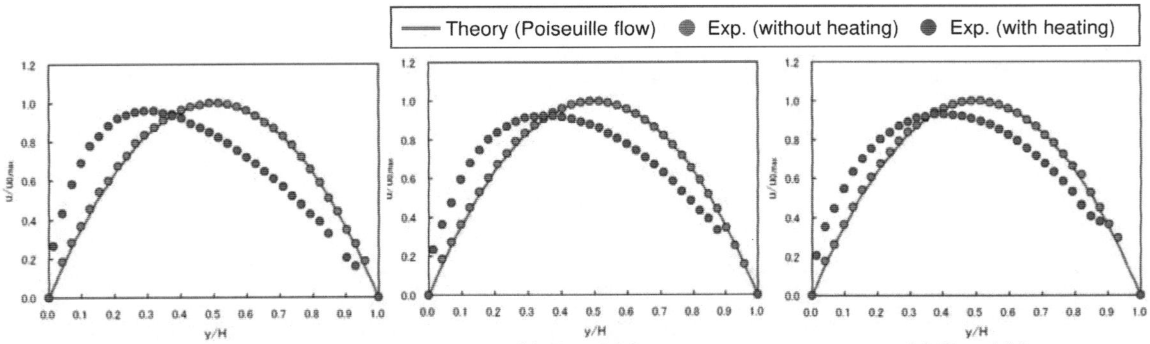

(a) Re=100 (b) Re=125 (c) Re=150

Fig. 6 Profiles of liquid mean velocity at different Reynolds numbers result that the flow with heating is the mixed-convection flow.

5.1.2 Local Heat Transfer Coefficient

Figure 7 shows the relationship between the Reynolds number and the local heat transfer coefficient at different measurement positions. At each measurement position, the local heat transfer coefficient increases with the Reynolds number. This originates from an increase in the streamwise liquid mean velocity near the heated wall, which is shown in Fig. 6. Moreover, at each Reynolds number, the local heat transfer coefficient decreases with an increase in the x. This is because the liquid flowing downstream is gradually heated by the heated plate. In Section 5.2.1, the local heat transfer coefficient obtained here is used as h_{x0}.

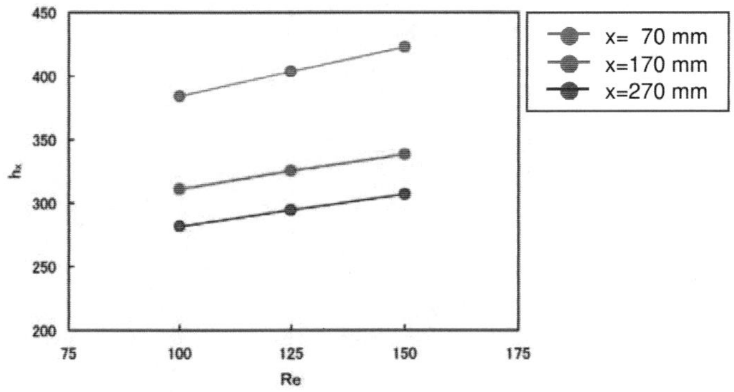

Fig. 7 Relationship between Reynolds number and local heat transfer coefficient at different measurement positions

5.2 Flow and Heat Transfer Characteristics with Sub-Millimeter-Bubble Injection

5.2.1 Heat Transfer Coefficient Ratio

Figure 8 shows the relationship between the Reynolds number and the heat transfer coefficient ratio h_x/h_{x0} at different measurement positions. The heat transfer coefficient ratio ranges from 1.23 to 1.35. This clearly demonstrates that sub-millimeter-bubble injection enhances laminar mixed-convection heat transfer for water. In addition, the heat transfer coefficient ratio decreases as the Reynolds number increases. This means that sub-millimeter-bubble injection for the laminar mixed convection is effective under relatively lower Reynolds number conditions. Furthermore, the heat transfer coefficient ratio decreases with an increase in the x. This tendency was also seen in our previous paper regarding laminar natural convection flow with sub-millimeter bubbles (Kitagawa et al, 2008).

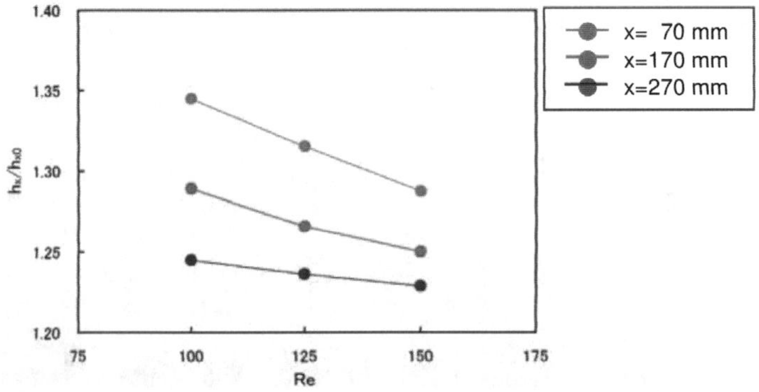

Fig. 8 Relationship between Reynolds number and h_x/h_{x0} at different measurement positions.

5.2.2 Heat Transfer Gain

Figure 9 shows the relationship between the Reynolds number and the heat transfer gain η. The heat transfer gain is the ratio of the heat transfer rate obtained by the sub-millimeter-bubble injection to the power consumption of the sub-millimeter-bubble generation P and is defined as

$$\eta = \frac{q_w A_H \left(\overline{h_x / h_{x0}} - 1\right)}{P}, \quad (5)$$

where A_H is the surface area of the heated wall and is set to 500 mm×90 mm for convenience. $\overline{h_x / h_{x0}}$ is obtained by averaging h_x/h_{x0} at three measurement positions. At each Reynolds number, the heat transfer gain is much higher than 1, and the maximum value of the heat transfer gain is approximately 2.65. It is therefore considered that injection of sub-millimeter bubbles is a promising technique for enhancing laminar mixed-convection heat transfer for water.

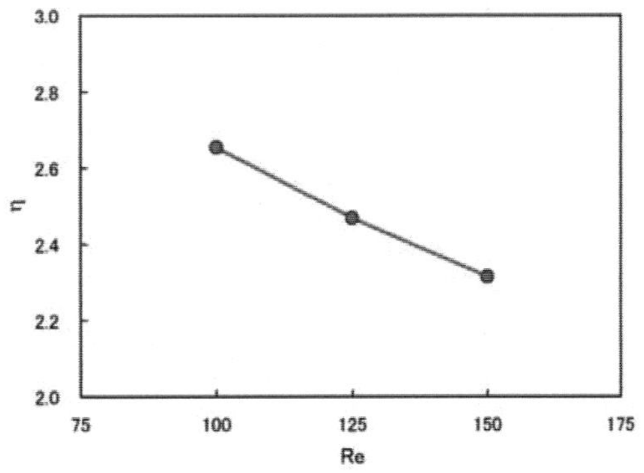

Fig. 9 Relationship between Reynolds number and heat transfer gain

5.2.3 Bubble Diameter

Figure 10 shows the probability distribution of the bubble equivalent diameter d at different Reynolds numbers. In this figure, n is the number of bubbles for each equivalent diameter, and N is the total of n. At each Reynolds number, the bubble diameter ranges from 0.3 to 0.8 mm. Additionally, at each Reynolds number, the probability distribution has a peak at d=0.52 mm. Therefore, the bubble diameter is independent of the Reynolds number.

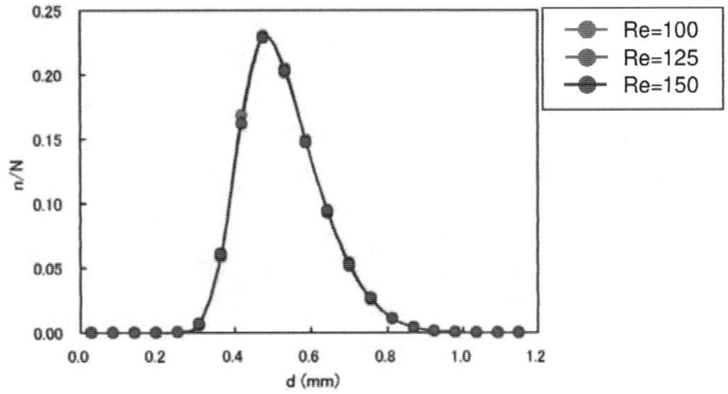

Fig. 10 Probability distribution of bubble equivalent diameter at different Reynolds numbers

5.2.4 Bubble Projection Area Ratio

When many bubbles concentrate near the wall, overlaps of bubble images taken by a camera occur significantly. In the image processing, this overlap leads to a decrease in the number of bubbles detected near the wall. Therefore, in this study, the bubble projection area ratio is used to discuss the probability distribution of the bubble location and is defined as

$$A_P = \frac{A_G}{A_L + A_G}, \quad (6)$$

where A is the projection area and the subscription L and G are, respectively, the liquid and the bubble. Figure 11 shows the profiles of the bubble projection area ratio at different Reynolds numbers. The vertical axis is normalized by $A_{p,sum}$ (sum of all A_p). Each datum is averaged over a 0.2 mm interval. At each Reynolds number, bubbles are located in the region of y/H=0–0.2. That is, the dependency of the bubble location on the Reynolds number is weak.

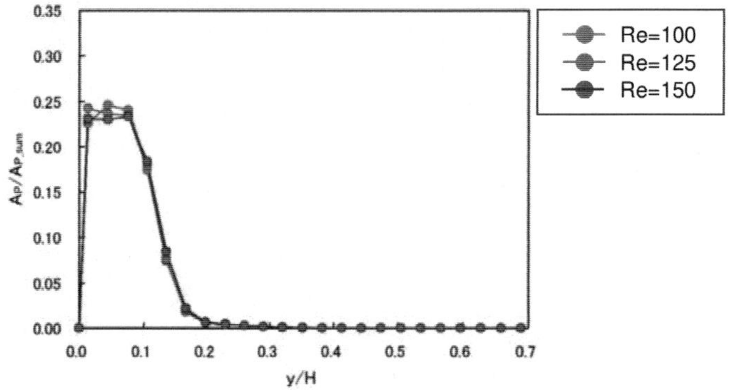

Fig. 11 Profiles of bubble projection area ratio at different Reynolds numbers

5.2.5 Liquid Mean Velocity

Figure 12 shows the profiles of the liquid mean velocity in the streamwise direction at different Reynolds numbers. In this figure, the results for more than 1000 samples are presented. Here, a major reason for the lack of data in the region of y/H=0–0.08 is that in the liquid velocity measurements, the intensity of the red light sheet decreases remarkably because of bubble concentration. At all the Reynolds numbers, the liquid mean velocities with sub-millimeter-bubble injection are higher than those without injection. In particular, the increase in the liquid mean velocity is significant near the heated wall. This is deeply related to the high bubble concentration near the heated wall. The increase in the liquid mean velocity leads to advection in the warm liquid near the heated wall and consequently contributes heat transfer enhancement. In the following, this is called the "bubble advection effect".

Figure 13 shows the relationship between the Reynolds number and the rate of increase in the maximum liquid velocity. In this figure, $u_{max,T}$ and $u_{max,S}$ are, respectively, the maximum liquid mean velocity with and without bubble injection, which are obtained from Fig. 12. The rate of increase in the maximum liquid velocity becomes lower as the Reynolds number increases. This result means that

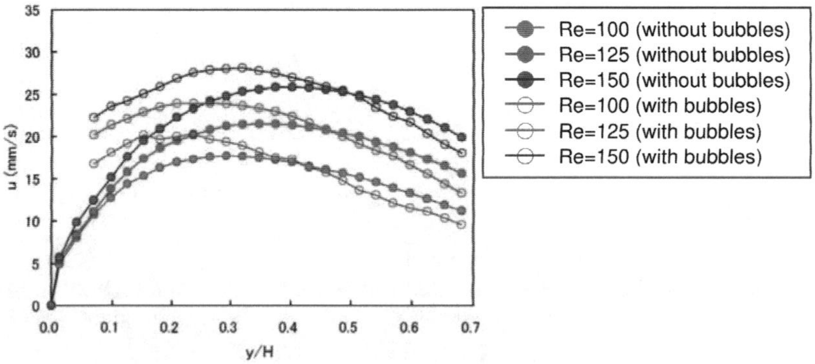

Fig. 12 Profiles of streamwise liquid mean velocity at different Reynolds numbers

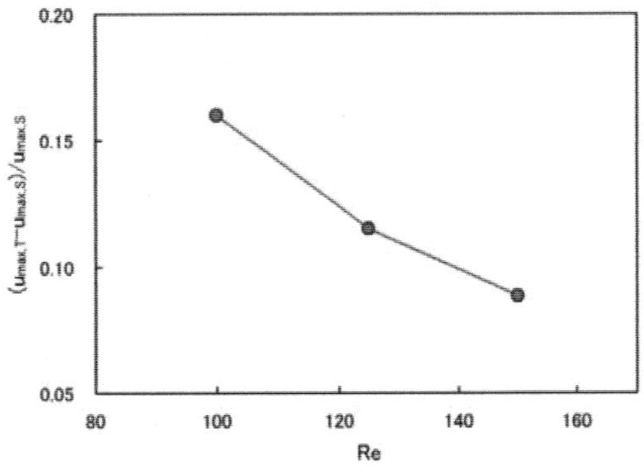

Fig. 13 Relationship between Reynolds number and rate of increase in the maximum liquid velocity

the "bubble advection effect" decreases with an increase in the Reynolds number.

5.2.6 RMS Values of Liquid Fluctuation Velocities

Figure 14 shows the profiles of the RMS values of the liquid fluctuation velocities at different Reynolds numbers. In this figure, u_{rms} and v_{rms} are the RMS values of the liquid fluctuation velocities in the streamwise and wall-normal directions, respectively. In this figure, the results for more than 1000 samples are presented, similar to Fig. 12. At all Reynolds numbers, the RMS values of the liquid fluctuation velocities in both directions are increased by the bubble injection. Particularly, the increase in the RMS values near the heated wall is significant. This is the same tendency as the profile of the liquid mean velocity. This increase near the heated wall activates the mixing of warm and cool liquids and as a result, contributes heat transfer enhancement directly. Hereafter, this is called the "bubble mixing effect".

Figure 15 shows the relationship between the Reynolds number and the RMS values of the liquid fluctuation velocities near the heated wall (which are obtained from Fig. 14). In this figure, $u_{rms,av}$ and $v_{rms,av}$ are obtained by averaging u_{rms} and v_{rms} over the range of y/H=0.08−0.2, respectively. $u_{rms,av}$ and $v_{rms,av}$ tend to increase as the Reynolds number increases. The tendency of increase in the $v_{rms,av}$ with increasing Reynolds number is clear compared with that in the $u_{rms,av}$. These results mean that the "bubble mixing effect" increases with the Reynolds number.

From Figs. 13 and 15, it was clear that the "bubble mixing effect" increases but the "bubble advection effect" decreases as the Reynolds number increases. Therefore, we conclude that the decrease in the heat transfer coefficient ratio with increasing Reynolds number, shown in Fig. 8, is mainly due to a decrease in the "bubble advection effect".

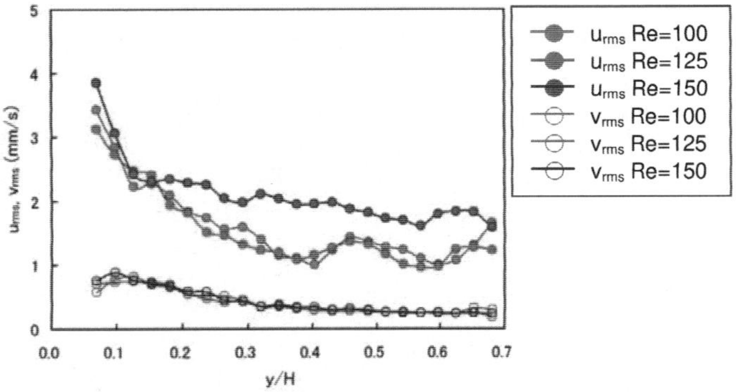

Fig. 14 Profiles of RMS values of liquid fluctuation velocities at different Reynolds numbers

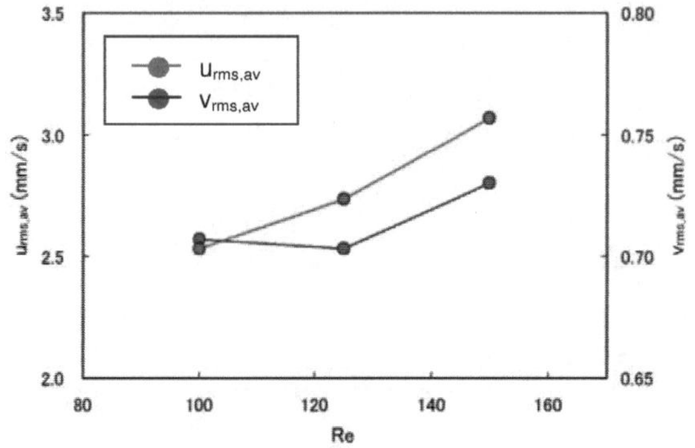

Fig. 15 Relationship between Reynolds number and RMS values of liquid fluctuation velocities near the heated wall

6. CONCLUSIONS

We experimentally investigated flow and heat transfer characteristics of the laminar mixed convection of water with sub-millimeter bubbles in a vertical channel. The Reynolds number of the main flow ranges from 100 to 150. The conclusions are as follows.

(1) The ratio of the heat transfer coefficient with sub-millimeter-bubble injection to that without injection (i.e., the heat transfer coefficient ratio) ranges from 1.23 to 1.35. Moreover, at each Reynolds number, the heat transfer gain is much higher than 1, and the maximum value of the heat transfer gain is approximately 2.65. It is therefore considered that injection of sub-millimeter bubbles is a promising technique for enhancing laminar mixed-convection heat transfer for water.

(2) The heat transfer coefficient ratio decreases with as the Reynolds number increases. This means that sub-millimeter-bubble injection for the laminar mixed convection is effective under relatively lower Reynolds number conditions. According to the liquid velocity measurements, the "bubble mixing effect" increases but the "bubble advection effect" decreases as the Reynolds number increases. Therefore, we conclude that the decrease in the heat transfer coefficient ratio with increasing Reynolds number is mainly due to a decrease in the "bubble advection effect".

NOMENCLATURE

A_H	surface area of heated plate
A_{st}	overall surface area of stainless-steel foil
d	bubble equivalent diameter
F	Faraday constant
h_x	heat transfer coefficient with bubble injection
h_{x0}	heat transfer coefficient without bubble injection
H	channel height
I	electrical current
l	total length of stainless-steel foil
n	number of bubbles
N	sum of all n
p	pressure
P	power consumption of bubble generation
Q	bubble flow rate
q_w	wall heat flux
R_0	universal gas constant
Re	Reynolds number
S	cross-section area of stainless-steel foil

T	temperature
u	liquid mean velocity in the streamwise direction
$u_{0,max}$	maximum value of streamwise liquid mean velocity without bubble injection and heating
U	streamwise mean velocity of liquid in the cross-section at z=0 mm
v	liquid mean velocity in the wall-normal direction
x, y, z	streamwise, wall-normal and spanwise directions

Greek Letters

η	heat transfer gain
ν	kinematic viscosity
ρ'	specific electrical resistance of stainless-steel foil

Subscripts

G	bubble
L	liquid
w	wall condition
∞	ambient condition

REFERENCES

Abdel-Wahed, R.M. et al. (1976). "Mixed convection on a vertical plate with an unheated starting length", Trans. ASME, J. Heat Transfer, 98, pp. 576-580

Celata, G.P. et al. (1999). "Heat transfer enhancement by air injection in upward heated mixed-convection fow of water", Int. J. Multiphase Flow, 25, pp. 1033-1052.

Gryzagoridis, J. (1975). "Combined free and forced convection from an isothermal vertical plate", Int. J. Heat Mass Transfer, 18, pp. 911-916.

Ishikawa, M. et al. (2000), "A novel algorithm for particle tracking velocimetry using the velocity gradient tensor", Exp. Fluids, 29, pp. 519-531.

Kitagawa, A. et al. (2005), "Flow structure of microbubble-laden turbulent channel flow measured by PIV combined with the shadow image technique", Exp. Fluids, 38, pp. 466-475.

Kitagawa, A. et al. (2008), "Heat transfer enhancement for laminar natural convection along a vertical plate due to sub-millimeter-bubble injection", Exp. Fluids, 45, pp. 473-484.

Kitamura, K. et al. (1987). "Turbulent heat and momentum transfer of combined forced and natural convection along a vertical flat plate (aiding flow)", Int. J. Heat and Mass Transfer, 30, pp. 23-41.

Murai, Y. et al. (2006). "Turbulent shear stress profiles in a bubbly channel flow assessed by particle tracking velocimetry", Exp. Fluids, 41, pp. 343-352.

Oosthuizen, P.H. et al. (1973). "A numerical study of laminar combined convective flow over flat plates", J. Heat Transfer, 95, pp. 60-63.

Takehara, K. et al. (1999), "A study on particle identification in PTV", J. Visualization, 1, pp. 313-323.

Tokuhiro, A. et al. (1998). "Turbulent flow past a bubble and an ellipsoid using shadow-image and PIV techniques", Int. J. Multiphase Flow, 24, pp. 1383-1406.

Yamamoto, F. et al. (1996), "Discussion of the cross-correlation methods for PIV", J. Flow Visual Image Processing, 3, pp. 65-78.

Journal of Physics: Conference Series **147** (2009) 012018 doi:10.1088/1742-6596/147/1/012018

A method for bubble volume calculating in vertical two-phase flow

H Y Wang, F Dong

Tianjin Key Laboratory of Process Measurement and Control, School of Electrical Engineering & Automation, Tianjin University, Tianjin, 300072, CHINA

fdong@tju.edu.cn

Abstract. The movement of bubble is a basic subject in gas-liquid two-phase flow research. A method for calculating bubble volume which is one of the most important characters in bubble motion research was proposed. A suit of visualized experimental device was designed and set up. Single bubble rising in stagnant liquid in a rectangular tank was studied using the high-speed video system. Bubbles generated by four orifice with different diameter (1mm, 2mm, 3mm, 4mm) were recorded respectively. Sequences of recorded high-speed images were processed by digital image processing method, such as image noise remove, binary image transform, bubble filling, and so on. then, Several parameters could be obtained from the processed image. Bubble area, equivalent diameter, bubble velocity, bubble acceleration are all indispensable in bubble volume calculating. In order to get the force balance equation, forces that work on bubble along vertical direction, including drag force, virtual mass force, buoyancy, gravity and liquid thrust, were analyzed. Finally, the bubble volume formula could be derived from the force balance equation and bubble parameters. Examples were given to shown the computing process and results. Comparison of the bubble volume calculated by geomettic method and the present method have shown the superiority of the proposed method in this paper.

1. INTRODUCTION

In recent years, gas-liquid two-phase flow has been widely used in biological fermentation, polymer polymerization, industrial wastewater treatment, environmental protection and so on. The movement of bubble is a basic subject in two-phase flow research as bubble in water always plays an important role in solving problems of experiments and projects. The state and motion of bubble are closed with the operating conditions, the nature of liquid and the form of ventilation. If bubble state could be got by visualized method, that will inspirit the development of chemical industry and its relevant filed.

High-speed visual is a non contact measurement which dose not interfere with the movement of bubble in flow. Many researchers have done to learning the characters of bubble using visualized method. Maurus et al.(2002) used digital imaging and analysing techniques tostudy the subcooled flow boiling and got the bubble characters and local void fraction. Rodríguez-Rodríguez et al.(2003) applied the particle tracking velocimetry approach of high-speed video images to a novel tracking and bubble break-up detection. Zaruba and Krepper et al.(2005) have used high-speed video observation to make study on bubble motion in a rectangular column and an individual bubble in the gas swarm could be tracked.

In addition, theoretic analysis of the force work on bubble has been studied in recent years. Marco et al.(2003) compared the theories of force balance along the vertical with the measured bubble rising velocities of gas bubble in a still liquid to explain the phenomena of sensible steady oscillation of velocity. A dynamic model of bubble based on pressure balance, force balance and mass transfer considerations has been developed to investigate the dynamics and mass transfer of a single bubble of carbon dioxide rising in a strong alkaline solution hydroxide by Madhavi et al.(2006). The total drag coefficient of single bubble rising freely with high viscosity was calculated from the accelerating motion to the steady motion with the added mass (virtual mass) and history force included by Zhang et al.(2008).

Bubble volume is important in studying bubble motion for it is closed related to void fraction. However, volume of bubble is difficult to get directly, especially for distorted bubble. A method for bubble volume calculation is proposed in this article. First of all, a suit of visualized device is designed and set up to make experiments. Signal bubble rising in stagnant water in a rectangular tank. Secondly, continues image of bubble rising in water was recorded by high-speed video camera. Thirdly, these images need to be processed by digital image processing algorithm. Several useful parameters could be acquired from the processed images. Such as, bubble area, equivalent diameter, bubble velocity and acceleration, and so on. Finally, analysis the force that bubble endured along vertical direction. The calculation formulas of every force have been summarized for the volume calculation. The computing formula of bubble volume was derived from force balance equation and bubble parameters. Examples have shown the analytic steps and calculated results run though the paper to verify the feasibility of this method for bubble volume calculation.

2. EXPERIMENTAL METHOD

High-speed video measurement is one of the important measuring technique of the movement of bubble in air-water two-phase flow. It's very important to get sharp, undistorted, continuous images to observe the flow character of two-phase flow. Therefore, the property of gas and liquid, the refractive index of fluid and channel and the gas speed must be considered during the system design. The experimental facility contains two parts: two-phase flow filed simulating and high-speed image recording.

2.1 Two-phase Flow Simulating

A schematic view of the visualized experimental test facility is present in Fig. 1. The vertical rectangular tank that made of Plexiglas has a cross-section of 20cm×20cm and is 125cm in height. As the tank is rectangular and the refractive index of Plexiglas is matched with that of liquid.

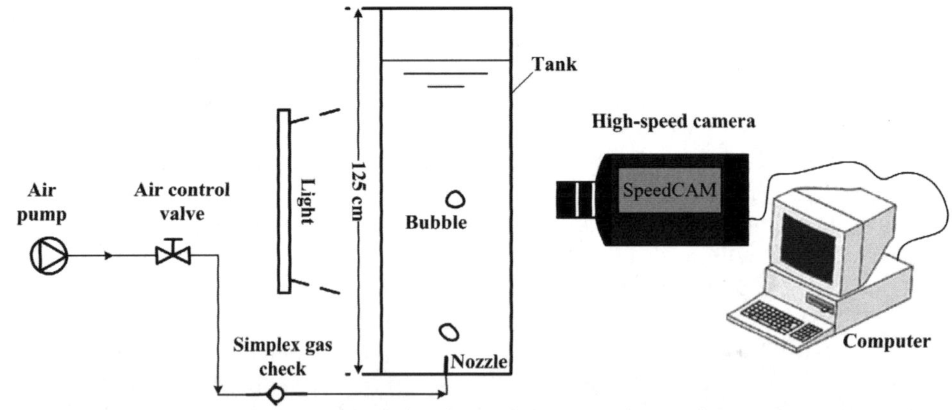

Fig. 1 Schematic view of the experimental setup

There are 4 orifices with different diameter (1mm, 2mm, 3mm, and 4mm) at the bottom centre of the channel. They could generate bubbles in liquid. Under each orifice, there is a simplex gas check to ensure that the liquid can not leak from the orifice. An air control valve which connected with the simplex gas check is the master switch. It controls the rising rate, frequency, size and stability of bubble by regulating the flow of air. A air pump connected to the air control value supplies air to the tank. At the bottom of the whole device, a square corrosion resistant plate connected with four column stainless steel tube was used to support the tank to stabilize the two-phase flow system.

2.2 High-speed Image Recording

A high-speed video camera consists of an advanced CMOS sensor was used to record images of the rising bubble. Its frame frequency is up to 32000fps and its highest resolution is 1280×1024. However, higher spatial resolutions result in longer recording times and fewer video frames per unit time. A compromise between recording speed and image resolution has to be found. Experimentally, 640×480 special resolution and 500fps frequency were enough to meet the requirement in the experimental process and was adopted in all experiments records inthis study. Auxiliary backlighting with 5400K color temperature was also used to outline the bubbles. As gas, water and tank are all transparent, back light was adopted in the experiment. The whole experiment platform was shown in fig. 2.

Fig. 2 High-speed image recording system

2.3 Experimental Steps

First, connect the camera to computer and turn on the camera. Second, adjust focus of high-speed camera untill the image is clear. Third, turn on the light and pump, turn the air control value to make bubbles rising from the orifice. Fourth, record continuous images of bubble rising. Each orifice should be measured independently. It's also necessary to change bubble rising rate and move the camera and the light to record bubbles rising images at different working condition. The following section select the single bubble images of the part from 14cm to 25.6cm from the tank bottom as the research object.

3. IMAGE PROCESSING AND PRAMETER CALCULATING

3.1 Image Processing

In order to abstract the character information from high-speed images the original images should be processed by digital image processing. First of all, transform RGB image into gray image. As digital images have a variety of noises, wavelet filter was adopted to emphasize some features or remove noises (Lv et al. 2007). Second, a brightness threshold t is applied to transform gray image into binary image as follows:

if $k(i,j) < t$, then $k(i,j) = 0$

if $k(i,j) \geq t$, then $k(i,j) = 255$

For the reverberation of bubble, the center of bubble has the same gray value as the background in binary image. It's necessary to fill the center of bubble to calculate the area or other prameters of bubble in two-phase flow. Infected algorithm which used to fill bubble of binary image is adopted in this paper. First, set a original infective point which is labeled value a. Second, scan the image matix begain at the original infective point. If the point in binary image is effected, then the points in it 8-neighborhood which is 0 will be labeled to number p. The formulation of infected algorithm is:

$$\begin{cases} h(i,j) = p, & [\bigcup_{m=i-1}^{i+1} \bigcup_{n=j-1}^{j+1} h(m.n) \cap \overline{k(i,j)}] = 1 \\ h(i,j) = 0, & else \end{cases} \quad (1)$$

Finaly, creat a zero matrix as large as the processed binary image matrix. Migrate all the infected points into the zero matrix to get the filled image.

Fig.3 has shown four typical images cut from a sequence of image when the diameter of the orifice is 4mm. and the processed results of them were also shown out.

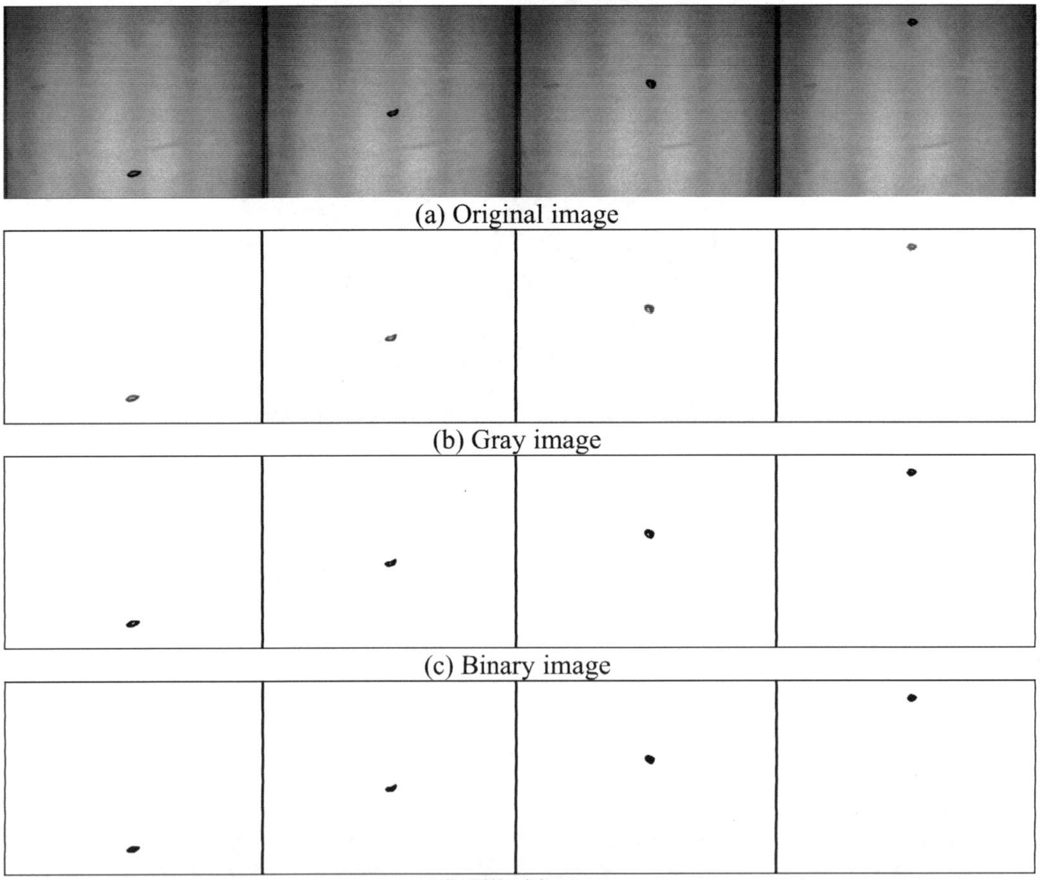

(a) Original image

(b) Gray image

(c) Binary image

(d) Filled image

Fig. 3 Example of high-speed image and the processing results

3.2 Parameters Calculating

Parameters that abstracted from processed image play a very important role in bubble rising study. Area, equivalent diameter, rising velocity and acceleration of bubble are all necessary in bubble volume calculation.

(1) Area of Bubble

Area of bubble is a useful parameter in obtaining void fraction. For simply, the area value of the bubble is defined as the total pixel number of a consecutive area in the bubble filled image. It is calculated according to:

$$A_i \cong \sum_j n_j \quad (2)$$

(2) Equivalent Diameter

In two phase flow, the diameter of bubble or particle is a important parameter. As bubble in two phase flow is not normally spherical, the diameter of sphere that has the same volume with bubble is defined as equivalent diameter of the bubble (Darton, 1977). The measurement of high-speed image is based on 2-D image, so equivalent diameter is the diameter of circle which has the same area with the bubble shadow. The formula of the equivalent diameter is:

$$d = \sqrt{\frac{4A_i}{\pi}} \quad (3)$$

(3) The Center of the Bubble

Under the press of water, bubble is distorting during the rising movement. So the position of bubble is difficult to be found. However, get the position of bubble in image is a pre-requisited work to do in analysing the movement of bubble (Zhang et al. 2006). Gray-scale centroid methold is applied to calculate the coordinate of bubble center based on gray image. The computing formulation is shown as follows:

$$x_c = \sum_{i,j\in\Omega} \frac{k(i,j)j}{M}, \qquad y_c = \sum_{i,j\in\Omega} \frac{k(i,j)i}{M} \quad (4)$$

where $M = \sum_{i,j\in\Omega} k(i,j)$ is the grand total of gray value of image. It can be seen that weighted average of the gray value in the image is the center of bubble.

(4)Bubble Velocity and acceleration

Bubble velocity is a key parameter in two-phase flow measurment, which infect other parameters directly. The velocity was calculated by measuring the distance of a bubble in two successive images for the time interval Δt between two relative image is known. Compute out the coordinate of bubble center (x_c, y_c) in every image. Get the displacement of bubble by subtracting bubble center of two sucessive images. Suppose the bubble centers are $(x_{c1}, y_{c1})$and(x_{c2}, y_{c2}). Bubble velocity v_x, v_y is a calculated by the formula asfollows:

$$v_x = (x_{c2} - x_{c1})/\Delta t, \quad v_y = (y_{c2} - y_{c1})/\Delta t \quad (5)$$

As the frequency of high-speed is very high, the displacement of bubble and time interval between two images are so small that the velocity can be regarded as instantaneous speed. And then, acceleration a of bubble can be computed out by defination as follows:

Journal of Physics: Conference Series **147** (2009) 012018 doi:10.1088/1742-6596/147/1/012018

$$a_x = (v_{x2} - v_{x1}) / \Delta t, \quad a_y = (v_{y2} - v_{y1}) / \Delta t \quad (6)$$

a_x and a_y is the acceleration of horizontal direction and vertical dircetion respectively. a_y is a usefull parameter in bubble volume calculating.

Table 1 gives the calculted results of four kinds parameter of which bubbles generate from orifices with different diameter. 20 images have be chosen to be processed in each situation. However, table 1 only shown the parameters of 6 images which is cut from the 20images for your information only.

Tab. 1 Calculated results of parameters

Diameter of orifice	Parameter	Parameter value of 6 images					
1mm	A(mm^2)	9.05	8.52	8.23	8.41	8.35	7.88
	d(mm)	3.39	3.29	3.24	3.27	3.26	3.17
	v_y(m/s)	0.152	0.311	0.132	0.248	0.228	0.152
	a_y(m/s^2)	7.935	-8.939	5.770	-0.959	-3.832	6.234
2mm	A(mm^2)	20.62	15.89	12.32	13.84	18.22	14.25
	d(mm)	5.12	4.50	3.96	4.20	4.82	4.26
	v_y(m/s)	0.188	0.2448	0.204	0.225	0.171	0.188
	a_y(m/s^2)	2.813	-2.016	1.069	-2.713	0.846	2.318
3mm	A(mm^2)	15.07	12.97	15.54	23.07	22.19	19.86
	d(mm)	4.38	4.06	4.45	5.42	5.32	5.03
	v_y(m/s)	0.265	0.177	0.216	0.192	0.215	0.199
	a_y(m/s^2)	-4.431	1.960	-1.219	1.183	-0.818	0.911
4mm	A(mm^2)	26.63	25.29	29.08	27.62	20.38	19.62
	d(mm)	5.82	5.67	6.08	5.93	5.094	5.00
	v_y(m/s)	0.216	0.221	0.243	0.247	0.195	0.206
	a_y(m/s^2)	0.274	1.094	0.204	-2.616	0.536	-0.301

4. BUBBLE VOLUME CALCULATION

Bubble volume is a very important parameter in two phase flow, which is difficult to measure. In this paper, we proposed a method based on 2-D high-speed image analysis to estimate volume of bubble. First, analysis the forces which bubble subjected in vertical direction. Then, get the force balance formulation based on Newton's second law.

4.1 Force Analysis in Vertical Direction

Usually, there are 5 kinds of force in vertical direction. They are drag force, virtual mass force, buoyancy, gravity and liquid thrust (Sherwood et al., 1999, Zhang et al., 2001).

4.1.1 Drag Force

Drag is the resistance of bubble moving in the water(Krepper et al., 2007). It can be expressed as:

$$F_D = -\frac{1}{2}\rho_l C_D |u_l - u_g| (u_g - u_l) s \quad (7)$$

s is the front face area of bubble. For bubble keep rolling during the experiment, s was approximated to the average area A of the side projection area of bubble. C_D is the Drag coefficient, which is related to Reynolds number.

By the definition, Reynolds number is the ratio of inertial force and viscous force. It is expressed as follw:

$$\mathrm{Re} = \frac{F_g}{F_m} = \frac{\rho_g S u_g^{\ 2}}{\frac{S}{d}\eta u_g} = \frac{u_g}{\eta}\rho_g d = \frac{u_g d}{v} \quad (8)$$

where u is average velocity of bubble. d is the equivalent diameter of bubble. v, η are the coefficient of kinematics viscosity and dynamic viscosity of the flow in motion state. When the temperature is 20 degrees Celsius, the dynamic viscosity of air is 18.09×10-6Pa· s, the kinematics viscosity of air is 14.8×10-6m2/s. Combining the working condition and formula (8), the Reynolds number could computed out.

In 1979, Ishii proposed Ishii formula. When 20<Re<260, C_D could be calculated by:

$$C_D = \frac{24}{\mathrm{Re}}(1 + 0.1\mathrm{Re}^{0.75}) \quad (9)$$

Table 2 have computed the Reynolds number and drag coefficient of the images refer to table 1. It's obviously that the smaller the diameter of the orifice is, the bigger the two parameters of the same bubble varying.

Tab. 2 Reynolds number and drag coefficient of bubbles

Name of coefficient	Diameter of orifice	Coefficient value of 6 images					
Reynolds number (Re)	d=1mm	34.94	69.24	28.93	54.75	50.33	32.49
	d=2mm	65.02	74.17	54.54	63.86	55.62	54.05
	d=3mm	78.51	48.51	64.88	70.13	77.28	67.55
	d=4mm	84.98	84.91	100.05	99.14	67.15	69.51
Drag coefficient (C_D)	d=1mm	1.085	0.606	1.26	0.744	0.800	1.151
	d=2mm	0.641	0.570	0.747	0.651	0.734	0.752
	d=3mm	0.542	0.825	0.642	0.599	0.550	0.620
	d=4mm	0.505	0.505	0.436	0.439	0.623	0.604

4.1.2 Virtual Mass Force

Liquid that surrounding the bubble will accelerate with the relative acceleration of bubble rising. An additional mass called virtual mass which caused by bubble acceleration will the give an additional force called virtual mass force. Virtual mass will increase the effective inertial mass (Chuag M. et al., 2001).

In the momentum equation of dispersion flows, virtual mass force is used to abrupt accelerate the continuous phase surround the dispersed phase. If the density of the dispersed phase is far outweighs that of the continuous phase, then the influence of virtual mass can be neglected. However, the density of dispersed phase air in this paper is much smaller than that of continuous phase water. Therefore, virtual mass force could not be neglected in this study. The expression of virtual mass force is:

$$F_V = \frac{1}{2} V_g \rho_l \frac{d}{dt}(u_l - u_g) \quad (10)$$

4.1.3 Buoyancy

Object in the liquid will suffer an upward force named buoyancy produced by the surrounding liquid or gas in which it is fully or partially immersed, due to the pressure difference of the fluid between the top and bottom of the object. According to Archimedean principle, the upward buoyancy force F_B is equal to the magnitude of the weight of fluid displaced by the body. The computing formula is:

$$F_B = G_l = \rho_l g V_g \quad (11)$$

Where g is the acceleration of gravity and g=9.8m/s2.

4.1.4 Gravity

Gravity is proportional to the mass of the body. So it was expressed as follow:

$$G = -m_g g = -\rho_g V_g g \quad (12)$$

4.1.5 Liquid Thrust

If liquid is flowing and the flowing direction of liquid is the same as bubble moving to, then bubble will endure the thrust of liquid. The thrust is:

$$F_P = \rho V_g \, {du_l}/{dt} \quad (13)$$

For the liquid is static in this subject, the flow rate of water u_l is 0. Thus, the liquid thrust dose not work on bubble moving.

4.2 Bubble Volume Calculating

4.2.1 Formula for volume

According to Newton's Second Law, the acceleration of an object as produced by a net force is directly proportional to the magnitude of the net force, in the same direction as the net force, and inversely proportional to the mass of the object. In other words, the net force is equated to the product of the mass times the acceleration a. It can be expressed in equation form as follows:

$$F_c = ma \quad (14)$$

where F_c is the net force. The direction of the net force is in the same direction as the acceleration.

Considering the above analysis of force, the net force of the rising bubble in static water is:

$$F_c = F_B + F_D + F_V + G \quad (15)$$

Combine formula (7) to (15), the force balance equation can be expressed as follows:

$$m_g a_y = -\frac{1}{2}\rho_l C_D u_g^2 A - \frac{1}{2} V_g \rho_l a_y + (\rho_l - \rho_g) g V_g \quad (16)$$

The finishing formula for volume is:

$$V_g = \frac{\rho_l C_D u_g^2 A}{2(\rho_l - \rho_g)g - (\rho_l + 2\rho_g)a_y} \quad (17)$$

Table 3 have calculated the volume of bubble refered to table 1 and table 2. And the average volume of 20 images were also shown in table 3.

Tab. 3 Volume of bubbles

Diameter of orifice	Volume value of 6 images						Average value
d=1mm	19.60	17.52	13.19	18.68	14.88	15.68	17.03
d=2mm	27.81	25.00	20.64	20.49	20.86	21.92	22.28
d=3mm	23.94	18.97	22.35	27.58	27.70	26.06	24.54
d=4mm	32.51	33.921	38.721	33.46	25.38	25.27	31.81

It is worth while to note that the drag coefficient C_D in formula (17) is related to the equivalent diameter d which supposed bubble to be a normal sphere. If we supposed the bubble is a normal sphere, bubble volume can be computed by $V_d = 4/3\pi(d/2)^3$. But bubbles deform during the rising course, the volume that only depends on equivalent diameter d will result in large error. However, formula (17) considered the forces endured on bubble in hydronamics, which combined many factors to analyse bubble motion and weakened the effect of the assumption that bubble is normal sphere in drag coefficient C_D computation. So, force analysis method for bubble volume calculation could be considered as an improved method which is much more accurate than direct calculation.

4.2.2 Comparison of Bubble Volume

Bubble volume also can be calculated by geometric methold from direct image measurement. In order to vertify the validity of the calculated results for the method metioned above, geometric method was introduced. And the standard deviation of bubble volume computed by this two method were compared with each other.

Under the press of water, bubble is distorting during the rising movement. However, the horizontal forces that worked on the bubble in the same high level are equal to offset each other. So, the cross section is quasi-circular. Therefore, the whole bubble could be described as an ellipsoid.

The geometric method is computed based on edge detected images. Suppose the background of edge detected image is white and the bubble edge is black. The geometric method of bubble volume computing is described as follows:

(1) Scan the edge detected image from the top down, left-to-right. To find the coordination of point with value 0.

(2) Compute the distance between the edge point (i, j) and the Center Coordinate. The distance r is calculating by:

$$r = \sqrt{(i - x_c)^2 + (j - y_c)^2} \quad (18)$$

(3) Determine the size of r and d/2. The mean value of the bigger is regarded as the major semi-axis a; but that of the smaller is regarded as the minor semi-axis b.

(4) Construct a ellipsoid with semi-axis: a, b, and c. where c=b. Compute the volume of the ellipsoid with volume formula:

$$V = \frac{4}{3}\pi abc \quad (19)$$

Table 4 has shown the comparison of the standard deviation for the volume calculated by geometric methold and the method proposed in this paper. It's obverse that the result of bubble volume calculated by the method proposed in this paper is much more accuracy than that of geometric method from table 4.

The geometric method for geometric volume computing of bubble is only based on 2-D high-speed image processing. For the shape changing during bubble rolling rising, the geometric volume has a larger fluctuation and compute error than that of the volume computed by the proposed method. However, the force analysis method based on a powerful theory produce stabilized results and small error. It is superior to geometric method in bubble volume calculating.

Tab. 4 Bubble volume standard deviation

Diameter of orifice	Standard deviation of geometric volume	Standard deviation of presented method
d=1mm	1.3742	0.4133
d=2mm	1.5685	0.4562
d=3mm	1.3507	0.4129
d=4mm	1.4590	0.6126

5. CONCLUSIONS

An experimental apparatus was set up and operated to study gas bubble motion. In order to get high-quality image of rising bubble by high-speed camera, two-phase flow filed was established by injecting gas bubbles into a rectangular tank through orifices with different diameter. Sequential images which acquired by the visualized system needed to be processed by digital image processing algorithm. Affective algorithm was applied to fill bubble in binary image. The area, equivalent diameter, center coordinate, velocity and acceleration of bubble were calculated by the processed images.

Additionally, when the contributions of drag force, virtual mass force, buoyancy and gravity acting on a single bubble were combined, the volume could be derived using Newton's second law combines with the experimental measurements of parameters of rising single bubbles released from orifices.

Experimental examples of bubble parameters calculation were shown in the paper. As bubble rolling in water, the same parameter of the same bubble in different image is different. Average value of volume could be viewed as the final rough estimated results. However, the accuracy of the results calculated method is much better than that got from direct image measurement after the standard deviation comparing.

ACKNOWLEDGEMENTS

The author appreciates the support from National Natural Science Foundation of China (No. 50776063) and Program for New Century Excellent Talents in University of China (No. NCET-06-230).

NOMENCLATURE

A	area value of bubble	[mm^2]

a	the acceleration of bubble	[m/s^2]
C	drag coefficient	
d	the equivalent diameter of bubble	[mm]
F	force	[N]
G	gravity	[N]
g	acceleration of gravity	[m/s^2]
h	gray value of image	
(i, j)	coordinate of pixel point	
K	gray value of image	
m	mass	[kg]
n	the number of pixels	
M	grand total of gray value of image	
Re	Reynolds number	
s	front face area of bubble	[mm^2]
t	brightness threshold	
u	flow velocity	[m/s]
V	volume	[mm^3]
v	bubble velocity	[m/s]
x	coordinate of x-axis	
y	coordinate of y-axis	

Greek Letters

Ω	the gather of the pixels of the same bubble	
ρ	density	[kg/m^3]
ν	kinematics viscosity coefficient	[m^2/s]
η	dynamic viscosity coefficient	[Pa·s]
Δt	time interval between image	[ms]

Subscripts

B	buoyancy
C	centre of bubble
D	drag force
I	the i-th bubble
j	the j-th pixel value
l	liquid
g	gas
V	virtual mass
x	parameter of horizontal direction
y	parameter of vertical direction

REFERENCES

Chuag, M. S. et al. (2001). "Effect of interfacial pressure jump and virtual mass terms on sound wave propagation in the two-phase flow," J. Sound. Vib., 244(4), pp. 717-728.

Darton, R. C. et al. (1977) "Bubble growth due to coalescence in fluidised beds," Trans. Instn. Chem. Engrs., 55(4), pp. 274-280.

Ishii, M. et al. (1979). "Drag coefficient and relative velocity in bubbly, droplet or particulate flows," Aiche. J., 25(5), pp. 843-855.

Krepper, E. et al. (2007). "Experimental and numerical studies of void fraction distribution in rectangular bubble columns," Nucl. Eng. Des., 237(4), pp. 399-408.

Lv, J. et al. (2007). "Wavelet denoising analysis based on image edge detection," J. Tianjin Polytechnic Uni., 26(5), pp. 64-67.

Madhavi, T. et al. (2007). "Studies on bubble dynamics with mass transfer," Chem. Eng. Sci., 128(2-3), pp. 95-104.

Maurus, R. et al. (2002). "Study of the bubble characteristics and the local void fraction in subcooled flow boiling using digital imaging and analysing techniques," Exp. Therm. Fluid. Sci., 26(2-4), pp. 147-155.

Marco, P. D. et al. (2003). "Experimental study on rising velocity of nitrogen bubbles in FC-72,"Int. J. Therm. Sci., 42(5), pp. 435-446.

Rodríguez-Rodríguez, J. et al. (2003). " A novel particle tracking and break-up detection algorithm: application to the turbulent break-up of bubbles," Meas. Sci. Technol., 14(3), pp. 1328~1340

Sherwood, J. D. (1999). " The force on a growing bubble in potential flow," Int. J. Multiphas. Flow, 25(4), pp. 705-713.

Zaruba, A. et al. (2005). "Experimental study on bubble motion in a rectangular bubble column using high-speed video observations," Flow Meas. Instrum., 16(5), pp. 277-287.

Zhang, M. L. et al. (2006). "Vapor void fractions for forced convection with subcooled boiling at low flow rates," Chinese Q. Mech., 25(2), pp. 208-212.

Zhang, L. et al. (2001). " A periodic formation from a submerged orifice," Chem. Eng. Sci., 56(18), pp. 5371-5381.

Zhang, L. et al. (2008). "Unsteady motion of a single bubble in highly viscous liquid and empirical correlation of drag coefficient," Chem. Eng. Sci., 63(8), pp. 2099-2106.

Journal of Physics: Conference Series **147** (2009) 012019 doi:10.1088/1742-6596/147/1/012019

Measurement of pressure fluctuation in gas-liquid two-phase vortex street

Zhiqiang Sun [1], Wenhui Sang [1], Hongjian Zhang [2]

[1] School of Energy Science and Engineering, Central South University, Changsha 410083, Hunan, CHINA
[2] Department of Control Science and Engineering, Zhejiang University, Hangzhou 310027, Zhejiang, CHINA

zqsun@mail.csu.edu.cn (Z. Sun, Corresponding author)

Abstract. The pressure fluctuation in the wake is an important parameter to characterize the shedding process of gas-liquid two-phase Karman vortex street. This paper investigated such pressure fluctuations in a horizontal pipe using air and water as the tested fluid media. The dynamic signal representing the pressure fluctuation was acquired by the duct-wall differential pressure method. Results show that in the wake of the gas-liquid two-phase Karman vortex street, the frequency of the pressure fluctuation is linear with the Reynolds number when the volume void fraction is within the range of 18%. Moreover, the mean amplitude of the pressure fluctuation decreases with the volume void fraction, and the mean amplitude is larger at higher water flowrates under the same volume void fraction. These findings contribute to an in-depth understanding of the gas-liquid two-phase Karman vortex street.

1. INTRODUCTION

The Karman vortex street may exist in some gas-liquid two-phase flow under certain fluid conditions (Hulin et al., 1982; Inoue et al., 1986). The gas-liquid two-phase vortex street is as the same as the single-phase vortex street in generating pressure fluctuations in the wake flow (Sun et al., 2006). The positive aspect of this kind of pressure fluctuation is that it can be used for measuring or analyzing partial parameters of the gas-liquid two-phase flow (Li et al., 1998; Shakouchi et al, 2001, 2002), and the negative aspect is that the pressure fluctuation may induce vibrations which is dangerous to the piping system (Yokosawa et al., 1986a, 1986b). Therefore, it is of importance to study the dynamic property of the pressure fluctuation induced by a gas-liquid two-phase vortex street.

In this paper, investigations were carried out on the fluctuating property of the pressure in the wake of the gas-liquid two-phase vortex street developed in a horizontal pipe. Experiments were conducted using air and water as the tested fluid media, and the dynamic signal representing the pressure fluctuation was acquired by the duct-wall differential pressure method (DDPM). Analyses and discussion were focused on the change of frequency and amplitude of the pressure fluctuation with the Reynolds number and the volume void fraction.

2. EXPERIMENT

Experiments were conducted in a horizontal 50mm-diameter pipe using air and water as the tested fluid media. As shown in Fig. 1, the water in the pool and the air from the compressor were first pressed into surge tanks before their flowrates were measured respectively by an electromagnetic flowmeter and a rotameter. After the water and air were mixed by a two-phase mixer, they entered in the test section where a bluff body was placed in the middle of the pipe perpendicular to the flow direction. The Karman vortex street was generated behind the bluff body and was shed in the wake flow. The bluff body was a trapezoidal cylinder with a blockage ratio of $b = d/D = 0.28$, where d was the width of the cylinder and D was the diameter of the pipe. The upstream and downstream straight pipe before and behind the bluff body were $30D$ and $20D$, which ensured the Karman vortex street fully developed.

From the two pressure tags on the pipe wall downstream of the front face of the bluff body, dynamic differential pressure signals were detected by a Honeywell 24PC sensor, were conditioned by an analogue electric circuit scope and then processed by a computer (Sun et al., 2004, 2007). The sampling rate used was 1000 Hz throughout the experiments. Each data set contained 10000 points. Signals were acquired only when the vortex street was stable. During the experiments, the air flowrate was 0.2–4.0 m^3/h with 1.5% accuracy degree, and the water flowrate was 5.0–19.0 m^3/h with an accuracy of 0.5%. Additionally, pressures and temperatures of air and water at the flowmeters and the test section were also measured.

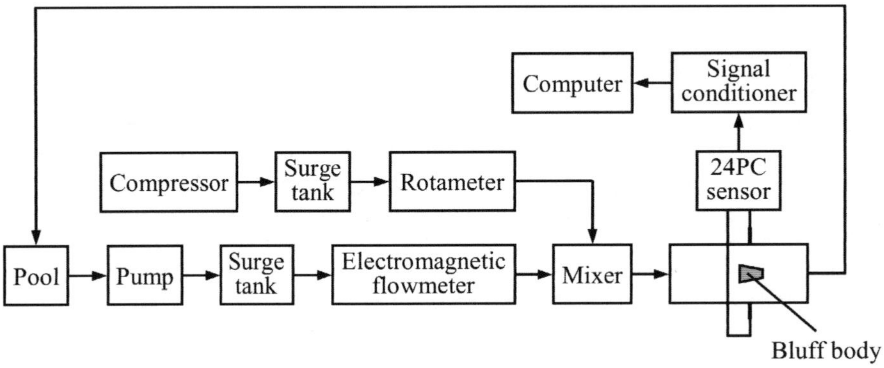

Fig. 1 Experimental set-up.

3. RESULTS AND DISCUSSION

3.1 Dynamic Differential Pressure Signal

Fig.2 shows the raw pressure fluctuation signal in vortex street wake respectively obtained from single water and gas-liquid two-phase flows. The two cases presented are featured with the same Reynolds number Re, which is defined as follows:

$$Re = v_{TP}\rho_{TP}D/\mu_{TP} \quad (1)$$

$$v_{TP} = (Q_G + Q_L)/S \quad (2)$$

$$\rho_{TP} = \rho_G\beta + \rho_L(1-\beta) \quad (3)$$

$$\mu_{TP} = \mu_G\beta + \mu_L(1-\beta) \quad (4)$$

where v_{TP} is the mean flow velocity of the two-phase mixture, Q_G and Q_L are the flowrates of the gas-phase and liquid-phase components, S is the cross-sectional area of the pipe, ρ_G, ρ_L and ρ_{TP} are the density of the gas-phase component, the liquid-phase component and the two-phase mixture, μ_G, μ_L and μ_{TP} are the dynamic viscosity of the gas-phase component, the liquid-phase component and the two-phase mixture.

It is seen that the pressure fluctuation in Fig.2(b) is quite stable as in the single water flow, which reveals that there still exist regular vortex shedding in the gas-liquid wake behind the bluff body. Within a certain range of the volume void fraction β ($\beta=Q_G/(Q_G+Q_L)$), the introduction of gas into the liquid component has little impact on the Karman vortex shedding. The corresponding power spectral densities (PSDs) of the raw signals were also calculated and were shown in Fig.2.

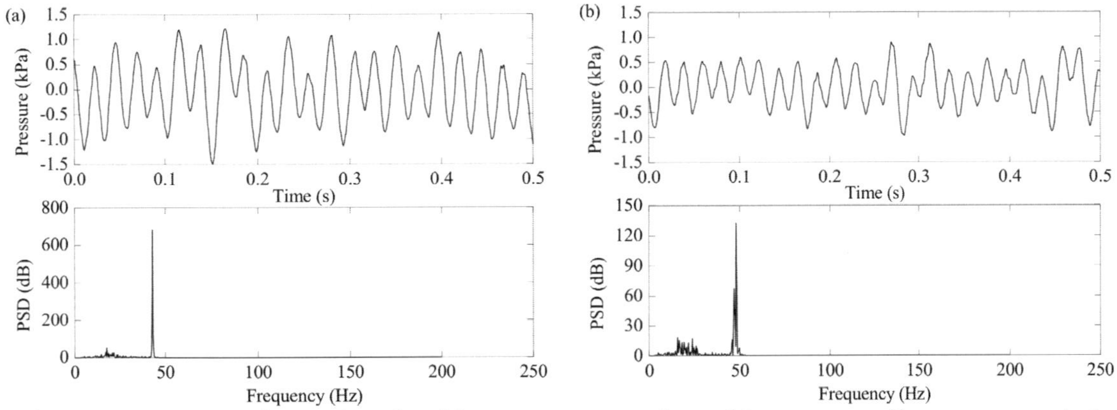

Fig. 2 Raw pressure fluctuation signal in vortex street wake and its corresponding power spectral density: (a) Single water; (b) Two phase.

3.2 Fluctuating Frequency

As mentioned above, the frequency f of gas-liquid two-phase vortex street could be calculated by analyzing the dynamic differential pressure signal in the frequency domain. Fig. 3 shows the relationship between the vortex frequency and the Reynolds number. It demonstrate that f increased with Re linearly regardless of the volume void fraction, which was just the same as that of single-phase flows. However, the linearity between f and Re maintained until the volume void fraction β was up to 18%, that is to say, when β was greater than 18%, the fluctuating differential pressure at vortex shedding frequency was not detectable.

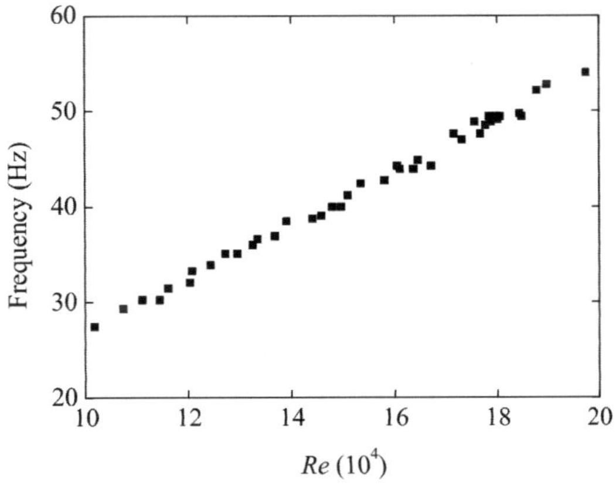

Fig. 3 Frequency of pressure fluctuation versus Reynolds number.

3.3 Fluctuating Amplitude

The amplitude of the fluctuating pressure in the vortex street wake has a close connection to the vortex shedding process and it is an important parameter to study the characteristics of the Karman vortex street. The relationship between the mean amplitude A of the fluctuating differential pressure and the volume void fraction are presented in Fig. 4. The calculation of A is as follows: first the number N of complete periods in an experiment data set was computed; then within each complete period the maximal and minimal values were selected respectively, and here we denote A^i_{max} and A^i_{min} for the maximal and minimal values in the ith period; finally the mean amplitude A was calculated by Eq. (5).

$$A = \frac{1}{2N}\left(\sum_{i=1}^{N} A_{\max}^{i} - \sum_{i=1}^{N} A_{\min}^{i}\right) \tag{5}$$

The signal amplitude is usually used to characterize the intensity or energy of the vortices generated from the view of fluid dynamics. Fig. 4 shows a decreasing trend of A with β at various water flowrates Q_L. The curve at lower Q_L is flat, which is nearly linear. When Q_L increases, the reduction in A turns to be a little faster and eventually it has the trend to be in a power function form at the higher Q_L. These results reveal that the energy of gas-liquid two-phase vortex decreases with the water flowrate.

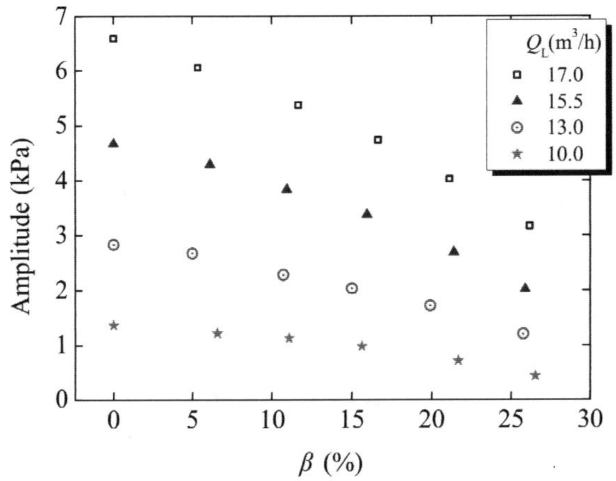

Fig. 4 Amplitude of pressure fluctuation versus volume void fraction.

4. CONCLUSIONS

This paper studied the pressure fluctuation in the wake of gas-liquid two-phase Karman vortex street in a horizontal pipe. Major findings are summarized as follows.

1) The frequency of the pressure fluctuation in the wake of gas-liquid two-phase Karman vortex street is linear with the Reynolds number when the volume void fraction is within the range of 18%.

2) The mean amplitude of the pressure fluctuation in the wake of gas-liquid two-phase Karman vortex street decreases with the volume void fraction, and under the same volume void fraction the mean amplitude is larger at higher water flowrates.

ACKNOWLEDGEMENTS

This work is financially supported by the China Postdoctoral Science Foundation Funded Project (200801346), the Hunan Postdoctoral Scientific Program (2008RS4022), and the Postdoctoral Science Foundation of Central South University.

NOMENCLATURE

A	signal amplitude	[kPa]
b	blockage ratio	
D	inner diameter of pipe	[m]
d	width of bluff body	[m]
f	vortex frequency	[Hz]
N	period number	
Q	volume flowrate	[m^3/h]
Re	Reynolds number	
v	mean flow velocity in pipe	[m/s]

Greek Letters

β	volume void fraction	
μ	dynamic viscosity	[N·s/m^2]
ρ	density	[kg/m^3]

Subscripts

G	gas-phase component
L	liquid-phase component
max	maximal value
min	minimal value
TP	two-phase flow

REFERENCES

Hulin, J. P. et al. (1982). "Experimental study of vortex emission behind bluff obstacles in a gas liquid vertical two phase flow," International Journal of Multiphase Flow, 8(5), pp. 475-490.

Inoue, A. et al. (1986). "Studies on two phase cross flow part I: characteristics around a cylinder," International Journal of Multiphase Flow, 12(2), pp. 149-167.

Li, Y. G. et al. (1998). "A study of measuring gas-liquid two-phase flow rate and component with gas-liquid two-phase vortex street characteristics," Proceedings of the 2nd International Symposium on Measurement Techniques for Multiphase Flows, Beijing, China, August 30-September 1, pp. 100-104.

Shakouchi, T. et al. (2001). "Measurement of flow rates of gas-liquid two-phase flow by Karman vortex," Proceedings of the 3rd International Symposium on Measurement Techniques for Multiphase Flows, Fukui, Japan, August 1-3, pp. 83-89.

Shakouchi, T. et al. (2002). "Behavior of vertical upward gas-liquid two-phase flow past obstacle in rectangular channel," JSME International Journal Series B: Fluids and Thermal Engineering, 45(3), pp. 686-693.

Sun, Z. Q. et al. (2004). "Research on detection of vortex flowmeter signal with differential pressure," Chinese Journal of Sensors and Actuators, 17(3), pp. 420-423. (in Chinese)

Sun, Z. Q. et al. (2006). "On measurement property of vortex flowmeter in bubbly two-phase flows," Proceedings of the 5th International Symposium on Measurement Techniques for Multiphase Flows, Macau, China, December 10-13, pp. 246-250.

Sun, Z. Q. et al. (2007). "Investigation of the pressure probe properties as the sensor in the vortex flowmeter," Sensors and Actuators: A. Physical, 136(2), pp. 646-655.

Yokosawa, M. et al. (1986a). "Studies on Two Phase cross Flow Part II: Transition Reynolds Number and Drag Coefficient," International Journal of Multiphase Flow, 12(2), pp. 169-184.

Yokosawa, M. et al. (1986b). "Studies on two phase cross flow part III: characteristics of unsteady flow behavior," International Journal of Multiphase Flow, 12(2), pp. 185-202.

Journal of Physics: Conference Series **147** (2009) 012020 doi:10.1088/1742-6596/147/1/012020

Frictional Drag reduction by wavy advection of deformable bubbles

Yoshihiko OISHI, Yuichi MURAI, Yuji TASAKA, Takeda YASUSHI

Hokkaido University, Sapporo, JAPAN

oishi@eng.hokudai.ac.jp

Abstract. Bubbles can reduce frictional drag in wall turbulence, and its effect is expected to use for ships and pipelines to save their power consumptions. A number of basic experiments have been carried out to date for finding out the best condition for enhancing the drag reduction. One issue that remains at present is the difference of the performance between steady and unsteady status in terms of bubble concentration. All the experiments in the past deal with the steady effect, i.e., the drag reduction is evaluated as a function of mean void fraction or given gas flow rate of continuous injection. Despite to this, the actual phenomena highly depend on local interaction between two phases upon unsteady manner. We focus on this point and elucidate the influence of time-fluctuating void fraction on the total response to the drag reduction. This view is in fact important to estimate the persistency of the bubble-based drag reduction in the flow direction since bubbles formulate wavy advection during their migration.

Our experiments are designed to measure the above-mentioned effect from laminar, transitional, and turbulent flows in a horizontal channel. For avoiding the contamination effect that worsens the reproducibility of the experiment, Silicone oil is used as carrier fluid. The oil also simulates the high Weber number bubble condition because of low surface tension. The unsteady interaction between the wavy advection of bubbles and the local skin friction, a synchronized system is constructed to connect the high-speed camera with the shear transducer, which can evaluate the interaction at 1000 fps. From the results, we confirm that the drag reduction is provided at Re>3000 in the turbulent flow regime, and also the total drag reduction is enhanced by the presence of the waves.

1. INTRODUCTION

Drag reduction is one of the most important fields of research in engineering since it directly contributes to energy saving in any facilities and devices. Use of bubbles is considered as feasible method recently because of several advantages in comparison with other methods. For instance, the engineers in shipbuilding are keen to apply this technique for the drag reduction of large ships. It is known that there are several important parameters that characterize the drag reduction by bubbles; void fraction, thickness of bubbly layer, and bubble size. In dimensionless expression, we need to investigate the dependency of Reynolds number, Weber number, and Froude number. On the mechanism of the drag reduction, which has been studied a lot to date, we need to conclude that it is not explained so simply. The drag reduction would be caused not only by the decreases in density and viscosity of mixture fluid but also by the dynamic two-phase interaction in the boundary layer. **Even in laminar flow**, the effective viscosity is significantly altered with deformation of bubbles which are subjected to shear stress, as reported by Rust and Manga (2002). Namely, the shear response of bubbles, or Rheological properties of bubbles plays an important role in modifying **the surrounding turbulent flow field**. At high Reynolds number, the interaction **process** is further complex than that at the laminar flow. Kawamura (2005) performed a DNS study for investigating the bubble deformation in a turbulent flow and found out the role of the bubble deformation. A different type of DNS study by Lu et al. (2005) has shown that the bubble deformation is one of the next key factor that can promote the sensitivity of the drag reduction. On the other hand, microbubbles that resist against the shear stress around have different effect of the drag reduction. Xu et al. (2002), Ferrante and Elghobashi (2004) investigated this and clarify that the advection of microbubbles relative to the near wall structures shifts the position of the momentum exchanging eddies far from the wall.

The many reports mention that there is a condition for the amount of injecting bubble in order to provide the highest performance of the drag reduction. We assess this as gain factor of the drag reduction, which is the ratio of the power-saving realized by the drag reduction to the power required for bubble generation. Air-film method so-called can separate the liquid from the wall to totally cut out the shear stress. However, the gain factor of this method is close to unity; 1% of void fraction provides 1% of drag reduction. To maximize the gain factor, we need to use mm-order bubbles that are easily generated with a variety of techniques. Using such a normal size bubble can obtain the gain factor from 4 to 10 (Murai et al. (2007)). This amplification of the gain factor happens when adequate interaction is realized between two-phases in boundary layer. Kitagawa et al. (2005) investigated the interaction between bubbles and turbulent events on structures in Reynolds stress distributions, including bubble deformation. They concluded that 'pseudo-turbulence' caused by the slip of bubbles relative to the liquid modifies the structure and characteristics of real-turbulence. Winkel et al. (2004) studied influence of bubble size and behavior using seawater, surfactant and tap water as carrier fluids, and concluded that bubble behaviors such as deformation and slipping are strongly affected by contaminants in the water such as particulates and ionic impurities. Therefore, it is plausible that theoretical prediction is hardly possible for explaining the drag reduction. The present experiment uses silicone oil and it has three advantages: 1) the gas-liquid interface is not susceptible to contamination due to the non-polar liquid, 2) bubble deformation can be achieved even in low Reynolds number turbulent flow because of lower surface tension than tap water, and 3) wide range of the Reynolds number covering laminar flow to turbulent flow can be achieved due to the high viscosity of silicone oil.

In this paper, we describe measurement techniques to investigate the relationship between the local void fraction and the local wall shear stress, and then discuss how the drag reduction is realized in the case of wavy advection of bubbles. With the data obtained, we reduce the factor that amplifies the drag reduction performance by bubbles.

2. EXPERIMENTAL APPARATUS

A schematic diagram of the experimental apparatus is shown in Fig.1. The apparatus comprises a flow controller, rectifier controller, test section, and deaeration unit. Silicone oil and air are used as working fluids under laboratory temperature. The silicone oil was driven by a pump (Ebara Pump, 80VNP2.2) that is controlled by an inverter (Showa Instrument Information Co. Ltd., FT-03070). The test section is a horizontal rectangular channel made of transparent acrylic resin, 20 mm in height, 160 mm in width and 6000 mm in length.

The outline of a bubble injection device is drawn in Fig.2. Air bubbles are injected into the horizontal channel from the top surface through 140 capillary tubes whose inner diameter is 0.1 mm. Air is supplied by a compressor (Amadera pneumatics co., Handicon T05P5S17), and the flow rate is measured by an air flowmeter (Tokyo Keiso Co. Ltd., F04-108478). The bulk mean liquid velocity ranges from 0.2 to 2.0 m/s. The injector is located 875mm downstream from the end of the diverging-converging channel.

Bubbles are completely removed by the double chamber structure of the tank. The tank utilizes the swirling effect in the inner chamber and buoyant separation effect in the outer chamber.

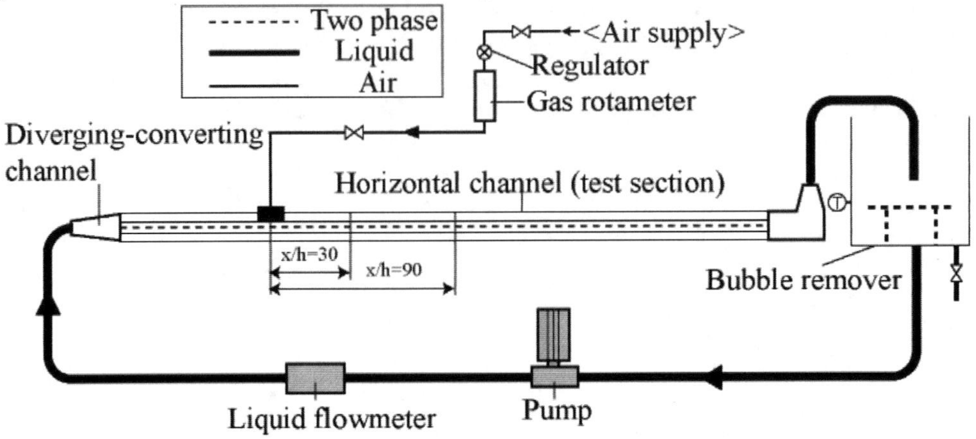

Fig. 1 Schematic diagram of the experimental facility.

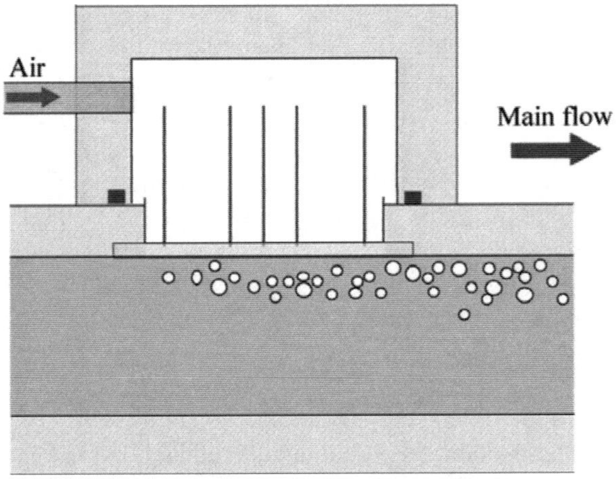

Fig. 2 Schematic of bubble injection device; air goes through array of capillary needles

Table 1 shows the experimental conditions. The skin friction coefficient C_f is defined by the following equation,

$$C_f = \frac{2\tau}{(1-\alpha)\rho U^2}, \tag{1}$$

where τ, α, U and ρ are respectively the local wall shear stress measured by the shear transducer, the bulk mean void fraction, the bulk mean velocity in the channel and the mean density of two-phase flow. α and U are given by,

$$\alpha = \frac{Q_g}{Q_l + Q_g}, \tag{2}$$

$$U' = \frac{Q_l + Q_g}{HB}, \qquad (3)$$

where Q is the volume flow rate, H (= $2h$) is the channel height and B is the channel width. The subscripts g and l indicate the gas and liquid phases, respectively. The bulk mean velocity U' increases due to the volumetric increase in the flow as the air bubbles are injected into the channel. (?) Therefore, to evaluate the skin friction coefficient in a two-phase flow, C_{f0} is modified by the following equation (Kodama et al. (2000)).

$$C'_{f0} = C_{f0}\frac{\tau_{lU}}{\tau_{lU'}}, \qquad (4)$$

where, τ_{lU} is the wall-shear stress estimated by the Blasius formula, and the subscript 0 indicates the bubble-free condition. In this study, the skin friction ratio, i.e. the ratio of the skin friction coefficient with bubbles to that without bubbles, is employed to evaluate the modification of the drag.

The steady geometry of the bubbles is solely a function of Weber number, which is calculated by

$$We = \frac{\mu G d}{2\sigma}, \qquad (5)$$

where d is the non-deformed bubble radius, G is the velocity gradient, μ is the suspending fluid viscosity, and σ is the surface tension.

Table 1 Experimental conditions

Main conditions			
Channel size		40 × 160 × 6000	[mm]
Channel half height	h	20	[mm]
Bulk void fraction	α	0.0 … 2.5	[%]
Mean bubble diameter	D	0.7	[mm]
Bulk liquid velocity	U	0.6, 1.0, 1.5, 2.0, 2.5	[m/s]
Bulk Reynolds number	$Re = 2hU/\nu$	1103 … 4425	[-]
Frictional Reynolds number	$Re\tau = hu\tau/\nu$	70 ... 140	[–]
Silicone oil			
Density	ρ	935	[kg/m^3]
Kinematic viscosity	ν	1.07×10^5	[m^2/s]
Temperature	t	20	[deg.C]
Surface tension	σ	21×10^{-3}	[N/m]
Camera conditions			
Measurement point	x/h	30, 90	[-]
Frame rate		1000	[fps]
Shutter speed		$1/2.0\times10^4$	[s]
Resolution		1024 × 1024	[pixels]
Sampling time	T	60	[s]

3. MEASUREMENT OF BUBBLY TURBULENT FLOW

3.1 Flow Visualization Technique

We carried out a synchronous technique to study the relationship between the bubble motion and wall friction. Images of bubbles are recorded by a high-speed digital video camera (Photron, Fastcam-Max). To obtain a backlight projection of the gas-liquid interface, a metal halide light (Photron, CVF-SL 150W) is used as the light source. The frame rate is 1000 fps and the shutter speed 1/20,000 s. The depth of field for the camera lens (Nikon, Micro-Nikkor105mm f/2.8) is 450 mm. A pulse generator synchronizes the trigger timings of the shear transducer and the high-speed camera. The calibration and image processing provide dimensionless values for the shear stress and bubble concentration, which are the friction coefficient and local projection of the void fraction. The measurement positions are x/h = 30 and 90, which correspond to 300 and 900 mm downstream of the air injection device, respectively. The measurement domain is 53.8 × 53.8 mm and the magnification factor is 0.0526 mm/pixel.

3.2 Direct Measurement of Wall Shear Stress

Measurement of the wall frictional drag is carried out as a direct measurement of the wall shear stress by a shear stress sensor (SSK Co. Ltd., S10W-4, capacity 3.9mN, temporal resolution 20Hz). Figure 3 shows the sensing part of the device, which is 10 mm in diameter. The flatness of sensitive area relative to the inner surface of the channel is carefully controlled using a lead plate. The shear stress signal is received and converted into electric signal by an analog amplifier (SSK Co. Ltd., M-1101). The measurement data is recorded by a PC as digital data via a data recorder (Keyence Co., NR-500).

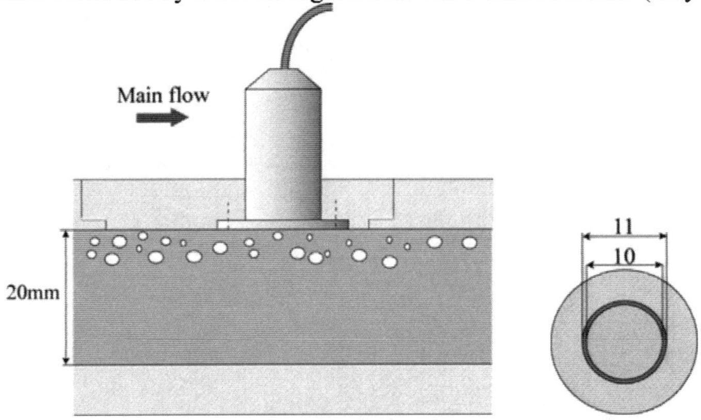

Fig. 3 Schematic of shear stress sensor mounted at the upper wall of the channel

3.3 Construction of Synchronous Measurement System

Figure 4 is a schematic diagram of the synchronized measurement system. A pulse generator synchronizes the trigger timings of the camera and the data logger for the shear stress sensor. The image acquisition position is 80 mm downstream from the measurement position of the wall-shear stress because the devices cannot be physically located at the same position. Hence there is a time lag between the data obtained by each device. The time lag is removed using the local mean advection velocity of the bubble interface, which is measured by particle image velocimetry (PIV). The advection velocity has a relative velocity to the mean liquid velocity in the channel. In the case of high void fraction, the bubbles become separated air films and move faster than the liquid. In the case of low void fraction, small bubbles appear and move slightly slower than the liquid because of the additional friction induced. The mean bubble diameter is 0.99 mm from the image analysis. The Weber number, which is the ratio of the shear stress to the surface tension, is 0.07 to 0.2.

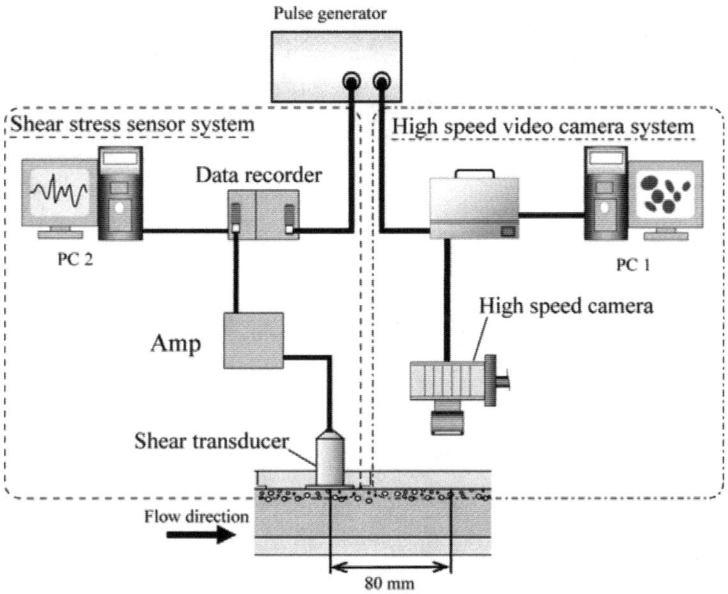

Fig. 4 Schematic diagram of the void-friction simultaneous measurement system

4. RESULTS AND DISCUSSION

4.1 Flow Pattern of Bubbles

In this experiment, we carried out particle tracking velocimetry (PTV). The center of gravity for individual bubble is extracted by image processing via edge detection, binarization, thinning and labelling. We obtained the velocity vectors using PTV (four-time steps PTV). Figure 5 shows a snapshot of bubbly flow with a velocity of 0.5 m/s and void fraction of 2.0 %. The injected bubbles extensively distribute in the spanwise direction and cluster in the streamwise direction. As a result, the bubble distribution becomes a stripe pattern for all conditions.

Figure 6 shows the original images of bubbles. The shape of a bubble is spherical in laminar flow as shown in Fig.6 (a). The bubble size irregularly expands in the streamwise direction with high Reynolds number. The expanded bubbles regain their sphere, and some bubbles break up.

Fig.5 Snapshot of bubbles that cover with channel wall ($U = 0.5$ m/s, $\alpha = 2.0$ %).

(a) Re = 1103, α = 0.5% (c) Re = 1798, α= 0.5% (e) Re = 2678, α= 0.5%

(b) Re = 1103, α = 2.5% (d) Re = 1798, α= 1.5% (f) Re = 2678, α= 1.5%

Fig. 6 Snapshot of bubbles near the upper wall (x/h = 90)

Figure 8 shows the skin friction ratio as a function of void fraction. When the liquid velocity U is slow, the ratio takes a larger value than unity but should be smaller than two. This ratio corresponds generally to the ratio of turbulent friction to laminar friction as shown before. However, its value decreases as the void fraction increases. On the other hand, the skin friction ratio takes a value smaller than unity when both the liquid velocity and the void fraction are high. In the case of the liquid velocity being equal to or larger than 2.0 m/s (Re=3734 to 3811), a frictional reduction of 9 % is obtained in the maximum case. This tells us the bubbles play a role in reducing the friction in such a low Reynolds number region.

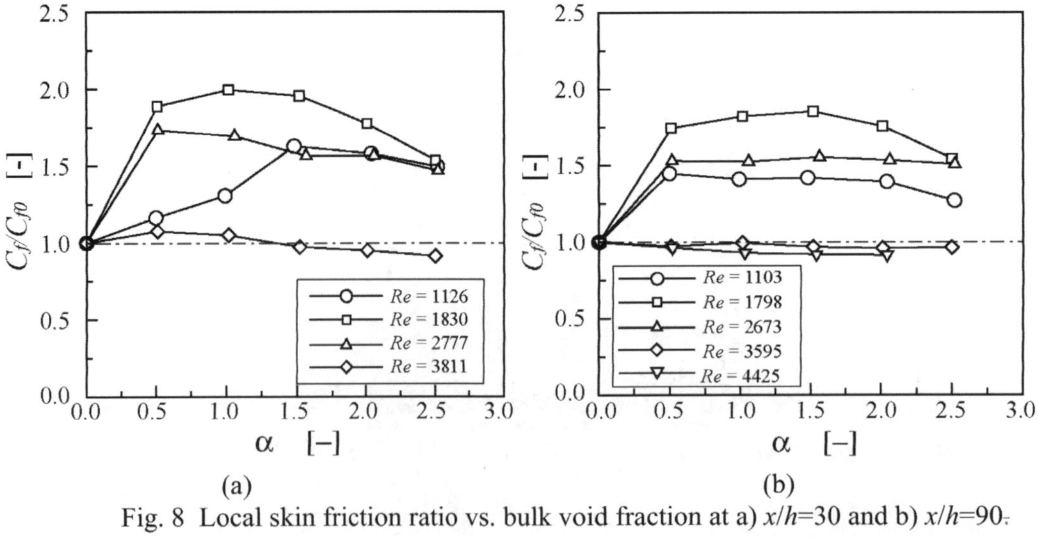

(a) (b)

Fig. 8 Local skin friction ratio vs. bulk void fraction at a) x/h=30 and b) x/h=90.

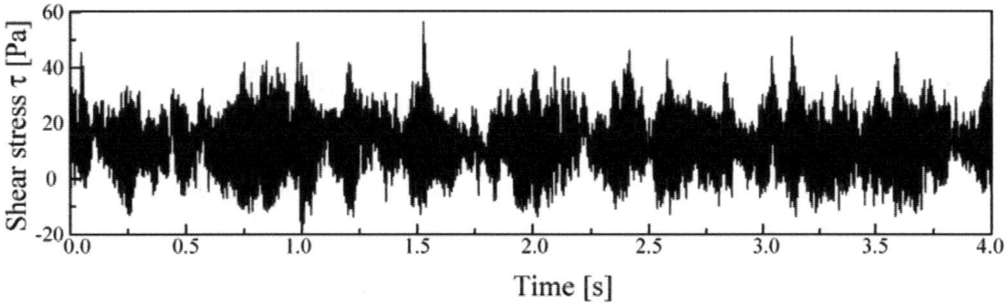

Fig. 9 Wall shear stress obtained by shear transducer

Figure 9 shows a sample of real data for the shear stress in the bubbly flow under these conditions. The fluctuation is significant relative to the average value. The relationship between the fluctuation frequency of the shear stress and fluctuation frequencies of the bubbles is of importance. Figure 10 shows the spectrum of real data. This fluctuation frequency of the shear stress is dependent on three factors of length. Frequencies in the range $5\times10<f<1\times10^2$ are dependent on the liquid velocity divided by the half width of the channel, those in the range $1\times10^2<f<2\times10^2$ are dependent on the bubble-bubble averaging spacing, and those in the range $2\times10^2<f<3\times10^2$ are dependent on the bubble size. In the latter case, the response to bubble size is high due to the size ranging from 1 to 5 mm. Although this relationship does not have a specific point, this frequency region is of interest in that it is the region of interference between frictional drag and bubble-bubble average spacing.

4.3 Equivalent Bubble Diameter

Figure 11 shows the probability density distribution of the bubble diameter in the drag reduction condition at x/h=30. The bubble diameter is an equivalent diameter extracted from the number of pixels in image processing. The mean bubble diameter is 988μm and the standard deviation is 410μm. The bubbles, which have larger Weber number than one, drastically deform in the turbulent flow. The range of the deformed bubble is between spherical microbubble and fragmental bubble.

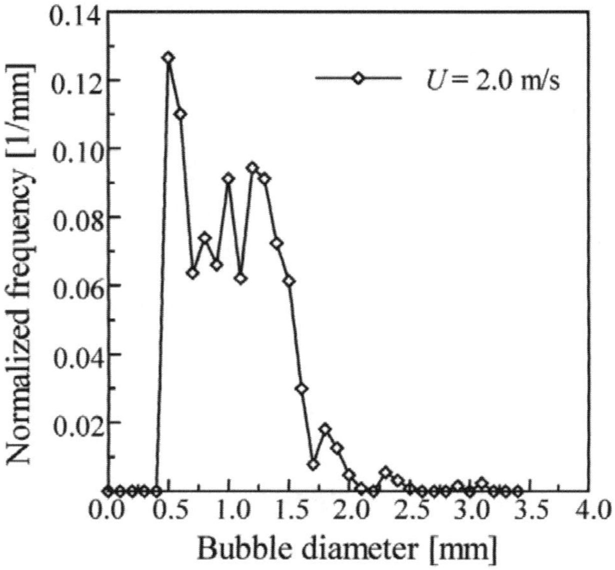

Fig. 11 Frequency of bubble diameter in the condition of drag reduction

4.4 Velocity of Bubbles

Figure 12 shows the bubble velocity vectors obtained using PTV. Figure 13 shows the velocity distribution of bubbles. The mean velocity of bubbles is 1.6m/s. In this experiment, bubbles of diameters smaller than 100μm (3 pixels) are ignored. Figure 13 shows the probability density distribution of the streamwise velocity of bubble at x/h=30. The results of this experiment show the individual bubble velocity remains constant, regardless of bubble size.

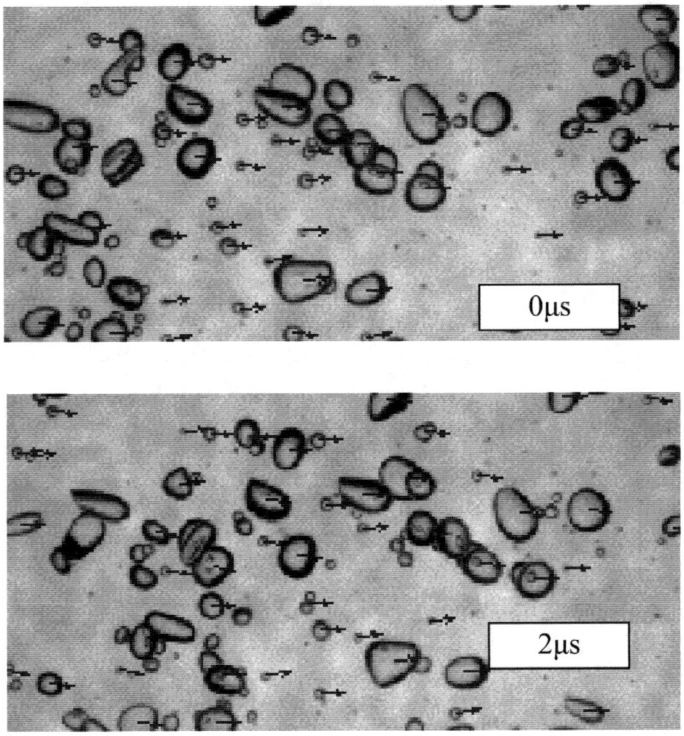

Fig. 12 Velocity vectors of bubble centroids

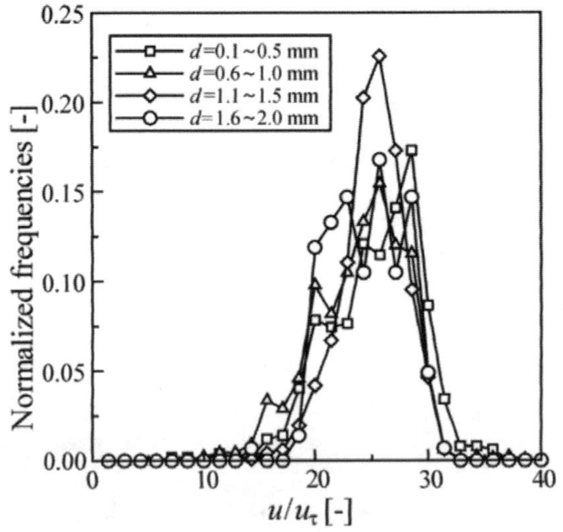

Fig. 13 Frequency of bubble velocity in the streamwise direction

Figure 14(a) shows an original image of the synchronized measurement experiment at x/h=30. We measured the projection void fraction from the binary image (Fig. 14(c)). Figure 15 shows the power spectrum of the skin friction (a) and void fraction (b).

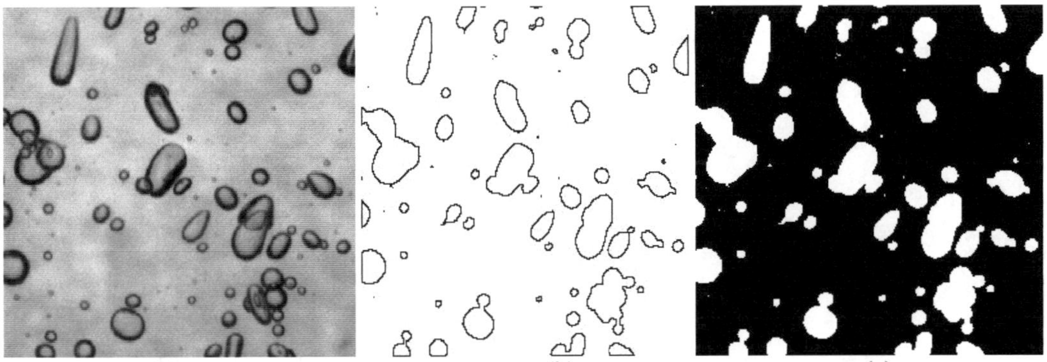

(a) (b) (c)

Fig. 14 Image analysis of the backlit projection of bubbles: (a) original image, (b) thinning image, (c) binary image to measure projection void fraction

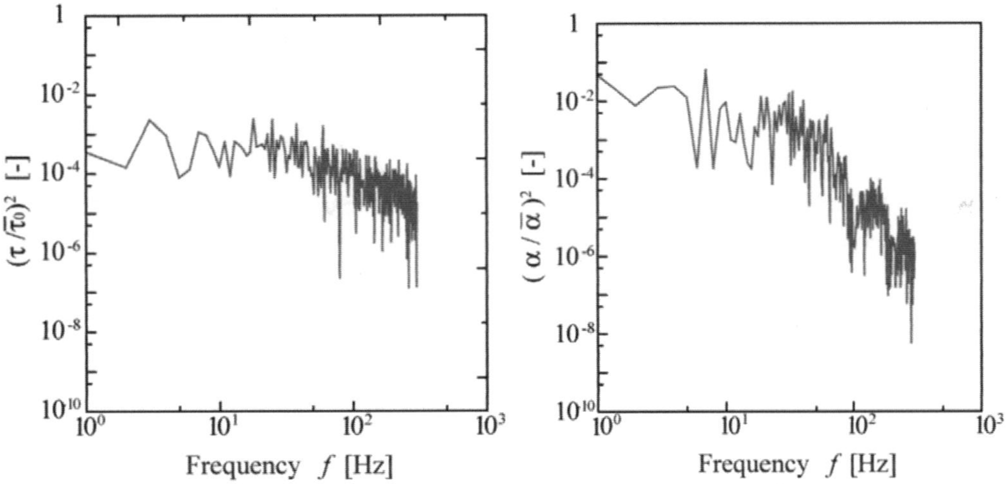

(a) (b)

Fig. 15 Power spectrum vs. frequency of (a) shear stress and (b) void fraction

4.5 Correlation of Bubble and Shear Stress

Figure 16 shows the waveforms of the shear stress and projection void fraction using the synchronized measurement system. The upper picture is the corresponding bubble image at the same sampling period. The mean void fraction, Reynolds number, and Weber number are α=2.0%, $Re=Uh/\nu$=1800, and $We=2\rho Uh/\sigma$=1870, respectively. The skin friction in the graph is normalized by that of the single-phase condition. The waveform of the skin friction (solid curve) fluctuates with large amplitude as the local projection void fraction (dotted curve) varies in time. The ratio of the fluctuation amplitude is roughly the same for both quantities; however, there is a certain phase shift between them. That is, the moment of highest void fraction does not match that of the largest drag reduction.

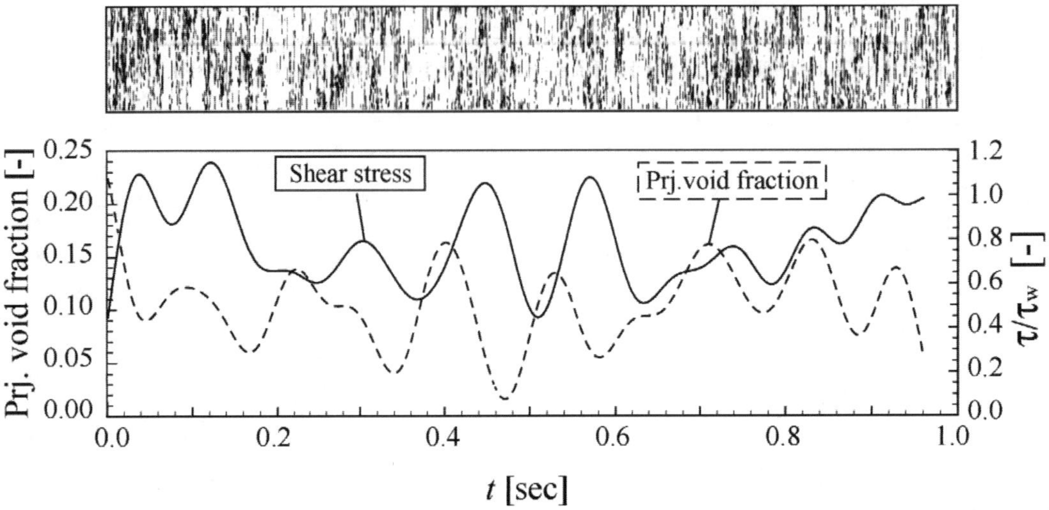

Fig. 16 Variation of local skin friction and projection void fraction obtained by synchronized measurement.

Figure 17 shows a two-dimensional trajectory of the two quantities. The trajectory is generally correlative relationship when the bubbles directly reduce a skin frictional drag. This is a historical effect of the bubbles on the skin friction, and furthermore the skin friction decreases greatly as bubbles begin to leave. This trend indicates the turbulent boundary layer is gradually altered in the streamwise direction by bubbles. Such a transient behavior was analyzed numerically by Xu et al. (2002) and Ferrante and Elghobashi (2004). A problem in numerical studies before this alteration was known was that the time-average skin friction hardly decreased as the time elapsed. Drag reduction occurred only in the initial stage of bubble-liquid interaction. However, the present experimental result has newly shown that the time average does decrease when the local/instantaneous value has a large fluctuation accompanying a phase shift.

Figure 18 shows the relationship between the time-averaged drag and instantaneous drag. The level of reduction estimated by this factor is something like the gain in drag reduction. The gain factor of the skin friction reduction per void fraction is calculated. The gain factor considered here is defined by the following equation, which indicates the magnitude of the drag reduction effect per void fraction.

$$\overline{G} = \frac{1}{\overline{\alpha}}\left(1 - \frac{C_f}{C'_{fo}}\right) = \frac{1}{\alpha}\overline{\left(\frac{\Delta\tau}{\tau_0}\right)}. \tag{5.6}$$

When drag reduction occurs, the gain factor takes a positive value. If the factor is greater than unity, it means the drag reduction being amplified and enhanced by the bubbles more than the density effect. On the other hand, the instantaneous gain factor considered here is defined by the following equation, which indicates the magnitude of the instant drag reduction effect per instant void fraction.

$$G' = \frac{1}{\alpha'}\left(\frac{\Delta\tau'}{\tau_0}\right). \tag{5.7}$$

Here, we define the instantaneous drag reduction ($\Delta\tau'$) and instantaneous void fraction (α') in terms of the maximum $\Delta\tau_{\max}$ and $\alpha_{\max}$ values. The time average value is obtained as 1.8, but the instantaneous

Journal of Physics: Conference Series **147** (2009) 012020 doi:10.1088/1742-6596/147/1/012020

maximum value is obtained as 3.2. As a result, the effect of the drag reduction is higher than the time-averaged drag when we focus on instantaneous behavior.

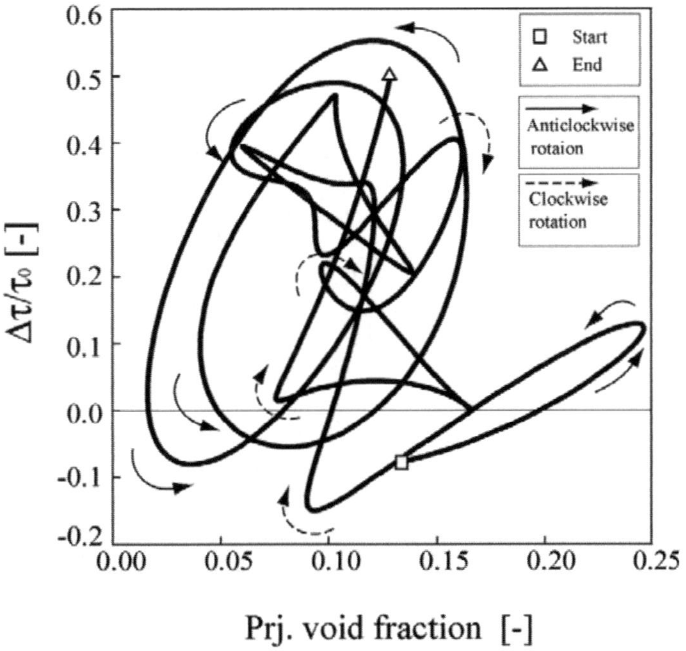

Fig. 17 Trajectory of two quantities: the drag reduction and the projection void fraction

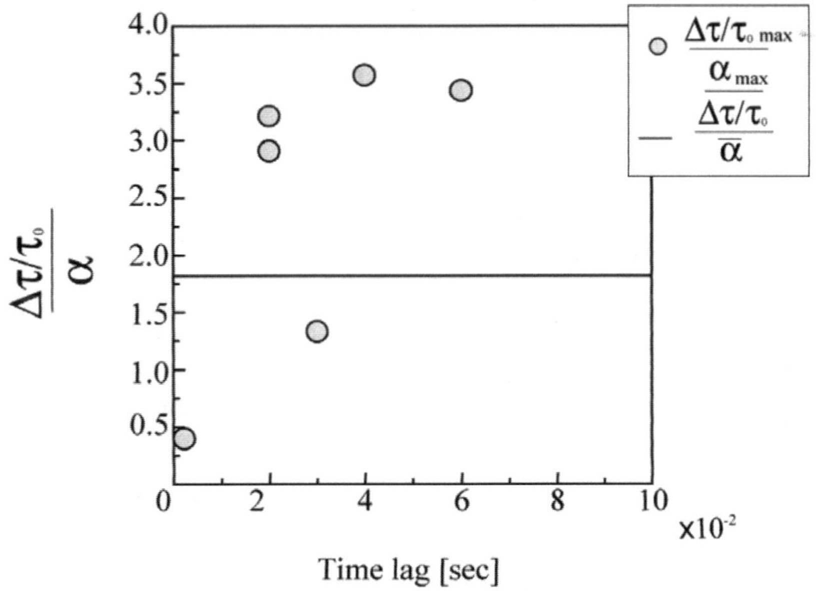

Fig. 18 Instantaneous gain factor vs. time lag between the peak of the ratio of drag reduction and the peak of void fraction

5. CONCLUSIONS

Bubble-induced drag modification is experimentally investigated using a horizontal channel. We use silicone oil as the carrier fluid since it enables us to stabilize the interfacial properties, to adjust low Reynolds number flow, and to control the condition of high capillary number. From the channel flow experiments, the following results are obtained.

- The injection of bubbles increases the friction coefficient by less than 50% in the case of laminar channel flows. In the transition region from laminar to turbulent flows, the friction coefficient increases up to two fold because bubbles activate the turbulent flow transition. The increase in the ratio of the friction coefficient matches that of the ratio of turbulent to laminar friction coefficients.
- After the transition to turbulent flow, the increased ratio of the skin frictional coefficient is reduced gradually by an increase in the void fraction. For the case when the flow is originally turbulent, the mixing of bubbles produces a certain drag reduction.
- The present experimental result has shown that the time average of the friction decreases when the local/instantaneous value has a large fluctuation. This indicates that the local drag reduction is promoted by unsteady interaction between two phases rather than the steady effect of the drag reduction. The wavy advection of the bubbles in our experiment is naturally created along the long spatial development of the two-phase boundary layer. Artificial supply of the wave, or intermittent injection of bubbles might cause a further enhancement of the total mean drag reduction (which is reported in another paper in this conference by H. Park et al).

ACKNOWLEDGEMENTS

This work was supported by JSPS (Japan Society for Promotion of Science) Research Fellowship from 2006 to 2007 by the name of Y. Oishi, and also funded by NMRI (National Maritime Research Institute of Japan) under ONR-NICOP project for fundamental research on microbubble drag reduction from 2003 to 2006. The authors thank to these supports as well as to repeated scientific discussions with research stuff associated to the project; Dr. Y. Kodama (NMRI), Dr. T. Kawamura (Univ. Tokyo), Dr. K. Sugiyama (Univ. Twente), Prof. F. Yamamoto (Univ. Fukui) , Prof. K. Hishida (Keio Univ.), and Prof. H. Kato (Toyo Univ., Japan).

REFERENCES

Ferrante, A. and Elghobashi, S. (2004). "On the physical mechanism of drag reduction in a spatially developing turbulent boundary layer laden with microbubbles," J. Fluid Mech., 503, pp. 345-355.

Lu, J., Fernandez, A. and Tryggvason, G. (2005). "The effect of bubbles on the wall drag in a turbulent channel flow, " Phys. Fluids, 17, 095102, pp. 1-12.

Kawamura, T. (2005). "Numerical and experimental investigation the mechanism of the microbubble drag reduction," Proc. 6th Symp. Smart Control of Turbulence – Tokyo, pp. 77-84.

Kitagawa A., Hishida K., Kodama Y. (2005). "Flow structure of microbubble-laden turbulent channel flow measured by PIV combined with the shadow image technique," Exp. in Fluids, 38, pp. 466-475.

Kodama Y., Kakugawa A. (2000). Takahashi T., Kawashima H., "Experimental study on microbubbles and their applicability to ships for skin friction reduction, " Int. J. Heat and Fluid Flow, 21, pp. 582-588.

Murai Y., Fukuda H., Oishi Y., Kodama Y., Yamamoto F., “Skin fricion reduction by large air bubbles in a horizontal channel flow,” Int. J. Multiphase Flow, (2007), 33, 2, 147-163.

Rust A.C., Manga M., (2002). "Bubble shapes and Orientations in low Re simple shear flow, " J. Colloid and Interface Sci., 249, pp. 476-480.

Rust A.C., Manga M., (2002). "Effects of bubble deformation on the viscosity of dilute suspensions," J. Non-Newtonian Fluid Mech., 104, pp. 53-63.

Sugiyama, K., Kawamura, T., Takagi, S., Matsumoto, Y. (2005). "Recent progress of microbubble flow simulation for elucidating drag reduction mechanism," Proc. 6th Symp. Smart Control of Turbulence – Tokyo, pp. 191-202.

Winkel E.S., Ceccio S.L., Dowling D.R., Perlin M., (2004). "Bubble-size distributions produced by wall injection of air into flowing freshwater, saltwater and surfactant solutions," Exp. in Fluids, 32, pp. 802-810.

Xu, J., Maxey, M.R. and Karniadakis, G.E. (2002). "Numerical simulation of turbulent drag reduction using micro-bubbles." J. Fluid Mech., 468, pp. 271-281.

Journal of Physics: Conference Series **147** (2009) 012021 doi:10.1088/1742-6596/147/1/012021

Study on Development of Ejector of Bubble Jet Engine (BJE) -Measurement of Thrust -

B Ono, K Nakashima, T Shigematsu and K Morishita [1]

[1] Sasebo National College of Technology1-1, Okishin-cho, Sasebo City, Nagasaki Pref., 857-1193 JAPAN

ono@post.cc.sasebo.ac.jp

Abstract. The AUV (Autonomous Under-water Vehicle), which is used for the present seabed investigations, has obtained the thrust with the screw driven by the battery. However, it has a disadvantage because of its size and cost. Therefore, this research is carried out to propose the Bubble Jet Engine (BJE) as an alternative propulsion device. It can directly transform combustion energy into kinetic energy, so it is expected that BJE can also rise the level of propulsion efficiency. This research aims at measuring exhaled mass flow rate and thrust to design ejectors, which become the core of BJE, and exploring practical possibility of BJE. Vertical type gas-water ejector experimental apparatus for measuring water entrainment was employed in order to understand the characteristics of operation conditions, such as inlet distance, air pressure of nozzle, diameter of nozzle, and so on. In addition, experiments for measuring the thrust in the condition of ejector were executed with horizontal type apparatus in water. However, the influence of the ejector to improve thrust can't have been recognized with high-pressure air at room temperature yet.

1. INTRODUCTION

The idea of using high-pressure gas as a driving force has been proposed by Witte (1969) and Mottard et al. (1961) since the old days. However, there has been no report in which the idea is applied to the practical use.

On the other hand, water jet promotion obtains the thrust by spouting the water, which is pumped up from a bottom of ship or from a rear nozzle, directly with high-pressure pump or screw. Recently, it has actually been used widely, such as the water motorcycle, the high-speed patrol boat, the passenger ship and so on. It is conceivable that the range of its application will spread out more if high-pressure gas can be used instead of the high-pressure pump and the screw. However, it is still difficult to transmit kinetic energy of the high-speed gas to water efficiently. Tsutahara et al. (2000) has actively been working on the research of the nozzle for the ship with high-speed gas used. It is also reported that efficiency is raised when steam is used instead of the high-pressure gas(Tsutahara et al., 2004). Thereupon, BJE aims at the high increase in its efficiency by transforming energy of the high temperature gas into kinetic energy of water using the high-pressure combustion gas with an ejector.

Balamurugan et al.(2007) show the behavior inside the ejector where a liquid mixes with the gas with simulation, and they investigate pump ability of the high-pressure gas experimentally in water pumping experiment.

This research examines the ejector performance in various experimental conditions empirically and aims at the commercialization of BJE.

2. Model of BJE

2.1 Structure

Ejector is the element in which gas is mixed with water, and the improvements in the propulsion efficiency can be realized by taking as much water as possible into the ejector. Figure 1 shows the schematic of BJE model. BJE is composed of fuel, a combustion vessel and an ejector. Hydrocarbon such as propane and butane is used as fuel. First, oxygen and fuel are burned in the combustion vessel. The combustion gas introduces seawater into an ejector through seawater intake. For this study, the ejector is the most important element where mixture of seawater and conversion gas form gas-liquid phase called bubble jet, and gives thrust by spouting.

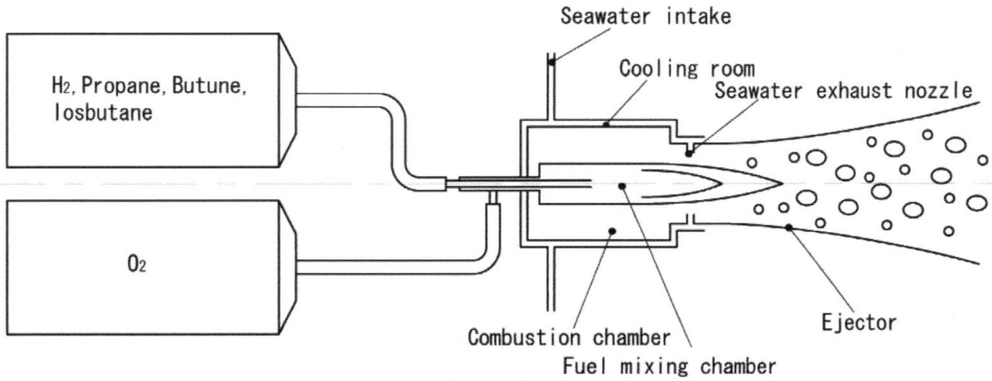

Fig. 1 Schematic of BJE model

2.2 Theory of BJE

This bubble jet engine system is expected to have high efficiency to transform combustion energy into the kinetic energy directly, reducing mechanical loss. The characteristics may be classified into three advantages under the following heads; (1) Improvement of thrust; (2) Improvement of propulsive efficiency; and (3) Transformation from heart energy to kinetic energy;

The first advantage is that improvement of trust to reduce loss diffused by attaching an ejector. The second is that the propulsive efficiency becomes predominance when the velocity of the gas spouted from nozzle is the same as the navigation speed of hull. The ejector can fully slow down the high-speed gas as bubble jet flow. The third is the possibility that thermal energy changes into the kinetic energy when the gas of 500℃ or higher is mixed with water in ejector.

This paper, however, discusses just (1) the improvement of thrust. Figure 2 shows schematic of jet from the nozzle without ejector and with one. Without ejector, the high-speed combustion gas is spouted out directly into water, therefore the law of conservation momentum is:

$$\dot{m}_{g0}v_{g0} = \dot{m}_{g1}v_{g1} + \Delta L_1 \qquad (1)$$

and with ejector:

$$\dot{m}_{g0}v_{g0} = \dot{m}_{j2}v_{j2} + \Delta L_2 \qquad (2)$$

where $\dot{m}_g$: mass flow rate of gas[kg/s], v_g: velocity of gas[m/s], $\dot{m}_w$: mass flow rate of water[kg/s], v_w: velocity of water[m/s], $\dot{m}_j$: mass flow rate of bubble jet[kg/s], v_j: velocity of bubble jet[m/s],

$$\dot{m}_{j2}v_{j2} = \dot{m}_{w2}v_{w2} + \dot{m}_{g2}v_{g2} \qquad (3)$$

The thrust of the purity net which can be taken out is $\dot{m}_{g1}v_{g1}$ and $\dot{m}_{j2}v_{j2}$, and it can be thought that the loss ΔL_1 without the ejector grows bigger than the loss ΔL_2 with the ejector. In other words, regarding the reduction of loss, the result is more favourable when the bubble jet hits water than when the light gas hits directly heavy water. Therefore, the possibility of improvement of thrust can be expected.

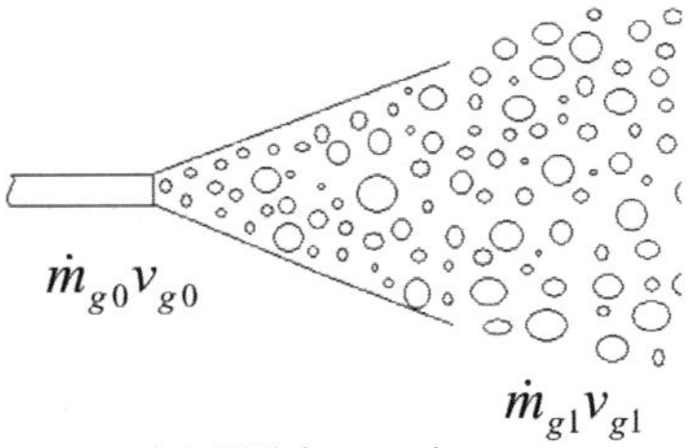

(a) Without ejector

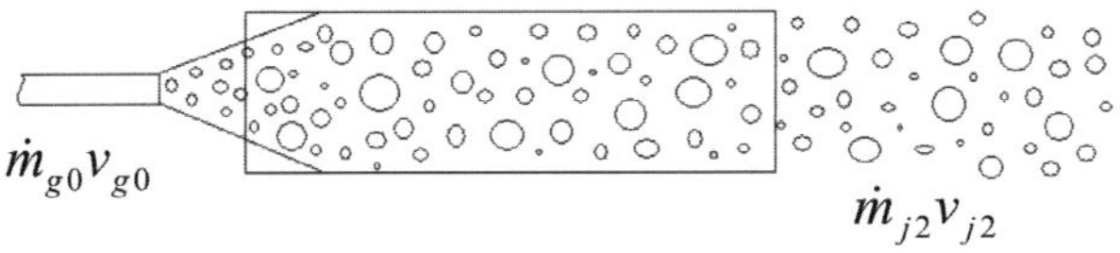

(b) With ejector

Fig. 2 Schematic of jet

3. EXPERMENT

3.1 WATER PUMPING TEST

3.1.1 Vertical type gas-water ejector experimental apparatus

Vertical type gas-water ejector experimental apparatus, which is adopted to understand the characteristics of the ejector, is shown in Fig. 3. That is composed of a water tank (diameter: 800 mm, height: 500 mm), a nozzle and an ejector. The air compressed with a compressor instead of the combustion gas is used in this experiment. After it spouts from the nozzle and passes inside the ejector with water, it is sent to void fraction measurement part (Fig.4). The void fraction measurement pipe has two valves which open and close simultaneously.

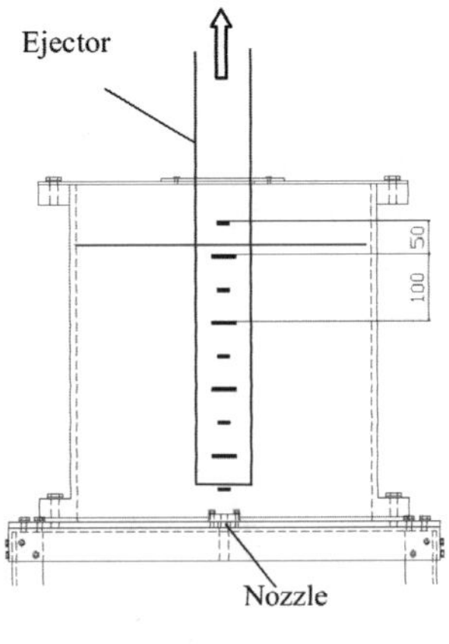

Fig. 3 Vertical type gas-water ejector experimental apparatus.

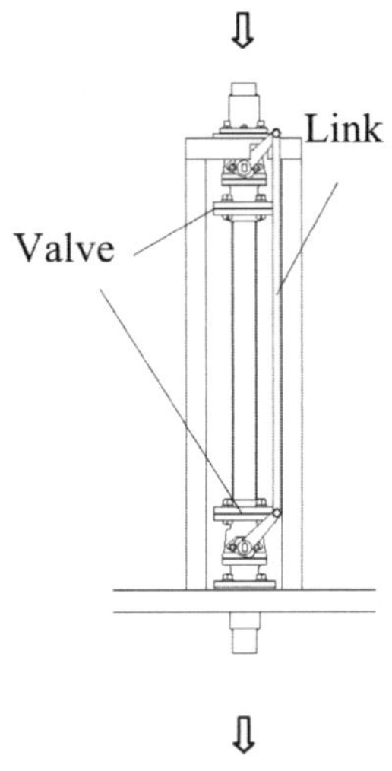

Fig. 4 Void fraction measurement pipe

3.1.2 Experimental method and condition

The detail of the inlet of the ejector is illustrated in Fig. 5. The experimental conditions are summarized in Table 1.

To investigate the performance of the ejector, the mass flow rate and velocity of the water introduced into the ejector have to be measured. In fact, however, the measuring the mass flow rate and the velocity of water is very difficult. Therefore, the void fraction was adopted, and it enabled us to do our estimation in this experiment.

(1) Estimation of mass flow rate

In this apparatus, because water to keep water level is not supplied, the mass flow rate that passes inside ejector is not steady, and it changes depending on level of water tank. However, because relation between mass flow rate and water level is linear as shown in Fig. 6, the mass flow rate at optional water level can be gained easily.

(2) Estimation of velocity

Flowing water was trapped by closing the valve of void fraction measurement pipe when water level passed at the water level 450 mm(at 50 mm underneath from water surface) of the water tank, and the void fraction can derivate from space volume. Therefore the water level 450 mm is considered criterion.

Velocity of water can be written by the mass flow rate and the void fraction as next equation. That is:

$$v_w = \dot{m}_w / A_\Phi \rho_w (1-\alpha) \qquad (4)$$

$\dot{m}_w$:mass flow rate of water[kg/s] A_Φ : cross section area of flow pipe[mm^2]
ρ_w :density of water[kg/m^3] α :void fraction

where, the influence of the gravity must be considered when water runs down in flow pipe or when it rises [Appendix 1]. However, the influence of this gravity can be ignored by positioning the void fraction measurement pipe to the same height of ejector inlet. Then, the friction loss between flow water and inside wall of the pipe is not considered at all.

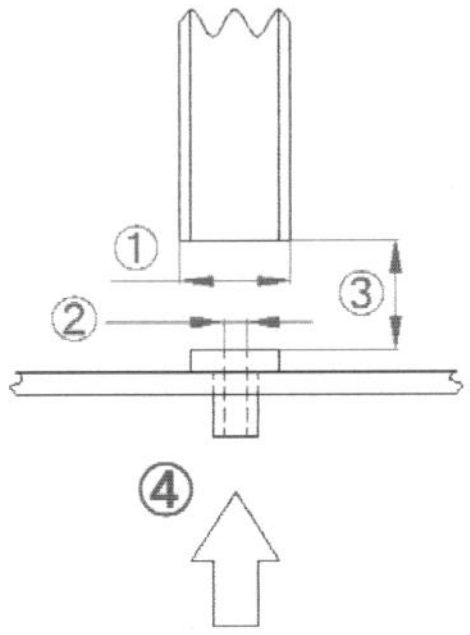

Fig. 5 Detail of the inlet of the ejector.

Table.1 Experimental condition

①	Nozzle diameter d[mm]	5, 8, 10
②	Ejector diameter Φ[mm]	20, 40, 60
③	Ejector Inlet distance, H[mm]	5, 10, 20, 30
④	Air pressure, p[MPa]	0.3, 0.4, 0.5

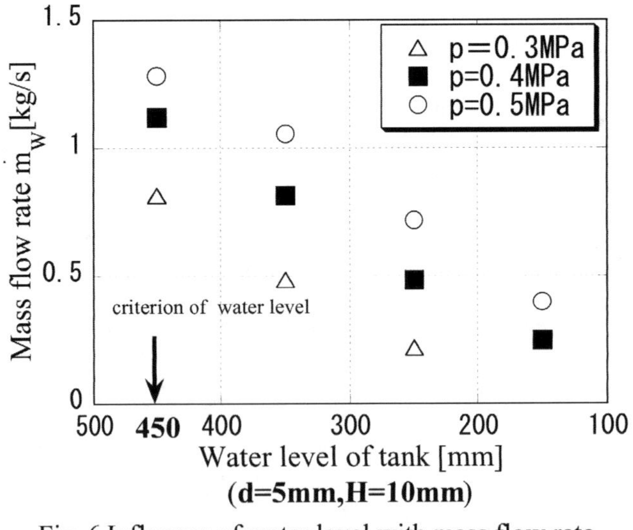

Fig. 6 Influence of water level with mass flow rate

3.2 THRUST MEASUREMENT

3.2.1 Horizontal type thrust measurement apparatus

The schematic diagram of horizontal type thrust measurement apparatus in water is shown in the Fig. 7. This system is composed of the chamber of the cylinder type to which the high-pressure air of the compressor is being supplied, the nozzle and the ejector, and they are submerged in water. The high-pressure air, which is mixed with water in the ejector, spouted from the nozzle, and it is exhausted as a bubble jet outside the ejector. Load cell installed on the guide rail detects the thrust.

3.2.2 Experimental method and condition

Figure 8 shows the enlarged view of the ejector inlet. Experiment parameters are ① nozzle diameter d, ② inlet distance H, ③ ejector length L, and ④ air pressure p, and they are shown in table 2 as the details. The experimental method is to measure the thrust in different conditions. In this experiment, the ejector diameter Φ used is only Φ 60 mm, which was conducted from the result of water pumping test.

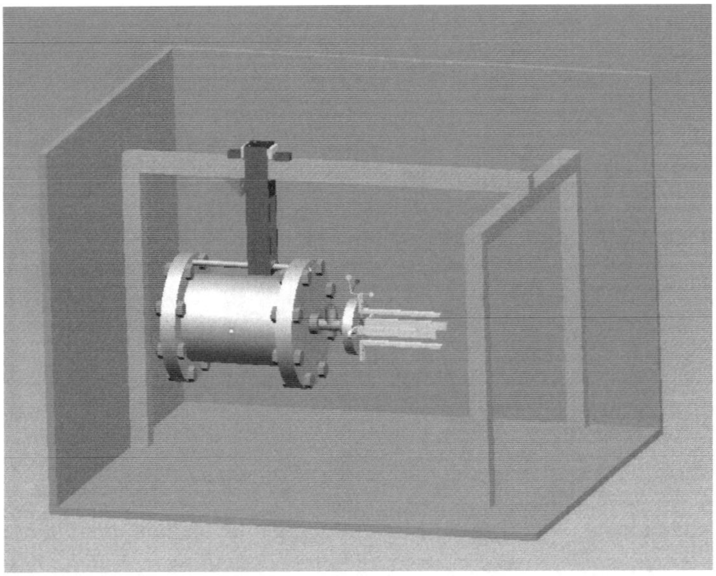

Fig. 7 Horizontal type thrust measurement apparatus

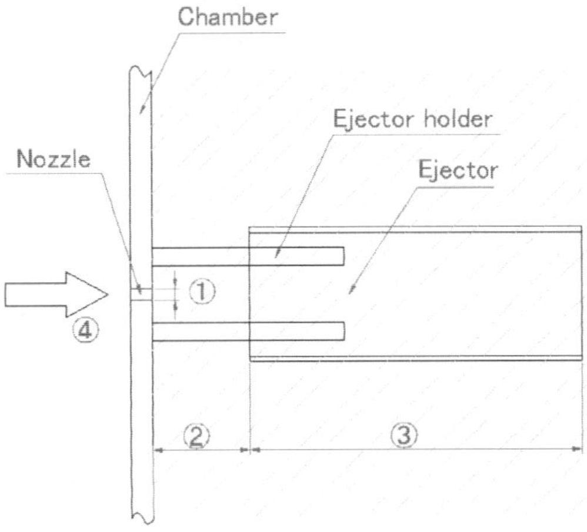

Fig. 8 Enlarged view of the ejector inlet

Table 2 Experimental condition

①	Nozzle d[mm]	3, 4, 5, 8, 10
②	Ejector inlet distance H[mm]	0, 10, 20, 30, 40, 50
③	Ejector length L[mm]	100, 200, 300, 400, 500
④	Air pressure p[MPa]	0.3, 0.4, 0.5

4. RESULTS AND DISCUSSION

4.1 Results of water pumping test

(1) Mass flow rate of water

The relation of mass flow rate versus inlet distance H is shown in Fig. 9(d=5mm). Each data shows the average value of three experiments in the same condition. The mass flow rate increases with increasing the inlet distance below H=20, and doesn't almost change in the range of above H= 20mm. There was the same trend in other nozzle diameters. It also increased according to the increase of pressure with all nozzle diameters.

(2)Void fraction α

A change in void fraction with a nozzle diameter d=10mm is shown in Fig. 10. Major changes could not be observed regardless of each nozzle diameter, and it is about 0.8～0.9. The void fraction decreased accordance to the increase of pressure with all nozzle diameters.

(3) I/O ratio η

The role of ejector is to give water energy, decreasing velocity of gas spouted from nozzle fully. Now, The input force F_{in} that acts on the ejector and the trust F_{out} of water exhausted from ejector are defined as follows:

$$F_{in} = A_d p \quad (5)$$

$$F_{out} = \dot{m}_w V_w \quad (6)$$

where, A_d ; cross section area of nozzle[mm^2], p ;air pressure[Pa],

Relations between F_{in} and F_{out} in various experimental conditions are shown in the Fig. 11. It is found that output F_{out} increases with the rise of pressure in any nozzle diameter. The result plotted this output F_{out} as I/O ratio $\eta = F_{out}/F_{in}$ is shown Fig. 12. The I/O ratio in a nozzle diameter 5mm is very high with every inlet distance condition. Especially, it is the highest with inlet distance 20mm at 0.5MPa in air pressure.

Thus, viewed in the above result, the high I/O ratio could be realized if the nozzle diameter 5mm in the range of this experiment condition was adopted.

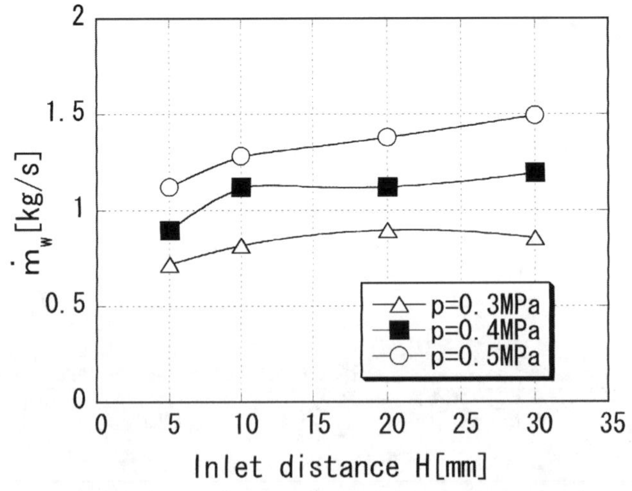

Fig. 9 Mass flow rate versus inlet distance H(d=5mm)

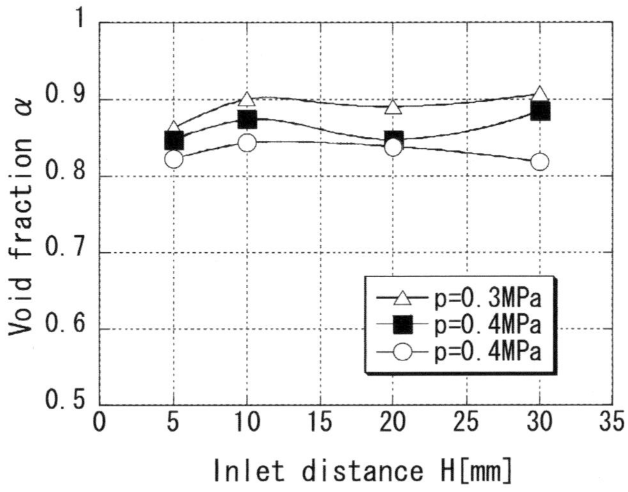

Fig. 10 Void fraction versus inlet distance H(d=10mm)

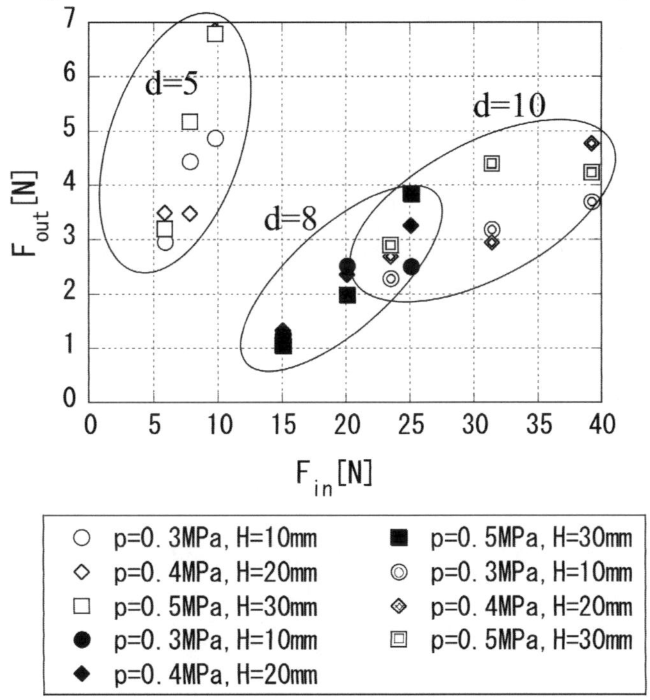

Fig. 11 Relations between F_{in} and F_{out} in various experimental conditions

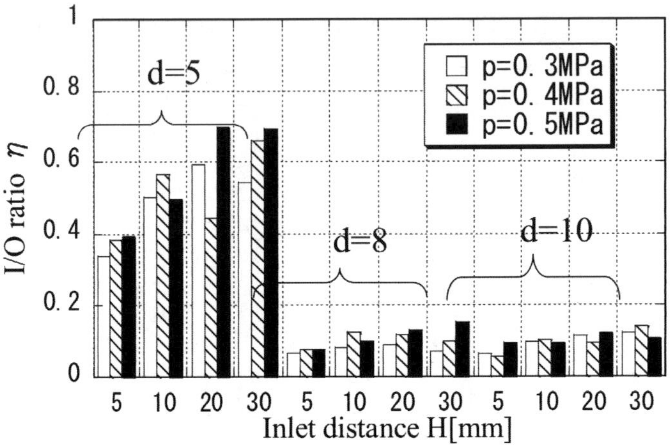

Fig. 12 I/O ratio η in water pumping test.

4.2 Results of thrust measurement

(1) Without ejector

Thrust measurement in water without installing an ejector was conducted to understand nozzle performance. The thrust F versus nozzle diameter d in case of air pressure p=0.5MPa without ejector is shown in Fig. 13. Table 3 summaries volume flow rate in each nozzle. The thrust of the nozzle of d= 5mm is the biggest corresponding to the volume flow rate of air. This data supports the results of water pumping test. Figure 14 shows influence of nozzle pressure on thrust. The thrust increases with increasing air pressure.

(2) With ejector

The relation of thrust for inlet distance H is shown the in Fig. 15(air pressure=0.5MPa, ejector diameter Φ60m). The thrust in the range of below inlet distance H=20mm increases. On the contrary, it is steady above H=20mm and for example when L=200; F=9.6N. This trend agree with the result of water pumping test; the mass flow rate of exhaust water from ejector didn't increase in the rage of above H=20mm.

Now, Figure 15 replotted the trust at H=30mm in Fig.16. Horizontal axis shows ejector length L. The trust is about 9.8N below L=200, and especially at L=0, so without ejector, it is the highest 9.9N. Also it rapidly falls right down above L=200mm because of friction loss and buoyant of air in ejector. Following what has been found, it can be said that improvement of thrust regarding ejector could not be achieved.

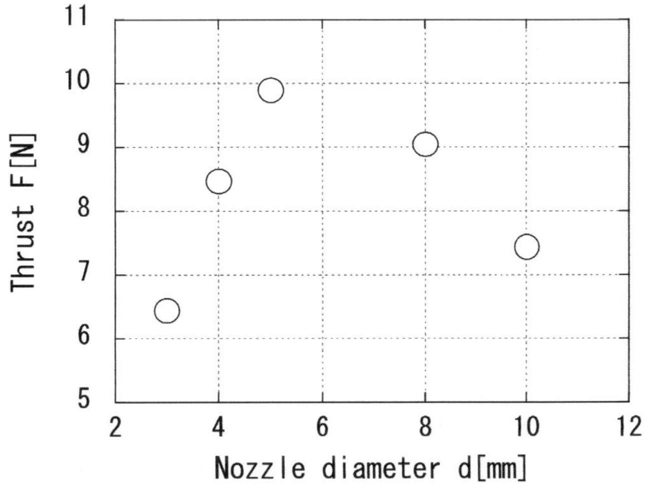

Fig. 13 Influence of nozzle diameter on thrust(p=0.5MPa)

Table.3 Volume flow rate of each nozzle

Nozzle diameter [mm]	3	4	5	8	10
Volume flow rate Q[l/s]	4.85	8.28	12.47	10.44	9.43

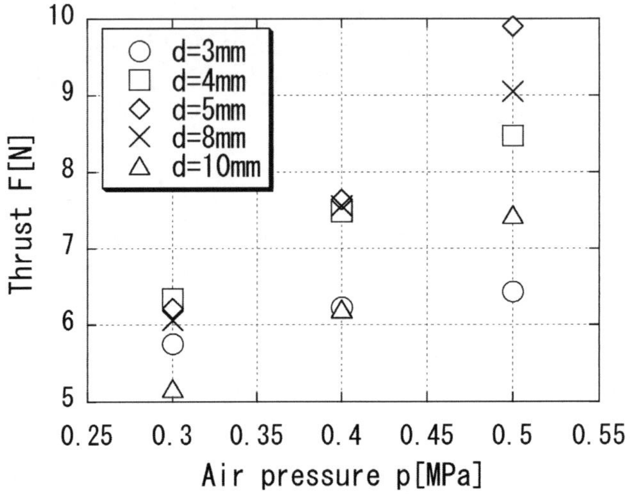

Fig. 14 Influence of air pressure on thrust

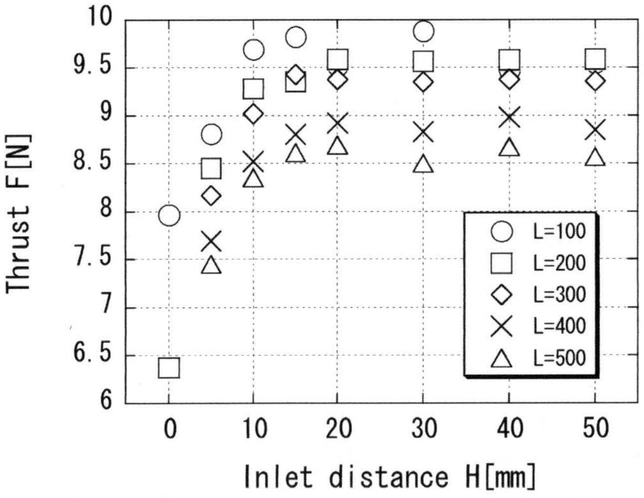

Fig. 15 Influence of ejector inlet distance on thrust

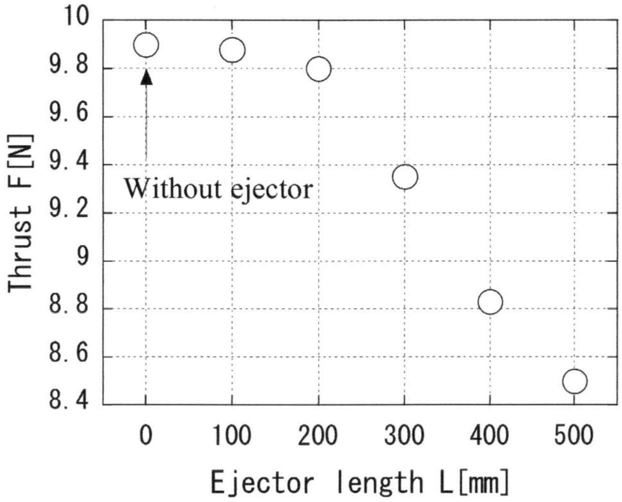

Fig. 16 Thrust at inlet distance H=30mm in various ejectors

5. CONCLUSIONS

In order to develop the bubble jet engine that is a new promotion system, the fundamental experiment was conducted and its practicality was estimated. The exhausted mass flow rate and velocity of water were achieved in water pumping test, and that output could be found.
A horizontal type thrust measurement apparatus was done, and the influence of the ejector on thrust was investigated in various experimental conditions. As a result, at present, improvement in the thrust by putting an ejector cannot be gained.

REFERENCES

Balamurugan, S., Gaikar V.G and Patwardhan, A.W. (2007). "Effect of ejector configuration on hydrodynamic characteristics of Gas-liquid ejectors," Chemical Engineer Science, doi:10.1016/j.ces. 10.002

Mottard, E. J. and Shoemaker, C. J., (1961), "Preliminary Investigation of an Underwater Ramjet, "NASA Tech. Note, D-991, pp.1-36.

Tsutahara M . Kimura T., Ogawa K., and Umeda S., (2000)," Performance Improvement of Ship Propulsion Equipment Driven by Injecting Moist Air," JSME Int. J. Series B. Vol 43, No.3 pp.427-433.

Tsutahara M. et al., (2004), "The Experimental Study of Flow in the Nozzle for Ships Population Directly Driven by High Pressure Gus," Mechanical Engineering Congress 2004 Japan, pdf.

Witte, J. H. (1969), "Predicted Performance of Large Water Ramjets", AIAA Paper, No.69-406, pp.1-10

Journal of Physics: Conference Series **147** (2009) 012021 doi:10.1088/1742-6596/147/1/012021

APPENDIX 1

Calibration of influence of the gravity on mass flow rate

Figure 17 indicates the difference in the velocity in case that gravity acts and doesn't. When the mass flow rate is found with void fraction measurement pipe, it is estimated lower than actual because of gravity. Specifically, the water flowed down through flow pipe due to the gravity is accelerated (Fig. 18(b)). The formula in acceleration in consideration of the influence of the gravity is:

$$v_a = v_0 + gt_1 \qquad (7)$$

$$h = v_0 t_1 + \frac{g{t_1}^2}{2} \qquad (8)$$

where, v_a : velocity in consideration of the gravity, v_0 : velocity when gravity isn't taken into consideration, g : gravity acceleration, : h :length of void measurement pipe.

Since v_a and h are well known here by equation (3), v_0 and t_1 can be looked for by the coalition. Next, $t_2 = h / v_0$, which is time to be necessary for water to fall in uniform velocity, is substituted for equation (8). Thus:

$$h_a = v_0 t_2 + \frac{g{t_2}^2}{2} \qquad (9)$$

Therefore, the mass flow rate in acceleration is modified to the mass flow rate in uniform velocity by finding out h_a.

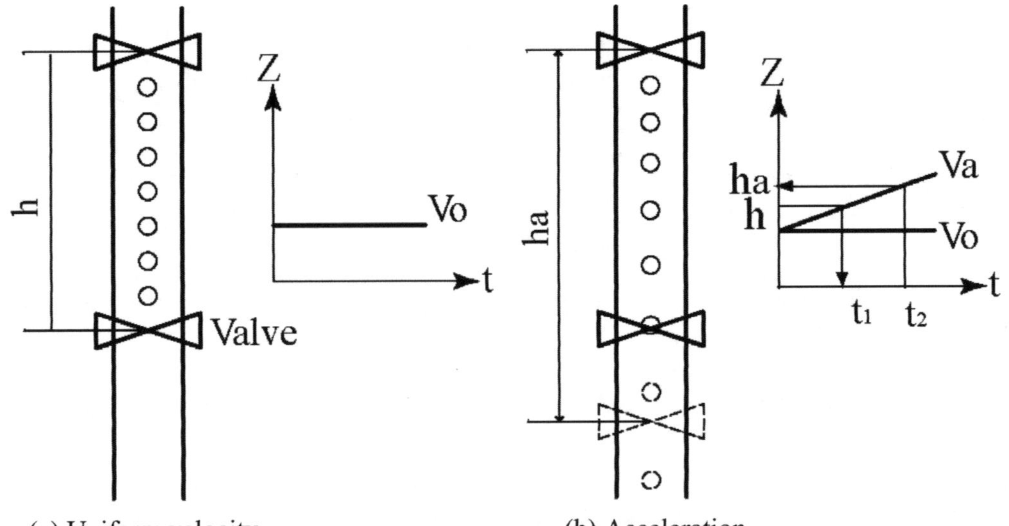

(a) Uniform velocity (b) Acceleration

Fig. 17 difference from the mass flow rate in uniform velocity and in acceleration

Journal of Physics: Conference Series **147** (2009) 012022 doi:10.1088/1742-6596/147/1/012022

The Multiphase Flow System Used in Exploiting Depleted Reservoirs: Water-based Micro-bubble Drilling Fluid

ZHENG Li-hui [1], HE Xiao-qing [1], FU Li-xia [2], Wang Xiang-chun [1]

[1] MOE Key Laboratory of Petroleum Engineering, China University of Petroleum, Beijing 102249, CHINA
[2] Production research institute, Huabei petroleum branch company, Renqiu CHINA

hexiaoqing2006@126.com

Abstract. Water-based micro-bubble drilling fluid, which is used to exploit depleted reservoirs, is a complicated multiphase flow system that is composed of gas, water, oil, polymer, surfactants and solids. The gas phase is separate from bulk water by two layers and three membranes. They are "surface tension reducing membrane", "high viscosity layer", "high viscosity fixing membrane", "compatibility enhancing membrane" and "concentration transition layer of liner high polymer (LHP) & surfactants" from every gas phase centre to the bulk water. "Surface tension reducing membrane", "high viscosity layer" and "high viscosity fixing membrane" bond closely to pack air forming "air-bag", "compatibility enhancing membrane" and "concentration transition layer of LHP & surfactants" absorb outside "air-bag" to form "incompact zone". From another point of view, "air-bag" and "incompact zone" compose micro-bubble. Dynamic changes of "incompact zone" enable micro-bubble to exist lonely or aggregate together, and lead the whole fluid, which can wet both hydrophilic and hydrophobic surface, to possess very high viscosity at an extremely low shear rate but to possess good fluidity at a higher shear rate. When the water-based micro-bubble drilling fluid encounters leakage zones, it will automatically regulate the sizes and shapes of the bubbles according to the slot width of fracture, the height of cavern as well as the aperture of openings, or seal them by making use of high viscosity of the system at a very low shear rate. Measurements of the rheological parameters indicate that water-based micro-bubble drilling fluid has very high plastic viscosity, yield point, initial gel, final gel and high ratio of yield point and plastic viscosity. All of these properties make the multiphase flow system meet the requirements of petroleum drilling industry. Research on interface between gas and bulk water of this multiphase flow system can provide us with information of synthesizing effective agents to enlarge the application of micro-bubble in petroleum industry.

1. INTRODUCTION

Mr Sebba (1987) found a kind of special biliquid foams called Aphron. It was Brookey (1998) who first introduced aphron into petroleum drilling industry and renamed aphrons as "micro-bubbles". Up to now, water-based micro-bubble drilling fluid (WMDF) has been drilled thousands of depleted reservoirs without any lost circulation problem. Laboratory experiments indicate that WMDF has very little spurt loss (Tom Brookey, 1998), has very high yield stress at extreme low shear rate, and can accumulate energy (C.D. Ivan et al., 2002). Many researchers think that micro-bubbles have the same structure as aphron (F.B. Growcock et al., 2003). But

Journal of Physics: Conference Series **147** (2009) 012022 doi:10.1088/1742-6596/147/1/012022

the recent research on the structure of "micro-bubble" indicates that it is slightly different from the structure of aphron because gas, water, oil, polymer, surfactants and solids exist in WMDF at the same time which would affect the structure. And it is the real structure of micro-bubble that explains mechanism for lost circulation controlled by WMDF and enables the drilling fluid system with good rheological properties to meet the requirements of petroleum drilling industry.

2. STRUCTURE

According to Sebba (1987), aphron is a soppy shell absorbing some surfactants staying in bulk water, as fig.1 shows. Different from aphron, except surfactants there are still some linear high polymers (LHP) and solids in bulk water, so micro-bubble in WMDF should have the structure as fig.2 shows.

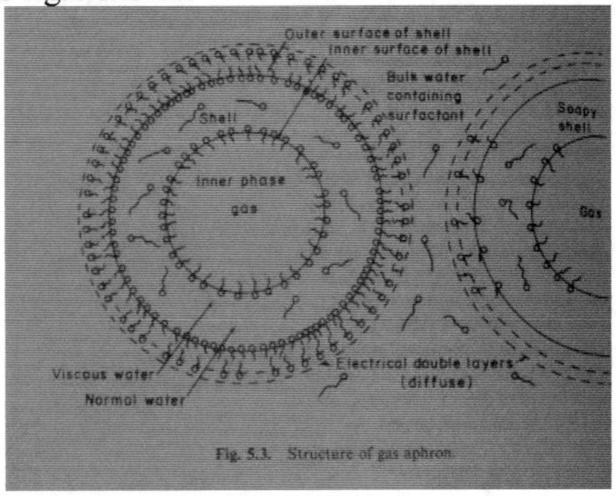

Fig. 1 Structure of aphron

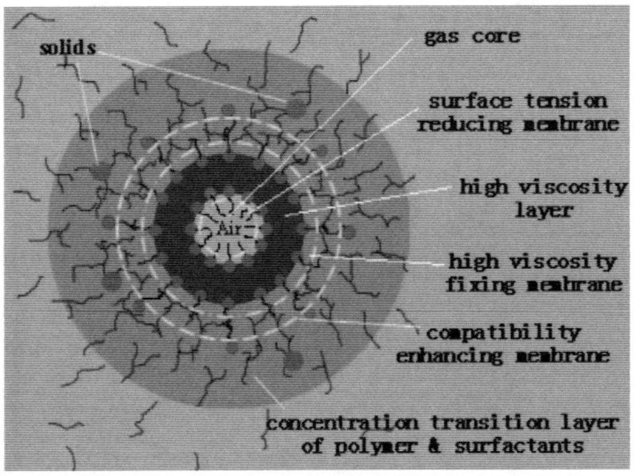

Fig. 2 Structure of micro-bubble

Fig.2 reveals a picture on which gas is wrapped by an aqueous coat that holds surfactants on both sides. The inner side surfactants point their hydrophobic ends to the gas core while hydrophilic ends residing within the aqueous coat. By contrast, the outer side surfactants have their hydrophilic ends reside within the aqueous coat while the hydrophobic ends stretching to the bulk water. LHP and surfactants in bulk water adhere outside the outer surfactants closely through van der Waals force forming a layer which has much higher concentration. With the attractive force reducing, concentration of LHP and surfactants decreases gradually. Little by little, concentration goes back to the same level as it is in bulk water after a few distances from the outer surfactants. In addition, some solids mixed with LHP appear in the higher concentration layer.

The gas, which is wrapped by aqueous coat and lying in the center of the micro-bubble, looks like a core. We name the core "gas core" and call the aqueous coat "high viscosity layer", for it is a special phase which has much higher viscosity than bulk water. The inner side surfactants, which is used to reduce the surface tension between gas and water, is named as "surface tension reducing membrane", while the outer side surfactants, which is used to keep the high viscosity of aqueous coat, is called "high viscosity fixing membrane". The closely adsorbed LHP and surfactants form a layer which is called "compatibility enhancing membrane", for they can enhance the compatibility of micro-bubble and bulk water, while the loosely adsorbed LHP and surfactants, which have changeable concentration and indefinite radium, are dimmed as "concentration transition layer of LHP & surfactants". "Gas core", "surface tension reducing membrane", "high viscosity layer" and "high viscosity fixing membrane" often work as a unit for "surface tension reducing membrane", "high viscosity layer" and "high viscosity fixing membrane" connect with each other through hydrogen bond, which is a relatively strong bond and will not be destroyed easily. We name the unit "air bag". However, it is the LHP and surfactants that form "compatibility enhancing membrane" and "concentration transition layer of polymer & surfactants". They connect with each other through van der Waals force and sensitization, which are relatively weak bond forces. So, the radium of the two parts can be changed easily and even these two parts can be destroyed by the outside force like diffusive force which makes the boundary between two parts indistinct. For the sake of decscribing easily, we merge the latter two parts as another unit and name it "incompact zone".

3 CHARACTERS

Aphron is a double layer micro-bubble which is weakly bonded by a layer of surfactants while the micro-bubble in WMDF is an "air-bag" wrapped by an "incompact zone". On the one hand, "incompact zone" wraps the "air-bag" and makes it water wetted, and keeps the whole system stable for a period of time. On the other hand, LHP and surfactants in "incompact zone" can diffuse into the bulk water while the opposite process is also realizable. Hence, "incompact zone" can separate from "air-bag" easily by the outside force and that will diminish air-bag's water wettability but enhance its hydrophobic ability. The dynamic changes of "incompact zone" lead the WMDF system to have its macro characteristics changing. These changesable characters will meet the requirements of petroleum drilling.

3.1 Micro-bubble's Existence and Its Wetting Behavior

Observing the micro-bubbles in WMDF through microscope, Growcock (2003) found that micro-bubbles can separate into bulk water automatically. He also discovered that several micro-bubbles may assemble together but perform as an individual micro-bubble. Experiments also show that micro-bubbles can form a bubble band if there is a pressure difference (F.B. Growcock, A. Belkin, M. Fosdick et al., 2006). By directly examining the wetting behavior of the oil and WMDF, F.B. Growcock and A. Belkin (2006) have found that whether the oil is applied to the drilling fluid or the fluid applied to the oil, the applied phase spreads over the substrate phase, i.e. the contact angle is very low(<5degrees).

These phenomena have close relations to the structure of micro-bubble. As fig. 2 shows, the hydrophilic ends of surfactants fixed in the "compatibility enhancing membrane" often have charges, which will cause electrostatic repulsion. Additionally, the viscoelasticity of "incompact zone" will not allow micro-bubbles in the WMDF to attach together but separate into bulk water automatically. The micro-bubbles were only made up of "air-bags" if the "incompact zones" were lost. That so many hydrophobic tails point to the outmost side will make the WMDF compatible with oil. Sometimes these bubbles can adsorb other micro-bubbles of the same structure and form congeries in different sizes. If the congeries only consist of several micro-bubbles, they can adsorb some LHP and surfactants in the bulk water to reconstruct the "incompact zones", then perform as an individual micro-bubble. Although the congeries have more than one "gas core", we still say that they are of the structure of "one core, two layers and three membranes" for the sake of describing conveniently. However, if the congeries consist of quite a large number of micro-bubbles and the pressure differences exist, congeries will expand and aggregate quickly toward the direction of pressure drop, thus forming a bubble band in the front of the fluid.

The micro-bubbles in WMDF will lose their "incompact zones" in the following three conditions. First, when WMDF moves at a very high speed, "incompact zones" will be held up in the bulk water for they have very good solubility and can be easily separated from "air-bags". this can be seen when WMDF flows through drill pipes. Second, when passing the small openings which are smaller than micro-bubbles' size, micro-bubbles will change their shapes or be cut into pieces and then lose "incompact zones". When WMDF passes through fine screen, lineal split or narrow pore throat belongs to the case mentioned above. Third, if micro-bubbles expand sharply in a split second, concentration of LHP and surfactants in "incompact zones" will shrink to the level as it is in bulk water. This will happen when WMDF encounts large fractures or big caverns.

3.2 High Yield Stress Shear Thinning Property

Compared with bulk water, "incompact zones" hold a much higer concentration of LHP and surfactants, so they have more complicated networks. If networks are destroyed, reconstruction of "incompact zones" will be much quicker than in bulk water. WMDF has very high yield stress at an extreme low shear rate since networks in "incompact zones" will be damaged first, then reconstructed. With the shear rate increase, networks will be broken up more quickly until the rate of broke and reconstruction reach a balance in "incompact zones" and bulk water. If the shear rate is high enough, the WMDF will have very good fluidity because the rate of destruction of network is far quicker than reconstruction and LHP will improve the fluidity. What's more, motived micro-bubbles will decrease viscosity of fluid (Sebba, 1987). Tom Brookey (1998) had determined the viscosity of WMDF and base fluid at different shear rates and had found the phenomenon described above.

4. MECHANISM FOR CIRCULATION CONTROL

Micro-bubbles can change their structures when passing openings of different sizes and that's why WMDF can control circulation loss in different conditions. Micro-bubbles in WMDF are composed of "air-bags" and "incompact zones" at surface rightly after preparation. When the WMDF is pumping from mud pit to well, "incompact zones" of micro-bubbles will be damaged by high shear rate caused by flow, and will leave "air-bags" alone in bulk water. However, there are still a number of LHP and surfactants in bulk water, these "air-bags" will not attach to each other to form aggregation but to keep independently and flow with the WMDF if there is no pressure difference. With the depth increasing, "air-bags" will be compressed and will accumulate more energy because of the increase of pressure and temperature. Professor F.B. Growcock (2006) found that micro-bubbles can exist at 3000psi for a long time. Compressed micro-bubbles will expand to normal size and reconstruct their "incompact zones" when coming back to surface through annular if no lost circulation zones are encountered during the whole process. Otherwise, micro-bubbles will perform quite differently in accordance with different conditions as following description.

4.1 Encounter Large Fracture or Big Cavern

Take the large fracture for example, considerable pressure differences exist between annular and lost circulation zones and that will lead the micro-bubbles to expand sharply and move quickly to lower pressure areas, causing micro-bubbles to lose their "incompact zones" completely and aggregate together as a large unit of bubbles band blocking between annular and large fracture.
Bubbles' flow appears to follow conventional Navier-Stokes theory (L.D. Landau ex tl., 1995). For a rigid sphere in a fluid is only affected by one-dimensional pressure gradient, $\triangle P/L$, the relative velocity of the bubble in an infinitely wide conduit is $V = 0.23\ r^2/\eta * \triangle P/L$, where r is bubble radius and η is the fluid viscosity. In this theory, large size bubbles will move faster and take up the front area. According to Laplace equation, the frontal bubbles will expand till there is a pressure balance between the front bubbles and formation. Anterior bubbles will restrain the latter bubbles' expanding and reduce their moving velocity gradually. Little by little, bubbles near annular will not need to expand any more, and pressure between these bubbles and WMDF in annular will reach a balance. At that point, WMDF accomplishes the task of controlling circulation lost, and then goes upward to surface as normal. However, if the frontal bubbles are broken into pieces in the process of expanding, latter bubbles will go on expanding and blocking in the front until no bubbles need to expand any more. Just as fig. 3 shows.

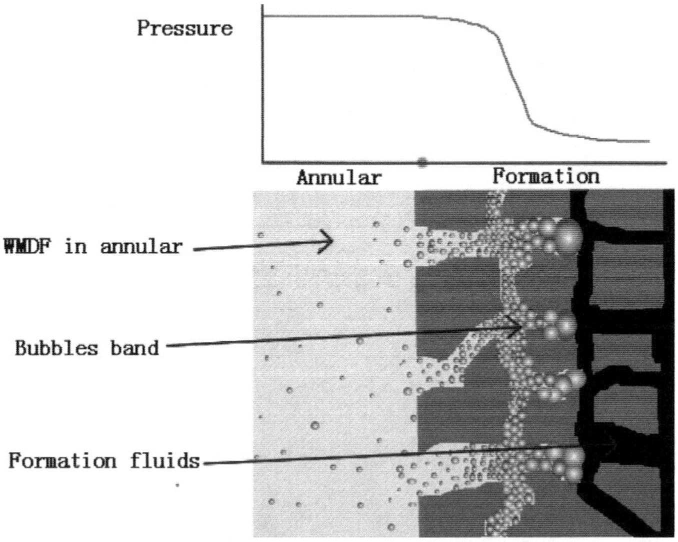

Fig. 3 Schematic of micro-bubble bridging macro-fracture

Actually, it is the "air-bags" that block the leakage zone while the WMDF is circulating. Because most of the formation rocks are water wetted, these hydrophobic "air-bags" will not stick to rocks. So, water in "high viscosity layer" will not enter into formation or enter in at a very low rate. Thus WMDF is able to circulate normally while micro-bubbles are blocking in the zone.

The first application of WMDF in West Texas (Tom brookey, 1998), where the bit dropped one foot, is very successful because the circulation can still be regained. Some laboratory experiments had been done to confirm the correctness of the explanation above. First of all, we design a specific formula using multiple regression method (Zheng lihui, 2008). Then test its sealing ability by measuring the filtration rate in 10 minutes conduct with sands of 80 screen mesh. All the results are put into table.1. From table.1 we can see that compared with the ordinary drilling fluid, WMDF can minimize the filtration rate effectively. With the pressure difference increase, its sealing ability will improve.

Table.1 sealing effect of WMDF

Pressure difference (MPa)	0	0.3	0.6	0.9	1.2
Ordinary drilling fluid (mL)	160	8000	10500	12000	13400
WMDF (mL)	0	4	3	0	0

Fig.4 is a real picture of the sealing experimental result of WMDF. We can clearly see that there is a departure of sands between bubble bands sealing region and remainder sands after the pressure difference is released.

Fig. 4 A real picture of sealing test

4.2 Encounter Rimula or Narrow Throat

Micro-bubbles will lose their "incompact-zones" while expanding rapidly under positive differential pressure. According to Navier-Stokes' theory, the larger-sized micro-bubbles will come to the leakage zone first. These bubbles will change their shapes and sizes properly in order to enter into the throat. Sometimes, micro-bubbles can enter into the throat directly if the positive differential pressure is not large enough to destroy the bubbles. Hydrophobic "air-bags" in throat will cause capillary force which will prevent following micro-bubbles from expanding and will slow down their velocity. Accumulation of the capillary force will create great residence that will stop WMDF from coming into the zone. When the rate slows down, "incompact zones" will be reconstructed and viscosity of the system will increase sharply. High yield stress at an extremely low shear rate will also prevent WMDF from entering, as fig. 5 shows.

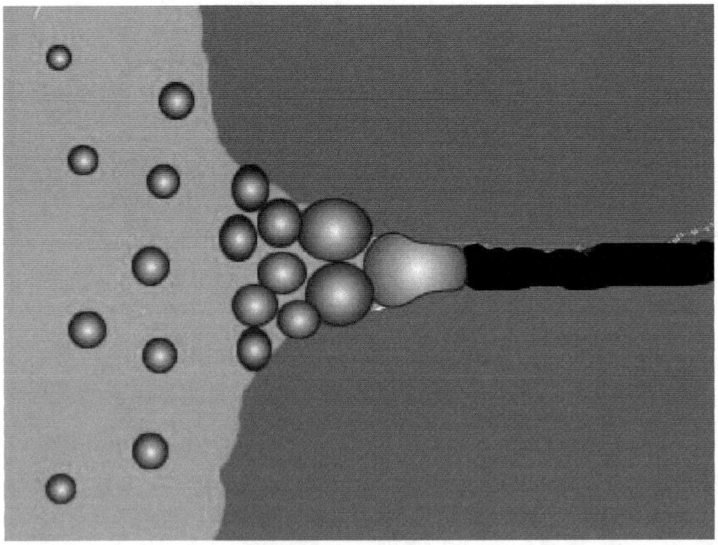

Fig. 5 Schematic of Micro-Bubble sealing narrow pore throat

However, if the positive differential pressure is very large, micro-bubbles will be destroyed or cut into fines before entering into the formation until the size is suitable to seal the throat. The mechanism for sealing is also the process of using capillary force mentioned above.

4.3 Encounter Percolation Zone

Though micro-bubbles can expand for bridging large fracture and sealing narrow throat, the expanded micro-bubbles can not enter into the percolation zone because the bubble size is too large compared with the percolation zone aperture. Fortunately, a great amount of polymer that creates high viscosity exists in the system. On the one hand, polymer slows down the rate of penetration by high viscosity. On the other hand, polymer will form the mud cake which effectively prevents WMDF from further entering.

In summary, micro-bubbles will adjust the size, shape or structure automatically according to the slot width of fracture, height of cavern or the aperture of openings to control lost circulation and help formation bearing larger pressure. While drilling a well with multiple pressured sequences, we can choose the WMDF system for the highest pressure zone. Bubbles will firstly seal the leakage zone which owns highest pressure difference with drilling fluid hydrostatic column. For the sealing bubbles are compressible and deformable, they can help sealed zones to bear higher pressure difference. Then, bubbles will automatically seal other leakage zones with less pressure differences until all of them are sealed, and the whole well will be drilled safely.

5. RHEOLOGICAL PROPERTY

The special structure of micro-bubble also makes WMDF have very good rheological property to meet the requirements of petroleum drilling industry. Table.2 is a series of formulas of WMDF designed by multiple regression and these systems' rheological parameters which are measured by rotary viscometer.

Table.2 rheological parameters of WMDF with different density

Density(g/cm^3)	Rheological parameter				
	PV (mPa s)	YP(Pa)	YP/PV	initial gel	Ten mins gel
0.81	17.00	23.51	1.38	8.18	10.73
0.91	25.00	30.66	1.23	8.18	10.73
1.02	31.00	39.35	1.27	9.71	13.29
1.12	22.00	32.19	1.46	10.73	13.80
1.18	20.00	38.33	1.92	10.99	16.35

These data reflect some important information. First of all, WMDF has very high viscosity no matter how high its density is. Plastic viscosity varies from 17 mPa s to 31 mPa s while the ordinary solids free water-based drilling fluids have the plastic viscosity ranging from 4 to 16 mPa s if they are of the same density of WMDF. Secondly, yield stress is extraordinarily high. Thirdly, the ratio of yield stress and plastic viscosity is much higher than any other drilling fluids. Table.2 shows each of them is over than 1.2 but the ratio of ordinary drilling fluids is controlled between 0.36 and 0.48. Fourthly, initial gel strength and final strength are very high. All of these are important for petroleum drilling.

For acceptable drilling fluids they must have abilities to clean the well, suspend the cuttings and weighting materials, keep the well wall stable and make sure it's safe while tripping. What's more, a good drilling fluid system should also increase the drill speed but not destroy the environmental balance. WMDF is the very system that meets these requirements.

In general, it is better to have a lower velocity of drilling fluids in annular to carry cuttings out of the well, and this will make sure that the drill bits have enough hydraulic horsepower to drill and enable the well wall to have stability because the returning rate will be not high enough to destroy the mud cake which sticks to the well wall. Reynolds number is often used to calculate the highest velocity which will not destroy the stability of well wall. It turns out that with the increase of PV and YP/PV the critical velocity will also be increased. So, from rheological view, high PV and YP/PV will prevent the WMDF system from turbulent flow and then ensure the stability of well. On the side, the high viscosity of WMDF makes it possible to carry drill cuttings out of well at low velocity. Additionally, high YP/PV will help the system turn to flat plate laminar flow which is in favor of carrying debris. Once the drilling process is stopped by the climate reasons or etc., network will be formed immediately and the gel strength will be strong enough to suspend the debris and weighting material to avoid sticking because WMDF has very high initial gel strength. The only thing that should be paid attention to is that if the WMDF stands still for some time, recirculation should begin with a small fraction of the system then the whole fluid system begins to recirculate. However, once the bubbles move, they will help to destroy the network structure formed by LHP and decrease the viscosity. Additionally, liner high polymers, the main agent of WMDF, are good flow pattern modifier which makes the system sheared thinning very well. On the one hand, this shear thinning system can avoid accident while tripping or pumping. On the other hand, it will increase the penetration rate in a great amount for it owns stronger impact force but less resistance while breaking the rocks.

6. CONCLUSION

Micro-bubble in WMDF has the structure of “one core, two layers and three membranes”. According to the different bond stress between membranes and layers, micro-bubble can be divided into two parts, structure stabled “air-bag” and dynamic changed “incompact zone”. Dynamic change of “incompact zone” is the radical reason why micro-bubbles can exist in WMDF independently or assemble together, why WMDF has very high yield stress at an extremely low shear rate but has good fluidity at an high shear rate, and why WMDF can wet both hydrophilic and hydrophobic surfaces.

It is the special structure that explains the mechanism for lost circulation control. However, if the formation rock is oil wetted, WMDF will lose its function. Fortunately, oil wetted rock is really rare. Some measurements indicate WMDF has high plastic viscosity, yield stress, YP/PV, initial gel and final gel which are important to meet the requirements of petroleum drilling.

What’s more, we can change the bond strength between different layers and membranes by choosing or synthesizing different agents to improve micro-bubbles’ stability, to enhance the system’s ability of antipollution and bearing high temperature. Thus, WMDF can meet requirements of different petroleum tasks like drilling, well completion, workover and so on. The range of application of WMDF will be enlarged.

ACKNOWLEDGEMENTS

The authors thank LihuiLab for permission to publish this paper.

REFERENCES

C.D. Ivan, F.B. Growcock, J.E. Friedheim. (2002). "Chemical and Physical Characterization of Aphron-Based Drilling Fluids" SPE 77445.

F.B. Growcock, A.M. Khan, G.A. Simon. (2003). "Application of Water-Based and Oil-Based Aphrons in Drilling Fluids" SPE 80208.

F.B. Growcock, A. Belkin, M. Fosdick et al. (2006). "Recent Advances in Aphron Drilling Fluids" SPE 97982.

L.D. Landau, and E.M. Lifshitz. (1995). Fluid Mechanics (Course of Theoretical Physics, Vol. 6), 2nd Edition, Butterworth-Heinemann.

Sebba. (1987). "Foams and Biliquid Foams-Aphrons," John Wiley and Sons, New York, pp. 46-61, 62-78, 102-127.

Tom Brookey. (1998). "'Micro-Bubbles': New Aphron Drill-in Fluid Technique Reduces Formation Damage in Horizontal Wells" SPE 39589.

ZHENG Li-hui. (2008). "Optimization of rheological parameter for micro-bubble drilling fluids by multiple regression experimental design" J. Cent. South Univ. Technology.

Journal of Physics: Conference Series **147** (2009) 012023 doi:10.1088/1742-6596/147/1/012023

Experimental Study on Bubble Induced Velocity Fluctuations in the Boundary Layer at Transitional Reynolds Numbers

J Huang [1], T Suzuki [2], Y Murai [3], F Yamamoto [2]

[1] Graduate School of Engineering, Fiber Amenity Engineering Course, The University of Fukui, 3-9-1 Bunkyo, Fukui, 910-8507, JAPAN
[2] Graduate School of Engineering, The University of Fukui, Fukui, JAPAN
[3] Division of Energy and Environment System, School of Engineering, Hokkaido University, Sapporo, JAPAN

huangj@mech.fukui-u.ac.jp

Abstract. Our previous study showed that the frictional drag decreases with increasing void fraction at Re > 1300, while it increases at Re < 1000. Decomposition of the Reynolds shear stress also implied that bubbles induce isotropy of turbulence. In order to confirm our previous analysis and to further investigate flow fields in the vicinity of bubbles, we analyze velocity fluctuations on the quadrant space in the streamwise and transverse direction ($u' - v'$ plane). Here, we focus on two specific Reynolds numbers (at Re ≈ 900 and ≈ 1410, which are close to the laminar-to-turbulent transition region) and discuss bubble effects on the sweep ($u' > 0$, $v' < 0$) and ejection ($u' < 0$, $v' > 0$) events as a function of the Reynolds number. We also illustrate the velocity fluctuations in the vicinity of an individual bubble and a swarm of bubbles on the $u' - v'$ coordinates. The result shows that a bubble swarm suppresses the velocity fluctuations at Re ≈ 1410. We then rearrange the scatted PTV vectors and compute power spectra of the kinematic energy of each velocity component. It only indicates that the elongated bubble suppresses low frequency E_{vv} at Re ≈ 1410.

1. INTRODUCTION

In order to explore the mechanism of drag reduction owing to microbubbles, many researchers have recently measured or simulated bubble-induced velocity fluctuations to discuss their effects on velocity distribution, vorticity, Reynolds shear stress, turbulent kinetic energy, power spectra, and other flow quantities. Xu *et al.* (2002) simulated the effect of different bubble sizes on the Reynolds shear stress and concluded that small bubbles tend to reduce the frictional drag. Kawamura and Kodama (2002) developed new computational methods to investigate interaction between bubbles and turbulence. Their results agreed well with the experimental studies on fully developed turbulent channel flows. Ferrante and Elghobashi (2004 and 2005) simulated microbubble-laden boundary layers over a horizontal flat plate and analyzed the bubble effects on the sweep and ejection events. Skudarnov and Lin (2007) also simulated the microbubble-laden flow. After investigating the velocity profiles and the distributions of the turbulent kinetic energy, they found that microbubbles decrease drag with increasing void fraction. Gabillet *et al.* (2002) experimentally studied the influence of bubbles on velocity fluctuations in the boundary layer of channel flows. They emphasized that the additional turbulent energy is caused by the relative motion of bubbles and the shear-induced turbulence. Murai *et al.* (2006) measured the velocity vectors using PTV in a horizontal channel and decomposed the Reynolds shear stress. Huang *et al.* (2008a) also experimentally investigated velocity fields on the streamwise and spanwise planes in boundary layers of horizontal channel flows. They discussed the bubble effect on vorticity and the power spectra. Lu *et al.* (2005) and Kitagawa *et al.* (2005) have focused on deformed bubbles and their oscillation in the boundary layer. They have investigated changes of flow fields in the vicinity of these bubbles.

All the studies mentioned above are performed under the fully developed turbulent flows at the Reynolds number of the order of 10^4. We also believe that one of the keys to revealing the drag-reduction mechanism is to investigate flow conditions from laminar to turbulent Reynolds numbers. In our previous study (Huang *et al.* 2008b), we confirmed that bubbles decrease the wall shear stress with increasing void fraction at Re > 1300, while they increase the drag at Re < 1000. We then used an approach derived by Murai *et al.* (2006) to investigate the bubble effect in detail. One of the results implied that bubbles contribute to isotropy of turbulence.

Our objective of this study is to further investigate bubble effects on the surrounding flow fields based on the quadrant analysis. This method expresses the turbulent diffusion and the relative motion of fluid elements in boundary layers. Therefore, we can estimate the contribution of velocity fluctuations from each quadrant, i.e. in the u' - v' coordinates, to the Reynolds shear stress, $-\rho\overline{u'v'}$, in detail. We first observe distributions of sweep ($u' > 0$, $v' < 0$) and ejection ($u' < 0$, $v' > 0$) events as a function of the transverse coordinate for both single- and two-phase flows. The results indicate that bubbles suppress the sweep and ejection events near the wall at Re ≈ 1410. We then explore the influence of an individual bubble and a swarm of bubbles on the velocity fluctuations in comparison with those for single-phase flows. The bubble swarm suppresses velocity fluctuations at Re ≈ 1410. After rearranging those scatted PTV vectors, there is no obvious influence from the elongated bubbles on power spectra of kinematic energy compared with those of single-phase flows. However, those bubbles enhance the kinetic energy of transverse velocity component, E_{vv}, at the high frequency at Re ≈ 900, while they suppress the low frequency E_{vv} at Re ≈ 1410.

2. EXPERIMENTAL APPARATUS AND CONDITIONS

Experiments are conducted on a horizontal channel, similar to our previous study (see Huang *et al.* 2008b in detail). The experimental conditions are summarized in Table 1. The half height of the channel, $h = 5mm$, is adopted as the length scale. Two specific Reynolds numbers, at Re ≈ 900 and ≈ 1410, are chosen, which belong to the drag increase range and the decrease one, respectively. They are also close to the laminar-to-turbulent transition regime. Silicone oil (Shin-Etsu, KF96, 10cSt) is adopted as the working fluid; therefore, we can ignore the contamination disturbance and affirm the experimental reproducibility due to its stable surface property. Compressed air is injected into the channel through a needle nozzle array at $\overline{\alpha} = 0.2\%$.

The test position is located at 250mm ($X/h = 50$) downstream of the injector as schematized in Fig. 1. In order to measure the flow fields surrounding bubbles in the boundary layer, a mirror is mounted

Table 1: Experimental conditions. Here h denotes the half-channel height, and u_τ is the friction velocity

Test position, X/h	50	[-]
Bulk velocity of silicone oil, $\overline{U}$	2.0, 3.0	[m/s]
Bulk void fraction, $\overline{\alpha}$	0.2	[%]
Temperature of silicone oil , T	21 ± 1	[°C]
Density of silicone oil, ρ	938 – 939	[kg/m^3]
Kinematic viscosity of silicone oil, ν	10.6–10.9×10^{-6}	[m^2/s]
Reynolds number, $\mathrm{Re} = \overline{U}h/\nu$	900, 1410	[-]
Friction Reynolds number, $\mathrm{Re}_\tau = u_\tau h/\nu$	53 – 131	[-]
Surface tension of silicone oil, σ_S	20.1	[mN/m]

on the channel window at an angle of 45º to the upper wall to reflect images of bubbles and tracer particles. Its effects on the target flow area are neglected. A high-speed video camera (Photron, FASTCAM-MAX 120 KC) is focused on the mid-span of the channel.

Using particle tracking velocimetry based on the shallow depth-of-field (Huang *et al.* 2008a), sufficient numbers of velocity vectors in the target depth (Δz = 0.3mm) have been detected around bubbles in the boundary layer, as shown in Fig. 2. The displacement thickness is evaluated as

$$\delta^*(x) = 1.72\sqrt{\frac{x\mu}{\bar{U}\rho}}, \tag{1}$$

where, x is the distance from the inlet of the channel (875mm, x/h = 175) and μ is the dynamic viscosity. The displacement thickness is 0.77 y/h (or 40 y^+) at Re ≈ 900 and 0.62 y/h (or 58 y^+) at Re ≈ 1410 ($y^+ = u_\tau y/\nu$, where, u_τ is the friction velocity, ν is the kinematic viscosity). The spurious vectors in the PTV measurement are removed by a statistical method.

Elongated bubbles are considered to strongly relate to the shear stress (Rust and Manga, 2002). At Re ≈ 900, as shown in Fig. 2 (a), the axial ratio of the bubble is about 3.4 (l/b, where l and b are the semi-major and semi-minor axes of the sheared bubble, respectively) with 31° of the inclination angle,

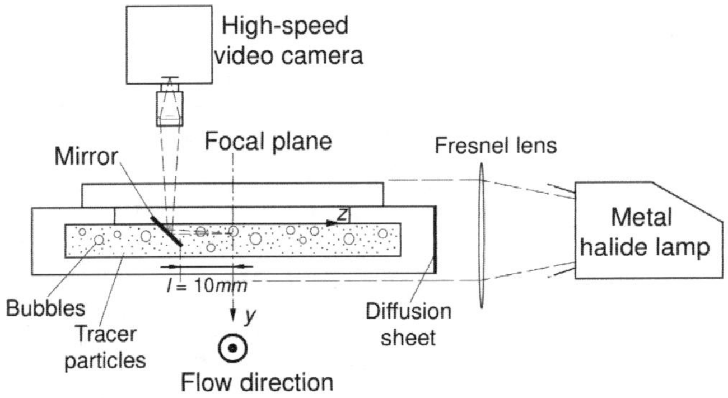

Fig. 1 Cross section of the flow visualization system for PTV. *Y* is the transverse direction, and *z* is the spanwise direction.

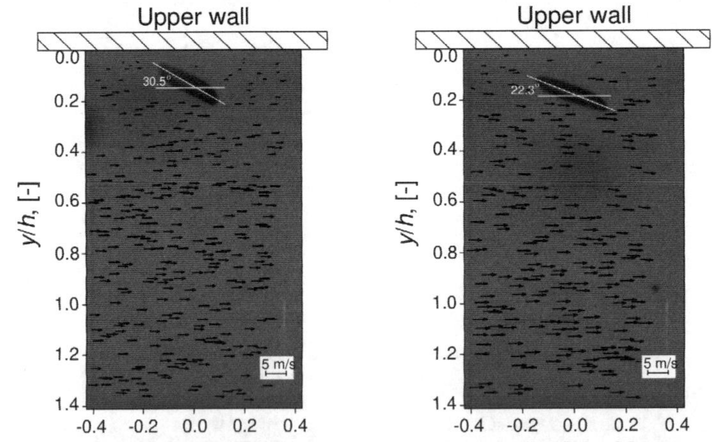

Fig. 2 The PTV velocity fields around an elongated bubble.
(a) at Re ≈ 900 ($\bar{U}$ = 2.0 m/s); (b) at Re ≈ 1410 ($\bar{U}$ = 3.0 m/s)

θ, while the axial ratio is 4.1 and θ = 22° at Re ≈ 1410, as shown in Fig. 2 (b). Their capillary numbers are about 0.24 at Re ≈ 900 and 0.80 at Re ≈ 1410, both of which are less than unity. At these *Ca* numbers, bubbles maintain steady elongated shapes. The *Ca* number is defined as

$$Ca = \frac{\mu U_b}{\sigma_s} , \qquad (2)$$

where, U_b is the velocity of bubbles and σ_s is the surface tension. It represents the ratio of the shear stress, which deforms bubbles, to the surface tension.

3. RESULTS AND DISCUSSION

3.1 Our Previous Results

In our previous studies (see Huang *et al.* 2008b in detail), we measured the wall shear stress, τ_w, with a shear force transducer (SSK Co., LTD. S10W-05) at three test positions X_n/h = 50, 200, 800 downstream of the bubble injector. The bulk void fraction, $\overline{\alpha}$, ranged from 0.2% to about 5%. After converting the wall shear stress to the skin friction coefficient, C_f, we compared the modified skin friction ratios of f two-phase flows, C_f, to single-phase flows, C_{f0}, C_f/C_{f0}, to minimize the uncertainty induced from injected bubbles and temperature variation. Fig. 3 displays the skin friction ratio at the test positions verse the bulk void fraction at different Reynolds numbers. The data points can be categorized into two types based on the Reynolds number. One is that the frictional ratios are greater than unity at Re ≈ 900, which indicates the drag increase for two-phase flows; the other is that the ratios are less than unity at Re ≈ 1410 indicating drag reduction. The ratios are monotonically decreased with increasing void fraction to about 20% when $\overline{\alpha} \approx 5.0\%$ at X_1/h = 50. According to the response frequency of the transducer (f_n = 30 Hz), noise at the higher frequency is filtered to ensure the measurement accuracy.

We also decomposed the Reynolds shear stress of bubbly flows into three components ($\tau_1 = -\rho \overline{f} \overline{u'v'}$, $\tau_2 = -\rho \overline{f'v'}\overline{u}$, $\tau_3 = -\rho \overline{f'u'v'}$, where f is the volume fraction of the continuous phase) based on the two-fluid model (Murai *et al.* 2006). Fig. 4 displays the normalized stress components together with the Reynolds shear stress, $-\rho\overline{u'v'}$, for single- and two-phase flows at the two Reynolds numbers. The injected bubbles decrease the stress components, τ_1, by reducing the mixture density. τ_2 generates negative momentum exchange near the wall at Re ≈ 1410 and dissipates the fluid energy. τ_3 maintains small positive stress across the upper half of the channel. This phenomenon implied that the

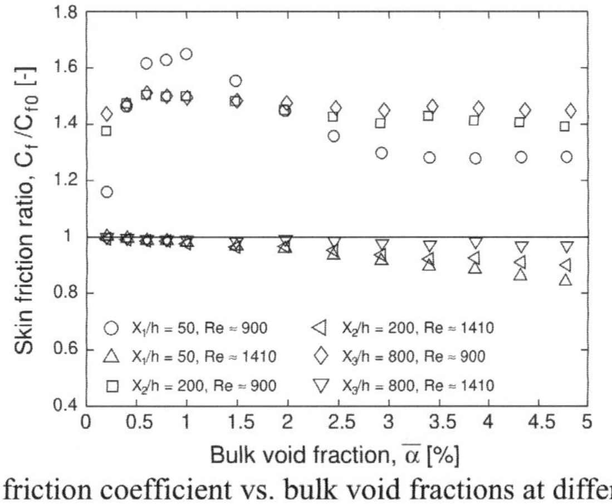

Fig. 3 Ratios of the skin friction coefficient vs. bulk void fractions at different Reynolds numbers.

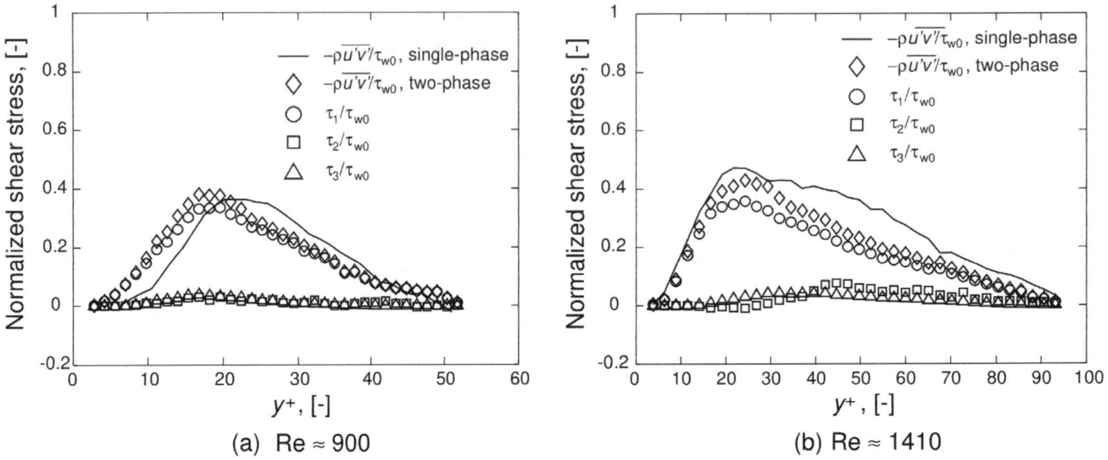

Fig. 4 Profiles of the normalized Reynolds shear stress and the decomposed shear stress components across the upper half channel for $\bar{\alpha} \approx 0.2\%$.

positive $u'v'$ of the continuous phase in the vicinity of the bubbles is increased, which contributes to the isotropy of the turbulence. However, in order to confirm our analysis, it is still important to further investigate instantaneous flow fields in the vicinity of bubbles.

3.2 Quadrant Analysis

In this study, we first analyze velocity fluctuation components in the streamwise and transverse directions, referred to as the *quadrant analysis*, to discuss the bubble-induced velocity fluctuations in detail. The streamwise and transverse components of the velocity fluctuations are defined as

$$u' = u - \bar{u} \quad \text{and} \quad v' = v - \bar{v}, \tag{3}$$

where, the over-bars denote the time-averaged variables and the primes indicate the fluctuation components. The quadrant analysis divides $u' - v'$ into four quadrants according to the sign of u' and v'. The first quadrant, $u' > 0$ and $v' > 0$, contains outward motion of high-speed flow; the second one, $u' < 0$ and $v' > 0$, contains the motion associated with ejection of low-speed flow away from the wall; the third, $u' < 0$ and $v' < 0$ contains inward motion of low-speed flow; and the fourth quadrant, $u' > 0$ and $v' < 0$, contains an inrush of high-speed flow, generally referred as a sweep event (Wallace *et al.*

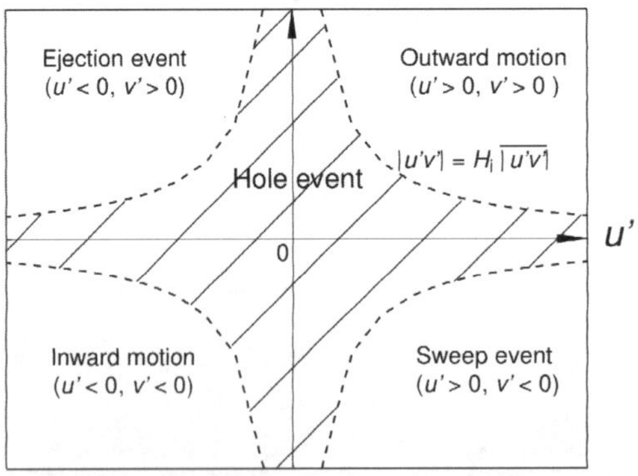

Fig. 5 The schematic of the division of $u' - v'$ coordinates and the hole event.

1972; Kim *et al.* 1987), as schematized in Fig. 5 together with the "hole event", which will be discussed in the next section. The sweep and ejection events contribute to the positive Reynolds shear stress, $-\rho\overline{u'v'}$, while the other two events corresponds to the negative part.

Fig. 6 shows the distribution of the normalized $u'-v'$ from each quadrant as a function of y-location. In each flow condition, the intensity of the ejection event, $|u'v'|_2$, (the second quadrant, $u' < 0$ and $v' > 0$) dominates at $y^+ \approx 20$, while that of the sweep event, $|u'v'|_4$, (the forth quadrant, $u' > 0$ and $v' < 0$) is greater than other events away from the wall, at $y^+ \approx 20$. Here, the subscripts 2 and 4 indicate the quadrant numbers. At Re $\approx$ 900, as shown in Fig. 6 (a), injected bubbles greatly enhance the intensity of the sweep and ejection events in the vicinity of the wall, $y^+ < 25$. The location where the intensities of the two events become equivalent ($|u'v'|_2 = |u'v'|_4$) also shifts from $y^+ \approx 25$ for single-phase flows, to the wall, at $y^+ \approx 20$ for two-phase flows. These events induced by bubbles have the potential to increase the shear stress on the wall (as derived in Gabillet *et al.* 2002). This is also consistent with our previously measured wall shear stress as shown in Fig. 3. In contrast, as shown in Fig. 6 (b), bubbles slightly suppress the intensity of the ejection and sweep events, especially the ejections, while they enhance the outward motion in the first quadrant ($u' > 0$ and $v' > 0$) and the inward motion in the third quadrant ($u' < 0$ and $v' < 0$). Therefore, the Reynolds shear stress can be attenuated due to isotropy of turbulence.

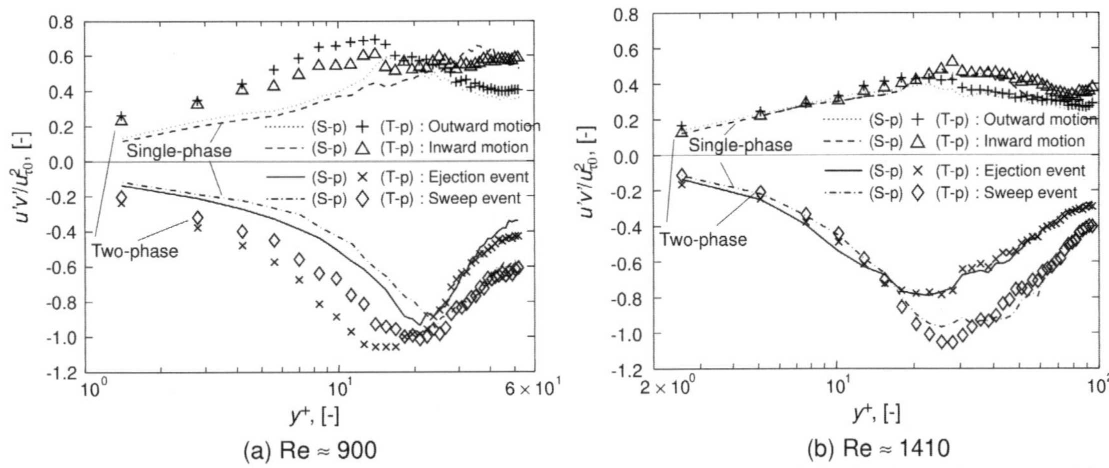

Fig. 6 $u'-v'$ from each quadrant normalized by the frictional velocity of single-phase flows, $u_{\tau0}$. (a) at Re $\approx$ 900; (b) at Re $\approx$ 1410. Lines express the experimental data for single-phase flows (S-p); symbols (+, ◊, ×, Δ) represent those for two-phase flows (T-p).

3.3 Fractional Contribution of the Four Quadrants

Besides displaying the distribution of different quadrant events, it is also important to quantitatively evaluate their relative importance in generating the Reynolds shear stress across transverse planes in the boundary layer. Here, we utilize the "hole event" (Lu and Willmarth, 1973; Hurther and Lemmin, 2000; Mazumde and Ojha, 2007), as illustrated by the cross-hatched region in Fig. 5, to separate the strength of turbulent diffusion. With this scheme, large contributors to Reynolds shear stress, $\overline{u'v'}$, can be extracted by leaving the smaller fluctuating $u'v'$ signal in the "hole region". We compute the contribution of each quadrant, I, to the Reynolds shear stress as a function of the threshold level of the hyperbolic hole region, H, as follows:

$$\langle u'v' \rangle_{i,H} = \lim_{t\to\infty} \frac{1}{t} \int_0^t u'(t)v'(t)\delta_{i,H}(u',v')\,dt, \qquad (4)$$

where, the angle brackets denote a conditional average. The indicator function $\delta_{i.H}$ obeys

$$\delta_{i,H}(u',v') = \begin{cases} 1 \\ 0 \end{cases}, \tag{5}$$

namely, only when (u', v') is in the ith quadrant and $|u'v'| \geq H|\overline{u'v'}|$, $\delta_{i.H}$ becomes unity. H can be selected as H = 0.5, 1.0, 1.5, 2.0, etc. The stress fraction of the i-th quadrant is normalized by the averaged intensity of $u'v'$, as

$$S_{i,H} = \langle u'v' \rangle_{i,H} \Big/ |\overline{u'v'}|. \tag{6}$$

Their summation also satisfies

$$\sum_{i=1}^{4} |S_{i,0}| = 1, \tag{7}$$

at H = 0. Figs. 7 and 8 respectively display the fractional contribution of each quadrant at the two Reynolds numbers at different y-locations.

As shown in Fig. 7 (a), the total fractions of the inward motion and ejection event ($|S_{1,0}|$ and $|S_{2,0}|$ at H = 0) are suppressed in two-phase flows, while that of the sweep event, $|S_{4,0}|$, is greatly enhanced at Re ≈ 900, y^+ = 10. Furthermore, for strong fluctuating $u'v'$ signals, such as at H > 1, the relative fraction of the sweep event becomes greater than that of the ejection ($|S_{4,H}| > |S_{2,H}|$, H > 1). This phenomenon is counter to single-phase flows, in which the ejection dominates the fractional contribution to the Reynolds shear stress ($|S_{2,H}| > |S_{4,H}|$, $H \geq 0$). Therefore, the velocity in this layer is enhanced by the injection of bubbles. However, the result is a little different at y^+ = 20, as shown in Fig. 7 (b). Bubbles enhance the total fractions of the ejection and sweep events and their fractional contributions are almost identical with each other ($|S_{2,H}| \approx |S_{4,H}|$, $H \geq 0$).

As shown in Fig. 8, bubbles slightly suppress the fraction of the sweep event ($|S_{4,H}|$, $H \geq 0$) at Re

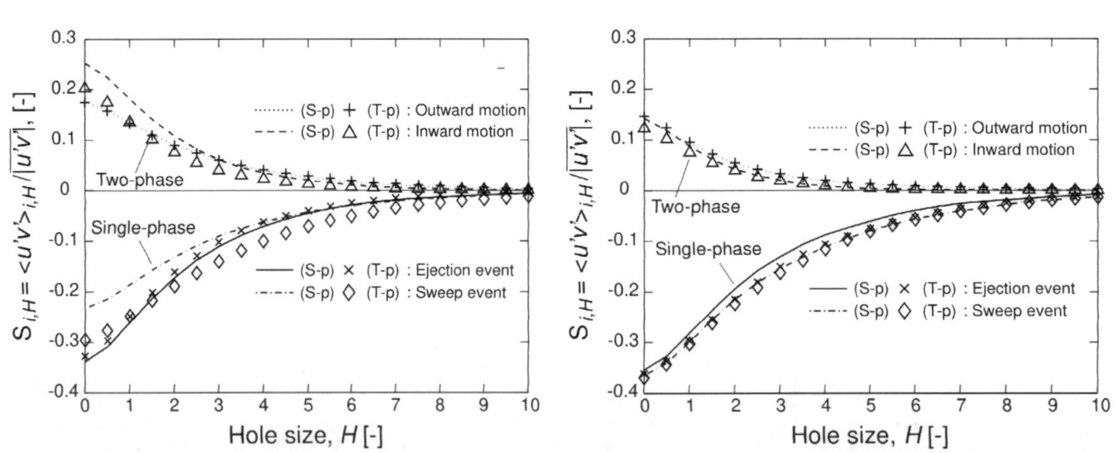

(a) Re ≈ 900, y^+ = 10

(b) Re ≈ 900, y^+ = 20

Fig. 7 The fractional contribution of each quadrant on the Reynolds shear stress against the hole size, H, at Re ≈ 900. (a) at y^+ = 10; (b) at y^+ = 20. Lines and symbols are the same to Fig. 6.

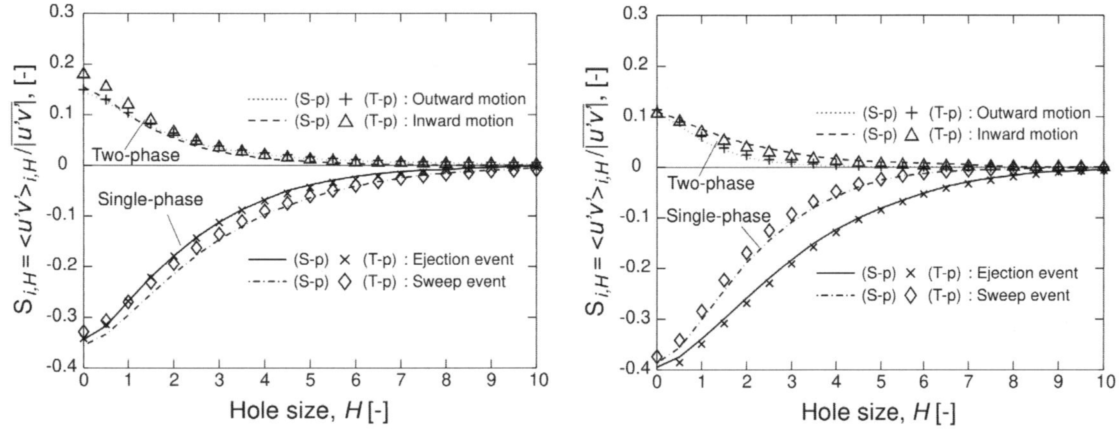

(a) Re ≈ 1410, $y^+ = 10$ (b) Re ≈ 1410, $y^+ = 30$

Fig. 8 The fractional contribution of each quadrant on the Reynolds shear stress against the hole size, H, at Re ≈ 1410. (a) at $y^+ = 10$; (b) at $y^+ = 30$.

≈ 1410. However, it is difficult to distinguish their influence on the other three quadrants due to the low bulk void fraction ($\overline{\alpha} = 0.2\%$), especially at $y^+ = 30$. At this y-location, the total fractions of the sweep and ejection events are almost the same, about 40% ($|S_{2,0}| \cong |S_{4,0}| \approx 40\%$, $H = 0$). But, the intensity of $u'v'$ for strong fluctuations, such as that at $H \geq 1$, is essentially contributed from ejection events ($|S_{2,H}| > |S_{4,H}|$, $H \geq 1$).

3.4 Velocity Fluctuation in the Vicinity of Bubbles

From the consecutive frames, we observe that there exist series of individual bubbles and bubble swarms in the boundary layer whose effects on the surrounding flows appear to be different. Therefore, we analyze the contribution of these types of bubbles to the quadrant events in details. For individual bubbles, the distance to the nearest bubble is generally much greater than its diameter and the bubble-bubble interaction is negligible. A swarm of bubbles consists of several bubbles and the typical distance between them is comparable with their sizes. Therefore, we should consider the influence of the bubble-bubble interaction on the surrounding flow fields.

Fig. 9 displays the velocity fluctuations of the continuous phase (silicone oil) in a thin shear layer including the peak Reynolds shear stress at each Reynolds number ($y/h = 0.28$ or $y^+ = 15$ at Re ≈ 900 and $y/h = 0.2$ or $y^+ = 20$ at Re ≈ 1410) for two-phase flows as well as for single-phase flows. The focused layer is nearly inside the free-stream rather than the bubble location ($0.05 < y/h < 0.2$) at Re ≈ 900. In contrast, it is closer to the bubble centroid ($y/h \approx 0.18$) at Re ≈ 1410. Individual bubbles are elongated, as displayed in Fig. 2.

At Re ≈ 900, velocity fluctuation in single-phase flows is weak, and data points are dense around $u' = v' = 0$ (in the range of hole event, $H < 1$). However, the injected bubbles promote turbulence. Individual bubbles induce the sweep event, while bubble swarms generate the ejection. At Re ≈ 1410, the magnitude of velocity fluctuations induced by the individual bubble is comparable with that of single-phase flows. Individual bubbles generate the ejection, opposed to the sweep event in the single-phase case. In contrast, bubble swarms suppress velocity fluctuations by decreasing both ejection and sweep events.

3.5 Rearranged Velocity Field and Power Spectrum of Kinetic Energy

According to the discussion on the relation between elongated bubbles and the shear stress (Rust and Manga 2002), we subsequently rearrange the scatted PTV vectors with the Spatio-temporal

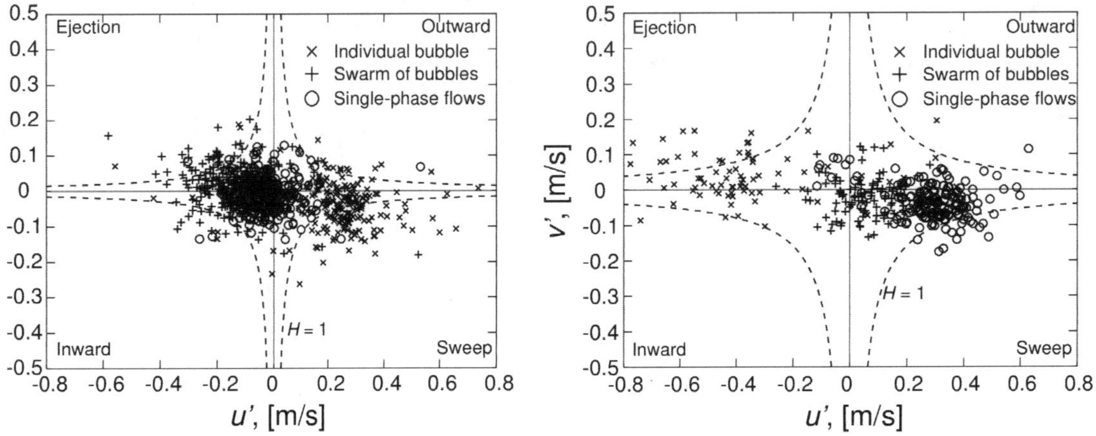

(a) Re ≈ 900 ($\overline{U}$ = 2.0 m/s), y^+=15 (b) Re ≈ 1410 ($\overline{U}$ = 3.0 m/s), y^+= 20

Fig. 9 Distribution of velocity fluctuations of continuous phase (silicone oil), *u', v'*, in the quadrant. indicates the velocity fluctuations around the individual bubbles shown in Fig. 2.

Biquadratic-ellipsoidal Equation Rearrangement (ST-BER) method developed by Ido *et al.* (2002) to obtain instantaneous flow fields around bubbles, which retain the third-order spatial continuity.

Fig. 10 shows the contours of the velocity magnitude at the two Reynolds numbers overlaying the relative velocity vectors of continuous phase to the elongated bubble illustrated in Fig. 2. The silicone oil flows from left to right. White parts express bubble images. Although these bubbles are elongated at each Reynolds number as described in Sec. 2, we find no remarkable effect on flow fields. We only observe that the velocity magnitude is slightly enhanced behind the bubble at Re ≈ 900 and velocity vectors indicate that flow elements move along the major axis of the bubble toward free-stream, which generates clockwise vorticity in front of bubbles. At Re ≈ 1410, the velocity magnitude behind the bubble attenuates. The relative velocity of the continuous phase to the bubble also directs along the major axis but toward the wall, especially in the range of $y^+ < 10$. These phenomena are consistent with the previous study by Kitagawa *et al.* (2005).

Subsequently, averaged power spectra of the kinetic energy for each velocity component ($E_{uu} = \overline{u^2}/2$ and $E_{vv} = \overline{v^2}/2$) are plotted in Fig. 11 as a function of y^+. The velocity in the bubble occupied area is interpolated by the Laplace equation. The interpolation is applied only inside the bubble area; therefore, the original measurement information of the continuous phase is unchanged. The abscissa is the frequency normalized by the frame rate of the high-speed camera. In both figures, the magnitudes of the power spectra increase with distance away from the wall. Their magnitudes keep constant in the low frequency, while decline greatly at the high frequency region. The slope at the high frequency part roughly matches -5/3 of Kolmogorov's law. We also notice that at Re ≈ 900, as shown in Fig. 11 (a), the kinetic energy of E_{vv} on $y^+ = 5$, close to the centroid of the bubble, is enhanced at $K^* > 0.25$. Its magnitude is comparable with that of the E_{uu} there. In contrast, as shown in Fig. 11 (b), E_{vv} on $y^+ = 20$ at Re ≈ 1410, where is also close to the centroid of the bubble, slightly decreases at $K^* < 0.2$. The decrease indicates that the elongated bubble suppresses the low frequency kinetic energy of the transverse velocity component and relatively large eddies, whose wavelength has the inverse proportion to frequency.

4. CONCLUSIONS

In this study, we experimentally investigated the bubble-induced velocity fluctuation at two specific Reynolds numbers (Re ≈ 900 and ≈ 1410), which belonged to the drag increase range and the decrease one, respectively. They were also close to the laminar-to-turbulent transition regime. From our

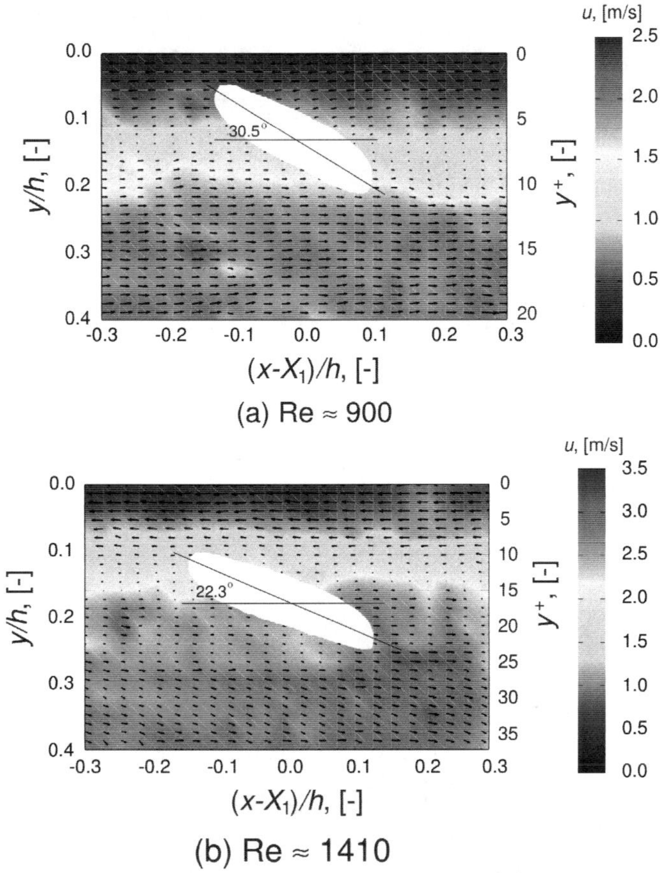

Fig. 10 Rearranged instantaneous contours of velocity magnitude for the streamwise component overlaid with relative velocity fluctuations to bubbles.

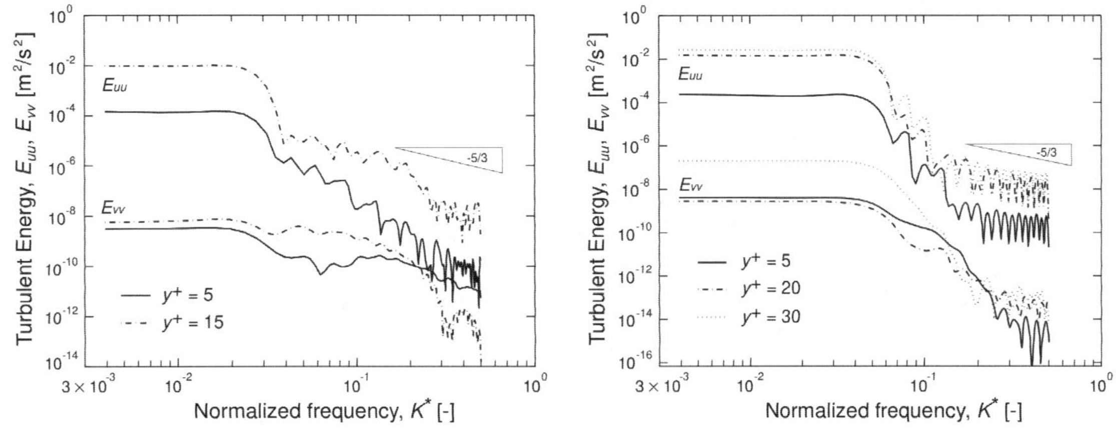

Fig. 11 Power spectra of the kinetic energy for each velocity component (u and v).

experiment, the following remarks were concluded:

1) Based on the quadrant analysis for the continuous phase, we found that the intensity of the ejection (the second quadrant, $u' < 0$ and $v' > 0$) and sweep (the forth quadrant, $u' > 0$ and $v' < 0$) events is greatly enhanced by bubbles at Re ≈ 900; in contrast, bubbles suppress those events at Re ≈ 1410.

2) For the hole event, whose threshold value is $H \geq 1$, the absolute stress fraction for the sweep events, $|S_{4,1}|$, in two-phase flows becomes greater than that for the ejections, $|S_{2,1}|$, on $y^+ = 10$ at Re ≈ 900. This phenomenon is counter to that for single-phase flows, in which the stress fraction for the ejections is greater than the sweep events.

3) By comparing the velocity fluctuations in the vicinity of individual bubbles and bubble swarms at Re ≈ 900 and ≈ 1410, we observe that the swarm of bubbles suppresses the ejection and sweep events at Re ≈ 1410.

4) But, after rearranging the instantaneous flow fields around the elongated bubbles, we found that those bubbles enhance the magnitude of the power spectrum for E_{vv} at the high frequency at Re ≈ 900, while they suppresses the low frequency E_{vv} at Re ≈ 1410.

ACKNOWLEDGEMENTS

We thank Mr. M. Koukawa, Mr. H. Fukuda, Mr. Q.Y. Yang and Mr. H. Ji at the University of Fukui for their great help during the experiments.

NOMENCLATURE

b	semi-minor axis of the elongated bubble	[m]
Ca	capillary number	[-]
C_f	skin-friction coefficient	[-]
E_{uu}, E_{vv}	kinematic energy for each velocity component per unit mass	[m^2/s^2]
f	volume fraction of the continuous phase	[-]
f'	fluctuations of volume fraction for the continuous phase	[-]
f_n	response frequency of the transducer	[Hz]
H	threshold level of the hyperbolic hole region	[-]
h	half height of the channel, used as the length scale	[m]
K^*	normalized frequency	[-]
l	semi-major axis of the elongated bubble	[m]
Re	Reynolds number	[-]
Re_τ	friction Reynolds number	[-]
S	Normalized fraction contribution of each quadrant	[-]
T	temperature of silicone oil	[°C]
t	time period	[s]
$\overline{U}$	bulk velocity of silicone oil	[m/s]
U_b	bubble velocity	[m/s]
u, v	streamwise and transverse velocity components	[m/s]
$\overline{u}$, $\overline{v}$	time-averaged streamwise and transverse velocity components	[m/s]
u', v'	streamwise and transverse components of velocity fluctuations	[m/s]
u_τ	friction velocity	[m/s]
X	test position from the bubble injector	[mm]
x	distance from the inlet of the channel	[mm]
y	transverse distance from the upper wall of the channel	[m]
y^+	wall unit	[-]
z	spanwise distance	[mm]
Δz	target depth	[mm]
	Greek Letters	
$\overline{\alpha}$	bulk void fraction	[%]
δ^*	displacement thickness	[mm]

$\delta_{i,H}$	indicator function of the hole event	[-]
μ	dynamic viscosity of silicone oil	[Pa·s]
v	kinematic viscosity of silicone oil	[m^2/s]
θ	inclination angle of bubbles to the streamwise direction	[°]
ρ	density of silicone oil	[kg/m^3]
σ_s	surface tension of silicone oil	[mN/m]
τ_w	wall shear stress	[Pa]
τ_1, τ_2, τ_3	decomposed Reynolds shear stress components for two-phase flows	[Pa]

Subscripts

i	i-th quadrant, i = 1, 2, 3 and 4
n	n-th test position downstream of the bubble injector, n= 1, 2, 3
0	parameters for single-phase flows

REFERENCES

Ferrante, A. and Elghobashi, S. (2004). "On the physical mechanisms of drag reduction in a spatially developing turbulent boundary layer laden with microbubbles," J. Fluid Mech., 503, pp. 345 – 355.

Ferrante, A. and Elghobashi, S. (2005). "Reynolds number effect on drag reduction in a microbubble-laden spatially developing turbulent boundary layer," J. Fluid Mech., 543, pp. 93–106.

Gabillet, C. et al. (2002). "Experimental study of bubble injection in a turbulent boundary layer," Int. J. Multi. Flow, 28, pp. 553–578.

Huang, J. et al. (2008a). "Shallow DOF-based particle tracking velocimetry applied to horizontal bubbly wall turbulence," Flow Measure. Instru., 19, pp. 93–105.

Huang, J. et al. (2008b). "Experimental study of bubbly channel flows in the laminar-to-turbulent transition regime," submitted to Int. J. Multi. Flow.

Hurther, D. and Lemmin, U. (2000). "Shear stress statistics and wall similarity analysis in turbulent boundary layers using a high-resolution 3-D ADVP," IEEE. J. Ocean. Eng., 25(4), pp. 446 – 457.

Ido, T. et al. (2002). "Post-processing algorithm for particle tracking velocimetry based on ellipsoidal equations," Exp. Fluids, 32, pp. 326 – 336.

Kawamura, T. and Kodama, Y. (2002). "Numerical simulation method to resolve interactions between bubbles and turbulence," Int. J. Heat and Fluid Flow, 26, pp. 627 – 638.

Kim, J. et al. (1987). "Turbulence statistics in fully developed channel flow at low Reynolds number," J. Fluid Mech., 177, pp. 133–166.

Kitagawa, A. et al. (2005). "Flow structure of microbubble-laden turbulent channel flow measured by PIV combined with the shadow image technique," Exp. Fluids, 38(4), pp. 466–475.

Lu, J. et al. (2005). "The effect of bubbles on the wall drag in a turbulent channel flow," Phys. Fluids, 17, pp. 1–12.

Lu, S.S. and Willmarth, W.W. (1973). "Measurements of the structure of the Reynolds stress in a turbulent boundary layer," J. Fluid Mech., 60(3), pp. 481 – 511.

Mazumde, B.S. and OJHA, S.P. (2007). "Turbulence statistics of flow due to wave current interaction," Flow Measure. Instru., 18, pp 129 – 138.

Murai, Y. et al. (2006). "Turbulent shear stress profiles in a bubbly channel flow assessed by particle tracking velocimetry," Exp. Fluids, 41(2), pp. 343–352.

Rust, A.C. and Manga, M. (2002). "Bubble shapes and orientations in low Re simple shear flow," J. Colloid Interface Sci., 249, pp. 476 – 480.

Skudarnov, P.V. and Lin, C.X. (2006). "Drag reduction by gas injection into turbulent boundary layer: Density ratio effect, " Int. J. Heat and Fluid Flow, 27, pp. 436–444.

Wallace, J.M. et al. (1972). "The wall region in turbulent shear flow," J. Fluid Mech., 54(1), pp. 39 – 48.

Xu, J. et al. (2002). "Numerical simulation of turbulent drag reduction using micro-bubbles," J. Fluid Mech., 468, pp. 271–281.

Journal of Physics: Conference Series **147** (2009) 012024 doi:10.1088/1742-6596/147/1/012024

Effect of electrical potential of microbubbles on ozone dissolution

H Hasegawa [1], H Kataoka [2], K Asano [3]

[1] Akita University, 1-1, Tegatagakuen-machi, Akita-shi, Akita 010-8502, Japan
[2] Graduate School of Akita University, Akita, JAPAN
[3] Akita National College of Technology, Akita, JAPAN

hhasegaw@mech.akita-u.ac.jp

Abstract. Microbubbles make ozone water generation effective due to the high dissolution rate of gas in contrast to a conventional generating method. Therefore, it is presumable that ozone water generation using microbubbles can be achieved by the low concentration ozone gas. In our previous study, a compact and low power microbubble generator was developed. The microbubbles are generated by the local shear stress in the flow through a pipe with slits. In the present study, in order to investigate the relationship between the electrical potential of the gas-water interface and the cleaning of cloth using ozone microbubbles, two models with different slit angles (θ=30 and 60 deg) were installed. High concentration ozone water is produced for θ = 60 deg in contrast to the θ=30 deg case. When a cloth is washed in the θ=60 deg case, the soiled cloth can be cleaned easily in comparison with the θ=30 deg case, because the zeta potential of microbubbles for θ=60 deg is larger than that for θ=30 deg.

1. INTRODUCTION

Microbubbles are very small air bubbles with an average diameter of about 50μm. Microbubbles have various unique chemical and physical characteristics in contrast to bubbles on the order of several tens of microns. For example, water quality can be improved by increasing the air supply with microbubble dissolutions. Furthermore, the electrical properties of the gas-water interface promote interaction with other materials, such as solid particles and oil droplets, to provide a basis for technical application in many fields, such as foam fractionation, food processing, and purification processes.

Ozone is a powerful oxidizing agent and is also unstable at high concentrations. Ozone is capable of bleaching substances and can be used for sterilization in place of the more common chlorine. Moreover, ozone is widely used to improve the taste and color of drinking water as well as to remove organics and inorganic compounds in drinking water and wastewater. Up until now, many factors have hindered the use of ozone, including its relatively low solubility and stability in water and the high cost of production. Ozone dissolution efficiency is improved by the use of microbubbles to generate ozone water, a method that is easy in comparison with other conventional methods. Furthermore, it was reported that microbubble technology was applied to increase the mass transfer rate of ozone as well as to enhance the ozone oxidation of synthetic wastewater containing azo dye (Li-Bing Chu ,

2007). In our previous study, high ozone concentrations were achieved using our proposed microbubble generator and cold water temperatures.

However, the effective microbubble characteristics that enhance ozonation efficiency have yet to be completely clarified. In the present study, the relationship between the electrical properties of microbubbles and the effect of ozonated water generated by ozone microbubbles on the cleaning of cloth is investigated.

2. EXPERIMENTAL APPARATUS AND METHOD

2.1 Microbubble generator

Our proposed microbubble generating system has advantages such as simplicity, ruggedness and low cost. Figure 1 shows the schematic diagram of the proposed microbubble generator developed in our previous study (Hasegawa, 2007). The microbubbles are generated by the local shear stress in the flow through a pipe with slits. The slit angle θ is defined as the angle of the slit against the main-flow direction in the pipe, and the same slit angle is set in the slit array. There are ten slits in the microbubble generating device, each of width 0.6 mm. The pipe water flow is discharged solely from the slits into the water tank. This changes the flow direction and produces a shearing force at the corner of the slits. That is, the flow separates at the corners of the slits and a shearing force at the gas-water interface is produced by the Kelvin-Helmholtz instability in the separated shear layer. Furthermore, the flow velocity is increased in the slit section due to the small cross-sectional area

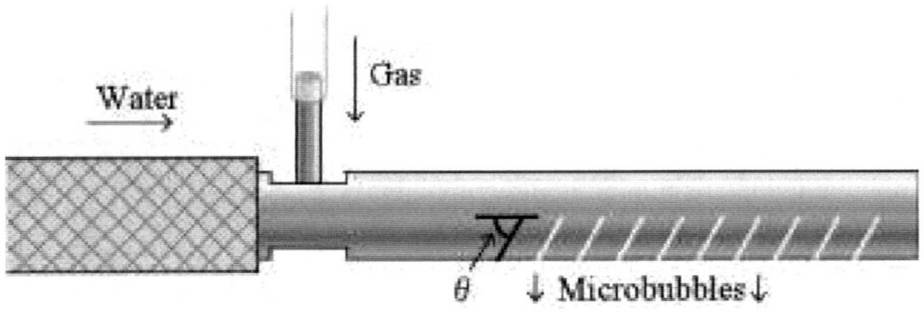

Fig.1 Microbubble generating device.

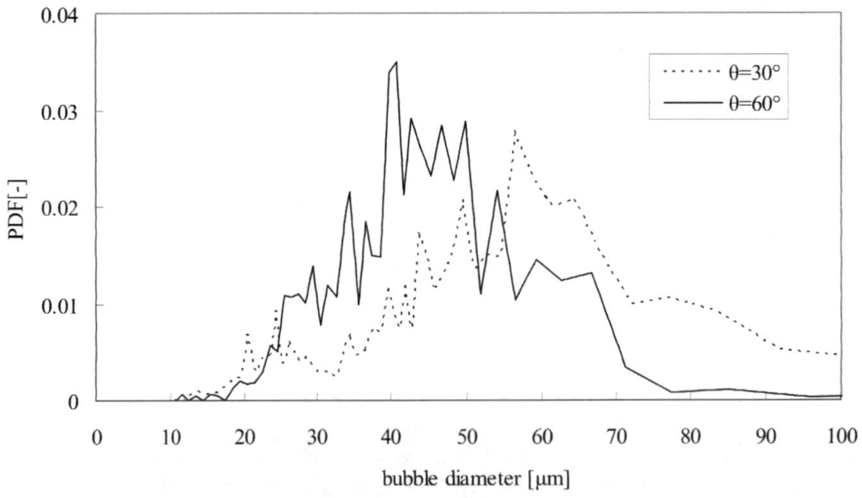

Fig.2 Bubble size distribution.

thereof. The shearing force is also produced by discharging water into the tank from the slit section with a large velocity gradient between the flow of water through the slits and the stationary surrounding water in the tank. This occurs because the accelerated flow of water through the slits quickly decreases when the water is discharged in the tank. In the present study, two models with different slit angles (θ = 30 and 60 deg) were installed.

Figure 2 shows the bubble size distribution. The bubble size distribution was measured by a light transmission method (Tabei, 2004). The principle method was based on changes in light transmission due to bubbles rising to the water surface. Small bubbles with a diameter of 50μm were generated for θ = 60 deg. On the other hand, the peak value in the distribution of the bubble diameter was 60μm for θ = 30 deg. In the θ=60 deg case, a large number of smaller bubbles can be produced in comparison with the θ=30 deg case.

2.2 Measuring system of zeta potential of microbubbles

Figure 3 shows the experimental apparatus used to measure the zeta potential of microbubbles with an enlarged view of the zeta potential measuring cell shown in Fig.4. In the present study, the surface charge of microbubbles was measured by an electrophoresis method. The zeta potential measurements

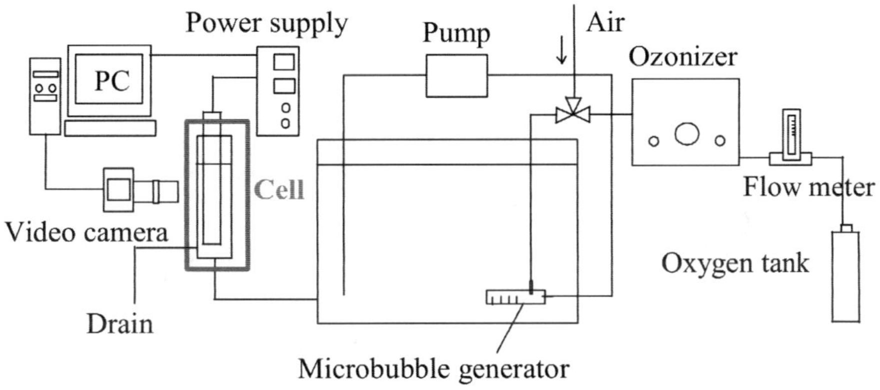

Fig.3 Experimental apparatus.

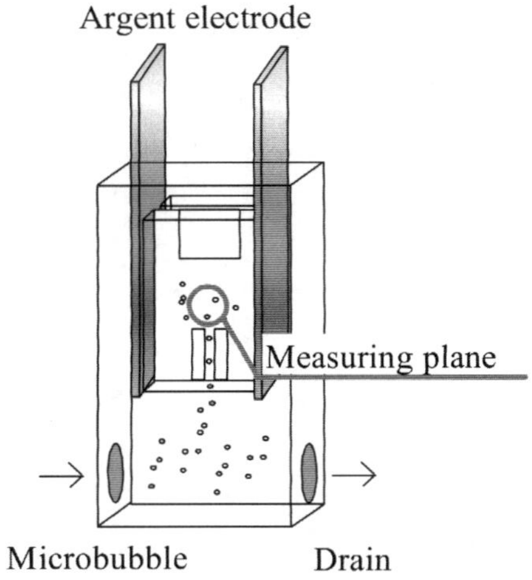

Fig.4 Enlarged view of zeta potential measuring cell.

were carried out by using a water reservoir with the microbubble generator, an electrophoresis cell consisting of two electrodes and a constant voltage power source (200 V), a pump, a camera system and a personal computer. The electrophoresis cell was composed of acrylic. In this measuring system, the direction of the flow of electrons in the measuring cell can be reversed per a constant time interval to clearly interpret the bubble movement in the horizontal direction. Microbubbles in the water tank were introduced to the electrophoresis cell and the movement of the bubbles in the horizontal direction was observed using a high speed camera. Light was scattered using a thin white sheet behind the cell for clear observation of the spherical bubbles in the cell. The horizontal movement of the microbubbles was analyzed under an electric field for the calculation of zeta potential using a graphical data processing method. The zeta potential was calculated from electrophoretic mobility in the horizontal direction using the Smoluchowski equation. (Takahashi, 2005)

3. RESULTS AND DISCUSSION

3.1 Microbubble Zeta potential

Figure 5 shows the zeta potential of microbubbles versus bubble diameter. The zeta potential of

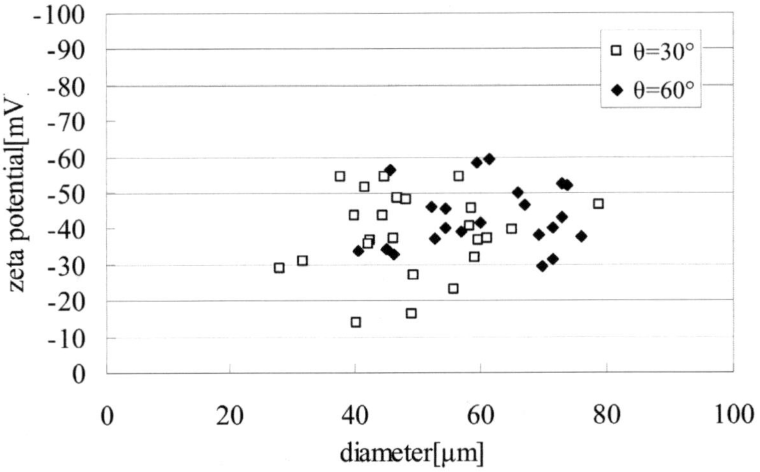

Fig.5 Zeta potential versus bubble diameter in purified water.

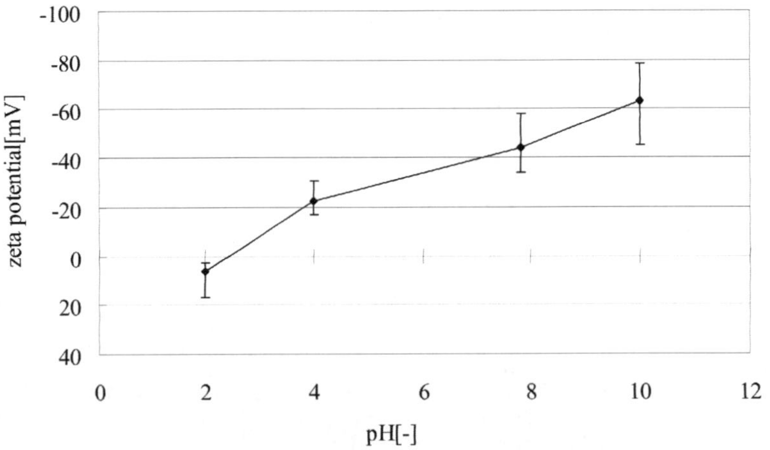

Fig.6 Averaged zeta potential versus pH.

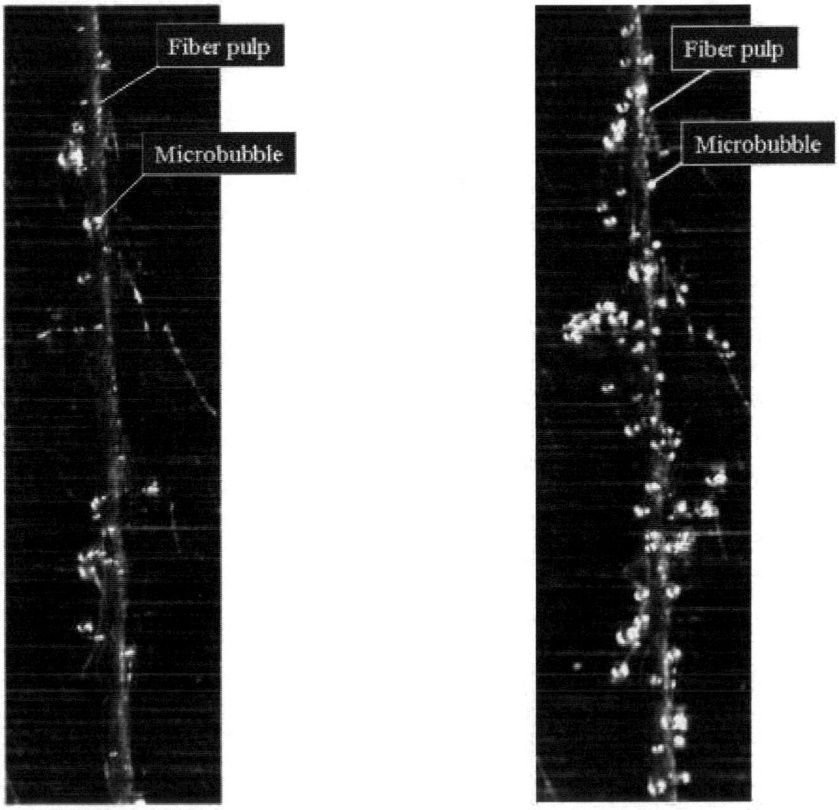

(a) pH 4 (b) pH 10

Fig.7 Enlarged view of adsorption of bubbles on a fiber.

microbubbles for θ=60 deg is larger than that for θ=30 deg. The microbubbles are negatively charged with an average zeta potential of approximately -38mV and -43mV in purified water (pH=7) for θ=30 and 60 deg, respectively. There was no relationship between the zeta potential of microbubbles and the bubble diameter, and the results suggest that the amount of electricity per unit surface area at the interface is independent of the bubble size. Figure 6 shows the zeta potential of microbubbles versus pH. The value of the pH was adjusted by the addition of HCl and NaOH to tap water. The zeta potential was negatively charged over a wide range of pH conditions and the negative potential decreased with decreasing pH. The changing mechanism has been explained as the adsorption of OH^- onto the interface by the difference in hydration energy between H^+ and OH^-. The hydrogen atoms point towards the water phase and oxygen atoms towards the gas phase, thereby attracting of anions to the interface. The adsorbed of H^+ and OH^- are crucial factors influencing the interface charge, and the electrolyte ions are attracted to the interface of the microbubbles. However, the zeta potential of the microbubbles is positively charged at pH=2, and it is difficult to explain the mechanism of a positively charged gas-water interface by the experimental result. To confirm theses phenomena, more detailed information regarding the gas-water interface charge is required.

In order to investigate the relationship between the zeta potential and adsorption of bubble onto the surface of a fiber, the behavior of microbubbles was observed using a high speed camera under different pH conditions. In this experiment, a fiber was centered vertically in the measurement cell and the microbubbles were discharged from the bottom of the cell. Bubble adsorption to the fiber is encouraged in the pH=10 case in contrast to the pH=4 case. The zeta potentials at pH=10 and 4 were measured to be approximately -50 mV and -10 mV, respectively (see Fig.6). The number of bubbles absorbed onto the surface of a fiber increased as the zeta potential of the microbubbles increased. That is, a huge amount of bubble adsorption was accomplished at large zeta potentials.

3.2 Ozone water generation using microbubbles

Figure 8 shows the variation in ozone water concentration after ozone microbubble generation. In the present study, measurements of ozone water concentration were performed by the colorimetric assay method. The ozone gas concentration of 80-100 ppm was used to generate ozone microbubbles. The ozone water concentration for θ = 60 deg becomes larger with elapsing time than that for θ = 30 deg. A high concentration of ozone water is achieved for the θ=60 deg case in contrast to the θ=30 deg case because the microbubbles with small size and large zeta potential are produced for θ=60 deg (see Figs.2 and 5). In other words, the result suggests that the ozone water generation is affected by the bubble size distribution and the electrical properties of microbubbles. In order to better understand ozone water generation, the relationship between the electrical potential of microbubbles and ozonation efficiency will be discussed later.

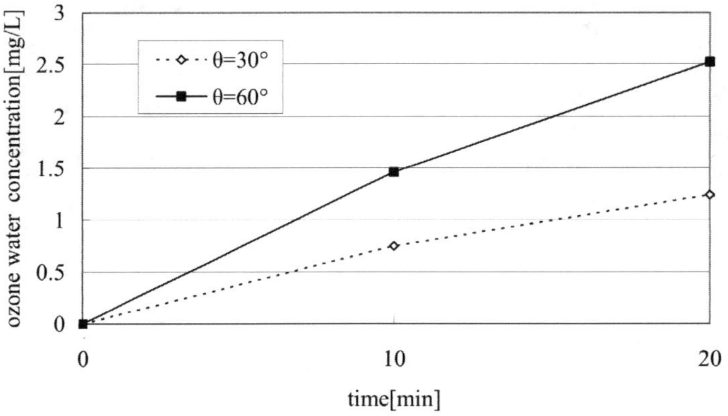

Fig.8 Ozone water concentration.

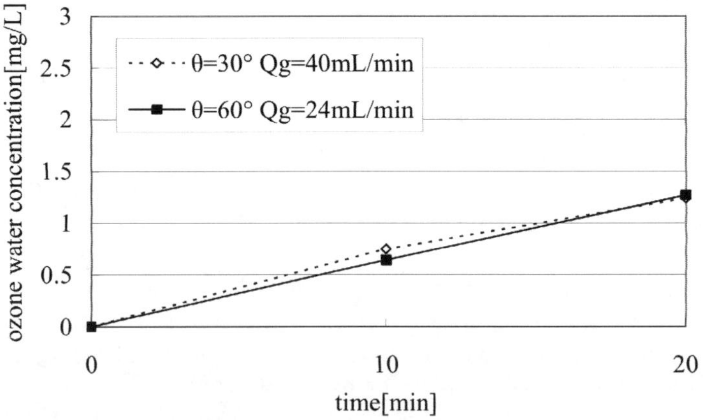

Fig.9 Ozone water concentration.

3.3 Soiled cloth cleaning using ozone microbubbles

In order to investigate the effect of the ozonated water generated by our proposed microbubble generator on cleaning power, experiments involving the cleaning of cloth were performed for the cases of θ= 30 and 60 deg. The cloths are artificially soiled by water including coffee powder and sauce. The soil was removed in the θ = 30 deg case after 40 minutes of generator operation. On the other hand, the same soil was removed in only 30 minutes in θ = 60 deg case. Furthermore, the laundered cloth in the θ = 60 deg case was cleaner than that for θ = 30 deg. In order to verify the effect of ozone water concentration from two generators on cleaning cloth, the same concentrations of ozone water were created by adjusting the ozone gas flow rate Q_g. Figure 9 shows the ozone water concentration for two

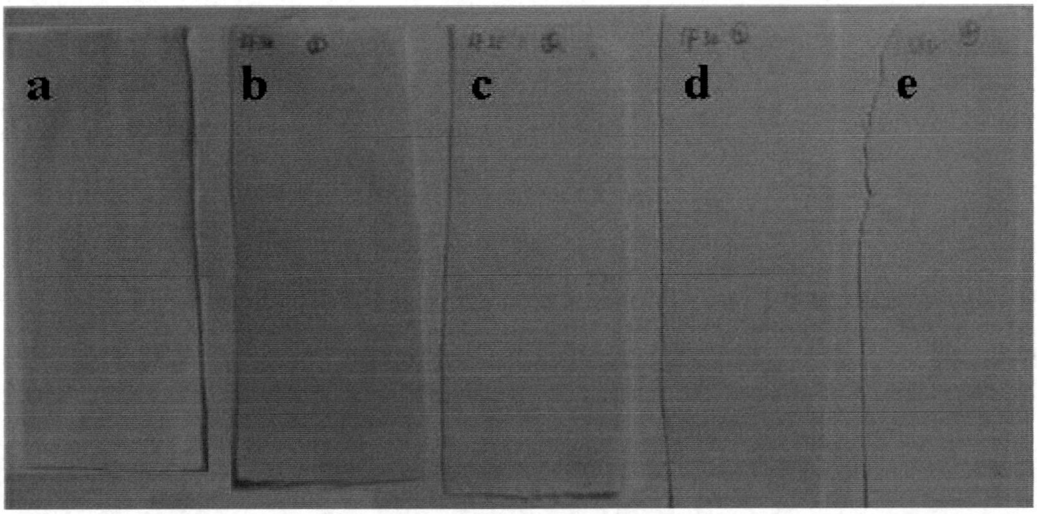

(a) θ =30 deg

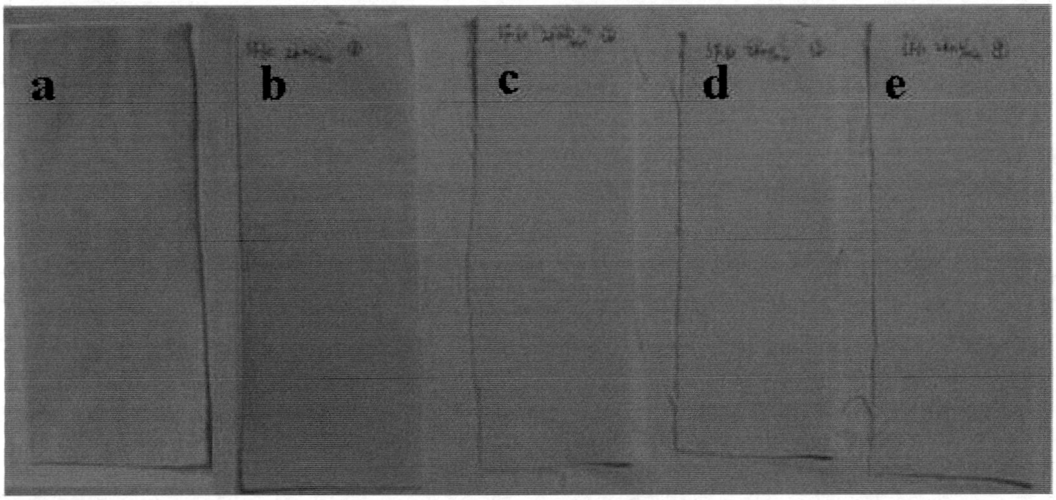

(b) θ =60 deg

Fig.10 Cleaning soiled cloths with coffee powder using ozone microbubbles. (a) Before washing; (b) 10 minutes washing; (c) 20 minutes washing; (d) 30 minutes washing; (e) 40

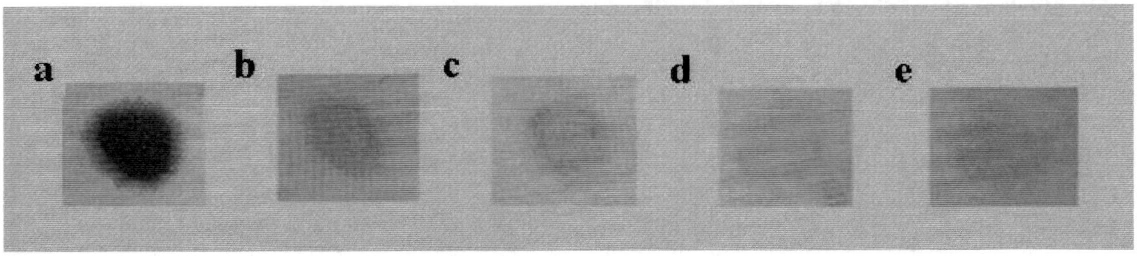

(a) θ =30 deg

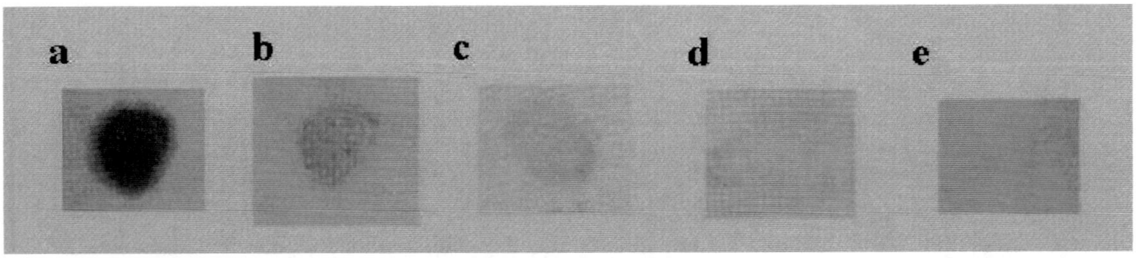

(b) θ =60 deg

Fig.11 Cleaning soiled cloths with sauce using ozone microbubbles. (a) Before washing; (b) 10 minutes washing; (c) 20 minutes washing; (d) 30 minutes washing; (e) 40 minutes washing.

generators in the case where the ozone gas flow rate was adjusted. Ozone water concentrations for θ = 30 and 60 deg show the same tendency. Figures 10 and 11 show the photographs of the soiled cloth during 40 minutes of the washing using ozone microbubbles under the same ozone water concentration., and Figs. 10 and 11 indicate the cases for soiled cloths with coffee powder and with sauce, respectively. Ozone microbubbles make soiled cloths clean effective for θ = 60 and 30 deg in both cases. In particular, the soiled cloth can be cleaned easily for the θ=60 deg case. That is, the θ=60 deg case shows superior performance in contrast to the θ=30 deg case for the washing using ozone microbubbles. The result suggests that washing using ozone microbubbles is affected by the electrical potential of microbubbles.

A bubble diameter of less than approximately 40 μm increases the zeta potential in the bubble shrinking process (Tokushige, 2008). It is presumable that a large number of small bubbles (less than 40 μm) exist in the bubble shrinking process for θ=60 deg because small bubble generation is achieved and the zeta potential becomes larger in contrast to the θ=30 deg case. Therefore, ozone microbubbles are easily absorbed onto the surface of the cloth for θ=60 deg, and the soil can be effectively removed from the cloth in comparison with the θ=30 deg case.

4. CONCLUSIONS

In the present study, the relationship between the electrical properties of microbubbles and the effect of ozonated water generated by ozone microbubbles on cleaning cloth was investigated. The findings of the present study are summarized as follows:

(1) In the proposed microbubble generator, the θ=60 deg slit angle exhibited superior performance due to its bubble size and electrical properties. That is, the microbubbles with small diameters and large electrical potentials are generated for θ=60 deg in comparison with the θ=30 deg case.

(2) The ozone water concentration with elapsing time for θ = 60 deg becomes larger than that for θ=30 deg.

(3) When a cloth is washed using the proposed microbubble generator with the θ=60 deg slit, the soiled cloth can be cleaned easily in comparison with the θ=30 deg case. That is, the θ=60 deg case shows superior performance in contrast to the θ=30 deg case for washing using ozone microbubbles.

(4) The zeta potential of microbubbles for θ=60 deg is larger than that for θ=30 deg. In cleaning cloth, the elapsed time during washing is improved by the microbubbles with large zeta potential, because the bubble adsorption onto the surface of a cloth is promoted due to the large electrical potential.

REFERENCES

Li-Bing, Chu (2007). "Enhanced sludge solubilization by microbubble ozonation", Chemosphere, 72 pp.205-212

Hasegawa, H (2007). "Micro-bubble Generation Using Shear Flow with Slits" The 1st International Colloquium on Dynamics, Physics and Chemistry of Bubbles and Gas-Liquid Boundaries, (ICBB2007), Niseko, Hokkaido, Japan

Takahashi, M (2005). "The ζ Potential of Microbubbles in Aqueous Solutions(Electrical property of the gas-water interface)" J. Phycs. Chem. B, 109, pp.21858-21864

Tabei, K (2004). "Study of Micro Bubble Generation by a Swirl Jet (Measurements of Bubble Distribution by Light Transmission and Characteristics of Generated Bubbles)", (in Japanese), Transactions of the Japan Society of Mechanical Engineers, Series B, 71-703, pp.848-853.

Tokushige, K (2008). "Characteristic of Electrical Potential of Microbubble", (in Japanese), J. JACT, Vol.13 No.2 pp.73-78.

Journal of Physics: Conference Series **147** (2009) 012025 doi:10.1088/1742-6596/147/1/012025

Study on Bubbly Flow Behavior in Natural Circulation Reactor by Thermal-Hydraulic Simulation Tests with SF_6-Gas and Ethanol Liquid

Yoshiyuki KONDO[*1], Keishi SUGA[*1], Koki HIBI[*1]

[*1] Mitsubishi Heavy Industries, Ltd.
Takasago Research and Development Center.
2-1-1, Shinhama, Arai-cho Takasago, Hyogo 676-8686, JAPAN
yoshiyuki_kondoh@mhi.co.jp

Toshihiko OKAZAKI[*2], Toshihiro KOMENO[*2]

[*2] The Japan Atomic Power Company

Tomoaki KUNUGI[*3], Akimi SERIZAWA[*4]

[*3] Kyoto University, [*4] Professor Emeritus of Kyoto University

Kimitoshi YONEDA[*5], Takahiro ARAI[*5]

Central Research Institute of Electric Power Industry

Abstract

An advanced experimental technique has been developed to simulate two-phase flow behavior in a light water reactor (LWR).

The technique applies three kinds of methods; (1) use of sulfur-hexafluoride (SF_6) gas and ethanol (C_2H_5OH) liquid at atmospheric temperature and a pressure less than 1.0MPa, where the fluid properties are similar to steam-water ones in the LWR, (2) generation of bubble with a sintering tube, which simulates bubble generation on heated surface in the LWR, (3) measurement of detailed bubble distribution data with a bi-optical probe (BOP), (4) and measurement of liquid velocities with the tracer liquid. This experimental technique provides easy visualization of flows by using a large scale experimental apparatus, which gives three-dimensional flows, and measurement of detailed spatial distributions of two-phase flow.

With this technique, we have carried out experiments simulating two-phase flow behavior in a single-channel geometry, a multi-rod-bundle one, and a horizontal-tube-bundle one on a typical natural circulation reactor system. Those experiments have clarified a) a flow regime map in a rod bundle on the transient region between bubbly and churn flow, b) three-dimensional flow behaviour in rod-bundles where inter-subassembly cross-flow occurs, c) bubble-separation behavior with consideration of reactor internal structures. The data have given analysis models for the natural circulation reactor design with good extrapolation.

1. INTRODUCTION

A power generation plant, such as nuclear and boiler, employs a high-temperature and high-pressure steam-water two-phase flow system for attaining high heat efficiency. An extensive effort has been directed to scale up of power plants. Though two-phase flow phenomena are investigated, those plants are difficult to design because of the complexity of flow characteristics, and their unknown behavior under different sets of design parameters such as temperature, pressure, and geometry.

For developing those systems, it is important to simulate phenomena in components constituting the systems. One of the important parameters on simulating is physical properties of fluids. Steam-water flow in a high-temperature and high-pressure system behaves quite different from in an atmospheric temperature and pressure system that is used for a basic hydraulic-simulation test. We have been developing an experimental technique using alterative fluids, sulphur hexafluoride (SF_6) gas and ethanol (C_2H_5OH) liquid.

This study focused on clarificaing steam-water bubbly flow behavior in a natural circulation reactor, as shown in Fig1.

Natural circulation is induced by hydraulic head difference between horizontal centres of a core and a heat-exchanger in the reactor vessel. The reactor vessel is pressurized by the steam from the boiled water itself. In this situation, there is free surface in the vessel. Steam and water phases will be separated at the surface unless any specific mechanistic device would be introduced. And a portion of steam will go up to a steam dome, while the rest of the steam will be carried under into a down-comer with the liquid. The steam is condensed and hot water is cooled down by heat-exchanger.

We have already carried out steam-water experiments under actual temperature and pressure conditions with a ϕ30mm-tube type apparatus from Japanese FY2001 to FY2004 as the phase-1 development, and have clarified that our natural circulation system is feasible. This study has also shown some issues to be solved;

1) two-phase flow with small bubbles around ϕ1mm keeps bubbly flow regime under void fractions less than 40%, in which the current data-base shows that flow regime goes into slag flow regime, commonly,
2) three-dimensional flow behavior in rod-bundles give large effects for the natural circulation system, because inter-subassembly cross-flow will occur in cases like that a subassembly with a small void fraction is loaded next a subassembly with large void fraction,
3) bubble-separation behavior must be evaluated with consideration of reactor internal structures, including the heat exchanger itself, around the free surface because of lack of data-base on the configuration.

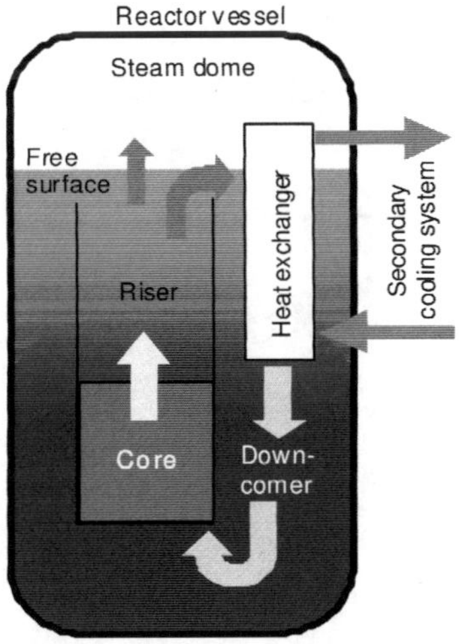

Fig. 1 Outline of integrated primary system reactor

As the phase-2 development, we have performed thermal-hydraulic simulation tests with an advanced experimental method from Japanese FY2005 to FY2007. The tests have simulated rod-bundles, core-internal structures in the riser and structures around the free surface.
This paper describes about the experiments and recommended evaluation methods for reactor designs of the natural circulation water reactor.

2. EXPERIMENTAL METHODS

2.1 Fluid Materials

Table 1 shows a comparison of physical properties of fluids. On the SF_6-C_2H_5OH system at 30°C and 0.7MPa, the gas density, gas-liquid density ratio and surface tension are close to those of the steam-water system at 270°C and 6MPa. For bubbly flow, those parameters have effects on a bubble behavior, such as terminal velocity, flow regime, and pressure drop.

The advantage of this experimental technique is that a test in the atmospheric temperature and a lower pressure easily simulates two-phase flow phenomena in a high-temperature and high-pressure steam-water system and detailed observations and measurements of two-phase flow data, such as void fraction, diameter, velocity, and inteference with a structure in a flow channel.

This experimental technique has been used for a study on a stable droplet in a fuel rod of a pressurized water reactor (PWR) under a loss of coolant accident (LOCA) [1].
In the study, the steady droplet diameter and the effects of system pressure and the gas velocity on it was clarified, and also the behavior of droplets was observed.

Table 1 Fluid Properties

Fluid	Unit	This study			Actual reactor	Conventional fluids	
		Sulfur hexafluoride gas (SF_6) Ethanol liquid (C_2H_5OH)			Steam Water	Air-Water	Freon (R123)
System pressure	[MPa]	0.3	0.5	0.7	8.7~15.5	0.1	0.77
Temperature	[° C]	30			300.9~344.8	30	97
Gas density	[kg/m³]	18.0	30.8	44.2	46.9~101.9	1.2	45
Liquid density	[kg/m³]	784			710~594	998	1214
Gas viscosity	[Pa s]	1.58×10^{-5}			$1.98\text{~}2.30\times 10^{-5}$	1.87×10^{-5}	1.3×10^{-5}
Liquid viscosity	[Pa s]	1.08×10^{-3}			$8.55\text{~}6.81\times 10^{-5}$	7.81×10^{-4}	1.96×10^{-4}
Surface tension	[mN/m]	22.2			14.2~4.7	71	7.1
Density ratio (L/G)	[-]	43.6	25.5	17.7	15.2~5.8	833	27
Density difference (L-G)	[kg/m³]	766	753	740	663~492	986.8	1169

2.2 Bubble Generation and Measurement

Our experiments was adopted a bi-optical probe (BOP) to measure void fractions, bubble diameters and interfacial velocities. Fig. 2 shows a tip of the BOP. A clearance between the sensor needles of the tip is about 1mm. The BOPs was installed in the test section and traversed over the cross section of the flow channels.

Bubble sizes to be measured are around 0.1mm described in the following. In this study, an applicability of the BOP for measuring such small bubbles was confirmed by comparing high-speed video, recording bubble contacts at the tip of the BOP, and measured BOP signals.

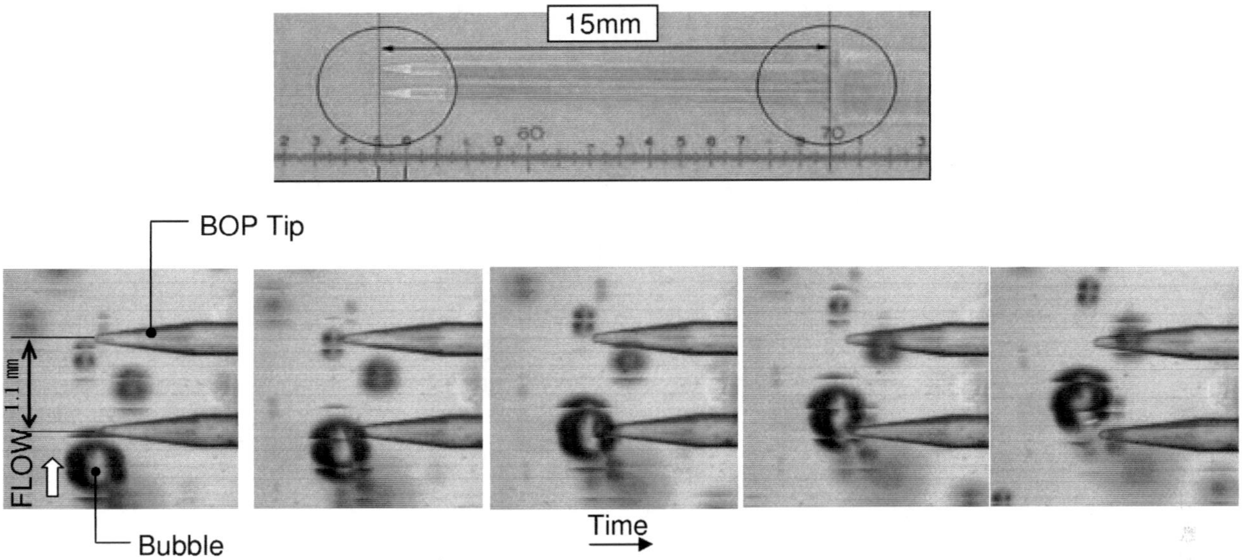

Fig. 2 Bubble Flow Measurement with Bi-Optical Probe (BOP)
(Air-Water System)

2.3 Liquid Flow Measurement by Tracer

The liquid cross-flow in the tube bundles was measured with a tracer-technique. Fig.3 shows a schematic of tracer concentration measurement. The tracer is a fluorescence dye, and fed from the one side of the bundle with a uniform concentration. The tracer is diffused by turbulent and cross-flow due to differential pressure. The tracer concentration is measured with an isokinetic suction method. The liquid cross-flow rate is calculated from the tracer concentrative distributions and the differences between the upstream and downstream. An averaged tracer concentration is calculated in each bundle measurement elevation.

2.4 Bubble Generation Method

This study has focused on bubble behavior in developing bubbly flow, collapse and/or coalescence, along the rod-bundle and the riser. Therefore, it is important to generate small-size bubbles as same as that in the actual reactors. We have established a method to generate such small bubbles with a sintering tube. The bubble sizes have been measured and compared with sizes evaluated by the empirical formula for departure diameter proposed by Levy [2] (see Fig. 4).

Estimated diameters of the departure bubbles are about 0.1~0.3mm under the actual conditions in the LWR core. Fig. 5 shows a bubble diameter distribution measured near the sintering tube surface. The distribution is ranging around 0.05~1mm and the peak is at 0.1mm. The bubble generation device with the sintering tube is confirmed to simulate bubble generation phenomenon from the heated wall in the actual reactors.

$$Y_B = C\left(\frac{\sigma D_H}{\tau_w}\right)^{\frac{1}{2}}\left[1 + C'\frac{g}{g_c}\frac{(\rho_L - \rho_V)D_H}{\tau_w}\right]^{-\frac{1}{2}} \qquad (1.1)$$

$$C=0.015, \qquad C'=0$$

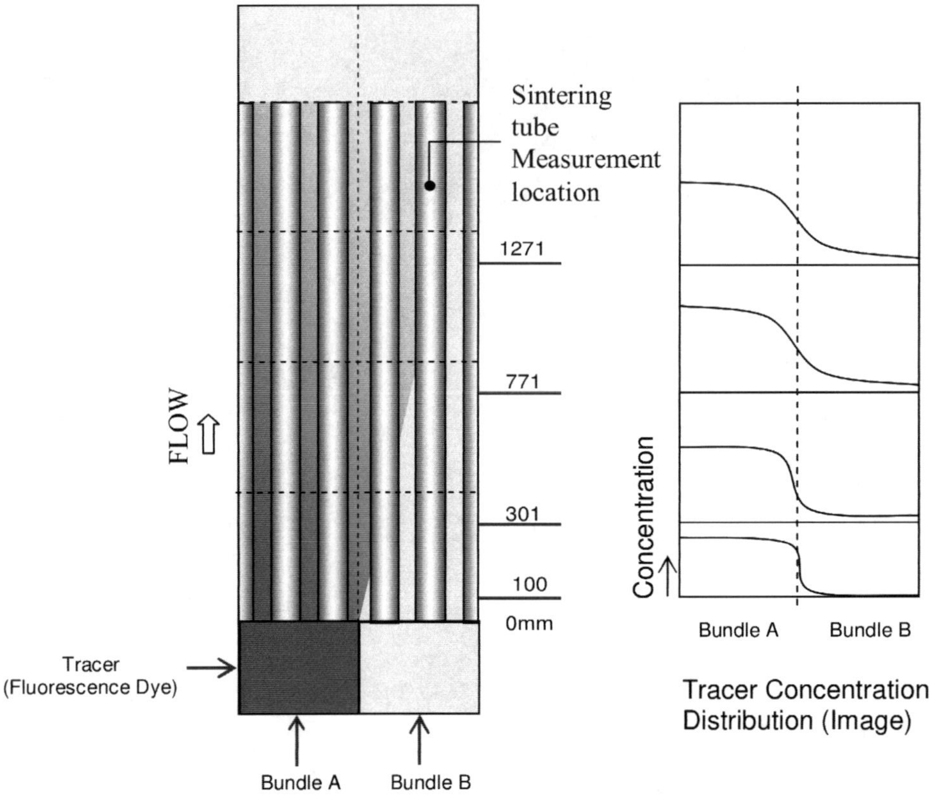

Fig. 3 Tracer concentration measurement

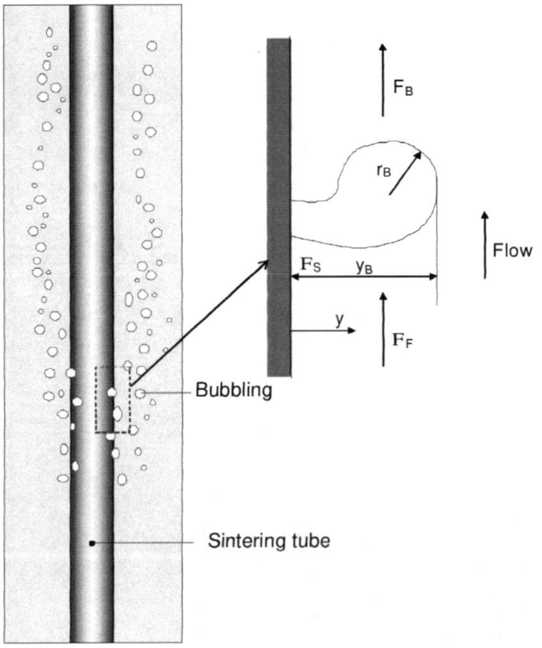

Fig. 4 Bubble prior to departure from heated surface

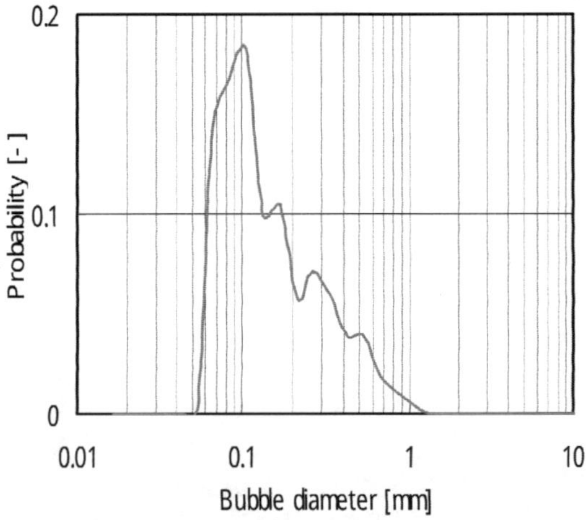

Fig. 5 Bubble diameter distribution measured near the sintering tube (P=0.7MPa[abs], T=30°C) .

3. TESTS and Results

3.1 Rod-Bundle Test

Rod-bundle tests were carried out in order to clarify pressure drop and inter-subassembly cross-flow performances in the core which consists from rod-bundles.

(a) Test Apparatus

Fig. 6 shows the schematic diagram of experimental apparatus for a rod-bundle simulation test. The gas was fed by a blower and the liquid by a pump. Each of flow rates was measured by the orifice flow meters.

The test section is shown in Fig. 7. The gas was injected at void-generation devices and recovered at each bundle exit. A gas buffer volume was placed at the apparatus top because pressure distributions give effects on inter-subassembly cross-flow phenomena of gas flow. A part simulating a core configuration consists from two bundles. The bundle has a 4x3 rod array. The outermost rods were half cylinders and four coner rods were quarter ones. The geometry reduces wall effects on flow channels near wall. Bubbles were not generate from the quarter cylinders in order highly to simulate two-phase flow phenomena in the actual rod bundles in the core.

The actual subassembly had a flat-plate grid to support the rod-bundle and the grid blocks out the cross-flow. Hence, we carried out tests with and without the grids to confirm grid effects. (See Fig. 7(d))

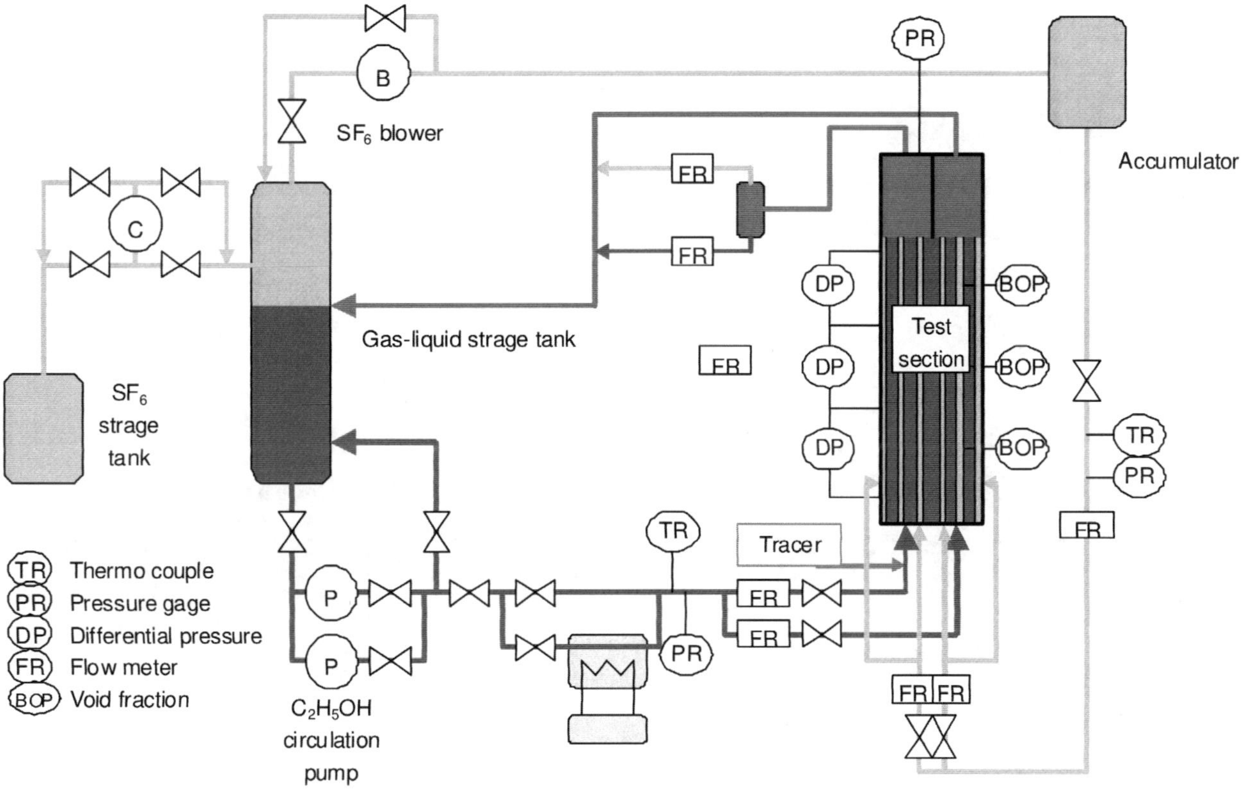

Fig. 6 Schematic diagram of experimental apparatus

Journal of Physics: Conference Series **147** (2009) 012025 doi:10.1088/1742-6596/147/1/012025

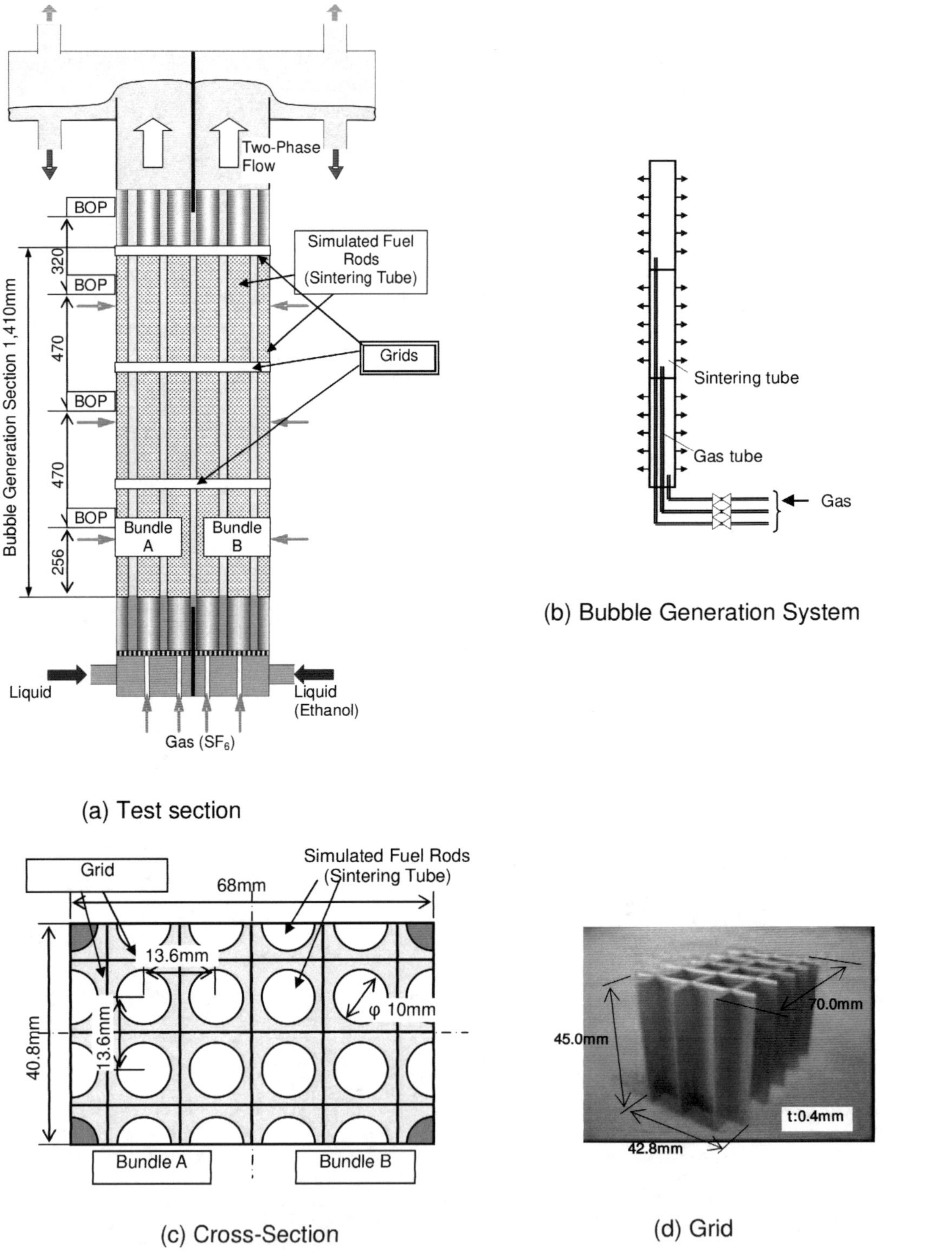

(a) Test section

(b) Bubble Generation System

(c) Cross-Section

(d) Grid

Fig. 7 Test section structure of rod-bundle simulation test

The liquid cross-flow in the tube bundles was measured with a tracer-technique. The probe was traversed over the cross section of the test section as shown in Fig. 8. The gas cross-flow rates were estimated by the measured void fraction and velocity.

We carried out simulation tests under several conditions shown in Table2. By using obtained data, we investigated pressure drop and inter-subassembly cross-flow performances.

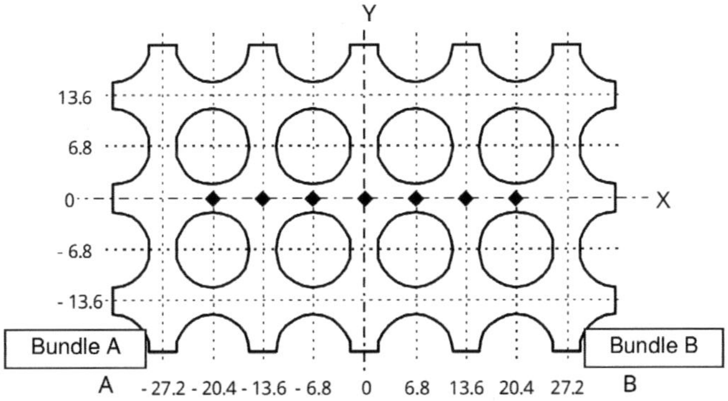

Fig. 8 Tracer concentration measurement location

Table 2 Experimental cases of rod-bundle simulation test

	No.	Superficial liquid velocity j_ℓ (m/s)		Superficial gas velocity j_g(m/s)	
		A	B	A	B
Without Grid	1	0.50	0.50	-	-
	2	1.01	1.01	-	-
	3	1.51	1.51	-	-
	4	2.01	2.01	-	-
	26	0.50	0.50	0.06	0.06
	12	1.00	1.01	0.12	0.13
	18	1.50	1.49	0.17	0.18
	41	0.51	0.51	0.48	0.48
	13	1.01	1.00	0.93	0.94
	33	1.52	1.51	1.41	1.40
	43	0.50	0.50	0.49	0.06
	49	1.00	1.00	0.97	0.12
	31	1.51	1.50	1.40	0.18
	39	1.01	1.00	1.00	0.12
With Grid	G-26	0.50	0.51	-	-
	G-27	1.01	1.01	-	-
	G-28	1.52	1.52	-	-
	G-29	2.02	2.02	-	-
	G-41	1.49	1.49	1.43	1.43
	G-43	1.01	1.01	0.96	0.96
	G-39	0.49	0.49	0.47	0.05
	G-45	1.00	1.00	0.95	0.12

(b) Results

1) Flow Pattern

Fig. 9 shows the bubble appearances observed at the outlet of the test section. Bundle A and B are high and low void fraction conditions of uniform inlet liquid velocity, respectively. Bubbles flowed without coalescence, even though, the bubbles were dense-packed and contacting each other, and the flow pattern was bubbly flow.

Fig. 10 shows a flow pattern map. A radius of a circle in the Fig.10 indicates a normalized standard deviation of differential pressure fluctuation in the measurement.

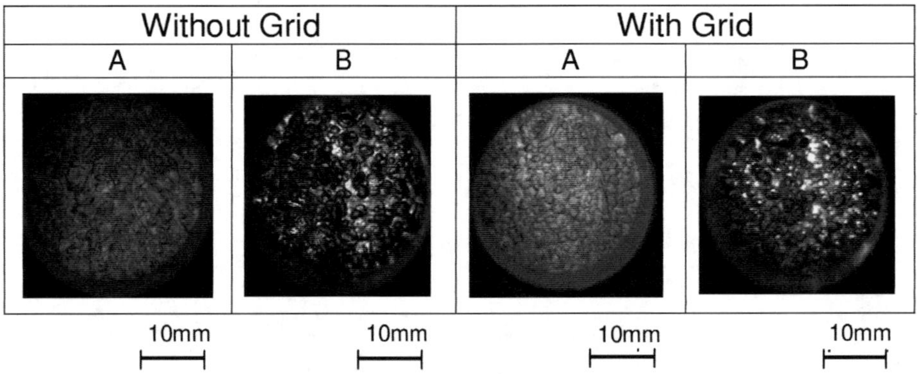

	A	B
j_g [m/s]	0.96	0.12
j_l [m/s]	1.0	1.0

Fig. 9 Bubbles observed at the outlet of the test section

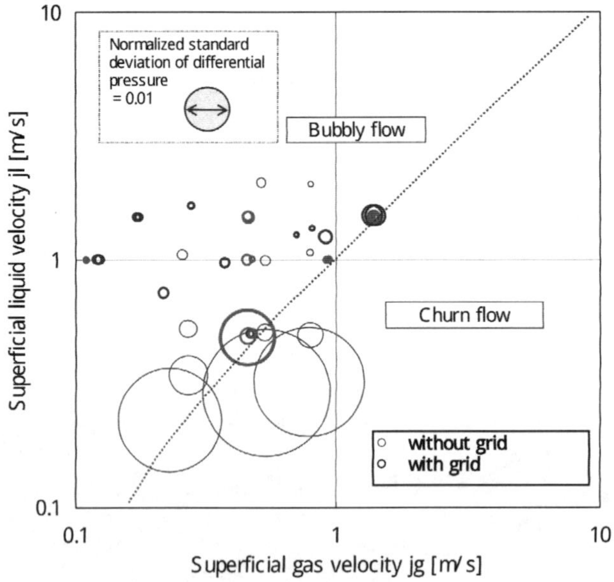

Fig. 10 Flow pattern in fuel bundle

The doted line is drawn for flow regime transition between bubbly and churn flow by the modified Mishima-Ishii's equations [3], which are as follows;

$$\frac{j_g}{\alpha} = C_0\left(j_g + j_l\right) + V_{gj} \tag{3.1}$$

$$C_0 = \left[C_\infty - (C_\infty - 1)\cdot\sqrt{\frac{\rho_g}{\rho_l}}\right]\cdot\left(1 - \exp(-18\cdot\alpha_{RB})\right) \tag{3.2}$$

$$C_\infty = 1.2 \tag{3.3}$$

$$V_{gj} = \sqrt{2}\cdot\left(\frac{\sigma\cdot g\cdot\left(\rho_l - \rho_g\right)}{\rho_l^2}\right)^{0.25}\cdot\left(1 - \alpha_{RB}\right)^{1.75} \tag{3.4}$$

$$\alpha_{RB} = f\cdot\alpha \tag{3.5}$$

$$\alpha = 0.3 \tag{3.6}$$

$$f = \frac{4/\pi\left(Pt/D_P\right)^2 - 1}{\left(\sqrt{2}\cdot Pt/D_P - 1\right)^2} \tag{3.7}$$

The equations are considered a geometrical effect on void fractions. Void fractions averaged over the cross-section area of rod bundle are higher than void fractions around the flow channel center and lower around the rod surface. The above equations apply the void fractions around the flow channel center by using Eq. (3.7).

2) Inter-Subassembly Cross-Flow

Fig. 11 shows horizontal distributions of the void fractions and the bubble daimeters. The experimental condition was j_g=0.96m/s for the BundleA, 0.12 for the BundleB, and uniform inlet liquid velocity j_l=1.0m/s, For H=796mm, the void fractions and bubble diameters in the bundle gap was higher than that in the flow center. For H=1196mm, the void fractions and bubble diameters in the flow center was higher than that in the bundle gap. This was because the generated bubbles flowed near the sintering tube coalesced and developed in the flow direction gathering around the flow center. The effect of the gird on the cross-flow was small.

Fig. 12 shows the tracer concentrations and those changing behavior along the tube bundle. Fig. 13 shows the comparison of cross-flow rate between the without and with grid. The effect of the gird on the cross-flow was small.

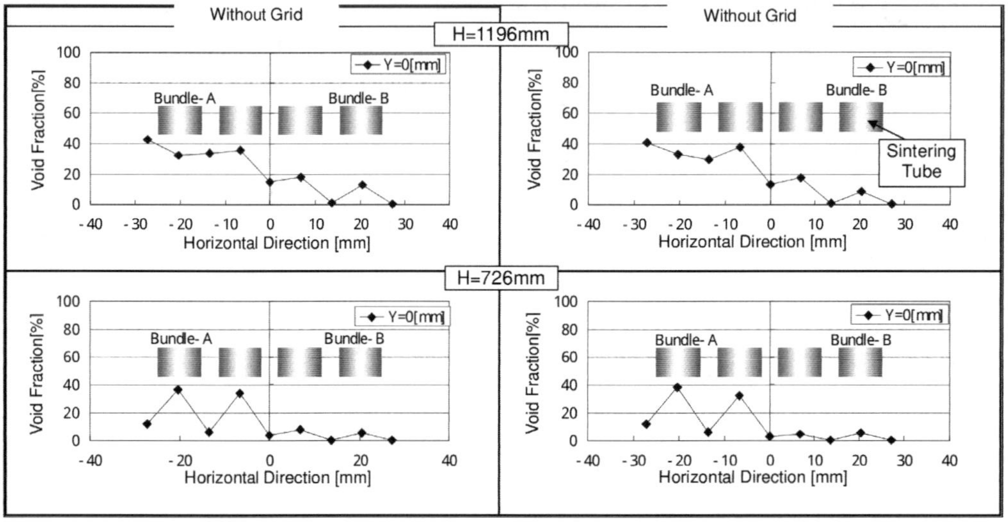

(a) Void fraction

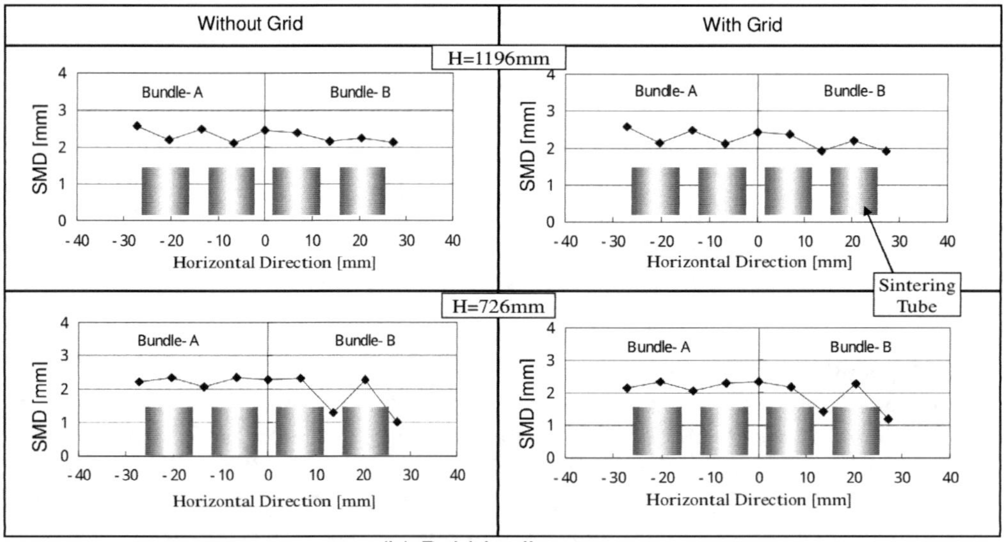

(b) Bubble diameter

Fig. 11 Void fraction and bubble diameter in rod bundle

Bundle-A: j_g : 0.96m/s, j_l :1.00m/s, Bundle-B: j_g : 1.00m/s
Measurement height : H=726mm, 1196mm

Journal of Physics: Conference Series **147** (2009) 012025 doi:10.1088/1742-6596/147/1/012025

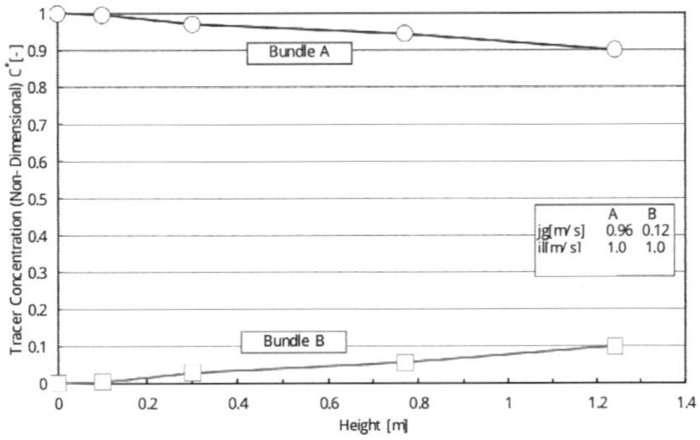

Tracer Concentration

Fig. 12 Liquid Cross-Flow Measured by Tracer Technique
(jg(A,B) = (0.96,0.12), jl(A,B) = (1.0,1.0), with grid)

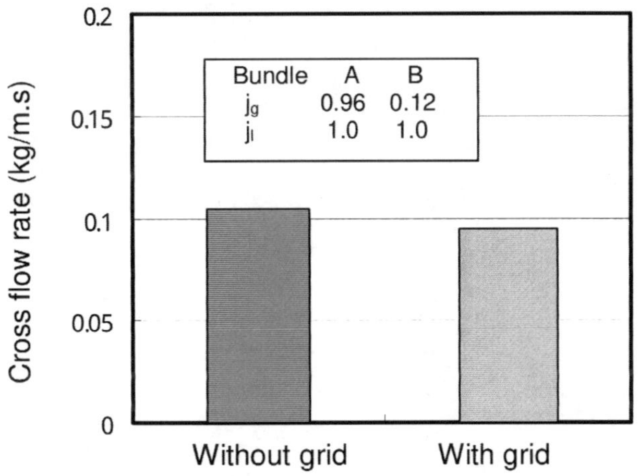

Fig. 13 Cross-flow between bundle A and B with/without grid
(jg(A,B) = (0.96,0.12), jl(A,B) = (1.0,1.0))

3.2 Gas-Separation Test

The gas separation test was carried out in order to clarify bubble separation performances at the liquid free surface and a herical-type steam generator under downward flow.

(a) Test apparatus

The test apparatus, as shown in Fig. 14, consisted of four parts, i.e a gas generation section, a riser simulation section, a down-comer simulation section, and a horizontal junction which combined the riser and down-comer simulation sections. SF_6 gas was injected at bubble generation devices, and recovered at the top of the riser simulation section without liquid and the bottom of the down-comer simulation section with liquid.

In the down-comer section, the horizontal tubes were placed to simulate a steam generator with helical-type tubes. The tube specifications were a 1/1-scale of a typical steam generator because a test fluid velocity range was similar to the actual reactor operation conditions. The bottom of the section is narrowed to simulate downward flow by heat removal around the steam generator.

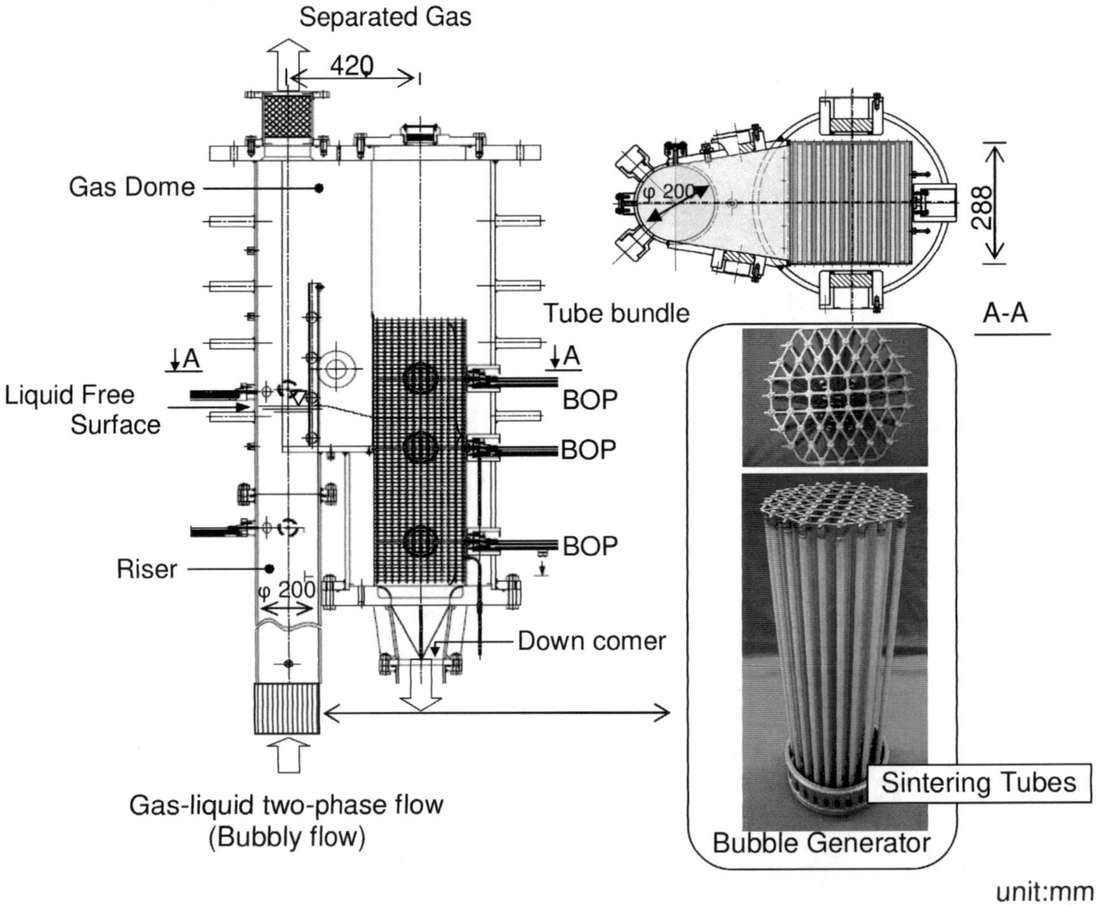

Fig. 14 Test apparatus structure of core-internal simulation test

Table 3 shows the experimental conditions. The parameters are the gas and liquid velocities, and the liquid free surface level.

Gas separation ratios from the liquid free surface were measured by the flow meter set at the outlet of the gas dome.

Bubble flow behavior near the liquid surface was observed though the observation port mounted on the side wall and the top of the test section.

Void fractions were measured with the BOP traversing it in the horizontal flow direction.

Table 3 Experimental Conditions

System Pressure	0.7MPa
System Temperature	30° C
Superficial Gas Velocity	0.2~0.4 m/s
Superficial liquid Velocity	0.35~0.55 m/s
Liquid Free Surface Level	100~300mm

(b) Results

Fig. 15 shows the observation of bubble appearances around free surface and horizontal tubes.

The bubble near the free surface coalesced and became larger, and finally, collapsed and separated at the surface. When putting riser-plate-slit, there existed a vortex at the down stream of the slit. The liquid free surface level was fluctuating.

In the horizontal tubes bundle, the bubble was absorbed near the horizontal free surface. The bubbles flowing in the tube bundle were trapped at the weak of the tubes, and then coalesced to become larger, sometimes slug. A number of large bubbles of which terminal velocity was higher than downward liquid velocity increased to the free surface and collapsed.

By using obtained data, we have investigated bubble separation performances.

Typical horizontal distributions of void fractions with and without the slit are shown in Fig. 18. The both of the void fraction distributions in the riser simulation section were convex upward and it was estimated that upward velocity distributions were bulging.

In the horizontal flow, the void fractions became smaller along downward, and it was estimated that bubble separation rates were almost constant in the case of the no-slit test. In the case of the slit test, the void fractions downstream of the slit were larger and bubble separation ratios were larger than the no-slit test. This was because the gas was drawn into a whirlpool in that area.

Hence, in the riser section, it is assumed that bubble separation ratios are proportional to ratios of void fractions around the free surface to ones in the riser inlet. The latter ratios were evaluated by the Ishii equation in the Drift-Flux Model [7] and the former are by the Wilson equation [8], which were constructed bubbly flow tests under natural circulation velocity.

< Bubble separation ratio >

$$\beta = 1 - \frac{\alpha_s}{\alpha_r} \tag{3.8}$$

< Ishii eq. [7]>

$$\frac{j_g}{\alpha_r} = C_o(j_g + j_l) + \sqrt{2}\left(\frac{\sigma g \Delta\rho}{\rho_l^{\,2}}\right)^{1/4} \tag{3.9}$$

< Wilson eq. [8]>

$$\alpha_s = a\left(\frac{\rho_g}{\rho_l - \rho_g}\right)^{0.12}\left(\frac{\Sigma}{D_H}\right)^{0.1} Fr^b \left(\frac{j_g}{j_g + j_l}\right)^{0.6} \tag{3.10}$$

$$\Sigma = \sqrt{\frac{\sigma}{g(\rho_l - \rho_g)}} \tag{3.11}$$

$$Fr = \frac{j_g}{\sqrt{\Sigma g}} \tag{3.12}$$

where, $Fr < 1.5, Fr > 10 : a = 0.564, b = 0.67$
$1.5 \leqq Fr \leqq 10 : a = 0.619, b = 0.47$

In the horizontal junction, the separation ratios was evaluated by a simple model, as shown in Fig. 19, by using a horizontal velocity U and a vertical terminal velocity V. A bubble with velocity of U and V is separated when a vertical migration length of the bubble is longer than dH_{si} before the bubble moves along a horizontal migration length dL. The V is evaluated by the Drift-Flux model with $C_0=1$, $j_l=0$, which values are in vertical open flow-channels, as follows;

< Bubble terminal velocity>

$$V = \sqrt{2}\left(\frac{\sigma g \Delta\rho}{\rho_l^2}\right)^{1/4}(1-\alpha_h)^{0.75} \tag{3.13}$$

The gas separation rate is obtained by integrating a product of bubble terminal velocity and its rising time in a differential section dL.

$$\begin{aligned} Q_{sep} &= \sum_i^N dH_{si} \cdot dA_{f\,i} \\ &= \sum_i^N \frac{dL_i}{U_i} \cdot V_i \cdot dA_{f\,i} \end{aligned} \tag{3.14}$$

The void fraction in each differential section is accounted for calculating terminal velocity. The void fraction is calculated with Smith's equation [9].

$$\alpha_h = \left[1 + \frac{\rho_g}{\rho_l} e\left(\frac{1}{x} - 1\right) + \frac{\rho_g}{\rho_l}(1-e)\left(\frac{1}{x} - 1\right) \cdot \left\{\frac{\rho_l/\rho_g + e(1/x - 1)}{1 + e(1/x - 1)}\right\}^{1/2}\right]^{-1} \tag{3.15}$$

where, $e = 0.4$

As shown in Fig. 20, we have confirmed that the above evaluation method gives good agreements within ±20%. The method is available in bubble separation ratio evaluations under actual thermal-hydraulic conditions in the core on the natural circulation reactors because of considering fluid properties.

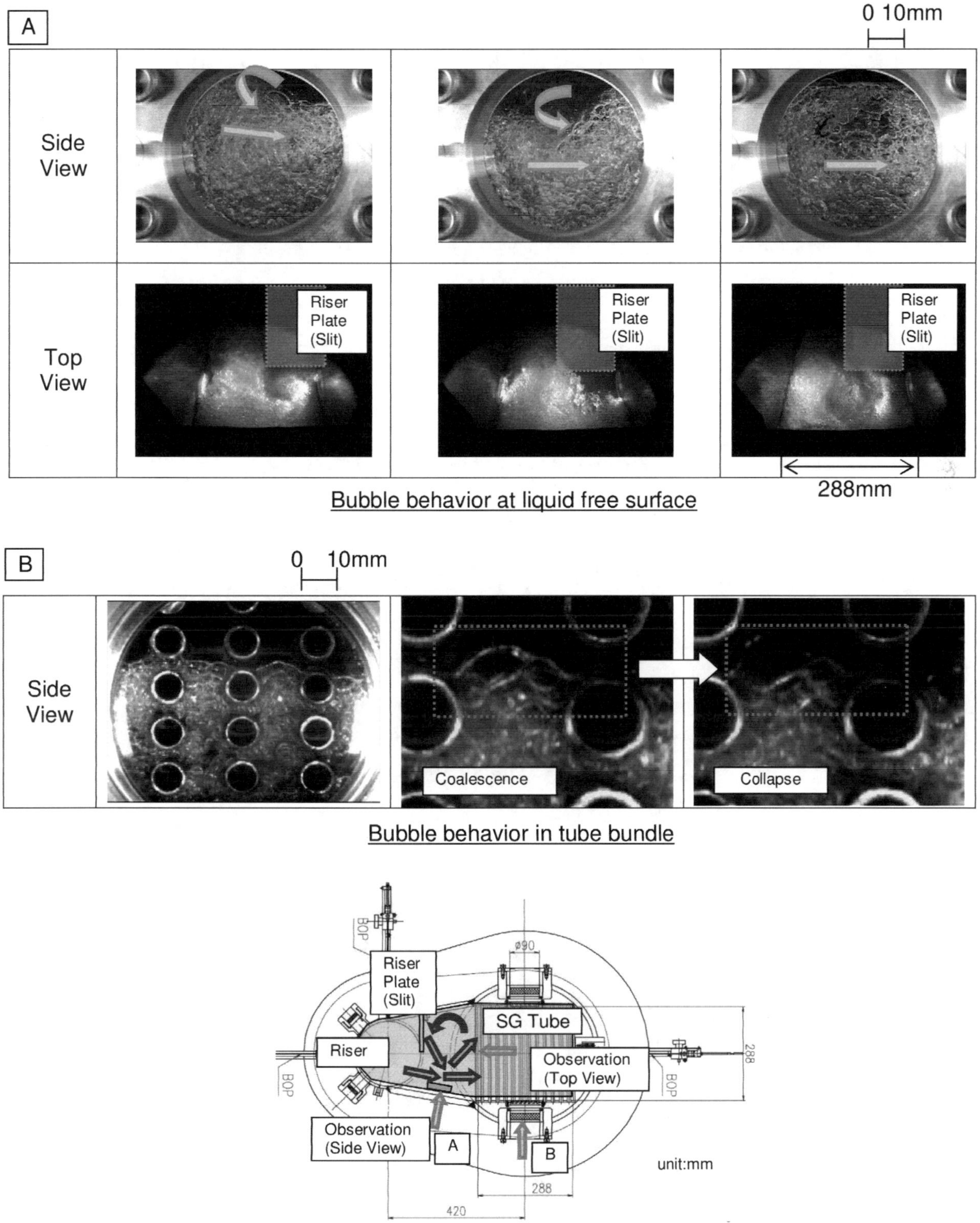

Fig. 15 Bubble behaviour at liquid surface and tube bundle

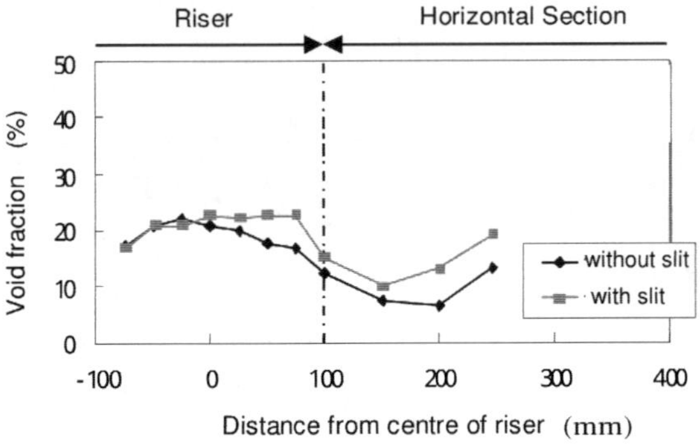

Fig. 16 Void fraction distribution around free surface

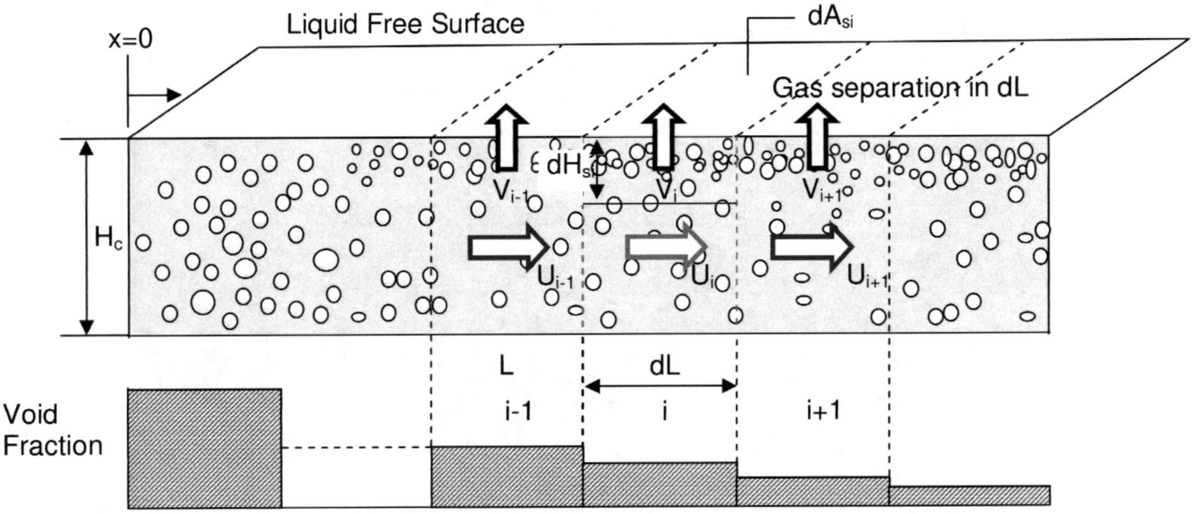

Fig. 17 Bubble separation model in horizontal junction

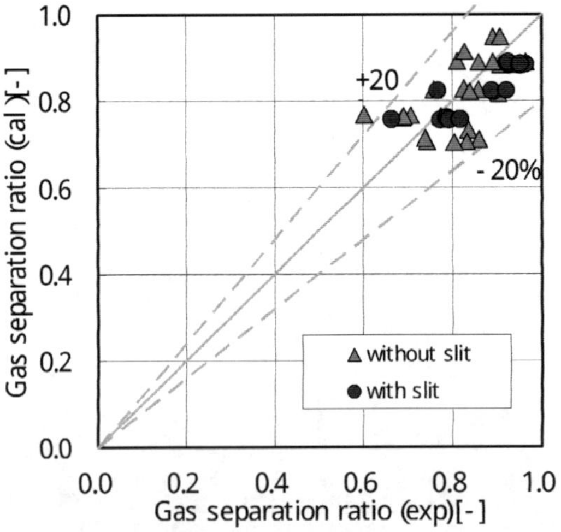

Fig. 18 Comparison of gas separation ratio between experiment and calculation

4. CONCLUSIONS

In order to obtain steam-water flow data in the natural circulation LWR, we have carried out gas-liquid two-phase flow tests with the advanced experimental technique, using SF_6-gas and ethanol liquid, measuring with BOP for gas phase, and tracer method for liquid phase. The tests simulated typical two-phase flow phenomena in the reactor vessel of the LWR. The tests have given very detailed and highly visualized data, and so we have established the empirical formulas for evaluating two-phase flow behavior.

ACKNOWLEDGEMENTS

The authors acknowledge the Institute of Applied Energy, and Ministry of Economy, Trade and Industry, Japan for their funding and support to this development since 2005. The authors are also grateful to other members of the project team in Kyoto University, Central Research Institute of Electric Power Industry (CRIEPI), The Japan Atomic Power Company (JAPC), and Mitsubishi Heavy Industries, Ltd. for their suggestions and discussions through this study.

NOMENCLATURE

A:	flow passage cross-section	[m^2]
A_f:	free surface area	[m^2]
A_T:	flow passage cross-section at orifice	[m^2]
Co:	distribution parameter	[-]
D_H:	hydraulic equivalent diameter	[m]
Fr:	Froude number	[-]
G:	total mass flux	[kg/m^2s]
g:	gravity acceleration	($9.8 m/s^2$)
j:	superficial velocity	[m/s]
Q:	volume flow rate	[m^3/s]
x:	quality	[-]
L:	distance from the origin	[m]
U:	horizontal velocity	[m/s]
V:	vertical velocity	[m/s]

(Greek letters)

α:	void fraction	[-]
μ:	viscosity	[Pas]
ρ:	density	[kg/m^3]
Σ:	Laplace length	[m]
σ:	surface tension	[N/m]
φ_L^2:	two-phase multiplier	[-]
χ_{tt}^2:	Lockhart-Martinelli parameter	[-]

(Subscripts)

g :	gas-phase
h :	horisontal
l :	liquid-phase
r :	riser
s :	surface
sep :	separation

Journal of Physics: Conference Series **147** (2009) 012025 doi:10.1088/1742-6596/147/1/012025

REFERENCES

1. Y.Kondo, "Experimental Study of Stable Droplet in Blowdown Phenomenon of Large Break LOCA," ICONE12-49349(2004)
2. S. Levy, "Forced Convection Subcooled Boiling Prediction of Vapour Volumetric Fraction", Int. J. Heat Mass Transfer. Vol.10, pp. 951-965. Pergamum Press Ltd. (1967)
3. D. Chisholm, "A Theoretical Basis for the Lockhart-Martinelli Correlation for Two-Phase Flow", Int. J. Heat Mass Transfer, 10, 1767-1778 (1967)
4. C. J. Baroczy, "A Systematic Correlation for Two-Phase Pressure Drop", Chem. Eng. Prog. Symp. Ser., 62, 232-249(1966)
5. R. W. Lockhart and R. C. Martinelli, "Proposed Correlation of Data for Isothermal Two-Phase, Two-Component Flow in Popes", Chem. Eng. Progress, 45-1, 39-4 (1949)
6. M. Ishii, "One-Dimensional Drift-Flux Model and Constitutive Equations for Relative Motion Between Phases in Various Two-Phase Flow Regimes", ANL-77-47 (1977)
7. M. Ishii, "One-Dimensional Drift-Flux Model and Constitutive Equations for Relative Motion Between Phases in Various Two-Phase Flow Regimes", ANL-77-47 (1977)
8. Wilson, S et al., "Preliminary Separation of Steam from Water by Natural Separation Part 1," Allis-Chalmers Atomic Energy Report No. ACNP-65002, (1965)
9. S.L.Smith, "Void Fractions in Two-Phase Flow, A Correlation Based upon an Equal Velocity Head Model", Proc. Instn. Mech. Engrs., 184 Pt 1, 36, 647-664, (1969-70)

Experimental investigation of laser-induced bubble dynamics near elastic/soft material in distilled water

S. Nakajima [1], Y. Yamamoto [1], M. Ota [2], K. Maeno [2]

[1] Graduate Student, Graduate School of Engineering, Chiba University, 1-33 Yayoi, Inage, Chiba, 263-8522, JAPAN
[2] Graduate School of Engineering, Chiba University, 1-33 Yayoi, Inage, Chiba, 263-8522, JAPAN

maeno@faculty.chiba-u.jp

Abstract. This study deals with an experimental investigation of the dynamics of laser-induced single bubble near the rigid material plate and near the elastic/soft material plate in the distilled water at room temperature under atmospheric pressure. A pulsed Nd:YAG laser was focused into the distilled water to make plasma and single bubble. The bubble repeated expanding and shrinking motion several times, and then collapsed. This behavior occurred on the sub-millisecond timescale. The solid wall near the bubble makes an asymmetric flow field. Many experiments on the behavior of laser-induced bubble near the rigid material have been reported. The bubble near the solid wall moves toward the rigid wall during its shrinking and rebounding process. The behavior of laser-induced bubble near the soft material, however, has not been well clarified. The soft material such as body tissue can deform and influence the behavior of the laser-induced bubble. Since the high peak power laser has been applied in the field of bioengineering and medical treatment, it is of great importance to clarify the effect of the soft material near the laser-induced cavitation bubble. In this research the behavior of laser-induced bubble near the elastic/soft material was visualized with schlieren method and investigated.

1. INTRODUCTION

Recently, the high power laser has been applied in the field of medical engineering, where optical breakdown is occasionally realized by laser beam focusing. Laser-induced optical breakdown generates shock waves and bubbles, and is used for medical applications (A. G. Doukas et al., 1991: Rok Petkovšek et al., 2007). Laser-induced shock waves and bubbles sometimes cause damage to tissues. Therefore, the propagation of laser-induced shock waves and the behavior of laser-induced bubble near the soft materials as body tissues need to be clarified. Many experiments on the behavior of laser-induced bubble near rigid material have been reported (Ed Zwaan et al., 2007). Rigid wall near the rebounding bubble makes an asymmetric flow field. Behavior of the single bubble near the solid wall is affected by the asymmetric flow field. It is well known that the bubble near the rigid wall moves toward the wall, and then unsteady movement of the bubble strongly depends on the positional relation between the bubble and rigid wall(s) (Ed Zwaan et al., 2007). The behavior of laser-induced bubble near the soft material, however, has not been clarified enough.

Our research has concentrated on the behavior of laser-induced bubble near a rigid wall and near an elastic/soft wall by visualization. We have analyzed the behavior of the bubble using the visualized bubble images, and then clarified that the behavior of laser-induced bubble is changed by the difference of material near the bubble. The observed bubble behavior is discussed in this paper.

2. EXPERIMENTAL APPARATUS AND METHOD

In our research a pulsed Nd:YAG laser beam was focused into the distilled water in order to make plasma and a single bubble. Bubble behavior was visualized by schlieren method and shadowgraph method. Experimental apparatus for schlieren method is shown in Fig. 1. This experimental apparatus consists of Nd:YAG laser, water vessel, test wall, lens system, flash lamp and digital still camera. A pulsed Nd:YAG laser (LOTIS TII, LS-2135, 532nm, 10-12ns pulse duration) oscillated at 1.0Hz. The emitted laser beam was shaped by the beam trimming parts, and converged into the distilled water. Distilled water was kept in acrylic water vessel at room temperature under the atmospheric pressure. Lens array was fixed on the vessel. Test wall was fixed near the focusing position of laser beam in water. The position of the test wall was adjusted with XYZ stage. We used flat plate of aluminum, foam rubber, sponge, and whetstone as a test wall in this research.

The laser beam was converged near the test wall to make plasma and a single bubble. At the same time, laser irradiation signal was sent from laser controller to delay generator (SUGAWARA Laboratories, RE-306), and then by the signal a flash lamp (SUGAWARA Laboratories, NPL-5) emitted a pulse light (180ns duration) after designated delay time. The emitted light passed through lens, water vessel, pinhole and R.B.filter (KAISER OPTICAL SYSTEMS, Rejection band filter #2891, 532nm) and entered into the digital still camera. A pinhole and R.B.filter were settled at a focusing position near the camera to weaken the laser light noise. We analyzed the behavior and shape of the bubble using these taken image data.

A high-speed camera (KEYENCE, Motion Analyzing Microscope VW-6000, Maximum 24000fps) was also used in order to observe the behavior of bubble with high-speed frame mode. Experimental apparatus for high-speed camera is shown in Fig. 2. Laser, water vessel, test wall were same as above-mentioned apparatus (Fig. 1). Light source was included in VW-6000, and light for visualization was supplied through optical fiber. The laser beam was converged near the test wall and made plasma and a single bubble. Light source for visualization emitted light continuously. Tracing paper was settled between light source and water vessel in order to diffuse the emitted light. An R.B.filter was settled between the water vessel and the camera to weaken the laser light noise. We observed the behavior of bubble and the deformation of test wall at 24000fps.

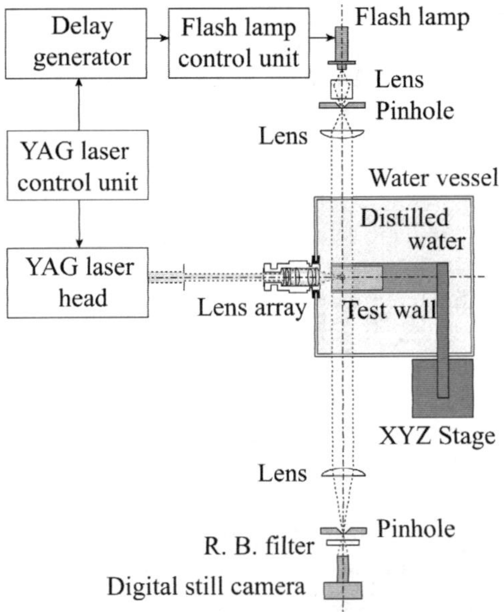

Fig. 1 Experimental apparatus for schlieren method

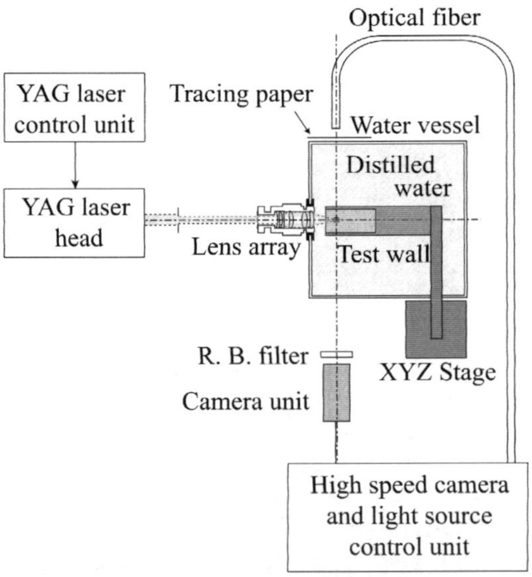

Fig. 2 Experimental apparatus for shadowgraph method

3. FUNDAMENTALS OF LASER-INDUCED BUBBLE BEHAVIOR

In this section, the fundamental aspects of laser-induced bubble behavior are described.

If only single bubble is in liquid, bubble behavior is as shown below;

(1) Pulsed laser beam focused into liquid makes plasma (Fig. 3 (a)).
(2) The single bubble is produced in liquid and rapidly expands by internal high temperature. At the same time, the strong pressure wave is generated by bubble inflation (Fig. 3 (b)).
(3) The bubble stops its growing. By inertial force the bubble grows beyond pressure equilibrium surface (described with broken line in Fig. 3). At this phase the pressure inside the bubble is lower than surrounding pressure (Fig. 3 (c)).
(4) The bubble starts to shrink because surrounding pressure is higher than the pressure inside the bubble (Fig. 3 (d)).
(5) During shrinking phase, the bubble shrinks beyond pressure equilibrium surface by inertia of bubble and surrounding. At the end of this shrinking phase, the bubble is at high pressure and high temperature (Fig. 3 (e)).
(6) The bubble re-expands. At this phase the strong pressure wave is generated by bubble inflation (Fig. 3 (f)).
(7) As above described, the bubble shows several expanding and shrinking, and collapses on sub-millisecond time scale.

The solid wall near the bubble makes asymmetric flow field. Behavior of the single bubble near the solid wall is affected by the asymmetric flow field. The flow field makes lower pressure region during shrinking phase as shown in Fig. 4. As the results, the bubble moves toward the wall, and sometimes makes micro-jet as shown in Fig. 5. It is known that the behavior of bubble in asymmetric flow field strongly depends on the distance between bubble and solid wall.

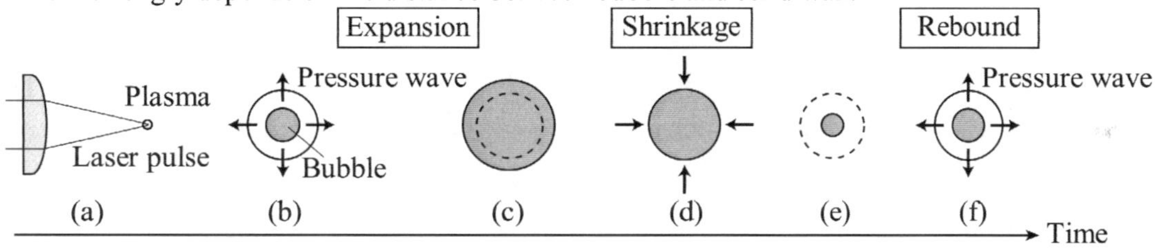

Fig. 3 Fundamentals of laser-induced bubble behavior

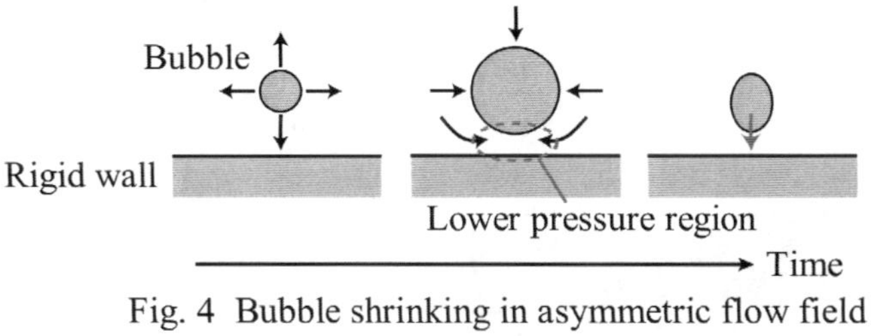

Fig. 4 Bubble shrinking in asymmetric flow field

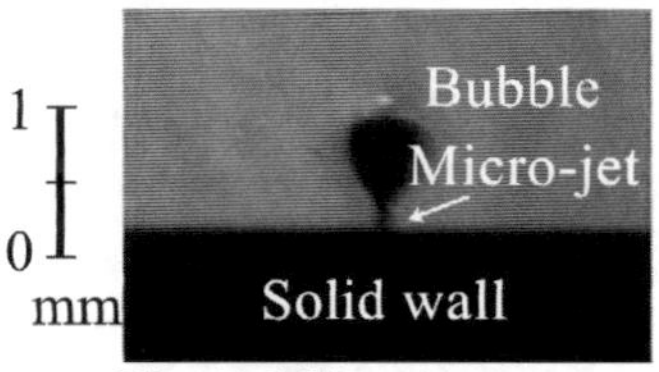

Fig. 5 Micro-jet

4. DEFINITION OF PARAMETERS

In this section the definitions of several parameters for analyzing bubble dynamics are described as follows.

(1) Elapsed time t: The pulsed Nd:YAG laser emits light at time $t = 0$. Elapsed time t can be changed by the delay generator at most 1μs temporal resolution.
(2) Coordinates and bubble position: The origin is test wall surface under the focusing position. Y-axis is taken perpendicular to wall surface, and "y_t" indicates position of bubble top, and "y_b" indicates bubble bottom.

(3) *Aspect ratio*: We introduce aspect ratio of the bubble in order to observe the change of bubble shape. *Aspect ratio* is defined by the following equation.

$$Aspect\ ratio = \frac{H}{W}$$

where $H = y_t - y_b$ is the height of the bubble and W is the width of the bubble.

(4) Non-dimensional focusing position γ : Bubble behavior is affected by the solid wall near the bubble. Relative distance between test wall and bubble is an important factor for bubble behavior. Non-dimensional focusing position γ is defined by the following equation.

$$\gamma = \frac{h}{R_{max}}$$

where h is focusing position (i.e. distance from the test wall surface to the focusing position) and R_{max} is maximum bubble radius.

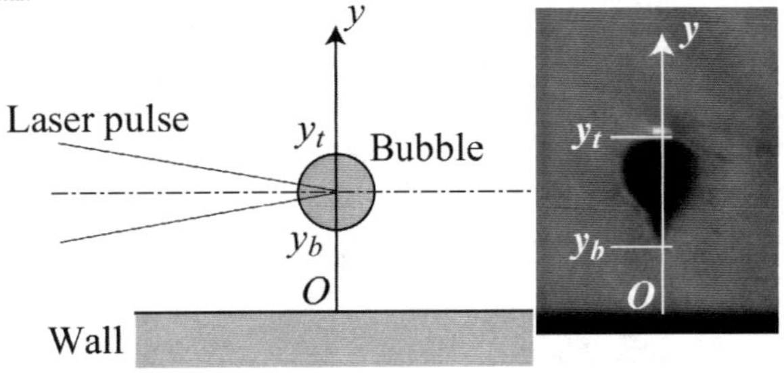

Fig. 6 Vertical coordinate

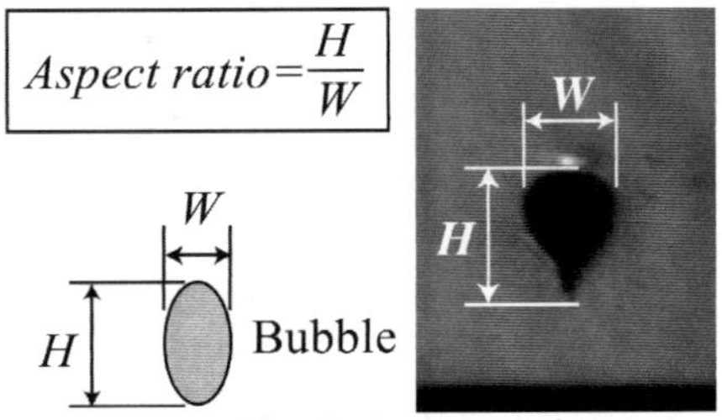

Fig. 7 Aspect ratio

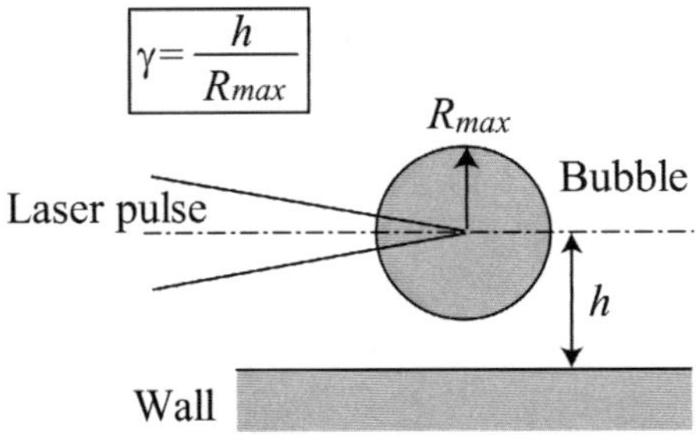

Fig. 8 Non-dimensional focusing position γ

5. EXPERIMENTAL RESULTS

In this section the experimental results are described. We performed four different types of experiments near an aluminum wall (h = 0.86mm), near a foam rubber wall (h = 0.85mm), near a whetstone wall (h = 0.93mm), and near a sponge wall (h = 0.85mm). The laser-induced single bubble visualized with schlieren method by digital still camera was observed as shown in Fig. 9.

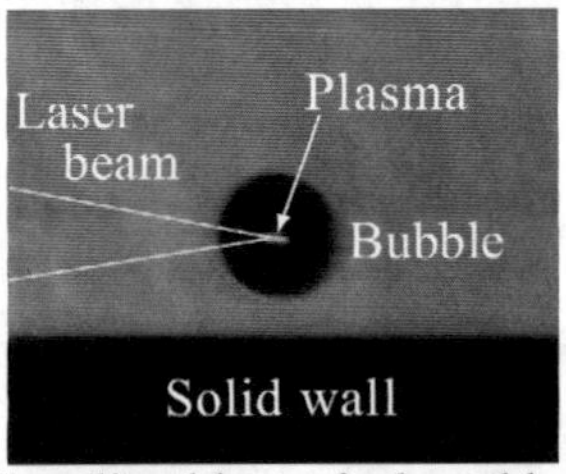

Fig. 9 Visualized laser-induced bubble

5.1 Behavior of the Bubble near an Aluminum Wall (h = 0.86mm, R_{max} = 0.69mm, γ = 1.24)

Behavior of laser-induced bubble near a flat aluminum wall was visualized and investigated. Temporal variation of bubble shape visualized with schlieren method is shown in Fig. 10. Elapsed time t is indicated under the each picture. Expansion of the bubble was observed until approximately t = 61.2μs. The bubble near an aluminum wall moves toward the wall and collides. Temporal variation of the bubble positions y_t and y_b are shown in Fig. 11 (a). Horizontal axis is the elapsed time t [μs] and vertical axis is y [mm]. Bubble positions (y_t and y_b) are described with averaged value and 2σ error bar from t = 1.2μs to t = 241.2μs and observation interval is 20μs. Broken line y = 0.87mm describes the focusing position. Expansion, shrinking and rebounding of the bubble can be observed. Lower bubble surface (indicated by y_b) hardly moves during its shrinking phase. Temporal variation of *Aspect ratio* is shown in Fig. 11 (b). Horizontal axis is elapsed time t [μs] and plotted open circles describe *Aspect ratio* of the observed individual bubbles. The bubble tends to keep spherical shape during the expanding phase. The bubble near an aluminum wall becomes slightly vertically elongated during the shrinking phase and horizontally elongated by collision with the aluminum wall.

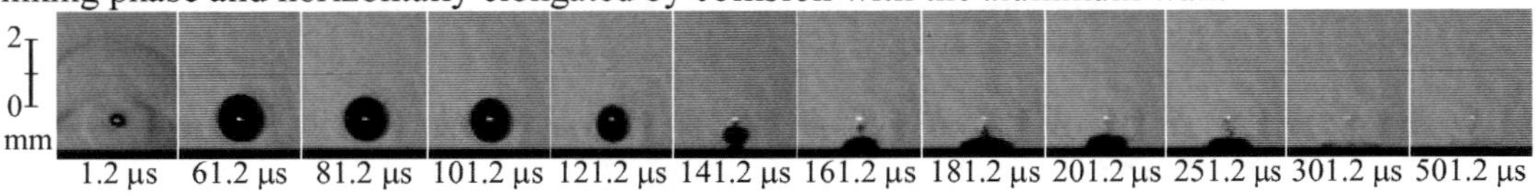

Fig. 10 Temporal variation of bubble shape near an aluminum wall

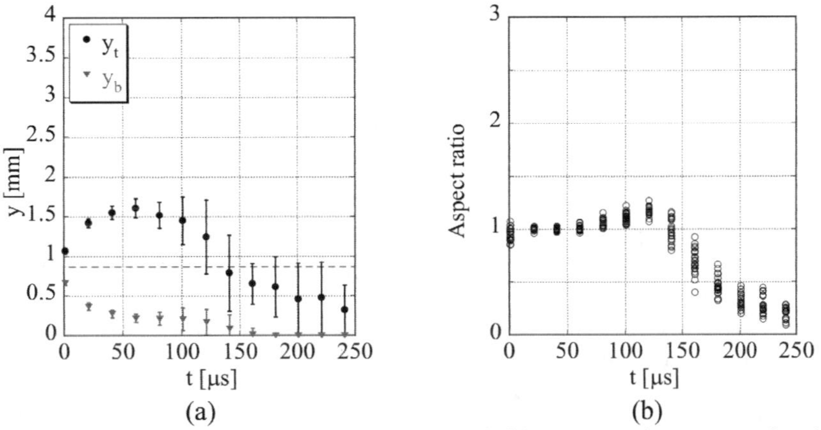

Fig. 11 Temporal variation of (a) bubble position and (b) aspect ratio near an aluminum wall

5.2 Behavior of the Bubble near a Foam Rubber Wall (h = 0.85mm, R_{max} = 0.65mm, γ = 1.32)

Behavior of laser-induced bubble near a flat foam rubber wall was visualized and investigated. Foam rubber is elastic and soft material. Water cannot flow through the foam rubber. Temporal variation of the bubble shape visualized with schlieren method is shown in Fig. 12. Expansion of the bubble was observed until approximately t = 61.2μs. The bubble near a foam rubber wall becomes vertically elongated and moves away from the wall after expansion. Temporal variation of the bubble position is shown in Fig. 13 (a). Expansion, shrinking and rebounding of the bubble can be observed. Rapid growth of upper bubble surface (indicated by y_t) is observed during the rebounding phase (approximately t = 120 ~ 180μs). Temporal variation of *Aspect ratio* is shown in Fig. 13 (b). The bubble near a foam rubber wall tends to become horizontally elongated during its shrinking phase and vertically elongated during the rebounding phase. The dispersion of *Aspect ratio* near a foam rubber is greater than that near an aluminum wall. Thus, behavior of the bubble near a foam rubber wall is obviously different from that near an aluminum wall.

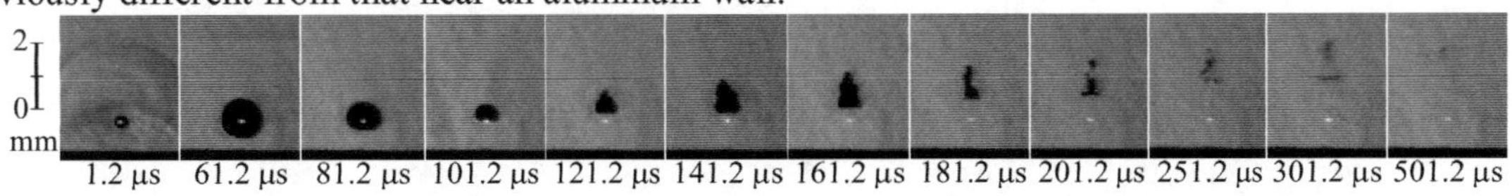

Fig. 12 Temporal variation of bubble shape near a foam rubber wall

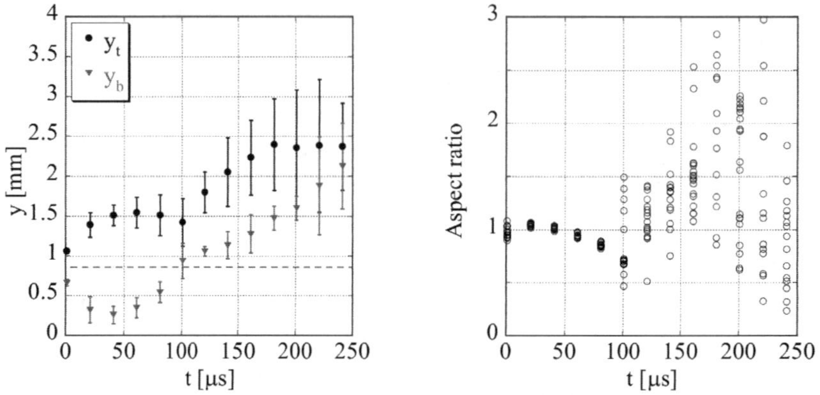

(a) (b)

Fig. 13 Temporal variation of (a) bubble position and (b) aspect ratio near a foam rubber wall

5.3 Behavior of the Bubble near a Whetstone Wall (h = 0.93mm, R_{max} = 0.64mm, γ = 1.47)

Behavior of laser-induced bubble near a flat whetstone wall was visualized and investigated. Whetstone is rigid material that has rough surface. Water cannot flow through the whetstone. Temporal variation of bubble shape visualized with schlieren method is shown in Fig. 14. Expansion of the bubble was observed until approximately t = 61.2μs. The bubble near a whetstone wall becomes vertically elongated and moves toward the wall and collided. Temporal variation of the bubble position is shown in Fig. 15 (a). Expansion, shrinking and rebounding of the bubble can be observed. This tendency is similar to that of the result near an aluminum wall. Temporal variation of aspect ratio is shown in Fig. 15 (b).The bubble tends to become spherical during its expanding phase. The bubble near a whetstone wall becomes slightly vertically elongated during the shrinking phase and horizontally elongated by collision with the whetstone wall. This tendency is also similar to the result near an aluminum wall.

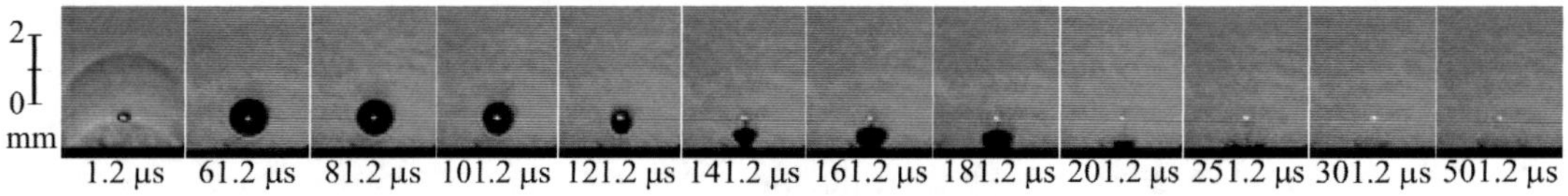

Fig. 14 Temporal variation of bubble shape near a whetstone wall

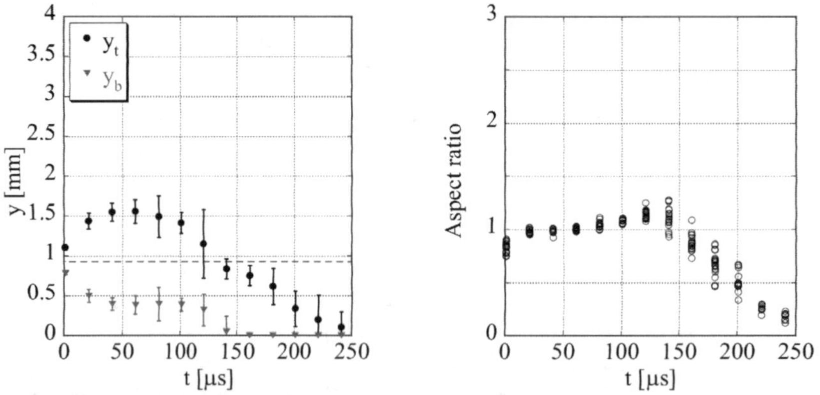

(a) (b)

Fig. 15 Temporal variation of (a) bubble position and (b) aspect ratio near a whetstone wall

5.4 Behavior of the Bubble near a Sponge Wall (h = 0.85mm, R_{max} = 0.74mm, γ = 1.14)

Behavior of laser-induced bubble near a flat sponge wall was also visualized. Sponge is porous, elastic and soft material. Water can flow through the sponge. Temporal variation of bubble shape visualized with schlieren method is shown in Fig. 16. Expansion of the bubble was observed until approximately t = 81.2μs. The bubble near a foam rubber wall shifts away from the wall after its rebounding phase. Temporal variation of the bubble position is shown in Fig. 17 (a). Expansion, shrinking and rebounding of the bubble can be observed. The growth of upper bubble surface (indicated by y_t) is smaller than the result near the foam rubber wall. Temporal variation of aspect ratio is shown in Fig. 18 (b). The bubble tends to keep spherical shape during the expanding and the shrinking phases. The dispersion of this value, however, suddenly increases when rebound starts.

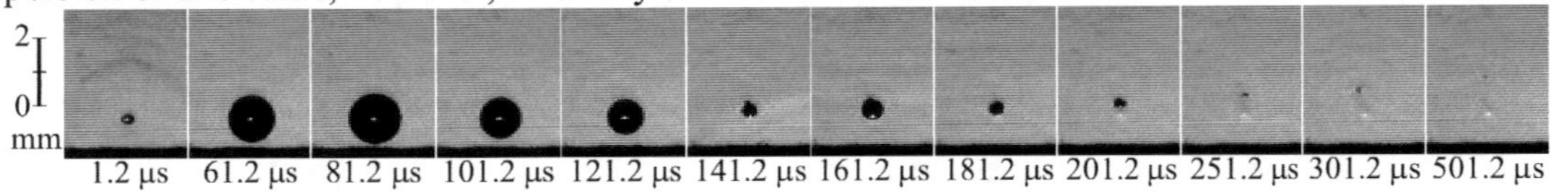

Fig. 16 Temporal variation of bubble shape near a sponge wall

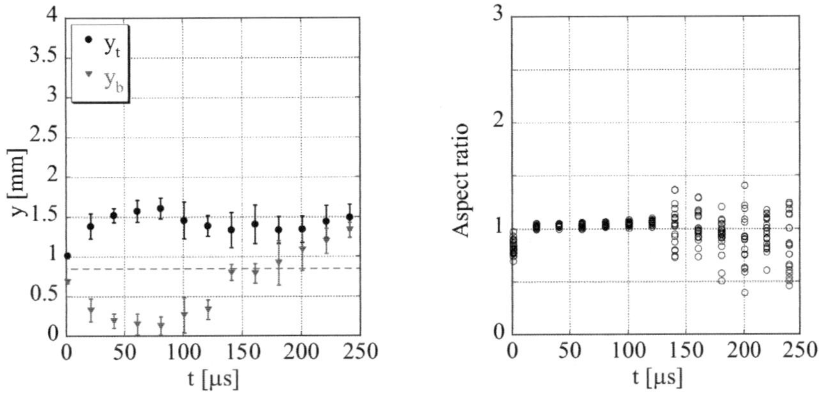

(a) (b)

Fig. 17 Temporal variation of (a) bubble position and (b) aspect ratio near a sponge wall

6. DISCUSSION

The behaviour of the laser-induced single bubbles near an aluminum wall and near a whetstone wall moved toward the wall and collided. On the other hand the bubble near a foam rubber wall and near a sponge wall moved away from the wall. The behavior of bubbles near a foam rubber wall and near a sponge wall cannot be explained by the fundamental scheme described in section 3. High-speed camera helps to clarify these phenomena. Slight deformation of the foam rubber wall was observed by high-speed camera visualization. Significant deformation of the sponge wall, however, was not observed. The deformation of solid wall makes higher pressure region between the bubble and the wall by its elasticity (Fig. 18). As a result, behavior of the bubble near an elastic and soft wall is inverted to the moving flow field near a rigid wall. According to the high-speed camera visualization, this wall deformation will be caused by primary pressure wave or bubble expansion, or both. These results imply that, in the case of medical laser application, the behavior of the laser-induced bubble near body tissue will be different from that near bone. Furthermore, it is shown that we can control behavior of the laser-induced bubble by changing material or position of wall near the bubble. Additional work is underway to clarify the effects of other materials for bubble behavior and shock waves.

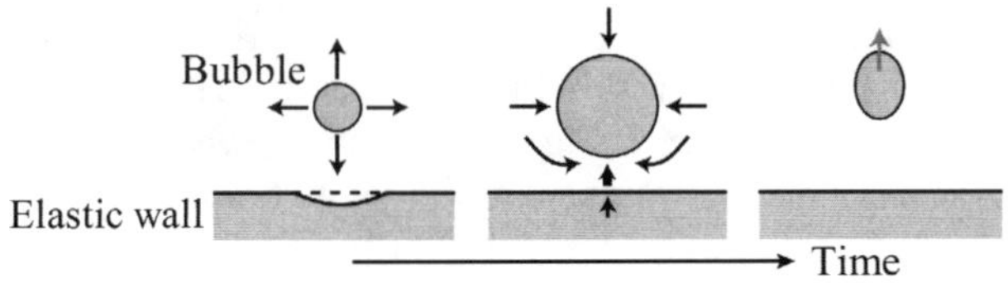

Fig. 18 The effect of deformed elastic wall

7. CONCLUSION

In this research the behaviors of laser-induced bubble near an aluminum wall (rigid material), near a foam rubber wall (elastic and soft material), near a whetstone wall (rigid material) and near a sponge wall (elastic and soft material) in distilled water have been investigated by visualization technique. The main results are indicated as follows:

1. Laser-induced bubble in the distilled water was able to be visualized clearly with schlieren method by using a rectangular acrylic water vessel.
2. High-speed camera helped us to observe the behavior of the bubble and the rapid deformation of the test wall continuously.
3. The behavior of the bubble was able to be measured with schlieren photograph. The video taken by the high-speed camera at 24000fps was not so enough for analysis of bubble behaviour itself.
4. The directions of the bubble movement were obviously different near the different materials; the bubbles near an aluminum wall and near a whetstone wall moved toward the wall, and the bubbles near a foam rubber wall and a sponge wall moved away from the wall.
5. Slight deformation of the foam rubber wall was observed.
6. Aspect ratio was introduced as an index of bubble shape. Aspect ratio tended to change drastically after the first expanding.

REFERENCES

A. G. Doukas, A. D. Zweig, J. K. Frisoli, R. Birngruber, and T. F. Deutsch (1991). "Non-Invasive Determination of Shock Wave Pressure Generated by Optical Breakdown," Appl. Phys. B53, pp. 237-245.

Ed Zwaan, Séverine Le Gac, Kinko Tsuji, and Claus-Dieter Ohl (2007). "Controlled Cavitation in Microfluidic Systems," Phys. Rev. Lett., 98, 254501.

Rok Petkovšek, Griša Močnik, and Janez Možina (2007). "Measurement of The High Pressure Ultrasonic Wave and The Cavitation Bubble by Optodynamic Method," Fluid Phase Equilibria, 256, pp. 158-162.

Journal of Physics: Conference Series **147** (2009) 012027 doi:10.1088/1742-6596/147/1/012027

Influence of Local Void Fraction Distribution on Turbulent Structure of Upward Bubbly Flow in Vertical Channel

T Ogasawara [1], S Takagi [1,2], Y Matsumoto [1]

[1] The University of Tokyo, 7-3-1 Hongo, Bunkyo-ku, Tokyo, 113-8656 JAPAN
[2] RIKEN, 2-1 Hirosawa, Wako, Saitama, 351-0198 JAPAN

ogasawara@fel.t.u-tokyo.ac.jp

Abstract. Turbulent structure of upward dilute bubbly flow with 1 mm bubbles in a vertical channel is investigated experimentally. Small amount of surfactant is added to water to avoid bubble coalescence and to control local void fraction distribution. Liquid phase velocity is measured using two-dimensional laser Doppler Velocimetry. In 1-Pentanol solution of 20 ppm, bubbles have half-slip surface and migrate strongly toward the channel wall due to the shear-induced lift force which leads to wall-peaked distribution of local void fraction. On the other hand, in Triton X-100 solution of 2 ppm, bubbles become fully-contaminated and do not migrate toward the wall or the channel centre due to near-zero lift force, causing uniform distribution of local void fraction in the wall-normal direction. Once bubbles accumulate near the wall, transport of turbulent energy produced by the wall shear towards the channel centre is blocked. Then turbulence induced by the bubble motion becomes dominant in a wide core region (so-called pseudo turbulence). By contrast, in the case of the uniform distribution of bubbles, a mechanism of a turbulent energy transport which is the same as that of a single-phase turbulence still exists and furthermore the bubble-induced turbulence is added on it.

1. INTRODUCTION

Bubbly flows are often encountered in nature. They are also observed in many industrial processes such as the flows in chemical reactors, aeration tanks and heat exchangers. Because of the wide application of these flows, a lot of investigations on their turbulent structure have been carried out over the past three decades.

The early studies on the measurement of bubbly flows were conducted by Serizawa et al. (1975) and Theofanous and Sullivan (1982), in which the local measurements of turbulent flow structure in pipe were made using a Hot-Film Anemometer and a Laser Doppler Anemometer (LDA or LDV), respectively. Following these studies, Wang et al. (1987), Lance and Bataille (1991), Serizawa et al. (1991), Liu (1993), Liu and Bankoff (1993a, 1993b), Moursali et al. (1995) and Marié et al. (1997) provided experimental information on: the local void fraction profile and its dependence on bubble size; the effect of initial bubble size and the bubble-induced liquid turbulence.

On the information of the phase distribution of bubbly flow in upward pipe flow at relatively lower void fraction, it is well known that the bubbles migrate toward the wall and the local void fraction has a wall-peak shape or a saddle shape (Serizawa et al., 1975; Wang et al., 1987). There are lots of likely causes of this transverse migration of bubbles; however, the mechanism has not been clarified. Most of existing experimental studies have been conducted with such factors as large deformable bubbles, no dilute bubbly flow and turbulence which complicate the bubbly flow structure. Kashinsky et al. (1993) investigated a laminar bubbly flow in a vertical pipe and reported the wall-peak distribution of the local void fraction. Single bubble behaviour in vertical channel flow and turbulent boundary layer with vertical plate have been investigated and the forces acting on the bubble was discussed (Žun, 1980; Felton and Loth, 2001).

So et al. (2002) investigated upward dilute bubbly flow in a vertical channel with mono-dispersed 1 mm spherical bubbles by the addition of surfactant into the tap water to avoid bubble coalescences. In 3-Pentanol solution of 20 ppm, these 1 mm bubbles strongly migrate toward the wall. It was observed that highly accumulated bubble near the wall formed crescent like shape horizontal bubble cluster. From their LDV measurement of bubbly flow, plug like flow structure and the bubble-induced turbulence in a core region were reported. Local void fraction distribution has a great influence on the bubbly flow turbulent structure. Therefore, clarification of the mechanism of the transverse bubble migration is important to analyse turbulent modification due to the presence of bubbles.

Here, contaminants in water such as surfactants greatly affect bubble behaviour. It is known that a bubble in contaminated water (surfactant solution) rises much slower than that in super-purified water (e.g. Duineveld, 1995). This phenomenon is explained by Marangoni effect (Frumkin and Levich, 1947). Non-uniform distribution of adsorbed surfactants on a surface of a rising bubble generate surface tension gradient and this causes the tangential stress. Thus, surfactants in water change gradually the boundary condition of the bubble surface from free-slip to no-slip. We (Takagi et al., 2008) reported that stronger Maragoni effect decreases a tendency of a lateral bubble migration of bubbles in an upward channel flow and the local void fraction distribution changes drastically from wall-peaking to uniform with more influential surfactants. This phenomenon is closely related to a shear-induced lift force on a bubble and its behaviour in surfactant solution. Fukuta et al. (2008) investigated numerically the influence of surfactants on the lift force acting on a spherical bubble in simple shear flow and indicated that the lift force decreases from that of a clean bubble, which has nearly the same lift coefficient as Auton's 0.5 (Auton, 1987), to that of a rigid sphere which is near-zero value by an immobilization of bubble surface due to Marangoni effect at bubble Reynolds number based on the bubble diameter and the relative velocity of the bubble of 100. This result qualitatively explains the lateral migration of bubbles in surfactant solution observed by the bubbly flow experiment (Takagi et al., 2008).

The local void fraction distribution is important information characterizing the bubbly flow structure. Despite strong dependence of surfactant on the local void fraction distribution, extent of contamination in water has not been evaluated in previous bubbly flow experiments in which tap water including some contaminants was used. Our objective is to investigate the effect of the local void fraction distribution on the bubbly flow turbulent structure. By controlling the surfactants in water, two different distributions of bubbles (wall peaking and uniform) are obtained. Liquid-phase velocity is measured using two-dimensional laser Doppler Velocimetry in both cases.

2. EXPERIMENTAL SET-UP

2.1 Experimental Apparatus

The experimental apparatus is shown in figure 1 and consists of a loop water tunnel. The tunnel volume is 270 litters, with a bubble separator at the top of the tunnel to eliminate bubble recirculation. The vertical channel has a thickness of $2H = 40$ mm and a spanwise width of 400 mm with an aspect ratio of 10. The channel height is 2500 mm. The distance from the contraction nozzle to the test section of 2000 mm is long enough for single-phase turbulent and laminar flow to be fully developed. The coordinate x denotes the streamwise direction, y denotes the wall-normal direction and z denotes the spanwise direction. Tap water in the tunnel is filtered through filter paper with a 10 micron pore size and an activated carbon filter, and is pumped vertically upwards through the channel. The output of the pump is controlled by a frequency converter.

In the bubbly flow experiment, air bubbles were injected from the bubble generator which is constructed with 474 stainless steel pipe of 0.1 mm inner diameter. The bubble generator is installed above the contraction nozzle of the tunnel, which is located at $x/H = 100$ downward from the test

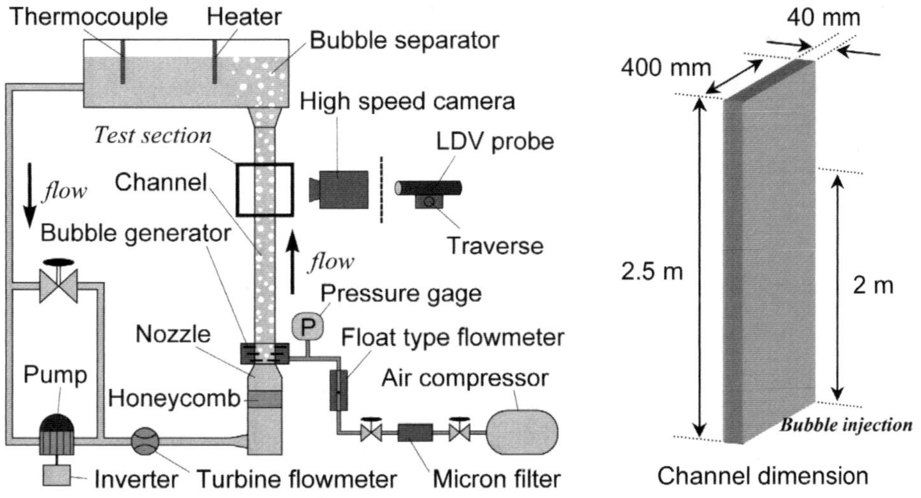

Figure 1 Experimental apparatus

section. The mean value of the bubble size is about 0.9 mm and the standard deviation is about 0.1 mm.

2.2 Experimental Conditions

The bulk Reynolds numbers Re ($=2U_bH/\nu$), based on the characteristic length of the channel width $2H$ (= 40 mm) and the bulk mean velocity of liquid phase U_b, is 10100 which corresponds to the friction Reynolds number Re_τ ($=u_\tau H/\nu$), based on the characteristic length of the half of the channel width H and the friction velocity of single-phase turbulent channel flow u_τ, of 300 and to liquid flow rate of 210 litters min^{-1}. The average void fraction is set to be 0.3% and 0.6% that are less than 1%. These amounts of small gas fraction are enough to modify the turbulent structure, lager gas fraction interferes with the optical measurement. Two kinds of surfactants, 1-Pentanol and Triton X-100, are used. The volumetric concentration of 1-Pentanol is 20 ppm for the wall-peaking local void fraction distribution, and of Triton X-100 is 2 ppm for the uniform one. These surfactant concentrations are quite less than the critical micelle concentration and the change of the water viscosity can be negligible.

2.3 Laser Doppler Velocimetry

To measure the liquid flow structure we applied a two-colour fibre LDV system previously described by So et al. (2002). The optical arrangement of the LDV system consisted of a commercial instruments (FLV 8835, Kanomax). Two components of the liquid phase velocity in *x*-*y* plain are measured. In order to measure the same velocity range by the different colour lasers, the optical probe is rotated approximately 45 degrees. In the present experiment, the Kolmogorov length and time scale of single-phase turbulence are estimated to be the order of 100 μm and 10 msec. Polystyrene particles of 8 μm in average diameter are added into the liquid phase as tracer particles.

3. RESULTS AND DISCUSSION

3.1 Surfactant Effect on Local Void Fraction Distribution

Figure 2 shows the snapshots of bubbly flow in different surfactant conditions: (a) Tap water, (b) 1-Pentanol solution of 20 ppm and (c) Triton X-100 solutions of 2 ppm. Left ones are view from the front of the channel (*x*-*z* plane) and right ones are view from the side of the channel (*x*-*y* plane). The bulk Reynolds number is 10100 and superficial gas velocity is 2.0 mm s^{-1} in every cases. The average void fraction is approximately 0.6%; however, it increases with the stronger Marangoni effect due to

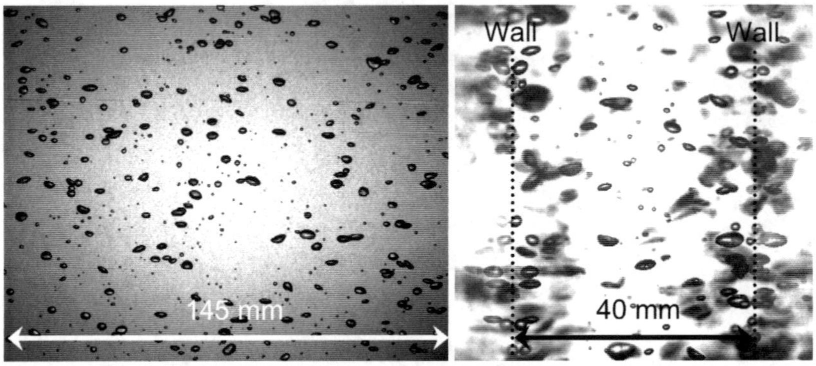

(a) tap water

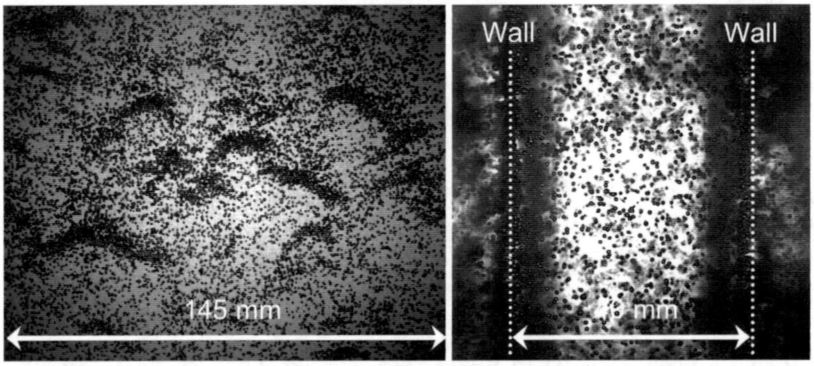

(b) 1-Pentanol 20 ppm

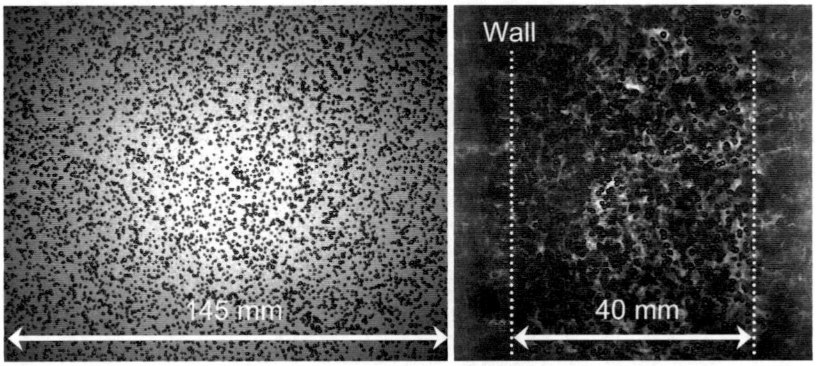

(c) Triton X-100 2 ppm

Figure 2. Photographs of upward bubbly flow in a vertical channel at Re=10100 and an average void fraction of 0.6%. The left photos show front-view and the right photos show side-view.

the presence of surfactant. Figure 2 (a) shows the snapshot of bubbly flow without any addition of surfactant into the tap water. The range of the bubble size becomes from 0.6 mm to 4 mm because of the coalescences up to the test section. The large deformed bubbles move upwards with zigzag or spiral motion. Figure 2 (b) and (c) are in the cases of 1-Pentanol solution of 20 ppm and Triton X-100 solution of 2 ppm, respectively. Mono-dispersed 1 mm bubbles are obtained by the prevention of the bubble coalescences. However, the spatial distribution of the bubbles is totally different. In the case of 20 ppm 1-Pentanol solution, the bubbles strongly accumulated near the wall and move upwards as they are sliding on the wall. At the same time, bubble clustering phenomenon can be observed. On the other hand, in the case of 2 ppm Triton X-100 solution, the bubbles distribute uniformly and do not migrate toward the wall.

The surfactant influence on a bubble can be explained by a Marangoni effect (Levich1962). Nonuniform surfactant adsorption produces the shear stress on the bubble surface. Regarding as this phenomenon from a continuum scale view point, a surface velocity gets lower and a drag on the bubble increases. That is to say, the Marangoni effect changes the boundary condition on the bubble surface from free-slip to no-slip. In 20 ppm 1-Pentanol solution, drag coefficient of an almost spherical bubble which is approximately 1 mm in diameter in stationary fluid takes the intermediate value between that of a clean spherical bubble and of a rigid sphere. On the other hand, in 2 ppm Triton X-100 solution, it becomes almost as same as that of a rigid sphere. The bubble surface in 20 ppm 1-Pentanol solution can still slip; the bubble surface in 2 ppm Triton X-100 solution almost takes no-slip condition.

Around Re_b=100, which is bubble Reynolds number and is defined using the bubble diameter and the relative velocity of a bubble, as is the case with our experiment, the lift coefficient of the rigid sphere is considerably less than that of the clean bubble (Legendre and Magnaudet 1998; Bagchi and Barachander 2002; Kurose and Komori 1999). In the case of 20 ppm 1-pentanol solution, it is assumed that the lift coefficient has an intermediate value and it decreases as the surface condition approaches no-slip condition due to the increase of the concentration. In the case of 2 ppm Triton X-100 solution, the lift coefficient might become as same as that of rigid sphere. Fukuta et al. (2008) reported numerical work on a shear-induced lift force on a spherical bubble in surfactant solution. With decrease of free-slip region on a bubble surface, basically C_D increases and C_L decreases. This decrease of the lift coefficient is a main reason why the tendency of the lateral migration of the bubbles toward the wall weakens with the effect of the stronger surfactant effect.

In this study, these two conditions, 1-Pentanol 20 ppm and Triton X-100 2 ppm, are set to be targets for LDV measurement. The bubble size is almost the same and the vocal void fraction distribution is quite different. In 1-Pentanol 20 ppm solution and Triton X-100 2 ppm solution, the local void fraction becomes wall-peaking and uniform distribution, respectively.

3.2 Surfactant Effect on Single-Phase Turbulence

To investigate the effect of added surfactant on the fully developed turbulent channel flow at the test section, the streamwise mean velocity and turbulent fluctuations in x and y directions are measured by the LDV system. Figure 3 (a) shows the mean velocity profiles in single-phase flow at Re=10100, which corresponds to Re_τ=303 in the cases of tap water, 320 ppm 1-Pentanol solution and 2 ppm

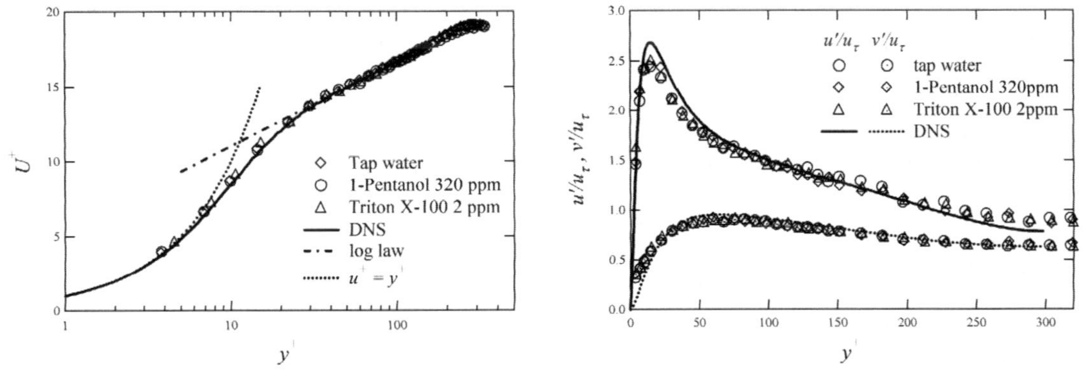

(a) mean velocity profile (b) RMS value of fluctuations

Figure 3. LDV results for single-phase flow at Re_τ=303 (Re=10100) - dependence of surfactant addition on single-phase turbulence. DNS result by Kasagi et al. (1992).

Triton X-100 solution. Figure 3 (a) also shows the comparison of the results with the law of the wall and that of Direct Numerical Simulation at Re_τ=300 (Kasagi et al. 1992). Regardless of the addition of surfactant, the present results show an excellent agreement with the DNS result. The turbulent fluctuations normalized by the wall friction velocity are shown in Figure 3 (b). These profiles do not change with surfactant condition and are also in good agreement with DNS result in both x and y directions. Thus, through the comparison of the results tap water and surfactant solutions, the surfactant effect on single phase turbulence was negligible in present study.

3.3 LDV Results of Bubbly Flow

To compare turbulent structure of two types of local void fraction distribution, wall-peaking and uniform distribution, bubbly flows with 20 ppm 1-Pentanol and 2 ppm Triton X-100 are measured using 2-D LDV system.

Figure 4 shows the LDV results of the bubby flows at Reynolds number of 10100. y/H = 0 and 1 means the channel wall and the centre, respectively. The circle and the triangle markers indicate the case of 1-Pentanol 20 ppm and Triton X-100 2 ppm with the average void fraction of 0.3 %, respectively. The plots with filled markers show the cases with a higher average void fraction of 0.6 % and the cross markers show the case of single-phase turbulence at Re = 10100 (Re_τ = 300).

Figure 4 (a) shows mean velocity profiles. In the case of 1-Pentanol 20 ppm, the mean velocity gradient near the wall becomes steep, and the mean velocity increases at the wall-near range around $0 < y/H < 0.4$. Then, at the channel centre around $0.4 < y/H < 1$, the velocity decreases and the plot flattens. Increasing the void fraction, this tendency becomes pronounced. In the case of f_g = 0.6 %, the velocity gradient becomes steeper, and the plot flattens over the wide region in channel core at $0.2 < y/H < 1$. In Triton X-100 2 ppm case, the mean velocity profile does not change from that of single-phase case at any void fraction. Once the local void fraction distribution takes the wall-peaking shape, the buoyancy force due to the accumulated bubbles lifts up the liquid-phase near the wall, and the velocity increases near the wall and decreases in the channel core region at the same time. On the other hand, in uniform distribution case, the acceleration of the liquid-phase does not occur in a particular region, because there is less nonuniformity of density as a mixture in the wall-normal direction.

Figure 4 (b) shows the plots of root-mean-square value of stream-wise velocity fluctuation. In the cases of 1-Pentanol 20 ppm, the stream-wise fluctuation decreases all over the channel. Especially, with higher void fraction condition (f_g = 0.6 %), it becomes more suppressed, and its plots flatten over the wide region in channel core at $0.2 < y/H < 1$. In contrast, in the cases of Ttiton X-100 2 ppm, it increases in whole channel region compared with that of single-phase turbulence. Here, it is interesting to say that its dependence on the average void fraction is opposite in these two kinds of local void fraction distribution. That is, increasing the average void fraction, the stream-wise fluctuation decreases in the case with wall-peaking distribution of the local void fraction or increases in the case with uniform distribution. This result is important information about the dominant factor which causes a turbulent fluctuation of wall bounded turbulent bubbly flows. Highly accumulate bubbles near the wall change the mechanism of turbulence transport and prevent it toward the channel centre. Stronger the bubbles accumulated near the wall, more difficult the turbulent transport toward the centre becomes. If the bubbles distribute uniformly, the turbulent transport mechanism, which is the same as that of single-phase turbulence, still exists at least in the condition of the present experiment. In this case, bubble induced turbulence is added furthermore, and this is the reason why the fluctuation increases the average void fraction.

Figure 4 (c) indicates the plots of root-mean-square value of wall-normal velocity fluctuation. This result shows an overall trend which is the same as that of stream-wise fluctuation. However, in the case of Triton X-100 2 ppm, the increments are less than that of stream-wise fluctuation. In Triton X-100 2 ppm solution, 1 mm bubbles move upright parallel to the wall and seem no to disturb the liquid-phase in the wall-normal direction. The wakes of the bubbles play a great role related to the turbulence

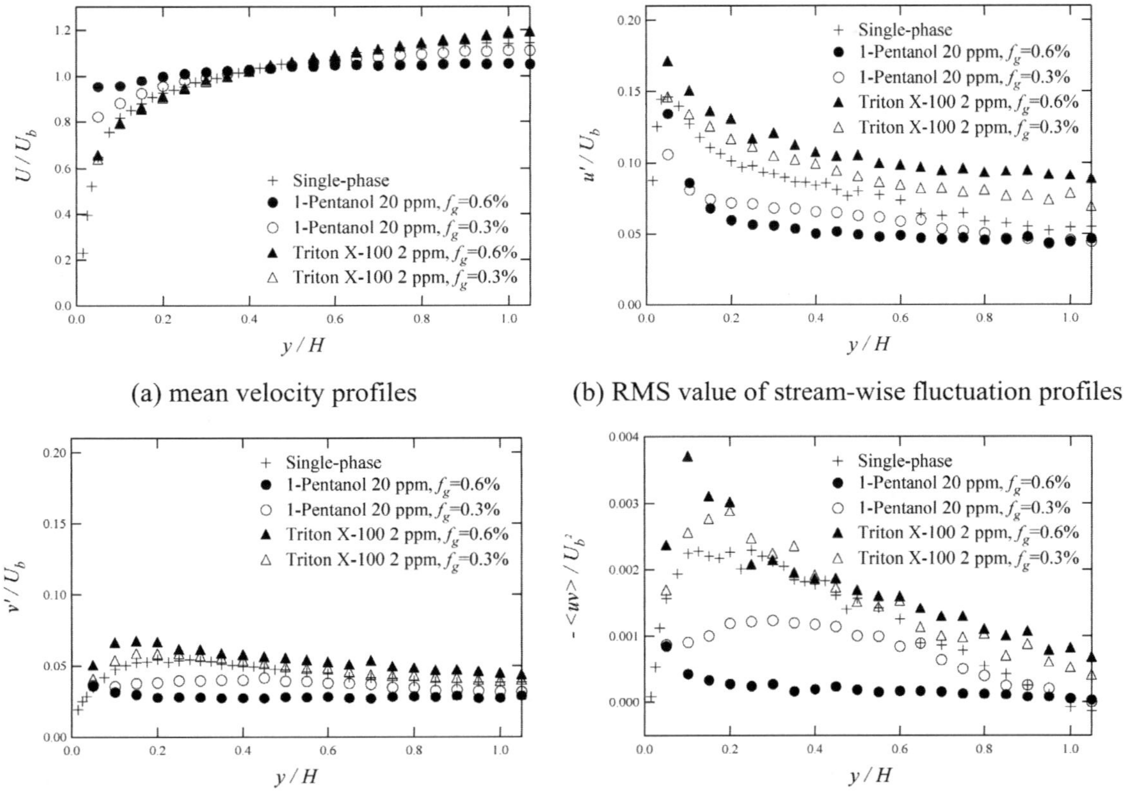

(a) mean velocity profiles

(b) RMS value of stream-wise fluctuation profiles

(c) RMS value of wall-normal fluctuation profiles

(d) RMS value of fluctuation profiles

Figure 4. LDV results of bubbly flow measurement (Re = 10100) - dependence of local void fraction distribution ('*wall-peaking*' for 1-Pentanol 20 ppm and '*coring*' for Triton X-100 2 ppm) and average void fraction.

induced by the bubbly motion, and the momentum of the bubble wake is dominant in the stream-wise direction not in the wall-normal direction.

Reynolds stress profiles are shown in Fig. 4 (d). In the case of 1-Pentanol 20 ppm, Reynolds stress decreases all over the channel. At higher void fraction, it becomes almost zero in the wide channel core region at $0.2 < y/H < 1$. In the case of Triton X-100 2 ppm, the profiles are almost the same as that in the case of single-phase turbulence. Here, we consider the stress balance of the turbulent channel flow of the mixture fluid. The total stress $\tau(y)$ is represented by summation of viscous shear stress and Reynolds stress.

$$\tau(y) = \mu \frac{dU}{dy} - \rho \langle uv \rangle \quad (1)$$

Local void fraction α as a function of y causes the buoyancy force and equation (2) is derived from the momentum equation.

$$\tau(y) + \int_0^y \Delta\rho g \left(\alpha(y) - f_g\right) dy = \tau_w \left(1 - \frac{y}{H}\right) \quad (2)$$

In the case of uniform distribution which means $\alpha(y) = f_g$, the second term of the left hand side of equation (2) becomes zero. Therefore, the stress balance is represented as the same as that of single-phase turbulence, summation of viscous shear stress and Reynolds stress should be balance with the right hand side of eq. (2). This corresponds to the case of Triton X-100 2 ppm. On the other hand, in the case of 1-Pentanol 20 ppm, $\alpha(y)$ is not uniform. So, total stress $\tau(y)$ does not balance with the right hand side of eq. (2). If the contribution of the second term of the left hand side of equation (2) due to the accumulation of the bubbles near the wall, total stress decrease in the channel core region.

4. CONCLUSIONS

In 1-Pentanol solution of 20 ppm, bubbles have half-slip surface and migrate strongly toward the channel wall due to the shear-induced lift force which leads to wall-peaked distribution of local void fraction. On the other hand, in Triton X-100 solution of 2 ppm, bubbles become fully-contaminated and do not migrate toward the wall or the channel centre due to near-zero lift force, causing uniform distribution of local void fraction in the wall-normal direction. Once bubbles accumulate near the wall, transport of turbulent energy produced by the wall shear towards the channel centre is blocked. Then turbulence induced by the bubble motion becomes dominant in a wide core region (so-called pseudo turbulence). By contrast, in the case of the uniform distribution of bubbles, a mechanism of a turbulent energy transport which is the same as that of a single-phase turbulence still exists and furthermore the bubble-induced turbulence is added on it. The contamination of water is a key factor to describe not only the individual bubble motions but also the large-scale turbulent structure of bubbly flow.

ACKNOWLEDGEMENTS

Special thanks are expressed to Mr. Teruyuki Nishihara and Mr. Hiroyuki Ushida for providing valuable experimental assistance.

NOMENCLATURE

f_g	average void fraction	[-, %]
g	gravity constant	[m/s^2]
H	half length of the channel width	[m]
Re	bulk Reynolds number	[-]
Re_τ	friction Reynolds number	[-]
U	mean velocity	[m/s]
u	instantaneous stream-wise velocity	[m/s]
U_b	bulk mean velocity	[m/s]
u_τ	friction velocity	[m/s]
v	instantaneous wall-normal velocity	[m/s]
x	coordinate (stream-wise direction)	[m]
y	coordinate (wall-normal direction)	[m]
z	coordinate (span-wise direction)	[m]
'	Root-Mean-Square value	[-]
	Greek Letters	
α	local void fraction	[-]
ν	kinematic viscosity coefficient	[m^2/s]

REFERENCES

Serizawa, A., Kataoka, I. & Michiyoshi, I., Turbulence structure of air–water bubbly flow–II Local properties. Int. J. Multiphase Flow **2**, 235–246. (1975)

Theofanous, T. G., Sullivan, J., Turbulence in two-phase dispersed flows. J. Fluid Mech. **116**, 343–362. (1982)

Wang, S. K., Lee, S. J., Jones, O. C. Jr & Lathey R. T. Jr, 3-D turbulence structure and phase distribution measurements in bubbly two phase flows. Int. J. Multiphase Flow **13**, 327–343. (1987)

Lance, M. & Bataille, J. Turbulence in the liquid phase of a uniform bubbly air–water flow. J. Fluid Mech. **222**, 95–118 (1991)

Serizawa, A., Kataoka, I., Gofuku, A., Takahashi, O. & Kawara, Z., Effect of initial bubble size on bubbly flow structure. Proceedings of the International Conference on Multiphase Flow, Tsukuba, Japan, 547–550. (1991)

Liu, T. J., Bubble size and entrance length effects on void development in a vertical channel, Int. J. Multiphase Flow **19**, 99-113. (1993)

Liu, T. J. & Bankoff, S. G., Structure of air–water bubbly flow in a vertical pipe –I. liquid mean velocity and turbulence measurements. Int. J. Heat Mass Transfer **36**, 1049–1060. (1993a)

Liu, T. J. & Bankoff, S. G., Structure of air–water bubbly flow in a vertical pipe–II. Void fraction, bubble velocity and bubble size distribution. Int. J. Heat Mass Transfer **36**, 1061–1072. (1993b)

Moursali, E., Marié, J. L. & Bataille, J., An upward turbulent bubbly boundary layer along a vertical flat plate. Int. J. Multiphase Flow **21**, 107–117. (1995)

Marié, J. L., Moursali, E. & Tran-Cong S., Similarity law and turbulence intensity profiles in a bubbly boundary layer at low void fractions. Int. J. Multiphase Flow **23**, 227–247. (1997)

Kashinsky. O. N., Timkin, L. S. & Cartellier, A., Experimental study of "laminar" bubbly flows in a vertical pipe. Exp. in Fluids **14**, 308-314. (1993)

Žun, I. The transverse migration of bubbles influenced by walls in vertical bubbly flow. Int. J. Multiphase Flow **6**, 583–588. (1980)

Felton, K. & Loth, E., Spherical bubble motion in a turbulent boundary layer. Phys. Fluids **13**, 2564-2577. (2001)

So, S., Morikita, H., Takagi, S. & Matsumoto, Y., Laser Doppler velocimetry measurement of turbulent bubbly channel flow. Exp. in Fluids **33**, 135-142. (2002)

Duineveld, P. C., The rise velocity and shape of bubbles in pure water at high Reynolds number, J. Fluid Mech. **292**, 325-332. (1995)

Frumkin, A. & Levich, V., On surfactants an interfacial motion. Zhur. Fiz. Khim. **21**, 1183-1204. (1947) (in Russian)
Takagi, S., Ogasawara, T. & Matsumoto, Y., The effects of surfactant on the multiscale structure of bubbly flows, Phil. Trans. R. Soc. A **366**, 2117– 2129. (2008)

Fukuta, M., Takagi, S. & Matsumoto, Y., Numerical study on the shear-induced lift force acting on a spherical bubble in aqueous surfactant solutions, Phys. Fluids 20, 040704 (2008)

Auton, T. R., The lift force on a spherical body in a rotational flow. J. Fluid Mech. **183**, 199-218. (1987)

Levich, V., Physicochemical Hydrodynamics. Prentice Hall. (1962)

Legendre, D. & Magnaudet, J., The lift force on a spherical bubble in a viscous linear shear flow. J. Fluid Mech. **368**, 81–126. (1998)

Bagchi, P. & Balachandar, S., Effect of free rotation on the motion of a solid sphere in linear shear flow at moderate *Re*. Phys. Fluids **14**, 2719–2737. (2002)

Kurose, R. & Komori, S., Drag and lift forces on a rotating sphere in a linear shear flow. J. Fluid Mech. **384**, 183–206. (1999)

Kasagi, N., Tomita, Y. & Kuroda, A., Direct numerical simulation of passive scalar field in a two dimensional turbulent channel flow. J. Heat Transfer **114**, 598–606. (1992)

Journal of Physics: Conference Series **147** (2009) 012028 doi:10.1088/1742-6596/147/1/012028

Measurement of Two-dimensional Bubble Velocity by Using Tri-fiber-optical Probe

Ruichang Yang, Rongchuan Zheng, Fanling Zhou, Ruolei Liu

Department of Thermal engineering, Key Laboratory for Thermal Science and Power Engineering, Tsinghua University, Beijing, 100084, China

Email: yangrc@mail.tsinghua.edu.cn

Abstract. In this study, an advanced measuring system with a tri-single-fiber-optical-probe has been developed to measure two-dimensional vapor/gas bubble velocity. The use of beam splitting devices instead of beam splitting lens simplifies the optical system, so the system becomes more compact and economic, and more easy to adjust. Corresponding to using triple-optical probe for measuring two-dimensional bubble velocity, a data processing method has been developed, including processing of bubble signals, cancelling of unrelated signals, determining of bubble velocity with cross correlation technique and so on. Using the developed two-dimensional bubble velocity measuring method, the rising velocity of air bubbles in gravitational field was measured. The measured bubble velocities were compared with the empirical correlation available. Deviation was in the range of ±30%. The bubble diameter obtained by data processing is in good accordance with that observed with a synchro-scope and a camera. This shows that the method developed here is reliable.

1. INTRODUCTION

Vapor/gas-liquid two-phase flows exist widely in the equipment of a lot of engineering area. To understand the two-phase flow behavior in the equipment is of great importance for its perfect design and safety analysis. The two-phase flow structures in some facilities such as in the rod bundles of nuclear reactors and in the tube bundles of steam generators are multi-dimensional and therefore are very complex. It is difficult to investigate the two-phase flow structures in the rod or tube bundles by experimental observation and measurements. With the development of computer technology, to use computational fluid dynamics (CFD) method to predict the two-phase flow characteristics in these facilities is considered. In order to verify the calculated results, it is necessary to obtain some information about two-phase flow in these devices such as the distribution of vapor bubble velocity, the vapor bubble size et al. Up till now, some advantage measurement methods, for example, particle image velocimetry (PIV) and particle tracking velocimetry (PTV) methods have been developed to measure directly the vapor bubble velocity fields and the distribution of vapor bubble size in multi-dimensional two-phase flows (Qu et al., 2004; Seol et al., 2007; Kawaguchi et al., 2005). However, these measuring devices are very complicated and expensive. Besides they are only suitable for the visual test facilities. Consequently the measuring systems consisting of many kinds of multi-sensor probes are rather simple and they can also be used to measure the parameters including the bubble velocities and the bubble size values of two-phase flows. The main problem is that a lot of probes available were only suitable for the measurement of local parameters of one-dimensional two-phase flows (Euh et al., 2005; El-Kamash et al., 2005; Ide et al., 2007; Chaumat et al., 2007). If ones try to use the multi-sensor probe to measure the behaviors such as the bubble velocities of multi-dimensional two-phase flow, they have to develop the technique of data processing of the signals from all sensors responding to the same bubble. Some researchers have used multi-sensor probes to measure the local time-averaged interfacial area concentration of the bubbles in multi-dimensional two-phase flow (Shen et al., 2005). Few measurements of multi-dimensional bubble velocity by using probes have been reported in the literature.

In this study, an advanced measuring system with a tri-single-fiber-optical-probe has been developed to measure two-dimensional vapor/gas bubble velocity. During the data processing, the technique of the rejection of non-correlation vapor/gas bubble signals has been developed. The measuring system has been used to measure the two-dimensional velocities of air bubbles with free upward flow in the stagnant water to verify the accuracy of the present measurement method.

2. MEASURING SYSTEM

An optical probe can distinguish liquid and vapor/gas phase by their different refractive index. Generally speaking, void fraction can be measured by a single optical probe, one-dimensional bubble velocity can be measured by a bi-optical probe. As for two-dimensional bubble velocity, it can be measured only by optical probes with at least three single-fiber sensors.

The optical probe developed by the authors consists of three sensors. Each sensor is a single optical fiber in itself 0.1mm in diameter, with the sensing tip molten into a sharp cone and is sleeved with a stainless steel tube 0.8 mm in diameter. The three sleeved fibers are assembled into one stainless tube 3 mm in diameter. The three sensing tips are arranged as a triangle in a plane parallel to the axis of the sleeve. The construction of the optical probe and the arrangement of the three sensors are shown in Fig.1. The measuring system of the optical probe is schematically shown in Fig.2.

In optical design instead of splitting lens which need accurate adjustment the beam splitting device is used. So this measuring system is simple and has the advantage of easy adjustment. According to the intensity required a He-Ne laser is used as the common light source for all three single-fiber sensors. Two types of splitters are used in the system: one-in three-out type and Y-type. Besides the tri-optical probe, the measuring system includes also photomultipliers, A/D converter and computer.

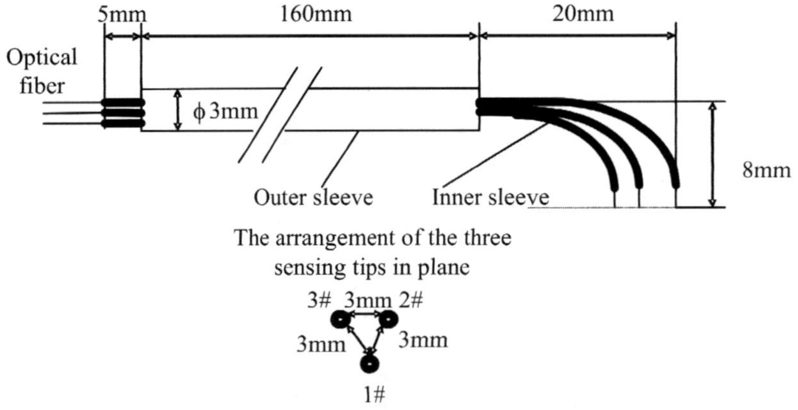

Fig. 1 The construction of the tri-fiber-optical probe and the arrangement of its sensors

3. DATA PROCESSING

The data processing is rather simple for one-dimensional bubble velocity measurement. Bubble velocity u can be obtained from the equation

$$u = L/\tau_m \qquad (1)$$

where L is the distance between two sensors and τ_m the "flying time" of the bubble. But the two-or three-dimensional velocity can not be obtained from eq.(1). As mentioned above, up to

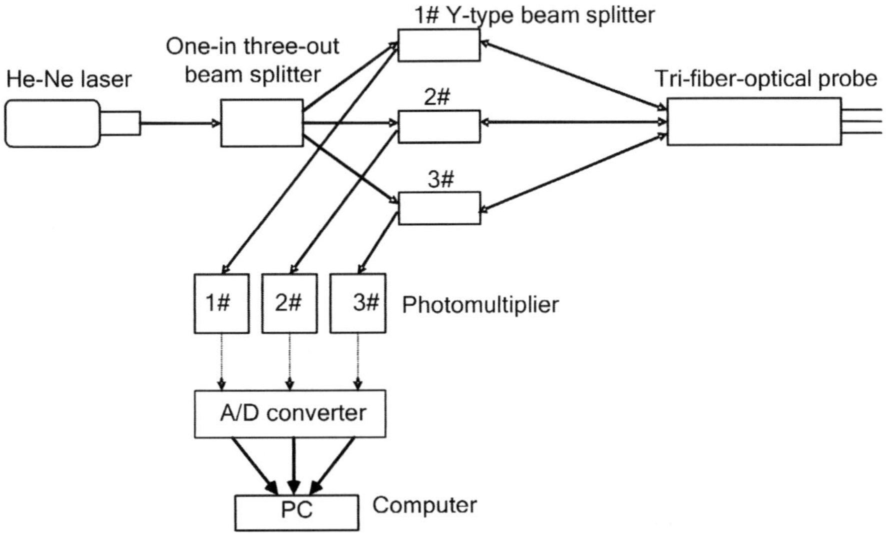

Fig. 2 Scheme of the measuring system of the tri-fiber-optical probe

now not so many papers related to measurement of two- or three-dimensional bubble velocity by using optical probe have been published, mainly because of the difficulties to develop an effective method for data processing. Firstly, it must be ensured that signals from different sensors are correlative with each other when measuring the bubble velocity, that is, signals from all sensors respond to the same bubble. All signals responding to bubbles which have not passed all the three sensors (unrelated signals) should be cancelled. Otherwise the measurement error will be great, that makes the results unreliable. In this sense, how to determine and cancel the unrelated signals becomes the key point of an effective measuring. Secondly, for measuring the two- or three-dimensional bubble velocity, it is necessary not only to determine the magnitude of the velocity, but also the movement direction of the bubble. The velocity direction is closely related to contact position of the bubble with the sensors and the flying time between the sensors. So another key point of realizing an effective measuring is to compose an algorithm with which the magnitude and direction of the velocity can be both determined. Having these two key points in mind, a data processing method for measuring two-dimensional bubble velocity has been developed. This method consists in following:

(1) Threshold for defining vapor/gas and liquid

The light signal reflected from the sensor is converted to electric signal and amplified. By means of A/D acquisition a set of varying with time digitized signals (voltages) can be collected in the computer. The ideal signals should be rectangular in shape. But in fact due to the surface tension there is some time delay when the sensor penetrates into the bubble and when the bubble leaves the sensor. So the signals obtained are not rectangular. The single threshold method is used to define whether the signal corresponds to vapor/gas phase or liquid phase. When the magnitude of the signal is larger than the threshold value, the sensor is located in vapor/gas phase. Otherwise the sensor is in the liquid phase. And by this all the signals are transformed to rectangular shape. Based on the characteristics of the experimental system, in the present study the threshold value V for the signals (voltages) is chosen as:

$$V = 0.3 \times (\text{max. signal value} - \text{min. signal value}) \quad (2)$$

(2) Cancelling of unrelated bubbles

In vapor/gas -liquid two-phase flow due to the ambiguity of bubble movement and the difference in bubble size there may be some cases when a bubble passes the probe (refer to Fig. 3).

(a) The bubble passes one sensor only;

(b) The bubble passes two sensors;
(c) The bubble passes three sensors, but some sensor just touches the edge of the bubble;
(d) The bubble passes all three sensors.

According to the correlation principle only case (d) is useful for the measurement. Bubbles passing one or two sensors are not only useless, but also cause errors in calculation. Case (c) also causes errors because signals from the sensor which only touches the edge of the bubble are too weak. So before doing the correlation analysis cases (a), (b) and (c) must be defined and cancelled. Only the signals of case (d) should be maintained. The method of cancelling unrelated bubble signals is as follows. First, estimate the approximate range of bubble velocity (u_{min} and u_{max}). Then estimate the corresponding maximum and minimum flying time of bubble between sensors.

$$t_{min} = L / u_{max} \qquad (3)$$

$$t_{max} = L / u_{min} \qquad (4)$$

Signal detected by one sensor is used as comparison base. If in the maximum and minimum flying time no signal is detected by the other two sensors, the case belongs to (a). If signal is detected by one of the other two sensors in the time period the case belongs to (b). If in this time period the other two sensors also give signals the case belongs to (c) or (d). Thus, cases (a) and (b) can be defined and cancelled easily. By comparing the impulse width, case (c) can be defined and then cancelled. The criteria are:

$$\varepsilon_1 < 2t_1 / (t_1+t_2) < \varepsilon_2 \qquad (5a)$$

$$\varepsilon_1 < 2t_1 / (t_1+t_3) < \varepsilon_2 \qquad (5b)$$

where $\varepsilon_1, \varepsilon_2$ are constant and are given ($\varepsilon_1<1$, $\varepsilon_2>1$) (Zheng, 1996), t_1, t_2 and t_3 are the impulse width of bubble signals from the three sensors accordingly, id est, the contact time of bubble with the sensor. When the signals meet the inequality (5), it is recognized as case (d). 0therwise the bubble signals belong to case (c) and should be cancelled.

(3) Determination of "flying time" by cross correlation techniques

After cancelling of the unrelated signals the cross correlation functions $R_{12}(\tau)$, $R_{13}(\tau)$ and $R_{23}(\tau)$ for each two sensors of the three sensors are calculated from the equation:

$$R_{xy}(\tau) = \frac{1}{T}\int_0^T x(t) \times y(t+\tau)dt \qquad (6)$$

and the "flying time" τ_{12}, τ_{13} and τ_{23} corresponding to the maximum value of the cross correlation functions $R_{12}(\tau)$, $R_{13}(\tau)$ and $R_{23}(\tau)$ are determined.

(4) Solving for the magnitude and direction of bubble velocity

Assume that the bubble is spherical, that the deformation of its shape when it touches and leaves the probe can be neglected or the bubble deformation when it touches and leaves the probe is the same and that the bubble velocity remains unchanged when it is in touch with the probe. As shown in Fig.4, the bubble passing through three sensors should have a velocity as L_{2c}/τ_{12} or L_{3b}/τ_{13} with a direction angle of θ. In order to calculate the velocity the distance L_{2c} or L_{3b} is required. The distance in turn is dependent on the bubble velocity, the arrangement of sensors in the plane and the bubble size. So the bubble velocity can be determined only by iteration. Assume the bubble velocity u_0, determine the bubble size according to "flying time", and then determine L_{2c} or L_{3b} in relationship with the geometry of probe and bubble. Thus, a new velocity can be obtained. Use the newly obtained velocity instead of u_0 and repeat the process until convergence is reached. The iteration process is as follows (refer to Fig. 4):

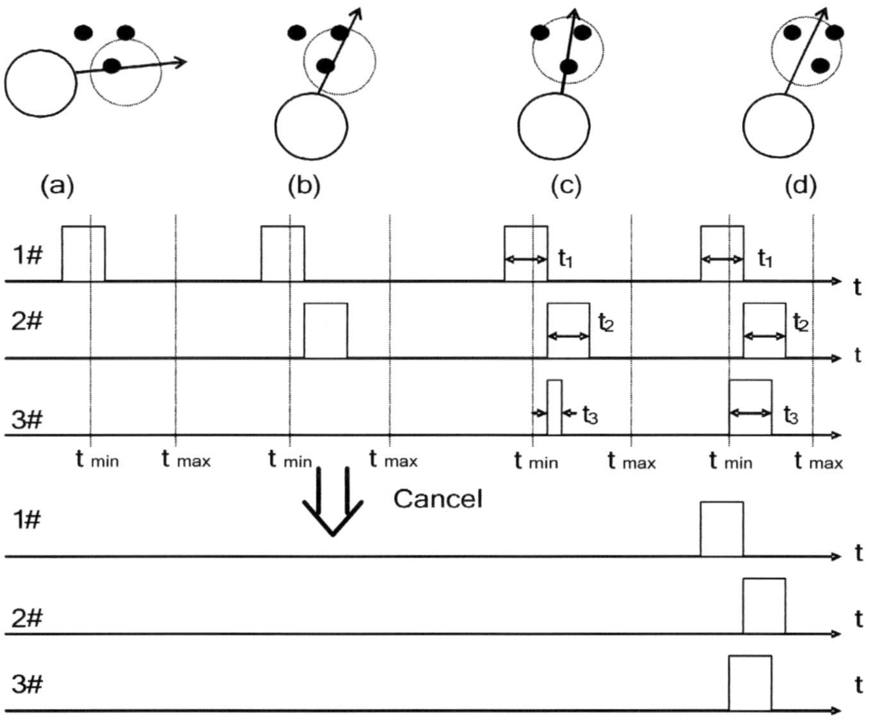

Fig. 3 Cancelling of unrelated bubble signals

(4.1) Calculate the chords of the bubble

$$L_{li} = t_1 \times u_0, \quad L_{cj} = t_2 \times u_0, \quad L_{bh} = t_3 \times u_0 \qquad (7)$$

where t_1, t_2, t_3 are the impulse width (the width of the rectangular wave) of bubble signal from sensor 1#, 2# and 3# respectively.

(4. 2) Calculate the transport distance

$$L_{2c} = \tau_{12} \times u_0, \quad L_{3b} = \tau_{13} \times u_0 \qquad (8)$$

(4. 3) Estimate the direction of bubble movement

As shown in Fig. 4a

$$L_{3d} = (0.5 \times L_{bh} + L_{3b}) - (0.5 \times L_{cj} + L_{2c}) \qquad (9a)$$

As shown in Fig. 4b

$$L_{2d} = (0.5 \times L_{cj} + L_{2c}) - (0.5 \times L_{bh} + L_{3b}) \qquad (9b)$$

If $L_{3d} \geq 0$, the bubble moves towards left to the sensor 3# as shown in Fig. 4a, and

$$\theta = \arcsin(L_{3d} / L_{23}) \qquad (10a)$$

If $L_{3d} < 0$, the bubble moves towards right to the sensor 2# as shown in Fig. 4b, and

$$\theta = \arcsin(L_{2d} / L_{23}) \qquad (10b)$$

where L_{23} is determined by the geometry of the probe.

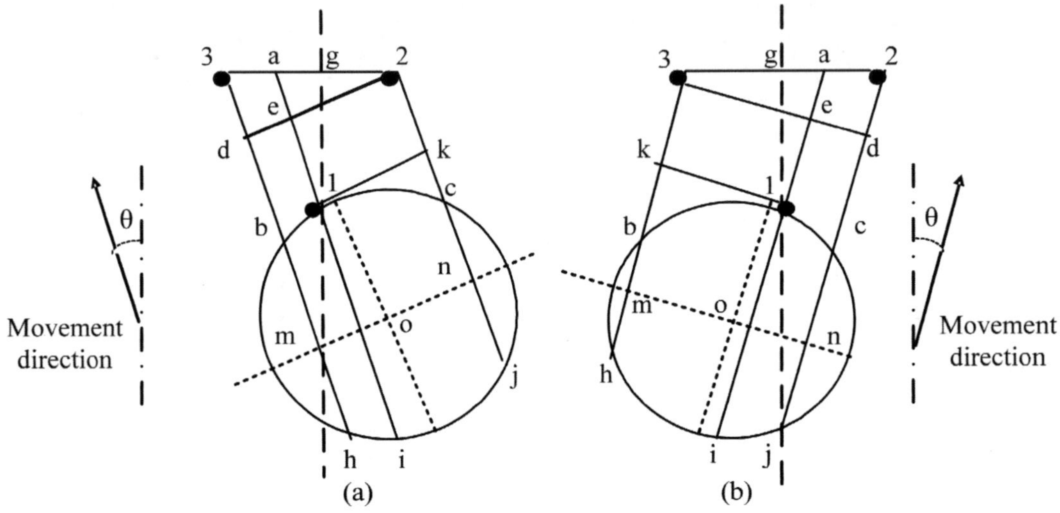

Fig. 4 Scheme of bubble touching with the probe

(4. 4) Calculate the new value of the velocity

Refer to Fig. 4a as an example:

$$L_{ag} = L_{1g} \times \text{tg}\theta \qquad (11a)$$

$$L_{2a} = L_{ag} + L_{2g} \qquad (11b)$$

$$L_{ae} = L_{2a} \times \sin\theta \qquad (11c)$$

$$L_{la} = L_{lg} \div \cos\theta \qquad (11d)$$

$$L_{le} = L_{la} - L_{ae} \qquad (11e)$$

$$L_{ck} = 0.5 \times (L_{li} - L_{cj}) \qquad (11f)$$

$$L_{2c} = L_{le} + L_{ck} \qquad (11g)$$

So the bubble velocity is

$$u = L_{2c} \div \tau_{12} \qquad (11h)$$

where L_{1g}, L_{2g} are determined by the geometry of the probe.

According to Fig. 4b, L_{3b} and the respective bubble velocity u can be determined in the same way.

(4. 5) Convergence discrimination

If $|u - u_0| \geq \varepsilon$, ε is the convergence factor, let $u_0 = u$, return to step (4.1) and continue iteration. If $|u - u_0| < \varepsilon$, end the iteration process. The bubble velocity is u, the included angle between the direction of movement and the vertical is θ.

(4. 6) Calculation of the bubble diameter

According to Fig. 4

$$L_{mn} = L_{23} \times \cos\theta \qquad (12a)$$

$$L_{mn} = L_{om} + L_{on} \qquad (12b)$$

$$L_{mn} = \left[(D/2)^2 - (L_{bh}/2)^2 \right]^{0.5} + \left[(D/2)^2 - (L_{cj}/2)^2 \right]^{0.5} \quad (12c)$$

The bubble diameter D can be obtained by solving the equations (12a) and (12c) simultaneously.

4. MEASUREMENT OF THE RISING BUBBLE VELOCITY

The test facility used for measuring the velocity of rising bubbles in stagnant water is shown in Fig. 5. Air is injected into distilled water through a syringe needle at the bottom of a plexiglass tube. In the upper part of the plexiglass tube the air bubbles rise freely. The triple-optical probe is located in the upper part and the plane of its three sensors is parallel to the axis of the tube. The visual observation shows that the bubble does not move vertically upwards, but spirals. That is, the upward movement of the bubble is S-shaped.

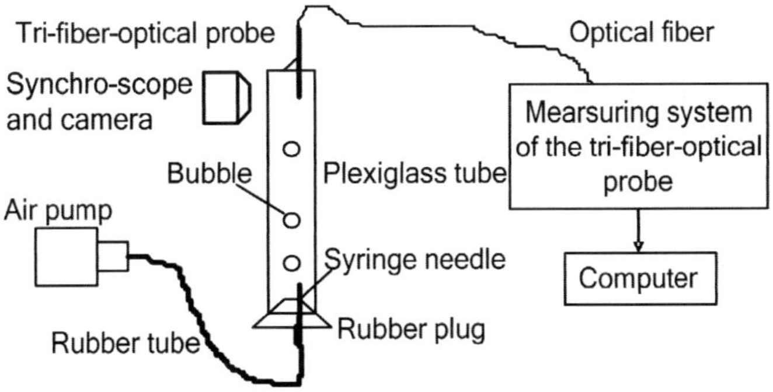

Fig. 5 The rising bubble velocity measuring facility

The output signals of the three sensors are collected simultaneously. For each velocity measurement 3×15000 data points are sampled. The time interval of sampling is 0.3ms. Bubble velocity is obtained by the data processing technique mentioned above, where ε_1=0.4, ε_2=1.6, u_{min}=0.044 m/s, u_{max}=0.273 m/s. The typical results obtained are listed in Table 1. In Table 1 the void fraction measured by each sensor is calculated as follows:

α = Data points of signal (voltage) with the value greater than the threshold value / Total data points of signal (voltage) (13)

It can be seen from Table 1 that the bubbles do not move vertically upwards. This is in accordance with the observation.

During the experiment the bubble diameter is observed with a synchro-scope and a camera. The bubble diameter obtained by data processing is in accordance with the investigated. This shows also that the method developed here is reliable for measuring two-dimensional bubble velocity. There were many experiments carried out for measuring the rising velocity of a bubble in gravitational field. The bubble rising velocity can be described by the following expression (Wallis, 1969):

$$u_b = 1.18\left[\sigma g(\rho_f - \rho_g)/\rho_f^2\right]^{0.25} \quad (14)$$

where σ is surface tension, g gravitational acceleration, ρ_f and ρ_g are liquid and vapor/gas density respectively. For the experiment condition presented in this paper the calculated with equation (14) bubble velocity is u_b=0.108 m/s. The comparison of measured and calculated vertical component of the velocity is shown in Fig. 6. The deviation is within ±30%. It should be noted that even in a stable two-phase flow field the buble velocity varies with time and eq.(14) gives the mean rising velocity of the bubble. The comparison shows that the velocity measured in the present system can be described by eq.(14).

If data are processed without cancelling of the unrelated signals, as shown in Table 1, 4 sets of signals will not converge in the iteration process and the velocity and its direction can not be obtained. From the other 5 sets of signals the magnitude and direction of the velocity can be obtained, but the velocity value is more discrepant, from 0.04 to 0.198 m/s. This shows the strong influence of unrelated signals. This also shows that the cancelling method presented here is successful.

Table .1 Measured velocities of rising bubbles

Measured void fractions

No	1	2	3	4	5	6	7	8	9
Void fraction, sensor 1#	0.245	0.233	0.295	0.244	0.234	0.286	0.290	0.169	0.264
Void fraction, sensor 2#	0.262	0.214	0.263	0.253	0.226	0.267	0.299	0.199	0.221
Void fraction, sensor 3#	0.178	0.327	0.356	0.189	0.256	0.292	0.297	0.276	0.326
Mean void fraction	0.288	0.258	0.305	0.229	0.239	0.282	0.295	0.215	0.270

Measured velocities without canceling of unrelated signals

Velocity magnitude, m/s	—	—	0.049	—	0.093	0.123	0.178	—	0.198
Direction angle, degrees	—	—	10.4 (a)	—	18.4 (b)	5.7 (a)	0.1 (a)	—	3.9 (b)

Measured velocities with canceling of unrelated signals

Velocity magnitude, m/s	0.148	0.085	0.117	0.137	0.138	0.119	0.122	0.091	0.095
Direction angle, degrees	20.8 (b)	3.7 (a)	13.7 (a)	6.6 (b)	15.5 (b)	17.8 (a)	3.7 (b)	21.8 (b)	0.0
Bubble diameter, mm	4.51	5.12	4.39	6.86	6.08	6.39	5.57	6.14	7.01

5. CONCLUSIONS

(1) The optical probe with three single-fiber sensors developed here can effectively get signals from the vapor/gas and liquid phases. The use of beam splitting devices instead of beam splitting lens simplifies the optical system, so the system becomes more compact and economic, and more easy to adjust. The sensor manufacturing method, melting the optical fiber tip to form a sharp corn is convenient. The sensor has little influence on the flow and has high dynamic responsibility. The system used here does not require large signal amplitude and is simple to handle.

(2) Corresponding to using triple-optical probe for measuring two-dimensional bubble velocity a data processing method has been developed, including processing of bubble signals, cancelling of unrelated signals, determining of bubble velocity with cross correlation technique and so on.

(3) Using the developed two-dimensional velocity measuring method, the rising velocity of air bubbles in gravitational field was measured. The results of bubble velocity were compared with the empirical correlation available. Deviation was in the range of ±30% . The bubble diameter obtained by data

processing is in good accordance with that observed with a synchro-scope and a camera. This shows that the method developed here is reliable.

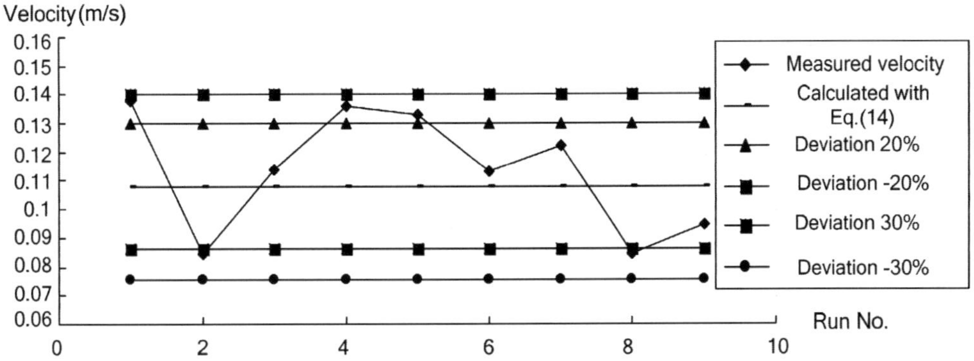

Fig. 6 Comparison of the measured and calculated velocity

ACKNOWLEDGEMENTS
This research was supported by the Natural Science Foundation of China (project number 50676051) and National Key Laboratory of Bubble Physics and Natural Circulation.

NOMENCLATURE

L	distance	[m]
t	impulse width of bubble signal (time) or flying time of bubble	[s]
u	velocity	[m/s]

Greek Letters

α	void fraction	
θ	direction angle of bubble movement	[degree]
τ	flying time of bubble	[s]

REFERENCE

Chaumat, H., Billet-Duquenne, A.M., Augier, F., Mathieu, C., Delmas, H. (2007). "On the Reliability of an Optical Fiber Probe in Bubble Column under Industrial Relevant Operating Conditions," Experimental Thermal and Fluid Science, 31(6), pp.495-504.

El-Kamash, M. K., Loewen, M. R., Rajaratnam, N. (2005). "Measurements of Void Fraction and Bubble Properties on a Stepped Chute Using a Fiber-optic Probe," Canadian Journal of Civil Engineering, 32(4), pp.636-643.

Euh, D. J., Yun, B. J., Park, W. M., Song, C.-H., Huh, B. G., Chung, C. H. (2005). "Investigation of the transport of the bubble parameters in air/water flow conditions," Proceedings of the American Nuclear Society - International Congress on Advances in Nuclear Power Plants, 5, pp.2831-2837.

Ide, H., Kimura, R., Kawaji, M. (2007). "Optical Measurement of Void Fraction and Bubble Size Distributions in a Microchannel," Heat Transfer Engineering, 28 (8-9), pp.713-719.

Kawaguchi, T., Maeda, M. (2005). "Measurement Technique for Analysis in Two-phase Flows Involving Distributed Size of Droplets and Bubbles Using Interferometric Method - Planar

Journal of Physics: Conference Series **147** (2009) 012028 doi:10.1088/1742-6596/147/1/012028

Simultaneous Measurement of Size and Velocity Vector Field," Multiphase Science and Technology, 17(1-2), pp.57-77.

Qu, J. W., Murai, Y., Yamamoto, F. (2004). "Simultaneous PIV/PTV Measurements of Bubble and Particle Phases in Gas-liquid Two-phase Flow Based on Image Separation and Reconstruction," Journal of Hydrodynamics, 16(6), pp.756-766.

Seol, D.G., Bhaumik, T., Bergmann, C., Socolofsky, S. A. (2007). "Particle Image Velocimetry Measurements of the Mean Flow Characteristics in a Bubble Plume," Journal of Engineering Mechanics, 133 (6), pp.665-676.

Shen, X., Saito, Y., Mishima, K., Nakamura, H. (2005). "Methodological Improvement of an Intrusive Four-sensor Probe for the Multi-dimensional Two-phase Flow Measurement," International Journal of Multiphase Flow, 31(5), pp.593-617.

Wallis, G. B. (1969). "One-dimensional Two-phase Flow," McGraw-Hill.

Zheng Rongchuan, (1996). "The Study of the Internal Circulation of Two-phase Flow among Horizontal Heated Tube Bundles," Ph.D. Thes

Journal of Physics: Conference Series **147** (2009) 012029 doi:10.1088/1742-6596/147/1/012029

Advances in Multiphase Flow Measurements Using Magnetic Resonance Relaxometry

Apostolos Kantzas, Sergey Kryuchkov, Blake Chandrasekaran

Perm Instruments, TIPM Laboratory
PO Box 61307, Brendwood PRO, Calgary, Alberta, Canada T2L 2K6

akantzas@ucalgary.ca

ABSTRACT. When it comes to the measurement of bitumen and water content as they are produced from thermally exploited reservoirs (cyclic steam stimulation or steam assisted gravity drainage) most of the current tools that are available in the market fail. This was demonstrated previously when our group introduced the first concept of a magnetic resonance based water-cut meter.

The use of magnetic resonance as a potential tool for fluid cut metering from thermally produced heavy oil and bitumen reservoirs is revisited. At first a review of the work to date is presented. Our recent approach in the tackling of this problem follows. A patented process is coupled with a patented pipe design that can be used inside a magnetic field and can capture fluids up to 260°C and 4.2MPa. The paper describes the technical advances to this goal and offers a first glimpse of field data from an actual thermal facility for bitumen production. The paper also addresses an approach for converting the current discrete measurement device into a continuous measurement system. Preliminary results for this new concept are also presented.

1. INTRODUCTION

Low field NMR relaxometry is a technology that offers significant benefits in reservoir characterization through the so-called magnetic-resonance logging tools that are offered by the oil and gas service companies (Coates et al., 1999). These tools can offer measurements of porosity, permeability, mobile and bound fluids and potentially saturations, if they are properly calibrated. Recently, the service tools also provide information about fluid viscosity and other fluid properties, again after careful calibration of the tools. These calibrations can be done through measurements in bench-top systems that perform the same measurements at much higher signal-to-noise ratio.

Over the past ten years the Tomographic Imaging and Porous Media (TIPM) Laboratory has focused on research, development and commercialization of low field NMR technology. Although the interest was primarily in heavy oil and oil sands applications, some technology advancements apply to conventional oils as well. Initially, much of the work has centred on core analysis. However, since it was realized that NMR could be used for determination of fluid content in cores, the jump to measurement of fluid content in fluid streams was natural. Fluid characterization in conventional oil and gas reservoirs can be done with a multitude of logging techniques including magnetic resonance. The term magnetic resonance induced fluid typing is used nowadays from the service companies to tackle this problem.

The fluid stream characterization work in our laboratory aimed at several potential applications, with original targets in heavy oil and oil sands. First the work was directed in the determination of fluid content in oil sand process plants that process oil sand mining ore and froth (Kantzas, 2004). The work was then extended to measurement of water cuts in oil/water streams (Mirotchnick et al., 2003, 2004). This was followed by work in the determination of fluid viscosities both in bulk streams, emulsions and in-situ (Bryan et al., 2005). Finally the research dealt with the measurement of fluid content, viscosity and mass transfer properties of heavy oil and bitumen mixtures with solvents (Goodarzi et al., 2007; Wen and Kantzas, 2004).

Journal of Physics: Conference Series **147** (2009) 012029 doi:10.1088/1742-6596/147/1/012029

In co-operation with Canadian Natural Resources Limited (CNRL) a water-cut meter utilizing low field NMR technology was operated for approximately eight years on two different cyclic steam injection heavy oil well pads (Allsopp et al., 2001; Wright et al., 2004).

The only other recently published technology that tries to measure fluid production at thermal conditions is the nuclear fraction metering technology offered by Schlumberger. This technology consists of the combination of a venturi meter and a gamma-ray source. It was field tested for a few hours in a SAGD operation in Western Canada (Hompoth et al., 2008a, 2008b). The technology shows promise but it cannot reach the specified temperature conditions imposed to us by the client operating companies.

2. PHYSICAL PRINCIPLES

The principle of proton NMR deals with the response of hydrogen bearing molecules to a sequence of magnetic field pulses. As the protons in hydrogen bearing molecules are exposed to a magnetic field they acquire energy. When the magnetic field stops the protons return to their (random) original state and emit this energy back to the environment. The released energy is recorded and amplified through specific hardware. This energy is detected as a decay curve that is then fitted to a multi-exponential spectrum of characteristic transverse relaxation times (T_2). The detailed physics of magnetic resonance can be found in numerous literature sources.

Figure 1 shows a spectrum of a heavy oil / water emulsion. This spectrum is typical of the types of spectra encountered in the analysis of NMR extracted fluid data. It is, therefore, used as an example to explain how the NMR analysis is done.

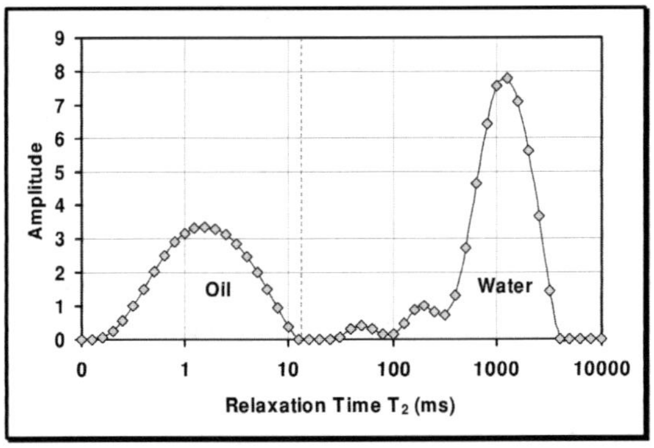

Fig. 1 Representative emulsion NMR spectrum indicating heavy oil and water peaks (from Kantzas, 2004)

The x-axis denotes the relaxation times T_2 and the amplitude denotes the signal strength at each relaxation time. The dashed vertical line indicates the characteristic cut-off point that is used to separate heavy oil signal from water signal and is based on theoretical and empirical considerations. In this case the slower relaxation is attributed to water and the faster relaxation is attributed to heavy oil. This interpretation was possible after numerous spectra of pure substances were collected. The smaller peaks between the dominant oil and water peaks are attributed to water droplets present in an emulsified state.

The information used for analysis of the NMR spectra often includes a weight or volume measurement, and calibration standards for comparison to the measured spectra. The properties of the spectra that are very useful are the amplitude (A) of the full spectrum or a part of the spectrum that corresponds to a given phase, the location of the spectra peaks as indicated by the relaxation time T_2, or in the case of complex spectra a characteristic average relaxation time. The most commonly used relaxation time for averaging purposes is the geometric mean relaxation time, T_{2gm} that is defined as:

Journal of Physics: Conference Series **147** (2009) 012029 doi:10.1088/1742-6596/147/1/012029

$$T_{2gm} = \exp\left(\sum \frac{A_i}{A} \cdot \ln(T_{2i})\right) \qquad (1)$$

A spectrum contains significant information that is qualitative until the amplitude is translated into a mass or volume number in which case the spectrum information becomes quantitative. In order to reach this point, we require calibrations that will offer the amplitude that corresponds to a unit volume or a unit mass. For most of our applications the amplitude per unit mass is defined through the amplitude index (*AI*) as follows:

$$AI = A/m \qquad (2)$$

As a result, if one knows the amplitude and amplitude index of a given compound, NMR spectra will be helpful to provide the actual amount of this compound as soon as we identify it in the spectrum. In many studies, instead of *AI*, it is customary to use another variable called the relative hydrogen index (*RHI*) that is defined as the ratio of *AI* values of a specific compound to that of water:

$$RHI = \frac{AI_i}{AI_w} \qquad (3)$$

Both *AI* and *RHI* have a sound scientific basis as they are related directly to the density of hydrogen atoms in the fluid. In practical applications they may be thought of as calibrations. The main difference is that *AI* values are instrument specific, but *RHI* values should be instrument independent and only depend on the composition of the stream at hand.

With respect to fluid analysis, the only parameter that will distort the NMR spectrum unexpectedly will be the presence of paramagnetic ions (such as iron or rust). Parameters that play havoc with conventional sensors such as salinity of brines and emulsion inversions are completely indifferent to NMR.

Temperature is an important factor that will shift the NMR peak locations of different fluids. Thus when measurements are made it is required that the temperature remains constant. Calibrations are then made to compare spectra at different temperatures to corresponding standards, and corrections are made to the predictive algorithms for fluid content and fluid property determination.

Before providing predictive equations for different circumstances, let us consider some relaxation times for different fluids. At room temperature bulk water relaxes at approximately 2300 ms whereas bitumen will relax with a T_2 of less than 10 ms.

In general, the more restricted the hydrogen bearing molecule, the faster its relaxation time. Restrictions to a molecule can occur either through an increase in viscosity, or through the confinement in small spaces (such as pores).

When dealing with emulsions of heavy oil or bitumen and water, it is often assumed that the T_2 of bitumen does not change much when the bitumen is emulsified because it is already quite fast, but the T_2 of water changes significantly when it is emulsified. This is taken into account when interpreting the spectra to conclude if we have a separated flow, oil in water or water in oil emulsion by the strong impact the location of the water signal has on the geometric mean of the spectra.

3. LABORATORY DATA

The first and most heavily studied NMR metering application was that of a two-phase (water-oil) compositional meter. Based on equation (2), if the total weight (or volume) of a sample is known, then given a constraint (or cut-off) for the separation of the two fluids, one can predict the water and oil cuts. In a simplified approach this can be done through:

$$\%w_c = \frac{A_w}{A_w + A_o(RHI)^{-1}} \qquad (4)$$

The cut-off approach separates the spectrum in two components, one corresponding to oil and one to water.

Fig. 2 shows some early laboratory results. The correlation between NMR results and Dean-Stark results is approximately 99%. Although it was not expected to achieve such precise results in the uncontrolled field environment, it was considered encouraging enough for field applications.

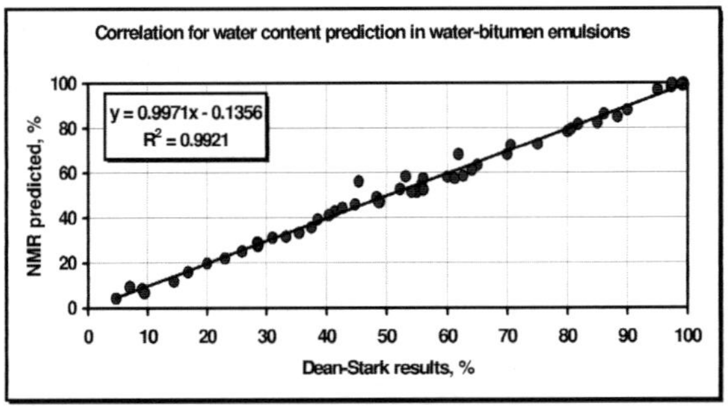

Fig. 2 Laboratory data for water-cut determination in water / oil streams (Allsopp et al., 2001)

The combination of equations (2) and (4) can be manipulated to provide the composition of a water / oil / gas stream. In this case the results can be plotted either for individual phases or in a ternary plot as it can be seen from Fig. 3 and Fig. 4 respectively.

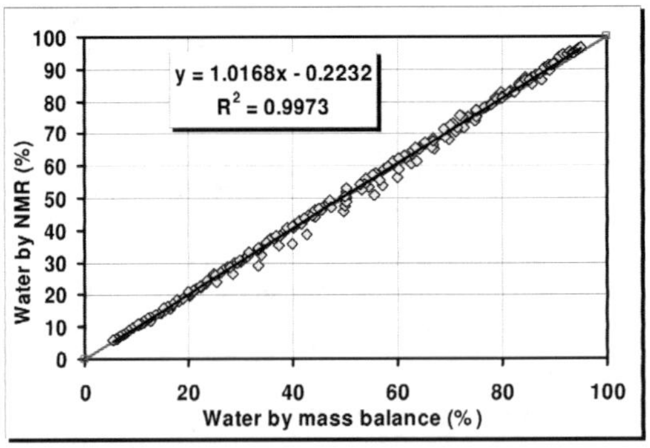

Fig. 3 Comparison of water content predictions from NMR and mass balance for mixtures of 2150 mPas oil with brine for a three-phase system (Kantzas, 2004)

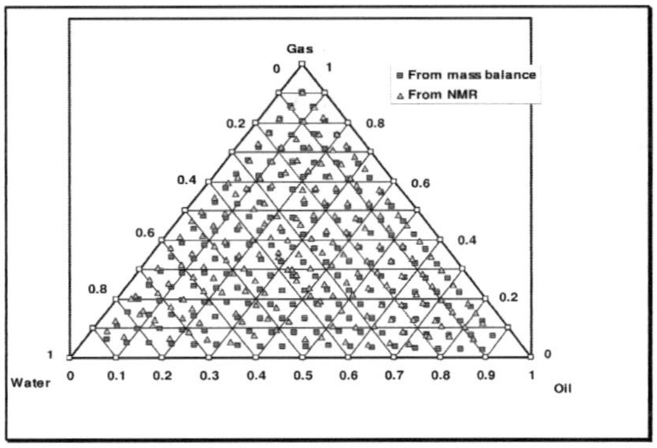

Fig. 4 Ternary diagram of composition of a three phase system extension of the data of Fig. 3 (Kantzas, 2004)

The next important development in fluid characterization was the determination of viscosity on-line. The results of this work have been extensively published (e.g. Bryan et al., 2005). It was found that NMR could predict viscosities of single fluids and emulsions with reasonable accuracy without tuning and with excellent accuracy if the algorithm is tuned to specific oils rather than the original generic approach.

The predictive models for viscosity measurements are of the form:

$$\mu = \frac{\alpha}{(RHI)^{\beta} T_{2gm}} \qquad (5)$$

The coefficients α and β vary with different systems and depending on whether the model is tuned or not to a specific reservoir.

Fig. 5 show the results of the predictive algorithm for viscosity measurements as it was applied in a wide variety of oils at different temperatures. For this model α=1.15 and β=4.55.

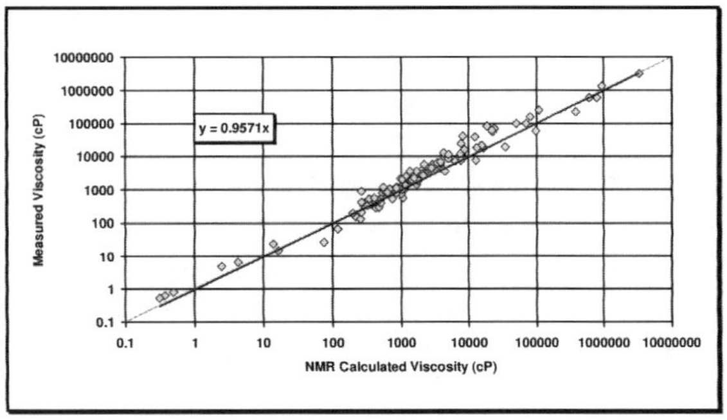

Fig. 5 Viscosity prediction algorithm for various oils (Kantzas, 2004)

In a parallel project, the value of NMR technology was investigated for bitumen-solvent mixtures. For processes like vapour extraction (VAPEX), it is anticipated that the operator would like to know the bitumen content and produced fluid viscosity for the evaluation of the performance of VAPEX. To this end laboratory work focused on using liquid solvents (both paraffin and cyclic based).

Journal of Physics: Conference Series **147** (2009) 012029 doi:10.1088/1742-6596/147/1/012029

Hundreds of mixtures of oils and solvents were created in different solvent to oil ratios. The NMR spectra were obtained over time and their trends were observed. Several predictive algorithms were designed that are tuned to different solvent / oil pairs. Fig. 7 shows a typical example of these plots that can be used for bitumen content determination from NMR derived parameters.

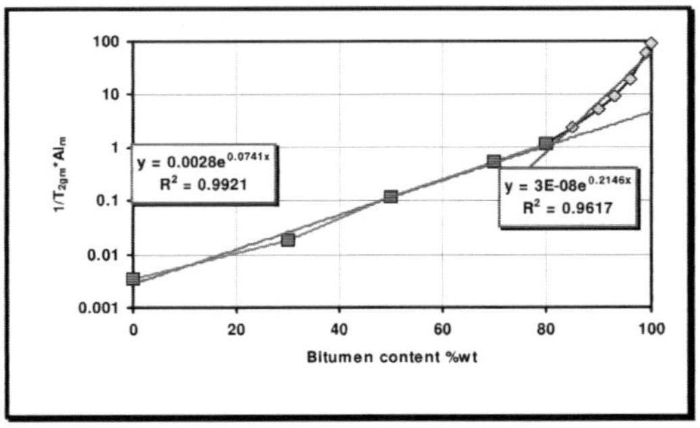

Fig. 6 Changes in NMR spectra properties with solvent addition (Wen and Kantzas, 2005)

It must be noted that the applications for NMR technology are numerous but only the water-cut metering technology has been field-tested to date. The specifics for field-testing and some preliminary results are discussed next.

4. SENSOR DESIGN

In order to apply the theoretical models for the creation of a field sensor, the standard laboratory designs to date were used as a starting point.

The sensor part consists of an assembly of permanent magnets and transmitter / receiver coils. The permanent magnets cannot be exposed to temperatures greater than 80°C without being damaged. Thus, we have developed a heat-shielded magnet that overcomes this drawback and at the same time increases the maximum process temperature to 260°C. The sampling pipe/tube itself has to be non-conductive and non-magnetic. Additionally, the pipe/tube also has to be able to handle the design conditions of 260°C at 4.2MPa (600psi) for the process fluid.

One of the most critical steps in the program was to define a set of criteria to use as both a guideline for the program and as a means to determine success or failure.

Based in large part on CNRL's extensive experience in the field the following criteria were set:

1. The sensor should give clear water-cut readouts in % on a local digital or computer display.
2. There should be no "tweaking" of the instrument by the end of the test.
3. The sensor should be operating to give water cuts consistently without interferences.
4. The apparatus results should be comparable to spun cuts within accuracy better than +/- 5 % of water cut readings.
5. The sensor should give repeatable results on the same sample (with +/- 5 % of the water-cut reading).
6. The sensor results should be comparable to Dean Stark extraction within accuracy of +/- 5 % of water cut reading.

Furthermore the instrument should be fitted with the hardware and software necessary for direct communication with the control system (PLC). The process of sampling and measuring should be automated. The measurement signal should be provided to the PLC by a 4-20 mA DC signal. Signals for measurement timing are passed between the instrument and PLC using 120

VAC on/off digital switches. Any other arrangement desired may also be used, as the proposed apparatus should be completely generic with respect to signals to or from the data logger or control systems. A special cabinet system to contain the instrument, its computer, and associated electronics for communication with the PLC was constructed. This cabinet had to protect the instrument, provide adequate ventilation and protect the environment from the instrument (Class 1, Div 2).

5. FIELD TESTING TO DATE

Example field data collected over a 24-hour period were presented elsewhere (Wright et al., 2004). It was demonstrated that the instrument sees changes in water cut with changes in well. Steady and reproducible results were obtained. It was further demonstrated that the test separator acts as a homogenizing buffer to the production data. Examination of the data shows that the tool is remarkably stable with measurements having a standard deviation of 2.2 %.

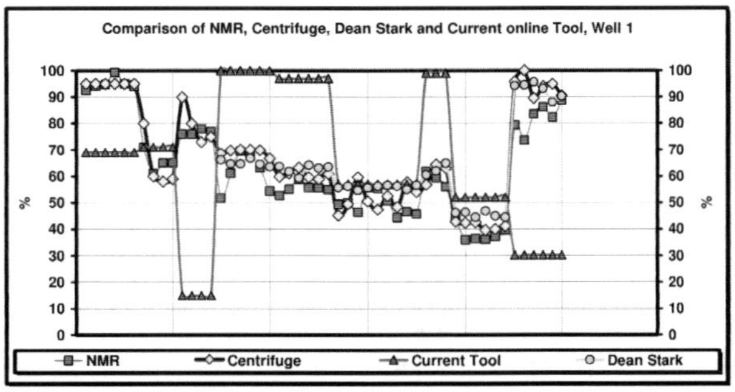

Fig. 7 Comparison of water cut measured by NMR, Centrifuge, Dean Stark and the current online tool (Wright et al., 2004)

Comparisons of the fit of the data with centrifuge cuts in the field and with subsequent Dean Stark extraction were considered satisfactory, considering the sample actually measured by the instrument is not the same as the sample sent to centrifuge, although it is as close as practically possible.

An attempt was made to compare the performance of the original online tool against NMR water cut measurements and against Dean Stark and centrifuge analysis of grabbed samples. A sample for such a comparison is shown in Fig. 7. The striking thing in this graph is that despite the sampling issues the NMR consistently tracks the discrete samples where as the competing instrument is only in range part of the time and is often reporting a number which is completely wrong. The pattern shown here, where the older tool consistently reports numbers which are different from the actual water cut, is typical of the operation of this well pad and is the major reason that the well pad was originally selected to test the NMR.

The first part of our work demonstrated that NMR based fluid content and fluid measurement technology is now feasible. Properties such as fluid viscosity, solvent content and water-cut can be measured on-line with remarkable accuracy and precision. The components that required intellectual property protection were patented (Mirotchnik et al., 2003, 2004). Additional field-testing will be required prior to full commercialization.

6. DESIGN IMPROVEMENTS AND NEW ALGORITHMS

A new company, Perm Instruments Inc., was created to commercialize patented and patentable technology developed in the Tomographic Imaging and Porous Media (TIPM) Laboratory. There has been a very strong interest by the oil companies that support our program to design a sensor that could measure oil, water, solvent, gas and solids in a flowing stream with immediate application in the heavy oil and oil sand development of Western Canada and at a cost comparable or less to that of three-phase test separators.

Fig. 8 is a simplified version of how the system works today. The fluid that has to be analyzed is conveyed to the sampling location inside the magnet (Fig. 8) where it is exposed to a relatively uniform magnetic field.

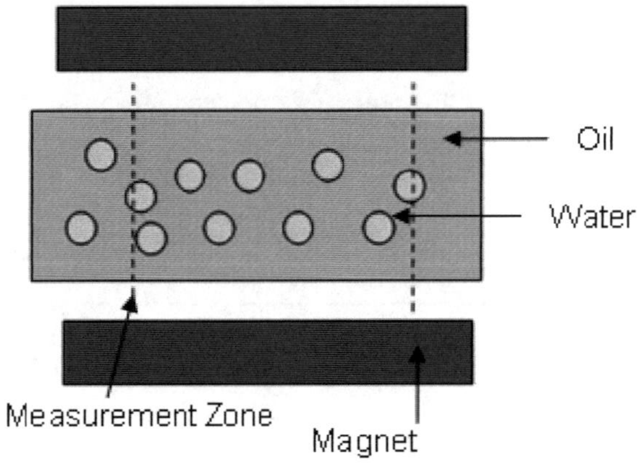

Fig. 8 Schematic of water-cut meter concept and approach

Any arbitrary emulsion in the measurement zone will yield a relaxation spectrum, which will compose of a water peak and an oil peak. Since the measurement volume is known we can calculate the amount of water present from the spectrum, normalize it to the measurement volume in the pipe and obtain the water cut. As long as there is no appreciable amount of gas the instrument will work very well as is also shown in Fig. 7, from the published data with Canadian Natural (Wright et al., 2004).

This idea has been implemented in an industrial type machine (MRWCM) that is shown in Fig. 9. The magnet, electronics and other support hardware is enclosed in a pressurized system that allows exposure to explosive environments under the Canadian Standard Association (CSA) Class 1 Div 2 classification. The cabinet and supporting structure have a minimal footprint of 1.5 x 1.5 m^2.

Fig. 9 Magnetic Resonance Water Cut Meter (MRWCM)

As mentioned earlier in Section 4, the MRWCM design concept meant that the pipe that conveys the process fluids to the point of measurement needs to be specially designed.

Journal of Physics: Conference Series **147** (2009) 012029 doi:10.1088/1742-6596/147/1/012029

Conventionally, pressure piping are designed as per ASME B31.3, which specifies the design requirements. Since, the pipe also needs to be transparent to a magnetic field and be highly resistive, the only materials that could be practically used were different kinds of plastics. Hence, an unconventional design approach was dictated by the meter, which further required extensive testing to demonstrate the performance of the new pipe. This process has since then been successfully completed with the necessary approvals from ABSA (Alberta Boilers and Safety Authority, which is the regulatory authority for Pressure Vessels and Pressure Piping designed and operated in Alberta). The effort has resulted in a one of a kind pipe design using plastic materials that can operate continuously at 260 C and 600 psig under the harshest fluid environments typical of the oilfields of Alberta. The detailed design of the pipe is currently being patented.

However, there are two areas of further improvement for this technology; (1) the sample has to be captured, thus there is no continuous monitoring; (2) the instrument is an add-on to the three-phase test separator and not a replacement. The first area is currently being addressed as follows:

The need to capture the sample within the MRWCM for a period of time is due to the fact that hydrogen spins need to reach equilibrium state when more of them are aligned along the direction of the magnetic field than in the opposite direction. The sample then remains within the same magnetic field for at least T_2 in order for the relaxation to be captured.

The complex population of spins that exist in hydrogen bearing molecules and are excited by the magnetic field of the MRWCM approaches equilibrium state after $T_{eq} \approx 5T_1$. Then the total time needed for the measurement procedure will reach $T_{meas} \approx 5T_1 + T_2 \approx 6T_1$. For bulk water with relaxation times close to $T_1 \approx T_2 \approx 2.5\,\text{sec}$, the above formula gives $T_{meas} \approx 15\,\text{sec}$.

With the current size of the uniform magnetic field area of about 10 cm along the axis of the pipe the flow speed is limited to not more than 10 cm / 15 sec ≈ 0.67 cm/sec. Therefore, a flow diameter of around 10 cm will translate to a production rate of 5 m^3/day, which in effect is an order of magnitude less than typical field production rates. Substantially larger magnets can be considered to resolve the issue, but this will noticeably increase the cost of the system. This approach is considered as a part of our broader development strategy. We also look at the possibility of sub sampling.

Alternatively, we suggest concentrating on the oil peak measurements instead and obtain water by difference. As the T_1 for heavy oil is typically in the range of 10 – 100 ms, enough time is available to accommodate realistic flow rates in the wells. For example the same flow diameter can accommodate in excess of 113 m^3/day flow rates for a 100 ms relaxing oil. The proposed approach could be accommodated with the current design for fluid flow measurements.

The planned steps for testing this approach are as follows:

- Modify our available high temperature loop for the purpose of this testing (accurate flow rate measurement, temperature control).
- Analyse flow regimes in the tube and assess the effects of viscosity, flow composition, turbulence and residence time distribution on accuracy of measurements.
- Test the capabilities of our digital fluoroscopy system (possibly combine two of them for getting stereo-effect) of delivering flow rate measurements based on appropriate image correlation analysis.
- With the known flow rate and flow velocity distribution analyse possibility of water content extraction from NMR signal.

- Look into the problem of proper continuous sampling (if a bypass loop will be needed in order to reduce flow rate of fluid passing through the meter).

A successful design of such a meter will require a wider diameter pipe and a wider magnet but of similar length. The potential incremental cost will have to be investigated but it is not considered prohibitive.

7. CONCLUSIONS

NMR based fluid content and fluid measurement technology is now feasible. Properties such as fluid viscosity, solvent content and water-cut can be measured on-line with remarkable accuracy and precision. Additional field-testing will be required prior to full commercialization.

ACKNOWLEDGEMENTS

The authors acknowledge the technical contributions and financial support of Petro-Canada and Laricina Energy Limited that went towards the development of the latest phase of this work. The work would not have been possible without the contributions of numerous staff and students in the TIPM laboratory, and in particular Florence Manalo, David Lastockin, Minghua Ding, Yuan Wen and Jonathan Bryan. Additional contributions were made by Kevin Allsopp, Ian Wright, Konstantin Mirotchnik. In addition, the support and contribution of Maureen E. Evers-Dakers of CNRL is gratefully appreciated.

NOMENCLATURE

A_o	Signal Amplitude of Oil	
A_w	Signal Amplitude of Water	
AI	Amplitude Index	
AI_w	Amplitude Index of Water	
DC	Direct Current	
m	Mass	[g]
NMR	Nuclear Magnetic Resonance	
PLC	Control System on Well Pad	
RHI	Relative Hydrogen Index	
T_1	Spin-Latice Relaxation Time	[ms]
T_2	Transverse Relaxation Time	[ms]
T_{2gm}	Geometric Mean T_2	[ms]
VAC	Volts Alternating Current	
w_c	Water-cut	
	Greek Letters	
α, β	Coefficients in Viscosity Models	
μ	Viscosity	[mPas]

Journal of Physics: Conference Series **147** (2009) 012029 doi:10.1088/1742-6596/147/1/012029

REFERENCES

Allsopp, K., Wright, I., Lastockin, D., Mirotchnik, K. and Kantzas, A. (2001). "Determination of Oil and Water Compositions of Oil/Water Emulsions Using Low Field NMR Relaxometry", J. Can. Pet. Tech., **40**(7), pp. 58-61.

Bryan, J., Kantzas, A. and Moon, D. (2005). "In-situ Viscosity of Oil Sands Using Low Field NMR", CIM 2003-107, J. Can. Pet. Tech., **44**(9), pp. 23-30.

Coates G., Xiao, L. and Prammer, M. (1999). *NMR Logging Principles and Applications, Halliburton Energy Services, Texas.*

Goodarzi, N., Bryan, J., Mai, A. and Kantzas, A. (2007). "Heavy Oil Fluid Testing with Conventional and Novel Techniques", SPE 97803, SPE Journal, **12**(3), pp. 305-315.

Hompoth, D., Pinquet, B.G. and Guerra, E. (2008). "Mulitphase Metering: Importance of Fluid Property Knowledge in SAGD Operations", Paper 2008-421, World Heavy Oil Congress, Edmonton, Alberta, Canada, March 10-12.

Hompoth, D., Khun, N., Pinquet, B.G. and Guerra, E. (2008). "Performance of Compbination of a Venturi and Nuclar Fraction Meter in SAGD Production Operations", Paper 2008-422, World Heavy Oil Congress, Edmonton, Alberta, Canada, March 10-12.

Kantzas, A. (2004) "On-line Measurement of Fluid Content and Fluid Properties for the Oil and Gas Industry with Low Field NMR Relaxometry", 2nd Canadian School of Hydrocarbon Measurement (C_SH_M), Calgary, Alberta March 24-25.

Mirotchnik, K., Allsopp, K., Kantzas, A. and Marentette, D.F. (2003). "Quantification of Bitumen Using NMR", Canadian Patent Office Serial #2,325,348. **US Patent 6,630,357**.

Mirotchnik, K., Allsopp, K., Kantzas, A. and Marentette, D.F. (2004). "Determination of Oil and Water Compositions of Oil/Water emulsions Using Low Field NMR Relaxometry", **US Patent 6,794,864**.

Wen, Y.W. and Kantzas, A. (2005). "Monitoring Bitumen-Solvent Interactions with Low-field Nuclear magnetic Resonance and X-ray Computer Assisted Tomography", Energy and Fuels, **19**, pp. 1319-1326.

Wright, I., Lastockin, D., Allsopp, K., Evers-Dakers, M.E. and Kantzas, A. (2004). "Field Trials of a Low Field NMR Water Cut Metering Device", CIM 2002-194 (Note), J. Can. Pet. Tech., **43**(5), pp. 17-21.

Journal of Physics: Conference Series **147** (2009) 012030 doi:10.1088/1742-6596/147/1/012030

Effect of pipeline thickness on electrical capacitance tomography

Yang Daoye[1], Zhou Bin[2], Xu Chuanlong[2], Tang Guanghua[2],Wang Shimin[2]

[1] Nanjing University of Technology, Nanjing, China
[2] Southeast University, Nanjing, China

yangdaoye@163.com, zhoubinde1980@gmail.com, chuanlongxu@seu.edu.cn, tgh164@163.com, smwang@seu.edu.cn

Abstract. For the purpose of analyzing the effect of pipeline thickness on ECT system include sensor and image reconstruction, radius-electrode ratio is defined as the ratio of inner and outer radius of pipeline. The rules of 8-electrode sensor parameters such as capacitance, capacitance change, and change rate of capacitance and sensitivity map are discussed by ANSYS and Matlab, which are combined to simulate sensor characteristic. With the increase of radius-electrode ratio, capacitance sensitivity to inner media increases, and the effect of pipeline thickness on capacitances of four typical inter-electrode pairs is much different under the same inner radius. LBP algorithm is adopted to reconstruct the ECT images of stratified flow, core flow and annular flow with different radius-electrode ratios. Simulation results show that image error under high radius-electrode ratio is small, and the image reconstructed from stratified flow is better than the ones from core flow and annular flow, especially in the condition of low radius-electrode ratio. As a result, ECT with low radius-electrode ratio is benefit to image reconstruction.

1. INTRODUCTION

Two-phase (multi-phase) flow is a commix flow mode which is widely existed, and its parameter measurement technology takes a more and more important role in science research and industrial production. Thus process tomography (PT) comes out with this trend. Electrical capacitance tomography (ECT) is one of the most mature and promising methods, which measures the capacitance change of multi-electrode sensor due to the change of dielectric permittivity being imaged, and then reconstructs the cross-section images using the measured raw data with a suitable algorithm. It has the characteristics such as low cost, fast response, non-intrusive method, broad application, safety (Li Haiqing et al.,2000; Yang W Q et al., 1995a, 1995b).

ECT system includes sensor, capacitance measuring circuit and imaging computer. And ECT sensor consists of insulating pipeline, measurement electrode, radial screen and earthed screen (Yang W Q et al., 1999). The measurement electrode is mounted symmetrically around the circumference of pipeline. Radial screen is fitted between the electrodes to cut the electro line external to the sensor pipeline and reduce the inter-electrode capacitance. The earthed screen surrounds the measurement electrodes to shield external electromagnetic noise. ECT sensor converts the permittivity distribution of multi-phase flow to inter-electrode capacitance, which is the ECT forward problem. Capacitance measuring circuit takes the capacitance data and transfers to imaging computer. Imaging computer reconstructs the distribution image with a suitable algorithm, which is called ECT inverse problem.

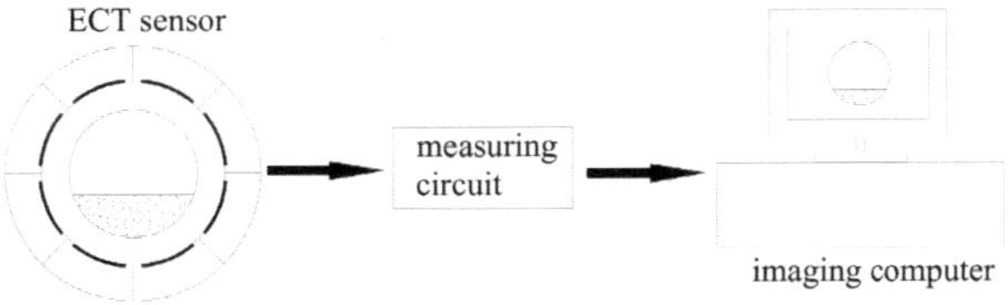

Fig.1 Sketch of ECT system

In most application, ECT electrode is mounted outside the ECT pipeline which is called external electrode ECT sensor (Yang W Q et al., 1997), so pipeline thickness is a very important parameter in ECT sensor design. The external electrode ECT sensor can avoid polluting and eroding by inner media, and affecting from the inner media static, so that non-intrusive measurement is realized.

External electrode ECT sensor is easy to manufacture and use, but the pipeline isolates the measured media from measurement electrode, therefore it brings some difficult to capacitance measurement and image reconstruction. Sensor pipeline results in the electro line aberrance at the boundary of different media. There are three factors which have great effect on electro line, e.g. pipeline material, inner dielectric permittivity and the ratio of pipeline thickness and diameter (A J Jaworski et al., 2000).

Finite element method (FEM) has been widely used to numerically solve mathematical problems in many engineering fields (Xie et al., 1992). Hence characteristic of 8-electrode ECT sensor with different pipeline thickness is studied in the method of numerical simulation, in which ANSYS is combined with Matlab. At the same inner diameter, ECT sensor models of different pipeline thickness are built in gas-solid two phase flow, and inter-electrode capacitance is simulated so that the sensitivity maps used to reconstruct image is taken. Then the effect of pipeline thickness on the imaging fidelity is investigated. The influence of pipeline thickness on ECT system is analyzed from aspect of forward problem and inverse problem.

2. EFFECT OF PIPELINE THICKNESS ON ECT SENSOR CHARACTERISTIC

2.1 Sensor Mathematical Model

For a sensor with N electrodes, there are N(N-1)/2 electrode pairs, hence N(N-1)/2 independent capacitance measurements. Fig. 2 is the cross section of 8-electrode ECT sensor, in which R_1 is inner pipeline radius; R_2 is outer pipeline radius; R_3 is earthed screen radius; and radial screen is connected to outer pipeline. High electrode covering ratio can get high image fidelity (A Martinez Olmos et al., 2007), because it can enlarge the inter-electrode capacitance, especially for adjacent electrode pairs, so electrode stretch angle θ should be chosen as large as possible. The axial length of electrodes is related to inter-electrode capacitance, signal bandwidth and measurement uncertainty of measured media. Longer electrodes produce an average signal over a greater axial length, which results in bad dynamic performance. Shorter electrode may result in capacitance too small to be measured accurately. Normally, the electrode length is chose as 10 cm (Peng et al., 2003).

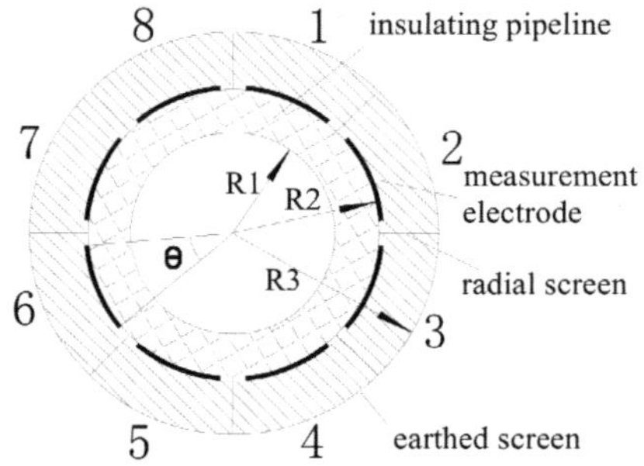

Fig.2 Cross section sketch of 8-electrode ECT sensor

For the purpose of making the describe of pipeline thickness easier, the concept of radius-electrode ratio $\rho(\rho=R_1/R_2)$ is put forward, which is the ratio of pipeline inner radius to electrode radius (i.e. pipeline outer radius). In ideal condition, pipeline thickness is zero, so ρ is equal to 1. The smaller radius-electrode ratio is, the more serious the effect of pipeline thickness becomes.

According to the conclusion of reference (Yan et al., 2003), when the ratio of axial length of electrode to outer radius of pipeline is larger than 1.5, the fringe effect of sensor can be ignored, namely, the sensor can be treated as two-dimensional field. The value of sensor described in the paper is much large than 1.5, so the sensor can be look on as simple 2-D structure. The operation frequency of capacitance measuring circuit is about 1 MHz, so the ECT sensor can be deal with electrostatic field theory.

If there exists no space charge inside the pipeline, the inner potential distribution can be described with 2-D Poisson equation (Xie et al., 1992; Jin Jianmin, 1998):

$$\nabla \cdot [\varepsilon_0 \varepsilon(x, y) \nabla \phi(x, y)] = 0 \qquad (1)$$

Where

$\phi(x, y)$ - spatial potential distribution

ε_0 - permittivity constant of free space

$\varepsilon(x, y)$ - spatial permittivity distribution

$\nabla \cdot$ - divergence operator

∇ - gradient operator

When electrode i is in excitation mode, its Dirichlet boundary condition is

$$\phi^{(i)}=\begin{cases} \mathrm{Vc} & (x,y)\subseteq\Gamma_i \quad (i=1,2,...,7) \\ 0 & (x,y)\subseteq\Gamma_j \quad (j=1,2,...,8,\, j\neq i) \quad \cup (x,y)\subseteq Q_s \end{cases} \tag{2}$$

Where Γ_1, Γ_2, …, Γ_8 are the spatial positions of the eight electrodes, Q_s is the earthed screen, Vc is the excitation potential. The forward problem is to take the inter-electrode capacitance at a certain dielectric distribution.

Solving Equ.1, the sensor potential distribution $\phi(x, y)$ can be obtained. For the structure irregularity of sensor electrode pair and the inhomogeneity of dielectric distribution, Equ.1 doesn't exist analytical solution, but it can be worked out through FEM. When using FEM to solve problem, the continuous field should be converted to a discrete form, namely, the solution domain should be divided into finite elements. For the convenience of numerical simulation, the FEM software ANSYS is used to build the geometry model of sensor and divide the cross section domain into triangle elements. The partition results such as node coordinates, boundary condition and relation between elements and nodes are extracted to calculate potential distribution in Matlab.

With potential distribution of sensor nodes, the inter-electrode capacitances can be obtained by following equation.

$$C_{i,j}=Q(\Gamma_j)/V_c=(\varepsilon_0/V_c)\int_{(x,y)\subseteq\Gamma_j}\varepsilon(x,y)\nabla\phi^{(i)}(x,y)\cdot d\Gamma_j \tag{3}$$

Where

$Q(\Gamma_j)$ - measured electric charge on electrode j, j= i+1, … , 8

$\phi^{(i)}(x, y)$ - potential distribution when electrode i is in excitation mode

The elapsed time of calculating a group of capacitances is determined by segmentation grade of grid, and node number and segmentation grade near electrodes are the main factors. Models of different pipeline thickness are built, and each electrode is divided into sixteen segments. Numerical simulation is implemented using Matlab on a PC with a Pentium IV 2.66G CPU and 512 Mbytes memory.

2.2 Effect of Pipeline Thickness on Sensor Capacitance

It is well known that the influence of a certain pipeline thickness to sensors with different inner radii is not the same. For the sake of scaling different sensors effectively, the concept of radius-electrode radio is defined (see section 2.1), so that it is more significant to evaluate the pipeline thickness.

The parameters of simulated ECT sensor, which used in dense-phase pneumatic conveying of pulverized coal, include inner radius R_1, outer radius R_2, earthed screen radius R_3, electrode stretch angle θ, axial length of electrodes L, gas permittivity ε_{gas}, solid permittivity ε_{solid}, insulating pipeline permittivity ε_{pl} and filling layer permittivity ε_{se}. The inner radius is constant, that is 5mm. The pipeline thickness (R_2-R_1) is selected as 5mm, 4mm, 3mm, 2mm, 1mm and 0mm respectively, so eight sensors are structured, whose radius-electrode radios are 0.50, 0.56, 0.64, 0.71, 0.77, 0.83, 0.91 and 1 respectively. Standing capacitance (when the pipeline is filled with gas), capacitance change and change rate of capacitance, which is the ratio of capacitance change to standing capacitance, are calculated, are investigated. Because of symmetry, 8-electrode sensor has four typical structures of inter-electrode pair; hence only four inter-electrode pairs (C1-2, C1-3, C1-4 and C1-5) are listed, as showing in Fig. 3. The electrode stretch angle θ is 40°, sensor axial length L is selected as 10cm, and permittivity (ε_{gas}, ε_{solid}, ε_{pl} and ε_{se}) is set to 1, 3.5, 4 and 2.5 respectively.

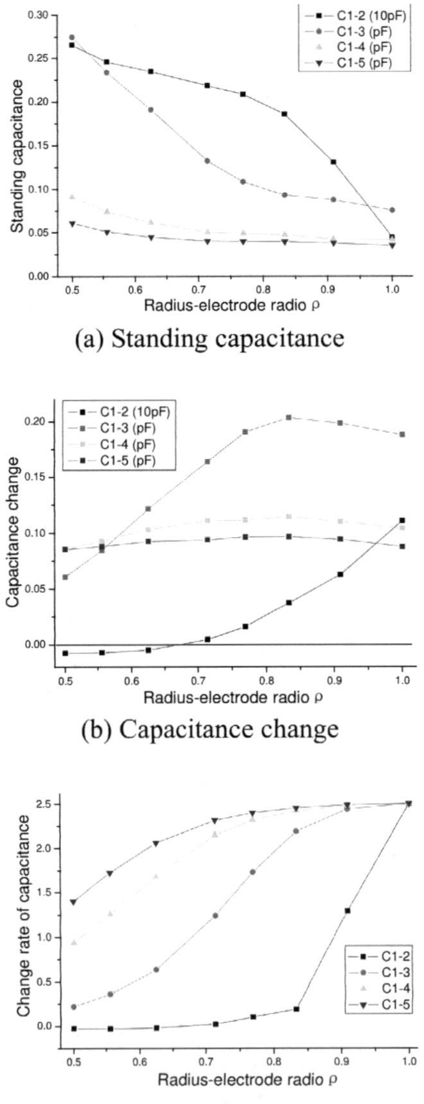

(a) Standing capacitance

(b) Capacitance change

(c) Change rate of capacitance

Fig. 3 Capacitance against radius-electrode ratio

For the particularity of position, capacitances of adjacent electrodes are much large than the others. For the convenience of plotting, standing capacitance and capacitance change of adjacent electrodes are reduced to 1/10.

Capacitance is proportional to permittivity, thus ε_{pl} is large than ε_{gas}. With the increase of radius-electrode radio, the gas phase acts a more and more important role, so the standing capacitances of four typical electrode pairs reduce (see Fig. 3(a)). But the standing capacitance tendency curve of adjacent electrode is different from the others. Capacitance of adjacent electrodes has significant change at high radius-electrode radio, but the non-adjacent ones have more obvious changes at low radius-electrode radio.

The tendency curve of capacitance change against ρ is shown in Fig. 3(b). In adjacent electrode pair, capacitance change increases monotonously with the increase of ρ, and turns from negative to positive. However, there exist maximum points in non-adjacent electrode pairs, and all the maximum values is at the same position where ρ = 0.85. So sensor with this ρ is more suitable to capacitance measurement.

Fig.3(c) is change rate of capacitance against ρ. Change rate of capacitance reflects how much capacitance of electrode pair is affected by the inner dielectric. In the ideal condition (ρ=1), change rate of capacitance is only related to permittivity, that is about 2.5 ((3.5-1)/1). At limit case of ρ=0, change rate of capacitance have no relationship with inner dielectric, namely, 0. Therefore, the effect of pipeline is little when ρ is near to 2.5, whereas the effect is great when ρ is near to 0. All of the change rate of capacitance rise with the increase of ρ. The effect of inner dielectric to opposite electrode pair is the largest, especially at high ρ. However it has little effect on adjacent electrode pair when ρ is little than 0.85. The influence of pipeline to electrode pairs decreases followed by C1-2, C1-3, C1-4 and C1-5.

2.3 Effect of Pipeline Thickness to Sensitivity map

Sensitivity distribution function $S_{i,j}(k)$ is defined as: relative capacitance change between electrode i-j due to permittivity change of *k*th element. When inner dielectric is gas and solid phase, it can be described as follows:

$$S_{i,j}(k) = \mu(k) \cdot \frac{C_{i,j}(k) - C_{gas}}{(C_{solid} - C_{gas})(\varepsilon_{solid} - \varepsilon_{gas})} \qquad (4)$$

Where

- k - serial number of element (k=1, 2, …, M)
- M - total number of finite elements inside the pipeline
- ε_{gas} - permittivity of gas phase
- ε_{solid} - permittivity of solid phase
- $C_{i,j}(k)$ - capacitance when permittivity of *k*th element is ε_{solid} and the others are ε_{gas}
- $\mu(k)$ - correlation factor of *k*th element related to area, which is the ratio of inner pipeline area to *k*th element area.

Based on Equ.4, the sensitivity matrix $S_{N\times M}$ can be taken, in which N is the total number of electrode pair in 8-electrode sensor. The calculated sensitivity maps with different ρ are shown in Fig. 4.

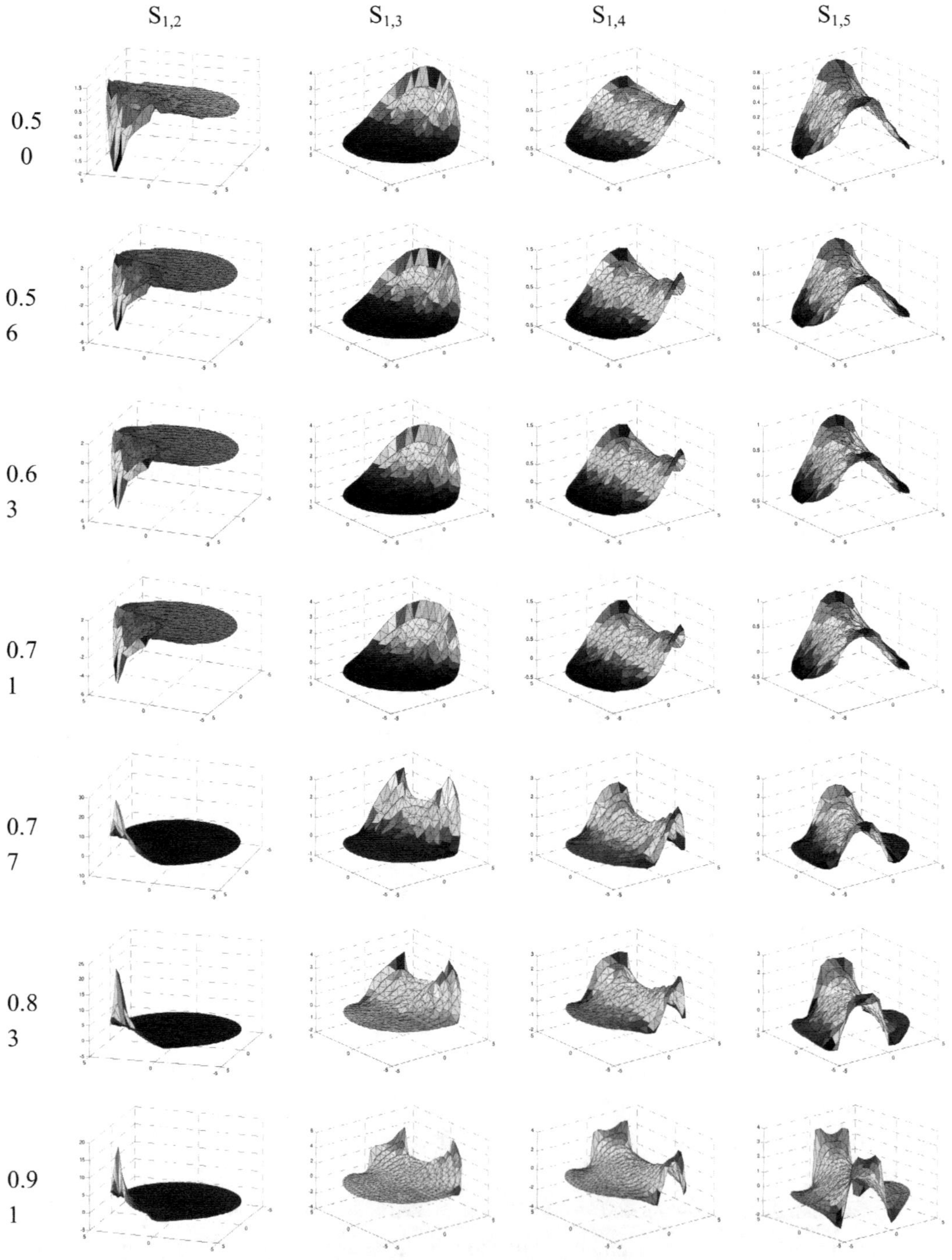

Fig. 4 Sensitivity map with different ρ

The largest sensitivity is nearby electrode 1 and electrode 2. The sensitivity is negative near to electrode 1 and 2 when $\rho \leqslant 0.71$. The shapes of sensitivity maps have no obvious variation except for intensity. With the increase of ρ, the maximum of sensitivity decreases. Nevertheless, when $\rho \geqslant 0.77$, sensitivity near to the two electrodes turns positive and the maximum of sensitivity decreases with the increase of ρ. Incidentally, capacitance change of adjacent electrodes turns positive at $\rho = 0.67$ (see Fig.3(b)), while maximum of sensitivity at $\rho = 0.71$ is still negative. It is not ambivalent because capacitance change reflects global variety and maximum of sensitivity represents local value. Fig. 5 is the variety tendency of maximum of sensitivity between adjacent electrodes. Maximum of sensitivity exists a broken point when ρ is between 0.71 and 0.77, but not a continuous monotonic curve.

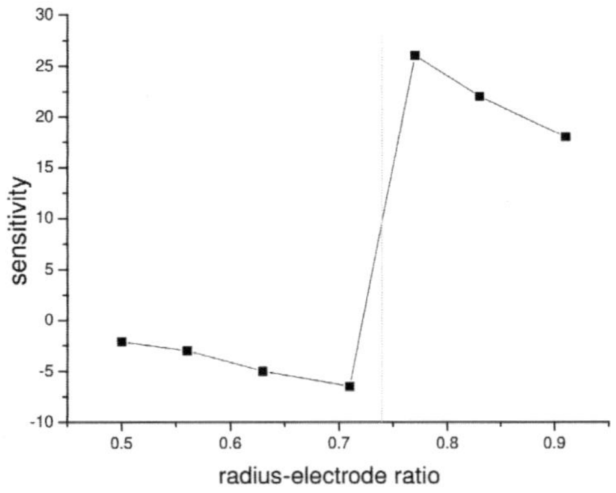

Fig.5 Sensitivity maximum of adjacent electrodes

$S_{1,3}$ is the sensitivity distribution between electrode 1 and 3. When $\rho < 0.71$, sensitivity peak is present between the two electrodes near to the pipeline, but not near to the electrodes. This is because the pipeline is so thick that inner pipeline wall is far from the electrode, which results in the two sensitivity peaks integrate into one. However, when ρ is large than 0.77, the effect of pipeline is weakened, so that there are two peaks of sensitivity map near to electrode 1 and 3.

$S_{1,4}$ is the sensitivity distribution between electrode 1 and 4. There exists two sensitivity peaks close to electrode 1 and 4 respectively. The distance between electrode 1 and 4 is far, the pipeline thickness can not conceal the characteristic that the sensitivity near electrode is high. Further more, the two peaks can be differentiated more obviously with the increase of ρ. When ρ is large than 0.77, negative sensitivity presents near to the two peaks.

$S_{1,5}$ is the sensitivity distribution between electrode 1 and 5. Shapes of sensitivity map in opposite electrode pair are similar to each other, look like saddles. The sensitivity peaks present near to electrode 1 and 5. Therefore the opposite electrode pair has minimum effect of pipeline. Similar to $S_{1,4}$, there also exist negative sensitivity near to the peaks when ρ is higher.

3. EFFECT OF PIPELINE THICKNESS ON IMAGE RECONSTRUCTION

Image reconstruction takes an important role in ECT, which is to reconstruct permittivity distribution inside the pipeline using measured capacitances of inter-electrode pairs. There are three principal problems presented in ECT image reconstruction. Firstly, the relation between permittivity and capacitance is not linear, and linear algorithm will result in imaging error; Secondly, number of capacitances is much less than number of image element; and thirdly, ECT inverse problem is ill-posed, which induces that the imaging result is much sensitive to measuring error and noise. Furthermore, sensor pipeline intensifies the difficult of image reconstruction

For the convenience of combination of forward and inverse problem, the partition of image reconstruction is the same as in forward problem. Then the image of inner pipeline is reconstructed using the capacitance data, which obtained by forward simulation (see Eqn.3) at a given permittivity distribution. The capacitance is normalized so that the measured data is dimensionless, which is easy to process image and reduce the effect of measurement error (Yang Lijian et al., 2002).

The normalized capacitance is defined as:

$$\lambda_{i,j}=(C_{i,j}^{m}-C_{i,j}^{gas})/(C_{i,j}^{solid}-C_{i,j}^{gas}) \qquad (5)$$

Where

$C_{i,j}^{m}$ - Measurement capacitance of electrode *i* and *j* at a given flow distribution

$C_{i,j}^{gas}$ - Capacitance of electrode *i* and *j* when the pipeline is empty

$C_{i,j}^{solid}$ - Capacitance of electrode *i* and *j* when the pipeline is full

Theoretically, normalized capacitances should be between 0 and 1, but it may be negative or large than 1 in adjacent electrode pair, especially when ρ is small, because of discontinuous permittivity distribution and radial screen.

LBP is the earliest algorithm used in ECT image reconstruction (Xie et al., 1989). Although mathematically not accurate and can only get qualitative image, LBP algorithm is still widely used in on-line image reconstruction because of its simplicity and celerity. If the effect of dielectric permittivity to sensitivity map can be ignored, the gray level *g(k)* of *k*th element in reconstructed image can be expressed as:

$$g(k)=\frac{\sum_{i=1}^{7}\sum_{j=i+1}^{8}\frac{C_{i,j}^{m}-C_{i,j}^{gas}}{C_{i,j}^{solid}-C_{i,j}^{gas}}S_{i,j}(k)}{\sum_{i=1}^{7}\sum_{j=i+1}^{8}S_{i,j}(k)} \qquad (6)$$

For the purpose of demonstrating the effect of pipeline thickness to image reconstruction, LBP algorithm is adopted to reconstruct the image of different flow patterns (stratified flow, core flow and annular flow) under the condition of ρ = 0.5 and ρ = 0.83. Image gray level is classified to five grades, and the reconstruction result is shown in Fig. 6 and Fig. 7.

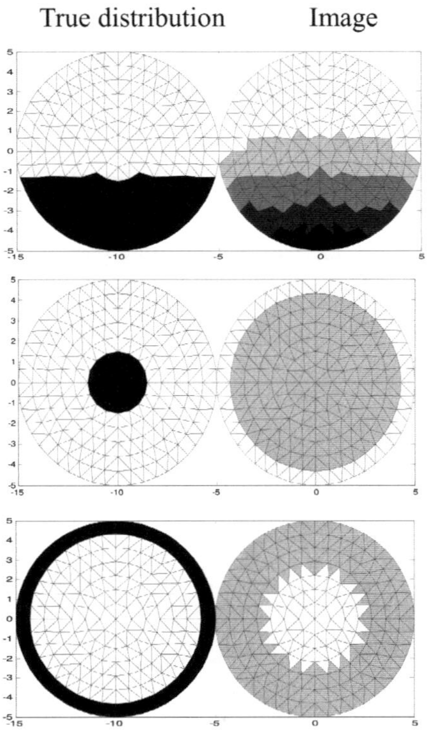

Fig. 6 Image Reconstruction at ρ=0.5

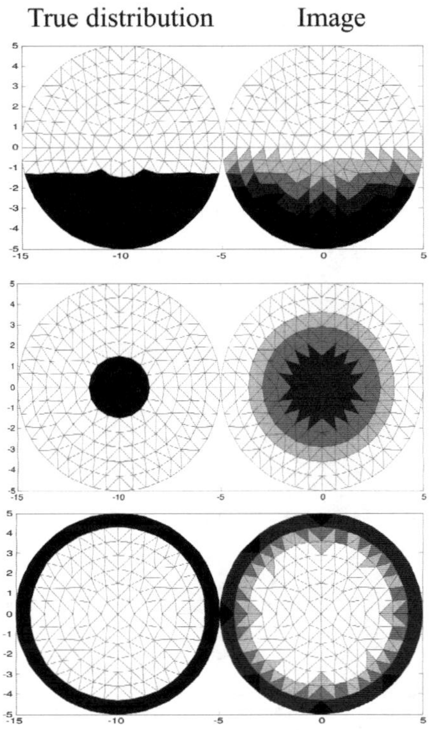

Fig. 7 Image Reconstruction at ρ=0.83

It is obvious that the reconstructed image at ρ=0.83 is better than the one at ρ=0.5 from the comparison of Fig.6 and Fig.7. Consequently, pipeline thickness has significant influence not only on capacitance but also on image reconstruction. The image fidelity is better in core and annular flow than in stratified flow when ρ is 0.83. But it is on the contrary when ρ is 0.5. In the condition of low radio-electrode radio, effect of pipeline thickness on opposite electrode pair is the least among these electrode pairs. The relation between opposite electrode pair and these flow patterns, which are symmetric with respect to circle center, is much less, so the reconstructed image fidelity under core and annular flow is not satisfactory. If an image filter threshold is introduced (Xie et al., 1993), image fidelity can be improved effectively.

4. CONCLUSION

There are two major computational problems in ECT: the forward problem and the inverse problem. The forward problem is to determine inter-electrode capacitance by solving the partial differential equations from permittivity distribution, and the inverse problem is to obtain permittivity distribution from measurement capacitance with a suitable algorithm.

The effect of pipeline thickness on ECT forward and inverse problem, such as capacitance, sensitivity and image reconstruction, is discussed completely by FEM numerical simulation. The influence of pipeline to electrode pairs decreases followed by 1-2, 1-3, 1-4 and 1-5. Maximum capacitance change is gotten at ρ=0.85, which is beneficial for capacitance measurement and image reconstruction. With the decrease of radius-electrode radio ρ, the distortion of ECT sensor sensitivity becomes more serious.

The sensor characteristic with low ρ is mainly manifested by opposite electrodes. However, opposite electrode pairs at low ρ can not reflect specialty of flow patterns which are symmetric with respect to circle center. Therefore, reconstructed image under flow patterns which are symmetric with respect to circle center is dissatisfactory, thus it is much better in these flow patterns such as stratified flow which is not symmetric with respect to circle.

ACKNOWLEDGEMENTS

This research was supported by Special Funds for National Key Basic Research of China (50836003) and Major State Basic Research Projects (2004CB217702). The authors wish to express their gratitude.

NOMENCLATURE

R_1	pipeline inner radius	[m]
R_2	pipeline outer radius	[m]
Vc	excitation potential	[F]
$C_{i,j}$	capacitance between electrode i and j	[F]
N	total number of electrode pair	

M total number of finite elements inside the pipeline
k serial number of element
$S_{i,j}(k)$ sensitivity of kth element between electrode i and j
$g(k)$ gray level of kth element
$C_{i,j}^{m}$ Measurement capacitance of electrode i and j [F]
$C_{i,j}^{gas}$ capacitance of electrode i and j when the pipeline is empty [F]
$C_{i,j}^{solid}$ capacitance of electrode i and j when the pipeline is full [F]

Greek Letters
ρ radius-electrode ratio
Φ spatial potential [V]
$\nabla \cdot$ divergence operator
∇ gradient operator
ε_0 permittivity constant of free space
$\varepsilon(x, y)$ spatial permittivity distribution
Γ_i spatial positions of electrode i
ε_{gas} permittivity of gas phase
ε_{solid} permittivity of solid phase
$\mu(k)$ correlation factor of kth element related to area
$\lambda_{i,j}$ normalized capacitance between electrode i and j

Subscripts
i i-th electrode
j j-th electrode
k k-th element

REFERENCES

A J Jaworski et al. (2000). The design of an electrical capacitance tomography sensor for use with media of high dielectric permittivity. Measurement Science and Technology, 11(6), pp. 743-757

A Martinez Olmos, et al. (2007). Fernandez Marron. Simulation design of electrical capacitance tomography sensors. IET Sci. Meas. Technol., 1(4), pp. 216, 216-223.

Jin.(1998). Finite Elements for Electromagnetic field. Xi'an: Xian University of Electronic Science and Technology Press, pp. 51-73

Li Haiqing et al. (2000). Special measurement technology and application. Hangzhou: Zhejiang University Press

Peng, et al. (2003). Image reconstruction algorithms for electrical capacitance tomography. Measurement Science and Technology, 14(1), pp. R1-R13

Xie, et al. (1992). Electrical capacitance tomography for flow imaging: system model for development of image reconstruction algorithms and design of primary sensors. IEE PROCEEDINGS-G, 139(1): 89-98

Xie, et al. (1989). 8-electrode capacitance system for two-component flow identification. Part 1: Tomographic flow imaging. IEEE Proceedings-A: Physical Science, 136(4), pp. 173-183

Xie, et al. (1993). Transputer-based electrical capacitance tomography for real-time imaging of oilfield flow pipeline. Tomographic Techniques for Process Design and Operation, pp. 333-346

Yan, et al. (2003). The simulation study of the electrical capacitance tomography. Journal of System Simulation, 15(11), pp. 1625-1627

Yang Lijian, et al. (2002). Normalization Approach for Capacitance Tomography Measurement Data . Chinese Journal of Scientific Instrument, 23(3), pp. 229-231, 225

Yang W Q et al. (1995a). Development of capacitance tomographic imaging systems for oil pipeline measurements. Review of Scientific Instruments,, 66(8), pp. 4326

Yang W Q, et al. (1995b). Electrical capacitance tomography –from design to applications. Measurement & Control, 28(9), pp. 261-266

Yang W Q, et al. (1999). New AC-based capacitance tomography system. IEE proceedings: Measurement Science and Technology, 146(1), pp.47-53.

Yang W Q, et al. (1997). Modelling of capacitance sensor. IEE proceedings: Measurement Science and Technology, 144(5), pp. 203-208.

Journal of Physics: Conference Series **147** (2009) 012031 doi:10.1088/1742-6596/147/1/012031

Imaging and Visualization of Multiphase Fluid Flow by different reconstruction method for Electrical Computer Tomography

Tong Zhao[1], Masahiro Takei[1], Kristian G. Basario[1]

[1] Nihon University, 1-8-14 Kanda Surugadai, Chiyoda-ku, Tokyo 101-8308, Japan

polomarslee@hotmail.com

Abstract. A new reconstruction method called Functional Weighted Inverse Matrix (FWIM) has been applied to an ill-posed inverse problem involving an electrical computer tomography for multiphase flow. Four image reconstruction algorithms——FWIM, Landweber (LW), Iterative Tikhonov regularization (ITR) and Generalized Vector Sampled Pattern Matching (GVSPM) methods——are tested using four types of pseudo particle distribution images with different dispersion concentration rates (DCR). The reconstructed images, as shown in figure, are compared in terms of capacitance residual, image error and image correlation. Overall, the accuracy is strongly dependent upon the image type and the dispersion concentration rates (DCR). However, the FWIM method proved superior to the LW, the ITR and the GVSPM methods in terms of the annular and bottom pseudo images.

1. INTRODUCTION

Recently, the application of non-invasive monitor, such as computed tomography (CT), in multiphase flow measurement has become increasingly popular. The electrical capacitance CT has been investigated as a visualization technique for the solid behavior in solid-air two-phase flow (Dyakowski et al. 1999; Halow et al. 1992; Huang et al. 1989). In capacitance CT, a sensor containing several electrodes is wrapped around the circumference of a pipeline, and the capacitances between the electrodes are measured. The particle concentration distribution, which is based on the permittivity distribution in a cross section, is obtained from the experimental capacitances. This is performed using an image reconstruction technique that is based on an ill-posed mathematical inverse problem. Because inverse problems are heavily dependent upon the system equation, a suitable image reconstruction technique for this capacitance CT is necessary. A variety of image reconstruction techniques have already been proposed (Isaksen 1996). Recently, the iterative techniques such as the Landweber method and the Iterative Tikhonov regularization method have been widely used because of the relatively high accuracy of their reconstruction images (Yang et al.2003). These conventional reconstruction techniques, however, have several drawbacks with regard to empirical value setting and convergence at an infinitive number of iterations. For example, the Landweber method requires an empirical gain value in order to converge upon the image, and it strongly depends upon the iteration number. It indicates that an image that has been processed over a suitable number of iterations will become extremely distorted (Liu et al. 2001). Therefore, the Landweber method requires advance knowledge of the empirical gain value and the empirical iteration number. The Iterative Tikhonov regularization method, on the other hand, needs an empirical value to use the singular value decomposition of the system equation to produce a pseudo inverse matrix. These conventional reconstruction methods are not able to accurately reconstruct an image with empirical values because they do not use an objective function to confirm the stability of the solution during the iterative process.

In order to overcome the drawbacks of these iterative methods, a novel solution strategy called the Generalized Vector Sampled Pattern Matching (GVSPM) method was proposed, and has already been applied to the optimization of electron beam dosing (Saotome et al. 1995; Yoda et al. 1997). However, as a kind iterative methods, the solve time of GVSPM method is too long for real time visualization. Therefore, a new non –iterative method called Functional Weighted Inverse Matrix (FWIM) method was put forward for the real time visualization of multiphase flow.

In this study, the new FWIM method is applied to the reconstruction of particle distribution images. This paper details the characteristics of this method that were examined using a simulation for pseudo particle concentration distribution images. The results of this simulation were then compared to those from the conventional Landweber, Iterative Tikhonov regularization and GVSPM methods. The results in three areas were compared: capacitance residual, image error and image correlation.

2. GOVERNING EQUATION FOR ECT

The capacitance CT sensor is shown in Fig. 1(A) and (B). The twelve sensor electrodes are separated by insulation material (Yang 1996). The relationship between capacitance and permittivity in a static-electro field is expressed by:

$$C_{i,j} = -\frac{\varepsilon_0}{V_c} \int_{r \subset \Gamma_j} \varepsilon(\mathbf{r}) \nabla V_i(\mathbf{r}) \cdot d\mathbf{r} \tag{1}$$

where i is the standard electrode number that ranges from 1 to 11, and j is the reference electrode number, which ranges from i+1 to 12. $C_{i,j}$ is the measured capacitance between the standard electrode i and the reference electrode j, ε_0 is the known vacuum permittivity, $\varepsilon(\mathbf{r})$ is the relative permittivity distribution on the cross section, $\mathbf{r}$ is a position vector on the cross section: $\mathbf{r}$ =(x, y), V_c is the known voltage to the ith electrode, Γ_j is the area affected by the electric line of force, and $V_i(\mathbf{r})$ is the potential distribution on the cross section between the ith and jth electrodes. Even though the values of $\varepsilon(\mathbf{r})$ and $V_i(\mathbf{r})$ in Eq. (1) are unknown, $\varepsilon(\mathbf{r})$ can be approximated by assuming that the electric charge's linear coupling at a position r with a weight of sensitivity in Γj area is the total capacitance. As detailed in a previous work (Polydorides et al. 2002; Williams et al. 1995), when a particle exists solely in a infinitely small $\Delta x \times \Delta y$ area at the center point $\mathbf{r}=\mathbf{r}_0$ between the i and j electrodes, and air exists at the remaining positions, the Laplace equation

$$\nabla \cdot [\varepsilon(\mathbf{r}_0) \nabla V(\mathbf{r}_0)] = 0 \tag{2}$$

is assumed to hold in the cross section. A Finite Element Method can be used to discritize Eq. (2), and the distribution of $V_i(\mathbf{r})$ can be obtained by substituting the boundary conditions into Eq. (1). Then, Eq. (1) can be rewritten as the following matrix expression:

$$\mathbf{C} = \mathbf{S_e E} \tag{3}$$

In other words, the capacitance CT can be used to obtain the permittivity distribution of particles **E** in the cross section from both the known sensitivity map matrix $\mathbf{S_e}$ and the measured capacitance matrix **C**. In the case of 12 electrodes and 32×32 = 1024 pixels in the pipe cross section, as in Fig. 1(C), the sensitivity matrix $\mathbf{S_e}$ in Eq. (3) is a 66×1024 matrix, the capacitance matrix **C** is a 66×1 column vector, and the permittivity distribution matrix **E** is a 1024×1 column vector. The mathematical method used to obtain the permittivity matrix **E** from both the capacitance matrix **C** and the sensitivity matrix $\mathbf{S_e}$ is an ill-posed inverse problem because the inverse matrix $\mathbf{S_e}^{-1}$ does not exist. The raw values of $\mathbf{S_e}$ have an extremely wide range; the sensitivity to an adjacent electrode pair is more than 100 times larger than that to an opposing electrode pair.

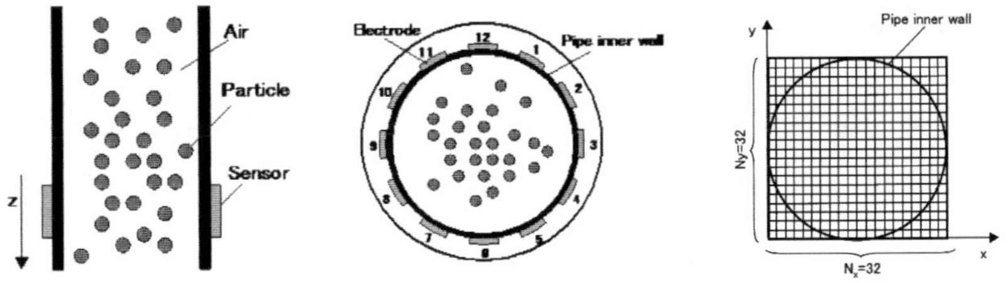

(A) Frontal view (B) Cross sectional view (C) Space resolution

Figure 1 Overview of capacitance tomography

3. FWIM METHOD THEORY FOR INVERSE PROBLEM

An inverse problem is the task that often occurs in solving the system where the values of some model parameter(s) must be obtained from the observed data. Inverse problems are typically ill posed, it means that the number of unknown is more than the number of formula. As addressed above, the ill posed problem need to be solve in this research is shown as Equation 3. Here, the solution vector **E** could be assumed to be the product of a weighted 1024×66 determinant **W** and a 66×1 column vector **S** as follow,

$$\mathbf{E}=\mathbf{WS} \tag{4}$$

If the determinant **CW** is the non-singular, Equation 3 can be reduced to

$$\mathbf{E}=\mathbf{W}[\mathbf{CW}]^{-1}\mathbf{Y} \tag{5}$$

In Equation 5, the key point is the way to decide the weight determinant **W**. If the solution e is the function of the solution space α, it could be given as,

$$e(\alpha)=s_0+s_1\cos[\alpha]+s_2\sin[\alpha]+s_3\cos[2\alpha]+s_4\sin[2\alpha]+\ldots \tag{6}$$

Equition 6 can be rewrite using complex numbers as,

$$e(\alpha)=s_0+s_1\operatorname{Re}[e^{i\alpha}]+s_2\operatorname{Im}[e^{i\alpha}]+s_3\operatorname{Re}[e^{i\alpha}]+s_4\operatorname{Im}[e^{i\alpha}]+\ldots \quad i=\sqrt{-1} \tag{7}$$

The solution vector **E** can be write as,

$$\mathbf{E}=\begin{pmatrix} e_1 \\ e_2 \\ \vdots \\ e_m \end{pmatrix}=\begin{pmatrix} 1 & 1 & 0 & 1 & \cdots \\ 1 & \cos[\alpha] & \cos[\alpha] & \cos[2\alpha] & \cdots \\ \vdots & \vdots & \vdots & \vdots & \vdots \\ 1 & \cos[(n-1)\alpha] & \sin[(n-1)\alpha] & \cos[2(n-1)\alpha] & \cdots \end{pmatrix}\begin{pmatrix} s_0 \\ s_1 \\ s_2 \\ \vdots \\ s_{n-1} \end{pmatrix}=\mathbf{WS} \tag{8}$$

Therefore, the weight determinant **W** can be given by,

$$\mathbf{W}=\begin{pmatrix} 1 & 1 & 0 & 1 & \cdots \\ 1 & \cos[\alpha] & \cos[\alpha] & \cos[2\alpha] & \cdots \\ \vdots & \vdots & \vdots & \vdots & \vdots \\ 1 & \cos[(n-1)\alpha] & \sin[(n-1)\alpha] & \cos[2(n-1)\alpha] & \cdots \end{pmatrix} \tag{9}$$

4. IMAGE RECONSTRUCTION AND DISCUSSION

4.1 Reconstruction conditions and example images

In this section, the Landweber (LW), Iterative Tikhonov regularization (ITR), GVSPM and FWIM methods are tested using pseudo particle concentration distribution simulation images of solid-air two-phase flow. The following guidelines are established to compare each method impartially.

1) The iterative number of LW, ITR and GVSPM method is 30.
2) The GVSPM's normalized sensitivity matrix is normalized with the norm of the column vector.
3) The first capacitance, which is obtained by multiplying the sensitivity matrix and the original pseudo image, is normalized with a minimum value of 0.0 and a maximum value of 1.0. The first

Journal of Physics: Conference Series **147** (2009) 012031 doi:10.1088/1742-6596/147/1/012031

capacitance is the measured capacitance in the real experiments.

4) In both the original pseudo images and the final reconstructed images, all of the pixels outside the pipe are ignored because the area outside the pipe is not relevant to the image comparisons.
5) After per iteration, the capacitances for the LW, ITR and GVSPM method are normalized with a minimum value of 0.0 and a maximum value of 1.0. The reconstructed image is not normalized by itself.
6) Because the results are normalized by the norm after each iteration, the final images of the GVSPM method provide only relative values. Therefore, the final images and final capacitances must be normalized with a minimum value of 0.0 and a maximum value of 1.0.
7) The gain α of LW method is fixed at 2.0 and the γ of ITR method is fixed at 0.01 for all of the pseudo particle images because of the relatively high correlation value.

Figure 2 shows the simulation images for four types of pseudo particle distributions. Because the actual solid-air two-phase flow image consists of main particle bulk and dispersion particles around it, many different dispersion concentration rates (DCR) between 0% and 100% are considered. For example, a 20% DCR correlates to a random white noise value between 0.0 and 0.2, which highlights the dispersion particles. These values are assigned to each pixel in the original 0% DCR image. Values over 1.0 replace all of the pixels with the original pseudo particle images.

Figure 2(a) displays the pseudo particle images with a 0% DCR and Figure 2(b) shows the representative images for a 100% DCR. In these figures, (*-1), (*-2), (*-3) and (*-4) are the annular, the center, the bottom and the four-bulk pseudo particles image within the pipe, respectively. In these pseudo images, the red pixels indicate the highest particle concentration, 1.0, and the blue pixels indicate air.

The representative reconstructed results of annular pseudo image that were obtained using each method are presented in Figure 3 as a reference. From these figures it is evident that the ring is reconstructed precision well using FWIM and GVSPM method, however, the reconstructed results using LW and ITR method are not ring and become circle.

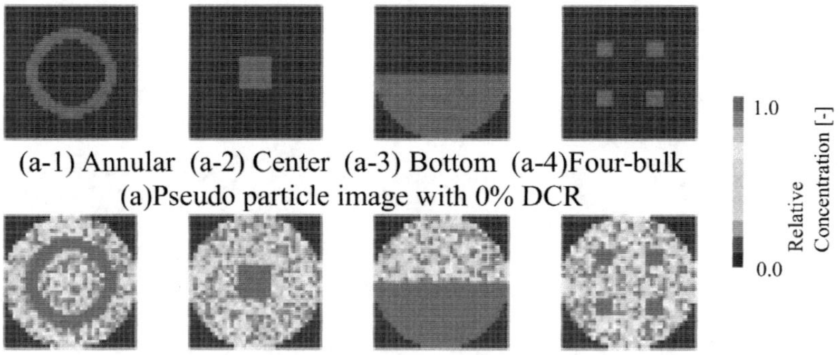

(a-1) Annular (a-2) Center (a-3) Bottom (a-4)Four-bulk

(a)Pseudo particle image with 0% DCR

(b-1) Annular (b-2) Center (b-3) Bottom (b-4)Four-bulk

(b)Pseudo particle image with 100% DCR

Figure 2 Pseudo images

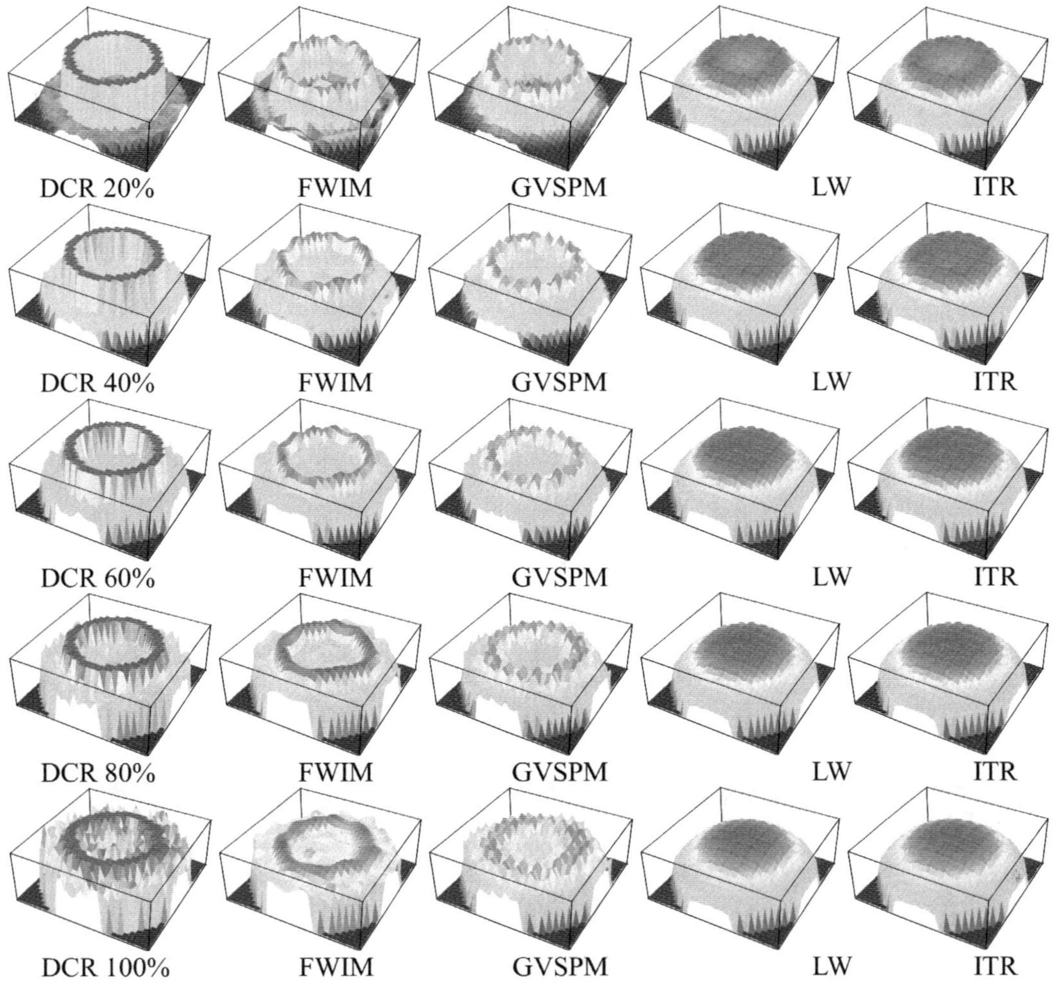

Figure 3 Reconstructed images of different DCR (Annular)

4.2 Estimation of reconstructed images

In order to estimate each method quantitatively, the capacitance residual C_R, the image error I_E and the image correlation I_C are calculated using the following:

$$C_R = \frac{\sqrt{\sum_{i=1}^{n}\left(c_i^{(k)} - c_i^{(\exp)}\right)^2}}{\sqrt{\left(\mathbf{C}^{(0)}\right)^2}}, \quad I_E = \frac{\sqrt{\sum_{i=1}^{m}\left(e_i^{(k)} - e_i^{original}\right)^2}}{\sqrt{\left(\mathbf{E}^{original}\right)^2}},$$

$$I_C = \frac{\sum_{i=1}^{m}\left[\left(e_i^{(k)} - \overline{\mathbf{E}^{(k)}}\right)\left(e_i^{original} - \overline{\mathbf{E}^{original}}\right)\right]}{\sqrt{\sum_{i=1}^{m}\left(e_i^{(k)} - \overline{\mathbf{E}^{(k)}}\right)^2}\sqrt{\sum_{i=1}^{m}\left(e_i^{original} - \overline{\mathbf{E}^{original}}\right)^2}} \tag{10}$$

In these equations, $e_i^{(k)}$ is ith element of the final reconstructed image $\mathbf{E}^{(k)}$, $\overline{\mathbf{E}^{(k)}}$ is the special mean pixel value of $\mathbf{E}^{(k)}$, $e_i^{original}$ is ith element of the original pseudo image matrix $\mathbf{E}^{original}$, and $c_i^{(k)}$ is ith element of the final capacitance $\mathbf{C}^{(k)}$ calculated from the final reconstructed image $\mathbf{E}^{(k)}$. The low values of C_R and I_E, the high value of I_C mean accurate reconstructed images.

Figure 4 to 7 show the estimation categories of the annular, the center, the bottom and the four-bulk pseudo particles images including the dispersion concentration rate DCR. In general, C_R and I_E increase and I_C decreases as DCR increases. The superior method is highly dependent upon the image type and the DCR. In the case of the annular image in Figure 4, the FWIM method proves superior to the LW, ITR and GVSPM methods. Figure 4(a) shows the I_C value for all the four methods is decrease as DCR increases. I_C for FWIM method is much higher than that for LW, ITR and GVSPM method, the I_C for LW, ITR method are nearly the same and always the lowest. Figure 4(b) displays that the I_E for FWIM method is lower than LW, ITR and GVSPM, specifically, when DCR<20% and DCR>80%, I_E for GVSPM is the lowest. The C_R value for FWIM and GVSPM are the same and the lowest when DCR>30% as shown in Figures 4(c), however, when DCR<30%, C_R for LW and ITR are the lowest.

The reconstructed images in the case of the center image of Figure 5 are highly dependent on the reconstructed method and the DCR. More specifically, in Figure 5(a), the I_C for GVSPM are higher than those of FWIM, LW and ITR in the case of DCR<20%; however, when 20%<DCR<80%, I_C for FWIM are the highest; when DCR>80%, I_C for LW and ITR become superior to FWIM and GVSPM, whose I_C value becomes extremely low. Figure 5(b) shows that the I_E for GVSPM is the lowest for all DCR. In terms of C_R as shown in Figure 5(c), LW and ITR is lower than FWIM and GVSPM, and when DCR>70%, C_R for all these four method nearly become the same.

Next, the estimation result of the bottom image is shown in Figure 6. Considering the three estimation categories as a whole, FWIM is also superior to LW, ITR and GVSPM method. As shown in Figure 6(a), the I_C for FWIM is nearly same as that for GVSPM, and always higher than LW and ITR. In Figure 6(b), it is clear that I_E for FWIM is the lowest and that for LW and ITR are the highest. In terms of C_R as shown in Figure 6(c), LW and ITR are the lowest when DCR<40%. When DCR>40%, C_R for FWIM and GVSPM become the lowest.

Finally, it is difficult to determine the best method from the estimation result of four-bulk reconstructed images as shown in Figure 7. In Figure 7(a), When DCR<20%, the I_C for GVSPM is superior to other method, however, when DCR>20%, I_C for FWIM become the highest. As shown in Figure 7(b), when DCR<20%, the I_E for GVSPM is the lowest; when 20%<DCR<90%, I_E for FWIM is lower than other method, however, when DCR>90%, it become to the highest. In terms of C_R as shown in Figure 7(c), LW and ITR is lower than FWIM and GVSPM, and when DCR>40%, C_R for all these four method nearly become the same.

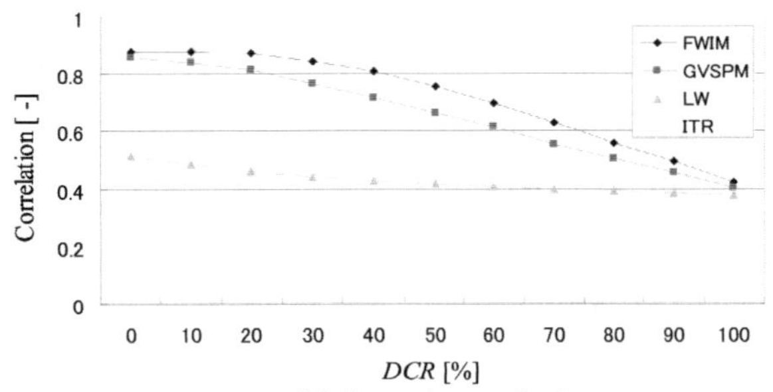

(a) Image correlation

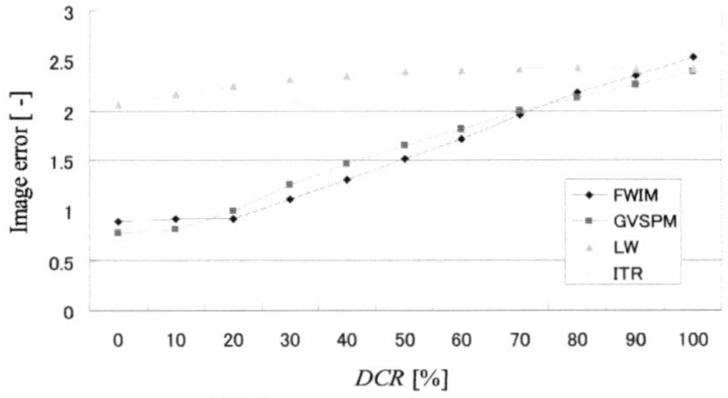

(b) Image error

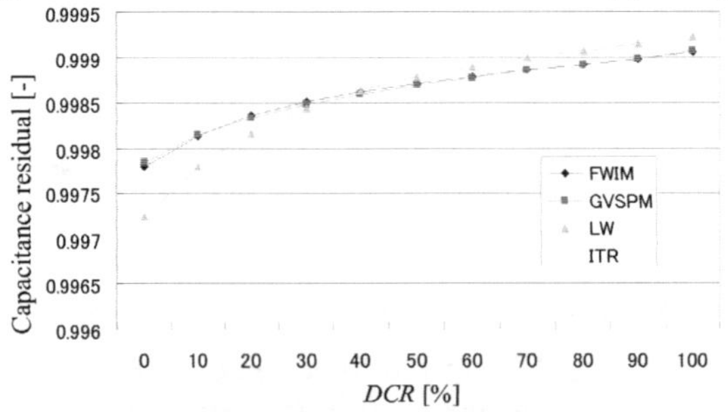

(c) Capacitance residual

Figure 4 I_C, I_E and C_R in the case of annular image

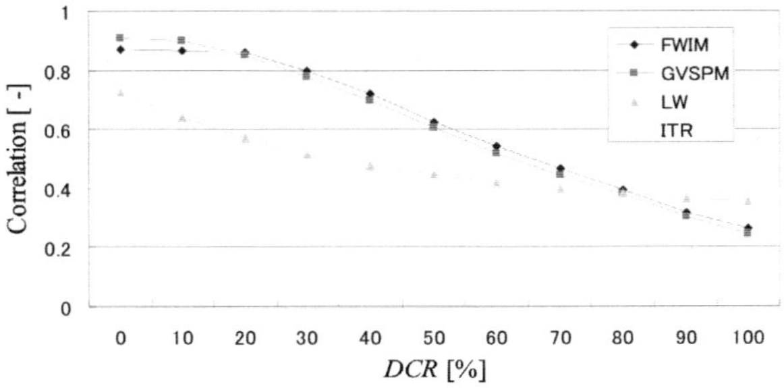

(a) Image correlation

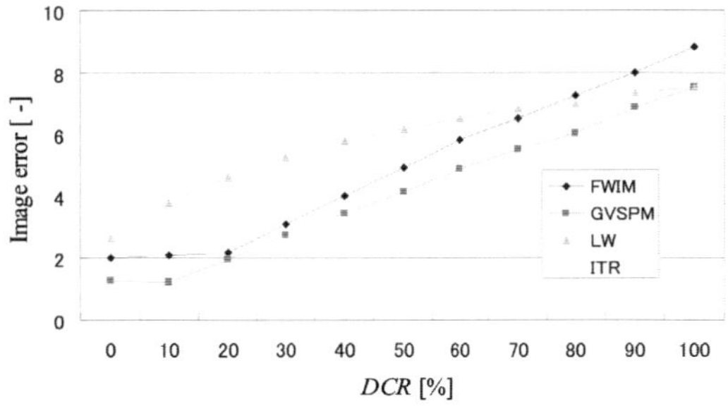

(b) Image error

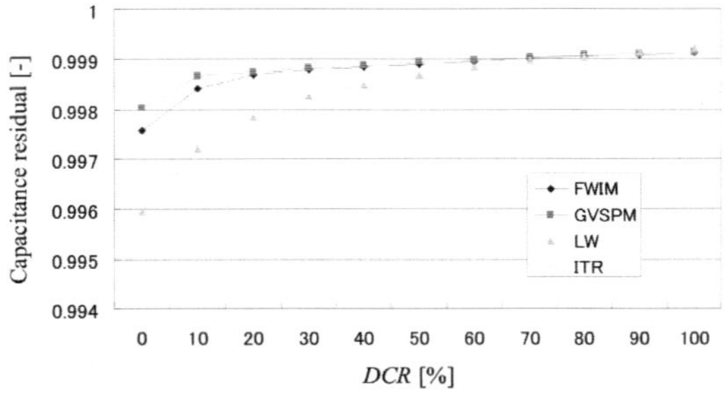

(c) Capacitance residual

Figure 5 I_C, I_E and C_R in the case of center image

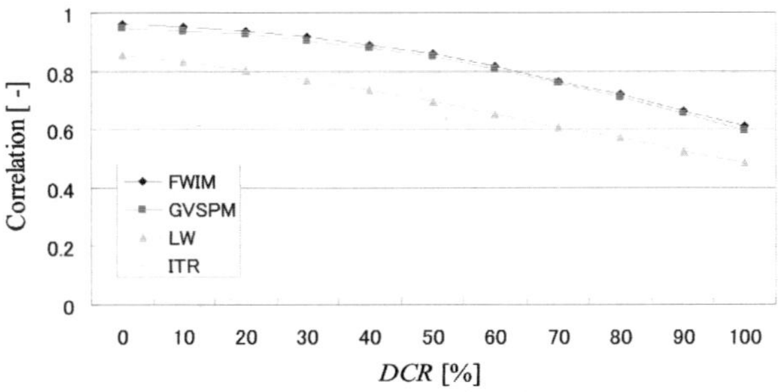

(a) Image correlation

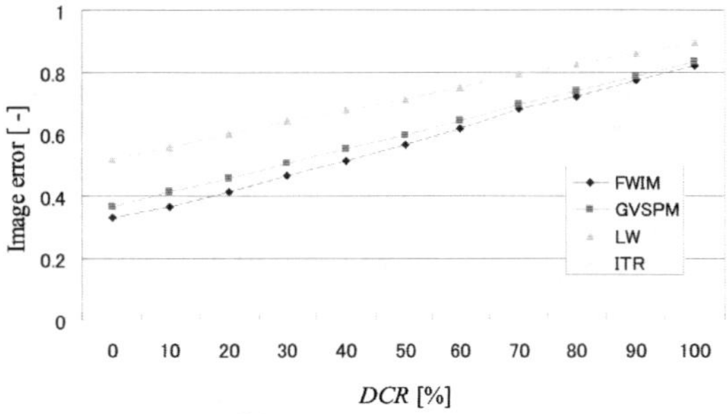

(b) Image error

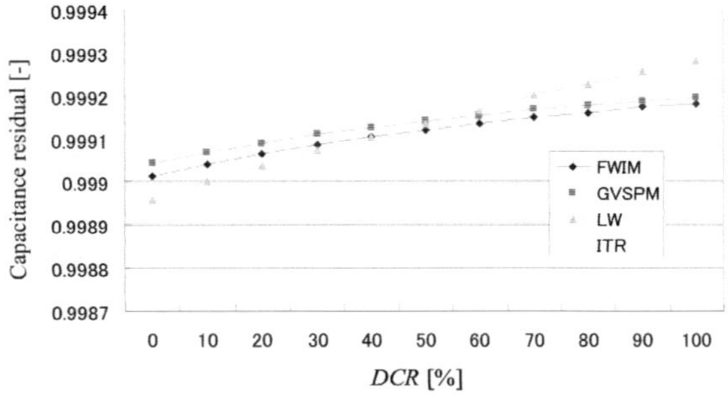

(c) Capacitance residual

Figure 6 I_C, I_E and C_R in the case of the bottom image

Journal of Physics: Conference Series **147** (2009) 012031 doi:10.1088/1742-6596/147/1/012031

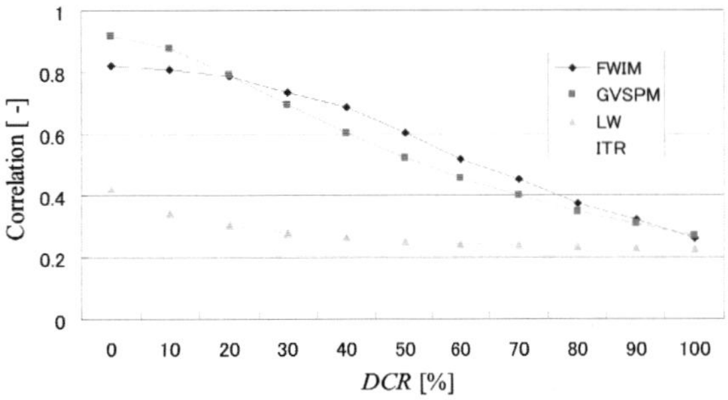

(a) Image correlation

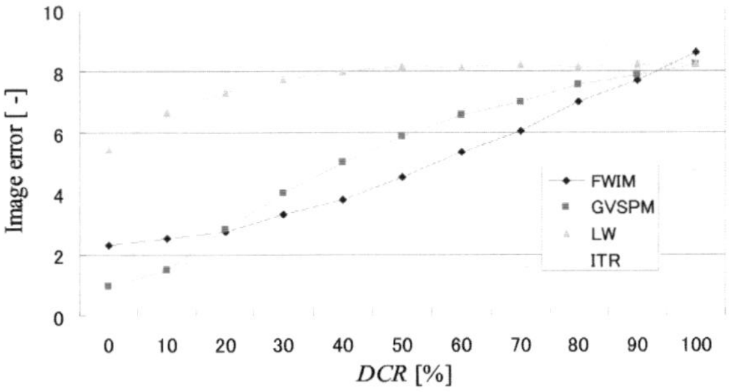

(b) Image error

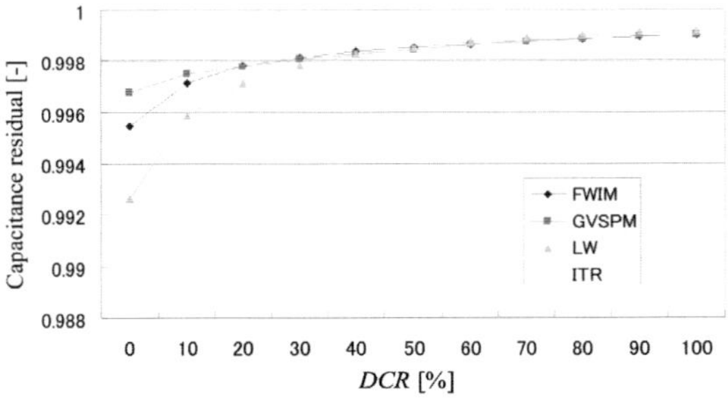

(c) Capacitance residual

Figure 7 I_C, I_E and C_R in the case of the four-bulk image

5. CONCLUSIONS

The Functional Weighted Inverse Matrix (FWIM) has been applied to an ill-posed inverse problem involving the electrical capacitance for multiphase flow. Four types of pseudo particle distribution images with dispersion concentration rates (*DCR*) were used to compare the FWIM method to the conventional Landweber (LW), Iterative Tikhonov regularization (ITR) and Generalized Vector Sampled Pattern Matching (GVSPM) method. The results of these comparisons are detailed below:

1) In terms of the annular pseudo particle images, the superior method depends on the *DCR*. However, considering the three estimation categories as a whole, FWIM method is a little better.
2) For the centre image cases, it is difficult to determine the best method.
3) FWIM is superior to the LW, ITR and GVSPM method for bottom pseudo particle image in terms of capacitance residual, image error, and image correlation, even though the FWIM method is not an iterative method.
4) For the four-bulk image cases, the superior method depends on the *DCR*. When 20%<*DCR*<90%, FWIM is proved superior to other method, however, when *DCR*<20% and *DCR*>90%, it is also difficult to determine the best method.

ACKNOWLEDGEMENTS

The authors wish to acknowledge the financial support provided by the Information Centre of Powder Technology Japan, and the Grant-in-aid for scientific research (C) from Japan Society for the Promotion of Science.

REFERENCES

Dyakowski,T. et al. (1999). "On-line monitoring of dense phase flow using real-time dielectric imaging", Powder Technology, 104, pp. 287-295.

Halow,J.S. et al. (1992). "Observations of fluidized bed coalescence using capacitance imaging", Powder Technology, 69, pp.255-277.

Huang,S.M. et al. (1989). "Tomographic imaging of two-component flow using capacitance sensors", J.Phys, E:Sci, Instrum, 22, pp.173-177.

Isaksen,O. (1996). "A review of reconstruction techniques for capacitance tomography", Meas. Sci. Technol, 7(3), pp.325-337.

Yang,W.Q. et al. (2003). "Image reconstruction algorithms for electrical capacitance tomography", Meas. Sci. Technol, 14, pp.R1-R13.

Liu,S. et al. (2001). "Comparison of three image reconstruction algorithms for electrical capacitance tomography", Proc. Second world congress on industrial process tomography, Hanover, Germany, pp.29-34.

Saotome,J., et al. (1995). "Magnetic core shape design by the sampled pattern matching method", IEEE Trans.Magn., 31, pp.1976-1979.

Yoda,K., et al. (1997). "Dose optimization of proton and heavy ion therapy using generalized sampled pattern matching", Phys.Med.Biol., 42, pp. 2411-2420.

Yang,W.Q., (1996). "Hardware design of electrical capacitance tomography systems", Meas. Sci. Technol, 7(3), pp.225-232.

Polydorides,N., et al. (2002). "A Matlab toolkit for three-dimensional electrical impedance tomography: a contribution to the Electrical Impedance and Diffuse Optical Reconstruction Software project", Meas. Sci. Technol., 13(12), pp.1871-1883.

Williams,R.A. et al. (1995). Process tomography principles, techniques and applications, Butterworth Heineman, Chapter 15, pp.301-312.

Gas-Liquid Flow Characterization in Bubble Columns with Various Gas-Liquid Using Electrical Resistance Tomography

Haibo Jin, Han Yuhuan, Yang Suohe

Department of Chemical Engineering, Beijing Institute of Petrochemical Technology, Beijing 102617, P.R. China

jinhaibo@bipt.edu.cn

Abstract. Electrical resistance tomography (ERT) is an advanced and new detecting technique that can measure and monitor the parameters of two-phase flow on line, such as gas-liquid bubble column. It is fit for the industrial process where the conductible medium serves as the disperse phase to present the key bubble flow characteristics in multi-phase medium. Radial variation of the gas holdup and mean holdups are investigated in a 0.160 m i. d. bubble column using ERT with two axial locations (Plane 1 and Plane 2). In all the experiments, air was used as the gas phase, tap water as liquid phase, and a series of experiments were done by adding KCl, ethanol, oil sodium, and glycerol to change liquid conductivity, liquid surface tension and viscosity. The superficial gas velocity was varied from 0.02 to 0.2 m/s. The effect of conductivity, surface tension, viscosity on the mean holdups and radial gas holdup distribution is discussed. The results showed that the gas holdup decrease with the increase of surface tension and increase with the increase of viscosity. Meanwhile, the settings of initial liquid conductivity slightly influence the gas holdup values, and the experimental data increases with the increase of the initial setting values in the same conditions.

1.INTRODUCTION

Bubble columns are widely used as gas–liquid contactors in many applications due to their simple construction, low operating cost, high energy efficiency and good mass and heat transfer rates (Kazakis et al., 2008), such as absorption, oxidation, fermentations, bio-reactions, coal liquefaction and waste water treatment etc.. During the past few years, special attention has been paid to local and global bubble flow characteristics as a key parameter that determines the overall reactor performance owing to the complexity of fluid mechanics in these systems (Dukovic et al, 2000; Fan et al, 1989). Developing qualities as well as quantitative understanding of the axial and radial holdup distribution in bubble columns turn to be essential (Jin et al., 2006).

Among all of those applications of process tomography as a robust noninvasive tool for direct analysis of the characteristics of multiphase flows, electrical resistance tomography (ERT) talent showing itself, owing to its noninvasive and fast data acquisition features, ERT might provide very useful tools for measuring and monitoring bubble column operation online (Wang et al, 1999; Jin et al., 2007). The speed of the technique to capture real-time data of highly fluctuating flow systems may be one of the most important concerns for the purpose of imaging multiphase flows in the process industries. And due to its high-speed capability, low construction cost, high safety and suitability for small or large vessels, ERT is considered to be the most powerful tool among other available tomography techniques.

In this study, the effect of conductivity, surface tension, viscosity on the mean holdups and radial gas holdup distribution was discussed using electrical resistance tomography with two axial locations (Plane 1 and Plane 2).

2. Experimental

2.1 Experimental setup

The experimental setup was a transparent perspex acrylic column with 2.5 m high and an inner diameter of 0.16m as shown in Fig. 1.

Two rings of ERT sensors, each composed of 16 rectangular electrodes, were mounted in the inner wall of the column in a non-invasive fashion. The electrodes were made of stainless steel with a contact area of 8 mm (w) by 16 mm (h). The data collection rate was 2 frames per second with an excitation signal frequency of 9.6 kHz. Meanwhile, two differential pressure sensors for measuring the differential pressure drop were placed along the column above the distributor. The differential pressure sampling frequency was 100 Hz. Two neighboring pressure-sampling ports were connected to the high end and low end of each different-pressure transducer, respectively.

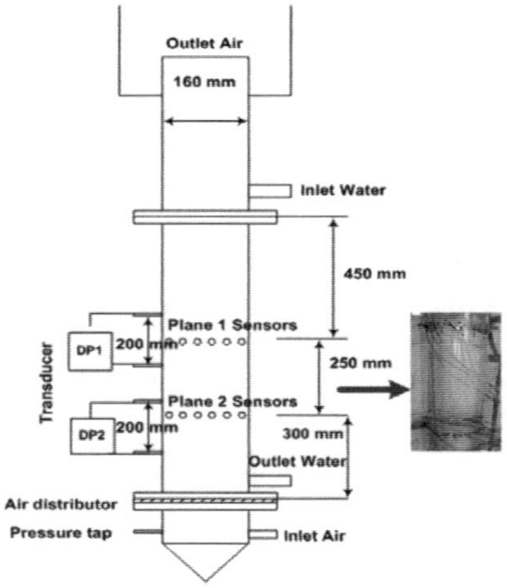

Fig. 1. Schematic diagram of bubble column

2.2. Experimental systems and operating conditions

All the experiments were carried out at ambient pressure and temperature conditions (20±1℃). Sodium oleate and Ethanol were used to reduce the liquid surface tension, Glycerin to change the liquid viscidity and KCl solution to change the liquid conductivity. The gas phase was atmospheric air for all runs, while several liquids were employed as liquid phase. Their physical properties are presented in Table 1. The superficial gas velocity varied from 0.02 m/s to 0.2 m/s.

Table 1 Physical property of liquids

Liquid phase	Conducivity, σ_l mS/cm	Viscosity, μ_L (mPas)	Surface tension σ_L(mN/m)
Tap water	0.96	1.0	71
KCl solution	1.10-1.90	1.0	71
Ethanol solution 0.2%(v/v)	0.96	1.0	69.5
Ethanol solution 0.6%(v/v)	0.96	1.0	68.6
Ethanol solution 5%(v/v)	0.96	1.0	59.7
Ethanol solution 10%(v/v)	0.96	1.0	53.4
Sodium oleate solution	0.96	1.0	33
Glycerin solution 8% (v/v)	0.96	1.29	71

2.3. Experimental procedures and calculation methods

The adjacent electrode pair strategy was adopted using 10mA injection current at 9.6 kHz for parameter measurement. Data collection rates for plane 1 and plane 2 were 50 ms, with time for image reconstruction and other overheads, yielding approximately two images per second. With conductivity data from ERT, the local gas volume fraction (ε_g) can be determined by applying the Maxwell equation (Maxwell, 1873).

$$\varepsilon_g = \frac{2\sigma_1 + \sigma_2 - 2\sigma_{mc} - \sigma_{mc}\sigma_2 / \sigma_1}{\sigma_{mc} - \sigma_2 / \sigma_1 + 2(\sigma_1 - \sigma_2)} \tag{1}$$

Where σ_1 is the conductivity of the first phase, σ_2 is the conductivity of the second phase, and σ_{mc} is the local value of mixture conductivity distribution. If the second phase is assumed to be nonconductive material, such as air in this study, the above equation can be simplified as follows.

$$\varepsilon_g = \frac{2\sigma_1 - 2\sigma_{mc}}{2\sigma_1 + \sigma_{mc}} \tag{2}$$

The local mixture conductivity (σ_{mc}) is determined from the pixel conductivity of ERT image.

The axial holdup can be measured from the two differential pressure signals. On the assumption that values of liquid acceleration term and the wall friction term were normally small and could be neglected, the mean gas holdup (ε_g) was calculated by equation (Jin et al., 2004).

$$\varepsilon_g = \frac{\Delta P}{\rho_l g \Delta H} \tag{3}$$

Where ΔP is the differential pressure between two pressure sensor points, ΔH the vertical distance between two pressure sensor points, and ρ_l liquid density.

3.RESULTS AND DISCUSSION

3.1 Effect of conductivity on the overall holdups

Fig.2 shows relationship between solution conductivity and overall gas holdups in the same condition by mean of ERT. From Fig.2, liquid conductivity behavior slightly influences the gas holdup values, and the gas holdup increases with the increase of the initial conductivity values. According to the results using differential pressure method in Fig.3, the gas holdup values under conditions of difference conductivity remains unchanged. This reason can be explained that resistance field is sensitive to process condition and system temperature, which result in nonlinear dependence on gas holdups from Eqn.(2). As a result , calibration techniques of ERT are required, and the system conditions should be monitors in a certain range.

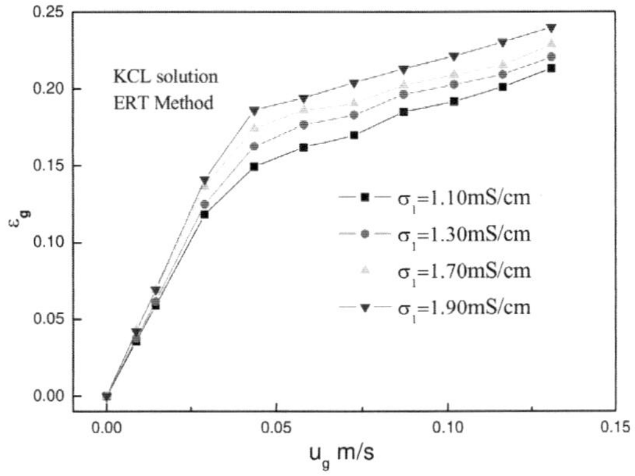

Fig.2 Comparison of different conductivity using ERT method

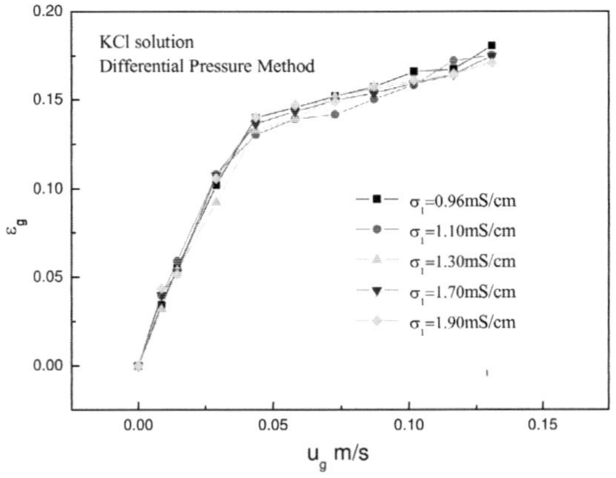

Fig.3 Comparison of different conductivity using differential pressure method

3.2 Effect of surface tension on the overall holdups

Surface tension of the liquid is an important physical parameter, but also an important factor that can affect gas holdup in gas-liquid column. The effect of surface tension on the gas holdups is studied by adding sodium oleate to change the surface tension in the air-water system and water-ethanol systems as shown Fig.4 and Fig.5. The results showed that the gas holdup increases with the decrease of surface tension in both two kinds of systems. However, the gas holdup in sodium oleate solution is much higher than one with tap water in the homogeneous regime, which is obvious different compared with the water-ethanol systems. It is considered that smaller bubbles are formed generally in tap water at first, and subsequently the coalescence phenomenon take place. However, the bubbles in a surfactant system have no tendency for coalescence (Bennett et al, 1999), which results in gas holdup for sodium oleate solution is much higher in the homogeneous regime.

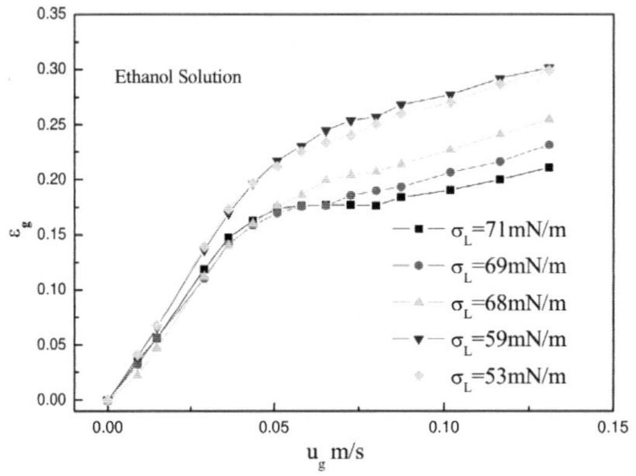

Fig.4 Comparison of different surface tension in Ethanol solution

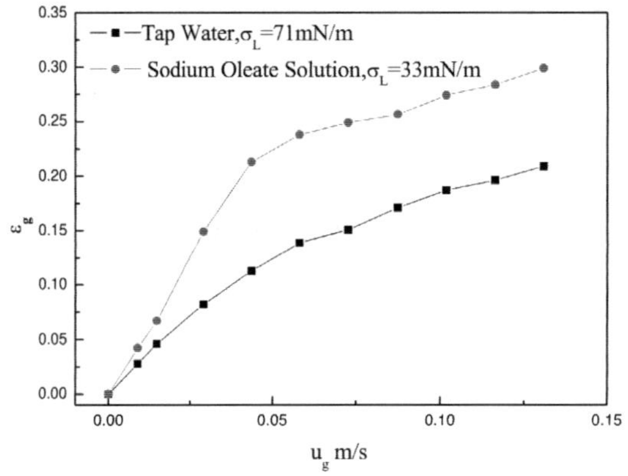

Fig.5 Comparison of different surface tension with surfactant

3.3 Effect of viscosity on the overall holdups

Viscosity of the liquid is an important physical property, and it also influence the performance of the flow. In this paper, experiment by adding glycerol to increase the viscosity of liquid (surface tension will remain basically unchanged) to study the impact of viscosity on the gas holdup, as it is shown in Fig.6, gas holdup increase with the increase of viscosity, the result is consistent with the analysis of Khare et al.(1990). Khare pointed out that with the increase of viscosity in the liquid, on the one hand, gas gathering strengthen, the average diameter of the bubble increase. On the other hand, increasing the viscosity makes the bubble up drag increased. So, in a certain range of viscosity, the gas holdup increase with the increase of liquid viscosity.

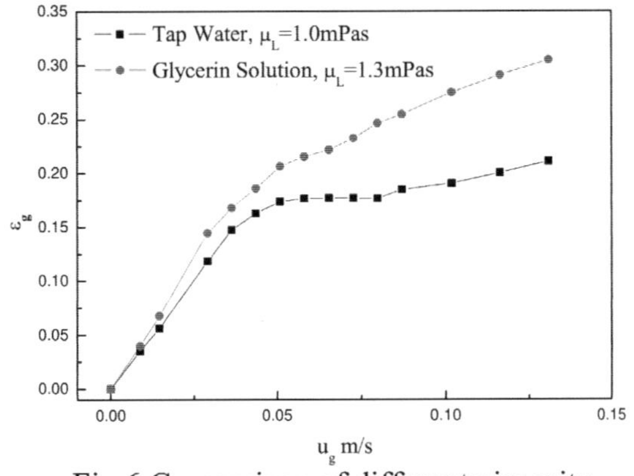

Fig.6 Comparison of different viscosity

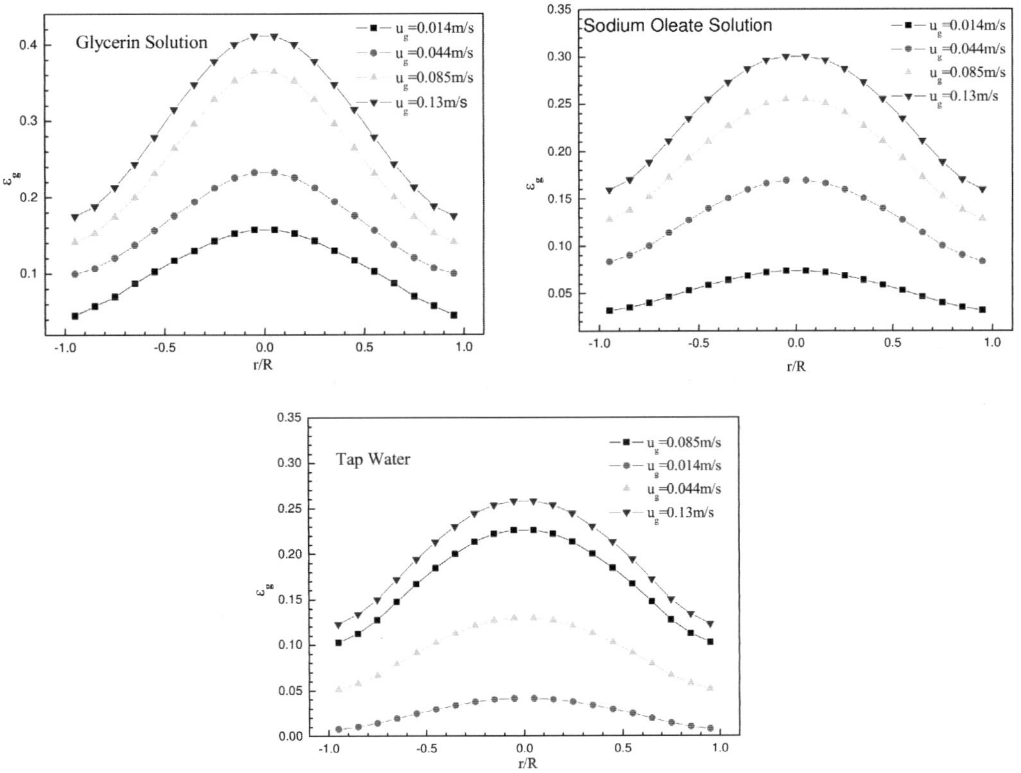

Fig. 7 The radial gas holdup profiles of different liquid systems.

3.3. The radial holdup profiles

Radial holdup profiles for different liquid systems as mentioned above are shown in Fig.7. The radial gas holdup increases with an increase of gas velocity. Meanwhile, the radial gas holdup profiles are steeper at the central region of the column with an increase of gas velocity, namely the higher gas holdups at the centre of the column and lower gas holdups at the wall of the column. According to these results, it is notified that the form of radial gas holdup distribution for different gas-liquid systems cannot change. But the form of radial holdups is different for various gas liquid system.

4.CONCLUSION

Electrical resistance tomography is used to study the overall and radial holdup in a series of different gas-liquid systems in bubble column. The effect of conductivity, surface tension, viscosity on the mean holdups and radial gas holdup distribution was discussed. The results showed that the gas holdup decrease with the increase of surface tension and increase with the increase of viscosity. Meanwhile, it is found that the settings of initial liquid conductivity slightly influence the gas holdup values, and the experimental data increases with the increase of the initial setting values in the same conditions. Different systems cannot affect the form of radial gas holdup distribution. So the measurement technique of electrical resistance tomography has verified the validity and effectiveness for gas liquid system in this operating condition. However, the conductivity of liquid phase should not be neglected as a impactful factor, more attention should be paid to measure gas-liquid flow behavior.

ACKNOWLEDGEMENTS

The research work was supported by National Natural Science Foundation of China under agreement number 20776018.

NOMENCLATURE

ΔH	The vertical distance between two pressure sensor points	[m]
ΔP	The differential pressure between two pressure sensor points	[Pa]

Greek Letters

ρ_l	Liquid density	[kg/m^3]
σ_1	The conductivity of the first phase	[mS/cm]
σ_2	The conductivity of the second phase	[mS/cm]
σ_{mc}	The local value of mixture conductivity distribution.	[mS/cm]
μ_L	Liquid viscosity	[mPas]
σ_L	Liquid surface tension	[mN/m]
ε_g	Overall gas holdups	[-]

REFERENCES

Bennett, M. A. et al. (1999). "Measurement and analysis of flows in a gas-liquid column reactor," Chem. Eng. Sci., 54, pp. 5003-5012.

Dukovic, M. P. (2000). "Opaque multiphase reactors: experimentation, modeling and troublesshooting," Oil Gas Sci. Technol., 55, pp.135–158.

Fan, L. S. (1989). Gas–liquid–solid fluidization engineering, Buttorworth, Boston.

Fransolet, E. et al. (2001). "Analysis of electrical resistance tomography measurements obtained on a bubble column," Meas. Sci. Technol., 12, pp. 1055-1060

Jin, H. et al. (2004). "Bubble behaviour of a large-scale bubble column with elevated pressure,"

Chem. Eng. Technol., 27, pp. 1007–1013.

Jin, H. et al. (2006). "The effect of sparger geometry on gas bubble flow behaviors using electrical resistance tomography," Chinese J. Chem. Eng., 14, pp. 127–131.

Jin, H. et al. (2007). "Measurement of gas holdup profiles in a gas liquid cocurrent bubble column using electrical resistance tomography," Flow Measurement and Instrumentation, 18, pp.191–196

Kazakis, N. A. et al. (2008). "Experimental study of bubble formation at metal porous spargers: Effect of liquid properties and sparger characteristics on the initial bubble size distribution," Chem. Eng. J., 137, pp. 265–281

Khare, A. S. et al. (1990). "Effect of fine particles on gas holdup in three-phase reactors," Chem. Eng. J., 44, pp. 11-25.

Maxwell, J. C. (1873). A treatise on electricity and magnetism, Clarendon Press, Oxford.

Wang, M. et al. (1999). "Electrical Resistance Tomography Sensing System for Industrial Applications," Chem. Eng. Comm., 175, pp. 49-70.

Ultra-filtration Measurement Using CT Imaging Technology

Junfeng Lu [1], Wenqiang Lu [2]

[1] Technical Institute of Physics & Chemistry, Chinese Academy of Sciences, No.2 Beiyitiao Street, Zhongguancun, Haidian District, Beijing, China, 100190
[2] Graduate University of Chinese Academy of Sciences, No.19A Yuquan Road, Shijingshan District, Beijing, China, 100049

junfenglu@mail.ipc.ac.cn

Abstract. As a functional unit in the hemodialysis process, dialyzer captured quite a few medical research interests since 1980s. In the design of dialyzer or in the ongoing hemodialysis process, to estimate the ultra-filtration amount of a dialyzer, the sideway loss of the running blood flow through hollow fibers or filtration channels should be measured. This further leads to the measurement of the blood flow inside the dialyzer. For this measurement, a non-invasive method is highly desired because of the high-dense bundled hollow fibers or packed channels inside the dialyzer. As non-invasive measurement tools, CT (Computed Tomography) technologies were widely used for tissue, bone, and cancerous clinical analyses etc... Thus, in this paper, a CT system is adopted to predict the blood flow inside a hollow fiber dialyzer. In view of symmetric property of the hollow fiber dialyzer, the largest cutting plane that parallels to the cylindrical dialyzer was analyzed by the CT system dynamically. And then, a non-invasive image analysis method used to predict the ultra-filtration amount is proposed.
Keywords. Ultra-filtration measurement, hemodialysis, CT imaging technology

1. INTRODUCTION

On failure of a biological kidney, the patient must be treated by hemodialysis therapy to evacuate metabolic wastes inside his body. The key functional unit in the process of hemodialysis is dialyzer. A cylindrical device shown in Fig. 1 describes an anatomic map of a hollow fiber dialyzer. This device

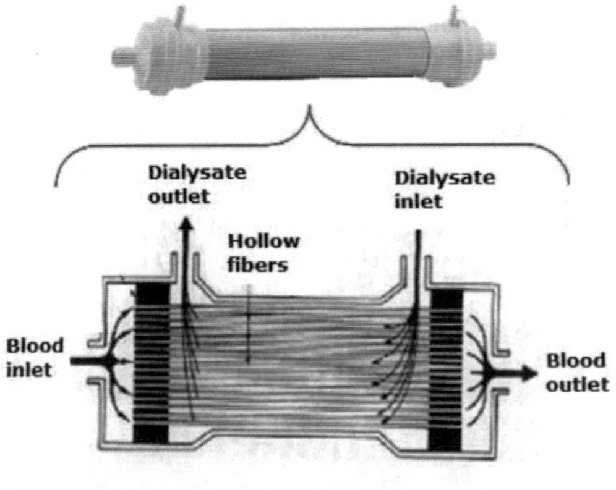

Fig. 1 Hollow fiber dialyzer

simulates the physiological functions of a biological kidney. Inside it, a bundle of hollow fibers dousing in dialysate solution perform the functions of kidney nephrons. As blood passing through these fibers, extra water and metabolic wastes such as urea and creatinine carried by plasma are filtered to the outside of the fibers due to the pre-setting of positive transmembrane pressure gradient between dialysate site and blood site. And then, the filtered wastes are removed by dialysate solution running in reverse direction to the blood. Because of its near perfect functionality, these sorts of devices are also called "artificial kidney". In recent years, due to the warm-up market in artificial biological devices, researchers have gradually joined in the projects of artificial organs. Artificial kidney stills captures big eye-balls in medical research field. And majority of research works were deployed in the design of high efficient more humanlike dialysis devices.

In the analysis of blood flow, dialysate flow or ultra-filtration flow inside the dialyzer, non-invasive measurement technologies are extremely useful because the intrusion of tiny flow measurement pins into a hollow fiber dialyzer is nearly impossible due to high-dense bundled fibers inside the dialyzer. Although Dr. Liao (2005) designed a method to make this measurement, this method is still in contact with blood and could not measure the ongoing blood during hemo-dialysis therapy. Non-invasive or intact measurement is a topic for years interested and studied by thousands of talented researchers. And it is found spreading use especially in medical research, clinical study, biology experimentation, and so on. In today's medical centers, CT, MRI, ultra-sound, or X-ray machines are diagnostic useful for physicians to make their judgment on patient disease. In view of these modern technologies, a choice of either MRI (Lu, 2005; Poh et al., 2003; Poh, 2003) or CT technology lights up a new way to analyze the flow behaviour inside the dialyzer.

In this paper, we present an image analysis method trying to measure ongoing ultra-filtration flow rate inside a hollow fiber dialyzer using a CT imaging system.

2. METHOD

The idea for the measurement of ultra-filtration flow rate of a hollow fiber dialyzer is based on an observation of the burst of pumped blood flow inside the dialyzer. As shown in Fig. 2, assuming that at time $t=T0+\Delta t$, the frontier of the burst blood flow inside the dialyzer reaches Y_i and the average

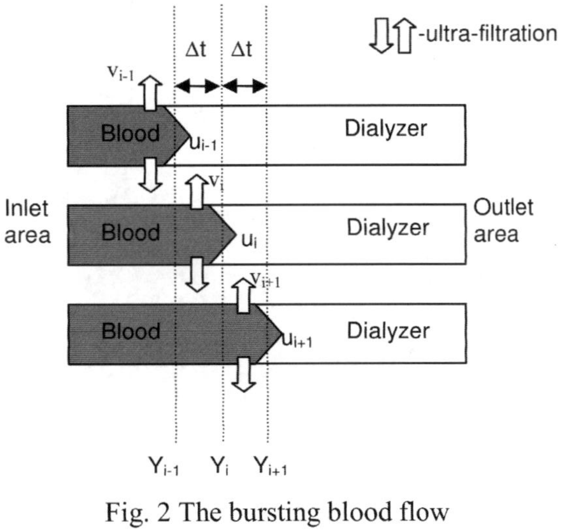

Fig. 2 The bursting blood flow

flow rate is u_i; after another time interval Δt it reaches Y_{i+1} with an average flow rate of u_{i+1}, then the blood flow rate of u_i can be determined by the following equation:

$$u_i = \frac{Y_i - Y_{i-1}}{\Delta t}, \quad i > 0 \tag{1}$$

Holding this equation, the side-way ultra-filtration flow v could be indirectly measured by the following equation:

$$v_i = \frac{(u_i - u_{i-1})N\pi R^2}{2N\pi R(Y_i - Y_{i-1})}, \quad i > 0 \tag{2}$$

where N is the bundle number (number of hollow fibers inside a dialyzer) of a hollow fiber dialyzer, R is the radius of a hollow fiber. One can further simplify equation (2) as follows:

$$v_i = \frac{(u_i - u_{i-1})R}{2(Y_i - Y_{i-1})}, \quad i > 0 \tag{3}$$

substitute u_i by equation (1), one can get:

$$v_i = \frac{(Y_i - 2Y_{i-1} + Y_{i-2})R}{2(Y_i - Y_{i-1})\Delta t}, \quad i > 0 \tag{4}$$

With the above equation, one can find that, to measure the ultra-filtration rate of the dialyzer, the only thing left is to find the frontier or "Y_i" series of the pumped blood flow inside the dialyzer.

The experimental setup for this measurement is shown in Fig. 3. The CT system is used as a camera to capture bursting blood flow pumped by a peristaltic pumping system. And the largest cutting plane that parallels to the cylindrical dialyzer is analyzed.

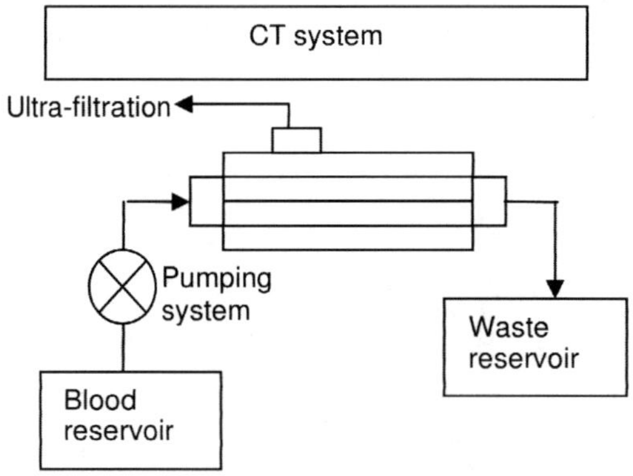

Fig. 3 Experimental setup

We capture flow image series between each pumping burst cycle. The time interval between each successive image in the series is set to be Δt. And this value is determined by the following equation:

$$\Delta t = \frac{T}{N} \tag{5}$$

where T is the pumping cycle of the pump and N is the number of images in the image series.

Fig. 4 shows the captured CT image series (N=9). In the collected images, the dialyzer is fixed on the table by an arm supporter. However, only from these images, one could not specify the blood flow frontier inside the dialyzer. Thus, the image series in Fig. 4 needs more delicate work to rip-off the useless background data. Notice that the background noises as well as support facilities for the dialyzer inside a captured CT image are inert to time, thus to remove these useless data, the histogram of the CT image is mostly helpful (Fig. 5).

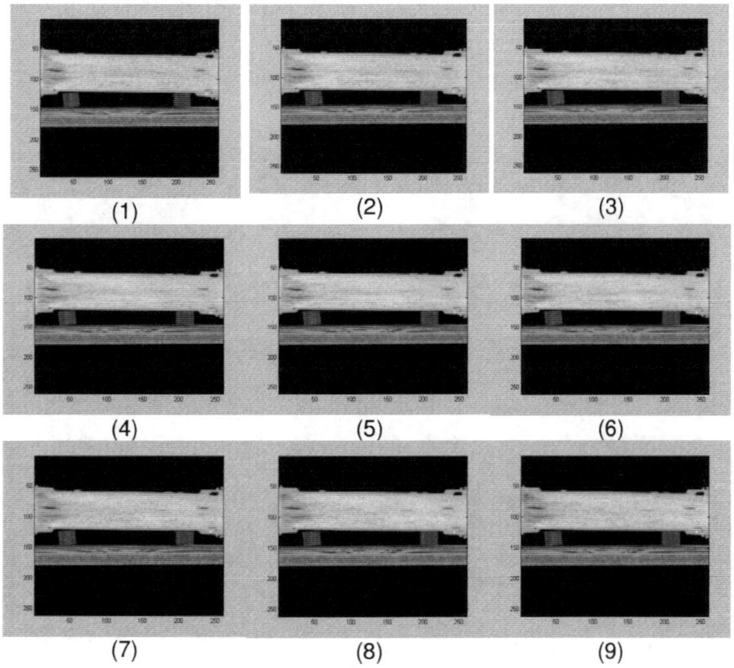

Fig. 4 The raw CT image data captured in a time interval Δt.

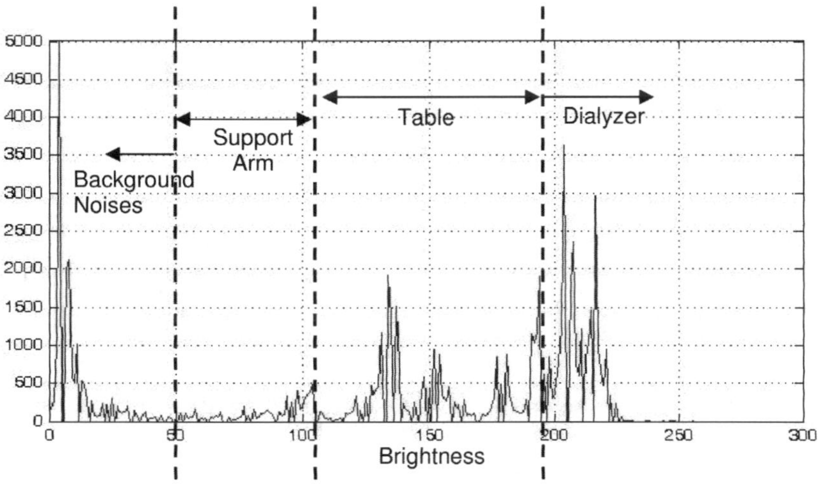

Fig. 5 Histogram of a CT image

As shown in Fig. 5, by employing multiple-threshold segmentation method (Gonzalez et al., 2003), three thresholds: 50, 108, and 190 (Due to the stability of the background data, these thresholds could be used for all those nine images in Fig. 4) are set to segment out four pattern areas (see Fig. 6) inside the CT image. Among these four patterns, only the area where the dialyzer locates is interested to us because we need to analyze the blood flow inside this area. We further assume the pixel set locking the dialyzer area as I_{DI}, as well the background area as I_B. Then the following morphological mechanism could determine the correct enclosing area I_D of the dialyzer:

$$I_D = ((I_{DI} \bullet E) \cap (!\, I_B)) \circ \mathrm{E} \tag{6}$$

where $\bullet$ is the morphological "closing" operation with structuring element E, $\cap$ is the morphological "and" operation, ! is the morphological "not" operation, as well, $\circ$ is the morphological "opening" operation. In the realization shown in Fig. 7, a circular structuring element E with a radius of 9 is chosen. In this figure, an enclosing solid structure is locked precisely where the dialyzer should be.

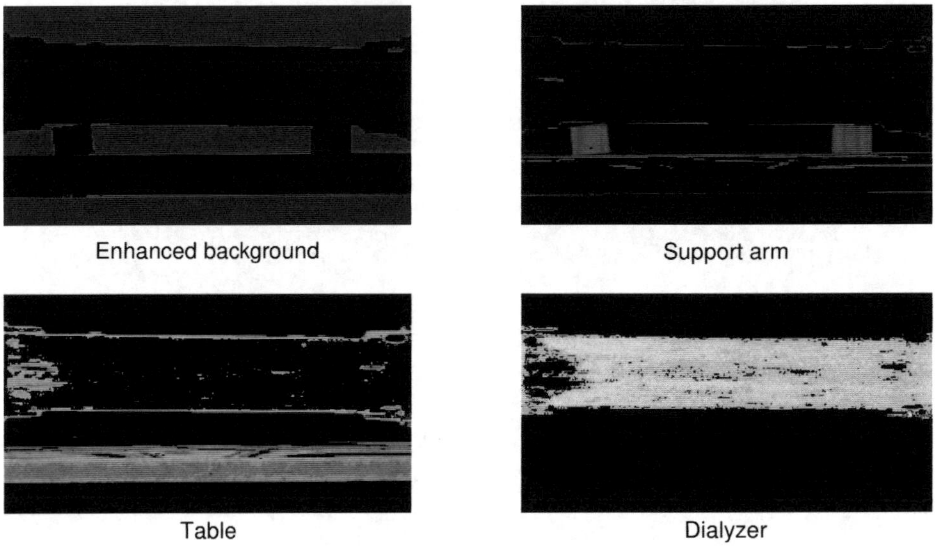

Fig. 6 Four areas inside a CT image

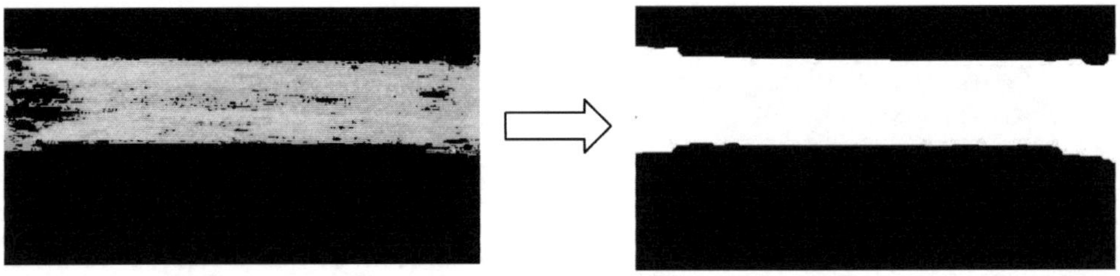

Fig. 7 Determination of dialyzer area by equation (6)

After the deletion of background noises and useless support patterns, next thing left to us is to pick out blood flow data inside the area determined by I_D. To do this, before the blood started to burst, a standard static pattern image is recorded by the CT machine as a baseline comparator. We name it as an image pixel set: I_C. Let's further define the image series as: I_i (i=1,2,...,9). Thus the following equation will help us to pick out useful data for blood flow inside the dialyzer:

$$I_{blood\_flow} = I_i \cap I_D - I_C \cap I_D \tag{7}$$

Substituting set images in Fig. 4 into equation (7), we get blood flow data as shown in Fig. 8.

At here, the data of blood flow still blurs one's vision. The layout boundary of the flow frontier cannot be specified even we can perceive a flying pattern (enclosed by a square in Fig. 8) in the image series. Moreover this flying pattern is compressed at the outlet area of the dialyzer due to the relatively lower volume flow rate.

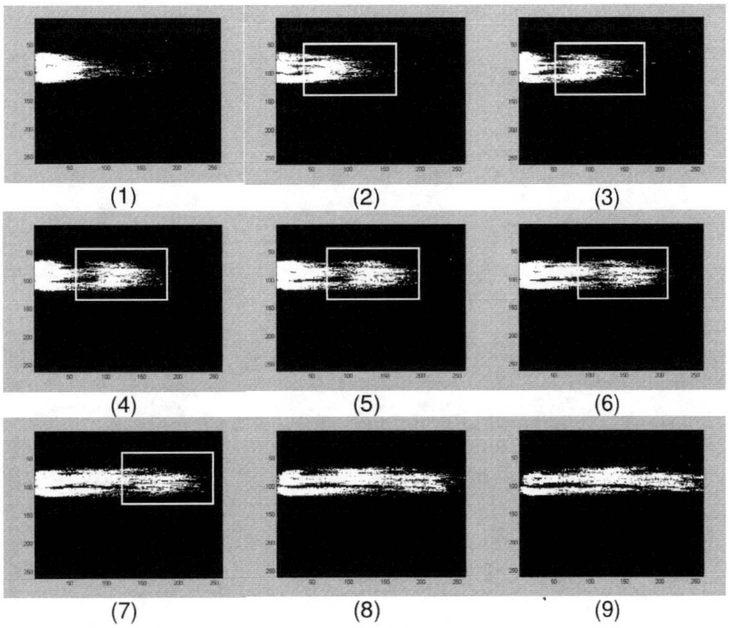

Fig. 8 Blood flow after ripping-off background noise

To project the frontiers of the bursting blood, a fast numerical interpolation method is desirable. Before we can move forward to interpolate the flow frontier, the frontier boundary points must be picked out. As shown in Fig. 9, we chose image No.1 as an example to clarify the process. Let's further define the set of these points as:

$$S = \{(x_i, y_i), i = 1,2,3,\ldots, N\} \tag{8}$$

where (x_i, y_i) is the screen coordinate and is in terms of pixel.

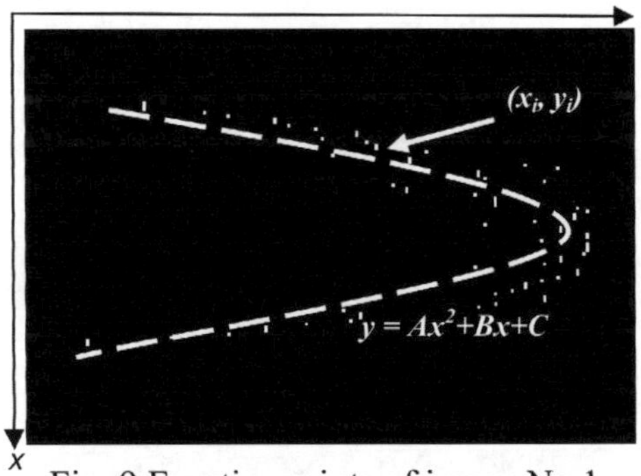

Fig. 9 Frontier points of image No.1

We assume that the layout boundary of the flow frontier is shaped as parabolic profile and can be described by the following equation:

$$y = Ax^2 + Bx + C \tag{9}$$

To interpolate the layout boundary, a least square parabola method (Mathews et al., 2004) is adopted in this paper. Thus, with those frontier points (x_i, y_i) ever recorded in Fig. 8, the values of A, B and C can be exactly determined by the following equations:

$$A\sum_{k=1}^{N} x_k^4 + B\sum_{k=1}^{N} x_k^3 + C\sum_{k=1}^{N} x_k^2 = \sum_{k=1}^{N} y_k x_k^2 \quad (10)$$

$$A\sum_{k=1}^{N} x_k^3 + B\sum_{k=1}^{N} x_k^2 + C\sum_{k=1}^{N} x_k = \sum_{k=1}^{N} y_k x_k \quad (11)$$

$$A\sum_{k=1}^{N} x_k^2 + B\sum_{k=1}^{N} x_k + CN = \sum_{k=1}^{N} y_k \quad (12)$$

3. RESULTS & DISCUSSION

Table 1 lists the parabolic interpolation results interpreted by the values of A, B and C.

Table 1. The values of A, B, and C

No.	1	2	3	4	5	6	7	8	9
A_i	-12.8	-13.7	-14.0	-13.6	-12.8	-13.2	-17.7	-18.1	-22.5
B_i	22.6	23.9	24.5	23.9	22.5	23.3	31.0	31.8	39.5
C_i	6.3413	6.4178	6.6683	7.1911	7.9829	8.6937	8.6953	9.5109	9.5658

Fig. 10 shows the normalized interpolation results for the frontier layout boundaries of the 9-image series. To determine the values of "Y_i" once defined in equation (1), one needs to further carry out a mean value for the frontier layout boundary. This is fulfilled by:

$$Y_i = \frac{\int_{x=.75}^{x=1} y_i(x)dx}{\Delta x} \qquad (i = 1,2,3,...,9)$$
$$= \frac{1}{\Delta x}\int_{x=.75}^{x=1} (A_i x^2 + B_i x + C_i)dx \quad (13)$$

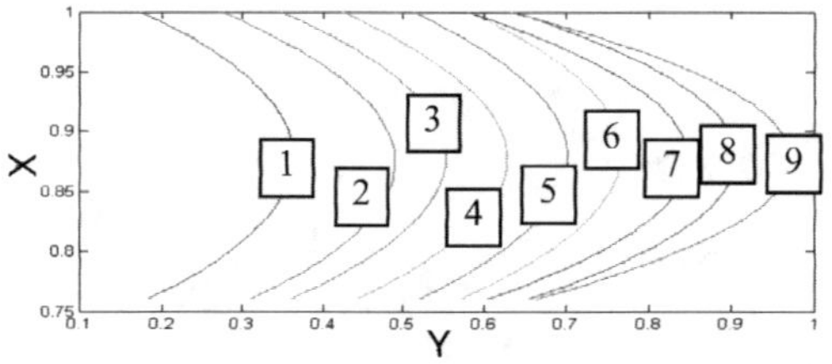

Fig. 10 Normalized frontier layout boundary of blood flow

Table 2 lists the results of Y_i series.

Table 2. Normalized Y_i series

No.	1	2	3	4	5	6	7	8	9
Y_i	0.5004	0.5299	0.5648	0.6059	0.6537	0.7071	0.7626	0.8193	0.8772

Substituting this Y_i series into equation (4), we finally get the normalized ultra-filtration rate as shown in Fig. 11. From this figure, one can find that the ultra-filtration on the run from dialyzer inlet to dialyzer outlet is generally decreased. An ideal shape of the ultra-filtration rate should follow the consine rule if the blood flow was going through a thin porous channel (Lu et al., 2008). The curve of the results shown in Fig. 11 also nearly follows this rule. Since in the experiment, the blood flow recorded by the CT system reflects a mass flow bundled by thousands of tiny flows that are detoured inside the hollow fibers (Fig. 12), and the burst begins at the in-let area of the dialyzer, thus, ultra-filtration mainly happens at the first 1/3 area of the dialyzer.

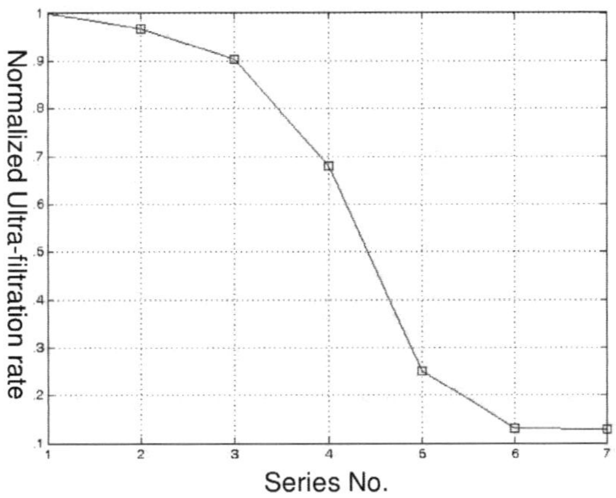

Fig. 11 The normalized ultra-filtration velocity

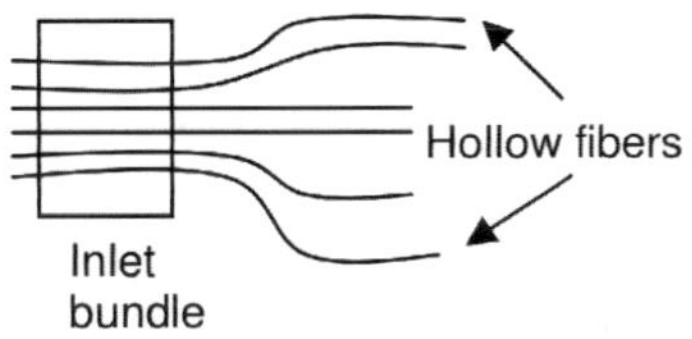

Fig. 12 The bundle of hollow fibers

4. CONCLUSION

The image processing method used in this paper predicted the ultra-filtration amount for a hollow fiber dialyzer, and also knocked another door for the non-invasive flow analysis inside a dialyzer. Even the work needs further delicate corrections (for example, in the experiment, a more sensitive high resolution imaging system with a smaller time interval to record image data could produce better resolution flow images), the results in this paper still generally proves the analytical work deduced by Dr. Lu (2008).

ACKNOWLEDGEMENTS

This research work is partially supported by the NSFC Grant No. 50776095.

REFERENCES

Gonzalez, R. C. et al. (2003). Digital Image Processing, Simplified Chinese Edition. ISBN 7-5053-8236-5, Beijing: Publishing House of Electronics Industry.

Liao, Z. J. et al. (2005). "Measurement of hollow fiber membrane transport properties in hemodialyzers," Journal of Membrane Science, Vol. 256, Issues 1-2, PP176-183

Lu, J. F. (2005). "Development of porous ceramic membranes for potential use in artificial kidney," PHD Thesis, University of Kentucky, Lexington, KY, USA

Lu, J. F. et al. (2008). "An approximate analytical solution to the ultrafiltration profile in a hemodialysis process between parallel porous plates", Chinese Science Bulletin, Vol. 53, No. 21: 3402-3408

Mathews, J. H. et al. (2004). Numerical Methods: Using Matlab, Fourth Edition, ISBN 0-13-065248-2, Prentice- Hall Pub. Inc., Upper Saddle River, NJ, USA.

Poh, C. K. et al. (2003). "Effect of spacer yarns on the dialysate flow distribution of hemodialyzers: a magnetic resonance imaging study," ASAIO Journal, 49(4):440-448.

Poh, C. K. (2003). "Effect of flow baffles on the dialysate flow distribution of hollow-fiber hemodialyzers: a nonintrusive experimenttal study using MRI," J Biomech Eng. Aug; 125 (4):481-9 12968572

Journal of Physics: Conference Series **147** (2009) 012034 doi:10.1088/1742-6596/147/1/012034

Particle Size Analysis by Transmission Fluctuation Spectrometry in the Two-Phase Flow Irradiated by a Rectangular Narrow Beam

J. Shen, B. Yu

University of Shanghai for Science and Technology
516 Jungong Rd, Shanghai 200093, CHINA

shenjq@online.sh.cn

Abstract. The theory of transmission fluctuation spectrometry (TFS) has been developed as a new method of particle analysis in the two-phase flow. In our earlier publications, a circular beam is used whose intensity is uniform or of a Gaussian profile. In this work, the TFS theory is studied for the case of a rectangular narrow beam. The signal process of the transmission fluctuations is performed in the time and frequency domains and the corresponding analytical expression expressed in terms of the expectancy of the transmission square (ETS) is obtained. In addition, the correlation of the fluctuating transmission signals is studied, expressed in terms of the expectancy of the transmission product (ETP). Numerical calculation shows that the transition function of the transmission fluctuation spectrum is sensitive to both the ratio of beam size to particle size and the shape of the beam cross section.

1. INTRODUCTION

The transmission signal measured in a flowing particle suspension with high spatial and temporal resolution shows significant fluctuations, which include complete information on the particle size distribution and particle concentration. In the last few years, the theory of transmission fluctuation spectrometry (TFS) has been developed as a new method of particle analysis (Breitenstein et al., 1999, 2001; Shen et al., 2003a, b, c, 2005a). After the transmission signal of a narrow beam has been subjected to a procedure of variable spatial and/or temporal averaging, the transmission fluctuations are expressed in terms of the expectancy of the transmission square (ETS). By changing the averaging parameter, the ETS is obtained as a spectrum, which is related to the physical properties of the suspension and the process of signal averaging (i.e. spatial averaging, temporal averaging and their combination).

The theory is based on the assumptions of geometric ray propagation and completely absorbent particles. This implies that the particles should be much larger than the wavelength and the transmission signal should be detected within the near field range so that effects from diffraction could be disaccounted for. In this sense, the TFS can also be realized with a wide parallel beam and a photo detector with a very small receiving area, which is equivalent to the optical arrangement of a narrow beam. The receiving area of the detector could be either circular or rectangular. In our earlier publications, the theory of TFS was developed for the circular beam whose intensity is uniform or of

Journal of Physics: Conference Series **147** (2009) 012034 doi:10.1088/1742-6596/147/1/012034

Gaussian profile and measurements were realized by using a convergent Gaussian beam with its beam waist being located in the middle of the measuring region (Shen et al., 2003b, 2005b; Yu et al. 2008). This work will discuss the case of the rectangular beam.

2. FORMULATION OF THE ETS

The transmission fluctuations through a 3-dimensional particle system can be derived from the transmission fluctuations through a thin suspension layer. As already applied successfully in the previous theories on the transmission of radiation through concentrated disperse systems (Riebel et al., 1994), the present model is based on a suspension monolayer, where the layer thickness corresponds to the particle diameter. Assuming geometric ray propagation of the radiation and perfectly absorbent (black) particles, the transmission behaviour of a monolayer can be described by the transparency function (see Fig.1):

$$T_{\mathrm{ML}}(\mathbf{r}) = 1 - \sum_{k=1}^{N} \mathrm{Heav}\left(\frac{d}{2} - |\mathbf{r} - \mathbf{r}_k|\right) \tag{1}$$

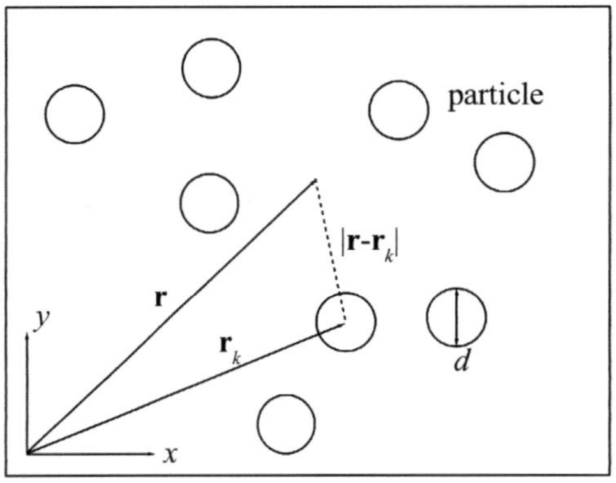

Fig. 1 Schematic of the monolayer

Here, $T_{\mathrm{ML}}(\mathbf{r})$ is the local value of the transmission, assuming a value of 1 or 0, respectively. d is the particle diameter and $\mathbf{r}_k$ is the centre coordinate of the particle. N is the number of particles in the monolayer. Heav is the Heaviside function describing the absorption of the radiation by the particle, which is defined as:

$$\mathrm{Heav}(u) = \begin{cases} u \geq 0: & 1 \\ u < 0: & 0 \end{cases} \tag{2}$$

According to the concept of the monolayer, a coincidence or superposition of particles is not possible within the layer.

In a first step, we shall consider the process of spatial averaging produced by the cross section of the beam. For a beam crossing the monolayer at $\mathbf{r}_0$ (coordinate of the beam centre), the transmission is equal to the average of the transparency function over the beam cross section:

$$T_{\mathrm{ML,B}}(\mathbf{r}_0) = \iint T_{\mathrm{ML}}(\mathbf{r}) B(\mathbf{r}_0, \mathbf{r}) \mathrm{d}\mathbf{r} \tag{3}$$

Here, $B(\mathbf{r}_0, \mathbf{r})$ is the normalized beam function describing the shape of the beam and its intensity profile. For a circular Gaussian beam, the beam function is:

$$B_{\mathrm{CG}}(\mathbf{r}_0, \mathbf{r}) = \frac{8}{\pi D^2} \exp\left[-\frac{8(\mathbf{r}-\mathbf{r}_0)^2}{D^2}\right] \tag{4}$$

It can be expressed as:

$$B_{\mathrm{CU}}(\mathbf{r}_0, \mathbf{r}) = \mathrm{rect}_{D/2}(\mathbf{r}-\mathbf{r}_0) \tag{5}$$

for a circular uniform beam and

$$B_{\mathrm{R}}(\mathbf{r}_0, \mathbf{r}) = \frac{1}{H \cdot W} \mathrm{rect}_{H/2}(x - x_0) \mathrm{rect}_{W/2}(y - y_0) \tag{6}$$

for a rectangular uniform beam. Here, $\mathrm{rect}_{L/2}(u)$ is the rect function with the width of L:

$$\mathrm{rect}_{L/2}(u) = \begin{cases} 1 & \text{if } |u| \le \frac{L}{2} \\ 0 & \text{else} \end{cases} \tag{7}$$

The variables in the beam functions are shown in Fig.2. The flow velocity of the particle suspension is v in y-direction. D is the diameter of the circular beam. H and W are the edge lengths of the rectangular beam perpendicular and parallel to the flow direction respectively.

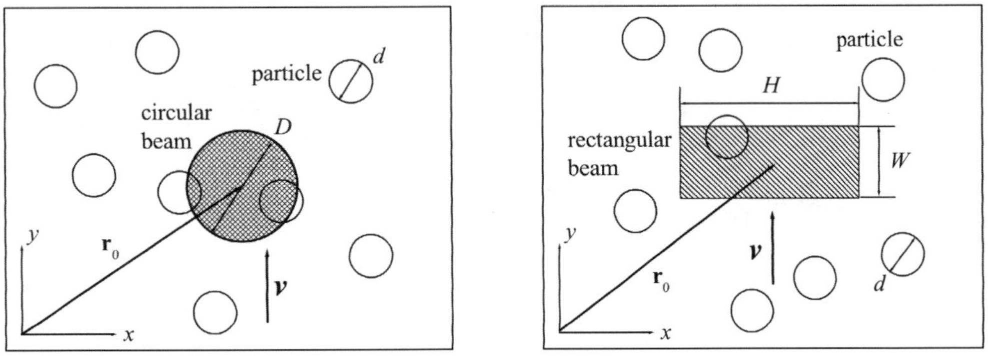

Fig. 2 Interaction between particles and the beam

So, Eq.(3) can be expressed as the convolution of the transparency function $T_{\mathrm{ML}}(\mathbf{r})$ and the beam function $B(\mathbf{r}_0, \mathbf{r})$:

$$T_{\mathrm{ML,B}}(\mathbf{r}_0) = (T_{\mathrm{ML}} \otimes B)_{\mathbf{r}_0} \tag{8}$$

Provided that the fluctuating transmission signal is averaged in the time domain and the time constant of temporal averaging is τ, the transmission averaged across the trajectory $L = v\tau$ of the beam is described as:

$$T_{\mathrm{ML,B},\tau}(\mathbf{r}_0) = \frac{1}{\tau}\int_{-\tau/2}^{\tau/2} T_{\mathrm{ML,B}}(\mathbf{r}_0 - \mathbf{e}_y \cdot vt)\,\mathrm{d}t \tag{9}$$

or equivalently

$$T_{\mathrm{ML,B},\tau}(\mathbf{r}_0) = \int_{-\infty}^{-\infty}\int_{-\infty}^{-\infty} T_{\mathrm{ML,B}}(\mathbf{r}_0 - \mathbf{r}) \cdot Z_L(\mathbf{r})\,\mathbf{dr} \tag{10}$$

where $Z_L(\mathbf{r})$ is defined as

$$Z_L(\mathbf{r}) = \frac{1}{L}\delta(x) \cdot \mathrm{rect}_{L/2}(\mathrm{y}) \tag{11}$$

Combining with Eq.(8), Eq.(10) can be written as the convolution of the functions $T_{\mathrm{ML}}(\mathbf{r})$, $B(\mathbf{r}_0, \mathbf{r})$ and $Z_L(\mathbf{r})$:

$$T_{\mathrm{ML,B},\tau}(\mathbf{r}_0) = (T_{\mathrm{ML}} \otimes B \otimes Z_L)_{\mathbf{r}_0} \tag{12}$$

The expectancy of the transmission square (ETS) is defined as

$$e\{T^2_{\mathrm{ML,B},\tau}\} = \lim_{A\to\infty}\frac{1}{A}\iint_{\mathbf{r}_0\in A}|T_{\mathrm{ML,B},\tau}(\mathbf{r}_0)|^2\,\mathbf{dr}_0 \tag{13}$$

Here, A is the area of the monolayer. Using Parseval's theorem of the Fourier transform, the ETS is obtained as:

$$e\{T^2_{\mathrm{ML,B},\tau}\} = \lim_{A\to\infty}\frac{1}{4\pi^2 A}\iint|\mathcal{F}\{T_{\mathrm{ML,B},\tau}(\mathbf{r}_0)\}|^2\,\mathrm{d}\boldsymbol{\omega} \tag{14}$$

The Fourier transform of the fluctuating transmission $T_{\mathrm{ML,B},\tau}(\mathbf{r}_0)$ is:

$$\mathcal{F}\{T_{\mathrm{ML,B},\tau}(\mathbf{r}_0)\} = \mathcal{F}\{T_{\mathrm{ML}}\}\cdot\mathcal{F}\{B\}\cdot\mathcal{F}\{Z_L\} \tag{15}$$

where $\mathcal{F}\{T_{\mathrm{ML}}\}$, $\mathcal{F}\{B\}$ and $\mathcal{F}\{Z_L\}$ are the Fourier transforms of the functions $T_{\mathrm{ML}}(\mathbf{r})$, $B(\mathbf{r}_0, \mathbf{r})$ and $Z_L(\mathbf{r})$ respectively.

$$\mathcal{F}\{T_{\mathrm{ML}}\} = 4\pi^2\delta(\boldsymbol{\omega}) - \frac{\pi d}{|\boldsymbol{\omega}|} J_1\left(\frac{d}{2}|\boldsymbol{\omega}|\right)\sum_{k=1}^{N}\exp(-\boldsymbol{\omega}\cdot\mathbf{r}_k)$$
$$\mathcal{F}\{Z_L\} = \mathrm{sinc}(\omega_y L/2) \tag{16}$$

J_1 is the 1st order Bessel function. The unnormalized sinc function is defined as

$$\mathrm{sinc}(u) = \frac{\sin(u)}{u} \tag{17}$$

The Fourier transform of the beam function depends on the beam function. The beam considered here is rectangular and is uniform in intensity. Thus the Fourier transform is:

$$\mathcal{F}\{B_{\mathrm{R}}\} = \mathrm{sinc}\left(\frac{\omega_x H}{2}\right)\mathrm{sinc}\left(\frac{\omega_y W}{2}\right) \tag{18}$$

The further solution of the integral given in Eq.(14) involves a large number of rather complicated mathematical operations (Shen, 2003d). The final result, which is obtained without any further simplifications or assumptions, is written as

$$e\left\{T^2_{\mathrm{ML,B},\tau}\right\} = 1 - \beta\cdot(2-\chi) + \beta^2\cdot(1+\varepsilon) \tag{19}$$

in which the analytical expressions of the transition functions χ and ε are

$$\chi = \int_0^\infty F(\Lambda_H, \Lambda_W, \Theta, u)\cdot\frac{2J_1^2(u)}{u}\cdot \mathrm{d}u$$
$$\varepsilon = \int_0^\infty F(\Lambda_H, \Lambda_W, \Theta, u)\cdot\frac{2J_1^2(u)}{u}\cdot G_{\mathrm{ML}}(u,\beta)\cdot \mathrm{d}u \tag{20}$$

$$F(\Lambda_H, \Lambda_W, \Theta, u) = \frac{2}{\pi}\int_0^{\pi/2}\left\{\begin{array}{l}\mathrm{sinc}(u\Lambda_H\cos\varphi)\\ \cdot\mathrm{sinc}(u\Lambda_W\sin\varphi)\\ \cdot\mathrm{sinc}(u\Theta\sin\varphi)\end{array}\right\}^2 \mathrm{d}\varphi \tag{21}$$

G_{ML} is a function describing the monolayer structure (Shen, 2003d). β is the monolayer density, which is defined as the ratio of the projected area covered by particles in the monolayer.

$$\beta = \frac{\pi d^2}{4}\cdot\frac{N}{A} \tag{22}$$

Λ_H, Λ_W and Θ are the dimensionless parameters, defined as:

$$\Lambda_H = \frac{H}{d}, \Lambda_W = \frac{W}{d}$$
$$\Theta = \frac{v\tau}{d} = \frac{L}{d} \tag{23}$$

The dimensionless sizes of the rectangular beam Λ_H and Λ_W describe the effect of spatial averaging of the transmission signal within the beam cross section and the dimensionless time interval Θ describes the effect of the temporal averaging.

The 3-dimensional particle suspension is modelled as a pile of independent monolayers of spherical particles. The transmission of the whole particle suspension is found to be the product of the transmissions through the individual monolayers and so is the expectancy of the transmission square (ETS) (Breitenstein et al., 1999):

$$e\left\{T^2_{\mathrm{B},\tau}\right\} = e\left\{T^2_{\mathrm{ML,B},\tau}\right\}^{N_{\mathrm{ML}}} \tag{24}$$

For a dilute particle suspension, the monolayer density is so small that the logarithm of the ETS of the particle suspension can be approximated to:

$$\ln e\left\{T^2_{\mathrm{B},\tau}\right\} \approx -N_{\mathrm{ML}}\beta\cdot(2-\chi) = -1.5\Delta Z\cdot C_V\cdot(2-\chi) \tag{25}$$

Here, N_{ML} is the number of monolayers of the particle suspension, ΔZ is the pathlength of the suspension along the beam propagation and C_V is the particle volume concentration (i.e. the ratio of particles' volume in the suspension).

3. NUMERICAL RESULTS

When the particle concentration is low, the transition function ε is not important so that only the function χ is considered in this work. As far as we know, the transition function χ can be calculated numerically only. Fig.3 shows some examples of the numerical results. It can be found that the transition function χ decreases gradually along with the increase of the parameter Θ, which means an increasing effect of the temporal averaging. The value of the transition function is between 0 and 1. The turning point is in the vicinity of $\Theta = v\tau/d = 1$, which gives the information of particle size as long as the flow velocity is known.

The limiting value of the transition function χ at large averaging temporal parameter Θ equals 0, i.e. $\lim\limits_{\Theta\to\infty}\chi = 0$. This corresponds to the traditional extinction measurement, in which the transmission signal is averaged in the low spatial and/or temporal resolutions. However, when Θ tends to 0 (corresponding to the highest temporal resolution), the limiting value depends on the spatial averaging parameters:

$$\mathrm{X}(\Lambda_H,\Lambda_W) = \lim_{\Theta\to\infty}\chi = \int_0^\infty F(\Lambda_H,\Lambda_W,\Theta,u)\cdot\frac{2J_1^2(u)}{u}\cdot \mathrm{d}u \tag{26}$$

$$\lim_{\Theta\to\infty} F\left(\Lambda_H,\Lambda_W,\Theta,u\right)=\frac{2}{\pi}\int_0^{\pi/2}\begin{Bmatrix}\operatorname{sinc}\left(u\Lambda_H\cos\varphi\right)\\ \cdot\operatorname{sinc}\left(u\Lambda_W\sin\varphi\right)\end{Bmatrix}^2 \mathrm{d}\varphi \tag{27}$$

Here, $\mathrm{X}\left(\Lambda_H,\Lambda_W\right)$ describes the stepheight of the transition function, i.e. the difference between the limiting values of χ, as is denoted in Fig.3. The numerical results are plotted in Fig.4. Similar to the temporal averaging, the increase of the spatial averaging parameters diminishes the fluctuation of the transmission signal.

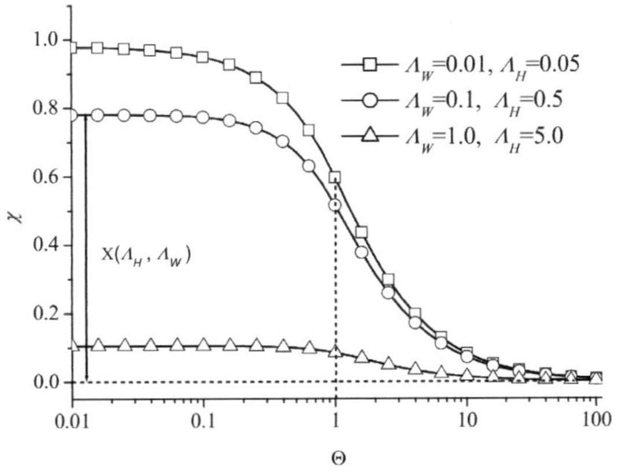

Fig. 3 The transition function χ at variant beam parameters Λ_H & Λ_W

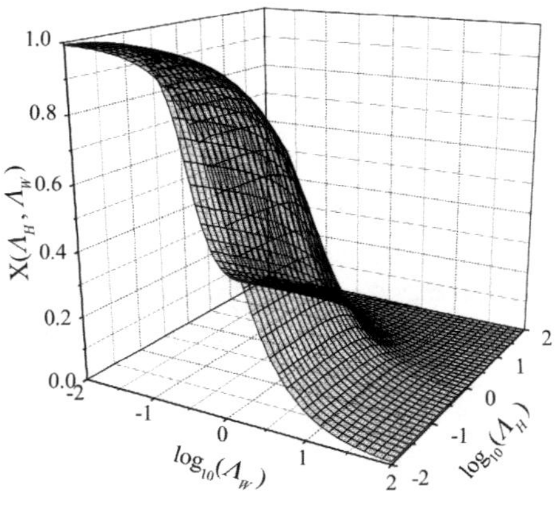

Fig. 4 The stepheight of the transition function $\mathrm{X}\left(\Lambda_H,\Lambda_W\right)$

To study the effect of the beam shape on the transmission fluctuation spectrum, the stepheight of the transition function $\mathrm{X}\left(\Lambda_H,\Lambda_W\right)$ is calculated via the dimensionless beam cross section, which is defined as the ratio of the beam cross section to the projected area of the spherical particle (i.e.

$\Sigma = W \cdot H \cdot 4/\pi d^2 = 4\Lambda_W \Lambda_H/\pi$) for the rectangular uniform beam. The beam shape is expressed with the ratio of the edge lengths $\eta = W/H = \Lambda_W/\Lambda_H$. Numerical results are shown in Fig.5. As a comparison, the stepheight of the transition function of a circular uniform beam is plotted too (Shen et al., 2003c). The dimensionless beam cross section is defined as the ratio of the beam cross section to the projected area of the spherical particle (i.e. $\Sigma = \Lambda_D^2$) for the circular uniform beam.

It can be found that the stepheight of the transition function is sensitive to the beam shape. For a fixed value of the dimensionless beam cross section, the stepheight is different depending on the ratio of the edge lengths η. The stepheight of the transition function reaches the maximum when $\Lambda_H = \Lambda_W$ (i.e. the case of a uniform beam in foursquare), which is close to that of the circular uniform beam. Therefore, in order to obtain the higher resolution of the spatial resolution in the transmission fluctuation spectrometry, it is better to use a narrow foursquare beam.

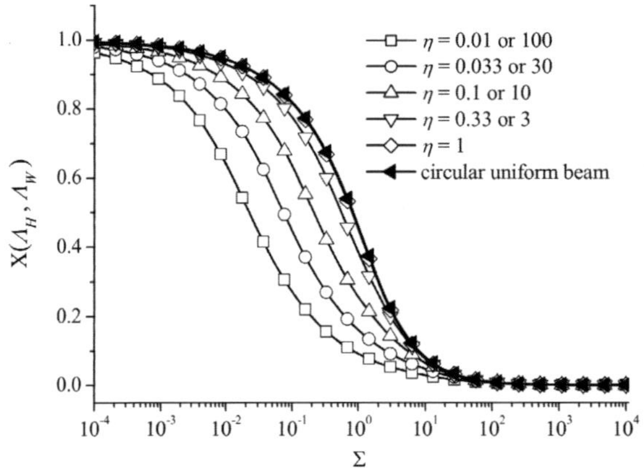

Fig. 5 The stepheight of the transition function $X(\Lambda_H, \Lambda_W)$ via the beam cross section

4. FURTHER DISCUSSIONS

The TFS theory presented in section 2 is developed in the time domain, i.e. the fluctuating transmission signal is averaged in a variable time interval. Similar to the work given by Shen et al. 2003a, the TFS theory can also be extended to the frequency domain, replacing the temporal averaging with the signal filtering. For example, as the 1st order low-pass filter is used, the function given in Eq.(21) is replaced with:

$$F(\Lambda_H, \Lambda_W, \Psi, u) = \frac{2}{\pi} \int_0^{\pi/2} \frac{\left[\text{sinc}(u\Lambda_H \cos\varphi) \cdot \text{sinc}(u\Lambda_W \sin\varphi)\right]^2}{1 + \left(\dfrac{u \cdot \sin\varphi}{\Psi}\right)^2} \mathrm{d}\varphi \tag{28}$$

Here, the dimensionless temporal averaging parameter Θ is replaced with the dimensionless cut-off frequency, which is defined as $\Psi = \pi d f_0/v$ (f_0 is the cut-off frequency of the 1st order low-pass filter). The deduction of the theoretical expression is very similar to that presented above and is hence not given here.

The theory may also be extended into the TFS with signal correlation. The temporal averaging expressed in Eq.(9) is replaced with the signal correlation and correspondingly the ETS defined in Eq.(14) is replaced with the ETP:

$$e\left\{T_{\mathrm{ML,B}}T_{\mathrm{ML,B},\tau}\right\}=\lim_{A\to\infty}\frac{1}{A}\iint_{\mathbf{r}_0\in A}\left|T_{\mathrm{ML,B},\tau}\left(\mathbf{r}_0\right)T_{\mathrm{ML,B},\tau}\left(\mathbf{r}_0+\mathbf{v}\tau\right)\right|\mathbf{dr}_0 \tag{29}$$

The flow velocity $\mathbf{v}$ is in y-direction (see Fig.2). After a large number of complicated mathematical operations, the function given in Eq.(21) is replaced with

$$F\left(\Lambda_H,\Lambda_W,\Psi,u\right)=\frac{2}{\pi}\int_0^{\pi/2}\begin{Bmatrix}\operatorname{sinc}\left(u\Lambda_H\cos\varphi\right)\\ \cdot\operatorname{sinc}\left(u\Lambda_W\sin\varphi\right)\end{Bmatrix}^2\cos\left(2\mathrm{u}\Gamma\sin\varphi\right)\mathrm{d}\varphi \tag{30}$$

Here, τ is the correlation time and the corresponding dimensionless correlation parameter is defined as $\Gamma=v\tau/d$.

5. CONCLUSIONS

In this work, we present the fundamentals of the transmission fluctuation spectrometry by using a rectangular uniform beam, which can be realized with a wide beam combining a tiny photo-detector.

An analytical expression of the expectancy of the transmission square (ETS) through a monolayer of spherical particles is obtained in the time domain and for a rectangular uniform beam. When the particle concentration is low, the transmission fluctuation spectrum can be approximately described by the monolayer density β and the transition function χ. The transition function χ depends on the spatial averaging parameters Λ_H and Λ_W and the temporal averaging parameter Θ. The maximum of the transition function $\chi=1$ is found at the highest spatial and temporal resolutions (i.e. the limiting case of $\Lambda_H\to 0$, $\Lambda_W\to 0$ and $\Theta\to 0$). The minimum of the transition function is $\chi=0$, corresponding to the case of the very low spatial or temporal resolution or both. Furthermore, the transition function χ is found to be related closely to the beam shape. For example, with a fixed value of the beam cross section, the stepheight of the transition function varies according to the ratio of the edge lengths of the rectangular beam and it reaches its maximum when the beam is foursquare. In addition, the TFS theory is also extended to the transmission fluctuation spectrometry in the frequency domain and that with signal correlation.

Based on the layer model, the theory presented here can be further extended to a 3-dimensional particle system and thus can be applied for real application of two-phase flow measurements.

ACKNOWLEDGEMENTS

The authors acknowledge the support from the National Natural Science Foundation of China (NSFC 50876069), the Key Scientific Research Project of Ministry of Education of the People's Republic of China (208041) and the Shanghai Municipal Education Commission (07ZZ88).

Journal of Physics: Conference Series **147** (2009) 012034 doi:10.1088/1742-6596/147/1/012034

NOMENCLATURE

A	area of the monolayer	$[m^2]$
B	beam function	
C_V	particle volume concentration	
D	diameter of the circular beam	[m]
d	particle diameter	[m]
$e\{\cdots\}$	expectancy of	
f_0	cutoff frequency of the filter	[1/s]
$\mathcal{F}\{\cdots\}$	Fourier transform of	
G_{ML}	function describing the monolayer structure	
H	edge length of the rectangular beam perpendicular to the flow direction	[m]
Heav	Heaviside function	
J_1	the 1st order Bessel function	
N	number of particles in the monolayer	
N_{ML}	number of monolayers in the particle suspension	
$\mathbf{r}$	coordinates in the monolayer	[m]
$\mathbf{r}_k$	coordinates of the particle center	[m]
$\mathbf{r}_0$	coordinates of the beam center	[m]
rect	rect function	
sinc	sinc function	
T	transmission	
$\mathbf{v}$	flow velocity	[m/s]
W	edge length of the rectangular beam parallel to the flow direction	[m]
Z_L	temporal averaging function	

Greek Letters

β	monolayer density	
Γ	dimensionless correlation time	
δ	Dirac delta function	
ΔZ	pathlength of the particle suspension	
η	ratio of the edge lengths of the rectangular beam	
Θ	dimensionless temporal averaging parameter	
Λ_H, Λ_W	dimensionless spatial averaging parameters	
Σ	dimensionless beam cross section	
τ	averaging time interval or correlation time	[s]
φ	integrand variable	
χ, ε	transition functions	
Ψ	dimensionless cutoff frequency of the filter	
ω	variable in the Fourier space	[1/m]

Subscripts

CG	circular Gaussian beam
CU	circular uniform beam
k	k-th particle in the monolayer
ML	monolayer
R	rectangular beam

REFERENCES

Breitenstein, M. et al. (1999). " The Fundamentals of Particle Size Analysis by Transmission Fluctuation Spectrometry. Part 1: A Theory on Temporal Transmission Fluctuations in Dilute Suspensions," Part. Part. Syst. Charact., 16(6), pp. 249–256.

Breitenstein, M. et al. (2001). " The Fundamentals of Particle Size Analysis by Transmission Fluctuation Spectrometry. Part 2: A Theory on Transmission Fluctuations with Combined Spatial and Temporal Averaging," Part. Part. Syst. Charact., 18(3), pp. 134-141.

Riebel, U. et al. (1994). " Extinction of Radiations in Sterically Interacting Systems of Monodisperse Spheres. Part 1: Theory," Part. Part. Syst. Charact., 11(3), pp. 212-221.

Shen, J. et al. (2003a). " The Fundamentals of Particle Size Analysis by Transmission Fluctuation Spectrometry. Part 3: A Theory on Transmission Fluctuations in a Gaussian Beam and with Signal Filtering," Part. Part. Syst. Charact., 20(2), pp. 94-103.

Shen, J. et al. (2003b). " Particle Size Analysis by Transmission Fluctuation Spectrometry: Experimental Results Obtained with a Gaussian Beam and Analog Signal Processing," Part. Part. Syst. Charact., 20(4), pp. 250-258.

Shen, J. et al. (2003c). " Fundamentals of Transmission Fluctuation Spectrometry with Variable Spatial Averaging," China Particuology 1(6), 242-246.

Shen, J. (2003d). Particle size analysis by transmission fluctuation spectrometry: Fundamentals and case studies, Cuvillier Verlag Göttingen, Germany.

Shen, J. et al. (2005a). " Transmission Fluctuation Spectrometry with Spatial Correlation," Part. Part. Syst. Charact., 22(1), pp. 24-37.

Shen, J. et al. (2005b): " Measurements on Particle Size Distribution and Concentration by Transmission Fluctuation Spectrometry with Temporal Correlation, " Optics Letters 30(16), pp. 2098-2100.

Yu, B. et al. (2008): " Measurements on Particle Size Distribution andConcentration by Transmission Fluctuation Spectrometry with Temporal Correlation, " Part. Part. Syst. Charact. 25(3), pp. 231-243.

Design of high-speed ECT and ERT system

Baoliang Wang, Zhiyao Huang, Haiqing Li

Zhejiang University, Hangzhou, CHINA

blwang@iipc.zju.edu.cn, zyhuang@iipc.zju.edu.cn, hqli@iipc.zju.edu.cn

Abstract. Process tomography technique provides a novel method to investigate the multi-phase flow distribution inside pipe or vessel. Electrical resistance tomography (ERT) and electrical capacitance tomography (ECT) are extensively studied in recent years. As the capacitance to voltage and resistance to voltage converters run faster, the speeds of other circuits in the system, such as MCU, A/D, D/A etc, have become the bottlenecks of improving the speed. This paper describes a new dual-modal, ECT and ERT, data acquisition system. The system is controlled by a digital signal processor. Both the ERT and the ECT systems use one platform to simplify the system design and maintenance. The system can work at high speed which is only limited by the capacitance to voltage converter or resistance to voltage converter. Primary test results show the speed of the new system is 1400 frames/second for 16-electrode ERT and 2200 frames/second for 12-electrode ECT.

1. INTRODUCTION

Electrical process tomography provides a powerful tool for visualize the phase distribution in two-phase flow. Our research group has developed ECT (Wang B.L. et al, 2005, Huang, Z. Y. et al, 2003) and ERT (Zhu J.P. et al, 2005) systems many years ago. They are both controlled by MCS-51 series MCU and use RS232 protocol to send data to PC. However, the two systems are designed in different circuit boards and very difficult to maintain. The two systems work independently and are difficult to precisely synchronize the sampling time. As the capacitance to voltage and resistance to voltage converters run faster, the other circuits in the system, such as A/D, D/A etc, have become the bottlenecks of improving the speed. In order to further improve the system speed, it is necessary to re-design the original ECT and ERT. The first target in the new system design is that the system can work at high speed which is only limited by the capacitance to voltage converter or resistance to voltage converter. The second target is to make the whole system flexible and easy to maintain.

This paper describes a new dual-modal, ECT and ERT, data acquisition system. The system is controlled by a digital signal processor. Both the ERT and the ECT systems use one platform to simplify the system design and maintenance. Primary test results show the speed of the new system is 1400 frames/second for 16-electrode ERT and 2200 frames/second for 12-electrode ECT.

2. SYSTEM DESIGN

2.1 System Platform

The block diagram of most ECT or ERT data acquisition system can be shown in Fig. 1. The difference of the two systems is the front-end signal processing circuit. In ECT, front-end signal processing circuit converts capacitance to voltage. In ERT, it converts resistance to voltage. The other circuits, such as A/D and D/A converter, processor, digital input and output, communication interface and power supply, are almost the same. The new system can be simplified based on the similarity between ECT and ERT.

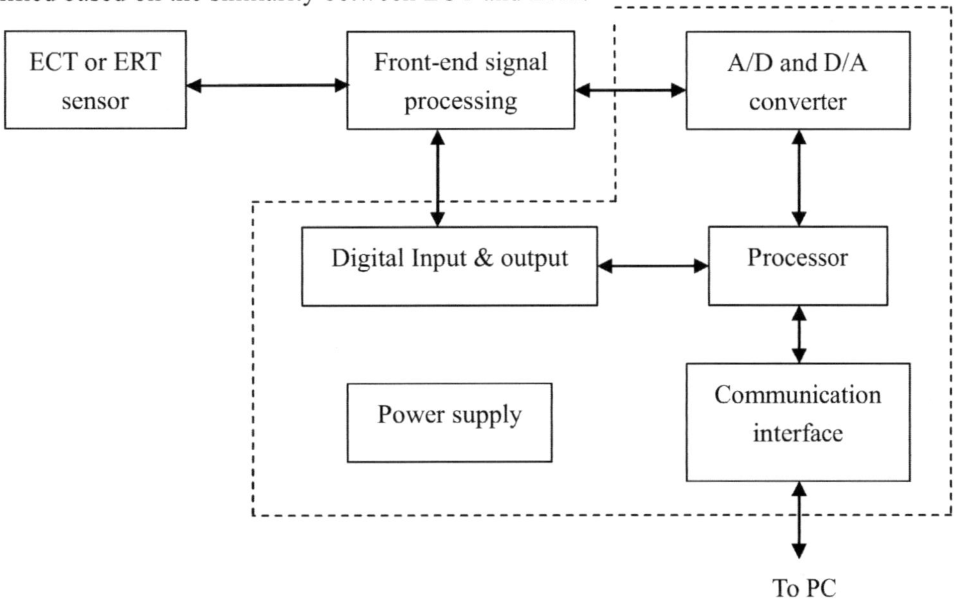

Fig.1 Block diagram of ECT and ERT system

The structure of the new system is shown in Fig. 2. The platforms, inside the dashed lines for both systems, are the same. The circuit modules inside dashed lines in Fig. 2 are corresponding to the circuits inside the dashed lines in Fig. 1.

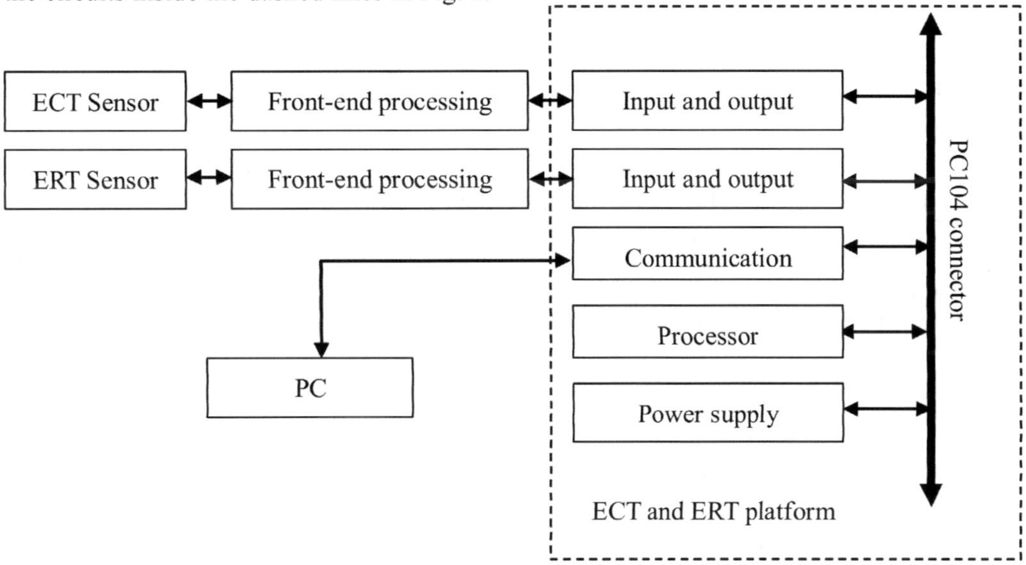

Fig. 2 Structure of new ECT and ERT

The platform consists of the following modules: power supply module, processor module, communication module and input and output (I/O) module.

(1) Power supply module: This module converts the 220V AC power to low voltage DC power for the system. The DC voltages used in the system are +5V, +3.3V, +15V, -15V and an isolated +24V. The DC voltages are distributed to the other modules through the PC104 connector.

(2) Processor module: The core of this module is a digital signal processor. The whole system is controlled by this module. This module includes DSP (ADSP-2188N), flash ROM and glue logic for the processor to peripherals. The DSP can run at 80 MIPS and integrate 48K words (24-bit) of program RAM and 56K words (16-bit) of data RAM. This chip provides enough processing power for the system.

(3) Communication module: This module provides USB2.0 interface (CY7C68013A) to the system. The front-end signal processing circuit used in the system runs very fast and thus the communication speed to PC should be very fast. USB2.0 interface can send data to PC at the speed up to, ideally, 480Mbps. Test results show the highest speed of the module is 22Mbps.

(4) I/O module: The I/O module provides the A/D (AD7472) and D/A (TLV5619) converter and digital input/output capability for the system. A/D converter converts the voltage from the front-end signal processing module to digital signal. The D/A can be used to balance the standing voltage when the measured area is empty. Digital output, implemented in a CPLD (XC9572), can be used to control the operation of front-end circuit and can also provide system status signal. If synchronization is required for multiple system, the digital input can be used to acquire the synchronize signal.

The modules of platform are connected by PC104 connectors. We use PC104 connectors only to connect each module and we defined the signals in each pin of connectors. The system is not compatible with PC104 PC module. The system can be extended easily by adding more new modules if more input signals are needed in the system.

The only difference in the new ECT and ERT system is in the front-end signal processing module. By adding front-end processing circuit for ECT or ERT, this system can operate as ERT system, ECT system or ECT and ERT dual-modal system.

2.2 ECT System Design

The main work of ECT system design based on the platform is the designing of front-end processing circuit. The core of the circuit is the capacitance to voltage (C/V) converter. The C/V circuit used in the system is based on differential sampling method (Wang B.L., 2005). Compared to other C/V circuits (Huang S.M., 1992, Yang W.Q., 1994), this circuit can measure the capacitance in one pulse. Fig. 3 shows the principle of the C/V circuit.

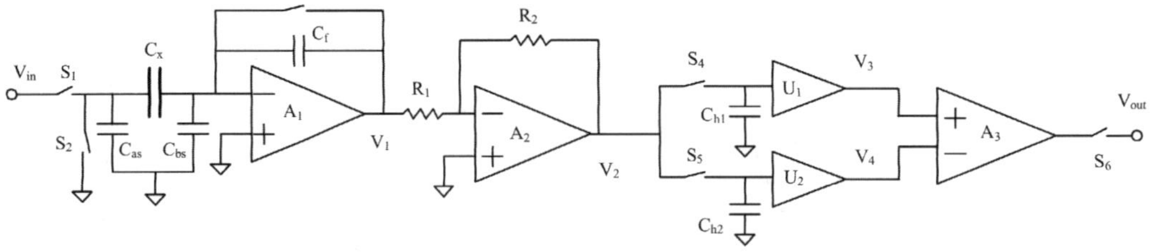

Fig. 3 Principle of differential sampling method

There are 12 electrodes in the ECT system. The front-end signal processing circuit for each electrode is designed in modular fashion. The circuit for each electrode is implemented in one print circuit board (PCB). Each PCB is mounted under the screen of ECT sensor. One flat cable connects all PCB boards inside the sensor to the platform. The block diagram of front-end module is shown in Fig. 4. The control signals, from the I/O module of platform, and the module address specified by binary-output rotary switch are connected to a decoder. The outputs of the decoder determine whether the module is in excitation mode or measurement mode. The control signals from the I/O module to each front-end module are identical. However the module address of each front-end module can be different. Therefore, modules can be set in different mode.

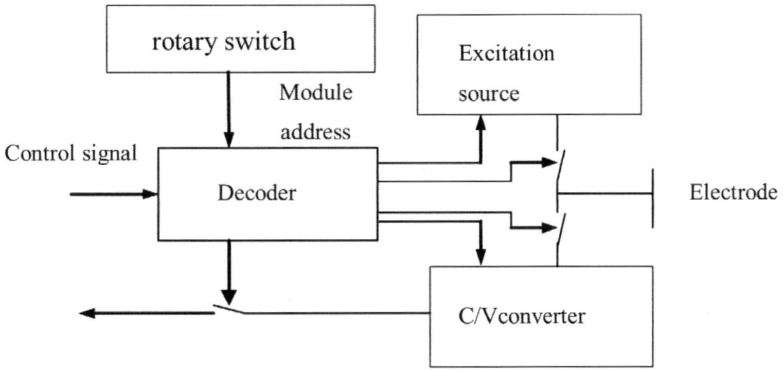

Fig.4 Block diagram of ECT front-end module

The modular design of front-end circuit simplified the system maintenance. By changing the number of front-end modules, the system can be use in 8 or 16 electrodes ECT system.

2.3 ERT System Design

The front-end processing circuits of ERT convert the resistance to voltage. The circuit is also designed in modular fashion. The module is designed based on bi-directional current pulse method (Andrew J. et al, 2005). The circuits for 2 electrodes are implemented in one PCB. For 16-electrode ERT system, 8 front-end modules are used. The modules are fixed on the wall of sensor. Each module is connected to a flat cable. One end of the flat cable is connected to the I/O module of the platform. Fig. 5 shows the principle of front-end module based on bi-directional current pulse method.

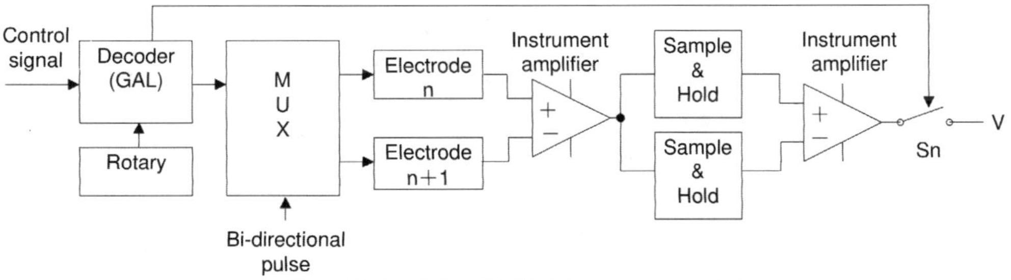

Fig. 5 Principle of ERT front-end module

The structure of ERT front-end module is similar to that of ECT front-end module. The control signals, from the I/O module of platform, and the module address specified by binary-output rotary switch are connected to the decoder. The outputs of decoder control the operation of the modules. All PCBs of ERT front-end modules are the same. Each module can be set to different address by the rotary switch.

3. PROTOTYPE SYSTEM AND TEST RESULTS

The new system is assembled after all modules are debugged. In order to test the system performance, the system is operated at ECT and ERT mode separately. First, the ECT system is tested. Fig. 6 shows the ECT system. By placing nylon stick at different positions inside the sensor, we can simulate different flow regimes of two-phase flow. Fig. 7 shows the reconstructed image for stratified flow, annular flow and core flow. Imaging results show that the reconstructed images match with the real flow regimes. The data acquisition speed can reach 2200 frames/s. The capacitance measurement resolution is 1fF.

Fig.6 ECT system

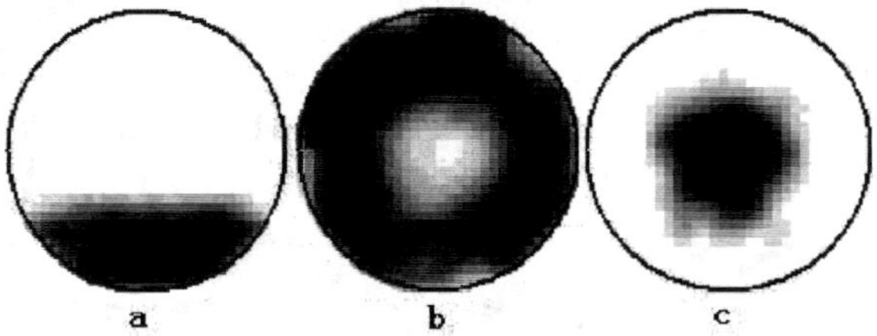

Fig.7 Imaging results of ECT

Fig. 8 shows the new ERT system based on the platform. The PCBs installed on the sensor pipe are the front-end processing modules for ERT. The low voltage signals induced from the sensor plate have been amplified before they are connected to the flat cable. Test results show that the system can capture the date at the speed of 1400 frames/second. To test the repeatability of the measurement electronics, we use the method in Wang M., 2005. The average standard deviation is 1.5 mV, which is less than 0.13% of typical maximum value in the measurement sequence.

The system performance are tested when the system is work as ECT or ERT. Further work is needed to test the system as dual-modal system. When the system works as dual-modal system we only need to rewrite the program. No more hardware design work is needed.

Fig.8 ERT system

4. CONCLUSIONS

A novel data acquisition system for ECT and ERT is developed. The system is designed in modular fashion which simplified the system maintenance. A common platform is proposed based on the similarity between ECT and ERT. The platform provides enough processing power for both ECT and ERT and can be extended easily by adding new modules in the platform. The front-end processing modules of ECT and ERT are also designed in modular fashion. All front-end modules in ECT and ERT are connected by a flat cable and are connect to the I/O module respectively. All the signals in the cable are active driven signals. The shielded cable is not needed. This structure simplified the system construction and maintenance. Because DSP, A/D, D/A and the communication interface are very fast, the speed of data acquisition is only limited by the front-end signal processing module. Test results show the ECT and ERT can work at speed of 2200 and 1400 frames/s respectively.

ACKNOWLEDGEMENTS

Project supported by the National Natural Science Foundation of China （No. 60532020,50576084）.

REFERENCES

Andrew J. et al. (2005). "A 1000-Measurement Frames/Second ERT Data Capture System With Real-Time Visualization," IEEE Sensors Journal, 5, pp.300-307.

Huang S. M. et al. (1992). "Design of sensor electronics for electrical capacitance tomography", IEE Proceedings-G, 139, pp.83-88.

Wang, B. L. et al. (2005). "A high-speed data acquisition system for ECT based on the differential sampling method, " IEEE Sensors Journal, 5(2):308-312.

Wang, M. et al. (2005). "A High-Performance EIT System, " IEEE Sensors Journal, 5, pp.289-299.

Yang W. Q. et al. (1994) "A high frequency and high resolution capacitance measuring circuit for process tomography, " IEE Proceedings -Circuits Device System, 141, pp. 215-219.

Huang, Z. Y. et al. (2003). "Application of electrical capacitance tomography to the void fraction measurement of two-phase flow, " IEEE Trans. Instrum. Meas., 52, pp.7-12.

Zhu, J. P. et al. (2005). "Design of ERT system", J Zhejiang Univ SCI, 6A(12), pp.1446-1448.

Journal of Physics: Conference Series **147** (2009) 012036 doi:10.1088/1742-6596/147/1/012036

Windows pollution problems of the dust concentration measurement based on scattering method

Yanjun Zhao [1], Yongtao Zhang [1], Xinyue Shi [1], Chuanlong Xu [2], Shimin Wang [2]

[1] Hebei Polytechnic University, China
[2] Southeast University, China

yjzhao@heut.edu.cn, zhyj_ts@sohu.com

Abstract. The windows are separated the measurement system from the dust space in the light Scattering dust concentration measurement system. The windows are polluted unavoidably by the dust and the measurement error is produced. Based on the Mie Scattering theory, the measurement error is researched in this paper. The numerical simulation results show that the measurement error is related to the particles diameter distribution and the refractive index, but is independent of the particles average diameter. A novel photoelectricity sensor is developed in this paper in order to solve the measurement error by the windows pollution. The calculated method is brought out which can amend the measurement errors by the windows pollution and improve the measurement accuracy.

1 Introduction

In China, dust pollution increases rapidly, and dust emission from all kinds of coal-fired boilers is one of major sources of atmospheric dust pollution. Over one hundred years ago, Ringelmann, a french engineer, proposed the eyeballing dust concentration measurement called Ringelmann blacken method(Zhao Y et al 2004, Yu Z et al 1990). This method is simple, but the measurement results are influenced largely by the operators' subjective factors and weather objective environment etc. In order to satisfy the need of dust pollution source monitoring and control, people present more stringent requirements for the on-line measurement of soot concentration. With the development of modern science and technology, people have developed several measuring methods for soot particles concentration based on different sensing principle, such as sampling(Zhao Y et al 2004, Yu Z et al 1990), Optics(Mu N et al 2004, Cai X et al 2001), electrostatic induction(Juliuse B.Gajewski 1999), ß-rays method(Li Q 2001) and ultrasonic method(Su M 2002), etc. The methods have their own characteristics and adaptability fields.

In recent years, Laser Scattering methods are widely applied to measuring soot concentration. But the optical measurement system is influenced easily by the windows pollution. As a certain distance between the windows and the protection of pipelines, a certain number of particles are in pipe connections. These particles are at a relatively static state, and the particles in the pipe are in motion, the signals measured by the photoelectric sensors are the mixed signals: the actual movement particle scattering signal and the relatively static particles scattering signals. The scattering intensity of the particles near the sensor is stronger than the scattering intensity of the particles in the pipe.The dust is absorbed on the windows surface by the electrostatic adsorption. The local multiple scattering is taken place so that the measured scattering intensity is smaller than the actual intensity. The measurement error is brought. Currently, the commonly and most effective method to solve windows pollution is using compressed air to purge the window glass, and regularly cleaning windows protection. The windows pollution dust can be cleanied by the method, but the impact of the measurement result can not be avoided entirely.

The windows pollution problem is studied deeply in this paper. A novel photoelectricity sensor is developed in order to solve the measurement error by the windows pollution.

2. Numerical Simulation Based On the Windows Pollution

The windows pollution problem is simulated based on Mie scattering theory(Ding w 2007). In simulations, the incidence light wavelength is 0.68μm and the range of the measuring angle is $0^{\circ} \sim 5^{\circ}$. The different average diameter can be obtained by adjusting the particles diameter distribution function and the largest diameter. The different indexes of 1.80 and 1.57-0.56i are considered respectively because the Mie scattering calculation is related to the refractive index m of the soot. Two types of the soot size distribution based on Johnson_SB function are chosen, and their distribution parameters are σ=2.5, μ=-2.0 ; σ=4.8, μ=0.0.

Table 1 shows measurement error of the particle whose refractive index is 1.8.

Table 2 shows measurement error of the particle whose refractive index is 1.57-0.56i.

Fig 1 shows the simulation result comparation of the different particle average diameter, different particle size distribution parameter and the different particle indexs.

distributing parameter	d =10 μm	d =15 μm	d =20 μm
σ =2.5 μ =-2	0.35%	0.36%	0.37%
σ =4.8 μ =0	0.89%	0.91%	0.95%

Table 1 the measurement error caused by windows pollution(particles refractive index is 1.8)

Distributing parameter	d =10 μm	d =15 μm	d =20 μm
σ =2.5 μ =-2	0.34%	0.35%	0.35%
σ =4.8 μ =0	0.84%	0.88%	0.90%

Table 2 the measurement error caused by windows pollution(particles refractive index is 1.57-0.56i.)

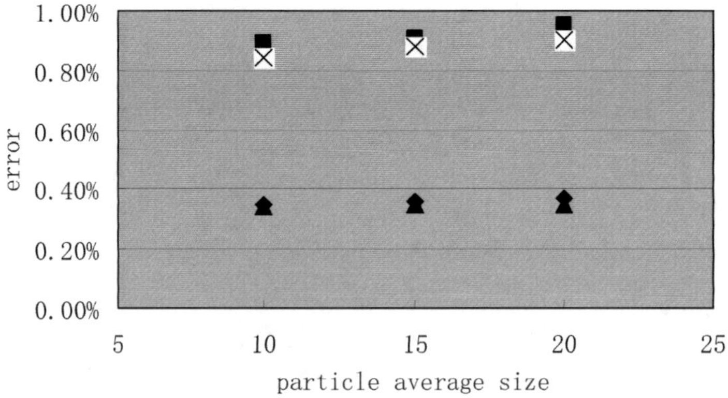

Fig 1 comparison of the numerical simulation result

■-the particle refractive index is 1.8, the size distribution is σ =4.8 μ =0;

×- the particle refractive index is 1.57-0.56i, the size distribution is σ =4.8 μ =0;

▲- the particle refractive index is 1.8, the size distribution is σ =2.5 μ =-2;

◆- the particle refractive index is 1.57-0.56i, the size distribution is σ =2.5 μ =-2.

From the table 1, table 2 and Fig 1, for the non-dissipativity and dissipativity particles,

1) Pollution particles on the windows would produce a certain measurement error.

2) For the different particles,refractive indexes, the the measurement error is different, but the difference comparatively small, so it can be regarded as approximately equality.
3) For the different particles size distribution parameters, the measurement error is different.
4) For the different particles average diameter, the measurement error is different, but the difference is comparatively small, so it can be regarded as approximately equality
Fig 2 shows the relation of the error with the pollution particle proportion while the particle index is 1.8 and the distribution parameter is σ=2.5, μ=-2.0.
Fig 3 shows the relation of the error with the pollution particle proportion while the particle index is 1.57-0.56i and the distribution parameter is σ=2.5, μ=-2.0.

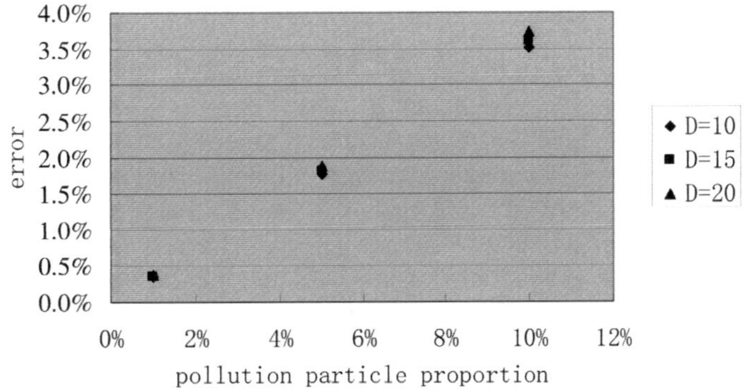

Fig 2 the relation of the error with the pollutionparticle proportion
(the particle refractive index is 1.8, the diameter distribution is σ=2.5 μ=-2)
◆-the particle average diameter is 10 μ m; ■-the particle average diameter is 15 μ m;
▲-the particle average diameter is 20 μ m;

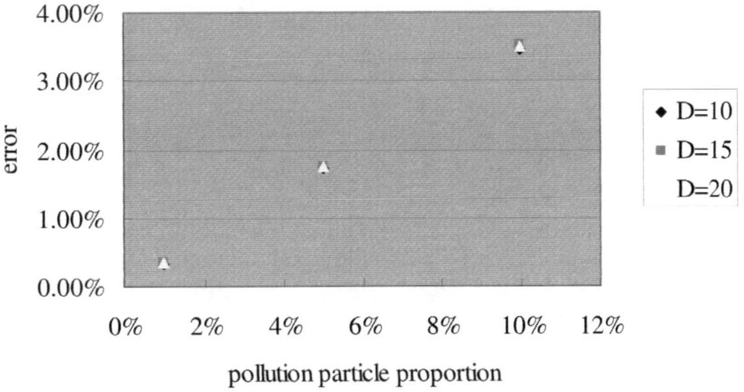

Fig 3 the relation of the error with the pollutionparticle proportion
(the particle refractive index is 1.57-0.56i, the diameter distribution is σ=2.5 μ=-2)
◆-the particle average diameter is 10 μ m; ■-the particle average diameter is 15 μ m;
▲-the particle average diameter is 20 μ m;

From the Fig 2 and Fig 3, for the non-dissipativity and dissipativity particles, the measurement error is increased along with the wingdows pollution degree is increased.
Regardless of the dissipation or non-dissipation of the particles, the conclusion can be gained from the table 1 to 2 and the Fig 1 to 2:
1) The window pollution have an impact for the measurement results. The measurement error has the

relationship with the particle size distribution, particle type and the concentration of particulate pollution, but the measurement error is independent of the particles average diameter.
2) For all type of particle, the measurement error is increased along with the windows pollution particles concèntraion increasing.
3) Under the condition of the particles same average diameter, if the particles diameter distribution is different, the measurement error has the different impact for the measurement results.
4) Under the condition of the same dianeter distribution of particles, if the particles average diameter is different, the measurement error is different. The measurement error increase along with the particle diameter increasing, but can be approximately considered the error as a constant .
5) Dissipation particles make the impact for the measurement results less than the non-dissipation particles.

3. The Solve Method of the Windows Pollution

Through the above analysis, regardless of the dissipation or non-dissipation particles, the measurement error is exist as long as the particle pollution exists. The measurement error is the relationship with the particle size distribution, the average diameter of the particle, particle type, and the concentration of pollution particulate. Through the analysis results, the measurement error caused by the same size distribution particles can be approximately seen as a constant. In most industrial environments, , if the production system works stably, the particle size distribution can be considered as unchanged, so the measurement error which made by pollution particles has the relationship with the concentration of particulate pollution.
A new multiple photoelectric sensor is designed, new multiple photoelectric sensor is shown in Figure3. The photosensitive measurement region is divided into five parts:
1) The centre of the sensor has a fixed diameter circle photosensitive region, used to measure the transmission and forward small angle scattering light intensity.
2) 2 to 5 fan-shaped photosensitive regions is four photoelectric sensors around the centre .Each photosensitive region is used to measure the scattering intensity within the definite angle range.

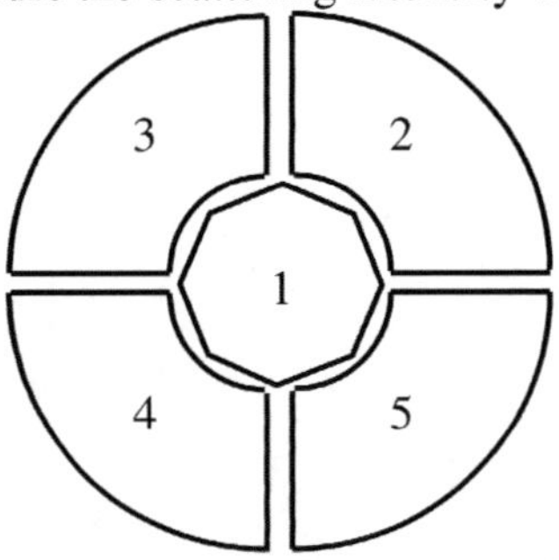

Figure 3 structure of the photoelectric sensor

Considering the movement's largest particle scattering angle α, the light scattering angle of pollution particle on the windows surface is greater than the α. The photosensitive region 1 measures the center and the forward small angle scattering light intensity. The photosensitive regions 2-5 measure the forward large angle scattering light intensity. Therefore, the scattering light intensity of the region 1 comprises the sports particle and the pollution particle scattering light intensity, the Scattering light intensity of region 2, 3, 4 and 5 is in the certain angle range of Scattering flux of the pollution particles.
For on-line particle concentration measurement system, the largest angle φ_1 of the flowing particle is determined. The scattering light intensity I_1 received by photoelectric sensor which angle is larger than φ_1 is produced by windows pollution particle, not including the the scattering light intensity in the pipe. Through I_1, the particle concentration C_{v1} of pollution particle can be calculated. The

pollution particles small angle scattering light I_1' can be calculated through Mie scattering theory. The small angle scattering light intensity I_0 is measured by the photoelectric sensor.

So

$$I_0' = I_0 - I_1'$$

I_0' is the actual scattering light intensity produced by the flowing particles in the pipe. So the actual flowing particles concentration C_{v0} can be calculated through the Mie scattering method, which can reduce or eliminate the measurement errors brought by the windows pollution particles.

4. Conclusion

The particle concentration measurement error is brought by the pollution particles. Numerical simulation shows that the measurement error has the relationship with the particle size distribution, the average diameter of the particle, particle type, and the concentration of particulate pollution. A novel photoelectricity sensor is developed in order to solve the windows pollution problem. Numerical simulation result based on the Mie theory shows that this method can amend the measurement errors by the windows pollution and improve the measurement accuracy.

REFERENCES

Wallis, G. B. (1969). One-dimensional Two-phase Flow, Mc-Graw Hill, Inc., New York, NY, USA, pp. 123-134.

Zhao Yanjun, et al. (2004). "The Study Of Light Scattering Average Value Method Online Monitoring Exhaust Dust Particle Concentration", Journal Of Electrical Engineering Of China, 11(24), pp. 217-221

Yu Zhengran et al. (1990). "Dust Testing Practical Skills", China Environmental Science Press, Beijing, China

Mu Ning (2004). "Continuous Measurement of Soot Concentration in the Exhaust Duct, " Southeast University

Cai Xiaoshu, et al. (2001). "The Study of Diagonosing the Running Condition of Pulverized Coal in Pipe, " Proceedings of the CSEE 21(7), pp. 83-86

Juliuse B.Gajewski, (1999). "Non-intrusive Solids Charge and Mass Flow Measurements with an electrostatic Flow Probe, " Journal of Electrostatics, 46, pp. 317-330

Li Qiping,(2001). "The Current Situation and Problem of Continuous On-line Testing Technology for Stack Dust in China," Sensor World, pp. 25-29

Su Mingxu, (2002). "The Ultrasonic Measurement Theory of the Particle Size Distribution and Concentration in the Gas-Solid Flow , " University of Shanghai Science and Technology

Ding Weiying,(2008). "Optical Monitoring System of the Soot Concentration in the Duct by the Scattering-Transmission method, " Hebei Polytechnic University

Journal of Physics: Conference Series **147** (2009) 012037 doi:10.1088/1742-6596/147/1/012037

Turbulent Shear Control with Oscillatory Bubble Injection

Hyun Jin Park, Yoshihiko Oishi, Yuji Tasaka, Yuichi Murai, Yasushi Takeda

Hokkaido University, Sapporo, JAPAN

hyun-jin@ring-me.eng.hokudai.ac.jp

Abstract. It is known that injecting bubbles into shear flow can reduce the frictional drag. This method has advantages in comparison to others in simplicity of installation and also in environment. The amount of drag reduction by bubbles depends on the void fraction provided in the boundary layer. It means, however, that certain power must be consumed to generate bubbles in water, worsening the total power-saving performance. We propose oscillatory bubble injection technique to improve the performance in this study. In order to prove this idea of new type of drag reduction, velocity vector field and shear stress profile in a horizontal channel flow are measured by ultrasonic velocity profiler (UVP) and shear stress transducer, respectively. We measure the gas-liquid interface from the UVP signal, as well. This compound measurement with different principles leads to deeper understanding of bubble-originated drag reduction phenomena, in particular for unsteady process of boundary layer alternation. At these experiments, the results have demonstrated that the intermittency promotes the drag reduction more than normal continuous injection for the same void fraction supplied.

1. INTRODUCTION

Frictional drag occupies 80 % of all drag and hence it affects significantly the fuel consumption rate of large ships like a tanker and a container carrier. McCormick (1973) found that the skin friction on the submerged body was reduced by injecting microbubbles around it. Madavan (1984) reported the efficiency of the microbubble drag reduction reaches as much as 80 %. After these reports, a number of researchers confirmed that this effect is feasible for practical use via various types of experiments. In comparison to other methodology for drag reduction, the use of bubbles has advantage in the both of simplicity and pollution free. By the latest experiments applied to a real ship, the net power efficiency of the ship was improved actually about 2 % (Kodama et al., 2006). However, this value is the result of subtraction of 5 % for generating bubbles, from the power-saving realized by drag reduction, 7 %. It obviously tells us that the power for generating bubbles worsens the net power-saving ratio, and we need to introduce a new designing concept to improve the net performance.

Reviewing a great deal of technical reports to date, how bubbles provide the frictional drag reduction can be classified into two types. One is quasi-steady effect that is characterized by change in fluid properties such as in average density and effective molecular/eddy viscosity (Murai et al., 2008). Another is unsteady effect in macroscopic sense, which takes place when two-phase boundary layer develops in streamwise direction before reaching fully developed state. The latter phenomenon is already deduced in several reports that deal with numerical simulations of bubbly shear flow. That is, the drag reduction occurs effectively when two phases exchanges their momentum before reaching the equilibrium state (Murai et al., 2006). On the other hand, experimental people quite often experience that local void fraction fluctuates naturally in time during drag-reducing situation. Even in such a case, they still evaluate the sensitivity of the drag reduction as only a function of time-average void fraction or mean gas flow rate whereas the physical mechanism is not explained by steady flow configuration.

Considering this background, we have investigated the effect of fluctuation in bubble flow rate. For this topic, we showed the waveform of local skin friction for a naturally provided fluctuation in void fraction in the previous report (Oishi et al., 2006). At that time, we mentioned the interesting behaviour of local gain factor, which was raised by amplifying the fluctuation in the local void fraction. This effect is obtained because the local drag reduction is non proportional to the local void fraction but is promoted exponentially. That is, a wavy supply of void fraction enhances the drag reduction more than constant one for the same gas flow rate. This can be called "non-linear effect" to the void fraction. On the other hand, we measured the liquid flow field around a single bubble with PTV in order to find out the effective region of the bubble in the streamwise direction (Murai et al., 2007). With this measurement, we concluded that the local skin friction was affected longer than the size of bubble, indicating the local unsteady effect of bubble-originated drag reduction. This effect can be called as "phase-lag effect" which also amplifies the drag reduction. As one of the method to combine these two effects; the non-linear effect and the phase-lag effect, we propose a new type of drag reduction by bubbles, that is, oscillatory bubble injection. This method utilizes the artificially provided oscillation in gas flow rate, but not the naturally provided voidage waves. In comparison to bubble injection at constant gas flow rate, i.e. continuous bubble injection, both the two effects are utilized in the case of oscillatory bubble injection, and thereby the net performance should be improved. In the present study, three types of bubble injection methods are designed and compared with changing the injection frequency.

2. EXPERMANTAL METHODS

2.1 Experimental Setup

The schematic diagram of the experimental facility is shown in Figure 1. The test section is a horizontal rectangular channel made of transparent acrylic resin and is 40 mm in height ($H = 2h$), 160 mm in width (W) and 6000 mm in length, respectively. Silicone oil is used to avoid the contamination effect on bubble interface, whose property is listed in Table 1. The silicon oil circulates in the channel, and the bubbles are mixed by capillary needle type of bubble injector. The bubbles are eliminated in the downstream region, by swirling the fluid in a bubble removable tank before returning to the inlet of the channel.

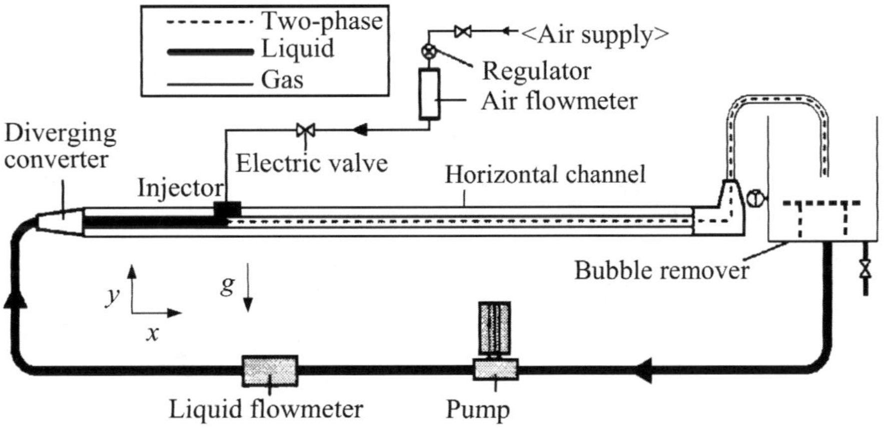

Fig. 1 Schematic diagram of experimental facility

Journal of Physics: Conference Series **147** (2009) 012037 doi:10.1088/1742-6596/147/1/012037

Table 1. Characteristics of silicone oil (KF96-10cs)

Temperature (T)	28	°C
Density (ρ)	932	Kg/m^3
Kinematic viscosity (ν)	7.7×10^{-6}	m^2/s
Surface tension (γ)	19.9×10^{-3}	N/m
Sound velocity (a_0)	970	m/s

Figure 2 and Figure 3 shows the detail schematic diagram of test section and the experimental system for control of bubble injection and simultaneous measurements. A bubble injector and a shear transducer are mounted on the upper wall of the channel at $x/H = 43.75$ and $x/H = 65$ from the channel inlet, respectively. The injector is connected with an electromagnetic valve so that the arbitrary fluctuation in gas flow rate is realized with a function generator. For measuring the internal flow structure with UVP, two ultrasound transducers are set on the bottom wall of the channel at $x/H = 68.75$ at different angles, ±8 degree. Two ultrasound absorber boards are mounted on the upper wall and the bottom wall of the channel near the measurement line of UVP. It absorbs ultrasound and prevents multi-reflected echo of ultrasound beam. Another function generator synchronizes two UVPs and a data logger. These measurement machines start to record the different data simultaneously, which are velocity, echo, shear stress and valve operation.

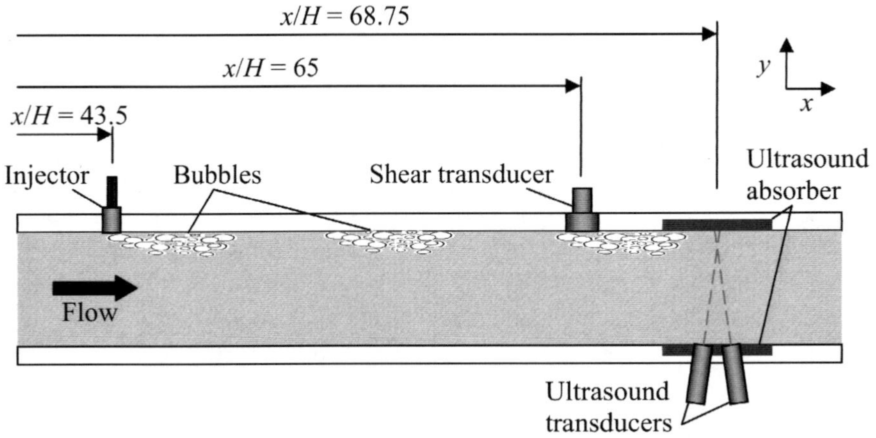

Fig. 2 Measurement instruments in test section of channel

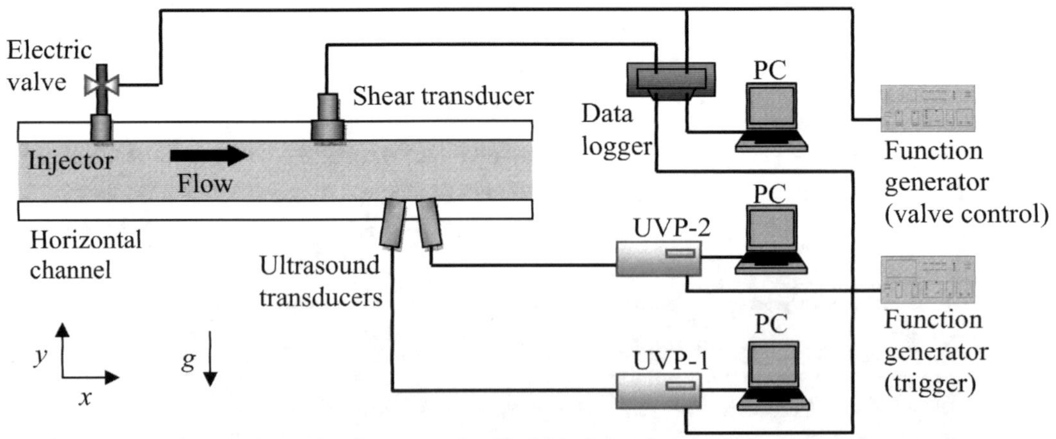

Fig. 3 Experimental system for control of bubble injection and simultaneous measurements

Figure 4 shows the definition of valve control parameters. The measurement conditions are shown in Table 2 and Table 3. The variable parameters in this experiment are, mean void fraction (α', Eq.(1); Q_g and Q_l are flow rates of gas and liquid phase), time of bubble injection pulse (t_{in}), and valve frequency (f). Local void fraction (α, Eq.(2)) is 1.5×10^{-2} and Re number defined by Eq.(3) is fixed at 2200. Here we treat (U) and ρ as invariable parameters because of sufficiently small value in bulk void fraction.

$$\alpha' = \frac{f\, t_{in}\, Q_g}{Q_l + f\, t_{in}\, Q_g} \tag{1}$$

$$\alpha = \frac{Q_g}{Q_l + Q_g} \tag{2}$$

$$Re = \frac{Uh}{\nu} \tag{3}$$

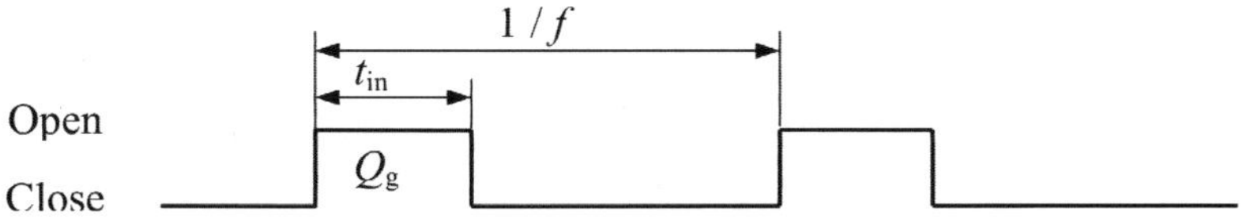

Fig. 4 Definition of valve operation for realizing intermittent injection

Table 2. Measurement conditions

Basic frequency of UVP	4	MHz
Particle density	930	Kg/m^3
Particle diameter	180	μm
Time resolution	10	ms
Velocity resolution	17	mm/s
Distance resolution	0.5	mm
Shear stress resolution	0.05	Pa

Table 3. Experimental conditions

Reynolds number (Re)	2200	-
Local void fraction (α)	1.5×10^{-2}	-
Mean void fraction (α')	$0 \sim 1.5 \times 10^{-2}$	-
Injection frequency (f)	0.2 – 0.5	Hz
Injection time (t_{in})	0.5 -1.5	s
Bubble diameter	5-50	mm

2.2 Detection of Gas-Liquid Interface

Presence of bubbles in the channel flow can be detected from the echo signal of the pulsed ultrasound. Murakawa (2008) used two types ultrasound transducers, 2 MHz and 8 MHz, to distinguish bubbles and solid particles suspended in liquid. Such a combination of different ultrasounds provides additional information of fluid flow in particular for multiphase fluids. In the case that the bubble size is sufficiently larger than the basic wavelength of ultrasound, just monitoring

Journal of Physics: Conference Series **147** (2009) 012037 doi:10.1088/1742-6596/147/1/012037

the echo intensity of monotonic ultrasound gives us the interfacial information of bubbles. Hence, we use a single ultrasound transducer for detecting of the gas-liquid interface in this study.

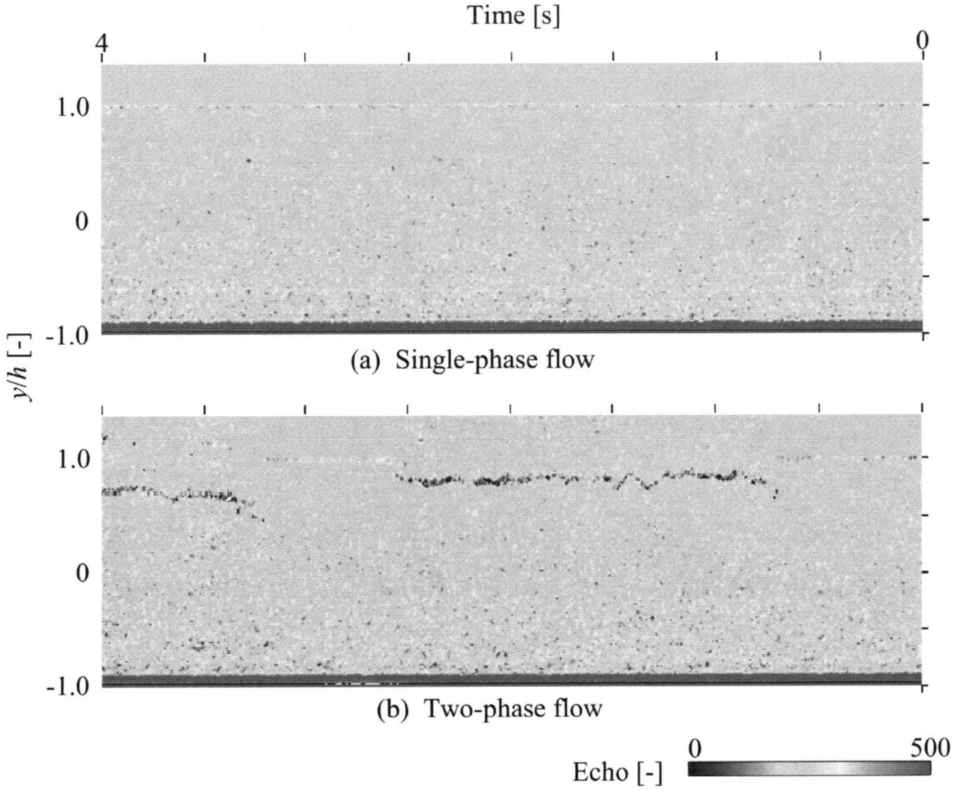

Fig. 5 Echo amplitude obtained by the UVP

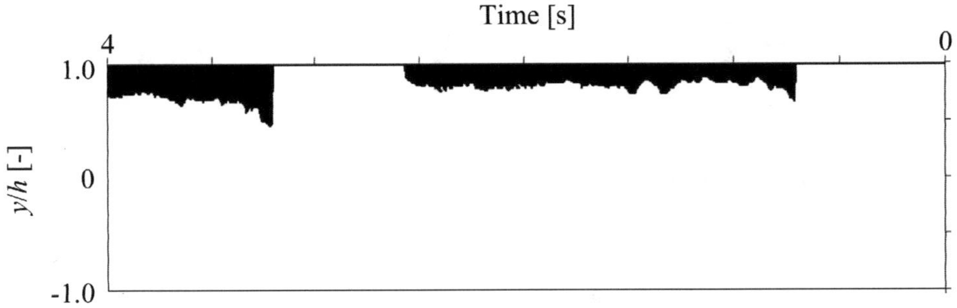

Fig. 6 The gas-liquid interface obtained by echo amplitude

Figure 5 shows the echo amplitude of single-phase flow and two-phase flow obtained by the current set of UVP. The green region in the figure indicates that the ultrasound transducer receives adequate echo intensity for tracer particles seeded in the flow. As seen in the red bottom layer in each figure, the transducer cannot receive significant echo near the bottom wall, $0 \leq y/h \leq 0.15$, because of near field characteristics of ultrasound pulse close to the head of the transducer. This problem, however, does not affect the detection of bubbles since bubbles exist in the upper half of the channel. In single-phase flow, it is possible to find the upper wall with the echo information by finding the maximum intensity at that place, $y/h = 1.0$. There is a single peak of the echo found in the vertical direction of the channel since ultrasound-absorbing boards are inserted. Without these boards, secondary peak takes place within the channel height due to multiple reflection of the pulse between the two parallel walls of the channel. In the case of two-phase flow, ultrasound pulse reflects on the gas-liquid interface, like a mirror, so the echo at the interface has the value higher or lower than that in surrounding region. The local high value in the echo intensity means that the ultrasound pulse reflects normal to the interface and returns straight to the transducer. The local low value comes up for oblique reflection of the pulse at the interface, with which the pulse does not return to the transducer. For these reasons, the outline of the bubble is identified as shown in Fig. 5(b), in which the interface is detected as a thin colourful layer in the channel.

Once the pulse reflects on the interface, it does not reach the upper wall to get the information deeper than the position of the interface. Although this is a demerit of the present technique, this full reflection of the pulse at the interface prevents from multiple reflection between the interface and the upper wall. Therefore, the position of the interface is uniquely captured. Figure 5 shows the gas-liquid interface measured by the present signal processing.

2.3 Two-Dimensional Velocity Vector Field

Utilizing two UVP systems with different measurement angles, velocity vector information can be obtained. For example, the same technique was reported by Taishi (2002). Figure 7 shows the relationship between the velocity components measured by UVPs and velocity vector of flow. In our system, two ultrasound transducers having same angle with opposite direction, $\theta = \pm 8°$, are used to get the velocity vector. The velocity vector (**u**) is then calculated by Eq. (4), where ξ_i is the velocity component along the ultrasound beam direction and (u, v) are the streamwise and the vertical velocity components of flow. Here, u_1 and u_2 are the velocity components which two UVP systems output as streamwise velocity.

$$\mathbf{u} = (u, v) = \left(\frac{\xi_1 - \xi_2}{2\sin\theta}, \frac{\xi_1 + \xi_2}{2\cos\theta} \right) = \left(\frac{u_1 - u_2}{2}, \frac{u_1 + u_2}{2} \tan\theta \right) \tag{4}$$

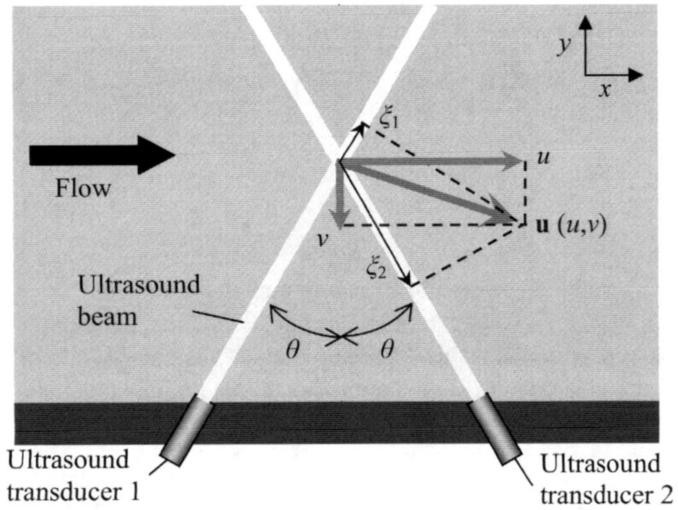

Fig. 7 Relationship among measured velocities and flow velocity vector

Figure 8 represents a sample of the velocity vector field inside the channel that contains bubbles, provided that the number of vectors measured is five times as that shown in the figure. Integrating the vector information in time, we can evaluate the turbulent shear characteristics of the flow and the role of bubbles, which will be mentioned later.

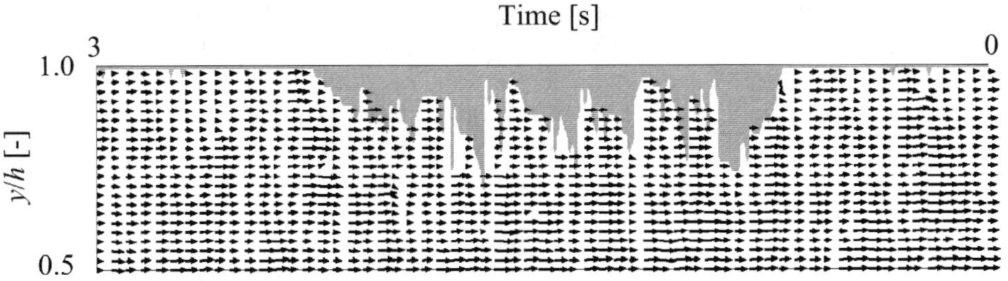

Fig. 8 Velocity vector field of liquid in the channel
(Gray region stands for gas bubble which is also detected by ultrasound pulse.)

3. RESULTS AND DISCUSSIONS

3.1 Bubble Shape by Oscillatory Injection

Figure 9(a) is a sample of the bubble shape measured by the echo information, which is averaged by phase of the valve operation (see (b)). At all the conditions, the time lag between the valve operation and the bubble measured is constantly obtained at given flow rate of liquid. The drifting speed of the bubble in the streamwise direction takes constant value, which is calculated by the time lag and the distance between the injector and the ultrasound transducer.

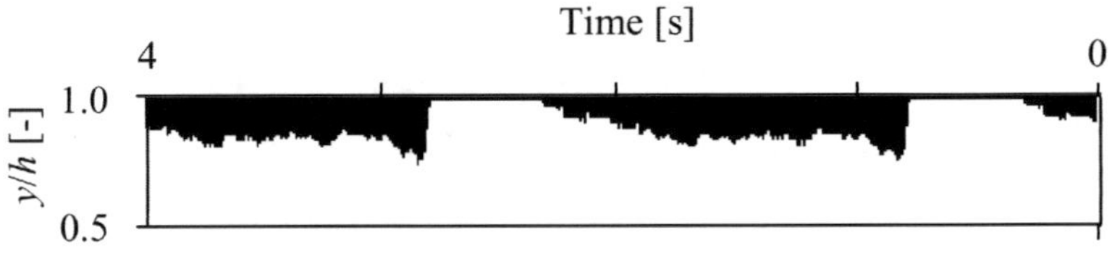

(a) Bubble shape measured by ultrasonic echo information

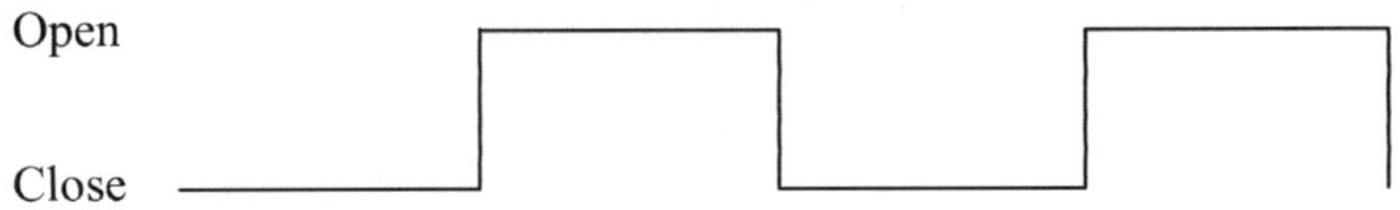

(b) valve operation for injecting bubbles intermittently

Fig. 9 Phase-averaged bubble shape and valve operation (f = 0.5 Hz, t_{in} = 1.0 s)

Figure 10 shows the measured relation between the injection time (the period of open status of the valve) and the passing time (the period of bubbles detected). As the linear trend seen in the result, the passing time or the length of individual bubble can be regularly controlled by the injection time as we designed so. When the valve operation has the highest frequency and the longest injection time, f = 0.5 Hz, t_{in} = 1.5 s, the bubbles are continuously injected since the passing time is longer than period of the bubble injection. It is noted that the individual bubble shape changes randomly within a certain range of the fluctuation because Weber number of the bubble is high enough to allow the turbulent deformation of the interface. We are going to take phase-statistic values to focus on the effect of the oscillatory bubble injection rather than investigating the local behaviour close to the interface.

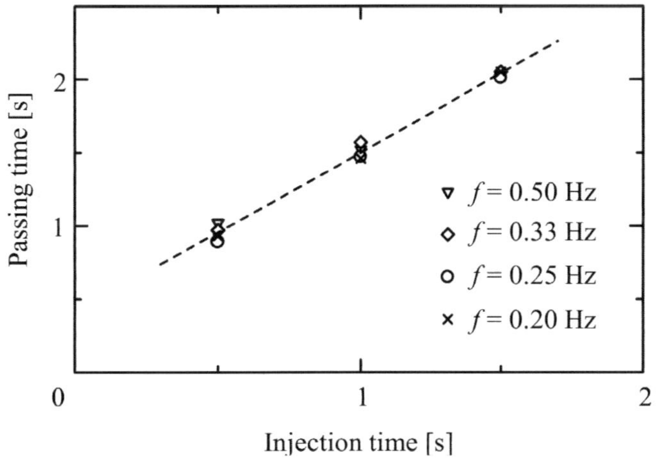

Fig. 10 Relation between the passing time and the injection time

3.2 Shear Stress on Upper Wall

The time-averaged shear stress at the upper wall is shown in Figure 11. The ordinate is normalized by the shear stress of the single-phase condition. With the data, we can confirm that the shear stress of oscillatory bubble injection provides the value smaller than that of continuous injection at the same void fraction. Namely, the drag reduction is promoted by injecting bubbles in the flow intermittently. The dotted lines in the graph are linearly fitted trends of the data for the two different types of the bubble injection. In the case of continuous injection, the drag is rather increasing by injecting bubbles at low void fraction. To obtain the drag reduction, the continuous injection requires to inject bubbles more than 0.5% in void fraction. To the contrary, the oscillatory injection immediately provides drag reduction at zero-limit of void fraction. Furthermore, the drag reduction by the oscillatory injection always has higher performance than that of continuous injection at the range tested.

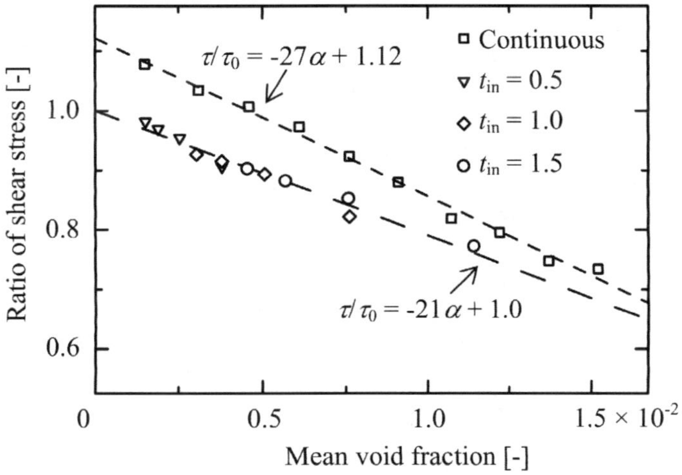

Fig. 11 Ratio of shear stress at the upper wall for two different bubble injections

A gain factor of skin friction reduction is calculated from the measured wall shear stresses, and its trend against mean void fraction is shown in Figure 12. The gain factor is defined by

$$G = \frac{1}{\alpha}\left(1 - \frac{\tau}{\tau_0}\right) \quad , \tag{5}$$

where α is the mean void fraction inside the channel. τ and τ_0 are the wall shear stress of two-phase and single-phase conditions, respectively. The gain factor, hence, indicates the magnitude of the drag effect per unit void fraction. This is to say, the sensitivity of drag reduction by bubbles. The gain factor takes a positive value when the drag reduction realizes. The gain factor at unity ($G = 1$) means that 1% of void fraction provides 1% of drag reduction. As seen in the results, the gain factor for the continuous injection reaches around 15 at enough void fraction supplied in the flow. This means that 1% of drag reduction realizes 15% of drag reduction. However, this high performance is not kept for low void fraction, by decreasing the gain rapidly to negative. The oscillatory injection overcomes this problem. The gain factor of the oscillatory injection is kept around 20 in considerably wide range of void fraction, $\alpha > 0.2 \times 10^{-2}$.

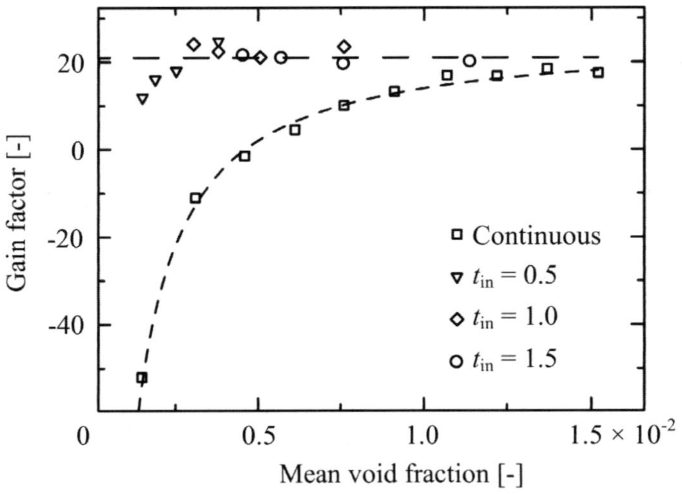

Fig. 12 Gain factor of skin friction reduction for two types of bubble injection

In order to discuss how the drag reduction is effectively enhanced by the oscillatory bubble injection, the waveform of the local skin friction around a single bubble is investigated. Fig. 13 shows the conditional averaged waveform of the shear stress at the upper wall; each the dashed line represent the average of the shear stress in the single-phase flow and continuous injection and the gray area represents the area existing bubble. The time lag between the waveform and the bubble area is corrected with the drifting speed of bubble. According to the results, before the bubble passes the position of the shear transducer, the shear stress is a little increased (see the left hand side of each bubble in the figure). This is explained by two-phase interaction in bubble-front region. When it enters the bubble existing area, the stress is decreased rapidly in a short time. The drag reduction achieves at the twice of the continuous injection. After that, the drag reduction is kept to be the same as the continuous injection when a long bubble is passing. Namely, the large drag reduction occurring in the first short period of bubble-entrance provides the improvement of the mean drag reduction in the case of oscillatory injection. In the later half of the bubble, the local shear stress restores smoothly to get the value of single-phase flow. This region has less drag reduction in comparison to the continuous injection, however, the volume of the gas is a little as well (see the shape of the bubble in Fig. 9). Thereby, the gain factor is not worsened by the phenomenon occurring in the later half of the bubble. From the compassion among these three cases, it is clear that the bubble with a shorter period (or a shorter length) enhances the drag reduction per the same volume fraction. It is worth noting that further shortening the bubble length will lead to the drag reduction performance lower. Our group investigated on the dependence on bubble length in the past and concluded that there was an optimum bubble size to maximize the drag reduction (Murai et al., 2007). This is so because small bubbles, which are comparable to the boundary layer thickness, will activate the momentum transfer, resulting in increment of drag. Therefore, the improved effect by the oscillatory bubble injection would be unavailable for the bubble injection at too high frequency, which exceeds the speed of bubble divided by the boundary layer thickness.

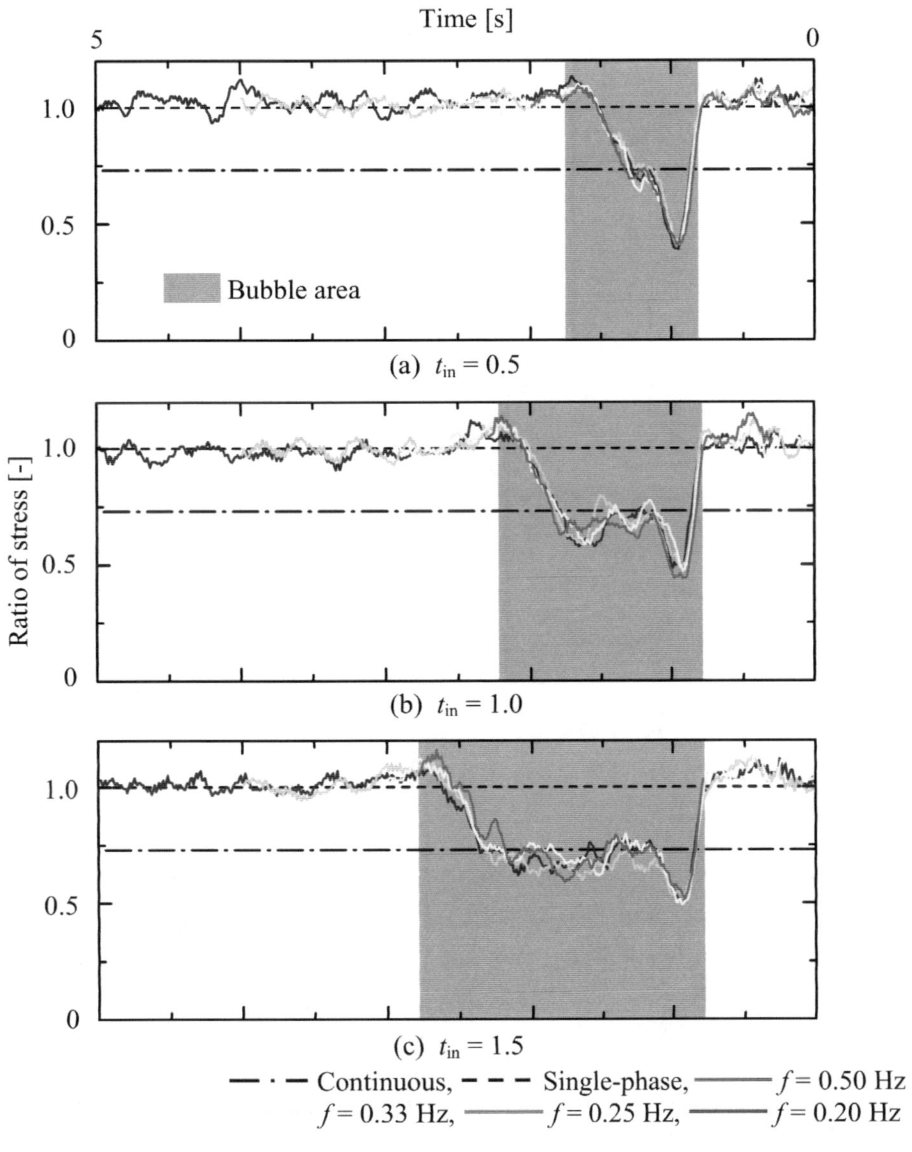

Fig. 13 Waveforms of skin friction caused by intermittent bubble injection

Journal of Physics: Conference Series **147** (2009) 012037 doi:10.1088/1742-6596/147/1/012037

3.3 Reynolds Shear Stress

In order to deepen the understanding on the effect of the oscillatory bubble injection, Reynolds shear stress is assessed from the UVP data.

Figure 14 shows the time averaged Reynolds shear stress for five different flow conditions. The results indicate that the flow is affected by bubbles at the upper half of the channel while the profiles are kept the same as that of single phase flow in the lower half, $y/h < 0$. Here we note that the Reynolds shear stress is not perfectly measured by two synchronized UVPs because of the limited spatial resolution of UVP. That is, the turbulence having the scale smaller than the resolution is uncounted in the data. Therefore, the measured value gets always smaller than the true one. However, we can still evaluate the measured value in relative sense since the energy cascade is known to occur obeying the universal law. The present results have shown that the Reynolds shear stress is reduced largest by the continuous injection, but smallest by the oscillatory injection at the shortest injection period. Since the Reynolds shear stress is defined only inside the liquid phase, this trend does not immediately correspond to the drag reduction performance obtained. Moreover, the time-averaged Reynolds shear stress cannot validly express the unsteady effect of the drag reduction, which is the key phenomenon of the oscillatory bubble injection.

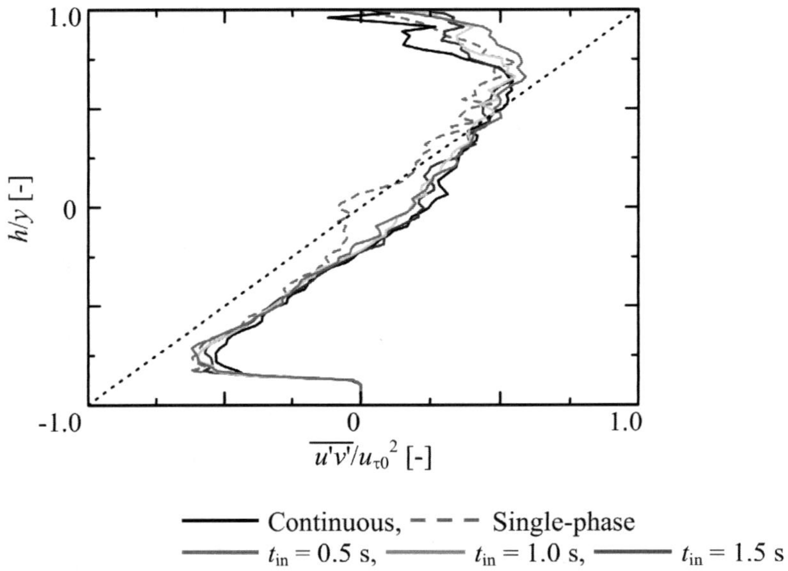

Fig. 14 Reynolds shear stress profiles for different bubble injections ($f = 0.33$ Hz)

Figure 15 shows the local Reynolds shear stress distribution around the bubble. This figure represents the moving-averaged Reynolds shear stress for a single shot of bubble, which is non-dimensionalized by friction velocity (u_τ). Since UVP cannot measure the velocity above the bubble interface, no information is obtained there. Such a region is also painted black. The time series of the Reynolds shear stress includes the one created by the original shear-induced turbulence in liquid flow. The result clearly shows that the local shear stress is reduced in the vicinity of bubble interface. This matches the general knowledge, i.e. the gas-liquid interface hardly keeps turbulence as well as the shear stress. In addition, the Reynolds shear stress in the upper half of the channel is broken to be wavy distribution beneath the bubble. This is a response of shear turbulence against suddenly unloaded stress at the upper wall. Hence it is explained that the effect of the oscillatory bubble injection is created with this unsteady response of liquid flow, which occurs periodically with passing of bubbles.

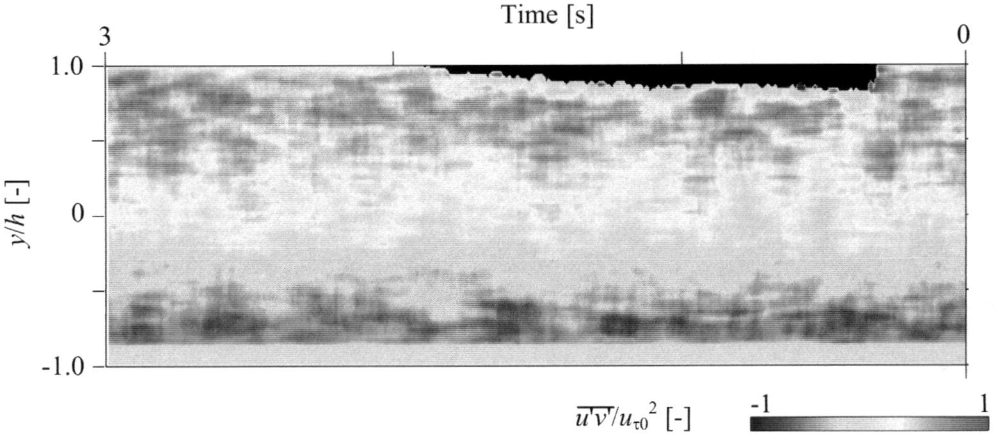

Fig. 15 Time series of Reynolds shear stress around bubble (f = 0.33 Hz, t_{in} = 1.0 s)

4. CONCLUSIONS

The effect of oscillatory bubble injection in frictional drag reduction is investigated experimentally, using a controlled bubble injection system. Information of liquid-area and the gas-liquid interface are successfully obtained from the echo amplitude of ultrasound. Hence, simultaneous measurement for the velocity and the phase distributions is realized by signal processing of ultrasound pulse. In addition, the wall shear stress and the Reynolds shear stress distributions are also successfully measured by compounding the two ultrasonic transducers and one shear transducer under synchronized control. With the present measurement, it is confirmed that the oscillatory bubble injection enhances the drag reduction significantly, and can save the power to generate bubble or can raise the gain factor of drag reduction. This improvement is realized by unsteady response of the shear flow against bubbles that pass intermittently. This effect is totally different from the one caused by the conventional continuous bubble injection. The oscillatory bubble injection amplifies the mean drag reduction by the repeated effect of unsteady drag reduction.

REFERENCES

Kodama, Y. et al. (2006). "A full scale microbubble experiment using a cement carrier," European Drag Reduction and Flow Control Meeting, POL1_2

Madavan, N.K. et al. (1984). "Reduction of turbulent skin friction by microbubbles," Phys. Fluids, Vol. 27, pp.356-363

McCormick, M.E. and Bhattacharyya, R. (1973). "Drag reduction of a submersible hull by electrolysis," Naval Engineers J., Vol. 85, pp.11-16

Murakawa, H. et al. (2008). "Application of ultrasonic multi-wave method for two-phase bubbly and slug flows," Flow Measurement and Instrumentation, Vol. 19, pp. 205-213.

Murai, Y. et al. (2006). "Turbulent shear stress profiles in a bubbly channel flow assessed by particle tracking velocimetry," Exp. Fluids, Vol. 41, pp.343-352.

Murai, Y. et al. (2006). "Turbulent Bubbly Channel Flow Investigated by Ultrasound Velocity Profiler," JSME Journal of Fluid Science and Technology, Vol.1, pp.12-23.

Murai, Y. et al. (2007). "Skin Friction Reduction by Large Air Bubbles in a Horizontal Channel Flow," Int. J. Multiphase Flow, Vol. 33, pp.147-163

Murai, Y. at al. (2008). "Frictional drag reduction in bubbly Couette-Taylor flow," Phys. Fluids, 10-3 (Available online) DOI: 10.1063/1.2884471.

Oishi, Y. et al. (2006). "Waveforms of local frictional drag modified by bubbles," European Drag Reduction and Flow Control Meeting, POL1_4

Taishi, T. et al. (2002). "Effect of the measurement volume in turbulent pipe flow measurement by the ultrasonic velocity profile method (mean velocity profile and Reynolds stress measurement)," Exp. Fluids, Vol. 32, pp.188-196.

Journal of Physics: Conference Series **147** (2009) 012038 doi:10.1088/1742-6596/147/1/012038

Flow Visualization and Characteristics of Vertical Gas-Liquid Bubbly Flow Around a Rectangular Cylinder (Bubble Size Effect)

A Voutsinas [1], T Shakouchi [1], K Tsujimoto [1], T Ando [1]

[1] Graduate School of Engineering, Mie University, Kurimamachiya-cho 1577, Tsu-shi, Mie, 514-8507, JAPAN

shako@mach.mie-u.ac.jp

Abstract. The present study deals with the effect of the bubble size, from small bubble scale to normal scale (d_b=0.25~2.6 mm), on the flow passing through a rectangular cylinder in an upward gas-liquid bubbly flow. Extensive visualization experiments are conducted and a digital camera and a high-speed camera analyzed the flow, while PIV analysis by the volume cross-correlation method is conducted to observe the differences in the flow pattern. In order to further understand the effect of bubble size, the pressure distribution along the pipe and the cylinder surface are measured. From the results taken, the drag force is calculated and compared to the case of single phase-flow. Furthermore, the fluctuation phenomena generating from the Karman vortex street downstream the cylinder are investigated, and how the intensity and frequency are affected by the bubble size and gas fraction is presented. The experiments are conducted under two different Reynolds number *Re*, and volumetric gas fraction ranging from α_v=0~5%, giving valuable information regarding the changes that occur due to bubble size differences and the relation it has with volumetric gas fraction.

1. INTRODUCTION

The emission of Karman vortex street of two vortex rows with opposite circulations for a gas-liquid two-phase bubbly flow past an obstacle(s), which is encountered in many industrial components, has been widely investigated and several studies exist.

These studies found in bibliography, are conducted under various parameters such as Reynolds number, gas fraction effect or cylinder's geometry effect (Dian, 2002; Hara et al., 1982; Hulin et al., 1982). In the recent years advanced personal computers are used and several models are suggested in order to conduct numerical analysis of the flow mechanism around an obstacle(s) (Shakouchi et al., 2004; Sugiyama et al., 1999). From these results, valuable information and deeper understanding on the flow pattern and mechanism has been achieved through the years.

The vibration phenomenon of a circular cylinder in cross flow was studied extensively by Hara et al. (1982), while the two-phase flow-induced vibration was introduced by Pettigrew (1994) and the characteristics of the unsteady behavior of two-phase flow is mentioned on Yokosawa et al. (1986) work.

The vortex emission behind bluff bodies, wake and drag of were studied and presented by Hulin et al. (1982), Roshko (1955) and Shakouchi et al. (2004). The results presented refer to various conditions of Reynolds number and void fraction. Furthermore, extensive numerical analysis on flow characteristics around cylinder, is presented by authors like Inoue et al. (1986) while numerical analysis has been conducted with different methods like Uchiyama et al. (1996) or Sugiyama et al. (1999).

From the above-mentioned studies, some include photographic and visualization material showing the flow pattern of the generated flow.

As seen, most if not all the work done so far, do not mention the effect of bubble size. Separate studies for micro-bubbly flows, or agitated bubbly with cap form bubble flows can be seen in bibliography. A study dealing with the same flow conditions but under different bubble sizes from small scale to normal scale d_b=0.27~2.6 mm however, have not been conducted, at least to the authors knowledge.

In the present study, the effect of the bubble size mean diameter under constant flow conditions and different gas fraction values are investigated. The differences on flow characteristics; such as flow pattern and vortex shedding frequency, with special attendance on visualization experiments with high speed camera and PIV analysis are introduced. The results will help to understand and explain how studies from individual authors, even conducted under similar conditions have shown different results. Moreover, the visualization material shows the undeniable changes occurring due to bubble size difference to the flow system. In addition, it is well known that in the industrial field, depending the application used or seen, a variation of bubble sizes into the gas-liquid system exists. By this, it will help understand and explain several situations and conditions that were not made clear so far.

2. EXPERIMENTAL CONDITIONS

2.1 Appartus

Figure 1 shows the experimental set-up. A rectangular test section was used and made of a transparent acrylic resin channel in order to conduct visualization experiments. The device has L=2400 mm in length, D=90.0 mm in width and H=45.0 mm in depth. A test cylinder of d=30.0 mm made of brass is mounted at the center of the channel perpendicular to the two-phase flow rigidly. The distance from the bottom of the test section to the front side of the test cylinder is 1520 mm and the coordinates are shown.

A pump gives the required amount of water flow rate, where water was storage in a tank. The water temperature was maintained constant at T=20 °C. The mixture of water and gas were separated at the top pf the device and water was re-circulated at the water tank. The water and gas flow rates were measured by float type flow meters. For the experiments, CO_2 gas was used as a driving gas because of the controllability of bubble size and to

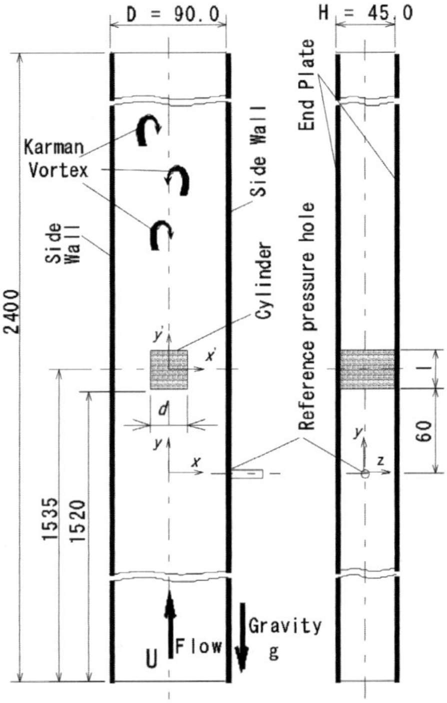

Fig.1 Test section

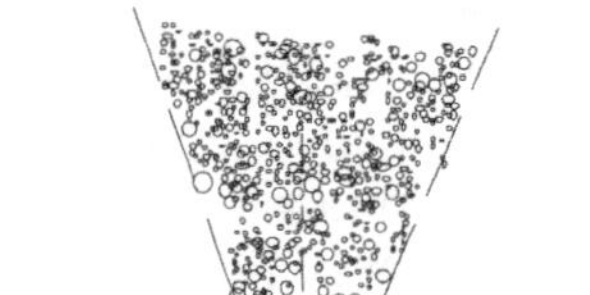

Fig.2 Bubble generation orifice nozzle

reduce collision phenomena before the test section and was provided by a pressurized tank as shown.

The bubble generator (Shakouchi et al., 2006) in the form of a simple orifice nozzle is given in Fig.2. The pipe diameter is of 5.0 mm and the orifice diameter is 2.0 mm, which gives a contraction area ratio of A_r=0.16. The shear stress generating due to the orifice jet effect was used for breaking up the induced gas from the two stainless steel pipes of diameter 1.0 mm, attached on the side of the orifice head as shown. The gas is induced at the sides and is entrapped into the jet, while changing the water flow rate of the nozzle under a given gas flow rate, manages to control the generated bubble size, due to the shear stress force change occurring. The flow rates are measured by float type flow rates like before.

2.2 Experimental Conditions and Procedure

In the following the condition under which experiments were conducted are presented. Two different Reynolds numbers were used at $Re=0.5\times 10^4$ and 1.0×10^4 of which Reynolds number is defined as follows:

$$Re=U_{wm}d/\nu \quad (1)$$

where, U_{wm} is the water mean velocity in the flow passage and the cylinder diameter is taken as a characteristic length.

The experiment was subject to the bubbly flow regime. Although there is some slip velocity between phases, for simplicity and convenience we assume that the relative velocity between phases in the axial direction can be neglected for the time being. Thus, we apply the homogeneous model and the void fraction which express the volume ratio of the gas phase in the flow, was varied from α_v=0~5% in the experiments and is given form the following expression:

$$\alpha_v=\frac{Q_g}{Q_g+Q_w}\times 100 \quad (2)$$

where Q_g and Q_w are the volume flow rates of the gas and the liquid phases.

It was mentioned above that the bubble size control was conducted by varying the volumetric water flow rate exiting from the orifice nozzle. Adjusting it from Q_w'=0~3.5 l/min per single nozzle allowed a mean bubble size variation of d_b=0.27~2.6 mm, while the gas value range was Q_g=0~2.4 l/min in total. Below in Table 1, the relation between gas fraction, volumetric water flow and mean bubble size is given.

Pressure holes of 0.7 mm diameter are drilled along the sidewall of the test section and a water manometer measures the pressure in respect to the reference point. In addition, ten pressure holes of same diameter were drilled at the front and rear side of the test cylinder and the pressure difference from the reference point, for different bubble size was also measured.

Table 1 Mean bubble size variation under different flow conditions

Water jet flow rate (per single water nozzle) Q_w' [l/min]	Mean bubble size d_b [mm]	
	α_v =1.5%	α_v =5%
0	2.6	2.6
1	0.7	0.9
3.5	0.27	0.27

A semi-conductor type pressure transducer measures the pressure fluctuation downstream at y/d=14. From the pressure transducer results, through the oscilloscope (Kanomax, TDS 2002 series), FFT analysis was conducted and the fluctuation amplitude was calculated. The pressure fluctuation was recorded for continues 2.5 s and 2.5 KHz. For each flow condition, data were recorded for 30 times and the mean value was considered and presented in the present study.

It is noted at this point that the experiments were carried out after the CO_2-water mix was re-circulated enough time within the loop system in order to bring it in a saturated state. This was done to make sure that the generated bubbles do not dilute within the system, which would cause the decrease of bubble size and reduction of gas fraction.

2.3 Visualization Experiments

A 5 mm slit of light provided by three halogen lights was used for the flow passage illumination, while an extra light from the back of the flow passage was used in some cases. The bubble size distribution was measured by the photographic images taken by a NIKKON D70 at 3600 FPS. For each bubble size cases, more than five hundred bubbles from several different photographs were measured. Since there were cases that the bubbles were not round but more spherical, the axial and vertical diameter was accounted and divided by two ($d_b=d_x+d_y/2$).

The overall flow pattern under different flow conditions was observed and recorded by a digital video camera with. Plastic beads of which density is approximately the same as that of water and mean diameter of 0.3 mm were used as tracers for single-phase flow (DIAION HP21). For the case of CO2-water two-phase flow the bubbles were used as tracers since they reflect the light and give good traceability.

A high speed photographing system (CCD camera, PHOTRON, FASTCAM M3PIV3/100K), with a Micro-NIKKOR 105mm lens was used for the image processing, with the shutter speed set at 1/2000 s and frame rate at 500FPS. The velocity vectors were extracted by PIV using the volume cross-correlation method to reproduce the complicated two-phase flow pattern and give the differences occurring due to bubble size effect. For the PIV analysis, two patterns were extracted. In the first, real time of 0.2s was extracted and the local instantaneous bubble velocity distribution was extracted in order to see the flow deviation. In the second, a total of more than 7s was taken, which allowed obtaining the mean velocity distribution.

3. RESULTS and DISCUSSIONS

3.1 Pressure Distribution and Drag Force

The pressure distribution along the sidewall vertical axis is measured under constant flow conditions and varying the bubble mean size. Results for $Re=0.5\times 10^4$ and $\alpha_v=1.5\%$ is illustrated in Fig.3. The vertical axis denotes the pressure coefficient which is a non dimensional value given from the equation shown below:

$$Cp=2(P_i-P_0-P_e)/(\rho_t U_{tm}^2) \quad (3)$$

where, ρ_t is the two-phase flow density and U_{tm} is the mean velocity of the two-phase flow upstream. P_i expresses the pressure value at a given point and P_0 is the pressure measured at the reference point and P_e the vertical elevation pressure.

The cylinder is illustrated for better understanding and is located between y/d=2~3, while the single-phase pressure distribution is also illustrated. The pressure decreases due to the friction effect and at the front corner of the cylinder an even more rapid decrease begins due to the increase of velocity. After it passes the narrow region, a further slight decrease continues from the generation of large-scale vortices that eventually generate the Karman vortex street. A slight recovery is observed right after and finally it follows its decrease due to frictional factors again. The pressure recovery differs clearly depending on the bubble size. For d_b=2.6 mm, the pressure recovery is the highest. That is due to the fact that larger bubbles tend to flow towards the center of the flow passage and less bubbles concentrate to the sidewall. On the other hand for smaller bubbles, d_b=0.9mm, where the size is small enough, bubbles flow throughout the whole area and near the sidewall, which leads to these kind of results. However, it is noted that the pressure gradient is almost the same for different bubble sizes, and the gradients is affected more by flow conditions rather bubble size, as shown in Fig.4.

Again, the pressure distribution pattern is the same like the previous condition. Naturally the pressure values obtained vary since the flow conditions are different. It is shown that the bubble size effect becomes smaller than for lower Reynolds number, which leads us to a first conclusion that the bubble size affects the pressure values more, under low velocity conditions.

Although it was considered that small bubbles would have the minimum pressure loss, it was found that the large bubbles that flow into the middle region show it instead. In addition, the uniform size bubbles of d_b=0.9 or 1.2 mm, had the maximum pressure decrease along the whole measured area.

After the measurements at the sidewall, the pressure distribution was measured at the front and rear surface of the cylinder itself, in order to investigate if the bubble size affects it. Since measurements on a two-phase flow are very difficult and unstable, each point is measured several times and the mean value is given on the graphs.

Figures 5 (a), (b) illustrate the effect of the bubble size on the cylinders front and rear surface pressure distribution for the same flow conditions mentioned above. The vertical axis denotes the non dimensional pressure coefficient and is given from the following equation:

$$C_p'=2(P_i-P_0)/(\rho_l U_{tm}^2) \quad (4)$$

It is expressed in the same way as above (Eq.3) with the difference that the elevation effect is not considered in this case. The pressure has the highest value at the center of the front side of the cylinder, Fig.5 (a), and decreases gradually in the direction of both sides. As

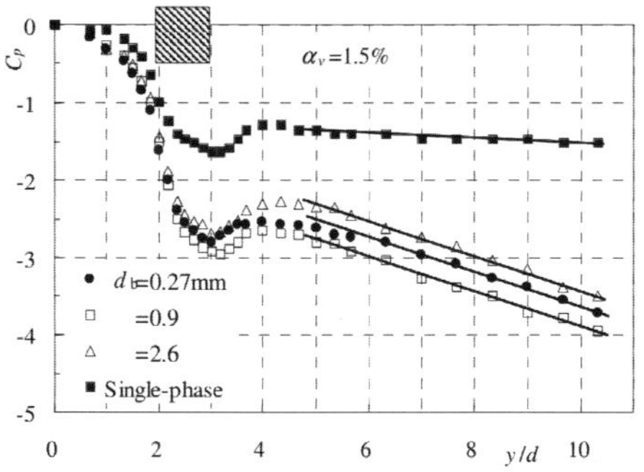

Fig.3 Pressure distribution along pipe (Re=0.5× 10^4)

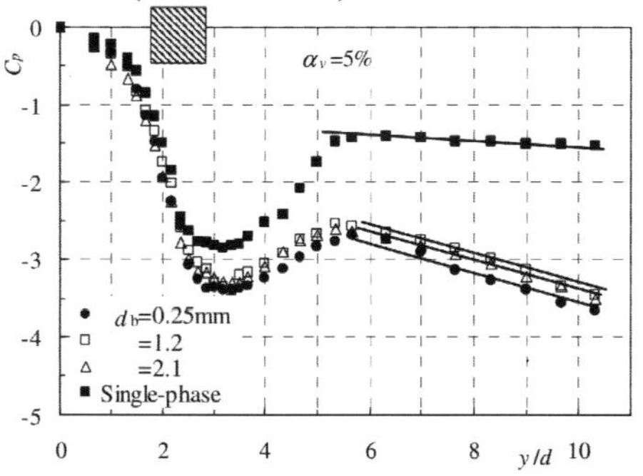

Fig.4 Pressure distribution along pipe (Re=1.0× 10^4)

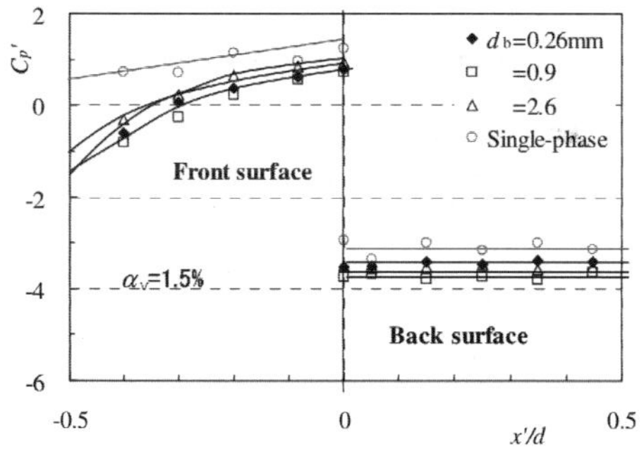

(a) Re=0.5× 10^4

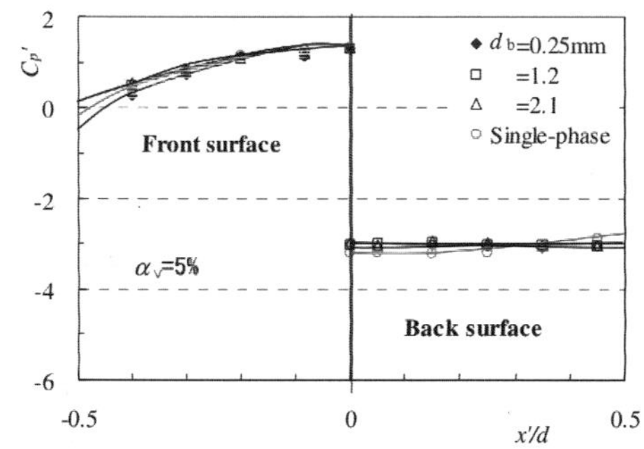

(b) Re=1.0× 10^4

Fig.5 Cylinder pressure distribution and effect of bubble size mentioned, the pattern is similar but a change on the pressure values is seen depending on the mean bubble size.

On the front side, the overall pressure value is affected by the mean bubble size, and the maximum pressure is obtained when d_b=2.6 mm, while the minimum is obtained when small size bubbly flow exists. That happens since large bubbles flow towards the center and more bubbles hit directly the cylinder's surface.

On the rear side, the pressure is almost the same under certain α_v. The maximum pressure is obtained when d_b=2.6 mm, while the minimum is obtained when small size bubbly flow exists. For that, the balance between the vortex generation area and bubble entrainment is considered. If the case of single-phase flow is considered as the standard value, then by inserting a bubble flow, although the mechanism remains the same the values change due to the interaction effect and thus the bubble entrainment into that region. Diameter bubbles of d_b=0.9 mm, are still easily entrapped into the vortex region and their density difference generates an even lower pressure core. When the flow is consisted by small bubbles, the flow characteristics show closer similarities to that of single-phase as it will be shown in a later section. For the case of large bubbles, even though that in our experiment the mean value is d_b=2.6 mm, it was mentioned that large bubbles exist inside the flow passage as it can be seen from Fig.8 as well. These sizes are entrapped more difficultly and wider gas free zone exists at the rear point of the cylinder, providing the results seen in this graph. From the pressure distribution measurements it was made apparent that the bubble size affects the flow pattern and characteristics in a flow past an obstacle.

In Fig.5 (b), where the Reynolds number is increased, it is shown that the bubble size effect is almost zero. That, in combination with the sidewall results, leads to the conclusion that the bubble size affects the flow in relatively low velocity values, at least concerning pressure phenomena.

From the pressure distribution and their differences due to mean bubble size effect on the front and rear surface of the rectangular cylinder the drag is considered. The drag D_r, on the cylinder's surface is calculated according to the C_{pi}' and is defined as the difference force exerted on the front and rear surfaces:

$$D_r = \frac{\int_{-d/2}^{d/2} C_{pi}'dx'}{d} \qquad (4)$$

From the given results of pressure coefficient from the front and rear side of the cylinder, the drag D_r, acting on it can be retrieved easily, as illustrated in Fig.6. For low gas fraction, the effect of bubble size is relatively smaller than the higher gas fraction case. A minimum is seen when the size is at small bubble scale, and slightly higher values are taken for the two other mean sizes. It is also shown that the values are slightly smaller than the case of single-phase flow. When increasing the gas fraction, the effect of the bubble size becomes considerably larger, where except the small bubble

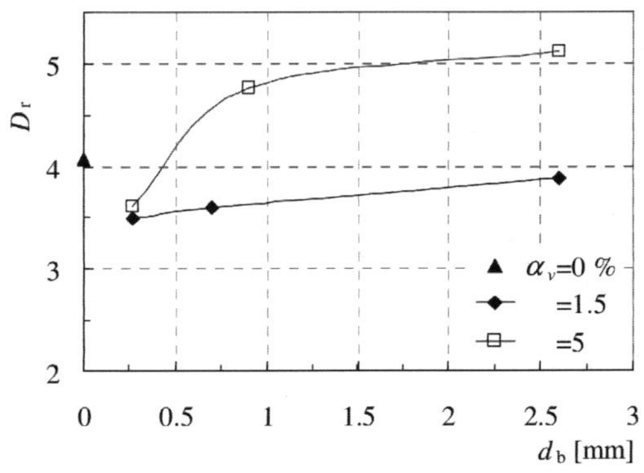

(a) Re=0.5× 10^4

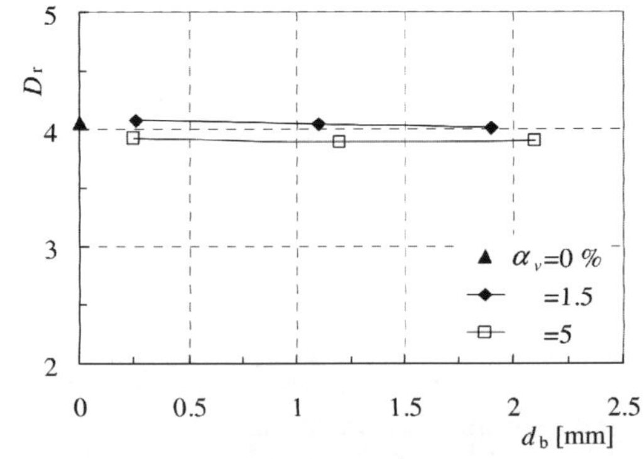

(b) Re=1.0× 10^4

Fig.6 Bubble size effect on drag, D_r

size case, the drag increase considerably more compared to the single-phase flow.

For an order of α_v=1.5%, the amount of larger bubbles is significantly small, as verified through visualization as well. Decreasing the bubble size, small bubbles spread throughout the whole flow passage, while medium size bubbles have a tendency to flow more near the wall region than in the middle. Larger bubbles, as known, flow towards the middle of the flow passage and hit directly the cylinders surface at low rates. This is why at α_v=1.5%, the drag takes the lowest value when d_b=0.7mm, since a larger amount of bubbles pass through the cylinder than hit directly to it.

When the gas fraction increases at α_v =5%, bubbles exist through the whole flow passage, even still maintaining their flowing characteristics. Increasing the mean bubble size, the drag increases, and the maximum value is seen when d_b=2.6 mm, since large bubbles flowing in the middle of the flow passage hit directly the front surface of the cylinder. For higher Reynolds number of $Re=1.0\times 10^4$ the bubble size effect is considerably decreased and shows that the differences occur due to gas fraction difference and not bubble size. That is also in good agreement with the results obtained from the wall side pressure distribution.

As it shows, the values are very close to that of single-phase flow with α_v=5%, having slightly lower values.

In the next section, the effect of bubble size distribution on fluctuation phenomena is introduced and discussed.

3.2 Pressure Fluctuation

The vortex shedding frequency f is studied under different bubble sizes. Figure 7 (a) illustrates the oscillation frequency results when FFT analysis is applied for $Re=0.5\times 10^4$, α_v=1.5%, at vertical position downstream of y/d=14. The horizontal axis denotes the oscillation frequency, and the vertical denotes the oscillation amplitude, or differently said, the intensity. On the horizontal axis range is set at 10Hz, since at higher frequencies the values were very lower and was not in the interest of the specific study.

Three main peaks are shown for all the cases, while the most intense oscillation is seen at f=0.6Hz for the three different bubble sizes, with a clear difference on the amplitude when the flow is consisted of small bubbles.

For d_b=2.6 mm, another two peaks with very close frequency are measured at f=1.6-2.2Hz. On the other hand, when d_b=0.7mm, the second peak has almost the same intensity as the first and its frequency is at f=2.2Hz, almost the same as before. When the bubble size decreases to small scale, relatively small peaks appear at lower frequencies than the other two cases of 1.4 and 3.2 Hz. The amplitude is drastically lower and the differences from larger bubbles are obvious. A clear explanation has not yet been found at the present stage and more analysis and investigation is required.

For α_v=5%, Three main peaks are shown for all cases of bubble size. The first dominant peak, which is the stronger as well, has the same frequency value at f=0.6Hz at all cases but the amplitude differs depending on the bubble size. It is clear that when the bubbles have a small size the amplitude is quite lower than the other two cases and this result comes to agreement with the previous condition and makes it even more certain that is caused due to the cylinder effect.

The second and third peak have different frequencies for d_b= 0.9 and 0.27 mm, with a value at f=2.4 and 5 Hz, while when the bubble size increases at d_b=2.6 mm the oscillation frequency tends to take lower values at f=2 and 4.6 Hz. It is pointed that both frequencies have a difference of 0.4 Hz and that fact

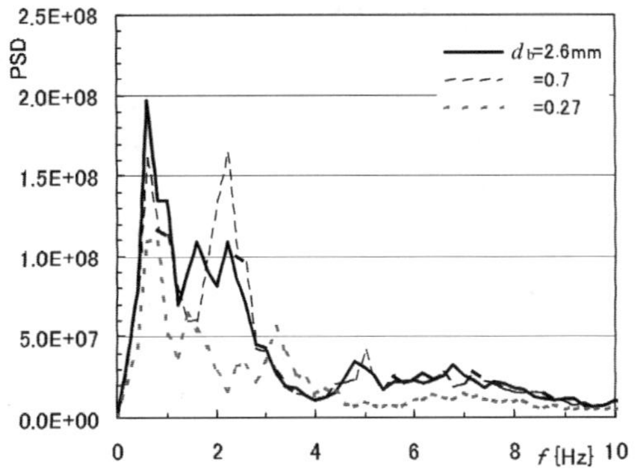

Fig.7 Bubble size effect on pressure fluctuation phenomena

leads us to the assumption that these oscillations might be related to the bubble size. Moreover, the fluctuation amplitude decreases when the bubble size decreases, showing that it affects the intensity of the fluctuation as well.

The first peak of high amplitude and low frequency is considered to be the effect of the Karman vortex street, since the frequency has been steady and independent from the bubble size variation. Even under different sizes, the effect of the cylinder and the Karman vortex street is considered to be similar. The difference on the amplitude can be explained considering the bubble size effect and the results given from the velocity distributions above. For the other two peaks, a theoretical explanation would be that the oscillation factor is composed of a mixture of frequencies, generated by the obstacle existence, bubble existence and small eddies on the side surface, since the amplitude is measured there.

It is noted that in both gas fractions the bubble size affects significantly the oscillation amplitude, showing that the smaller the bubbles the less the intensity power acting on the sidewall downstream.

3.3 Visualization and PIV Measurements

Figure 8 (a), shows the instantaneous visualized flow pattern and the equivalent PIV analysis results of the two-phase bubbly flow, for $Re=0.5\times 10^4$ and $\alpha_v=5\%$ with d_b=0.27mm. The white colored parts denote the bubbles in the flow passage. The flow passage is filled with the existence of small bubbles, and the dead water region that was observed behind the obstacle for larger bubbles, as it will be shown later, has been totally disappeared as shown in the photographic material. The small-scale size bubbles are easily entrapped into the region, since the buoyancy force acting on them is significantly small and shows very close flow characteristics with that of single phase. The velocity vector scale at the downstream region does not indicate the entrapment of bubbles inside the region just behind the obstacle as the picture, but that is due to the fact that the bubbles besides of being too small, the movement was difficult to be tracked from the PIV analysis, since they were extremely slow. However, from visualization results it was clear that a difference depending on the bubble size exists at the rear point region. The maximum bubble velocity has obviously decreased at the side area of the minimum

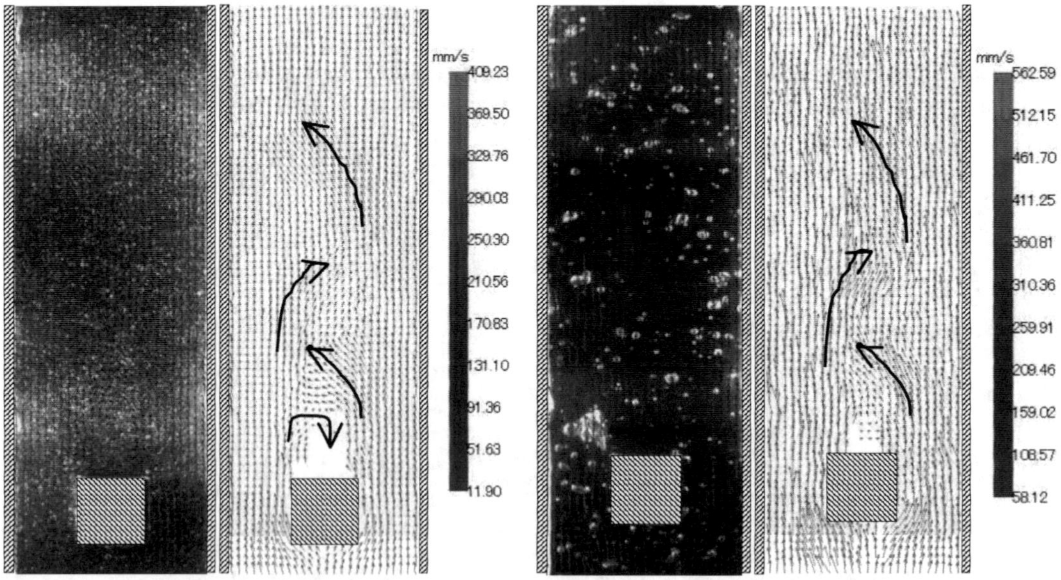

(a) d_b=0.27 mm (b) d_b=2,6 mm

Fig.8 Visualized flow pattern and PIV analysis (Effect of bubble size) flow passage and the effect of the bubble size is considered as the reason for that result. It is also noted that the velocity value expressed is very close to that of single phase traced.

Figure 8 (b), shows the results at the same flow conditions but bubble mean diameter at d_b=2.6mm. The gas-liquid bubbly flow separates at the front corners of the cylinder and forms a high local void fraction area on the two sides of the cylinder as well as the front side, where the separation area exists. A mix of relatively large bubbles of approximately 3~4mm and small ones can be observed at the illustration taken. A dead water region, which contains no bubbles, or few bubbles occasionally just behind the cylinder is observed clearly through the picture. The bubbles swarm in this region, which oscillates and develops into Karman vortex at the down stream reaches to the sidewalls of the flow passage. The velocity vector given by PIV is shown and the increase of bubble velocity between the cylinder and the sidewall caused due to the flow area decrease is observed. The generation of the Karman vortex street is clear. Again, a clear path of the Karman vortex street is illustrated.

From the PIV analysis results the local bubble velocity distributions at y'/d=0, 1.5, 6 are extracted and the differences due to bubble size are shown in Fig.9 (a~c) for two different Reynolds number and α_v=5%. The results present the mean velocity value of a total recording time of 8 seconds, in order to secure that several passes have occurred during that time and present the average values and not the instantaneous ones.

At y'/d=0, which is at the center of the obstacle, the increase of velocity due to the decrease of flow passage is shown. In all three cases, the maximum velocity appears at the center of the space between the sidewall and the cylinder wall. The velocity decrease at the sidewall of the flow passage is considered to be due to the wall friction. Although the velocity pattern is similar for all cases, a clear difference on the velocity itself can be seen, depending the mean bubble size.

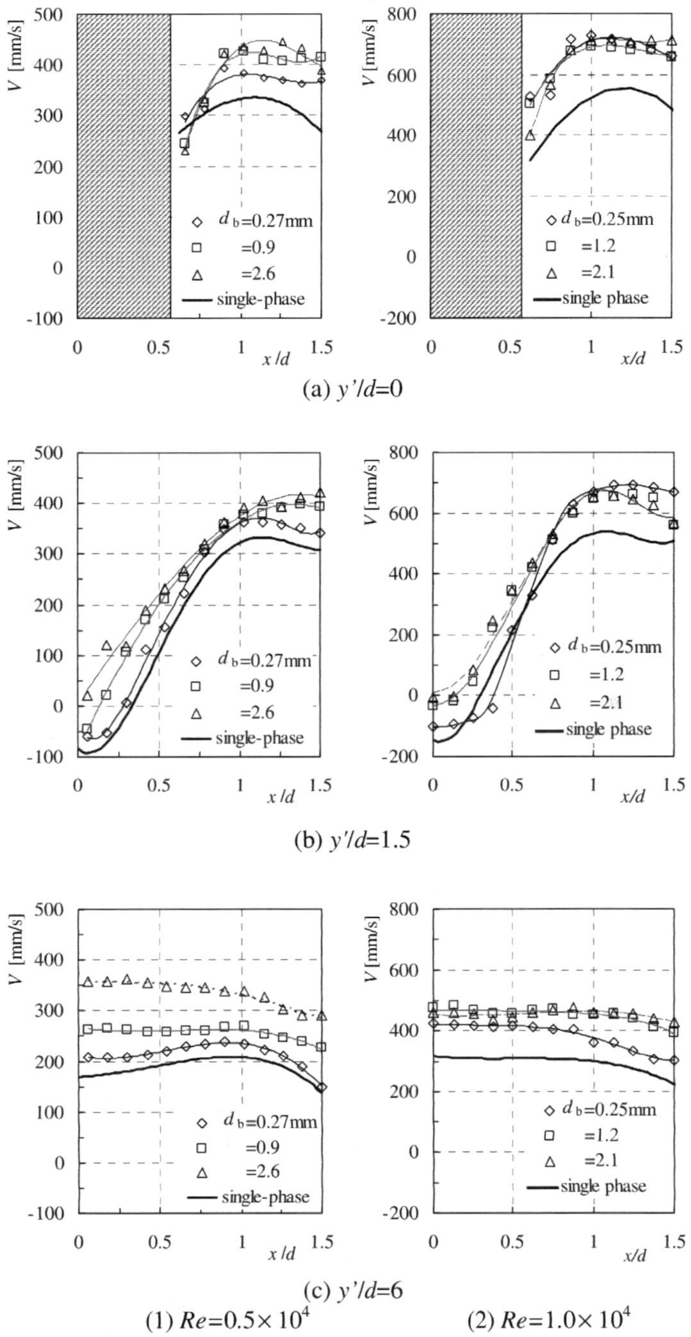

Fig.9 Bubble size effect on local velocity distribution (α_v=5%)

Journal of Physics: Conference Series **147** (2009) 012038 doi:10.1088/1742-6596/147/1/012038

The bigger the bubbles the higher the overall velocity becomes, with a maximum difference of approximately 80 mm/s.

The next graph illustrates the velocity distribution at position y'/d=1.5. The distribution pattern is the same for all the conditions, with the maximum velocity still obtained at x/d= -1,1 which is the center of the flow passage between obstacle and wall. Clear differences however can be seen in the velocity values. As mentioned above, the smaller the bubbles the lower the velocity. Especially at the region just behind the obstacle the effect of the bubble size is clearly shown. When d_b=0.27mm, negative values of velocity are extracted. That denotes a downward flow, which are the small bubbles entrapped into the vortex region generated there. When the bubble size is d_b=0.9mm, the negative velocity seen previously does not exist anymore. Although the velocity is very low, meaning that the downward force due to vortex acting on the bubbles and buoyancy force are nearly close to equilibrium, positive values are seen. For the bigger size of d_b=2.6mm, even higher velocity is seen and the effect of the bubble size is clear.

At an even downstream location of y'/d=6, the velocity differences are even greater. For the case of small bubbles the influence of the Karman vortex street still exists, showing a slight decrease of velocity at the center of the flow passage and having maximum values at the same position from the previous two graphs. Increasing the bubble size the velocity becomes almost the same with a decrease shown at the sidewall area where the wall friction exists. For the largest bubble size, opposed to the other two cases, maximum velocity is obtained at the center of the flow passage. That means that the effect of the Karman vortex street is not that strong and that the bubbles tend to flow into the center.

From visualization analysis it was made apparent that the bubble size affects the overall flow pattern as well the local velocity distribution. The smaller the bubbles the more they tend to have similarities with that of single phase flow and more they get affected by the Karman vortex street generated behind the obstacle. Increasing the bubble size, the overall velocity increases and effect of vortex becomes less.

4. CONCLUSIONS

Investigation on flow pattern and characteristics past a rectangular cylinder in a gas-liquid bubbly flow were conducted. The effect of the bubble size, with a variation from small scale to relatively large was conducted, using a laboratory developed control mechanism. From the experiments the following results are given.

The pressure distribution on the sidewall of the flow passage was measured. The maximum pressure drop varies with bubble size for low Reynolds number, with the medium size bubbly flow showing maximum decrease, while for higher *Re* the effect becomes very small. The pressure gradient remains the same even when the bubble size differs.

The pressure values at the front and rear side of the cylinder surface showed an obvious effect due to the bubble size for low Reynolds number. On the other hand, increasing the water velocity, the C_p' values where not affected by the bubble size, showing that a relation between *Re*, gas fraction and bubble size exists.

FFT analysis was conducted and three main fluctuations were obtained. For d_b=2.6mm the frequency values are lower than the other two cases and the effect of buoyancy force is considered. Moreover, the amplitude takes its minimum value when small bubbles flow exists and it was made clear that the bubble size affects the fluctuation intensity.

Extensive visualization experiments were conducted and differences on bubble entrapment into the vortex region behind the obstacle, as well as at the downstream was made apparent.

From PIV analysis it was found that the bubble size affects the total and local velocity distribution and value. In general, the smaller the bubbles (micro bubbles) the velocity shows a resemblance to that of the theoretical value of single-phase flow. The larger the bubbles become, the higher the local velocities at certain locations get affecting the overall flow pattern downstream and Karman vortex street.

NOMENCLATURE

A_r	Contraction area ratio of orifice nozzle	[-]
Cp, Cp'	Pressure coefficient	[-]
C_D	Drag coefficient	[-]
D_r	Drag of cylinder	[-]
D	Width of test section	[mm]
d	Width of cylinder	[mm]
d_b	Bubble diameter	[mm]
f	Vortex shedding frequency	[Hz]
g	Gravity	[m/sec^2]
H	Height of experimental section	[mm]
L	Length of experimental section	[mm]
P	Pressure	[Pa]
$Q_{w/g}$	Flow rate of water or gas	[m^3/s]
Re	Reynolds number ($=U_{wm}d/\nu$)	[-]
x/d, y/d	Distance ratio	[-]
T	Water temperature	[$^\circ C$]
U_{tm}	Mean velocity of two-phase flow	[m/s]
U_{wm}	Mean velocity of water flow	[m/s]

Greek Letters

α_v	Volumetric gas flow rate ratio	
ν	kinematic viscosity of water	[m^2/s]
ρ	density	[kg/m^3]

REFERENCES

Dian, T. (2002). "An Investigation on a Vertical Upward Gas-Liquid Bubbly Two-Phase Flow Past Obstacle," Doctoral degree thesis dissertation, Mie University.

Hara, F. et al. (1982). "Vibration of circular cylinder in cross two-phase flow, (1st report, Karman vortex shedding and pressure fluctuations)," Trans. of JSME, 48 (431), pp. 962-971.

Hulin, J-P. et al. (1982). "Experimental study of vortex emission behind bluff obstacles in a gas liquid vertical two-phase flow," Int. J. Multiphase Flow, 8 (5), pp. 475-490.

Inoue, A. et al. (1986). "Studies on two-phase cross flow, part I: flow characteristics around a cylinder," Int. J. Multiphase flow, 12 (2), pp. 149-167.

Pettigrew, M.J. (1994). "Two-phase flow-induced vibration: An Overview, J. Pressure Vessel Technology," ASME, 116, pp. 233-253.

Roshko, A. (1955). "The wake and drag of bluff bodies," J.of Aeronautical Sciences, 22, pp. 124 - 132.

Shakouchi, T. et al. (2004). "Flow Characteristics Around and Drag of Obstacle in Vertical Upward Gas-Liquid Bubbly Pipe Flow," Proceedings of the 3rd International Symposium on Two-Phase Flow Modeling and Experimentation, CD-ROM, Pisa, Italy.

Shakouchi, T. et al. (2007). "Gas-Liquid Two-phase Flow Passing Through Single Raw Tube Bundle with Non-Equal Spaces," (In Japanese), Proceddings of Annual Meeting of JSME-FMD, No.07-16,

CD-ROM.

Shakouchi, T. et al. (2006). "Flow and Aeration Characteristics of Micro Bubble Jet Flow," Japanese Journal of Multiphase Flow, Development of Multiphase Research II, pp.33-38.

Sugiyama, K. et al. (1999). "Three dimensional analysis of bubbly flow around a circular cylinder," Trans. of JSME, Ser. B, 65 (638), pp.3260-3267.

Uchiyama, T. et al. (1996). "Numerical analysis of incompressible two-fluid model by finite element method (Numerical method and calculation of flow around a cisrcular cylinder)," Trans. of JSME, Ser. B, 64 (595), pp.977-984.

Yokosawa, M. et al. (1986). "Studies on two-phase cross flow, part III: characteristics on unsteady flow behavior," Int. J. Multiphase Flow, 12 (2), pp. 185-202.

Journal of Physics: Conference Series **147** (2009) 012039 doi:10.1088/1742-6596/147/1/012039

Measurement of Falling Film Thickness around a Horizontal Tube Using Laser-induced Fluorescence Technique

Xiaofei Wang, Maogang He, Hualiang Fan, Ying Zhang

State Key Laboratory of Multiphase Flow in Power Engineering, Xi'an Jiaotong University,Xi'an 710049, P. R. China

mghe@mail.xjtu.edu.cn (Maogang He)

Abstract. An optical method for the non-intrusive measurement of falling film thickness on the perimeter of the horizontal tube is described. This non-intrusive measurement method is based on the visualization of the falling film flowing around the horizontal tube and uses the laser induced fluorescence and image processing techniques. The falling film thickness around a Turbo-CII horizontal tube and a plain horizontal tube was measured when water flushed the tube vertically with a stable velocity. The results for the Turbo-CII tube are compared with those for the plain tube, and show that the film thickness around the Turbo-CII tube varies with the same trend as that around the plain tube, and the minimal values of the film thickness tend to locate at different angular position of 95°-120° under different conditions. The film thickness around the enhanced tube changes within a smaller range with time than that around the plain tube. The corresponding effects on flow uniformity have also been discussed.

1. INTRODUCTION

Falling-film horizontal-tube evaporator used in the refrigeration has many clear advantages over flooded tube bundles, such as high heat transfer coefficient, low refrigerant charge and so on. It represents the superiority in energy saving and environment protection. However, There are some difficulties in liquid distribution and tube alignment, which affect flow uniformity and dryout, especially in deep bundles, (Ribatski G., Jacobi A. M. 2005). So the falling-film horizontal-tube evaporator has not been widely used in refrigeration system and air-conditioning at present.

The process of falling film flowing around a horizontal tube is the main carrier of heat exchanging in the falling-film horizontal-tube evaporator and has a significant impact on heat transfer. Clearly, when the liquid flush the horizontal tube vertically, the flow uniformity is the key factor for the study of the heat transfer. The falling film thickness around a horizontal tube can be used to reflect the flow uniformity directly and predict dryout, so the research on the falling film thickness around a horizontal tube is an effective method to improve liquid distribution and tube alignment. Here it's should be declared, for a fully developed laminar flow, the actual thickness of the film is also a sensitive factor for falling film heat transfer, which is inversely proportional to the local heat transfer coefficient. Thus, the film thickness is an important parameter for mechanism research and design of falling-film horizontal-tube evaporator.

The traditional expression of the falling film thickness was introduced by Nusselt, which was induced by assuming momentum effects on the falling film to be negligible and the intertube flowing mode to be a continuous sheet, as follows:

$$\delta = \left(\frac{3\eta_L \Gamma}{\rho_L (\rho_L - \rho_G) g \sin\beta} \right)^{1/3} \tag{1}$$

Where δ is the falling film thickness, β is the angle around the perimeter from the top of a horzontal tube, Γ is the flow rate on one side of the tube and 2Γ is the total flow rate on the tube, and the film Reynolds number is defined as follows (Nusselt, W. 1916):

$$Re = \frac{4\Gamma}{\eta_L} \tag{2}$$

The experimental research on the falling film thickness has also been developed and several different test techniques have been introduced, which are classified into intrusive and non-intrusive methods.

For the intrusive method, one popular technique is using parallel-wire conductance probes (Coney, M.W. E. 1973; Brown, R. C. et al., 1978; Koskie, J. E. et al., 1989). Another popular technique is by means of a micrometer to measure the falling film thickness. With this method, Thomson gave the values of the water film thickness at the angular position of β=90°, which is in good agreement with Nusselt's theory (Thomson, A. K. G. 1937). In literatures (Rogers, J. T. et al., 1989; Xu Li. et al., 2003), the water film thickness at 45°, 90° and 135° location were measured by means of a micrometer, the experimental results of Rogers were about 30% greater than theoretical values of Nusselt. As we know, the intrusive method may perturb the film flow and raise the surface at the insertion location due to surface tension effects.

For the non-intrusive method, the capacitance method is adopted to measure the liquid film thickness, which is a function of the electrical capacitance of an electrical condenser. For example, Dukler and Bergelin used this method to measure the liquid film thickness on a vertical plate (Dukler, A. E. et al. 1952). Recently, optical methods have become more and more widely used. Zhang et al. used the images from a camera to measure the average film thickness on a vertical tube, while a laser and a photo diode were used to measure the transient amplitude of waves (Zhang, J. T. et al., 2000). Desevaux et al. presented an interface measurement technique for a liquid film flowing inside small grooves by laser-induced fluorescence, which could provid a high resolution in the detection of the liquid-gas interface (Desevaux et al., 2002). D. Gstoehl et al. used the same technique as Desevaux to measure the falling film thickness around a plain horizontal copper tube in the sheet intertube flow mode. The test results showed relatively good agreement with Nusselt's theory around the upper perimeter of the tube but much poorer agreement on the lower perimeter (D. Gstoehl et al., 2004).

Although there are so many methods which can be used to measure the falling film thickness outside a horizontal tube, none tested results of the film thickness around any kinds of enhanced surface tubes has been reported by now as far as we know. While for the falling-film horizontal-tube evaporator or other kinds of heat exchanger, the enhanced surface is widely used and may be more significant. So the film thickness distribution around the enhanced tube is the topic of our present interest. Furthermore, for enhanced surface, the simulation computation is more complex and hardly achieved, so it is more practical to measure the film thickness around some enhanced surface horizontal tube.

In this work, the film thickness of water flowing around the outside of horizontal enhanced tube was measured using laser-induced fluorescence, and compared with that of the plain tube. The variation trend of the thickness with time and tube perimeter position has been recorded by a high-speed digital camera with a high temporal and spatial resolution.

2. EXPERIMENTAL METHOD

2.1 Theory and Apparatus

The laser-induced fluorescence technique is adopted to investigate the film thickness outside the horizontal tube. When the fluorescent substance is illuminated by the laser, it will emit certain wavelength fluorescence. In aqueous solution, the gas-liquid interface could be identified by the fluorescence, which is captured by the high-speed digital camera. The key process in this experiment is to adjust the position of the sheet light and the camera to ensure the sheet light pass through the central axes of the horizontal tube and keep the camera at a position perpendicular to the sheet light, see fig.1. With the appropriate relative position of the sheet light and the camera, the film thickness obtained in the experiment could be just in the radial direction. The theory of the measurement can be described simply as follows. Firstly, capture the image of the test position on the tube surface without test fluid flowing around, which characterizes the geometry of the solid-gas interface, and then capture the image of the test position on the tube surface with test fluid flowing around, which characterizes the geometry of the liquid-gas interface. Finally, the falling film thickness can be obtained by comparing the interface of the liquid-gas with that of the solid-gas, as shown in fig. 2.

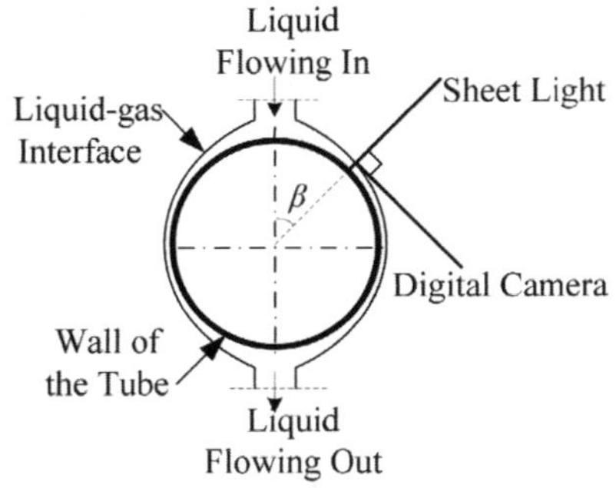

Fig. 1 Relative location of the light sheet and the digital camera

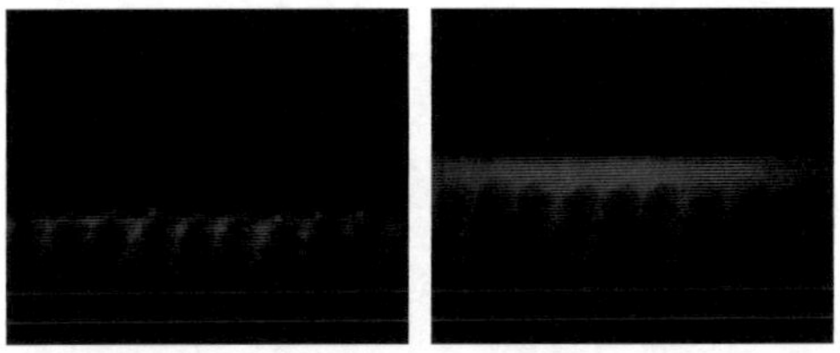

Fig. 2 The interface of the liquid-gas and that of the solid-gas

The mainly equipments used in this technique were introduced as follows.

The light source: A continuous solid-state laser emitting in the green light of 532 nm wavelength with a power of 0.3 W, the beam diameter is 2.0 mm.

The laser sheet generator: It can provide a sheet light beam which could vary in light density, light width and length with a limit range of 0.2×10 mm (minimum width × maximal length).

The position adjustor: It could be used to move the sheet light around the central axis of the horizontal tube with a maximal range from 30° to 150° and at intervals of 6°. It also ensures the digital camera keep vertical to the incident direction of the sheet light.

The fluorescent dye: The fluorescent substance used in the experiment is Rhodamine B, the concentration of the aqueous solution is $0.2 g \cdot L^{-1}$, which has no significant modification of the physical properties of water.

The image acquisition: The flow images are obtained by the high-speed digital camera which has a resolution of 1280×1024 pixels. A stop band filter (λ=532 nm) placed in front of the camera eliminates the spurious reflections and the incident light in order to perceive only the fluorescent light. So the interface can be easily detected because of the sharp contrast, resulting in an error of less than 25μm in the film thickness.

The falling film fluid circuit subsystem (as shown in Fig. 3): It is used to supply different kinds of flowing condition to ensure a steady falling film around the horizontal tube. It consists of the liquid distributor, the tested tube, the liquid trap, the pump, several values, the filter and the flow meter. The liquid distributor provides a uniform flow distribution along the tube, it is 300×80×16 mm（inside Length × Height × Width）, with the liquid entering at the top through two holes and leaving at the bottom through a flat plate with 3 mm diameter hole spaced 5 mm apart center-to-center. A grid with foam is added in the liquid distributor to increase the homogenization effect. Three tubes are arranged vertically and the second one is chosen for the measurement. The distance between the first tube and the liquid distributor is 3mm and the intertube spacing is defined as *s* which is variable according to the different experimental condition. The tested tubes are plain tubes and enhanced tubes. The plain tube is 19.05 mm-diameter copper tube whose surface is carefully cleaned. The enhanced tube is 19.05 mm-diameter Turbo-CII copper tube with the fin height of 0.559 mm and the fin density of 43 fpi. Fig. 4 depicts the photograph of the Turbo-CII tube. The changes of the flow rate could read directly from the flow meter by varying the valve opening.

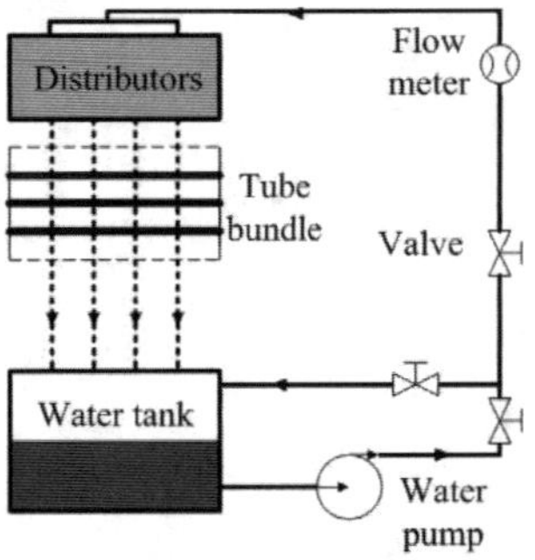

Fig.3 The falling film fluid circuit subsystem

Fig.4 The surface of the Turbo-CII tube

2.2 Experimental Procedure

Before starting a test, the liquid distributor would be leveled carefully. Tubes were aligned in the test section at a stated tube spacing. Since a tube inclination would cause a nonuniform distribution of the flow, the tubes also were leveled before a test, and leveling was checked again after the liquid started circulating in the system. The liquid flow rate was adjusted to obtain a column mode, if the column fell at fixed sites without drifting towards either end of the test tube, the tubes were level and its horizontal axes were considered in the middle of the hole center in the liquid distributor. The position of the sheet light and the camera was adjusted carefully for the test and the solid-gas interface were recorded by the digital camera. After that, the test liquid was circulated through the system for several hours to ensure that the tubes were carefully wetted, at last the liquid-gas interface was captured and the flow rate was read.

An image processing program written in MATLAB was developed specifically for extracting the interface of liquid-gas and solid-gas. In this program, the original image was digitized. An appropriate threshold of the gray levels permitted the removal of the background noise to obtain a clear image. The edge detection was then achieved by identifying sequence number of the pixel in which the gray levels have a step change, and the film thickness is obtained by comparing the numbers of the solid-gas pixels with that of the liquid-gas pixels and multiplying pixel size. It should point out that, for the enhanced tube, because the enhanced surface is accidented, the solid-gas interface for the measurement of the film thickness is not the real one captured by the camera, but a reconstructed plane which was fixed according to the adjacent fins. Therefore, the experimental film thickness around the root section of the enhanced tube was thinner than its real value. Although, it would not affect the study result of the flow uniformity and the distribution trend of the film thickness around the tube.

Several tests were performed to estimate the reliability of the measurement technique. The average observed difference between two measurements under the same condition for the plain tubes was ±7% while for the Turbo-CII tube was ±5%.

3. EXPERIMENTAL RESULT AND DISCUSSTION

The experimental liquid was pure water with a very minor amount of Rhodamine B under the normal temperature (about 27℃) and pressure (about 98 kPa) conditions. The intertube spacing of the measurements was 3 mm, 6 mm, 9mm, 12mm, and 19mm. All test conditions were below the Reynolds number for transition to turbulent flow (typically cited to

Journal of Physics: Conference Series **147** (2009) 012039 doi:10.1088/1742-6596/147/1/012039

be 1600-1800), so the wave influence on film thickness could be neglected.

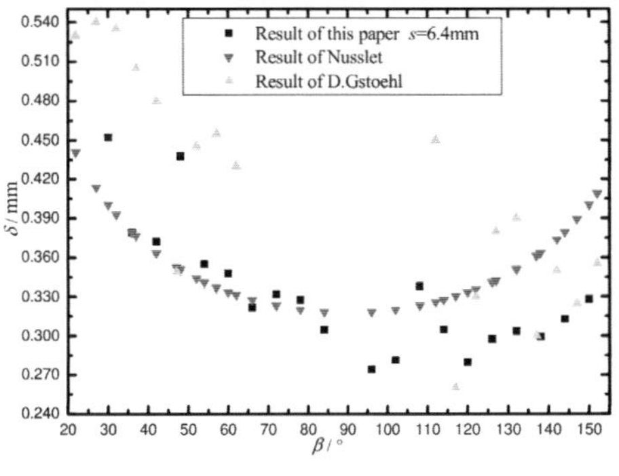

Fig.5 Film thickness versus angle of plain tube at *Re*=574

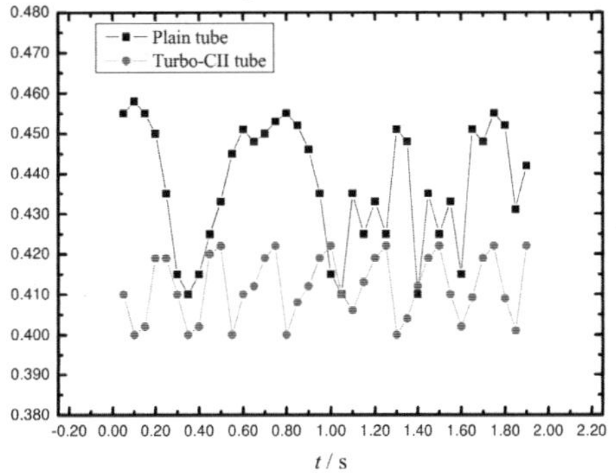

Fig.6 Film Thickness at certain position along the tube with the time interval of 0.05s

Fig. 5 shows the variation trend of the film thickness around the plain tube with the angle at *Re*=574 and the intertube spacing was s=6.4mm. Compared with Gstoehl's experimental results attained under the same condition, the maximal film thickness is about 9% smaller and the minimal film thickness is much closed to Gstoehl's values (D. Gstoehl et al., 2004). From the figure, the experimental results at β<84° approach to Nusselt's theory value, while the experimental results at β>96° are smaller. This variation trend validates that Nusselt's theory gives a reasonable prediction of film thickness on the upper perimeter of the tube and tends to significantly overestimate values on the lower perimeter. This variation trend is also in accord with the results of the previous numerical simulation on the film thickness reported by our group (X. F. Wang et al., 2008).

The temporal change of the film thickness is recorded in fig. 6. The time interval is 0.05s. The flow is under the condition of *Re*=225 with tube spacing of 12mm at the angular position of β=48°. The film thickness around the plain tube and the Turbo-CII tube all change randomly and the film thickness around the Turbo-CII tube vary with a more narrow range than that around the plain tube. The differences between the maximum film thickness and the minimum film thickness are 0.070 mm for the plain tube and 0.020 mm for the Turbo-CII tube, which are respectively about a rate of 15% and 6% to the average film thickness.

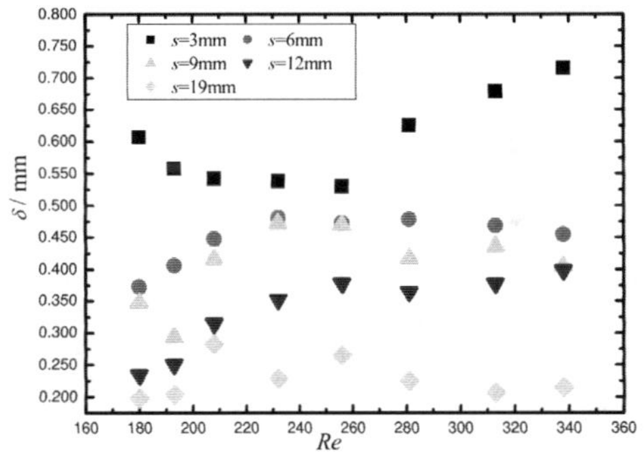

Fig.7 Film thickness versus flow rate of Turbo-CII tube at β=48°

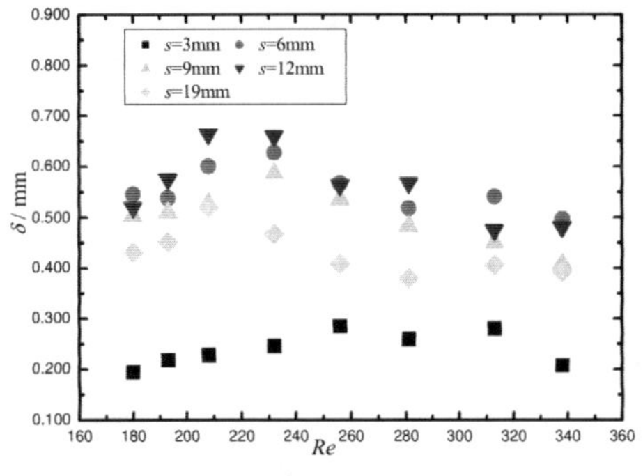

Fig.8 Film thickness versus flow rate of plain tube at β=48°

Fig.7 and fig.8 show the film thickness versus flow rate of the Turbo-CII tube and the plain tube with *s*=3 mm, 6 mm, 9 mm, 12 mm, and 19 mm at β=48°, respectively. The flow is in steady column flow mode at the Reynolds number from 180 to 340, which is also within the *Re* range for column flow mode of Roques (J. F. Roques., 2002). In these two figures, the film thickness change slowly with an inflection point. The increasing flow rate may induce the liquid accumulated between the adjacent coming liquid column and result in the increasing of the film thickness, and the liquid will be accumulated to a limit as the flow rate increasing,

Journal of Physics: Conference Series **147** (2009) 012039 doi:10.1088/1742-6596/147/1/012039

then it will be outspread along the tube. The location where the liquid is accumulated varies with different flow rate. And the change trend of film thickness is different due to the relative location of the tested area and the film accumulated area. For the plain tube, the inflection point is more obvious, which may be due to block effect of the fin of the enhanced surface on the liquid accumulation.

The film thickness decrease with the increase of the tube spacing because of the flow accelerated, while the film thickness of the plain tube at *s*= 3 mm is smaller than others, as shown in fig.8. The same phenomenon for *s*=3.2 mm is also observed by D.Gstoehl. The reason may be the effect of the surface tension on the film flow. For a small tube spacing (smaller than a certain limit) where the liquid could not spread between the adjacent tubes completely, a large amount of the flowing liquid are accumulated on the bottom of the upper tube on the function of the surface tension. Little liquid flowing down around the tested tube results in a smaller film thickness.

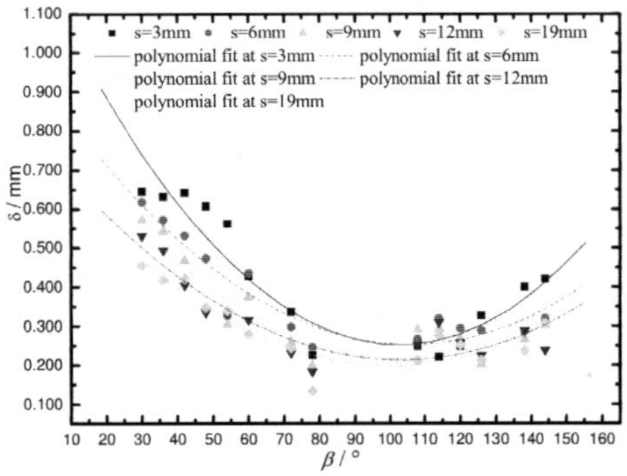

Fig.9 Film thickness versus angles of Turbo-CII tube at *Re*=180

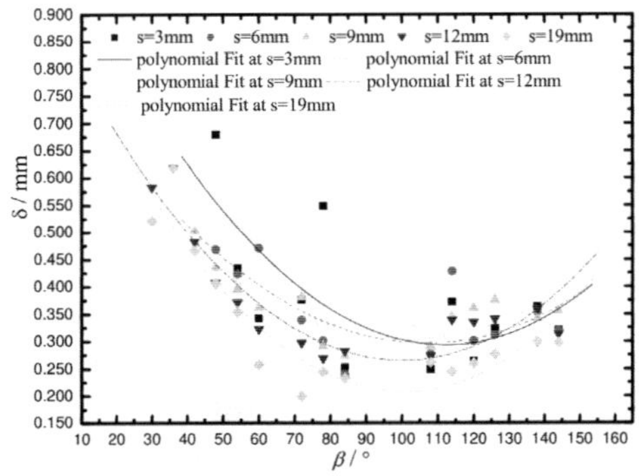

Fig.10 Film thickness versus angles of Turbo-CII tube at *Re*=260

Fig.9 and fig.10 show the film thickness of the Turbo-CII tube versus angles at *Re*=180 and *Re*=260, respectively. Values of the film thickness on the upper tube are larger than that on the lower tube, which is similar to that of the plain tube. In these two figures, the minimal values of the film thickness are not just at angle of 90°. According to all our test results, the minimal values of the film thickness tend to locate at different angular position of 95°-120° under different conditions.

Generally, if the film thickness gets thinner or the flow become unstable under some flow conditions, the local dryout on the surface of the tube would occur. The area where the local dryout probably occurs could be found from all these tested results. With regard to falling film evaporation with nucleate boiling in the film, the thinner film at the angle from 95° to 120° on the low perimeter of the tube tend to suppress nucleation and facilitate the formation of dry patches. And by the observation, the most unsteadily flowing may occur with the liquid accumulating which accompanied by the local film getting thicker, so it is significant to study the area and condition of the liquid accumulating. Compared with the plain tube, the fin of the Turbo-CII tube may store a little liquid and steady the film flowing, thus, the Turbo-CII tube provides a good performance for the flow uniformity.

CONCLUSIONS

An optical method based on laser-induced fluorescence and image processing of high-speed digital camera has been developed to measure the falling film thickness on the outside of a horizontal tube. The film thickness was measured for the plain tube also for the Turbo-CII tube under different flow rate condition with different intertube spacing, respectively. The measurement provides the variation of the thickness with time, angular position from image processing of the image sequences.

The present measurements show that the film thicknesses around the Turbo-CII tube vary with the same trend as that around the plain tube, which validates Nusselt's theory giving a

Journal of Physics: Conference Series **147** (2009) 012039 doi:10.1088/1742-6596/147/1/012039

reasonable prediction of film thickness on the upper perimeter of the tube and significantly overestimate values on the lower perimeter. The minimal values of the film thickness tend to locate at different angular position of 95°-120° under different conditions, and the angular position of the thinner film may be the main position for dryout detection. For the enhanced tubes, the falling film thickness change within a smaller range with time than that around the plain tube.

NOMENCLATURE

g acceleration due to gravity (9.81), [m·s^{-2}]
Re film Reynolds number, $4\Gamma/\eta$
s intertube spacing, [mm]
t test time, [s]

Creek symbols

β angular position on a horizontal tube measured from the top, [°]
Γ film mass flow rate on one side per unit length of cylinder, [kg·m^{-1}·s^{-1}]
η dynamic viscosity, [Pa·s]
δ film thickness, [mm]
λ wave length of the light, [nm]

Subscripts

G gas
L liquid

REFERENCES

Brown, R. C. et al. (1978). "The Use of Wire Probes for Measurement of Liquid Film Thickness in Annular Gas–Liquid Flows," Can. J. Chem. Eng., 56, pp. 754–757.

Coney, M.W. E. (1973). "The Theory and Application of Conductance Probes for the Measurement of Liquid Film Thickness in Two-Phase Flow," J. Phy. E, 6, pp. 903–910.

Desevaux, P. et al. (2002). "Interface Measurement Technique for Liquid Film Flowing Inside Small Grooves by Laser Induced Fluorescence," App. Thermo. Eng., 22, pp. 521-534.

D. Gstoehl et al. (2004). "Measurement of Falling Film Thickness Around a Horizontal Tube Using a Laser Measurement Technique," Heat Transfer Eng., 25(8), pp. 28-34.

Dukler, A. E. et al. (1952). "Characteristics of Flow in Falling Liquid Films," Chem. Eng. Progress, 48(11), pp. 557-563.

J. F. Roques. et al. (2002). "Falling Film Transitions on Plain and Enhanced Tubes," J. Heat Transfer, 124, pp. 491-499.

Koskie, J. E. et al. (1989). "Parallel-Wire Probes for Measurement of Thick Liquid Films," Int. J. Multiphase. Flow, 15(4), pp. 521–530.

Nusselt, W. (1916). "Die Oberfl¨achenkondensation des Wasserdampfes," Zeitschr. Ver. Deut. Ing., 60, pp. 541–546, 569–575.

Ribatski G. et al. (2005). "Falling-film evaporation on horizontal tubes a critical review," Int. J. Refrigeration, 28, pp. 635-653.

Rogers, J. T. et al. (1989). "Experimental Laminar Falling Film Heat Transfer Coefficients on

Large Diameter Horizontal Tubes, " Can. J. Chem. Eng., 67, pp. 560–568.

Thomson, A. K. G. (1937). "Heat Transmission in Film-Type Coolers," J. Society Chem. Industry, Japan, Supplemental Bind, 40, pp. 380T–384T.

Wang X. F., He M. G., Zhang Y. (2008). "Numerical Simulation of the Liquid Flowing Outside the Tube of the Horizontal-tube Falling Film Evaporator", J. Eng. Thermophysics, 29(8), pp. 1347-1350.

Xu Li et al. (2003). "Flowing State in Liquid Films over Horizontal Tubes," Desalination, 156, pp.101-107.

Zhang, J. T. et al. (2000). "Falling Liquid Fin Thickness Measurement by an Optical-Electronic Method," Review of Scientifique Instruments, 71(4), pp. 1883-1886.

Journal of Physics: Conference Series **147** (2009) 012040 doi:10.1088/1742-6596/147/1/012040

The behavior of bouncing jet

Naoto Oka [1], Toshiyuki Gomi [1], Ichiro Ueno [2]

[1] Graduate School, Tokyo University of Science, 2641 Yamazaki, Noda, Chiba 278-8510, JAPAN
[2] Tokyo University of Science, Noda, JAPAN

uenolab@gmail.com

Abstract. A series of experimental study was carried out with a special interest on a behavior of impinging liquid jet to a flowing bath of the same liquid. It has been known that the jet bounces from the bath against the gravity even after the penetrating to the liquid bath under a narrow range of condition. This unique behavior is often called as 'the bouncing jet.' Entrainment of ambient gas around the jet results in forming a thin gas layer between the jet and the liquid bath to realize a non-coalescnece jet in the bath. This is a key phenomenon to understand the dynamics of the impinging liquid jet. In the present study the authors focus on this ambient gas entrainment in the system that the impinging jet penetrates to coalesce into the flowing liquid bath; especially on the evaluation of the thickness of the entrained gas around the penetrated jet.

1. INTRODUCTION

Impinging liquid jet on a flowing liquid bath can been seen in a variety of industrial applications, for example, prevention of air bubble mixing in casting process and improvement of mixture efficiency of gas-liquid/liquid-liquid system. Entrainment of ambient gas around the jet results in forming a thin gas layer between the jet and the liquid bath [Lorenceau et al, (2004); Eggers, (2001); Zhu et al, (2000)] in such a system. This ambient-gas entraiment realizes a non-coalescnece jet in the bath. This is a key phenomenon to understand the dynamics of the impinging liquid jet. It has been known that the jet bounces off the bath surface against the gravity even after the penetrating to the liquid bath under a narrow range of condition on impinging jet velocity and flow rate of the bath. This unique behavior is often called as 'the bouncing jet' for Newtonian liquid[Thrasher et al. (2007)]. In the present study the authors focus on this ambient gas entrainment in the system that the impinging jet penetrates to coalesce into the flowing liquid bath; especially on the evaluation of the thickness of the entrained gas around the penetrated jet.

2. EXPERIMENT

Figure 1 shows the experimental apparatus. The apparatus consists of major three parts: jet generation system, flowing liquid bath, and observation system. Linear jet of 100-cSt silicone oil was produced from a nozzle attached to an overflow tank. This system enabled us to generate a jet of constant flow rate. The jet was impinged to the flow of the same fluid in a bath. Inner width, depth and height of the bass were of 320 mm, 20 mm and 50 mm, respectively. The flow rate in the bath was controlled by regulating valve. The flow was confirmed to have a linear profile at the test region. The jet generation system was placed above the bath at a designated height from the fluid surface in the bath. The impact velocity of the jet to the fluid in the bath was varied by varying the height of the jet generation system. The fluids of the jet and in the bath were recirculated by two pumps after the filter. The behavior of the impinged jet to the flowing fluid in the bath was observed with a high-speed camera of 1024×1024 pixcel^2 at a frame rate up to 2000 fps. The jet and the fluid near the impinging jet was visualized under backlighting condition.

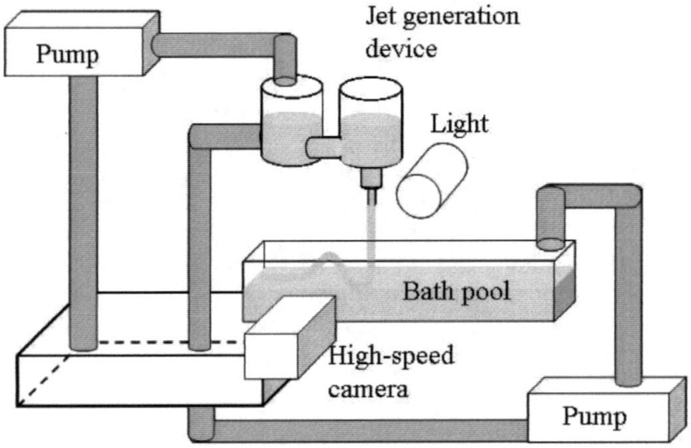

Fig. 1 Experimental apparatus

3. RESULTS AND DISCUSSION

3.1 Behavior classification of liquid jet in bouncing jet generation process

Four major behaviors of impinged jet were observed by changing the velocities of the jet and the flow in the bath. Typical examples of the major behaviors are shown in Fig. 2. In case of large difference between the jet velocity, V_{jet}, and the bath flow velocity, V_{bath}, the jet penetrates into the bath and bends towards the downstream direction in the bath as shown in Fig. 2-1. Entrained-air sheath wraps the impinged jet to realize that the jet never coalesces with the fluid in the bath until the air sheath breaks down. Tiny bubbles are formed and detach at the tip of the sheath. One can see an 'air pocket' does exist beneath the deformed free surface of the fluid in the bath to sustain the penetrating jet. If one decreases the difference $V_{jet} - V_{bath}$, the penetration length of the jet in the bath becomes shorter, and the air sheath breaks down right after the sustaining air pocket as shown in Fig. 2-2. In this case frequency of tiny bubble generation becomes higher. Noted here that the tiny bubbles detach at the bottom of the tip end. This is because the tip of the impinged jet is pinned by the boundary of the air pocket. In further decreasing the difference $V_{jet} - V_{bath}$, the bouncing jet [Thrasher et al.(2007)] is formed as shown in Fig. 2-3; that is, the impinged jet is bounced off the surface of the fluid flowing in the bath without any penetration into the bath fluid. The jet is sustained by the deformed free surface of the fluid in the bath with an air cushion between them. The bounced jet is bent due to the gravity and re-entry to penetrate into the bath. Further decreasing the difference $V_{jet} - V_{bath}$, the jet crawls on the free surface of the fluid in the bath after bouncing off as shown in Fig. 2-4.

The authors especially focus on the thickness of the entrained air sheath formed around the jet in the cases as shown in Figs. 2-1 and 2-2.

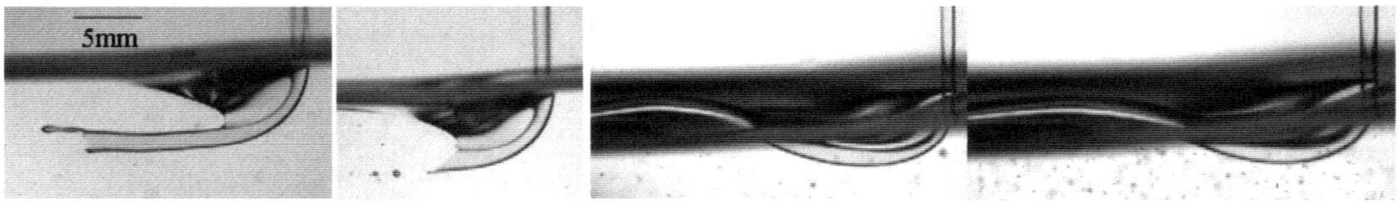

2-1 2-2 2-3 2-4

Fig. 2 Typical examples of the behaviors of the impinged jet to the fluid flowing in the bath. (1): (V_{jet}, V_{bath}) [mm/s] = (1.36×10^3, 2.37×10^2), (2): (1.04×10^3, 2.37×10^2), (3): (8.31×10^2, 3.04×10^2) and (4): (5.44×10^2, 4.00×10^2)

3.2 Measurement of thickness of air film

The authors try to evaluate the thickness of entrained air wrapping the penetrated jet. The thickness is evaluated at a position right beyond the air pocket sustaining the jet. As aforementioned tiny bubbles grow and detach from the tip of the jet coalescing to the bulk fluid. One can evaluate a mean gas flow rate around the jet by measuring a growth rate of the detaching bubble. Figure 3 indicates a successive snapshots of a growing bubble at the jet tip observed from the side. Assumptions are made for the bubble growth and for the air flow in the sheath around the jet; (i) the bubble at the jet tip grows in an axisymmetric manner, and (ii) the jet and the air sheath are co-axial cylinders, (iii) the velocity of the air in the sheath corresponds to the mean value of $(V_{jet}+V_{bath})/2$. The diameter of the periphery of the jet and the air sheath is evaluated from the recorded images of the jet in the bath from the side.
The images around the jet tip and around the bath surface where the jet is penetrating to the bath are detected with synchronized two high speed cameras. Field of view for each camera is of about 10 mm $\times$10 mm. Evaluated thickness of the air sheath around the impinged liquid jet in the case of the jet flow rate of 0.83 cm^3/s and the test fluid of 100-cSt silicone oil as functions of V_{jet} and V_{bath} is presented in Fig. 4.

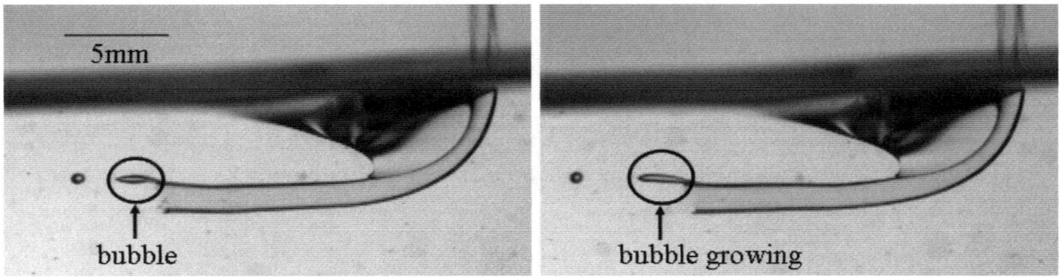

$t = t_l$ [ms] t_l+6.0

Fig. 3 Succesive snapshots of a growing bubble at the jet tip observed from the side: V_{jet} = 1.4×10^3 mm/s and V_{bath} = 2.4×10^2 mm/s.

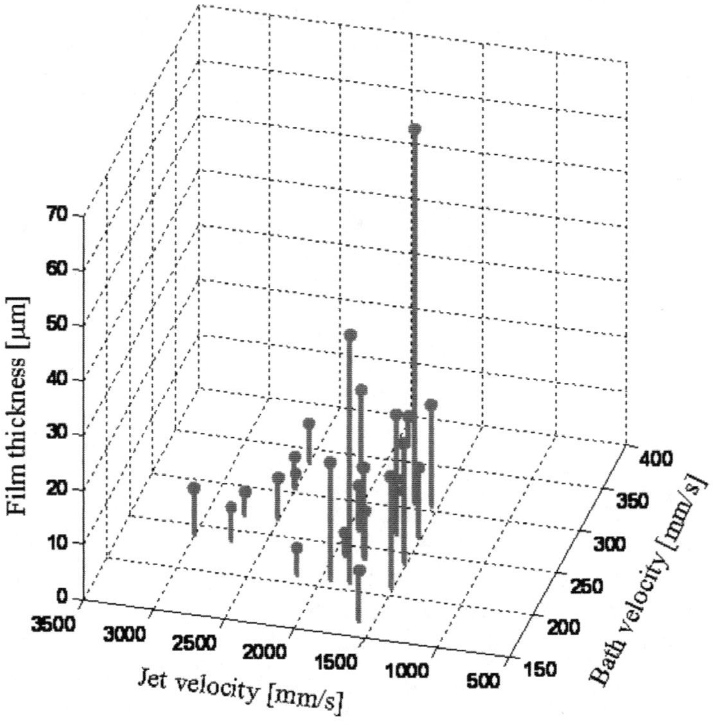

Fig. 4 Evaluated thickness of the air sheath around the impinged liquid jet as functions of V_{jet} and V_{bath}. Flow rate of the jet: 0.83 cm^3/s, test fluid: 100-cSt silicone oil.

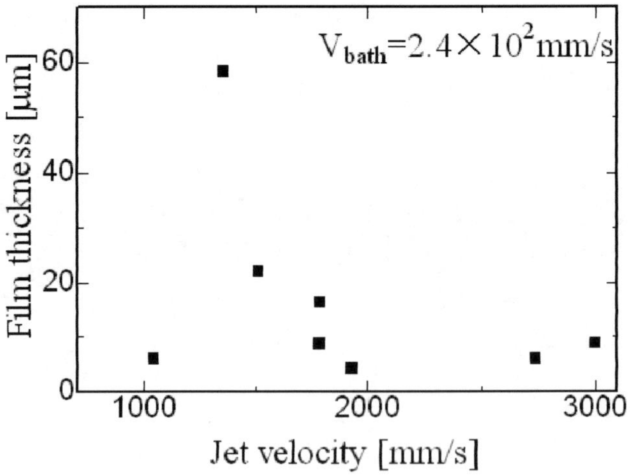

Fig. 5 Evaluated thickness of the air sheath around the impinged liquid jet as a function of V_{jet}. Flow rate of the jet: 0.83 cm^3/s, test fluid: 100-cSt silicone oil, $V_{bath} = 2.4\times10^2$ mm/s.

4. CONCLUSIONS

Averaged air sheath thickness is about ten to several tens micrometer, which is rather less than the case of impinging jet to a pool [Lorenceau et al, (2004)]. The distribution of the thickness has a profile that exhibits a maximum in a range of the jet velocity under a constant V_{bath}. Figure 5 indicates the air sheath thickness against the jet velocity under the conditions of 0.83 cm^3/s of jet flow rate and of $2.4x10^2$ mm/s. The thickness increases as the jet velocity increases in the case of lower V_{jet}. The thickness abruptly decreases after the maximum point; this can be explained by considering shape of the free surface of the fluid in the bath sustaining the impinged jet. It can be explained by considering as follows; The momentum of the flowing fluid in the bath results in the jet bending toward the downstream direction, and this causes a pocket of the deformed free surface of the fluid in the bath. The entrained air around the jet snakes away toward the pocket because of the pressure gradient between the air sheath around the jet and the ambient pressure in the pocket, then the net amount of the entrained air around the jet penetrated into the bath fluid decreases.

Concluding Remarks
Entrainment of ambient gas around an impinging jet to a flowing fluid in the bath is examined. This ambient gas entrainment results in forming a thin gas layer between the jet and the liquid bath, which realizes a non-coalescence jet in the bath. The authors evaluate the thickness of the entrained gas around the penetrated jet as functions of the jet velocity and the velocity of the fluid flowing in the bath.

REFERENCES

Lorenceau, E. Quere, D. Eggers, J. (2004). "Air Entrainment by a Viscous Jet Plunging into a Bath ," Phys.Rev.Lett.93,254501.

Thrasher, M. Jung, S. Pang, Y. K. Chuu, Chih-Piao. Swinney, H. L. (2007). "Bouncing jet: A Newtonian liquid rebounding off a free surface," Phys.Rev.Lett.76,056319.

Journal of Physics: Conference Series **147** (2009) 012041 doi:10.1088/1742-6596/147/1/012041

Entropy feature extraction on flow pattern of gas/liquid two-phase flow based on cross-section measurement

J Han [1,2], F Dong [1], Y Y Xu [1]

[1] Tianjin Key Laboratory of Process Measurement and Control, School of Electrical Engineering & Automation, Tianjin University, Tianjin, 300072 CHINA
[2] Department of Educational Technology, Capital Normal University, Beijing, 100037 CHINA

fdong@tju.edu.cn

Abstract. This paper introduces the fundamental of cross-section measurement system based on Electrical Resistance Tomography (ERT). The measured data of four flow regimes of the gas/liquid two-phase flow in horizontal pipe flow are obtained by an ERT system. For the measured data, five entropies are extracted to analyze the experimental data according to the different flow regimes, and the analysis method is examined and compared in three different perspectives. The results indicate that three different perspectives of entropy-based feature extraction are sensitive to the flow pattern transition in gas/liquid two-phase flow. By analyzing the results of three different perspectives with the changes of gas/liquid two-phase flow parameters, the dynamic structures of gas/liquid two-phase flow is obtained, and they also provide an efficient supplementary to reveal the flow pattern transition mechanism of gas/liquid two-phase flow. Comparison of the three different methods of feature extraction shows that the appropriate entropy should be used for the identification and prediction of flow regimes.

1. INTRODUCTION

The gas/liquid two-phase flow is very typical in nature and industrial applications, and the accurate measurement of its parameters is always an urgent issue that needs to be solved in engineering and science (Baker, O., 1954; Dong, F. et al., 2003). Research on two-phase flows is of great importance to scientific development, national economy and human living. The meausrement of two phase flows presents a great challenge and fascination to researchers due to its inherit complexity and nonlinearity. Flow regime, a most characteristic spatial distributional parameter of the two phase mixture flows in pipes, impacts flow characteristic and the performance of heat and mass transfer. Moreover, it has a great influence on the accuracy when measuring other two-phase flow parameters. Therefore, the feature extraction and analysis, the identification and prediction of flow regimes are important.

Horizontal pipe flow is an important flow in the two-phase flows. The cross-sectional measurement based on Electrical Resistance Tomography (ERT) is a kind of process tomography that can be widely used in the measurement of multi-phase flow parameters (Fabre, J. et al., 1992; Dong, F. et al., 2005). The cross-sectional measurement system has a great potential in process applications. As a kind of Process Tomography (PT), this technique extracts spatial and temporal information of process parameters by using multiple sensors mounted around the region of interest. The characteristics of cross-sectional measurement are non-intrusive, non-radioactive, visible and etc (Dong, F. et al., 2006).

There are many methods to process the experimental data for feature extraction, such as the power spectral density functions (PSDF), signal to noise ratio (SNR), autocorrelation functions (ACF), probability density functions (PDF), fractal and chaos time series analysis, entropy time series analysis and etc. Entropy is a common concept in many fields, mainly in signal processing. In information theory, entropy is used to measure the concept of information laws. Classical entropy-based criteria matches these conditions and describes information-related properties for an accurate representation of a given signal. So the entropy is adopted to signal analysis and feature extraction for cross-sectional measured resistance information of the gas/liquid two-phase flow in this paper. Five main entropies are adopted to analyze the experimental data which is acquired by a cross-sectional measurement system in different flow regimes, and the five methods are compared in three different perspectives.

In spite of the technological development and efforts in the literatures for feature extraction of the gas/liquid two-phase flow, providing a more reliable and accurate feature is still considered a challenge. In this paper, the general information of cross-sectional measurement based on a ERT system and the notion of entropy is introduced, the five entropies and their definitions of the method are presented. The method of five entropy features analysis and extraction from the three different perspectives which cover the characteristic of cross-section measurement is given, and the three different perspectives are also analyzed. The three different perspectives of entropies feature extraction are sensitive to the flow pattern transition in gas/liquid two-phase flow, they can fully reflect the dynamics structure of gas/liquid two-phase flow and reveal the flow pattern transition mechanism of gas/liquid two-phase flow. Three different methods of feature extraction are compared to show that the appropriate entropy should be used.

2. CROSS-SECTIONAL MEASUREMENT BASED ON ERT AND ENTROPY

2.1 Cross-Sectional Measurement Based on ERT

The cross-sectional measurement system based on ERT, which is developed from the principle that the conductivity of different medium varies, is one of the process tomography techniques. By using the cross-sectional measureing technique, the medium distribution of measured field can be identified if the conductivity distribution of sensing field is obtained.

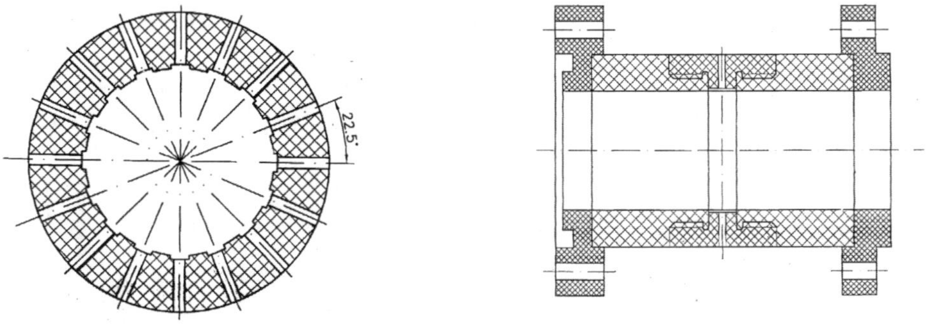

Fig. 1 The structure of electrode array

Sensitive electrode array using embedded structure is composed of 16 rectangular electrodes which are made from titanium alloy and evenly mounted in the pipe wall. The drift δ of measuring voltage in the cross-sectional measurement system is less than 1%, it meets the requirements of the qualitative and quantitative analysis. The drift δ is defined as the root-mean-square error of the 208 measurement voltage between the N-th cross-section and M-th cross-section (Dong F. et al., 2005).

$$\delta = \sqrt{\frac{1}{208}\sum_{i=1}^{208}(V_{Ni} - V_{Mi})^2} \times 100\% \quad (1)$$

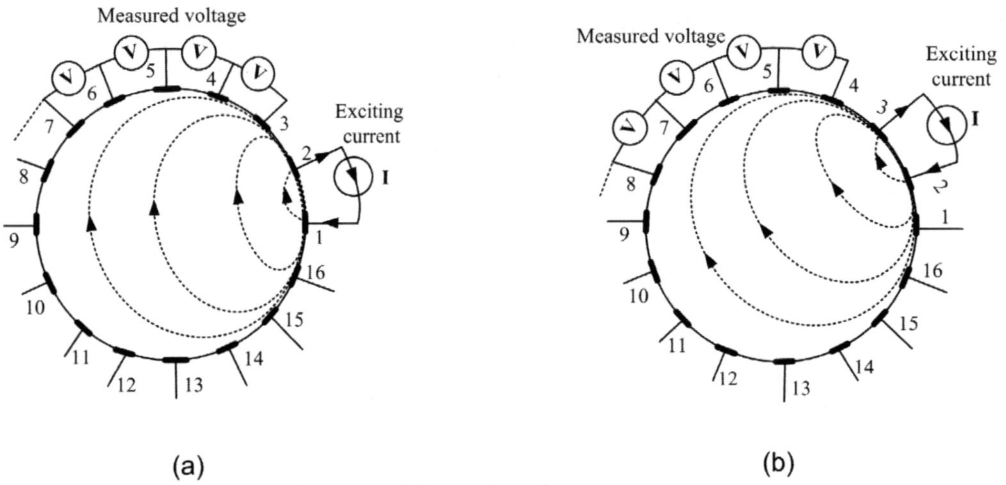

Fig. 2 Exciting and measurement strategy

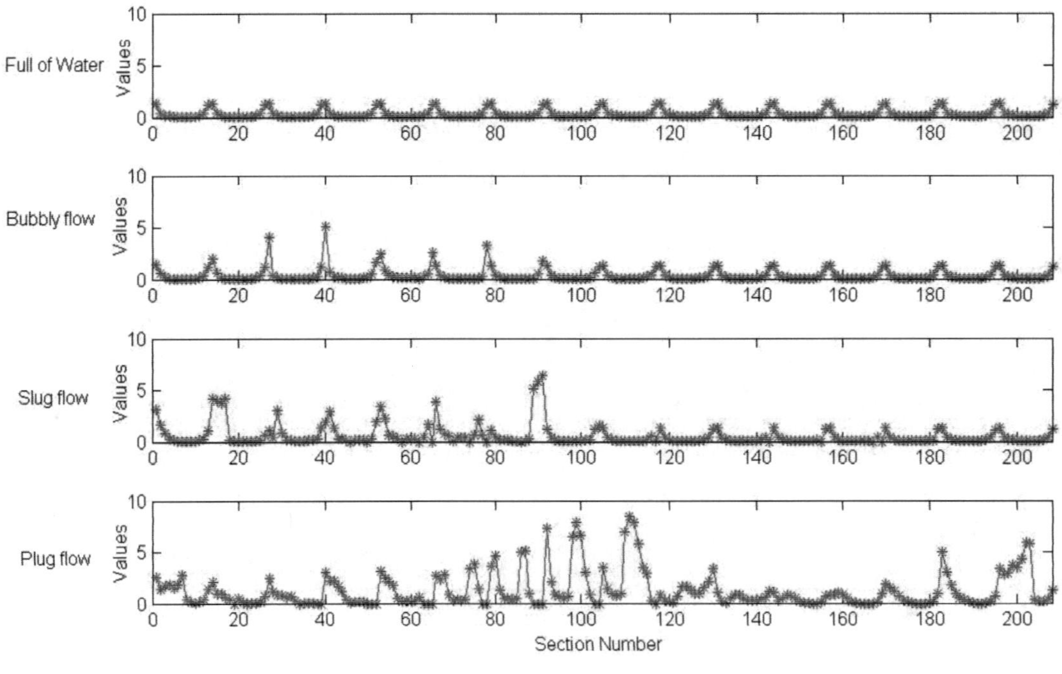

Fig.3 The curves of one frame data under four flow regimes

Normally, the operating principle of the cross-sectional measurement system based on ERT is current-excitation and voltage-measurement, as shown in Fig.2. The exciting current is applied into the measured section through a pair of electrodes and establishes the sensing field. The sensing field varies with the conductivity distribution and results in the change of the distribution of electric potential, and correspondingly the change of boundary voltage of sensing field. Therefore, the measured voltage contains the information of the conductivity of sensing field, and the internal flow status can be obtained from further information processing.

The 16-electrode sensors were adopted in the cross-sectional measurement system, shown as in

Fig.2. The adjacent strategy was used in this research. According to the operating procedure of this strategy, gathering data of one frame needs 16 excitations and 13 voltage data acquires under each excitation. The data of one frame which is also known as one section will form a 16×13 array, so 208 measured voltages from 16 excitations for one frame of cross-sectional measurement are measured. When the pipe is full of water, the 208 collected voltage data are in the trend of 16 U-shape. The trends of one section data in four kinds of flow regimes are shown as Fig.3.

2.2 Concept of Entropy and Five Entropies from Information Viewpoint

In thermodynamics (a branch of physics), entropy is a measure of the unavailability of a system's energy to do work (Coifman, R.R. et al., 1992; John. A., 2003). It is a measure of the randomness of molecules in a system and is central to the second law of thermodynamics and the fundamental thermodynamic relation, which deals with physical processes and whether they occur spontaneously. Spontaneous changes, occur with an increase in entropy in isolated systems, tend to smooth out differences in temperature, pressure, density, and chemical potential that may exist in a system, so entropy is thus a measure of how far this smoothing-out process has progressed and it is one of the factors that determines the free energy of the system. In terms of statistical mechanics, the entropy describes the number of the possible microscopic configurations of the system and is a quantitative measure of disorder in a system. The statistical definition of entropy is a more fundamental definition, from which all other definitions and all properties of entropy follow. The term "entropy" is generally interpreted in three distinct, but semi-related, ways, i.e. from macroscopic viewpoint (classical thermodynamics), a microscopic viewpoint (statistical thermodynamics), and an information viewpoint (information theory).

In the wavelet packet framework, compression and de-noising ideas are identical to those developed in the wavelet framework. The only new feature is a more complete analysis that provides increased flexibility. A single decomposition using wavelet packets generates a large number of bases. The following five entropies are the different entropy criterias in the wavelet packet framework. Many others are available and can be easily integrated. On the concept of entropy from the information viewpoint (information theory), there are many definitions of the method and the selection must be based on their applied occasions. Most of them use the five main entropies: the (nonnormalized) Shannon entropy, the concentration in norm entropy, the "log energy" entropy, the threshold entropy and the "SURE" entropy. In the following expressions, s is the signal and $(s_i)_j$ the coefficients of s in an orthonormal basis. The entropy E must be an additive cost function such that $E(0)=0$ and $E(s)=\sum_i E(s_i)$.

(1) The (nonnormalized) Shannon entropy

$$\mathrm{E1}(s_i) = s_i^2 \log(s_i^2) \quad (2)$$

So

$$\mathrm{E1}(s) = -\sum_i s_i^2 \log(s_i^2) \quad (3)$$

(2) The "log energy" entropy with the convention

$$\log(0)=0 : \mathrm{E2}(s_i) = \log(s_i^2) \quad (4)$$

So

$$E2(s) = \sum_i \log(s_i^2) \quad (5)$$

(3) The threshold entropy

$E3(s_i) = 1$ if $|s_i| > p$ and 0 elsewhere so $E3(s) = \#\{i \text{ such that } |s_i| > p\}$ is the number of time instants when the signal is greater than a threshold p.

(4) The "SURE" entropy

$$E4(s) = n - \#\{i \text{ such that } |s_i| \le p\} + \sum_i \min(s_i^2, p^2) \quad (6)$$

(5) The concentration in norm entropy

$$E5(s_i) = |s_i|^P \quad (7)$$

So

$$E5(s) = \sum_i |s_i|^P = \|s\|_P^P \quad (8)$$

3. ENTROPY-BASED FEATURE EXTRATION

According to the operating procedure of the cross-sectional measurement system, gathering data of one frame needs 16 excitations with 13 voltage data measured under each excitation. The 208 measured voltages in one frame of measured cross-section can be formed. The multi-section measurement data is formed by continual measurement.

For a cross-section of data, a variety of data-processing methods can be adopted. The first method is the entropies feature extraction based on the 208 meausrement data from one cross-section. The second method is the entropies feature extraction based on different direction excition electrodes, i.e. the feature is extracted from the 13 data under each pair of excitation to form 16 sets of data. The third method is the entropies feature extraction based on measurement electrodes. The 13 data which is acquired by each pair of measurement electrode is analyzed and processed to form 16 sets of data in a cross-section.

In the following sections, frive entropies are extracted from 60 sequential frames' data with three data-processing methods repectively in four flow regimes of the gas/liquid two-phase flow in horizontal pipe flow.

3.1 Entropies Feature Extraction Based on Section

In this perspective, the 208 data in each section is analyzed with the results shown in Fig.4. The x-coordinate denotes the sequential sections and y-coordinate denotes the values of entropy feature analysis in the subgraph. The analyzed results are also formed according to the four flow regimes of the gas/liquid two-phase flow in horizontal pipe flow.

When the pipe full of water, the Shannon entropy, the SURE entropy as well as the NORM entropy have the good characteristic, but the Log entropy and the Threshold entropy are easily influenced by disturbance.

In Bubbly flow, the Log entropy, the SURE entropy and the NORM entropy well manifest the bubble characteristic. The Shannon entropy has also manifested the bubble characteristic, but is inferior to the former three entropies. The Threshold entropy does not obviously manifest the bubble characteristic in bubbly flow, so that the threshold value must be selected carefully.

All the five entropies manifest slug flow in the characteristic. And the performances of Log entropy, the Threshold entropy, the SURE entropy and the NORM entropy are more obvious, and manifest the appearance of bubble in the wave crest and the trough blends.

In plug flow, the Log entropy, the Threshold entropy, the SURE entropy, and the NORM entropy perform better than others, and may also manifest the wave ridge and the trough blends.

In this kind of data processing of the characteristic analysis and extraction method, the Log entropy, the SURE entropy and the NORM entropy have better performance, and well manifest the flow quality. Through proper selction of threshold value, the Threshold entropy can be used in characteristic analysis and the extraction of the flow under particular conditions. The results indicate the Log entropy may be more suitable for the identification of the four flow regimes of the gas/liquid two-phase flow in horizontal pipe flow.

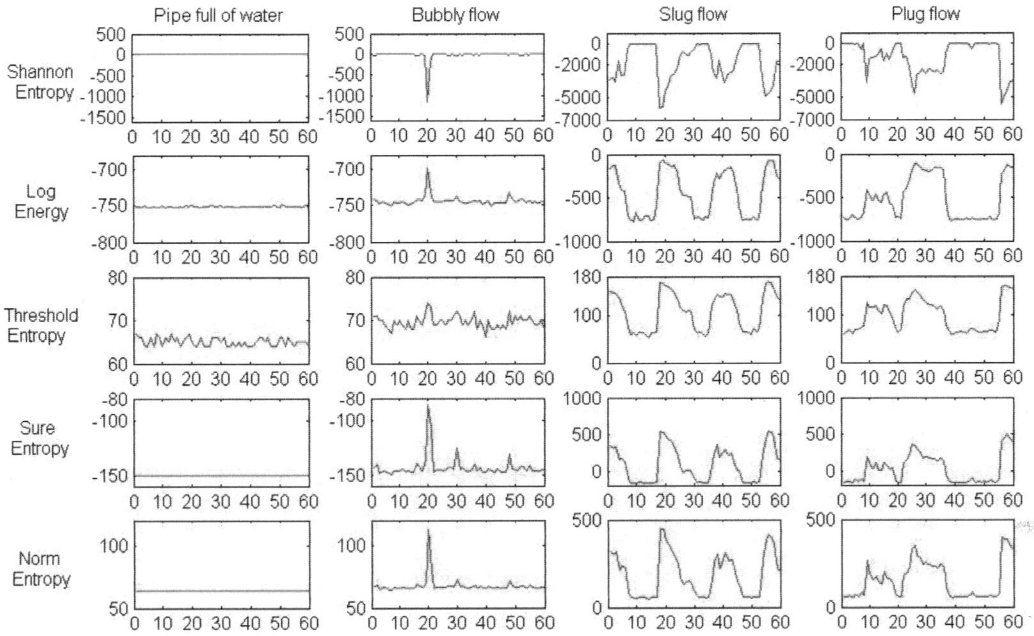

Fig.4 Entropy feature analysis based on 208 data measured in one section

3.2 Entropies Feature Extraction Based on Different Direction Exciting Electrodes

In this perspective, the 13 data measured by each excitation is analyzed, the 16 analyzed results are formed in each section and the analyzed results of the 60 sequential sections are presented. The results are shown in Fig.5, Fig.6, Fig.7, Fig.8 and Fig.9. The coordinate definition of the figures refers to the nomenclature part. The x-coordinate denotes 16 excitations, y-coordinate denotes the sequential sections and z-coordinate denotes the values of entropy feature analysis in the sub graph. According to the four flow regimes of the gas/liquid two-phase flow in horizontal pipe flow and five entropies, the analyzed results are presented in 3 dimensional (3D) cordinataes, and the cross-section space better reflects the characteristics of the flow.

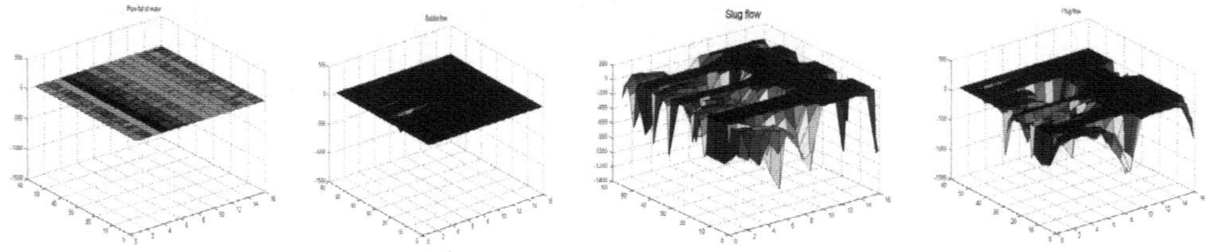

(a) (b) (c) (d)

Fig.5 Shannon Entropy feature based on the 13 measurement data by Each Excitation in each section: (a) Pipe full of water, (b) Bubbly flow, (c) Slug flow, (d) Plug flow.

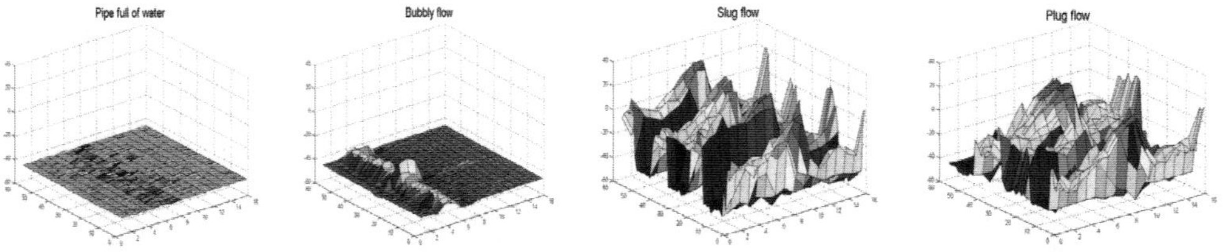

(a) (b) (c) (d)

Fig.6 Log Entropy feature based on the 13 measurement data by Each Excitation in each section : (a) Pipe full of water, (b) Bubbly flow, (c) Slug flow, (d) Plug flow.

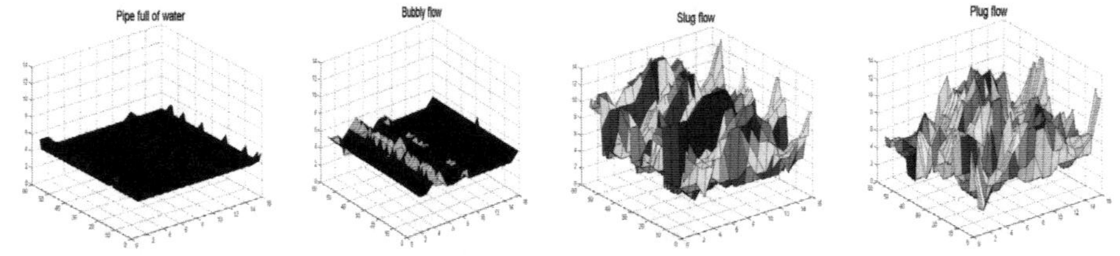

(a) (b) (c) (d)

Fig.7 Threshold Entropy feature based on the 13 measurement data by Each Excitation in each section : (a) Pipe full of water, (b) Bubbly flow, (c) Slug flow, (d). Plug flow.

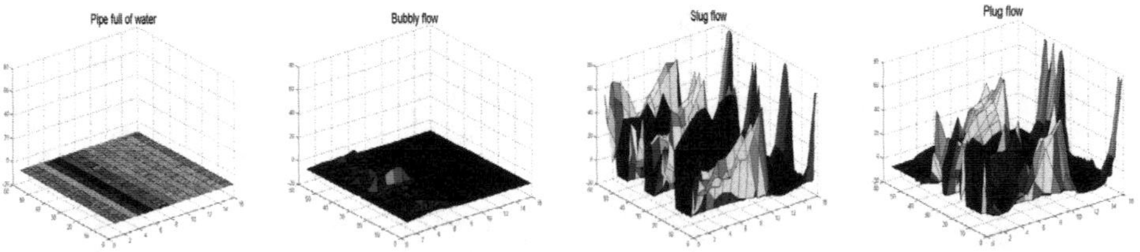

(a) (b) (c) (d)

Fig.8 SURE Entropy feature based on the 13 measurement data by Each Excitation in each section : (a) Pipe full of water, (b) Bubbly flow, (c) Slug flow, (d) Plug flow.

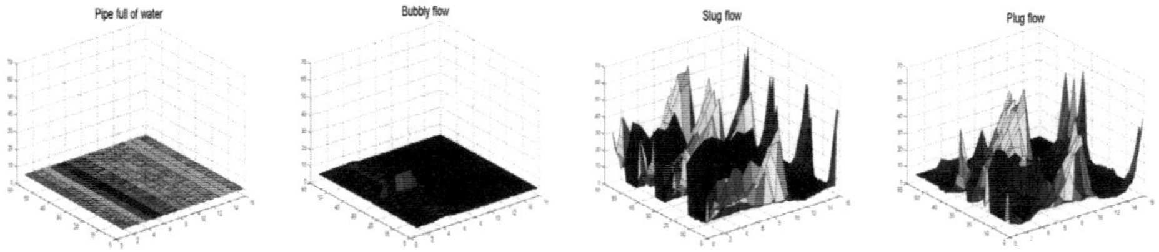

(a) (b) (c) (d)

Fig.9 Norm Entropy feature based on the 13 measurement data by Each Excitation in each section : (a) Pipe full of water, (b) Bubbly flow, (c) Slug flow, (d) Plug flow.

Regarding the pipe full of water, it is similar with the above conclusion that the Shannon entropy, the SURE entropy and the NORM entropy have better performance, but the Log entropy and the Threshold entropy is easy affacted by disturbance. For the bubbly flow, the five entropies well manifest the bubble characteristic. The Log entropy effectively manifests the characteristics of the flow of bubble in the process and the cross-sectional space. The threshold value of Threshold entropy must be selected suitably. The five entropies manifest the characteristic of slug flow and plug flow. In slug flow, the performance of the Log entropy, the Threshold entropy is more obvious over others, and better manifest the appearance of bubble in the wave crest and the trough blends. On the other hand, data processing of the Log entropy and the Threshold entropy is smooth, while the others are incisive.

In this kind of data processing of the characteristic analysis and extraction method, the Log entropy is considered better, and can well and smoothly manifest the flow characteristic and quality in process and the cross-sectional space of the flow.

3.3 Entropies Feature Extraction Based on Measurement Electrodes

In the perspective, the 13 data measured by each pair of the measurement electrodes is analyzed, the 16 sets of analyzed results of data measured by 16 pairs of the measurement electrodes are achieved in each section and the analyzed results of the 60 sequential sections are also presented. The results are shown in Fig.10, Fig.11, Fig.12, Fig.13 and Fig.14. The coordinate definition of the figures also refers to the nomenclature part. The x-coordinate denotes 16 excitations, y-coordinate denotes the sequential sections and z-coordinate denotes the values of entropy feature analysis in the subgraph. Corresponding to the four flow regimes of the gas/liquid two-phase flow in horizontal pipe flow and five entropies, the analyzed results and comparison groups are presented in 3D cordinates.

When the pipe is full of water and in bubbly flow, it is similar with the above analysis. For the slug flow and plug flow, the five entropies also well manifest the characteristic of the flow. This method is more capable of distingushing slug and plug flow , but the characteristics of plug in flow are not obvious enough.

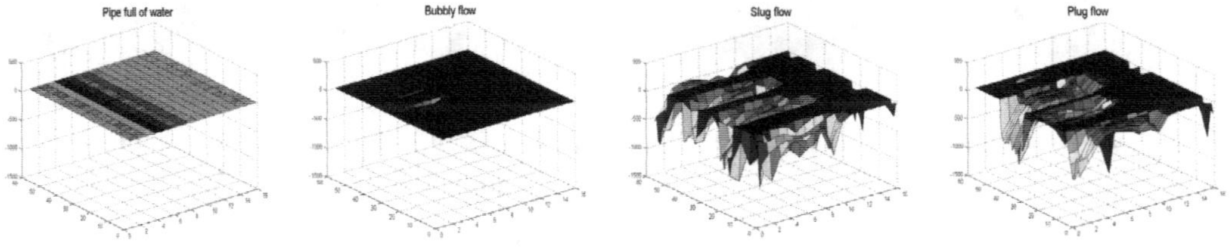

(a) (b) (c) (d)

Fig.10 Shannon Entropy feature based on the 13 data measured by each of the measurement electrodes in each section : (a) Pipe full of water, (b) Bubbly flow, (c) Slug flow, (d) Plug flow.

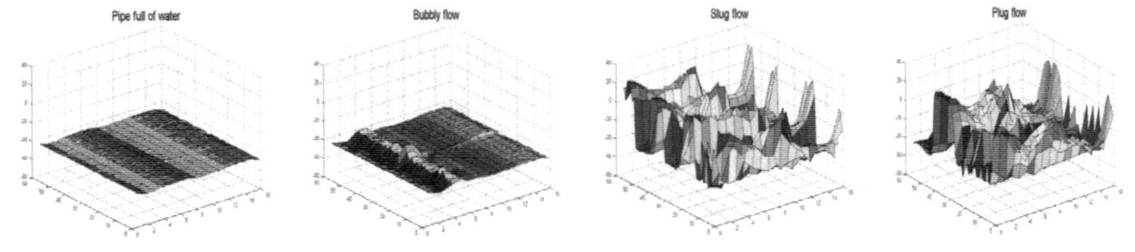

(a) (b) (c) (d)

Fig.11 Log Entropy feature based on the 13 data measured by each of the measurement electrodes in each section : (a) Pipe full of water, (b) Bubbly flow, (c) Slug flow, (d) Plug flow.

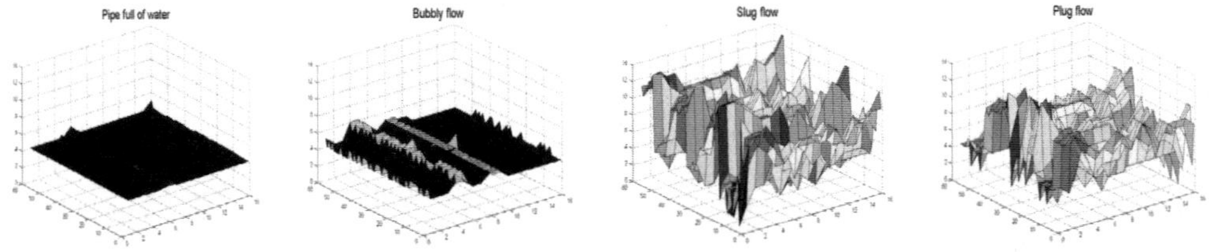

(a) (b) (c) (d)

Fig.12 Threshold Entropy feature based on the 13 data measured by each of the measurement electrodes in each section : (a) Pipe full of water, (b) Bubbly flow, (c) Slug flow, (d) Plug flow.

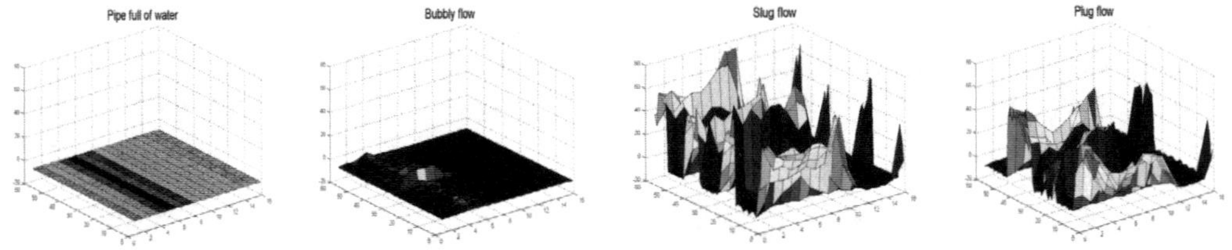

(a) (b) (c) (d)

Fig.13 SURE Entropy feature based on the 13 data measured by each of the measurement electrodes in each section : (a) Pipe full of water, (b) Bubbly flow, (c) Slug flow, (d) Plug flow.

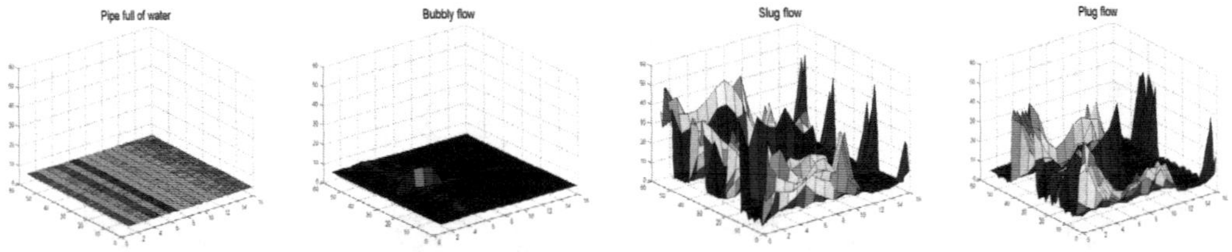

(a) (b) (c) (d)

Fig.14 Norm Entropy feature based on the 13 data measured by each of the measurement electrodes in each section : (a) Pipe full of water, (b) Bubbly flow, (c) Slug flow, (d) Plug flow.

4. CONCLUSIONS

This paper introduces the fundamental of the cross-sectional measurement based on a ERT system. The measured data of four flow regimes of the gas/liquid two-phase flow in horizontal pipe flow were obtained. For the measured data, five entropies are extracted to analyze the experimental data in different flow regimes, and the analysis method is introduced in three different perspectives. From the different method of feature extraction, comparison groups are formed and they show that the appropriate entropy should be used for the identification and prediction of flow regimes. In the first method, the Log entropy, the SURE entropy and the NORM entropy have better performance in representing flow characteristic, and well manifest the flow quality. The results indicate the Log entropy is more suitable for the identification of the four flow regimes of the gas/liquid two-phase flow in horizontal pipe flow. In the second method, the Log entropy is also considered better and smoothly manifest the flow characteristic in the process and cross-sectional space of the flow. In the third method, it is not obvious between each method and all the five entropies well manifest the flow characteristic.

The results indicate that the three different perspectives of entropy feature extraction are sensitive to the flow pattern transition in gas/liquid two-phase flow. By analyzing the rules of three different perspectives with the changes of gas/liquid two-phase flow parameters, the dynamic structures of gas/liquid two-phase flow can be obtained, and an efficient supplementary to reveal the flow pattern transition mechanism of gas/liquid two-phase flow can also be provided. From three different perspectives of feature extraction, comparison groups are formed and they show that the appropriate entropy should be used for the identification and prediction of flow regimes.

ACKNOWLEDGEMENTS
The author appreciates the support from National Natural Science Foundation of China (50776063) and Natural Science Foundation of Tianjin (08JCZDJC17700).

NOMENCLATURE

δ	The drift of measuring voltage
s	The signal
$(s_i)_j$	The coefficients of signal s in an orthonormal basis
E	Entropy value
$\sum_i$	Summation symbol
N	the number of the cross-section
M	the number of the cross-section
i	*i*-th coordinates
j	*j*-th coordinates

Coordinates definition:

2D Coordinates The example is shown in Fig. A (a). The x-coordinate denote the sequential cross-section number and y-coordinate denote the values.

3D Coordinates The example is shown in Fig. A (b). The x-coordinate denote the electrode number which can be exciting electrodes or measuring electrodes, y-coordinate denote the sequential cross-section number and z-coordinate denote the values.

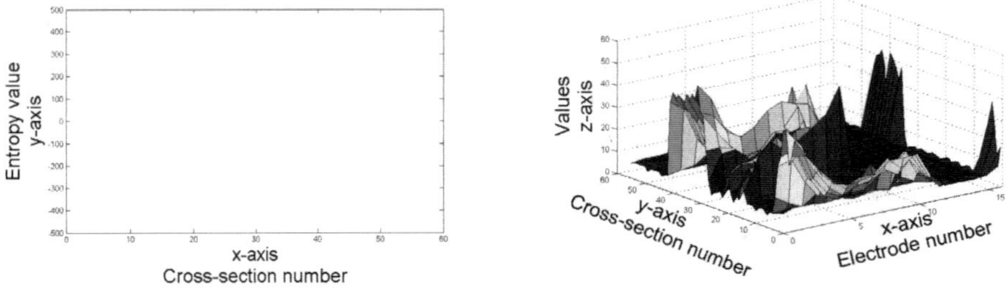

(a) (b)

Fig.A. The example of coordinates definition

REFERENCES

Baker, O. (1954). "Simultaneous flow of oil and gas," Oil Gas J. 53, pp. 184–195.

Bendat, J.S. et al. (1971). Random Data Analysis and Measurement Procedures. Wiley Interscience, New York.

Bergh, H. et al. (1965). "Theoretical and experimental results for the dynamic response of pressure measuring systems," Report NLR-TR F. pp. 238, National Aeronautical and Astronautical Research Institute, Amsterdam.

Bertola, V. (2003). "Experimental characterization of gas–liquid intermittent sub-regimes by phase density function measurements," Exp. Fluids 34(1), pp. 122–129.

Broadhurst, T.E. et al. (1976). "Measurement and spectral analysis of pressure fluctuations in slugging beds," In: Keairns, D.L. (Ed.), Fluidization Technology, Vol. 1. Engineering Foundation, New York, pp. 63-85.

Chatfield E. (1991). The Analysis of Time Series, New York, USA, Chapman and Hall, Fourth edition.

Chigier, N. (2005). "The future of atomization and sprays," In: ILASS-Europe, Orleans, France, September 5–7.

Coifman, R.R. et al. (1992). "Entropy-based Algorithms for best basis selection," IEEE Trans. on Inf. Theory, 38(2), pp. 713-718,.

Coleman, J.W. et al. (1999). "Characterization of two-phase flow patterns in small diameter round and rectangular tubes," Int. J. Heat Mass Transfer, 42, pp. 2869–2881.

Daw, C.S. et al. (1991). "Characterization of voidage and pressure signals from fluidized beds using deterministic chaos theory," Proc. Eleventh Int. Conf. Fluidized Bed Combustion. American Society of Mechanical Engineers, New York, pp. 777-786.

Dong F. et al. (2001). "Identification of two-phase flow regimes in horizontal, inclined and vertical pipes, " Measurement Science and Technology, 12(8), pp. 1069-1075.

Dong F. et al. (2003). "Application of electrical resistance tomography to two-phase pipe flow parameters measurement," Flow Measurement and Instrumentation, 14(4-5), pp. 183-192.

Dong F. et al. (2005). "Application of dual-plane ERT system and cross correlation technique to measure gas-liquid flows in vertical upward pipe," Flow Measurement and Instrumentation, 16(2-3), pp. 191-197.

Dong F. et al. (2006). "Two methods for measurement of gas-liquid flows in vertical upward pipe using dual-lane ERT system," IEEE Transaction on Instrumentation and Measurement,.55(5), pp.1576-1586.

Drahos, J. et al. (1996). "Deterministic chaos analysis of pressure fluctuations in an horizontal pipe at intermittent flow regime," Chem. Eng. J. 64, pp. 149–156.

Dumouchel, C. (2005). "Experimental analysis of a liquid atomization process at low Weber number," In: ICHMT International Symposium, 6th–10th, Antalya, Turkey, June.

Dumouchel, C. et al. (2005) "On the role of the liquid flow characteristics on low Weber number atomization processes," Exp. Fluids 38, pp. 637–647.

Elman J. L. (1990). "Finding structure in Time," In Cognitive Science, 14, pp. 179-211.

Fabre, J. et al. (1992). "Modelling of two-phase slug flow," Ann. Rev. Fluid Mech. 24, pp. 21–46.

Faeth, G.M. et al. (1995). "Structure and breakup properties of sprays," Int. J. Multiphase Flow 21, pp. 99–127.

Farabee, T.M. et al. (1991). "Spectral features of wall pressure fluctuations beneath turbulent boundary layers," Phys. Fluids A 3, pp. 2410-2420.

Frisch, U. et al. (1981). "Intermittency in nonlinear dynamics and singularities at complex times," Physical Review A 23, pp. 2673-2705.

John, A. (2003). Information Theory and Evolution. World Scientific. ISBN 981-238-399-9.

Kawahara, A. et al. (2002). "Investigation of two-phase flow pattern, void fraction and pressure drop in a microchannel". Int. J. Multiphase Flow 28, pp. 1411–1435.

Kohn, R. (1980). "On the spectral decomposition of stationary time series using Walsh functions," I & II. Adv. Appl. Prob. 12, pp. 183–199.

Lirag, R.C. et al. (1971). "Statistical study of the pressure fluctuations in a fluidized bed," Chem. Eng. Prog. Symp. Series 67, pp. 11-22.

Monin, A.S. et al. (1975). Statistical Fluid Mechanics. MIT Press, New York.

Morettin, P.A. (1981). "Walsh spectral analysis," SIAM Rev. 23, pp. 279–291.

Newby, R.A. et al. (1986). "Test of the scaling relationships for fluid-bed dynamics," In: Ostergaard, K., Sorensen, A. (Eds.), Fluidization V. Engineering Foundation, New York, pp. 31-38.

Paglianti, A. et al. (1996). "Characterization of subregimes in two-phase slug flow," Int. J. Multiphase Flow, 22, pp. 783–796.

Proakis, J.G. et al. (1989). Introduction to Digital Signal Processing. Macmillan, New York.

Wu, M.M. et al. (2006). "Feature extraction method for gas/Liquid two-phase flow based on wavelets transform," Proceedings of the Fifth International Conference on Machine Learning and Cybernetics, Dalian, China, August13-15, pp.1422-1427.

Journal of Physics: Conference Series **147** (2009) 012042 doi:10.1088/1742-6596/147/1/012042

A numerical boundary integral method for cross-sectional distribution of gas/water two phase flows

Y Y Xu, F Dong, C Tan

Tianjin Key Laboratory of Process Measurement and Control
School of Electrical Engineering and Automation, Tianjin University, Tianjin 300072, CHINA

fdong@tju.edu.cn

Abstract. A clearer understanding of two-phase flows requires getting the distribution of gas/water two-phase flows non-invasively. A numerical method to get the cross-sectional distribution of gas/water two-phase flows, via acquiring the common boundary of the two domains, is presented. This approach is based on the boundary integral equations reduced from Green's formula and their derivatives, considering that the gas phase is isolating and the conductivity of water phase is known. The forward problem is solved by boundary element method, which shows considerable accuracy but much smaller matrix size compared with finite element method. The boundary integral equations with non-smooth boundaries are discreted, and their Frechet derivatives are linearized. An iterative method is adopted, and the initial guess is estimated with the aid of the Linear Back Projection Algorithm for electrical resistance tomography. Numerical results show that the method can locate the gas bubbles and determine their shape effectively. Residual error is defined to investigate the convergence processes. In the anterior steps, the processes converge more quickly for fewer elements; while in the posterior steps more mesh elements are needed to get better results. Element increment contributes to decrease the final residual errors, but it tends to be faint when the number of elements is considerably large.

1. INTRODUCTION

Gas/water two-phase flows are widely encountered in nature and modern industry. A clearer understanding of two-phase flows requires getting the distribution of gas/water two-phase flows non-invasively, since it has an influence up on the characteristics and behaviors of the mixtures of the two fluids. Numerous techniques have been developed for this purpose over the years: radiation radiography techniques (Uchimura, 1998; Takenaka, 1998), X-ray computed tomography (Toye, 1998; Harvel, 1999), ultrasonic tomography (Xu, 1997), electrical tomography techniques (Bennett, 2002; Kim, 2002), optical techniques (Ursenbacher, 2004) and so on.

Electrical resistance tomography (ERT) is a novel method to determine the electrical conductivity distribution of a cross-section by applying currents at the boundary and measuring the resulting voltages at the electrodes. Comparing to computerized tomography (CT), which is a standard tool in medical diagnostics and non-destructive testing of materials, ERT has many advantages in cost, speed and safety, which makes it more suitable in two-phase or multi-phase flow context. Applications of ERT on two-phase flows have been presented by Dong (2001), Fransolet (2005) and Jin (2007).

ERT system operates in the following way: A number of electrodes are distributed around the cross-section at equal intervals. A constant current of low frequency is applied to an adjacent pair of electrodes, and potentials are measured on the electrodes. It then switches the current to another pair of electrodes. When the distribution of gas differs, the measured voltages vary. Therefore, the measured potentials contain the information of the gas distribution. After the excited pair of electrodes makes a rotation, a set of potential data is collected for further processing.

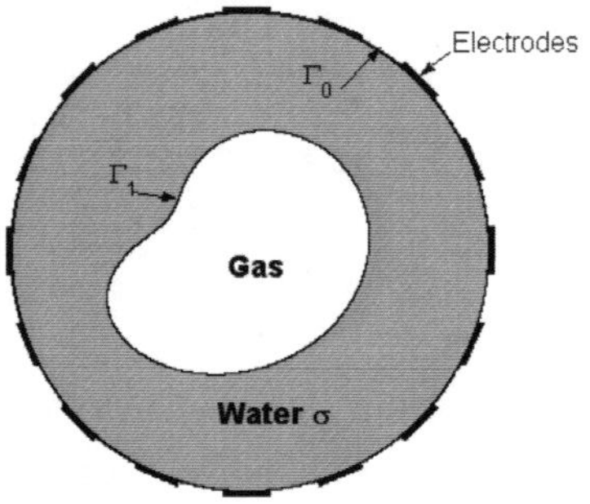

Fig.1 A sketch map of ERT sensors

The following difficulty makes it hard to apply ERT to online quantitative multi-phase flow measurement. As the adjacent excitation mode is generally adopted, electrical currents mainly travel from one excited electrode directly to the other. As a result, variation in the conductivity near the centre of the cross-section has little effect on the boundary potentials measured. And concerning the inverse problem, small measurement errors easily cause vibration of the conductivity distribution in the interior of the body. Furthermore, total independent measurements are much fewer than the number of conductivity mesh due to capability constraints of the hardware system. Mathematically speaking the electrical impedance tomography is a nonlinear and severely ill-posed problem, which makes it hard to get a precise image of the electrical conductivity distribution. So it is important to incorporate as much a priori knowledge about the object as possible.

As to the gas/water two-phase flow, the discrete gas phase is isolating, and the conductivity of water phase is constant and can be measured using additional equipment. We only have to find out the interior boundary which separates out the gas and water. Thus priori knowledge is fully utilized, and makes the problem much easier to approach and more likely to be eventually solved numerically.

This paper is partly based on recent work of Kress (2005), Ivanyshyn (2006) and Eckel (2007) where a new integral equation method has been developed to recover inclusions with piecewise constant conductivities, inclusions, and perfectly conducting, respectively. The main work of this paper is to generate and linearize the boundary integral equations (BIEs), and also their derivatives, of 2D laplace equation with non-smooth boundaries, concerning that non-smooth points exist in realistic ERT systems. Another reason is that the discretization of boundary into some set of elements is often used for numerical solutions, which inevitably makes the boundary unsmooth. Thus it can be regard as a supplement to their theory.

2. BOUNDARY INTEGRALS

Suppose $D \in R^2$ is a simply connected bounded domain which represents the conducting medium, the electric potential u satisfies the Laplace equation and boundary conditions:

$$\begin{cases} \Delta u = 0 & \text{in } D \\ \int_{E_i} \sigma \dfrac{\partial u}{\partial n} ds = g_i & \text{on electrode } i \ (i = 1, 2, ..., 16) \\ u = const & \text{on each electrode} \\ \sigma \dfrac{\partial u}{\partial n} = 0 & \text{on other points of } \partial D \end{cases} \quad (1)$$

where g_i represents the imposed current on the i-th electrode, n the outward unit normal to the boundary ∂D, and σ the electrical conductivity in D.

Introduce the well-known fundamental solution to the Laplace equation in R^2:

$$u^*(x,y) = -\frac{1}{2\pi} \ln|x - y| \quad x \neq y \quad (2)$$

Applying Green's integral theorem, get the integral equation directly:

$$\alpha(y)u(y) = \int_{\partial D} \left(u^*(x,y) \frac{\partial u}{\partial n} - u \frac{\partial u^*(x,y)}{\partial n} \right) dSx \qquad y \in \partial D \quad (3)$$

where $\alpha(y) = \theta(y)/2\pi$, and $\theta(y)$ is the internal angel at point y.

For analytical convenience, the boundary of the domain D is divided into two curves: Γ_0 (exterior) and Γ_1 (internal), and denote the unit normal of Γ_1, n, orient to the interior D. And rewrite equation (3) to be:

$$\alpha(y)u(y) + \int_{\Gamma_0} u \frac{\partial u^*(x,y)}{\partial n} dSx - \int_{\Gamma_0} u^*(x,y) \frac{\partial u}{\partial n} dSx = \int_{\Gamma_1} u \frac{\partial u^*(x,y)}{\partial n} dSx \qquad y \in \Gamma_0 \quad (4)$$

$$-\alpha(y)u(y) + \int_{\Gamma_1} u \frac{\partial u^*(x,y)}{\partial n} dSx = \int_{\Gamma_0} u \frac{\partial u^*(x,y)}{\partial n} dSx - \int_{\Gamma_0} u^*(x,y) \frac{\partial u}{\partial n} dSx \qquad y \in \Gamma_1 \quad (5)$$

Now define the following functions:

$$w_0(f,g)(y) = \alpha(y)u(y) + \int_{\Gamma_0} f \frac{\partial u^*(x,y)}{\partial n} dSx - \int_{\Gamma_0} u^*(x,y) g \, dSx \qquad y \in \Gamma_0 \quad (6)$$

$$w_1(f,g)(y) = \int_{\Gamma_0} f \frac{\partial u^*(x,y)}{\partial n} dSx - \int_{\Gamma_0} u^*(x,y) g \, dSx \qquad y \in \Gamma_1 \quad (7)$$

$$K_0 h(y) = \int_{\Gamma_1} h \frac{\partial u^*(x,y)}{\partial n} dSx \qquad y \in \Gamma_0 \quad (8)$$

$$K_1 h(y) = -\alpha(y)h(y) + \int_{T_1} h \frac{\partial u^*(x,y)}{\partial n} dSx \qquad y \in \Gamma_1 \quad (9)$$

where f, h and g is $u| \, T_0$, $u| \, T_1$, $\frac{\partial u}{\partial n}| \, T_0$ respectively.

Equations (6)-(10) are not difficult to calculate, given an explicit description of the interior boundary and boundary conditions. However conditions given in equation (1) are not strictly Neumann nor Dirichlet forms, the integrals can be converted to algebraic equations via linearization which have a unique solution. Thus, if find a Γ_1, and the corresponding h satisfy following equations, then the inverse problem is solved.

$$K_0 h = w_0(f, g) \quad (10)$$

$$K_1 h = w_1(f, g) \quad (11)$$

Linear boundary elements are adopted to carry out above calculations and the forward problem. The potentials on the boundaries are considered to be continuous, but the derivatives of them are discontinuous on ends of electrodes. Fig. 2 compares the boundary potentials computed by two numerical methods: linear boundary element method (BEM) and finite element method (FEM) using commercial software *COMSOL*. The interior boundary is determined by $r = 0.5$ in the polar coordinate system, the radius of the exterior circle is 1, and electrode coverage rate is 0.5. Imposed current on the electrodes are 1 and -1 for the first and second electrode respectively, and 0 for others. For FEM, the mesh size is up to 18720, while the boundaries are meshed with only 174 elements for BEM, which shows considerable accuracy, but much smaller matrix size and higher speed.

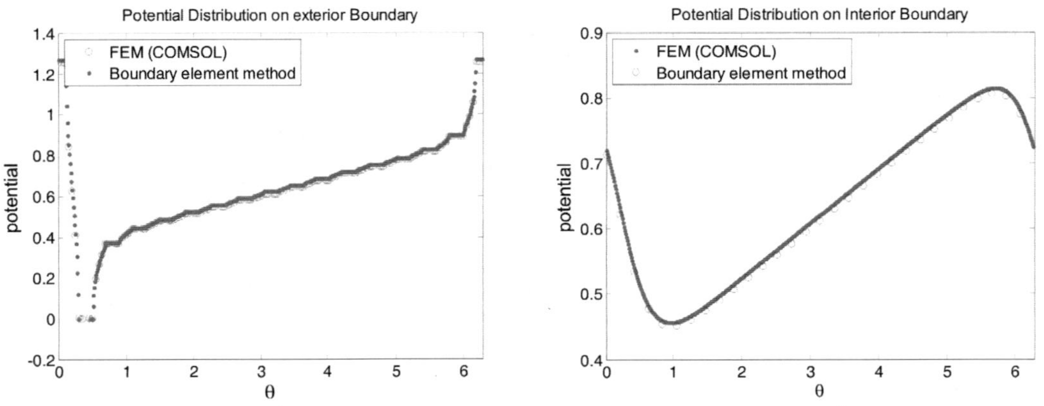

(a) (b)

Fig. 2 Comparison of potential distributions on the boundaries with two numerical methods; (a): on the exterior boundary; (b): on interior boundary

If equations (10) and (11) don't hold, the Frechet derivatives of the operators $K_0 h$ and $K_1 h$, with respect to Γ_1, is needed to get iterative equations. Suppose the interior boundary is meshed into line segments:

$$\Gamma_1 : \ = \left\{ \cup z_1^e(t), \quad e = (1, 2, 3, \ldots, n1), t \in [0, 1] \right\} \quad (12)$$

where $n1$ is the amount of segments. The derivative of $K_0 h$, $\mathrm{d}K_0 h$, can be calculated by the following formula:

$$\mathrm{d}K_0 h^e = \frac{1}{2\pi}\int_0^1 \left\{ \frac{2\left[(z_1^e)'(\tau)\right]^{\perp}\cdot\left[z_1^e(\tau)-y\right]\ \left[z_1^e(\tau)-y\right]\cdot\zeta(\tau)}{\left|z_1^e(\tau)-y\right|^4} - \frac{\left[(z_1^e)'(\tau)\right]^{\perp}\cdot\zeta(\tau)+\left[\zeta'(\tau)\right]^{\perp}\cdot\left[z_1^e(\tau)-y\right]}{\left|z_1^e(\tau)-y\right|^2}\right\} h^e(\tau)d\tau \qquad y\in\Gamma_0 \tag{13}$$

where $\zeta(\tau)\quad \tau\in[0,1]$ is the small spatial increment on z_1^e. It is much more complicated to get the derivative of $K_1 h$, which requires consideration of both the variation of internal angel and Γ_1 in the second part of equation (9).

$$dK_1 h = h d\alpha + d\widetilde{K}_1 h \tag{14}$$

where $\mathrm{d}\widetilde{K}_1 h$ may be calculated by:

$$d\widetilde{K}_1 h^e = \begin{cases} \frac{1}{2\pi}\int_0^1 \left\{ \frac{2\left[(z_1^e)'(\tau)\right]^{\perp}\cdot\left[z_1^e(\tau)-y\right]\ \left[z_1^e(\tau)-y\right]\cdot\zeta(\tau)}{\left|z_1^e(\tau)-y\right|^4} - \frac{\left[(z_1^e)'(\tau)\right]^{\perp}\cdot\zeta(\tau)+\left[\zeta'(\tau)\right]^{\perp}\cdot\left[z_1^e(\tau)-y\right]}{\left|z_1^e(\tau)-y\right|^2}\right\} h^e(\tau)d\tau & y\in\Gamma_1^j,\ e\neq j \\ \frac{1}{2\pi}\sum_j\int_0^1 \left\{ \frac{2\left[(z_1^j)'(\tau)\right]^{\perp}\cdot\left[z_1^j(\tau)-y\right]\ \left[z_1^j(\tau)-y\right]\cdot\zeta(\tau)}{\left|z_1^j(\tau)-y\right|^4} - \frac{\left[(z_1^j)'(\tau)\right]^{\perp}\cdot\zeta(\tau)}{\left|z_1^j(\tau)-y\right|^2}\right\} h^j(\tau)d\tau & y\in\Gamma_1^e,\ e\neq j \end{cases} \tag{15}$$

Thus get the iterative equations:

$$K_0 h + K_0 \mathrm{d}h + \mathrm{d}K_0 h = w_0(f,g) \tag{16}$$

$$K_1 h + K_1 \mathrm{d}h + \mathrm{d}K_1 h = w_1(f,g) + grad\ w_1(f,g) \tag{17}$$

3. NUMERICAL EXPERIMENTS AND RESULTS

A few numerical examples are presented to illustrate the feasibility of the method as described in the previous sections. In all these examples the exterior boundary was chosen according to real equipment, which consists of a circular Plexiglas pipe and 16 rectangular electrodes. We tested the algorithm with one unknown interior boundary, which is confined to: (a) circular, (b) elliptical, (c) heart-like and (d) dumbbell-like. Differential voltages on adjacent electrodes, which constitute a 13*18 matrix, are simulated by the finite element tool *COMSOL*. As to the inverse problem, for the sake of rapidity, the amount of discrete line segments is confined to 25, and iteration steps to less then 10. Results are shown in Fig. 3. The initial circular guess is roughly estimated according to the image of conductivity distribution, which is acquired by using the Linear Back Projection (LBP) Algorithm of ERT. It is shown that the method can locate the gas bubbles and determine their shape effectively. Comparing the figures, the accuracy and the convergence speed are in connection with the shape of the gas phase. Since the common boundaries are discretized for approximation, the geometric coefficients on nodes usually don't equal to 0.5, and hence affect the iterative process. If the geometric coefficients are uniform and close to 0.5, a comparatively accurate result is easily acquired; Otherwise, the accuracy and convergence speed will be little constrainted where they pulse. In point of the electric field, the local electrical field strength where exists bigger reconstructive error, is so weak, and hence variation of this part of the boundary has little effect on the boundary measurements.

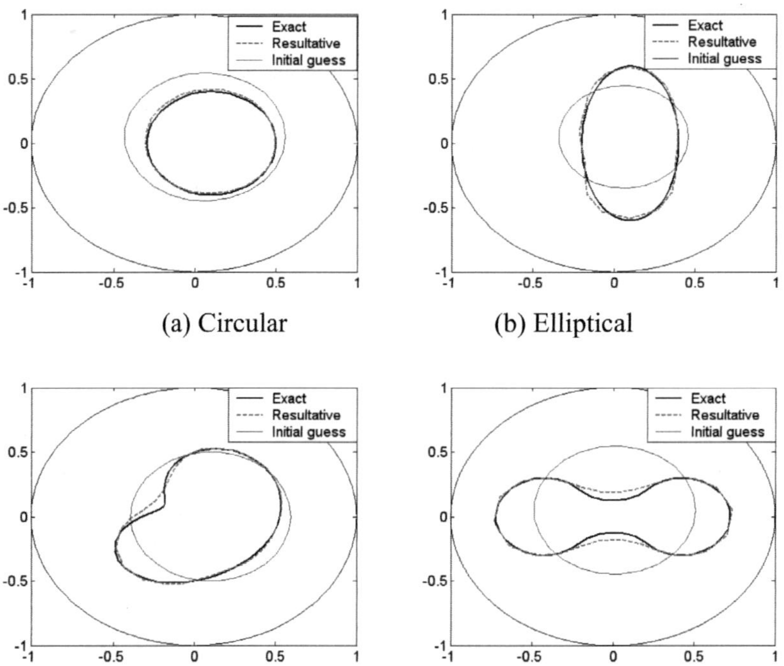

(a) Circular (b) Elliptical

(c) Heart-like (d) Dumbbell-like

Fig. 3 Reconstructions of 4 kinds of interior boundaries

In order to investigate the convergence process, the residual error is defined to be:

$$RE = \sum_{i=1}^{16} \sum_{j=1}^{13} \left| V_i^j - \overline{V_i^j} \right| \quad (18)$$

where V_i^j is the j-th voltage simulated of i-th excitation for current computed interior boundary, and $\overline{}$ denotes the given data. Fig. 4 shows the decrease of residual error while the number pf iterative steps increases for each kind of interior boundary in fig. 3. Three mesh sizes (20, 50, and 80 elements) are adopted to discuss the effect on the reconstructions.

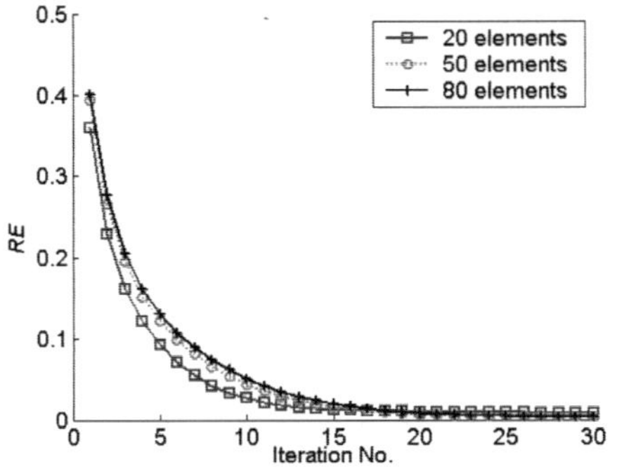

(a) Circular case (b) Elliptical case

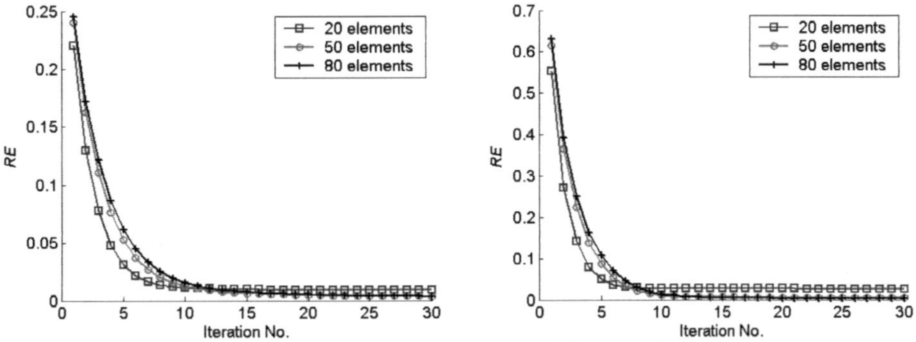

(c) Heart-like case (d) Dumbbell-like case

Fig. 4 Convergence processes for the four kinds of interior boundaries

The iterative processes converge more quickly in the anterior steps (less than 10 approximately) for relatively less elements, while in the posterior steps more mesh elements are favourable. The residual errors in the final iterations decrease as the number of elements increases, but this effect tends to be faint when the number of elements is large enough. The final residual errors differ little for the three mesh sizes in fig. 4(a), where the curve has a constant function of curvature. While in other subgraphs where the curvatures are similar in total to fig. 4(a), but partially with big bending, the final residual errors vary. Curvature of the boundaries has relation to the required mesh size. If the curve curls sharply we should refine the mesh and vice versa. Further research on mesh algorithm would be carried out.

4. CONCLUSIONS

Gas/water two-phase flows are widely encountered in nature and modern industry. A clearer understanding of two-phase flows requires getting the distribution of gas/water two-phase flows non-invasively. The ill-posed problem makes it hard to apply ERT to online quantitative multi-phase flow measurement. This approach is based on the BIEs of 2D laplace equation. BIEs, and also their derivatives, of non-smooth boundaries are deduced, and hence an iterative method is adopted. Four kinds of interior boundaries are tested which illustrate the feasibility and effectivity of the method. It shows that the accuracy and the convergence speed are in connection with the shape of the gas phase. The accuracy and convergence speed are little constrainted where the geometric coefficients vary sharply along the common boundary. The residual error is defined to investigate the convergence process. In the anterior steps, the processes converge more quickly for fewer elements; while in the posterior steps more mesh elements is required to get better results. Element increment contributes to decrease the final residual errors, but it tends to be faint when the number of elements is large enough.

ACKNOWLEDGEMENTS
The author appreciates the support from National Natural Science Foundation Committee of China (No. 50776063) and Program for New Century Excellent Talents in University of China (No. NCET-06-230).

NOMENCLATURE

D	simply connected bounded domain
R	real-space
u	electric potential
g	current density
n	unit normal
x	global 2-D coordinates
y	global 2-D coordinates
f	electric potential on Γ_0
h	electric potential on Γ_1
n1	amount of segments of Γ_1
z	curve function
w	curvilinear Integral on Γ_0
K	curvilinear Integral on Γ_1
RE	residual error of boundary data
V	boundary voltage
	Greek Letters
Γ	boundary
α	geometric coefficient on interval (0,1)
θ	the internal angel at point y
σ	electrical conductivity
ζ	natural coordinates
	Subscripts
i	*i*-th coordinates
0	exterior boundary
1	internal boundary
	Superscript
e	*e*-th coordinates
^{2}	2-D
^{j}	*j*-th coordinates

REFERENCES
Bennett, M.A. et al. (2002). "The investigation of bubble column and foam processes using electrical capacitance tomography", Miner. Eng., 15(4), pp: 225-234

Dong, F. et al. (2001). "Identification of two-phase flow regimes in horizontal, inclined and vertical pipes", Meas. Sci. Tech., 12(8), pp:1069-1075.

Eckel, H. et al. (2007). "Nonlinear integral equations for the inverse electrical impedance problem", Inverse Probl., 23(2), pp: 475-491.

Fransolet, E. et al. (2005). "Analysis of gas holdup in bubble columns with non-Newtonian fluid using electrical resistance tomography and dynamic gas disengagement technique", Chem. Eng. Sci., 60(22), pp: 6118-6123.

Harvel, G. D. et al. (1999). "Cross-sectional void fraction distribution measurements in a vertical annulus two-phase flow by high speed X-ray computed tomography and real-time neutron radiography techniques", Flow Measure. Instrum., 10(4), pp: 259–266.

Ivanyshyn, O. et al. (2006). "Nonlinear integral equations for solving inverse boundary value problems for inclusions and cracks", J. Integr. Equat. Appl., 18(1), pp: 13–38.

Jin, H. et al. (2007). "Analysis of bubble behaviors in bubble columns using electrical resistance tomography", Chem. Eng. J., 130(2-3), pp: 179-185.

Kim, M. et al. (2002). "An experimental study of electrical impedance tomography for the two-phase flow visualization", Int. Commun. Heat Mass Transf., 29(2), pp: 193-202

Kress, R. et al. (2005). "Nonlinear integral equations and the iterative solution for an inverse boundary value problem", Inverse Probl., 21(4), pp: 1207-1223.

Takenaka, N. et al. (1998). "Three-dimensional visualization of void fraction distribution in steady two-phase flow by thermal neutron radiography". Nucl. Eng. Des., 184(2-3), pp:203-212

Toye, D. et al. (1998). "Local measurements of void fraction and liquid holdup in packed columns using X-ray computed tomography", Chem. Eng. Process., 37(6), pp: 511-520

Uchimura, K. et al. (1998). "An image processing approach for two-phase interfaces visualized by a real time neutron radiography technique", Flow Measure. Instrum., 9(4), pp: 203-210

Ursenbacher, T. et al. (2004). "Interfacial measurements in stratified types of flow.Part I: New optical measurement technique and dry angle measurements", Int. J. Multiph. Flow,30(2), pp: 107-24

Xu, L.J. et al. (1997). "Gas/liquid two-phase flow regime identification by ultrasonic tomography". Flow Meas. Instrum. 8(3-4), pp. 145–155

Observation of the fluid pulsation behind a prism in gas-liquid annular flow

Zhiqiang Sun [1], **Wenhui Sang** [1], **Hongjian Zhang** [2]

[1] School of Energy Science and Engineering, Central South University, Changsha 410083, Hunan, CHINA
[2] Department of Control Science and Engineering, Zhejiang University, Hangzhou 310027, Zhejiang, CHINA

zqsun@mail.csu.edu.cn (Z. Sun, Corresponding author)

Abstract. The fluid pulsation behind a prism is closely concerned with the design and operation of gas-liquid two-phase flow systems. This paper explored the property of such fluid pulsations in a horizontal pipe using air and water as the tested media. The pressure difference between the upstream and downstream of the prism was adopted to represent the characteristics of the fluid pulsation. Based on the experimental results, it is found that there exit obvious pulsations in the fluid behind the prism in annular flows, and this pulsation is more marked at low gas mass qualities than those at high gas mass qualities. Furthermore, the average pressure basically decreases with the gas mass quality and increases with the gas-liquid total mass flowrate. These findings could be useful for the analysis of flowing characteristics in gas-liquid annular flow across a prism.

1. INTRODUCTION

Annular flow is a major kind of regime of gas-liquid two-phase flow, which is seen widely in the industry such as power, chemical and petroleum engineering etc. Generally, the annular flow happens at high gas phase velocity and low liquid phase velocity (Zaidi et al., 1998). In this flow regime, the liquid phase component flows along the pipe wall and forms a layer of liquid film, whereas the gas phase component entraining some small liquid drops flows along the pipe center (Han et al., 2006; Kim et al., 2006). Due to this special fluid structure of annular flow, fluid pulsation is a common phenomenon that occurs (Inoue et al., 1986; Yokosawa et al., 1986). The consequence of the fluid pulsation may result in piping vibration and it is dangerous to the equipment and the production processes. Especially when there exist some obstacles in the flow passage, the intrinsic pulsation of annular flow will be superimposed upon additional pulsations arising from the block of the obstacles (Sun et al., 2008a). However, investigations on the complex fluid pulsation behind a prism in annular flow are seldom reported so far.

This paper contributes to the study on the fluid pulsation property of gas-liquid annular flow across a horizontal quasi-trapezoidal prism. The pressure difference between the upstream and downstream of the prism was adopted as the characteristic parameter representing the fluid pulsation. Based on the experiments using air and water as the test fluid media, the relationship of average pressure, gas mass quality and total mass flowrate were discussed.

2. EXPERIMENT

Experiments were conducted on the same apparatus as described in the authors' previous work (Sun et al., 2008b). The tested media were air and water both flowing horizontally across a quisi-trapezoidal prism which was placed at the center of the pipe perpendicular to the flow direction. Fig.1 shows the layout of the experiment system. The flowrates of air and water were measured respectively by a vortex flowmeter and an electromagnetic flowmeter, and were regulated by the valves before they entered into a gas-liquid mixer. The air flowrate Q_G at the entrance was 30–120 m^3/h and the water flow rate Q_L was 0.5–2.5 m^3/h. Additionally, pressures of air and water at this part were also measured to calculate their mass flowrates G_G and G_L. After passing through the gas-liquid mixer, the mixture of air and water reached the test section, in which the experiments were carried out and signals were acquired. The upstream and downstream straight pipe were 120D (D is the inner diameter of the pipe, D = 50 mm) and 30D long respectively. The water was recirculated to the storage pool while the air was exhausted to the atmosphere. The flow regime was recorded manually via a transparent observation window in the test section.

The blockage ratio b ($b = d/D$, d was the width of the prism, d = 14 mm) of the quasi-trapezoidal prism used was 0.28. The duct-wall differential pressure method (DDPM) (Sun et al., 2007) was adopted to acquire the fluid pulsation signal. From the two pressure tags on the pipe wall which were respectively 1D upstream and 0.2D downstream of the prism, the dynamic differential pressure was detected by a Honeywell 24PC sensor, then sampled and processed by a computer, as shown in Fig.2. Signals were acquired only when the annular flow regime appeared.

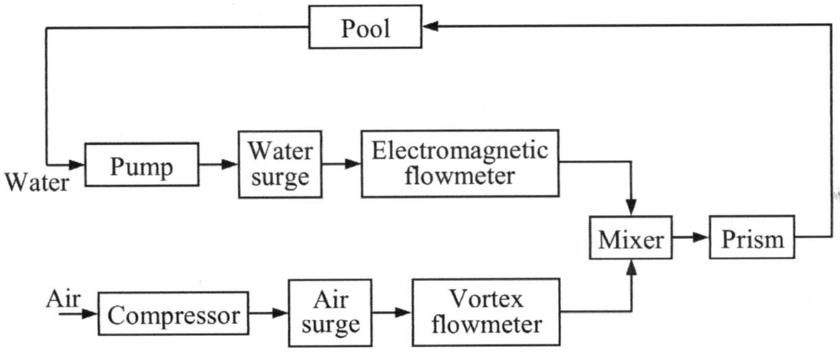

Fig. 1 Experimental set-ups

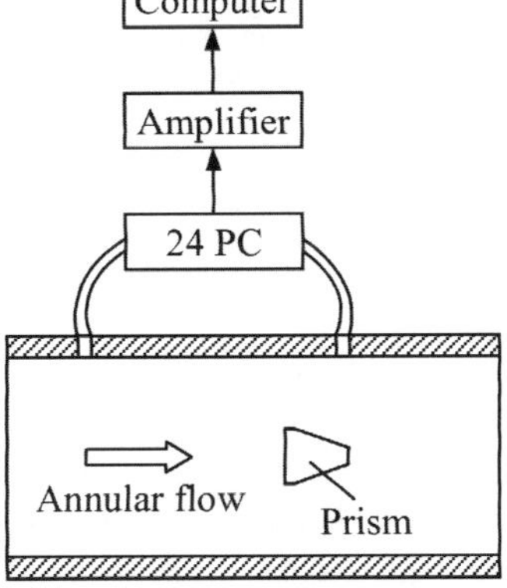

Fig. 2 The prism and signal acquisition system

Journal of Physics: Conference Series **147** (2009) 012043 doi:10.1088/1742-6596/147/1/012043

3. RESULTS AND DISCUSSION

3.1 Fluid Pulsation Induced Pressure Fluctuation

Under the annular flow regime, the flowrates of air and water were adjusted randomly to produce various flow conditions. The pressure difference between the upstream and downstream of the prism was adopted as the characteristic parameter representing the fluid pulsation. A series of experiments were conducted at different gas mass qualities x ($x = G_G/(G_G + G_L)$). Due to the restriction of the laboratory condition, the gas mass quality of the annular flow can not exceed 0.31. The sampling rate of 1 000 Hz was adopted in the experiments and 10000 points were included in each data set. Fig.3 presents the fluid pulsation induced pressure fluctuations under some typical gas mass qualities. It is seen that there are obvious fluctuations in the measured pressure differences, which reveals the fluid pulsation behind the prism in annular flows. Moreover, this pulsation is more marked at low gas mass qualities than those at high gas mass qualities. When x is less than 0.1, the maximum amplitude of the pressure fluctuation is basically greater than 5 kPa. However, when x is from 0.1 to 0.31, the amplitude of the pressure fluctuation is basically within 3 kPa. In addition, the power spectral densities of the pressure fluctuation were also calculated, and results show that the major frequency components were concentrated on the frequency band from 1–10 Hz.

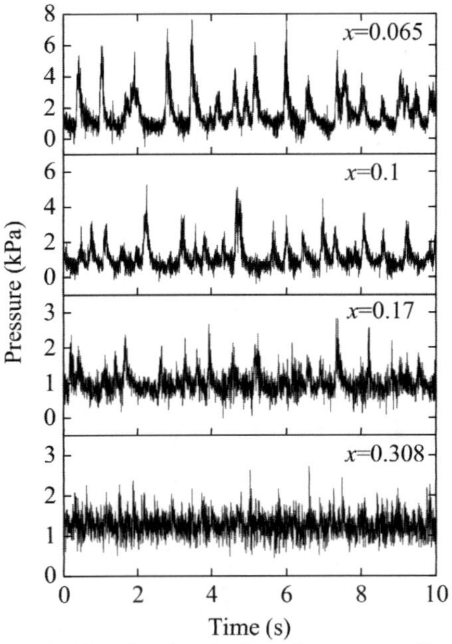

Fig. 3 The fluid pulsation induced pressure fluctuations

2.2 Average Pressure Fluctuation

The pressure fluctuation induced by the fluid pulsation is strongly affected by the gas and liquid component contents of the two-phase flow which is a random and nonlinear process. To analyze the whole property of the pressure fluctuation, the time average value of the pressure fluctuation were calculated. The time average pressure fluctuation can represent the integral energy of the fluid pulsation.

Fig.4 shows the relationship between the average pressure fluctuation and the gas mass quality. It is obvious that the average pressure fluctuation is closely concerned with the gas mass quality. When x = 0.05–0.15, the average pressure distributes from 1.2 to 2.3 kPa; when x = 0.15–0.2, the average pressure assembles around 1 kPa; and when x = 0.2–0.31, the average pressure distributes mainly from 0.6 to 1 kPa. On the whole, the average pressure decreases with the gas mass quality.

Fig.5 presents the relationship between the average pressure fluctuation and the gas-liquid total mass flowrate. The average pressure increases basically with the gas-liquid total mass flowrate. It is interesting that the average pressure differs from each other under the same gas-liquid total mass flowrate. This difference mainly results from the different gas mass quality of the annular flow in which the signal was acquired.

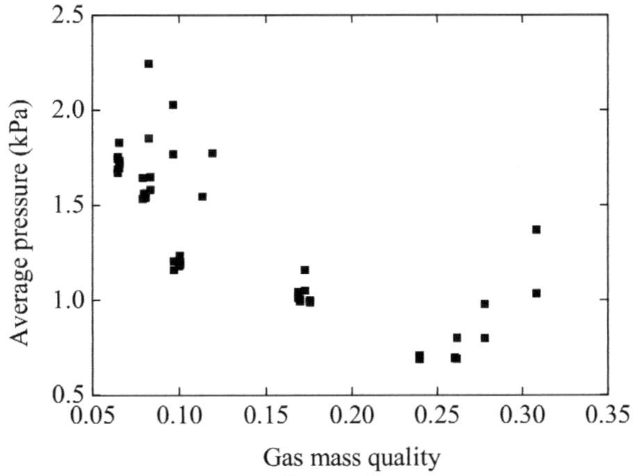

Fig. 4 Average pressure fluctuation versus gas mass quality

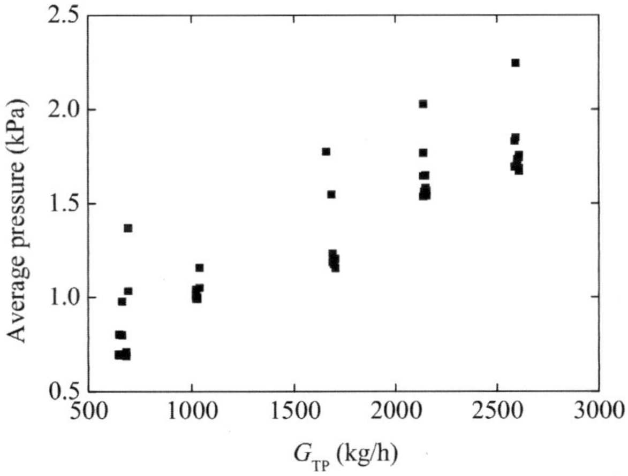

Fig. 5 Average pressure fluctuation versus total mass flowrate

3. CONCLUSIONS

The fluid pulsation behind a prism in gas-liquid annular flows was explored in this paper. The pressure difference between the upstream and downstream of the prism was used to represent the characteristics of the fluid pulsation. Based on the experimental results, it is found that there exit obvious pulsations in the fluid behind the prism in annular flows, and this pulsation is more marked at low gas mass qualities than those at high gas mass qualities. Furthermore, the average pressure decreases basically with the gas mass quality and increases with the gas-liquid total mass flowrate.

ACKNOWLEDGEMENTS

This work is financially supported by the China Postdoctoral Science Foundation Funded Project (200801346), the Hunan Postdoctoral Scientific Program (2008RS4022), and the Postdoctoral Science Foundation of Central South University.

Journal of Physics: Conference Series **147** (2009) 012043 doi:10.1088/1742-6596/147/1/012043

NOMENCLATURE

b	blockage ratio	
D	inner diameter of pipe	[m]
d	width of prism	[m]
G	mass flowrate	[kg/h]
Q	volume flowrate	[m^3/h]
x	gas mass quality	

Subscripts

G	gas-phase component
L	liquid-phase component
TP	two-phase flow

REFERENCES

Han, H. et al. (2006). "A study on the effect of gas flow rate on the wave characteristics in two-phase gas-liquid annular flow," Nuclear Engineering and Design, 236(24), pp. 2580-2588.

Inoue, A. et al. (1986). "Studies on two phase cross flow part I: characteristics around a cylinder," International Journal of Multiphase Flow, 12(2), pp. 149-167.

Kim, N. et al. (2006). "Two-phase flow distribution of air-water annular flow in a parallel flow heat exchanger," International Journal of Multiphase Flow, 32(12), pp. 1340-1353.

Sun, Z. Q. et al. (2007). "Investigation of the pressure probe properties as the sensor in the vortex flowmeter," Sensors and Actuators: A. Physical, 136(2), pp. 646-655.

Sun, Z. Q. et al. (2008a). "Neural networks approach for prediction of gas-liquid two-phase flow pattern based on frequency domain analysis of vortex flowmeter signals," Measurement Science and Technology, 19(1), pp. 015401(8pp).

Sun, Z. Q. et al. (2008b). "Dynamic pressure drop property of gas-liquid two-phase flow around a horizontal cylinder," Journal of Chemical Engineering of Chinese Universities, 22(3), pp. 378-383. (in Chinese)

Yokosawa, M. et al. (1986). "Studies on two phase cross flow part III: characteristics of unsteady flow behavior," International Journal of Multiphase Flow, 12(2), pp. 185-202.

Zaidi, S. H. et al. (1998). "A comparative study of phase Doppler and laser diffraction techniques to investigate drop sizes in annular two-phase flow," Chemical Engineering Journal, 71(2), pp. 135-143.

Journal of Physics: Conference Series **147** (2009) 012044 doi:10.1088/1742-6596/147/1/012044

Comparison of differential pressure model based on flow regime for gas/liquid two-phase flow

F Dong, F. S Zhang, W Li, C Tan

Tianjin Key Laboratory of Process Measurement and Control, School of Electrical Engineering and Automation, Tianjin University, Tianjin 300072, CHINA

fdong@tju.edu.cn

Abstract. Gas/liquid two-phase flow in horizontal pipe is very common in many industry processes, because of the complexity and variability, the real-time parameter measurement of two-phase flow, such as the measurement of flow regime and flow rate, becomes a difficult issue in the field of engineering and science. The flow regime recognition plays a fundamental role in gas/liquid two-phase flow measurement, other parameters of two-phase flow can be measured more easily and correctly based on the correct flow regime recognition result. A multi-sensor system is introduced to make the flow regime recognition and the mass flow rate measurement. The fusion system is consisted of temperature sensor, pressure sensor, cross-section information system and v-cone flow meter. After the flow regime recognition by cross-section information system, comparison of four typical differential pressure (DP) models is discussed based on the DP signal of v-cone flow meter. Eventually, an optimum DP model has been chosen for each flow regime. The experiment result of mass flow rate measurement shows it is efficient to classify the DP models by flow regime.

1. INTRODUCTION

Research work of gas/liquid two-phase flow in horizontal pipe receives great attention in the research field of multi-phase flow. It was found that because of the complexity and variability of two-phase flow, the real-time parameter measurement becomes a difficult issue in the field of engineering and science (Hetsroni, 1982).

The flow regime, which ties up with the pressure, temperature and flow rate, could reflect the flow paten and inner structure of two-phase flow. Other parameters of two-phase flow could be measured more easily and correctly in case of the flow regime has been correctly recognized. The mass flow rate measurement is achieved by ultilyting a multi-sensor data fusion system which contains temperature sensor, pressure sensor, cross section system and v-cone flow meter. It was found that multi-sensor data fusion combines the data from multiple sensors and relates the information from associated databases to achieve improved accuracies and more specific inferences than that could be achieved by the use of a single sensor (Llinsa et al., 1998). Multi-sensor data fusion has already been widely used in the field of military and robot but just comes to a start in the field of multi-phase flow.

Cross-section information technology is an on-line measurement technique with the advantages of non-invasion, non-radiation. it could provide us with two/three dimensional information which is far more than other conventional techniques. The measured voltage contains rich information of the cross-section and could be used to make the flow regime recognition and cross fraction calculation. D-S evidence theory is adopted to recognize the flow regime based on the measured voltage.

Journal of Physics: Conference Series **147** (2009) 012044 doi:10.1088/1742-6596/147/1/012044

V-cone flow meter is a new kind of throttling set which appeared in the middle of 1980s. it was found that because of the special shape, it has the advantages of small pressure drop, high repeatability, good linear ability, wide measurement scale and so on (Stewart et al., 2002). In the field of single-phase flow, v-cone flow meter has achieved pretty high accuracy, at the mean time it has become a popular device for two-phase flow measurement by combine with differential pressure (DP) models. Steven (2003) found its efficient use in the measurement of wet gas.

A method to measure the mass flow rate of gas/liquid two-phase flow in horizontal pipeline is put forward. According to the flow regime, a comparison of DP models has been made based on the DP signal from v-cone flow meter. The most suitable DP models for v-cone flow meter have been respectively adopted under different flow regime which has been recognized by the cross-section information system. The experiment result shows this method is acceptable.

2. MULTI-SENSOR FUSION SYSYTEM

The multi-sensor system contains temperature sensor, pressure sensor, cross-section information system and v-cone flow meter. Fig 1 shows the simplified structure of the multi-sensor data fusion system. The cross-section information system is used to recognize the flow regime while other sensors are used to make the mass flow rate calculation.

The physics fundamental of cross-section information system is different medium has different conductivity. Dong (2005) found that the medium distribution can cause the change of conductivity distribution which could be measured by the system. There are two electrod-plane in the cross-section information system in which 16-electrode and adjacent exciting mode are adopted. 16 electrodes are uniformly installed around the pipe in each plane. Each plane of electrode could be regarded as an independent sensor of the same kind which can be used in the data fusion. It was found that the adjacent exciting mode refers to the exciting current is injected into the measured area through a pair of adjacent electrodes, and the voltages are measured through successive pairs of neighboring electrodes except the exciting electrodes. After all the voltages have been measured, the current is injected through the next electrodes until all the independent voltage measurements have been done (Dong et al., 2001). In this mode every circle contains 16 excitations and each excitation contains 13 voltage data, in order to get one frame information of the cross-section 208 voltage data need to be gathered.

The dual-plane cross-section information system used in the experiment could provide us with 50 frames data per second in every plane. Differential pressure signal is obtained across a v-cone flow meter whose β is 0.65, front angle is 45°, and back angle is 65°.

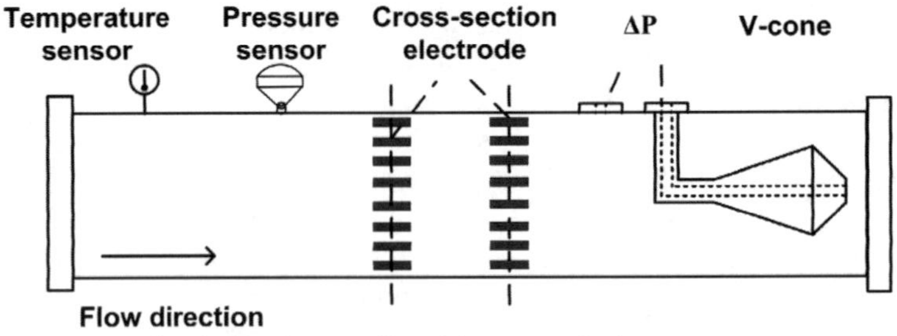

Fig. 1 Structure of multi-sensor fusion system

The experiment is carried out in Tianjin University Multi-phase Flow Laboratory. The horizontal pipe is made of organic glass, so that the flow regime could be observed more clearly in the experiment. The pipe is about 40m long and the inner diameter is 50mm. The flow rate of water is about 0.1-16.0m^3/h, and gas is 0.06-82.0m^3/h. Different flow regime could be formed by controlling the valve of water and gas pipe. Of these flow regimes bubble flow, short plug flow, long plug flow and annular flow are objective flow regimes to be recognized.

3. FLOW REGIME RECOGNITION

3.1 Feature Extraction

In the adjacent exciting and measure strategy, one frame information of the pipe cross-section is consists of 208 measured voltage data. It was found that it is difficult to make the flow regime recogonition directly and data pretreatment is needed in order to reduce the data dimension, R_i is defined as follows (Tan et al., 2007):

$$R_i = \frac{1}{13}\sum\left(V_{ij} - V_{ij0}\right) \quad (1)$$

Where V_{ij} is the j-th voltage data at the i-th excitation, V_{ij0} is the V_{ij} vale when the pipe is full of water.

An array $V_{Ri\times n}$ (where $n = 16m$) could be formed through the time axis, In order to get a better known about how well $V_{Ri\times n}$ represents the void fraction at different flow regime on a large scale, set $m = 500$. three features: $\overline{X}$, SD, SK are extracted from the above R_i series in order to reduce the data demension at a further step:

$$\overline{X} = \frac{1}{n}\sum_{k=1}^{n} R_{ik} \quad (2)$$

$$SD = \left(\frac{1}{(n-1)}\sum_{k=1}^{n}\left(R_{ik} - \overline{R}\right)^2\right)^{1/2} \quad (3)$$

$$SK = \frac{n}{(n-1)(n-2)}\sum_{k=1}^{n}\left(\frac{R_{ik} - \overline{R}}{SD}\right)^3 \quad (4)$$

The experiment results show the three features have good capability of differentiating (shown in Tab.1), so they could be used to identify the flow regime separately.

Tab.1 Feature extraction result

Flow Regime	$\overline{X}$	*SD*	*SK*
	0.828	8.511	2.491
Bubble Flow	1.192	8.768	1.982
	1.047	9.119	2.272
	8.345	20.521	0.511
Short Plug Flow	7.832	19.898	0.414
	8.9259	20.850	0.5295
	12.5947	32.578	0.797
Long Plug Flow	11.674	34.176	0.631
	10.889	28.729	0.615
	5.743	50.434	0.894
Annular Flow	5.650	49.438	0.824
	4.870	50.535	0.861

3.2 D-S Evidence Theory

Dempster-Shafer (D-S) evidence theory is a statistical-based data fusion classification algorithm. It was found that D-S evidence theory is based on a non-zero set Θ, which is called Frame of Discernment (Shafer, 1976).

3.2.1 Basic Probability Assignment Function

In the theory a basic probability assignment over Θ is defined as m: $2^{\Theta} \rightarrow [0,1]$, such that:

$$\begin{cases} m(\phi) = 0 \\ \sum_{A \subset \Theta} m(A) = 1 \end{cases} \quad (5)$$

If $A \subseteq \Theta$, and $m(A) > 0$, A is called focal element and $m(A)$ is an evidence which represents how well A is supported.

3.2.2 Combination Rule

The key point of D-S evidence theory is that the single evidence BPAF may be combined to an integrated BPAF, this is the evidence combination. The combination rule is as follows:

$$m(A) = \begin{cases} 0, A = \phi \\ K \sum_{A_i \cap B_j \cap \cdots = A} m(A_i) m(B_j) \cdots, A \neq \phi \end{cases} \quad (6)$$

where $K = (1 - \sum_{A_i \cap B_j \cap \cdots = \phi} m_1(A_i) m(B_j) \cdots)^{-1}$.

3.3 Recognition Process and Result

Features extracted from both planes of cross-section information system are used to get the basic probability assignment function $m_i(A_j)$ ($i = 1,2,\cdots,6$, $j = 1,2,3,4$) which represent how well the *ith* feature supports the *jth* flow regime. The features $\overline{X}$, *SD* and *SK* from both planes are respectively defined as S_i, $i = 1,2,\cdots,6$ and the four objectivel flow regimes: bubble flow, short plug flow, long plug flow and annular flow, which appointed as the elements of the frame of discernment in D-S evidence theory, are respectively defined as A_j, $j = 1,2,3,4$. The concepts of Union Core, Distance and proportion coefficient K are introduced to get the value of $m_i(A_j)$:

$$C_i = \frac{1}{N} \sum_{k=1}^{N} S_i, i = 1,\ldots 6 \quad (7)$$

$$D_i = \sqrt{(S_i' - C_i)^2}, i = 1,\ldots 6 \quad (8)$$

$$K_i = \frac{D_i}{C_i}, i = 1,\ldots 6 \quad (9)$$

In the equation, S_i is the sample data, $S_i^{'}$ is the recognized object.

The sample data S_i is used to get the Union Core C_i and then the basic probability assignment function $m_i(A_j)$ could be acquired through normalizing the four K from these four flow regimes. Through utilizing the combination rule with $m_i(A_j)$, the recognition result is obtained (shown in Tab.2). In this experiment, 30 sets of data of each flow regime are used as the sample data to get the Union Core C_i (six Cores of each regime), and another 30 sets data of each flow regime are used as the recognizing objects with the regime recognition method based on D-S evidence theory.

Tab.2 Recognition result under the method based on D-S evidence theory

Flow regime	Data sets	Right recognized	Success rate
Bubble Flow	30	29	96.7%
Short Plug Flow	30	26	86.7%
Long Plug Flow	30	27	90%
Annular Flow	30	30	100%

Only one set data of bubble flow is taken as short plug flow, totally 7 sets data of short plug and long plug flow are mistaken as each other, and all the annular flow have been right recognized. The experiment results shows the flow regime can be recognized with high accuracy by cross-section information system, and the recognition method based on D-S evidence theory is acceptable. If a better way is found to get the BPAF in evidence theory, better recogniton result would be achieved.

4. DIFFERENTIAL PRESSURE MODEL

It was found that in the field of gas/liquid two-phase flow measurement based on differential pressure, there are already some differential pressure (DP) models, such as Homogeneous model, Murdock model, Chisholm model, Lin model and so on (Murdock, 1962; Chisholm, 1977; Lin, 1982). However the above DP models are deduced from the orifice plate meter and venturi tube not v-cone flow meter, the direct use of these DP models to v-cone flow meter would cause considerable larger measurement error. Since every model has its own adaptive condition, an optimum DP model from them could be found under a certain flow regime for v-cone flow meter.

4.1 Homogeneous Model

In this model, gas/liquid two-phase flow is considered as well-mixed single-phase flow with a mixed density, so single-phase flow DP model could be used. The measurement error of this model is considerable big especially when the gas mass quality x is small and pressure is low.

$$W = \frac{\varphi CA\sqrt{2\rho_G \Delta P}}{\sqrt{1-\beta^4}\sqrt{\frac{\rho_G}{\rho_L}+x\left(1-\frac{\rho_G}{\rho_L}\right)}} \qquad (10)$$

4.2 Murdock Model

Murdock model is derived from separated model to which a coefficient 1.26 has been added. The experiment condition for Murdock model is that gas mass quality: 0.11-0.98 and the pressure: 0.1-4 MPa.

$$W = \frac{\varphi CA\sqrt{2\Delta P_{TP}\rho_G}}{\sqrt{1-\beta^4}[x+1.26(1-x)\sqrt{\rho_G/\rho_L}]} \quad (11)$$

4.3 Chisholm Model

Slip model is taken into consideration in Chisholm model. When the experiment pressure is 1.0-7.0 MPa and the gas mass quality $x \le 0.1$, the prediction error of this model is about 20%. This error reduces with the increase of experiment pressure and gas mass quality.

$$W = \frac{\varphi CA\sqrt{2\rho_G \Delta P_{TP}}}{x\sqrt{1-\beta^4}\sqrt{1+\left[\left(\frac{\rho_L}{\rho_G}\right)^{0.25}+\left(\frac{\rho_G}{\rho_L}\right)^{0.25}\right]\chi+\chi^2}} \quad (12)$$

Where $\chi = \sqrt{\frac{\Delta P_L}{\Delta P_G}} = \left(\frac{W_L}{W_G}\right)\sqrt{\frac{\rho_G}{\rho_L}} = \left(\frac{1-x}{x}\right)\sqrt{\frac{\rho_G}{\rho_L}}$.

4.4 Lin Model

On the basis of a modified separated flow model, a simple and rational relationship for the flow rate is put forward by introduction of a corrective coefficient θ in this model. θ is a function of ρ_G/ρ_L and is derived from the experiment of high ρ_G/ρ_L condition across a sharp edged circular orifice. Lin model performs well especially when the ρ_G/ρ_L is high or the gas mass quality x is large.

$$W = \frac{\varphi CA\sqrt{2\Delta P_{TP}\rho_G}}{\sqrt{1-\beta^4}[x+\theta(1-x)\sqrt{\rho_G/\rho_L}]} \quad (13)$$

5. EXPERIMENT RESULT

The DP models mentioned above have been applied to each objective flow regime using the differential pressure signal acquired by the v-cone flow meter. Fig. 2 shows the mass flow rate error of the models under each flow regime.

The first part of this figure show the DP models' experiment error at bubble flow. The gas mass quality x is small and the pressure is low in bubble flow and Murdock model has done a good job at this flow regime, the experiment error mostly shakes within 5% while others models beyond 10%. When the gas mass quality x and pressure increses, bubble flow turns into short plug flow, Murdork model continues its good performance in short plug flow, the mean value and standard deviation is the smallest of the four models. Chisholm model becomes the best behaved model at long plug flow, the error curve is smooth and within 5%, while the error of others is variable and big. At annular flow, the standard deviation of these models is samller than other flow regimes, while the mean value is biggest of all the flow regimes. Gas/liquid two-phase flow becomes more mixed when the pressure is high and gas mass quality is large, so Homogenous model suits anuular flow much better than other flow regimes. However the mixture could not becomes so well mixed that it could be seen as single-phase flow or rather Homogenous model is an ideal model, the experiment error of it is considerable big. The last part of the figure shows Lin model is supperier to others.

Journal of Physics: Conference Series **147** (2009) 012044 doi:10.1088/1742-6596/147/1/012044

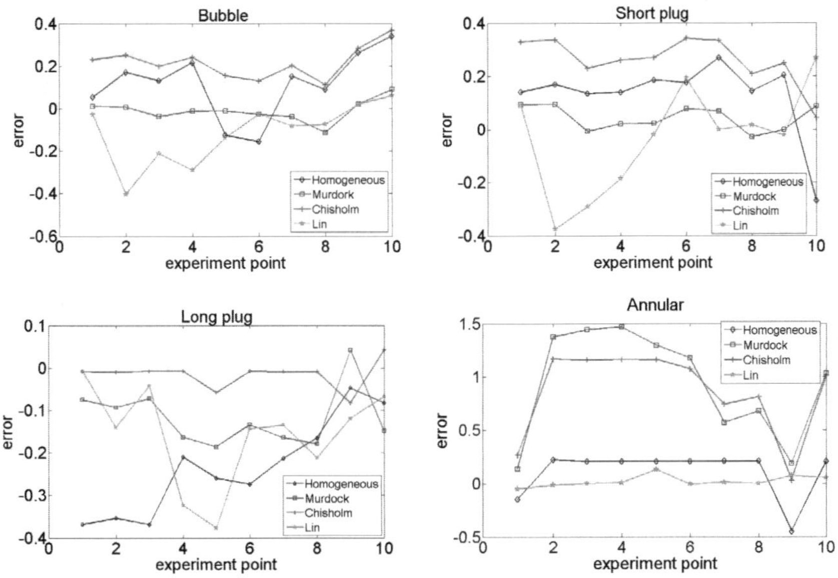

Fig. 2 Experiment error of DP models at objective flow regime

After the error comparison of the models at certain flow regimes, in this experiement Murdock model is the most suitable model for bubble flow and short plug flow, or rather Murdock model well worked in the condition of low pressure and small gas mass quality, Chisholm model is the best for long plug flow, and Lin model for annular flow.

The optimum model would be applied to certain flow regime which has been recognizied after the error comparison of the models.

Based on the flow regime recognition and optimum model selection, the mass flow rate result is given by every 10 sets measured data of each flow regime which have been right recognizied. Fig. 3 shows the mass flow rate error.

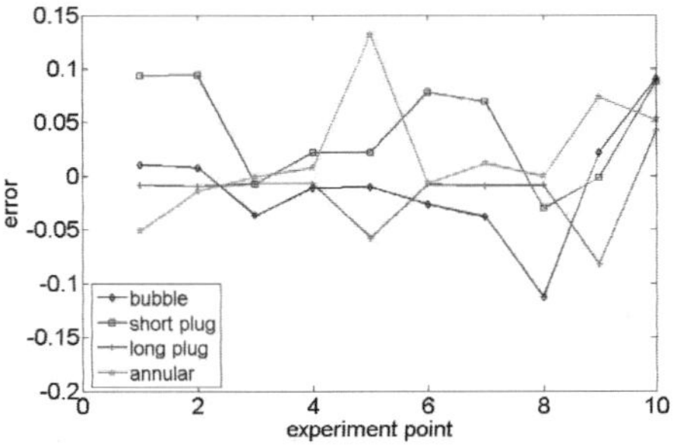

Fig. 3 Mass flow rate measurement error

The experiment error of long plug flow with Chisholm is smallest while the error of others is more variable but mostly centralized within 10%. Only one experiment point of bubble flow and one point of annular flow is beyond 10%. The error is acceptable for the flow rate measurement of gas/liquid two-phase flow. A higher accuracy could be achieved if these DP models have been modified.

6. CONCLUSION

A method of mass flow rate measurement for gas/liquid two-phase flow based on flow regime recognition and differential pressure model is put forward in this paper. The experiment result is achieved by a multi-sensor fusion system which is mainly consists of cross-section information system and v-cone flow meter. Firstly the flow regimes have been recognized by a dule-plane cross-section information system. D-S evidence theory as an important fusion method in decision-fusion level has been used to recognize the objective flow regimes. The high recogniton accuracy indicates the feasibility of this recognition method. After the flow regime recognition, an error comparision among four typical DP models has been made according to objective flow regimes. Finally a optimum DP model has been selected for each objective flow regime which has been recognizied, that is Murdock model for bubble flow, Chisholm model for long plug flow, and Lin model for short plug flow and annular flow. By utilizing the optimum DP model, the mass flow rate error is within 10%, it is shown that the method based on multi-sensor fusion is acceptable for flow rate measurement of gas/liquid two-phase flow.

Because of the complexity and variability of multi-phase flow, it is needed to observe and measure the flow process from multi-aspect in both time and space. This provide us a good opportunity to apply multi-sensor data fusion to the field of multi-phase flow measurement. With the research work get deeper, more fusion methods and fusion strategies will be applicated in this research area.

ACKNOWLEDGEMENTS

The author appreciates the support from National High Technology Research and Development Program of China (2006AA04Z167) and Natural Science Foundation of Tianjin (08JCZDJC17700).

NOMENCLATURE

x	gas mass quality	
m	frame number	
n	length of the pre-treatment data series	
A	minimum cross sectional area of v-cone	[m^2]
C	discharge coefficient of the V-cone flow meter	
N	sample data number	
R	pre-treatment data series	
V	measured voltage	
W	mass flow rate of gas/liquid two-phase flow	[kg/s]
	Greek Letters	
β	diameter ratio of the v-cone flow meter	
φ	thermal expansion factor	
ρ	density of gas and liquid	[kg/m^2]
θ	corrective coefficient.	
ΔP	differential pressure of two-phase flow	[Pa]
	Subscripts	
i	i-th excitation	
j	j-th voltage data	
G	gas phase	
L	liquid phase	
0	the pipe is full of water	

REFERENCES

Chisholm, D. (1977). "Two-phase Flow through Sharp-edged Orifices," J. Mech. Eng. Sci., 19(3), pp.128-130.

Dong, F. et al. (2001). "Identification of Two-phase Flow Regimes in Horizontal, Inclined and Vertical Pipes," Meas. Sci. Techno., 12(8), pp. 1069-1075

Dong, F. et al. (2005). "Void Fraction Measurement for Two-Phase Flow Using Electrical Resistance Tomography," Can. J. Chem. Eng., 83 (1), pp.19-23.

Hetsroni, G. (1988). Handbook of Multiphase Systems, New York, Hemisphere, USA.

Llinsa, J. et al. (1998). "An Introduction to Multi-sensor Data Fusion, " International Symposium on Circuits and System, pp. 537-540. 31 May-3 June.

Lin, Z. H. (1982). "Two-phase Flow Measurements with Sharp-edged Orifices," Int. J. Multiphase Flow., 18(6), pp.683-693.

Murdock, J. W. (1962). "Two-Phase Flow Measurement with Orifices," J. Basic. Eng., pp.419-433.

Shafer, G. (1976). "A Mathematical Theory of Evidence," Princeton, New Jersey, USA.

Steven, R. (2003) "Research Developments in Wet Gas Metering with V-Cone Meters, " North Sea Flow Measurement Workshop, October 28-31.

Stewart, D. G. et al. (2002). "Wet Gas Metering with V-Cone Meters," North sea flow measurement workshop, October 22-25.

Tan, C. et al. (2007). "Identification of Gas/liquid Two-phase Flow Regime through ERT-based Measurement and Feature Extraction," Flow. Meas. Instrum., 18(5), pp.255-261.

Journal of Physics: Conference Series **147** (2009) 012045 doi:10.1088/1742-6596/147/1/012045

Experimental Study on Occurrence-Time Ratio Measurements of Air Entrainment in a Suction Sump

K. Inagaki[1], J. Funaki[1], K. Hirata[1]
[1] Department of Mechanical Engineering, Doshisha University
Kyoto 610-0321, Japan.

dti0316@mail4.doshisha.ac.jp

Abstract. In order to get accurate measurements of air entrainment in a suction sump, we design some new simple bubble sensors, which can detect the existence of air bubbles inside a suction pipe with no disturbances by the sensors and with a fine spatial resolution. We force on an intermittency factor γ, that is, an occurrence-time ratio of the air entrainment, and compare the result by the present sensor with those by conventional two methods; namely, visual and auditory ones. As a result, we show the criteria which specify lower-accuracy conditions in the conventional methods. By the visual method, the accuracy of the γ becomes low, when γ is less than 0.05. By the auditory method, the accuracy of γ becomes low, when the submergence depth S of the suction pipe is close to the critical one S_c.

1. INTRODUCTION

In power generation plants, irrigations, drainages and so on, we have increasingly required the innovations in suction-sump designs for low cost, compact size and high efficiency. In such situations, air entrainments into suction pipes become to appear more frequently. The air entrainments often induce vibration, noise and low efficiency, and sometimes result in pumps' collapses at the worst.

In the present study, our purpose is to get accurate measurements of the air entrainment into a suction pipe inserted vertically down in a suction sump, which is called as the "vertical wet-pit pump" configuration. Concerning the air entrainment in such a suction sump, there have been several studies (see Iversen, 1953; JSME standard, 1984; Hirata et al., 2003). Recently, in order to reveal the air-entrainment mechanism, the authors have conducted flow-velocity measurements by a UDM (Ultrasonic Doppler Method) and shown three-dimensional time-mean velocity distributions and equi-vorticity contours (Funaki et al., 2007).

Until now, we have usually judged the air-entrainment occurrence by means of a visually-based method (hereinafter, referred to as a visual method). This conventional method has been used prevailingly, because of its practicability and simplicity. However, as this visual method inherently means just the observation of free surfaces (will be described later), we do not identify the air entrainment itself. Then, we should require an accurate assessment on the visual method.

So, we design a new and simple conductance-type electric bubble sensor, which can detect the existence of air bubbles inside a suction pipe with no disturbances by sensor probes and with a fine spatial resolution. We focus upon an occurrence-time ratio of the air entrainment, and compare the result by the bubble sensor with those by conventional two methods; namely, the visual and auditory ones. The auditory method is an auditorily-based method where we use a sound-level meter as a sensor. Because of its simplicity and handiness, the auditory method has been prevalent as well as the visual method, despite the lack of direct relations with the air entrainment.

2. EXPERIMENTAL METHOD

2.1 Experimental apparatus

Figure 1 shows the present model, that is, a simple system with a suction sump and a suction pipe, having a configuration of the vertical wet-pit pump. Here, D is an outside diameter, B a breath of the suction sump, X a back clearance which is the distance from a suction-pipe centre to a back wall, Z a bottom clearance, and H a height level.

In order to specify a dynamical state, we define the three kinematic non-dimensional governing parameters; that is, the Froude number Fr, the Reynolds number Re and the Weber number We, as follows.

$$Fr = V_b / (gD)^{0.5} \ . \qquad (1)$$

$$Re = V_b \, D / \nu \ . \qquad (2)$$

$$We = V_b \, (\rho D / \sigma)^{0.5} \ . \qquad (3)$$

g, ν, ρ and σ denote the gravitational acceralation, kinetic viscosity, fluid density and water-to-air surface tension, respectively. The mean flow velocity V_b at the suction-pipe intake is defined as

$$V_b = 4\,Q\,/\,(\pi D^2), \qquad (4)$$

where Q is the flow rate into the suction pipe. As a supplementary parameter, we use the Bond number such as

$$Bo = \rho g D^2 / \sigma \ . \qquad (5)$$

When we define governing parameters, we can consider various options in the choice of characteristic scales. In the present study, we choose D and V_b as length and velocity scales, respectively. In Table 1, we summarise the present values of experimental parameters.

Figure 2 shows a schematic diagram of the present experimental apparatus, which is similar with those in references (Hirata et al., 2003; Funaki et al., 2007). A turbo pump B (No. 2 in the figure) feeds working fluid (water) to a suction sump (No. 9) from a reservoir tank. We control the flow rate from the pump A by a valve, and then control the water level in the suction sump. In the upstream of the suction sump, we put a strainer (No. 10) in order to get uniform flow. A bend-type jet pumps (No. 7) pumps up water in the suction sump into a suction pipe (No. 8). Here, the jet pump has less swirling component, than ordinary pumps. The water from the suction pipe and the jet pump falls into a container tank on a platform scale (No. 11).

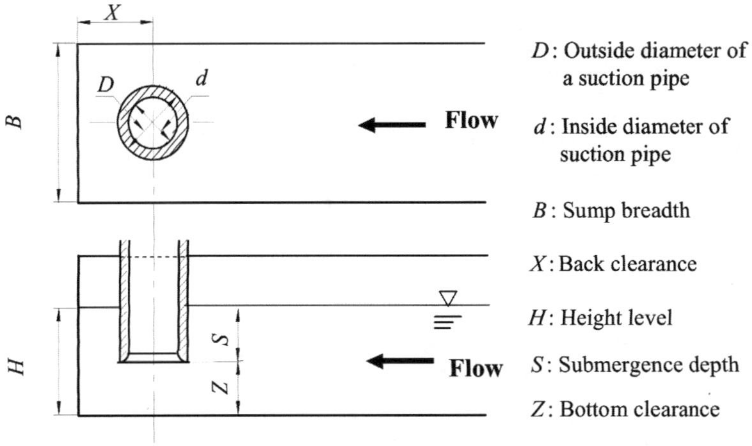

Fig.1 Suction sump and suction pipe, with a configuration of the vertical wet-pit pump.

Journal of Physics: Conference Series **147** (2009) 012045 doi:10.1088/1742-6596/147/1/012045

Table1 Experimental parameters.

D [mm]	38
d [mm]	34
B/D	3.15
X/D	1.71-2.32
Z/D	0.71
S/D	0.8-2.25
V_b [m/s]	1
Fr	1.6
Re	3.8×10^4
Bo	200
We	28

In the visual method, we observe free surfaces near the suction pipe, using a camcorder with a frame rate of 60frames/sec, which is fixed downstream outside the suction sump. In the auditory method, a sound-level meter is placed near the outer wall of the suction pipe and at 0.2m above the suction-pipe intake. We have to determine an adequate threathold value for the sound-level meter, as well as the bubble sensor. The present threathold value is $1.16V_0$ in the output voltage V from the sound-level meter, where V_0 denotes the output voltage when the spatially-continuous air core (see later for its definition) does not appear. The threathold values depend upon the sensor location and so on. We will discuss the threathold values of the sound-level meter and the bubble sensor, in the following subsections.

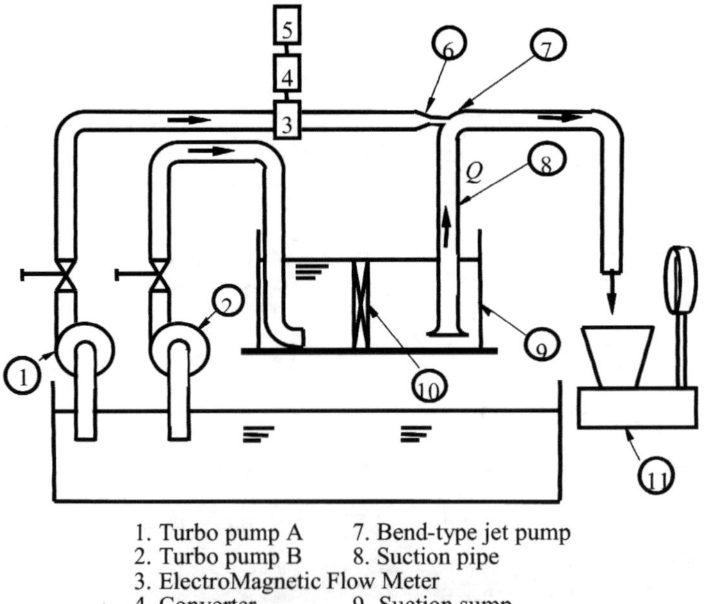

1. Turbo pump A
2. Turbo pump B
3. ElectroMagnetic Flow Meter
4. Converter
5. Digital multimeter
6. Nozzle
7. Bend-type jet pump
8. Suction pipe
9. Suction sump
10. Strainer
11. Platform scale

Fig.2 Experimental apparatus.

2.2 Air-entrainment-occurrence judgment by the visual measurement

The visual method is the most common method to judge the air-entrainment occurrence. However, due to ambiguous definitions of the air-entrainment occurrence, there has existed the confusion among researchers. (The main reason for the confusion, of course, comes from the lack of direct relations with the air entrainment itself.) In the present study, the judgment of the air-entrainment occurrence by the visual method are given by the following manner, which is the same as the conventional manner as in reference (Hirata et al., 2003). When we observe any spatially-continuous air cores from a free surface to a suction-pipe intake, namely, any underwater air bulks accompanied with so-called "the fully-developed entraining vortices even at a short instant," we recognise the air-entrainment occurrence (see Fig. 3(a)). On the other hand, when we observe some spatially-intermittent air cores, namely, some underwater air bulks accompanied with so-called "the intermittently-developed entraining vortices," we do not recognise the air-entrainment occurrence (see Fig. 3 (b)).

2.3 Definition of critical submergence S_c

We often use a critical submergence S_c as a prime indicator for the air entrainment. The definition of S_c is as follows. When the submergence depth S (see Fig.1) decreases from a sufficiently large value to zero, we can first observe the spatially-continuous air cores at $S = S_c$. In the present experiments, we gradually decrease S, with a step of 0.005m. At each step, we search for the spatially-continuous air cores during 300sec. In most cases, duration times of the air entrainment are much shorter than 300sec. Of course, the appearance of the spatially-continuous air cores does not mean the air-entrainment occurrence, exactly.

Incidentally, such a judgment manner is more objective and keeps a good reproductivity which has been confirmed by many preparatory experiments, although it lacks a thoroughly-theoretical background. (In fact, we have also confirmed that the results obtained with much longer sampling times than 300sec almost coincide with the present results.) The present judgment manner is the same as reference (Hirata et al., 2003). Even if we use a somewhat-deferent judgment manner from the present one, the obtained results are considered to be qualitatively the same as the present results (also see Hirata et al., 2003).

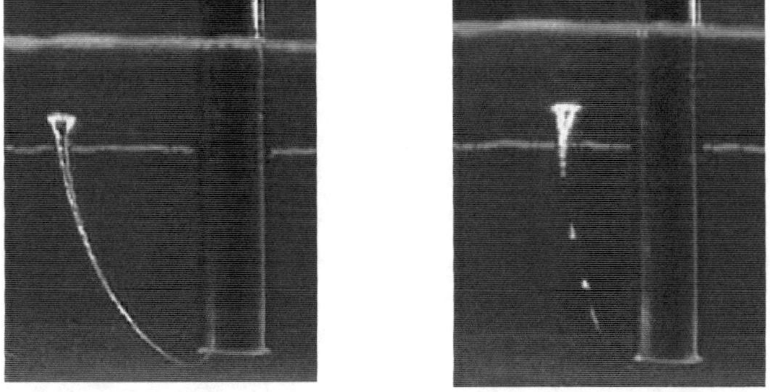

(a) Spatially-continuous air core from free surface to a suction-pipe intake

(b) Spatially-intermittent air core from free surface to a suction-pipe intake

Fig.3 Types of air cores, for the judgment of air-entrainment occurence.

Journal of Physics: Conference Series **147** (2009) 012045 doi:10.1088/1742-6596/147/1/012045

2.4 Bubble sensor

For an accurate detection of the air entrainment into the suction pipe, we develop a bubble sensor. The bubble sensor utilises the same principle as the conductance-type void-ratio meters (for example, see Hewitt, 1978); namely, it measures the electrical resistance which increases with increasing air volume inside the suction pipe. So, we can instantaneously know the occurrence of the air entrainment by means of monitoring the output voltage.

Figure 4 shows a schematic diagram of the bubble sensor. Two copper ring-shaped electrodes are flush-mounted inside a suction pipe, at 0.2m downstream from the suction-pipe intake. The gap between the two electrodes is fixed to 0.002m, and we impress a constant voltage between them. If air bubbles exist between them, the conductance value changes. Then, we measure the voltage impressed to another resistance in a bubble-sensor series circuit as a real-time output corresponding to this conductance change.

2.2.2 Definition of intermittency factor

We usually record the output voltage from the bubble sensor for 300sec. Using this raw time-series data, we get an intermittency factor γ. γ represents the occurrence-time ratio of the air-entrainment to the total recording time. Its definition is given by

$$\gamma = \frac{\sum_{i=1}^{n} \tau_i}{t_0}. \qquad (6)$$

As shown in Fig. 5, during a duration time τ_i, the i-th air entrainment occurs. Here, t_0 denotes the total recording time which is equal to 300sec. And, n denotes the total number of the air-entrainment occurrence for $t = 0 - t_0$. Supplemently speaking, one main reason why we consider γ is that the concerning phenomenon has a very-weak periodicity, and that γ tends to could have a constant value with enough long t_0.

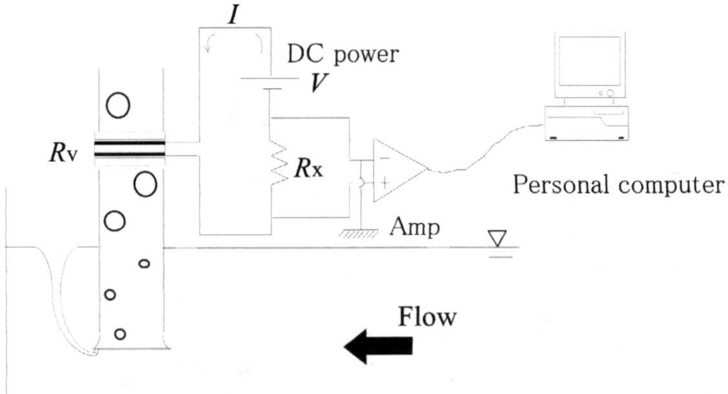

Fig.4 Bubble sensor.

3. RESULTS AND DISCUSSION

3.1 Comparison on raw time-series data

Figures 6-8 show the raw time-series data of the simultaneous measurements by the three methods; namely, the bubble-sensor method, the auditory method and the visual method. In each figure, the abscissa denotes time. The ordinate denotes the output voltage from the bubble sensor in figure (a), the sound-level meter in figure (b), or the appearance/disappearance of the spatially-continuous air core in figure (c). Figures 6, 7 and 8 are the results at $S/D = 0.80$, 1.65 and 2.25 ($\approx S_c/D$), respectively.

Although figures (a), (b) and (c) are simultaneously-measured data, the data are not perfectly synchronised with one another. For example, because the bubble sensor is located somewhat downstream from the suction-pipe intake, such a time lag as about 0.2sec occurs in comparison with the air entrainment at the suction-pipe intake. However, time scales of the abscissa of Figs. 6-8 are much longer than the time lag. Then, we have no corrections on the data, because the smaller time lag does not bring us any miss-understandings. Furthermore, the time lag intrinsically does not appear, when we consider the intermittency factor γ.

The threathold values for the bubble sensor and the sound-level meter should be determined in advance. To begin from the conclusion, we cannot find out the perfectly-accordable values among the three methods, for all the cases tested. However, but for such exceptional cases as the air entrainment occurs rarely (for example, see Fig. 8), we can find out the accordable threathold values among the three methods. In fact, both in Figs. 6 and 7, we can find out the well-accordable threathold values for the bubble sensor and the sound-level meter. In such situations, S/D is smaller than S_c/D. So, because the air entrainment occurs frequently, the occurrence-time ratios of the spatially-continuous and spatially-intermittent air cores without the actual air entrainment are considered to become relatively small. Furthermore, the sound-pressure perturbations induced by the air entrainment are also considered to become large.

Now, we temporarily focus our attention upon a single bubble on a suction pipe, in order to consider the threathold value of the bubble sensor. (Of course, it is difficult to suppose a single bubble as an approximation for the actual air entrainment.) Incidentally, because the threathold value of the sound-level meter strongly depends upon its location and its surrounding sound environments we can hardly reveal its physical relations. On the other hand, it is easier to consider the physical relations concerning.

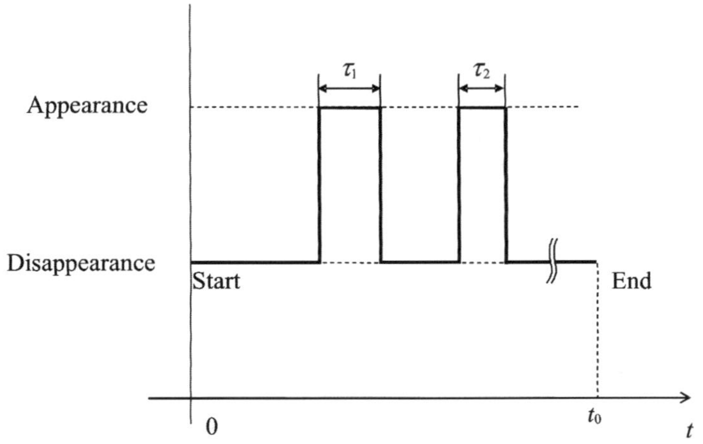

Fig.5 Intermittency factor γ.

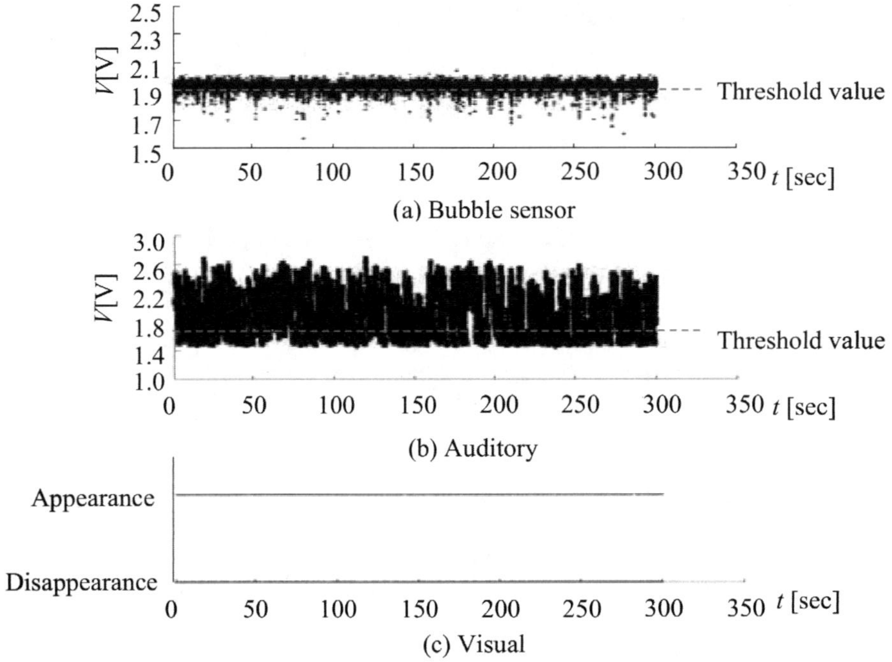

Fig.6 Time histories (X/D=1.71, S/D=0.80).

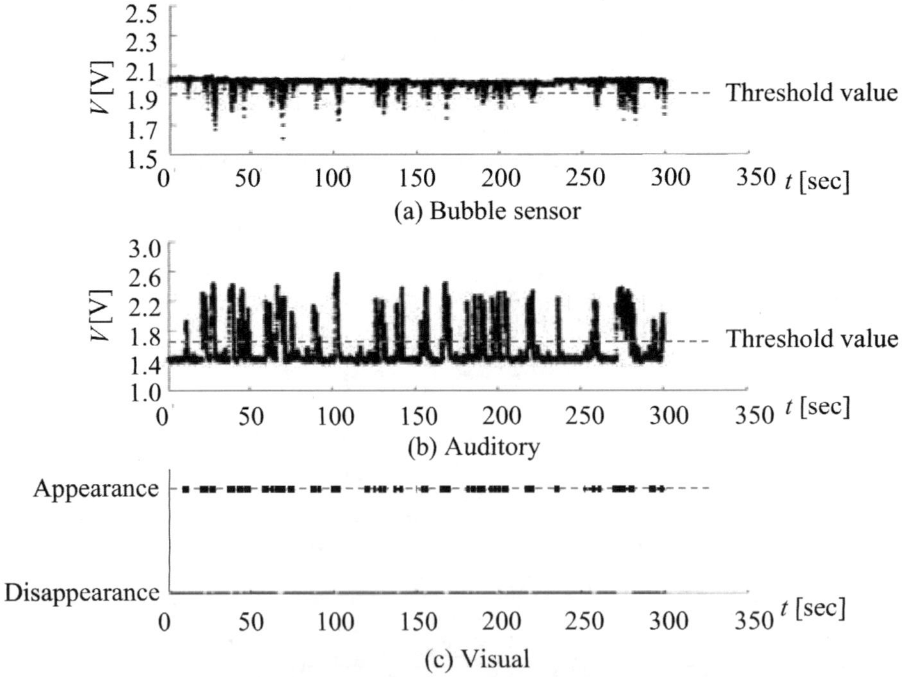

Fig.7 Time histories (X/D=1.71, S/D=1.65).

Figure 9 shows the relation between the threathold value V of the bubble sensor and the diameter ϕ of a single air bubble between the two electrodes. When we choose 1.9V as the threathold value, we regard the passage of a single air bubble with a diameter $\phi > 0.15D$ (= 5.7mm) as the air entrainment. (Strictly speaking, we also regard the passage of a group of air bubbles with $\phi < 0.15D$ as the air entrainment.) Incidentally, as ϕ increases, the error on V tends to increase. This is considered to be related with the air-bubble deformation and the air-bubble eccentricity from the suction-pipe centre with a timely-random fluctuation.

Figure 8 shows different features from Figs. 6 and 7. At first, we compare the bubble-sensor method in figure (a) with the visual method in figure (c). At $t \approx 10$sec, the air entrainment is not detected by the bubble-sensor method, but detected by the visual method. In addition, at $t \approx 90$sec, the duration time of the air entrainment measured by the bubble-sensor method is much smaller than that by the visual method. The above two discrepancies can be explained by frequent appearances of the spatially- continuous air core without the actual air entrainment.

Second, we compare the bubble-sensor method in figure (a) with the auditory method in figure (b). At $t \approx 90$sec, the air entrainment is detected by the bubble-sensor method, but not by the auditory method. This can be explained by the small sound-level component of the air entrainment at $S/D \approx S_c/D$, where the amount of entrained air and the corresponding sound level become small. Then, the relatively-small sound related with the air entrainment tends to be drowned out by the back-ground sound noises due to a pumps operation, water flow and so on.

3.2 Intermittency factor of air entrainment

In Fig. 10, we summarise Figs. 6-8. The ordinate and the abscissa denote the intermittency factor γ and the reduced submergence depth S/D, respectively. For all the three methods, as S/D increases, γ decreases monotonously and linearly. In addition, all the values of S/D at $\gamma = 0$, which are obtained by extrapolations, almost coincide with S_c/D, which is determined by the visual method. However, the values of γ obtained by the three methods at $S/D \approx S_c/D$ do not always coincide with one another.

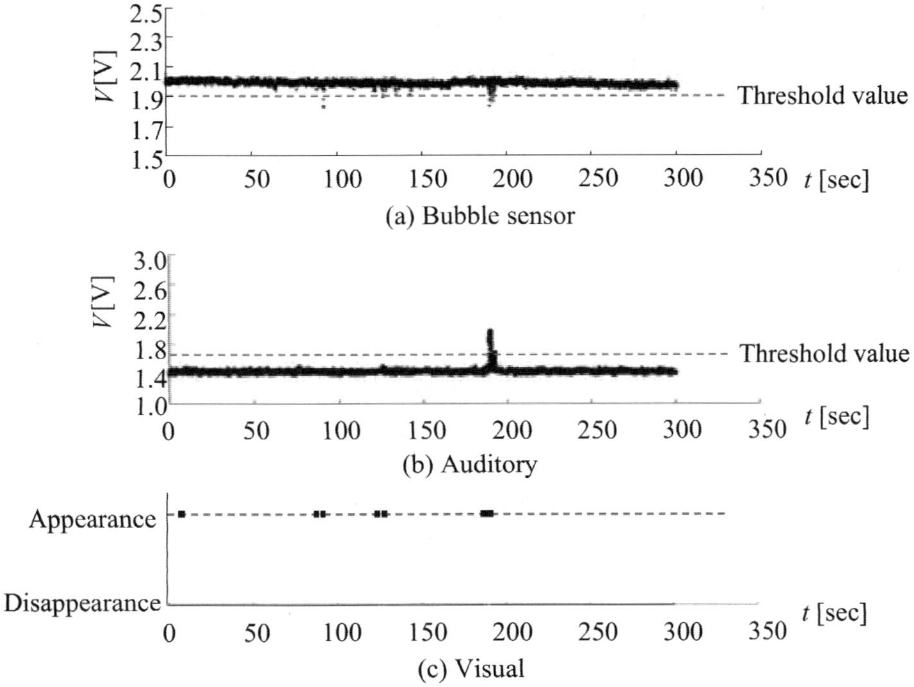

Fig.8 Time histories (X/D=1.71, S/D=2.25).

Journal of Physics: Conference Series **147** (2009) 012045 doi:10.1088/1742-6596/147/1/012045

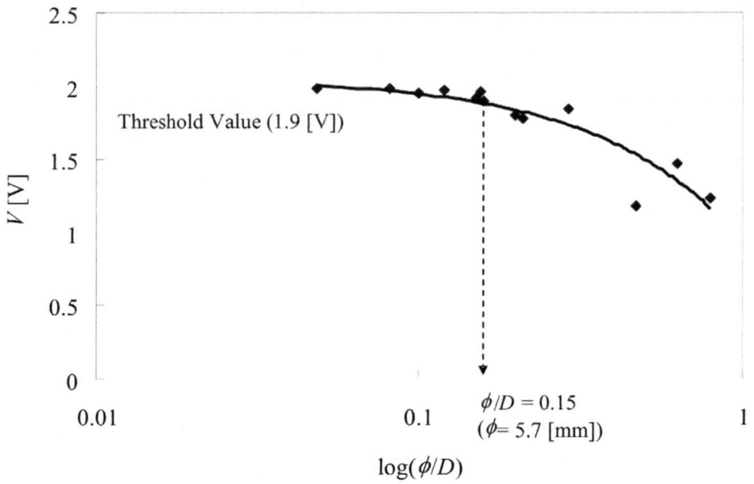

Fig.9 Threshold value and bubble diameter ϕ.

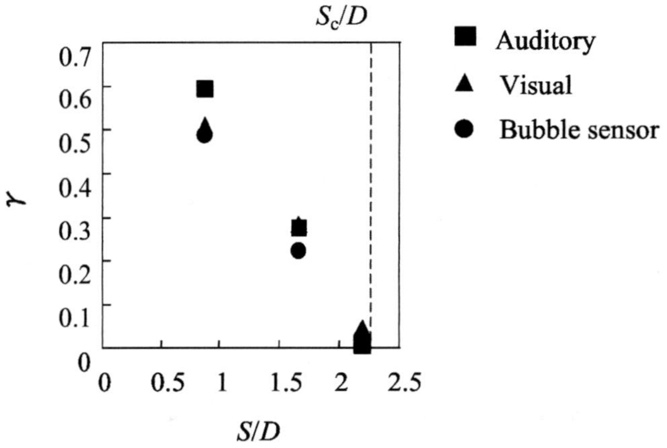

Fig.10 Intermittency factor γ versus S/D (X/D=1.71).

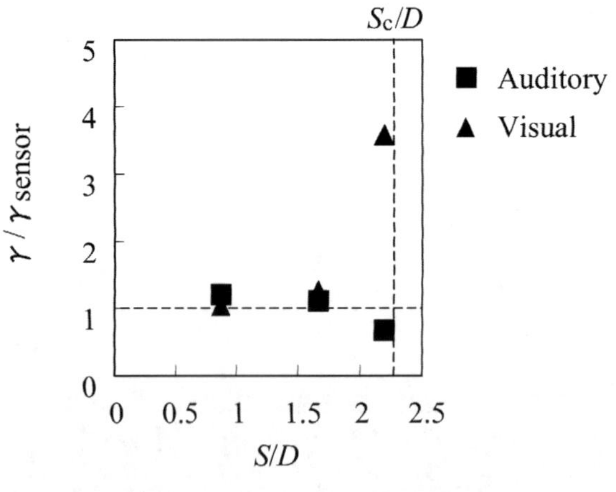

Fig.11 Reduced intermittency factor (X/D=1.71).

Table 2 Summary of comparison among bubble sensor, auditory and visual.

Auditory	Not accurate at $S/D \approx Sc/D$
Visual	Not accurate at $\gamma < 0.05$

In Fig. 11, we re-plot Fig. 10 in order to see the differences of the auditory and visual methods from the bubble-sensor method. Specifically speaking, the ordinate denote the reduce intermittency factor γ / γ_{sensor} instead of γ, where γ is normalized by the intermittency factor γ_{sensor} by the bubble-sensor method.

In Fig. 11, most of γ / γ_{sensor} are approximately unity except for those at $S/D \approx S_c/D$. However, at $S/D \approx S_c/D$, γ by the visual method is much larger, and γ by the auditory method is slightly smaller than γ_{sensor}. These results well correspond to Fig. 6-8.

Concerning the other many results with different values of X/D, we have finally obtained such a criteria as Table 2. The criteria specifies lower-accuracy conditions in the conventional auditory and visual methods. Specifically speaking, by the auditory method, the accuracy of γ becomes low, when submergence depth S of the suction pipe is close to the critical one S_c. On the other hand, by the visual method, the accuracy of γ becomes low, when γ is such a small value as less than 0.05.

Incidentally, the value of S_c/D determined by the visual method is considered to be accurate, when we refer to the results by the bubble-sensor method.

At $S/D \approx S_c/D$, we cannot anticipate accurate measurements of γ by both the auditory and visual methods, because γ tends to have a small value. At $S/D \ll S_c/D$, we can anticipate accurate measurements of γ by the auditory method, but cannot always anticipate accurate ones by the visual method.

4. SUMMARY

Until now, we have conventionally judged the air-entrainment occurrence in suction sumps by means of the visual method or the auditory method. These conventional methods have been used prevailingly, because of their practicability and simplicity. However, as the methods inherently mean just the observation of free surfaces or noise level, we do not identify the air entrainment itself. Then, we have required accurate assessments on the methods.

In order to get accurate measurements of air entrainment in a suction sump, we heve designed a new and simple conductance-type electric bubble sensor. And, we have focused on the intermittency factor γ (namely, the occurrence-time ratio of the air entrainment), and have compared the result by the bubble sensor with those by the conventional two methods; namely, the visual and auditory ones. As a result, we have shown the criteria which specify lower-accuracy conditions in the conventional methods.

ACKNOWLEDGEMENTS

We appreciate useful advice on void-ratio measurements by Prof. Takenaka (Kobe University).

NOMENCLATURE

B	Sump breadth	[mm]
Bo	Bond number	
D	Outside diameter of a suction pipe	[mm]

d	Inside diameter of a suction pipe	[mm]
Fr	Froude number	
H	Height level	[mm]
Re	Reynolds number	
S	Submergence depth	[mm]
t	time	[s]
V_b	Mean flow velocity	[m/s]
We	Weber number	
X	Back clearance	[mm]
Z	Bottom clearance	[mm]

Greek Letters

γ	Intermittency factor
τ	Air entrainment occurs

Subscripts

c	critical

REFERENCES

Funaki, J., Neya, M., Hattori, M., Tanigawa, H. and Hirata, K. (2007). “Flow Measurements in a Suction Sump by UVP,” Journal of Fluid Science and Technology of the Japan Society of Mechanical Engineers, series B, Vol.73, No.726, pp531-537.

Hewitt, G.. F., (1978). Measurement of Two Phase Parameters, Academic Press, London.

Hirata, K., Funaki, J. and Nakajima, N. (2003). “On the Critical Submergence for Air-Entraining Vortices in a Suction Sump,” Journal of Hydraulic, Coastal and Environmental Engineering, Proceeding of the Japan Society of Civil Engineers. No.747, pp.61-69 (in Japanese).

Iversen, H. W. (1953). “Studies of Submergence Requirements of High-Specific-Speed Pumps,” Transactions of the American Society of Mechanical Engineers, Vol.75, No.4, pp.635-641.

JSME Standard, (1984).JSME S 004. “Standard Method for Model Testing the Performance of a Pump Sump,” Japan Society of Mechanical Engineers.

AUTHOR INDEX

AUTHOR INDEX

AUTHOR INDEX

Institute of Physics
Dirac House, Temple Back
Bristol BS1 6BE UK

CURRAN ASSOCIATES INC
proceedings
.com

6th International Symposium on Measurement Techniques for Multiphase Flows 2008

Journal of Physics: Conference Series Volume 147

Naha, Okinawa, Japan
15-17 December 2008

Volume 2 of 2

ISBN: 978-1-61567-331-5

IOP Institute of Physics

6th International Symposium on Measurement Techniques for Multiphase Flows 2008

Journal of Physics: Conference Series Volume 147

Naha, Okinawa, Japan
15-17 December 2008

Volume 2 of 2

ISBN: 978-1-61567-331-5

Printed from e-media with permission by:

Curran Associates, Inc.
57 Morehouse Lane
Red Hook, NY 12571

Some format issues inherent in the e-media version may also appear in this print version.

Printed by Curran Associates, Inc. (2009)

For permission requests, please contact the Institute of Physics at the address below.

Institute of Physics
Dirac House, Temple Back
Bristol BS1 6BE UK

Phone: 44 1 17 929 7481
Fax: 44 1 17 920 0979

techtracking@iop.org

Additional copies of this publication are available from:

Curran Associates, Inc.
57 Morehouse Lane
Red Hook, NY 12571 USA
Phone: 845-758-0400
Fax: 845-758-2634
Email: curran@proceedings.com
Web: www.proceedings.com

TABLE OF CONTENTS

VOLUME 1

VOLUME 2

Journal of Physics: Conference Series **147** (2009) 012046 doi:10.1088/1742-6596/147/1/012046

Optical measurement of the instantaneous wavy surface structure on a water plane jet

K Itoh [1], D Maruwaka [2], H Kumamaru [1]

[1] University of Hyogo, 2167 Shosha, Himeji, Hyogo, 671-2201, Japan
[2] NTN Corporation, Osaka, 550-0003, Japan

itoh@eng.u-hyogo.ac.jp

Abstract. Laser beam refraction technique for measurement of free surface waves on a high-speed water jet has been developed in the present work. The new method employs two pulse laser diodes and the single light-spot detector. Through detecting the two dimensional trajectory of laser beams refracted at free surface, the local slope-angle fluctuation at two measuring locations 1.2 mm apart on a jet free surface is evaluated. The experiments are conducted for several locations along the jet center axis within the range of average velocity up to 5 m/s. The wave speed is evaluated from dominant time lag of cross-correlation coefficient for individual waves, which are discriminated by applying the zero-up-crossing method to the slope-angle time series data. The shape of waves is also calculated by integrating the free-surface slope angle data numerically. The ensemble averaged wavelength increases with the distance from nozzle exit almost monotonically. On the other hand, the mean wave height first increase with distance, reaches a maximum and then decreases. The maximum value of wave height is less than 0.23 mm for the tested velocity range. The steepness of free surface wave, that is the ratio of wave height and wavelength, takes a maximum at 20 ~ 30 mm downstream nozzle exit. This suggests that the wave breaking occurs on the jet free surface for higher velocity case.

1. INTRODUCTION

The occurrence of waves on a liquid free surface has been known to affect the absorbing and diffusion phenomena through the interface between different materials and phases. Then, many researches have been performed to measure the free surface waves during the last several ten years. In the recent proposals for neutron source facilities (Katsuta et al., 1996), the evaluation of the wave height on the high-speed (~15m/s) liquid lithium jet becomes important design issue because the occurrence of free-surface fluctuation may induce no only an increase in vaporization and dispersion of lithium into high vacuum environment but also an insufficient absorption of high deuteron beam energy (40 MeV, 125mA × 2).

Several measurements have been conducted on jet free-surface waves on lithium jet (Horiike et al., 2003; Kondo et al., 2006) and water as stimulant of lithium (Nakamura et al., 1998; Itoh et al., 2004); however only restricted data have been obtained because of the difficulties in measuring waves on high-speed liquid surface. Non-intrusive, high-response methods have been strongly preferred for such measurement.

Then, the aim of present study is to develop the measurement technique for the waves on a high-speed liquid jet. Present optical technique makes it possible to measure the local slope angle fluctuations at adjoined two locations on a wavy surface. High-speed light spot detector is used to detect the trajectories of laser beams refracted at jet free surface. The wave speed is derived from the dominant time lag of cross correlation coefficients between slope-angle fluctuation data at two measuring location. The spatial shape of waves is also reconstructed by integrating the free-surface slope angle data.

Since this study focuses the development of measurement technique for liquid free surface, the experiments is concerned with a lower velocity case (≤ 5 m/s) rather than of proposed beam target flow. However, the present method will be applied for higher velocity case with adjusting the geometry of two lasers (that is, spatial interval between two lasers, and their diameters) and with improving the time response of the laser blinking. Furthermore, the measured data in the present work will be useful in verifying the wave generation mechanism, such as free surface instability theory, even if the data obtained from water jet in simulating lithium flow (Kondo et al., 2005).

2. EXPERIMENT

2.1 Experimental Facility

Figure 1 shows a schematic view of test section, which is made of transparent acrylic resin. The facility uses water as the working fluid at room temperature and atmospheric pressure. Water is used since its kinematic viscosity is not too much different from those of lithium at the operating temperature (523K) of the proposed beam target (Itoh et al., 1999). The nozzle-exit height is 10 mm and the width is 100 mm. A plane water jet is generated by a two-dimensional convergent nozzle and flows along a horizontal flat wall. The jet width is fixed by sidewalls and the jet free surface is open to atmosphere to allow optical measurement and visual observation.

The optical arrangement for measurement of free surface waves is also shown in Fig. 1. Two laser beams are shot vertically on the water surface. We employ pulse laser diodes (Global laser, Premier LC, 655 nm, 1 mW) for the light sources. This diode is possible to blink by 300 kHz. The control of blinking of lasers will be mentioned in the next section with a topic of detection of two laser's loci by using an optical sensor. A focusing lens (150 mm focal length) and cylindrical collimator (1 mm dia., 3 mm length) reduces the beam diameter to be ≤ 24 μm at the water surface. The interval of two lasers L was adjusted to be 1.2 mm. The control signals for blinking of lasers were synchronized carefully with the data acquisition of output signals of the optical sensor. The lasers were alternately blinked at the frequency of 40kHz, and the output signals including laser control signals were sampled at the rate of 800 kHz in the present experiment.

The experiments are conducted for cross-sectional average velocity at the nozzle exit $U = 3 \sim 5$ m/s. Optical measurement of free surface is achieved for the distance $x = 80$ mm from nozzle exit as the optical system and sensor are moved along the center axis of the test section by using fine motion stages.

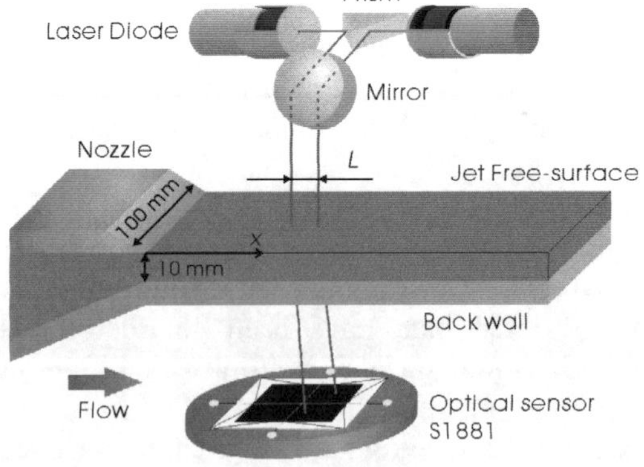

Fig. 1 Schematic view of test section and optical system.

2.2 Measurment of Free-surface Slope Angle

Present approach estimates the fluctuation of free surface slope angle from measurement of refracted laser beam loci. The relationship between the displacement of the beam spot on the sensor and the local slope angle θ of free surface is shown in Fig. 2.

The beam is refracted by the local slope of the water surface and passes through the water jet, the transparent back wall and air before hitting the optical sensor (Hamamatsu Photonics, S1881). The displacement r_x in streamwise x-direction is represented by local slope angle

$$r_x = D_w \tan(\theta_x - \theta_w) + D_b \tan(\theta_b) - D_a \tan(\theta_a) , \quad (1)$$

where D_w and D_b are thickness of jet and transparent back wall (10 mm), and D_a is distance from backwall to the sensor (20 mm). θ_w, θ_b and θ_a are the angle of refracted beam at each interface, which is obeyed the Snell's refraction law,

$$\frac{\sin\theta_x}{\sin\theta_w} = \frac{n_w}{n_a}, \frac{\sin(\theta_x - \theta_w)}{\sin\theta_b} = \frac{n_b}{n_w}, \frac{\sin\theta_b}{\sin\theta_a} = \frac{n_a}{n_b} , \quad (2)$$

where n_w, n_a and n_b are refractive indexes of water, acrylic resin and air respectively.

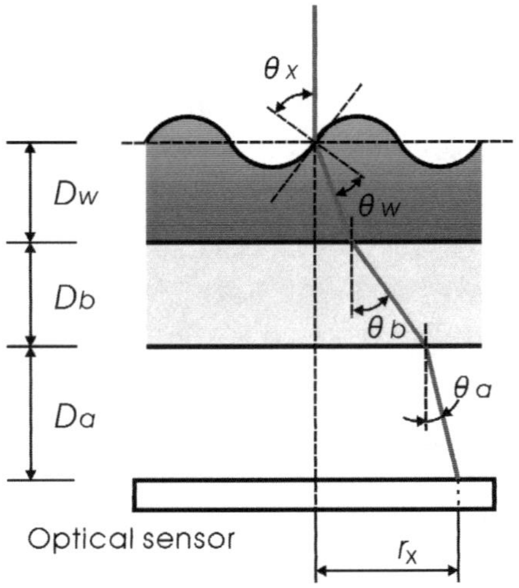

Fig. 2 Relation between free surface slope angle and displacement on optical sensor.

The instantaneous free surface slope angle θ_x is calculated by substituting the detected displacement r_x into eqs. (1) and (2). Although the solution θ_x of eqs. (1) and (2) can be obtained by using numerical iteration such as bisection method, a lot of time will be needed for data processing. Then, preliminary evaluated relation between r_x and θ_x (least-square fitted with 5th order algebraic equation) are used for the instantaneous calculation of θ_x in the present work.

The theoretical curve calculated from eqs. (1) and (2) is shown in Fig. 3 in comparison with the calibration data. The experimental slope-angle data is obtained by inclining the thin cover glass on the filled water surface in the present test section. The deviation of theoretical curve from the calibration data is less than ± 0.02 rad. We also checked the fluctuation in the jet thickness D_w affects only little the measured value of θ_x. For example, a 10% (= 1 mm) increase in the jet thickness (that is quadruple as large as the typical mean amplitudes of waves) would shift the theoretical curve by only 0.01 rad for present experimental condition. Then, the overall uncertainly in this measurement is evaluated to be less than ± 0.03 rad. Moreover, the maximum slope angle measurable with this method is determined by the thickness of light path, the refraction indexes of materials and the size of the optical sensor. It is ± 0.80 rad for the present geometry.

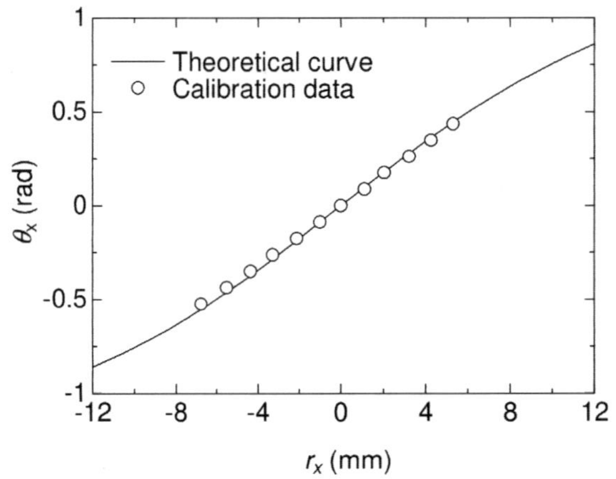

Fig. 3 Relation between displacement on optical sensor and free-surface slope angle.

The optical sensor S1881 consists of single p-i-n photo diode and can detect the two dimensional displacement of the beam spot on the sensitive area. As shown in Fig. 4, the beam spot induces photocurrent signals X_1 and X_2 in streamwise, and Y_1 and Y_2 in spanwise direction by photovoltaic effect. These signals are amplified by high responsive op-amp and are sampled independently. Then, the instantaneous displacements r_x and r_y is calculated from

$$r_x = \frac{L_x}{2}\frac{(X_2+Y_1)-(X_1+Y_2)}{X_1+X_2+Y_1+Y_2}, \quad r_y = \frac{L_y}{2}\frac{(X_2+Y_2)-(X_1+Y_1)}{X_1+X_2+Y_1+Y_2} \quad (3)$$

where $L_x = L_y = 26$ mm are the length of the each side of sensitive area (including non-active area) of S1881.

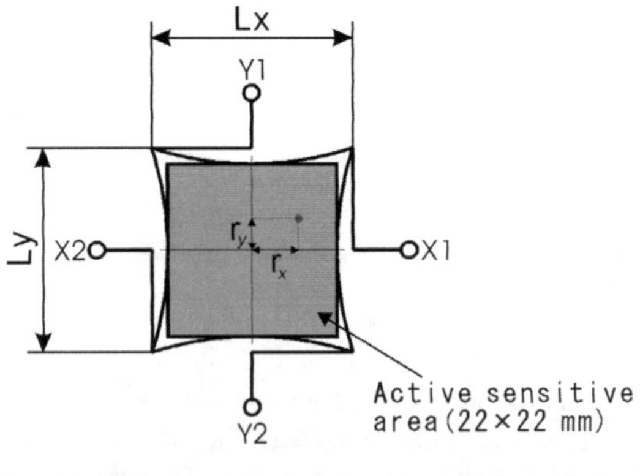

Fig. 4 Output signals of optical sensor, S1881.

The present single-photodiode sensor is suitable for a measurement of fluctuation on high-speed jet free surface since its maximum response frequency is 300 kHz for continuous beam. However, if more than two beams are hit on the sensitive area at the same time, it only detects the median point of each beam because of the overlap of induced photocurrents. Therefore, we have the two laser-diodes switched by turns in order to prevent the two beams from existing on sensitive area simultaneously. Although this switching operation fragments the obtained cross correlation coefficient, the dominant time lag can be determined by examining appropriate blinking frequency and spatial interval of two lasers.

2.3 Reconstruction of Wave Shape

The cross correlation coefficient $R(\tau)$ is represented by the slope angle fluctuation θ_{x1} and θ_{x2},

$$R(\tau)=\frac{\overline{\theta_{x1}(t)\theta_{x2}(t+\tau)}}{\sqrt{\overline{\theta_{x1}^2}}\sqrt{\overline{\theta_{x2}^2}}}, \quad (4)$$

where τ is time lag at spatial interval of measuring location. The wave speed c_x is evaluated from the dominant time lag τ_a, corresponding with the maximum cross-correlation coefficient, as $c_x \equiv L/\tau_a$. The waves with different wavelengths are considered to propagate with different wave speed. Then, in the present work, τ_a is calculated for individual waves, which are separated from time series data of θ_{x1} by using zero-up-crossing method (Bazargan et al., 2007).

In what follows, we explain the reconstruction of spatial wave shape from measured slope angle data. When the shape of free surface is denoted as $z = \eta_x(x)$, the local slope angle θ_x in streamwise direction can be written as

$$\tan\theta_x = \frac{\partial \eta_x}{\partial x} = \frac{\partial \eta_x}{\partial t}\cdot\frac{\partial t}{\partial x}, \quad (5)$$

where $\partial x/\partial t$ is the speed of free surface wave which passes through the measuring location. If waves are assumed to be traveling with frozen profiles across the measurement location, the wave shape η_x is obtained by integrating eq. (5) with respect to time as

$$\eta_x = \int_{wave}\left(\frac{\partial x}{\partial t}\right)\cdot \tan\theta_x dt = \int_{wave} c_x \cdot \tan\theta_x dt \ . \quad (6)$$

Thus, η_x is obtained from numerical integration of eq. (6) by substituting the evaluated wave speed c_x for individual waves. The wave height h is calculated from the deviation between maximum and minimum value of η_x. The wavelength λ is also calculated as the product of c_x and wave period, which is separated from time series data of θ_x by using zero-up-crossing method.

3. RESULTS AND DISCUSSIONS

3.1 Typical Slope Angle Data

Typical data of displacement r_x of beam spots are compared with the control signals in Fig. 5. While control signal takes positive value, laser 1 (LD1) is switched on (and laser 2 (LD2) is switched off). The free surface fluctuation at two measuring location is well captured, alternatively. However, it is found that the transitional signals are included in the moment that the laser interchanges. The transitional motion of temporal signals is clearly quicker than the wave period. And such signals were also observed in the preliminary test when still water (no wave) surface were measured. Therefore, these temporal outputs are considered to be related to change of beam intensity during the switching operation. Then, only one signal at the intermediate time while each laser is switched on is employed as responsible data for slope angle analysis. The responsible data are indicated by red or blue symbols in Fig. 5.

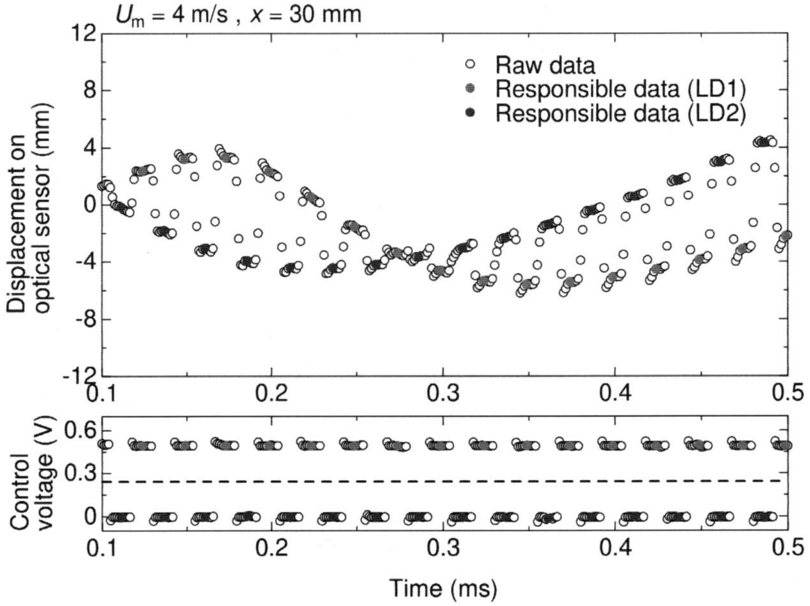

Fig. 5 Typical streamwise displacement data obtained by two lasers.

The traces of streamwise slope angle θ_1 and θ_2 are shown in Fig. 6. Although the deletion of transitional data restricts the practical sampling rate to 40 kHz, meaningful data are obtained at two measuring location. Because of laser switching operation, the time of data acquisition of θ_{x2} lags 0.025 ms behind that of θ_{x1}.

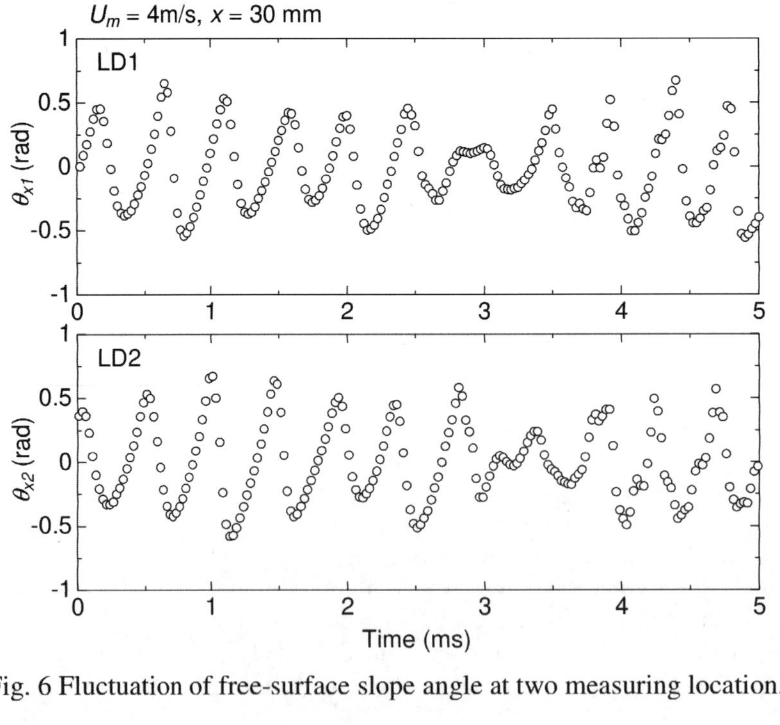

Fig. 6 Fluctuation of free-surface slope angle at two measuring location.

After dividing time series data of θ_x into individual waves by using zero-up-crossing method, the cross correlation coefficient eq. (4) is calculated. The wave speed c_x is evaluated from the dominant time lag τ_a corresponding maximum cross-correlation coefficient R_{max}. In the present work, the threshold value of R_{max} is chosen to be $R_{max} \geq 0.90$, and the target wave is assumed to be lost at downstream measuring location when the value of R_{max} is lower than the threshold value. It is confirmed that the average value of wave height and wavelength are almost independent of the threshold vale through the calculations by using another threshold value, that is $R_{max} \geq 0.95$ and 0.98, whereas a number of population decreases with an increase in the threshold value.

3.2 Visual Observation of Jet Free Surface

Typical microflash (4 μs) picture of the jet free surface is shown in Fig. 7 for U_m = 5 m/s. Three distinct regions can be identified with respect to the wave development in the flow direction. The jet starts with a smooth region with almost no visible waves. It is followed by the second region, which indicates periodic two-dimensional (2D) waves with dominant wavelength of 0.5 to 1.0 mm and wave amplitudes becoming greater with the distance x from the nozzle exit. Finally, the 2D structure of waves breaks down into less regular 3D-wave patterns.

As the jet velocity in raised, the smooth and 2D-wave regions become shorter and eventually disappear. The linear analysis of the instability modes of the velocity shear layer underneath the free surface implies that shear-mode instabilities are responsible for periodic wave generation on high-speed liquid jet with an initially-laminar shear layer (Itoh et al., 2007). Therefore, the decrease and disappear in the smooth and 2D-wave region may be related to the relaxation distance of shear layer underneath free surface. Furthermore, these structures of jet free surface were observed in the liquid lithium jet (Hassberger, 1983; Kondo et al., 2005) and water jet in simulating beam target flow (Itoh et al., 1999).

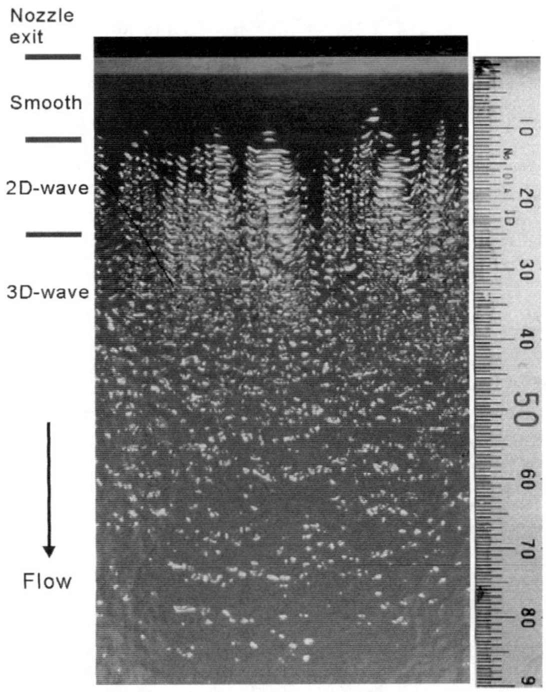

Fig. 7 Typical structures of jet free surface (U_m = 5m/s).

Journal of Physics: Conference Series **147** (2009) 012046 doi:10.1088/1742-6596/147/1/012046

3.3 Wavelength and Wave height

The ensemble average of measured wavelength λ_{ave} are plotted in Fig. 8 against the distance x from nozzle exit for various values of U_m. The mean wavelength increases with x for the tested whole velocity range. The absolute value of wavelength decreases with increase in the jet average velocity U_m. The ranges of smooth, 2D and 3D-wave regions, defined from the still photograph as shown in Fig. 7, are indicated in this figure by horizontal arrows. It is found that the wavelength increases quickly in the smooth region, while they increase at a smaller rate in the 2D and 3D-wave region. Furthermore, the data of θ_x indicated the occurance of intermittent wave packets in the smooth region.

In order to check the validity of present technique, the wavelength was also measured from the luminance profile of the photograph. Five profiles of streamwise luminance were extracted from the photograph shown in Fig.7, and the wavelength was measured. The dominant wavelength, derived by ensemble averaging the wavelength on the luminance profile, is compared with optical measurement data for average jet velocity of U_m = 5m/s. The error bars indicate the ranges from the observed maximum wavelength to the minimum value. It is confirmed that the mean wavelength acquired from optical measurement agrees well with the photograph observation.

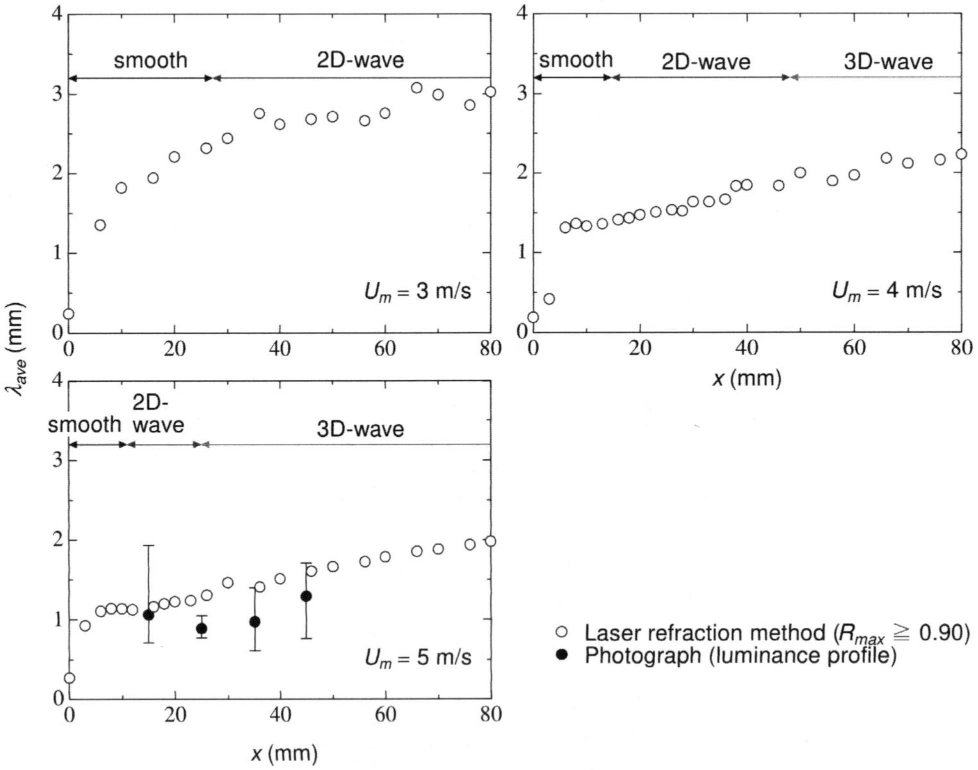

Fig. 8 Variation of mean wavelength with the distance from nozzle exit; comparison between optical measurment and photgraph observations.

The mean wave height is plotted against x in Fig. 9. Wave height increases with the distance from nozzle exit and reaches ~ 0.2 mm at the most downstream measuring location x = 80 mm. For $U_m \geq 4$ m/s, the wave height takes a peak at x = 20 to 30 mm from the nozzle exit. The location of the peak height approaches to the nozzle exit with increase in jet velocity, and it corresponds approximately to the end of 2D wave region. On the other hand, the lowest-velocity case (U_m = 3 m/s), without 3D-wave region for the measured whole location, shows no clear peak.

The wave steepness as defined to be the ratio of the wave height to the wavelength is plotted in Fig. 10. The ensemble average of the ratio h/λ for individual waves is indicated. For higher velocity case $U_m \geq 4$ m/s, mean steepness first increases with x, reaches a maximum at 2D-wave region and then decreases with further increase in the distance from nozzle exit. The peak is higher and clearer for higher exit velocities. The non-linear water wave theory suggests that the shape of wave collapses when the wave develops sharper and its steepness exceeds a limit value. For example, the breaking limit of Stokes waves is known to be 0.142 (Whitham, 1974). Then, the changes of steepness may indicate that the periodic 2D-wave becomes unstable at the end of growth and breaks down into less regular 3D-waves.

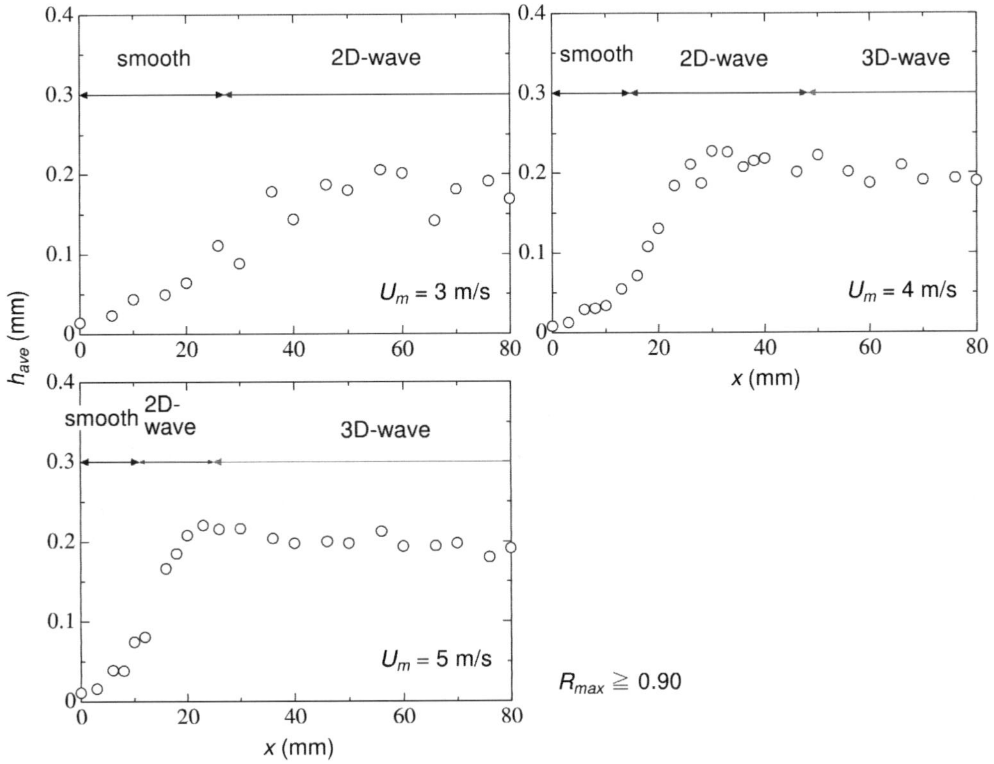

Fig. 9 Growth of mean wave height with the distance from nozzle exit.

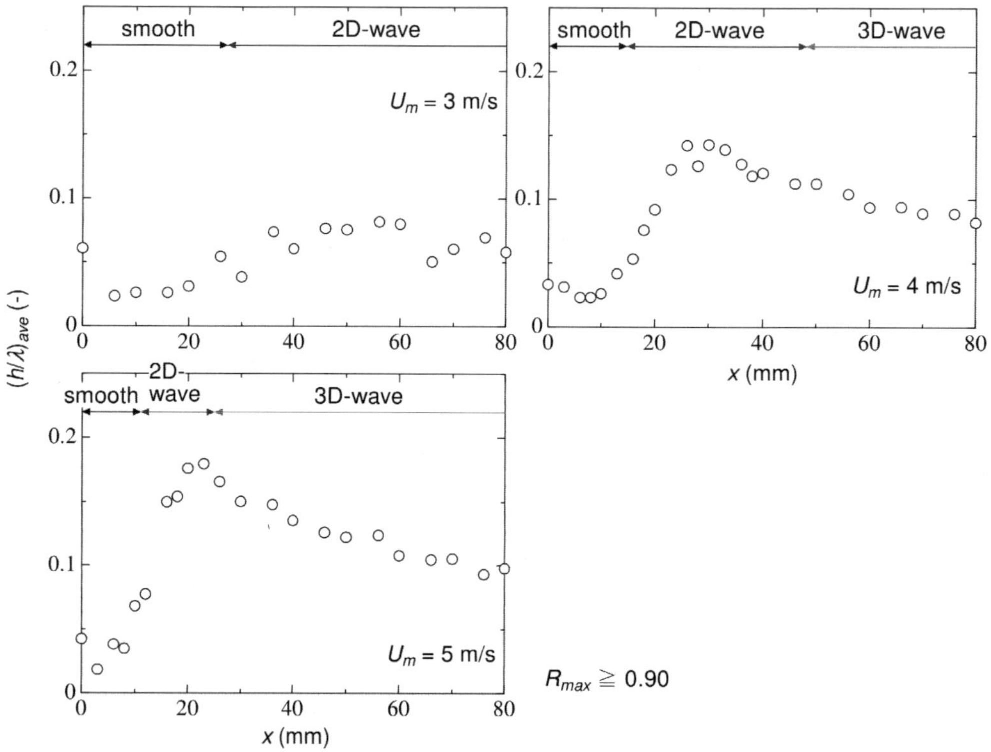

Fig. 10 Variation of mean wave steepness with the distance from nozzle exit.

4. CONCLUSIONS

Optical measurements of free surface waves on a water jet have been conducted for the jet average velocity of $U_m \leq$ 5m/s. A new technique employs high-speed optical sensor to detect the displacements of two pulse laser beams refracted at neighboring locations on jet free surface. The shape of free surface wave is reconstructed by integrating slope angle data by substituting wave speed evaluated from cross-correlation of neighboring free-surface fluctuation.

The high-speed photograph demonstrates that the periodic 2D-wave occurs and subsequently grows into 3D-wave pattern on the jet free surface. The mean wavelength measured by using present optical technique agrees with the value evaluated from the luminance profile of photograph. It is found that the wave amplitude increases with the distance from the nozzle exit and takes a maximum at the 2D-wave region for higher velocity case $U_m \geq$ 4 m/s. In the present experimental range, the maximum value of mean wave height is less than 0.23 mm. The wave steepness also indicates a local peak in the 2D-wave region. This suggests that the change of periodic 2D-wave into less regular 3D structure may relate to the wave breaking on the jet free surface.

REFERENCES

Katsuta, H., et al. (1996), "Conceptual design study of IFMIF target facility", Fusion Technol., 30(3), pp.1152-1160.

Horiike, H., et al. (2003), "Lithium Free Surface Flow Experiment for IFMIF", Fusion Eng. Des., 66-68, pp.199-204.

Kondo, H., et al. (2006), "Experimental Study of Lithium Free-Surface Flow for IFMIF Target Design, " Fusion Eng. Des., 81(1-7), pp.687-693

Nakamura, et al. (1998), "Experimental and Analytical Studies on High-Speed Plane Jet along Concave Wall Simulating IFMIF Li Target Flow," J. Nucl. Mater., 258-263(1), pp.440-445.

Itoh, K. et al. (2004), "Internal-Shear Mode Instabilities on High-Speed Liquid Jet, (II) Experimental Analysis of Curved Target Flow," J. Nucl. Sic. Technol., 41-8, pp.809-816.

Kondo, H., et al. (2005), "Surface wave on high speed liquid lithium flow for IFMIF," Fusion Eng. Des., 75-79, pp. 865-869.

Itoh, K. et al. (1999), "Initial Free Surface Instabilities on a High-Speed Water Jet Simulating a Liquid-Metal Target," Fusion Technol., 36, pp.69-84.

Bazargan, H. et al. (2007), "Simulation of the mean zero-up-crossing wave period using artificial neural networks trained with a simulated annealing algorithm," J. Marine Sci. Technol. 12(1), pp. 22-33.

Itoh, K. et al. (2007), "Linear Stability Analysis on Free-Surface Liquid Jet with Different Simplification of Velocity Profile," J. Fluid Sci. Technol., 2(2), pp.417-428.

Hassberger, J. A. (1983), "Stability of the FMIT High Speed Free Surface Liquid Jet Flowing Along a Curved Back Wall," Proc. 10th Symp. Fusion Eng., Philadelphia, USA, December 5-9, Vol. 2, p. 1849

Whitham, G. B. (1974). Linear and Nonlinear Waves, John Wiley & Sons, New York, p.471.

Journal of Physics: Conference Series **147** (2009) 012047 doi:10.1088/1742-6596/147/1/012047

Online recognition of the multiphase flow regime and study of slug flow in pipeline

GUO Liejin [1], BAI Bofeng[1], ZHAO Liang[1], Wang Xin[1], Gu Hanyang[1]

[1] Xi'an Jiaotong University, Xi'an 710049, CHINA

lj-guo@mail.xjtu.edu.cn

Abstract. Multiphase flow is the phenomenon existing widely in nature, daily life, as well as petroleum and chemical engineering industrial fields. The interface structure among multiphase and their movement are complicated, which distribute random and heterogeneously in the spatial and temporal scales and have multivalue of the flow structure and state[1]. Flow regime is defined as the macro feature about the multiphase interface structure and its distribution, which is an important feature to describe multiphase flow. The energy and mass transport mechanism differ much for each flow regimes. It is necessary to solve the flow regime recognition to get a clear understanding of the physical phenomena and their mechanism of multiphase flow. And the flow regime is one of the main factors affecting the online measurement accuracy of phase fraction, flow rate and other phase parameters. Therefore, it is of great scientific and technological importance to develop new principles and methods of multiphase flow regime online recognition, and of great industrial background.

In this paper, the key reasons that the present method cannot be used to solve the industrial multiphase flow pattern recognition are clarified firstly. Then the prerequisite to realize the online recognition of multiphase flow regime is analyzed, and the recognition rules for partial flow pattern are obtained based on the massive experimental data. The standard templates for every flow regime feature are calculated with self-organization cluster algorithm. The multi-sensor data fusion method is proposed to realize the online recognition of multiphase flow regime with the pressure and differential pressure signals, which overcomes the severe influence of fluid flow velocity and the oil fraction on the recognition. The online recognition method is tested in the practice, which has less than 10 percent measurement error. The method takes advantages of high confidence, good fault tolerance and less requirement of single sensor performance.

Among various flow patterns of gas-liquid flow, slug flow occurs frequently in the petroleum, chemical, civil and nuclear industries. In the offshore oil and gas field, the maximum slug length and its statistical distribution are very important for the design of separator and downstream processing facility at steady state operations. However transient conditions may be encountered in the production, such as operational upsets, start-up, shut-down, pigging and blowdown, which are key operational and safety issues related to oil field development. So it is necessary to have an understanding the flow parameters under transient conditions. In this paper, the evolution of slug length along a horizontal pipe in gas-liquid flow is also studied in details and then an experimental study of flowrate transients in slug flow is provided. Also, the special gas-liquid flow phenomena easily encountered in the life span of offshore oil fields, called severe slugging, is studied experimentally and some results are presented.

1. INTRODUCTION

Multiphase flow is a phenomenon which exists widely in petroleum, chemical engineering industrial fields, as well as in natural, life living. The interfacial structure between phases and its movement is very complicated, which has the randomness and heterogeneity distribution in the spatial and temporal scales, therefore, the flow structure and state of multiphase mixture is in multivalue. Flow pattern is defined as the macro feature about the multiphase interfacial structure and their distribution, which is an important feature to describe multiphase flow. When gas and liquid two-phase flow in a pipe, several flow patterns can be observed, including typical stratified and slug flow. The energy and mass transport mechanism differs much for different flow patterns. It is necessary to solve the recognition of flow patterns to get a clear understanding of the physical phenomena and their mechanism of multiphase flow. And the flow regime is one of the main factors affecting the online measurement accuracy of phase fraction, flow rate and other phase parameters. Therefore, it is of great scientific and technological importance to develop new principles and methods of multiphase flow regime online recognition, and of great industrial background as described in Guo (2002, 2005).

The one of the earliest research work with online recognition of flow regime was presented by Hubbard and Dukler (1966), in which they proposed a method based on pressure fluctuation probability density function. Until now, the flow regime recognition methods can be classified into two categories. One is on basis of flow structure image directly, and the other is the indirect method, that is, the recognition with the analysis of fluctuation signal representing the multiphase flow character. The indirect method is becoming main approach because it can quantitatively describe the flow regime. The fluctuating signals about flow regime have many feature extraction methods, such as time series analysis, wavelet, time-frequency joint analysis, fuzzy, chaos and fractal by Tutu 1984; Matsui (1986); Wambsganss & Jendrzejczyk (1991); Cai, Wambsganss & Jendrzejczyk (1996),; and Langford, Beasley & Ochterbeck (1998). The artificial neural network has being developed to a conventional method to recognize the flow regime by Mi, Ishii &Tsoukalas (1998); Monji & Matsu (1998); Wu & Zhou (2001); and Mahvash & Ross (2008).

Until now all the online recognition methods proposed in previous publications still could not give a satisfactory solution for the industry flow due to their little reliability and repeatability as pointed out by Bai, Guo & Zhao (2001); Li (2002); Yang et al (2005), and Bai et al (2008). The main reasons are as follows:

(1) The fluid physical properties and flow velocity have great influence on the recognition accuracy.

(2) The feature parameter intersects for different flow regime, which make it is unable to identify accurately basing on one single technology.

The physical mechanism resulted in pressure fluctuation can explain the two reasons. The pressure fluctuation is caused by the phase interface movement and its induced turbulence, the bilateral differential pressure, as well the liquid turbulence by Samways, Bradbury & Bruun (1997). The oil-gas-water multiphase flow as the example, its great flow rate and great oil fraction will certainly reduce the sensitivity of the pressure fluctuation to the phase movement and distribution.

In this paper, the prerequisite is analyzed to realize the online recognition of multiphase flow regime firstly. Secondly the recognition rules for partial of flow pattern are discussed and given basing on the feature analysis of pressure signal and differential pressure signal. Then the self-organization cluster algorithm is proposed to design the standard templates for every flow pattern feature. At last, the multi-sensor data fusion method is proposed to recognize the multiphase flow regime, which is tested in the practice.

Among various flow patterns of gas-liquid flow, gas-liquid slug flow occurs frequently in the petroleum, chemical, civil and nuclear industries. It is characterized by a sequence of elongated bubbles separated by liquid slugs that may contain small bubbles. It can be divided into two main groups: hydrodynamic and terrain slugging. Hydrodynamic slugging is the normal slugging pattern encountered in straight flow lines. Terrain slugging results from variable flow line topography which lead to local dips and peaks in the flow. An extreme case of this form of slugging occurs when a slightly downwardly inclined flow line is connected to a vertical riser which connects the flow line to the platform-mounted separator. Severe slugging often results in large backpressure accumulation, as well as widely varying liquid and gas production levels. For its transient and intermittent nature, it is difficult to predict the flow characteristics correctly. Therefore, reliable flow models are crucial for the safe and cost efficient design and operation of industrial facilities. This paper describes the experimental and numerical investigations on the slug flow in pipeline in our laboratory.

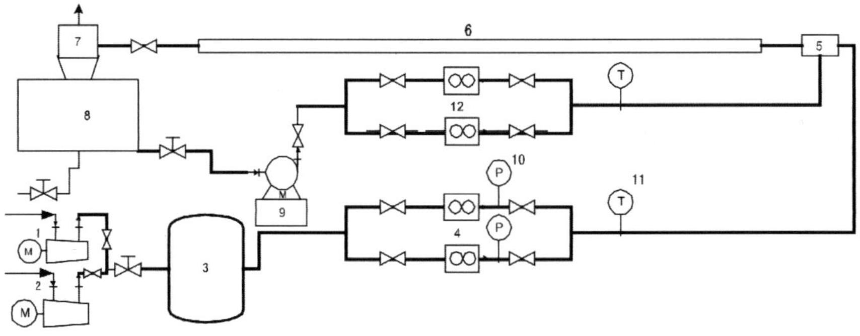

Fig. 1 Schematic of the experimental system
1, 2: Air Compressor; 3: Air tank; 4: Air flowmeters; 5: Mixer; 6: Test section; 7: Separator; 8: Water tank; 9: Centrifugal pump; 10: ROSEMOUNT pressure transducer; 11: Thermocouple; 12: Liquid flowmeters

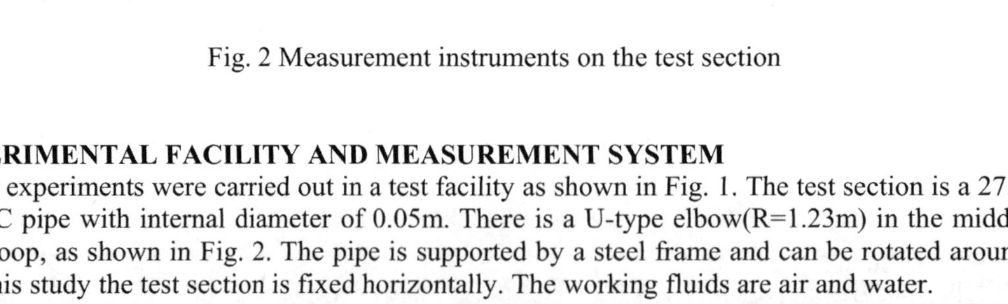

Fig. 2 Measurement instruments on the test section

2. EXPERIMENTAL FACILITY AND MEASUREMENT SYSTEM

The experiments were carried out in a test facility as shown in Fig. 1. The test section is a 27.33m long PVC pipe with internal diameter of 0.05m. There is a U-type elbow(R=1.23m) in the middle of the test loop, as shown in Fig. 2. The pipe is supported by a steel frame and can be rotated around its axis.In this study the test section is fixed horizontally. The working fluids are air and water.

Fig. 2 gives the locations of the instruments installed on the test section. There are 14 KELLER PA23e pressure transducers, which are sequenced from P1 to P14 along the tube. The frequency response of the KELLER PA23e pressure transducers is 5kHz, which is high enough to record the transient characteristics of slug flow. The analog output signals from pressure transducers and flowmeters were sampled by a 12-bit PCI-6071e data acquisition card produced by NATIONAL INSTRUMENTS company running on a personal computer. The sampling frequency of each channel is setting at 2500Hz.

The slug flow parameters in the pipe were detected by the fast response pressure transducers. This method was presented in the Ph.D. dissertation of He (2002). In this paper the signal of pressure transducer was compared with that of the double parallel wire conductivity probes. It is found that the sharp increasing edge of the pressure is corresponding to the front of the liquid slug, as shown in Fig.3, 4. However, when the tail of the liquid slug leaves the pressure transducer there is no clear mark in the pressure signal. But if we subtract the downstream pressure from the upstream pressure we can get the differential pressure and the end of the sharp decreasing edge is corresponding to the moment of the tail of the liquid slug leaving the downstream pressure transducer. The beginning point of the sharp increasing edge of the differential pressure trace is the moment that the liquid slug front arrives at the upstream pressure transducer. Therefore liquid slug translational velocity and its length can be measured from three neighboring pressure transducers. In this study, ten pressure transducers from P5 to P14 were installed on the test section and all the separation distances between two neighboring taps were 1m. Fig.4 shows the differential pressure traces from the subtractions of three close pressure transducer signals. The slug translational velocity is:

$$V_t = \frac{L}{t_1} \quad (1)$$

where t1 is the time spent by the liquid slug front moving from the upstream pressure transducer to the downstream one. L is the separation distance between two neighboring pressure taps. The slug length could be calculated by:

$$L_s = V_t t_2 - L \quad (2)$$

where t2 is the difference between the time at which the liquid slug front contacts the upstream pressure tape and the time at which the liquid slug tail contacts the downstream pressure tape. It is assumed that the liquid slug length is constant in this time interval.

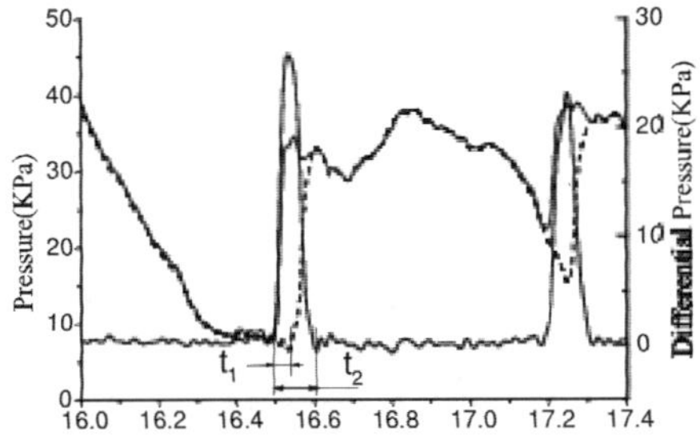

Fig. 3 Pressure and differential pressure vs. time

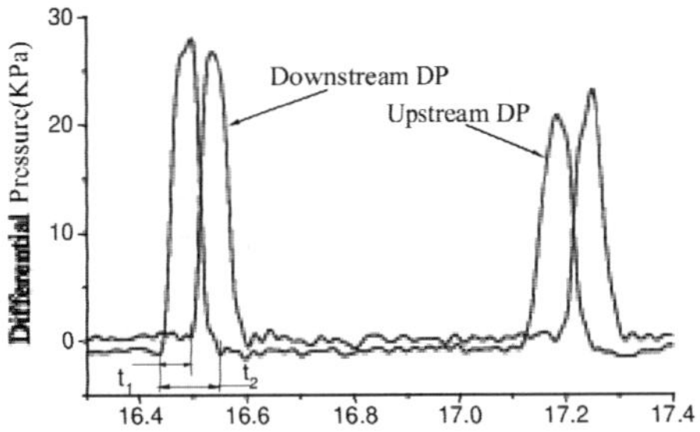

Fig. 4 Differential pressure vs. Time

3. FLOW PATTERN ONLINE IDENTIFICATION

3.1 Basic Principles of Flow Pattern Online Identification

The online recognition of flow regime need to measure real-time information of parameters, such as pressure, differential pressure, phase fraction and so on, at first, then to extract and analyze the feature of these, finally to make decision of what flow pattern it belongs to based on these feature. Choice of flow regime feature strongly affects the strategy of flow regime recognition and its accuracy. The flow regime recognition method that is constituted of significantly different features will has a better performance as shown in Bai (1999), Huang & Zhong (1991). In fact we often don't know what feature is the most important, and we also can't make sure weather the pre-determined feature combination is appropriate. So feature extraction and analysis of signals is needed to find the most effective features to form a low-dimensional mode vector.

The basic principles of flow regime recognition are as followed:

(1) Sample sets which represent the flow regime feature should exist in the flow feature space of flow regime information,

$$\{X_{ij}\} \subset \{\varpi_k\} \quad (3)$$

where, i =1,2,…,n is the kind of flow regime; j=1,2,…,N is the flow regime samples composed of feature parameters; k=1,2,…,M is the number of flow regime, M=n is desirable. The principle includes the following meanings. The flow regime samples composed of feature parameters must be representative; Flow pattern samples should have the attribute property of a certain type of flow pattern, and it can express physical significance of the flow pattern to a certain extent; Feature parameters which react sensitively with flow regime can be chosen as pattern samples.

(2) The feature space should meet the principle of close cohesion for a certain flow regime mode, and far separation for different flow regimes mode. The less the distance in the same mode and the greater the distance among the different modes, the easier to separate different flow patterns will be.

According to this principle, feature extraction and analysis is conducted on the signals of pressure and pressure difference over a wide test range, which can be acquired easily in the industry. At the sane time the clustering and separation characteristics of the feature parameter are studied. Then find the most effective feature to form the low-dimensional flow regime simple space and finally choose the concrete strategy to recognize the flow regimes on-line.

3.2.Feature Analysis and Flow Regime Recognition Rules

3.2.1 Feature Analysiss

Experiment was conducted on oil-gas-water multiphase flow loop. The test section was made up of plexi glass pipe with the diameter of 40 mm and the length of 20 m. The measurement points of pressure and differential pressure located at 15 m section from the entrance. 40 # diesel was used during the experiment. The experimental range of oil fraction is from 0.11 to 0.75, the absolute pressure from 0.1 to 0.3Mpa, fluid temperature was 4 to 15□, oil superficial velocity from 0.05 to 0.51 ms^{-1}, water superficial velocity from 0.05 to 1.51 ms^{-1}, and gas superficial velocity from 0.02 to 50.6 ms^{-1}. The experiment basically covered the range of industrial flow parameter under the medium velocity. Reference to the dividing method of two-phase flow of Mandhane, Gregory & Aziz (1974) and Acikgoz, Franca & Lahey (1992), and according to both visual observation and the fluctuating pressure and differential pressure, the oil-gas-water multiphase flow regimes in horizontal tube can be classified into four kinds: bubbly flow, intermittent flow, stratified flow and annular flow.

Since pressure and differential pressure signals contain a variety of noise, de-average value processing and filtering processing was done before feature extraction and analysis. Figure 5 shows a group of experimental results at oil superficial velocity of 0.05 m/s and water superficial velocity of 0.10 m/s. To bubble flow, the pressure and differential pressure fluctuations was small; To plug flow, the pressure signal was similar to rectangular wave and the differential pressure signal have a periodic rise; To slug flow, the rise frequency of differential pressure signal increase, whereas the periodicity of

Journal of Physics: Conference Series **147** (2009) 012047 doi:10.1088/1742-6596/147/1/012047

pressure signal decrease; To annular flow, energy of pressure and differential pressure signal increase, the pressure and differential pressure signal don't have periods phenomenon no longer.

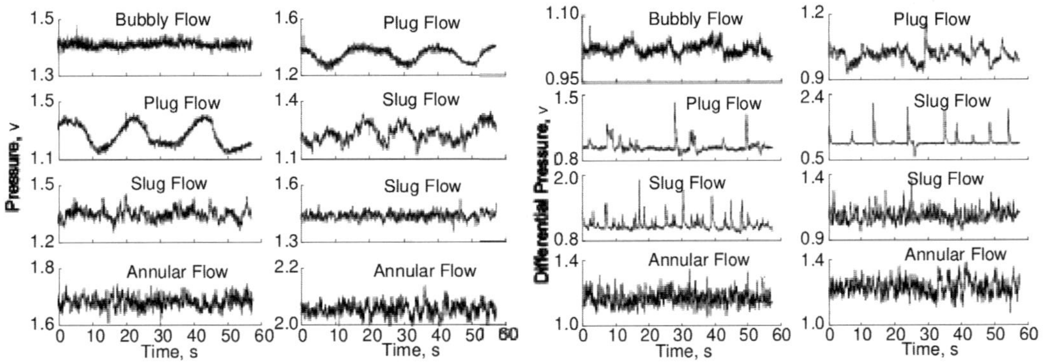

Run	Gas velocity	Flow regime	Run	Gas velocity	Flow regime
1	0.68	Bubbly	5	20.44	Slug
2	2.53	Plug	6	25.61	Slug
3	2.54	Plug	7	36.68	Annular
4	15.27	Slug	8	42.15	Annular

Fig. 5 Results at different pressure and differential pressure (see table as well)

Figures 6 and 7 respectively show the root mean square (RMS) features of the pressure signal and the variance features of differential pressure signal. Noting that the differential pressure signal is processed dimensionless, that is, divided by the differential pressure of the full tube water in order to remove the impact of diameter:

$$\Delta P = (\Delta P)_m / (\rho_w g D) \quad (4)$$

Where, ΔP, $(\Delta P)_m$, ρ_w, g, D are respectively the dimensionless differential pressure, average differential pressure (kPa), the density of water (kgm^{-3}), acceleration of gravity (ms^{-2}), and the diameter of the tube (m).

The change rule of pressure RMS is the same with that by Wambsganss, Jendrzejczyk, France (1991). Before the appearance of annular flow, RMS increases with gas velocity, but the liquid velocity has a non-monotonic influence. The RMS drops suddenly when the annular flow appears after the slug flow and increases with the liquid velocity. The RMS of differential pressure signal increases with the increment of gas velocity and liquid velocity. In addition, other feature parameters of pressure signal and differential pressure signal, such as the average deviation, the peak factor, the energy of every scale, fractal dimension, the biggest power spectrum, these all can't reflect the changes of flow regime. Increase the oil fraction, the relative velocity between gas and liquid will reduce, it can also restrain the bubble combination, the occurrence and development of interfacial wave, so it cause the above parameters even less sensitive to the change of flow regime.

3.2.2 The Recognition Rule of Flow Pattern

The flow regime surely exists in a certain feature space according to the basic rules of multiphase flow dynamics. The differential pressure fluctuation energy of annular flow and high speed intermittent flow is higher than that of bubbly flow, stratified flow, etc. According to the dimensionless variance (total energy) of differential pressure signal, the flow regimes can be divided into two classes:

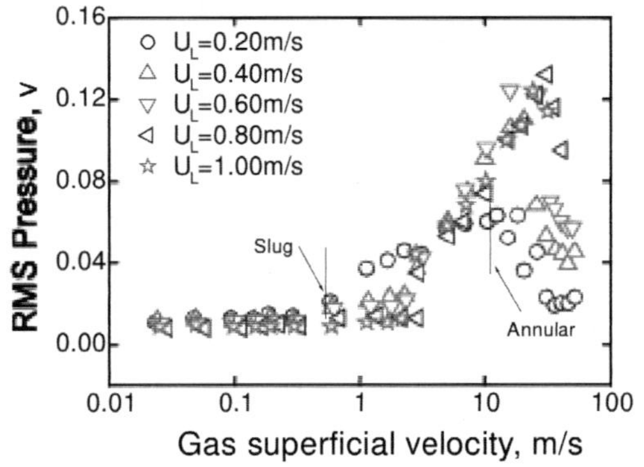

Fig. 6 RMS characteristin of the pressure signal(oil fraction is 0.2)

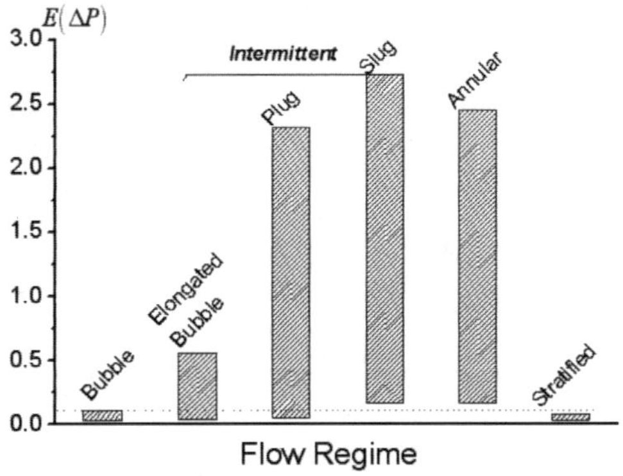

Fig. 7 Dimensionless variance of the differential pressure signal

$E(\Delta P) \geq E(\Delta P)_c$, for all annular flow and part of intermittent flow (5)

$E(\Delta P) < E(\Delta P)_c$, for all bubble flow, stratified flow and part of intermittent flow (6)

where, $E(\Delta P)_c = 0.12$.

Because the differential pressure can be measured directly, all feature parameters are unfolded on the dimensionless time-averaged differential pressure domain, and the law of these combined features is analyzed. Then the feature space of special flow pattern is established. Figure 8 indicates that the feature parameter of annular flow and part of intermittent flow (mainly slug flow and few plug flow) clusters together. As well, the feature parameter of bubble flow, stratified flow and part of intermittent flow (mainly plug flow and a few slug flows) gathers. The annular flow and intermittent flow is easy to be separated. The flow regime recognition rules can be established according to dividing these features space:

$(\Delta P)_m \geq (\Delta P)_1, E(\Delta P)_c < E(\Delta P) \leq E(\Delta P)_1$, annular flow (7)

$(\Delta P)_m > (\Delta P)_2, E(\Delta P) \geq \left[E(\Delta P)_c, E(\Delta P)_2\right]_{max}$, intermittent flow (8)

In above equations,

$E(\Delta P)_1 = 3.6 + 3.9\Delta P - 3.3(\Delta P)^2$, $E(\Delta P)_2 = -2.1 + 2.1\Delta P$ (9)

$(\Delta P)_1 = 3.2$, $(\Delta P)_2 = 2.8$ (10)

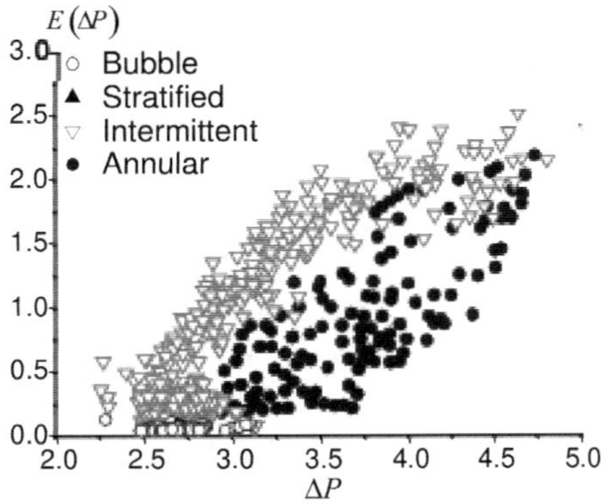

Fig. 8 Feature parameters unfolede on the differential pressure domain

3.3.Classification Machine of Flow Pattern

Although the flow regime recognition rule of the two regimes was put forward, the feature space of the two kinds still overlap. Therefore these flow regime rules are not suitable for all feature space. Moreover the feature spaces of other flow regimes overlap seriously. Other technique is needed to solve the problem. One of the solutions is to calculate the feature samples according to the experimental results with some algorithm, then compose feature space using these samples. The feature space can be used to identify the flow regime. Because the learning vector quantization (LVQ) classification machines have the character of self-organization, the present research chooses it to calculate the standard template of flow regimes. Suppose the euclidean distance between input sample x with the attribute ω_r and m_u which is the u-th sort is the nearest, then the iterative learning rule is as follows from Kohonen (1990):

$$m_i(k+1) = m_i(k) \qquad (i = 1,2,\cdots,D) \qquad \text{if } r = u \quad (11)$$

$$\begin{cases} m_r(k+1) = m_r(k) + \alpha(k)[x(k) - m_r(k)] \\ m_u(k+1) = m_u(k) - \alpha(k)[x(k) - m_u(k)] \end{cases} \qquad \text{if } r \neq u \quad (12)$$

Where, $m(k)$ is template, D is the amount of template, $0<a(k)<1$ is learning coefficient. The training samples $\{x_1, x_2, \ldots, x_N\}$, the initial template vector $\{m_1(0), m_2(0), \ldots, m_D(0)\}$ and the form of learning coefficient $a(k)$ are all given. Of template samples $x_j \in \omega_r$, calculate the distance between x_j and all types of template samples and then find the minimum:

$$u = \arg\min_i \left\{ \left\| x_j - m_i(k) \right\|^2 \right\} \quad (13)$$

If r=u, the template vector of each kind is invariable, according to equation (9). Otherwise iterative calculate the quantitative vector of each kind of template, according toequation (10). Then standard template $m_i(k)$ of different flow pattern can be obtained if the iteration is convergent.

Fourier spectrum of pressure signal processed uniformly is selected to be the input vector of the learning vector quantization (LVQ) classification machines. When apply these standard templates to recognition unknown flow pattern, all distances between standard template and Fourier spectrum of the pressure signal of unknown flow regime should be calculated:

$$Z(x, m_i) = \left\| x - m_i \right\|^2 \quad \text{i=1, \ldots, D} \quad (14)$$

Find the attribute of template quantization vector of the minimum distance, and the attribute is the sort of unknown flow regime.

3.4.Date Fusion Method of Flow Pattern On-line Recognittion

3.4.1 Flow Process of Data Fusion

The pressure and differential pressure signal constitute the feature space of the flow regime recognition. The application of spectral analysis, statistical analysis and nonlinear analysis are limited in the space, it can only provide several kind of supposition set of flow regime recognition in certain knowledge scope. If the pressure and differential pressure signal features realize supplementary, the supposition set may reduce and approach to the real state and obtain the uniform description and the forecast of multiphase flow regime. Therefore, the success flow regime recognition method must be integration and fusion of the multi-sensor multi-characteristics and the knowledge. This research proposed one kind of solution according to multi-sensor data fusion technology as from Kohonen (1990) and Hall (2001). For pressure signal, the homogenized Fourier power spectrum, the root-mean-square and the fractal dimension number are extracted. For differential pressure signal, the time-domain characteristic such as dimensionless average value and the variance are extracted.

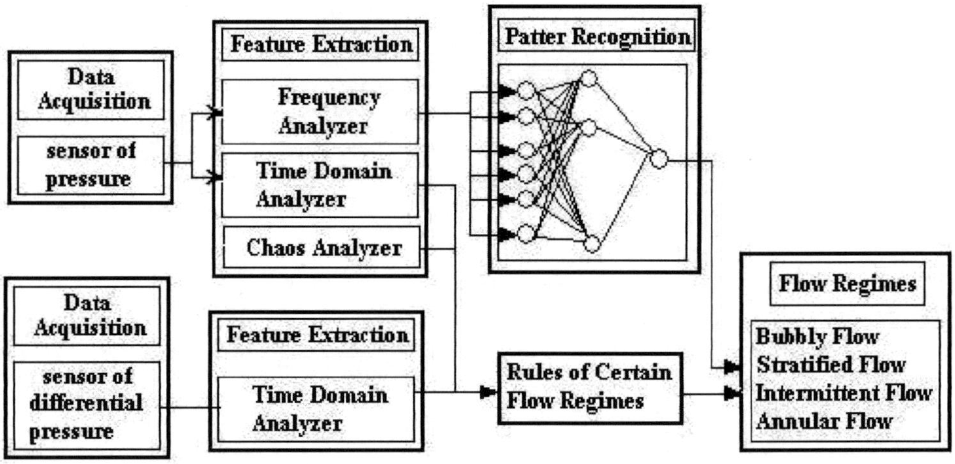

Fig. 9 Flow chat of the multi-sensot data fusion online recognition

Figure 9 shows the concrete flow chart of the multi-sensor data fusion online recognition of flow regime. First realize real-time measurement of pressure and differential pressure signal of multiphase flow, and then carry on the multi-feature extraction to the pressure and differential pressure signal, judge flow regime mode according to result of feature extraction and the recognition rules of specific flow pattern. If the flow regime feature does not satisfy the rule, take the learning vector quantization pattern classifier as the unknown flow regime classifier and complete the recognition according to standard template of the feature space.

The flow regime recognition method takes advantages of high confidence, good fault tolerance and less requirement of single sensor performance because of using the multi-sensor data fusion theory.

3.4.2 Flow Regime Online Recognition System and the Test Results and Analysis

According to the above flow regime online recognition theory, the flow regime recognition instrument is developed using the micro-processing device, together with pressure, differential pressure sensors that layout on the pipeline, as well as the concurrent voltage source, these composed the flow regime online recognition system. The layout of measurement sensor is shown in Figure 10. This system does not change flow structure in pipelines.

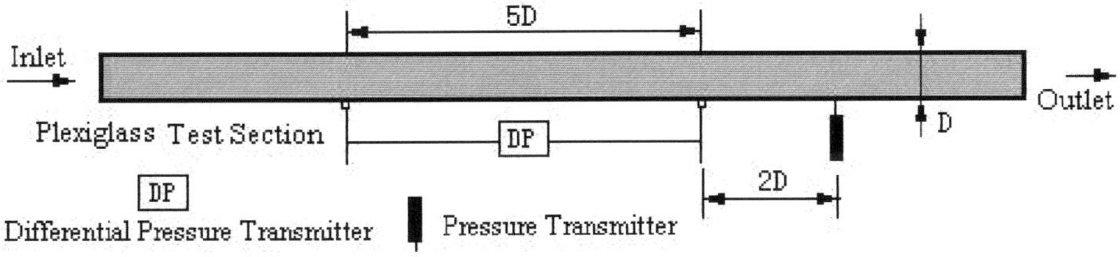

Fig. 10 Layout of measurement sensor in the flow pattern online recognition system

Carrying on the examination directly on the experiment system in laboratory, the test parameter area covered all scope of experiment flow parameter. To test the repeatability, we use three consecutive identify ways to the same condition. The test results are shown in Table1. In the wrong recognition, the annular flow was mistakenly identified as intermittent flow, the intermittent flow was mistakenly identified as annular flow, the bubbly flow and stratified flow, the stratified flow was mistakenly identified as intermittent flow and bubbly flow, the bubbly flow was mistakenly identified as intermittent flow and stratified flow. The total average flow regime correct recognition rate is 93.2%. In addition, the metrology institute measuring results show that the system has good duplication and good recognition accuracy rate, the recognition accuracy rate is as high as 96.7% except flow regime transition. With the experimental station of Daqing Oil Field Design Institute, using 80mm mixing pipeline and the crude oil-natural gas-water real fluid, the application results indicated that accuracy rate of flow regime online recognition was 91.3%.

Table. 1 Test result of oil-gas-water multiphase flow pattern recognition technology

Flow pattern	case number	Correct recognition	Correct rate
Annular flow	200	185	92.5%
Intermittent flow	400	377	96.8%
Stratified flow	50	46	90.2%
Bubbly flow	60	54	90.0%

The recognition method divided all flow regimes into four kinds, bubbly flow, stratified flow, intermittent flow and annular flow; it didn't define the flow regime transition as the new category. The test result indicated that accuracy rate of these 4 kinds of flow regime online recognition is higher than 90%. The wrongly distinguished flow regime mainly appeared at the transition region, the partial reasons are observer's subjective judgment, and the following research may consider the definition of flow regime transition. For the flow regimes under low air speed, such as bubbly flow, stratified flow, intermittent flow and so on, the pressure, differential pressure fluctuation energy is low, this cause high incorrect recognition rate, the following research might consider to layout differential-pressure sensor on the upper and lower sides of pipeline.

The flow regime recognition needs to obtain enough flow information. For the intermittent flow under low fluid velocity, its flowing periods may reach as high as above 10s. Therefore, the sampling time of the online recognition method of this research can't reduce at will.

4.AN EXPERIMENTAL AND MUMERICAL STUDY OF SLUG FLOW IN PIPE FLOW

Gas-liquid slug flow occurs frequently in the petroleum, chemical, civil and nuclear industries. It is characterized by a sequence of elongated bubbles separated by liquid slugs that may contain small bubbles. For its transient and intermittent nature, it is difficult to predict the flow characteristics correctly. Therefore, reliable flow models are crucial for the safe and cost efficient design and operation of industrial facilities.

In the offshore oil and gas field, the maximum slug length and its statistical distribution are very important for the design of separator and downstream processing facility at steady state operations. However transient conditions may be encountered in the production, such as operational upsets, start-up, shut-down, pigging and blowdown, which are key operational and safety issues related to oilfield development. So it is necessary to have an understanding the flow parameters under transient conditions. In this paper the evolution of slug length along a horizontal pipe in gas-liquid flow is studied firstly and then an experimental study of flowrate transients in slug flow is provided. Also, the special gas-liquid flow phenomena easily encountered in the life span of offshore oil fields, called severe slugging, was studied experimentally and some results are presented in this paper.

4.1 Slug Length Evolution of Gas-liquid Slug Flow inHorizontal Pipe

For steady state slug flow in horizontal pipe, Dukler & Hubbard (1975) presented the "unit cell" concept to predict average liquid fractions and pressure drops for given flow rates and pipe diameters. Several models based on this approach were developed later by other researchers to predict slug flow parameters. But all the models used constant slug length as the primary assumption and it is impossible to predict the slug length evolution along the pipe.

From visual observations, it is found that very short slugs generated at the entrance section of the pipe maybe have length of several diameters of the pipe. However the frequency of the slugs is relatively high. When the short slugs move downstream along the pipe, some elongated bubbles disappear and short slugs merge to form longer slugs. In this process frequency of slug reduces and reaches a stable one when most slugs are long enough and can flow out of the pipe. The surviving of slug is controlled by the pick-up process at the slug front, where the moving slug front scoops the liquid film under the large gas bubble, and by the liquid shedding process at the tail of the slug. The liquid shedding rate is determined by the translational velocity of the Taylor bubble of the slug unit, while the translational velocity is related to the maximum local velocity of the slug tail. At the tail of short slug, the velocity profile is not developed fully as the long one, therefore short slugs are unstable and are easily overtaken by the following Taylor bubble (Moissis & Griffith 1962). When a short slug disappeared, the liquid is scooped by the approaching slug and its length increases. Moissis & Griffith (1962) studied the rate of collapse of liquid slugs due to wake effect in upward vertical flow. Cook & Behnia (2000) studied this effect in near horizontal pipe. It is interesting that both their results can be regressed in the form of exponential expression.

From field data Brill et al. (1981) found the slug length followed a log-normal distribution and Nydal et al. (1992) also had the same conclusion from experimental data. Zheng et al. (1994) proposed a simple slug tracking model to simulate slug flow in hilly terrain pipeline. The model applied source and sink concepts to the elbows between pipe sections which have different inclinations. Barnea & Taitel (1993) also used a simple slug tracking model to calculate slug evolution along vertical pipe. In their model an assumption relation of the translational velocity between neighboring liquid slugs is used to consider the wake effect. Cook & Behnia (2000) compared the experimental slug length distribution in near horizontal pipe with this model and found the agreement was very well. Hout et al. (2001) studied slug length distributions in upward vertical flow. In the model presented by Barnea & Taitel, the film surrounding the Taylor bubble zone and the small gas bubbles in liquid slug are neglected.

In 1996, Nydal and Banerjee developed a Lagrangian slug tracking model, which is based on the integral balances of mass and momentum of the liquid slug and gas bubble in a slug unit. An important assumption is that there is no gas was entrained in the liquid slug. The program of this model is implemented with object oriented technique. The predicted results of hilly terrain slug flow from this model agreed well with measured data. Taitel and Barnea (1998, 1999) proposed a slug tracking model which based on the unit cell model presented by Dukler and Hubbard. This is a Lagrangian model and could record trajectories of every slug and Taylor bubble in the pipe. In this model, liquid film was assumed to have an averaged equilibrium height to simplify the calculation and the acceleration pressure drop, which is brought by the velocity increase of the liquid film scooped by the next liquid slug, is not considered.

In the present work, the Taitel and Barnea model is modified to considered the acceleration pressure drop and programmed in C++ with object oriented technique. The evolution of slug length along pipe of hydrodynamic slug flow was studied experimentally in a horizontal pipe and numerically by the simulator.

4.2 Slug Tracking Model

The model proposed by Taitel & Barnea(1998,1999) was adopted and modified to consider the acceleration pressure drop in this paper.

The slug tracking model was based on the mechanistic "unit cell" concept presented by Dukler and Hubbard(1975). A typical slug unit consists of a liquid slug zone of length Ls and an elongated bubble zone of length Lf. There is a liquid film under the long gas bubble. The main assumption of the "unit cell" model is that slug flow is comprised of a series of slug units in which every liquid slug has the equal length. However in the tracking model each liquid slug has its individual length and the velocities of liquid slug front and tail are calculated respectively. Therefore the lengths of all slugs in the pipeline can be determined. In the pipe some slugs might disappear when their lengths are shorter than a critical length and two neighboring elongted bubbles would be merged. To simplify the calculation a uniform equilibrium film height is used.

As shown in Fig. 11 the position of liquid slug front is Zfr and the tail is Zta. The velocities of liquid slug front and tail are Ufr and Uta respectively. These parameters as well as the fluid velocity, pressure and holdup in every liquid slug and liquid film and the gas velocity in elongated bubble are calculated. In the tracking model, the slug unit at the entrance is calculated by the steady state "unit cell" model and its slug length is specified, while in the pipe the lengths of liquid slug and liquid film are calculated by the velocities of liquid slug front and tail from the former time step. In this paper, some modifications of the Taitel and Barnea model are presented.

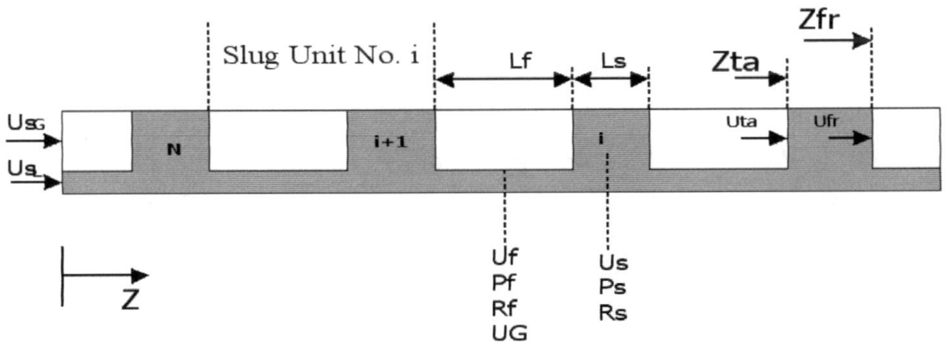

Fig. 11 Simplified diagram of slug flow structure in horizontal pipe

4.2.1 The Translational Velocity and the Wake Effect

The translational velocity of an elongated bubble following a developed liquid slug has been studied extensively. The well known relation proposed by Nicklin et al. (1962) has been applied for all pipe inclinations. In this study the translational velocity was also analyzed and it was found that the following correlation can be used:

$$V_t = 1.2V_m + 0.54\sqrt{gD} \qquad (15)$$

Due to the wake effect, the translational velocity of elongated bubble will increase as the liquid slug length in front of it decreases. Cook & Behnia (2000) experimentally studied it near horizontal pipe and obtained a correlation. The highest mixture velocity used in their experiments was 2.5m/s. In this work their correlation is applied to consider the wake effect:

$$\frac{V_{tb}}{V_{t\infty}} = 1.0 + 0.56\exp\left(-0.46\frac{L_S}{D}\right) \qquad (16)$$

where Vt∞ is the translational velocity of long bubble following a liquid slug of stable length, calculated from Eq.(? ? ?) Ls is the length of the liquid slug separated the neighboring elongated bubbles.

4.2.2 Pressure Drop Caluculation

In this work the acceleration pressure drop due to slug front scooping film liquid ahead of it is considered. The pressure drop is calculated by force balance. Bendiksen (1996) described that the mixing zone at the slug front is short when the mixture velocity is low, but when the mixture velocity is high the mixing zone is long and maybe has half the slug length. To simplify the problem, we assume that the acceleration pressure drop distributes over half of the liquid slug. By the force balance between the pressures of the point i and i+1, it gives:

$$\begin{aligned} P_{S\,i+1} &= P_{S\,i} + \tfrac{1}{4} f_{S\,i}\rho_{S\,i}L_{S\,i}\frac{S}{A}\left|U_{S\,i}\right|U_{S\,i} + \tfrac{1}{2}\rho_{S\,i}g\sin\varphi L_{S\,i} \\ &+ \tfrac{1}{4} f_{S\,i+1}\rho_{S\,i+1}L_{S\,i+1}\frac{S}{A}\left|U_{S\,i+1}\right|U_{S\,i+1} + \tfrac{1}{2}\rho_{S\,i+1}g\sin\varphi L_{S\,i+1} \\ &+ \left[\frac{\tau_f S_f}{A} + \frac{\tau_G S_G}{A} + \rho_f g\sin\varphi\right]_{f\,i} L_{f\,i} + \Delta P_{acc\ i+1} \end{aligned} \qquad (17)$$

where the acceleration pressure drop is (Sharma, Y. 1985) given by:

$$\Delta P_{acc} = \rho_L R_S\left(U_S - U_f\right)\left(V_t - U_S\right) \qquad (18)$$

At each time step, the variables $P_{S\,i+1}$, $P_{S\,i}$, $U_{S\,i}$, $U_{S\,i+1}$ are unknown and the other variables are taken from the previous step. The equation used to linearize last two equations is:

$$\left|U_{S\,i}^{new}\right|U_{S\,i}^{new} = 2\left|U_{S\,i}\right|U_{S\,i}^{new} - \left|U_{S\,i}\right|U_{S\,i} \qquad (19)$$

Therefor Eq. (18) can be written:

$$\Delta P_{acc\,i+1} = \rho_{L\,i+1} R_{S\,i+1} \left(U_{S\,i+1} - U_{f\,i}\right)\left(Uta_{i+1} - U_{S\,i+1}\right)$$
$$= \rho_{L\,i+1} R_{S\,i+1} \begin{bmatrix} \left(U_{f\,i} + Uta_{i+1}\right) U_{S\,i+1} - U_{f\,i} Uta_{i+1} \\ -2U_{S\,i+1} U_{S\,i+1}^{new} + U_{S\,i+1}^{2} \end{bmatrix} \quad (20)$$

Thus the discretization form of Eq. (17) is:

$$P_{S\,i}^{new} + \left[\tfrac{1}{2} f_{S\,i} \rho_{S\,i} L_{S\,i} \frac{S}{A} \left|U_{S\,i}\right|\right] U_{S\,i}^{new} + \begin{bmatrix} \tfrac{1}{2} f_{S\,i+1} \rho_{S\,i+1} L_{S\,i+1} \frac{S}{A} \left|U_{S\,i+1}\right| \\ -2\rho_{L\,i+1} R_{S\,i+1} U_{S\,i+1} \end{bmatrix} U_{S\,i+1}^{new}$$
$$- P_{S\,i+1}^{new} = -\tfrac{1}{2} \rho_{S\,i} g \sin\varphi L_{S\,i} - \tfrac{1}{2} \rho_{S\,i+1} g \sin\varphi L_{S\,i+1}$$
$$+ \tfrac{1}{4} f_{S\,i} \rho_{S\,i} L_{S\,i} \frac{S}{A} \left|U_{S\,i}\right| U_{S\,i} + \tfrac{1}{4} f_{S\,i+1} \rho_{S\,i+1} L_{S\,i+1} \frac{S}{A} \left|U_{S\,i+1}\right| U_{S\,i+1}$$
$$- \left[\frac{\tau_f S_f}{A} + \frac{\tau_G S_G}{A} + \rho_f g \sin\varphi\right]_{f\,i} L_{f\,i}$$
$$- \rho_{L\,i+1} R_{S\,i+1} \left[\left(U_{f\,i} + Uta_{i+1}\right) U_{S\,i+1} - U_{f\,i} Uta_{i+1} + U_{S\,i+1}^{2}\right] \quad (21)$$

where ρ_f is defined by:

$$\rho_f = R_f \rho_L + \left(1 - R_f\right)\rho_G \quad (22)$$

The slug tracking model was abstractly analyzed with object oriented technique and programmed in C++ language. All the liquid slug and elongated bubble in the pipe are managed by a double linked list.

4.2.3 Results and Discussion

Liquid slug lengths were measured by the differential pressure method from the experiments covering a wide range of mixture velocities. The pressure signals to determine the slug lengths were measured from P11, P12 and P13. Fig. 10 shows the maximum, minimum and mean slug length versus to the local superficial mixture velocity. From the experimental data the liquid slug lengths were collected and the statistical result was obtained:

The mean slug length ranges from 8.3~19D; the averaged value is 13D.

The maximum slug length ranges from 22~36D; the averaged value is 31D.

The minimum slug length ranges from 2~11D; the averaged value is 4.3D.

The error bars show in Fig. 12 represents the standard deviation of the mean slug length. From Fig. 12 it is clearly shown that the slug length is not dependent on the mixture velocity, which agrees well with the work of many former researchers.

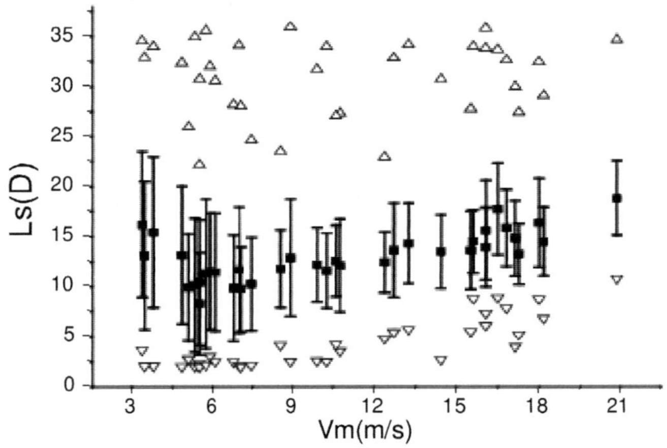

Fig. 12 Measured maximum(Δ), minimum(∇) and mean(■) slug length. The error bars represent the standard deviation of the mean slug length

In order to get accurate statistical distribution of liquid slug length, some experiments run for a very long time and several hundreds of liquid slugs were recorded. Fig. 13 presents the observed liquid slug length distributions which were measured at 12.83m, 16.83m and 18.83m from the pipe entrance respectively. The air and water superficial velocities are 7.3m/s and 1.52m/s respectively. The liquid slug length distributions calculated by the simulator are also shown in Fig. 14 and represented by shadow filled columns. The model used a normal distribution of liquid slug length between 2D and 3D entering the pipe. Comparison with the measurements in Fig. 13, it is clear that the predicted slug length distributions are very good and both Fig. 13 and Fig. 14 show a right-skewed distribution for slug length. The maximum and the averaged slug length and its standard deviation are given in both figures for each measured results. The numbers of slugs involved in the calculation process are also shown in Fig. 14.

As shown in Fig. 13, both the averaged and the maximum slug length increase along the pipe. In Fig. 14, the number of simulated liquid slugs reduced 53 from 12.83m to 16.83m after the pipe entrance, i.e. averaged 13.3 per meter, while from 16.83m to 18.83m the slugs reduced 20, i.e. averaged 10 per meter. It was distinctly shown that short slugs disappeared during the liquid slugs flowing along the pipe and the slug lengths tended to stabilized at a longer length. When some short slugs disappeared, the other slugs would scoop the liquid shed by the decaying slug and their lengths increase, therefore both the mean and the maximum slug length increased along the pipe. The calculation agrees well with the measurements. The agreement between the predicted maximum slug length and the measured is also quite good.

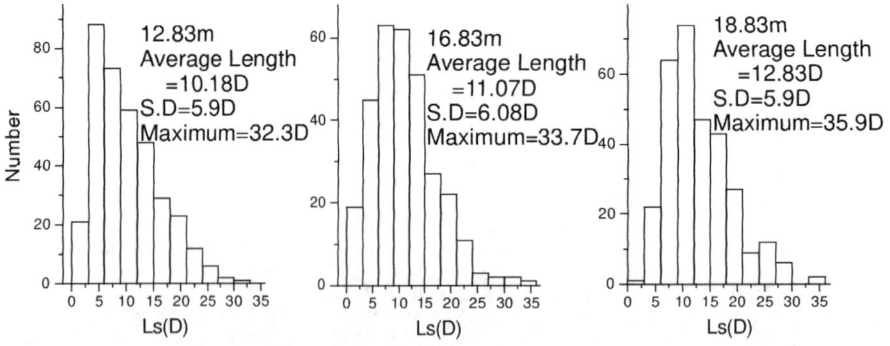

Fig. 13 Measured slug length distributions at 3 positions along the pipe (V_{SL}=1.52m/s, V_{Sg}=7.3m/s)

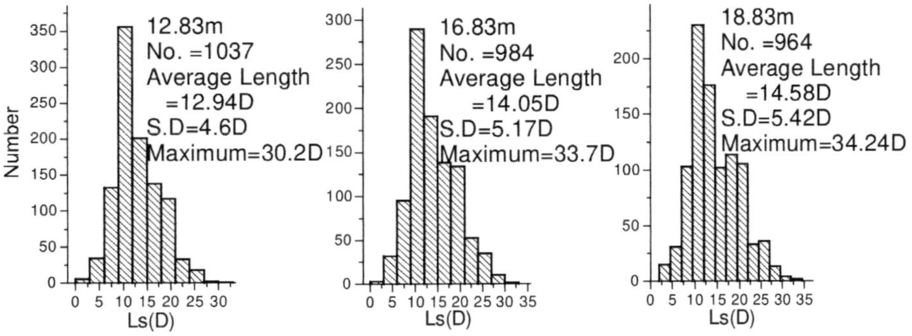

Fig. 14 Predicted slug length distributions at 3 positions along the pipe (V_{SL}=1.52m/s, V_{Sg}=7.3m/s)

But the predicted averaged slug length is higher than the measured length and the mean difference is 2.5D. This difference may be due to the separation effect of the U-type elbow (Fig. 2. When a slug is passing in the elbow pipe, in addition to the gravitational force there is centrifugal force put on the liquid slug. For the stronger volume force on the slug in the elbow liquid, liquid in the slug will be distributed to the film zone thereby the slug length will decrease. For short slugs this effect would be more effective because both the front and tail of liquid slug locate in the elbow pipe. Whereas for long slugs, such as the maximum slug, their lengths are longer than the elbow pipe length (about 25D) and it means that at least one of the ends of slug is not in the elbow pipe section and this will prevent these liquid slugs in the elbow section from decreasing in length. Hence the separation effect of the elbow on the maximum slug is very little. On the other hand, the start of slug flow from stratified flow need a small distance from the entrance, but it is not considered in the simulator and this also might increase the predicted mean slug lengths. From 12.83m to 16.83m after the entrance, the increase of the measured mean slug length is 2.65D, while the predicted is 1.64D, so it is clear that the elbow blocks development of the slug flow.

Fig. 15 and 12 are the measured and the predicted slug length distribution when the air and water velocities are 8.93m/s and 0.98m/s respectively. The predicted mean slug length is 2.57D longer than the measured, which is similarly with the former case. The difference of the maximum slug lengths between the observed and the predicted is only 1.34D. Fig. 15 and 12 also present the shape fitted to the measured distributions of the liquid slug lengths. It was found that both measured and predicted distributions follow a log-normal distribution. The both fit equations are given:

Measured liquid slug lengths:

$$\bar{y}\left({L_S}/{D}\right) = 0.00577 + 0.255\exp\left(-\frac{\ln^2\left({L_S}/{(11.472D)}\right)}{2\times 0.238^2}\right) \tag{23}$$

Predicted liquid slug lengths:

$$\bar{y}\left({L_S}/{D}\right) = 0.0154 + 0.28498\exp\left(-\frac{\ln^2\left({L_S}/{(12.945D)}\right)}{2\times 0.15334^2}\right) \tag{24}$$

To understand the disappearing and merging process of the liquid slugs and the elongated bubbles close to the entrance, the moving trajectories of 8 slugs and 8 bubbles were recorded in about 0.35s

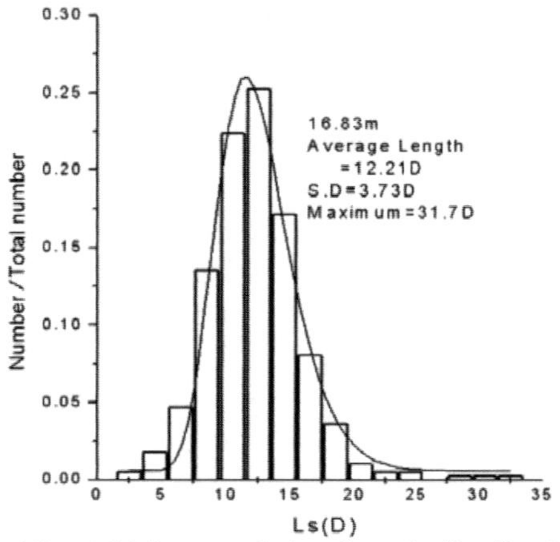

Fig. 15 Measured slug length distribution, (—)log normal fit (V_{SL}=0.98m/s, V_{Sg}=8.93m/s,)

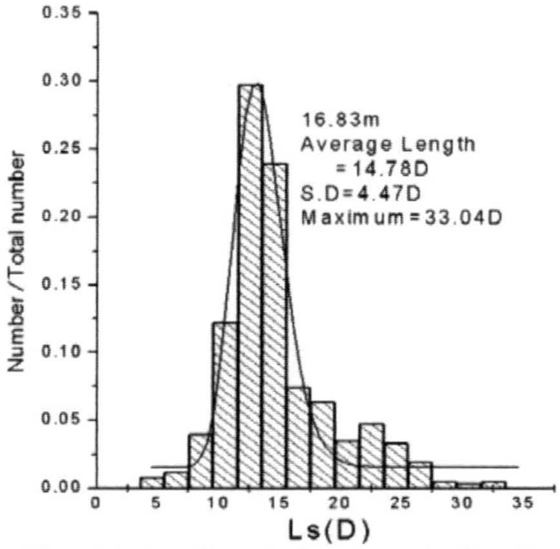

Fig. 16 Predicted slug length distribution, (—)log normal fit (V_{SL}=0.98m/s, V_{Sg}=8.93m/s)

from one calculated case as shown in Fig. 12. Fig. 17 shows the trajectories of all slug s and elongated bubbles vs. time in 0.35s. There were 8 slugs entering the pipe, but only the third and the eighth liquid slug flowed out of the pipe successfully. Four liquid slugs preceding the eighth liquid slug disappeared and the shed liquid is collected by the eighth liquid slug in the process, thus its length increased and the stable length was reached after the velocity of front was equal to the tail velocity. At the same time, five bubbles merged and formed a long one before the eighth liquid slug. The third liquid slug collected the liquid from the first and the second liquid slug, therefore its length was shorter than the eighth liquid slug.

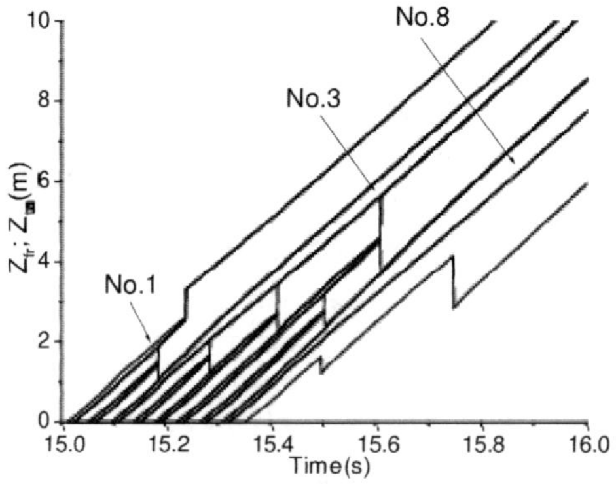

Fig. 17 Predicted slugs and elongated bubbles trajectories(V_{SL}=1.52m/s, V_{Sg}=7.3m/s)

4.3 An Experimental Study of the Flowrate Transients in Slug Flow

Slug flow is of crucial significance as a great number of multiphase transport lines are operated in this flow regime. The flowrate transient with slug flow is widely met in oil and gas pipelines. For steady state operations, parameters such as holdup and pressure gradient are important, whereas under transient conditions, an understanding of additional parameters such as maximum slug length and peak pressure is also required.

Transient conditions include operational upsets, start-up, shut-down, pigging and blowdown, which are key operational and safety issues related to oilfield development. Flowrate transient phenomena mean that the mass flux of two-phase at inlet or outlet changes with time. They may be classified into slow transient and fast transient condition depending on whether the state of fluids is in steady state. Under slow transient, the state of fluid at every location of the pipeline is in steady or quasi-steady state so that the fluid flow can be modeled by a steady state or quasi-steady state approach. Under fast transient, the fluid flow and its state parameters in every location of pipeline can not be modeled by a steady state or quasi-steady state approach and have to be dealt with by a special method. In a pipeline for oil and gas transportation, the flowrate change caused by the fluctuations of pressure beneath the well and the liquid height of downstream separator or storage tank is called slow transient, whereas the flowrate change caused by man-controlled fast operations and other accidents is called fast transient. Usually, the flowrate transient is accommodated with the changes of liquid and gas fluxes simultaneously, but, for convenience of research, the experiment was designed and operated in such a way that the change of mass flux was in one phase first, then in two phases simultaneously. We examined the effect of a step change in one of fluid flowrates of two phases. Five types of transients, i.e., "Up-gas" transient, "Down-gas" transient, "Up-liquid" transient, "Down-liquid" transient and simultaneously change of gas and liquid mass flux, were simulated in our experiments.

The effects of flowrate transients in horizontal pipes have been studied for a number of years. Sakaguchi et al (1973) looked, in particular, at the formation of slugs in flowrate transients between two steady states of stratified flow. Taitel et al (1978) conducted a more detailed study of air and water flowrate transients in a pipe with 38.1 mm internal diameter. Caussade et al (1987) conducted an air-water two-phase flow experiments in an 89.9 m long pipeline of 53 mm internal diameter, in which the void fractions and pressures were recorded at five locations along the test section. Their experimental results showed the propagation of both pressure and holdup waves along the pipe. For the gas flow transients, a pressure overshoot beyond the value of the new steady state was observed at the entrance to the test section, but not apparent at the exit. Minami (1991) conducted a flowrate transients experiment in slug flow regime using air and kerosene in a 3-inch diameter pipeline of 420 m length. He pointed out that an around 30 kPa pressure overshoot was identified in the entrance of test section for a gas flowrate increase transient and associated with an increased liquid discharge rate corresponding to a period of intense slugging, and about 500 s was needed to establish a new steady state. In the case of gas flowrate decreasing transient, the pressure undershoot below the new steady state was around 25 kPa and the liquid build-up in the separator appeared to be consistent with the existence of a period of stratified flow.

More recently, King et al (1996) presented preliminary findings for flowrate transients within the slug flow regime in WASP(Water, Air, Sand, Petroleum) facility of Imperial College. They concluded that gas transients exhibit pressure and holdup effects and that decreasing gas transients caused the temporary formation of a stratified flow. Liquid transients caused changes in slug frequency with smooth pressure changes between the steady states. King et al (1998) considered the large-scale or ensemble-averaged response of a slug flow to inlet flowrate changes in the same facility and observed the response of individual slugs to flowrate changes using a series of conductivity probes along the test section. This allowed a detailed study of the growth and collapse of individual slugs under transient conditions to be undertaken.

All of above-mentioned researchers observed the change of characteristic parameters in the flowrate transients, but did not give out any quantitative relationship between the changes of characteristic parameters and flowrate.

In this work, we investigate the large-scale or ensemble-averaged response of a slug flow and the response of individual slug to inlet flowrate changes using a series of pressure transducers and differential pressure transducers along the test section. The quantitative relationships between the changes of characteristic parameters and flowrates are obtained from the experimental results.

4.3.1 Test Loop and Ewperimental Method

A schematic diagram of the facility located at the University of Petroleum(East China) is shown in Fig.18. The rig can be operated at pressure up to 1200 kPa using air-oil-water three-phase flow as working fluid, in which air was powered by a frequency modulation compressor, water was supplied by a centrifugal pump from water tank, and the oil was driven by a screw pump. The compressor was controlled by the mode of frequency modulation according to the fluctuation of gas pressure in buffer tank to maintain the pressure constant. The regulating valves were used to maintain the flow velocity constant. A vessel slug catcher and a multiple-pipe slug catcher were installed in the end of test section and all were controlled by regulating valves and computer. The present experiment was limited to air-water flow under 100 kPa to 200 kPa at the exit of system. The test section is a stainless steel pipe with 378 m length and 80mm inner diameter.

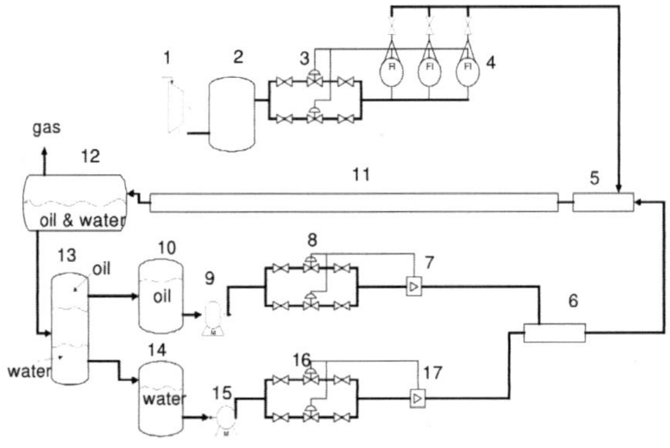

Fig. 18 A schematic of the experimental system
1—frequency modulation compressor; 2—gas buffer tank; 3—gas regulator; 4—metal rotameter; 5—gas & liquid mixer; 6—oil-water mixer; 7—ROSEMOUNT mass flowmeter; 8—oil regulator; 9—screw pump; 10—oil tank; 11—experimental section; 12—gas-liquid separator; 13—oil-water separator; 14—water tank; 15—centrifugal pump; 16—water regulator; 17—ROSEMOUNT mass flowmeter

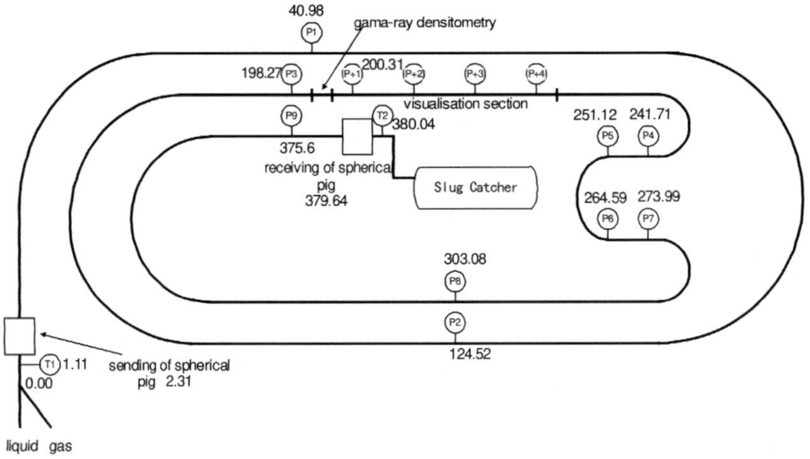

Fig. 19 Location of measuring instruments

Two ROSEMOUNT mass flowmeters (model CFM100 M328NU and CFM100 M328NU, accuracy±0.1 %) and three Metal Rotameters (model H54, accuracy class 1) were used for the flowrates measurement of every fluid phase before entering the mixers. Then the fluids were respectively introduced into mixers to form two or three phase flow. A serial of KELLER PA23e pressure transducers and a gama-ray densitometer were installed in the test section for identification of flow pattern and measurement of pressure and liquid holdup as shown in Fig.19. Ten KELLER PA25e pressure transducers $P_0 \sim P_9$ were arranged respectively at nine locations along the test section, P_0 was located at the inlet of the gas flowmeter to measure the gas pressure and four KELLER PA25e pressure transducers $P_{+1} \sim P_{+4}$ were used for the measurement of slug velocity and slug length. The changes of pressure associated with flowrate transient were measured using above mentioned pressure transducers of KELLER PA25e model. Their frequency response is 5000 Hz, which is high enough to pick up the characteristics of pressure change. The signal from the flowmeters and pressure transducers were sampled by a model PCI-6071 data acquisition system from NATIONAL INSTRUMENTS. The sampling frequency of each channel is 2500Hz.

As described by King et al. (1998), the upstream compressible volume has an effect on the flowrate transients. To minimize this effect the gas and liquid flows were throttled through the regulator valve before flowing through the part for flowrate measurement of test loop.

4.3.2 Gas Flowrate Transients

We study the effect of a step increase or decrease in the flowrate of gas (called "up-gas" transients or "down-gas" transients) on the flow regime and characteristic parameters of slug flow first. Although the flow regime transition can occur during the transient process the new steady state is also in the slug flow regime after a flowrate increases or decreases. During the gas flowrate transients, the pressure of gas buffer tank in the loop is maintained at 800 kPa and the liquid regulator is at rest. According to change of gas flowrate, the gas regulator was adjusted and controlled by a computer program to make the flow reach a new steady state.

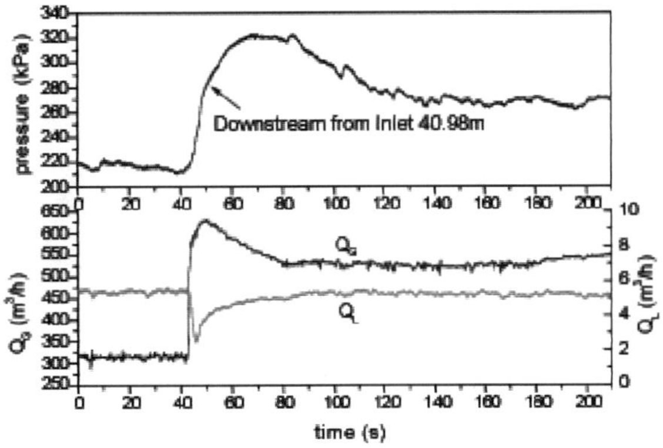

Fig. 20 Flowrate and pressure in "Up-gas" transient

4.3.2.1 Analysis of Pressure Fluctuation

(1) "UP-GAS" transients

Fig.20 shows a typical result of pressure profile with time at the location of 40.98 m away from the entrance in "up-gas" transient. A peak of gas flowrate was found between the initial steady state flowrate 315.3 $Nm^3 \cdot h^{-1}$ and final steady state flowrate 523.1 N $m^3 \cdot h^{-1}$. The peak of gas flowrate was 632.2 $Nm^3 \cdot h^{-1}$. At the moment of 40.6 s, the gas flow rate increased rapidly to reach the peak, then decreased slowly and continued for 36.8 s to reach the final steady state. As the gas flowrate increased, the liquid flowrate decreased from 5.35 $m^3 \cdot h^{-1}$ to 2.46 $m^3 \cdot h^{-1}$ rapidly, though the liquid regulator did not take any action. This is not desirable but inherent characteristics in our system.

At the beginning of the "up-gas" flowrate transient, the pressure in pipeline increased rapidly almost at the same time. As soon as the peak value reached, the pressure decreased slowly from the peak value and continued for 136.3 s before reaching the new steady state. This phenomenon is called pressure overshoot. Under the conditions of 315.3 m^3h^{-1} for gas flowrate and 5.35 m^3h^{-1} for liquid flowrate, the value of pressure overshoot is 56 kPa at the location 40.98 m away from entrance.

Journal of Physics: Conference Series **147** (2009) 012047 doi:10.1088/1742-6596/147/1/012047

The reason for this feature is that there exist different response speeds of the pressure build-up and the acceleration of liquid slugs. According to the idea of Nicklin et al. (1992), the slug velocity is always represented as a sum of two terms as follows:

$$V = C_0 V_m + V_\infty \quad (25)$$

where V_∞ is the drift velocity and V_m is the homogeneous mixture velocity. Therefore, the slug velocity is directly proportional to the homogeneous mixture velocity which equals to the sum of liquid superficial velocity V_{SL} and gas superficial velocity V_{SG}. The increase of gas flowrate results in an increase of slug velocity. However, inertia of liquid slugs has a baffle to its acceleration. This must induce a larger acceleration pressure drop and cause a pressure overshoot above the new steady state. This additional pressure drop is directly proportional to the change of the homogeneous mixture velocity.

The value of pressure overshoot is a function of the gas flowrate change, gas slug volume and others. The more the gas flowrate change is and the less the downstream gas slug volume is, the more the value of pressure overshoot is.

Table. 2 The pressure overshoot at different location from entrance

Diatance, m	Pressure overshoot, kPa
-22	53
40.98	56
124.5	67.8
198.3	76.3
241.7	83.7
251.1	83.1
264.6	85.1
274	86.9
303.1	80.4
375.6	66.4

Table 2 presents the relationship of the value of pressure overshoot and the distance from the entrance. The value of the pressure overshoot increases with the distance but decreases at the location close to the exit. The reason is that with the decrease of pressure along the pipeline, the gas volume in slug and the slug velocity increase, so that the pressure drop for mixture acceleration increases to overcome the inertia of liquid slugs. Because there may be no liquid slug or only one liquid slug at the location near the exit of pipeline, the value of pressure overshoot is lower than that at upstream locations. At the upstream location of the pipeline exit, the interaction between two nearby liquid slugs is not negligible.

Journal of Physics: Conference Series **147** (2009) 012047 doi:10.1088/1742-6596/147/1/012047

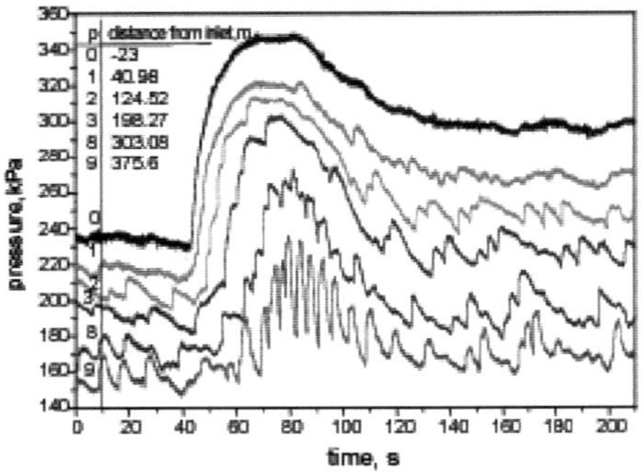

Fig. 21 Pressure profile of different locations

The pressure in test section is a function of the number of slugs, length of slug, slug velocity and the distance between measuring location and the exit of pipeline. The increase of gas flowrate results in a pressure increase . However there exists a time delay for the pressure increase between two different locations due to the damp of gas slug. Fig.21 shows the pressure profiles at different locations in pipeline. The "up-gas" flowrate transient is often associated with the formation of a larger slug. This is consistent with that the increased gas flowrate promotes the formation of a slug which sweeps down the pipeline with a higher gas driving velocity behind the slug, which forcing it over a relatively high holdup in front of the slug. A new steady state slug is being effectively pushed over a higher film holdup than the old steady state one. Under these circumstances, the possibilities for extremely rapid slug growth are increased.

Table 3 presents the spread speed of pressure wave obtained using an approximate method. The spread speed of pressure wave decreases along the pipeline.

Table. 3 Spread speed of pressure wave at different locations

Distance from entrance, m	Pressure wave speed, $m.s^{-1}$
82.74	16.22
161.4	10.87
220	9.528
257.9	9.32
288.6	9.25
339.4	7.85

(2)"DOWN-GAS" transients

Figure 22 shows a typical pressure profile at the location 40.98 m away from the entrance under "down-gas" transients. The phenomena occurred during "down-gas" transient are different from those during "Up-gas" transient. With decreasing gas flowrate, the pressure at different location of the pipe reduced almost simultaneously and reached a valley below the value of final steady state. This phenomenon is called pressure undershoot. The main reason is that as the gas flowrate decreases, the velocity of liquid slug reduces to one below that of downstream slugs, then the volume of gas slug between two nearby liquid slugs increases, gas expands and the pressure reduces to one less than that in the new steady state.

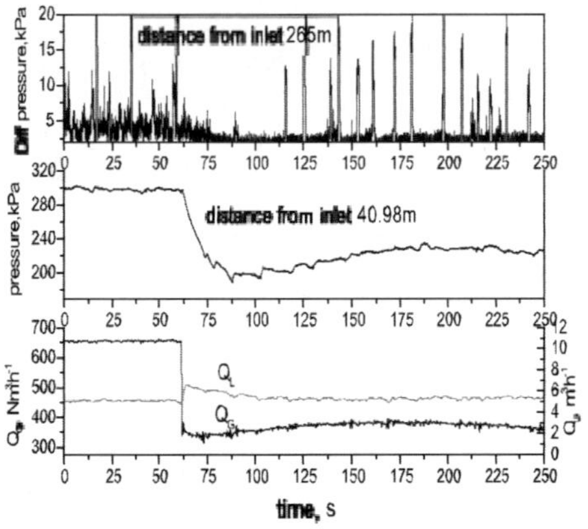

Fig. 22 Flowrate and pressure in "Down-gas" transient

The value of pressure undershoot is sensitive to the flow condition and magnitude of gas flowrate change. Except that, the value of pressure undershoot is related to the locations of pipeline Figure 23 and Table 4 show the change of pressure undershoot with the distance away from the entrance. As the distance away from pipeline entrance increased, the values of pressure undershoot increased at first, then decreased linearly.

During the "down-gas" transient, the slug velocity decreased rapidly. Since these slugs were already existed in the test section when the transient initiated, the liquid layer in front of these slugs retained the low thickness feature corresponding to the previous higher gas velocity. These slugs picked up less liquid from the thin layer than when they shed from their tails, so that they collapsed rapidly and led to a short period of stratification. As shown in Figure 23 and Table 4, the period of stratified flow is a function of the distance away from the entrance. The further the distance is, the longer the period of stratified flow is.

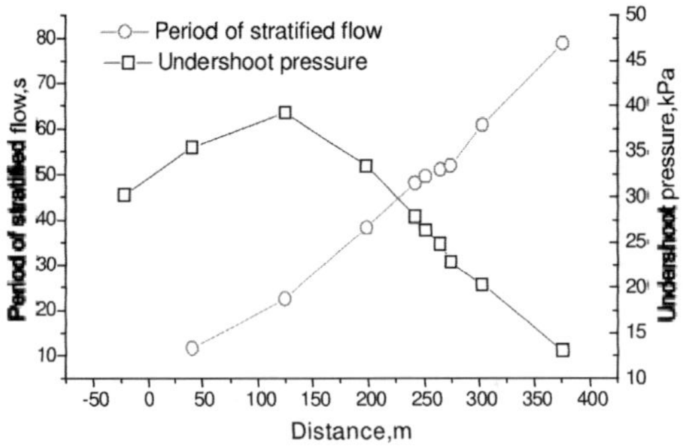

Fig. 23 The value of pressure undershoot and theperiod of stratified flow versus distance

Table. 4 The value of pressure undershoot and the periodof stratified flow at different locations of pipeline

Distance from entrance,m	Pressure undershoot, kPa	Stratified Period,s
-22	30.2	11.71
40.98	35.5	22.54
124.5	39.28	38.25
198.3	33.41	48.126
241.7	27.86	49.53
251.1	26.33	51.05
264.6	24.81	51.87
274	22.86	60.79
303.1	20.29	78.84
375.6	13.08	11.71

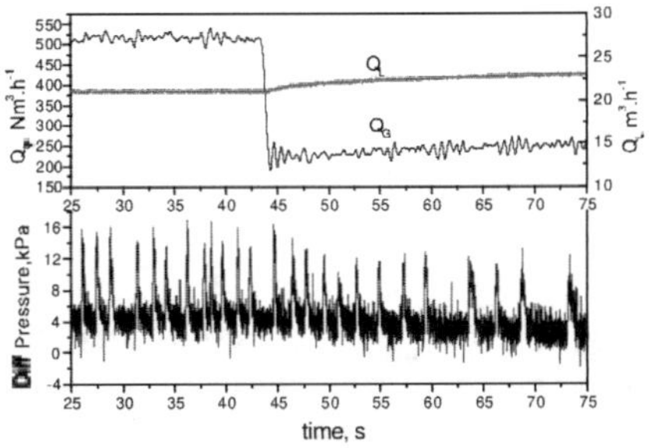

(a) Q_L=20.6m^3·h^{-1}, Q_G=484~181N m^3·h^{-1}

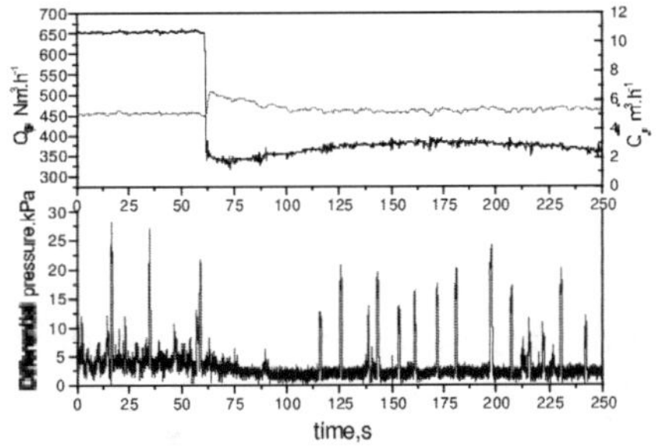

(b) Q_L=4.98 m^3·h^{-1}, Q_G=639.98~341.86N m^3·h^{-1}

Fig. 24 Differential pressure fluctuation at same point

Except the distance away from the entrance, the liquid superficial velocity has an important influence on the period of stratified flow. The greater the liquid superficial velocity is, the shorter the period of stratified flow appears to be. This is an extension of the natural slug frequency cycle; a higher superficial liquid velocity corresponds to a higher slug frequency and thus the liquid layer recovers more quickly to the new equilibrium value corresponding to the inlet holdup (1977). Fig.24 shows the differential pressure profiles at different liquid superficial velocities. In Fig.22a, the period of stratified flow is not obvious.

(3) Empirical correlation of pressure overshoot

In order to describe quantitatively the pressure overshoot, a new regression model is built up by analyzing the reason for producing the pressure overshoot.

The reason causing pressure overshoot is that the response speeds of pressure at different locations along the pipeline are different and there exist the acceleration effects of liquid slugs. The value of pressure overshoot is associated with the pressure drop for accelerating two-phase mixture. The equation for the acceleration pressure drop of two-phase flow may be a reference to establish a regression model for pressure overshoot prediction.

With the assumption that the liquid slug is free of gas bubble, the slug velocity and mass flux at initial steady state are U_{S1} and ρU_{S1} respectively, these at new steady state are U_{S2} and ρU_{S2} respectively. From momentum conservation, the acceleration pressure drop of liquid slug can be written as follows (1992),

$$\Delta p_a = \rho U_{S2} U_{S2} - \rho U_{S1} U_{S1} = \rho \left(U_{S2}^2 - U_{S1}^2 \right) \quad (26)$$

From Nicklin et al. (1962), the slug velocity is proportional to the homogeneous mixture velocity. Assume that the drift velocity V_∞ equals zero, considering the effects of gas compressibility, the lengths of liquid and gas slug and keeping meet the condition that the pressure overshoot must be zero if the change of gas flowrate is zero, the regression model can be gotten as follows,

$$\Delta p_{os} = a\rho \left(U_{m2}^2 - U_{m1}^2 \right) \quad (27)$$

Where α is the regression coefficient.

Since the homogeneous mixture velocity U_m equals the sum of liquid superficial velocity V_{SL} and gas superficial velocity $V_{SG,}$ and that the liquid superficial velocity is constant during the gas flowrate transient, Eq.(27) can be rewritten as follows:

$$\Delta p_{os} = a\rho \left[\left(V_{SG2} + V_{SL} \right)^2 - \left(V_{SG1} + V_{SL} \right)^2 \right] \quad (28)$$

24 groups effective data were obtained in present experiment. Based on these data, a correlation for the pressure overshoot was finally obtained as follows,

$$\Delta p_{os} = a\rho \left[\left(V_{SG2} + V_{SL} \right)^2 - \left(V_{SG1} + V_{SL} \right)^2 \right] \quad (28)$$

By T-test(test for averaged value analysis) and F-test(Fisher distribution function analysis test for root mean square deviation), Eq.(29) and the coefficient of equation is remarkable at remarkable level 0.0001.

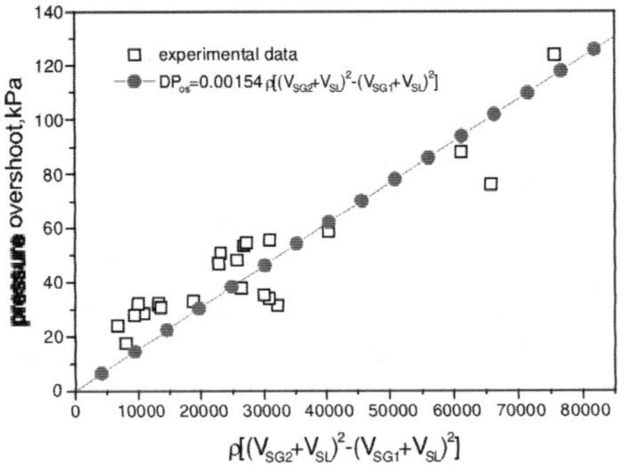

Fig. 25 The comparison between regression equation and experimental data

Fig.25 shows a comparison of the regression equation to the experimental data. From Eq.(29) and figure 25, it can be concluded that the value of pressure overshoot is directly proportional to a square of the difference of the homogeneous mixture velocities between in initial and final steady states. When the change of gas flowrate equals zero, the pressure overshoot is equal to zero

4.3.3 Liquid Flowrate Transients

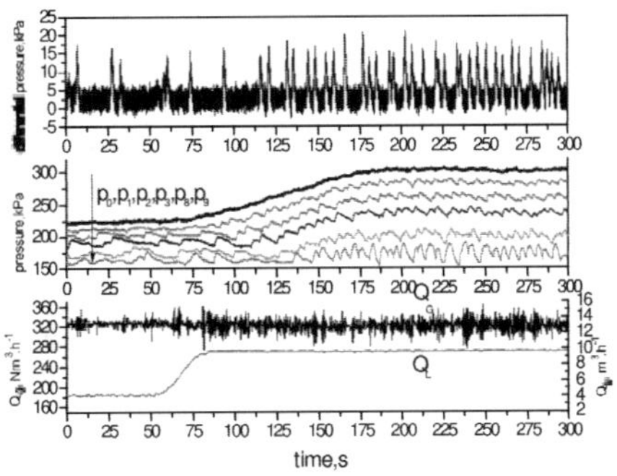

Fig. 26 "Up-liquid" transient

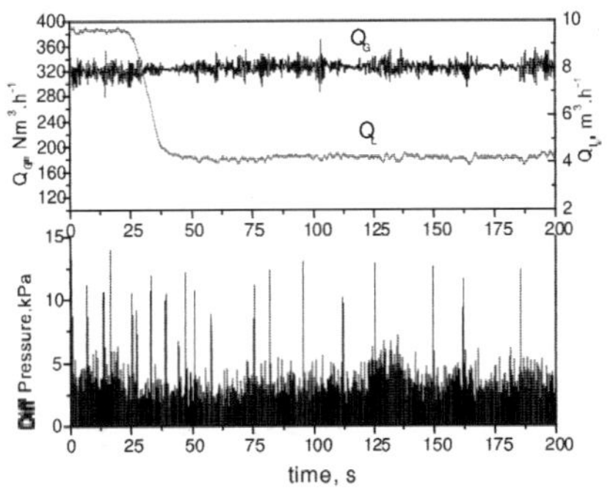

Fig. 27 "Down-liquid" transient

Fig. 26 shows the pressure and differential pressure profiles at the location of 245 m away from the entrance under "Up-liquid" transient. When the gas flowrate was kept constant, and liquid flowrate changed from a steady state to a new steady state. The flow regime of mixture is still slug flow during this transient.

Fig. 27 shows the differential pressure profile at the location of 245 m away from the entrance under "Down-liquid" transient.

During the liquid transients, the pressure changed smoothly to the new steady state value. The far the location of measuring point is from the entrance, the smaller the pressure change is. There was no pressure overshoot and no pressure undershoot in test section and no flow pattern transition occurred. Only the slug frequency appeared rapid change in this transient, because the slug frequency at a steady state depends strongly on the superficial liquid velocity. The new steady state was established quickly with two or three slugs to bridge the period. This suggests that the liquid level near the inlet responded quickly to the change of liquid flowrate, hence the slug formation at new steady state was established in a similar mechanism (1977).

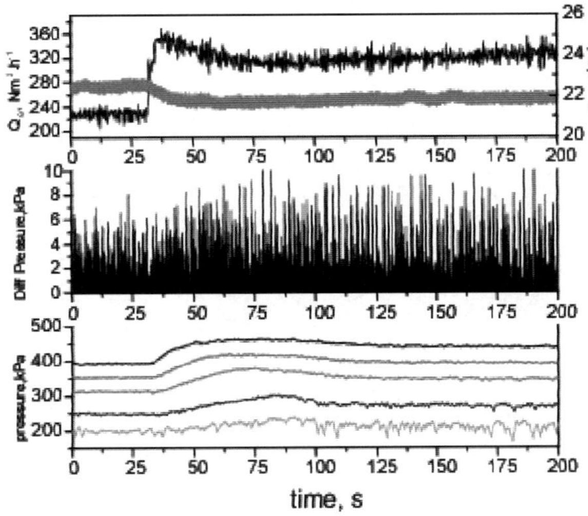

Fig. 28 Pressure and differential pressure changes of simultaneous change of gas and liquid flowrate

4.3.4 Simultaneous Change of Gas and Liquid Flowrate

In our experimental system, the change of gas flowrate had an effect on the liquid flowrate. The raising of the gas flowrate brought on a falling of liquid flowrate consequentially. In order to complete more tests in wider range, the gas regulator and liquid regulator were used to adjust the gas and liquid flowrate.

There are four cases of simultaneous transients, i.e., gas and liquid flowrates increase or decrease simultaneously, gas flowrate increases and liquid flowrate decreases, or otherwise. In our experiment, the third case is considered in which gas flowrate increases and liquid flowrate decreases. Fig. 28 shows the pressure and differential pressure profiles. There were a pressure overshoot and a decrease of slug frequency in this case. The feature of pressure overshoots and slug frequency change consisted with that of single-phase flowrate transient.

4.4 An Experimental Study of Severe Slugginig in Offshore Pipeline/riser System

In the offshore oil and gas industry, liquid and gas are frequently transported in pipeline simultaneously. When the subsea pipeline is connected to the platform or onshore facility through a riser, there may be a special flow phenomenon, called severe slugging, occurring in the pipeline and riser configuration. Compared with normal hydrodynamic slug flow, the intermittent characteristics of severe slugging are stronger and regularly, such as the length of the liquid slug may be several times of the riser length and the pressure changes periodically and its fluctuation magnitude may be equal to the hydrostatic head of the riser full of liquid. Also, the liquid slug would be impelled with a higher velocity leaving the riser than the liquid velocity at the pipe inlet when the compressed gas in the pipeline expanded and flowed out of the riser exit. This phenomenon makes it very difficult to design the production facilities economically. Farghaly et al. (1987) reported that the liquid carry over of some separators at average production rate was only half of its nominated capacity. Also the operation of the downstream processing equipments would be very difficult when the severe slugging was occurring.

Some researchers have studied the severe slugging phenomenon in real field or laboratory. Yocum (1973) was the first to discuss the severe slugging encountered in several oil fields although he did not differentiate it from the conventional slug flow. Farghaly et al. (1987) also reported that the severe slugging was occurred and abnormal long liquid slugs were generated in offshore pipeline-riser systems. The impact of the severe slugging on the field operation was discussed comprehensively and a hydrodynamic model was presented. Whereas in most laboratory studies, small scale test rigs were used to study the severe slugging, such as reported by Schmidt et al. (1979,1985), Pots et al. (1987), Taitel et al. (1990), Tengesdal (2002). In their test rigs, a buffer vessel before the liquid and gas mixer was used to simulate the extra length of the pipeline. In this paper, tests have been carried out in a large scale experimental facility that the length of the real pipeline was 127m and no buffer vessel was used. Therefore the effect of the severe slugging on the flow in the upstream pipeline could be studied rightly and vice versa.

4.4.1Experimental Facility and Method

An experimental facility was designed and constructed for this study in the State Key Laboratory of Multiphase Flow at Xi'an Jiaotong University. A schematic diagram of the experimental facility is shown in Figure 29.

Clean and filtered tape water was driven by a multi-stage centrifugal pump and then measured by the electromagnetic flowmeters. Oil was supplied by a gear pump from an oil tank and then measured by a CMF050 or CMF200 Micro Motion mass flowmeter. Air was compressed by a screw compressor and flowed into a buffer vessel. Then the air was filtered and measured by an orifice plate or a digital vortex flowmeter before entering the mixer. A horizontal plate was fixed in the mixer pipe to get stratified flow regime at the initial part of the test section, as shown in Figure 4.1. The gas was injected above and the liquid below the segregation plate. After the mixer, liquid and air flowed into the test loop with stratified flow pattern.

The test loop was consisted of a 108m long horizontal section, a 19m long downward inclined section and a 16m high riser. The total length of the test loop is 143m. The downward section could be inclined between 0° and -5° from the horizontal and this section was connected to the riser at its lower end through a 30cm radius bend. In this study, the inclination angles were limited to 0°, -2° and -5°. Most segments of the test loop were consisted of stainless steel pipe, but several segments located at the horizontal, downward and vertical parts of the loop were made of PVC pipe and were clear and visualization. The inner diameter of the test loop was 50mm. At the end of the riser, the gas-liquid mixture was separated in a vertical separator which was operated under atmospheric condition. The air was vented to atmosphere and the liquid was returned to a liquid-liquid separator tank. Oil and water were separated and flowed back to the storage tanks and recirculated respectively.

In this study, air and water were used as the test fluids and the water was metered by the Micro Motion mass flowmeters for their high precision.

The test rig was instrumented with many pressure and differential pressure transducers along the profile of the pipeline and riser. The liquid holdup near the outlet of the riser was measured by a gamma ray densitometer which was aligned horizontally through the vertical pipe center. A video camera was equipped to record the flow configuration in the horizontal, downward or vertical transparent pipe sections. The instruments also included a series of conductance probes mounted in the horizontal and downward PVC pipe sections to identify the liquid slugs and elongated bubbles. The temperatures of every phase before the mixer and the mixture at the outlet of the riser were measured by 4 thermocouples.

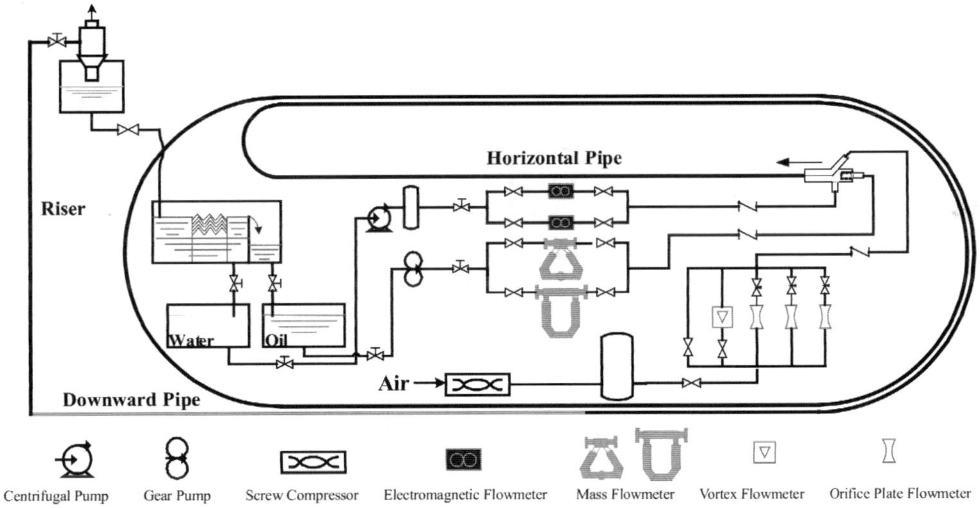

Fig. 29 Schematic of the test facility for severe slugging flow

The hardware of the data acquisition system consisted of a PC and a high speed multifunction board PCI-6071E made by National Instruments Inc. The data acquisition software, also called virtual instruments, was developed in the graphical programming system of LabVIEW6.0. The analog signals were acquired and stored into binary files and also could be displayed on the screen in waveform simultaneously. All the concerned signals were sampled at 500Hz.

The liquid and air superficial velocities varied in the range of 0.01-2m/s and 0.01-5m/s respectively for this study. A 54600-Series oscilloscope, produced by Agilent Inc., was used to acquire and display two absolute pressure signals from the transducers located on the horizontal and vertical pipe sections. The pressure signals were analyzed and the cycle time could be acquired if there was cyclic process occurring in the facility. Moreover, determined from the real time information of the system pressures, if the constant operational conditions were reached, the acquisition program was run to collect transient data for about several periods of severe slugging if this flow pattern was occurring in the test configuration. If the other stable flow patterns occurred, data was collected for 10 minutes. While the acquisition process was initiated, visual observations of the flow in the horizontal, downward inclined and vertical pipe sections were performed and the flow patterns were identified and recorded, together with the transition process of the flow patterns in a cycle time. Combined with the signals of pressure, differential pressure and liquid holdup displayed on the computer screen, it could be determined whether the flow in the pipeline-riser system was severe slugging or not.

4.4.2Results and Discussion

(1)Flow pattern map

Figure 30 presents the experimental flow pattern maps that illustrate the occurring region of severe slugging as a function of the flow rates. The inclination angles of the downward pipe section of the three maps are 0°, -2° and -5° respectively. Note that the inlet gas liquid superficial velocity is normalized to the atmospheric condition. Except the severe slugging, the other flow patterns are classified as steady flow, which means that the flow patterns in the pipeline or the riser are fixed and do not change periodically, for example occurring of slug flow in pipeline and churn flow in the riser simultaneously.

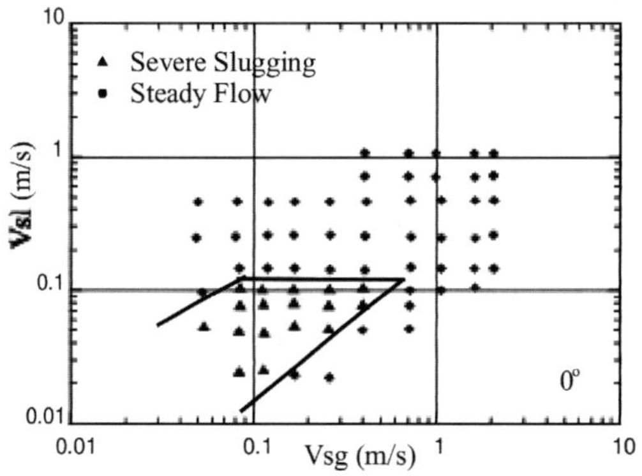

(a) Downward pipe section angle of

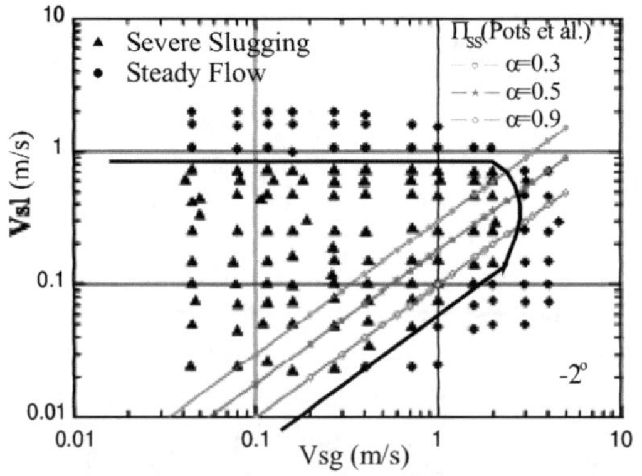

(b) Downward pipe section angle of

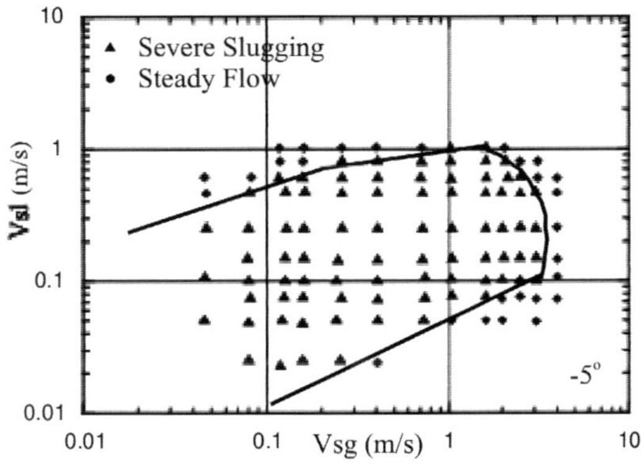

(c) Downward pipe section angle of -5°

Fig. 30 Flow pattern map for various downward pipe section angles: 0°, -2° and -5°.

From these figures it can be seen that severe slugging is easier occurrence in the pipeline/riser system with a downward inclined pipe section before the riser. Compared with the maps of -2° and -5° inclination, Figure 30(a) shows that severe slugging only occurs in a very small area when the inclination of the downward pipe is 0°, i.e. the whole pipeline is horizontal. Also, the transition boundaries move down along both the liquid and the gas flow rates axes when the inclination angle is decreased to 0°, while the flow pattern in the pipeline is limited to stratified flow. However, when the inclination is increased to -2° or -5°, severe slugging is easily occurring in the system and there is no distinct difference between the flow pattern maps of the -2° and -5° inclination, as shown in Figure 30(b) and 30(c). Figure 31 shows all the experimental severe slugging cases of -2° inclination downward pipe section on the Mandhane horizontal flow pattern map. It can be seen that the flow pattern in the horizontal pipe section may be stratified flow, bubble flow or slug flow, which was also observed in the experiments. So severe slugging is easily encountered in field pipeline/riser system at low liquid and gas rates. From the experiments, it was observed that in most time of the severe slugging period the flow pattern in the downward pipe section was stratified flow though there might be several liquid slugs passing in the gas blowout process. Therefore this study agrees with the conclusion of Schmidt et al.(1985), they suggested that the stratified flow in the downward pipe section is a precondition for severe slugging. In their experiment the air accumulator before the mixer was used to simulate a long pipeline, however it was different from the real pipeline used in this study and the facility having the long pipeline can simulate the field offshore pipeline/riser system comprehensively. From the analysis, it can be concluded that stratified flow in the pipe section ahead of the riser, which is horizontal or downward, is the necessary condition for severe slugging occurrence. On the other hand, it was observed in the experiments that normal slug flow in the upstream horizontal pipe section and severe slugging in the whole system can coexist and both flow patterns are not mutually exclusive, which was also shown in Figure 30(b) and Figure31

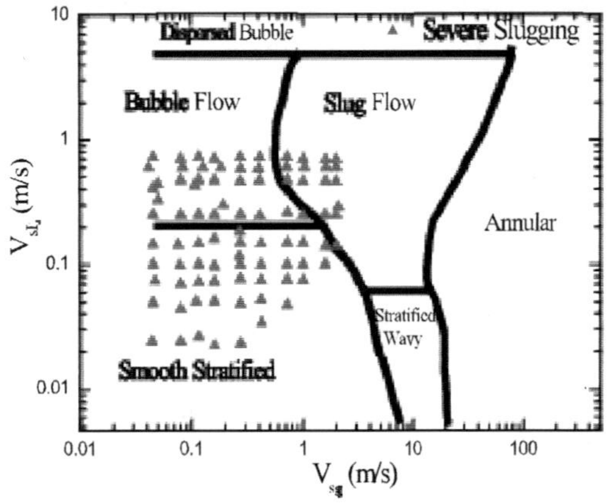

Fig. 31 Experimental severe slugging data of -2° downward pipe section shown on the flow pattern map for horizontal flow (Mandhane et al. 1974)

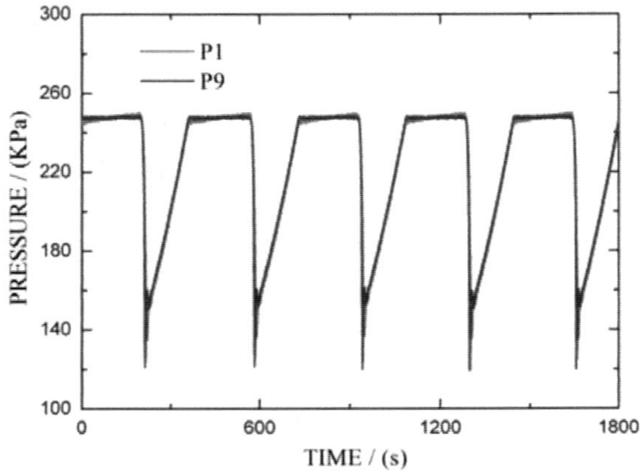

Fig. 32 Inlet pressure P1 and riser base pressure P9 vs. time , -2° inclination, Vsg=0.12m/s, Vsl=0.15m/s

Boe (1981) proposed a mathematical criterion to predict the severe slugging region. Later Pots et al. (1987) proposed a similar criterion considering that the rate of hydrostatic head buildup of the liquid in the riser caused by the growth of the slug must exceed the rate of gas pressure buildup in the pipeline. The ratio between the pressure buildup rates can be expressed as:

$$\Pi_{SS} = \frac{zRT/M}{\mathrm{g}L_{\mathrm{p}}\alpha}\frac{W_g}{W_L} \qquad (30)$$

where W_g and W_L are gas and liquid mass flow rates respectively. α is the void fraction in the pipeline. L_p is the length of the pipeline. If $\Pi_{SS}<1$, severe slugging could occur.

In Figure 28(b), three curves $\Pi_{SS}=1$ calculated for the test facility of this study are shown. The assumption value of the void fraction in the pipeline was 0.3, 0.5 and 0.9 respectively. It is expected that severe slugging would occur in the left region of these curves. It is shown that only the right curve with $\alpha = 0.9$ is close to the experimental boundary. Therefore this criterion does not predict the transition boundary of the flow patterns successfully.

(2) Pipeline pressure

Some experimental cases are presented in this paper and the flow characteristics of severe slugging are analyzed. The inclination angle of the downward section is limited to -2°.

Figure 32 shows the transient effect of the absolute pressures in the pipeline during several periods of the typical severe slugging, in which the liquid slug length is longer than the riser length. P1 is the inlet pressure of the system located 0.5m after the mixer and P9 is the pressure at the bottom of the riser. The air and liquid superficial velocities for this case were 0.12m/s and 0.15m/s respectively. From this figure it is shown that the pressures fluctuate periodically and regularly in severe slugging and have a period up to 360 seconds. The fluctuation magnitude of the inlet pressure P1 is 110KPa, while it is 124KPa for the riser bottom pressure P9. Also, it is found that both traces almost superpose each other in the cyclic process, though the distance is 127m between P1 and P9 transducers. Therefore the severe slugging has significant effect on the horizontal pipeline and the flow in that could not be treated separately from the whole system.

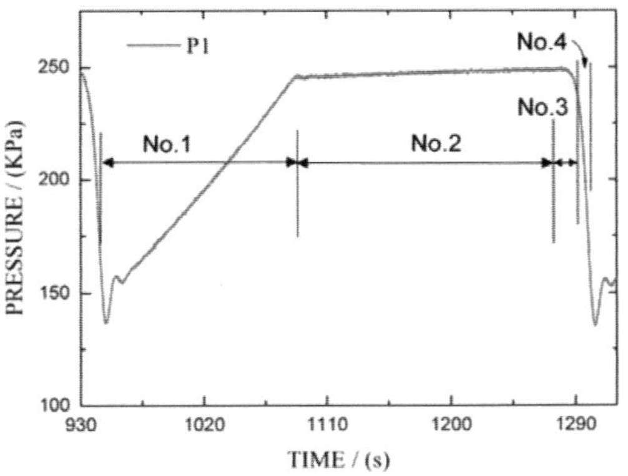

Fig. 33 Inlet pressure P1 vs. time in a severe slugging cycle, -2° inclination, Vsg=0.12m/s, Vsl=0.15m/s

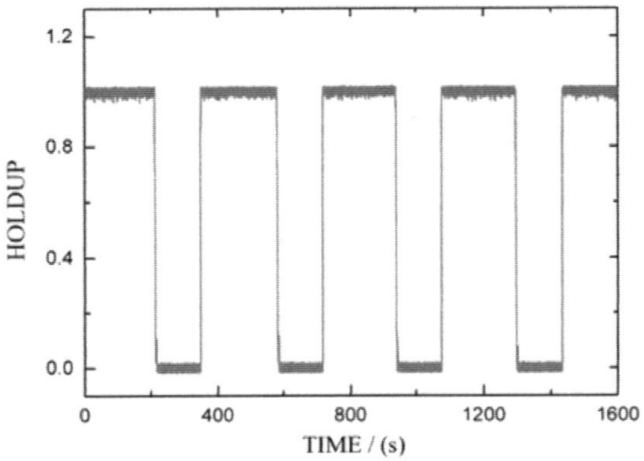

Fig. 34 Liquid holdup of riser vs. time, -2° inclination, Vsg=0.12m/s, Vsl=0.15m/s

Figure 33 shows the inlet pressure P1 in a severe slugging cycle time. As reported by Schmidt et al. (1985), the cycle of severe slugging can be divided into four stages:

(1) Slug generation stage;
(2) Slug production stage;
(3) Bubble penetration stage;
(4) Gas blowout stage.

From the experimental observation, it was found that during the gas blowout stage there were several liquid slugs passage in the downward pipe section in most cases and their lengths were longer than the length in normal slug flow. These liquid slugs were produced in the horizontal pipeline caused by the sudden expansion of the compressed gas for the pressure decrease and moved rapidly to the riser base and blocked the gas passage into the riser together with the liquid fallback in the vertical riser. After the blockage of the riser base, the following liquid slugs stopped moving in the downward pipe and filled it with the carrying liquid. The slug generation stage started again. Because of the slugging phenomenon occurring in the pipeline during gas blowout process, the length of liquid penetration into the downward pipeline in this study is longer than that only formed from the liquid fallback in the vertical riser pipe, such as reported by Schmidt (1979).

(3) Riser liquid holdup

Figure 34 presents the liquid holdup measured at the top of the riser from the gamma ray densitometer during severe slugging cycles. It is shown that the pipe area of the measured cross section had two conditions, "full" or "empty", in most time and the transition time is very short. When the pipe was empty, it was during the slug generation stage and the liquid did not reach the top of the riser pipe. The holdup increased to unity when the liquid slug reached the top of the riser and the liquid began to flow into the separator. At the end of the bubble penetration stage and the beginning of the gas blowout stage, gas bubble reached the top of the riser pipe and gas flowed out violently. When the remaining liquid fell down the riser and blocked the bottom, the liquid holdup decreased sharply to zero and there was no fluids flow out of the riser. This process occurred in a very short time comparing with the whole severe slugging period, as shown in Figure 34. From the video recorded at the transparent vertical pipe section, it is found that the gas flowed in 16 seconds in a cycle. So it meant that the gas volume entering the pipeline in a severe slugging cycle time was compressed and released rapidly in a very short time.

(4) CLiquid slug length

In order to measure the liquid production, a special cylinder vessel was designed optimally and small liquid level changes could be indicated by the differential pressure transducer. A T-junction equipped with an electromagnetic quick valve was used to switch the liquid stream into either the cylinder vessel or the liquid-liquid separator tank. The length of liquid slug flowing out of the riser could be calculated from the differential pressure between the bottom and the liquid level in the vessel.

Figure 35 shows the length of liquid slug produced in severe slugging vs. the liquid superficial velocity with -2° inclination of the downward pipe section. The riser length is also plotted in Figure 35. It is indicated that the liquid production in a severe slugging cycle time increases as the liquid flow rate increases when the gas flow rate is constant. However the liquid slug length decreases sharply as the gas flow rate increases. Therefore increasing the inlet gas flow rate is a feasible method to eliminate the severe slugging. From Figure 35, it is found that the maximum of the liquid slug length is about 7 times of the riser length.

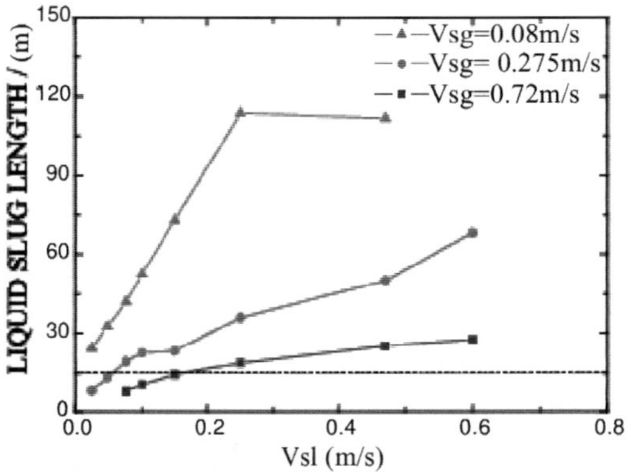

Fig. 35 Liquid slug length vs. liquid superficial velocity, -2° inclination

5.SUMMARY

5.1 Conclusions to the Flow Pattern Online Identification

The flow velocity of fluid and the oil fraction influence strongly the signal feature of pressure and differential pressure. The feature parameters for every flow pattern intersect seriously. To overcome the online recognition of industrial multiphase flow is very difficult only based on a single type of flow pattern information and processing method.

The recognition method based on feature space needs to meet two basic prerequisite. One is that the feature space about flow regime information should have sample set which can represent the feature of flow regime. The other is that the feature space should meet the principle of close cohesion for a certain flow regime, and far separation for different flow regimes.

The recognition rules for partial of flow pattern are proposed on the basis of decollating the feature space composed by the signal features of pressure and differential pressure. For every flow pattern feature the self-organization cluster algorithm is proposed to design the standard templates, which can be taken as the basis to recognize flow regime.

Based on the pressure and differential pressure signals, the multi-sensor data fusion method is proposed to realize the online recognition of multiphase flow regime, which can overcomes the severe influence of flow velocity and the oil fraction of fluid on the recognition. The online recognition method is tested and confirmed in the practice, which has less than 10 percent measurement error. The method takes advantages of high confidence, good fault tolerance and less requirement over single sensor performance.

5.2 Conclusions to Slug Flow Tracking Model

Systematical experiments two-phase slug flow of air and water mixture were conducted on a 27.33m long horizontal pipe. The measurements of hydrodynamic parameters of slug flow evolution along the pipe were completed by the differential pressure method. In this method Two close pressure transducers were used to measure the translational velocity of elongated bubble and the liquid slug length. From the experimental results, it was found that the averaged liquid slug lengths are not dependent on the mixture velocities, and the maximum and the averaged liquid slug length increase along the pipe.

A modified slug tracking model was presented in our laboratory which considered the affections of acceleration pressure drop and the wake effect. This model was programmed in C++ language with OOP technique. Compared with the data of observed liquid slugs, the model correctly predicted the length distributions with quite good agreement. Both the predicted and the measured liquid slug lengths followed a log-normal distribution. The model can predict the maximum slug lengths correctly. However the predicted averaged liquid slug lengths were longer than the experimental data, which was considered due to the separation effect of the elbow which decreased the length of the short slugs

5.3 Conclusions to the Experimental Study of the Flowrate Transients in Slug Flow

The results obtained in our laboratory demonstrate that the "Up-gas" transient would produce pressure overshoot. The magnitude of overshoot is related to the magnitude of change of gas flow rate and the distance away from the entrance. The further the distance is, the higher the magnitude of pressure overshoot. Near the exit, the pressure overshoot begins to decrease and is close to zero in the downstream slug catcher. Pressure waves exist in the pipeline, which spread from the entrance to the exit. The spreading speed of pressure wave decreases along the pipeline. At the same time, the increase of gas flow rate is often associated with the formation of larger slug.

The decrease of gas flow rate is associated with pressure undershoot and temporary stratified flow. The magnitude of pressure undershoots and period of stratified flow are sensitive to change of gas flow rate and the distance away from the entrance. With increasing the distance away from the entrance, the magnitude of pressure undershoots increase at first, then decrease linearly, but the period of stratified flow increase linearly.

Under liquid transients, the obvious feature is the change of liquid slug frequency.

Based on the experimental results an empirical equation for the calculation of pressure overshoot was proposed in which the pressure overshoot was correlated with the acceleration pressure drop caused by the increase of gas flow rate.

5.4 Conclusions to Severe Slugging Flow

An experimental facility was designed and constructed to study the severe slugging in SKMFPE. A series of flow pattern maps were obtained and the effect of the inclination angle of the downward pipe was experimentally studied and analyzed. It was found that the stratified flow in the downward pipe is a precondition for the occurrence of severe slugging. However, the normal slug flow can be occurring in the horizontal pipe with the severe slugging occurring in the whole system. The hydrodynamic characteristics of severe slugging were reported, such as the pressure, the liquid holdup and the length of the liquid slug. Figure 35 shows that the liquid slug length decreases sharply as the gas flow rate increases at same liquid flow rate. Therefore, increasing the inlet gas superficial velocity is a feasible method to eliminate the severe slugging

REFERENCES

Acikgoz M, Franca F and Lahey R T Jr. (1992). An Experimental Study of Three-Phase Flow Regimes. Int J of Multiphase Flow. 18:327-336

Andritsos N. and T.J. Hanratty (1987). Interfacial instabilities for horizontal gas-liquid gas-liquid flows in pipelines. Int.J. Multiphase flow, .13(5):583-603.

Bai B F. (1999). Flow Parameters Fluctuations and Online Automatic Identification of Flow Regime for Steam (Gas)-Water Two-Phase Flow. Ph.D Dissertation. Xi'an: Xi'an Jiaotong University, 1999: 3-15

Bai B F, Guo L J, Zhao L. (2001). Development Of One-Line Identification Of Steam (Gas)-Liquid Two-Phase Flow Regimes. Advances in Mechanics, 31(3): 437-446

Bai B F, Zhang S J， zhang X M ，Zhao L., Guo L J. (2008). Online recognition of the multiphase flow regime, Sci China Ser E-Tech Sci, Aug. 2008, 51(8):1-9.

Banjamin T.B. (1959). shearing flow over a wavy surface, J.Fluid Mech. Vol.6, pp.161-205.

Barnea, D. (1991). On the effect of viscosity on stability of stratified gas-liquid flow-application to flow pattern transition at various pipe inclinations. *Chem. Engng. Sci.*, 46, 2123-2131.

Barnea, D. & Taitel, Y. (1993). Kelvin-Helmholtz stability criteria for stratified flow: viscous versus non-viscous (inviscid) approaches, *Int. J. Multiphase Flow,* 19(4), 639-649.

Barnea, D.A., Taitel, Y (1993). A model for slug length distribution in gas-liquid slug flow. Int. J. Multiphase Flow. 19, 829-838

Barnea D., Taitel Y. (1994). Interfacial and structural stability of separated.flow, Int. Multiphase Flow 20 (Suppl.) 387–414.

Bendiksen, K., Espedal, M. (1992). Onset of slugging in horizontal gas-liquid pipe flow, Int. J. Multiphase Flow, 18, 237~247.

Bendiksen, K. H. et al. (1996). On the modeling of slug flow. Chem. Eng. Comm. V141-142, 71-102

Boe, A. (1981). Severe slugging characteristics; Part 1: Flow regime for severe slugging; Par 2: Point model simulation study. Presented at Selected Topics in Two-Phase Flow, Trondheim, Norway, March 1981.

Brauner N., Maron D.M. (1992). Stability analysis of stratified liquid–liquid flow, Int. J. Multiphase Flow 18, 103–121.

Brill, J.P. et al. (1981). Analysis of two-phase tests in large-diameter flow lines in Prudhoe Bay Field. SPEJ, 363-378

Cai, Y, Wambsganss M W, Jendrzejczyk J A. (1996). Application of Chaos Theory in Identification of Two-phase Flow Patterns and Transitions in a Small, Horizontal, Rectangular Channel. ASME J Fluid Engineering, 118: 383-390

Caussade, B., Fabre, J., The'ron, B. (1987). Unsteady gas-liquid flow in horizontal pipes, European Two-Phase Flow Group Meeting, Trondheim, Norway

Chu, K.J. and Dukler, A.E. (1974). Statistical characteristics of thin, wavy films: II Studies of the substrater and its wave structure. AIChE J. 20, 695-706.

Chu, K.J. and Dukler, A.E. (1975). Statistical characteristics of thin, wavy films: III Structure of the large waves and their resistance to gas flow. AIChE J. 21, 583-593.

Cook, M.&Behnia, M. (2000). Slug length prediction in near horizontal gas-liquid intermittent flow.

Chem. Eng. Sci. 55, 2009-2018

Crowley C.J., Wallis G.B., Barry J.J. (1992). Validation of a one-dimensional wave model for the stratified to slug flow regime transition, with consequences for wave growth and slug frequency, Int. J. Multiphase Flow 18 (2) 249–271.

Dukler, A. E., Hubbard, M. G (1975). A model for gas-liquid slug flow in horizontal and near horizontal tubes. Ind. Engng Chem. Fundam. 14, 337-347

Farghaly, M.A. (1987). Study of severe slugging in real offshore pipeline riser-pipe system. SPE 15726, Presented at SPE Middle East Oil Show, Manama, Bahrain, March, 1987.

Funada, T., Joseph, D.D. (2001). Viscous potential flow analysis of Kelvin–Helmholtz instability in a channel. *Int. J Multiphase Flow* 28(9), 263–283.

Giovine D. (1991). Stability of liquid flow down an inclined tube, Int. J. Multiphase Flow 17,485–496.

Guo L,J, Li G,J, Chen X,J (2002). A linear and non-linear analysis on interfacial instability of gas-liquid two-phase flow through a circular pipe.Int,J. Heat and Mass transfer,45,1525-1534.

Guo L J. (2002). Two-phase and Multi-phase Flow Dynamics. Xi'an: Xi'an Jiaotong University Press, 2002

Guo Liejin, Gu Hanyang, Wang Xin, Chen Xuejun (2005). Interfacial waves and slug flow in pipeline, in Proceeding of 5th int. symp on multiphase flow, heat mass transfer and energy conversion, Xi'an, 2005.

Hall, David L. (2001). Handbook of Multisensor Data Fusion. New York: CRC Press, 2001

Hanratty, T.J., Engen, J.M. (1961). Interaction between a turbulent air steam and a moving water surface, AICHE J.7, pp488-497

He Limin (2002). An investigation of the characteristics of oil-gas two-phase slug flow in horizontal and near horizontal pipes. Ph.D. Dissertation. Xi'an Jiaotong Univ.

Huang R, Zhong B L. (1991). Monitor and Failure Diagnosis during Machine Manufacture. Xi'an: Xi'an Jiaotong University Press, 1991: 48-70

Hubbard M G, Dukler A E. (1966). The Characterization of Flow Regimes for Horizontal Two-phase Flow. In: M.A.Saad & J.A.Moller eds. Proc. of the 1966 Heat Transfer and Fluid Mechanics Institute, Stanford University Press, 100-121

Jeffreys, H. (1925). On the formation of water waves by wind, Proc.R.Soc. A , 107,189.

Jurman L.A., M.J. (1989) Study of waves on thin liquid film sheared by turbulent gas flows, Phys. Fluids A 1,522–536.

King, M. J. S., Hale, C. P., Mendes-Tatsis, M. A., Hewitt, G. F. (1996). An experimental investigation of the effect of flowrate transients within the slug flow regime in horizontal pipes, 7th International Conference on Pressure Surges and Fluid Transients in pipelines and Open Channels, Harrogate, UK.

King, M. J. S., Hale, C. P., Lawrence, C. J., Hewitt, G. F. (1998). Characteristics of flowrate transients in slug flow, Int J Multiphase Flow,24, 825-854.

Kohonen T. (1990). The Self-Organizing Map. Proceedings of the IEEE. 1990, 178 (9): 1464-1480

Langford H M, Beasley D E, Ochterbeck J M. (1998). Chaos Analysis of Pressure Signals in Upward Air-water Flows. Third International Conference on Multiphase Flow, ICMF'98, Lyon, France, June 8-14

Lay. C.E. and Dukler A.E.(1994) Flooding in vertical tubes-II A film model for entry region flooding. Int.J. Multiphase flow, Vol.20, No.2, pp.235-247.

Li, G.J. (1996). Ph. D Thesis School of Energy and power Engineering Xi'an jiaotong University, Xi'an (1996)

Li H. (2002). Application Of Wavelet Multi-Resolution Analysis To Pressure Fluctuations Of Gas–Solid Two-Phase Flow In A Horizontal Pipe. Powder Technology, 2002, 125:61-73.

Lin P.Y., Hanratty T.J. (1986). Prediction of the initiation of slugs with linear stability theory, Int. J. Multiphase Flow 12,79–98.

Lin, Zonghu, (1992). Characteristics of gas-liquid two-phase flow in pipelines and their engineering application, Xi'an Jiaotong University Press, 57~59.

Mahvash A, Ross A. (2008). Two-phase Flow Pattern Identification Using Continuous Hidden Markov Model. Int J of Multiphase Flow, 34 (1): 303–311

Mandhane J M, Gregory G A and Aziz K. A. (1974). Flow Pattern Map for Gas-Liquid Flow in Horizontal Pipes. Int J of Multiphase Flow, 1: 537-553

Matsui G. (1986).Automatic Identification of Flow Regimes in Vertical Two-phase Flow Using Differential Pressure Fluctuations. Nuclear Engineering and Design, 95:221-231

Mi Y, Ishii M, Tsoukalas L H. (1998). Vertical Two-Phase Flow Identification Using Advanced Instrumentation and Neural Networks. Nuclear Engineering and Design, 184:409-420

Miles J.W. (1959). On the generation of surface wave by shear flows. J.Fluid Mech. Vol.3, pp.185-204.

Minami, K. (1991). Tansient flow and pigging dynamics in two-phase pipelines, PhD Thesis, University of Tulsa, Tulsa, OK, USA.

Moissis, R., Griffith, P. (1962). Entrance effects in a two-phase slug flow. J. Heat Transfer. 84, 29-39

Monji H, Matsui G. (1998). Flow Pattern Identification of Gas-liquid Two-phase Flow Using a Neural Network. Third International Conference on Multiphase Flow, ICMF'98, Lyon, France, June 8-12,

Nicklin, D. J., Wilkes, J. O., Davidson, J. F. (1962). Two-phase flow in vertical tubes, Tans Inst Chem Eng, 40, 61~68.

Nydal, O. J. et al.(1992). Statistical characterization of slug flow in horizontal pipes, Int. J. Multiphase Flow. 18, 439-453

Nydal, O. J. et al. (1996). Dynamic slug tracking simulations for gas-liquid flow in pipelines. Chem. Eng. Comm. V141~142, 13~39

Pots, B.M., Bromilow, I.G., Konijn, M.J. (1987). Severe slug flow in offshore flowline/riser systems. Soc. Pet. Eng. J., SPE13723, , 319-324.

Sakaguchi, T., Akagawa, K., Hamaguchi, H., Ashiwake, N. (1973). Transient behaviour of air-water two-phase flow in a horizontal tube, Winter Annual Meeting of Heat Transfer Division ASME, Detroit, MI, USA, November.

Samways A L, Bradbury L J S, Bruun H H.(1997). Pressure Measurements and Convection Velocity Evaluations in Two-Phase Flow. Int J of Multiphase Flow, 23(6):1007-1029

Schmidt, Z., Brill, L.P., Beggs, D. (1979). Experimental study of severe slugging in a two-phase flow pipeline-riser-pipe system. SPE8306. Presented at SPE Annual Technical Conference and Exhibition, Las Vages Nevada.

Schmidt, Z., Doty, D.R., Dutta Roy, K. (1985) Severe slugging in offshore pipeline riser-pipe system. Soc. Pet. Eng. J., SPE12334, 27-38.

Sharma, Y. (1985). Modeling transient two-phase slug flow. Ph.D. dissertation. Univ. Tulsa.
Shi J. and Kocamustafaogullari,G., 1994, Interfacial measurements in horizontal stratified flow patterns, Nuclear Engineering and Design,149,81

Taitel, Y. and Dukler, A.E. (1976). A model for predicting flow regime transitions in horizontal and near horizontal gas-liquid flow. AIChE J. 22, 47-55.

Taitel, Y., Lee, N., Duckler, A. E. (1978). A model for slug flow frequency during gas-liquid flow in horizontal and near horizontal pipes, Int J Multiphase Flow, 3, 585-596.

Taitel Y, Vierkandt, S, Shoham O, Brill JP. (1990). Severe Slugging in a riser system: Experiments and Modeling Int. J. Multiphase Flow. 16(1): 57-68.

Taitel, Y. et al. (1998). Effect of gas compressibility on a slug tracking model. Chem. Eng. Sci. 53, 2089-2097

Taitel, Y. & Barnea, D. (1999). Slug tracking in hilly terrain pipelines. SPE Ann. Tech. Conf. Exhib. Prod. Oper. and Eng. Houston, 321-330

Takahama, H. and Kato, S. (1980). Longitudinal flow characteristics of vertical falling liquid films without concurrent gas flow. Int. J. Multiphase Flow 6, 203-215.

Tengesdal, J.O. (2002). Investigation of self-lifting concept for severe slugging elimination in deep-water pipeline/riser systems. Ph.D. Thesis, Pennsylvania State University.

Tutu N K. (1984). Pressure Drop Fluctuations and Bubble-Slug Transition in a Vertical Two-phase Air-water Flow. Int J of Multiphase Flow, 10 (2):211-216

van Hout et al. (2001). Evolution of statistical parameters of gas-liquid slug flow along vertical pipes. Int. J. Multiphase Flow. 27, 1579-1602

Wambsganss M W, Jendrzejczyk J A, France D M. (1991). Two-phase Flow Patterns and Transitions in a small Horizontal, Rectangular Channel. Int J of Multiphase Flow, 17 (3):327-342

Woods, B.D., Hanratty, T.J. (1996). "Relation of slug stability to shedding rate", Int. J. Multiphase Flow, 22,809~828.

Wu Haojiang, Zhou Fangde, Wu Yuyuan. (2001). Intelligent Identification System of Flow Regime of Oil-Gas-Water Multiphase Flow. Int J of Multiphase Flow, 27: 459-475

Yang G, Wang Y T, Lu Z X, and et al. (2005). Application of Multi-sensor Data Fusion in Multiphase Flow Parameter Measurement. Instrument Technique and Sensor, No.11: 51-53

Yocum, B. T. (1973). Offshore riser slug flow avoidance: mathematical models for design and optimization. SPE4312, Presented at SPE European Meeting, London, April, 1973.

Zheng, G.H. et al. (1994). Slug flow behavior in a hilly terrain pipeline. Int. J. Multiphase Flow. 20, 63-79

Journal of Physics: Conference Series **147** (2009) 012048 doi:10.1088/1742-6596/147/1/012048

Digital image processing based mass flow rate measurement of gas/solid two-phase flow

Ding Song[1], Lihui Peng[1], Geng Lu[1], Shiyuan Yang [1], Yong Yan [2]

[1] Tsinghua National Laboratory for Information Science and Technology, Department of Automation, Tsinghua University, Beijing, 100084, China
[2] University of Kent, Canterbury, Kent CT2 7NT, U.K.

lihuipeng@tsinghua.edu.cn

Abstract. With the rapid growth of the process industry, pneumatic conveying as a tool for the transportation of a wide variety of pulverized and granular materials has become widespread. In order to improve plant control and operational efficiency, it is essential to know the parameters of the particle flow. This paper presents a digital imaging based method which is capable of measuring multiple flow parameters, including volumetric concentration, velocity and mass flow rate of particles in the gas/solid two phase flow. The measurement system consists of a solid state laser for illumination, a low-cost CCD camera for particle image acquisition and a microcomputer with bespoke software for particle image processing. The measurements of particle velocity and volumetric concentration share the same sensing hardware but use different exposure time and different image processing methods. By controlling the exposure time of the camera a clear image and a motion blurred image are obtained respectively. The clear image is thresholded by OTSU method to identify the particles from the dark background so that the volumetric concentration is determined by calculating the ratio between the particle area and the total area. Particle velocity is derived from the motion blur length, which is estimated from the motion blurred images by using the travelling wave equation method. The mass flow rate of particles is calculated by combining the particle velocity and volumetric concentration. Simulation and experiment results indicate that the proposed method is promising for the measurement of multiple parameters of gas/solid two-phase flow.

1. INTRODUCTION

Pneumatic conveying as a tool for the transportation of a wide variety of pulverized and granular materials has become widespread in many industrial fields. A well known example is pulverized coal in electrical power generation where the dilute-phase flow conditions give a typical solids/air mass ratio of 0.5:1 kg/kg to 5:1 kg/kg (equivalent to volumetric concentration of 0.05% to 0.5%) in the pneumatic pipelines (Hu et al., 2006). However, it is the mass flow rate of solids that is of primary interest to operators of a pneumatic conveyor (Yan et al., 1996).

It has long been recognized that the on-line measurement of gas/solid two-phase flow is a technically challenging area. Much work has been carried out in the field of gas/solids two-phase flow measurement over the past three decades. In general, the measurement methods can be classified into two categories, direct method and indirect method. The direct methods such as the thermal methods (Moriyama et al., 1985), the active charging and detecting method (Zhang, 1988), and the passive charge detecting method (Masuda et al., 1994), are capable of providing the mass flow rate directly.

However, besides the mass flow rate, the indirect method also provides other parameters such as particle velocity and volumetric concentration, which are useful for the monitoring of pneumatic conveying systems.

Recently, optical methods have attracted a lot attention because of their non-intrusive nature in the sensing process. Carter and Yan (2003) use a CCD camera for online particle size distribution and concentration measurement. Song et al. (2006) proposed a digital image processing based method for the measurement of particle velocity. This paper focuses on the mass flow rate measurement by using a single cost-effective digital camera.

2. MEASUREMENT PRINCIPLE

As depicted in Fig.1, the measurement system consists of a cost-effective solid state laser for illumination and a CCD camera for particle image acquisition. The principle of the proposed method is to capture particle images with adjustable exposure time. Clear images with shorter exposure time are used to determine the volume concentration, and the blurred images with longer exposure time are used to derive the particle velocity. The mass flow rate of particles is then deduced according to the following equation:

$$M_s(t) = \rho_s A V_s(t) \beta_s(t) \qquad (1)$$

Where $V_s(t)$ and $\beta_s(t)$ are the particle velocity and cross-section averaged volumetric concentration, respectively, ρ_s is the true density of the solids material, and A is the cross sectional area of the pipe.

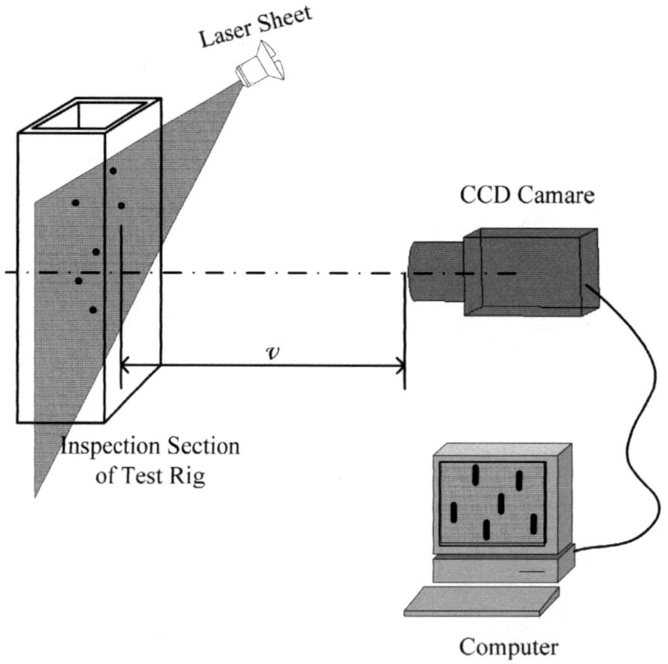

Fig.1 Principle of the measurement system

2.1 Velocity Measurement

To deduce the velocity of particles, the key point is to obtain from the blurred image the motion length due to the exposure time of the camera. A method based on the point spread function (PSF) has been reported (Song et al., 2006). However, this method is sensitive to the noise in the image. The travelling wave equation (TWE) can be used to solve this problem (Cai et al., 2003). In addition, the travelling wave equation (TWE) method is easy to compute because the image processing is in the space domain. Without losing generality, we use $w(x, y, t)$ representing the gray scale of pixel (x, y) at time t in a moving image. Suppose the image keeps a linear uniform motion at a velocity of V_p pixels per second in x direction and its gray scale does not change during the motion, therefore,

$$\frac{\partial w(x,y,t)}{\partial x}\cdot\frac{\partial x}{\partial t}+\frac{\partial w(x,y,t)}{\partial y}\cdot\frac{\partial y}{\partial t}+\frac{\partial w(x,y,t)}{\partial t}=0 \qquad (2)$$

Assume that the image moves only in x direction at a velocity of V_p pixels per second, so $\frac{\partial x}{\partial t}=V_p$ and $\frac{\partial y}{\partial t}=0$. Equation (2) now becomes

$$\frac{\partial w(x,y,t)}{\partial t}+V_p\frac{\partial w(x,y,t)}{\partial x}=0 \qquad (3)$$

Suppose $f(x, y)$ is the initial value (image) of $w(x, y, t)$ when $t = 0$, i.e.

$$f(x,y)=w(x,y,0) \qquad (4)$$

Equations (3) and (4) can be modelled with a 1-D travelling wave equation as follows

$$\begin{cases}(\frac{\partial}{\partial t}+V_p\frac{\partial}{\partial x})w(x,y,t)=0\\ w(x,y,0)=f(x,y)\end{cases} \qquad (5)$$

The D'Alembert solution to equation (5) (Zhang et al., 2003) is

$$w(x,y,t)=f(x-V_pt,y) \qquad (6)$$

According to the aforementioned analysis, $g(x, y, \tau)$, the so-called motion blurred image, is the total exposure of moving image on the CCD chip and can be obtained by integrating its instantaneous image $w(x, y, t)$ over the exposure time τ during which the camera shutter is open. In other words,

$$g(x,y,\tau)=\int_0^\tau w(x,y,t)dt \qquad (7)$$

Multiplying the Equation (7) by V_p and differentiating both sides yields

$$\begin{aligned} V_p \frac{\partial}{\partial x} g(x,y,\tau) &= \int_0^\tau V_p \frac{\partial}{\partial x} w(x,y,t)dt \\ &= \int_0^\tau -\frac{\partial}{\partial t} w(x,y,t)dt = -w(x,y,t)\Big|_0^\tau \\ &= f(x,y) - f(x - V_p\tau, y) \\ &= f(x,y) - f(x-\sigma, y) \end{aligned} \quad (8)$$

where $\sigma = V_p \tau$ is defined as the motion blur length and can be obtained by calculating the autocorrelation function of the row vectors of $\frac{\partial}{\partial x} g(x,y,\tau)$. The lens model (Song et al., 2006), which links the motion blur length σ to the particle velocity V_s, can be described as:

$$V_s = \frac{(u-f)\sigma e}{f\tau} \quad (9)$$

where u, v and f are the object distance, image distance and focal length of the lens system, respectively, τ stands for the exposure time of the CCD camera, and e is the size of each pixel.

2.2 Concentration Measurement

The volumetric concentration of solids represents the total quantity of moving solids within a pneumatic pipeline and is defined as the cross sectional area occupied by the moving solids normalized with respect to the pipe cross sectional area. The key task is to identify the particles from the acquired clear images under short exposure time. In fact, the acquired images are far less clear than their static counterparts because of the motion of particles and the non-uniformity of the illumination. For this reason, the background image is first obtained and then used to subtract from the original image. A simple thresholding method (Otsu, 1979) is used for the image segmentation, which is the key step to identify the particles from the dark background.

Otsu's algorithm assumes the image contains two classes of pixels (e.g. foreground and background) and finds the optimum threshold separating the two classes so that their combined spread (within-class variance) is minimal (Liu et al., 1993). As a method to select a threshold automatically from a gray level histogram, the Otsu method directly deals with the problem of evaluating the goodness of thresholds. An optimal threshold is selected by the discriminant criterion, namely by maximizing the discriminant measure (or the measure of separability of the resultant classes in gray levels).

Once a binarized image is available the sizes of particles can be determined. The system software calculates A_s which represents the sum of the area of all particles in the binary image. The volumetric concentration of β_s is calculated from:

$$\beta_s = \frac{A_s}{A} \quad (10)$$

where A represents the total area of the CCD camera scope.

3. RESULTS

Experimental tests were performed on a pneumatic conveying test rig (Carter et al., 2003; Peng et al., 2008) as depicted in Fig.2. The inner diameter of the pipeline is 40mm. A 1.3Mega pixel progressive scan CCD camera was installed on the inspection section to capture particle images. The camera works in two modes - short exposure time mode and long exposure time mode. Motion blurred images for particle velocity measurement were captured under the long exposure time mode. On the other hand, clear images for the volume concentration measurement were captured under the short exposure time mode. The superficial air velocity and mass flow rate of particles were varied by controlling the power of the vacuum system and the frequency of the vibrating feeder, respectively.

Sago particles with the diameter of approximately 3.2 mm were used as a test material. The true density of the particle is 1.329×10^3 kg/m^3. Fig.3 and Fig.4 show some examples of particle images captured for different exposure times. The short exposure time (50us) was set to obtain the average volume concentration. Fig.3 shows the clear image and its thresholded version which was processed using the Otsu method. Long exposure time (1ms) was set to obtain motion blurred image for particle velocity measurement. The blurred images are illustrated in Fig.4. The Mass flow rate of solids was deduced using equation (1). Results obtained are summarized in Table 1.

Glass particles of 0.4mm in diameter and 2.213×10^3 kg/m^3 true density were also tested. Table2 outlines the results.

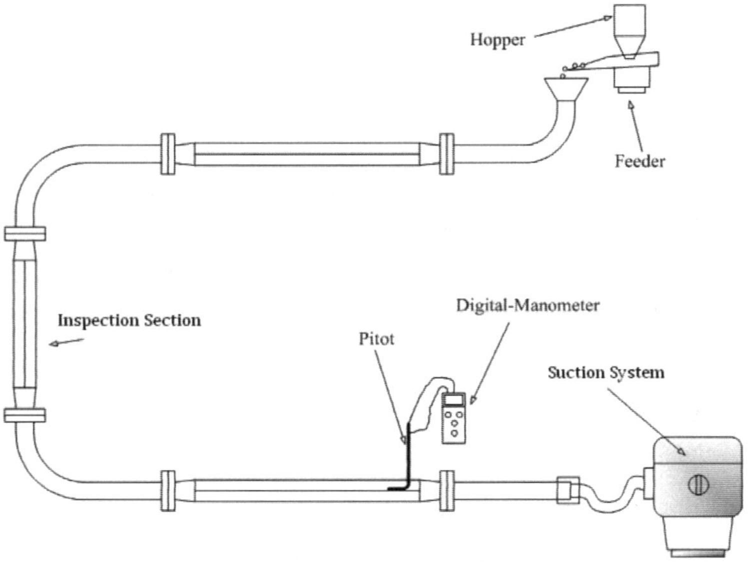

Fig.2 Schematic diagram of the test rig

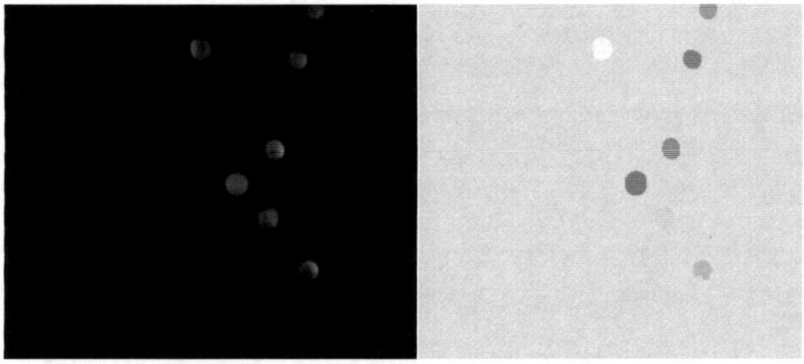

Fig.3 Clear image and its thresholded version

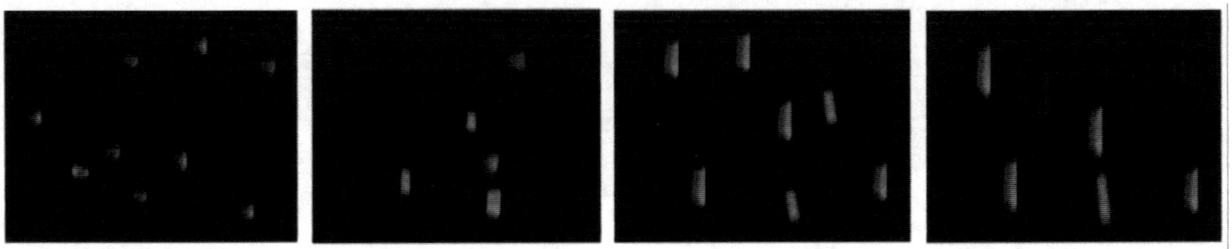

Air velocity=7*m/s* Air velocity=9*m/s* Air velocity=11*m/s* Air velocity=13.3*m/s*

Fig. 4 Motion blurred images for different air velocities (CCD exposure times=1ms)

Table 1. Mass flow rate measurement results using sago particles as a test material

Total mass (kg)	Measurement time (s)	Air velocity (m/s)	Measurements (kg)							
			1	2	3	4	5	6	Mean value (kg)	Standard deviation (kg)
2	120	7	1.835	1.965	2.253	1.796	2.092	1.781	1.946	0.173
2	120	9	2.182	2.096	1.865	1.968	2.163	1.802	2.041	0.163
2	120	11	2.223	2.152	1.873	1.902	1.825	1.854	1.953	0.163
2	120	13.3	1.914	1.963	2.123	2.208	2.107	1.907	2.063	0.134
2	180	7	2.172	2.302	1.875	1.869	1.924	2.197	2.078	0.181
2	180	9	2.249	1.853	1.864	1.832	2.092	1.867	1.958	0.157
2	180	11	2.155	1.896	1.874	2.106	1.909	1.865	1.947	0.135
2	180	13.3	1.874	2.166	1.968	2.201	2.136	1.985	2.032	0.135
2	300	7	2.091	2.135	2.093	1.833	1.891	2.103	2.037	0.122
2	300	9	1.826	2.271	1.864	1.934	2.202	1.921	2.024	0.159
2	300	11	2.123	1.865	1.832	1.968	2.168	2.244	2.044	0.157
2	300	13.3	1.909	2.124	2.214	1.934	1.936	2.21	2.026	0.154

Table 2. Mass flow rate measurement results using glass particles as a test material

Total mass (g)	Measurement time (s)	Air velocity (m/s)	Measurement value (g)							
			1	2	3	4	5	6	Mean value (g)	Standard deviation (g)
500	120	7	550.57	545.85	519.16	481.36	500.93	463.27	510.19	34.94
500	120	9	532.58	510.51	439.91	445.59	481.37	500.45	485.07	36.75
500	120	11	528.69	465.93	494.82	479.20	505.50	556.81	505.16	33.29
500	120	13.3	533.71	491.74	462.42	542.99	531.15	540.37	517.06	32.61
500	180	7	553.44	477.08	519.78	446.03	553.04	517.28	511.11	42.56
500	180	9	551.46	464.34	550.94	474.80	499.99	518.72	510.04	37.13
500	180	11	533.47	491.61	452.93	451.56	524.94	508.19	493.78	35.24
500	180	13.3	505.76	505.15	480.61	528.87	515.74	485.02	503.53	18.26
500	300	7	520.20	539.60	495.37	482.20	488.42	461.49	497.88	27.95
500	300	9	466.21	480.13	504.31	476.59	455.92	511.82	482.50	21.67
500	300	11	528.66	484.93	547.67	458.69	497.99	552.50	511.74	37.31
500	300	13.3	475.53	494.47	535.61	471.27	462.31	490.26	488.24	26.11

4. CONCLUSIONS

A method for the mass flow rate measurement of pneumatic conveyed particles through digital imaging is proposed. By controlling the exposure time of a CCD camera, both a clear image and a motion blurred image are obtained, respectively, and these images are then used to derive the particle velocity, the volume concentration, and the mass flow rate. The experimental tests on the pneumatic conveying test rig have produced favourable results supporting that the proposed digital imaging method is a viable approach to mass flow rate measurement. Future work will focus on the enhancement of the real-time performance of the system and the practical design and implementation of a prototype system for field trials.

ACKNOWLEDGEMENTS

The authors wish to acknowledge the National Natural Science Foundation of China for providing financial support under the framework of two-base research project between Tsinghua University, China and University of Kent, U. K. (contract number 60440430641). The authors are also grateful to Tsinghua Basic Research Foundation for supporting the project under the contract No. JC2007026.

NOMENCLATURE

A	the cross-section area of the pipe	[m^2]
A_s	sum of the area of all particles in the binary image	[m^2]
e	the size of CCD pixel	[m]
f	focal length of the lens system	[m]
$f(x,y)$	original image	[-]
$g(x,y,\tau)$	motion blurred image during the exposure time	[-]
$M_s(t)$	mass flow rate	[kg/s]
u	object distance of the lens system	[m]
V_s	particles velocity	[m/s]
V_p	particle image velocity	[pixel/s]
$V_s(t)$	solids velocity of the two-phase flow	[m/s]
$w(x,y,t)$	instantaneous exposure of the moving object at time t	[-]
x, y	dimensions of image array	[-]
$\beta_s(t)$	volumetric concentration	[-]
ρ_s	density of the solids material	[kg/m^3]
σ	motion blur length	[-]
τ	exposure time of the CCD camera	[s]

REFERENCES

Cai, L. D. (2003). "Travelling wave equations and restorations of motion blurred images," Acta Automatica Sinica, 29, pp. 466-471.

Carter, R. M., Yan, Y. (2003). "On-line particle sizing of pulverized and granular fuels using digital imaging techniques," Measurement Science & Technology, 14(7), pp. 1099-1109.

Hu, H. L., Xu, T. M., Hui, S. E., Zhou, Q. L. (2006). "A novel capacitive system for the concentration measurement of pneumatically conveyed pulverized fuel at power stations," Flow Measurement and Instrumentation, 17(2), pp. 87–92.

Liu, J. Z. et al. (1993). "The automatic thresholding of gray-level picture via 2D Otsu method," Acta Automatica Sinica, 19(1), pp. 101-105.

Masuda, H., Matsusaka, S., Nagatani, S. (1994). "Measurement of powder flow rate in gas–solids pipe flow based on the static electrification of particles," Advanced Powder Technology, 14(5), pp. 5241-5254.

Moriyama, T., Fujii, S., Abe, K., Kobayashi, M, (1985). "Mass flowmeter using heat transfer for dense phase solid gas two-phase flow, " 'Int. Conf. on Fluid Control and Measurement, Tokyo, 1985, September.

Peng, L. H., Zhang, Y., Yan,Y. (1993). "Characterization of electrostatic sensors for flow measurement of particulate solids in square-shaped pneumatic conveying pipelines," Sensors and Actuators A: Physical, 141(1), pp. 59-67.

Otsu, N. (1979). "A threshold selection method from gray-level histograms," IEEE Transactions on Systems, Man, and Cybernetics, 9(1), pp. 62-66.

Song, D., Peng, L. H., Zhang, Y., Yang, S. Y. (2006). "Simulation studies on the on-line concurrent measurement of velocity and size distribution of pneumatically conveyed particles through digital imaging," 'The 5th International Symposium on Measurement Techniques for Multiphase Flows, Macau, China, December 10-13.

Yan, Y. (1996). "Mass flow measurement of bulk solid in pneumatic pipelines," Meas. Sci. Technol., 7(12), pp. 1687-1760.

Zhang, J., Chen, H. R., Rong, G. (2003). "Motion blurred image restoration based on wave equation," Journal of Tsinghua Science and Technology, 45(7), pp. 1003-1007.

Zhang, X. L. (1988). "The development and test of FG-1 type flowmeter for pulverized materials," 'Proc. First Int. Conf. on Measurement and Control of Granular Materials, Shenyang, China, September.

Journal of Physics: Conference Series **147** (2009) 012049 doi:10.1088/1742-6596/147/1/012049

Effects of Velocity Ratio on turbulent Mixing Layer at high Reynolds number

Fude Guo, Bin Chen*, Liejin Guo, Ximin Zhang

State Key Laboratory of Multiphase Flow in Power Engineering, Xi'an Jiaotong University, Xi'an 710049, China

chenbin@mail.xjtu.edu.cn

Abstract. In this paper the effect of velocity ratio on plane mixing layer was experimentally investigated by Particle Image Velocimetry (PIV). The velocity ratio between low and high speed side are 0.25 and 0.50, respectively. The Reynolds number based on the velocity difference of two streams and hydraulic diameter of the channel is 66000. The results indicate that the maximum mean Reynolds stress on the same cross section is decreasing with the velocity ratio increasing. The maximum dimensionless mean vorticity is decreasing according to an exponential law along the stream-wise direction and the decreasing speed of dimensionless mean maximum vorticity is increasing with the velocity ratio increasing. The dimensionless vorticity caused by plate wake is decreasing sharply in a very short distance along the down stream wise direction and the decreasing speed is faster when the velocity ratio is larger.

1. INTRODUCTION

Mixing layer flow can be found widely in engineering and nature with many applications ranging from chemical reaction to combustion, where the turbulent mixing of two flowing streams is of primary significance to the efficiency of such devices. As a result, many researchers have studied mixing layer flow for very long time, and many important results have been obtained on the effect of parameters on the evolution of mixing layer. Brown and Roshko (1974) concluded that the large coherent vortical structures were the intrinsic features of the mixing layer. Almost at the same time, Winant and Brownand (1976) showed that the growth of a mixing layer is governed by the pairing mechanism of these vortical structures. Ho and Huang (1982) showed the initial condition had a long-lasting effect on the development of the mixing layer. Bernal and Roshko (1986) indicated that the critical Reynolds number of the instability decreases as the velocity ratio increases. Yet the distributions of turbulent parameters in the mixing layer are sensitive to many factors even small changes in initial conditions may significantly affect the Reynolds stress distributions in the near-field (Plesniak et al. 1993). Mehta (1991) found the splitter wake played a very dominant role, in some cases, a lasting role in the development of the mixing layer. Mehta (1991) pointed, for velocity ratios between 0.5 and 0.7, self-similarity of the mixing layer was observed with the asymptotic states comparable with the Reynolds number range from 686~1301. Wiecek and Mehta (1998) found that the peak mean stream wise vorticity levels decrease with increasing velocity ratio at all the stream wise locations investigated when one side of the mixing layer was set to 12 m/s and the other side was varied from 6 to 10.8 m/s in wind tunnel. Azim (2003) showed that with the velocity ratio increasing the development distance of mixing layer was increased when the Reynolds number ranges from 300~1500. M. D. Slessor et al.

(1998) reported the influence of inflow conditions on turbulent mixing layer with chemical reaction under Reynolds number $2*10^5$.
Although many researchers have studied the factors which effect the distributions of turbulent parameters and development of the mixing layer and some useful conclusions are given, the understanding of the turbulent mechanism of mixing layer especially effect of velocity ratios is still lack. So the motivation of this paper is to investigate the effect of velocity ratio on turbulent plane mixing layer by Particle Image Velocimetry (PIV), and try to get some quantitative conclusions..

2. EXPERIMENTAL SET-UP

The experiments were conducted in a vertical mixing layer water tunnel, which is specifically designed for free-shear flow experiments with multi velocity ratios. As shown in Fig.1, the main test section of the system consists of a glass vertical water channel to visualize the development of the turbulent mixing layer. Three different splitters are used to generate two streams of flow with designed velocity ratio between low and high side of the mixing layer following the principles proposed and discussed by previous researchers (Paule Dimotakis etc.1976, Daniel Bernard Lang 1985, E. Loth etc. 1995, B. Ford etc. 2000), which are made by 3mm thickness plastic-glass curved into "s" shape with two sharp splitters at the up- and down-flow end parts separately. In order to visualize the mixing layer, a LaVision GmbH PIV system is adopted, which integrates a double pulsed, 15 Hz, 532 nm, 200 mJ Nd-Yag lasers including sheet optics to generate a light sheet from the laser beam. In present paper, the velocity ratios between the low and high speed side r are 0.25 and 0.5, respectively. The Reynolds number based on the velocity difference of two steams and hydraulic diameter of the channel is 66000.

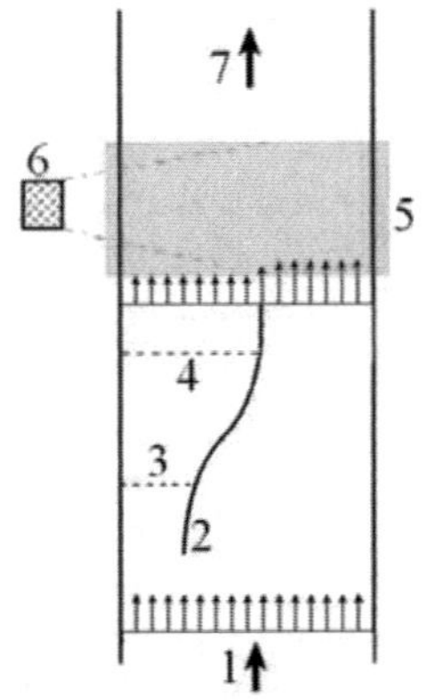

Fig.1 Sketch of the experimental system

1. In Flow 2. Splitter Plate
3. Turning Perforated Plate
4. Calming Material 5. Laser Sheet
6. PIV CCD 7. Out Flow

In this section, we analyze and discuss the variations of turbulent quantities with different velocity ratios. In the following figures, the origin is at the starting point of the mixing layer (center of the splitter plate edge down-stream). The x axis is along the down stream-wise direction, while y axis is from the low to the high speed side cross the mixing layer. Both x and y are normalized by the actual extracted half width of the channel d as $x^*=x/d$ and $y^*=y/d$.

3.1 Velocity

Fig.2 shows the mean stream-wise velocity profiles in different cross-section of the mixing layer with Reynolds numbers 66000 for different velocity ratio. In these figures the dimensionless mean stream-wise velocity u+ is normalized by the low speed side velocity $\overline{u}_l$ ($u^+ = \overline{u}/\overline{u}_l$). It can be seen that the average velocity ratios between the low and high speed side r are consistent with the designed parameter 0.25 and 0.50, which verifies that the design of our experimental setup is good enough to study the effect of velocity ratio on the development of mixing layer. It can be noticed that there are deficits of the average velocity profile at the low speed side at $x^*=0$ (the location just near the end of the splitter plate) because of the influence of the splitter plate wake. The impact of the splitter plate wake disappears rapidly along the down-stream direction and the mean velocity at high and low speed sides both become flat at $x^*=1$. In other words, the effect of the splitter wake on the mean velocity only exists at a very short distance along the stream-wise direction.

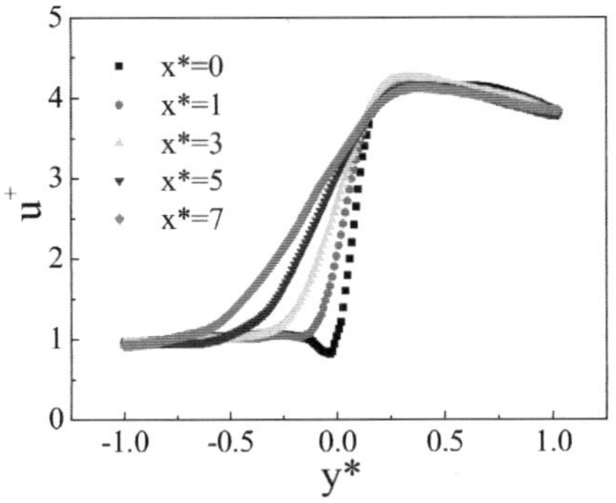

a) *r*=0.25

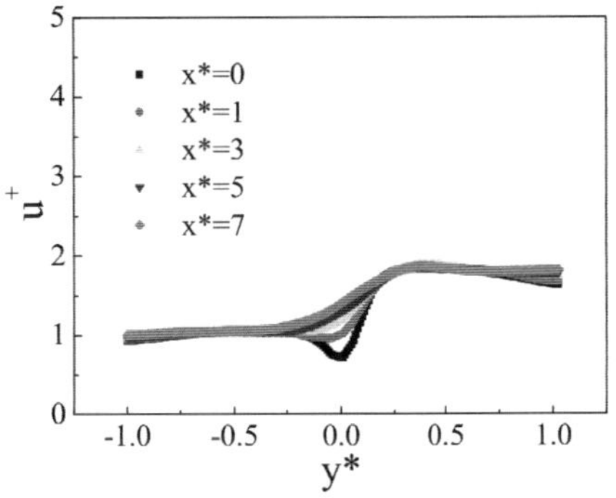

b) *r*=0.50

Fig. 2 Profiles of mean velocity at different stream-wise cross sections

3.2 Reynolds stress

The instantanenous distributions of Reynolds stress for different volocity ration are shown in Fig. 3 combined with the instantaneous velocity fluctuation vectors u' and v'. The negative values of Reynolds stress mainly concentrate at the central region of the mixing layer as the blue contours shows in Fig.3, while the small red regions indicate the positive Reynolds stress. Compared with Fig.3.(a) and (b), we can find that the intensity of velocity fluctuations in Fig.3(a) almost distributes homogeneously in the whole flow field when velocity ratio is 0.5. Different with Fig.3(a), it fluctuates more tempestuously at the central part than the other place of the flow field in Fig.3(b) when velocity ratio is 0.25.

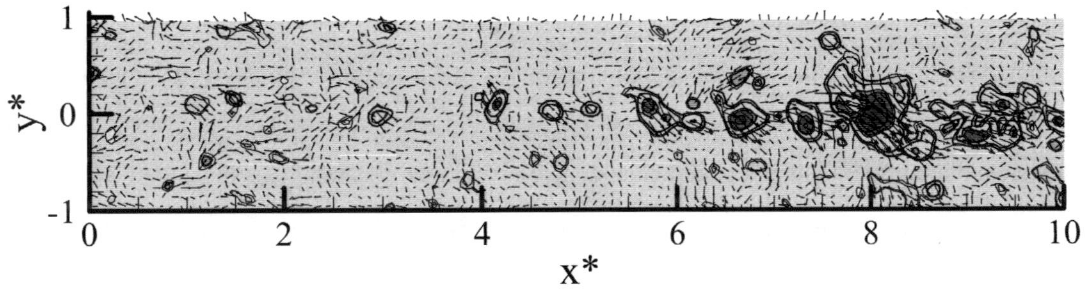

a) r=0.5

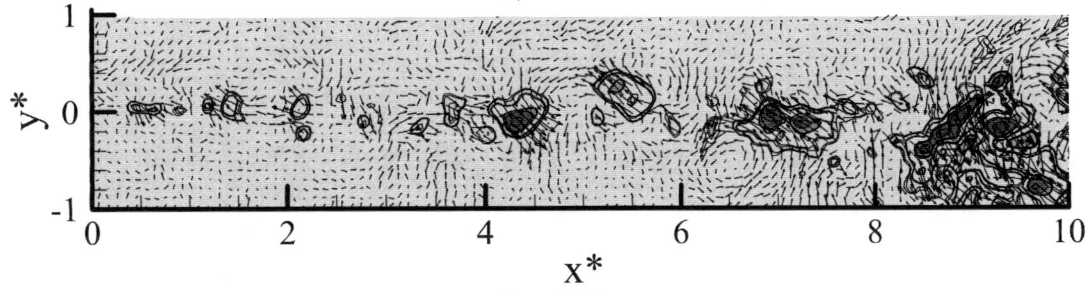

b) r=0.25

Fig. 3 The instantaneous distribution of Reynolds stress

Fig.4 gives the mean Reynolds stress with the velocity ratios r=0.5 and 0.25 at different cross sections to investigate the effects of velocity ratios on the distributions of Reynolds stress quantitatively. Previous researchers had investigated the Reynolds stress distribution along the stream-wise direction for a long time. However, there exists considerable scatter in the published data. Naka (2006) found that the Reynolds stress decreased along down stream-wise when the Reynolds numbers of high and low speed side of the wind tunnel are 610 and 260. But numerical simulation results of Cheng (2005) showed that the peak values of Reynolds shear stress $\overline{(-u'v')}/(u_1-u_2)^2$ of the free shear layer (Mc=0.64) increased along the down stream-wise direction. The results of our experiments in Fig.4 indicates that the mean Reynolds stress increases along the down streamwise at the same velocity ràtio but decreases with the velocity ratio increasing at the same cross section. So not only the Reynolds numbeer but also the the velocity ratios has important effects on the distribution of mean Reynolds stress..

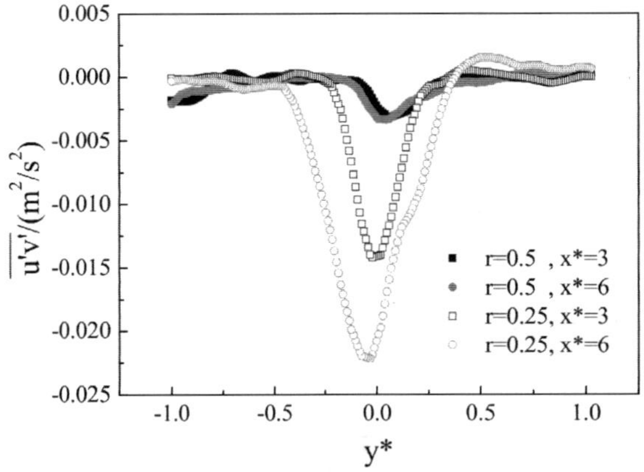

Fig.4 Distributions of Reynolds stress along the stream-wise direction under different velocity ration

3.3 Vorticity

Fig.5 shows the distributions of mean vorticity on different cross-section under different velocity ratio along the stream-wise in mixing layer. It indicates that the mean vorticity will decrease with the development of mixing layer. There are deficits of the mean vorticity profiles in Fig.5 a) and b) at the location x*=0 of the low speed side because of the splitter wake. Compared Fig.5 a) and b) we can find that the value of deficit at the plane edge is increasing with the velocity ratio. But the splitter wakes only have effects for a short distance downstream. The negative vorticity almost vanishes when the normalized x* is equal and greater than 1 for both velocity ratios.

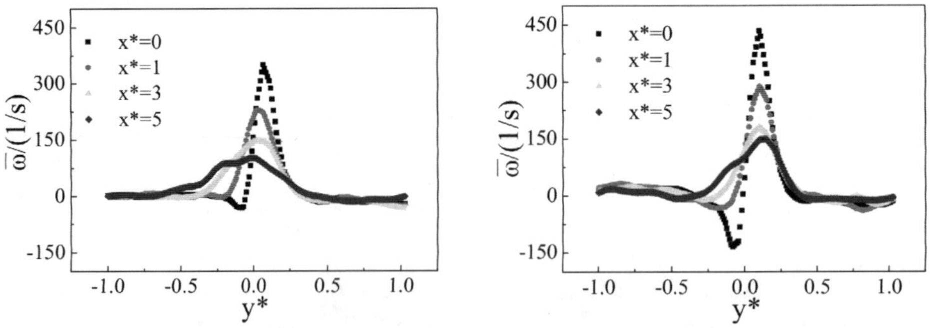

a) r=0.25 b) r=0.50

Fig5. Distributions of vorticity on different cross-section

The peak value of normalized statistics vorticity ω^+ ($\omega^+ = (\overline{\omega})_{max}/(\overline{\omega}_{x^*=0})_{max}$) is an important parameter for the development of mixing layer, which indicates the decreasing speed of the peak values of mean vorticity compared to the maximum mean vorticity at the place of the mixing layer merging. The distributions of dimensionless vorticity ω^+ are plotted in the Fig. 6. This Figure shows that the dimensionless vorticity decreases along the stream wise according to an exponential law $\omega^+ = 0.75\exp(-x^*/(4.2-4.4/r))+0.23$. It is obviously that the decreasing speed of the dimensionless vorticity is faster when the velocity ratio is larger.
In order to investigate the role of wake vorticity in the mixing layer development, the dimensionless vorticity of plate wake ω^*_{wp} ($\omega^*_{wp} = \overline{\omega}_{\max wp}/\overline{\omega}_{\max wp(0)}$) along the streamwise is shown in Fig. 7. It is shown that the dimensionless vorticity of the wake decreases sharply in a short distance the mixing layer beginning and the decreasing speed is faster when the velocity ratio is smaller. The results of Fig. 7 also say that the dimensionless vorticity is increasing with the velocity ratio at the same croccsection.

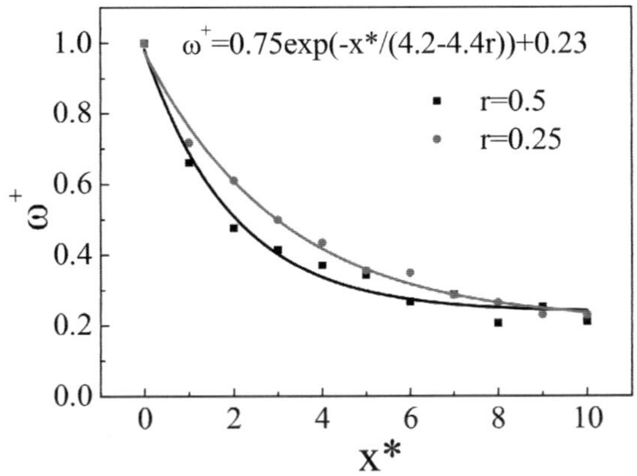

Fig. 6 Distribution of vorticity along the stream-wise(Re=66000)

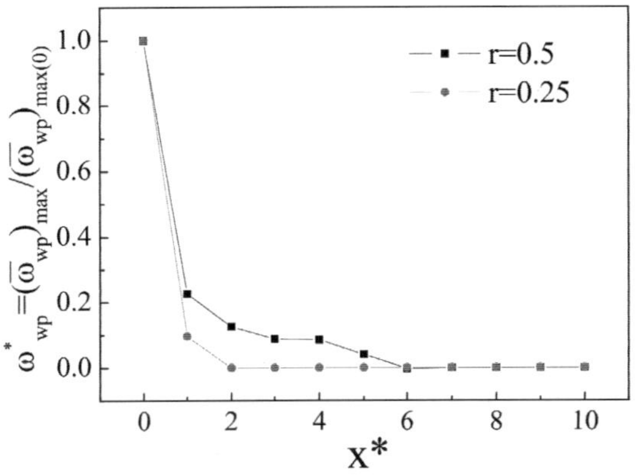

Fig. 7 Distribution of dimensionless vorticity of plate wake

4. CONCLUSION

With two different velocity ratios 0.25 and 0.5, our PIV experiments indicate that the mean Reynolds stress increases with the velocity ratio decreasing on the same cross section and with the distance increasing along the stream wise direction at the same velocity ratio.
The mean vorticity will decrease along the stream wise and the dimensionless mean vorticity ω^+ also decreases along the stream wise according to an exponential law. The decreasing speed is faster when the velocity ratio is larger.
The plate wake makes the mean velocity and vorticity have deficits in a short distance where the mixing layer began merging. The peak value of dimensionless vorticity of plate wake is decreasing sharply to almost zero. The decreasing speed is faster when the velocity ratio is larger. However, the value of this dimensionless vorticity is increasing with the velocity ratio at the same Reynolds number.

ACKNOWLEDGEMENTS

We should thank the financial support of Natural Science Foundation of China (Project No.50536020, 50676079), Program for New Century Excellent Talents in University (NCET-07-0661) and Key Project of Chinese Ministry of Education (No. 107101).

REFERENCES

Azim M. Abdul et al. (2003). “Plane mixing layers from parallel and non-parallel merging of two streams,” Experiments in Fluids, 34(2), pp.220-226.

Bernal L.P., et al. (1986). "Streamwise vortex structure in plane mixing layer" J. Fluid Mech., 170, pp.499-525.

Brown G.L. et al. (1974). "On density effects and large structure in turbulent mixing layers," J. Fluid Mech., 64, pp. 775-816.

Browand F.K. et al. (1976). "Large scales in the developing mixing layer," J. Fluid Mech., 76(1), pp.127-144.

Chih-ming Ho et al. (1982). "Subharmonics and vortex merging in mixing layer," J. Fluid Mech., 119, pp.443-473.

Dimotakis Paule et al. (1976). "The mixing layer at high Reynolds number: large-structure dynamics and entrainment," J. Fluid Mech., 78(3), pp.535-560.

Ford B. et al. (2000). "Ellipsoidal Bubble Diffusion in a Turbulent Shear Layer," Int. J. Multiphase Flow, 26(3), pp.503-516.

Guo Fude et al. (2007). "PIV Experimental Investigation of a Turbulent Mixing Layer (I): Single Phase Flow," 13th China Symposium for Multiphase Flow, pp.164-171.

Lang Daniel Bernard (1985). "Laser Doppler velocity and vorticity measurements in turbulent shear layer," PHD Thesis, California Institute of Technology Pasadena, California, USA.

Loth E. et al. (1995). "Modulation of shear layer thickness due to large bubbles. Int. J. Multiphase Flow," 21(5), pp.919-927.

M. D. SLESSOR et al. (1998). "Turbulent shear-layer mixing at high Reynolds numbers: effects of inflow conditions". Journal of Fluid Mechanics, 376, pp. 115-138.

Mehta R.D. (1991). "Effect of velocity ratio on plane mixing layer development: influence of the splitter plate wake," Experiments in Fluids, 10, pp.194-204.

Naka Yoshitsugu, et al. (2006). "Simultaneous measurements of fluctuating velocity and pressure in a turbulent mixing layer". International Journal of Heat and Fluid Flow, 27(4), pp737-746.

Plesniak M.W. et al. (1993). "Effects of small changes in initial conditions on mixing layer three-dimensionality," Experiments in Fluids, 14(4), pp.286-288.

Wiecek Kevin C. et al. (1998). "Effects of velocity ratio on mixing layer three-dimensionality," Experimental Thermal and Fluid Science, 16, pp.165-176.

Journal of Physics: Conference Series **147** (2009) 012050 doi:10.1088/1742-6596/147/1/012050

PIV Measurement of Turbulent Mixing Layer Flow with Polymer Additives

T. NING, F. GUO, B. CHEN*, X. ZHANG

State Key Laboratory of Multiphase Flow in Power Engineering, Xi'an Jiaotong University, Xi'an 710049, China

chenbin@mail.xjtu.edu.cn

Abstract. Turbulent mixing layer flow with polymer additives was experimentally investigated by PIV in present paper. The velocity ratio between high and low speed is 4:1 and the Reynolds number for pure water case based on the velocity differences of two steams and hydraulic diameter of the channel ranges from 14667~73333. Flow field and turbulent quantities of turbulent mixing layer with 200ppm polymer additives were measured and compared with pure water mixing layer flow. It is shown that the dynamic development of mixing layer is greatly influenced by polymer addictives. The smaller vortices are eliminated and the coherent structure is much clearer. Similar with pure water case, Reynolds stress and vorticity still concentrate in a coniform area of central part of mixing layer and the width will increase with the Reynolds number increasing. However, compared with pure water case, the coniform width of polymer additives case is larger, which means the polymer additives will lead to the diffusion of coherent structure. The peak value of vorticity in different cross section will decrease with the development of mixing layer. Compared with pure water case, the vorticity is larger at the beginning of the mixing layer but decreases faster in the case with polymer additives.

1. INTRODUCTION

After the phenomenon that flow friction of polymer solution was smaller than pure solvent under the unsteady turbulence state had been found by Toms(1948), resistance reduction with polymer additives has been widely applied in many fields, such as ship transportation, petroleum transportation and fluids pipeline conveying because it is very important to save energy and protect ecological environment. After almost 60 years' study on this phenomena, the research direction has changed a lot. At the very beginning, researches were concentrated on the aspect that they conjecture the turbulent influence with polymer additives by studing motion characteristics of polymer molecular in shear flows and elongational flows (Lumley J L, 1969; Rabin Y et al., 1989). Later, Virk P S. (1967) found that onset of drag reduction occurs at a well-defined wall shear stress related to the random-coiling effective diameter of the polymer, and the maximum drag reduction possible is limited by an asymptote that is independent of polymer and pipe diameter. Gordons R J. (1970) showed the effects of turbulent coherent structures with addition of polymer, and he thought that the intensity of bursting process of turbulent boundary layer was weaken with polymer additives. Then, several researchers also put forward their theories such as pseudo-plastic behavior, turbulent fluctuation inhibition, and viscoelasticity, etc(Shao, 2001). However, the conclusion on the mechanism of drag reduction by polymer additive wasn't clear enough.

It is well known that the mixing layer is a classical free shear flow with large eddy coherent structure..

It will be very helpful to study the influence of polymer additives on turbulent fluctuation and coherent structure. Hibberd (1982) measured the mixing layer with polymer additives by LDA, and found that small-scale structures were decreasing and turbulent coherent structures were more clear. Riediger (1989) found that the survival time of coherent structure was longer and the increasing of momentum thickness was inhibited. SHAO (1998) reported the influence of polymer addition on coherent structure and turbulent fluctuation, and found that the incresing of momentum thickness was slower.). However, no quantitative conclusions were reported up to now.

In present paper Particle Image Velocimetry (PIV) will be used to measure the distribution of velocity, Reynolds stress and average vorticity to get the quantitative information of the turbulent mixing layer with polymer additives and compared with pure water case to study the drag reduction mechanism by polymer additives in turbulent flow.

2. EXPERIMENTAL EQUIPMENT

All the tests were done in the vertical channal of mixing layer experimental system. By using a special designed splitter, a vertical inflow was seperated into two streams of flow with designed different velocity to form the mixing layer at the end of the splitter (Guo F D. et al., 2007). In order to visualize the mixing layer, a PIV system supplied by La Vision GmbH is adopted. The PIV system integrates a double pulsed, 15 Hz, 532 nm, 200 mJ Nd-Yag lasers including sheet optics to generate a light sheet from the laser beam. The focus and the thickness of the laser sheet can be adjusted via the laser-sheet optics and a thickness of approximately 1mm is used throughout the experiment. Silver-coated hollow glass balls with the diameter size distribution of 8-12μm and a density of 1.1g/cm^3 are used as tracer particles. Each illumination of the measurement area is recorded using two (2D) ImagerPro4M CCD cameras (8 bit, 15 Hz) with a resolution of 2048 × 2048 pixels.

In present paper, the velocity ratio between high and low speed is 4:1 and the Reynolds number for pure water case based on the velocity difference of two steams and hydraulic diameter of the channel ranges from 14667~73333. Odium salicylate (NaSal) was added into the pure water and the concentration is 200ppm. The typical test conditions were shown in Tab 1.

Tab 1. Test conditions

u_h (m/s)	u_l (m/s)	$\Delta u = u_h - u_l$ (m/s)	Water $\mathrm{Re} = (\Delta u d)/v$
0.33333	0.08333	0.25	14667
0.66667	0.16667	0.50	29333
1.33333	0.33333	1.00	58667
1.66667	0.41667	1.25	73333

3. RESULTS AND DISCUSSION

In this section, we analyze the typical test condition in Tab 1 and discuss the variations of turbulent quantities. In the following figures, the origin is the starting point of the mixing layer (centre of the splitting plate end). The x axis is along the downstrem direction, and y axis is from the low speed side to the high speed side along the vertical flow direction. x^* and y^* are the non-dimensional length normalized by half width of the channel.

3.1 Velocity

At first, we checked the average velocity profiles along the dowmstream direction in different cross-section of the mixing layer shown in Fig. 1(Reynolds number of pure water case is 73333). It can be seen that the velocity ratio between high speed side and low speed side is 4:1. Besides, it was well consistent with the designed ratio for both of the pure water case and polymer addition case. Through the comparison between the data of two cases, we can find that the wake caused by the splitter plate are stronger for the case with polymer additives.

Fig. 2 shows the velocity vector distribution in the mixing layer which is given by $U = u - (\overline{u}_h + \overline{u}_l)/2$, and the velocity difference between high speed and low speed is 0.5m/s. Apparently, we can see the coherent structures in both of two conditions. Compared with pure water case, the vortex structure of the polymer case was enhanced, so it was more clear and larger. Further more, the large eddy also rolled up earlier. With the increasement of velocity difference, these influence is more obvious. And there are similar results for other conditions with different Reynolds numbers.

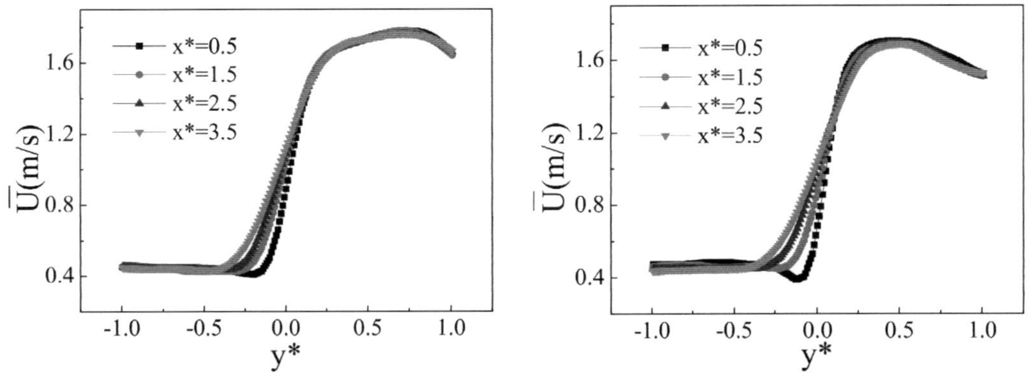

(a) Pure water (b) with polymer additives

Fig. 1 Profile of the stream wise velocity in different cross-section

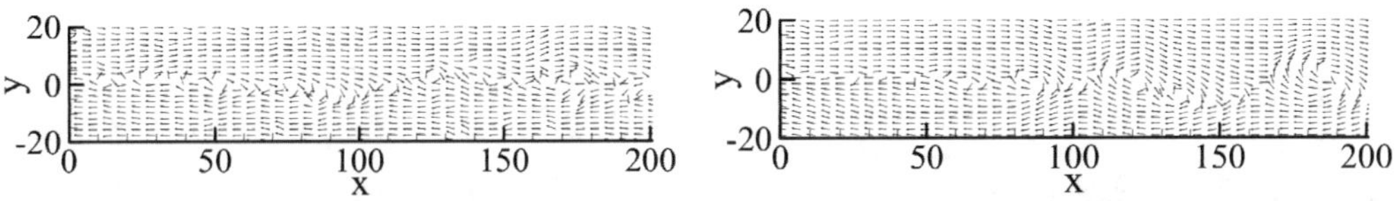

(a) Pure water (b) With polymer additives

Fig. 2 Downstream velocity vector (Δu=0.5 m/s)

3.2 Reynolds Shear Stress

Fig. 3 gives the Reynolds shear stress $-\overline{u'v'}$ with various velocity differences in the mixing layers. It is obvious that Reynolds shear stress with polymer additives appeares a peak value near the central line of the mixing layer. Moreover, similar with pure water case, it was concentrating in a coniform area of central part of mixing layer and the width of this area will increase with increasing of velocity

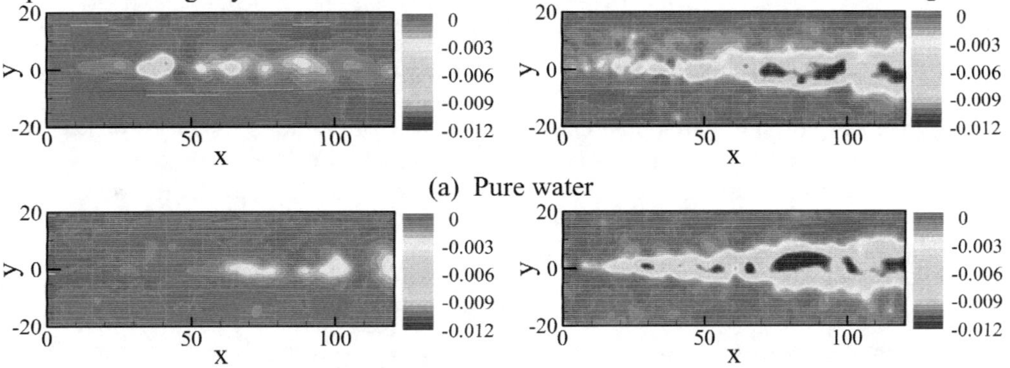

(a) Pure water

(b) Polymer additives

Δu=0.5 m/s Δu=1.25 m/s

Fig. 3 Reynolds shear stress field

difference. Further more, Reynolds shear stress increases along the down stream direction but decreases with the increasement of Reynolds number, and there is a trend that it is developing to the low speed side. Compared with pure water case, the coniform width of polymer additives case is larger, which means the polymer additives will lead to the diffusion of coherent structure. These rules can be reflected more directly in Fig. 4. Profile of Reynolds shear stress between pure water and polymer at the same cross-section and the same velocity difference can be compared in this picture. It was clear that the coniform width and the value for polymer additives case are larger.

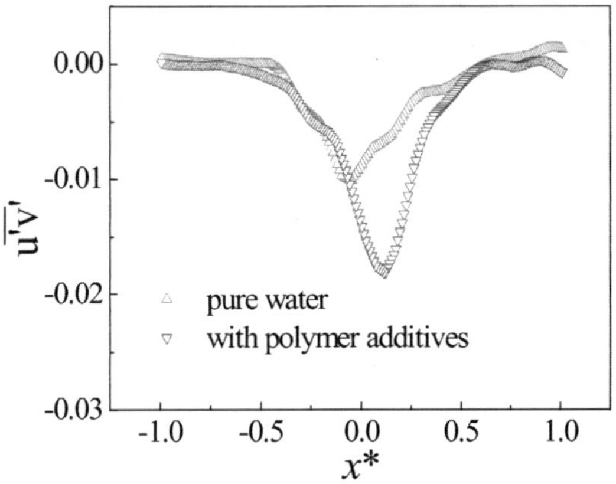

Fig. 4 Profile of Reynolds shear stress (Δu=1.25 m/s , x*=1)

3.3 Vorticity

The flow field distribution of average vorticity was shown in Fig. 5. It indicates that the average vorticity in the flow field increases with the velocity difference and Reynolds number increasing, while it decreases with the development of the mixing layer along the streamwise. The peak value of average vorticity concentrated on the central area of mixing layer along streamwise. Different with pure water case, coniform width of polymer additives case is larger, and the value of vorticity is greater, too.

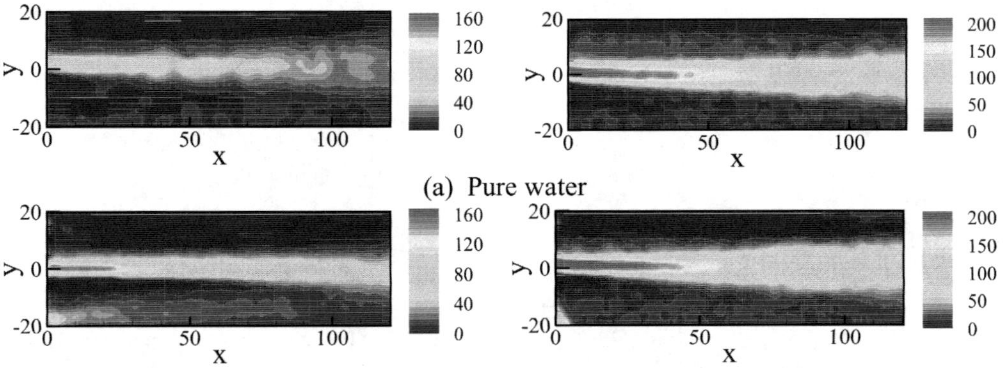

(a) Pure water

(b) Polymer additives

Δu=0.5 m/s Δu=1.25 m/s

Fig. 5 Average vorticity field

To reveal the average vorticity variations more clearly, the peak value of average vorticity along the main flow direction was given in Fig. 6. For the pure water case, the average vorticity decreases directly. But in the polymer case, the peak value of average vorticity increases at the very beginning of the mixing layer, and then it decreases. Compared with pure water case, the average vorticity is larger at the beginning of the mixing layer but decreases faster in the case with polymer additives. It means that attenuation rate of polymer case is higher. The reason is that the viscosity of polymer fluid is greater than pure water, which will result in the reduction of velocity fluctuation in y direction. Then, the velocity gradient in flow field will decrease and there will be a faster decay of vorticity in polymer case. For the other velocity difference not shown here, the peak value of average vorticity are similar with Fig 6.

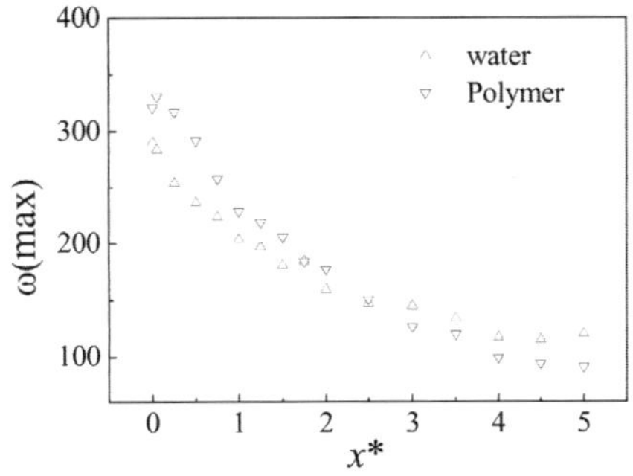

Fig. 6 Peak value of average vorticity (Δ u=1.25 m/s)

4. CONCLUSIONS

In order to reveal the mechanism of drag reduction by polymer addition, turbulent mixing layer flow with polymer additives was experimentally investigated by PIV in present paper. Flow field and turbulent quantities of turbulent mixing layer with 200ppm polymer additives were measured and compared with pure water mixing layer flow. Our experiment indicates that large coherent vortex structures existe in both of pure water case and polymer additives case, and they are enhanced when polymer was add to the water. We found that Reynolds stress and vorticity are concentrated in a coniform area of central part of mixing layer and the width of peak value area will increase with increasing of velocity difference. And values of vorticity and Reynolds stress are increased when polymer was added into the water. The peak value of average vorticiy of polymer additives case decreased faster than pure water one.

ACKNOWLEDGEMENTS

The author would like to acknowledge the joint support of National Natural Science Foundation of China (Grant No. 50676079), Program for New Century Excellent Talents in University (NCET-07-0661) and Key Project of Chinese Ministry of Education (No. 107101).

REFERENCES

Toms BA. (1948). “Some Observation on the Flow of Linear Polymer Solutions through Straight Tubes at Large Reynolds Numbers,” Proc. 1st Int. Cong. Rheol., North Holland, Amsterdam, pp. 135-141

Virk P S. et al. (1967). “The Toms phenomenon: turbulent pipe flow of dilute polymer solutions,” J. Fluid Mech., 30. 305

Lumley J L. (1969). “Drag reduction by additives,” Ann. Rev. Fluid Mech., 1. 367

Gordons R J. (1970). “Mechanism for turbulent drag reduction in dilute polymer solutions,” Nature, 227. 599

Hibberd M, Kwade M, Scharf R. (1982). “Influence of drag reducing additives on the structure of turbulence in mixing layer,” Rheologica Acta., 21(4-5), pp. 582-586

Rabin Y, Zielinska B J A. (1989). “Scale development enhancement and damping of vorticity disturbance by polymers in elongational flow,” Phy. Rev. Lett., 63. 512

Riediger S. (1989). “Drag Reduction in Fluid Flow,” Chichester: Ellis Horwood, pp. 303-310

Shao Xueming et al. (1998). "Experimental Research for the Influence of Polymer Additives on the Characteristics of Mixing Layer Flow," Journal of Experimental Mechanics, 13(4), pp. 520-525

Guo Fude et al. (2007). "PIV Experimental Investigation of a Turbulent Mixing Layer (I): Single Phase Flow, " 13th China Symposium for Multiphase Flow. Daqing China, pp. 164-171

Jiao Lifang et al. (2008). "Review of Turbulent Drag-Reducing Flow and Heat Transfer with Additives," Advances in Mechanics, 38(3), pp. 339-357.

Journal of Physics: Conference Series **147** (2009) 012051 doi:10.1088/1742-6596/147/1/012051

Measurement about the Interaction between Two Falling Droplets

N Ninomiya [1], **Y Takanashi** [2]

[1] Utsunomiya University, 7-1-2 Yoto, Utsunomiya, 321-8585 JAPAN
[2] Yamagata Fujitsu Ltd., Yamagata, JAPAN

nino@utmu.jp

Abstract. Although the phenomena related to the multiphase flow can be found in many kinds of industrial and engineering applications, the physical mechanism of the multiphase flow has not been investigated in detail. The major reason for the lack of data in the multiphase flow lies in the difficulties in measuring the flow quantities of the multiple phases simultaneously. The difference in the refractive indices makes the visualization in the vicinity of the boundary of the multiple phases almost impossible. In this study, the refractive index of the aqueous phase has been equalized to that of the oil phase by adjusting the concentration of aqueous solution. Presently, the simultaneous visualization and the PIV measurement have been carried out about the both phases of the liquid-liquid two-phase flow. The measurement has been carried out for the flow field around and inside of two falling droplets interacting each other while they travel.

1. INTRODUCTION

Multiphase flow has been one of the most difficult and, at the same time, one of the most practically important phenomena in the fluid mechanics. Many studies for the multiphase flow have been conducted in order to investigate the details about the momentum and material transfer at the interface. Numerical studies, such as Dandy and Leal (1989), have revealed many kinds of features about the both phases. Even though the measurements about the surrounding fluid, such as Cieslinski et al. (2005), have been extensively carried out, only a few results have been reported about the flow near the boundary and the flow inside of a bubble or a droplet. This is mainly because the discontinuity of the medium at the interface of the multiphase flow makes the measurement or even the visualization almost impossible.

Recent developments in the visualization technique and the improvements in the imaging devices make the great progress in the measurement techniques. Especially, the success by the particle image velocimetry (PIV) and by the particle tracking velocimetry (PTV) is outstanding. Nevertheless, the PIV or PTV measurements in the multiphase flow are still limited to the examination of outer phases. The reason of the lack of data for the inner phases is because the visualization itself is still very difficult with the refraction at the interface of the multiphase flow. Yamauchi et al. (2000) have succeeded in measuring the flow field inside of the water droplet falling through the stationary oil by compensating the complex distortion caused by the refraction at the boundary of a droplet assuming that the droplet is perfectly spherical and Eötvös number and Reynolds number are both small. Ninomiya and Yasuda (2006) have carried out the flow visualization and the PIV measurements of the flow around and inside of a falling droplet simultaneously under various flow conditions by using the index matching technique, as was used by Knapp and Bertrand (2005). Presently, the flow field around and inside of two falling droplets interacting each other has been extensively examined by PIV measurement with the moving camera. As a result, many features about the flow related to the two falling droplets have been investigated by the results of the PIV measurements.

2. MEASUREMENT PROCEDURE

In order to visualize the flow around and inside of a droplet simultaneously, the refractive indices of the inner and the outer phases should be matched. As the refractive index of silicon oil at the room temperature is 1.3999, whereas that of water is 1.3330, the refractive index of water phase might be matched by introducing some miscible substance of higher refractive index. In this study, the glycerol, whose refractive index is 1.4716 and is also immiscible to silicon oil, is chosen for the refractive index matching substance.

Figure 1 shows the change of the refractive indices of two grades of the silicon oil and of the glycerol solution of various concentrations. It is obvious that the glycerol solution of higher concentration takes higher refractive index. The refractive index of ordinary liquid gets smaller with higher temperature, but its temperature sensitivity is different with liquid. Thus the refractive index of the silicon oil can be matched with that of glycerol solution of adequate concentration and temperature. Presently, 52 % glycerol solution is chosen so that its refractive index matches that of 20 cSt silicon oil, i.e., 1.4026, at room temperature of 25 deg.

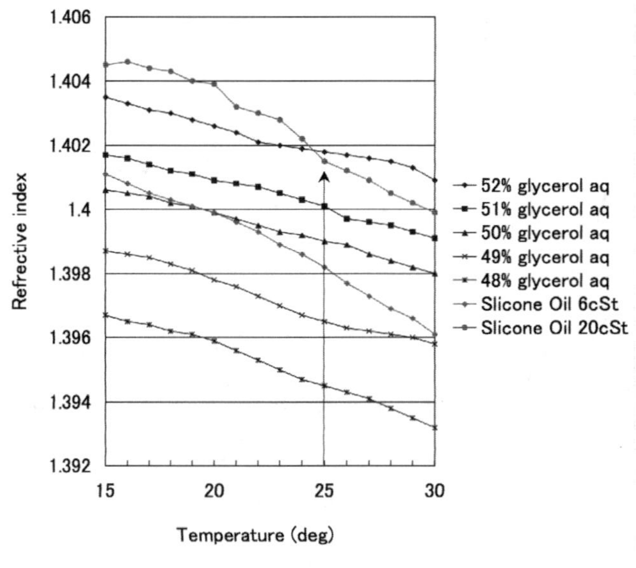

Fig. 1 Refractive indices of fluids

Table 1. Physical properties of fluids at 25 deg

	Silicon Oil 20 cSt	52% Glycerol aq.
refractive index	1.4026	1.4026
viscosity	21.8 m Pa s	6.70 m Pa s
specific density	0.925	1.130
tracer particle	Polyethylene ρ=0.95,15μm	Nylon 6 ρ=1.13, 20μm

The physical properties of the working fluids used in this study are summarized in Table 1. As for the oil phase, silicon oil of 20 cSt is chosen so that the terminal velocity of the falling droplets comes into the appropriate range for the PIV measurement. As shown in Table 1, the refractive indices of 52 % aqueous solution of glycerol and silicon oil of 20 cSt are exactly matched. As for the tracer particle for each phases, nylon 6 and polyethylene is selected whose specific densities are almost the same as those of fluids. Thus, the motions of each fluid can be visualized properly.

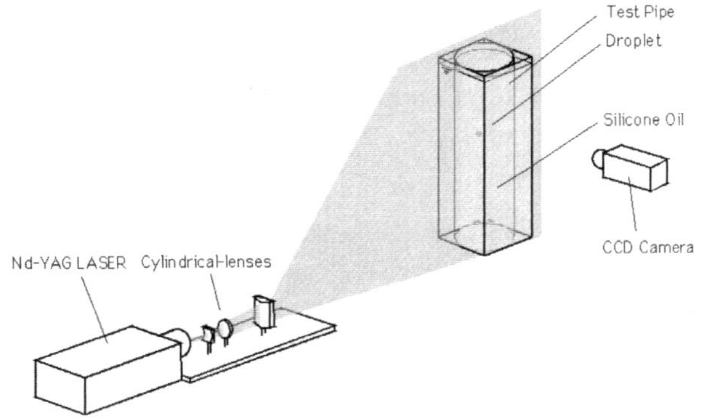

Fig. 2 Experimental apparatus

Figure 2 shows the schematic view of experimental apparatus. Silicon oil of 20 cSt is filled in the reservoir of 100 x 100 x 500 mm, inside of which thin cylindrical wall of Plexiglas is placed in order to avoid the effect of corner of the container. To illuminate the tracer particles, double-pulsed Nd-YAG laser is used and the beam is expanded by the cylindrical lens and then focused into thin sheet by the sheet forming optics as shown in Fig. 2. The positions of the droplet injecting nozzles are carefully aligned with the laser light sheet so that the exact center plane of the two droplets can be visualized.

As the refractive indices of both phases are perfectly matched, the boundary of the droplet is no more visible. Thus, in this study, the droplet is slightly colored by Rhodamine B so that the shape of the droplet can be distinguished by the laser-induced fluorescence (LIF).

The image of the droplet is captured by moving CCD camera that has the resolution of 1376 x 1040 at the frame rate of 30 fps and then the images are directly transferred to PC. The CCD camera is mounted on the slider that travels at the constant speed in order to capture the images of the flow field around and inside of the falling droplets as long as possible. The terminal velocities of the two falling droplets have been measured in the preliminary experiment to determine the traveling speed of the slider.

Before measuring the velocity field around and inside of a falling droplet, the size and the position of the droplet are calculated from the images. As the droplet is slightly colored by the fluorescent dye, the droplet can be identified as the slightly bright region. In order to eliminate the images of the tracer particles, the smoothing is applied to the original image. Then the derivative of the brightness is taken in order to enhance the outline of the droplet. Finally, by fitting the equation of ellipse to the outline by the least square method, the center and the radii in major and minor axis of the droplet are obtained.

The typical size of a droplet, which is almost spherical under the condition of present study, is about 4.5 mm in diameter. The terminal velocity is calculated from the displacement and the time interval and then the non-dimensional parameters, such as the aspect ratio of the droplet, drag coefficient C_D, Reynolds number Re, Morton number Mo and Eötvös number Eo or Bond number Bo. The typical terminal velocity of a droplet is 66 mm/s and this value ends up with Re = 12.9 and C_D = 2.56. According to the Grace's diagram, excerpted from Clift et al. (1978), the present droplet is plotted in "spherical" zone and the measured aspect ratio of 0.95 agrees well with the diagram.

3. PIV RESULTS

Even though many kinds of feature about the two-phase flow can be examined by the index matching technique, presently the attention has been paid upon the interaction between two falling droplets at the same height. Similar works have been done by Kitagawa et al. (2004) and by Murai et al. (2006), but the flow field inside of the dispersed phase has not been investigated. In this study, two water droplets have been simultaneously injected into Silicone oil from two nozzles of some horizontal distances. This distance has been changed to investigate the details of the effect of the initial distance to the interaction between two droplets. In order to investigate the details of the interaction between two droplets at the same height, the terminal velocities of two droplets have been measured in the preliminary tests so that the CCD camera mounted on a slider can capture the motions of two droplets as long as possible.

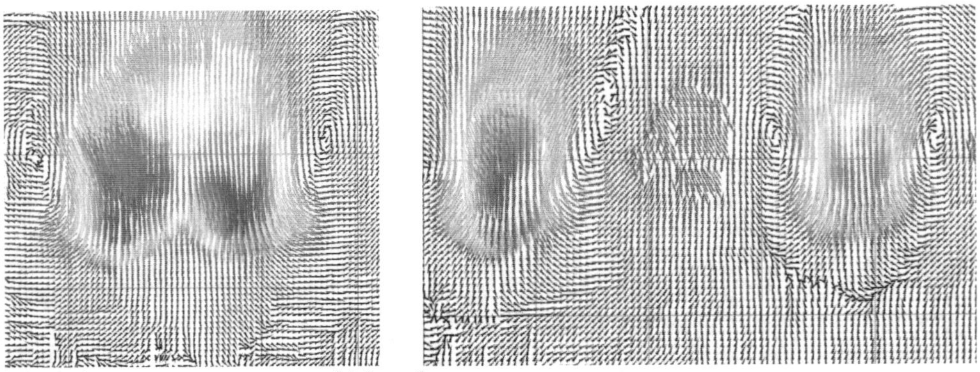

(a) Lx = 4 mm (b) Lx = 9 mm

Fig. 3 PIV results (flow at fixed coordinates)

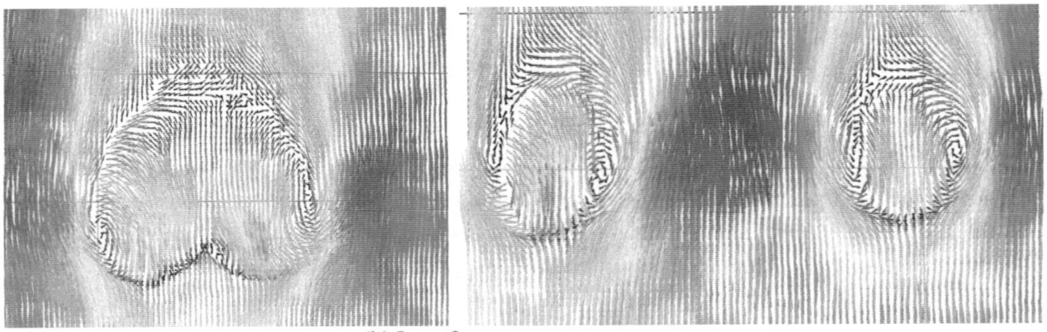

(a) Lx = 4 mm (b) Lx = 9 mm

Fig. 4 PIV results (flow relative to droplets)

The typical results of the PIV measurements in this study are shown in Figs. 3 to 6. These results are for the cases that the initial horizontal distance between two droplets is Lx = 4 mm for case (a) and is Lx = 9 mm for case (b). Figures 3 and 4 designate the flow field around and inside of two falling droplets at their early stage, which is just after the injection of the droplets. Figures 3(a) and 3(b) show the flow field at the fixed coordinates. The strong downward motions of two droplets are evident and the fluids behind the droplets are drawn by the droplets. It is also seen that the fluids in front of the droplets are pushed aside by the droplets when they penetrate the surrounding fluid. As for the case (b) of large initial distance, the two droplets seem to travel almost independently and the upward motion, which may be the result of superposition of the effects of two droplets, is found in the midst of the two droplets. On the contrary, as for the case (a) of small initial distance, the fluid between two droplets is drawn by the two droplets and moves downward.

Figures 4 are the plot of the flow relative to the droplets. The average of the terminal velocities of two droplets is subtracted from the results in Figs. 3 to highlight the motion inside of the droplets. As for the case (b), the almost symmetric two counter-rotating vortices are seen inside of both of the two droplets. This implies that there is only very little interaction between two droplets for the case (b) of large initial distance. But for the case (a) of small initial distance, as the fluid between two droplets is pulled by two droplets and travels with them, the shear on the facing side of the droplet gets very weak and thus rotating motion inside of the droplet is not seen on this side of two droplets. It is also seen that the wakes of two droplets merge into single motion, which may look like a wake of a single object. But it should be noticed that the two droplets themselves do not merge into single under this condition.

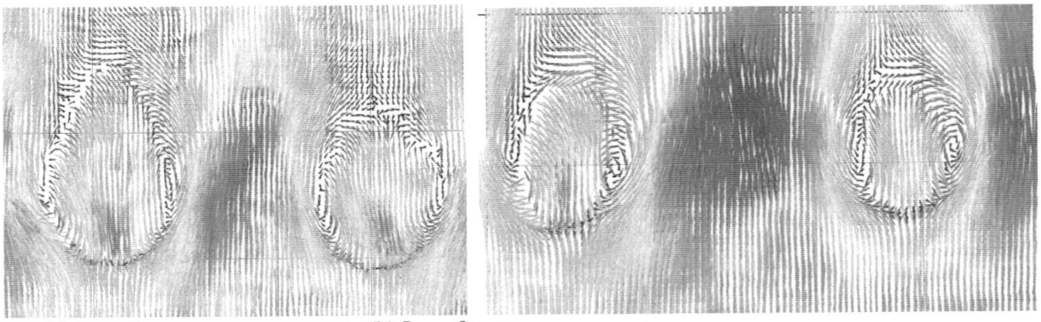

(a) Lx = 4 mm (b) Lx = 9 mm

Fig. 5 PIV results (flow relative to droplets)

Figures 5 shows the flow fields at the vertical position of 90 mm from the nozzle. Beyond the vertical position of 50 mm from the nozzle, the velocities of the falling droplets reach their terminal velocities of 66 mm/s. Here again, Figs. 5 are the results in the moving coordinates. The horizontal distances between two droplets have been increased to 10.4 mm for case (a) and 13.6 mm for case (b). The increase for the small initial distance case (a) is much larger as the interaction between two droplets is very strong in the early stage. Now the surrounding fluid flows into the gap of two droplets and the shear on the both side of the droplets produces the counter-rotating vortices inside of the droplets. Even though that a slight asymmetry is seen for the counter-rotating vortices and a little outward motions inside of the droplets are found for case (a), now the flow fields inside of two droplets are almost symmetric for case (b).

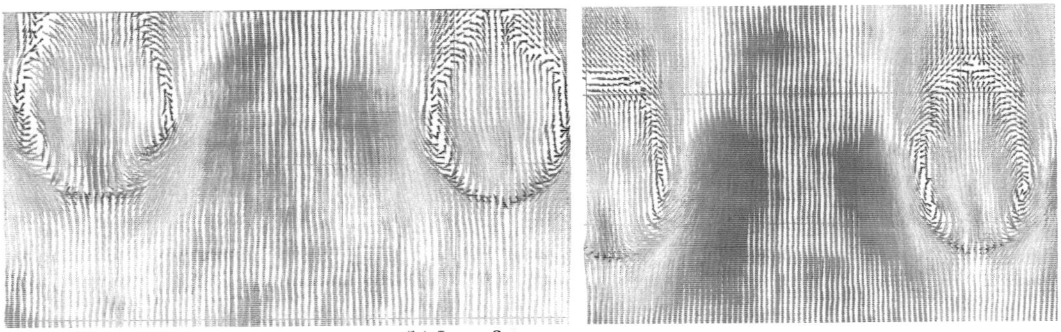

(a) Lx = 4 mm (b) Lx = 9 mm

Fig. 6 PIV results (flow relative to droplets)

In Figs. 6, the flow fields around and inside of two falling droplets when the two droplets travel far downstream to 200 mm are shown. The horizontal distances are 12.8 mm for case (a) and 14.4 mm for case (b). Though there still are a little bit of increases in horizontal distances for both cases, the flow field looks like a superposition of two flow fields of single falling droplet. This is evident from the facts that the counter-rotating vortices inside of a droplet is symmetric and two shear layers around a droplet on the facing side of each droplets do not collapse and flows almost independently.

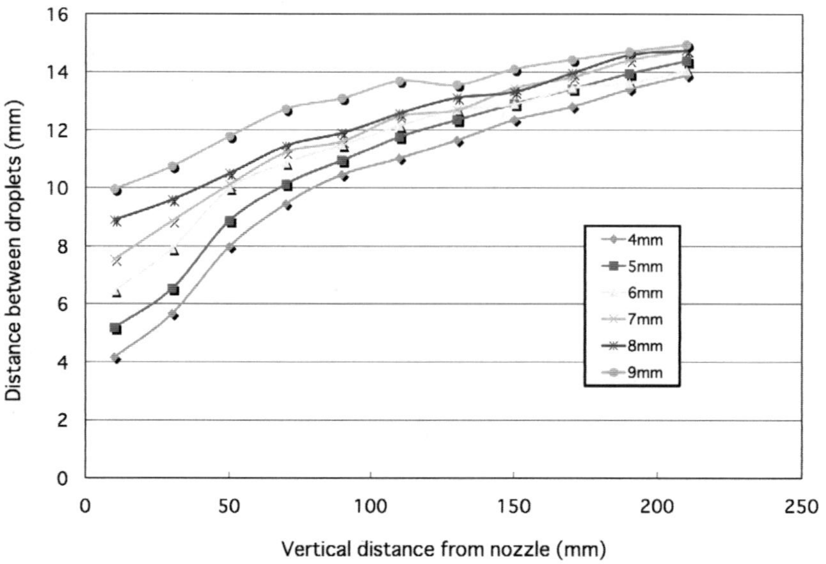

Fig. 7 PIV results (flow relative to droplets)

Figure 7 shows the plots of the horizontal distances between two droplets while they travel for the cases of different initial distances between two droplets. For any initial distance between two droplets, the distances between two droplets increase monotonously. It is also obvious that the smaller the initial distance, the faster two droplets separate. The rate of increase of the distance between two droplets is very large before they reach their terminal velocities at 50 mm downstream and gets quite small after the distance between droplets is larger than 14 mm. It is quite straightforward to conclude that the closer the initial distances, the stronger the interaction between two falling droplets.

3. CONCLUSIONS

In order to investigate the details of the flow around and inside of the falling droplets simultaneously, the refractive index of the water phase is perfectly matched to that of the oil phase and thus the refraction at the boundary of water droplets falling in oil is completely avoided. By introducing the tracer particles, which are neutrally buoyant to each phases, and by slightly coloring the droplet by the fluorescent dye, the flow around and inside of the falling droplets and the positions of the droplets are clearly visualized. As a result of the PIV measurement using this index matching technique, the details of the flow fields around and inside of the two falling droplets injected at the same height have been investigated in detail.

Thus, the followings are the conclusions of this study:

1) The index matching technique is very effective in visualizing the both phases of the multiphase flow. As for the PIV measurement, the difference in the background brightness of the slightly LIF-colored droplets and the non-colored surrounding fluid helps in distinguishing flow fields in each phases.

2) When the droplets fall, the fluids in front of the droplets are pushed aside and then coincide in the midst of two droplets. This may be the driving force of the separating motions of two droplets. But, for the case of small initial distance 4 mm, at the early stage after injection the fluid cannot penetrates

into the gap of two droplets and thus the two droplets fall as though the lump of two droplets is a single object.

3) When the falling speeds of droplets reach their terminal velocities at the vertical distance beyond 50 mm, even though slight asymmetry is still observed for the counter-rotating vortices inside of droplets and the droplets keep on leaving from each other, the interaction between two droplets gets very weak and the flow field looks like a superposition of two flow fields of single falling droplet.

4) According to the plots of the horizontal distances between two droplets, the smaller the initial distance between two droplets, the faster the two droplets separate. The rates of increase of the distances between two droplets are larger before they reach their terminal velocities.

ACKNOWLEDGEMENTS

The authors greatly acknowledge the support by the Venture Business Laboratory of Utsunomiya University.

REFERENCES

Dandy, D. S. and Leal, L. G. (1989). "Buoyancy-driven motion of a deformable drop through a quiescent liquid at intermediate Reynolds numbers", Journal of Fluid Mechanics, 208, pp. 161-192.

Cieslinski, J. T., Polewski, J. and Szymczyk, J. A. (2005). "Flow field around growing and rising vapour bubble by PIV measurement", Journal of Visualization, 8(3), pp. 209-216.

Yamauchi, M., Uemura, T. and Ozawa, M. (2000). "Velocity distributions inside and outside of a water drop in oil", Application of Laser Techniques to Fluid Mechanics, CD-ROM, pp. 1-6.

Ninomiya, N. and Yasuda, K. (2006). "Visualization and PIV measurement of the flow around and inside of a falling droplet", Journal of Visualization, 9(3), pp. 257-264.

Knapp, Y. and Bertrand, E. (2005). "Particle imaging velocimetry measurements in a heart simulator", Journal of Visualization, 8(3), pp. 217-224.

Clift, R., Grace, J. R. and Weber, M. E. (1978). Bubbles, Drops and Particles, Academic Press.

Murai, Y., Qu, J. and Yamamoto, F. (2006). "Three-Dimensional Interaction of Bubbles at Intermediate Reynolds Numbers", Multiphase Science and Technology, 18(2), pp. 175-197.

Kitagawa, A., Sugiyama, K., and Murai, Y. (2004). "Experimental Detection of Bubble-Bubble Interactions in a Wall-Sliding Bubble Swarm", International Journal of Multiphase Flow, 30(10), pp. 1213-1234.

Rigorous description of holograms of particles illuminated by an astigmatic elliptical Gaussian beam

Y. J. Yuan, K. F. Ren, S. Coëtmellec and D. Lebrun

UMR 6614/CORIA, CNRS & Université et INSA de Rouen
Avenue de l'Université BP 12,
76801 Saint Etienne du Rouvray, France

email : fang.ren@coria.fr

Abstract The digital holography is a non-intrusive optical metrology and well adapted for the measurement of the size and velocity field of particles in the spray of a fluid. The simplified model of an opaque disk is often used in the treatment of the diagrams and therefore the refraction and the third dimension diffraction of the particle are not taken into account. We present in this paper a rigorous description of the holographic diagrams and evaluate the effects of the refraction and the third dimension diffraction by comparison to the opaque disk model. It is found that the effects are important when the real part of the refractive index is near unity or the imaginary part is non zero but small.

1. INTRODUCTION

It is essential for the laboratory research to study the combustion engine in an automobile or aircraft and industrial processes in order to improve the efficiency and reduce the pollutant emission. The digital holography is a non-intrusive optical method well adapted for the measurement of the size and 3D velocity fields of particles in a spray of fluid.

The scalar theory of diffraction is often used to describe the recording and reconstruction of particle holograms. The mathematical tools like wavelet transformation or fractional Fourier transformation give a useful tool to perform the reconstruction. However, the particle is often considered as an opaque disk and the laser beam used is non-polarized (Nicolas et al, 2005; Nicolas, 2007). As a result, the effects of the third dimension diffraction and the polarization of the incident beam are not taken into account.

A study based on a rigorous theoretical tool – Generalized Lorenz-Mie Theory (GLMT) (Goesbet et al 1988; Ren 1995) has been conducted to predict the holographic image of a spherical particle illuminated by different forms of laser beam – plane wave, circular Gaussian beam and astigmatic elliptical Gaussian beam (AEG beam). The influences of the third dimension diffraction, the complex refractive index, the position of the particle, and the polarization and the nature of the incident beamare evaluated by comparison between the holograms obtained by this method and the images of an opaque disk predicted by diffraction theory.

The paper is organized as follows. Section 2 recalls the principles of the two models used in the paper to describe the hologram – opaque disk and rigorous electromagnetic scattering of a

Journal of Physics: Conference Series **147** (2009) 012052 doi:10.1088/1742-6596/147/1/012052

homogeneus sphere. The simulated results by the two models are presented and compared in section 3 to evaluate the effects of the refraction and the third dimension difraction. Section 4 is devoted to conclusions.

2. PRINCIPLES OF HOLOGRAM SIMULATION

2.1 Experimental setup of digital in-line holography

The system of digital holography under study is composed of a laser, a beam collimator and a camera, as shown in Fig. 1. A particle located at a distance z from the camera is illuminated by the laser beam and the hologram is recorded by the camera. To obtain a collimated beam, a lens of focal length $f_\xi = f_\eta = 200$ mm is added between the laser and the particle. A astigmatic elliptical Gaussian beam is obtained by a cylindrical lens of focal length $f_\varepsilon \to \infty$ and $f_\eta = 200$ mm. The obtained beam has a diameter of $2w_0 \approx 14$mm.

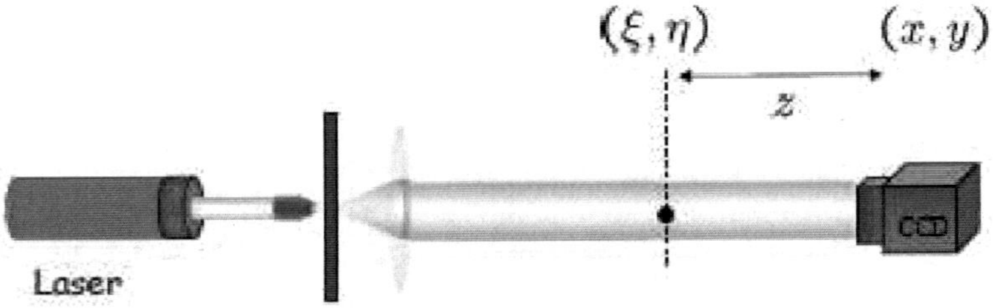

Fig. 1. Experimental setup for in-line hologram recording.

2.2 Model of opaque disk - scalar theory of diffraction

Consider an opaque disk of diameter d located at the center of the plane (ξ,η) with a distance z from the CCD camera as shown in Fig. 1. According to the Huygens-Fresnel integral, the amplitude of the electromagnetic field in the (x,y) plane is given by :

$$A_z(x,y) = \frac{\exp\left(i\frac{2\pi}{\lambda}z\right)}{i\lambda z}\int_{-\infty}^{\infty}\int_{-\infty}^{\infty} E(\xi,\eta)\left[1-\text{circ}(\xi,\eta)\right]\exp\left\{i\frac{\pi}{\lambda z}\left[(\xi-x)^2+(\eta-y)^2\right]\right\}d\xi d\eta \quad (1)$$

where $E(\xi,\eta)$ is the scalar field of the incident beam in the (ξ,η) plane, λ is the wavelength and $\text{circ}(\xi,\eta)$ the transmittance function in the object plane defined by :

$$\text{circ}(\xi,\eta) = \begin{cases} 1 & \text{if} \quad \sqrt{\xi^2+\eta^2} < d/2. \\ 0 & \text{else.} \end{cases} \quad (2)$$

In the cases of circular Gaussian beam and astigmatic elliptical Gaussian beam on-axis illumination, the integration (1) can be carried out analytically (see part 4.2 in Nicolas 2007). For example, when the disk is illuminated by a plane wave, the amplitude in the image plane is given by :

$$A_z = \frac{\exp\left(i\frac{2\pi}{\lambda}z\right)}{i\lambda z}\int_{-\infty}^{\infty}\int_{-\infty}^{\infty}\left[1-\text{circ}(\xi,\eta)\right]\exp\left\{i\frac{\pi}{\lambda z}\left[(\xi-x)^2+(\eta-y)^2\right]\right\}d\xi d\eta \quad (3)$$

In far-field case ($\lambda z \gg 1$), the intensity distribution on the image plane can be calculated by:

$$I(x,y)=A_z(x,y)A_z^*(x,y)\simeq 1-\frac{2}{\lambda z}\sin\left[\frac{\pi}{\lambda z}(x^2+y^2)\right]F(x,y) \tag{4}$$

where

$$F(x,y)=\frac{\pi d^2}{2}\frac{J_1\left(\frac{\pi d\sqrt{x^2+y^2}}{\lambda z}\right)}{\frac{\pi d\sqrt{x^2+y^2}}{\lambda z}} \tag{5}$$

J_1 is the first kind Bessel function of the first order. In the more general cases the analytical expressions given by F. Nicolas (Nicolas, 2007) are used to predict the holographical diagrams of an opaque disk.

2.3 Model of sphere – Generalized Lorenz - Mie theory

The generalized Lorenz-Mie theory (GLMT) is rigorous theory to describe the scattering of shaped beam by a spherical, homogeneous and isotropic particle(Gouesbet et al 1988; Ren, 1995). All the particle parameters and the incident beam characteristics, such as refractive index, position of particle and polarization of incident beam, are taken into account. The particle can be located at any point in the beam (Ren, 1995; Ren et al, 1998).

In the framework of GLMT, the holographic diagram of a sphere is obtained by the superposition of the indent field and the scattered one (Ren, 1995; Girasole et al, 2000). The total intensity on the image plane is given by

$$I_{total}=\left|E_{total}\right|^2 \tag{6}$$

where the total electric field is the summation of the incident beam and the scattered wave: $E_{total}=E_{incident}+E_{scattered}$.

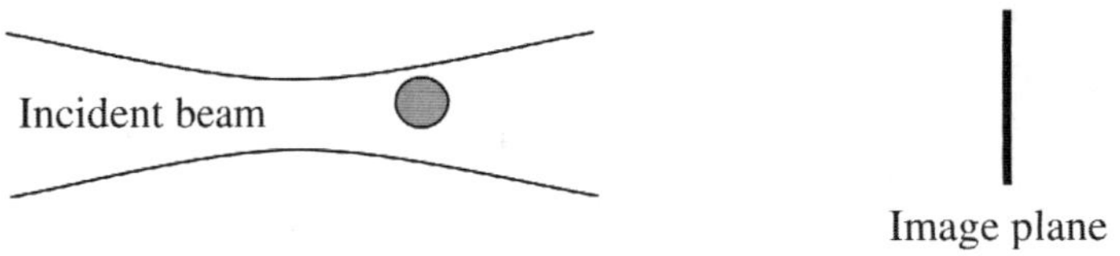

Fig. 2. Geometry of scattering problem by GLMT

3. SIMULATION DIAGRAMS AND DISCUSSION

We present in this section the holographic diagrams predicted by the two methods. By comparison of the diagrams we investigate the effects of the refraction and the third dimension diffraction of the particle according to its optical properies such as the refractive index of the particle, the polarization and the form of the beam.

3.1 Holograms of different beam

3.1.1 Case of a plane wave

Suppose an opaque disk of diameter d = 100μm located at a distance z = 80 mm from the CCD camera is illuminated by a plane wave of wavelength λ=632.8nm. The intensity distribution predicted by the opaque disk model is illustrated in Fig. 3. The images simulated by the GLMT as function of complex refractive index are presented in Fig. 4-5.

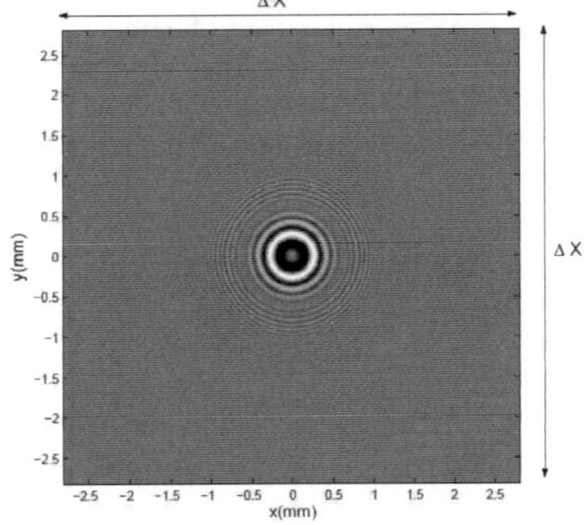

Fig. 3. Intensity distribution observed at a distance z = 80mm of an opaque disk of diameter d = 100µm illuminated by a plane wave of wavelength λ=632.8nm.

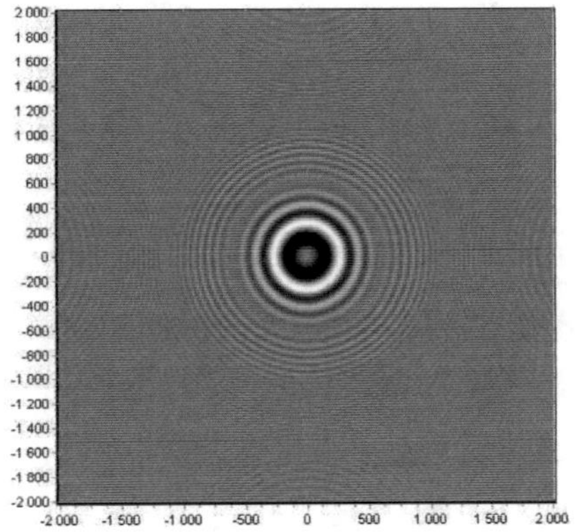

Fig. 4. Holographic images of a sphere simulated by the GLMT with the refractive index m= (1.33, 0) or m= (1.33, 0.1), in the same configuration as Fig. 3.

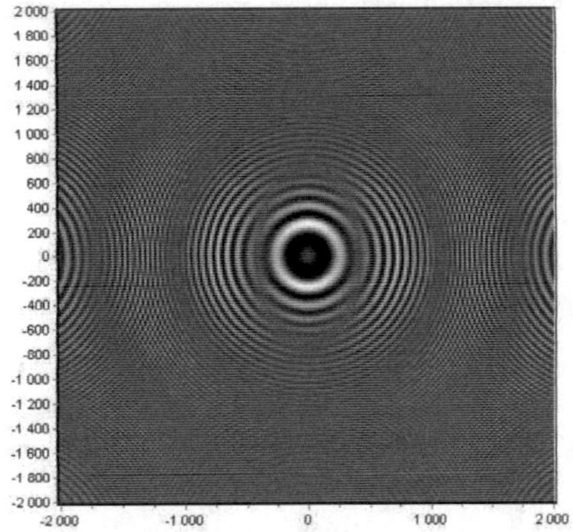

Fig. 5. The same case as Fig. 4. except m= (1.33, 0.01).

By comparison of the Figs. 4 and 5 to Fig. 3, we find that the form of holographic images simulated by the GLMT is similar to that of the opaque disk. But in the case $m = (1.33,\ 0.01)$, an obvious asymmetry is observed. This asymmetry is due to the polarization of incident beam. To confirm this hypothesis, a study on the polarization influence has been carried out and will be presented in section 3.3.

To quantify the difference between the two models, we compared the intensity along x axis in the image plane in Fig. 6 for different refractive index m = (1.33, 0.1), m = (1.33, 0.005) and m = (1.33, 0.01). We find that when m = (1.33, 0.1), the results of two models are in good agreement. In other words, in this case, a spherical particle can be considered as an opaque disk. In the contrary, when m = (1.33, 0.005) and m = (1.33, 0.01), the difference between the two models is evident. That is to say, in these two cases, we can not simplify a spherical particle as an opaque disk. The refraction and diffraction of third dimension must be taken into account.

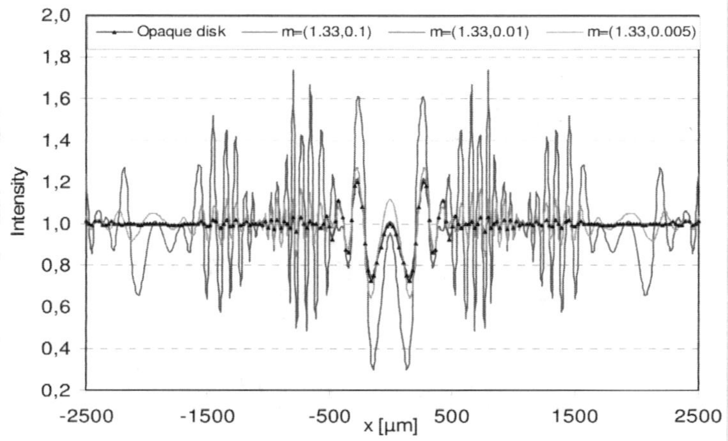

Fig. 6. Comparison of intensity distribution along x axis obtained by the two models with different refractive index in the same configuration as Fig. 3.

3.1.2 Case of a circular Gaussian beam

The diffraction pattern predicted by the scalar theory of diffraction is shown in Fig. 7. for an opaque disk of a diameter d = 150µm illuminated by a circular Gaussian beam and located at z = 120mm from the CCD camera. The local waist radii and curvatures of the Gaussian beam in the object plane are $w_\varepsilon = w_\eta$=1.75mm and $R_\varepsilon = R_\eta$=50mm. The holographic diagram simulated by the GLMT in the same case is shown in Fig. 8. We find that the form of holographic images simulated by the GLMT is similar to that obtained by the model of an opaque disk. Simmilar as in the plane wave case, the difference of diagrams obtained by the two models is also observable.

3.1.3 Case of an astigmatic elliptical Gaussian beam

In the case of a sphere illuminated by an AEG beam, the diffraction patterns are no longer symmetry since the beam is asymmetry. The diffraction pattern obtained by opaque disk model in the same case as Fig. 7 is shown as Fig. 9. but disk is located at z =107mm after the beam waist. Conversely, Fig. 10. presents the diffraction pattern observed at z =154mm where the disk is before the beam waist. By comparison to the diagrams simulated with GLMT given in Figs. 11 and 12 for the same cases, similar comments to the circular Gaussian beam illumination case can be made for the effects of the refractive index on the holographic digrams.

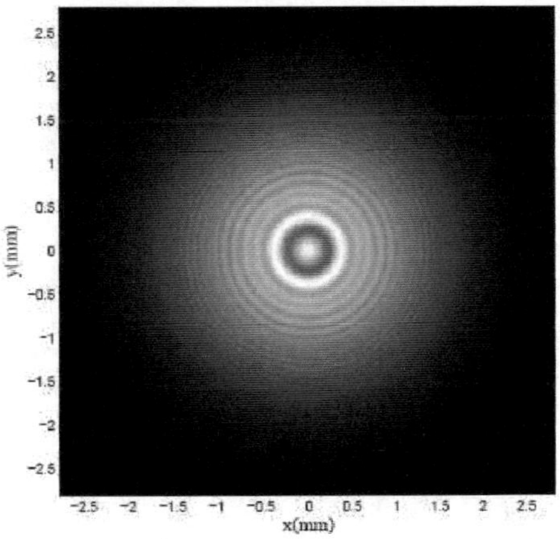

Fig. 7. Diffraction pattern of an opaque disk d=150μm illuminated by a circular Gaussian beam w_ε= w_η=1.75mm, R_ε= R_η=50mm and z =120mm.

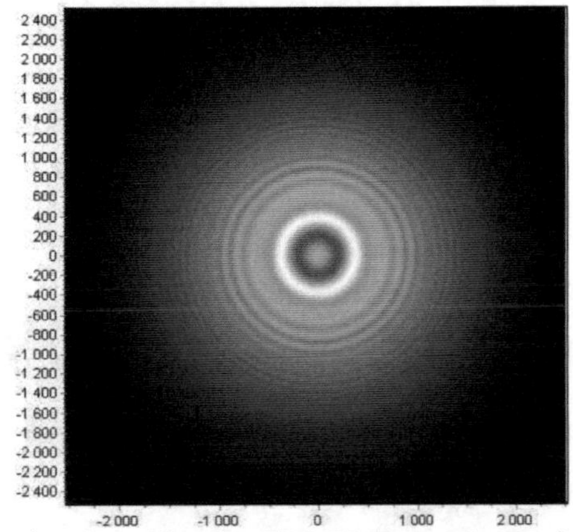

Fig. 8. Holographic images simulated by the GLMT for a sphere of diameter d=150μm and refractive index m= (1.33, 0) or m= (1.33, 0.1) located at the distance z=120mm illuminated by a circular Gaussian beam of wavelength λ=632.8nm.

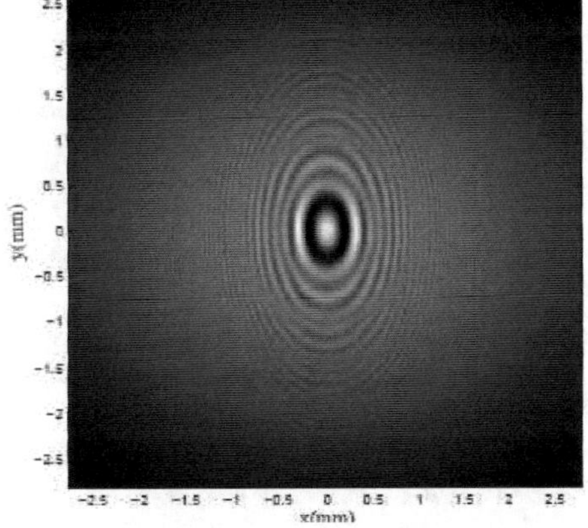

Fig. 9. Diffraction pattern of an opaque disk illuminated by an AEG beam d=150μm, w_ε=7mm, w_η=1.75mm, R_ε= – 2.36×10^{5}m, R_η= – 50mm et z=107mm.

Journal of Physics: Conference Series **147** (2009) 012052 doi:10.1088/1742-6596/147/1/012052

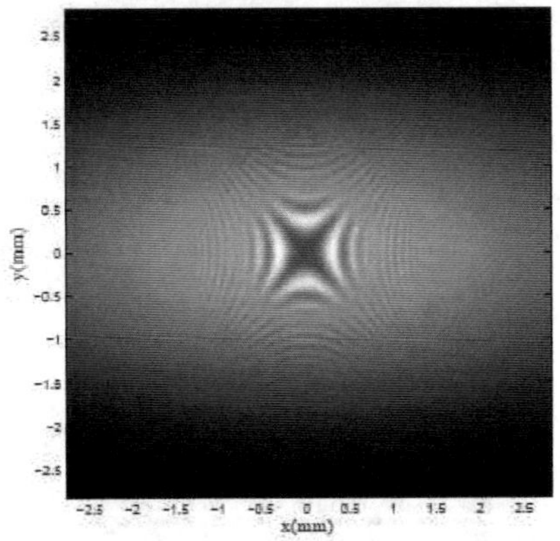

Fig. 10. Diffraction pattern of an opaque disk illuminated by a AEG beam D=150μm, w_{ε}=7mm, w_{η}=2.27mm, $R_{\varepsilon}= -4.38\times10^{5}$m, R_{η}=65mm and z=154mm.

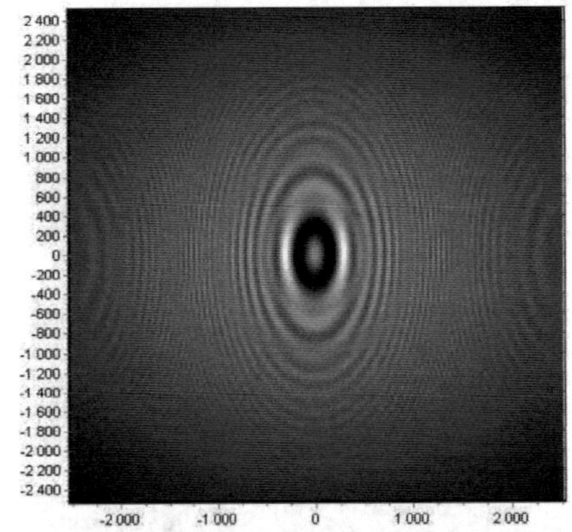

Fig. 11. Same case as in Fig. 9 simulated by the GLMT.

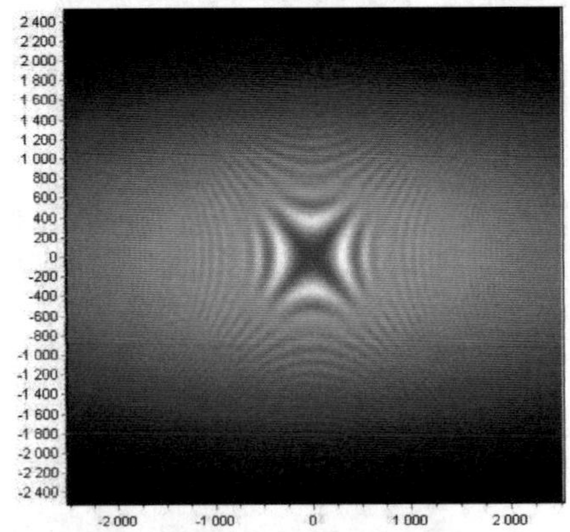

Fig. 12. Same case as in Fig. 10 simulated by the GLMT.

3.2 Effect of refractive index

In the previous section, we have chosen certain values of refractive index for the simulation and compared the holographic diagrams of two models. In this section, the effects of refractive index will be assessed systemically. Fig. 13. displays the intensity at the centre of image as function of imaginary part of refractive index with two values of real part of refractive index.

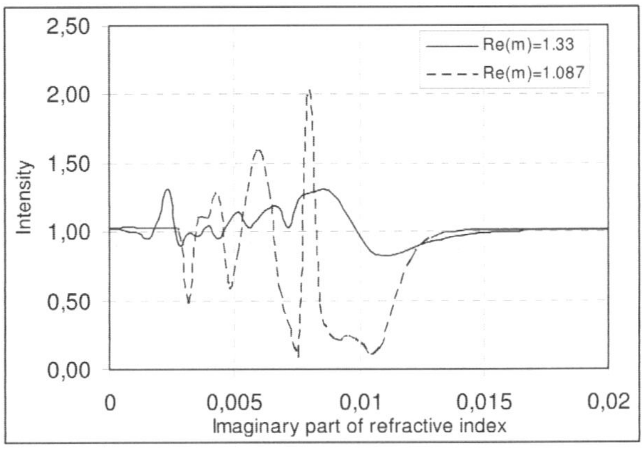

Fig. 13. Intensity at the centre of image as function of the imaginary part of refractive index.

We find that the intensity is very sensitive to the imaginary part of refractive index, espetially when the real part of index equals to 1.087. That is to say, when the real part of refractive index is near unity, a spherical particle can not be considered as an opaque disk.

We have also examined the effects of real part of refractive index and we can conclude that if the imaginary part of refractive index is nul or relatively important (0.1 for example), the image remains the same whatever the real part of refractive index between 1 and 1.8. But if the imaginary part is relatively small but not nul (0.01 or 0.005 for example), the real part of refractive index plays an important role in the form of the diagram (see Fig. 14).

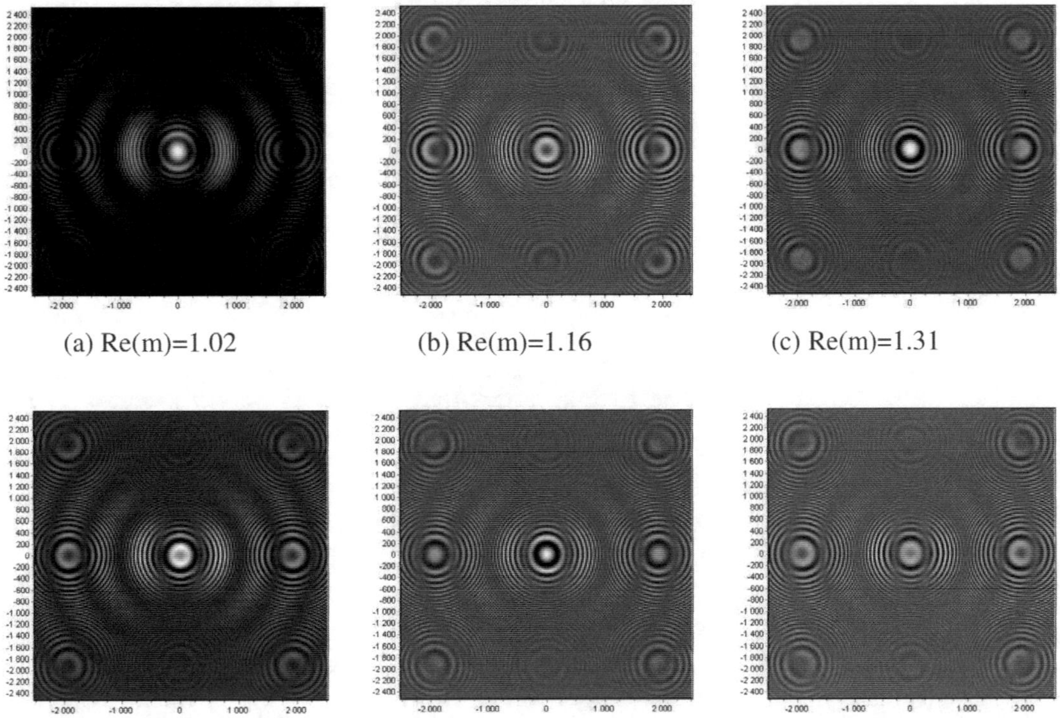

(a) Re(m)=1.02 (b) Re(m)=1.16 (c) Re(m)=1.31

(d) Re(m)=1.47 (e) Re(m)=1.62 (f) Re(m)=1.77

Fig. 14. Images simulated by the GLMT with imaginary part of refractive index 0.01 in the same configuration Fig. 3.

Fig. 15. represents the intensity at the centre of the image as function of real part of refractive index with four values of imaginary part. We find that when the imaginary part is 0 and 0.1, the effects of the real part of refractive index on holograms are not obvious at all. But when the imaginary part equals to 0.005 and 0.01, the real part of refractive index plays a role in holograms, especially when the real part of index is close to 1.

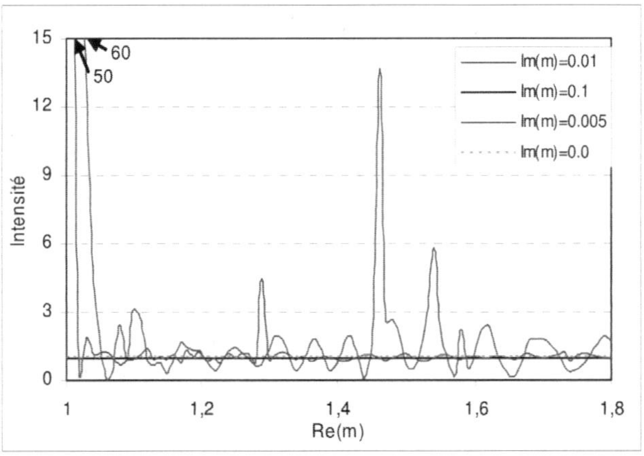

Fig. 15. Intensity evolution at centre of image in function of different real part of refractive index.

3.3 Effect of light polarization

Fig. 16-18. illustrate the simulated holographic images of a sphere illuminated by a plane wave in three polarization states. The intensity distributions along x axis for this three cases are compared in Fig. 19. Naturally, the image of polarization in y corresponds to the image of polarization in x but turned by 90 degrees, and the image in non-polarized case is perfectly axisymmetric.

In the cases of a sphere illuminated by a circular Gaussian beam and an AEG beam, there is also the polarization influence for absorbing particles (Re(m)=0.003), but the influence is less obvious relative to the case of a plane wave.

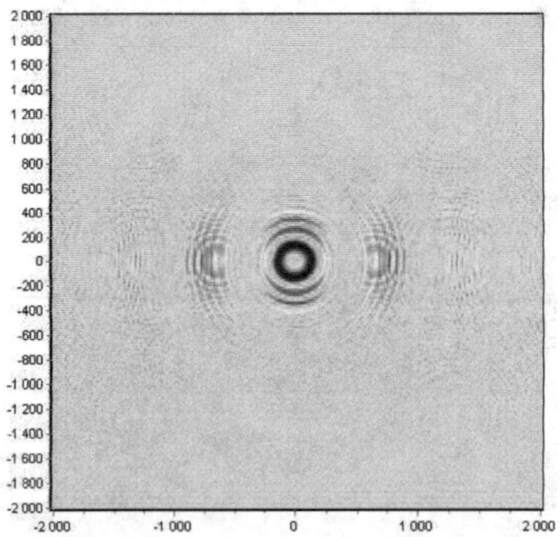

Fig. 16. Image simulated by the GLMT at z=80mm for a sphere of diameter d=100μm and refractive index m= (1.33, 0.01) illuminated by a plane wave polarized in the x direction of wavelength λ=632.8nm.

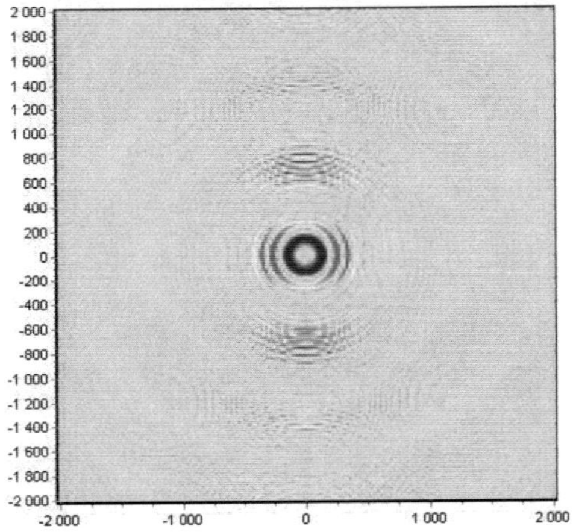

Fig. 17. The same case as Fig. 16. but the incident wave is polarized in the *y* direction.

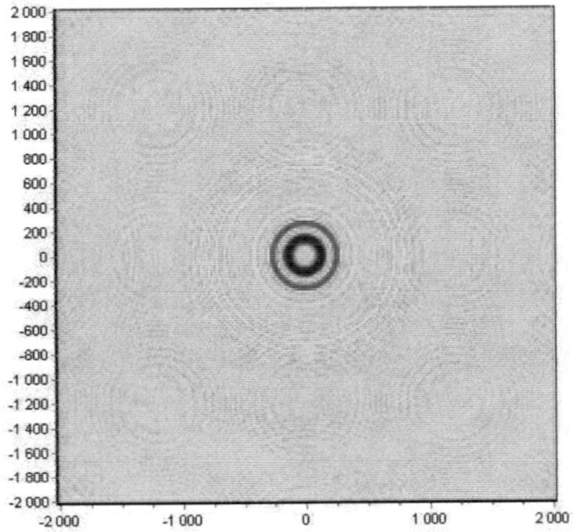

Fig. 18. The same case as Fig. 17. but the incident wave is non-polarized.

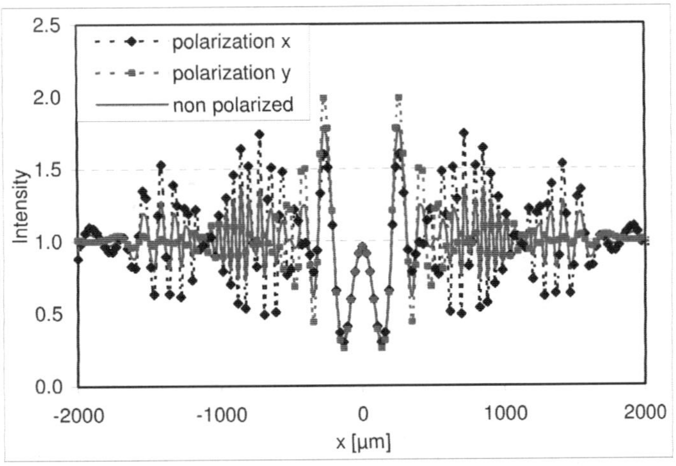

Fig. 19. Intensity distributions along the axis x with the three polarization states.

3.4 Effect of particle position

Since the AEG beam used in the simulation is much more focused in *y* direction than in *x* direction, we just move the particle along *y* axis. We find that the figure structure remains the same but the images moved in the opposite direction (Fig. 20).

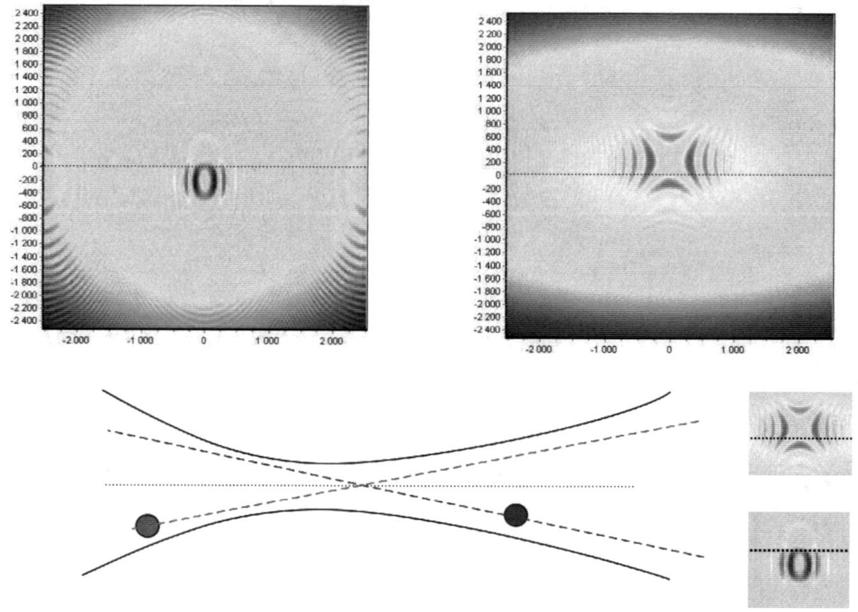

Fig. 20. Images simulated by GLMT for a sphere of d=150μm and m= (1.33, 0.003) located off optical axis (x_0=0, y_0=100μm) and illuminated by a AEG beam of λ=632.8nm.

4. CONCLUSIONS

In this paper, the holograms of a homogeneous sphere are simulated by a rigorous model based on the generalized Lorenz-Mie theory and compared with those obtained by a simplified model of an opaque disk in order to evaluate the various effects of the refraction and third dimension difraction of a particle, such as the refractive and the position of particle, the form and the polarization of the incident beam. It is found that the effects of the third dimension (diffraction and refraction) is obvious when the imaginary part of refractive index is moderate (not zero nor too large) or the real part is near unity. We find also that the polarization effects is more important in the plane wave case than in the circular or astigmatic elliptical Gaussian beam illumination cases.

NOMENCLATURE

f	focal length of lens	[mm]
A	complex amplitude of EM field	[V/m]
E	electrical field	[V/m]
d	particle diameter	[mm]
I	intensity	[V^2/ m^2]
m	complex refractive index of particle	
Im(m)	imaginary part of refractive index	
Re(m)	real part of refractive index	

R	curvatures of the wave front	[mm]
w	waist radius	[mm]
	Greek Letters	
λ	wavelength	[nm]
	Subscripts	
ξ	coordinate in the image plane	
η	coordinate in the image plane	

REFERENCES

Girasole, T. et al (2000). "Particle imaging : GLMT simulation, J. of Visualization, 3(2), pp195-202.

Gouesbet, G. et al (1988). "Light scattering from a sphere arbitrarily located in a Gaussian beam, using a Bromwich formulation," J. Opt. Soc. Am. A 5, pp1427-1443.

Nicolas, F. et al (2005). "Application of the fractional Fourier transformation to digital holography recorded by an elliptical, astigmate Gaussian beam," J.Opt. Soc. Am. A. 22(11), pp2569-2577

Nicolas, F. (2007). "Holographie numérique femtoseconde : Modélisation et restitution par transformation de Fourier fractionnaire," PhD thesis, University of Rouen, France.

Ren, K. F. (1995). "Diffusion des Faisceaux Feuille Laser par une Particule Sphérique et Application aux Ecoulements Diphasiques," PhD Thesis, University of Rouen, France.

Ren, K. F. et al (1998). "The integral localized approximation in generalized Lorenz-Mie theory," Appl. Opt., 37(19), pp.4218-4225.

Journal of Physics: Conference Series **147** (2009) 012053 doi:10.1088/1742-6596/147/1/012053

Measurement technique for solid-liquid two-phase flow using a Normal-line Hough Transform method

Xiaoran Yu, Dongchang Xing, Tatsuya Hazuku, Tomoji Takamasa [1], Takashi Ishimaru, Yuji Tanaka [2], Tatsuro Akiba [3]

[1] Tokyo University of Marine Science and Technology, 2-1-6 Etchujima, Koto-ku, Tokyo, 135-8533 JAPAN
[2] Tokyo University of Marine Science and Technology, 4-5-7 Konan, Minato-ku, Tokyo, 108-8477 JAPAN
[3] Research Institute for Cell Engineering Advanced Industrial Science and Technology, Central 6, 1-1-1 Higashi, Tsukuba, Ibaraki, 305-8566 JAPAN

yuxr@kaiyodai.ac.jp, m073065@kaiyodai.ac.jp

Abstract. Image processing method is proved to be an available and important approach for the measurement of multi-phase flow in the past researches. But there still remain some difficult technical issues such as to separate and position the overlapped particle phase in a dispersed two-phase flow, and to carry it out in an efficient way. In our research, we firstly proposed a Normal-line Hough Transform (NHT) method which can solve the problems in an efficient manner than the past ones. Then, we observed the measurement ability and validated the accuracy of the proposed method via a simulation experiment using a pair of pseudo particles. The result revealed that our proposed method could be a practicable one with a high efficiency of 114 times than usual Hough Transform method. What's more, we also compared the usability of our method with other existed ones using a Monte Carlo method, and the result reveals that our method is a robust one with a wider applicability.

1. INTRODUCTION

Researchs concerning the construction of interfacial area transport equation model have attracted much attention in the fields of multi-phase flow for many years. For which, detail interfacial information of two-phase flow under various of flow regimes, such as void fraction, interfacial velocity and interfacial area concentration, are necessary (Ishii et al., 2005). An image processing method, which can observe the interfacial information directly without disturbing the flow and provide us with a high time-space resolution, has become an important measurement technique along with the development of the computer image processing technology.

Many image processing methods for measuring a two-phase flow have been developed in the past. Among these methods, the most conventional method is the particle detection based on the edge tracking and labelling method as well as the template matching technique(Zabulis et al., 2007). Though they are relative efficient methods for the detection of particles and to reconstruct their interface in dispersed two-phase flows under a low particle density condition, however, the recognition and separation of the overlapped dispersed particles become very difficult when the particle density increases to a certain level.

On the other hand, Shen et al. (2000) proposed a method to divide a planar curve into circle arcs based on a curve-rotate breakpoint detection method, after the arcs belong to the same circle are identified and clustered using an area cross-correlation method, overlapped particles can be recognized in a certain approving accuracy. Unfortunately, the area cross-correlation method has a threshold of area correlation coefficient for about 0.85, thus, the proposed method would become invalidate when the center of two particles are adjacent to each other in an extreme manner.

Furthermore, Qu et al. (2004) employed a two-dimensional median filter together with a triangular Bézier patch method. The proposed method is proved to be an efficient one with a high accuracy in detecting and isolating the small-scale phases from those of the large ones, such as to isolate PIV tracer particles from bubble phase. But it still remains a problem in recognizing and separating overlapped particles that are similar in area fraction.

Besides, the other methods are base on the detection of interface curvature extrema (Fujitsuka et al., 2004). However, the edge gradient breakpoint cannot be obtained easily in practice owing to a low image resolution. Therefore, the development of an appropriate method, which can recognize particles and separate the overlapping ones from defective edge information are eagerly demanded.

In our research, firstly, we proposed a technique, Normal-line Hough Transformation method (NHT), that is able to detect the solid particle phase even though the particle edge was concealed and interrupted each other. We applied this method to solid-liquid two-phase flow images and confirmed that this method is more effective than the existing ones. Secondly, we compared the detection result of NHT with the actual relative position information of a pair of pseudo particles that were adjusted by a traverse, and the accuracy of the proposed method was evaluated in detail. Moreover, a Monte Carlo method is carried out to clarify the measurement limit and the validity of the proposed method, which demonstrates the relationship between the average volume rate of uniform particles and the extraction rate of each method.

2. NORMAL-LINE HOUGH TRANSFORM

The most conventional circle extraction method was the circular arc detection based on the Hough transformation (Duda et al., 1972; Yu et al., 2006). Supposing we are to extract a circle of radius r from an image use a Hough tranformation. Usually edge of the particle image is detected, then each edge pixel contributes a circle of radius r to an output accumulator space (Fig. 1a), which we call it voting parameter space. If the circle radius is unknown, the algorithm should be run for all possible radiuses to form a three-dimensional voting parameter space, in which, two dimensions represent the coordinate position information of the circle center, and the third dimension represents the variable radiuses. The voting parameter space will have a peak where these contributed circles cross at the center of the detecting circle.

However, in practice, a large number of points have to be accumulated in the parameter space in order to construct peaks. Furthermore, the Hough transformation also suffers from the storage problem (Yuen et al., 1996). In this paper, we present an innovative method to recognize overlapped circle particles in images. It makes use of the property that the normal-line of every edge point on a circle consequentially passes through the circle center. We combine the normal-line gradient algorithm to a Hough transformation, which is proved to be an efficient one than the Hough transformation.

In our research, low density dispersed particles with the same radius are used. As a result, all the circles in an image are of identical known radius, here we suppose the radius are of length r, and the three-dimensional voting parameter space can be reduced to two-dimensional. The objective is to find the coordinates of the centers. Assume that all the circles are of radius r, the center of a circle will certainly fall on the extending line of gradient vectors with a distance of r from the edge pixel. Thus, each edge point contributes one point (Fig. 1a) instead of a circle (Fig. 1b).

The normal-line direction of the edge is obtained according to the gradient, which is calculated by the edge operator. Generally, an image is described as function of brightness, such as $f(x, y)$, in which, variable x and y represent the coordinates of two dimensions. The first order differential (gradient) of the image function is shown in form of vector in equation (1).

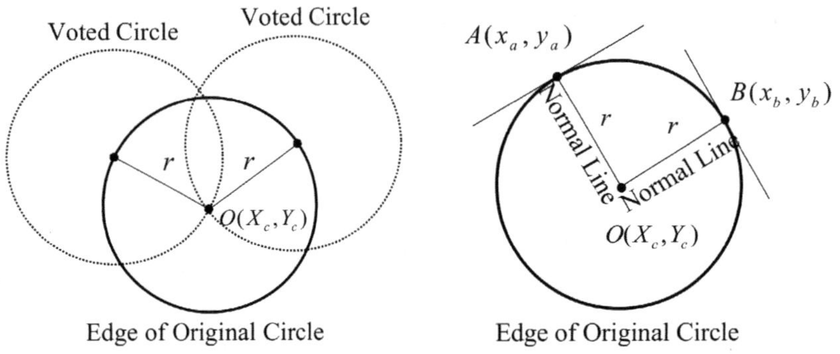

(a) Hough transform (b) NHT method

Fig. 1 Circle detection using a Hough transform and NHT method.

$$\nabla f(x,y) = \frac{\partial f(x,y)}{\partial x} u_x + \frac{\partial f(x,y)}{\partial y} u_y \qquad (1)$$

Here, we use u_x, u_y to indicate x and y axial direction unit vector. By equation (1), the gradient can be calculated by the space differentiates of brightness of the image, and the direction of the gradient can be obtained by equation (2),

$$\theta = \tan^{-1}\left(\frac{f_y(x,y)}{f_x(x,y)}\right) \qquad (2)$$

where θ stands for the vertical direction (normal-line direction) on the circle edge point. Therefore, the direction of the normal-line is obtained from the gradient on each edge point of the particle. Afterwards, we make use of the property that the center of circle is on the normal-line of edge point, and draw all the normal-lines in the voting parameter space. As a result, the intersecting point of all normal-lines will be the circle center.

In this research, we utilized both the Hough transform method and NHT method under the same experimental condition (CPU: Pentium M 1.8*GHz*, Memory: 1*G*, Display Memory: 64*M*, Image resolution: 640*480) to process about 800 particles with a 13 pixels radius in 56 piece of solid-liquid two-phase flow images. The number of average votes of each method in detecting one particle is $V_{HT} = 4538$, $V_{NHT} = 55$, and the processing time is $t_{HT} = 15748ms$, $t_{NHT} = 138ms$. The NHT method improves the processing speed by decreasing the amount of votes dramaticly, and obtained a remarkable efficiency as a result.

3. VERIFICATION OF MEASUREMENT ACCURACY

To verify the accuracy of the proposed NHT method, we carry out an additional verification experiment using a pair of pseudo particles (particle diameter: 2.381*mm*, test tube diameter: 9.0*mm*). Firstly, we fix the position of particle 1 to the center of the test tube, and move particle 2 in horizontal direction with a unit distance of 250μm using a traverse. Secondly, we compare the actual distance between the two particles with the measured one using NHT method, and then, the measurement accuracy of the NHT method is investigated according a quantitative analysis. Figure 2 shows the extraction result of the pseudo particles in the accuracy verification experiment. Figure 3 shows the result of the accuracy evaluation. The result shows that the standard deviation of the actual position and the measured position is about 0.014, accordingly, the proposed method is proved to be able to detect the overlapped solid particle with a relative accuracy.

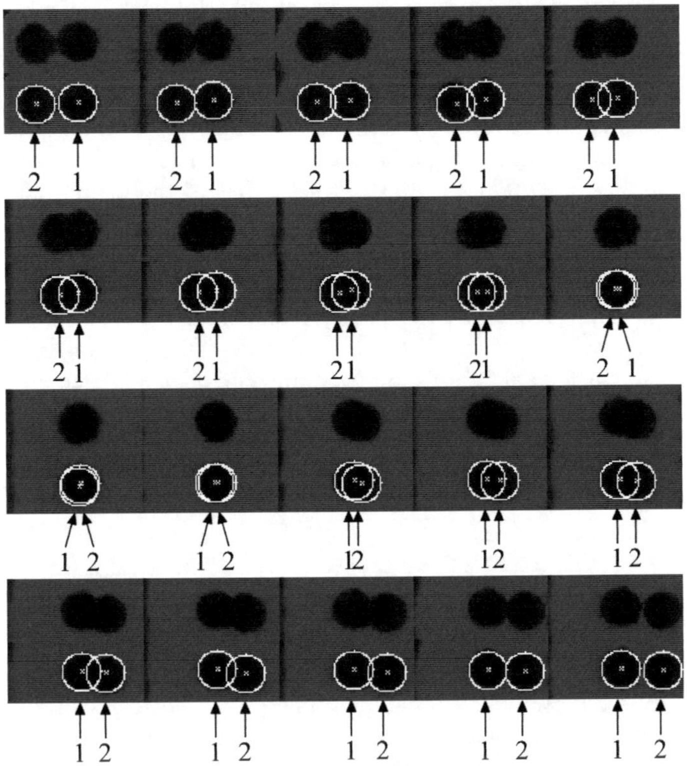

Fig. 2 Extraction result of normal-line Hough method in the accuracy evaluation experiment.

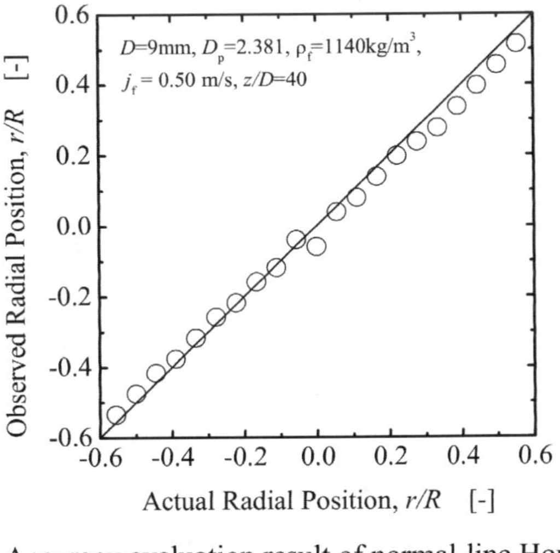

Fig. 3 Accuracy evaluation result of normal-line Hough method.

4. TRANSITION OF EXTRACTION RATE

To investigate the transition of the extraction rate of NHT and the past methods under various of particle phase volume rate, a simulation experiment using a pair of pseudo particles with a 13 pixels radius was carried out. In addition, a simulation experiment surveying the extraction rate of each method at various of particle phase volume rates was conducted. In detail, firstly, in order to investigate the measurement limit of the proposed method as well as the past ones, we detach the two totally overlapped pseudo particles from each other, with an one-pixel step in center distance. The least center distance between the two particles when they can be distinguished using an visual observation is considerd to be the visual measurement limit. Moreover, the measurement limit of interface curvature extrema method appears when the centers of two particles are close enough so as to lost the curvature extrem. Figure 4 shows the simulation result for surveying measurement limit of each technique. For easy observation, black particles of 13 pixels and a white background was utilized. The result indicates that the measurement limit of the visual method, interface curvature extrema method and the proposed NHT method are 2.3 pixels, 5 pixels and 1 pixel, respectively.

In the simulation experiment that evaluates the transition of extraction rate at a range of average particle phase volume rates, three-dimensional coordinates of particle centers are generated at random in the test tube in certain solid phase volume rate (by adjusting the quantity density of particles). Assume that two particles with a center distance closer than the measurement limit can not be recognized. The theoretical detectable ratios of each method is calculated and illustrated, which concerns the extraction rate on average volume rate of uniform particles. Here, in order to simulate the dispersed two-phase flow in a realistic manner, the randomly generated coordinates of the particles are assigned to meet the additional restrictions of both equation (3) and equation (4).

$$\sqrt{(x_n - x_i)^2 + (y_n - y_i)^2 + (z_n - z_i)^2} \geq 2r \qquad (3)$$

$$\sqrt{(x_n - X_o)^2 + (y_n - Y_o)^2} \leq R - r \qquad (4)$$

Where (x_i, y_i, z_i)stands for the three-dimensional coordinates of a particle center that has already been generated. r and R are radii of the particle and test tube, respectively. (X_o, Y_o) indicates the tube center coordinates on the tube section. In each particle volume rate, 1000 cases of simulations are conducted. The simulated extraction rate of each method according to the average dispersed particle phase volume rate is shown in Fig. 5.

From Fig. 5, it is understood that the extraction rate of NHT method achieves 98.7% when the average particle volume rate is about 0.3, and it is proved to be more accurate than the past methods.

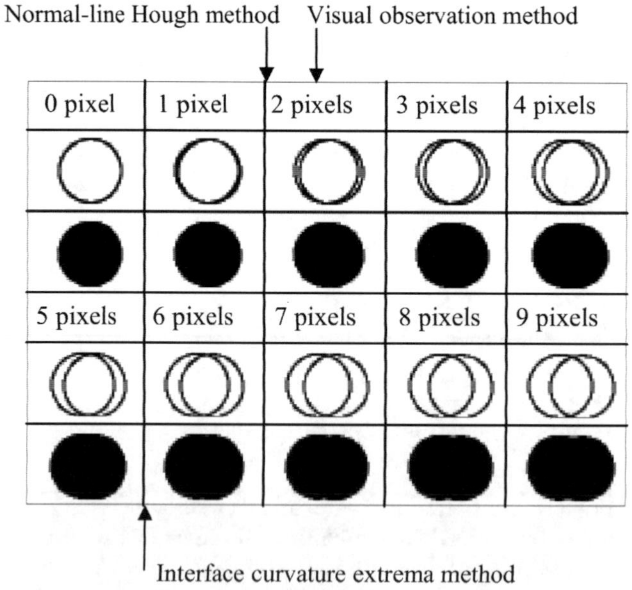

Fig. 4 Measurement limit of each technique.

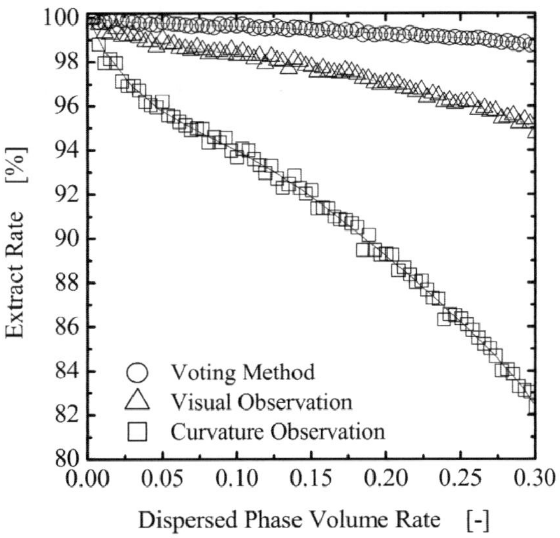

Fig. 5 Transition of extraction rate of each technique.

Moreover, it is confirmed that overlapping occurs more frequently according to the increasement of particle volume rate, and a decrease in extraction rate consequently. Since the extraction rate of interface curvature extrema method and the visual observe method decreases remarkably in the case of a high particle volume rate, a miss-evaluating is considered to be possible.

5. EXPERIMENT RESULT

Subsequently, to measure the three-dimensional interfacial information of the flow, we utilize our new method to the solid-liquid two-phase flow images which were taken using the Takamasa's Stereo Image Method (SIM) (Takamasa et al., 1993; Takamasa et al., 1997), and the interfacial transportation information of the disperse phase in the tube axis direction and the tube transect direction, such as the interfacial area concentration and interfacial velocity, etc, can be built up efficiently.

Figure 6 and 7 show one example of the measurement result that applies the NHT method to the

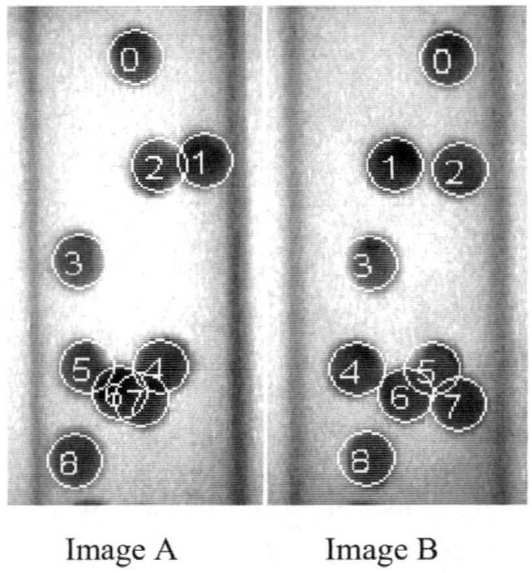

Fig. 6 One example of processing result using normal-line Hough method.

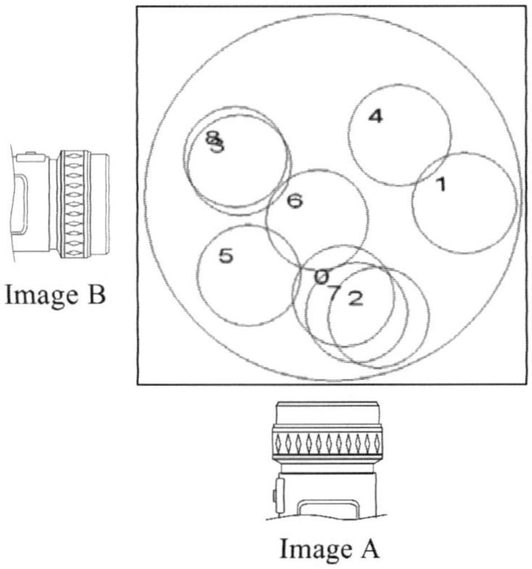

Fig. 7 Particle distribution reconstruction.

solid-liquid two-phase dispersed flow images. Firstly, three-dimensional coordinates are extracted and reconsturcted from the position information of the particles in the stereo images, as being shown in Fig. 6, the particles are identified to be the identical one by comparing and pairing the tube axial coordinates in image A and image B. Then, the particle distribution information in the tube transect is reconstructed with the use of three-dimensional coordinates information consequently, and the result is shown in Fig. 7.

6. CONCLUSION

This paper describes a new method to detect particles using NHT in order to recognize and separate the overlapped particles and observe the interfacial distribution and velocity analysis in flow images. It makes an improvement in efficiency by combining a normal-line detection technique with the Hough transform method on the base of SIM method. And it is proved to be an applicable and efficient method for the separation and position measurement of overlapped disperse particles.

Moreover, we conducted a simulation experiment using a pair of pseudo particles to verify the measurement accuracy of the proposed technique. The standard deviation between the actual center distance of the two pseudo particles and the observed value of our method is very small. Thus, it comes to the conclusion that the present measurement method is suitable for detecting overlapped disperse particles with a high accuracy.

On the other hand, a Monte Carlo method to clarify the measurement limit and the validity of the proposed method as well as the previous methods is carried out. The result illustrates a comparison of the extraction rate of each method under a range of average volume rate of uniform particles, and clarify the proposed NHT method to be an available and promising technique.

Furthermore, in practice, we utilized the NHT method to the stereo images of solid-liquid two-phase flow taken under micro-gravity circumstance, and reconstructed the three-dimensional interface distribution information of dispersed particles. The result reveals that individual particles can be recognized and separated in the stereo flow images regardless of two or more of them are overlapping each other. It can also be expected to be utilized to the gas-liquid two-phase flow under micro-gravity circumstance, when the dispersed bubbles are assumed to be approximate circles in shape considering of the agravic effect.

REFERENCES

Duda, R. et al. (1972). "Use of the Hough Transformation To Detect Lines and Curves in Pictures", Communications of the Association for Computing Machinery, 15, pp. 11-15.

Fujitsuka, T. et al. (2004). "Study of Bubble Motion in Gas-Liquid Babbly Flow in Vertical Pipes", Proceedings of Japanese Society for Multiphase Flow, D110, pp. 155-156(in Japanese).

Ishii, M. et al. (2005). "Thermo-fluid Dynamics of Two-Phase Flow", Springer.

Qu, J. et al. (2004). "Simultaneous PIV/PTV Measurements of Bubble and Particle Phase in Gas-liquid Two-phase Flow Based on Image Separation and Reconstruction", Journal of Hydrodynamics, Ser. B, 16(6), pp. 756-766.

Shen, L. et al. (1999). "A method for recognizing particles in overlapped particle images", Pattern Recognition Letters 21, pp. 21-30.

Takamasa, T. et al. (1997). "Measurement of Interfacial Configuration of Bubbly Flow under Normal and Microgravity Conditions using Stereo Image- Processing Method", Transaction of JSME, 63(606), pp. 40-47 (in Japanese).

Takamasa, T. et al. (1993). "Measurements of Bubble Interface Configurations in Vertical Bubbly

Flow using Image-Processing Method", Transaction of JSME, 59(564), pp. 63-69 (in Japanese).

Yu, X. et al. (2006). "Measurement of Solid-liquid Tow-phase Flow Using a New Robust Normal-line Hough Transform Method", Proc. 5th ISMTMF, 1, pp. 335-340.

Yuen, P.C. et al. (1996). "A novel method for parameter estimation of digital arc", Pattern Recognition Letters, 17, pp. 929-938.

Zabulis, X. et al. (2007). "Detection of densely dispersed spherical bubbles in digital images based on a template matching technique Application to wet foams", Colloids and Surfaces A: Physicochem. Eng. Aspects 309, pp. 96–106.

Journal of Physics: Conference Series **147** (2009) 012054 doi:10.1088/1742-6596/147/1/012054

Development of Two-Color Digital Holographic PTV for Dispersed Two-Phase Flow

M Naito [1], M Inoue [1], M Ichiyanagi [1], Y Sato [1], K Hishida [1]

[1] Department of System Design Engineering Faculty of Science and Technology, Keio University, 3-14-1 Hisyoshi, Kohoku-ku, Yokohama, 223-8522, Japan

naito@mh.sd.keio.ac.jp

Abstract. Bubbly flows have been widely used in various industrial fields such as the drag reduction of ships and the purification of water. It is expected to investigate the three-dimensional mechanism of the two-phaes flows which involve bubbles with various sizes. Digital holographic particle tracking velocimetry (DHPTV) is an effectual tool for making the flow structure clear, because DHPTV enables to measure instantaneous three-dimensional time series velocity fields of bubbles without complicated optical systems. In a holographic image, microbubbles are recorded as fringe patterns and a millibubble is recorded as a shadow image. The objective of the present study is to establish the measuring method for the velocity of bubbles with micro- and milli-meter sizes by using two laser beams. Finally, the method was realized by combining stereo holographic technique and stereo shadow imaging technique. A millibubble was measured by stereo shadow imaging technique and microbubbles were measured by stereo holographic technique.

1. INTRODUCTION

Bubbly flows are used in various industrial fields such as the frictional drag reduction for ships (Kitagawa et al., 2005) and the purification of water. For the enhancement of the transport efficiency and the improvement of the industrial water treatment, the mechanism of the bubbly flows which contain bubbles with various sizes should be investigated. Recently, several researchers have developed measurement systems and examined the structures of bubbly flows. Murai et al. (2000) showed that the energy spectrum of liquid phase has a strong relation to the bubble-bubble interval distance by measuring the motions of both liquid phase and rising bubbles with particle tracking velocimetry (PTV). Lindken and Merzkirch (2002) examined the turbulence modification in the liquid phase with bubbles by utilizing digital particle imaging velocimetry (DPIV) system, laser induced fluorescence (LIF) and shadowgraphy. Kitagawa et al. (2005) revealed the effect of bubble deformation on the turbulence modification around microbubbles by utilizing PTV and shadow image technique (SIT). Their efforts have contributed to clarify the mechanism of the bubbly flows. In actual, bubbly flows have three-dimensional structures. The present study focuses on experimental measurement of instantaneous three-dimensional velocity of bubbles.

There are several measuring techniques for three-dimensional velocity fields such as scanning PTV, computed tomography (CT) or DHPTV. Among these systems, DHPTV is an effectual tool for making the bubbly flow structures clear, because DHPTV enables to detect the velocity of bubbles without complicated optical systems, compared with scanning PTV or computed tomography (CT).

Journal of Physics: Conference Series **147** (2009) 012054 doi:10.1088/1742-6596/147/1/012054

The objective of the present study is to establish the measuring method for the velocity of bubbles with micro- and milli-meter sizes. In a recording plane for hologram, microbubbles are recorded as fringe patterns which have three-dimensional information of positions of them and millibubbles are recorded as two-dimensional shadow images. In order to analyze fringe patterns and shadow images in the same system, the method is proposed by combining holographic technique and stereo shadow imaging technique with two laser beams.

2. Experimental procedure

2.1 Principle of holography

The digital holographic technique consists of digital recording and numerical reconstruction procedures. In the recording process, an objective beam and a reference beam are interfered. The objective beam is formed by reflecting, scattering and diffracting from the test volume and the reference beam have the constant phase profile. Images of fringe patterns are obtained in the recording plate as shown in Fig. 1(*a*). The images are known as holograms which have information of the intensity and phase of light in three-dimensional whole field. The holograms are numerically reconstructed by utilizing fresnel diffraction theory. In numerical simulation, the reconstructed wavefront $g(X, Y, Z)$ in a certain plane at a distance Z from the recording plane is obtained by using the Fresnel Kirchhoff equation defined as eq. (1).

$$g(X,Y,Z) = \frac{-i}{\lambda}\iint \frac{f(X',Y')}{Z} e^{ik\sqrt{Z^2+(X'-X)^2+(Y'-Y)^2}}\,dxdy \quad (1)$$

where (X',Y') denote the pixel coordinates in the recording plane, (X,Y) denote the pixel coordinates in the reconstructed plane, Z is the optical distance from the CCD recording plane to the ambient test volume X-Y plate, $f(X',Y')$ is the intensity profile of fringe patterns, $g(X, Y, Z)$ is the reconstructed wavefront, λ is wavelength, k is wave number and i is the square root of -1. Fig. 1(*b*) shows the reconstructed image of (*a*) at Z = 218 mm from the CCD recording plane. The images of particles which locate near Z = 218 mm are formed. One of them is marked in a red circle in Fig. 1(*b*). Those which do not locate near Z = 218 mm are reconstructed as interference fringes in concentric patterns. All of the obtained holograms are analyzed along the following procedures as described in section 2.3. In the present study, (X, Y, Z) are defined as pixel coordinates and (x, y, z) are defined as real space coordinates.

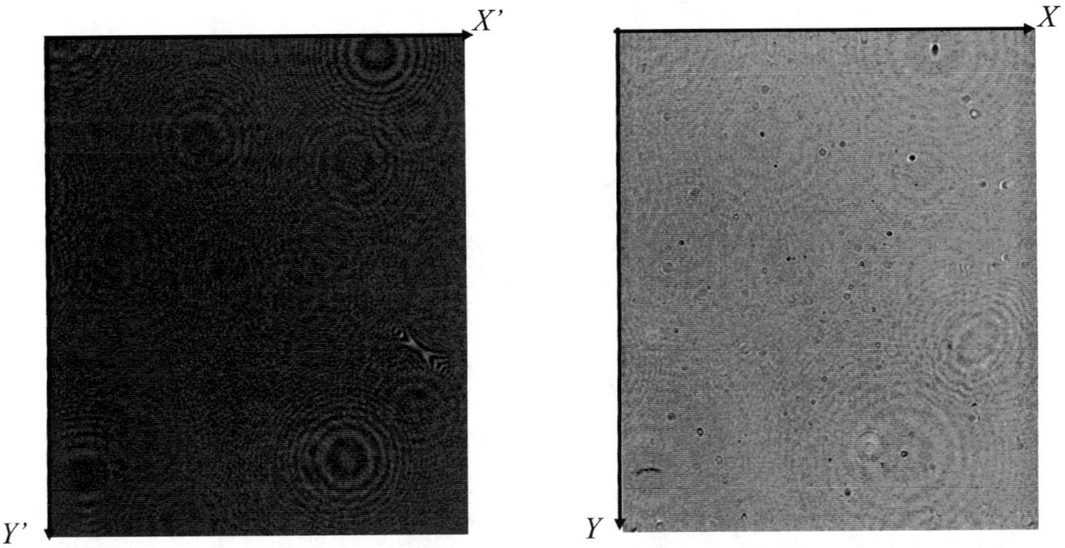

Fig. 1 (*a*) Recorded hologram by He-Ne laser (*b*) Reconstructed image of (a) at Z = 218 mm

Journal of Physics: Conference Series **147** (2009) 012054 doi:10.1088/1742-6596/147/1/012054

2.2 Experimental setup

Fig. 2 illustrates the schematic of the experimental setup and Fig. 3 shows the schematic of the test section which locates at the area where two laser beams are crossing. A channel was made of acrylic, whose length, width and depth were 1000 mm, 30 mm and 22 mm, respectively. The test section was only made of glass to keep the coherency of laser beams. The center of test section was located at 800 mm from the upstream of the channel. Working fluid was water and it was circulated by a pump. CW YAG laser ($\lambda = 532$ nm) and CW He-Ne laser ($\lambda = 632.8$ nm) were used as light sources. Each laser beam was expanded, reflected by mirrors and crossed at the center of test section. Images were recorded by a 12-bit CCD camera (Imperx IPX-2M30H-L, 33 fps) which has 1920 × 1080 resolution with the pixel size of 7.4 μm × 7.4 μm and the exposure time was set to 50 μs. Both two laser beams were vertically introduced to the camera. The left side of CCD array recorded interference fringe patterns by He-Ne laser. The other side of CCD array recorded fringes by YAG laser. For the validation of the developed system, spherical polystyrene particles with diameter of 50 μm were suspended as tracers in the flow. Furthermore, for the applications to the two-phase flows, two kinds of bubbles were prepared that were microbubbles and milibubbles. Microbubble flows were injected at the upstream of the tank. Their diameters were from 50 μm to 80 μm. A milibubble, whose shape was ellipsoidal, was generated by the injector needle at the test section.

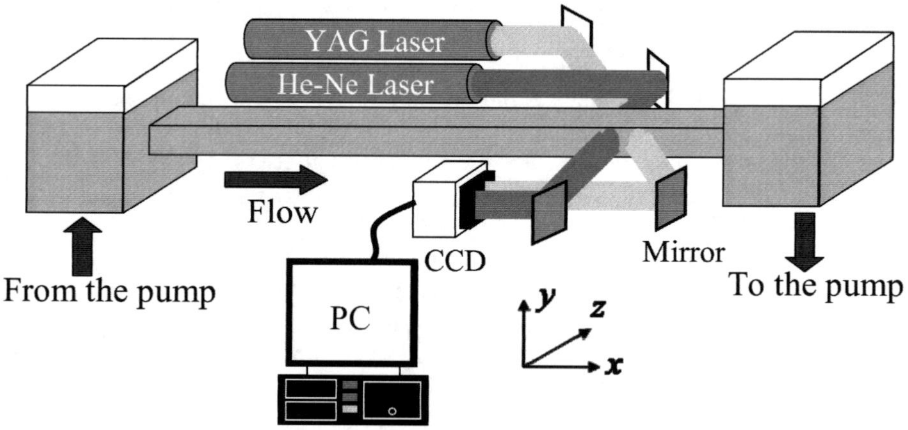

Fig. 2 Schematic of the experimental setup

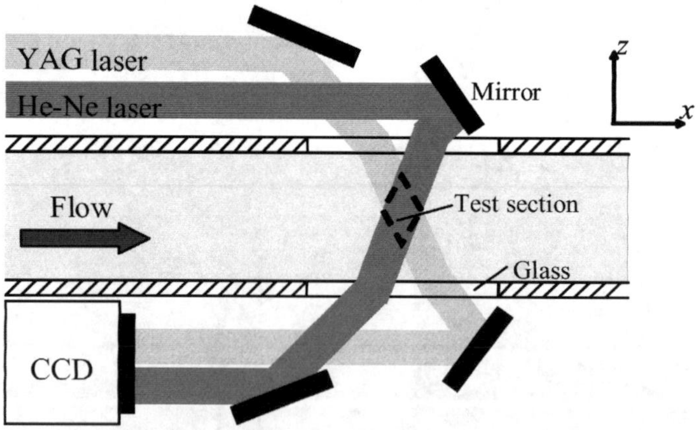

Fig. 3 Schematic of the test section which locates at the area where two laser beams are crossing

2.3 Image processing

For the investigation of the three-dimensional bubbly flow structures, the presesnt study captured the holographic images. Fig. 4 shows the flow chart of the method for obtaining the position of a millibubble and velocity-vectors of microbubbles. In a hologram, a millibubble is recorded as a two-dimensional shadow image and microbubbles are recorded as fringe patterns which contain three-dimensional information of them. Firstly, the millibubble is extracted from the holographic image and its position is determined by stereo shadow imaging technique. Secondly, microbubbles are analyzed by stereo holographic technique as described in section 2.1. Details of the procedures are described in section 2.3.1 and 2.3.2, respectively.

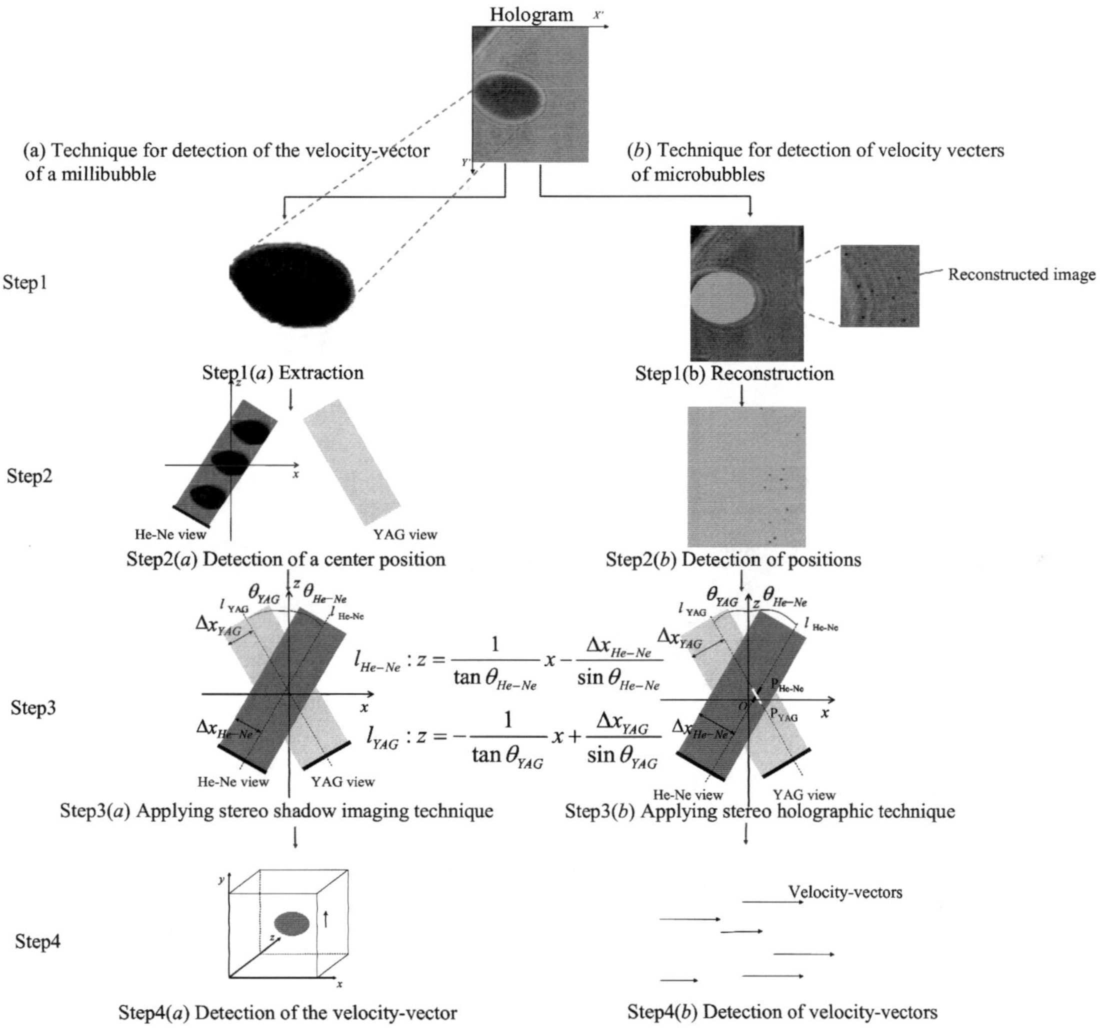

Fig. 4 Flow chart of the image processing of the simultaneous measuring technique for (a) the three-dimensional position of a millibubble and (b) velocity-vectors of microbubbles

2.3.1 Detection technique for the velocity-vector of a millibubble

Fig. 4(*a*) shows the procedures of the measurement of the position of a millibubble. Each procedure is explained below (Step 1(*a*) - 4(*a*)).

Step 1(*a*) Extraction of a millibubble

When black portions of a hologram image occupy over 100 times pixels of a microbubble image, they are regarded as a millibubble. The portions are extracted from the image.

Step 2(*a*) Detection of a center position in each laser side view

The center position, (X, Y), is calculated from the extracted image in each laser side view.

Step 3(*a*) Applying stereo shadow imaging technique and detection of a three-dimensional position

The position of a millibubble has the uncertain range in x-z direction with only one laser as shown in step 2(*a*). To improve the problem, the present study developed the stereo shadow imaging technique by utilizing two laser beams. In the pixel coordinates, the length of the major axis of the millibubble and that of the minor axis are calculated, and the center position is detected. In the consideration of the refraction between air and water, the pixel coordinates, (X, Y, Z), are translated to the real space coordinates, (x, y, z). After translation, the matched center position of the millibubble, (x, z), is obtained from the function of He-Ne view, (2), and that of YAG view, (3),

$$l_{He-Ne} : z = \frac{1}{\tan\theta_{He-Ne}} x - \frac{\Delta x_{He-Ne}}{\sin\theta_{He-Ne}} \quad (2)$$

$$l_{YAG} : z = -\frac{1}{\tan\theta_{YAG}} x + \frac{\Delta x_{YAG}}{\sin\theta_{YAG}} \quad (3)$$

where θ_{He-Ne} is crossing angle between z-axis and He-Ne laser, θ_{YAG} is crossing angle between z-axis and YAG laser, Δx_{He-Ne} is the distance between the center position of x and original point in the x-direction in He-Ne view and Δx_{YAG} is the distance in YAG view, respectively. θ_{He-Ne}, θ_{YAG}, Δx_{He-Ne} and Δx_{YAG} are described in step 3(*a*) of Fig. 3. The center position of y is calculated by multiplying pixel size 0.0074 mm due to laser beam introducing parallel to the y-axis. In this way, the position of the millibubble is detected.

Step 4(*a*) Detection of the velocity-vector

The instantaneous velocity-vector is obtained by calculating the difference between the center position at time t_1 and that at t_2.

2.3.2 Detection technique for velocity-vectors of microbubble

Fig. 4(*b*) shows the procedures of the measurement of the position of the microbubbles. Each procedure is explained below (Step 1(*b*) to 4(*b*)).

Step 1(*b*) Reconstruction in each laser side view

The average brightness value of the image is substituted for the brightness value of the portions of the extracted millibubble. The image with 1mm interval along Z-axis is reconstructed by utilizing the Fresnel-Kirchhoff equation as described in eq. (1) in each laser side view.

Step 2(*b*) Detection of three-dimensional position of microbubbles

(Detection of X-Y position)

In the intensity profile of ambient X position along Y-direction, the averaged brightness value, a, and root-mean-square deviation value, δ, are calculated. The threshold value is defined as a–4δ. When the intensity value at ambient (*X*, *Y*) position is less than the threshold value, it is defined as the particle position.

(Detection of *Z* position)

By comparing the intensity of the detected *X*-*Y* positon along *Z*-direction, the position of the lowest value is determined as the optical distance, *Z*.

Step 3(*b*) Applying stereo holographic technique

$P_{He\text{-}Ne}$ and P_{YAG}, which are described in step 3(*b*) of Fig. 3, are the ranges of position of the microbubble from He-Ne view and from YAG view, respectively. By using only one laser, *z*-position of a microbubble has about 1 mm uncertainty due to depth of focus (DOF). However, the stereo shadow holography proposed in the present study yielded the uncertainty of less than 1 mm by matching two laser views.

Step 4(*b*) Detection of velocity-vectors

In this step, in order to detect velocity-vectors of microbubbles, multiple time series PTV is applied to flow fields. For example, the principle of 3 time series PTV is mentioned in the follwing statements.

Any point in the flow is assumed to be moving in a straight line due to the minuteness of the scale of flow modulation. Fig. 5 illustrates the schematic of 3 time series PTV. Blue circles, red circles and yellow circles are positions detected at t_1, t_2 and t_3, respectively. $\vec{D}_{12}$ denotes the displacement vector detected between time t_1 and t_2, and $\vec{D}_{23}$ denotes the displacement vector detected between time t_2 and t_3. Fig. 5(a) illustrates the schematic of the detection of $\vec{D}_{23}$. At first, $\vec{D}_{12}$ vectors are obtained from the one position detected at t_1 to several positions detected at t_2. Next, the vectors with the same length and direction as $\vec{D}_{12}$ vectors are obtained from positions detected at t_2. Due to the above assumption, the position detected at t_3 is expected to be near the termination of the vector. When the error ratio of vector, given as eq. (4), is less than 0.1, $\vec{D}_{23}$ is determined as the velocity-vector in three-dimensional flow field. In Fig. 5(*a*), circles surrounded by dotted lines are the ranges of position of yellow circles based on eq. (4) and red vectors are detected vectors. The procedure is applied to all of the positions detected at t_1 and velocity vectors, $\vec{D}_{12}$ and $\vec{D}_{23}$, are detected as shown in Fig. 5(*b*).

$$\left| \frac{\vec{D}_{23}}{\vec{D}_{12}} - 1 \right| \le 0.1 \qquad (4)$$

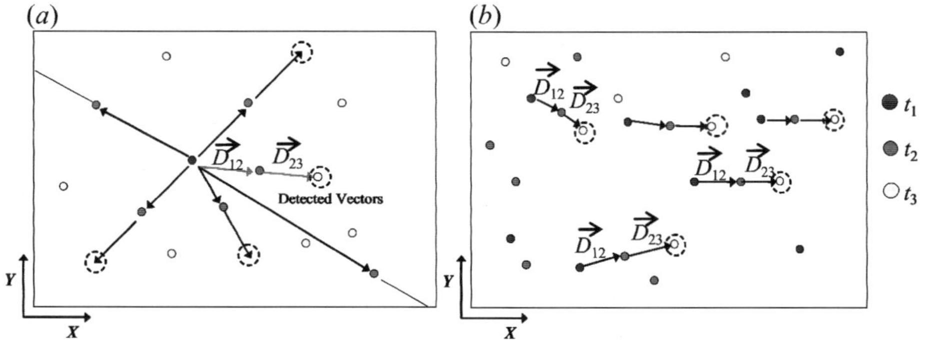

Fig. 5 Schematic of 3 time series PTV
(*a*) Schematic of the detection of $\vec{D}_{23}$ (*b*) Velocity-vectors obtained from 3 time series PTV

3 Results and Discussion

3.1 Application to single-phase flow

For the validation of the developed system, it was firstly applied to the laminar single-phase flow whose Reynolds number, *Re*, is 465. The result of the velocity-vectors is shown in Fig. 6. 864 vectors were detected from 150 subsequent images and the number density of tracer particles was 0.09 mm^{-3}. It can be seen that the velocity-vectors were going toward downstream along *x*-axis. The velocity-vectors within the region at $-0.5 < y < 0.5$ mm are compared with the theoretical value as shown in Fig. 7. The experimental value agreed well with the theoretical values. It is indicated that the proposed system is applicable for obtaining three-dimensional and three- components velocity-vectors of single-phase flow.

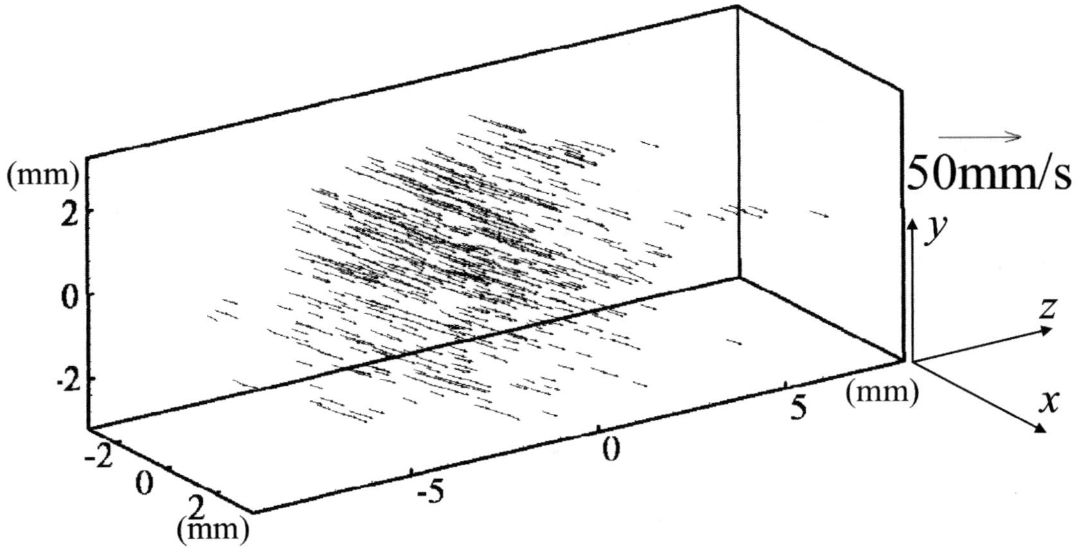

Fig. 6 Velocity-vectors in a laminar single-phase flow

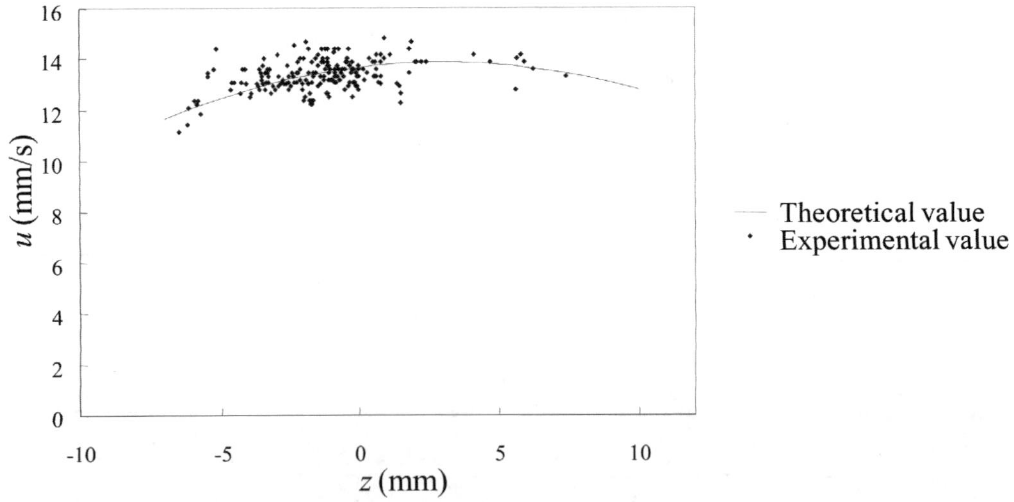

Fig. 7 *u*-*z* plots of the detected velocity within the region between $y = -0.5$ mm and $y = 0.5$ mm

3.2 Application to two-phase flows

3.2.1 Measurement of microbubble flows

Fig. 8(*a*) shows intensity profiles in terms of *Z*-position of a tracer particle in the reconstructed image. It usually has the peak, which shows the lowest intensity value. Its position is the infection point and has zero value of the differential coefficient. It is determined as the position in *Z*-direction. On the other hand, Fig. 8(*b*) and (*c*) show the intensity profiles of microbubbles. One microbubble has the same intensity over 2 mm interval and the other microbubble has two peaks. Their irregular intensity profiles are marked in red circles. When a microbubble which has irregular intensity profile is detected, its position has uncertainty from 2 mm to 5 mm in the depth direction. With only one laser, it is difficult to detect the accurate position of microbubbles due to the uncertainty in *Z*-direction. The irregular intensity profiles should be caused by the internal diffraction of microbubbles. In order to break through the problem of the uncertainty of detecting positions of microbubbles, stereo holographic technique is proposed. Developed system was applied to the flow fields with microbubbles and the obtained velocity-vectors are presented in Fig. 9. 157 velocity-vectors are detected from 150 subsequent images and the number density of microbubbles is 0.05 mm^{-3}. It can be seen that stereo holographic technique with two laser beams contributes to detect accurate positions of microbubbles.

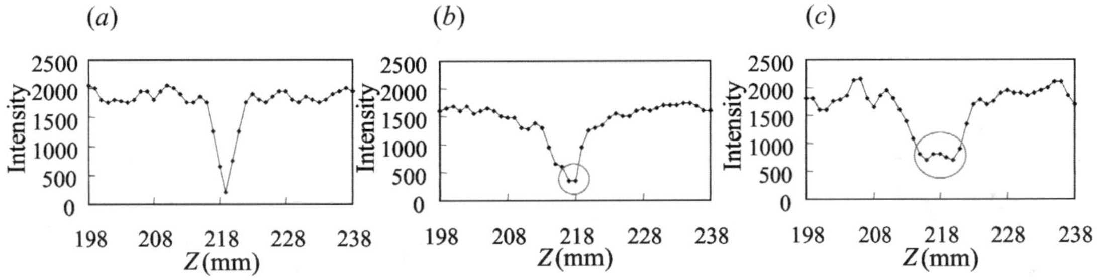

Fig. 8. Intensity profiles along the *Z*-axis
(*a*) Intensity profile of a tracer particle (*b*), (*c*) Intensity profiles of some microbubbles

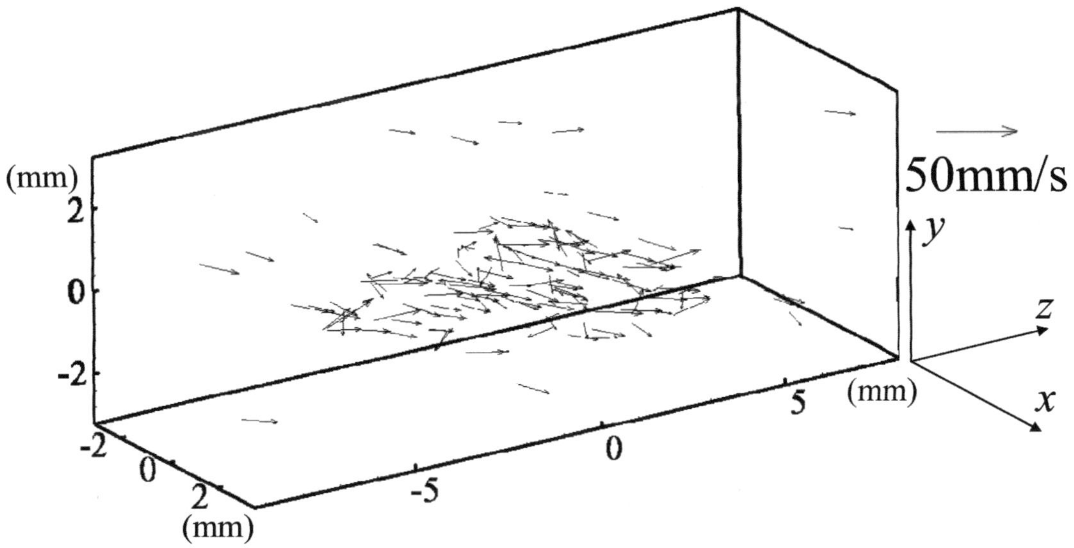

Fig. 9. Velocity-vectors of microbubbles

3.2.2 Simultaneous measurement of a millibubble and microbubbles

Fig. 10 (*a*) shows the reconstructed image of a millibubble and microbubbles at He-Ne laser side view, (*b*) shows the intensity profile of the microbubble which is marked in the red circle in (*a*) and (*c*) shows the intensity profile of the millibubble along *Z*-axis. In Fig. 10(*b*), the microbubble has the intensity peak as described in section 3.2.1. However, in Fig. 10(*c*), it is found that the image of the millibubble is recognized as a shadow image and its intensity does not have a peak in *Z*-direction. It can be confirmed that the position of a millibubble cannot be detected by holographic technique.

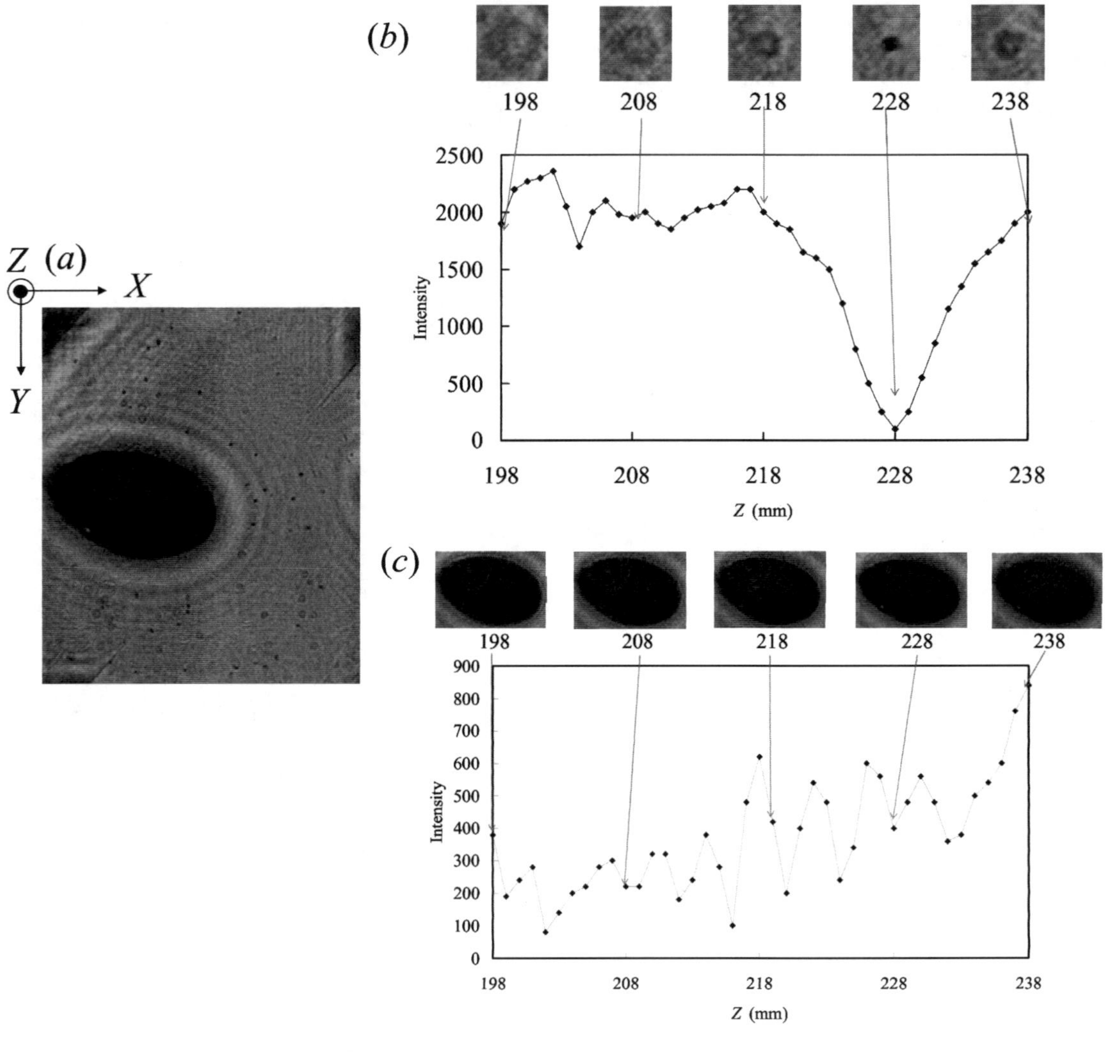

Fig. 10.(*a*) Reconstructed image of a millibubble and microbubbles at He-Ne laser side view, (*b*) intensity profile of the microbubble which is marked in an red circle along *Z*-axis and (*c*) intensity profile of the millibubble along *Z*-axis

In the present study, the position of the millibubble was estimated by stereo shadow imaging technique with two laser beams described in section 2.3.1. Three-dimensional velocity-vectors of microbubbles were calculated by PTV with stereo holographic technique. Fig. 11 shows the result of the instantaneous three-dimensional position of the millibubble, which was indicated by the dotted ellipsoid, and velocity-vectors of microbubbles obtained from 10 subsequent images. Fig. 11(*a*) and (*b*) show the results from different views based on the same data. The center position of the millibubble was $(x, y, z) = (0.39, 0.36, 2.00)$ where the cross is marked. White circles indicate instantaneous positions of microbubbles and 58 velocity-vectors of microbubbbles were detected. The result reveals that the proposed system can be applied to the two-phase flows involving bubbles with micro- and milli- meter sizes.

(*a*)

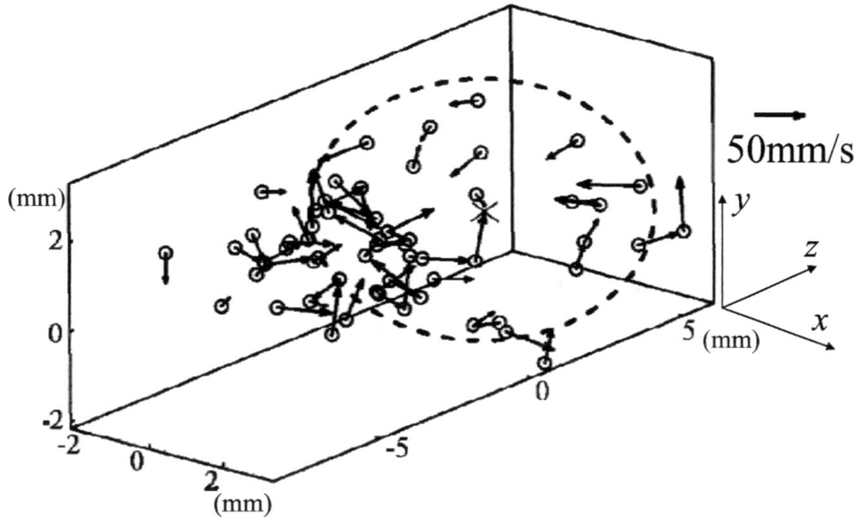

(*b*)

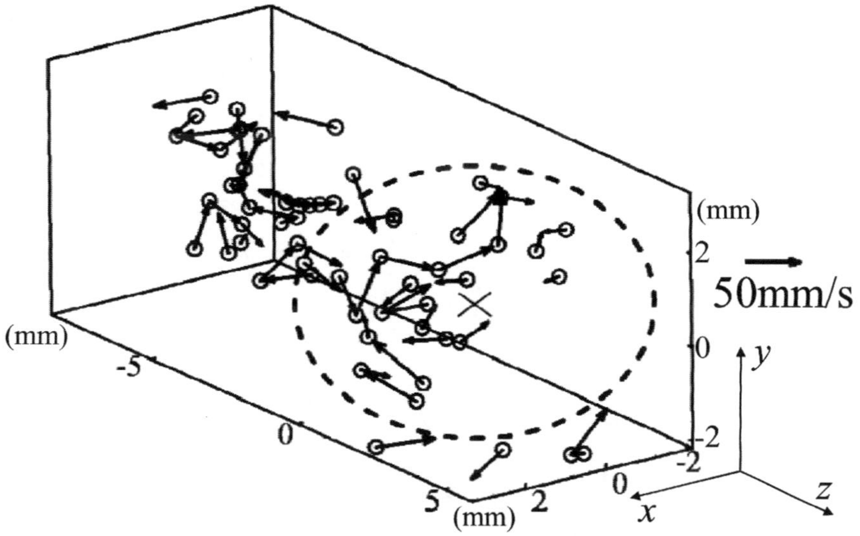

Fig. 11. Instantaneous position of the millibubble and velocity-vectors of microbubbles (*a*) and (*b*), the results from different views based on the same data

4. Conclusion

A new method is proposed in order to measure the three-dimensional positions and velocity-vectors of bubbles with various sizes. The method is realized by combining holographic technique and stereo shadow imaging technique. A millibubble is measued by stereo shadow imaging technique and microbubbles are measured by stereo holographic technique. The important conclusions obtained from this work are summarized below.

1. To validate the proposed system, it was applied to the laminar flow in the rectangler channel. The mainstream velocity component agrees with the theoretical value and the profile of vectors in the stream-wise direction, u, and the channel width-direction, z, showed parabolic shape.

2. The system was applied to the measurement of the microbubbles whose sizes are from 50 μm to 80 μm. Although many microbubbles have the position uncertainty from 2 mm to 5 mm in Z-axis, accurate positions are detected from two laser beam view. It is revealed that the system can be applied not only to the single-phase flow but also to the two-phase flows.

3. The system was applied to the simultaneous measurement of a milibubble and microbubbles. In the present study, with two laser beams, a millibubble and microbubbles are measured by shadow imaging technique and stereo holographic technique, respectively. The instantaneous position of a millibubble is evaluated and instantaneous three dimensional velocity-vectors of microbubbles are detected. It is revealed that the system can be applied to the two-phase flows which involve bubbles with micro- and milli- meter sizes.

The proposed DHPTV method is a promising technique to measure motions of bubbles with various sizes. In the near future, the measurement system will contribute to investigate the mechanism of bubbly flows.

ACKNOWLEDGEMENTS

This work was subsidized by Grant-in-Aid for Scientific Research (No. 18206024) from Ministry of Education, Culture, Sports, Science and Technology in Japan. The third author is supported by Research Fellowships for Young Scientists (No. 19-8799) from Japan Society for the Promotion of Science.

NOMENCLATURE

(X, Y, Z) pixel coordinates

(x, y, z) real space coordinates

i	the square root of -1	[-]
k	wave number	[-]
a	averaged brightness value	[-]
$\vec{D}$	displacement vector of the position	[-]
t	time	[s]

Greek Letters

λ	wavelength	[nm]
θ	crossing angle between z-axis and the laser beam	[rad]

Δx	distance between center position of x and original point in the x-direction	[mm]
δ	root-mean-square deviation value	[-]

Subscripts

He-Ne	He-Ne side view
YAG	YAG side view

REFERENCES

Kitagawa, A. et al. (2004). "Two-phase turbulence structure in a microbubble channel flow," Proc. 5th Symp. Smart Control of Turbulence.

Kitagawa, A. et al. (2005). "Flow structure of microbubble-laden turbulent channel flow measured by PIV combined with the shadow image technique," Exp. Fluids, 38, pp. 466-475.

Linden, R. and Merzkirch, W. (2002). "A novel PIV technique for measurements in multiphase flows and its application to two-phase bubbly flows," Exp. Fluids, 33, pp. 814-825.

Zhang, J. et al. (1997). "Turbulent flow measurement in a square duct with hybrid holographic PIV," Exp. Fluids, 23, pp. 373-381.

Murai, Y. et al. (2000). "Inverse energy cascade structure of turbulence in a bubbly flow (PIV measurement and results)," JSME International Journal Series B, 43, pp. 188-196.

Murai, Y. et al. (2007). "Skin friction reduction by large air bubbles in a horizontal channel flow," International Journal of Multiphase Flow, 33, pp. 147-163.

Seok, K. et al. (2008). "Effect of particle number density in in-line digital holographic particle velocimetry," Exp. Fluids, 44, pp. 623-631.

Journal of Physics: Conference Series **147** (2009) 012055 doi:10.1088/1742-6596/147/1/012055

Research and application of flow visualization technique in large-scale hydraulic models*

Wen Cheng[1, 2], Bao-wei Hu[2, 3], Tian Wan [2], Zhang-bin Wang [2], Fujio YAMAMOTO[4]

[1] State Key Laboratory of Multiphase Flow in Power Engineering; Xian Jiao tong University, Xian 710049, China
[2] Key Laboratory of Northwest Water Resources and Environmental Ecology of Education Ministry; Xian University of Technology, Xi'an 710048, China
[3] ShaoXing University, ShaoXing,312000, China
[4] Fiber Amenity Engineering Course, Graduate School of Engineering, University of Fukui, Bunkyo 3-9-1, Fukui 910-8507, Japan

wencheng@xaut.edu.cn; emiwbp22@163.com; hbw@ zscas.edu.cn

Abstract. With the rapid development of computer and optical technique, as a powerful measuring method in hydrodynamics, flow visualization technique is able to get much flow field information in practical engineering. It also has an important meaning to projects and environmental areas. The image processing is the key to flow visualization for gaining flow field information. In terms of the PIV principles, the algorithm of PIV based on Fast Fourier Transformation (FFT) is studied, and the image mosaic program based on genetic algorithm has been compiled. The flow field of seabuckthorn flexible dam has been calculated and analyzed by using those methods as above-mentioned, and in the meanwhile, the application of flow visualization technique to measure the outdoor flow field has been studied exploringly, which proves that it is feasible to apply this technique to measure the large-scale outdoor flow field.

1. INTRODUCTION

Flow visualization is the major method to study the field of fluid dynamics. Its basic principle is that some maters are released in transparent or semitransparent fluid medium, and then make the flow field visualizable by optical function. It is the most intuitionistic and effective tool to study the basal flow phenomena, to realize characteristics of flow, and then to deeply explore its physical mechanism.(X. Liu. 2006) This technique attracts many human eyes and plays an important role in the study of fluid dynamics. Long before 1914, Hot Wire Film Anemometer was invented, which made great contributions to fluid measurement. However, this method interferes with flow field for its contacted measuring. Laser Doppler Velocimeter developed in 1960s has achieved contact free less measurement to flow field, but it is a single point measuring technique which is applicable for steady-state or unsteady-state flow field, but not for the study of time-variant one(L. Xu,et al. 2003).
Since 1980s, quantitative flow visualization technique has been developed rapidly with the noticeable development of electronic technique, image processing methods and signal analyzing theory. Particle

* Project supported by the National Natural Science Foundation of China(Grant No.: 50679071); and China Postdoctoral Science Foundation (Grant No.: 20070410378)

Image Velocimetry (PIV) was developed this time. Based on fluid visualization, it is the newest fluid measurement means developed by introducing the research achievements of modern computer technique, optical technique and image analysis technology (Y.Murai, 2008). PIV can not only show the physical pattern of flows, but also supply quantitative information of instantaneous whole flow field, which makes the study of flow visualization produce a qualitative leap forward.

As the most advanced technique of flow visualization, PIV can be widely used in engineering and environmental applications, especially in studying flow velocimetry of large-scale experimental models in irrigation projects and natural waters. The velocimetry theory of single-phase two-dimensional flow field in lab has been developed maturely by 30 years' research. However, for the limitation of its experimental setup and conditions, so far, many domestic researchers have been major in studying the application of PIV/PTV to indoor mini-type experiment table, and the application of visualization technique to outdoor flow field observation is singular. In order to extend this advanced velocimetry technique to outdoor and explore its effect in sea buckthorn flexible dam, this paper studies the changing modality of flow field in field observing sea buckthorn flexible dam by applying the PIV.

2. METHOD OF PARTICLE IMAGE VELOCIMETRY

The basic principle of PIV can be described in the following operations (Y.Murai, et al. 2007): the nearly noise-free tracer particles with good traceability and dispersion are seeded to the flow field, particle velocity is used to represent the flow velocities at the corresponding positions in the flow field, and the powerful lighting system is used to illuminate a measurement plane in the flow field. The photographing method is used to record the positions of those particles which get exposure twice or more times, then the displacement of each particle are available by the image analysis techniques. As a result, the velocity vector of each point in the flow can be obtained from the displacement and exposure time interval, and other motion parameters can also be calculated, whereby the local motion laws of fluid can be derived (X. Wang, et al. 1998; Wen Cheng, et al. 2005).It can be expressed as the following equation:

$$\begin{cases} u(x,y) = \dfrac{S_x \times k}{dt} \\ v(x,y) = \dfrac{S_y \times k}{dt} \end{cases} \quad (1)$$

The performance of the whole flow field is evaluated by Eq. (2):

$$C = \frac{\sum(\phi_{th} \cdot \phi_R)}{\sqrt{\sum \phi_{th}^{\ 2} \cdot \sum \phi_R^{\ 2}}} \quad (2)$$

The PIV technique includes three steps as follows: (1) Acquisition of the digital images; (2) Compiling the PIV programs; (3) Data post-processing. Based on the principle of PIV, using CCD cameras to constantly shoot the flow field, and then the digital images can be obtained. The FFT-based Gray level cross-correlation method is adopted in the images, and then, the tracing images can be divided up from the measured ones in experiment by compiled programs. Finally, the algorithms can be achieved by using MATLAB to compile programs.

3. THE APPLICATION OF FLOW VIZUALIZATION IN PRACTICAL ENGINEERING

The sea buckthorn flexible dam mainly makes use of the principles in fluid dynamics that the drag reduced flow formed by flow obstacle preventing water, flow rate reduction and silt settlement caused by flood sediment retardation, and pressure head going through the interspaces between main branches of sea buckthorn. Planting sea buckthorn in gully between mountains, and when flood bringing with silt goes through the sea buckthorn, the silt will be lagged in the upper or middle reaches of sea buckthorn cluster, which can control water and soil loss of gullies and intercept silt(Ri-jun Zhang,2006) .

Explored and studied for more than 20 years, many qualitative results have been obtained. In order to deeply realize the watertight effect of sea buckthorn, and explore the flow property and the effect low of sea buckthorn to flow rate alteration, this paper observes the flow field changing form of sea buckthorn in field with recently advanced PIV technique. In order to get the data of basic images, a flow rate observation experiment was carried out successfully in experimental base of sea buckthorn in Hua County, Shaanxi Province, China in November, 2006. This paper mainly uses 2# sea buckthorn dam as the research object to explain that it is feasible to observe the outdoor flow field by applying the PIV technique.

3.1 Experimental Setup

This test uses daylight as light sources. The tracing particles are circle white foams with good light rain reflecting characteristics and small density, which diameter is about 5mm. Fig.1 shows the test system.

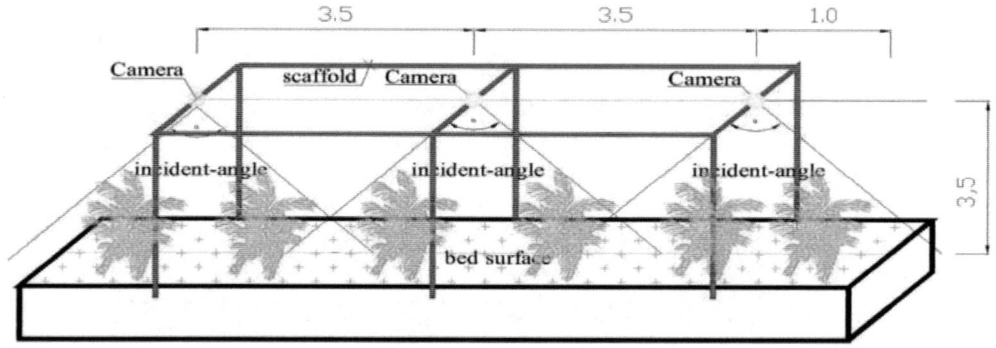

Fig. 1 The layout for the test facilities （m）

The frequency of camera used in experiments is 30frame/s. The images in this paper employ one frame every 3 frames. Two continuous moments at some state in test are intercepted to shoot the flow field. The whole length of 2# sea buckthorn dam is 9.64m. The camera is fixed on the 3.5m-high scaffold. With the limitations of experimental conditions such as the performance and objective distance of camera, three cameras are used to cover the whole bed in this test, and in the meanwhile, the spillway face is divided into three sections, which is upper、middle and bottom, respectively. Fig 2 shows the flow field image of each section.

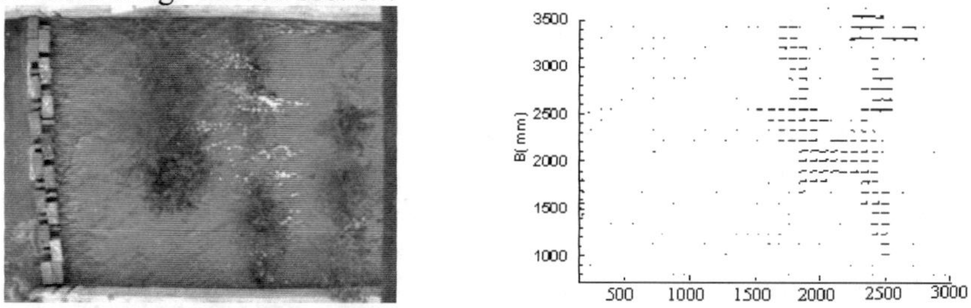

(a)The instantaneous image and flow velocity vector graph of the dam in the upriver

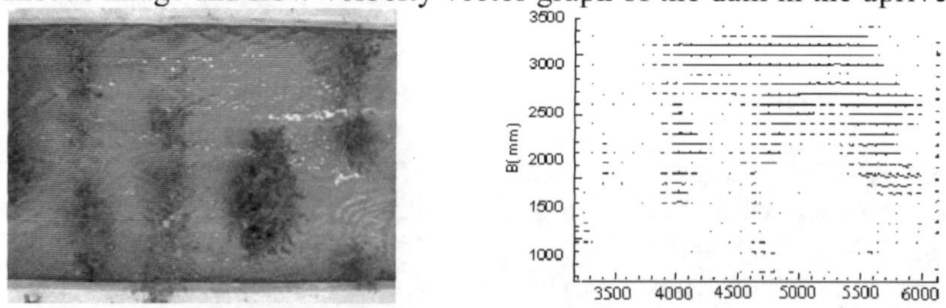

(b) The instantaneous image and flow velocity vector graph of the dam in the middle reach

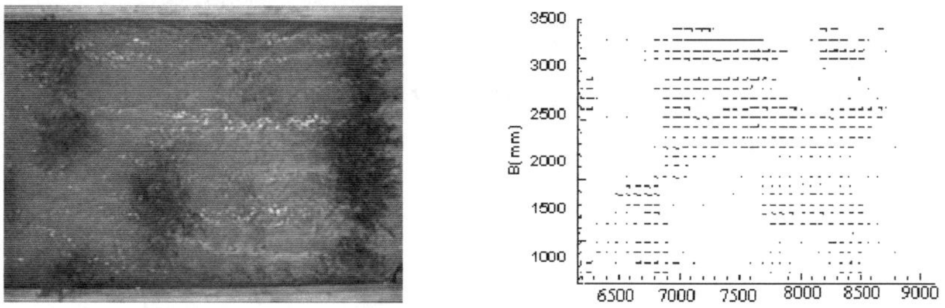

(c)The instantaneous image and flow velocity vector graph of the dam in the backward section

Fig. 2 Flow field images in each section of 2# bed

3.2 Experimental results and Analysis

3.2.1 Calculation of flow field

The flow changes in flexible dam for test process are simulated by the water level variation of the impounding reservoir which supplies water to the dam. Fig.3 shows the time variation of discharge in impounding reservoir.

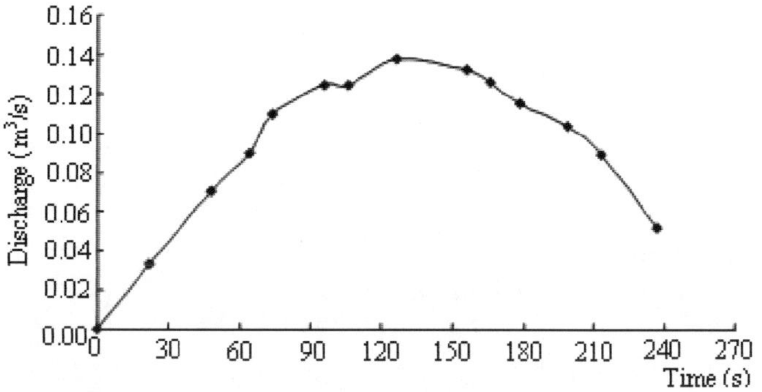

Fig. 3 The time variation of discharge in impounding reservoir

In order to calculate the whole flow flied of the flexible dam, this paper studies the method of image mosaics in the pre-processing stage. The basic principle of this method is used for reference, and the flow field is calculated. On the basis of this data, the digital flow field is mosaic. Fig.4 shows the result. The discharge is $0.11 m^3/s$. The water is flowing from left to right. L represents the length of the dam bed, while B is width. The arrows in the field vector graph represent the velocity of tracing particles, and the length expressed the size of velocity (1mm=0.34m/s), while the arrows represent the direction. Comparing to the real flow field, it is shows that the field vector graph obtained by PIV can fundamentally reflect the flow movement characteristics such as the field states and the size of velocity.

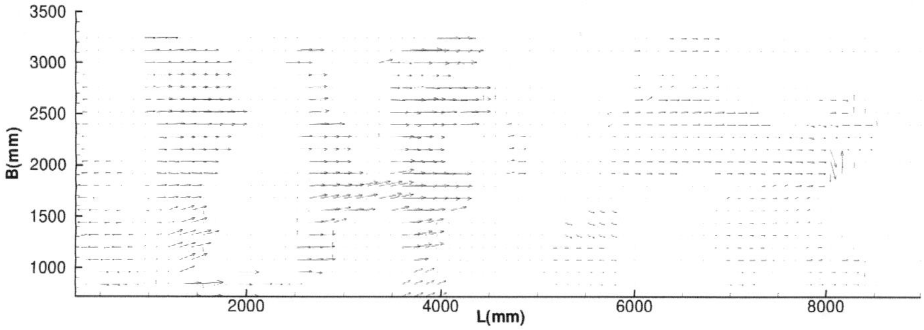

Fig.4 The whole flow field shown as vector graph in 2# dam

3.2.2 The analysis of the results

Through analyzing the distribution of the flow field, the energy dissipation produced by the flexible dam can be reflected.

(1) **The velocity distribution along the flexible dam bed for the same discharge**

In order to explore the change in the velocity along the flexible dam bed for the same discharge, on the basis of the embarking time and order of the tracing particles in the image sequence, the discharge of the tracing particles embarked each time and its corresponding velocity in each cross section can be deduced by combining the outflow processing curve of impounding reservoir. Fig. 5 shows the flow velocity changing process along the 2# dam bed in 5 types of different discharges.

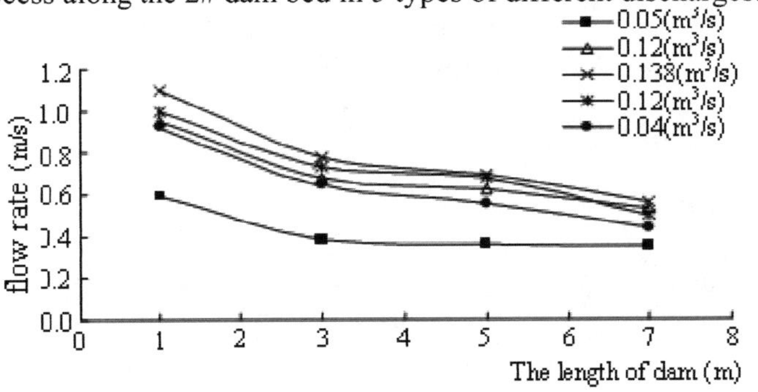

Fig.5 The relationship between flow velocity and the length of dam along 2# bed for different discharges

Fig.5 shows that the variation in the velocity for each discharge along the dam bed is decreasing. When the water flow comes into the sea buckthorn flexible dam, the velocity curve presents descending trend, especially in the first row. The changing characteristic of Fig.5 shows that the water flow velocity is continuously reducing because of the existence of the sea buckthorn, which explains the preventing water effect of the sea buckthorn. This conclusion is in concordance with the results obtained in the indoor lab flume in Sin kiang Agricultural University, China. (Y. Cheng,et al 2003, 2004)

Observing from the discharge angle, the effect of the sea buckthorn on the flow is presenting certain laws because of the different orders the flow coming into the sea buckthorn dam. As the discharge is small in the beginning stage of the test, the sea buckthorn is not lodging and the backwater phenomenon doesn't appear in the flexible dam, which shows that the flood detention action of the dam is not remarkable. With the discharge becoming heavier, the detention function will come to be very obvious. Fig.5 shows that the velocity in the discharge of $0.12m^3/s$ is as same as that in the discharge of $0.04m^3/s$ in the later stage, which explains that the flood peak discharge is delayed by sea buckthorn flexible dam. This result is the same as the conclusion obtained by Wang Cun (2003) in Hohai University, China who studies the effect of the plants on the flow.

(2) **The variations of velocity in the same monitoring cross section**

In order to know the flow rate variation in the same cross section of the whole outflow process, the outflow discharge of impounding reservoir is calculated by volume method, and the locations that 1m downstream of entrance region and 3m、5m、7m away from the dam site are set as 4 monitoring cross sections. After measurements are made once per 30s in these sections, and 8 times each section, then the average flow rate is calculated. The result is shown in Fig.6.

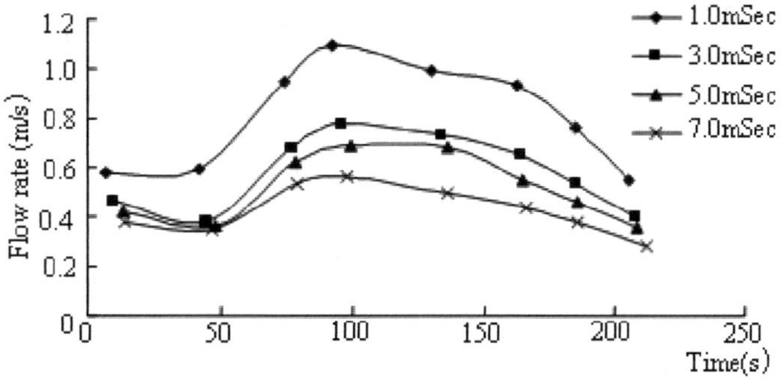

Fig.6 The time variation of flow rate in different monitoring cross-section of 2# dam bed

The velocity at 1.0m section is drafted as inlet velocity. Fig. 6 shows that the flow rate variation trend at each section is the same as the outflow changing trend of reservoir. The velocity in the first cross-section takes a maximum value of 1.10m/s, at which the sea buckthorn doesn't have remarkable influence on the entrance region water flow. Compared with the first cross-section, the flow rate at the middle section is reducing obviously. It can be explained that according to the principle of the continuity of flow movement, the backwater region appears in the dam for the existence of the sea buckthorn. Because of the interactions of the backwater flow before the dam and the flexible features of the dam body, the flow at the last monitoring cross-section in the middle and lower parts reaches has changed to be steady, which is beneficial to the sediment disposition. Every monitoring cross-section has undergone the same outflow process; however, because of the existence of the sea buckthorn, the flow rate at each cross-section seems to be reduced step by step, which explains that it is feasible to employ the PIV to observe the flow field of flexible dam(Zhang-bin Wang et al. 2008).

3.3 Effect of the sea buckthorn flexible dam on distribution of soil grain size

In order to know that when the flow conditions in channal are changed by sea buckthorn flexible dam, whether the soil can be sorted and intercepted naturally, a certain research about the distribution of soil grain size in spacial and time have also be done in this paper.

Three sections of 2# dam in upstream, middle reaches and downstream are employed to correspoud with the above-metioned research. And five soil layers in each section are used to sample soil, which is 0～20、20～40、40～60、60～80、80～100cm respectively. The soil samples are graded by sieving method after drying. Fig.7 shows the result.

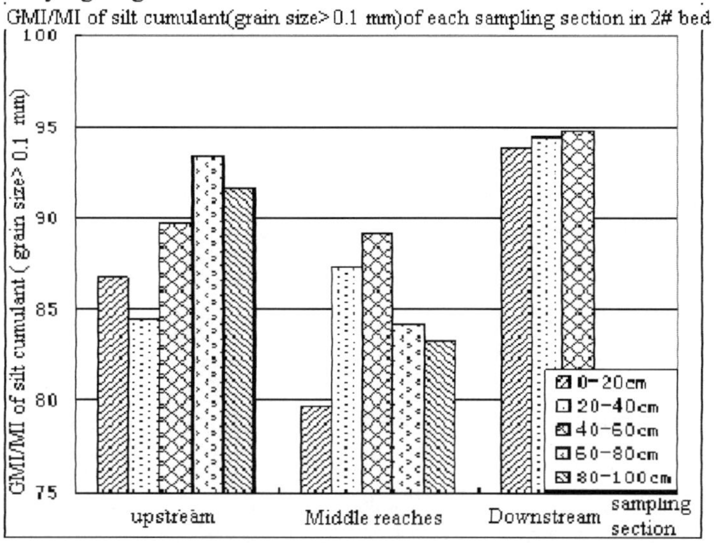

Fig.7 The cumulative mass percentage of sediment grain size bigger than 0.1mm in each layer of different sampling sections

The content of soil particles whose grain sizes are larger than 0.1mm is increasing layer by layer. The main reason is that the water flow is held back by the sea buckthorn, and the flow rate is decreasing along the dam, then the coarse particles cannot move with water (Tao Zhang. 2008). The existence of the sea buckthorn has changed the flow pattern, then the distribution of the soil grain size can also be changed, which has greatly improved the erosion resistant of the soil.

5. CONCLUSIONS

The basic principle of flow visualization is studied in this paper. The flow field measurement with flow visualization technique is used in the practical engineering, and the following conclusions have been obtained.

(1) Through analyzing the digital flow field measured by PIV, it can be seen that the sea buckthorn has remarkable influence on the flow characteristics, especially on the velocity, and the flow rate in the dam is obviously smaller than that before the dam. It means that the water flow energy loss caused by the resistance of the sea buckthorn flexible dam to flow is mainly concentrated in the water-saturated stage, and it is most possible for silt depositing in this area, which provides the theoretical basis for controlling soil erosion of gully channel with sea buckthorn flexible dam.

(2) By analyzing the distribution of soil grain size in the channel with the existence of the sea buckthorn, it also can be obtained the sand-blocking effect of the flexible dam.

(2) The feasibility of the PIV technique to be used in field test site has been verified, which provides the technological supports for outdoor flow field observation in the future.

ACKNOWLEDGEMENTS

The authors thank Prof. Huai-en Li of Xi'an university of Technology and Prof. Yamamoto of University of Fukui for useful discussion.

NOMENCLATURE

S_x	Displacements of particles in water flow direction	[m]
S_y	Displacements of particles in the direction normal to water flow direction	[m]
k	Scale	
dt	exposure time.	[s]

Greek Letters

Ø	target variable

Subscripts

th	theoretical value
R	reconstruction value

REFERENCES

X. Liu.(2006). "Study on particle image velocimetry and inverse analysis in aeration tank,"M.D. Xi'an University of Technology ,Shaanxi,CHINA

L. Xu,et al.(2003) "Study on development of Particle Image Velocimetry" J. Advance of Dyn. 33(04),pp. 533-540

Yuichi Murai.(2008) "Increase of effective viscosity in bubbly liquids from transient bubble deformation,"Fluid Dynamics Research. 40(7-8):565-575

X. Wang, et al. (1998) "The Initial exploration of particle image velocimetry (PTV-PIV) in two-phase flow,"J. Dyn. 30(01),pp. 121-125

Yuichi Murai, Kazumasa Inaba, Yasushi Takeda(2007) "Skin friction reduction by large air bubbles in a horizontal channel flow,"19(2),pp.223-229

Wen Cheng, et al.(2005) "Bubble velocity measurement with recursive cross correlation PIV technique," J.Fluid Measurement and Instrumentation. 16,pp. 35-46

R. Zhang(2006). "The water flow experiments of seabuckthorn flexible dam," M.D. Xi'an University of Technology ,Shaanxi,CHINA

Y. Cheng,et al. (2003) "Experimental study on Hydrodynamic behavior of channel vegetation," J. Sin kiang Agricultural University,26 (2),pp.59-64

Y. Cheng. (2004) "Hydrodynamic characteristics of plant flexible dam based on experiment," M.D. Sin kiang Agricultural University, Sin kiang, CHINA

C. Wang. (2003) "The water flow test study of river channel with vegetations," M.D.Hohai University, Jiangsu, CHINA

Z. Wang, et al.(2008) "Research and application of flow visualization technique in the fluid measurement of Sea buckthorn plant flexible dam," J. Hydro. Eng.27(1),pp.28-31

Z.Wang. (2008) "Research and application of flow visualization technique in Sea buckthorn plant flexible dam outdoor fluid measurement," M.D. Xi'an University of Technology, Shaanxi, CHINA

T.Zhang. (2008) “Experimental research on sand-blocking and ecological benefit of sea Sea buckthorn plant flexible dam," M.D. Xi'an University of Technology, Shaanxi, CHINA

Journal of Physics: Conference Series **147** (2009) 012056 doi:10.1088/1742-6596/147/1/012056

Phase Separation Techniques in Two-Phase Microchannel Flow

K Hishida [1], M Ichiyanagi [1], Y Sato [1]

[1] Keio University, 3-14-1 Hiyoshi, Kohoku-ku, Yokohama, 223-8522 JAPAN

hishida@sd.keio.ac.jp

Abstract. Two kinds of phase separation techniques in a microchannel have been developed to advance microfluidic devices. One is the technique to combine acoustic radiation and electrostatic forces, and these forces act on particles in aqueous solution perpendicular to the flow direction simultaneously. The acoustic radiation force trapped both 9.6 μm and 4.0 μm diameter particles at nodal positions in the standing wave field, and the electrostatic force moved only 4.0 μm diameter particles towards the anode. The magnitudes of each force were evaluated by using particle images and particle tracking velocimetry for particle separation in the flow field. On the other hand, another technique is to utilize bilayered acoustic radiation forces. The ultrasonic transducers with 5 and 2.5 MHz were equipped parallel to the flow direction. Both 15.0 μm and 4.0 μm diameter particles in the aqueous solution were trapped at the nodes of the upstream in 5 MHz sound field, and 2.5 MHz transducer was radiated to move only 15.0 μm diameter particles toward a nodal position of its sound field. The exposure time and time cycle of 2.5 MHz transducer were determined to be 0.1 s and 1 s from the migration times of large and small particles transfer to the nodal positions, respectively. It is confirmed that the continuous and selective separation based on particle diameter was accomplished by the present techniques.

1. INTRODUCTION

Material separation and dispersion control are significant techniques for product quality enhancement in various fields including engineering, medical science and sitology (Petersson *et al.* 2004; Scampicchio *et al.* 2004). These techniques are used for pre-processing, sample dilution or extracting impurities. The recent development of microfabrication technology has yielded microfluidic devices, which allows to possess numerous functions on a single miniature device (Auroux *et al.* 2002; Reyes *et al.* 2002; Sinton 2004). The conventional techniques using a centrifugal separation or membrane separation have a number of difficulties in implementing within these miniaturized devices due to their larger apparatus in comparison with the devices. Furthermore, many are contact-type techniques with the subject matter, which may cause the disruption of subject and is difficult to handle samples of very small quantity. Thus a non-intrusive and continuous separation technique is required to implement in microfluidic devices.

A number of non-intrusive techniques such as the magnetic force (Relle and Grant 1998; Ueno *et al.* 1990) and laser radiation pressure (Ashkin *et al.* 1986; Ito *et al.* 2002) were applied to microfluidic devices, and the separation technique utilizing acoustic radiation force gives a significant attention in the recent years due to their possession of numerous unique properties (Hawkes *et al.* 2002; Petersson *et al.* 2004). The acoustic radiation force affects particles in proportional to physical parameters such as particle diameter, density and compressibility, and the force moves particles toward the nodal positions of the standing wave field perpendicular to the flow direction. Furthermore, the electrostatic force is one of the most non-intrusive separation techniques in microfluidic devices (Harrison *et al.* 1993; Woolley *et al.* 1998), and the electrophoresis is commonly used for material transport and separation. However, since this technique depends on the electrical properties of the subjects, its applicability is limited. For further enhancement of the separation efficiency in microfluidic devices, it is expected to develop an innovative separation technique.

The objective of the present work is to develop two kinds of novel separation techniques based on particle diameter. One is the technique to combine acoustic radiation and electrostatic forces. The acoustic radiation force traps both large and small size particles at nodal positions in the standing wave field, and the electrostatic force moves only small size particles towards the anode. The magnitudes of each force were evaluated by using particle images and particle tracking velocimetry for particle separation in the flow field. On the other hand, another technique is to utilize bilayered acoustic radiation forces. This technique utilizes two ultrasonic transducers with 5 and 2.5 MHz and depends on only physical parameters such as the particle diameter, density and compressibility. Two kinds of particles with the different diameters are selected, and both the migration times of particle transfer to the nodal positions and the acoustic radiation force are estimated from the particle velocity data. 5 MHz ultrasonic sound traps both the large and small particles at the nodal positions and subsequently only large particles are moved toward the nodal position of 2.5 MHz sound field, because the 2.5 MHz transducer is radiated at the pulse time to trap only large particles which is determined by the migration times of particles transfer to the nodal positions.

2. PRINCIPLES OF PHASE SEPARATION TECHNIQUE

2.1 Effect of Acoustic Radiation Force on Particles

When an acoustic standing wave is set up in fluid, particles within the standing wave are affected by acoustic radiation force, F_{ac} (Laurell *et al.* 2007). The particles are translated to either nodes or anti-nodes of the standing wave depending on their physical properties relative to the ones of the suspended fluid. Acoustic radiation force is given by:

$$F_{ac} = -\frac{4}{3}\pi r^3 k E_{ac} A \sin(2kl), \qquad (1)$$

where r [m] is the particle radius, k [-] is the wave number, E_{ac} [J/m^3] is the time-averaged acoustic energy density, l [m] is the distance from the nearest node, and A [-] is the acoustic constant factor which is derived from both the particle density and its compressibility. Figure 1 illustrates the schematic of the particle motion affected by the acoustic radiation force within the standing wave. Particles with different diameter are suspended in a channel as shown in Figure 1 (*a*). Within the standing wave field, nodes with small acoustic pressure amplitude and anti-nodes with large amplitudes exist one after another as represented in Figures 1 (*b*) and (*c*). The direction of the force is determined by the sign of A factor of equation 1; a positive A factor results in movement toward a pressure node and a negative in movement toward a pressure anti-node. Generally, solid particles in aqueous solutions are transported to nodes. Furthermore, the acoustic radiation force is proportional to third power of particle radius as described in equation 1. When the acoustic radiation force is constant, firstly large size particles are moved toward the nodes of the sound in shorter time than small size particles as exhibited in Figure 1 (*b*), and subsequently small size particles are collected at nodes as shown in Figure 1 (*c*).

2.2 Phase Separation Technique Utilizing Acoustic and Electrostatic Force

The acoustic standing wave field was yielded by equipping the ultrasonic transducer and reflector plate perpendicular to the flow direction. Electrodes were located at four corners of the microchannel.

When applying the acoustic radiation force, particles are collected at the nodes as shown in Figure 2 (*a*). When applying the electrical field, particles with a negative charge are moved towards the positive electrode by electrostatic force, F_e. Thus, both the acoustic radiation and electrostatic forces act on the particles simultaneously. When the acoustic radiation force is larger than the electrostatic force as shown in Figure 2 (*b*), large and small size particles are confined at nodal positions in the standing wave field. When the electrostatic force is dominant as shown in Figure 2 (*c*), large and small size particles are transferred towards the positive electrode. Finally, it is observed in Figure 2 (*d*) that large size particles were collected at the node of the sound field and small size particle were transferred towards the positive electrode. This is caused by controlling the balance of the acoustic radiation and electrostatic force, that is, $F_{ac} > F_e$ for large size particles and $F_{ac} < F_e$ for small size particles, which yields to move only small size particles towards the positive electrode, because the acoustic radiation force is proportional to third power of particle radius. Thus it is possible to separate the large size particles from small size particles by setting appropriate acoustic radiation and electrostatic forces which are important parameters of the present separation technique.

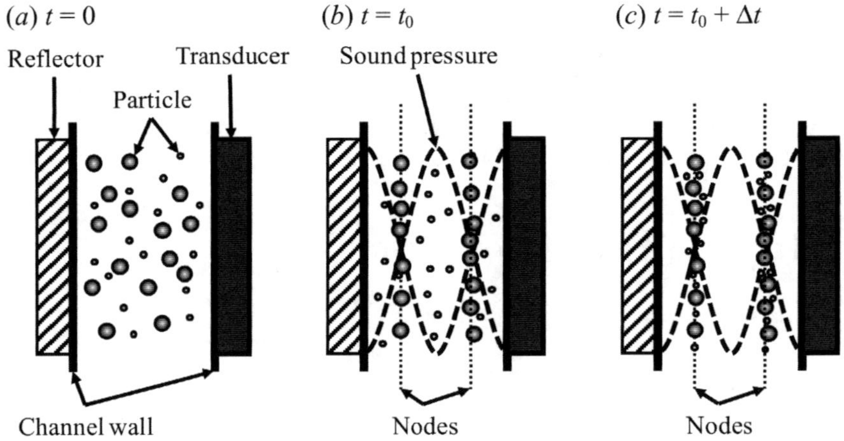

Figure 1. Schematic of particle motion affected by acoustic radiation force in standing wave field. (*a*) Particles are suspended in a channel prior to an application of acoustic radiation force ($t = 0$). (*b*) Only large size particles are collected at nodes of a sound field at $t = t_0$ after an application of acoustic radiation force. (*c*) Both large and small size particles are collected at $t = t_0 + \Delta t$.

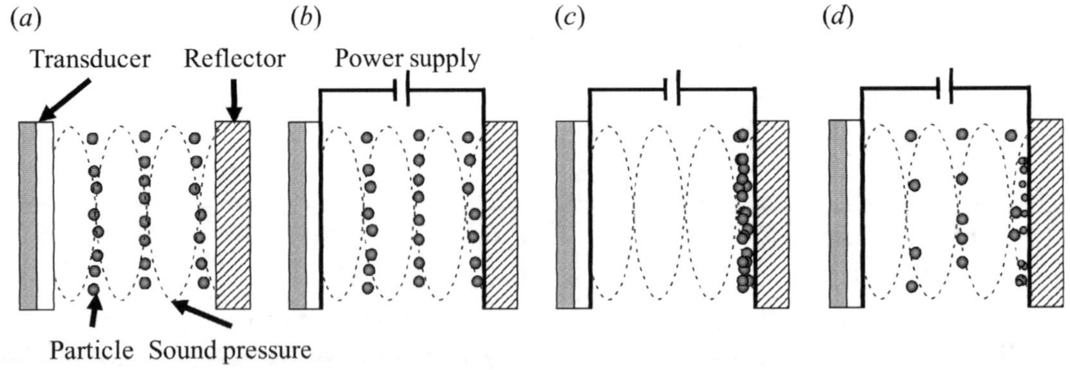

Figure 2. Concept of separation technique utilizing acoustic radiation and electrostatic forces. (*a*) Particles are trapped in the standing wave field. (*b*) Particle motion when $F_{ac} > F_e$ for both large and small size particles. (*c*) Particle motion when $F_{ac} < F_e$ for both large and small size particles. (*d*) Particle motion when $F_{ac} > F_e$ for large size particles and $F_{ac} < F_e$ for small size particles.

2.3 Phase Separation Technique Utilizing Bilayered Acoustic Force

Figure 3 represents the concept of the present separation technique. Acoustic standing wave fields were generated by using two ultrasonic transducers and a reflector plate, which were placed parallel to each other in the span-wise direction of the channel as illustrated in Figure 3 (*a*). Both large and small size particles were randomly suspended within in a flow field. The oscillating frequency of ultrasonic transducer at upstream was two times as large as that at downstream, because two and one nodes were generated at the upstream and downstream, respectively. Figure 3 (*b*) indicated that when entering the upstream sound field, both particles were collected at the two nodes of standing wave because the upstream transducer was constantly radiated. On the other hand, it is observed at the downstream of Figure 3 (*c*) that large size particles were collected at the node of the downstream sound field and small size particle were kept at the nodal positions of upstream sound field. This is caused by controlling the exposure time of downstream ultrasonic transducer which yields to move only large size particles towards the node, because the acoustic radiation force is proportional to third power of particle radius and the migration time to move large size particles is shorter than that to move small size particles as shown in Figure 1. Thus it is possible to separate the particles into three threads by setting an appropriate exposure time of downstream transducer. Since the exposure time is determined from the migration time to move particles towards the node, the migration time, t_{steady}, is an important parameter of the present separation technique. The present work defined the following equation as the separable condition of particles:

$$t_{trap} \geq t_{steady} \text{ for large size particles },$$
$$t_{trap} \leq t_{steady} \text{ for small size particles },$$
$$t_{trap} = \frac{Y_l}{v} \times \frac{t_{ex}}{t_{per}}, \tag{2}$$

where Y_l [m] is the length of downstream transducer, v [m/s] is the bulk velocity and t_{ex} [s] and t_{per} [s] are the exposure time and time cycle of exposure of the downstream transducer, respectively. Y_l /v is

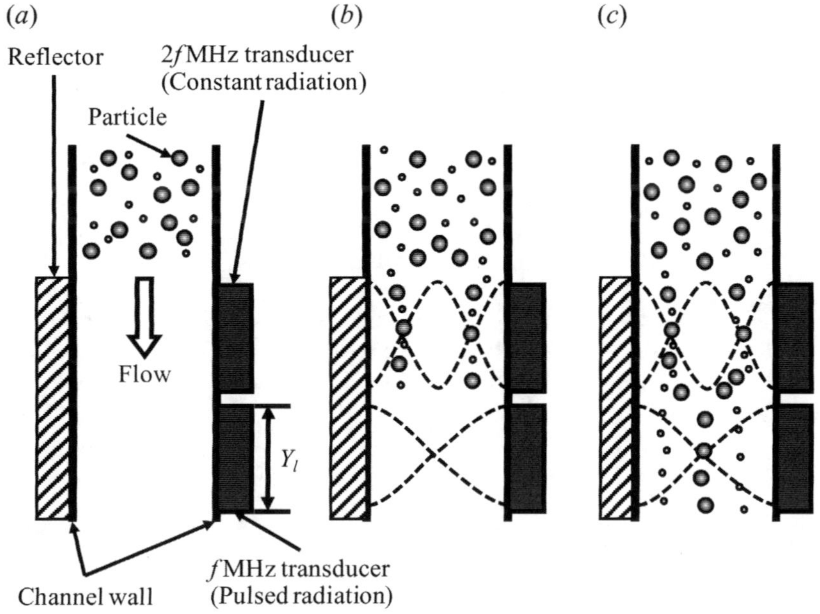

Figure 3. Concept of separation technique utilizing bilayered acoustic radiation forces.

the time for particles to pass through the downstream transducer, and t_{ex}/t_{per} is the duty ratio for the exposure of the downstream transducer.

2.4 Evaluation Technique of Acoustic Radiation Force and Migration Time

The conventional measurement techniques for the acoustic radiation force such as the hydrophone method are not suitable for measuring within micro-scale field, because they require the insertion of probes and involve sensitive axis alignments for accurate results. In the present work, since it is difficult to directly measure the acoustic radiation force, they were obtained by measuring the particle velocities affected by the acoustic force. Within the Stokes region, the relaxation time of particles, τ_p, is given by Clift *et al.* (1978):

$$\tau_p = \frac{r^2(2\rho_p + \rho_m)}{9\mu}, \tag{3}$$

where ρ_m and ρ_p [kg/m^3] are the density of fluid and particles, respectively, and μ [Pa·s] is the fluid viscosity. The relaxation time of the largest particle used in this work was O (10^{-6}) s, while the relaxation time of fluid flow, τ_f, was O (10^{-1}) s which was derived from the channel width divided by the bulk velocity. Since $\tau_p << \tau_f$, the particle motion is assumed to follow the flow field. In addition, when the volume concentration of particles is less than 2 %, the fluid motion generated by the particle motion is negligible (Probstein 1994). Thus, the acoustic radiation force described in equation 1 can be equal to the drag force, F_d, that particles receive from the fluid. The drag force is provided by:

$$F_d = 6\pi\mu r u, \tag{4}$$

where u is the particle velocity, as the experiments were performed in a stationary fluid. The effects of gravity and buoyancy can be ignored because the direction of sound is perpendicular to the direction of gravity. The particle velocity was obtained by particle tracking velocimetry (PTV), and the acoustic radiation force was evaluated by substituting the particle velocity to equation 4.

Furthermore, particles are separated by trapping at the nodal positions within the acoustic standing wave fields, so that the migration time to move particles towards the nodes, t_{steady}, is an important parameter. The present work evaluated the migration time by using PTV, and defined as the steady-state condition when the averaged span-wise velocity is less than 0.6 μm/s, which is the measurement uncertainty in 95 % confidence level of PTV.

3. EXPERIMENT APPARATUS

3.1 Microchannel

Figure 4 shows the schematic of the I-shaped flow channel for the particle separation with combining the acoustic radiation and electrostatic forces. The channel is composed of acrylic and glass, and is 2 mm width and 25 mm length with a depth of 1 mm. The ultrasonic transducer with oscillating frequency of 8.125 MHz was equipped at the center of the channel perpendicular to the flow direction. The oscillating frequency was set by considering transmission rate of sound, D, through surface boundaries (Takeda and Kikura 2002), which is given as:

$$D = 1\Big/\sqrt{1+\frac{1}{4}\left(m-\frac{1}{m}\right)^2 \sin^2\frac{2\pi d}{c}f}, \text{ when } m = Z_1/Z_2 \tag{5}$$

where d (= 2 mm) is the thickness of acrylic wall, c is the sound velocity within acrylic, f is the sound

Journal of Physics: Conference Series **147** (2009) 012056 doi:10.1088/1742-6596/147/1/012056

frequency, and Z_1 and Z_2 are the acoustic impedance of the materials at the interface. Table 1 compiles the acoustic impedance of each material. The transmission rate between water and glass is 0.23, and the transmission rate between water and acrylic is 0.76, so that the ultrasonic wave oscillated by the transducer penetrates the acrylic wall and fluid in the channel, and is reflected by the glass wall across the channel. In addition, four platinum electrodes with 100 μm diameter were placed along the microchannel wall, and those at the glass wall were set to be positive, and those at acrylic were set to be negative.

Figure 5 shows the schematic diagram of the I-shaped microchannel for the particle separation with bilayered acoustic radiation forces. The channel is composed of acrylic and glass, and is 300 μm width and 55 mm length with a depth of 1.5 mm. Ultrasonic transducers with oscillating frequency of 5 and 2.5 MHz were equipped perpendicular to the flow direction. The oscillating frequency was derived from equation 5. The point of origin was set at the center of transducer, and *X*, *Y*, and *Z*-axis were positioned at span-wise, stream-wise and depth-wise direction, respectively. 5 MHz sound field of upstream transducer provides two nodes at $x = 75$ and $x = 225$ μm, and the oscillating frequency of 2.5 MHz yields one node at the center of the channel ($x = 150$ μm).

Table 1. Acoustic impedance of various materials

	Acoustic impedance [$\times 10^6$ kg/m^{2}s]
Acrylic	3.26
Glass	12.8
Water at 20 degrees	1.48

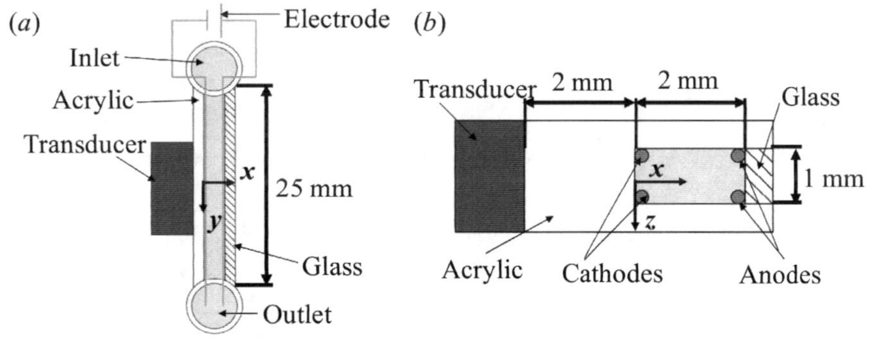

Figure 4. (*a*) Top view and (*b*) cross-sectional view of a microchannel for the particle separation with acoustic radiation and electrostatic forces.

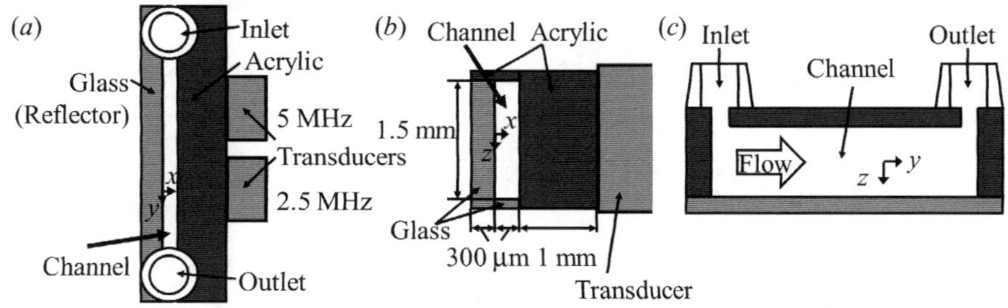

Figure 5. (*a*) Top view, (*b*) cross-sectional view and (*c*) side view of a microchannel for the particle separation with bilayered acoustic radiation forces.

3.2 Flow Visualization and Control System

Figure 6 (*a*) illustrates a schematic of the optical measurement system. A transmitted light by LED lamp (Ohm Electric Inc, LL-10W) was set below the stage and emitted towards the channel to project the shadow of particles. The particle images were captured by the CCD camera (SONY Co., Ltd, XCDL-X710CR, 1024 × 768 pixels, 24 bits, 30 fps) through the 10× magnification objective lens. This set-up corresponds to a measurement area of 433 × 325 μm^2. Figure 6 (*b*) shows the control system for the particle separation with the acoustic radiation and electrostatic forces. The transducer was oscillated at a sinusoidal wave by a function generator (Micronix Co., Ltd, MFG206). Platinum electrodes were connected to a DC voltage which was generated by high-voltage power supply (KEPCO Inc. BOP1000M). Figure 6 (*c*) exhibits the control system for the particle separation with the bilayered acoustic radiation forces. Each of the transducers was oscillated at a sinusoidal wave by a function generator. Relay switch driven by pulse generator (Quantum Composers Inc, 9300 series) was equipped to the 2.5MHz transducer because of cutting the signal from the function generator at intervals set on the pulse generator.

3.3 Particle and Fluid Properties

Table 2 lists the properties of the polystyrene particles. Three kinds of particles with 2.0, 4.0 and 15 μm diameters (Invitrogen Corp., FluoSpheres Sulfate Microspheres) were fluorescent, and the particle with 9.6 μm diameter (Interfacial Dynamics Co., White Sulfate Latex) was non-fluorescent. The particles were suspended in ion-exchanged water and were negatively charged because the particle surface was modified by sulfo groups ($-SO_3H$).

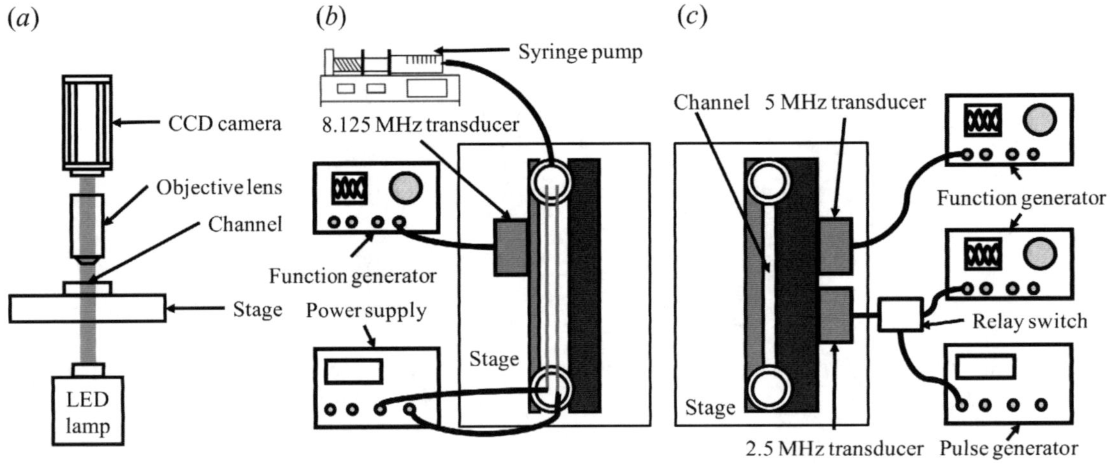

Figure 6. (*a*) Schematic of optical measurement system. (*b*) Control system to generate acoustic radiation and electrostatic forces. (*c*) Control system to generate bilayered acoustic radiation forces.

Table 2. Properties of polystyrene particles

Diameter	[μm]	2.0	4.0	9.6	15
Density	[g/cm^3]	1.055	1.055	1.055	1.055
Volume concentration	[%]	0.020	0.020	0.168	0.336
Absorption wavelength	[nm]	505	505	–	540
Emission wavelength	[nm]	515	515	–	560

4. RESULTS AND DISCUSSION

4.1 Particle Separation Utilizing Acoustic Radiation and Electrostatic Forces

4.1.1 Evaluation of Acoustic Radiation and Electrostatic Forces

The aqueous solutions with 2.0 µm, 4.0 µm and 9.6 µm diameter particles were injected through the channel inlet, and the acoustic radiation force acting on particles was evaluated by measuring particle velocities in a liquid phase flow and substituting particle velocities to equation 4. The particle velocities were averaged by 5 pixel intervals along *X*-axis. Figure 7 shows the profiles of drag force acting on the particles. It is found that the sinusoidal pattern with twice the oscillating frequency was observed and the acoustic radiation force was increasing with an increase in the particle diameter.

The electrostatic force acting on particles was evaluated by measuring particle velocities in a liquid phase flow and substituting particle velocities to equation 4. It is observed that the velocity distribution in the measurement area is uniform and the standard deviation of the velocity measurement was approximately 2.5 %. The relationship between the applied voltage and drag force is plotted in Figure 8. The particle velocities are summation of electrophoretic and electroosmotic flow velocities. The electroosmotic flow velocity was estimated to be approximately 10 µm/s at 15 V/cm by using the numerical simulation. Thus, it was evaluated that around 20 to 30% of the force shown in Figure 8 was due to electroosmotic flow.

The applied voltage has been determined from the relationship between the acoustic radiation and electrostatic forces. For 2.0 µm diameter particles, the amplitude of drag force and applied electric field, which the electrostatic force is equal to the acoustic radiation force, are 0.5 pN and 12 V/cm, respectively. This is the point where the dominant force of the particle motion changes from acoustic radiation force to electrostatic force. For 4.0 µm diameter particles, 15 V/cm is the point of equilibrium. For 9.6 µm diameter particles, the acoustic radiation force is dominant factor, because the acoustic radiation force is five times larger than the electrostatic force. Thus, their balances of the acoustic radiation and electrostatic forces enable to separate large size particles of 9.6 µm from small size particles of 2.0 and 4.0 µm.

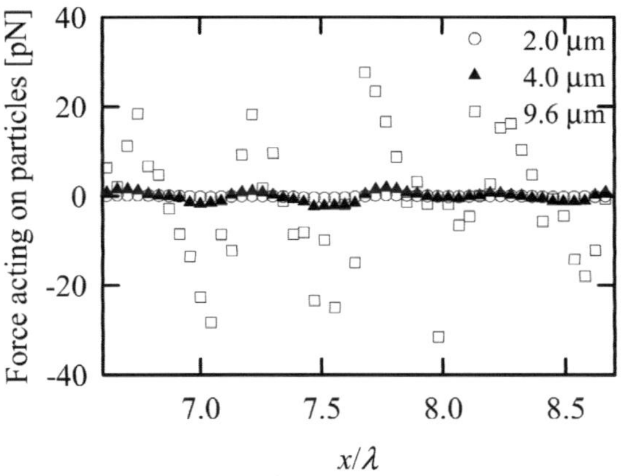

Figure 7. Profile of force acting on particles induced by acoustic radiation force.

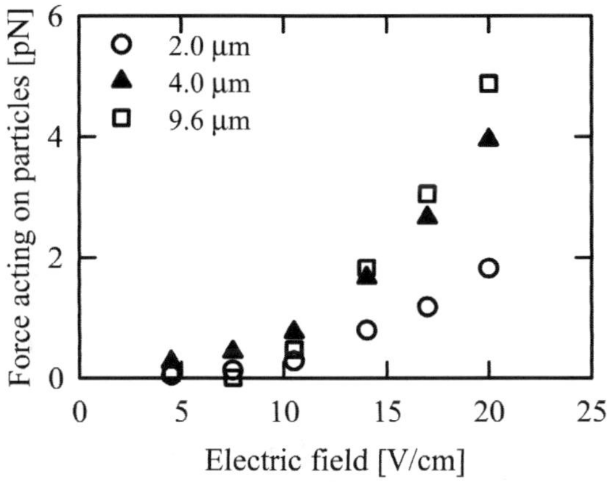

Figure 8. Relationship between electric field and force acting on particles induced by electrostatic force.

4.1.2 Demonstration of Particle Separation Utilizing Acoustic Radiation and Electrostatic Forces

For the demonstration of the particle separation, the flow field was formed by injecting the aqueous solutions with 2.0 μm and 9.6 μm diameter particles in the I-shaped channel as shown in Figure 4. The acoustic radiation force with the frequency of 8.125 MHz and the electric field of 14 V/cm were simultaneously applied. Bulk velocity was set at 45 μm/s by a pressure-driven flow which was generated by the syringe pump. Figure 9 shows the instantaneous particle images with velocity-vectors which were obtained by the PTV measurement. It is observed that 2.0 μm diameter particles were moved towards the span-wise direction, while 9.6 μm diameter particles were trapped at the nodal position of the standing wave field. The averaged span-wise velocity of 2.0 μm diameter particles was 23.1 μm/s, so that these particles were transferred to the location of half channel width in approximately 45 s. Thus, it is found that the particle separation utilizing the acoustic radiation and electrostatic forces is realized by evaluating the acoustic radiation and electrostatic forces. The separation efficiency of this technique will be improved by optimizing several factors, that is, the channel design, the transducer positioning, the flow velocity of continuous phase and the amplitude of applied voltage.

4.2 Particle Separation Utilizing Bilayered Acoustic Radiation Forces

4.2.1 Evaluation of Migration Time of Particles

In the separation technique utilizing the bilayered acoustic radiation forces, particles are selectively separated by using the difference of migration time to move towards the node of 2.5 MHz standing wave. Migration time of particles from nodal positions of 5 MHz standing wave to nodal position of 2.5 MHz was evaluated by the PTV measurement. Instantaneous images of 15 μm diameter particles in 2.5 MHz standing wave were presented in Figure 10, and black circles indicate particles. The present work defined as t = 0 s, when the acoustic radiation force of 2.5 MHz standing wave was applied to the channel as shown in Figure 10 (*a*). When the span-wise velocity becomes less than 0.6 μm/s as shown in Figure 10 (*b*), the particles were assumed to be steady-state. Figures 11 (*a*), (*b*) and (*c*) give profiles of brightness intensity of 4.0 μm, 9.6 μm and 15 μm diameter particles, respectively, which were obtained after each 10 iterative measurements. Black and white circles indicate brightness intensity when applying the acoustic radiation force by the sound field of 2.5 MHz and when reaching

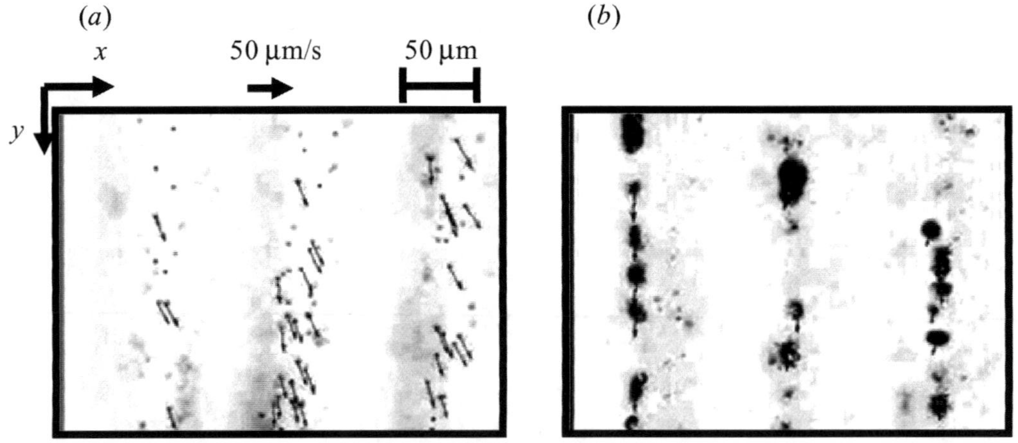

Figure 9. Instantaneous images of (*a*) 2.0 and (*b*) 9.6 μm diameter particles with velocity-vectors which were obtained by the PTV measurement.

the migration time to move particles towards the nodes, respectively. Figure 12 summarizes the migration time in terms of particle diameter, and exhibits that the migration time of 15 μm and 9.6 μm is approximately fifth part of that of 4.0 μm. Their difference of migration time enables to separate large size particles of 15 μm and 9.6 μm from small size particles of 4.0 μm.

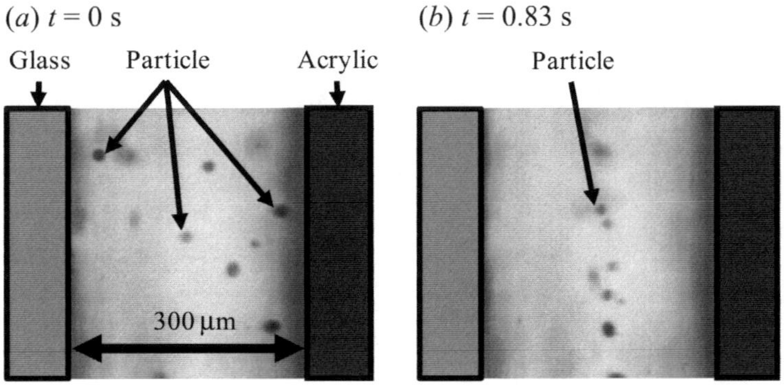

Figure 10. Instantaneous images of 15 μm diameter particles in sound field of 2.5 MHz.

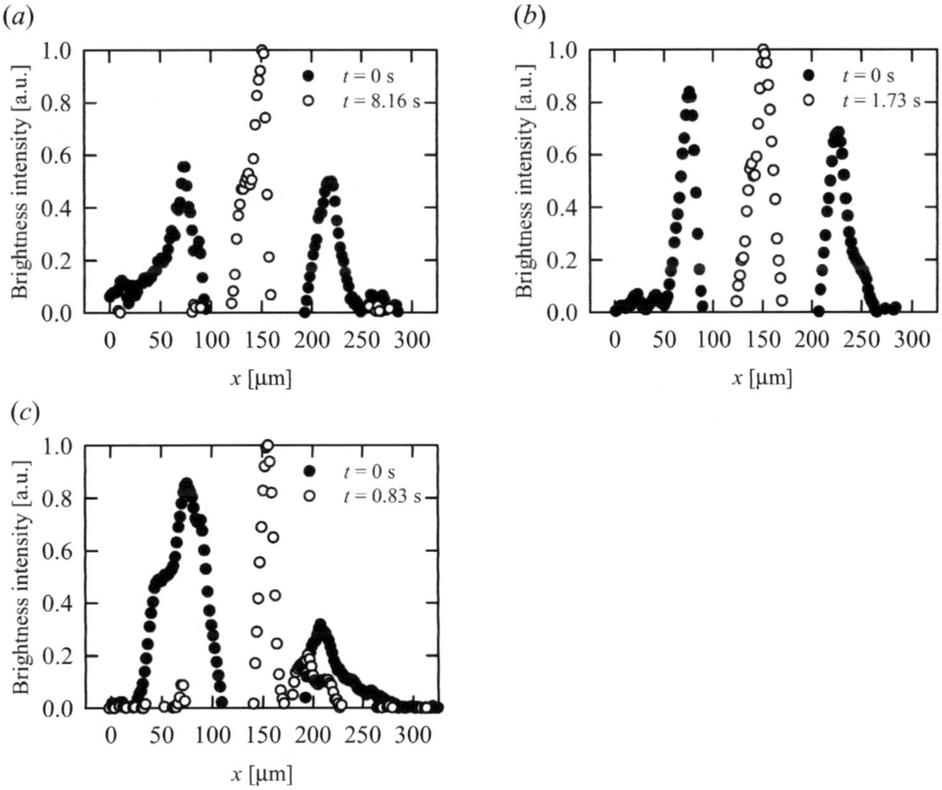

Figure 11. Profiles of brightness intensity of (*a*) 4.0 μm, (*b*) 9.6 μm and (*c*) 15 μm diameter particles. Black and white circles are brightness intensity when applying the acoustic radiation force by the sound field of 2.5 MHz and when reaching the migration time to move particles at the nodes, respectively.

4.2.2 Demonstration of Particle Separation Utilizing Bilayered Acoustic Radiation Forces

For the application of the separation technique, the flow field was formed by injecting the aqueous solutions in the I-shaped channel as shown in Figure 5. The experiments were performed under two conditions, that is, a flow with 15 μm and 4.0 μm diameter particles and a flow with 9.6 μm and 4.0 μm diameter particles. Bulk velocities, v, were set at 750 μm/s, 450 μm/s, 250 μm/s and 100 μm/s by a pressure-driven flow which was generated by the difference of the water surface in both reservoirs, and the Reynolds number, Re, were 0.37, 0.22, 0.12 and 0.05, based on the hydraulic diameter of 0.5 mm, respectively. The parameters in equation 2 were set below. The length of downstream transducer, Y_l, is 11 mm, and the exposure time, t_{ex}, and time cycle of the exposure of the downstream transducer, t_{per}, are 0.1 s and 1 s, respectively. The bulk velocity of 750 μm/s, 450 μm/s and 250 μm/s qualifies as the separable condition as described in equation 2, and that of 100 μm/s does not qualify.

Figure 13 shows the spatial distributions of particles, which were obtained after each 10 iterative measurements and located at the center of 2.5 MHz transducer (x = 0 mm). The black and white circles indicate large and small size particles, respectively. As typical results, Figures 13 (*a*) exhibit the separation of 15 μm and 4.0 μm diameter particles with bulk velocities of 750 μm/s, and Figures 13 (*b*) plot the separation of 9.6 μm and 4.0 μm diameter particles with bulk velocities of 100 μm/s. From Figures 13 (*a*) which qualifies as the separable condition, large particles were collected at the nodal

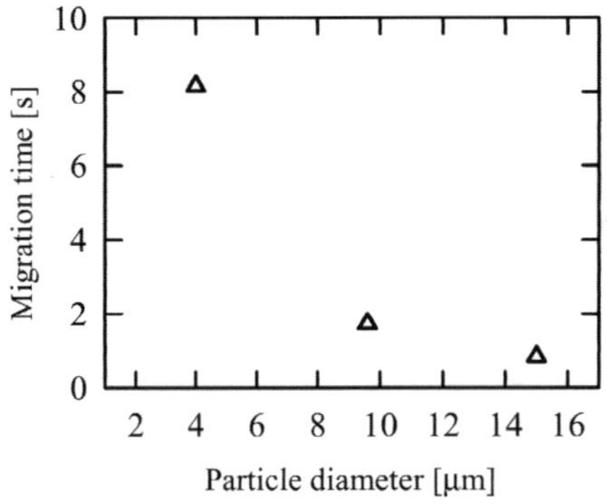

Figure 12. Summary of migration time in terms of particle diameter.

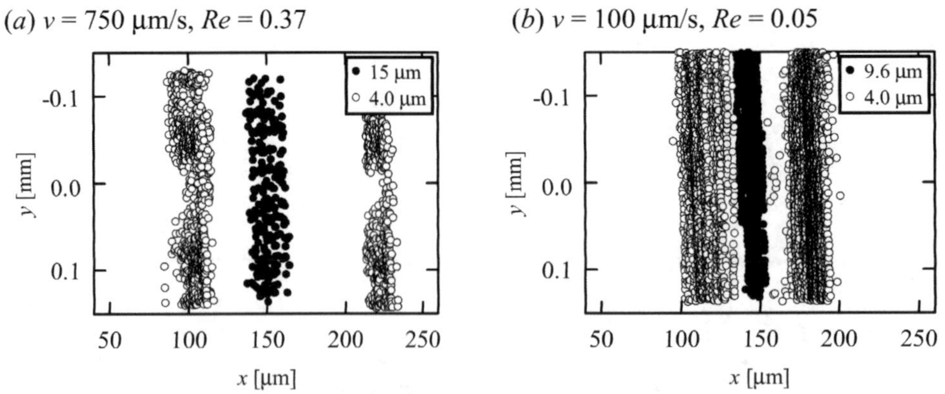

Figure 13. Experimental results of particle separation by utilizing two acoustic radiation forces. Black and white circles indicate large and small size particles, respectively. The experiments are performed by injecting (*a*) 15 μm and 4.0 μm diameter particles in flow field with bulk velocity of 750 μm/s and (*b*) 9.6 μm and 4.0 μm diameter particles in flow field with bulk velocity of 100 μm/s.

position of 2.5 MHz sound field and small particles were existed at the nodes of 5 MHz, and the particle separation by size was successful. On the other hand, the experiments with 100 μm/s show that both large and small particles were accumulated at the span-wise position from 100 μm to 200 μm. Thus, it is obvious that the particle separation utilizing the acoustic radiation force with two ultrasonic transducers is advanced by evaluating the migration time of particles as shown in Figure 12 and qualifying the separable condition of equation 2.

5. CONCLUSIONS

Two kinds of novel non-intrusive and continuous separation techniques for suspended particles in a microchannel were proposed for the application to the microfluidic device. One is the technique to combine acoustic radiation and electrostatic forces, and another technique is to utilize bilayered acoustic radiation forces. The important conclusions obtained from this work are summarized below.

1. The separation technique utilizing the acoustic radiation and electrostatic forces was applied to a flow field with 9.6 μm and 2.0 μm diameter particles. The transducer and electrodes were equipped at the location perpendicular to the flow direction. 2.0 μm diameter particles were transferred towards the side wall of the microchannel by the electrostatic force, and 9.6 μm diameter particles were trapped at the nodal position of the acoustic field. This is caused by evaluating the acoustic radiation and electrostatic forces acting on particles and investigating the relationship where the dominant force of each particle changes from the acoustic radiation force to the electrostatic force.

2. The separation technique utilizing the bilayered acoustic radiation forces was applied to a flow field with 15 μm and 4.0 μm diameter particles. 5 MHz and 2.5 MHz ultrasonic transducers were equipped at upstream and downstream, respectively. The acoustic radiation force acts on particles in proportion to third power of particle radius and moves particles towards the node of standing wave field. So that the large size particles have short migration time of transfer to the nodal positions in comparison with the small size particles. By controlling the exposure time of downstream ultrasonic transducer, large size particles were collected at the node of the downstream sound field and small size particles were kept at the nodal positions of upstream sound field. The migration times of each particle were evaluated by using the PTV measurement, and the exposure time and time cycle of 2.5 MHz transducer was determined to be 0.1 and 1 s, respectively, which were derived from the relationship between their difference of migration time and the bulk flow velocity. Finally, it is observed that the small size particles were trapped at the nodal position of upstream transducer, and large size particles were moved at the nodal position of downstream transducer.

ACKNOWLEDGEMENTS

The authors would like to thank Messrs. H. Ishida and H. Nakanosono at Keio University for performing their experiments. This work was subsidized by Grant-in-Aid for Scientific Research (No. 18206024) from Ministry of Education, Culture, Sports, Science and Technology in Japan. The second author is supported by Research Fellowships for Young Scientists (No. 19-8799) from Japan Society for the Promotion of Science.

NOMENCLATURE

c sound velocity
d wall thickness
D transmission rate of sound

E_{ac}	time-averaged acoustic energy density
f	oscillating frequency of sound
F_{ac}	acoustic radiation force
F_d	drag force
F_e	electrostatic force
k	wave number
l	distance from the nearest node
r	particle radius
Re	Reynolds number
t_{ex}	exposure time
t_{per}	time cycle of exposure
t_{steady}	migration time to move particles towards nodes
u, v	velocity
Y_l	length of transducer
Z_1, Z_2	acoustic impedance
	Greek Letters
μ	fluid viscosity
ρ_m	fluid density
ρ_p	particle density
τ_f	relaxation time of fluid flow
τ_p	relaxation time of particle

REFERENCES

Ashkin A; Dziedzic JM; Bjorkholm JE; Chu S (1986) Observation of a single-beam gradient force optical trap for dielectric particles. Opt Lett 11: 288–290

Auroux PA; Iossifidis D; Reyes DR; Manz A (2002) Micro total analysis systems. 2. Analytical standard operations and applications. Anal Chem 74: 2637–2652

Clift T; Grace JR; Weber ME (1978) *Bubbles, Drops, and Particles.*, New York: Academic Press

Harrison DJ; Fluri K; Seiler K; Fan Z; Effenhauser CS; Manz A (1993) Micromachining a miniaturized capillary electrophoresis-based chemical analysis system on a chip. Science 261: 895–897

Hawkes JJ; Coakley WT; Gröschl M; Benes E; Armstrong S; Tasker PJ; Nowotny H (2002) Single half-wavelength ultrasonic particle filter: Predictions of the transfer matrix multilayer resonator model and experimental filtration results. J Acoust Soc Am 111: 1259–1266

Ishida H; Sato Y; Hishida K (2006) Continuous separation technique of suspended particles by utilizing acoustic radiation and electrostatic force. Trans JSME Ser B 72(722): 2473–2480

Ito S; Yoshikawa H; Masuhara H (2002) Laser manipulation and fixation of single gold nanoparticles in solution at room temperature. Appl Phys Lett 80: 482–484

Laurell T; Petersson F; Nilsson A (2007) Chip integrated strategies for acoustic separation and manipulation of cells and particles. Chem Soc Rev 36: 492–506

Petersson F; Nilsson A; Holm C; Jonsson H; Laurell T (2004) Separation of lipids from blood utilizing ultrasonic standing waves in microfluidic channels. Analyst 129: 938–943

Probstein RF (1994) *Physicochemical Hydrodynamics, 2nd edn.*, John Wiley & Sons

Reyes DR; Iossifidis D; Auroux PA; Manz A (2002) Micro total analysis systems. 1. Introduction, theory, and technology. Anal Chem 74: 2623–2636

Relle S; Grant SB (1998) One-step process for particle separation by magnetic seeding. Langmuir 14: 2316–2328

Scampicchio M; Wang J; Mannino S; Chatrathi MP (2004) Microchip capillary electrophoresis with amperometric detection for rapid separation and detection of phenolic acids. J Chromatography A 1049: 189–194

Sinton D (2004) Microscale flow visualization. Microfluid Nanofluid 1: 2–21

Takeda Y; Kikura H (2002) Flow mapping of the mercury flow. Exp Fluids 32: 161–169

Ueno S; Shimohata K; Yamada T; Iwamoto M (1990) Magnetic separation of fine particle dusts using a superconducting magnet and wire meshes. J Appl Phys 67: 5904–5906

Woolley AT; Lao K; Glazer AN; Mathies RA (1998) Capillary electrophoresis chips with integrated electrochemical detection. Anal Chem 70: 684–688.

The 6th International Symposium on Measurement Techniques for Multiphase Flows IOP Publishing
Journal of Physics: Conference Series **147** (2009) 012057 doi:10.1088/1742-6596/147/1/012057

Correlations between the Two-Phase Gas/Liquid Spray Atomization and the Stokes/Aerodynamic Weber Numbers

Mohammad A. Rahman [1], Ted Heidrick [2], Brian A. Fleck [3]

[1] The Department of Mechanical Engineering, University of Alberta, Edmonton, Alberta, Canada

[1] marahman@ualberta.ca

Abstract. The The effects of air-to-liquid ratio (β) and void fraction (α) on Sauter mean diameter (SMD or D32), arithmetic mean diameter (D10), surface mean diameter (D20), volume mean diameter (VMD or D20), and radial velocity profiles were experimentally investigated for a two-phase gas/liquid (TPGL) nozzle with a hybrid design of classical twin-fluid and effervescent nozzles. Radial spray profiles were measured using a Phase-Doppler-Particle-Anemometer (PDPA) system on 15Dn, 30Dn, 60Dn, 120Dn;(Dn represents nozzle diameter = 3.10,mm) axial distances. In addition, the effects of spray break-up patterns were analyzed with changing axial distances. The average void fraction in the feeding conduit (FC) was measured by a pneumatic controlled quick-closing-valve (QCV). The experiments were performed using mixtures of air with water at water flow rates of 1.50 to 7.50 kg/min and air-to-liquid mass ratios (β) of 0.30 to 10;%. The length and diameter of the FC was 36.8,cm and 6.35,mm, respectively. Result indicates that as the St number reaches the value of one, no more break-up continues, thus the droplets start to coalesce each other forming bigger droplets (higher D10 values) with increasing radial distances. Knowledge from this study will provide better understanding that ensures an increase in plant efficiency and product yield in oil sands bitumen upgrading.

1. INTRODUCTION

In fluid coking, the gas (steam) and liquid (bitumen) mix well upstream of the feed nozzles where the mixture is atomized to form a spray. One of the drawbacks found with the fluid coking nozzles is the development of instabilities in the spray caused by the *TPGL* flow patterns formed inside or upstream the nozzle at the higher β (Ariyapadi et al., 2005; Maldonado, 2006). A stable spray is demonstrated by a good dispersion of the liquid phase. It is desirable to produce a stable spray with minimum *SMD* (D_{32}) and well-dispersed liquid droplets. A desired bitumen drop size in contact with a given coke particle is one with the same nominal diameter as the coke particle. This ensures that the coke is adequately and sufficiently coated with a thin layer of bitumen (Ejim, 2008). On the other hand, an unstable spray is characterized by intermittency/pulsation in its flow regime with the random formation of fine and coarse droplets in the spray. These pulsations are attributed to the *TPGL* fluids conditions, such as β, α, the mixing pressure, P_m, the design of the mixing chamber, and the geometry of nozzle (Tafreshi et al., 2002). Previous studies (Roesler et al., 1989; Whitlow et al., 1993) showed that as the β is increased, for a constant operating pressure, at a certain transition point (e.g. β >1.0% in the large-scale nozzle) the spray becomes unstable. A homogeneous mixture of the gas-liquid

Journal of Physics: Conference Series **147** (2009) 012057 doi:10.1088/1742-6596/147/1/012057

entering the nozzle would maximize the effect of the decompression of the gas phase, resulting in a stable spray. On the other hand, a heterogeneous flow entering into the nozzle cause an unstable spray formation (Barker et al., 1991).

Enhanced heat and mass transfer can be achieved from a spray, which is composed of dispersed droplets with larger spread rates. Moreover, as the droplet sizes are reduced, the energy of the droplets is more readily transferred to the surrounding fluid (MacGregor, 1991). This would ensure proper mixing with the surrounding fluids. Furthermore, in processes where the feed needs to be injected into a cross-flowing stream, the droplets in the spray must have enough momentum to penetrate the cross-flowing fluid stream (Ariyapadi et al., 2000). Continuous and fine spray characteristics are desirable in the feeding nozzle. This feed nozzle is used in the heavy oil process industry. Preheated bitumen and steam is mixed upstream of the nozzle and subsequently injected into fluid bed coker reactors via feed nozzles. One of the drawbacks of this spray characteristic is the pulsation within the spray and in the feeding conduit, which is highly undesirable to yield high productivity. These pulsations result in poor atomization and in most instances, a slug of liquid is ejected out of the nozzle.

It is convenient to work with mean drop sizes instead of complete drop size distributions in the *TPGL* atomization characterization. The mean drop size distribution is generalized as follows (Lefebvre, 1989; Mugele et al., 1951):

$$(D_{ab})^{a-b} = \frac{\int_{D_0}^{D_m} D^a (DN / dD) dD}{\int_{D_0}^{D_m} D^b (DN / dD) dD} \tag{1}$$

The values of a and b can be found in Lefebvre (1989) and the mean diameter expression is presented in Table 1.

Table 1. Mean diameters (adapted from Lefebvre, 1989).

Symbol	Name of mean diameter	Expression
D_{10}	Length	$\frac{\sum N_i D_i}{\sum N_i}$
D_{20}	Surface area	$\left(\frac{\sum N_i D_i^2}{\sum N_i}\right)^{1/2}$
D_{30}	Volume	$\left(\frac{\sum N_i D_i^3}{\sum N_i}\right)^{1/3}$
D_{32}	Sauter mean (*SMD*)	$\frac{\sum N_i D_i^3}{\sum N_i D_i^2}$

Generally D_{32}/SMD provides a good indication of the drop size dispersion characteristics and used for mass transfer application. The D_{10} diameter is used for comparison purpose, the D_{20} diameter is used for surface area controlling application, the D_{30} diameter is used for volume controlling application, the D_{43} diameter is used for combustion equilibrium application. In the droplet motion, the *Stokes number (St)* is a very important parameter. The *St* is defined as the ratio of the particle momentum response time over a flow system time, defined as: $St = \tau_p / \tau_c = \rho_p d_p^2 u_c / 18\mu_c L$. Two types of situations can be observed for particles (bubbles/droplets) suspended in fluid, namely: a) if the $St<<1$, the particles will have ample time to respond to changes in flow velocity, b) if $St>>1$, then the particle will have essentially no time to respond to the fluid velocity changes and the particle velocity will be little affected by fluid velocity change (Crowe, 2006).

A spray breaks up further downstream from the tip of a nozzle. A typical *TPGL* atomization process involves a) primary atomization (*PA*), b) secondary atomization process (*SA*). The dominant forces involve in the atomization process are: (i) hydrodynamic or inertial force attributed to undulations/perturbations, (ii) aerodynamic force attributed to drag/shearing effect, (iii) viscous force attributed to oppose a change in liquid geometry, and (iv) surface tension forces attributed to a minimum surface energy (Nasr et al., 2002). The first two forces are disruptive in nature and second two forces are cohesive in nature. Due to the interaction of internal forces such as: a) turbulence, b) inertial effects, c) momentum transfer between transverse layers of a jet are mainly responsible for the *PA* (McCarthy et al., 1974). At this stage the disruptive forces exceed the consolidating forces resulting the oscillations on the liquid surface and subsequently disintegrate the bulk liquid into drops (Liu, 1999; Shavit et al., 1996). The *SA* in a spray occurs when larger droplets or liquid ligaments break down into smaller droplets. The breakup of a single droplet in a gas is caused by either greater relative velocity, or turbulence (Crowe, 2006). The *SA* occurs due to two force ratios acting on the drop (Pilch et al., 1987). Firstly, if the aerodynamic forces overcome the forces due to the surface tension, the droplet will further deform (Low et al., 1982). The ratio of this two forces can be represented by the *Weber number*, which can be defined as: i) aerodynamic/gas *Weber number*: $We_g = \rho_g u_{rel}^2 d_d / \sigma$ or ii) liquid *Weber number*: $We_l = \rho_l u_l^2 d_d / \sigma$ .Secondly, the *Ohnesorge number/Laplace number*, denoted by: $Oh = \sqrt{We_l} / \mathrm{Re}_l = \mu_l / \sqrt{\rho_l d_d \sigma} = 1/\sqrt{Lp}$, which is the ratio of liquid viscous force to surface tension force on the drop, also plays an important role in the *SA*. Here Re_l is the liquid *Reynolds number* defined: $\mathrm{Re}_l = \rho_l u_l d_d / \mu_l$. Mathematically, if the We_g exceeds the We_{crit}, the *SA* occurs. For low-viscosity liquids, We_{crit} = 6 to 13 for $Oh < 0.1$, $We_{crit} \approx Oh^2$ for $Oh > 0.1$ (Faeth, 1990). In the *TPGL* spray characteristics measurement, the *PDPA* system and digital image analysis techniques has been using as an advanced experimental techniques. The *PDPA* measurement techniques to measure the *TPGL* spray can be found in literature (Ariyapadi et al., 2000; Ariyapadi et al., 2001; Ariyapadi et al., 2003; Copan et al., 2001; Ziesenis et al., 2002). Recently developed digital image analysis techniques potentially are also capable of sizing particles of arbitrary shape and size and with a wide dynamic range (Ariyapadi et al., 2005; Kashdan et al., 2003; Zama et al., 2004).

Measurement techniques of the *TPGL* flows/sprays are a challenge. Due to highly non-uniform volumetric flow distributions and intermittency in the flow, it is extremely difficult to predict accurate droplet sizes (d_d) and flow pattern distributions in this type of flow. This uncertainty of the d_d prediction is exaggerated if the *FC* length of the nozzle is short, as in the present study (36.8 cm long); since the *TPGL* flow cannot be fully developed within this short pipe length. The nozzles currently used in this study do not belong specifically to any of the nozzle classes for the *TPGL* flows existing such as twin-fluid nozzles and effervescent nozzles. Moreover, in

Journal of Physics: Conference Series **147** (2009) 012057 doi:10.1088/1742-6596/147/1/012057

recent years Nasr et al. (2002), Hsiang et al. (1992), Lefebvre (1989), Lefebvre et al., (1988), and Roesler et al. (1989) conducted atomization studies in the feed nozzles. However, these studies were based on the equilibrium flow condition, in vertical nozzle configurations and in a larger length scale set-up. In contrast to the above studies, the current study deals with the non-equilibrium flow condition (in which the *FC* is very short), in a horizontal nozzle configuration, and in a smaller length scale set-up.

The objectives of this study are: i) to predict correlations between the spray disintegration process and the upstream *FC* flow characteristics, ii) to understand the fundamentals of the *TPGL* flow patterns, and spray characteristics. The proposed study will contribute to the fundamental knowledge of the *TPGL* flows and make concrete headway in the design of an industrial nozzle used in a large-scale high impact operation.

2. EXPERIMENTAL SET-UP

In this study, a laboratory scale nozzle assembly was implemented. The dimension was at one-quarter scale of a patented full-scale design (US Patent #: 6003789) employed in a fluidized bed coker for heavy oil upgrading. A feeding conduit of 36.8 cm in length and 6.35 mm ID was used upstream of the nozzle. The nozzle diameter (D_n) was 3.10 mm. Gas (air) was supplied from a compressor and liquid was supplied from a reciprocating pump. These two fluids mixed together at a T-mixer prior to the *FC*. This nozzle assembly was mounted on a 3-D automated traversing rig. The experimental schematic diagram is presented in Fig. 1.

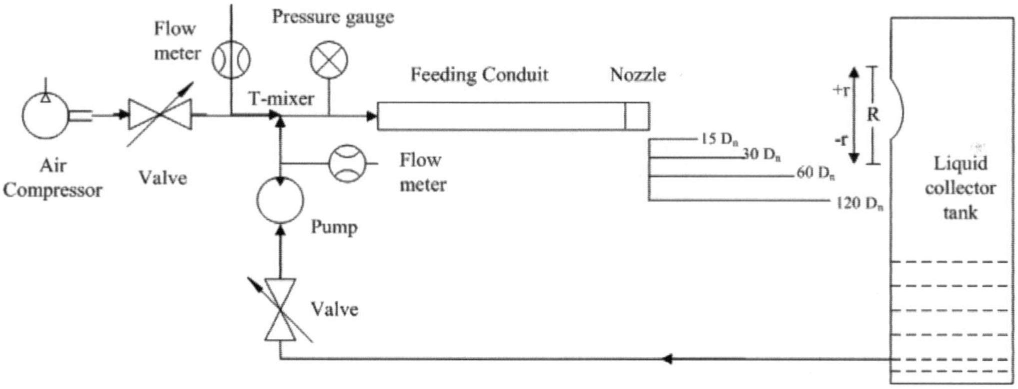

Fig. 1 Schematic of the experimental set-up (adapted from Rahman et al., 2008)

Mean drop size was measured using a 2D-*PDPA* from the Dantec Dynamics specifications (Ejim et al., 2005). The focal lengths of the *PDPA* transmitter and receiver lenses were 400 mm and 310 mm, respectively. During data collection, the *PDPA* was operated in refraction and forward-scatter mode, and the receiver was set to a scattering angle (ϕ) of 30° for the air-water tests. Dantec (Dantec Dynamics, 2003) specified that first order refraction was the most dominant scattering mode at $\phi = 30°$ for water droplets in air (about 10^2 orders of magnitude higher compared to the backward scattering mode).

This technique can measure simultaneously velocity and particle size of known refractive index. This method is termed as "2D-*PDPA*" as this technique can measure velocity of two orthogonal axes. The optical setting of the 2D-*PDPA* is presented in Table 2.

Table 2. Optical settings of the PDA unit used in the study

Parameter	Unit	Values
Scattering angle	(-)	30°
Beam spacing	mm	38.0
Beam expansion ratio	(-)	1.0
Receiver focal length	mm	310
Beam diameter	mm	1.35
Transmitter focal length	(mm	400
Receiver slit width or aperture	mm	0.1
He-Ne laser wavelength	nm	632.8
Nd-YAG laser wavelength	nm	532

The green Nd-YAG and red He-Ne laser power are 200 mW and 20 mW, respectively. Their wavelengths are 532 and 632.8 nm, respectively. Each laser split in two laser beams using the unit's Bragg cell. The resulting four beams (two pairs for each laser) are aprt of 90° each other and converge at the focal length of the transmitter lens to form a control volume. Droplet size was measured by detecting the incident droplets on the receiver detectors. The size of a droplet is directly proportional to the phase shift of scattered light in the control volume. The velocity of droplets can be found also by detecting the incident droplets on the receiver detectors. However, in this case the droplets passing through the control volume transmit Doppler frequencies or signals that are directly proportional to their velocity. These Doppler frequencies are detected by the receiver.

Radial spray profiles were measured using the *PDPA* on the $30D_n$, $60D_n$, and $120D_n$;axial distances downstream of the nozzle orifice. Measurements were taken varying the radial positions (r) by the 3-D traversing rig. The values of D_{32}, D_{10}, and u_x were measured with the changing β, r, and x positions. In Fig.1Fig. 1, the '*R*' indicates the radius of the spray.

3. RESULTS AND DISCUSSIONS

The *SMD* (D_{32}) provides a good indication of the drop size dispersion characteristics in the *TPGL* spray. From Fig. 2 it is evident that D_{32} values are greater at the center of the *TPGL* spray due to higher non-spherical droplet density persists around this zone. However, further away from the center of the spray, the D_{32} values significantly decrease as around this zone droplets are more disperse. At the periphery of the spray, the D_{32} values are flattened out and in some cases a bit increase as around this region droplets may coalesce to each other.

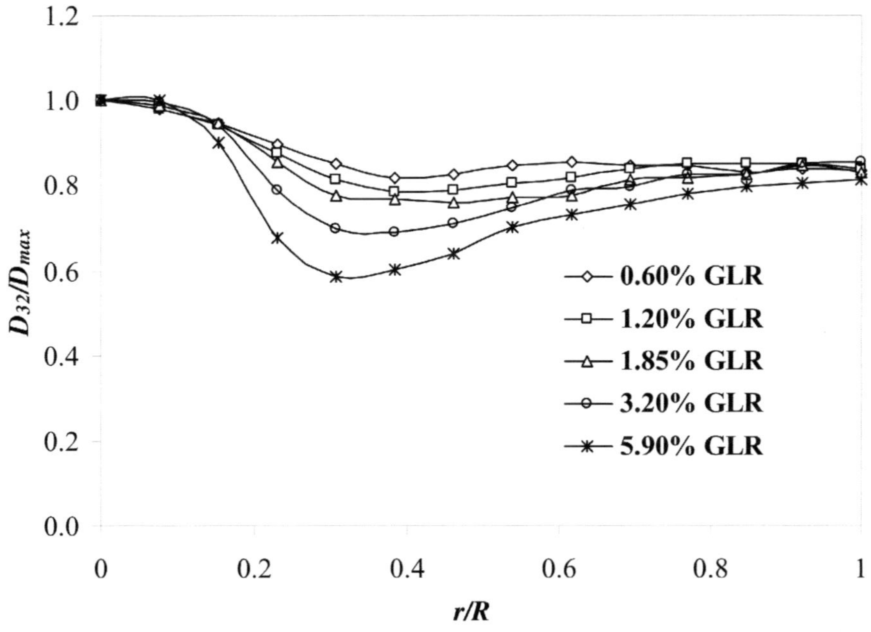

Fig. 2 Effects of the *GLR* (gas-to-liquid-ratio, β, by mass) on the SMD/D_{32} profiles with changing r at a fixed axial distance of $60\ D_n$ and 620 kPa mixing pressure.

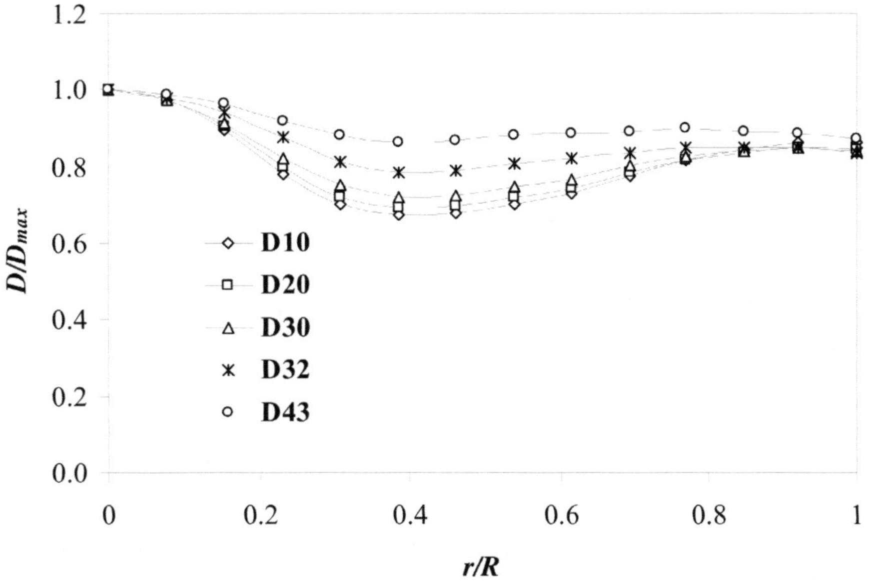

Fig. 3 Profiles of D_{10}, D_{20}, D_{30}, D_{32}, D_{43} with changing r at a fixed axial distance of $60\ D_n$ and 620 kPa mixing pressure.

In Fig. 3, D_{10}, D_{20}, D_{30}, D_{32}, and D_{43} profiles with changing r at a fixed axial distance of $60\ D_n$ and 620 kPa mixing pressure (P_m) is depicted. All the profiles indicate that the droplet sizes are greater at the center of the spray. However, further away from the center of the spray, the droplet size decreases significantly and finally fatten out at the periphery of the spray.

The *PDPA* provides the two-component droplet velocity (u_x, u_y) and droplet diameter (d_d). However, it cannot provide the continuous phase gas (air) velocity data in the *TPGL* sprays. Experimentally it is difficult to obtain the continuous phase gas velocity data in the *TPGL* atomization process. However, Crowe (2006) provided a correlation for how to obtain the continuous phase gas velocity by solving a particle motion equation and knowing the droplet velocity from the *PDPA*. This correlation was subsequently defined: $U_d/U_g \cong 1/(1+St)$. Here, *St* was the *Stokes number*. If the *St* number tends to be zero, there would be no slip between the two phases of fluid. In this study, the U_g value was obtained from the *St* number and by knowing the U_d and D_{10} values from the *PDPA* data.

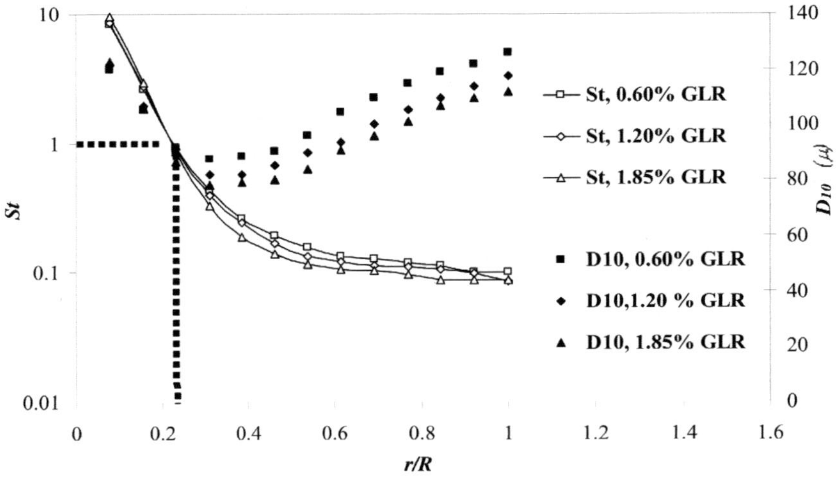

Fig. 4 Effects of the *GLR* (gas-to-liquid-ratio, β, by mass) on the *St* and D_{10} profiles with changing r at a fixed axial distance of $60D_n$ and 620 kPa mixing pressure.

In the atomization process, the *St* number is a very important parameter. The *St* number is defined as the ratio of the droplet momentum response time over a flow system response time, defined as: $St = \tau_p/\tau_c = \rho_l D^2_{10} U_l / 18\mu_l r$. In Fig. 4, the droplet *St* number and D_{10} profiles with changing r and β at a fixed axial distance of $60\ D_n$ are depicted. The P_m was maintained at a constant pressure of 620 kPa and the flow rates of gas/liquid were varied to obtain different β of 0.60%, 1.20%, and 1.85% at the constant pressure of 620 kPa. Here, R is the radius of the spray which is 65 mm. First, it is observed that if the 'r' increases, the *St* number decreases remarkably due to the smaller droplet response time at the periphery of the spray. If the $St \ll 1$, the droplets will have ample time to respond to changes in continuous phase flow velocity. However, if $St \gg 1$, then the droplets will have essentially no time to respond to the continuous phase fluid velocity changes and the droplet velocity will be affected very little by the fluid velocity change (Crowe, 2006). In our study similar observations for droplets suspended in the atmospheric air were noticed. The *St* number reached a value of one at r/R of 0.25. Thus, at the center of the spray the droplet response time is much higher than that of the continuous air phase. At the center of the spray, a remarkable amount of slip exists between the liquid and gaseous phases. However, at the spray outer region, the droplet response time followed the continuous phase response time. In addition to the above fact, it is also observed that the *St* number profiles do not differ significantly in the radial direction for the β values of 0.60%, 1.20%, and 1.85% at the constant pressure of 620 kPa until the point where the *St* number is equal to 1. However, after the point where the *St* number is equal to 1, the higher β exhibits the lower *St* numbers in the radial direction due to the reduction of the jet half width.

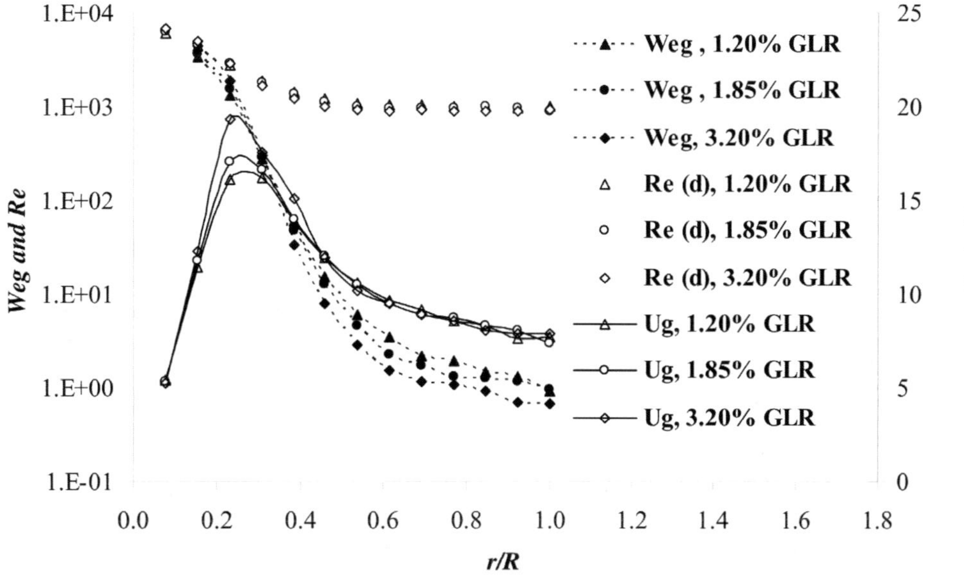

Fig. 5 Effects of the *GLR* on the We_g, Re_d and U_g profiles with changing *r* at a fixed axial distance of $60\ D_n$ and 620 kPa mixing pressure.

Droplet response with the *St* number can be further demonstrated by the D_{10} profile data set. In Fig. 4, it is initially observed that if the β values are increased, the D_{10} values are decreased. Secondly, until the $St = 1$ at the radial distance of $r/R = 0.25$, the D_{10} values are decreased as the droplets continue to break-up up to this point. However, as soon as the *St* reached 1, droplet break-up began to cease, thus the droplets started to coalesce forming bigger droplets (higher D_{10} values) with increasing radial distances.

The *TPGL* atomization behavior can be fully understood by the aerodynamic *Weber* number (We_g) and droplet *Reynolds* number (Re_d). The We_g is the measure of the relative importance of the fluids' inertia compared to its surface tension, which can be defined as $We_g = \rho_l (U_l - U_g)\ r/\sigma$. Whereas the Re_d is the measure of the relative importance of the droplet's inertia compared to its viscosity, which is defined as $Re_d = \rho_l\ U_l\ r/\mu$. A spray breaks up further downstream from the tip of a nozzle. A typical *TPGL* atomization process involves primary atomization (*PA*) and secondary atomization (*SA*). The dominant forces involved in the atomization process are (Nasr et al., 2002): a) hydrodynamic or inertial force attributed to undulations/perturbations, b) aerodynamic force attributed to drag/shearing effect, c) viscous force attributed to an oppositional change in liquid geometry, and d) surface tension forces attributed to a minimum surface energy. The first two forces are disruptive in nature and the second two forces are cohesive in nature. Another mechanism for the *TPGL* atomization mechanism is the '*bubble energy*' explosion (Lefebvre et al., 1988; Roesler et al., 1989). It was postulated that jet break-up occurs when the bubbles within the bulk liquid possess enough energy to overcome the surface tension forces that hold the liquid jet together. When the bubbles have enough energy, the droplets further atomize through bubble explosions.

In Fig. 5, the droplet break-up mechanism was investigated by the We_g, Re_d and U_g profiles with changing r and β at a fixed axial distance of $60\ D_n$. Here, R was the radius of the spray which was 65 mm. The P_m was maintained at a constant pressure of 620 kPa and the flow rates of gas and liquid were varied to obtain the β of 1.20%, and 1.85%, and 3.20% at that constant pressure. An interesting observation is that the U_g has higher values between the center and periphery of the spray in the radial direction (r). Due to this peak of the U_g value, the droplets break-up further downstream radially and axially, which is termed as the *SA*. In literature, several correlations were hypothesized to predict the *SA* in terms of the We_g. The interaction of internal forces such as turbulence, inertial effects, and momentum transfer between transverse layers of a jet, are mainly responsible for the *PA* (McCarthy et al., 1974). At this stage the disruptive forces exceeded the consolidating forces which resulted in oscillations on the liquid surface and subsequently disintegrated the bulk liquid into drops (Liu, 1999; Shavit et al., 1996). The *SA* in a spray occurs when larger droplets or liquid ligaments break down into smaller droplets. The breakup of a single droplet in a gas is caused by either greater relative velocity, or turbulence (Crowe, 2006). In Fig. 5, the greater relative velocity exists at this peak point of U_g profile. The *SA* occurs due to two force ratios acting on the drop (Pilch et al., 1987). If at first the aerodynamic forces overcome the forces due to the surface tension, the droplet will further deform (Low et al., 1982). The two dimensionless numbers can represent the ratio of these two forces. The We_g plays a vital role in the first stage of the *SA* process. Secondly, the *Ohnesorge number*, denoted by $Oh\ (\mu_l/\sqrt{\rho_l d_d \sigma})$, which is the ratio of liquid viscous force to surface tension force on the drop, also plays an important role in the *SA* process. It was postulated that if the We_g exceeds the We_{crit} ('*crit*' denotes for critical), the *SA* occurs. For low-viscosity liquids, $We_{crit} = 6$ to 13 for $Oh<0.1$, $We_{crit} \approx Oh^2$ for $Oh>0.1$ (Faeth, 1990). To find out the We_{crit}, in Fig. 5 a perpendicular line was drawn from the peak value of the U_g profile that intersects the We_g number profile at a point. The value of the We_g number at this point can be termed as the We_{crit}, which is approximately 500. In Fig. 5, it is also observed that the We_g and Re_d decreased remarkably with the radial distances up to the We_{crit} limit. However, after a few radial distances from the We_{crit} limit, the We_g and Re_d decreased slightly with the variation of the β values.

In Fig. 5, it was also observed that the Re_d varied from 700 to 7000 due to the variation of the r/R values. If the $Re_d \ll 1$, the *TPGL* flow can be termed as *Stokes* flow. In the *Stokes* flow regime, viscous droplets remain spherical. The wake formed behind the droplets became stronger as the Re_d increases and the inertia of the flow around the droplets overcame the viscosity effects on the surface of the droplets (Crowe, 2006). It is also observed that the Re_d value decreased with the radial distances. However, the U_g still has enough momentum, which can induce a slip between the two phases. Thus, although the R_d value decreases with the r, the higher relative velocity between the phases is still responsible for the *SA* process.

In Fig. 6, the droplet *St* and D_{10} profiles with changing axial distances (x of $30D_n$, $60D_n$, $120D_n$) and radial distances (r) are depicted. Here, R is the radius of the spray which was 40 mm with the β of 2% at a P_m of 520 kPa. It was initially observed that if the 'r' and 'x' increase, the *St* number decreases remarkably due to smaller droplet response time compared to that of the continuous phase gaseous medium. Secondly, the *St* number reaches the value of 1 at r/R of 0.40 for axial distances of $30D_n$, $60D_n$, $120\ D_n$, respectively. Droplet break-up characteristics with the *St* number can also be verified with the D_{10} data. It was observed that at $120D_n$ downstream of the nozzle orifice, the droplets are more disperse compare to other two cases. Thus, near the center (r=0) of the spray at $120D_n$ downstream from the nozzle orifice, the D_{10} values are lower than at $30D_n$ and $60D_n$ downstream D_{10} values at the similar radial position (r=0). In the later cases, the droplets were not fully atomized, which can be further explained with the We_g and Re_d profiles. However, as soon as the *St* value reaches 1, further away from the center, the D_{10} values are higher at $120D_n$ downstream from the nozzle orifice than at $30D_n$ and $60D_n$ downstream values at the similar radial position. This is due to the fact of coalescence of droplets around this zone.

In Fig. 7, the droplet break-up mechanism was examined by the We_g, Re_d and U_g profiles with changing axial distances ($30\ D_n$, $60\ D_n$, $120\ D_n$) and radial distances (r). Here, R is the radius of the spray which was 40 mm. The β for this condition was maintained at 2% with a P_m of 520 kPa.

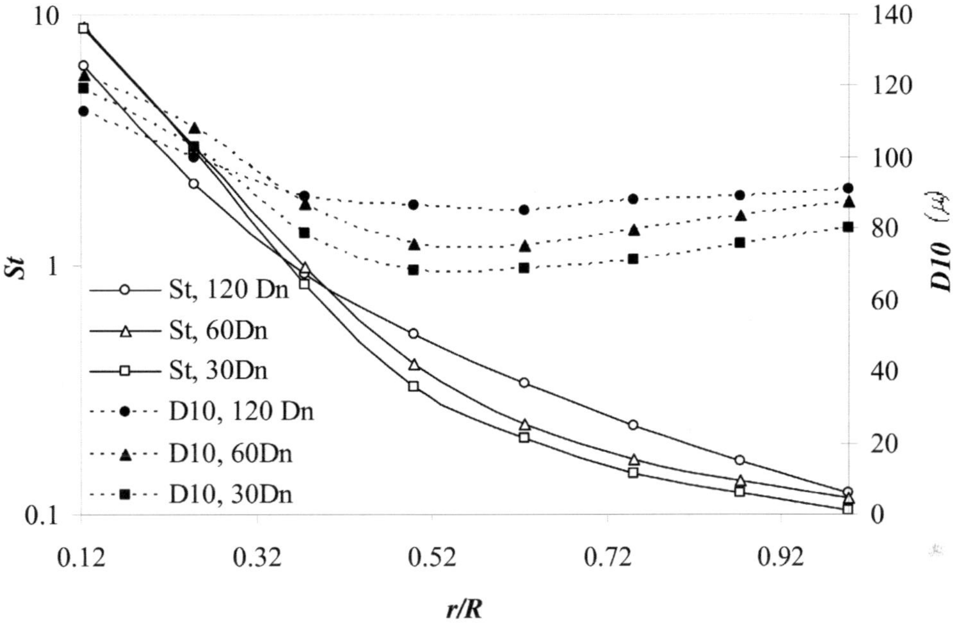

Fig. 6 Effects of the axial distances on the droplet *St* and D_{10} profiles with changing radial distances (r) at 2% *GLR*.

An interesting observation in Fig. 7 is that the U_g had higher values between the center and periphery of the spray in the radial direction. Until the highest U_g value, the droplets had a tendency to breakup further. Thus, in Fig. 7, again the We_{crit} value was obtained by drawing a perpendicular line from the U_g profile that intersects the We_g profile at a certain point. The value obtained at this point is termed the We_{crit}. In Fig. 7, it was also observed that the We_g and Re_d decreased remarkably with axial distances and radial distances. If the We_g exceeds the We_{crit}, droplets will have the tendency to break-up further due to higher momentum transfer between the phases. Thus, it was observed that further downstream of the spray ($120D_n$), the We_{crit} value was less compared to the upstream condition ($30D_n$). However, the reverse case was observed for the U_g values. At $120D_n$ downstream from the orifice of the nozzle, the U_g had greater values compared to $30D_n$ downstream of the spray, indicating the bubble explosion imparted higher momentum than the gas phase further downstream.

In Fig. 7, it is also observed that the Re_d varied from 800 to 6500 in the radial direction. Downstream of the spray ($120D_n$), the momentum of the droplet decreased remarkably compared to the upstream condition ($30D_n$). However, for similar positions the gas phase still had enough momentum as it had just been exploded at the nozzle orifice. This also provides the basis for why the U_g exhibits higher values at $120D_n$ compared to $30Dn$ and $60D_n$ downstream from the nozzle orifice. However, it seems that the Re_d values are lower further downstream as the droplets lose momentum. Thus, the Re_d values are not fully capable of explaining the droplet atomization behavior further downstream of the spray.

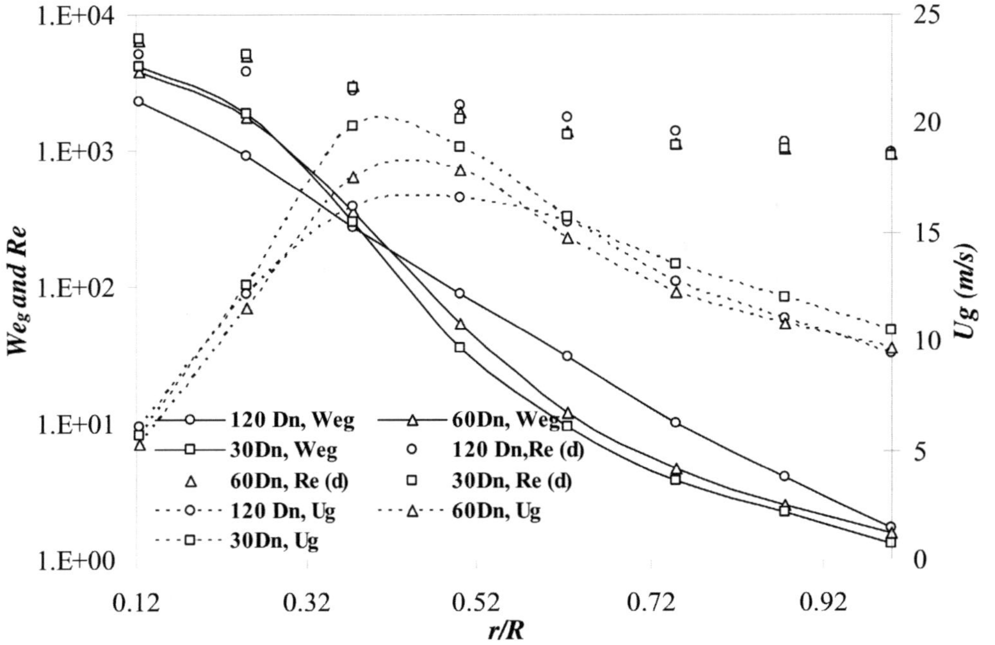

Fig. 7 Effects of the axial distances on the We_g, Re_d and U_g profiles with changing radial distances (r) at 2% *GLR*.

4. CONCLUSIONS

The fundamental knowledge of the *TPGL* flow/atomization process in nozzles is important for many industrial and chemical reactions. The outcome of this research will help in optimization of commercial process conditions and provide a comprehensive means of improving the design conditions of the *TPGL* flow/atomization process. Specifically, this project will assist to optimize the operating range of the existing steam/bitumen fluid cooking nozzles used in the heavy oil upgrading process. Knowledge obtained will contribute to the development of a new series of nozzles heavy oil industry is currently bringing to market. Atomization from the nozzle likely depends on flow patterns, void fractions, and bubble size distribution in the upstream *FC*. Thus, it is essential to have a good understanding and a reasonable estimate of the bubble size, flow regime, void fraction, pressure drop, and subsequent droplet size distribution in turbulent *TPGL* flow. A wide variety of operating conditions need to be introduced in the experiment to obtain a better correlation of the flow conditions and spray characteristics.

The *St* number is useful to evaluate the U_g values using a particle motion equation. Eventually, the U_g value is required to obtain the We_g and Re_d values, which are very useful dimensionless numbers used to predict the *TPGL* spray atomization and specially the droplet *SA* commencement. In this study, two types of conditions were studied. The value of β was first varied from 0.60%, 1.20%, and 1.85% at the constant pressure of 620 kPa at a fixed axial downstream of $60D_n$. It was observed that the *St* number reached the value of 1 at the *r/R* of 0.25 distance. Thus, at the center of the spray the droplets response time was much higher than that of a continuous phase gaseous medium. However, at the spray outer region, the droplet response time followed the continuous phase response time. As soon as the *St* reached 1, break-up stopped, thus the droplets started to coalesce forming bigger droplets (higher D_{10} values) with increasing radial distances. It was also observed that the We_g and Re_d decreased remarkably with the radial distances (r) before the We_{crit} limit. However, the We_g and Re_d decreased slowly with the variation of the β values after the We_{crit} limit.

In the second set of studies, the droplet atomization behavior was examined with changing axial distances (x of $30D_n$, $60D_n$, $120D_n$) and radial distances (r). In this case the β was 2% at a P_m of 520 kPa. It was observed that if the 'r' and 'x' was increased, the *St* number decreased remarkably due to smaller droplet response times compared to the continuous phase response time. The *St* number reached the value of 1 at the r/R of 0.40 for axial distances of $30D_n$, $60D_n$, $120\ D_n$, respectively. Thus, near the center ($r = 0$) of the spray at $120D_n$ downstream, the D_{10} values were lower than that of $30D_n$ and $60D_n$ at the similar radial positions ($r = 0$) due to greater droplets break-up in further downstream.

5. ACKNOWLEDGMENTS

The authors wish to acknowledge the financial support used to carry out this study provided by Syncrude Canada Ltd., CRD grant, Alberta Ingenuity, and NSERC.

6. LIST OF SYMBOLS, NOMENCLATURE, OR ABBREVIATIONS

Abbreviations

GLR gas to liquid ratio by mass
FC feeding conduit
PDPA Phase-Doppler-Particle-Analyzer
PA primary atomization
SA secondary atomization
TPGL two-phase gas/liquid flow

Symbols

D_n nozzle diameter (m)
D_{10} arithmetic mean diameter (m)
D_{32} sauter mean diameter (m)
D diameter (m)
d_d droplet diameter (m)
d_b bubble diameter (m)
m_g mass flow rates of the gas phase (kg/sec)
m_l mass flow rates of the liquid phase (kg/sec)
N number of samples (-)
P_m mixing pressure (Pa)
Q_g volume flow rate of the gas phase (m^3/sec)
Q_l volume flow rate of the liquid phase (m^3/sec)
r radial position (m)
R radius of the spray (m)
U_g velocity of the gas phase phases (m/sec)
U_l velocity of the liquid phases (m/sec)
U_d velocity of the droplet (m/sec)
x axial distance (m)

Greek Letters

ϕ scattering angle (^{o})
β gas to liquid ratio by mass (%)
ρ density (kg/m^3)
ρ_l density of liquid phase (20°C pure water, 998 kg/m^3)
μ viscosity $(kg/m.s)$
μ_l viscosity of the carrier fluid (20°C pure water, 1.00×10^{-3} kg/m.s)
μ_g viscosity of the gas phase (20°C air, 1.82×10^{-5} kg/m.s)
σ surface tension (20°C pure water, 72.8 N/m)
τ_p particle momentum response time (s)
τ_c flow system time (s)

Subscripts

c, l, w continuous/liquid/water phase
p ,d, g, a particle/dispersed/gas/air phase
d droplet

Dimensionless numbers

Oh Ohnesorge number
St Stokes number
Re Reynolds number
We Weber number
We_{crit} Critical Weber number

7. REFERENCES

Ariyapadi, S. et al. (2000). "Effect of Cross-Flow on the Spray Characteristics of an Industrial Feed Nozzle", paper 75b Proc. AIChE Spring Annual meeting, Atlanta, GA, 2000.

Ariyapadi, S. et al. (2000). "Effect of Cross-Flow on the Spray Characteristics of an Industrial Feed Nozzle", paper 75b Proc. AIChE Spring Annual meeting, Atlanta, GA.

Ariyapadi, S. et al. (2001). "Characterizing Spray Pulsations Using a Phase-Doppler Particle Analyzer," Ind. Eng. Chem. Res. 40, pp.5282-5290.

Ariyapadi, S. et al. (2005). "Stability of horizontal gas–liquid sprays in open-air and in a gas–solid fluidized bed," Powder Technology 155 pp.161 – 174.

Ariyapadi, S. et al. (2003). "Effect of Crossflow on the Spray Characteristics of an industrial Feed Nozzle," Chem. Eng. Comm. 190: 1, pp.681-1704.

Barker, C.L. et al. (1991). "Acoustic Monitoring of Two-Phase Feed Nozzles," pp.US Patent 5 004 152.

Copan, J. et al. (2001). "Droplet Size-Velocity Characteristics of Sprays Generated by Two-Phase Feed Nozzles," Chem. Engg. Comm. 184 pp.105- 124.

Crowe, C.T. (2006). "Multiphase flow handbook", CRC : Taylor & Francis, Boca Raton, FL.

Dantec Dynamics, A.S. (2003). "BSA Flow Software", Version 2.1: Installation and User's guide, Skovlunde.

Ejim, C.E. (2008). "Effect of Liquid Viscosity, Surface Tension and Nozzle Size on Atomization in Two-Phase Nozzles", PhD thesis, Mechanical Engineering, University of Alberta.

Ejim, C.E. et al. (2005). "A Scaling Study of the Atomization of a Two-Phase Industrial Nozzle: Part 1 - Effect of Surface Tension and Viscosity on Mean Drop Size Profiles," Proceedings of the 20th ILASS - Europe Meeting, Sept. 5-7, Orléans, France.

Faeth, G.M. (1990). "Structure and atomization properties of dense turbulent sprays", 23rd Symposium on Combustion, The Combustion Institute, Pittsburgh, 1990, pp. 1345–1352.

Faeth, G.M. (1990). "Structure and atomization properties of dense turbulent sprays," 23rd Symposium on Combustion, The combustion Institute, Pittsburgh, pp. 1345–1352.

Hsiang, L.P. et al. (1992). "Near-limit drop deformation and secondary breakup," Int. J. Multiphase Flow 18, pp.635–652.

Kashdan, J.T. et al. (2003). "Two-Phase Flow Characterization by Automated Digital Image Analysis. Part 1: Fundamental Principles and Calibration of the Technique," Part. Part. Syst. Charact. 20

Lefebvre, A.H. (1989). "Atomizations and Sprays", Hemisphere, New York.

Lefebvre, A.H. et al. (1988). "Spray characteristics of aerated-liquid pressure atomizers," AIAA Journal of Propulsion and Power 4 (6), pp.293-298.

Liu, H. (1999). "Science and Engineering of droplets - Fundamentals and Applications," William Andrew Inc./Noyes Publications.

Low, T.B. et al. (1982). "Collision, coalescence and breakup of raindrops," J. Atmos. Sci. 39, pp.1591–1618.

MacGregor, S.A. (1991). "Air Entrainment in Spray Jets," Int. J. Heat and Fluid Flow 12, pp.279–283.

Maldonado, S. (2006). "Improving the Stability of Gas-Liquid Sprays by Modifying the Two-Phase Flow Entering The Nozzle," MASc. Thesis, University of Alberta.

McCarthy, M.J. et al. (1974). "Review of stability of liquid jets and influence of nozzle design," Chemical Engineering Journal 7, pp.1-20.

Mugele, R. et al. (1951). "Drop size distributions is sprays," Ind. Eng. Chem. 43 (6), pp.1317-1324.

Nasr, G.G. et al. (2002). "Industrial sprays and atomization : design, analysis and applications," Springer London ; New York.

Pilch, M. et al. (1987). "Use of breakup time data and velocity history data to predict the maximum size of stable fragments for acceleration-induced breakup of a liquid drop," Int. J.

Multiphase Flow 13, pp.741–757.

Rahman, M.A. et al. (2008). "Characterizing the Two-Phase, Air/Liquid Spray Profile Using a Phase-Doppler-Particle-Analyzer," proceedings of the 14TH International Symposium on Applications of Laser Techniques to Fluid Mechanics, Lisbon, Portugal.

Roesler, T.C. et al. (1989). "Studies on aerated-liquid atomization," International Journal of Turbo Jet Engines 6, pp.221-230.

Shavit, U. et al. (1996). "Development and evaluation of a new turbulence generator for atomization research," Experiments in Fluids 20 (4), pp.291-301.

Tafreshi, Z.M. et al. (2002). "Improving the efficiency of fluid cokers by altering two-phase feed characteristics," Powder Technology 125, pp.234– 241.

Whitlow, J.D. et al. (1993). "Effervescent atomizer operation and spray characteristics," Atomizations and Sprays 3, pp.137–155.

Zama, Y. et al. (2004). "Simultaneous Measurement of Droplet Size and Three-Components of Velocity in Spray," Optical Review 11 (6), pp.358–364.

Ziesenis, J. et al. (2002). "Absorption and Scattering of Light by Highly Concentrated Two-phase Flows," Part. Part. Syst. Charact. 19 pp.195-202.

Measurement of Emulsion Flow in Porous Media: Improvements in Heavy Oil Recovery

J Bryan, J Wang and A Kantzas

TIPM Laboratory, University of Calgary
2500 University Dr NW, Calgary, AB Canada

akantzas@ucalgary.ca

Abstract. Many heavy oil and bitumen reservoirs in the world are too small or thin for thermal enhanced oil recovery methods to be economic. In these fields, novel methods of less energy intensive, non-thermal technologies are required. Previous experience has shown that the injection of low concentrations of aqueous alkali-surfactant solutions into the reservoir can significantly improve the oil recovery, beyond that of waterflooding. This is due to the *in-situ* formation of emulsions, which plug off the water channels and lead to improved sweep efficiency in the reservoir. The proper control of these floods requires methods for monitoring the formation and effect of these emulsions. In this paper, the results of laboratory core floods are interpreted to demonstrate how the pressure and flow response can be related to the formation of these emulsions. A new technique (low field NMR) is also used to directly measure W/O emulsions in porous media. Finally, a numerical study is performed in order to demonstrate how the *in-situ* formation of emulsions can be simply represented in simulation software.

1. INTRODUCTION

Several countries in the world contain significant resources of heavy oil and bitumen, which is characterized by having high viscosity, and density similar to that of water. For countries such as Canada, Venezuala and China, this oil resource constitutes a significant part of the economy, and as oil demand rises worldwide this unconventional crude oil could be could an important source of global energy in years to come.

In Canada alone, the total bitumen resource base is estimated at approximately 270 billion m^3 of oil in place (AEUB, 2007). From this incredible resource, the oil volumes that can be recovered under current technology (i.e. the oil reserves) are approximated at 10% of the total oil in place (Radler, 2006). A fraction of this oil can be recovered through surface mining, but over 90% of the oil in place has to be recovered through *in-situ* technologies (AEUB, 2007). This presents a tremendous challenge for increasing the amount of oil that can be economically recovered from these reservoirs.

In most cases, bitumen is found in unconsolidated, high porosity and permeability sand deposits. Ease of flow is therefore not a major concern in these reservoirs; rather the high oil viscosity is the single greatest impediment that must be overcome in order to produce the oil. As a result, most enhanced oil recovery schemes tend to focus on some variation of thermal injection, whereby the oil is heated to reduce its viscosity and induce flow. In order for these thermal methods to be applied, it is generally accepted that the reservoir must be more than 10 m thick (AEUB, 2007) or heat loss to the overburden and underburden will be too significant.

Heavy oil is a special subset of these bitumen deposits, whereby the oil is slightly less viscous and therefore has some limited mobility at reservoir conditions. The oil may initially contain some solution gas, so approximately 5% of the oil may be recovered through primary solution gas drive mechanisms. This leaves significant volumes remaining for improved oil recovery. In many cases, however, the reservoirs are relatively thin (under 10 m in thickness) and may have been disturbed to an unknown extent during primary production. Many reservoirs also may contain bottom water aquifers, or gas caps. All of these conditions make them poor candidates for thermal oil recovery processes, and less-energy intensive non-thermal methods are required to produce this remaining oil.

Chemical flooding is a relatively old technology (Nutting, 1925; Reisberg and Doscher, 1956), and has been applied in enhanced recovery schemes for conventional oil for several decades. In this technique, a surface-active agent (surfactant) is injected along with water to reduce the interfacial tension between oil and water. This in turn enables displacement of discontinuous oil ganglia that are present after waterflooding. Chemical flooding is therefore a tertiary recovery process in conventional oil, and is performed with the goal of reducing the residual oil saturation in waterflooded zones. Alkali flooding is a special subset of surfactant flooding, whereby the surfactants are generated *in-situ* through the reaction between the injected alkali and organic acids from the oil. This recovery technique is highly sensitive to factors such as the salinity of the aqueous phase (Nelson *et al.*, 1984), thus surfactants are often injected along with the alkali to stabilize the system. This procedure, known as alkali-surfactant flooding, provides similar stability as conventional surfactant floods, while substantially reducing the cost of the required surfactant.

In heavy oil systems, chemical flooding has also been shown to yield improved oil recovery compared to waterflooding (Scott *et al.*, 1965; Dranchuk *et al.*, 1974; Farouq Ali *et al.*, 1979; Dong *et al.*, 2007). A common observation made in these floods was that the injection of alkali and/or surfactant appeared to lead to the generation of W/O emulsions. These emulsions were expected to be more viscous than the constituent oil, thus the proposed recovery mechanism was that of mobility ratio enhancement by displacing heavy oil by a higher viscosity W/O emulsion phase (Farouq Ali *et al.*, 1979; Dong *et al.*, 2007). However, it is not clear how the emulsion phase could be displaced in a stable fashion by the aqueous solution at the injection end of the flood, nor is it clear where the emulsions are forming and from where oil is being displaced.

In waterflooding, oil is significantly by-passed by water due to the disparity between the oil and water viscosity. Chemical injection still involves the displacement of viscous oil by a low viscosity aqueous solution, so the improved oil recovery observed must be somehow related to the generation and movement of emulsions in the porous media. In order to properly understand these recovery mechanisms, techniques need to be developed to interpret the response of the system to chemical injection. Specifically, the presence and effect of emulsions must be quantified before any attempt can be made to model this improved oil recovery response. Once methodologies are present for measuring the generation and effect of these emulsions, then it will be possible both to predict their behaviour in both laboratory and field scale projects. Thus, the development of techniques for measuring emulsions is key to the design and control of chemical floods in heavy oil systems. In this work a suite of core floods was performed, and the *in-situ* formation of emulsions was identified through measurements of pressure and produced fluids, and through spectra acquired using low field nuclear magnetic resonance. The observations made allowed for an improved understanding of how heavy oil is recovered through alkali-surfactant injection.

2. MATERIALS AND PROCEDURES

In previous work (Bryan and Kantzas, 2007), core floods were performed at ambient temperature (23°C) on a heavy oil in sand packs of varying permeability. The oil used in this study comes from a field in Saskatchewan, Canada. It has a viscosity of 11,500 mPa·s at 23°C, and a density of 982 kg/m^3. Although all the oil came from the same field, it was still heterogeneous so some of the core floods were performed with oil that had a viscosity of 15,000 mPa·s. The surfactant is a commercial anionic sodium alkane sulphonate (Bio-Terge PAS-8S) and was supplied by Stepan Company. The alkali used was sodium carbonate, Na_2CO_3, which was chosen over other alkali agents due to its buffering capacity and the fact that it is less sensitive to variations in water chemistry than NaOH (Martin and Oxley, 1985). Measurements of oil-water interfacial tension showed that under relatively low chemical concentrations (0.1 wt% surfactant and < 1 wt% alkali in water), the interfacial tension could be reduced by approximately three orders of magnitude. Thus, ultra-low interfacial tensions are possible even in viscous oil systems. This is a necessary requirement for emulsion formation, but it does not explain which emulsion (O/W or W/O) will form. Through phase behaviour testing, it was observed that when the aqueous phase contained 2 wt% NaCl, only viscous W/O emulsions would be generated, even for systems containing significantly more water than oil (Bryan and Kantzas, 2007). This is due to the fact that oil is much more viscous than water, thus in an emulsion system oil will more easily become the continuous phase (Garret, 1972; Isaacs and Chow, 1992). In chemical floods performed using brine (2 wt% NaCl in water), it was therefore expected that only W/O emulsification would occur.

Table 1 lists the dimensions of the core floods performed with brine being the aqueous phase. In all tests, the cores were first saturated with brine, and the permeability to brine was measured. The cores were then flooded to irreducible water saturation (S_{wi}) by injecting heavy oil at a fixed rate. Subsequently, the cores were then either flooded by brine, alkali-surfactant (AS) solution, or a combination of the two. All the chemical floods were conducted using a fixed concentration of 0.1 wt% surfactant and 0.5 wt% alkali in brine.

Experiments Q – T, Y and Z were conducted as primary AS floods, meaning that in these systems the AS solution was injected directly into the core that contained oil and irreducible water saturation. In these tests, approximately five pore volumes of fluid were injected. In floods U – X and AA – AC, brine injection was first carried out at the same fixed rate for five pore volumes, followed by an additional 1 – 2 pore volumes of AS injection at varying rates.

Table 1. Chemical Core Floods Performed Using Brine as the Aqueous Phase

Core ID	D (m)	L (m)	Porosity	k (D)	S_{wi}	μ_o (mPa s)	WF rate (m/d)	AS rate (m/d)
Q	0.011	0.103	0.45	15.98	0.163	11,500		0.51
R	0.011	0.103	0.45	11.04	0.255	11,500		0.51
S	0.011	0.103	0.45	15.9	0.159	11,500		0.51
T	0.011	0.103	0.45	11.74	0.250	11,500		0.51
U	0.011	0.103	0.45	17.94	0.194	11,500	0.51	0.51
V	0.011	0.103	0.45	11.55	0.243	11,500	0.51	0.51
W	0.011	0.103	0.45	11.81	0.150	15,000	0.51	0.26
X	0.011	0.103	0.45	10.24	0.135	15,000	0.51	1.02
Y	0.011	0.105	0.49	1.48	0.109	11,500		0.51
Z	0.011	0.105	0.49	1.19	0.203	15,000		0.51
AA	0.011	0.105	0.49	0.92	0.105	15,000	0.51	0.51
AB	0.011	0.105	0.49	0.97	0.165	15,000	0.51	0.26
AC	0.036	0.211	0.45	9.05	0.115	11,500	0.42	0.42

In all of the tests, the pressure gradient across the core was monitored, as was the ratio of water and oil produced. Specifically, the relationship between the pressure gradient and the oil recovery was compared between chemical flooding and waterflooding, in order to identify the influence of the W/O emulsions, and to determine the efficiency of the oil displacement. Several core floods were performed in high peremeability cores (values on the order of 11 – 15D), while other tests were performed in cores with permeabilities on the order of 1D. In systems where O/W emulsions formed preferentially to W/O emulsions, it had been observed that different recovery profiles were generated for high vs. low permeability sand (Bryan and Kantzas, 2007). In the high permeability cores, chemical injection led to the generation of a pressure build-up in the system, with a corresponding

Journal of Physics: Conference Series **147** (2009) 012058 doi:10.1088/1742-6596/147/1/012058

drop in the produced water cuts. Despite the observed alteration in the behaviour of the system, oil and water appeared to be produced as separate phases. In contrast, when O/W emulsions were generated in low permeability sand, these emulsions were actually produced, with a corresponding loss in the flood efficiency (Bryan and Kantzas, 2007). In these tests, therefore, one of the objectives was to determine if the same differences in flood response would be observed in the production and displacement of viscous W/O emulsions.

Finally, several core floods were performed in a TECAPEEK™ core holder, which is NMR-transparent. The sand pack measured was approximately 0.5 m in length, and spectra were acquired at four locations within the sand pack. The spectra were then interpreted to separate the oil and water signals, and the water spectra were used to identify the presence of *in-situ* W/O emulsions. These tests all allowed for improved visualization of the effect of emulsion generation on improved oil recovery.

3. PRIMARY AND SECONDARY CHEMICAL FLOODING

In reservoir flooding applications, chemical flooding could be carried out either as a primary process (i.e. directly into the sand that contains oil and irreducible water saturation), or after waterflooding. Figure 1 compares the oil recovery response waterfloods and primary AS floods. Both configurations are qualitatively similar: oil is recovery is high during the first 1 – 2 PV injected, and the slope of the recovery profile is less at later times. In general, chemical flooding recovers more oil than waterflooding.

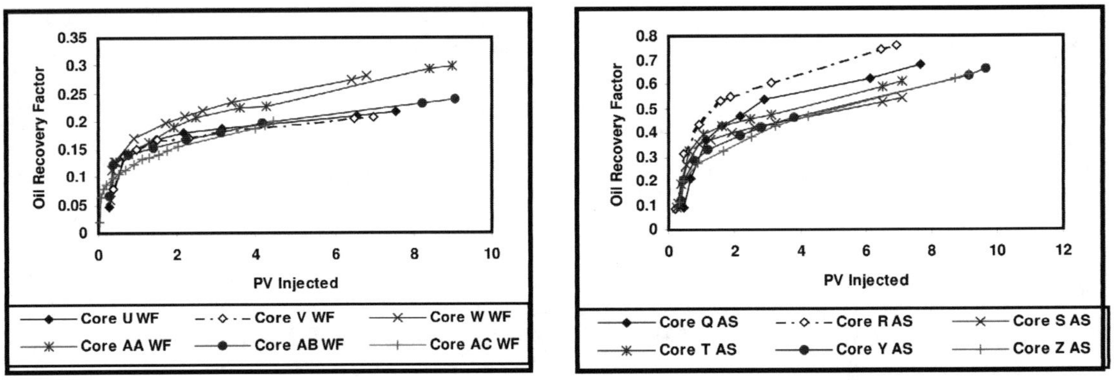

(a) (b)

Fig. 1 Oil recovery profiles from constant rate waterflooding (a) and primary AS flooding (b)

The viscosity of water and AS solution are essentially the same, thus the improved oil recovery response in Figure 1 is not a reflection of a higher viscosity aqueous phase, as is the case in polymer flooding. Rather, the low interfacial tension and shear due to flow through rock pores leads to the generation of emulsions, and these alter the flow properties of oil and water in the pores of the rock. A key to understanding the flood response is to compare the oil recovery up to the point of breakthrough, and after breakthrough. These results are summarized in Table 2.

In heavy oil flooding, the actual point of breakthrough is difficult to determine accurately, since it happens very quickly, and only discrete samples of produced oil and water are measured so it is likely that the actual point of breakthrough does not correspond to a sampling time. However, it is evident from Figure 1 that the change in slope of the oil recovery (indicative of a continuous aqeous phase) occurs within the first PV of fluid injection. Therefore, in Table 2 the recovery values are summarized for the first PV of fluid injection and for the subsequent injected volumes separately.

On average, the waterflood recovery was 0.16 for the first PV of water injection, compared to 0.33 during the first PV of primary AS injection. The recovery is therefore significantly higher during the chemical floods. Even after water breakthrough, it appears that when W/O emulsions are formed the flood still behaves more efficiently than a waterflood. However, the difference in flood efficiency (oil

Journal of Physics: Conference Series **147** (2009) 012058 doi:10.1088/1742-6596/147/1/012058

recovery/PV injected) is less significant at later times. The important contribution of primary chemical flooding appears to be in improving the breakthrough oil recovery.

Table 2. Oil Recovery Values for Waterflooding and AS Flooding

Core ID	WF Recovery (1st PV Inj)	WF Recovery (last 4 PV)	Primary AS (1st PV Inj)	Primary AS (last 4 PV)	Secondary AS (1st PV inj)
Q			0.315	0.276	
R			0.445	0.236	
S			0.356	0.128	
T			0.382	0.156	
Y			0.306	0.198	
Z			0.280	0.214	
U	0.153	0.046			0.143
V	0.151	0.044			0.150
W	0.172	0.084			0.300
X	0.156	0.055			0.199
AA	0.151	0.088			0.123
AB	0.144	0.061			0.169
AC	0.128	0.088			0.405

AS flooding is more expensive than waterflooding, and is much more difficult to design at the field scale. It is therefore unlikely that chemical floods will be carried out without first recovering some oil by less-expensive waterflooding. In the waterfloods, continuous channels of water form quickly due to the adverse mobility ratio between oil and water. After breakthrough, these water channels act as low-resistance pathways for flow, therefore most of the injected water simply flows through these fingers and by-passes the oil. This is evidenced by the decline in the oil production for both waterfloods and primary chemical floods in Figure 1. Therefore, a inference that could be made is that for secondary AS flooding, the injected chemical will simply flow through these channels and not contact significantly more oil. In reality, however, it was observed that secondary AS injection still led to an improvement in the recovery response. Figure 2 shows the oil production and water cut profiles for a case where a constant rate waterflood was conducted for 4.5 PV, followed by AS injection at the same rate. The response of the system to chemical injection is evident in both the reduction in the water cut and the increase in recovery.

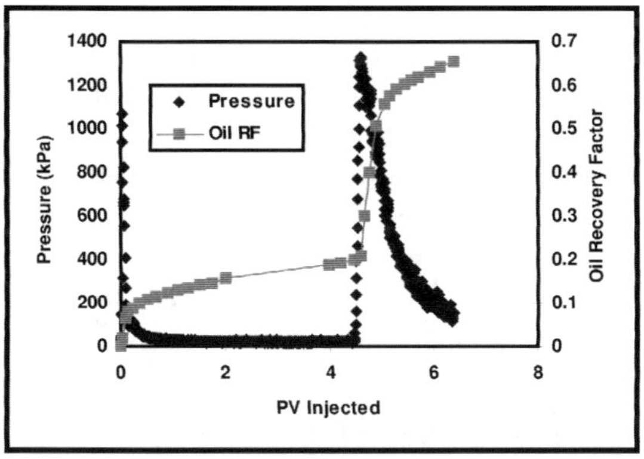

(a)

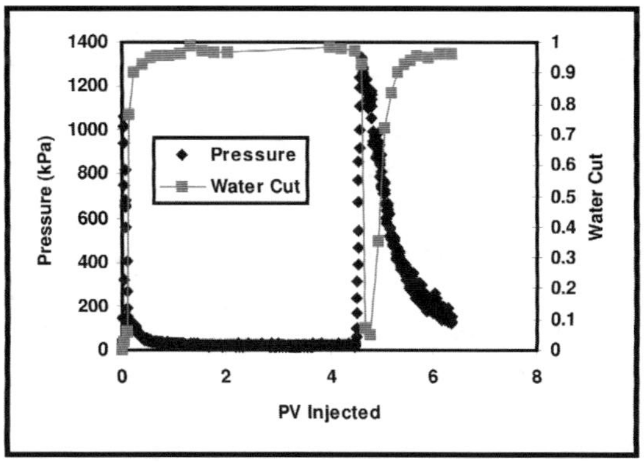

(b)

Fig. 2 Pressure with oil recovery (a) and water cut (b) profiles for secondary AS flooding

Even though continuous water channels were already present across the core at the start of AS injection, the presence of surfactants still led to an increase in the oil recovery compared to waterflooding. The response shown in Figure 2 is that of Experiment AC in Table 2; most of the core floods did not exhibit such a high secondary flooding recovery enhancement. On average, secondary chemical flooding leads to an additional recovery of 21% more oil. This is similar to the recovery efficiency of the primary chemical floods in the last four PV of fluid injection (0.2 Oil RF/PV Injected). Therefore, the formation of emulsions seems to be effective as a heavy oil recovery mechanism even in reservoirs that have been previously waterflooded.

One observation that could be made from Figure 2 is the presence of very high maximum pressures for both waterflooding and AS flooding. The fact that pressure builds up in the system could be evidence not of emulsions, but rather the simple formation and subsequent displacement of an oil bank. This is the common interpretation of such a response in enhanced oil recovery projects (Green and Willhite, 1998). If this were the actual displacement mechanism, then the improved oil recovery is simply a reflection of an increase in the pressure gradient that is applied across the core. Thus, in order to properly understand the enhanced oil recovery mechanism, this interpretation must be validated. Figure 3 compares the ratio of the AS/waterflood pressure gradients to their recoveries. In all the cores tested, the recovery of oil for the same pressure gradient is higher under chemical flooding. Therefore, AS injection does not simply lead to mobilization of oil and the subsquent formation of an oil bank. Rather, for the same pressure gradient applied across the core, more oil is displaced by the chemical than by water.

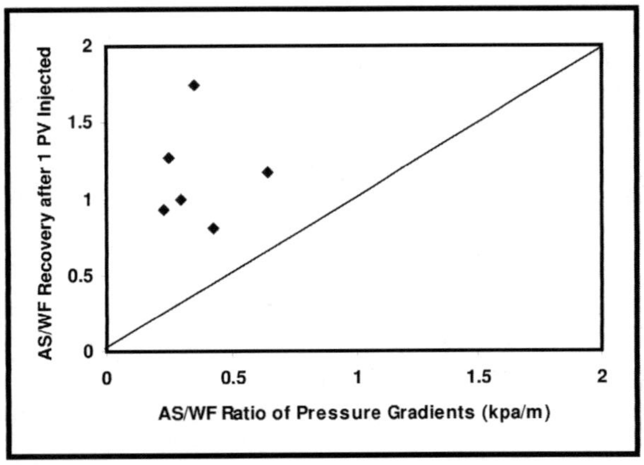

Fig. 3 Correlation between recovery and max pressure for waterflooding vs. AS flooding

The results from Figures 1 to 3 show how simple measurements of pressures and produced fluids can be used to make important observations regarding the nature of heavy oil displacement by chemical solution. Even with no additional information available, it is possible to determine that AS injection results in changes in the flow properties of oil and water in porous media, and that this is responsible for improved oil recovery. Thus, in both laboratory and field studies, measurements of the pressure vs. production response are key to identifying that a change has actually occurred. The mechanism proposed is that of the formation of W/O emulsions, as observed in bulk liquid phase behaviour studies. However, simple pressure measurements cannot definitively identify the presence of the emulsions. Therefore, a more detailed study was also performed using low field NMR.

4. NMR MEASUREMENTS OF *IN-SITU* WATER EMULSIFICATION

Low field nuclear magnetic resonance (NMR) has been applied in the petroleum industry for several decades (Brown and Gamson, 1960). NMR logging tools are commonly used to predict rock properties such as porosity and permeability (Straley *et al.*, 1997; Coates *et al.*, 1999). In recent years, applications of NMR technology have also been identified in heavy oil and bitumen formations, for the determination of *in-situ* oil viscosity (Bryan *et al.*, 2007) and fluid content in oil sand samples (Bryan *et al.*, 2006). Recently, methodologies were developed (Mai, 2008; Bryan *et al.*, 2008) to determine the *in-situ* distribution of water in porous media. In this work, these methodologies were used to provide quantitative evidence of W/O emulsification during chemical flooding.

NMR measures the response of hydrogen in the presence of magnetic fields. Hydrogen has a property known as "spin", causing protons to behave essentially like tiny bar magnets, lining up in the direction of magnetic field lines. In an NMR experiment, radio-frequency pulses are applied to tip the protons onto a perpendicular plane, known as the transverse plane. The time for them to return to their equilibrium position (lined up with the external magnetic field lines) is recorded as a decaying signal on the transverse plane. Subsequently, this data is converted into a spectrum of amplitudes and associated relaxation time constants and all analysis is performed on these spectra.

Relaxation rates (and therefore relaxation times) depend on both the physical properties of the fluid being measured and the location of the fluids in the porous medium. The total relaxation is the summation of bulk relaxation – proportional to the fluid viscosity, and surface relaxation – proportional to the size of the pore containing the fluid (Straley *et al.*, 1997; Coates *et al.*, 1999). Figure 4 shows the spectra of fluids at four length locations in the unconsolidated sand pack in the NMR-transparent core holder.

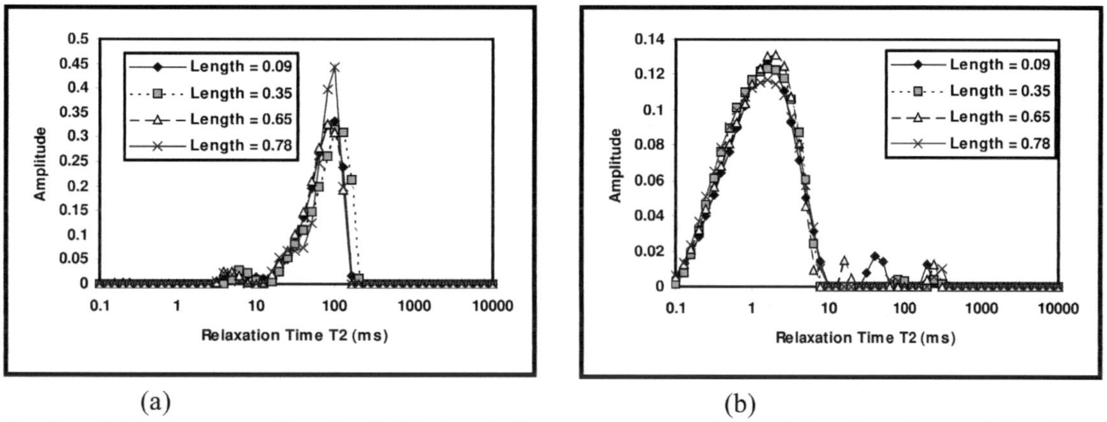

Fig. 4 NMR spectra of *in-situ* fluid signals for water-saturated sand (a) and sand containing heavy oil and connate water (b)

For low viscosity fluids like water, the bulk relaxation is significantly slower than the surface relaxation, so the spectra of water in porous media are analogous to the pore size distribution of the sand (Straley *et al.*, 1997; Coates *et al.*, 1999). In the sand pack shown in Figure 4(a), there is a low fraction of small pores with relaxation times between 5 – 10 ms, but the majority of the pores have relaxation times between 10 – 200 ms. All four length fractions show similar T_2 peak locations, so the pore sizes are homogeneous along the length of the core. The core was then flooded down to irreducible water saturation (S_{wi}) using the same viscous heavy oil as in the other core floods, and the resulting spectra at the same four locations are presented in Figure 4(b). These spectra all have a large, fast-relaxing peak under 10 ms. This is the signal from the viscous oil; for high viscosity fluids the relaxation times are governed by the fluid viscosity, thus the relaxation times of the oil are essentially independent of any influence from the porous medium (Bryan *et al.*, 2007). The signal from the oil therefore does not provide any information about where the oil is located, but the signal from any water remaining is still a reflection of the physical location of the water. In Figure 4(b), the signal after the oil peak is the contribution from connate water; most of the amplitude between 10 – 200 ms has disappeared, so oil is physically present in the larger pores.

The sand pack was first waterflooded for approximately 5 PV's at a constant rate, followed by AS injection at the same rate. The resulting spectra after the AS flood are shown in Figure 5, for the same four locations in the sand pack. When comparing the amplitude of the first peak of the spectra in Figures 5 and 4(b), it is evident that the amplitude is lower after waterflooding and AS flooding. This is therefore an indication that there is less oil remaining to give off an NMR signal, so the spectra validate the fact that oil has been displaced out of the core. It is significant, however, that there does not appear to be an obvious difference in the oil amplitudes with length. This shows that oil was displaced all along the sand pack, and the chemical flood did not lead to the generation of a high oil saturation bank anywhere in the core.

Any signal after the first peak is the contribution from water. During the waterflood, as the oil amplitude decreased, a water peak formed between 10 – 200 ms. This was water relaxing in the pores of the sand, and corresponds to injected water that has displaced oil completely out of pores. This signal is therefore that of the continuous water channels formed during the waterflood (Mai, 2008; Bryan *et al.*, 2008). In Figure 5, however, there is additional signal relaxing after 200 ms. There were no pores present to give such a slow relaxing signal, so this water is still physically in the same size of pore. The slower relaxation times indicate that protons in the water are no longer in contact with the sand, so the surface relaxation effects are reduced and the water is now relaxing closer to its bulk values. This is direct and quantitative evidence of W/O emulsions that have formed within the core.

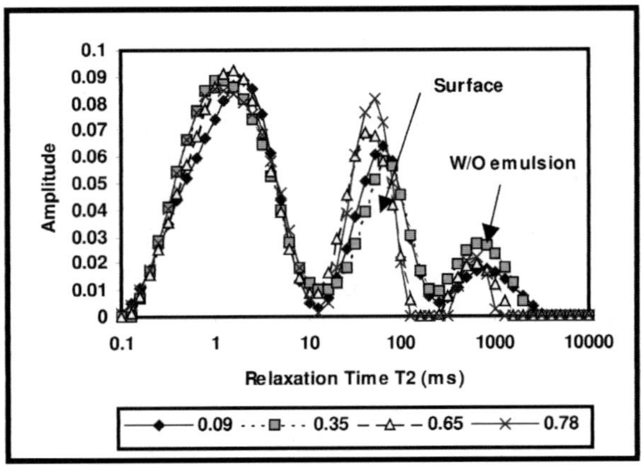

Fig. 5 Possible distribution of water during a chemical flood

In Figure 5, it is also evident that W/O emulsions form all along the length of the sand pack; once again there was no significant variation in the emulsion amplitude with length. The water saturation in the pores and in emulsions was therefore averaged for the four locations, and the mean values of saturation are plotted in Figure 6, for time of the AS flood. The water in the pores decreases from 40% saturation down to 20%. As this happens, the saturation of W/O emulsions increases from 10% up to 50%. The total water saturation increases during the chemical flood, indicating that additional oil is displaced out of the core and this pore space is now occupied by chemical solution.

The majority of emulsions are formed within the first half a PV of secondary AS injection. During this time, pressure would build up in the core, as shown in Figure 2. The pressure build up was accompanied by a reduction in the produced water cut, and correspondingly an increase in the oil recovery. These results present an important implication of chemical flooding: as the majority of W/O emulsions are forming, this leads to a decrease in the water production in the system. The improved recovery response from chemical flooding is therefore at least partially a reflection of the formation of these W/O emulsions all along the length of the system.

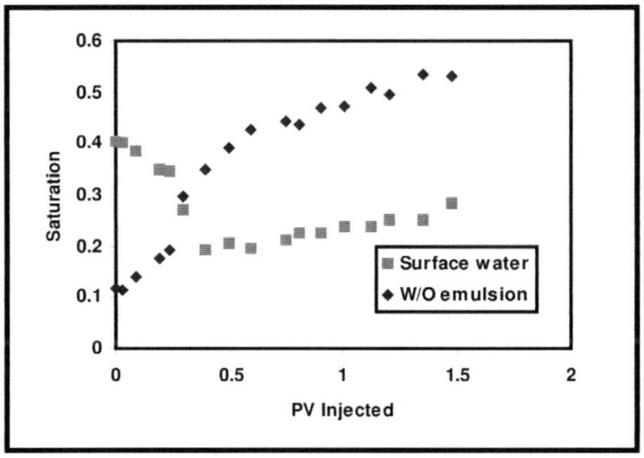

Fig. 6 Change in distributin of water during a chemical flood

The utilization of low field NMR for measurements of *in-situ* fluid distributions therefore provides a valuable contribution to the pressure and recovery data measured during the core floods of Tables 1 and 2. Specifically, the spectra validate the hypothesis that emulsification is somehow responsible for improved oil recovery, compared to the concept of an oil bank formation. Thus, NMR technology can be used at least at the laboratory scale to monitor the formation and presence of *in-situ* W/O emulsions.

5. PREDICTIONS OF HEAVY OIL RECOVERY BY W/O EMULSIFICATION

With the combined methods for measuring emulsion formation and its effect on the produced fluids, a mechanism can now be proposed to explain how chemical injection can lead to enhanced heavy oil recovery. This proposed mechanism is then tested through a simple reservoir simulation model.

5.1 Proposed AS Flooding Recovery Mechanism

In Figure 7, a waterflood was conducted for approximatly ten PV's of fluid injection, followed by an additonal three PV's of AS injection. The oil recovery profile is compared against the predicted recovery, which was made by following the same trend of the waterflood for many more PV's of constant rate injection.

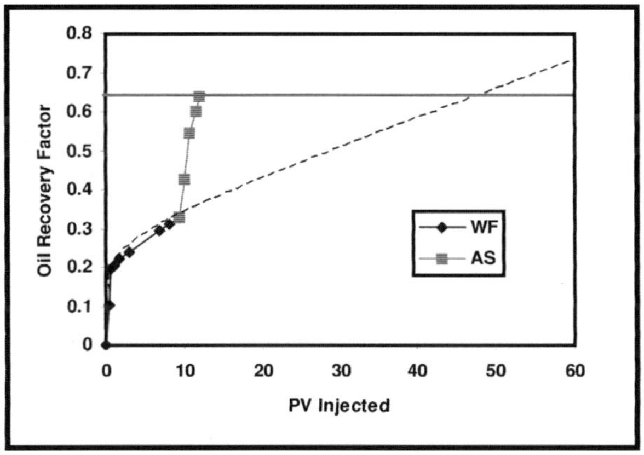

Fig. 7 Recovery profile for secondary chemical flood and an extended waterflood in heavy oil

It is possible that with very long times left for the waterflood, similar oil recoveries may be possible for waterflooding and chemical flooding. The chemical flooding appears to provide an acceleration to the oil recovery, thus the value of chemical injection is mostly in the fact that oil can be recovered

Journal of Physics: Conference Series **147** (2009) 012058 doi:10.1088/1742-6596/147/1/012058

much more quickly due to the build-up of pressure in the system. Moreover, after the pressure has once again declined, there is still 30 – 40% of oil remaining in the sand pack. Therefore, chemical flooding does not necessarily yeild lower residual oil saturations, as is expected in conventional oil surfactant flooding (Green and Willhite, 1998), but the oil is simply recovered much more quickly than if waterflooding were continued.

The data from the experiments can be summarized in several key observations. First, the injection of chemical, whether as a primary or secondary agent, leads to a more efficient displacement of oil for any given pressure gradient. Emulsions are formed all along the length of the system, and correspondingly oil is produced along the length of the core instead of as an oil bank. During the time when most of the emulsions are forming, pressure builds up in the core and correspondingly water production decreases and oil recovery increases. By focusing on measurements of flow of multiple fluid phases and *in-situ* emulsification, it is now possible to propose a mechanism by which the injected chemical is displacing the oil.

AS solution mixes with oil, and the combination of low interfacial tension and shear due to flow through rock pores leads to the formation of W/O emulsions. During secondary AS injection, these emulsions are forming in the continuous water channels, as evidenced by the decrease in surface water saturation in the NMR spectra. When these emulsions form, they effectively block off the water channels, so the produced water cuts decrease. These emulsions cannot be easily propagated through the rock pores, so they will instead block off the water channels and lead to improved sweep efficiency of the flood. In this manner, oil is recovered much more quickly, compared to waterflooding. In primary AS flooding, the same mechanisms are present – instead of water fingering through the oil, emulsions form and block off the water channels. This in turn leads to improved breakthrough recovery of the oil. After continuous emulsion channels form along the length of the core, the plugging of these channels is much more difficult and subsequent chemical injection will not be as efficient. Therefore, AS injection will result in maximum oil recovery within the first pore volume; at later times pressure once again declines in the core and the oil rates decline as well.

In order to model emulsion flow in porous media, mathematical formulations have been proposed in the literature (Alvarado and Marsden, 1979; Abou-Kassem and Farouq Ali, 1995), incorporating emulsion viscosity as a function of shear rate. However, in systems where emulsions are formed *in-situ*, the total multi-phase flow condition involves emulsion formation, breaking, plugging and flow, along with flow of un-emulsified fluids. Therefore, a simplier approach is to incorporate all of these unknown parameters into the effective relative permeability of the oil and water. If the mechanism of improved oil recovery is really that of plugging of the water channels and improved sweep efficiency, then a simple way to model and predict the chemical flooding response is simply to reduce the apparent relative permeability to water. Using the data from the core floods in Tables 1 and 2, relative permeability relationships were developed by assuming a simplistic Darcy's Law relationship. These relative permeability values are apparent values, since the true relative permeability behaviour cannot be determined in unstable flooding configurations. However, for these systems, the apparent oil and water relative permeability behaviour could be summarized in the following expressions:

$$k_{ro} = 1.015(1 - S_e)^{0.2997} \tag{1}$$

Where k_{ro} = the apparent relative permeabilty to oil
S_e = normalized water saturation.

For waterflooding and AS flooding, the corresponding relative permeability relationships are:

$$k_{rw} = 0.0148(S_e)^{2.0161} \tag{2}$$

And

$$k_{rw} = 0.0266(S_e)^{5.3851} \quad (3)$$

In these expressions, the normalized water saturation is defined as:

$$S_e = \frac{S_w - S_{wi}}{(1 - S_{or}) - S_{wi}} \quad (4)$$

The correlations for Equations (1) to (3) were developed empirically, using the normalized water saturations for all of the core floods performed. The resulting water saturations are shown in Figure 8.

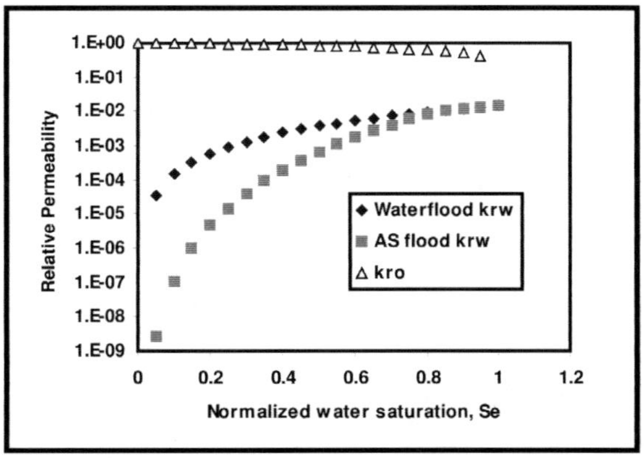

Fig. 8 Predicted oil and water relative permeability curves

The apparent water relative permeability values are much lower than those of oil, due to the significant viscosity difference between the oil and water. It should be noted that for the data obtained in this study, there was no distinct difference in the oil relative permeability behaviour for waterflooding vs. chemical flooding, so only one relationship was developed for the oil. This is in contrast to what other researchers have found (Wang *et al.*, 2008). The main contribution from the chemical injection is a reduction in the relative permeability to water, which should theoretically reduce the production of water. This was tested in a simple reservoir simulation study.

5.2 Numerical Simulation of AS Flooding

A simple model for simulating waterflooding and chemical flooding in a heavy oil sand pack was constructed using CMR Stars™ (2006.10). This model is a laboratory scale simulation, containing dead oil and aqueous phase (either water or AS solution) as the two components in the reservoir. The model was run isothermally at 25°C, using the input parameters summarized in Table 3.

Table 3. General Input Parameters for Laboratory Scale Simulation

Input Parameter	Value
Length of model	10.3 cm
Cross-sectional area	0.933 cm²
Gridblocks in "i" direction	20 (0.515 cm long)
Gridblocks in "j" direction	1 (0.966 cm long)
Gridblocks in "k" direction	10 (0.0966 cm long)
Porosity	0.45
Oil viscosity	11,500 mPa·s
Irreducible water saturation	0.187

The reservoir consists of two layers: the first eight rows have a permeability of 6 D in all directions. The bottom two layers have a permeability of 21 D. Overall the permeability of the model is 9 D; the two layers are present in order to simulate unequal displacement through the higher permeability layer. This leads to early breakthrough of water, which occurs during viscous fingering in heavy oil waterfloods. Figure 9 shows a schematic of this system, with a single injector at the inlet face of the model and the producer at the outlet.

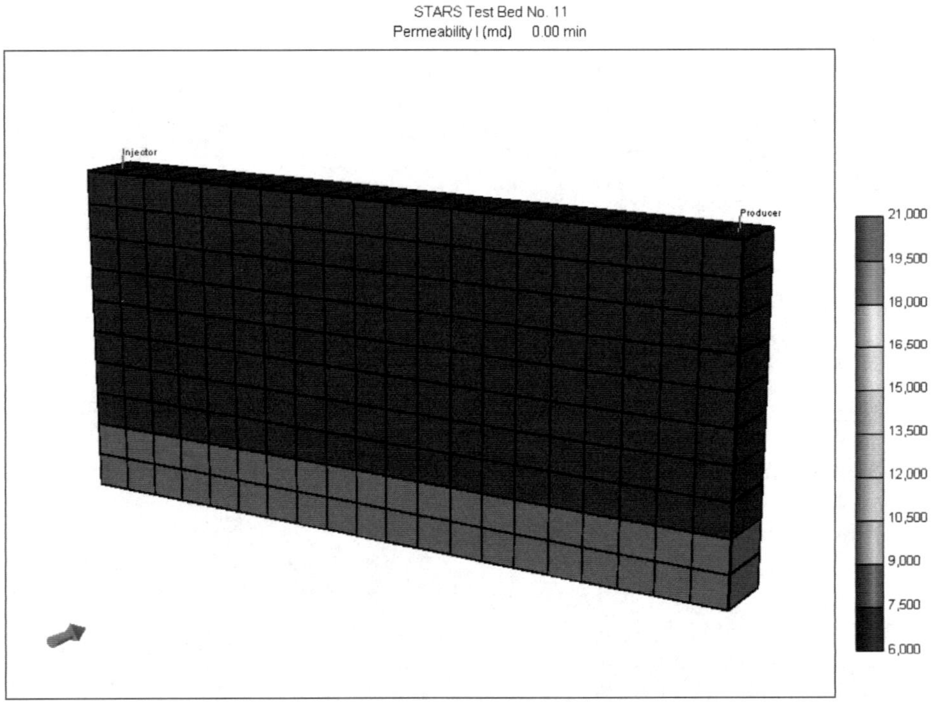

Fig. 9 3-D reservoir model, showing variation in permeability

Fluids were injected at 2 mL/hr, giving a frontal velocity of 0.51 m/day, which is the same as what was used in the core floods in Table 1. In one case, waterflooding was carried out at a fixed injection rate, using the relative permeability to water given by Equation (2). In a separate run, AS solution was injected at the same fixed rate, by using the relative permeability to water given by Equation (3). The oil recovery profiles for waterflooding vs. primary AS flooding are shown in Figure 10. It is evident from this plot that the main contribution from the primary AS flood is a higher breakthrough recovery. This is similar to the behaviour shown in Figure 1.

After water breakthrough the slope of the oil recovery from the AS flood is still higher than that of waterflooding, indicating that even at later times when the pressure has already declined across the core, chemical flooding is still more efficient due to plugging of some of the water channels. As expected, therefore, the improved recovery of the primary AS flood can be modeled by a simple change in the apparent relative permeability to water.

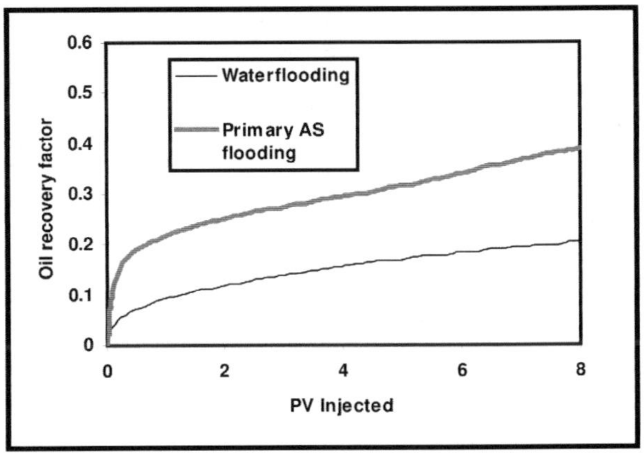

Fig. 10 Simulated results for waterflooding vs. primary AS flooding

Figure 11 shows the pressure gradient with its associated production and water cut response to a waterflood with a secondary AS flood. Waterflooding was carried out for approximately four PV, followed by another three PV of AS injection. Approximately 23% of the oil was recovered during the waterflood, followed by another 20% of oil that was mainly recovered during the first PV of chemical injection. The profiles in Figure 11 are qualitatively very similar to what was observed in the laboratory experiments exemplified in Figure 2, but no attempt was made in this study to actually history match the production data obtained.

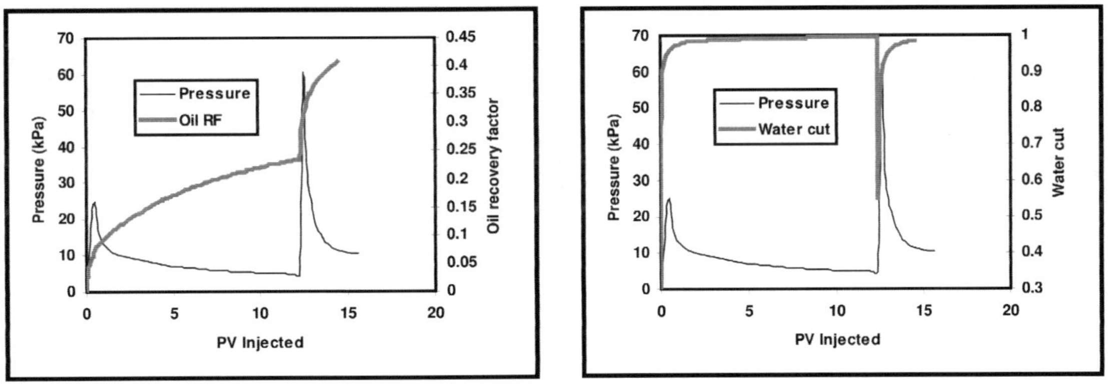

(a) (b)

Fig. 11 Simulated oil recovery (a) and produced water cut (b) profiles for secondary AS flooding

The results of Figures 10 and 11 demonstrate how a simple change in the apparent relative permeability to water can be used to generate the flood responses observed in the laboratory core floods. It is also important to note that in both primary and secondary AS flooding systems, the same water relative permeability relationship was used. The injected chemical therefore behaves similarly in both flooding configurations, which is expected. The oil relative permeability is also unaffected by the injection fluid, since the main influence of the AS solution is to plug off the water channels through the formation of emulsions.

6. CONCLUSIONS

By properly monitoring the response of the system, both through quantitative NMR spectra and even through simple measurements of pressure vs. produced fluids, an improved understanding of chemical flooding was achieved. This understanding was validated through numerical simulation.

Journal of Physics: Conference Series **147** (2009) 012058 doi:10.1088/1742-6596/147/1/012058

Upon injection of chemical, pressure builds up again in the system, and this corresponds to lower water cuts and improved production of oil. In this study, measurements were made of these multi-phase systems in order to identify the mechanism by which chemical injection can be used to recover oil. Using a combination of pressure measurements, collection of produced oil and water, and low field NMR of *in-situ* water distributions, the recovery mechanism has been identified. Alkali-surfactant solutions reduce interfacial tensions and result in the formation of W/O emulsions. These emulsions plug off the water channels, leading to improved sweep efficiency in the system.

The consequence of this recovery mechanism is that chemical injection in heavy oil may not necessarily lead to lower residual oil saturations, as is the case in conventional oil surfactant flooding. The presence of emulsions simply accelerates the oil recovery, thus alkali-surfactant injection behaves almost like a polymer flood in heavy oil applications. In order to model the response from chemical injection, the apparent relative permeability of the aqueous phase is reduced. With this simple change in the relative permeability behaviour, the response from both primary and secondary chemical flooding can be modeled.

ACKNOWLEDGEMENTS

The authors wish to acknowledge the technical contributions of Jun Gao, Michael Erath and Xiao Dong Ji of TIPM Laboratory in helping to set up the experimental systems used in this study. The advice from Dr. An Mai is also gratefully acknowledged. The oil used in these core floods was generously provided by Nexen Inc., and the surfactant was donated by Stepan Company. Financial support for this project was received from NSERC, COURSE, ISEEE and the Canada Research Chair in Energy and Imaging with its Industrial Affiliates (Shell, Nexen, Devon, PetroCanada, Canadian Natural, ET Energy, Suncor, Schlumberger, CMG Foundation, Laricina, ConocoPhillips and Paramount).

NOMENCLATURE

AS	Alkali-surfactant flooding	
D	Diameter	[m]
k	Permeability	[D, μm^2]
k_{ro}	Apparent relative permeability to oil	
k_{rw}	Apparent relative permeability to water	
L	Length	[m]
NMR	Nuclear magnetic resonance	
PV	Pore volume	
RF	Recovery factor	
S_e	Normaliezed water saturation	
S_{or}	Residual oil saturation	
S_{wi}	Irreducible water saturation	[fraction]
T_2	NMR transverse relaxation time	[ms]
W/O	Water in oil	
WF	Waterflood	
	Greek Letters	
μ_o	Oil viscosity	[mPa·s]

REFERENCES

Abou-Kassem, J.H. and Farouq Ali, S.M., "Modeling of Emulsion Flow in Porous Media", *J. Can. Pet. Tech.*, **34** (6), 30 – 38, Jun 1995.

AEUB, *ST9802007: Alberta's Energy Reserves 2006 and Supply/Demand Outlook 2007-2016*, Farhood Rahnama, coord., Alberta Energy and Utilities Board, Calgary, AB Canada, Jun 2007, http://www.ercb.ca/docs/products/STs/st98_current.pdf.

Alvarado, D.A. and Marsden, S.S. Jr., "Flow of Oil-in-Water Emulsions Through Tubes and Porous Media", SPE 5989, *SPE Journal*, 369 – 377, Dec 1979.

Brown, R. and Gamson, B., "Nuclear Magnetism Logging", *AIME Petr. Trans.*, **219**, 1960.

Bryan, J., Mai, A., Hum, F. and Kantzas, A., "Oil- and Water-Content Measurements in Bitumen Ore and Froth Samples Using Low Field NMR", SPE 97802, *SPE Res. Eval. & Engg.*, 654 – 663, Dec 2006.

Bryan, J., Kantzas, A., Badry, R., Emmerson, J. and Hancsicsak, T., "In Situ Viscosity of Heavy Oil: Core and Log Calibrations", *J. Can. Pet. Tech.*, **46** (11), 47 – 55, Nov 2007.

Bryan, J. and Kantzas, A., "Enhanced Heavy Oil Recovery by Alkali-Surfactant Flooding", SPE 110738, 2007 SPE Annual Technical Conference and Exhibition, Anaheim, CA USA, Nov 11 – 14, 2007.

Bryan, J., Mai, A. and Kantzas, A., "Investigation into the Processes Responsible for Heavy Oil Recovery by Alkali-Surfactant Flooding", SPE 113993, 2008 SPE Improved Oil Recovery Symposium, Tulsa, OK USA, Apr 19 – 23, 2008.

Coates, G., Xiao, L. and Prammer, M., *NMR Logging Principles and Applications*, Halliburton Energy Services, TX USA, 1999.

Dong, M., Liu, Q. and Li, A., "Micromodel Study of the Displacement Mechanisms of Enhanced Oil Recovery By Alkaline Flooding", SCA 2007-47, International Symposium of the Society of Core Analysts, Calgary, AB Canada, Sept 10 – 12, 2007.

Dranchuk, P.M., Scott, J.D. and Flock, D.L., "Effect of the Addition of Certain Chemicals On Oil Recovery During Waterflooding", *J. Can. Pet. Tech.*, Jul – Sept 1974.

Farouq Ali, S.M., Figueroa, J.M., Azuaje, E.A., and Farquharson, R.G., "Recovery of Lloydminster and Morichal crudes by caustic, acid and emulsion floods", *J. Can. Pet. Tech.*, 53 – 59, Jan – Mar 1979.

Garrett, H.E., *Surface Active Chemicals*, Pergamon Press, New York, NY USA, 1972.

Green, D.W. and Willhite, G.P., *Enchanced Oil Recovery*, SPE Textbook Series Vol. 6, Society of Petroleum Engineers Inc., Richardson, TX USA, 1998.

Isaacs, E.E. and Chow, R.S., "Practical Aspects of Emulsion Stability", *Emulsions Fundamentals and Applications in the Petroleum Industry*, L.L.Schramm ed., Advances in Chemistry Series 231, 51 – 78, American Chemical Society, 1992.

Mai, A., *Mechanisms of Heavy Oil Recovery by Waterflooding*, PhD Thesis, Department of Chemical and Petroleum Engineering, University of Calgary, Calgary, AB Canada, 2008.

Martin, F.D. and Oxley, J.C., "Effect of Various Alkaline Chemicals on Phase Behavior of Surfactant/Brine/Oil Mixtures", SPE Paper 13575, International Symposium on Oilfield and Geothermal Chemistry, Pheonix, AZ USA, Apr 9 – 11, 1985.

Nelson, R.C., Lawson, J.B., Thigpen, D.R. and Stegemeier, G.L., "Cosurfactant-Enhanced Alkaline Flooding", SPE/DOE 12672, SPE/DOE Fourth Symposium on Enhanced Oil Recovery, Tulsa, OK USA, Apr 15 – 18, 1984.

Nutting, P.G., "Chemical Problems in the Water-Driving of Petroleum Reservoirs", *Ind. And Eng. Chem.*, **17**, Oct 1925.

Radler, M., "Oil production, reserves slightly increase in 2006", *Oil & Gas Journal*, **104**, 20 – 23, Dec 2006.

Reisberg, J. and Doscher, T.M., "Interfacial Phenomena in Crude Oil – Water Systems", *Prod. Monthly*, 43 – 50, Nov 1956.

Scott, G.R., Collins, H.N. and Flock, D.L., "Improving Waterflood Recovery of Viscous Crude Oils by Chemical Control", *J. Can. Pet. Tech.*, **4** (10), October-December, 1965.

Straley, C., Rossini, D., Vinegar, H., Tutunjian, P. and Morriss, C., "Core Analysis by Low Field NMR", *The Log Analyst*, **38** (2), Mar – Apr 1997.

Wang, J., Dong, M. and Arhuoma, M., "Simulation of Chemical Flooding for Heavy Oil Recovery in Thin Reservoirs", PS 2008-159, 59th Annual Technical Meeting of the Petroleum Society, Calgary, AB Canada, Jun 17 – 19, 2008.

Journal of Physics: Conference Series **147** (2009) 012059 doi:10.1088/1742-6596/147/1/012059

Flow rate measurement of oil-water two-phase flow based on V-cone flow meter

F Dong, C Tan, W Li, F S Zhang

Tianjin Key Laboratory of Process Measurement and Control, School of Electrical Engineering and Automation, Tianjin University, Tianjin 300072, CHINA

fdong@tju.edu.cn

Abstract. In petroleum industry, oil-water two-phase flow widely exists in production wells, and the measurement of oil-water flow is significant in oil well logging. One method in common use is to build a theoretic model to connect differential pressures generated by a throttling set with flow-rate. Several measuring models were adopted and compared for model selection and modification. A V-cone with 0.65 diameter ratio was selected as the throttling set because of its special structure and intrinsic advantages. In the experiments, the differential pressures measured under different flow range were processed with a homogeneous flow model. And the results show that the error is small in the condition where the oil and water are well mixed, while the error is big when they are not well mixed. A polynomial of the error and the water percentage was fitted to improve the model accuracy and widen the measurement range.

1. INTRODUCTION

In the measurement of oil-water two phase flow, one method that is commonly used is getting flow rate information from the pressure drop of fluid flows. This measuring method in oil-water two flow field is usually sumarized as one measuring model which is called pressure drop model. And pressure drop model includes stratified flow model and homogeneous flow model. The basic idea of the separated model is to treat the two phases separately as a single fluid. Each phase is assumed to flow alone in the pipe with its own flow parameters and properties. Homogeneous flow model is the condition that water and oil mixes so well that the mixture can be seen as one single phase approximately. Unlike gas-water two phase flow, each phase of oil-water two phase flow has the similar flow characteristics. So the measuring model that suits for gas-water two phase flow can not be used in oil-water two phase flow. Charles (1961) was a earlier researcher who worked on the oil-water two phase flow regime. He laid some foundations for oil-water two phase flow. In 1980s, Arirachakaran (1989) did some experiments about stratified flow, mixed flow, annular flow, intermittent flow and homogeneous flow. And they came to the conclusion that take oil phase as one kind of high density "gas phase", the flow regime were much similar. Several years later, Hewitt (1997) studied and compared the flow regime of oil-water two phase flow and gas-water two phase flow, and concluded that stratified flow exists in both of the two kinds of two phase flow; though slug flow is very common in gas-water two phase, it is very rare in oil-water two phase flow; in gas water two phase flow, only as the gas percentage is less than 30% gas phase could disperse into water phase, but water phase never disperse into gas phase, while in the oil-water two phase flow, each can disperse into the other phase no matter what the phase percentage is; annular flow is a typical and common

Journal of Physics: Conference Series **147** (2009) 012059 doi:10.1088/1742-6596/147/1/012059

flow regime in gas-water two phase flow, but in oil-water two phase flow annular flow is hardly appeared. During this period, Nadler (1995) also reported some much similar conclusions. Trallero (1996) summarized the method about flow regime of oil-water two phase flow and that is widely used so far.

$$\begin{cases} Stratified\begin{cases} ST \\ ST\ \&\ MI \end{cases} \\ Homogeneous\begin{cases} Water\begin{cases} Do/w\ \&\ w \\ o/w \end{cases} \\ Oil\begin{cases} Dw\ \&\ Do/w \\ w/o \end{cases} \end{cases} \end{cases}$$

The throttling set is one of the primary elements used in the flow-rate measurement by introducing pressure drop. When fluid passes the throttling set, the cross section area of the fluid decreases, then the velocity of the fluid increases. Consequently the pressure difference between the upstream and downstream of the throttling set results in the differential pressure which represents the flow-rate of the fluid.

For the advantages of high reliability and accuracy, simple structure as well as easy installation, the throttling sets have gained wide appliactions in single phase flow measurement and in broad investigation on two-phase flow rate measurement. In the middle of 1980s, came out a new kind of V-cone flow meter, which is shown in Figure 1. The streamlined structure of V-cone overcomes the disadvantages of other throttling sets that can be easily jammed due to the high viscosity of oil flows. The noise signal created by the fluid flows through the V-cone is much lower than that generated by other throttling sets (Sun, 2004).

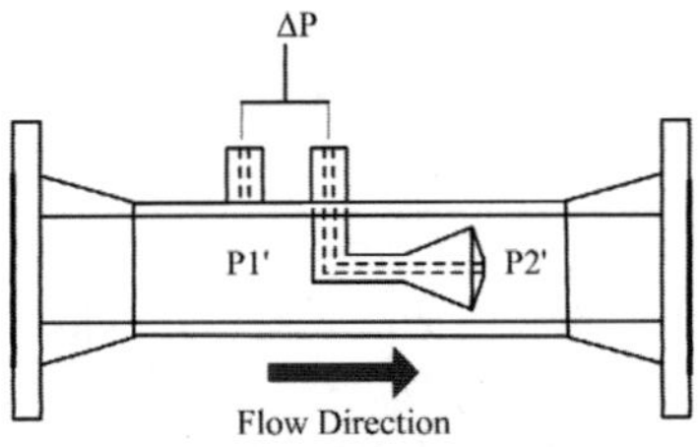

Fig. 1 Structure of V-cone flow meter

As the fluid flows through the V-Cone meter, the differential pressure ΔP is generatd from the shrinking area of the cross-section. With a wider measurement scale, it is more accurate and effective in flow rate measurement, for which, V-cone becomes a popular research field for two-phase flow measurement.

2. MEASUREMENT METHOD OF FLOW-RATE ON OIL-WATER TWO PHASE FLOW

Previous researchers had worked on the flow regime of the oil-water two phase flows, but their achievements never refer to the relationship of the flow rate and the flow regimes. However, the flow regime factor is very important for flow rate measurement of oil-water two phase flows. It also determines the measuring model adopted.

When the mixture of water and oil passes the V-cone, according to the Bernoulli equation

$$\frac{p_1'}{\rho_1}+\frac{u_1'^2}{2}=\frac{p_2'}{\rho_2}+\frac{u_2'^2}{2} \quad (1)$$

And the fluid continuity equation:

$$A_1 u_1 = A_2 u_2^{'} \quad (2)$$

$p_1^{'}, p_2^{'}, u_1, u_2^{'}, \rho_1, \rho_2, A_1, A_2$ represent the upstream pressure of V-cone, the downstream pressure of V-cone, upstream flow-rate, downstream flow-rate, upstream fluid density, downstream fluid density, upstream circulation area, downstream circulation area respectively. If the fluid is incompressible, then $\rho_1 = \rho_2 = \rho$, $A_2 = \mu A_0$, where A_0 is the circulation area of the V-cone, thus

$$u_2^{'} = \frac{1}{\sqrt{1-\mu^2}} \sqrt{\frac{2}{\rho}(p_1^{'} - p_2^{'})} \quad (3)$$

and $\mu^2 = \beta^4$, β is the equivalent diameter ratio. Because $p_1^{'}$ and $p_2^{'}$ are average pressures, the modified formula $\sqrt{p_1^{'} - p_2^{'}} = \sqrt{\psi}\sqrt{p_1 - p_2}$ is necessary, where $\sqrt{\psi}$ is called pressure tapping coefficient which changes as the mode of the pressure tapping changes. Introducing ε to modify the $u_2^{'}$ (considering the loss due to fluid viscosity), so

$$u_2 = \varepsilon u_2^{'} = \frac{\varepsilon\sqrt{\psi}}{\sqrt{1-\beta^4}} \sqrt{\frac{2}{\rho}(p_1 - p_2)} \quad (4)$$

The volume flow-rate is

$$q_v = u_2 \mu A_0 = \frac{\mu\varepsilon\sqrt{\psi}}{\sqrt{1-\beta^4}} A_0 \sqrt{\frac{2}{\rho}\Delta p} \quad (5)$$

Discharge coefficient is defined: $C = \mu\varepsilon\sqrt{\psi}$, then

$$q_v = \frac{C\varepsilon}{\sqrt{1-\beta^4}} A_0 \sqrt{2\Delta p \rho} \quad (6)$$

For $A_0 = \frac{\pi}{4}(D^2 - d^2)$ D is the diameter of pipe, d is the largest cross section area of V-cone, so the mass flow-rate is determined by:

$$q_m = \frac{C\varepsilon}{\sqrt{1-\beta^4}} \frac{\pi}{4}(D^2 - d^2)\sqrt{2\Delta p \rho} \quad (7)$$

Consider oil-water two-phase flow as one single kind of fluid, the density of the fluid ρ_m is expressed as

$$\frac{1}{\rho_m} = \frac{\omega_O}{\rho_O} + \frac{1-\omega_O}{\rho_w} \quad (8)$$

ω_O is mass fraction of oil, ρ_O is the density of oil phase, ρ_w is the density of the water phase. In fact, the density of the oil-water flow should be modified with parameter n. Substitute equation (8) into equation (7), we get

$$G = \frac{CA}{\sqrt{(1-\beta^4)}} \sqrt{2\Delta p \rho_w / [1 + (\frac{\rho_w}{\rho_O} - 1)\omega_O{}^n]} \quad (9)$$

G is the mass flow-rate of oil-water two-phase flow, and n represents the situation of oil-water flow. And the more mixture mixed, the more n close to unity (Ma ea al., 2007).

The relationship between mass fraction of oil ω_O and volume fraction of water φ_w is

$$\omega_O = [1 + \frac{\rho_w}{\rho_O}(\frac{1}{1-\varphi_w} - 1)]^{-1} \quad (10)$$

So

$$G = \frac{CA}{\sqrt{1-\beta^4}} K \sqrt{2\Delta p \rho_w} \quad (11)$$

K is modified coefficient of oil-water two-phase flow based on V-cone.

$$K = \left\{ 1 + (\frac{\rho_O}{\rho_w} - 1) / [1 + \frac{\rho_w}{\rho_O}(\frac{1}{1-\varphi_w} - 1)]^n \right\}^{-\frac{1}{2}} \quad (12)$$

Thus, mass flow rate of oil and water are

$$G_O = G / [1 + \frac{\rho_w}{\rho_O}(\frac{1}{1-\varphi_w} - 1)] \quad (13)$$

$$G_w = G - G_O \quad (14)$$

3. EXPERIMENT AND RESULTS

The test section is a Perspex pipe in a stainless horizontal pipe loop. The diameter of the pipe is 50mm. The Oil-water loop contains an oil-water separation tank, an oil tank, a water tank, a V-cone flow meter, an industrial control computer, several vortex flow meters, and several pumps and valves. The industrial control computer controls the opening degree of the valves in order to control the flow rate of oil phase and water phase. The flow range of water and oil in this work is about: 0 - 12 m^3/h and 0 - 50 m^3/h respectively. The density of the oil is 841.7 kg/m^3, and its viscosity is around 14.3 cp. As the mixture of water and oil flows through the V-cone flow meter with a diameter ratio of 0.65 for this work, there will be a certain differential pressure between the upstream and the downstream of V-cone flow meter. Thus groups of differential pressure are obtained with the flow rate of oil-water two phase flow changes. At last, the mixture of water and oil flows into the oil-water tank for separation, then the oil and water are pumped to oil tank and water tank separately.

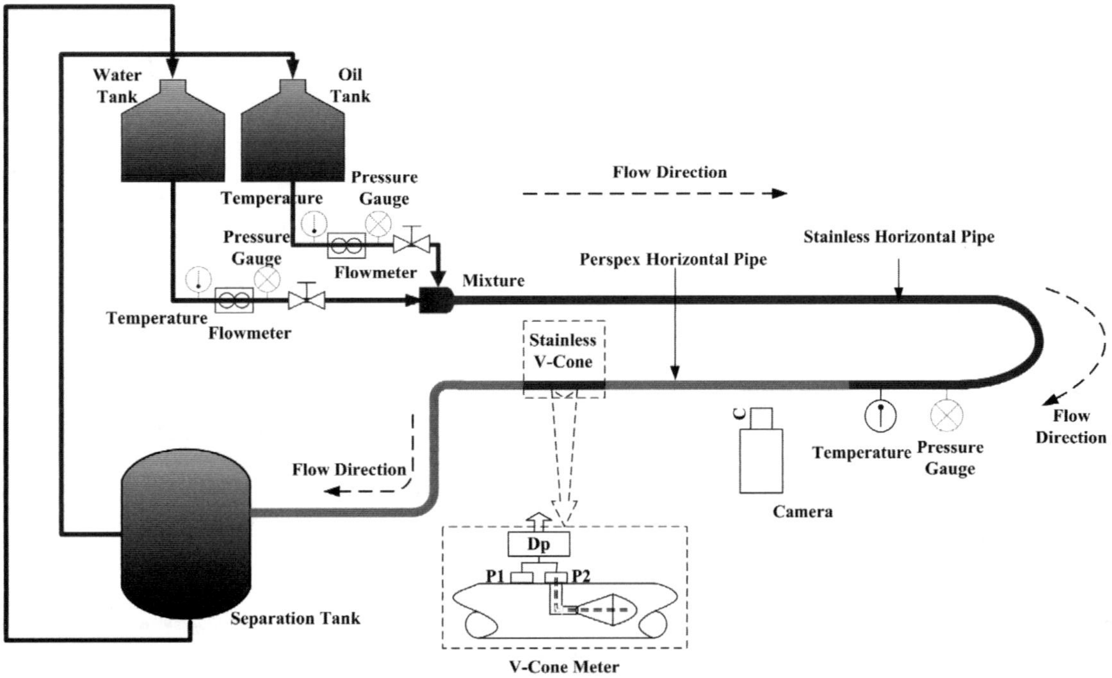

Fig. 2 Oil-water two phase flow loop

The V-cone meters are not standard flow meters, so they must be calibrated before experiments. The parameter that needs calibration is Discharge coefficient C. Ten groups of water experiments were conducted to calculate this parameter, as shown in Figure 3 that all the data points are stabilized around 0.8-0.9. An average value of 0.83 was used in the models adopted.

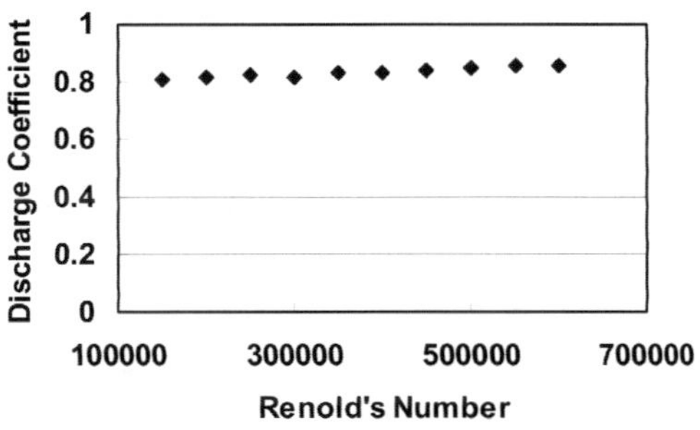

Fig. 3 Calibration of Discharge Coefficient

The procedure of the experiments is: in one experiment data set, water flow rate is fixed, then gradually increase the oil flow rate until reach the maximum; change the water flow rate to a bigger value, increase the oil in the same way to get next data set. The results of the experiments are shown in Figure 4:

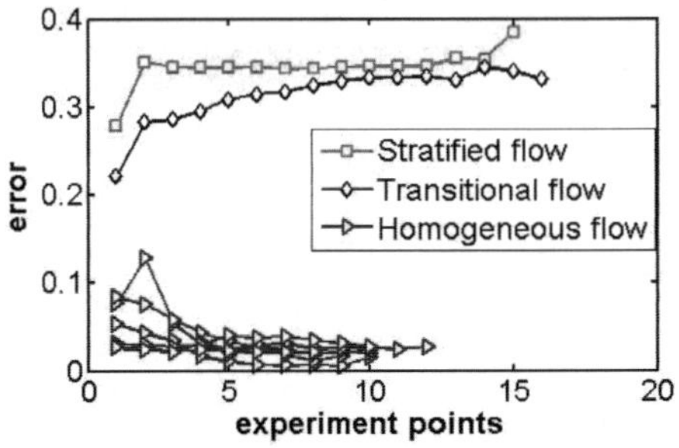

Fig. 4 Error comparison of each experiment point

where the error is defined as $\varepsilon = \frac{G_A - G}{G_A}$, and G_A is actual flow rate, G is the flow rate which is calculated through the measuring model.

When the oil-water two phase flow was in stratified flow, the measuring error was very big with the maximum error reached 40%. As the mixing degree increasing so that the oil-water two phase flow was in transitional flow, then the measuring error decreased a little but it was still too high to suit for the measurement. When the oil and water were well mixed that the oil-water two phase flow was approximately considered as one single phase, i.e. homogeneous flow condition, thus the error reduces to less than 10%.

In order to make the measuring model fit for a wider scale and improve the measuring accuracy, a certain modification is needed when the oil-water two phase flow is in stratified flow or transitional flow. And the repeatability of oil-water two phase flows when they are in homogeneous flow needs to be verified.

The basic idea of the first task is, when an oil-water two phase flow is in stratified flow or transitional flow, a relationship that is about error ε and its determinant will be determined and a correlation will be fitted. As known that the mixing degree is a decisive factor that affects the measuring error, while water percentage φ_W determines the mixing degree. Consequently, error ε is of the relation with water percentage φ_W. Thus, from the definition of error, the actual flow rate can be expressed as $G_A = \frac{G}{1-\varepsilon}$.

On the condition that the oil-water two phase flows were in stratified flow, a polynomial about error ε and water percentage φ_W was fitted:

$$\varepsilon = -0.1226\varphi_W{}^3 + 0.0975\varphi_W{}^2 - 0.0943\varphi_W + 0.3421 \quad (15)$$

When it was in stratified flow, applied equation (15) into $G_A = \frac{G}{1-\varepsilon}$ and found that the error was significantly reduced, as shown in Figure 5.

With the same method, a polynomial about error ε and water percentage φ_W was fitted when the oil-water two phase flow was in transitional flow, as expressed by equation (16).

$$\varepsilon = -0.7092\varphi_W{}^3 + 0.9968\varphi_W{}^2 - 0.3895\varphi_W + 0.3847 \quad (16)$$

The comparison results showed that the error decreased significantly as well. As shown in Figure 6, the calculated flow rate is very close to the actual flow rate. Thus the results based on the modified measuring model are meaningful under the condition that the oil-water two phase flow is in stratified flow or transitional flow.

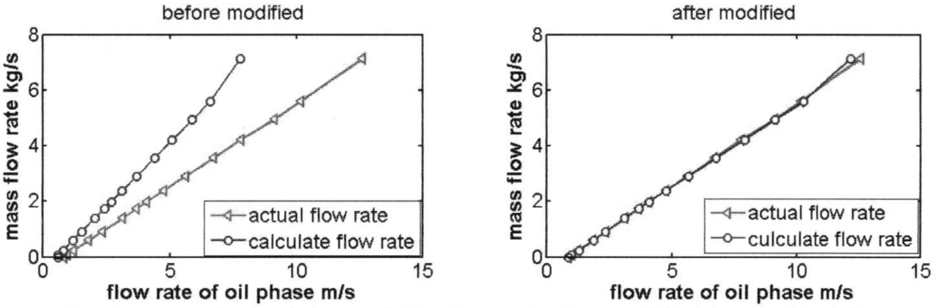

Fig. 5 comparison of the model before and after modified in stratified flow

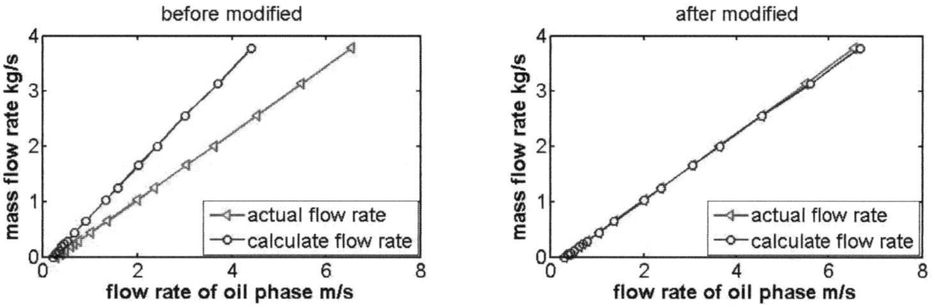

Fig. 6 Comparison of the model before and after modified in transitional flow

As discussed above, when the oil and water are well mixed, the error of measurement based on the model reduced. In order to make sure that the measuring model is repeatable for the measurement, this research involved many groups of repetitive experiments. By changing the ratio of water and oil, the results of homogeneous flow experiments and repetitive experiments are shown in Figure 7.

Obviously, under the similar conditions, the results of repetitive experiments are very close to the results of original experiments. The experimental results show that in oil-water two-phase flows, the measuring model adopted in this work can only be used in homogeneous flow regime unless it is modified. And the modification varies with flow regime. For verification, the experimental results are compared with the increasing water flow rate and with different oil fraction range. The results are seen in Figure 8.

The three groups of conditions in this work are oil percentage is bigger than 50%, oil percentage is bigger than 20% and smaller than 50%, oil percentage is smaller than 20%. As the flow rate of water phase increases, the measurement error based on the measuring model is decreasing. And the higher the flow rate of water is, the better the oil-water two phase flow mixed. So the experiments also show that the measuring model without modification is only suitable for the homogeneous flow, and when the oil-water two phase flow is stratified flow or transitional flow the modification is necessary.

In addition, when water and oil mixture flows in low flow rate, they tend to flow separately and usually are treated as two separate fluids flow alone, and the separate models rather than homogeneous models would be more appropriate in that case.

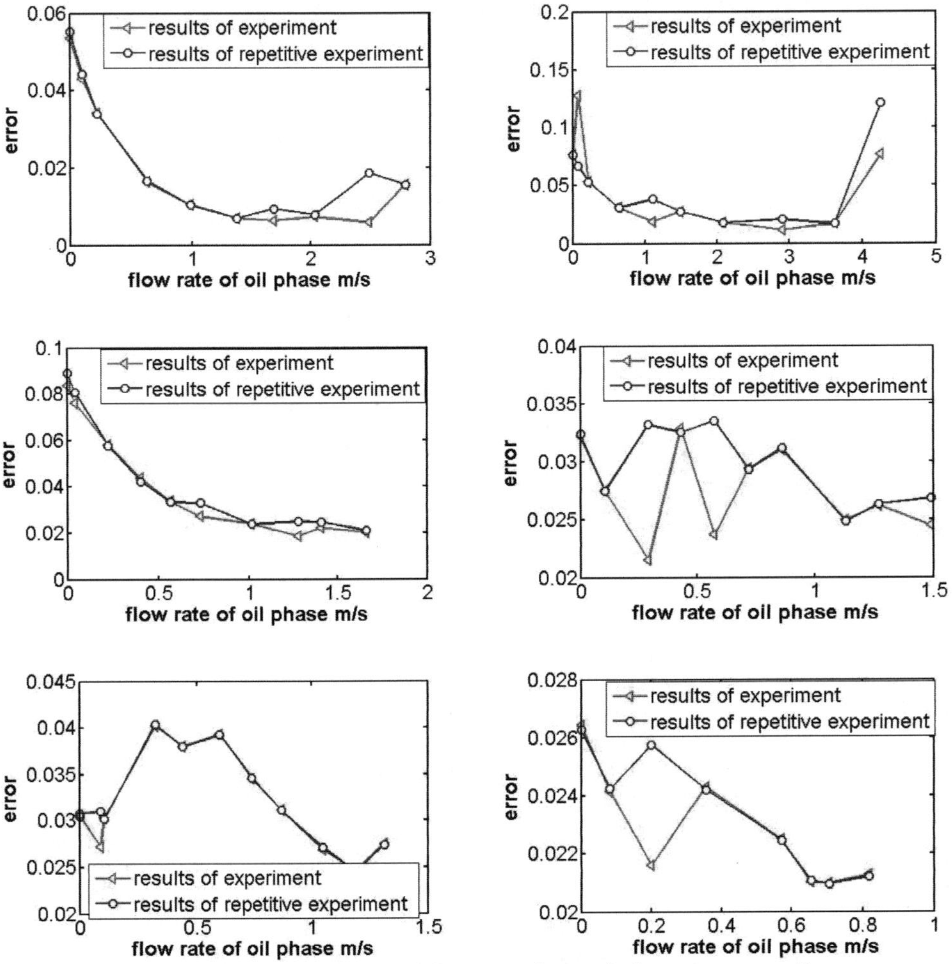

Fig. 7 Results of repetitive experiment in homogeneous flow

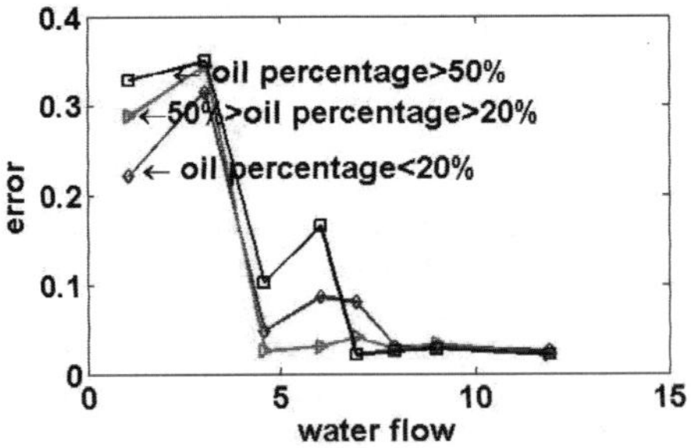

Fig. 8 Error comparison under different oil percentage

4. CONCLUSIONS

From a series of experiments of oil-water two-phase flows in a 50mm diameter Perspex horizontal pipe, the flow rates were measured with a V-cone flow meter with 0.65 diameter ratio. A measuring model is presented and proved efficient when the oil-water two phase flows are in homogeneous flow through repetitive experiments. However, the error of the measuring model grows beyond the acceptable limit, when oil-water mixtures were in stratified flow or transitional flow. In order to make the measuring model suitable for a wider scale and better accuracy, a certain modification is carried out. From the analysis that the factor which affects the measuring accuracy is the mixing degree of the oil and water, the modification correlation was established. The results show that the measurement accuracy was improved by the modified correlations. These experiments and results suggest that the better the oil and water mix, the lower the measuring error would be.

ACKNOWLEDGEMENTS

The author appreciates the support from National High Technology Research and Development Program of China (2006AA04Z167) and Natural Science Foundation of Tianjin (08JCZDJC17700).

NOMENCLATURE

p	pressure	[Pa]
u	flow velocity	[m/s]
A	circulation area	[m^2]
D	diameter of pipe	[m]
S	the largest cross section area of V-cone	[m^2]
q	flow-rate	[m^3/s]
C	discharge coefficient	
n	mixing degree coefficient	
K	modified coefficient of oil-water two-phase flow based on V-cone	
	Greek Letters	
ρ	fluid density	[kg/ m^3]
β	equivalent diameter ratio	
$\sqrt{\psi}$	pressure tapping coefficient	
ε	error	
	Subscripts	
1	upstream	
2	downstream	
0	V-cone flow meter	
o	oil	
w	water	

v volume

m mass

' absolute value

REFERENCES

Arirachakaran, S. et al. (1989) "An analysis of oil/water phenomena in horizontal pipes," SPE Production Operating Symposium, SPE Paper 18836, Oklahoma, pp.155–167

Charles, M. E. et al. (1961) "The horizontal pipeline flow of equal density oil-water mixture," Canadian Journal of Chemical Engineering, 39(17) pp.27-36

Hewitt, G. F.(1997), "Form gas-liquid to liquid-liquid two-phase flow: a difficult journey," Int. Symposium on Liquid-Liquid Two-Phase Flow and Transport Phenomema, Antalya, Turkey, Nov.3rd-7th

Ma, L.B. et al. (2006) "Flow rate measurement of oil-water two phase flow based on differential pressure method," Journal of Zhejiang University (Engieering Science), 41(2) pp.365-368

Nadle, M. et al. (1995) "The effect of gas injection on the flow of two-immiscible liquids in horizontal pipes". Chemical Engineering Technology, 18(3) pp.156-165

Sun, Y. Z. (2004). "V-cone flow meter," Natural Gas Industry, 24(3), pp.105-110

Trallero, J L. et al. (1997) "A study of oil- water flow patterns in horizontal pipes," SPE Production & Facilities, 12(3) pp. 165-172

Journal of Physics: Conference Series **147** (2009) 012060 doi:10.1088/1742-6596/147/1/012060

Effect of properties of pulverized coal on dense phase pneumatic conveying at high pressure

Chen Xiao-ping[1], Zhou Yun[1], Liang Cai[1], Meng Qing-min[1], Lu Peng[1], Pu Wen-hao[1], Xu Pan[1]

[1] School of Energy & Environment, Southeast University , Nanjing 210096, China

xpchen@seu.edu.cn

Abstract. Experiments of dense-phase pneumatic conveying of pulverized coal using nitrogen were carried out in an experimental test facility with the conveying pressure up to 4 MPa. The relationship between mass flow rate of pulverized coal with different mean particle size and gas injecting velocity were investigated. Also, the characteristics of pressure drop in the horizontal pipe, vertical pipe and bend with different mean particle size of pulverized coal were analyzed. Test results indicate that, with the increase in total gas flow rate enters into the feeding hopper, the mass flow rate of fine coals reach the peak value earlier than that of coarse coals, but the optimum injecting velocity is lower than that of coarse coals. In addition, it is showed that the pressure drop for coarse coals is higher than that of fine coals with the same mass flow rate, and when the coal concentration in the pipe is higher enough, the conveying pressure drop for fine coal is close to that of coarse coal. The pressure drop in vertical pipe is higher than that in horizontal pipe, and the pressure drop in horizontal pipe is lower than that in horizontal bend with the same mass flow rate.

1. INTRODUCTION

High-pressure dense-phase pneumatic conveying of pulverized coal has the advantages of high solid/gas ratio, low gas consumption, less pipe wear and stability. So, it is generally used as coal supply system in large-scale pressured coal gasification units(Shen et al., 2005). In recent years, extensive valuable achievements on dense-phase pneumatic conveying technology have been obtained(Konrad et al., 1986; Pan, 1999; Molerus, 1996; Hong et al., 1995; Herbreteau et al., 2000; Geldart et al., 1992). However, these researches mainly focused on low-pressure dense-phase pneumatic conveying; only few references on high-pressure dense-phase pneumatic conveying are available, especially for pneumatic conveying of pulverized coal with different mean particle size. An experimental setup of dense-phase pneumatic conveying of pulverized coal was developed by Southeast University with the conveying pressure up to 4 MPa and the solid/gas mass flow ratio up to 600kg/m^3. Up to now, researches on conveying characteristics and pipeline resistance properties have been done with this test facility, and some important results were obtained(Xiong et al., 2004; LIANG et al., 2007; LIANG et al., 2007). The gas–solid two-phase flow becomes very unsteady and complicated in high-pressure dense-phase pneumatic conveying due to its low conveying velocity and high solid concentration. And the theory of dense-phase pneumatic conveying under high pressure is still not mature. Further experimental research and theoretical research on this field is still needed.

In this paper, some new researches reported here were carried out at Southeast University.

Pulverized coal with different mean particle size was used. The conveying characteristics and pressure drop along the pipe for pulverized coal with different mean particle size are presented. Besides, an important different between this work and that of earlier researches is that the pressure drop across bend have been evaluated by us.

The results that we obtained are very important for the designing and operation of large-scale high- pressure dense-phase pneumatic conveying system.

2. Experimental apparatus and properties of coal used

The high-pressure dense-phase pneumatic conveying experimental facility is shown schematically in Fig 1. High pressure nitrogen from buffer tank is divided into fluidizing gas, pressurizing gas and supplement gas. Pulverized coal in bottom area of the feeding hopper is fluidized by fluidizing gas, and then enters the conveying pipeline driven by conveying differential pressure. During conveying process, pressurizing gas is mainly used to maintain the feeding hopper pressure, while supplement gas can effectively regulate the solid/gas mass flow ratio and maintain stable conveying. Nitrogen in receiving hopper is discharged through back pressure regulation valve while the contained pulverized coal is separated by a bag filter. The detailed description of the experimental facility is reported in Ref. (Shen et al., 2005).

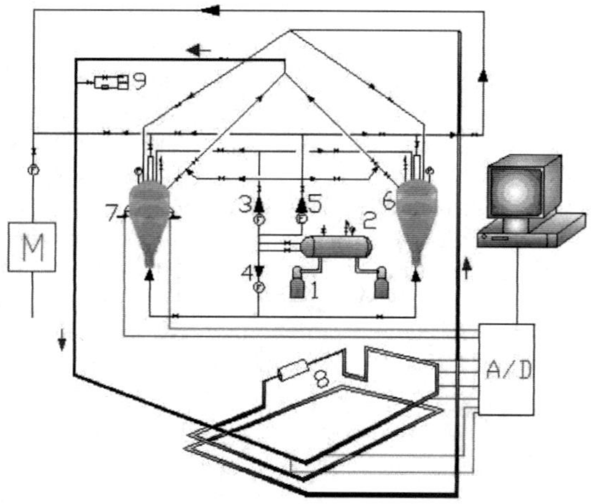

1- nitrogen cylinder 2- buffer tank 3- pressurizing gas 4- fluidizing gas 5- supplement gas
6- hopper 7- weigh cell 8- inspection window 9-water pump

Fig.1 Schematic diagram of high-pressure dense-phase pneumatic conveying test facility

Pulverized coal with the density of 1400 kg/m^3 and the mean particle size of 0.52×10^{-4}m, 1.15×10^{-4}m and 3.0×10^{-4} m are used as testing material.

3. Results and discussion

3.1 The effect of injecting velocity on solids mass flow rate

Using the fluidization properties of pulverized coal could make dense-phase pneumatic conveying of pulverized coal easier. In order to maintain stable conveying, it is important to seek the optimum flow pattern of pulverized coal in the feeding hopper. As shown in Fig.2, fluidizing gas was injected into the feeding hopper through gas distributor. Pulverized coal in area A was fully fluidized, and area B is the mass-flow region where pulverized coal is moving down. Area C is defluidized zone, where there is little nitrogen to flow through this area. And pulverized coal in the bottom of area C collapses to area B. The particle concentration in area B is larger than that of area C and smaller than that of area A. Pulverized coal with different mean particle size exhibit different fluidization behaviour. The effect of injecting gas velocity and particle size on conveying stability was researched by means of investigation on the relationship between injecting velocity and mass flow rate of pulverized coal with different mean particle size.

Tab. 1 Experimental data

d_p (μm)	V_i (m/s)	M_s (kg/s)	Q_f (m^3/h)	Q_p (m^3/h)
52	3.0	0.16	0.4	0.8
300	2.3	0.21	0.4	0.8

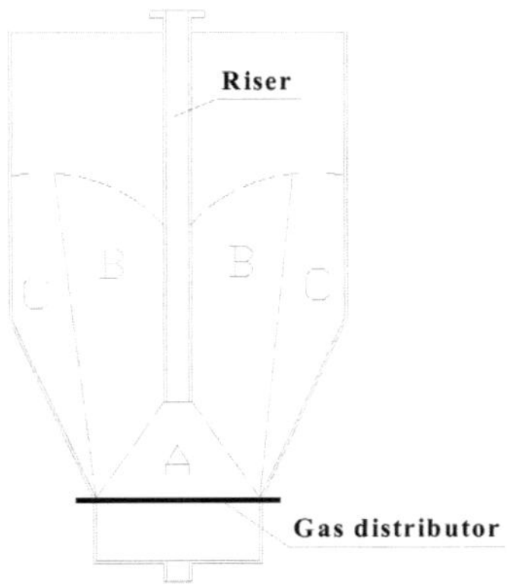

Fig. 2 Schematic diagram of feeding hopper

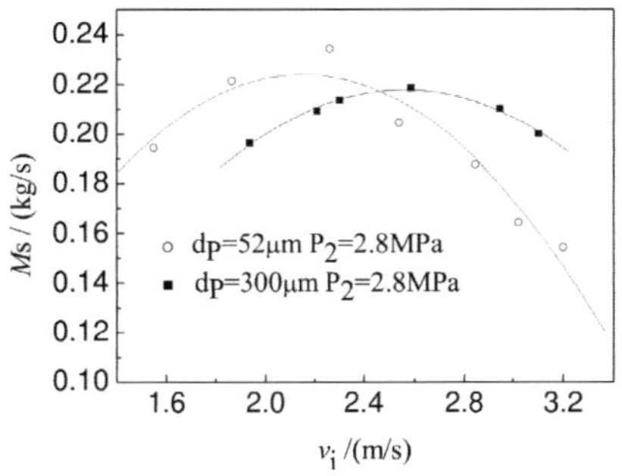

Fig. 3 Mass flow rate of coal vs. injecting velocity

While the nitrogen entering into the feeding hopper, a part of the nitrogen is to replace the coal and maintain the feeding hopper pressure, and the other part enters into the riser. Gas velocity in the riser is called as injecting velocity. During the experiments, the pressure of receiving hopper is kept constant. Tab.1 shows that, at this experimental condition, the injecting velocity for fine coal is larger than that for coarse coal and the mass flow rate of fine coal is less than that of coarse coal. With the same volume flow rates of fluidizing gas and pressurizing gas, higher injecting velocity means that less pulverized coal is replaced, which leads to less mass flow rate of pulverized coal.

Fig.3 shows that, with the increase in injecting velocity, the mass flow rate of pulverized coal increases firstly and then declines, and the mass flow rate of fine coal reaches the peak earlier than that of coarse coal. With the increase in injecting velocity, that is to say with the increase in gas volume rate entering into the feeding hopper, the fine coal reaches the fluidized state relatively earlier than coarse coal. And the particle concentration for fine coal is higher in the incipient fluidized state. Therefore, the mass flow rate of fine coal is larger than that of coarse coal at the earlier stage, and reaches the peak earlier. After the mass flow rate reaching its peak, it declines with the increase in gas volume rate. Also the mass flow rate of fine coal decreases faster than that of coarse coal and is less than that of coarse coal lately. The reason for this phenomenon is that when the pulverized coal is fully fluidized, further increasing in gas flow rate leads to decreasing in particle concentration. Also, the interaction between particles for fine coal is larger than that of coarse coal, resulting in particle aggregation easily. If the gas volume rate is large enough, the fine coal will be in poor fluidized state and the particle concentration becomes non-uniform, so the average particle concentration above the distributor decreases more quickly than that of coarse coal, which leads to the carrying capacity of unit volume gas for the fine coals lower than that of coarse coal.

Journal of Physics: Conference Series **147** (2009) 012060 doi:10.1088/1742-6596/147/1/012060

3.2 Phase diagram of dense-phase pneumatic conveying at high pressure

A phase diagram is usually a series of curves of pressure drop per unit length *versus* superficial gas velocity, and the solid mass flow rate Ms or solid/gas mass flow ratio μ is same in each curve. The phase diagram can accurately describe the pneumatic conveying characteristics and could be used to determinate the flow pattern for a given condition. It also could be used for the analysis of conveying stability and parameters optimization. The phase diagram for pneumatic conveying of pulverized coal under high pressure was given in this paper to research the conveying characteristics and conveying stability for pulverized with different mean particle size.

With high superficial gas velocity, the flow pattern is of homogeneous dilute suspension flow. The pressure drop is mainly caused by the gas phase and increases with the increase in superficial gas velocity. With the decrease in superficial gas velocity, the pressure drop for gas declines. But the particles begin to fall down to the bottom of the pipe, and the particle concentration in the bottom area increases gradually. So the pressure drop for solids increases. When the pressure drop increment for solids equals the pressure drop decrement for gas, the total pressure drop reaches minimum value. The superficial gas velocity value in this situation is defined as economic velocity (saltation velocity), and it is often used as a criterion to distinguish between steady flow and unsteady flow. If the superficial gas velocity is further decreased, a stationary or slow-moving particle layer will be formed on the bottom of the pipe, and the gas-solid two phase flow becomes unsteady and falls into dense phase flow, such as dune flow or plug flow. At this condition, the inter-particle collisions and particle-wall force are enhanced, and the pressure drop increases with the decrease in superficial gas velocity.

3.2.1 Phase diagram of horizontal pipe

Fig.4 is the phase diagram of horizontal pipe for the conveying of pulverized coal with mean particle size of 52 μm . The pressure drop decreases at first and then increases slightly with the increase in superficial gas velocity. Fig.4 also shows that, at the same total conveying pressure drop (pressure difference between feeding hopper and receiving hopper), the mass flow rate of coal increase with the increase in pipe pressure. And at the same pipe pressure, the coal mass flow rate and economic velocity increase with the increase in total conveying pressure drop.

Fig. 5 is the phase diagram of horizontal pipe for the conveying of pulverized coal with mean particle size of 52 & μm and 300 μm . The pressure drop for coarse coal is higher than that of fine coal with the same mass flow rate. While the coal particles flowing with the gas, the following feature of fine coal is better than that of coarse coal. So the fine coal has the tendency to aggregate at the centre of the pipe and easy to be conveyed. On the contrary, the coarse coal tends to fall down to the bottom of the pipe due to gravity, so the wall frictional pressure drop for coarse coal is higher than that for fine coal.

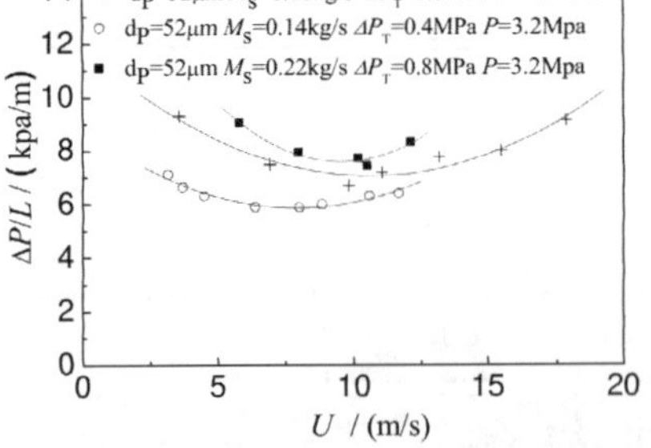

Fig. 4 Phase diagram of horizontal pipe

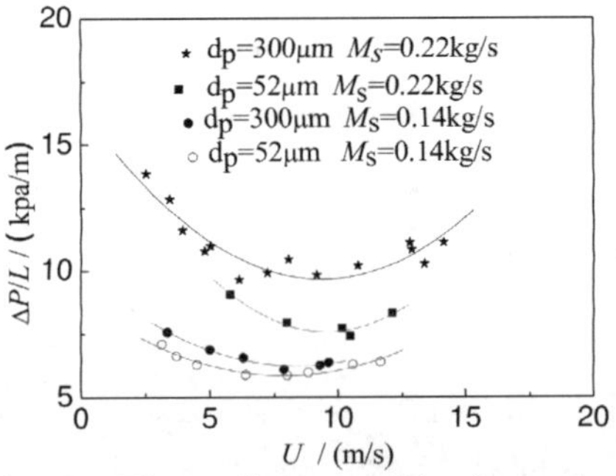

Fig. 5 Phase diagram of horizontal pipe for pulverized coal

Journal of Physics: Conference Series **147** (2009) 012060 doi:10.1088/1742-6596/147/1/012060

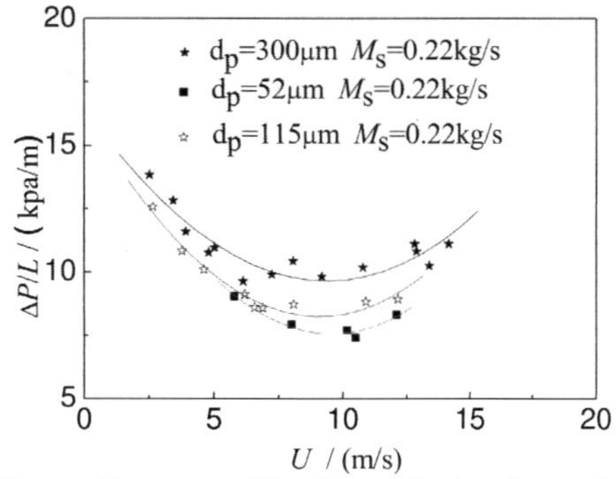

Fig. 6 Phase diagram of horizontal pipe for pulverized coal with mean particle size of 0.52×10^{-4}m, 1.15×10^{-4}m and 3.00×10^{-4}m

Tab. 2 Empirical formula of the pressure drop for per unit length and the economic velocity

d_p	Expression	U_S
$\times10^{-4}$m		m/s
52	$\frac{\Delta P}{L} = 17.26 - 2.02U + 0.106U^2$	9.53
115	$\frac{\Delta P}{L} = 16.48 - 1.46U + 0.08U^2$	9.12
300	$\frac{\Delta P}{L} = 16.39 - 1.79U + 0.095U^2$	9.42

Generally, when the coal concentration in the pipe is high enough, with the decrease in superficial gas velocity, the pressure drop of fine coal is close to that of coarse coal(Geldart et al., 1990). It is because that reducing the gas velocity has a greater effect on the deposition trend of fine coal than that of coarse coal. Decrease in gas velocity leads to a large number of fine coal falls down to the bottom of the pipe, so the wall frictional pressure drop increases. Meanwhile, the deposition of fine coal reduces the gas flow area. So the gas velocity is increased, which leads to the increase in the pressure drop for gas. It could be seen from Fig.5 that, to the left of the minimum point of the two curves with the coal mass flow rate of around 0.22kg/s, the pressure drop for fine coal indeed has the tendency to close to that of coarse coal when the superficial gas velocity is reduced, but further study is needed due to the limits of experimental condition.

Curve fitting analysis of the three pressure drop curves for pulverized coal with mean particle size of 52 μm, 115 μm and 300 μm in Fig.6 shows that the pressure drop for per unit length could be expressed with the following formula:

$$\frac{\Delta P}{L} = \frac{aU^2 + bU + c}{U} \qquad (1)$$

Let $\frac{d(\Delta P/L)}{dU} = 0$, the economic velocity could be obtained. With experimental data fitting, the empirical formulas of the pressure drop for per unit length and the economic velocities for the three kinds of pulverized coal are obtained, as shown in Table 2. The economic velocity for pulverized coal with mean particle diameter of 115 μm is lower than that of other two kinds of pulverized coal. It could be assumed that the fine coal is easy to be conveyed from the view of power consumption, whereas the medium-sized coal is easy to be conveyed from the view of conveying stability. Because just three kinds of coal were tested in this paper, in order to extend the above conclusions, further study is needed.

3.2.2 Comparison of Phase diagram of horizontal pipe and that of vertical pipe

As shown in Fig.7, the pressure drop for vertical pipe is higher than that of horizontal pipe with the same coal mass flow rate. According to Barth's theory, the total pressure drop of the pipe could be expressed as below:

$$\Delta p = \Delta p_{g} + \Delta p_{s} \qquad (2)$$

Where Δp_g is the pressure drop for gas and Δp_s is the pressure drop for solids. For horizontal pipe, Δp_g and Δp_s are due to gas-pipe wall friction and solids-pipe wall friction respectively. While for vertical pipe, they also include pressure drop due to gas and solids gravity.

With the same superficial gas velocity, the pressure drop for gas is almost the same for the horizontal pipe and the vertical pipe. Though there was a tendency towards higher solids friction due to coal deposition in horizontal pipe, the pressure drop due to gravity in vertical pipe is relatively larger, which finally leads to the pressure drop for vertical pipe larger than that of horizontal pipe.

3.2.3 Comparison of Phase diagram of horizontal pipe and that of horizontal bend

Particles enter into the bend can be expected to aggregate at the outer lateral wall and collide with outer lateral wall due to centrifugal force. They would be rebounded backwards and collide with particles arriving later. It would cause a net loss of kinetic energy. Then, the particles would be accelerated by the gas again and leads to a loss of gas momentum. So, the pressure drop across horizontal bend is much higher than that of horizontal pipe.

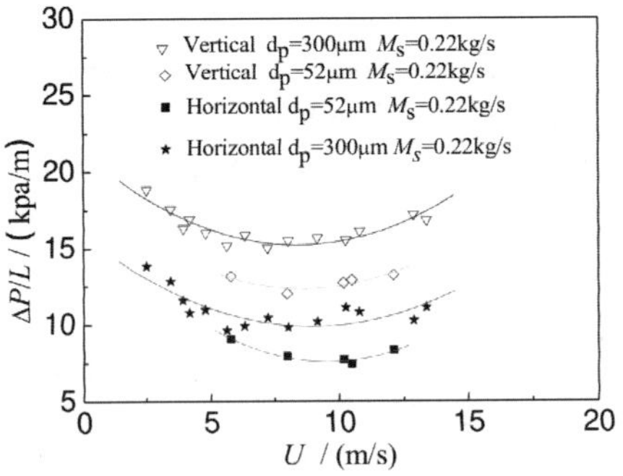

Fig. 7 Phase diagram of vertical pipe and horizontal pipe

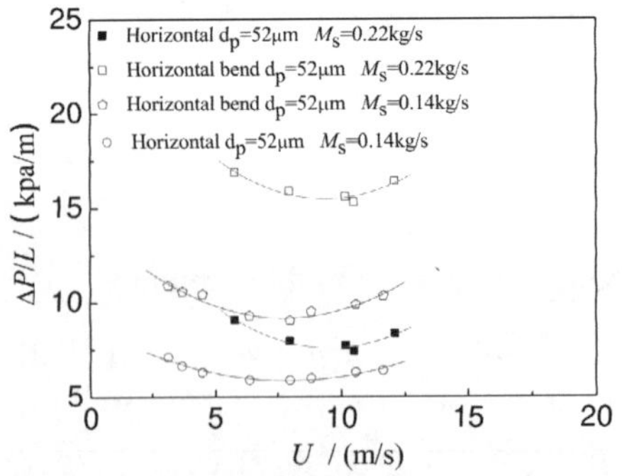

Fig. 8 Phase diagram of horizontal pipe and horizontal bend

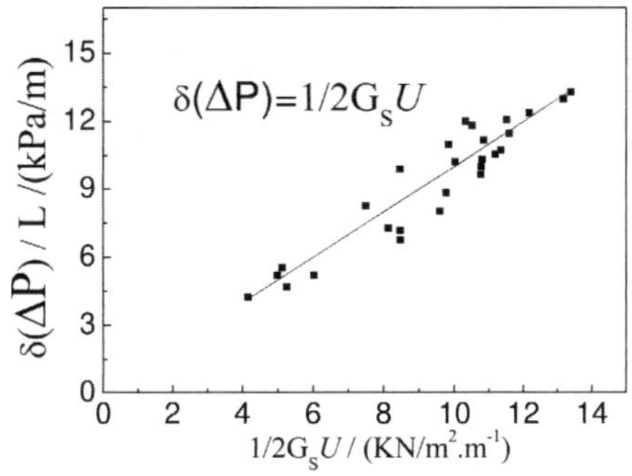

Fig. 9 Correlation of pressure drop due to solids kinetic energy losing across bend

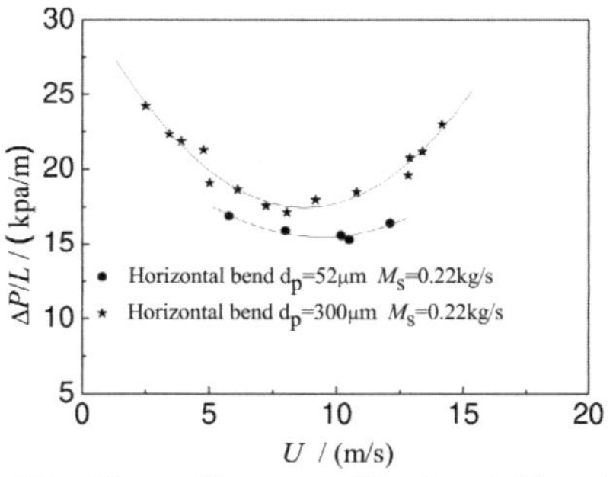

Fig. 10 Phase diagram of horizontal bend

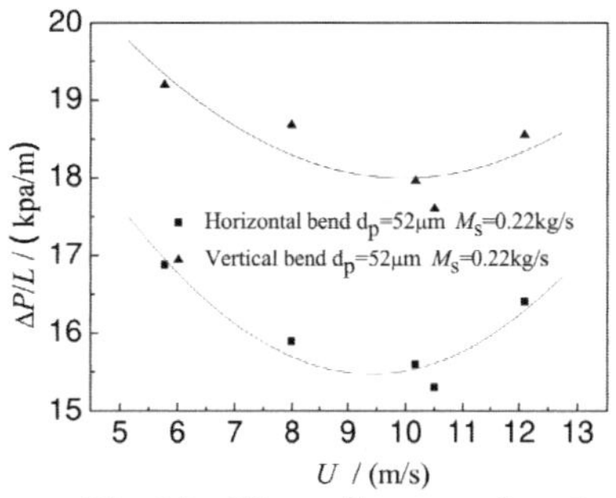

Fig. 11 Phase diagram of vertical bend and horizontal bend

As shown in Fig.8, with the same mass flow rate of coal, the pressure drop of horizontal bend is higher than that of horizontal pipe. The higher the mass flow rate, the larger the difference between pressure drop of horizontal bend and that of horizontal pipe. The frequency of collision between solids and wall increase with the increase in the mass flow rate, which lead to the increase in the pressure drop due to kinetic energy dissipation. Furthermore, with the increase in the mass flow rate, the pressure drop due to solids friction is also increase.

With experimental data fitting, it is found that the difference value between the pressure drop of bend and that of straight pipe is $1/2G_S U$, which can be seen from Fig. 9.

Fig. 10 illustrates that, for horizontal bend, the pressure drop for coarse coal is larger than that for fine coal with the same mass flow rate of coal. The centrifugal force of coarse coal is larger than that of fine coal, so the coarse coal tends to move to the outer lateral wall more seriously, and the coal concentration is higher near the outer lateral wall. Therefore, the pressure drop due to collision and wall friction are higher for coarse coal.

3.2.4 Comparison of Phase diagram of horizontal bend and that of vertical bend

The pressure drop for vertical bend is higher than that of horizontal bend with the same conveying condition, which can be seen from Fig.11. The pressure drop of vertical pipe is larger than that of horizontal pipe as mentioned in chapter 3.2.2. For the same reason, the pressure drop of vertical bend is larger than that of horizontal bend. Moreover, coals conveyed in the vertical bend would dissipate a part of kinetic energy due to gravity while redirecting coals from horizontal pipe to the vertical pipe. Coals would then be re-accelerating upwards, and the pressure drop due to kinetic energy of vertical bend increase as well as(Lee et al., 2004).

4. Conclusions

(1) With the increase in total gas flow rate enters into the feeding hopper, the mass flow rate of fine coal reaches the peak value earlier than that of coarse coal. The optimum injecting velocity for fine coal is 2.26m/s, while this value is 2.58m/s for coarse coal when the receiving pressure is 2.8MPa.

(2) At the same total conveying pressure drop, the mass flow rate of coal is increased with the increase in pipe pressure. And at the same pipe pressure, the coal mass flow rate is increased with the increase in total conveying pressure drop.
(3) The pressure drop decreases at first and then increases with the increase in superficial gas velocity. The pressure drop per unit length could be expressed with the formula as eqn.(1). The superficial velocity at which the total pressure drop reaches minimum is 9.12m/s for pulverized coal with mean particle size of 1.14×10^{-4} m. And it is lower than that of two other pulverized coal.
(4) The pressure drop in vertical pipe is higher than that in horizontal pipe, and the pressure drop in horizontal pipe is lower than that in horizontal bend at the same mass flow rate. The pressure drop for vertical bend is higher than that of horizontal bend with the same mass flow rate. With experimental data fitting, it is found that the difference value between the pressure drop of bend and that of straight pipe is $1/2G_S U$.
(5) For horizontal pipe and horizontal bend, the pressure drop for coarse coal is larger than that for fine coal with the same mass flow rate.

ACKNOWLEDGEMENTS
Financial support from the National Key Program of Basic research in China (2004CB217702) was sincerely acknowledged.

NOMENCLATURE

d_P	particle size	[μm]
G_S	mass flow rate of solids	[kg/(m^2.s)]
M_S	mass flow rate of solids	[kg/s]
P_1	feeding hopper pressure	[MPa]
P_2	receiving hopper pressure	[MPa]
ΔP_T	conveying differential pressure	[MPa]
$\Delta P/L$	total pressure drop per unit length across pipe	[KPa/m]
Q_f	volumetric flow rate of fluidizing gas	[m^3/h]
Q_P	volumetric flow rate of pressurizing gas	[m^3/h]
Q_s	volumetric flow rate of supplemental gas	[m^3/h]
U	superficial velocity of gas	[m/s]
U_S	saltation velocity	[m/s]
V_i	injecting velocity	[m/s]

Greek Letters

$\delta(\Delta P)$ the difference value between the pressure drop of bend and that of straight pipe
μ solids to gas mass flow rate ratio

REFFERENCE

Geldart D. et al. (1990). "Dense Phase conveying of fine coal at high total pressure,". J.Powder Technology, 62(3) pp.243-252.

Geldart D. et al. (1992). "Saltation velocities in high pressure conveying of fine coal," J.Powder Technology., 69(2), pp. 157-172.

Hong J. et al. (1995). "Phase diagrams in dense phase pneumatic transport," J.Powder

Technology., 84(3), pp.213-219.

Herbreteau C. et al. (2000). "Experimental study of parameters which influence the energy minimum in horizontal gas-solid conveying," J.Powder Technology., 112(3), pp.213-220.

Konrad K. (1986). "Dense-Phase Pneumatic Conveying: A Review," J.Powder Technology., 49(1), pp. 1-35.

Lee L., et al. (2004). "Pneumatic transport of granular materials through a 90° bend,". J.Chemical Engineering Science, 59(2) pp. 4637-4651.

LIANG Cai et al. (2007). "Dense-phase pneumatic conveying of pulverized coal under high pressure," J.Journal of southeast university, 37(3), pp. 427-431.

LIANG Cai et al. (2007). "Investigation on dense-phase pneumatic conveying characteristic of pulverized coal under high pressure," J.Proceedings of the CSEE, 27(14) pp. 31-35.

Molerus O. "Overview: Pneumatic transport of solids," J.Powder Technology., 88(3), pp. 309-321.

Pan R(1999). "Material properties and flow modes in pneumatic conveying," J.Powder Technology., 104(2), pp. 157-163.

SHEN Xiang-lin et al. (2005). "Experimental study on dense-phase pneumatic conveying of pulverized coal at high pressures," J.Proceedings of the CSEE., 25(24), pp. 103-107.

Xiong et al. (2004). "A Study on the resistance properties for vertical pipe in dense phase pneumatic conveying of pulverized coal under high pressures," J.Proceedings of the CSEE., 24 (9), pp. 248-251

The 6th International Symposium on Measurement Techniques for Multiphase Flows IOP Publishing
Journal of Physics: Conference Series **147** (2009) 012061 doi:10.1088/1742-6596/147/1/012061

Confocal Micro-PIV Measurement of Droplet Formation in a T-shaped Micro-junction

M Oishi [1], H Kinoshita[1], T Fujii[1], M Oshima[2]

[1] Institute of Industrial Science, The University of Tokyo, 4-6-1 Komaba Meguro-ku Tokyo, 153-8505, JAPAN
[2] Interfaculty Initiative in Information Studies, The University of Tokyo, Tokyo, JAPAN

oishi@iis.u-tokyo.ac.jp

Abstract. This paper aims to investigate a mechanism of microdroplet formation using "multicolor confocal micro particle image velocimetry (PIV)" technique. The present system can measure dynamical behavior of multiphase flow separately and simultaneously. It also enables to identify the interactions between two immiscible fluids. We have applied this system to measure the water droplet formation at a micro T-shaped junction. We have also succeeded in dispersing fluorescent tracer particles into both phases. The interaction between the internal flow of to-be-dispersed water phase and of continuous oil phase is measured as a liquid-liquid multiphase flow. As a result of PIV measurement and interface scanning, the relationship between flow structure of each phase and interface shape is clarified. It indicates that the gap between the tip of to-be-dispersed phase and capillary wall, and interface area play an important role in the flow structure and shear stress on the interface.

1. INTRODUCTION

For the last decade, a lot of microfluidic devices have been designed as micro total analysis systems (Micro-TAS) or lab-on-a-chip applications (Reyes et al., 2002; Auroux et al., 2002). These devices are used for many purposes such as mixing, separation, sensing, or chemical reactions and so on. Especially microdroplet is used as a smallest micro chamber to obtain fast reaction process and to make monodisperse microspheres or capsules.

There are some previous studies about droplet formation mechanism using micro T-shaped junction. Most of all uses Capillary number (*Ca*) in order to explain its phenomena. *Ca* is written in following equation (1).

$$Ca = \frac{\mu_c u_c}{\gamma} \qquad (1)$$

Ca means a ratio of shear stress and interfacial tension. The shear stress acts to tear off the tip of to-be-dispersed phase and the interfacial tension acts to keep the surface of to-be-dispersed phase to minimize its surface area.

Garstecki et al (2006) suggests that the mechanism changes from "squeezing" to "dripping" between $Ca \sim 10^{-2}$. Similarly, Tice et al (2003) and Nishisako (2006) suggest its border of change as $Ca \sim 1$, and Graaf et al (2005) suggests $Ca \sim 0.1$. This means it is inadequate to clarify this mechanism with only *Ca* number.

In this paper, we vary only flow rate of each phase as a parameter to control *Ca* number, because viscosity and interfacial tension affects each other.

2. MEASUREMENT METHOD

2.1 Multicolor Confocal Micro PIV System

In order to measure multiphase flow characteristics simultaneously, we developed a multicolor confocal micro PIV system (Oishi et al., 2006). Its configuration is shown in Fig. 1. Our system is based on a confocal micro PIV system, which was developed by Kinoshita et al (2005), and consists of a high-speed confocal scanner (CSU22, Yokogawa Electric Corp.), high-speed camera (Phantom v7.1, Vision Research Inc.) and an inverted microscope (DMIRE2, Leica Microsystems).

Our upgraded system allows multiphase flow measurement through additional devices and modules. To simultaneously measure multiphase flow phenomena, we added a second laser and camera, a laser combiner, and a multi-wavelength dividing unit to the system. The two illumination lasers (543-BS-A03, Melles Griot Inc. and Excel 1500, Laser Quantum Ltd.) are combined using a dichroic mirror in the laser combiner and led to the laser port of the confocal scanner using optical fiber. Lasers illuminate the target flow, and Stokes' shifted fluorescent light from the fluorescent particles dispersed in the flow are emitted back through confocal scanner. Then, the multi-wavelength emitted light is separated and filtered at the separation unit, which is shown in Fig.2, and recorded by two high-speed cameras. The present system can collect a sequence of images at a speed of up to 2000 frames per second (fps).

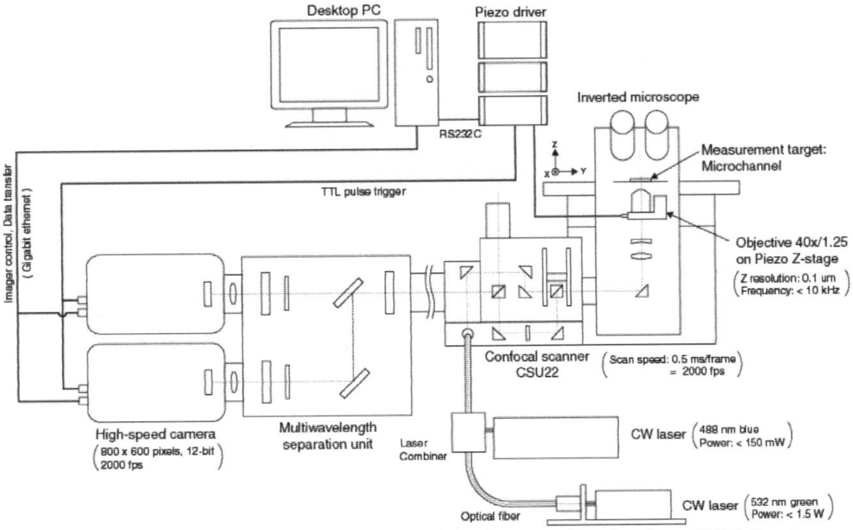

Fig. 1 Schematic diagram of the multicolor confocal micro PIV System

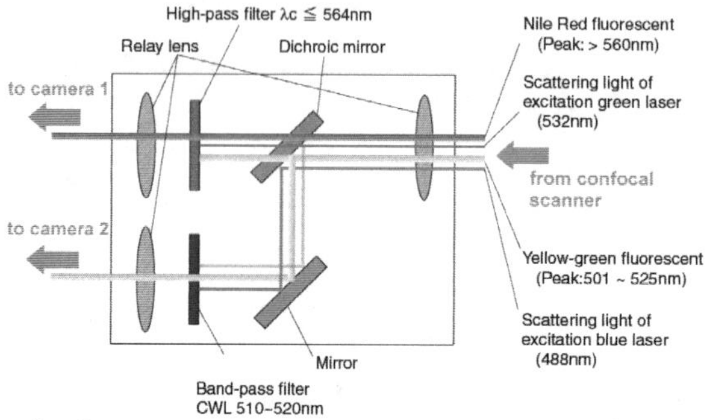

Fig. 2 Schematic diagram of the multicolor separation unit and its optical design

2.2 Target Microchannel

Figure 3 is a schematic illustration of a droplet formation device with a T-shaped junction which width of 100 μm and depth of 80 μm. The patterned microchannel is placed and sealed off on a cover glass where PDMS is spin-coated at a thickness of 10-20 μm using the fabrication method of Hong et al (2001). Two inlet ports and one outlet port are fabricated at the ends of the microchannel. The inlet ports are connected to separate 250 μL glass syringes (Gastight Syringes 1725LT, Hamilton Company, USA) on syringe pumps (KDS100, KD Scientific Inc., USA) through Teflon tubes, and the outlet port is connected to the drain. The oil phase flows in the straight main channel at a constant flow rate from inlet A, and the water phase flows from inlet B to the T-shaped junction at a constant flow rate too. Syringe pumps control both flow rates.

The oil phase flow intersects with the water phase flow at the T-shaped junction, where the water flow is diverted and transformed into droplets. The droplets are generated sequentially at a constant frequency. The droplet size and the generation frequency are dependent on the inlet flow rates, the geometry of the channel, driving pressure, interfacial tension, wettability on the channel wall, shear force at the oil–water interface and the viscosities of the working fluids.

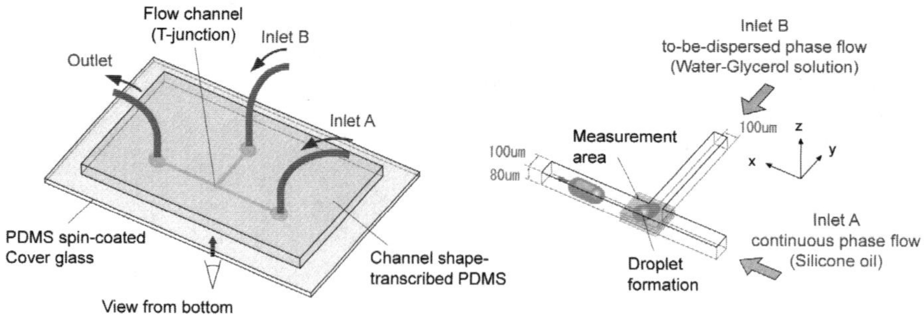

Fig. 3 T-shaped microchannel

3. RESULTS AND DISCUSSIONS

3.1 Preliminary Experiment

As a preliminary experiment, we searched critical *Ca* number which indicates the changing point of the droplet formation mechanism. We choose two immiscible fluids shown in Tab. 1. Although a small amount of flourescein is dissolved into to-be-dispersed phase to visualize interface, other ingredients are same as for PIV measurement in order to keep liquid and interfacial characteristics.

We varied the flow rate of continuous phase Q_c and to-be-dispersed phase Q_d. The range of flow rate and *Ca* number are also shown in Tab. 1.

We choose droplet length D_h as a characteristic parameter of mechanism shown in Fig. 4 and 5. Relatively slow continuous phase flow rate, it means low *Ca* number, the diameter of droplet become

Journal of Physics: Conference Series **147** (2009) 012061 doi:10.1088/1742-6596/147/1/012061

larger than the channel size. Then it becomes a plug shape. On the other hand, for fast continuous phase flow, it means high *Ca* number, droplet is torn off from the tip of to-be-dispersed phase before attaching opposite wall. Then it becomes smaller droplet than channel size.

Table. 1 Characteristics of two immiscible fluids

	Continuous phase (inlet A)	To-be-dispersed phase (inlet B)
Working fluid	Silicone Oil	Glycerol solution
Specific gravity [g/cc]	0.984	1.162
Viscosity [mPa·s]	68.6	10.58
Flow rate [ml/hr]	2.4 ~ 100.0	2.0, 8.0
Average flow velocity [mm/sec]	0.83 ~ 3.47	0.69, 0.28
Interfacial tension [mN/m]	11.7	
Capillary number	4.89×10^-4 ~ 2.04×10^-2	

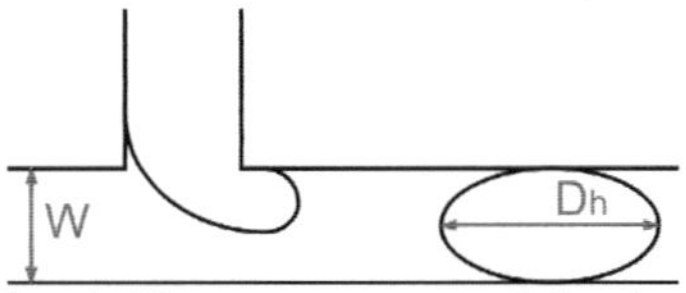

Fig. 4 Size parameter of the droplet

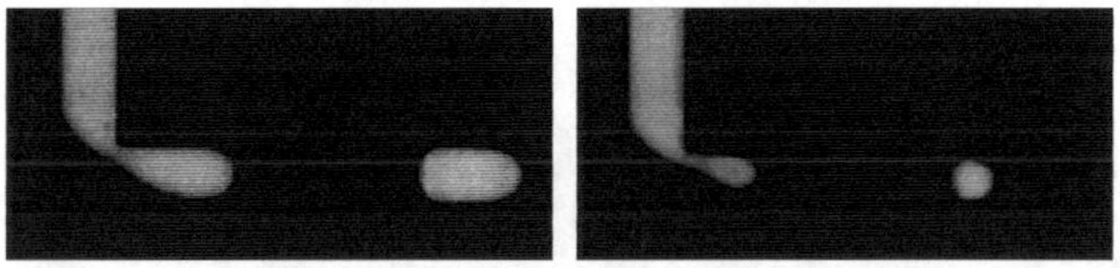
Fig. 5 Droplet formation images
Left: Low Ca number condition (squeezing), Right: High Ca number condition (dripping)

As a result of preliminary measurement, fig 6 shows the relation between *Ca* number and dimensionless droplet shape parameter, which is defined by dividing droplet length D_h by channel width W. This shape parameter indicates wall contact of droplet. When this parameter indicates over 1.0, the droplets attach to the channel wall and become bullet shape elongated streamwise direction. The inflection points for each Q_d plot are about $Ca = 3.0 \times 10^{-3}$. It is thought to be causally related to the three-dimensional gap between tip of to-be-dispersed phase and channel wall.

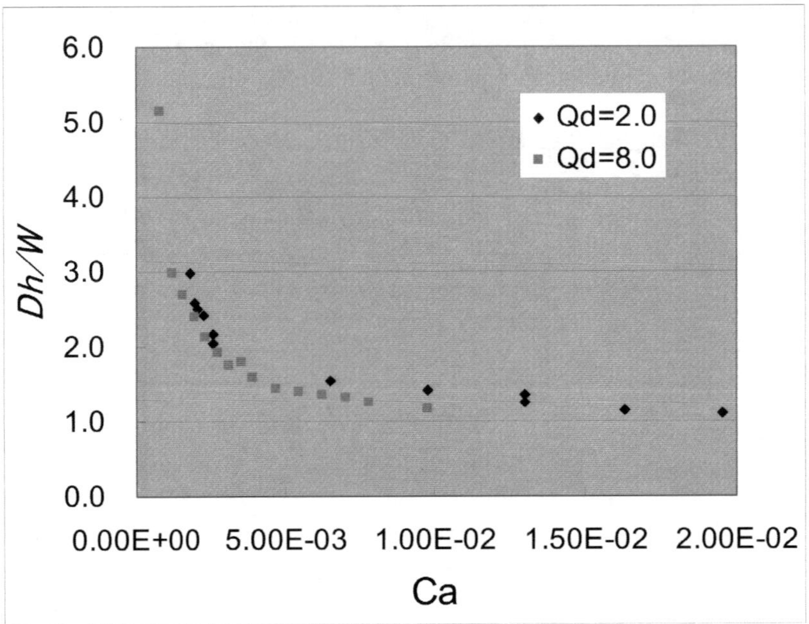

Fig. 6 Relation between *Ca* number and droplet shape

3.2 PIV Measurement

In order to clarify the droplet formation mechanism in more detail, we measured the velocity distribution of both phase using multicolor confocal micro PIV system. From preliminary experiment, the mechanism changes between $Ca = 3.0 \times 10^{-3}$ for our experimental condition. So we chose two condition of $Ca = 1.63 \times 10^{-3}$ and 8.14×10^{-3}. Other conditions of PIV measurement is written in Tab. 2. The measurement region is 229 x 172 μm with the in-plane spatial resolution of 0.286 μm/pixel, and measurement height is at its center height.

The refractive indices of the working fluids and channel material needed to match exactly to minimize the refraction and reflection of light at the interface. In this experiment, two immiscible liquids were used as working fluids for refractive index matching: silicone oil (KF-6002, Shin-Etsu Chemical Co. Ltd., Japan) as the oil phase and a mixture of dilute water and glycerol (Wako Pure Chemical Industries, Ltd., Japan) as the water phase. Silicone oil was chosen because it has a refractive index of 1.4122, which is almost equal to that of the PDMS material of the microchannel (refractive index: 1.412). Also, the mixture of dilute water and glycerol has a refractive index of 1.412 when mixed in the proportion of 43% dilute water and 57% glycerol.

To visualize the oil phase, we utilized oil-immersible microspheres made from porous silica particles (Godball E-2C, Suzuki Yushi Industrial Co. Ltd., Japan). These particles have high surface lipophilicity, and high traceability due to their low specific weight of porous structure. These particles were dyed using Fluorescein (Wako Pure Chemical Industries, Ltd., Japan) and dried in a vacuum desiccator. Clustered particles were separated using vibrating agitator. The distribution of particle diameter ranged between 0.9 to 1.4 μm. The ϕ 1 μm diameter red fluorescent particles (FluoSpheres F8821, Invitrogen-Molecular Probes Inc.) were dispersed in the water phase with a volume ratio of 0.4%. The measurement depths of each particle are measured and it results in about 3μm.

Table. 2 Conditions of PIV Measurement

	Continuous phase (inlet A)		To-be-dispersed phase (inlet B)
Working fluid	Silicone Oil		Glycerol solution
Flow rate [μl/hr]	8.0	40.0	2.0
Average flow velocity [mm/sec]	0.278	1.390	0.069
Capillary number	1.63E-03 (Squeezing)	8.14E-03 (Dripping)	←
Tracer particle	φ0.9 ~ 1.4μm Green Fluorescent silica		φ1.0μm Red fluorescent polystylene
Measurement Depth [μm]	3.10 ~ 3.84		3.34
Refractive index	1.412		
Interfacial tension [mN/m]	11.7		

(Experiment temperature : 25°C)

The results of multicolor PIV measurement at each phase are shown in Fig. 7. We compared each three periods, Period 1: after generating droplet, the remaining tip of to-be-dispersed phase reaches to the lower edge of the junction. Period 2: just before detachment of to-be-dispersed phase. Period 3: just after detachment. The left images are the results of low *Ca* condition and the right images are that of high *Ca* condition.

At the period 1, the tip of to-be-dispersed phase of low *Ca* condition is sharper than that of high *Ca* condition. It derives from differcence of shear force affected by the continuous flow near the interface. Strong shear force of high *Ca* condition drugs and elongates the tip surface of to-be-dispersed phase. This shear force also makes flow circulation inside the tip of to-be-dispersed phase. Only high *Ca* condition makes strong vortex which has vortex core.

At the period 2, the continuous phase is clogged up by to-be-dispersed phase at low *Ca* condition. Then, continuous phase squeezes the neck of to-be-dispersed phase. Furthermore, the velocity inside droplets is faster than that of near interface, it means this acceleration is not derived from shear stress by the continuous phase at the interface. On the other hand, at high *Ca* condition, the continuous flow is not clogged up and it accelerates flow near the interface of to-be-dispersed phase.

At the period 3, the continuous flow follows after the droplet at low *Ca* condition. In contrary, the continuous flow pushes the droplet at high *Ca* condition.

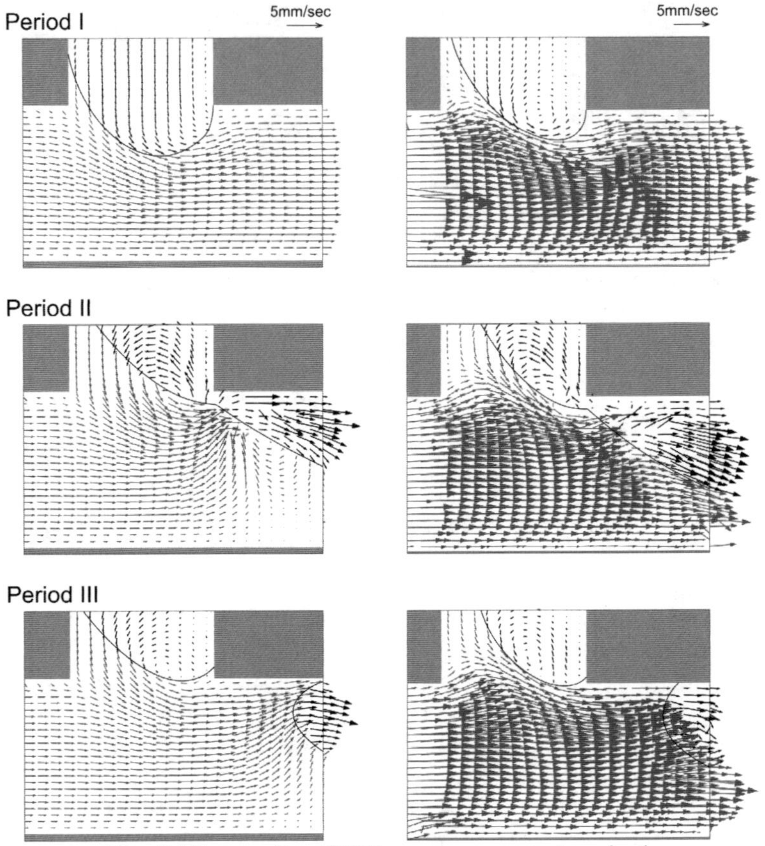

Fig. 7 The results of PIV measurement at each phase
Left: Low *Ca* number condition, Right: High *Ca* number condition

3.2 Three-dimensional Reconstruction of PIV Results

The piezoelectric device can control z-position of objective lens precisely. Each plane are measured every 3μm pitch in z-direction, and we can make three-dimensional velocity distribution map by piling up velocity datas of these planes. After that, in order to calcurate vertical velocity component, continuity equation was used.

In this study, we considered flow structure symmetric about the mid plane (center height). Therefore, vertical velocity component becomes zero at the mid plane, and we use this assumption for the boundary condition. The calculation proceeds from the mid plane to the near wall plane.

Figure 8 shows three-dimensional flow structures of each phase at squeezing condition. The timing period is about 0.45t from droplet generation.

About to-be-dispersed phase, flow path goes straight to its tip and there are no vortex inside it. In contrast, continuous phase have three-dimensional flow structure. The faster flow at the centerline is devided to two direction, one is to the gap between tip of to-be-dispersed phase and channel wall, the other is to upstream of inlet channel of to-be-dispersed phase. Since these gaps near the four channel corners are larger than that of mid plane, continuous flow moves to these corners, resulting in three-dimensional flow is generated.

From these results, although there are less interaction between two phases in the case of queezing condition and time period, it may change depends on *Ca* number and another parameters. We have to speed up the confocal scanning speed for further investigation.

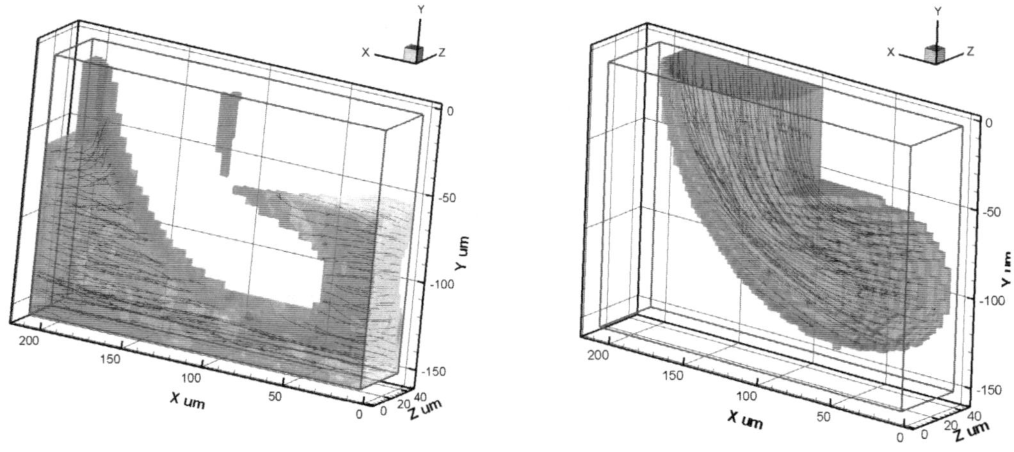

Fig. 8 Three-dimensional flow structures of each phase (half region)
Left: Continuous phase, Right: to-be-dispersed phase

4. CONCLUSIONS

From these results, it seems that the important parameter is a gap between the tip of to-be-dispersed phase and channel wall. If the gap is large, surrounding continuous fluid gives strong shear stress on the surface of to-be-dispersed phase and finally it tears off ("dripping") its tip. On the contrary, smaller gap can't put through the continuous fluid flow, and it leads to low shear stress, which is not enough to tear off small droplet, resulting in "squeezing"

In conclusion, the droplet formation mechanism depends on the *Ca* number and gap region that is determined by flow rate of to-be-dispersed phase and channel size. Since both parameters are dependent on time, it is not enough to decide the critical *Ca* number, which identifies change of formation mechanism from "squeezing" to "dripping". The present measurement also clarifies mechanism of the transient behavior.

We also constructed a multicolor confocal micro-PIV system to measure behavior of multiphase flow. The selection of lasers, optical components and fluorescent materials for ideal multicolor separation is a key point of this system. It is also important that the optical arrangement in the multicolor separation unit. Using the present system, we can remove unnecessary scattering light and obtain only particle fluorescent images in different phases. The present system was applied to measure the liquid-liquid multiphase flow in a microchannel. It clarifies the strong interaction between phases in detail.

NOMENCLATURE

W	channel width	[m]
D	channel depth	[m]
A	area of bubble surface	[m2]
Ca	Capillary number	[-]
Q	flow rate	[m3/sec]
u	flow velocity	[m/sec]
D_h	length of droplet	[m]
	Greek Letters	
α	thermal diffusivity	[m^2/s]
μ	viscosity	[mPa sec]
γ	interfacial tension	[mN/m]

Journal of Physics: Conference Series **147** (2009) 012061 doi:10.1088/1742-6596/147/1/012061

Subscripts

c fluid of continuous flow

d fluid of to-be-dispersed flow

REFERENCES

Reyes, D. R., Iossifidis, D., Auroux, P. A. and Manz, A. (2002). "Micro total analysis systems. 1. Introduction, theory, and technology," Anal. Chem., 74, pp. 2623–2636.

Auroux, P. A., Iossifidis, D., Reyes, D. R. and Manz, A. (2002). "Micro total analysis systems. 2. Analytical standard operations and applications," Anal. Chem., 74, pp. 2637–2652.

Garstecki, P., Fuerstman, M. J., Stone, H. A., Whitesides, G. M. (2006). "Formation of droplets and bubbles in a microfluidic T-junction—scaling and mechanism of break-up," Lab on a Chip., 6, pp. 437-446.

Tice, J. D., Song, H., Lyon, A. D. and Ismagilov, R. F. (2003). "Formation of Droplets and Mixing in Multiphase Microfluidics at Low Values of the Reynolds and the Capillary Numbers," Langmuir, 19, pp. 9127-9133.

Nishisako, T. (2006). "Formulation on Droplets and Particles Using Microfluidic Devices," Ph.D. thesis, The Univ. of Tokyo.

van der Graaf, S., Steegmans, M. L. J., van der Sman, R. G. M., Schro¨en, C. G. P. H., Boom, R. M. (2005). "Droplet formation in a T-shaped microchannel junction," Colloids and Surfaces A: Physicochem. Eng. Aspects, 266, pp. 106–116.

Oishi, M., Kinoshita, H., Oshima, M. and Kobayashi, T. (2006). "Multicolor conforcal micro PIV system for multiphase flow measurement," Proc of 12th ISFV, No. 223.

Kinoshita, H., Oshima, M., Kaneda, S. and Fujii, T. (2005). "Confocal micro PIV measurement of internal flow in a moving droplet," Proc of Micro-TAS, No. 0192.

Hong, J. W., Hosokawa, K., Fujii, T., Seki, M. and Endo, I. (2001). "Microfabricated Polymer Chip for Capillary Gel Electrophoresis," Biotechnology Progress, Vol. 17, Issue 5, pp. 958-962.

Journal of Physics: Conference Series **147** (2009) 012062 doi:10.1088/1742-6596/147/1/012062

Investigation of the mixture flow rates of oil-water two-phase flow using the turbine flow meter

Dong-hui Li, Fei-fei Feng, Ying-xiang Wu, Jing-yu Xu

Institute of Mechanics, Chinese Academy of Sciences, Beijing 100190 , China

yxwu@imech.ac.cn (Professor Y.-x. Wu)

Abstract. In this work, the mixture flow rate of oil-water flows was studied using the turbine flow-meter. The research emphasis focuses on the effect of oil viscosity and input fluids flow rates on the precision of the meter. Experiments were conducted to measure the in-situ mixture flow rate in a horizontal pipe with 0.05m diameter using seven different viscosities of white oil and tap water as liquid phases. Results showed that both oil viscosity and input oil fraction exert a remarkable effect on measured results, especially when the viscosity of oil phase remained in the area of high value. In addition, for metering mixture flow rate using turbine flow-meter, the results are not sensitive to two-phase flow pattern according to the experimental data.

1. INTRODUCTION

Flows of oil and water two-phase are encountered in a diverse range of processes and equipment. For example, in the oil industry, mixtures of oil and water are transported in pipes over long distances. Accurate oil and water flow rate measurements are essential in order to properly manage both the wells and reservoirs. The problem of how to meter oil/gas/water mixtures has been of interest to the petroleum industry since the early 1980s. Since then considerable research has been conducted into the development of a three-phase flow meter suitable for use in an industrial environment (Falcone, 2002). Ideally such an instrument needs to be reasonably accurate (typically 5% of rate for each phase), nonintrusive, reliable, flow regime independent, and suitable for use over the full component fraction range. Usually, the solution to the problem of metering multiphase flows is to separate the component of the gas phase first, and then measure the flow rate of each using conventional single-phase instruments. In recent publications (Johnson and Farroll, 1995; Luxhùj, 1998; Skea and Hall, 1999; ZHANG et al. 2005), some researcher tried to measure the mixture flow meter using the turbine flow-meter. However, a survey of the past literature shows that no experiment, to the best of the authors' knowledge, has been reported till date to investigate on the effect of oil viscosity on measure results at a fixed system. In this work, the mixture flow rate of oil-water flows was studied using the turbine flow-meter. The research emphasis focuses on the effect of oil viscosity and input oil fraction on the precision of the meter.

Journal of Physics: Conference Series **147** (2009) 012062 doi:10.1088/1742-6596/147/1/012062

2. EXPERIMENTAL SET-UP AND PROCEDURE

The experiments reported below were carried out on the multiphase flow facilities at Institute of Mechanics, Chinese Academy of Sciences. A schematic diagram of the experimental system was shown in Fig.1. All experiments were conducted using white oils and water at room-temperature and atmospheric outlet pressure. Measurements were made for water flow rates from input oil volume fractions from 5% to 95%. The system consisted of a steel frame supporting a transparent Perspex pipes. Water and oil were pumped from their respective storage tanks, metered, and introduced into pipes via a T-junction. The mixture flowed along a 12 m long horizontal pipe from the entry point, which provided sufficient entrance length to stabilize the flow, to the test section. Different flow structures can be obtained through adjusting the input flow rates of water and oil by the pumps themselves. When the steady state was reached under a flow condition, the input flow fluids rates were measured. The high-speed camera recorded the continuous flow process synchronously when the camera frequency was set at 0.5 kHz, which was high enough to capture the details of the flow process. The holdup measurement was taken after the mixture flowing along a 4.5 m long horizontal pipe from the bend. The rapid closing valves system was operated three times to obtain the averaged water holdup under the flow condition. Seven different viscosities white oil have been used to study the effect of the fluid properties on the measurement. The physical properties of liquids and test matrix have been listed in Table 1.

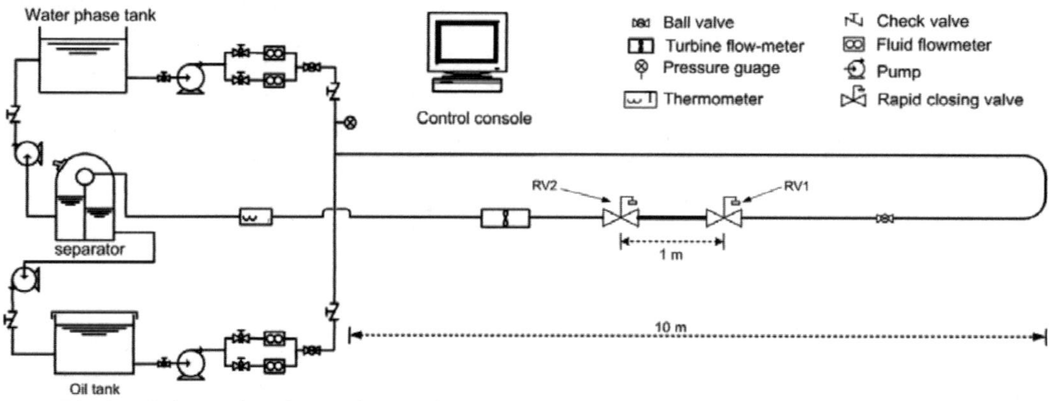

Fig.1 Schematic of experimental setup

Table 1 Experimental matrix

Oil viscosity (mPa·s)	50	160	225	400	700	1100	1450
Water flow rate, Qw (m^3/h)	0~15	0~15	0~15	0~15	0~15	0~15	0~15
Oil flow rate, Qo (m^3/h)	1.7~13	2.5~12	2.9~12	0.8~9.9	0.7~5.7	0.7~1.4	1.4~2.8
Test points	52	41	43	54	40	31	31

3. EXPERIMENTAL RESULTS AND DISCUSSIONS

In this work, input fluids flow rates (Q_W, Q_O) were measured separately under a fixed oil viscosity. Thus, the input oil and water volume fractions can be obtained by:

$$\beta_O = \frac{Q_O}{Q_O + Q_W} = \frac{Q_O}{Q_{M1}} \tag{1}$$

$$\beta_W = \frac{Q_W}{Q_O + Q_W} = \frac{Q_W}{Q_{M1}} = 1 - \beta_O \tag{2}$$

where Q_{M1} is the input mixture flow-rate, Q_{M2} is in-situ mixture flow rate which was measured by turbine flow-meter (shown in figure 1). Compared Q_{M1} with Q_{M2}, the relative error can be cauculated by:

$$Er = \frac{Q_{M2} - Q_{M1}}{Q_{M1}} \times 100\% \tag{3}$$

Figures 2 to 8 show the relative errors using seven different viscosities, respectively. In this work, ±5% relative error is used as the standard to define the work area, and a vertical line is used to distinguish the effective boundary of work area in figures. Relative error vs. input oil volume fraction is shown in figure 2 for water and oil flow with oil viscosity 50mPa·s. It can be found that most of errors are within ±5% during the entire range and turbine flow-meter works well at any given oil fraction. The experimental results with oil viscosity 160mPa·s in figure 3 shows that relative errors are to increase with the increase of input oil fraction. Relative errors can keep within ±5% during a large range of input oil fraction, which is consistent with oil viscosity 50mPa·s for two-phase flow. With oil viscosity increased further to 225mPa·s, similar trend can be observed as shown in figure 4. However, relative errors are within ±5% when βo is less than 70%. Error curves decline with increased βo. When βo reaches 70%, errors can reach to 10%. The failure area appears in the range of 70%~100% for input oil fraction. Relative error vs. input oil volume fraction with oil viscosities 400mPas and 700mPa·s are shown in figures 5 and 6 respectively. It can be seen that, when βo reaches 50~60%, the curves sharply decline and extend ±5%. Work of Turbine flow-meter goes beyond its effective area and the errors turn positive to negative.

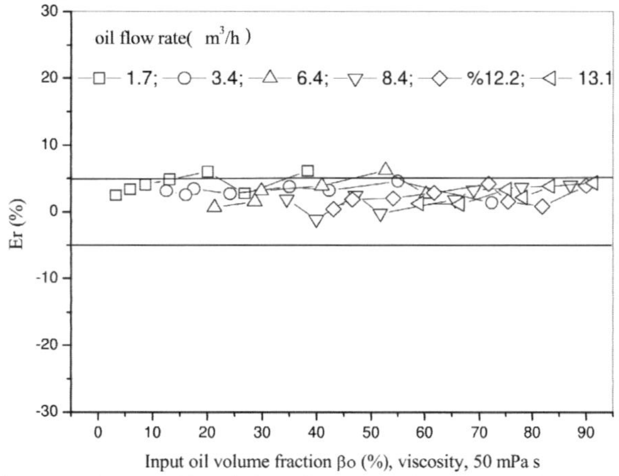

Fig.2 Relative error (Er) vs. input oil volume fraction (βo) with oil viscosity 50mPa·s

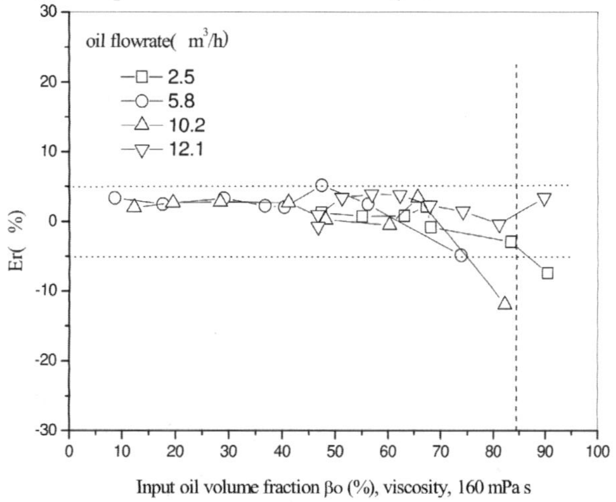

Fig.3 Relative error (Er) vs. input oil volume fraction (β_o) with oil viscosity 160mPa·s.

In order to further investigate of the effects high-viscosity oil on measurement precision, we observed the characteristics of turbine flow-meter when oil viscosities are 1100 mPa·s and 1450mPa·s respectively. The results are shown in figure 7 and 8 respectively. It can be seen that errors sharply increase when βo remains in the range of 30%~40%. Thus, the effective area of turbine flow meter is very narrow for super heavy oil.

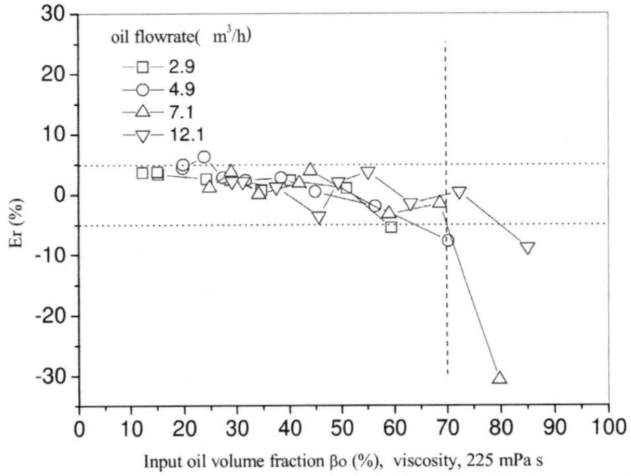

Fig.4 Relative error (Er) vs. input oil volume fraction (β_o) with oil viscosity 225mPa·s

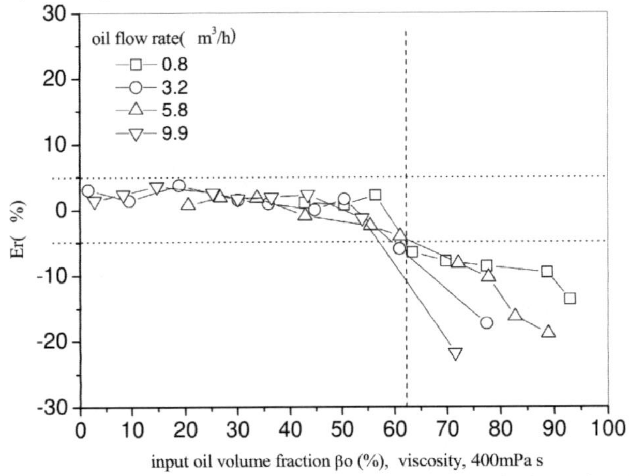

Fig.5 Relative error (Er) vs. input oil volume fraction (β_o) with oil viscosity 400mPa·s.

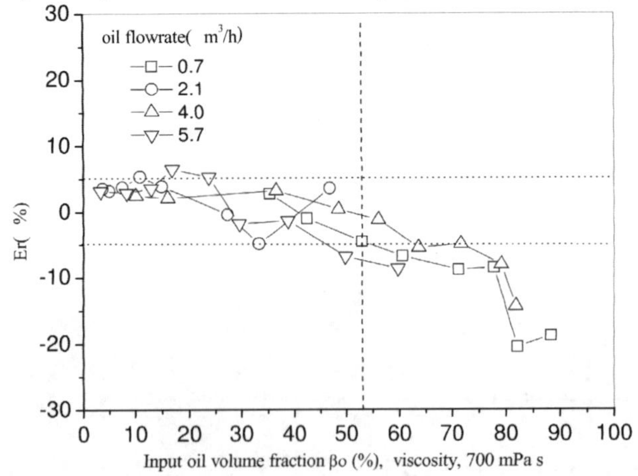

Fig.6 Relative error (Er) vs. input oil volume fraction (β_o) with oil viscosity 700mPa·s.

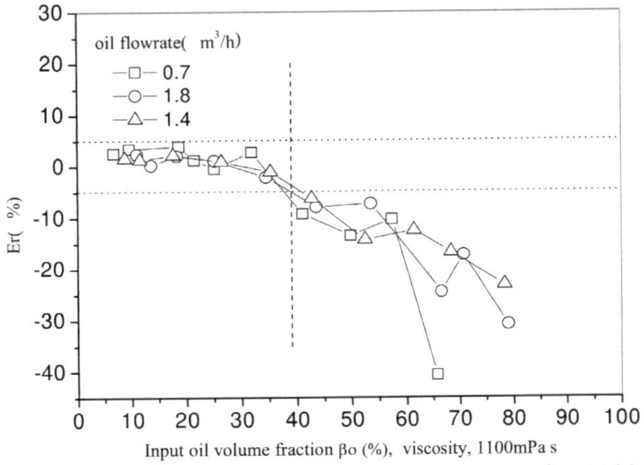

Fig.7 Relative error (Er) vs. input oil volume fraction (β_o) with oil viscosity 1100mPa·s.

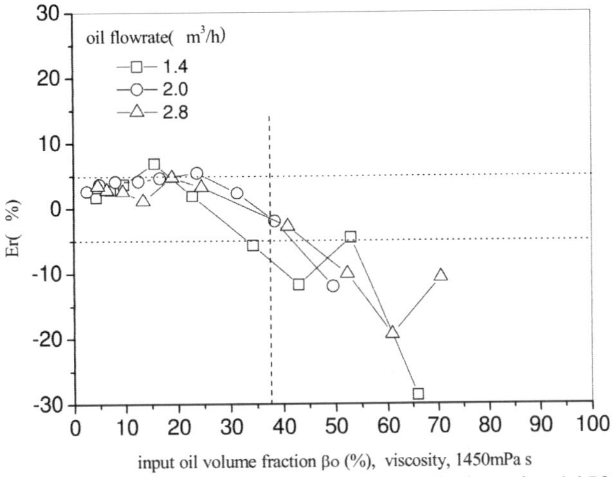

Fig.8 Relative error (Er) vs. input oil volume fraction (βo) with oil viscosity 1450mPa s.

Form above results, it can be concluded that measured errors tend to increase with the increase of input oil fraction (βo) and the effective area of turbine flow meter becomes narrow. For oil-water flow with high oil viscosity, the viscosity of fluid has great impact on measure precision. When oil viscosity goes beyond a certain value, the measure errors will be enlarged. In order to explain this, we study the effective viscosity of oil-water flow: regarding water-oil mixture as homogeneous flow, the mixture viscosity can be defined as:

Water as continuous phase: $$\mu_1 = \frac{\rho_o(1-\beta_w)+\rho_w\beta_w}{\frac{\rho_o(1-\beta_w)}{\mu_o}+\frac{\rho_w\beta_w}{\beta_w}} \tag{4}$$

Oil as continuous phase: $$\mu_2 = \frac{\mu_o}{1+\mu_w}\left(1+\frac{1.5\mu_w\beta_w}{\mu_o+\mu_w}\right) \tag{5}$$

In the present work, when water phase is continuous, the calculated effective viscosity is between 1.02 mPa ·s and 2.35mPa ·s and measured errors are within ±5%. When oil phase is continuous, calculated effective viscosity is between 283 and 370 mPa ·s and measured errors go beyond 10% and even reach to 21%. During the effective area of turbine flow meter, most of relative errors are positive with low oil viscosity, which means measured values are larger than real values. However, most of absolute errors are negative with high oil viscosity, which means measured values are smaller than real values.

In order to investigate the relationship between measure precision and flow pattern, flow pattern with oil viscosity 50mPa·s are presented and compared with the flow pattern map obtained by Lovick and Angeli (2004). The flow patterns have been observed visually as well as by photographic techniques. In the present work the flow structures of oil-water two-phase flow are distinguished, in general, as six basic flow structures according to the definition of Trallero et al (1997). They are: stratified flow (*ST*), stratified flow with mixing at the interface (*ST&MI*), dispersion of oil in water and water (*Do/w&w*), oil in water emulsion (*o/w*), dispersions of water in oil and oil in water (*Dw/o&Do/w*) and water in oil emulsion (*w/o*). Flow pattern map of Lovick and Angeli with corresponding superficial velocity (solid lines), showing oil-water two-phase flowing in horizontal pipes for the experiments reported in the present work, is presented in Figure 10. From boundary of different flow pattern we can find that flow pattern obtained in this experiment accord well with flow pattern obtained by Lovick and Angeli. According to the figure 3, we can see that errors between different patterns have no obvious difference, so we conclude that the characteristics of turbine flow meter is not sensitive to flow pattern for low viscosity oil and water flow.

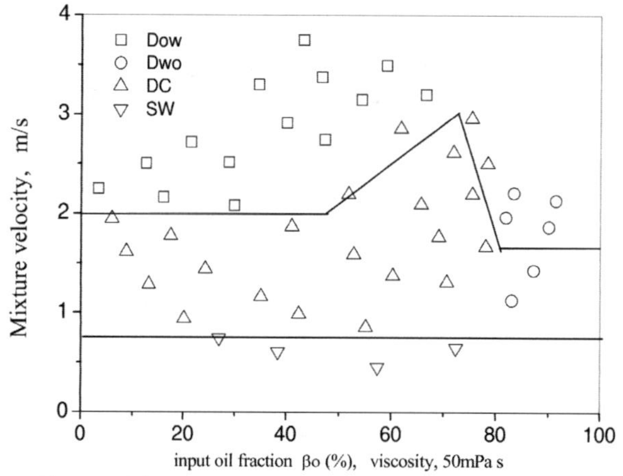

Fig.9 Oil/water flow pattern map in horizontal pipe

4. CONCLUSIONS

An experimental study of the mixture flow rate of oil-water flows using the turbine flow-meter has been conducted. The research emphasis focuses on the effect of oil viscosity and input fluids flow rates on the precision of the meter. Results showed that oil viscosity and input oil fraction exert a remarkable effect on measured errors, especially when the viscosity of oil phase remained in the area of high value. In addition, for metering mixture flow rate using turbine flow-meter, the results are not sensitive to flow-pattern for low viscosity oil and water flow according to the experimental data..

ACKNOWLEDGEMENTS

The authors gratefully acknowledge that the work described here was financially supported by both the National Natural Science Foundation of China (Grant No. 10572143) and the key project of "Eleventh Five-Year Plan" of the Chinese Academy of Sciences (Grant No. KJCX2-YW-L02).

REFERENCES

Falcone G., Hewitt G.F., Alimonti C. et al. (2002). "Multiphase flow metering: Current trends and future developments," Journal of Petroleum Technology, 54 (4), pp 77-84.

ZHANG H.-j., YUE W.-t., HUANG Z.-y. (2005). "Investigation of oil-air two-phase mass flow rate measurement using Venturi and void fraction sensor," Journal of Zhejiang University Science, 6A (6), pp601-606.

Luxhùj J.T. (1998). "An artificial neural network for nonlinear estimation of the turbine flow-meter coefficient," Engineering Applications of Artificial Intelligence, 11, pp 723-734.

Johnson M.W., Farroll S. (1995). "Development of a turbine meter for two-phase flow measurement in vertical pipes," Flow Measurement and Instrumentation, 6(4), pp279-282.

Skea A.F., Hall A.W.R. (1999). "Effects of water in oil and oil in water on single-phase flow meters," Flow Measurement and Instrumentation, 10, pp151-157.

Lovick J., Angeli P. (2004). "Experimental studies on the dual continuous flow pattern in oil-water flows," International Journal of Multiphase Flow, 30, pp139-157.

Trallero J.L., Sarica C., Brill J.P., (1997). "A study of oil/water flow patterns in horizontal pipes," SPE Production and Facilities August, SPE 36609.

Advances in Measurements and Simulation of Gas-Particle Flows and Coal Combustion in Burners/Combustors

L. X. Zhou

Tsinghua University, Haidian District, Beijing 100084, CHINA

zhoulx@mail.tsinghua.edu.cn

Abstract. Innovative coal combustors were developed, and measurement and simulation of gas-particle flows and coal combustion in such combustors were done in the Department of Engineering Mechanics, Tsinghua University. LDV/PDPA measurements are made to understand the behavior of turbulent gas-particle flows in coal combustors. Coal combustion test was done for the non-slagging cyclone coal combustor. The full two-fluid model developed by the present author was used to simulate turbulent gas-particle flows, coal combustion and NO_x formation. It is found by measurements and simulation that the optimum design can give large-size recirculation zones for improving the combustion performance for all the combustors. The combustion test shows that the non-slagging coal combustor can burn 3-5mm coal particles with good combustion efficiency and low NO emission. Simulation in comparison with experiments indicates that the swirl number can significantly affect the NO formation in the swirl coal combustor.
Keywords: PDPA Measurements, Simulation, Gas-Particle Flows, Coal Combustion

1. INTRODUCTION

Developing highly efficient and low pollutant coal burners/combustors remains to be important tasks of power engineering in China, since in many cases low-volatile and high-ash-content coals are used in power stations with serious problems of stabilizing coal flames at low load. In some cases coals are burned in small-size industrial and domestic furnaces with very low combustion efficiency and serious pollution, including high dust emission. Besides, low NO_x burners should be developed for utility boiler furnaces in power stations. To solve these problems, systematic studies on developing coal combustors, measurement and simulation of gas-particle flows and coal combustion in such combustors were carried out by the present author's group. Still in the years of 1983-1986, a coal combustor with high-velocity jets was developed for stabilizing anthracite and low-volatile coal flames, and measurements and simulation of gas-particle flows and simulation of coal combustion were done in optimization design of this kind of combustors. Then, in the years of 1989-1996 a non-slagging cyclone coal combustor was developed, and measurements and simulation of gas-particle flows were made for subsequent design of this kind of combustors and combustion experiments. Finally, in the years of 2000-2003, the development, measurements and simulation of gas-particle flows and coal combustion in a low-pollutant swirl coal combustor were conducted. The effect of fuel feeding location and swirl on NO_x formation was studied.

2. DEVELOPED COAL BURNERS/COMBUSTORS

There are many types of low-grade coals produced in China, and in many power stations, high-moisture, high-ash-content low-volatile coals, even anthracite coals have to be used. For burning these types of coals the problem of flame stabilization at low load is crucial, frequently, it was obliged to burn oil for stabilizing the coal flame. To solve this problem, a coal burner/combustor with high-velocity jets, as shown in Fig.1, was developed (Zhou et al., 1994). Its principle is to open some tiny holes at the front end wall of a sudden-expansion chamber, from these holes high-velocity jets with velocities about 200~300m/s are injected into the combustor, while the coal-particle-laden main stream velocity is about 20m/s. Due to the entrainment effect of the jets a large-size recirculation zone connecting the corner part and the near-axis part is induced, in which the particle concentration is sufficiently high, so it is favourable to coal flame stabilization. The measurements and simulation are aimed to give optimal jet-hole location and flow-rate ratio.

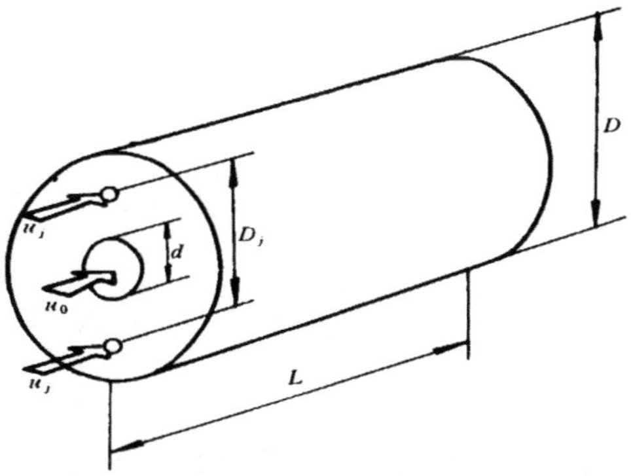

Fig.1 A coal Burner/Combustor with high-Velocity Jets

In China, more than 20% of coals are burned in small-size industrial and domestic furnaces with very low combustion efficiency and serious pollution, including high dust emission. It is expected to achieve highly efficient and low-pollutant combustion in these furnaces. To solve these problems, a two-staged non-slagging cyclone combustor, organized in perpendicularly swirling flows, called "spouting-cyclone combustor" (SCC) was proposed (Zhou et al., 1992). The principle of SCC is shown in Fig.2. There is a two-staged combustion in two perpendicularly swirling flows. In the lower part of the combustor, called "spouting zone", an opposed-jet system consists of one or a pair of primary-air jets laden with 3-5mm coal particles and one or a pair of bottom-air jets. The primary-air jet and bottom-air jet are in parallel and opposite directions. This opposite-jet system can create large-size recirculating flows around the horizontal axis as a primary combustion zone. The larger the particle sizes, the longer the residence time in which particles will multiply circulate in this zone. In the upper part of the combustor, called "cyclone zone", one or two pairs of tangentially introduced secondary-air jets can create cyclonic flows around the vertical axis Here small particles or particles not yet burnout in the primary zone will complete their combustion. The particles collected on the wall can be led to return to the primary zone and burn again. The spouting zone is a fuel-rich zone for reducing the NO_x formation. The secondary-air flow rate will keep the combustion temperature lower than the ash melting point, so as to maintain the coal combustion in a non-slagging mode. Subsequently it was found that the original shape of the spouting zone is not sufficiently good for particle recirculation. Then the shape of the spouting zone was changed from a cylinder to a rectangular cross-sectional channel (Wang et al., 1996), as shown in Fig.3, it is an improved spouting-cyclone combustor

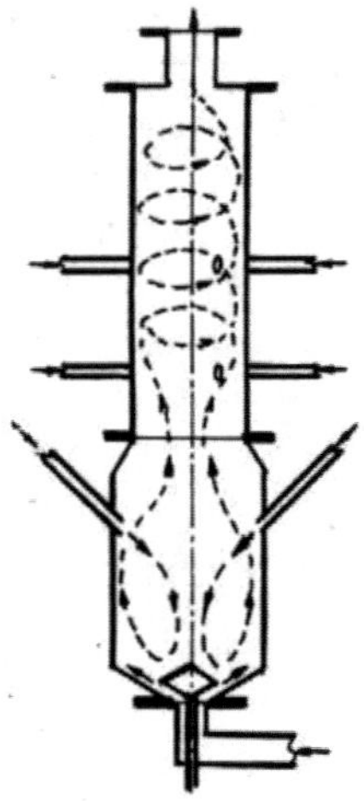

Fig.2 A Spouting-Cyclone Combustor

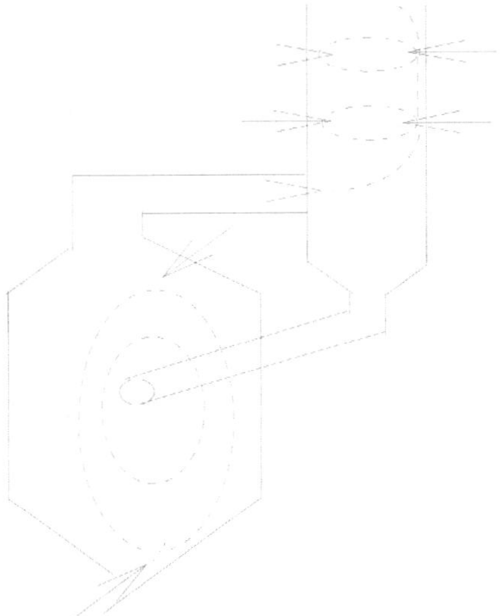

Fig.3 An Improved Spouting-Cyclone Combustor

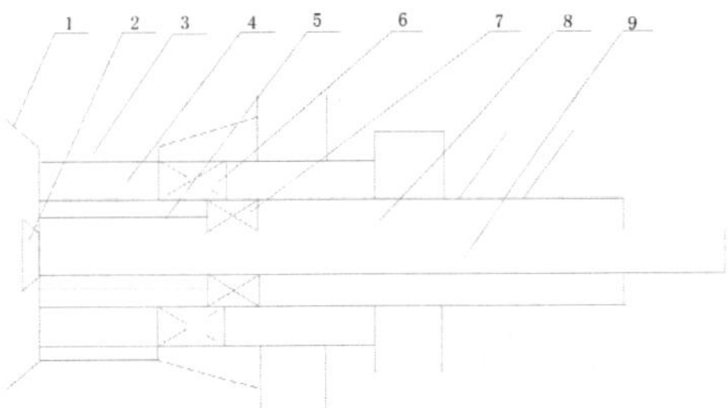

Fig.4 A Swirl Coal Combustor
1-Expansion; 2-Cone ; 3-Non-swirling secondary air ; 4-Swirling secondary air;
5-Annular space ; 6-swirler blade ;
7-Swirler blade ; 8-Primary-air ; 9-Central tube.

Furthermore, swirl coal burners are widely used in utility boiler furnaces of power stations in China. Its advantage is that it can stabilize the coal flame by itself. Recently, for designing such kind of combustors, not only the flame stabilization but also reducing pollution should be taken into account. It is expected that there is high coal concentration in the recirculation zone, but too high concentration will lead to reducing combustion efficiency. A typical swirl coal combustor was proposed (Li et al., 2000), as that shown in Fig.4. It includes a central tube, primary-air tubes for supplying coal-particle-laden flows, secondary-air tubes for supplying non-swirling and swirling flows. The measurements and simulation are aimed to study the effect of different flow rates and swirl angles of primary air and secondary air on the size of recirculation zone and coal concentration, in order to give good flame stabilization and simultaneous low NOx formation.

3. EXPERIMENTAL SET-UP & MEASUREMENT TECHNIQUE FOR ISOTHERMAL GAS-PARTICLE FLOWS

The experimental set-up for measuring isothermal gas-particle flows in Case 1 is shown in Fig.5. It consists of the test section 15, an air-supply system (compressor 9 and blower 12), a powder-supply system 10 and 11, and a LDV measurement system 1, 2, 3. The measurement instrumentation is a DISA-55X 2-D LDV system with forward scattering. The test section is given in Fig.1, where L=400mm, D=100mm, d=24mm, D_j=67mm and d_j=1mm. The inlet flow parameters are: u_0=15m/s; u_j=260m/s; G_p=60g/min; d_p=60μm.

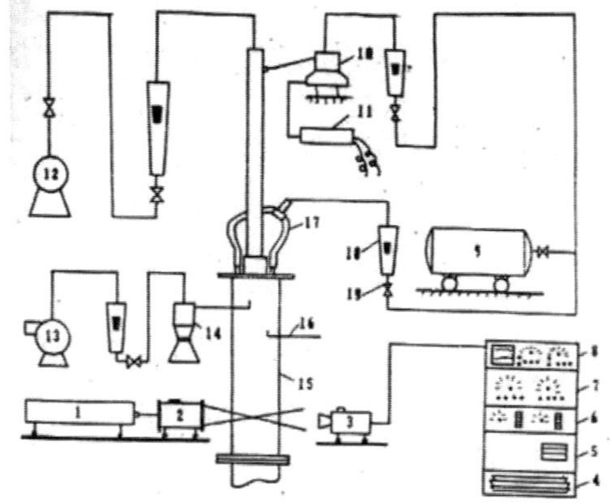
Fig 5 Experimental Set-Up for Case 1

The experimental set-up for measuring isothermal gas-particle flows in Case 2 is shown in Fig.6. It consists of blowers 1, pipe lines and valves 2, flow meters 3, air compressor 4, secondary-air pipe 5, atomizer 6, powder feeder 7, spouting zone 8, cyclone zone 9, laser 10, transmitter 11, receiver 12, bottom-air pipe 13 and primary-air pipe 14. A TSI 2-D LDV system was used to measure the gas and particle velocities separately. A TSI atomizer was used to produce 1-5μm liquid droplets tracking the gas flow and an electro-magnetically vibrating powder feeder was employed to supply 250-300μm glass beads as the particle phase. The cylindrical test section (Fig.2) of spouting zone and cyclone zone are made of plexiglass. There are one or two downward inclined primary-air inlets and one or two upward inclined bottom-air inlets in the spouting zone, and four tangential secondary-air inlets at two different heights of the cyclone zone. The test section was mounted on a traversing system for flow measurements at different cross sections.

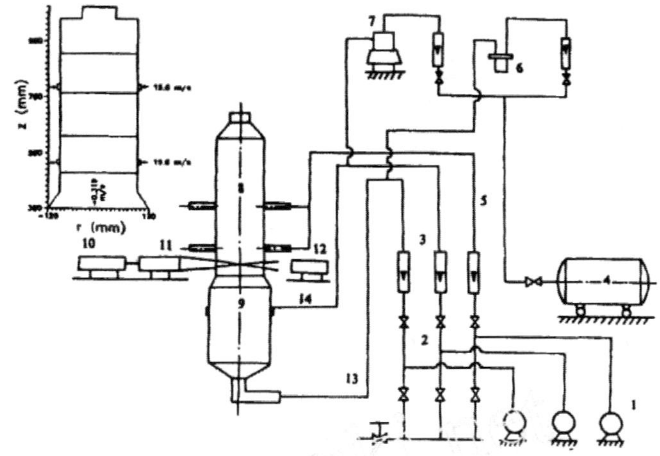
Fig.6 Experimental Set-Up for Case 2

The experimental set-up for measuring isothermal gas-particle flows in Case 3 is shown in Fig.7. It consists of air compressors 18, 19, a Powder feeder 15, a test section 12, a PDPA system 1-11, and a cyclone separator 14. The test section is given in Fig.4. A 3-D PDPA made by Dantec Inc was used to measure gas-particle flows, and a backward scattering arrangement was used. The particle velocity is measured based on the Doppler frequency shift. The particle size and concentration are measured based on the phase difference caused by Mie scattering and the number of particles passing the measuring volume during a certain time interval.

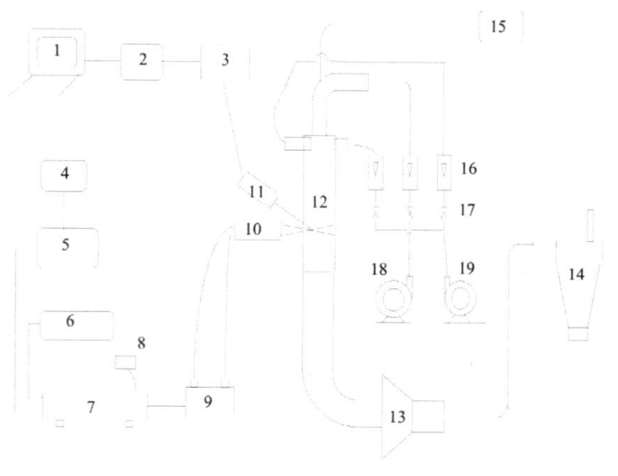

Fig.7 Experimental Set-Up for Case 3

The test section is a cold model of swirl burner (Fig.4) together with a cylindrical furnace of 800mm diameter and 1500mm height. The measurement system is a DANTEC 3-D phase Doppler particle anemometer (PDPA). Glass beads of 0~10μm are taken as the gas tracer and glass beads of 50μm represent the particle phase. The inlet flow parameters are shown in Tab. 1. For each case there are 5 measurement cross sections along the axial direction and for each cross section there are 21 measurement points along the radial direction. At each measurement point the data rate is 500 per minute and each time of sampling lasts 2 minutes.

Table 1 Flow Parameters for Case 3

Burner	Case	Primary Air	Swirler Angle	Q (m^3/h)	q_1(%)	q_{21} (%)	q_{22} (%)
Type I	3-1	No Coal Concentrator	45°	1179	20.1	54.3	25.6
Type II	3-2	With Coal Concentrator	45°	1174	19.2	53.3	27.5

where Q——total air flow rate, q_1—primary air ratio, q_{21}——swirling secondary-air ratio, q_{22}——non-swirling secondary air ratio, air temperature 291K.

4. NUMERICAL MODELS AND METHODS FOR SIMULATION

For simulating turbulent gas-particle flows and coal combustion, a full two-fluid model was proposed (Zhou et al., 2000). The time-averaged equations of turbulent reacting gas-particle flows can be written as:

$$\frac{\partial \rho}{\partial t}+\frac{\partial}{\partial x_j}(\rho v_j)=-\sum_p n_p \dot{m}_p = S \tag{1}$$

$$\frac{\partial}{\partial t}(\rho v_i)+\frac{\partial}{\partial x_j}(\rho v_j v_i)=-\frac{\partial p}{\partial x_i}+\mu\left(\frac{\partial v_j}{\partial x_i}+\frac{\partial v_i}{\partial x_j}\right)+\rho g_i$$
$$+\sum_p \frac{m_p}{\tau_r}\left[n_p(v_{pi}-v_i)+\overline{n_p v_{pi}}\right]+v_i S+F_{mi}-\frac{\partial}{\partial x_j}\left(\rho\overline{v_j v_i}\right) \tag{2}$$

$$\frac{\partial}{\partial t}(\rho h)+\frac{\partial}{\partial x_j}(\rho v_j h)=\frac{\partial}{\partial x_j}\left(\lambda\frac{\partial T}{\partial x_j}\right)-q_r+\sum_p n_p Q_p-\sum_k n_p \dot{m}_p C_{pp} T_p+$$
$$W_s Q_s-\frac{\partial}{\partial x_j}\left(\rho\overline{v_j h}\right) \tag{3}$$

Journal of Physics: Conference Series **147** (2009) 012063 doi:10.1088/1742-6596/147/1/012063

$$\frac{\partial}{\partial t}(\rho Y_s)+\frac{\partial}{\partial x_j}(\rho v_j Y_S)=\frac{\partial}{\partial x_j}\left(D\rho\frac{\partial Y_S}{\partial x_j}\right)-W_S-\alpha_S\sum_p n_p\dot{m}_p-\frac{\partial}{\partial x_j}\left(\rho\overline{Y_s v_j}\right) \tag{4}$$

$$\frac{\partial n_p}{\partial t}+\frac{\partial}{\partial x_j}(n_p v_{pj})=-\frac{\partial}{\partial x_j}\left(\overline{n_p v_{pj}}\right) \tag{5}$$

$$\frac{\partial(n_p m_p)}{\partial t}+\frac{\partial}{\partial x_j}(n_p m_p v_{pj})=-m_p\frac{\partial}{\partial x_j}\left(\overline{n_p v_{pj}}\right)+n_p\dot{m}_p \tag{6}$$

$$\frac{\partial(n_p m_c)}{\partial t}+\frac{\partial}{\partial x_j}(n_p m_c v_{pj})=-m_c\frac{\partial}{\partial x_j}\left(\overline{n_p v_{pj}}\right)+n_p\dot{m}_c \tag{7}$$

$$\frac{\partial(n_p m_w)}{\partial t}+\frac{\partial}{\partial x_j}(n_p m_w v_{pj})=-m_w\frac{\partial}{\partial x_j}\left(\overline{n_p v_{pj}}\right)+n_p\dot{m}_w \tag{8}$$

$$\frac{\partial}{\partial t}(n_p v_{pi})+\frac{\partial}{\partial x_j}(n_p v_{pj} v_{pi})=n_p g_i+\frac{1}{\tau_r}\left[n_p(v_i-v_{ki})-\overline{n_p v_{pi}}\right]+\frac{n_p\dot{m}_p}{m_p}(v_i-v_{pi})$$
$$-\frac{\partial}{\partial x_j}\left(n_p\overline{v_{pj}v_{pi}}+v_{pj}\overline{n_p v_{pi}}+v_{pi}\overline{n_p v_{pj}}\right) \tag{9}$$

$$\frac{\partial}{\partial t}(n_p h_p)+\frac{\partial}{\partial x_j}(n_p v_{pj} h_p)=\frac{n_p}{m_p}(Q_h-Q_p-Q_{rp})+\frac{n_p\dot{m}_p}{m_p}(h-h_p)$$
$$-\frac{\partial}{\partial x_j}\left(n_p\overline{v_{pj}h_p}+v_{pj}\overline{n_p h_p}+h_p\overline{n_p v_{pj}}\right) \tag{10}$$

where Eqs. (1) through (10) are gas continuity, gas momentum, gas energy, gas species, particle number density, total particle mass, particle raw coal, particle moisture, particle momentum and particle energy equations respectively, and W_s is the time-averaged reaction rate. All correlations and the time-averaged reaction rate in these equations need to be modeled. The particle mass changing rate $\dot{m}_p$ due to moisture evaporation, pyrolization and char combustion are determined by the algebraic expressions based on a surface evaporation model, one-equation pyrolyzation model and diffusion-kinetic model of char combustion. For two-phase turbulence modeling, a k-ε-kp model is used. For volatile and CO combustion in the comprehensive modeling of coal combustion, originally the conventional EBU-Arrhenius model is used (Zhou at al., 2000). The nomenclature for the above-stated equations can be found there.

The NO formed in coal combustion consists of mainly thermal NO and fuel NO. For the reaction kinetics of thermal NO formation, the well-known Zeldovich mechanism is used

$$N2+O \longrightarrow NO+N$$
$$N+O2 \longrightarrow NO+O$$

$$N+OH \longrightarrow NO+H$$

The total reaction rate is determined by

$$W = 8.39\times10^{16}\rho^{1.5}T^{-0.5}Y_{N_2}Y_{O_2}{}^{0.5}\exp(-564.4\times10^3/RT)$$

For the reaction kinetics of fuel NO formation, the DeSoete mechanism is used

$$HCN+O2 \longrightarrow NO \quad (a)$$

$$HCN+NO \longrightarrow N2 \quad (b)$$

$$C+NO \longrightarrow N2 \quad (c)$$

Table 2 gives the kinetic constants in $K = B\exp(E/RT)$, where K is the reaction coefficient.

Table 2. Kinetic Constants of fuel NO formation

Reaction	B	E (J/mol)
a	1×10^{11} (1/s)	2.803×10^5
b	3×10^{12} (1/s)	2.5105×10^5
c	41.8kg-mol/(atm.m^2.s)	1.425×10^5

For NO formation in turbulent flows, an algebraic second-order moment (ASOM) turbulence-chemistry model is proposed. The time-averaged reaction rate is (Zhou et al., 2003)

$$\overline{W_S} = B\rho^2[(\overline{Y_1}\,\overline{Y_2} + \overline{Y_1'Y_2'})\overline{K} + \overline{Y_1}\,\overline{K'Y_2'} + \overline{Y_2}\,\overline{K'Y_1'}] \tag{11}$$

$$K = \exp(-E/RT)$$

where, $\overline{K} = \int \exp(E/RT)p(T)dT$, $p(T)$ is the PDF of temperature. Assuming a top-hat PDF gives

$$\overline{K} = \{\exp[-E/R(T+g_T^{1/2})] + \exp[-E/R(T-g_T^{1/2})]\},\ g_T = \overline{T'^2} \tag{12}$$

The correlations $\overline{K'Y_1'}$, $\overline{K'Y_2'}$, $\overline{Y_1'Y_2'}$ in Eq.(11) can be determined by a generalized form of the following transport equation :

$$\frac{\partial}{\partial t}(\rho\overline{\psi\varphi}) + \frac{\partial}{\partial x_j}(\rho V_j\overline{\psi\varphi}) = \frac{\partial}{\partial x_j}\left(\frac{\mu_e}{\sigma_\varphi}\frac{\partial\overline{\psi\varphi}}{\partial x_j}\right) + c_{g1}\mu_T\frac{\partial\Phi}{\partial x_j}\frac{\partial\Psi}{\partial x_j} - c_{g2}\frac{\varepsilon}{k}\rho\overline{\psi'\varphi'} \tag{13}$$

where ψ, φ are the instantaneous species concentration and reaction coefficient K; Φ, Ψ are the mean species concentration and reaction coefficients.

In case of coal combustion, due to the complexity of the processes, for the sake of simplicity Eq.(13) is reduced to the following algebraic expressions by neglecting the convection and diffusion terms

$$\overline{Y_1'Y_2'} = C_1\frac{k^3}{\varepsilon^2}\frac{\partial\overline{Y_1}}{\partial x_j}\frac{\partial\overline{Y_2}}{\partial x_j};\ \overline{K'Y_1'} = C_1\frac{k^3}{\varepsilon^2}\frac{\partial\overline{K}}{\partial x_j}\frac{\partial\overline{Y_1}}{\partial x_j};\ \overline{Y_2'K'} = C_1\frac{k^3}{\varepsilon^2}\frac{\partial\overline{Y_2}}{\partial x_j}\frac{\partial\overline{K}}{\partial x_j} \tag{14}$$

where k, ε are the turbulent kinetic energy and its dissipation rate. The temperature or enthalpy fluctuation correlation g_T can be determined by:

$$\frac{\partial(\rho \overline{h'^2})}{\partial t}+\frac{\partial\left(\overline{\rho \bar{u} h'^2}\right)}{\partial x}+\frac{\partial\left(\overline{\rho \bar{v} h'^2}\right)}{r\partial r}+\frac{\partial\left(\overline{\rho \bar{w} h'^2}\right)}{r\partial \theta}=\frac{\partial}{\partial x}\left(\lambda_t\frac{\partial \overline{h'^2}}{\partial x}\right)+\frac{\partial}{r\partial r}\left(r\lambda_t\frac{\partial \overline{h'^2}}{\partial r}\right)$$

$$+\frac{\partial}{r^2\partial \theta}\left(\lambda_t\frac{\partial \overline{h'^2}}{\partial \theta}\right)+c_1\mu_t\left[\left(\frac{\partial \bar{h}}{\partial x}\right)^2+\left(\frac{\partial \bar{h}}{\partial r}\right)^2+\frac{1}{r^2}\left(\frac{\partial \bar{h}}{\partial \theta}\right)^2\right]+c_2\frac{\varepsilon}{k}\rho\overline{h'^2}$$

(15)

Equations (11), (12), (14), (15) constitute the algebraic unified second-order moment (ASOM) turbulence-chemistry model for NO formation in turbulent flows. The nomenclature in these equations is given by Zhou et al. (2003).

5. MEASUREMENT & SIMULATION RESULTS OF GAS-PARTICLE FLOWS

Figure 8 gives the predicted and measured gas and particle axial velocities for Case 1. It is seen that measurements and predictions are near to each other. Both of them show large-size reverse flow zones and particles keep forward motion in the central reverse flow zone. Figure 9 shows the measured and predicted particle mass fluxes. Predictions and measurements are in good agreement. Due to the particle inertia, a large amount of particles enter the central reverse flow zone, forming high particle concentration there, hence enhance flame stabilization at low loads and reduce NO formation. The measurements and simulation give the optimal jet-hole location and flow-rate ratio (not permitted to be reported here due to technical license) under which the size of recirculation zone and the reverse flow rate are maximal.

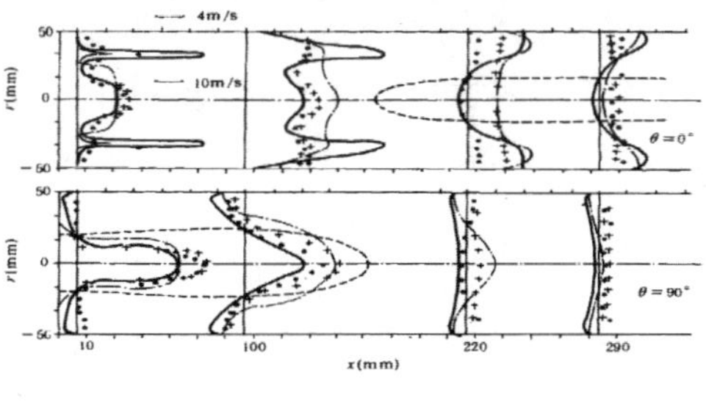

Fig.8 Two-Phase Velocities (•, Gas, Exp; +Particle,Exp, —Gas, Pred., -.-Particle, Pred)

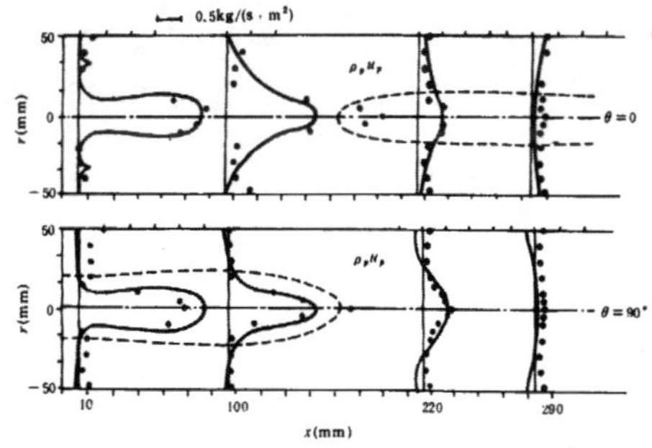

Fig.9 Particle Mass Fluxes (•Exp. , —Pred., --Reverse flow boundary)

Figure 10 gives the time-averaged gas and particle axial velocities for Case 2 in the spouting zone in the longitudinal plane perpendicular to the plane containing the primary-air and bottom-air inlets. Clearly, there are strongly recirculating flows in the primary combustion zone, i.e. spouting zone. The time-averaged axial and tangential gas and particle velocities in the cyclone zone for Case 2 are shown in Figs.11 and 12 respectively. The tangential velocity distribution shows a typical Rankine-vortex structure--solid-body rotation plus potential vortex. Near the bottom of the cyclone zone the potential vortex region is so small that it cannot be detected. Almost no recirculation takes place in the cyclone zone.

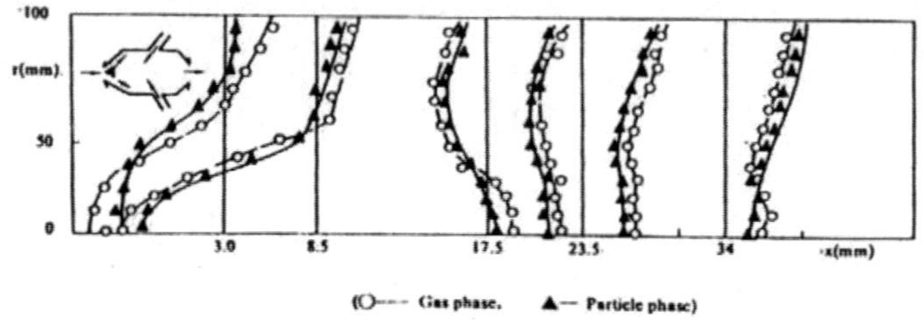

Fig.10 Gas and Particle Axial Velocities in the Spouting Zone of Case 2

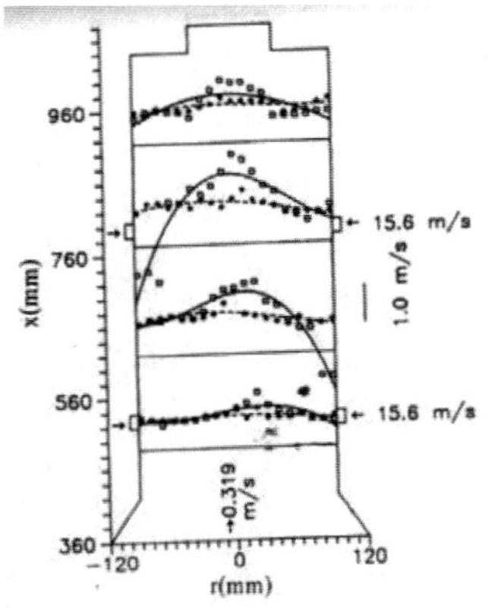

Fig.11 Two-phase Axial Velocities (Case 2)

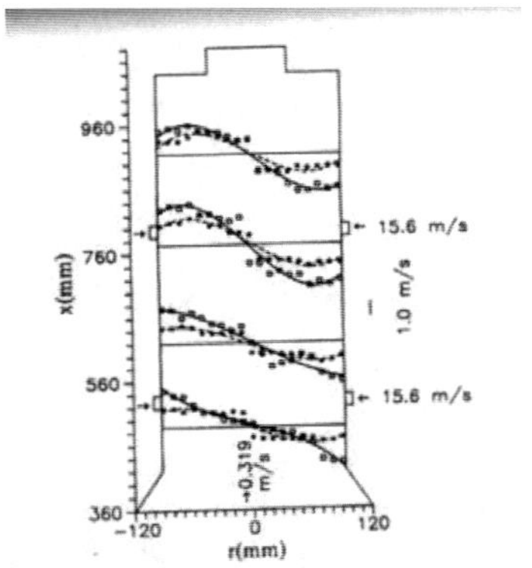

Fig.12 Two-phase Tangential Velocities (Case 2)

Figure 13 gives the predicted two-phase u-v velocity vectors in the spouting zone of the improved SCC for Case 2. It is seen that the gas and particle recirculation is much better than that in the original SCC, almost occupying the full space of the primary combustion zone

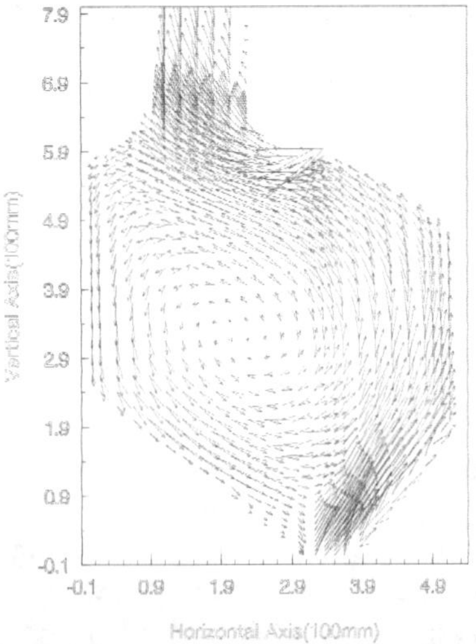

Fig.13. Two-phase Velocity Vectors (Spouting Zone of the Improved SCC)

Figure 14 shows the measured and predicted gas and particle axial velocities for Case 3 respectively. The agreement is quite good. It can be seen that a large reverse flow zone ranging from x=0 to x=1.5m is formed at the near-axis region of the burner exit. The maximum width of the near-axis recirculation zone locates at x=0.3 m, and highest negative velocities of about -

5m/s are found at the location x=0.1m within the near-axis recirculation zone. Figure 15 shows the measured and predicted gas and particle tangential velocities for Case 3 respectively. At first and second cross sections (x=10mm, x=110mm), the predicted tangential velocities are higher than those measured, and further in the down-stream region there still exists the discrepancy between predictions and experiments. Since the k-ε-k_p model can well predict weakly swirling flows, and predicted tangential velocities show reasonable Rankine-votex structure observed in swirling flows, it may be considered that the discrepancy is caused mainly by errors in measurement of tangential velocities. Because the radius of the furnace model is so large, it is impossible to guarantee the symmetry of measurements. The measured tangential velocities at the axis are not zero which is obviously wrong. Figure 16 gives the comparison between the predicted gas and the particle axial velocities for Case 3. It is found that the velocity slip between the gas and particle phases is evident only at the first, second and third cross sections. The particles initially have almost the same velocity as that of the primary-air flow. The secondary air flow has higher velocity and mass flow rate than those of the primary-air flow; therefore the velocity slip between gas and particle phases is large near the inlet region. On the other hand, due to higher inertia, the particles are not able to follow the rapid expansion and deceleration of the air flow, so that the particle velocity is higher than the gas one in some regions.

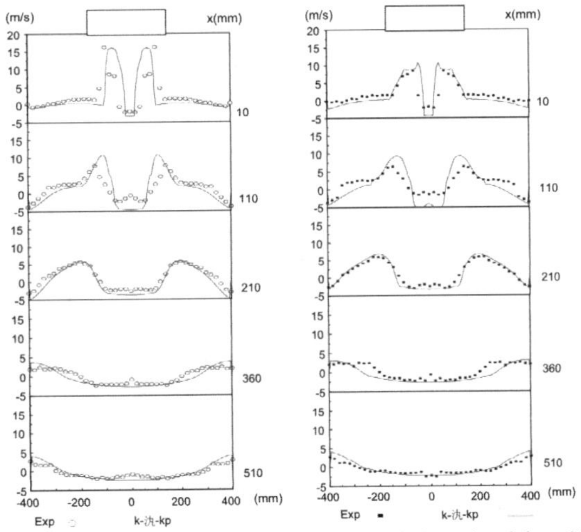

Fig.14 Gas (left) and Particle (right) Axial Velocities (Case 3)

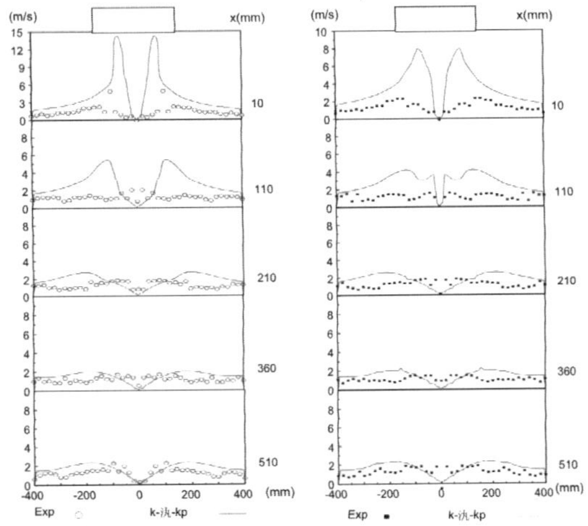

Fig.15 Gas (left) and Particle (right) Tangential Velocities(Case 3)

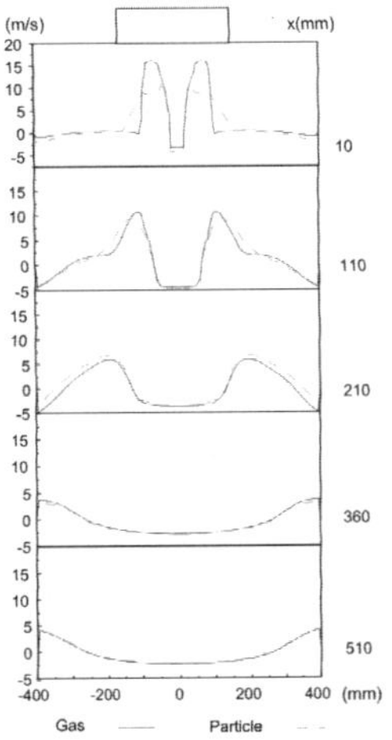

Fig.16 Comparison of Two-Phase Axial Velocities(Case 3)

Journal of Physics: Conference Series **147** (2009) 012063 doi:10.1088/1742-6596/147/1/012063

6. MEASUREMENT RESULTS OF COAL COMBUSTION

The combustion test for Case 2 was done with bituminous coal particles of 3-5mm. The proximate analysis of the Chinese Datong coal is: volatile 24.7%, ash 11.7%, moisture 3%, fixed carbon 63.3%. The temperature rise in the spouting zone and the cyclone zone is shown in Fig.17. The temperature in the spouting zone is in the range of 1273-1373K, which is below the ash melting point. The temperature in the cyclone zone is about 100K higher than that in the spouting zone. The NOx concentration is in the range of 100-320ppm (Fig.18), which is acceptable for a small-size furnace. The combustion efficiency reaches 92%, which is considered as a good performance for such small-size combustors burning ungrounded coal particles.

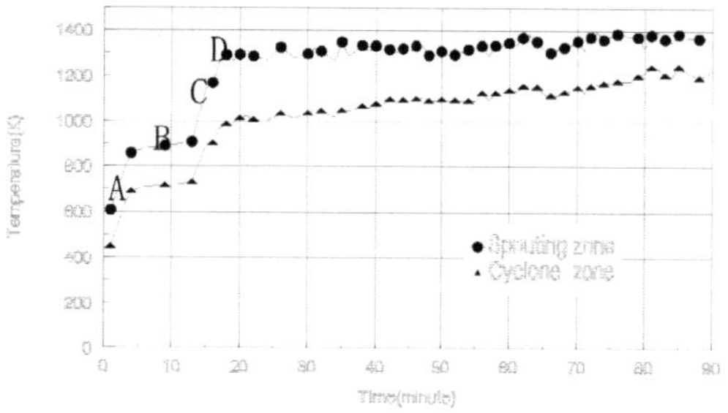

Fig.17 Temperature Rise

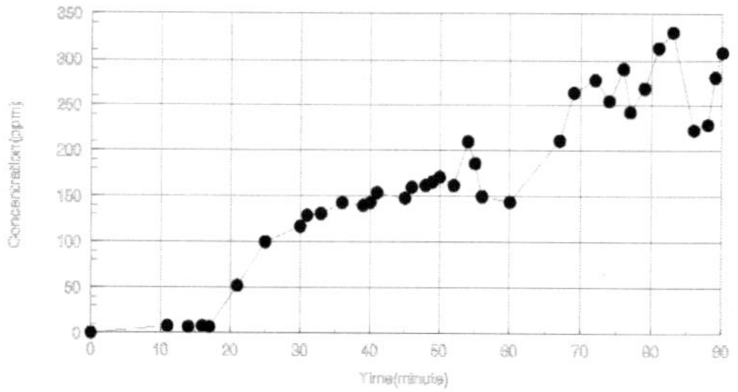

Fig.18 NOx Concentration Change

7. SIMULATION RESULTS OF COAL COMBUSTION

Figures 19 and 20 show the predicted gas and particle temperature maps for Case 1. It is seen that the high temperature is developed right at the stagnation point of the central reverse flow zone, and particle temperature is higher than the gas temperature. Figures 21 and 22 give the predicted thermal and total NO concentration for Case 1 respectively. Obviously, the total NO concentration is much higher than the thermal NO concentration. Simulation of coal combustion and NO formation was carried out also for a swirl combustor measured by Abbas et al. (1993), as shown in Fig.23. The geometrical sizes and inlet-flow conditions are given in Tables 3 and 4. The coal proximate analysis is given in Table 5. The pulverized coal is fed through the annular space "b", the secondary air is supplied through the annular space "c", and a bluff body is located in the central orifice "a". The average initial size of the pulverized coal is 45 μm. The prediction results for cold flows are compared with the PDPA measurement results and for coal combustion are compared with the coal combustion experimental results (Abbas et al., 1993).

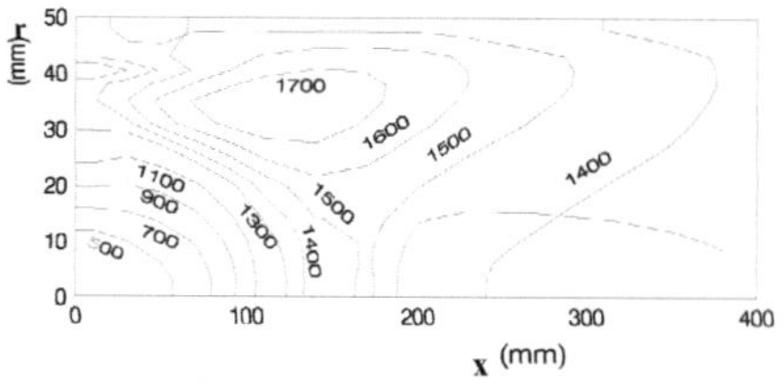

Fig.19 Gas Temperature Map

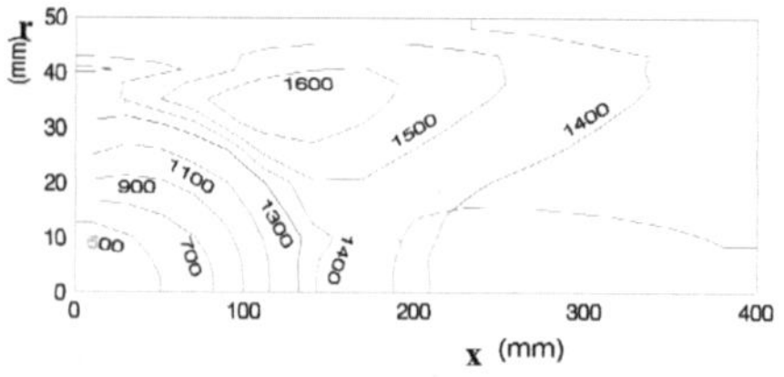

Fig.20 Particle Temperature Map

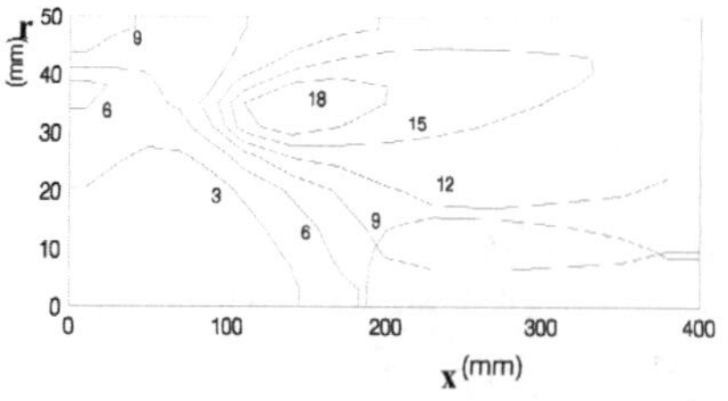

Fig.21 Thermal NO Concentration(ppm)

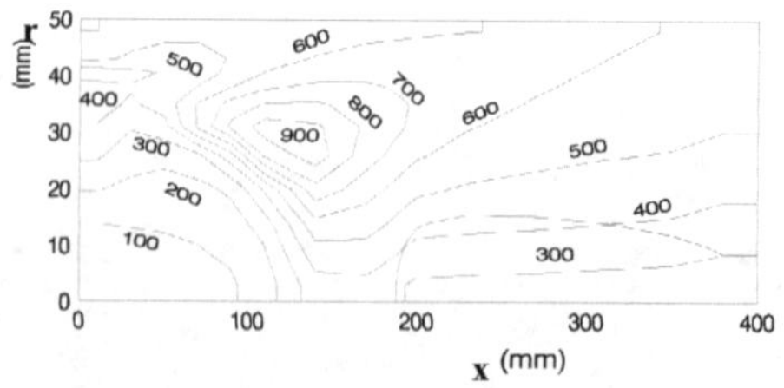

Fig.22 Total NO Concentration (ppm)

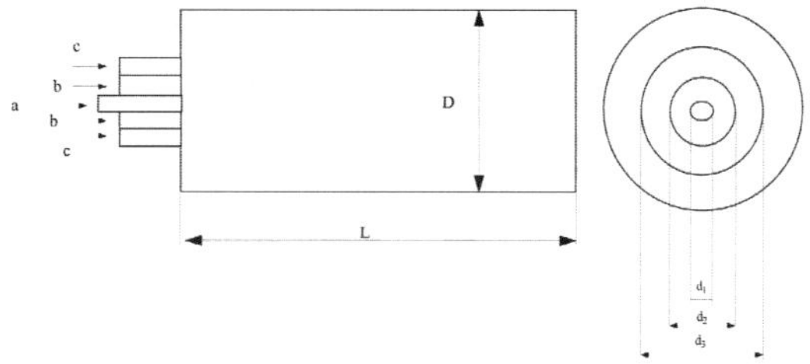

Fig.23 The Coal Swirl Combustor

Table 3 The Geometrical Sizes of the Swirl Combustor

D	L	d_1	d_2	d_3
0.6m	3.0m	0.022m	0.035m	0.104m

Table 4 Inlet-Flow Parameters of the Combustor

	Air Flow Rate (kg/h)	Coal Feeding Rate (kg/h)	Swirl Number
Primary Air	32.8	14	0.0
Secondary Air	120.4	0	0.5;0.8; 1.0;1.4

Table 5 Proximate Analysis of the Coal

Volatile (%)	Char (%)	Water (%)	Ash (%)
35.8	53.7	6.3	4.2

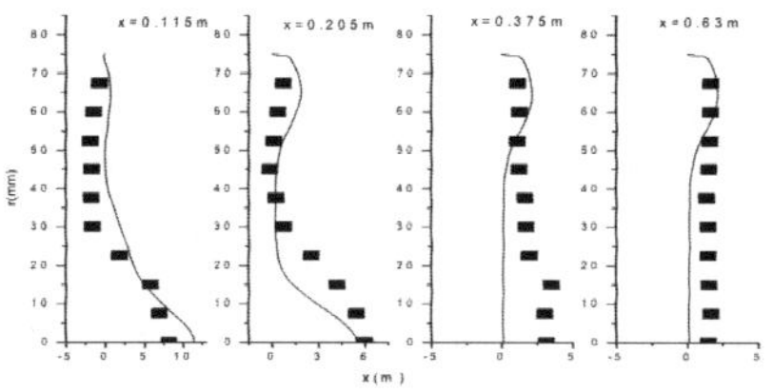

Fig.24 Gas Axial Velocity (—Pred ■ Exp)

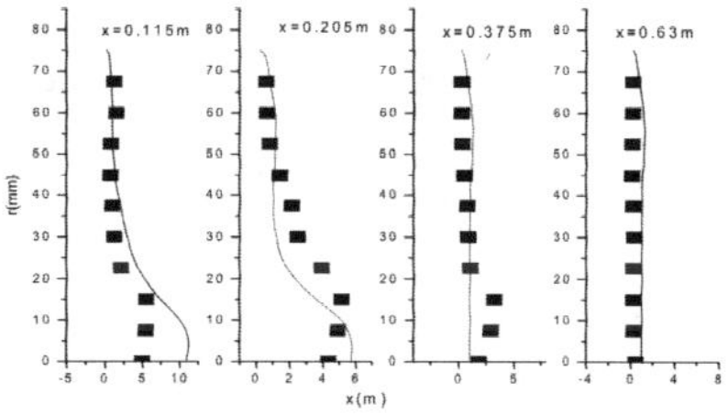

Fig.25 Turbulent Energy (—Pred ■ Exp)

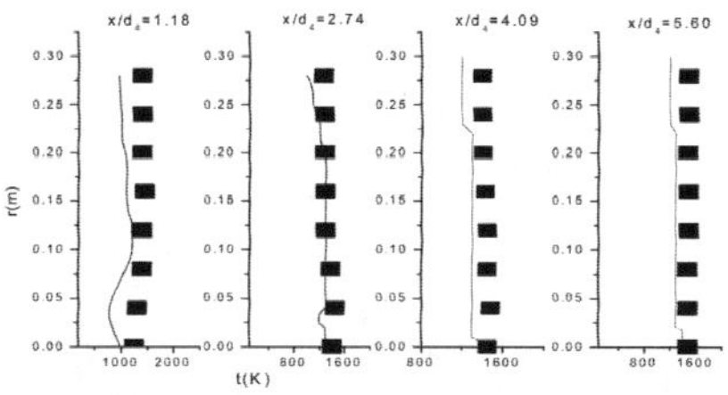

Fig.26 Temperature (— Pred. ■ Exp.)

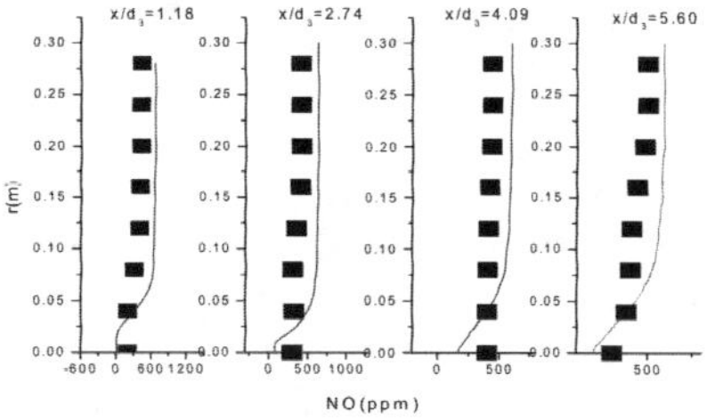

Fig. 27 NO Concentration (— Pred. ■ Exp.)

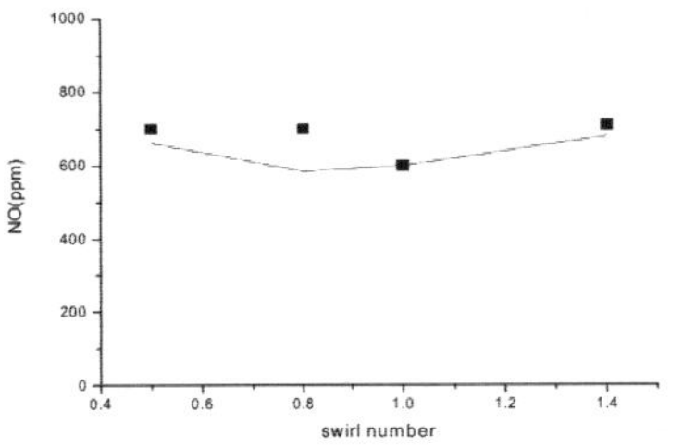

Fig.28 Average NO Concentration at the Exit

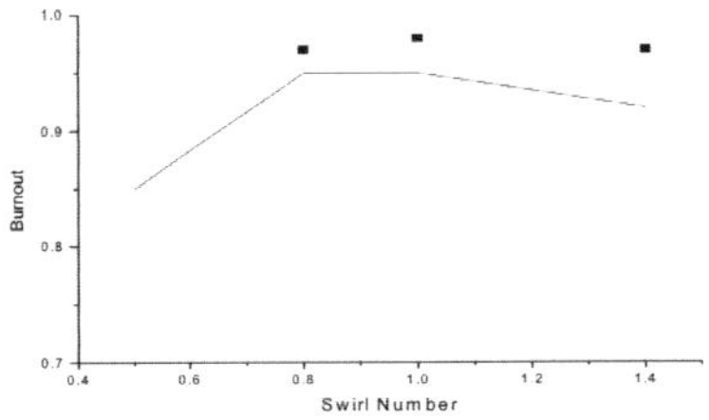

Fig.29 Coal Burn-out Rate

Figures 24 and 25 show the predicted and measured gas velocity and turbulent kinetic energy for cold flows and Figure 26 gives the predicted and measured temperature profiles, showing a good agreement between predictions and experiments. Figure 27 shows the predicted NO concentration profiles and their comparison with the experimental results. The agreement is also good. Figures 28 and 29 give the predicted averaged NO concentration at the exit and burnout rate vs. as the swirl number respectively and their comparison with experimental results. Both predictions and experiments show the common tendency: as the swirl number increases the NO concentration at first will decrease and then will increase, while the burnout rate at first will increase and then will decrease. There is a quantitative discrepancy between predictions and experiments. The predicted lowest NO emission and highest burnout rate occur at the swirl number of 0.8, but the measured ones occur at the swirl number of 1.0. This discrepancy may be caused by numerical errors and inaccuracies of the models. The overall NO formation in coal combustion should be determined by the temperature, coal concentration and turbulent fluctuation. With the increase of swirl number from 0.5 to 0.8, the temperature increases not so much, but the coal concentration in the inlet zone increases and the turbulent fluctuation decreases. Therefore, the NO formation decreases.

8. CONCLUSIONS

(1) LDV/PDPA measurements and simulation of isothermal gas-particle flows give the optimum design which can create large-size recirculation zones and high particle concentration for stabilizing coal flame and reducing NO formation in coal combustors.
(2) Combustion measurements show that a small-size non-slagging cyclone coal combustor can burn 3-5mm coal particles with high combustion efficiency and low NO emission.
(3) The full two-fluid model developed by the present author can well simulate turbulent gas-particle flows, coal combustion and NO_x formation.
(4) Simulation in comparison with experiments gives the effect of swirl on NO emission in a swirl coal combustor

ACKNOWLEDGEMENT
This study was originally sponsored by the National "Pandeng" Project of China under the Grant 85-06-1-2, the National Fundamental Research "973" Project of China under the Grant G1999-0222-07 and is now sponsored by the National Natural Science Foundation of China, under the Grants 50606026 and 50736006.

REFERENCES
Abbas, T., Costen, P., Hassan, M.A.(1993). "The effect of the near burner aerodynamics on pollution, stability and combustion in a PF-fired furnace", Combustion Science and Technology, 93:73-90

Li Zhiqiang, Li Rongxian, Zhou Lixing (2000). "Cold gas-particle flows in a new swirl pulverized-coal burner by PDPA measurement", Tsinghua Science and Technology, 5(1): 100-104

Wang, D.X., Wang, X.L., Zhou, L.X. (1996). "Experimental studies on gas-particle flows and coal combustion in a new-generation of spouting-cyclone combustor", J. Thermal Science, 5(2):132-137

Zhou, L.X., Guo, Y.C., Lin, W.Y. (2000). "Two-fluid models for simulation reacting gas-particle flows, coal combustion and NOx formation", Combustion Science and Technology, 150:161-180,

Zhou, L.X., Lin, W.Y., Zhang, J., Luo, W.W., Huang, X.Q. (1994). "Gas-particle flows and coal combustion in a burner/combustor with high-velocity jets", Combustion and Flame, 99:669-678

Zhou, L.X., Wang, F., Zhang, J. (2003). "Simulation of swirling combustion and NO formation using a USM turbulence-chemistry model", Fuel, 82: 1579-1586

Zhou, L.X., Zhou, B., Wang, X.L., Li, R.X. (1992). "Experimental studies on swirling gas-particle flows in a spouting-cyclone combustor", J. Thermal Science, 1(3): 203-207

Journal of Physics: Conference Series **147** (2009) 012064 doi:10.1088/1742-6596/147/1/012064

Determination of Mass Diffusion Coefficients of Oxygenated Fuel Additives in Air Using Digital Real-Time Holographic Interferometry

Maogang He, Ying Guo, Qiu Zhong, Ying Zhang

State Key Laboratory of Multiphase Flow in Power Engineering, Xi'an Jiaotong University, Xi'an 710049, P. R. China

mghe@mail.xjtu.edu.cn (Maogang He)

Abstract. In this work, an experimental system based on digital real-time holographic interferometry for measuring the mass diffusion coefficients of fluid is introduced. The method of processing interference fringe hologram is also introduced thoroughly. By uncertainties analysis and experimental verification, the accuracy of this system is validated. The experimental uncertainties in temperature and mass diffusion coefficient are estimated to be no greater than ± 0.16 K and ± 0.2 %, respectively. On this basis, the mass diffusion coefficients of three fuel additives, diethyl 1,6-hexanedioate (diethyl adipate, DEA), dimethyl carbonate (DMC) and diethyl carbonate (DEC) in air were measured at T = (278.15 to 338.15) K under atmospheric pressure, and polynomial was fitted by the experimental data.

1. INTRODUCTION

Recently, with the gradual reduction of conventional energy reserves such as coal and petroleum, the research and promotion of new energy are becoming more urgent (Ogawa et al., 2001; Alam et al., 2006). Among all of the research the use of mixed fuel is considered to be the most convenient and simple way to solve the problem of worldwide energy crisis. Mixed fuel is a mixture of conventional liquid fuel (gasoline, diesel oil, etc.) and some fuel additives. Research shows that the addition of oxygenated fuel additive to gasoline or diesel oil will accelerate the process of combustion, shorten the combustion time in the internal combustion engine, greatly improve the combustion performance and efficiency, and effectively reduce the emissions of harmful off-gas such as NO_x, CO_x, HC and soot etc. (Yoshimoto et al., 1999; Alfredo et al., 2003). Since the late 20th century, many thermophysical properties of oxygenated fuel additive and mixed fule have been researched a lot, such as the surface tension, viscosity, mutual solubility, stability and the optimal mixing ratio etc. Fully understanding and utilizing the thermophysical properties have promoted the application of mixed fuel in engineering (Poulopoulos et al., 2001; Shing et al. 2002).

The mass diffusion coefficient of the fuel additive is also an important thermophysical property for experimentally researching the spray, atomization, and combustion processes of the mixed fuel in a combustion engine, and it is also a key parameter for the numerical simulation of the combustion process (Wang et al., 2004). Therefore, the theoretical prediction and experimental research of mass diffusion coefficients of fuel additives are of great importance. A literature review indicated that there is a dearth of accurate data on mass diffusion coefficients of oxygenated fuel additives in air. The motivation of this article is to conduct a detailed experimental investigation on the diffusivity of three kinds of oxygenated fuel additives in air.

By now, the experimental methods of measuring mass diffusion coefficients mainly include: diaphragm cells (George et al., 1988), Taylor's dispersion (Matsunaqa et al., 2002), pulsed-field gradient nuclear magnetic resonance (Botton et al., 2001), dynamic light scattering (Hushcha et al., 2000) and holographic interferometry (Baranski et al., 2003; Charlotte et al., 2000). Compared with other methods, digital real-time holographic interferometry is the newest method with the highest accuracy of measuring the mass diffusion coefficient and is widely used recent years (Kato et al., 1997; Reyes et al., 2002; Richter et al., 2003). In this work, an experimental system based on digital real-time holographic interferometry which was constructed by our group (He et al., 2008; Guo et al., 2008) was used to measure the mass diffusion coefficients of fluid. The mass diffusion coefficients of three fuel additives, diethyl 1,6-hexanedioate (Diethyl adipate, DEA), dimethyl carbonate (DMC), diethyl carbonate (DEC) in air were measured at different temperatures from 278.15 K to 338.15 K under atmospheric pressure.

2. EXPERIMENTAL SYSTEM AND MEASURING THEORY

Figure 1 shows the optical system of digital real-time holographic interferometry. All the optical parts are required to be installed on the optical shockproof apparatus to reduce the influence of the environment shark. From Figure 1, we can see that the laser beam was emitted from a He–Ne laser and was reflected by a mirror, and it then went into a spatial filter in which it was expanded. Then, the expanded laser beam was collimated by an achromatic doublets lens into a parallel laser beam. The parallel laser-beam was split into a reference beam and an object beam by a beam splitter. The object beam passed through the diffusion cell and recorded the diffusion information. Finally, both the object beam and the reference beam interfered at the beam splitter to form the interference fringe, which was collected by a CCD camera and recorded on a computer.

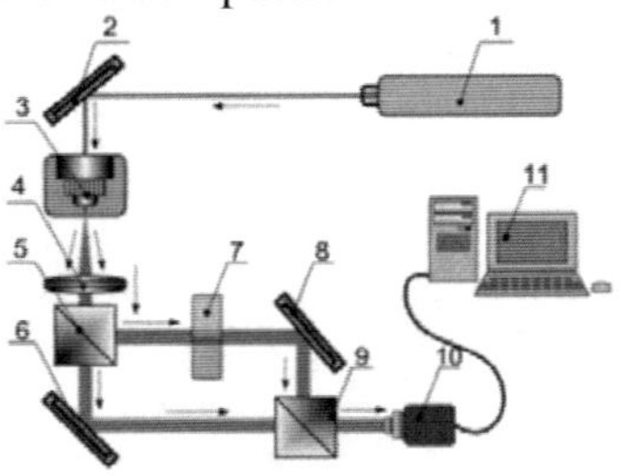

(1) He–Ne laser, (2) mirror, (3) spatial filter, (4) achromatic doublets lens, (5) beam splitter prism, (6) mirror, (7) diffusion cell, (8) mirror, (9) beam splitter prism, (10) CCD camera, (11) computer
Fig. 1 Optical system of digital real-time holographic interferometry

Figure 2 shows the diffusion cell used in this work, and it is an improved version of our previous apparatus reported before (He et al., 2008). In order to make sure the experiment goes on in a stable temperature environment, the diffusion cell is placed in the thermostatic water bath whose temperature could be controlled form (0 to 80) ℃ with an uncertainty of ± 0.2 ℃.

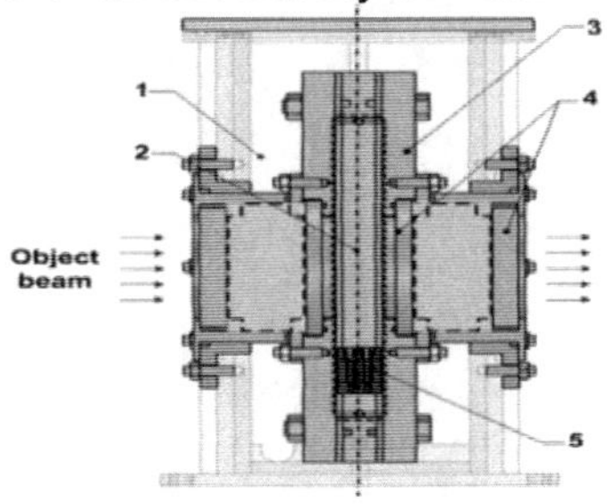

(1) thermostatic water bath, (2) diffusion zone, (3) diffusion cell, (4) optical glass, (5) honeycomb
Fig. 2 Diffusion system

In order to calculate the mass diffusion coefficient from the interference fringe image, the primary goal is to establish the relationship of mass diffusion coefficient and the phase information of object wave. Along with the development of the mass diffusion in the experiment, the refractive index of the solution changes continuously and the phase of object wave which pass through the solution changes responsively, and it leads to the continuous change of interference fringe on the hologram. So if we extract the phase of object wave from the interference fringe image and use the relationship established above, we can finally calculate the mass diffusion coefficient.

For one-dimensional mass diffusion, the mass diffusion coefficient D_{12} can be expressed as equation (1), which was introduced by the second Fick's law (Welty et al., 2001)

$$(\frac{\partial c}{\partial t}) = D_{12}(\frac{\partial^2 c}{\partial z^2}) \tag{1}$$

where, D_{12} is the mass diffusion coefficient, z is the direction of diffusion, t is the elapsed time of diffusion, and c is the concentration of the liquid, which is a function of z and t. From the linear relationship between phase difference of object beam and solution's concentration, the final equation for calculating D_{12} can be deduced as (Crank, 1972)

$$D_{12} = \Delta z_m^2 \frac{t_1/t_2 - 1}{8t_1 \ln(t_1/t_2)} \tag{2}$$

where, t_1 and t_2 are the times of getting two holograms, Δz_m is the vertical distances between the two extreme points of concentration difference. From equation (2), we can see that as long as t_1, t_2 and Δz_m are obtained, we can calculate the mass diffusion coefficient. t_1 and t_2 may be read from the computer timer, and Δz_m may be extracted from the interference fringe image by image processing method (will be introduced in the part 3).

The mass diffusion coefficient of KCl in aqueous solution at a temperatures of 298.15 K and condensation of 0.33 mol·L^{-1} was measured to verify the accuracy and reliability of the system. When the experimental results are compared with literature values (Gosting et al., 1950), the absolute average of relative deviations is within 1.20 %.

3. INTERFERENCE FRINGE PROCESSING METHOD

The processing method of holographic interference fringe image used in this paper includes following three steps: (1) wavelet de-noising; (2) frequency-domain filter and reconstruction of object wave; (3) determination of wrapping phase difference and its unwrapping. After performing the above operations we can finally get the value of Δz_m, the process is shown in Figure 3.

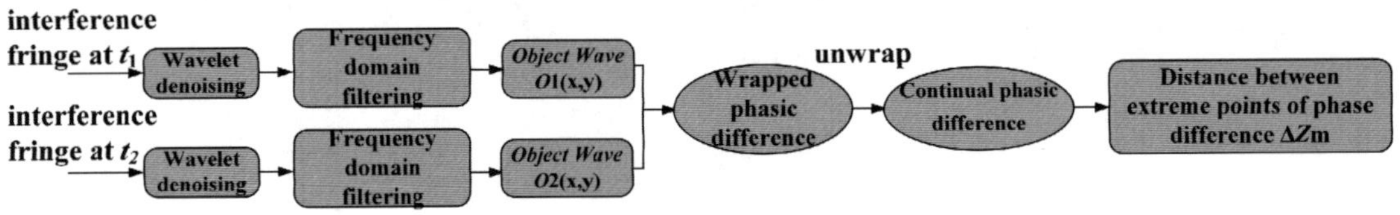

Fig. 3 Flowchart of interference fringe image processing

Step one: wavelet de-noising. In the process of gathering the interference fringe image, there are many factors that lead to the existence of errors on the interference fringe image such as the uncorrelated light, the dust in the path of the light, and the water fog on the glass wall of the diffusion cell, so the reduction of noise is necessary for the further processing of interference fringe image. In this paper, we de-noise the interference fringe image with the wavelet de-noising method: Perform three layers 'two-dimensional wavelet decomposition' to the interference fringe image with the 'db1' wavelet. The wavelet coefficient is calculated by NeighCoeff's algorithmic method with a 3×3 window (Cai et al., 2001). This method can reduce the noise amply and retain the useful information at the same time. The de-noising result is shown in Figure 4.

Original interference fringe image De-noised interference fringe image
Fig. 4 Original interference fringe image and de-noised interference fringe image

Step two: frequency-domain filter and reconstruction of object wave. After performing two-dimensional Fourier transform to the de-noised interference fringe image we obtained the frequency spectrum of the interference fringe image. Then filtered the frequency spectrum with a rectangular window to extract the first level frequency spectrum which contains the information of object wave, the inverse Fourier transform was performed to reconstruct the object wave in the spatial domain, as shown in Figure 5.

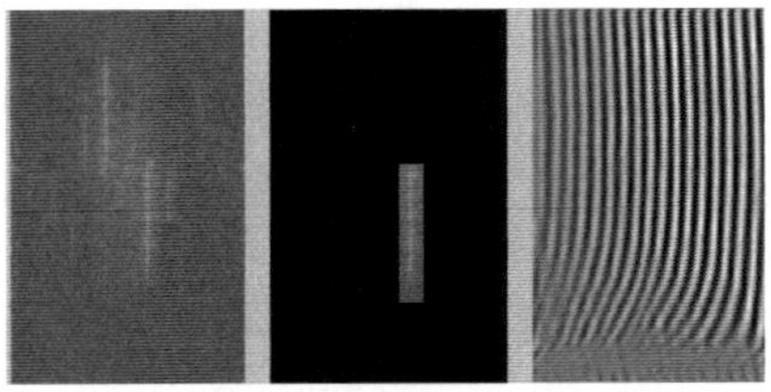

(a) (b) (c)
(a)- Frequency spectrum of interference fringe image
(b)- The first level frequency spectrum extracted by a rectangular window
(c)- Reconstruction of the object wave
Fig. 5 Frequency spectrum and reconstruction of object wave

Step three: determination of wrapping phase difference and its unwrapping. Performing division to the reconstructed object waves of different moments and we can get the wrapped phase difference, $\Delta\psi$, as shown in fig. 6-a. Adopting the least square phase unwrapping method to unwrap the wrapped phase difference, and we can get the continual phase difference, $\Delta\varphi$, as shown in fig. 6-b. Through calculating the mean value of each row vector of the continual phase difference matrix, we can obtain the line of continual phase difference, as shown in fig. 6-c. In the Figure, vertical coordinate represents for the number of pixel by which we can determinate the value of Δz_m. For the gas-liquid mass diffusion system which is different from the liquid-liquid mass diffusion system (He et al., 2008), we can only obtain the interference fringe in the up half of the diffusion cell. There is only one extreme point on the line of the continual phase difference on the gas side, and the distance between this point and the gas-liquid interface is half of Δz_m, so by doubling the distance, we can get Δz_m. Finally, putting Δz_m, t_1, and t_2 into equation (2), the mass diffusion coefficient D_{12} can be calculated.

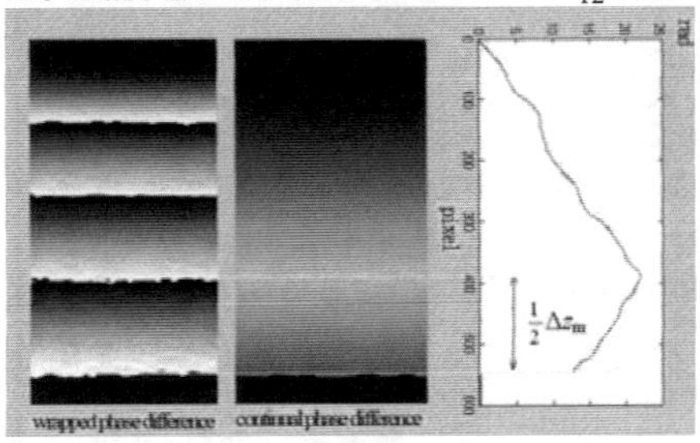

(a) (b) (c)
(a)-Distribution of wrapped phase difference $\Delta\psi(x,y)$
(b)-Distribution of continual phase difference $\Delta\varphi(x,y)$
(c)-Linear distribution of continual phase difference
Fig. 6 Phase difference of object waves

Journal of Physics: Conference Series **147** (2009) 012064 doi:10.1088/1742-6596/147/1/012064

4. EXPERIMENTAL RESULTS AND DISCUSSION

4.1 Experimental Results

The physical properties of the three oxygenated fuel additives, diethyl 1,6-hexanedioate (diethyl adipate, DEA), dimethyl carbonate (DMC), diethyl carbonate (DEC) are listed in Table 1.

Table 1. Physical Properties of the Oxygenated Fuel Additives

	DEA	DMC	DEC
CAS no.	141-28-6	616-38-6	105-58-8
molecular formula	$C_{10}H_{12}O_4$	$C_3H_6O_3$	$C_5H_{10}O_3$
molecular weight	202.25	90.07	118.13
relative density	1.026	1.0694	0.975
mass purity	≥99.5%	≥99.5%	≥99.7%

The measured mass diffusion coefficients of diethyl adipate (DEA) in air at T = (278.15 to 338.15) K under atmospheric pressure are listed in Table 2 and Figure 7.

Table 2. Experimental Mass Diffusion Coefficients $D_{12}/(\mathrm{cm^2 \cdot s^{-1}})$ of DEA in Air from T = 278.15 K to 338.15 K

T/K	$10^5 D_{12}/\mathrm{cm^2 \cdot s^{-1}}$	T/K	$10^5 D_{12}/\mathrm{cm^2 \cdot s^{-1}}$	T/K	$10^5 D_{12}/\mathrm{cm^2 \cdot s^{-1}}$
278.15	4.310	299.15	4.766	320.15	5.699
281.15	4.336	302.15	4.876	323.15	5.872
284.15	4.391	305.15	4.976	326.15	6.064
287.15	4.436	308.15	5.122	329.15	6.274
290.15	4.518	311.15	5.259	332.15	6.467
293.15	4.591	314.15	5.396	335.15	6.677
296.15	4.675	317.15	5.562	338.15	6.878

Uncertainties: D_{12}, ± 0.2 %; T, ± 0.16 K

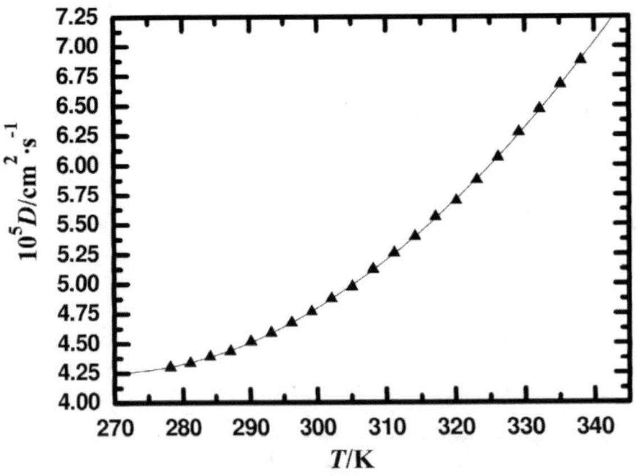

Fig. 7 Mass diffusion coefficients of DEA in air against temperature. ▲ , experimental data; —, polynomial fit of experimental data

The polynomial of the mass diffusion coefficients of DEA in air calculation was fit by the experimental data as follows

$$D_{12}/\mathrm{cm^2 s^{-1}} = 4.231\times10^{-4} - 2.846\times10^{-6}(T/\mathrm{K}) + 5.319\times10^{-9}(T/\mathrm{K})^2 \quad (3)$$

The measured mass diffusion coefficients of dimethyl carbonate (DMC) in air at T = (278.15 to 338.15) K under atmospheric pressure are listed in Table 3 and Figure 8.

Table 3. Experimental Mass Diffusion Coefficients $D_{12}/(\mathrm{cm^2 \cdot s^{-1}})$ of DMC in Air from T = 278.15 K to 338.15 K

T/K	$10^5 D_{12}/\mathrm{cm^2 \cdot s^{-1}}$	T/K	$10^5 D_{12}/\mathrm{cm^2 \cdot s^{-1}}$	T/K	$10^5 D_{12}/\mathrm{cm^2 \cdot s^{-1}}$
278.15	8.257	299.15	8.802	320.15	9.929
281.15	8.301	302.15	8.916	323.15	10.14
284.15	8.340	305.15	9.074	326.15	10.32
287.15	8.419	308.15	9.211	329.15	10.57
290.15	8.503	311.15	9.379	332.15	10.79
293.15	8.598	314.15	9.548	335.15	11.03
296.15	8.693	317.15	9.727	338.15	11.33

Uncertainties: D_{12}, ± 0.2 %; T, ± 0.16 K

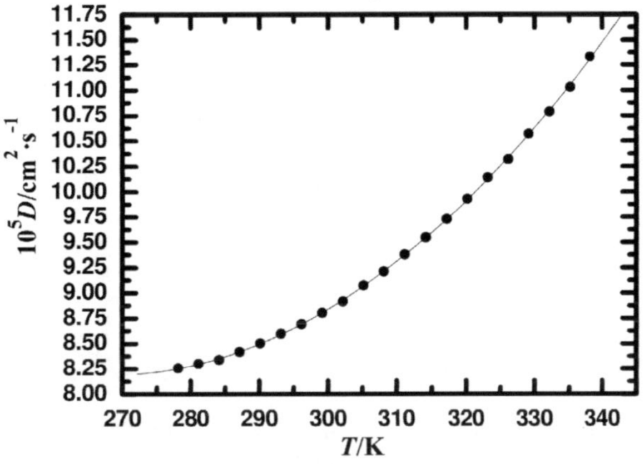

Fig. 8 Mass diffusion coefficients of DMC in air against temperature.
● , experimental data; —, polynomial fit of experimental data

The polynomial of the mass diffusion coefficients of DMC in air calculation was fit by the experimental data as follows

$$D_{12}/\mathrm{cm^2 s^{-1}} = 5.318\times10^{-4} - 3.364\times10^{-6}(T/\mathrm{K}) + 6.286\times10^{-9}(T/\mathrm{K})^2 \quad (4)$$

The measured mass diffusion coefficients of diethyl carbonate (DEC) in air at T = (278.15 to 338.15) K under atmospheric pressure are listed in Table 4 and Figure 9.

Table 4. Experimental Mass Diffusion Coefficients $D_{12}/(\mathrm{cm^2 \cdot s^{-1}})$ of DEC in Air from T = 278.15 K to 338.15 K

T/K	$10^5 D_{12}/\mathrm{cm^2 \cdot s^{-1}}$	T/K	$10^5 D_{12}/\mathrm{cm^2 \cdot s^{-1}}$	T/K	$10^5 D_{12}/\mathrm{cm^2 \cdot s^{-1}}$
278.15	6.791	299.15	7.492	320.15	8.467
281.15	6.852	302.15	7.612	323.15	8.634
284.15	6.951	305.15	7.738	326.15	8.811
287.15	7.029	308.15	7.885	329.15	8.968
290.15	7.147	311.15	8.013	332.15	9.157
293.15	7.247	314.15	8.163	335.15	9.314
296.15	7.394	317.15	8.302	338.15	9.558

Uncertainties: D_{12}, ± 0.2 %; T, ± 0.16 K

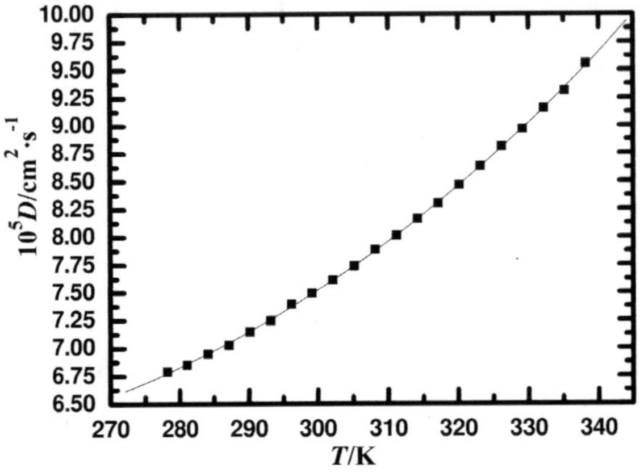

Fig. 9 Mass diffusion coefficients of DEC in air against temperature.
■ , experimental data; — , polynomial fit of experimental data

The polynomial of the mass diffusion coefficients of DEC in air calculation was fit by the experimental data as follows

$$D_{12}/\mathrm{cm}^2\mathrm{s}^{-1} = 2.327\times10^{-4}-1.458\times10^{-6}(T/\mathrm{K})+3.111\times10^{-9}(T/\mathrm{K})^2 \quad (5)$$

From Table 2, 3, 4 and Figure 7, 8, 9, we can see that within the same temperature range the mass diffusion coefficients of the above three fuel additives are about in the same order of magnitude, and the specific magnitude sequence is D_{12}(DMC)>D_{12}(DEC)>D_{12}(DEA). This difference is mainly determined by the molecular size of these three fuel additives. By Stokes-Einstein formula, $D_{12} = k_B T / (6\pi R_0 \eta)$ (Cussler, 2000), we can see that under the same thermophysical conditions, the smaller the diffusion molecular radius is, the bigger its corresponding mass diffusion coefficient is. With the other correlation equations we can also get the inversely-proportional relationship between mass diffusion coefficient and molecular volume. Figure 10 shows the molecular structure of these three fuel additives approximately, we can see that the sequence of molecular size is in accordance with the mass diffusion coefficient.

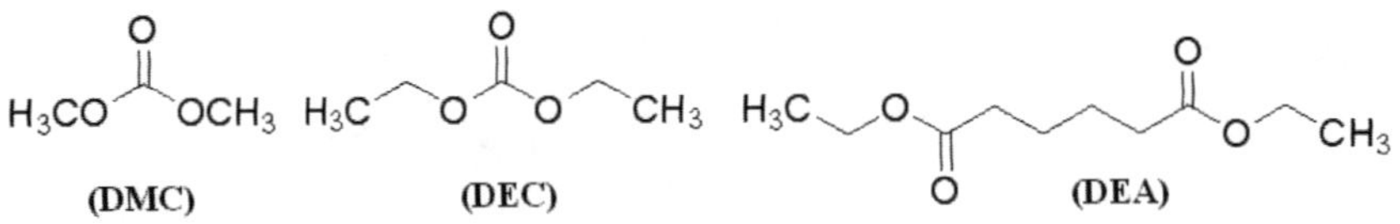

Fig. 10 Molecular structural formula of DMC, DEC and DEA

4.2 Experimental Uncertainty Analysis

The mass diffusion coefficient is described by temperature, the moment interference fringe was received, and the distance between two extreme points of concentration change. To determine the uncertainty in the measurement, we use the following uncertainty equation (Geneve, 1995).

$$U = k\cdot u_c = k\sqrt{\sum (u_i)^2} \quad (6)$$

where, subscript i is the influencing factor of mass diffusion coefficient. u_i is the uncertainty for each variable. u_c is the compound uncertainty composed by each variable, k is the confidence coefficient and it is usually taken to be 2 or 3. When k = 2, the degree of confidence is 95 %; When k = 3, the degree of confidence is 99 %. In this study, the confidence coefficient of the compound uncertainty is taken to be 2. The result shows that the uncertainties in temperature and mass diffusion coefficient in this work are estimated to be no greater than ±0.16 K and ±0.2 % (See Table 5), respectively.

Table 5. Experimental Uncertainty of Temperature and Mass Diffusion Coefficient

Temperature		
	platinum resistance thermometer, u_1	± 0.01 K
	data collection and process detector equipment, u_2	± 0.025 K
	temperature control system, u_3	± 0.05 K
	temperature stability of constant temperature cabinet, u_4	± 0.05 K
	combined standard uncertainty, u_c	± 0.08 K
Mass Diffusion Coefficient		
	time of getting hologram, u_1	± 2.22 ppm
	distance between two extreme points of concentration, u_2	± 0.1 %
	combined standard uncertainty, u_c	± 0.1 %

5. CONCLUSIONS

In the present work, an experimental system based on the theory of digital real-time holographic interferometry was introduced. With the experimental system, the mass diffusion coefficients of oxygenated fuel additives diethyl 1,6-hexanedioate (Diethyl adipate, DEA), dimethyl carbonate (DMC), diethyl carbonate (DEC) in air were measured at 21 different temperatures from 278.15 K to 338.15 K under atmospheric pressure. The values of these three mass diffusion coefficients were compared with each other and the differences were explained on the theory of molecular structure. The experimental uncertainties in temperature and mass diffusion coefficient are estimated to be no greater than ± 0.16 K and ± 0.2 %, respectively. Finally the polynomial was fitted by the experimental data which is convenient for engineering application.

ACKNOWLEDGEMENTS

We gratefully acknowledge the National Nature Science Fund Committee (NSFC nos. 50676069 and 50836004) for financial support of this work.

NOMENCLATURE

D_{12}	mass diffusion coefficient	[cm^2/s]
c	concentration of liquid	[mol/L]
t	moment of getting interference fringe hologram	[s]
z	the direction of one dimensional mass diffusion	[mm]
k_B	Boltzmann constant	[J/K]
R_0	the radius of diffusion molecule	[m]
η	the viscosity of diffusion substance	[Pa·s]
u	the uncertainty in the experimental measurement	
U	the expanded uncertainty in the experimental measurement	
k	the confidence coefficient of uncertainty	

Greek Letters

$\Delta\psi$	wrapped phase difference	[rad]
$\Delta\varphi$	continual phase difference	[rad]

Subscripts

i	influencing factor of mass diffusion coefficient

REFERENCES

Alam, M. et al. (2006). "Spray and combustion visualization of a direct-injection diesel engine operated with oxygenated fuel blends," Int. J. Energ. Res., 7(6), pp. 503-521.

Alfredo, A. et al. (2003). "Biodegradation of Potential Diesel Oxygenate Additives dibutyl maleate (DMB), and tripropylene glycolmethyl ether (TGME)," J. Chemosphere, 52, pp. 861-868.

Baranski, J. et al. (2003). "Determination of Binary Diffusion Coefficients of Gases Using Holographic Interferometry in a Loschmidt Cell," Int. J. Thermophys., 24(5), pp. 1207-1220.

Botton, V. et al. (2001). "Measurement of solute diffusivities. Part II .Experimental measurements in a convection- controlled shear cell Interest of a uniform magnetic field," Int. J. Heat. Mass Tran., 44(17), pp. 3345-3357.

Cai, T. T. et al. (2001). "Incorporating information on neighbouring coefficients into wavelet estimation," Indian Journal of Statistics B, 63(2), pp. 127-145.

Charlotte, M. et al. (2000). "Diffusion of lysozyme in gels and liquids A general approach for the determination of diffusion coefficients using holographic laser interferometry," J. Chromatogr. B, 743, pp.151-167.

Crank, J. (1972). The Mathematics of Diffusion, Oxford: Oxford University Press, pp. 92-114.

Cussler, E. L. (2000). Diffusion, Mass Transfer in Fluid Systems (2nd Edition), Cambridge University Press. pp. 70-74.

Geneve. (1995). Guide to the Expression of Uncertainty in Measurement, International Organization for Standardization: Switzerland.

George, M. et al. (1988). "In Vitro Measurement of Gastrointestinal Tissue Permeability Using a New Diffusion Cell," Pharmaceutical Research, 5(6), pp.372-376

Gosting, L. J. (1950). "A study of the diffusion of potassium chloride in water at 25℃ with the gouy interference method," J. Am. Chem. Soc., 72(10), pp. 4418-4422.

Guo, Y. et al. (2008). "Mass Diffusion Coefficients of Dimethyl Carbonate in Heptane and in Air at T = (278.15 to 338.15) K," J. Chem. Eng. Data., 53(12), pp.2861-2864.

He, M. G. et al. (2008). "Measurement of mass diffusion coefficients of saccharose solution and dimethyl ether in air using digital image holographic interferometry system," High Temp-High Press, 37(1), pp. 61-70.

He, M. G. et al. (2008). "A New Method of Processing Mach Zehnder Interference Fringe Gathered in Procedure of Diffusion," submitted to International Journal of Thermophysics, under review.

Hushcha, T. O. et al. (2000). "Conformation Changes of Albumin in its Interaction with Physiologically Active Compounds as Studied by Quasi-elastic Light Scattering Spectroscopy and Ultrasonic Method," Talanta, 53(1), pp. 29-34.

Kato, N. et al. (1997). "Numerical Simulation and Laser Holographic Study on Thermal Diffusion in Counter Flow with Different Temperatures," Energy Conversion and Management, 38(10-13), pp. 1197-1207.

Matsunaqa, N. et al. (2002). "Measurement of mutual diffusion coefficients of gases by the Taylor method: Measurements on H2-Air, H_2-N_2, and H_2-O_2 systems," Heat Transfer Asian Research, 31(3), pp. 182-193.

Ogawa, H. et al. (2001). "Approaches for Ultra Clean Diesel Combustion with Newly Developed Fuels." Proceedings of the International Symposium on Clean and High Efficiency Combustion in Engines, Tianjin, China, 18-24; Vol. A, pp. 22-35.

Poulopoulos, S. G. et al. (2001). "Regulated and Unregulated Emissions from an Internal Combustion Engine Operating on Ethanol-Containing Fuels," J. Atmospheric Environment, 35, pp. 4399-4406.

Reyes, L. et al. (2002). "Influence of Conformational Changes on Diffusion Properties of Bovine Serum Albumin: A Holographic Interferometry Study," Colloids and Surfaces B: Biointerfaces, 25(1), pp. 99-108.

Richter, J. et al. (2003). "Digital Image Holography for diffusion measurements in Molten Salts and Ionic Liquids-method and first results," J. Mol. Liq., 103-104, pp. 359-370.

Shing, T. L. et al. (2002). "Applicability of Gasoline Containing Ethanol as Thailand's Alternative Fuel to Curb Toxic VOC Pollutants from Automobile Emission," J. Atmospheric Environment, 36, pp. 3495-3503.

Wang, X. B. et al. (2004). "Estimation of DME Thermophysical Properties by Molecule Theory," Transactions of CSICE, 22(6), pp. 486-492.

Welty, J. et al. (2001). Fundamental of Momentum, Heat, and Mass Transfer (4th Ed), John Wiley & Sons Inc.: Singapore.

Yoshimoto, Y. et al. (1999). "Reduction of NOx and smoke, and BSFC in a diesel engine fuelled by biodiesel emulsion with used frying oil," Society of Automotive Engineers, 108(4), pp. 1913-1920.

Journal of Physics: Conference Series **147** (2009) 012065 doi:10.1088/1742-6596/147/1/012065

A Video Probe Measurement System for Coarse Water Droplets in LP Steam Turbine

Xueliang Fan, Zhihai Jia, Jingjing Zhang, Xiaoshu Cai

University of Shanghai for Science & Technology, Shanghai 200093, China
xlfanustc@gmail.com

Abstract. A video-probe measurement system has been developed at IPTFM. Designing process and test results of the system are described in detail in this paper. The measurement system mainly consists of camera module (XC-HR70), frame grabber (Matrox Meteor-II /Multi-Channel), function generator and image acquisition and processing interface. Static and dynamic experiments on water droplets were carried out on the spraying platform. During the static calibration testing, standard glass beads were used to obtain the parameters. And test results showed that this method has good repetition and credibility. During the dynamic testing, spraying experiments were carried out. The size and velocity of spray droplets were measured, which showed good agreement with the known empirical value. Both dynamic and static testing results show the measurement system is capable of online measuring the coarse water droplets in low-pressure steam turbine and obtain the size and velocity of water droplets from images snapped by the system.

1. INTRODUCTION

For the large power plant, the inlet steam of turbine is superheated, and mixed by fine and coarse water droplets in the LP cylinders and condensers, which will decrease thermal efficiency, cause dangerous corrosion of turbine blades and influence the operation safety. Researches on the water droplets measurement could enhance the basis investigation of the blade water corrosion model, as well as predicted the corrosion areas and provided information for precautionary measures.

For the fine droplet, the global method, known as the extinction method has been chosen to obtain its size distribution and concentration (wetness). Several optical probes based on this measurement principle have been developed(Cai Xiaoshu, et al., 2001; Ross, et al.,1994). Up to now, however, there is no universal measuring method available for CWD (coarse water droplet). EDF in France has developed a Holography probe and a Microvideo probe to measure the size and velocity of CWD (Kleitz A, et al., 1991). In Germany and Japan, the side scattering method is employed to obtain the size of CWD(Bohn D, 1997; Tanuma T, 1991). "Catchpot" probes were developed in Britain and Czech to measure the amount of CWD(Williams et al., 1976; Petr et al. 1995). However, it is still not very clear about the properties of CWD as well as its proportion in the total wetness. In order to better understanding the mechanism of erosion of blades and to determine the total wetness accurately, it is necessary to measure size, velocity, concentration and amount of CWD.

Digital imaging is becoming increasingly attractive because of the availability of low-cost high performance imaging devices and computing hardware. An advantage of using the imaging approach is that non-intrusive real-time measurement of many individual particles can be measured concurrently and continuously.

For measuring the CWD in steam turbine, A video-probe measurement system has been developed at IPTFM. Designing process and test results of the system are described in this paper.

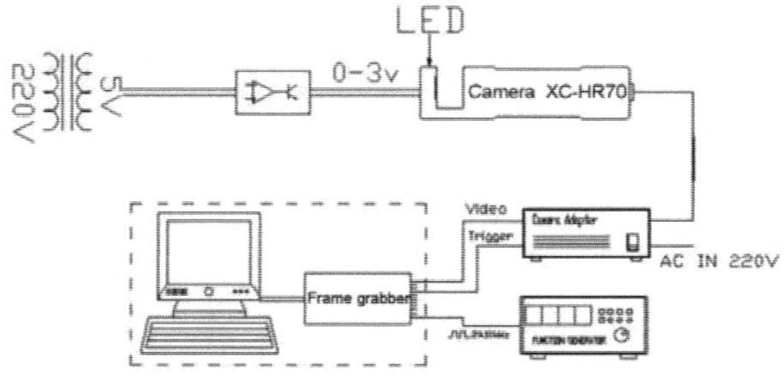

Fig.1 Schematic diagram of the video probe system

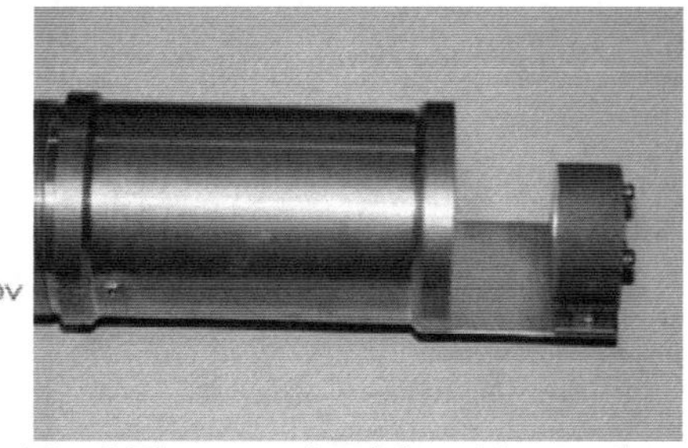
Fig.2 Photograph of the video probe

2. VIDEO PROBE MEASUREMENT SYSTEM COMPONENTS

The measurement system mainly consists of camera module (XC-HR70), frame grabber (Matrox Meteor-II /Multi-Channel), function generator and image acquisition and processing interface (shown in Fig.1 and Fig.2). XC-HR70 is an ultra-compact monochrome camera module ideal for high-resolution image capturing applications. A 1/3 type progressive scan CCD incorporated in the XC-HR70 allows the output of XGA resolution (1024 x 768) images at a rate of 30 frames/s. By inputting an external trigger signal, one can capture the information of single screens at arbitrary timing. In this system, the minimum exposure time can be set as short as 10μs. That is very important for capturing the fasting moving CWD.

3. SYSTEM CALIBRATION

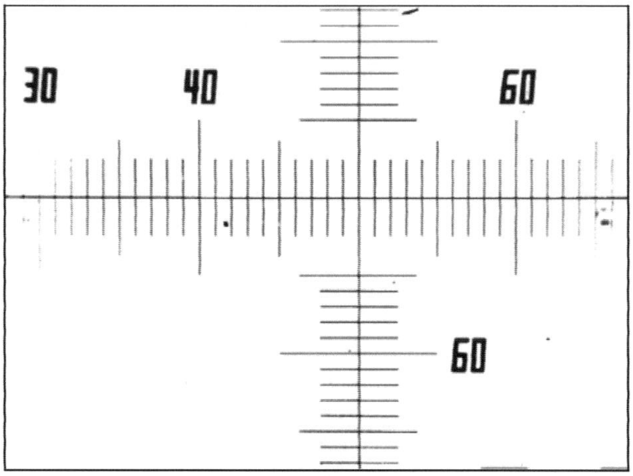

Fig. 3 Reticule image

System calibration is very straightforward and is realised off-line by the use of a low-cost precision lined reticule whose smallest division is 100μm. During the calibration, the video probe is set on the optical rail. The lined reticule is put in front of the camera and the reticule is perpendicular to the camera. Move the reticule forward and backward and get the corresponding reticule images. When the reticule image is sharpest (as shown in Fig. 3), the distance between the reticule and the lens is defined as the standard working distance. In this system, the standard working distance is 11mm. From Fig. 3, it can be found the field of view is 3.9mm by 2.95mm at the standard working distance. With the fact that the image size is 1024 x 768 pixels, it can be noted that one pixel corresponds to 3.8μm (scaling

Journal of Physics: Conference Series **147** (2009) 012065 doi:10.1088/1742-6596/147/1/012065

factor). The calibration is automatically accomplished in the software and the scaling factor obtained will be used in the process of measuring the size and velocity of droplets.

4. STATIC TESTING

After the system calibration, the static testing is carried out. Three kinds of standard particles whose mean diameter are 39.0±2.1μm (particle A), 56.4±2.2μm (particle B), and 77.2±4.2μm (particle C) respectively are chosen for the static testing. The particle images and the measurement results are shown in Fig.4-Fig.6. The statistics samples of three kinds of particles are 526, 574 and 223 respectively, and the corresponding measurements of mean diameter are 36.78μm, 55.26μm and 72.43μm respectively. So the measurement errors are -5.69%, -2.02% and -6.18% respectively. The measurement results agree well with the known values.

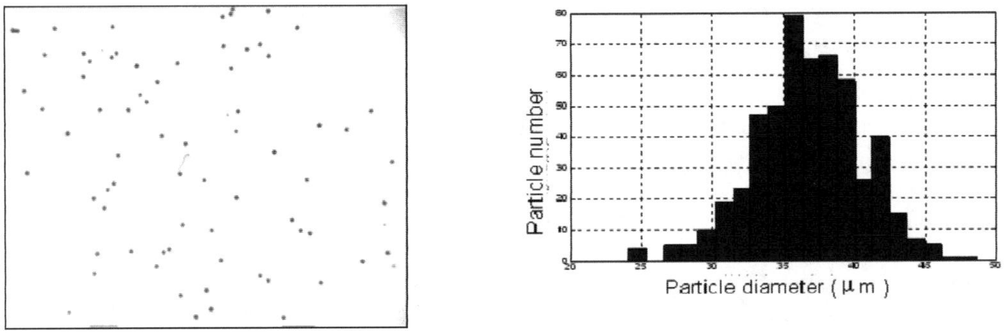

Fig.4 The image and the particle diameter distribution of particle A

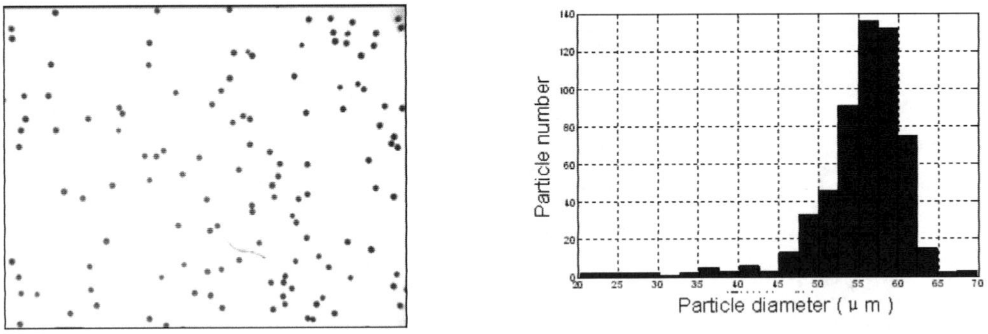

Fig.5 The image and the particle diameter distribution of particle B

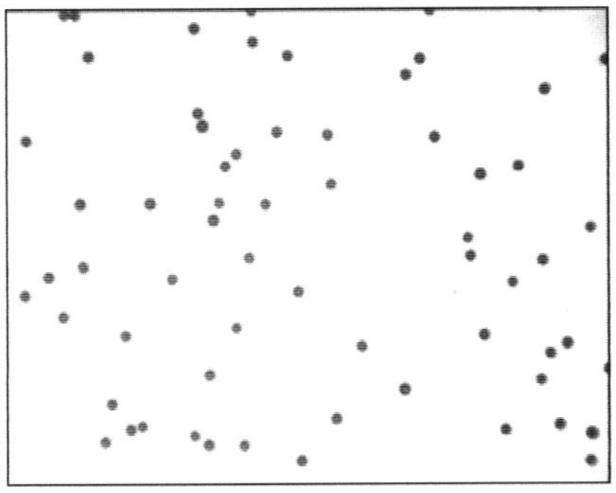

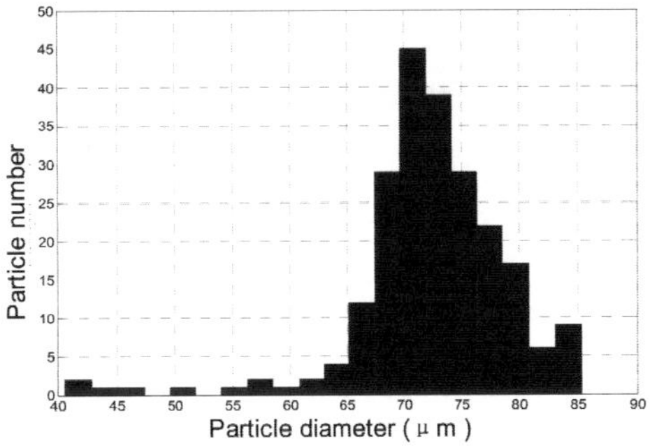

Fig.6 The image and the particle diameter distribution of particle C

4. DYNAMIC TESTING

4.1 Spray Testing Set-up

The spray testing system is composed mainly of video probe, nozzle, triplex plunger pump, and water tank (as shown in Fig.7). The water from the tank is thus pressurized as it goes through the plunger pump and atomized at the nozzle exit. The spray droplets size relates to flow rate, and water pressure. The larger water pressure yields smaller droplet and vice versa. The type of spray nozzle used in testing is Danfoss 45°S, and its design flow rate is 2.75gal/h. The flow pressure is 0.6 MPa.

4.2 Size Measurement of Spray Droplets

Figure 8 shows the measurement results and Figure 9 shows the images of droplets. The total number of water droplets is 971, and the mean diameter is 73.98μm. The diameter of water droplets range from 10μm to 400μm, but mainly distributed from 10μm to 50μm.

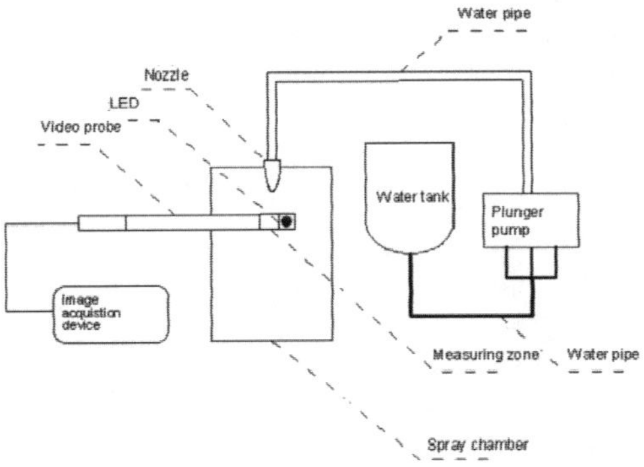

Fig.7 Schematic diagram of the spray testing system

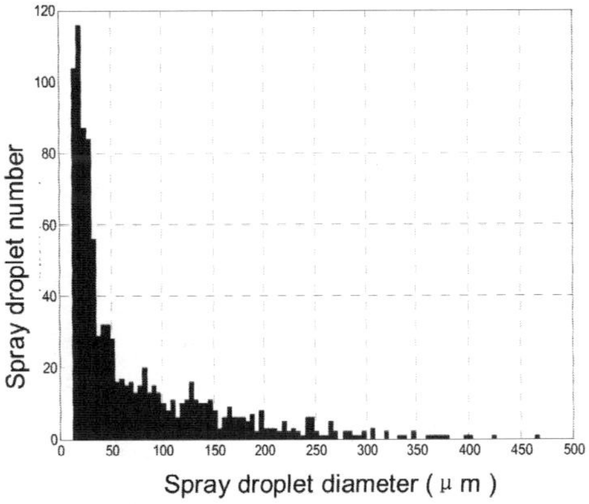

Fig.8 Particle diameter distribution of spray droplets

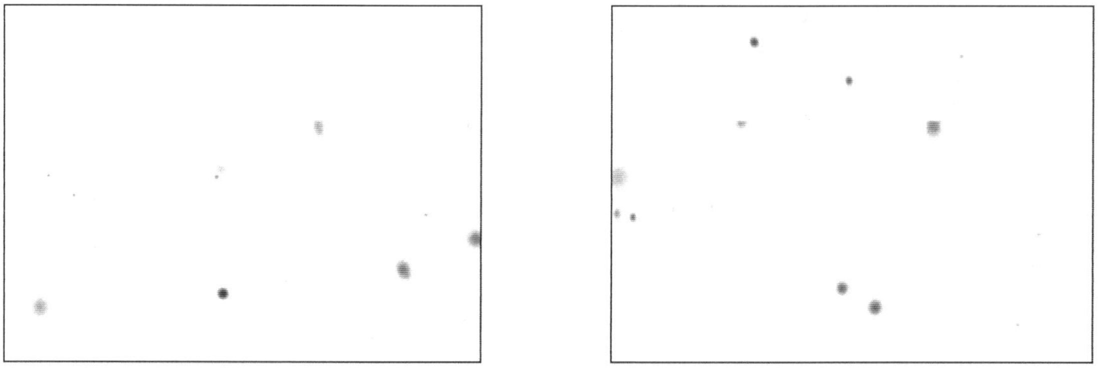

Fig. 9 Spray droplets images

4.3 Velocity Measurement of Spray Droplets Based on a Single Motion Blurred Image

When the exposure time of the camera becomes longer or the velocity of water droplet becomes faster, it is difficult to get the sharp image of droplets. The image becomes blurred and it shows the trajectory of the water travelling in the exposure time (as shown in Fig. 10).

Fig.10 Motion blurred image of fast moving spray droplet

Assuming the length of trajectory of the spray droplet is L, the diameter of the spray droplet is d, then the displacement of the spray droplet during the exposure time t can be expressed as $S = L - d$ and the velocity of water droplet can be given as

$$v = \frac{S}{t} = \frac{L-d}{t}$$

For the droplet shown in Fig.10, $t = 50\mu s$, $L = 95 \times 3.8 = 361\mu m$, $d = 35 \times 3.8 = 133\mu m$, so the velocity is

$$v = \frac{(361-133)\mu m}{50\mu s} = 4.56 m/s$$

Fig. 11 Motion blurred image of spray droplets when the flow pressure is 0.6 MPa.

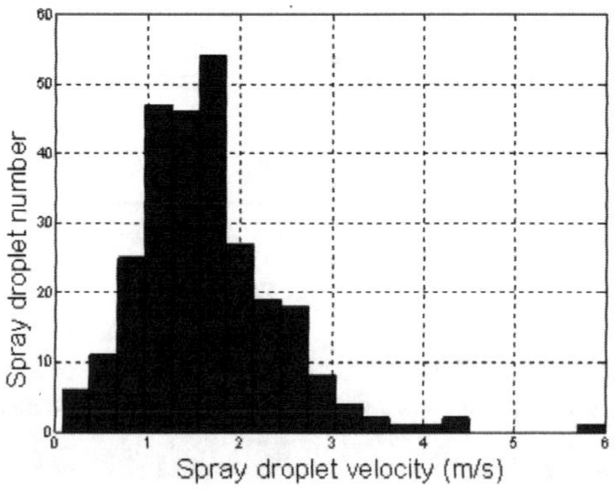

Fig. 12 Velocity distribution of spray droplets when the flow pressure is 0.6 MPa.

Fig. 11 shows the motion blurred image of spray droplets when the flow pressure is 0.6 MPa. By the analysis of the Fig.11, it can be determined that the total number of spray droplets is 272 and their average velocity 1.64m/s and their average diameter 67.87μm. The velocity distribution of spray droplets is shown in Fig.12. It can be seen from Fig.12 that the velocity of spray droplets mainly distributed in the range of 0.5m/s~3m/s.

5. CONCLUSIONS

In this paper, the designing and testing of a video-probe measurement system for coarse water droplets are described in detail. Experiments on static and dynamic water droplet were carried out on the spraying platform. During the static calibration tests, standard glass beads were used to obtain the parameters. And testing results show that this method has good repetition and credibility. During the dynamic tests, spraying experiments were carried out. The size and velocity of spray droplets were measured, which show good agreement with the known empirical value. Both dynamic and static test results show the measurement system is capable of online measurement of the coarse water droplets in low-pressure steam turbine and obtain the droplet size parameters from pictures snapped by the system.

ACKNOWLEDGEMENTS

The authors like to Shanghai special fund for choosing and cultivating young scholars (slg-07003).

REFERENCES

Bohn D, Holzenthal K. (1997). "Measurement of the Wetness Fraction and Droplet Size Distribution in a Condensation Turbine," Aachen: Aachen University of Technology.

Cai Xiaoshu, et al. (2001). "A Novel Integrated Probe System For Measuring The Two Phase Wet Steam Flow In Steam Turbine," Journal Of Engineering Thermophysics, 22(6),pp.52-27.

Kleitz A, et al. (1991). "Water droplet sizing in LP and HP wet steam turbines," Fluid Machinery Forum, New York, ASME, FFD-119.

Petr V, Stastny M, Turyna R.(1995). "International Collaboration on Turbine Steam Chemistry and Corrosion in the Phase Transition Zone," AP-108184, EPRI.

Ross M M, Spratling M W,Kirkland C B,et al. (1994). "Measurement of microfog wetness in a model steam turbine using a miniature optical spectral extinction probe," Proceedings of the Conference on Optical Methods and Data Processing in Heat and Fluid Flow, London .

Tanuma T, Sakamoto T. (1991). "The Removal of Water from Steam Turbine Stationary Blades By Suction Slots, " C432/022 IMechE.

Williams G J, Lord M J. (1976). "Measurements of coarse water distribution in the LP cylinders of operating steam turbines, " Proc. Inst. Mech. Engnrs., 190(4), pp.59-69.

Journal of Physics: Conference Series **147** (2009) 012066 doi:10.1088/1742-6596/147/1/012066

Ignition characteristic of blended coal in a drop-tube furnace

Tianyang Chi[1], Hongjian Zhang[1], Hongliang Zhou[1], Rui Kan[1]

[1]State Key Laboratory of Industrial Control Technology, Zhejiang University
Hangzhou 310027 ,China

hjzhang@ iipc.zju.edu.cn

Abstract. The ignition of pulverized coal is an important preliminary step in the coal combustion process, due to its influence on flame stability, the formation and emission of pollutants and flame extinction. In recent years there has been an increasing utilization of coal blends for combustion, but information on the changes of coal flame during ignition process is scarce. In this work the ignition behavior of a series of coal blends, made up from four coals of different rank was studied in a drop-tube furnace. A new method of oscillation frequency based on empirical mode decomposition is introduced. The Ignition feature of oscillation frequency was obtained from the coals studied; the feature proved well in distinguishing ignition mechanisms from homogeneous, hetero-homogeneous to heterogeneous in the ignition section of a drop-tube furnace. The different ignition mechanisms of coal blends were observed depending on the blends mass proportion and inclined to the characteristic of the coal consisted of higher volatile.

1. INTRODUCTION

The coal blending from multiple sources for power generation is a global practice in the commercial utilities. This is due to strategic, environmental reasons or in order to improve the combustion properties of the blended coal. The ignition of coal particles is an important preliminary step in the coal combustion process. The reactivity and ignition behaviors of coal particles are of considerable importance for designing the boiler and controlling the combustion process (Faundez el al., 2005; Su et al., 2001).

It is now generally accepted that coal particle ignition is a multistage process. The ignition of coal particles may take place in homogeneously, heterogeneously or hetero-homogeneously. In the homogeneous ignition, the initial step involves pyrolysis, subsequent ignition of volatiles which react in the gaseous phase surrounding the coal particles and followed by ignition of the char. The heterogeneous coal ignition involves the direct attack of oxygen on the whole coal particles. The hetero-homogeneous ignition is the simultaneous ignition of the volatile matter and the coal particle surface, which is the intersection of homogeneous and heterogeneous ignitions (Essenhigh et al., 1989; Zhang et al., 1994).

TGA techniques have been widely employed to determine the ignition behavior of coal, they can operate at very different conditions to those encountered in a pulverized coal combustor (Arenillas et al., 2004; Essenhigh et al., 1989;). Drop-tube furnaces (DTF) more closely simulate combustion conditions in industrial pulverized-coal combustors than thermogravimetric analyzers.

Journal of Physics: Conference Series **147** (2009) 012066 doi:10.1088/1742-6596/147/1/012066

This paper reports a novel method on judging the ignition behavior of coal blends at the ignition section in a drop-tube furnace. The characteristic parameter based on empirical mode decomposition of the flame signal is determined from a flame monitoring system. The ignition behavior of coal blends can be obtained from the characteristic parameter.

2. Methodology

2.1 Drop-dupe furnace and test conditions

A drop-tube furnace (DTF) is designed for laboratory experiments under similar conditions in temperature and residence time as those of industrial boilers and power plants. A schematic diagram of the drop-tube furnace utilized in the present study is depicted in Fig. 1. A heating rate of 10^4-10^5℃s^{-1} and maximum temperature of 1500℃ can be achieved in the DTF. In addition, the particles are in a dynamic, dilute phase, which allows for individual and cloud particle combustion.

The ignition section and burnout section are the main parts of the furnace. The auxiliary equipment consists of an air compressor, air preheater, flow metering device, pulverized fuel feeder, etc. The DTF is heated by silica-carbon rods which are perpendicular to the furnace axis. There are 15 quartz optical windows with a diameter of 7mm along the axial direction of the ignition section which is used to observe the ignition of pulverized coal. The data reported in this paper are obtained from 9 quartz optical windows. The ignition and burnout sections can be heated to maximum temperatures of 1000℃ and 1500℃, respectively. The latter is normally used for burnout and slagging experiments. The flow rate of air is measured using a spinner flow meter and the air preheater heats secondary air before it goes into the DTF. Thermocouples are used to measure the wall temperature of the air preheater and the temperatures of ignition section, burnout section and secondary air. The experimental process is monitored and controlled through an on-line computer data acquisition system. A screw feeder is used to feed pulverized coal. By changing the rotation speed of the electromotor, the desired feeding rate can be obtained.

In this study, the temperatures of the ignition and burnout sections were kept constant at 850℃ and 1350℃ respectively in all the experiments. The size of coal particles is below 180 micron which is sized through a screen of 80 meshes. The feeding rate of the feeder was 1.2kg/h. In the present research four coals of different rank with a mass ration of 70% to 30%, 60% to 40%, 50% to 50%, 40% to 60% and 30% to 70% were tested. All types of coal were sources from coal fired power stations. Table 1 summarizes the proximate analysis, ultimate analysis and gross calorific value. It is clear that single coals A, B, C and D have the higher volatile content (>25%). The volatile content of coal blends which affect the ignition behavior very much are between 25% and 30%. The property of the coal blends approach to the main coal as its mass proportion increases. It is important to note that two coals with the similar proximate analysis may not have the same ignition and flame stability characteristics, because the ignition and flame stability depend on early heat release, not necessarily early volatile release(Su et al., 2001).

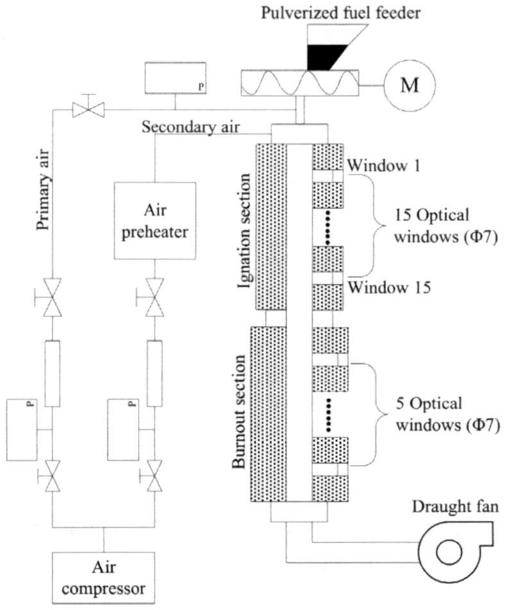

Fig.1 Experimental apparatus of the drop-tube furnace

Table 1. Properties of pulverized coals tested

Coal sample	Proximate analysis (wt% , ad)				Ultimate analysis (wt% , ad)					Gross calorific value (MJ/kg)
	M	A	V	FC	C	H	N	S	O	
A	7.24	12.13	30.57	50.06	60.61	4.45	0.98	0.47	14.12	25.72
A:B (70%:30%)	5.86	16.76	30.03	47.35	58.83	4.3	0.89	0.37	12.99	24.20
A:B (60%:40%)	5.38	18.10	29.44	47.08	57.86	4.38	0.92	0.39	12.97	24.82
A:B (50%:50%)	4.88	19.60	29.28	46.24	56.94	4.29	1.16	0.36	12.77	24.36
A:B (40%:60%)	4.54	19.91	29.26	46.29	56.72	4.26	0.88	0.44	13.25	24.47
A:B (30%:70%)	4.18	21.44	29.09	45.29	55.76	4.32	0.96	0.39	12.95	23.19
B	2.55	26.15	27.82	43.48	53.50	4.16	1.10	0.44	12.10	23.36
C	20.26	21.67	28.40	29.67	40.40	1.56	1.09	0.33	14.69	17.11
C:D (70%:30%)	11.09	28.15	28.21	32.55	42.74	1.77	1.05	0.44	14.76	18.24
C:D (60%:40%)	11.02	29.00	27.41	32.55	42.67	2.08	1.02	0.43	13.78	18.03
C:D (50%:50%)	9.68	30.28	27.51	32.53	42.32	1.78	1.01	0.46	14.47	18.13
C:D (40%:60%)	8.15	32.12	27.09	32.64	42.32	1.81	0.97	0.50	14.13	18.19
C:D (30%:70%)	6.53	34.03	28.83	30.61	42.21	1.63	0.94	0.53	14.13	18.83
D	5.49	36.21	26.27	32.03	42.25	2.21	0.81	0.56	12.47	18.16

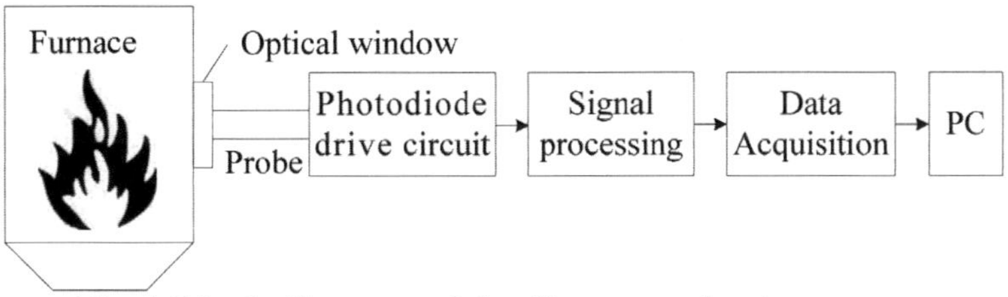

Fig.2 Block diagram of the flame monitoring system

2.2 Flame monitoring system

The system consists of a flame detecting probe, photodiode drive circuit, signal processing circuit, data acquisition unit and PC-based signal processing software. Fig.2 shows the structure and main components of the system. The flame detector has four photodiodes covering the visible and infrared bands. The detector produces four flame signals containing the dynamic characteristic information of the flame and hence the ignition characteristics of the fuel (Xu et al., 2004). In this study, only one flame signal which band is between 400nm and 1100nm is used to produce the characteristic parameters.

When monitoring, the flame monitoring probe whose diameter is the same as that of the optical window is pressed closely against the observation window of a DTF. The flame signal was sampled with a frequency of 1kHz and a data length of 20 seconds. At the whole experiment the combustion are stable on different observing window at different time due to an on-line furnace monitoring system.

2.3 Empirical mode decomposition

The Empirical mode decomposition (EMD) method which is proposed to analyze the complex signal (Huang et al., 1998) decomposes a time-series into a finite set of oscillatory functions called the intrinsic mode functions (IMF). An IMF function must satisfy two conditions: (1) the number of extreme and the number of zero crossings must either equal or different at most by one; (2) At any point, the mean value of the envelope defined by the local maxima and the envelope defined by the local minima is zero. Obviously, most original signals do not satisfy those requirements and need to be decomposed into some different IMF components. The process is called EMD. The procedure to decompose a signal $x(t)$ into intrinsic mode functions is as follows:

1. Identify all extreme of $x(t)$;
2. Interpolate between maxima to obtain an envelope $e_{\max}(t)$;
3. Interpolate between minima to obtain an envelope $e_{\min}(t)$;
4. Compute the running mean, i.e. $m(t)=[e_{\max}(t)+e_{\min}(t)]/2$;
5. Subtract $m(t)$ from $x(t)$ to extract the detail $h(t)=x(t)-m(t)$;
6. Examine if $h(t)$ is an IMF;
7. If not iterate on the detail $h(t)$; this is called the sifting process.

An IMF $C_1(t)$ is extracted once the detail signal $h_1(t)$ is zero-mean and the number of extreme with the number of zero crossings equal or differ at most by one. The first IMF is subtracted from the original signal $x(t)$; the difference called the first residue $r_1(t)$. This residue is treated as the new signal and the whole process is repeated. The sifting process ends when the last residue $r_N(t)$ is a constant or a monotonic function. The signal $x(t)$ is written as the sum:

$$x(t)=\sum_{j=1}^{N} C_j(t)+r_N(t) \quad (1)$$

Fig. 3 is the IMFs of typical flame signal. It can be seen that the coal flame signal can get N IMFs and a residue. C1~C10 are listed from high frequency to low frequency. At the same time IMF with high frequency and low amplitude have great influence on the result of oscillation frequency of a coal flame.

In the calculation of the oscillation frequency, the IMFs with high frequency (>50Hz) are taken as noise from the signal. For example, the coal flame signal in fig. 3, the frequency of C1~C3 are 377Hz, 192Hz and 82Hz which are not belong to the intrinsic fluctuation of coal flame, so they are cut from the original signal. The new signal is made up of the rest IMFs and residue.

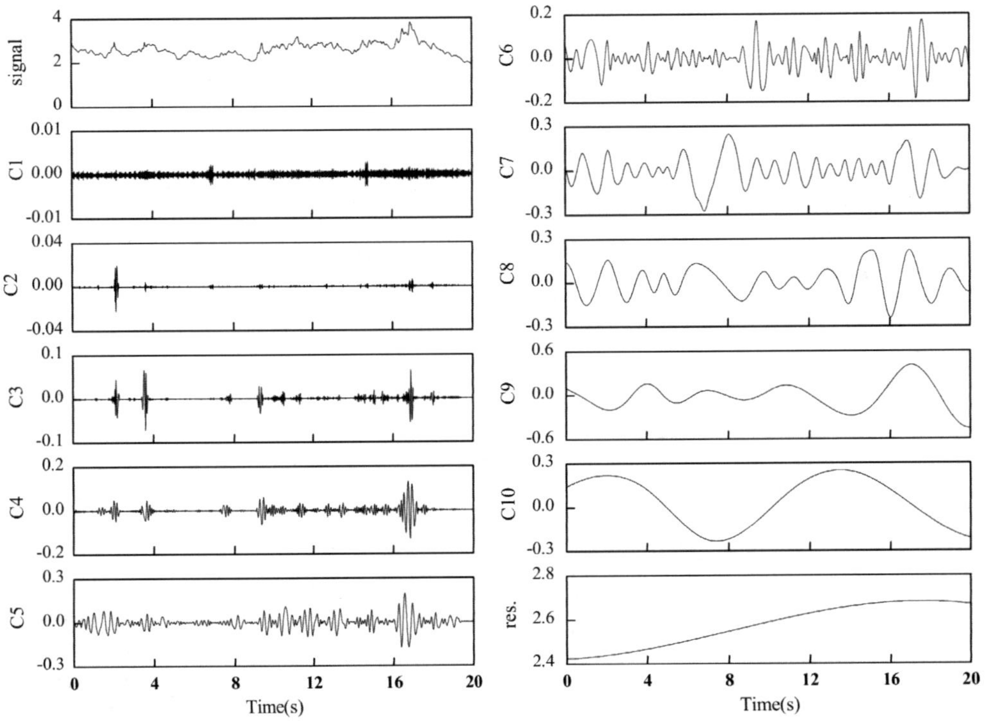

(a) (b)

Fig.3 the resulting empirical mode decomposition components from the coal flame (a) the original data and the components *c*1~*c*5; (*b*) the components *c*6~*c*10. Notice the last component res. is not an IMF; it is the trend.

2.4 Feature extraction-

The parameters of a flame can vary, depending upon the type and structure of the furnace and the targeted interests (Lu et al., 2008).It should be noted that the ignition section of a drop-tube furnace is the primary reaction zone of a coal combustion process in term of energy conversion and combustion products formation, the flame signal obtained from flame monitoring system is naturally fluctuating due to the inherent dynamic nature of the flame. According the characteristics of the furnace and the flame, the following parameter is measured:

Oscillation frequency (f_o) – the mean frequency at which the flame fluctuates, which is defined as the weighted average frequency of the IMFs of coal flame signal.

$$f_o = \sum_{i=1}^{n} p_i f_i \qquad (2)$$

Where n is the number of the IMFs which frequency is below 50Hz, f_i stands for the i th IMF which frequency is below 50Hz and p_i is the weighted factor of the i th IMF of the flame signal.

3. Result and discussion

Fig 4 shows that the oscillation frequency of coals in the process of ignition has three kinds of curve shape along with the flame axis. The one curve0 of coal A, coal A and B blends with different mixing proportion have two peaks; the first peak is the rapid release and subsequent ignition of volatiles, and the second peak is mainly the char ignition. Its ignition behavior is thought as homogeneously. The other curves of coal B, coal C, coal D and blends of coal C and D with different mixing proportion has one peak. This may be the joint of homogeneous ignition and heterogeneous ignition — hetero-homogeneous ignition behavior. The oscillation frequency curves of coal blends with one coal in higher volatile contents ((>25%) in mixing proportion of 70% to 30% and 60% to 40% are similar as well as that of 50% to 50%, 40% to 60% and 30% to 70%. The coal blends are in a higher oscillation frequency because of their higher volatile content. It can also be concluded that the oscillation frequency characteristic of coal blends is the combination of single coals and depends on the coal ranks greatly.

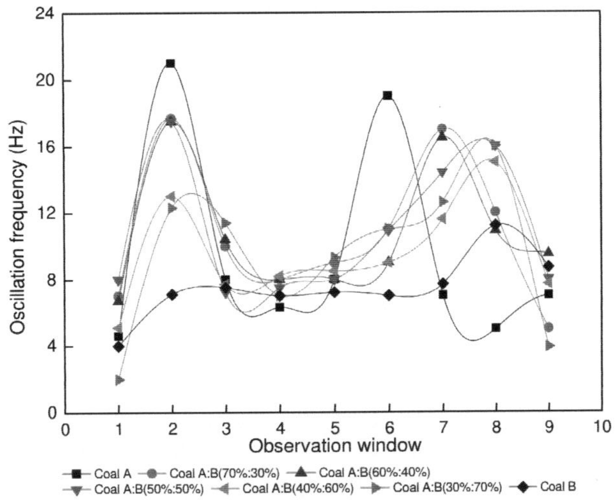

(a)Coal blends of coal A and B

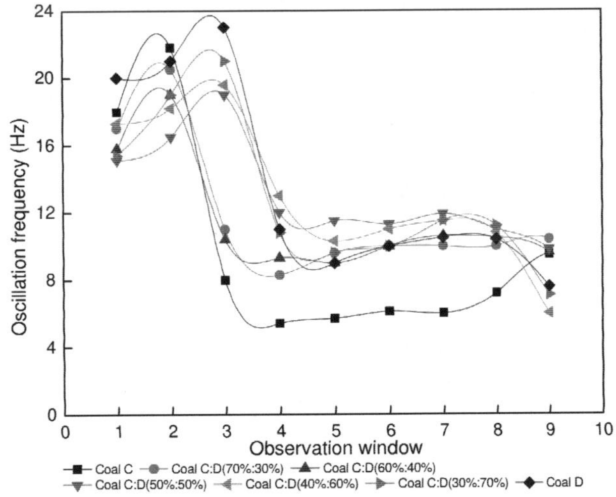

(b) Coal blends of coal C and D

Fig.4 Oscillation frequency of blended coals

4. CONCLUSIONS

A novel method on judging the ignition behavior of coals has been developed in a drop-tube furnace using a flame monitor. A series of tests has been conducted along the axial quartz observing windows of the ignition section during steady combustion. The feature parameter of oscillation frequency based on empirical mode decomposition has been measured from flame signal which is found most sensitive to coal types and therefore can be used as an indicator of the dynamic nature of flame. The results have suggested that ignition behavior can be determined from the feature. It has also been demonstrated that under the experimental conditions the mass ratio between the two coals in coal blends has a major impact on the ignition behavior. The measurement has been proved effective to judging the ignition behavior of coal which provided a new research field to study and understand pulverized coal ignition.

ACKNOWLEDGEMENTS

This work is financially supported by the Key Program of the National Natural Science Foundation of China with Grant No.60534030.

REFERENCES

Arenillas, A. et al. (2004). "A TG/DTA study on the effect of coal blending on ignition behavior," Journal of Thermal Analysis and Calorimetry, 76, p. 603-614.

Zhang , D.K. et al. (1994). "Igintion of coal particles : the influence of experimental technique, " Fuel, 73(7), p. 1114-1119.

Huang , N. E. et al. (1998). "The empirical mode decomposition and the Hilbert spectrum for nonlinear and non-stationary time series analysis, " Proceedings of Royal Society of London Series, A 454, p.903–995.

Lu ,G. et al. (2008). "Impact of co-firing coal and biomass on flame characteristics and stability," Fuel, 87, p. 1133-1140.

CARD , J.B. A. et al. (1995). "A Drop Tube Furnace Study of Coal Combustion and Unburned Carbon Content Using Optical Techniques," Combusiton and Flame, 101, p. 539-547.

Faundez , J. et al. (2005). "Ignition behavior of different rank coals in an entrained flow reactor," Fuel, 84, p. 2172-2177.

Xu , L.J. et al. (2004). "On-Line Fuel Identification Using Digital Signal Processing and Fuzzy Inference Techniques. IEEE Transactions On Instrumentation and Measurement, 53, p. 1316-1320.

Cloke , M. et al. (2002). "Combustion characteristics of coals using a drop tube furnac,"Fuel, 81,p. 727-735.

Essenhigh , R. H. et al. (1989). "Ignition of Coal Particles: A Review. Combusiton and Flame," 77, p. 3-30.

Kizgut , S. et al. (2003). "Characterization and non-isothermal decomposition kinetics of some Turkish bituminous coals by thermal analysis," Fuel Processing Technology, 85, p. 103-111.

Su , S. et al. (2001). "Techniques to determine ignition, flame stability and burnout of blended coals in p.f. power station boilers," Progress in Energy and Combustion Science, 27, p. 75-98.

Chen , Y. et al. (1996). "Studying the mechanisms of ignition of coal particles by TG-DTA," Thermochimica Acta, 275, p. 149-158.

Fan , Y.S. et al. (2008). "Ignition Characteristics of Pulverized Coal under High Oxygen Concentrations," Energy & Fuels, 22, p. 892-897.

Journal of Physics: Conference Series **147** (2009) 012067 doi:10.1088/1742-6596/147/1/012067

Study on airflow and inhaled particle deposition within realistic human upper respiratory tract

J Lin[1], G L HU[1], J R FAN[2], D Pan[2]

[1] School of Light Industry, Zhejiang University of Science and Technology, No. 318, Liuhe Road, Hangzhou, 310023, P. R. China

[2] State Key Laboratory of Clear Energy Utilization, Zhejiang University, No. 38, Zheda Road, Hangzhou, 310027, P. R. China

melinj@163.com

Abstract. Based on the CT (Computerized Tomography) scanned images of a 19-years-old healthy boy, a realistic geometric model of URT from nasal cavity to the upper six-generation bronchial is rebuilt. To investigate airflow and particle deposition in the obtained realistic human upper respiratory tract, RNG k-ε turbulence model was used to describe the primary flow and particle deposition under three breathing intensity such as 15 L/min, 30 L/min and 60 L/min. The particle is tracked and analyzed in the Lagrangian frame. The velocity fields of airflow under different airflow rates were computed and discussed. The trapping of particles with diameter 1μm on the wall surfaces was monitored, and the locations of trapping in different region were visualized.

In order to study the characteristics of particles movement and the effect of particles diameter on the deposition pattern, eleven kinds of sphere particles with different diameters are selected as research object. The diameters of selected particles as follows: 0.1μm, 0.5μm, 1μm, 2.5μm, 3μm, 3.5μm, 4μm, 4.5μm, 5μm, 6.5μm and 8μm. The variation of inhalable particles deposition in realistic human upper respiratory tract with respiratory intensity and particle size was researched and compared.

1. INTRODUCTION

Particles ranging from toxic particulate matter to drug aerosols may be inhaled into people's respiratory tract, and some of them can be either harmful or therapeutic to humans depending upon the aerosol material, deposition site and local concentration. In turn, these aspects and parameters are greatly determined by the airflow field, particle properties, breathing pattern and geometric airway characteristics. Thus, good knowledge of airflow structure, deposition of particles in the upper respiratory tract can provide credible theoretical foundation not only for preventing illness of respiratory tract effectively. On the other hand, it is desirable to deliver pharmaceutical aerosols to the targeted part accurately, and thus effectively reduce drug dosage and pain of patients (Balásházy et al., 2003).

Hitherto, especially in the last decade, many research literatures about computation fluid dynamics modeling of airflow and particles movement in the respiratory tract are available all over the world. Most models employ low-Reynolds k-ω model and standard k-ε model to describe gas flow,

Journal of Physics: Conference Series **147** (2009) 012067 doi:10.1088/1742-6596/147/1/012067

track particles movement in the frame of Lagrange. Using a mouth-to-trachea replica plus three generations of the Weibel A bronchia, Zhang et al. (2005) as well as Zhang et al. (2006) investigated numerically steady inhaled particle transport and deposition. Employing a Weibel-type bifurcation and considering chronic obstructive pulmonary disease (COPD), Yang et al. (2006) and Liu et al. (2003) compared the resulting airflow structures under different inlet conditions. Tang et al. (2004) presents a brief summary of flow features in the human respiratory system and simulates an airflow field based on a 3D real-anatomical geometry of the human nasal cavity. The dispersed phase in the flow field is solved by Lagrangian particle-tracking approach. Hofmann et al. (2001) computed the inspiratory deposition efficiencies of ultrafine particles with 6500 nm in airway Generations 3 and 4 for different inlet flow rates. Just recently, the LES method is employed to the study of airflow in the human respiratory tract (Jin et al., 2007). For the geometric model, it can be found that most simulation is based on the idealized, i.e., planar and symmetric, lung airway model published by Weibel (1963). Recently, modern imaging techniques allowed for even more detailed mapping of the human respiratory system (Cebral and Summer, 2004). CFD simulations based on the CT scanned images are conducted (Xi and Longest, 2007; Shi et al., 2007). Nowak et al. (2003) conducted computational fluid dynamics (CFD) simulations of airflow and particle deposition in geometries representing the human tracheobronchial tree. Two geometries were used in this work; one of them is based on a CT scan of a cadaver lung cast. Flow conditions used included both steady-state inhalation and exhalation conditions as well as time-dependent breathing cycles. Particle trajectories were calculated in each of these models by solving the equations of motion of the particle for the deterministic portion of particle displacement. The trapping of particles on the wall surfaces was monitored, and the locations of trapping in each generation were recorded.

The objectives of this paper is to rebuilt a three-dimensional geometric model of realistic human upper respiratory tract including nasal cavity, pharynx, larynx, trachea and the upper six-generation bronchial based on the CT scanned images. On the other hand, RNG k-ε turbulence model is employed to study the airflow and particles deposition in this realistic human upper respiratory tract.

2. DESCRIPTION OF MATHEMATICAL MODEL

2.1 Geometric model

Transverse CT images were obtained from a healthy 19-year-old normal adult male (the volunteer was awake and reclining). The subsequent CT images were then transferred to a 3-dimensional structure of the airway by image and graphics recognition technologies. The obtained 3-dimension upper airway is shown in figure 1. The geometric structure of the upper airway is very complex, especially the nasal cavity, which is different from most previous model. This will cause complicated airflow fields in the upper airway, and complicated particle transport and deposition principles along the airway. The more real flow characteristics can be obtained using this realistic geometrical model.

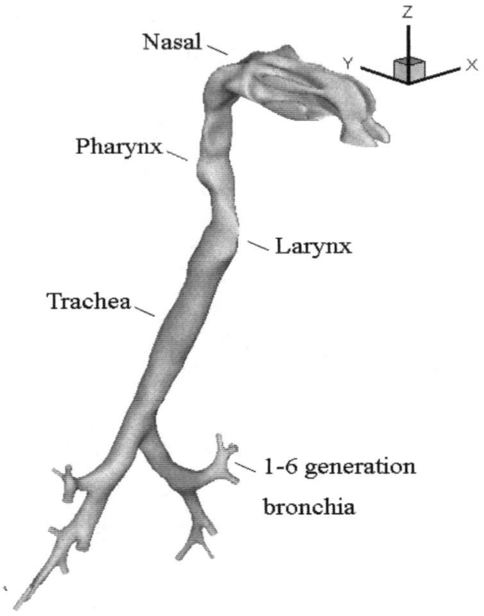

Fig.1 The three-dimensional geometrical model of realistic human upper respiratory tract rebuilt from the CT scanned images

2.2 **Governing equation**

The airflow structure will be very complex because sharp shrink and expand segment exist in the pharynx-larynx, bifurcate structure exist in the bronchia and the nasal cavity with more complex structure. Thus, the RNG $k-\varepsilon$ turbulent model is adopted in this paper, which provides an option to account for the effects of swirl or rotation by modifying the turbulent viscosity appropriately. Furthermore the RNG $k-\varepsilon$ model is more responsive to the effects of rapid strain and streamlines curvature than the standard k- ε model. The continuity and movement equations:

$$\frac{\partial \rho}{\partial t}+\frac{\partial}{\partial x_i}(\rho u_i)=0 \tag{1}$$

$$\frac{\partial(\rho u_i)}{\partial t}+\frac{\partial(\rho u_i u_j)}{\partial x_j}=-\frac{\partial p}{\partial x_i}+\mu\frac{\partial^2 u_i}{\partial x_j \partial x_j}+\frac{\partial}{\partial x_j}\left[\mu_t\left(\frac{\partial u_i}{\partial x_j}+\frac{\partial u_j}{\partial x_i}\right)\right]+\rho g_i \tag{2}$$

The turbulence kinetic energy, *k*, and its rate of dissipation, ε, are obtained from the following transport equations:

$$\frac{\partial(\rho k)}{\partial t}+\frac{\partial(\rho k u_i)}{\partial x_i}=\frac{\partial}{\partial x_j}\left(\alpha_k \mu_{eff}\frac{\partial k}{\partial x_j}\right)+G_k+\rho\varepsilon \tag{3}$$

$$\frac{\partial(\rho\varepsilon)}{\partial t}+\frac{\partial(\rho\varepsilon u_i)}{\partial x_i}=\frac{\partial}{\partial x_j}\left(\alpha_\varepsilon \mu_{eff}\frac{\partial \varepsilon}{\partial x_j}\right)+\frac{C_{1\varepsilon}^{*}\varepsilon}{k}G_k-C_{2\varepsilon}\rho\frac{\varepsilon^2}{k} \tag{4}$$

$$\left.\begin{aligned} &\mu_{eff} = \mu + \mu_t ; \mu_t = \rho C_\mu \frac{k^2}{\varepsilon} \\ &C_{1\varepsilon}{}^* = C_{1\varepsilon} - \frac{\eta(1-\eta/\eta_0)}{1+\beta\eta^3} ; \eta = \left(2E_{ij}\cdot E_{ij}\right)^{1/2}\frac{k}{\varepsilon} \\ &E_{ij} = \frac{1}{2}\left(\frac{\partial u_i}{\partial x_j} + \frac{\partial u_j}{\partial x_i}\right) \end{aligned}\right\} \tag{5}$$

Where i、j=1、2、3, ρ is fluid density, t is time, u is fluid velocity vector, x denotes the spatial coordinate, p stands for average pressure, μ is fluid viscosity, μ_t is turbulent viscosity, g_i is the gravity in the i direction, k is turbulence kinetic energy, ε is turbulence dissipation rate, G_k represents the generation of turbulence kinetic energy due to the mean velocity gradients, The quantities α_k and α_ε are the inverse effective Prandtl numbers for k and ε, β is thermal expansion coefficient, E_{ij} implies main time-average strain rate. C_μ and η_0 are model constants. The model constants $C_{1\varepsilon}$ and $C_{2\varepsilon}$ have values derived analytically by the RNG theory.

The particle phase in the air can be considered as dilute phase, and neglecting the interaction between particles. The Lagrangian dispersed phase model is used for the prediction of the trajectory of a particle. This is done by integrating the force balance on the particle. In a Lagrangian reference frame, movement vector of a single particle can be written as:

$$m_{\mathrm{p}}\frac{dv_i}{dt} = \frac{\pi d_{\mathrm{p}}^2}{8}C_D\rho\left|\bar{u}_i - v_i\right|\left(\bar{u}_i - v_i\right) + m_{\mathrm{p}}g_i + 1.6(\mu\rho)^{1/2}d_{\mathrm{p}}\left|\bar{u}_i - v_i\right|\left|\frac{\partial \bar{u}_i}{\partial x_i}\right|^{\frac{1}{2}} \tag{6}$$

Where, m_p is particle mass, d_p is particle diameter, v_i represents particle velocity vector, C_D stands for particle resistance coefficient, g acceleration of gravity. The three terms on the right hand are Stokes resistance, gravity force and Saffman's lift force respectively, which are considered in this paper.

The particle deposit efficiency (DE) is introduced to quantificationally analyze particle deposit in different region of the respiratory tract:

$$\mathrm{DE} = \frac{N_{\mathrm{d}}}{N_{\mathrm{t}}}\times 100\% \tag{7}$$

Where N_d denotes the number of particle deposited on wall of some region in the respiratory tract, N_t represents the total number of particle inhaled into the model.

2.3 Initial condition and boundary condition

First of all, three-cycle pre-simulation are conducted to eliminate initial effect because the inlet velocity is stable during stable breath. The pressure outlet boundary is imposed on the outlet of bronchial, i.e., the pressure is zero and gradient of other variables is zero. When the particles impact with model wall, it is marked as deposition and the track computation is over, which accords with the real instance well. Different from many papers, the pressure inlet condition is selected for the computation of this paper. One box extending from nasal inlet is added to consider the inlet effect because the pressure at nasal inlet is not equal one atmosphere but the boundary of the box.

3. RESULTS AND DISCUSSIONS

3.1 Gas phase

Three breathing intensities Q=15L/min, Q=30L/min, Q=60L/min are computed respectively.

Figure 2 depicts the typical velocity vectors and pressure contours during inhalation phase for the airflow rate Q=15L/min. The outside air is inhaled into nasal cavity by the pressure difference between interior lung and outer circumstance. The air stream arrive a relatively high velocity after entering the nasal cavity due to the constrictive structure in the inlet segment. At the cross-section A-A (Figure 2a), the maximum velocity is located at the lower wall when airflow through the syphon structure. Then the velocity value decrease from the lower wall to the upper wall till little-value reverse velocity appears close to the upper wall. Corresponding to Figure 6b, we can find that the domain with reverse velocity is the negative pressure region, which is the first negative pressure region after the gas stream entering the nasal cavity and there exists one relatively faint vortex. After the syphon structure, the maximum velocity value of gas stream is gradually transferred to the upper wall of olfactory airway of nasal cavity, which is directly eroded by gas stream. The possibility of particle deposition in the vortex region is very low for the low velocity in this domain. However the particle is prone to deposit at the upper wall of syphon structure and partial olfactory airway behind that.

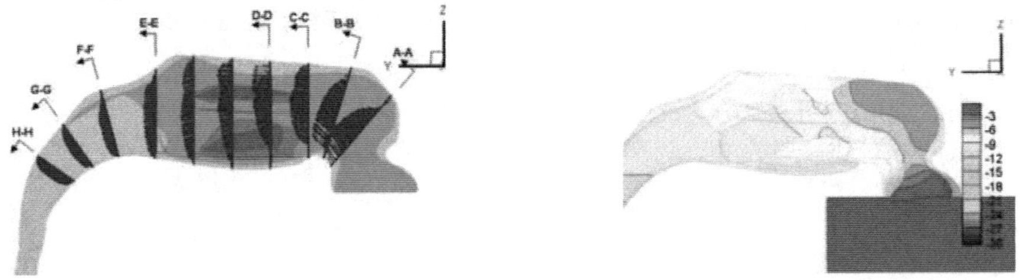

(a) Velocity vector at selected cross-sections in the nasal cavity (b) Pressure contours in the right nasal cavity

Fig.2 The typical velocity vectors and pressure contours during inhalation phase for the airflow rate Q=15L/min

Figure 3 illustrates the velocity contours in the pharynx-larynx part and bronchia bifurcation for breath intensity Q=60L/min during inhalation phase. The maximum velocity of the whole model with Vmax=3.82 m/s lies in the pharynx. There has one shrink structure in the larynx; the maximum velocity at the cross-section in the larynx exhibits close to the posterior wall. After entering into bronchia, the maximum velocity at the cross-section in the bronchia gradually moves towards the center of the tube, and the gradient of pressure and velocity is small. The velocity of bifurcation is plotted in figure 3b. It can be seen from figure that the gas stream erodes the carinal ridge directly, the carnial ridge of every generation bronchia and its daughter tube will be the hot spot for particles deposition.

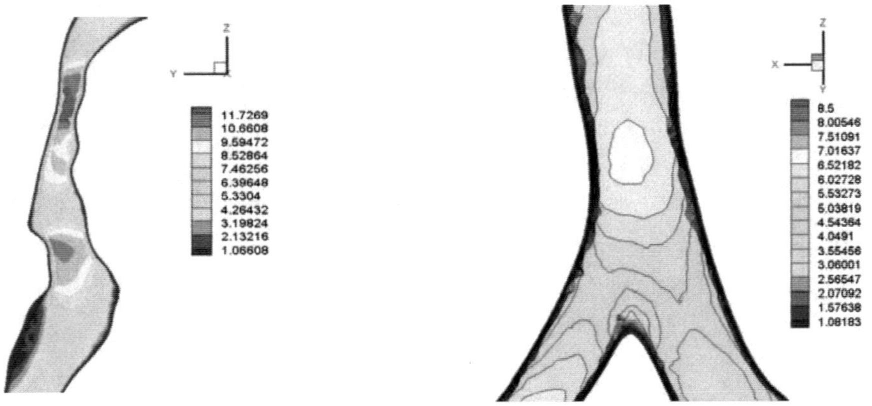

(a) Velocity contours in the pharynx-larynx (b) Velocity contours in the bronchia bifurcation

Fig. 3 Velocity contours in the pharynx-larynx and bronchia bifurcation for breath intensity Q=60L/min during inhalation phase

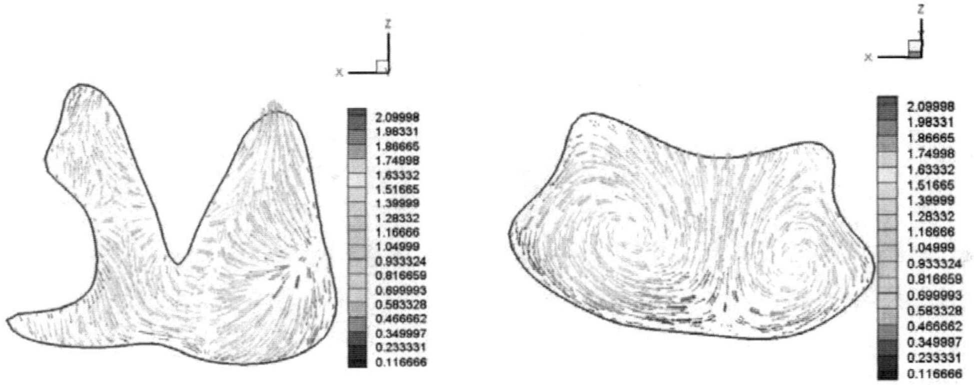

(a) E-E (b) F-F

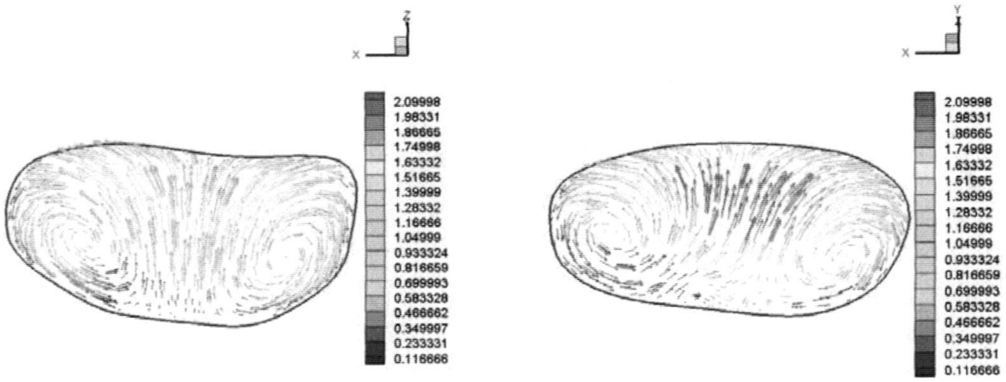

(c) G-G (d) H-H

Fig. 4 Secondary velocity vectors at selected cross-sections in the nasal cavity during exhalation phase for the breath intensity Q=15L/min

Figure 4 plots the velocity vectors at selected cross-sections in the second half nasal cavity with breath intensity Q=15L/min during exhalation phase. It is different from inhalation process that the maximum velocity is located at the center of section and there have two distinct and symmetry vortex in both sides of the section when the gas stream flows through section H-H. Then, when the gas stream flows through the G-G and F-F section, the vortexes in two sides of section move towards center, the maximum value of velocity at the center position decrease gradually and the corresponding velocity gradient. After that, when the gas stream flows through the E-E section, around which the gas stream is divided into the left and right nasal channels, the vortexes disappear.

3.2 Particles deposition

3.2.1 Map of particle deposition

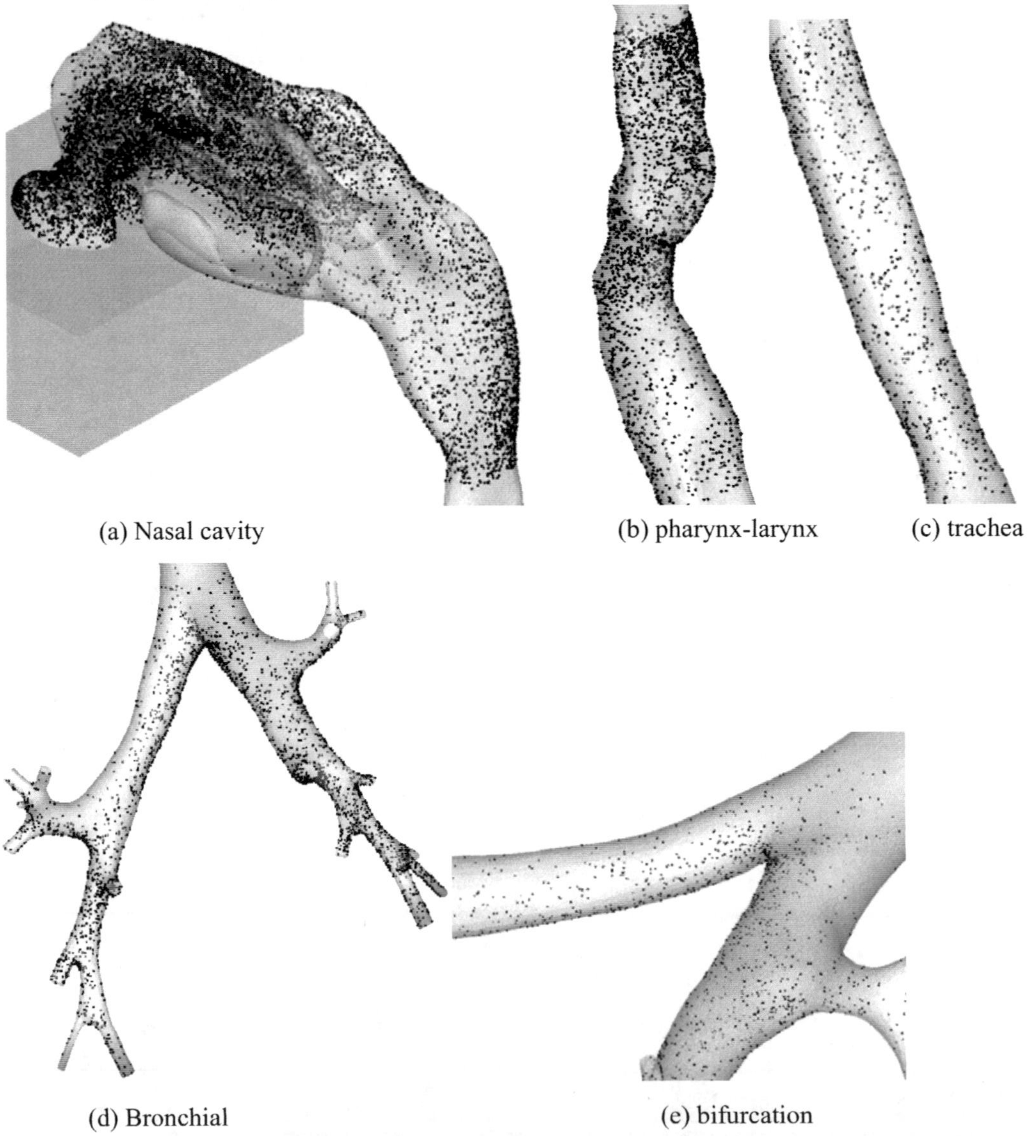

Fig. 5 Particles deposition in different part of the respiratory tract with d=1μm and Q=15 L/min during inhalation phase, the darkish particle indicates deposition at the rear wall

In order to investigate the particle deposition position and hot spots, Lagrange method is implied to track the particle movement. The particle deposition at different parts and under different breath intensities are visualized in figure 5. The figure a plots the deposition in the bronchia, from which we can see that the particle preferred deposition in the carinal ridge and downstream at the inner tube of the daughter tubes for the effect of impact between particle and wall. The deposition in the nasal cavity, pharynx-larynx, and trachea is shown in figure. It can be seen from figure that the deposition in the front of the pharynx-larynx is small for the air flow affected by the geometry structure.

The deposition efficiencies of particle with diameter 1μm in the different parts under different breath intensities are plotted in Figure 6. It can be seen from figure that the total deposition efficiency in the whole model slackly increases as the breath intensity increases. The deposition efficiency in the nasal cavity decreases as the breath intensity increases; however, the deposition efficiency in the pharynx-larynx, trachea and bronchia slackly increases as the breath intensity increases. Among all the parts, the deposition efficiency in the nasal cavity is the largest; almost half particle is deposited in this region. The next sequence of deposition efficiency is bronchia, pharynx-larynx and trachea. The deposition efficiency in the trachea is the smallest for the relatively smooth tube structure in the trachea position.

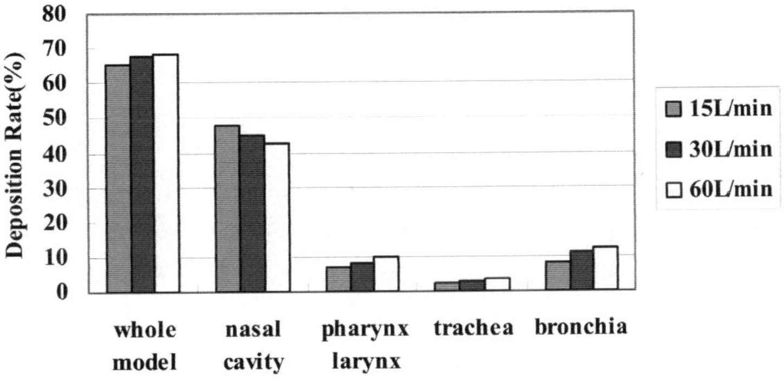

Fig. 6 Numerically predicted total and regional deposition efficiencies of 1 μm diameter particles under three breath intensities

3.2.2 Variations of deposition rate with inhaled particles diameter and airflow rate

In order to study the effect of particles diameter and airflow rate on the deposition characteristics, eleven kinds of monodisperse particles with diameter 0.1μm, 0.5μm, 1μm, 2.5μm, 3μm, 3.5μm, 4μm, 4.5μm, 5μm, 6.5μm, 8μm and three airflow rates Q=15, 30 and 60L/min are simulated.

The variation of total particles deposition rate in the whole model with breath intensity is illustrated in figure 7. For the breath intensity with 15 L/min, the deposition rate faintly decreases as increase of the diameter smaller than 6.5μm, increase as increase of the diameter larger than 6.5μm. For the breath intensity with 30 L/min, the deposition rate decreases as increase of the diameter which is smaller than 4μm and increases as increase of the diameter which is larger than 4μm. For the breath intensity 60 L/min, the deposition rate decreases as increase of the diameter smaller than 2.5μm, and increase as increase of the diameter which is larger than 2.5μm. The deposition in the nasal cavity occupies more than half deposition in the whole model and directly determines the total deposition characteristics. Be similar to the deposition rate in the nasal cavity, the diameter corresponding to the minimum deposition transfers towards small size as the breath intensity increase.

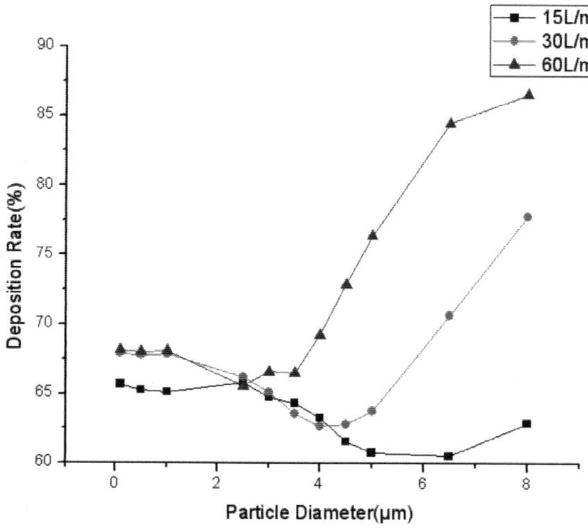

Fig. 7 Variation of total particles deposition rate in the whole computation domain with breath intensity

4. CONCLUSIONS

A three-dimension geometric model of a real human upper respiratory tract including nasal cavity, pharynx, larynx, trachea and upper six generations bronchial based on the CT scanned images is rebuilt in this paper. RNG k-ε turbulence model is used to compute airflow, particle movement and deposition efficiency in the obtained complex structure geometrical model. The three-dimensional flow structure under three different breathing intensity are computed and analyzed. Then particle of 1μm diameter in every part of respiratory tract are tracked and visualized under Lagrange frame. The result indicates that the total deposition efficiency in the whole model slackly increases as the increase of breathing intensity. The deposition efficiency of different particles with diameter 0.1μm、0.5μm、1μm、2.5μm、3μm、3.5μm、4μm、4.5μm、5μm、6.5μm、8μm in the realistic human respiratory tract is simulated in this paper.

ACKNOWLEDGEMENTS

The authors wish to express their thanks to Natural Science Foundation of P.R. China (granted No 50776082) and Keynote Natural Science Foundation of Zhejiang Province (P.R. China, granted No Z505050). It is appreciated that professor Shen Huahao, director of institute of respiratory disease, second affiliated hospital, Zhejiang university college of medicine has provided lots of information about the respiratory tract structure and respiratory principles, and his doctoral candidate Huang Huaqiong, who has provided the CT data images under volunteer's consent.

REFERENCES

Balásházy, I. et al. (2003). "Local particle deposition patterns may play a key role in the development of lung cancer," J. of Applied Physiology, 94, pp. 1719–1725.

Cebral, J. R. et al. (2004). "Tracheal and central bronchial aerodynamics using virtual bronchoscopy and computational fluid dynamics," IEEE Transactions on Medical Imaging, 23(8), pp. 1021-1033.

Hofmann,W.et al. (2001). "Environmental tobacco smoke deposition in the human respiratory tract: Differences between experimental and theoretical approaches," Journal of Aerosol Medicine, 14(3), pp. 317–326.

Jin H. H. et al. (2007). "Large eddy simulation of inhaled particle deposition within the human upper respiratory tract," J. of Aerosol Science, 38, pp. 257–268.

Liu, Y. et al. (2003). "Modeling the bifurcating flow in an asymmetric human lung airway," Journal of Biomechanics, 36(7), pp. 951-959.

Nowak N. et al. (2003). "Computational fluid dynamics simulation of airflow and aerosol deposition in human lungs," Annals of Biomedical Engineering, 31, pp. 374-390.

Tang, H. et al. (2004). "Dynamic Analysis of Airflow Features in a 3D Real-Anatomical Geometry of the Human Nasal Cavity," 15th Australasian Fluid Mechanics Conference The University of Sydney, Sydney, Australia, December 13-17.

Xi, J. et al. (2007). "Transport and deposition of micro-aerosols in realistic and simplified models of the oral airway," Annals of Biomedical Engineering, 35(4), pp. 560–581.

Yang, X. L. et al. (2006). "The effect of inlet velocity profile on the bifurcation COPD airway flow," Computers in Biology and Medicine, 36(2), pp. 181-194.

Zhang, Z. et al. (2006). "Water vapor transport and its effects on the deposition of hygroscopic droplets in a human upper airway model," Aerosol Science and Technology, 40, pp. 1-16.

Zhang, Z. et al. (2005). "Comparison of micro- and nano-size particle depositions in a human upper airway model," Journal of Aerosol Science, 36(2), pp. 211–233.

Journal of Physics: Conference Series **147** (2009) 012068 doi:10.1088/1742-6596/147/1/012068

Effect of Characterization of Powder on Particle Velocity in Dense-phase Pneumatic Conveying

Zongming Liu[1], Bin Du[1], Weilin Zhao[1], Xiansong Li[1], Jinkai Li[1] , Congling Bu[2], Hua Yi[2]

[1] School of Materials Science and Engineering, University of Jinan, Jinan 250022, China
[2] Shandong Zhanggu Pneumatic Conveying Engineering Co.Ltd, Jinan 250200, China

kjclzm@ujn.edu.cn

Abstract. Particle velocity is an important characteristic parameter in the process of evaluating dense-phase pneumatic conveying system. It relates to the properties of conveying materiel, system performance, operating conditions, physical properties of conveying gas etc, and has some noticeable effect on the conveying capacity, energy consumption, wear between particles and pipe. Of all influencing factors, characterization of powder is crucial to impact particle velocity. In this paper, pneumatic conveying experiments of four kinds of powdery material whose different solid physical properties are tested and taken are carried out using compress air in long distance pipe. The trend of particle velocity was achieved by the experimental data gained from Groups of differential pressure transmitters. Dimension analysis and mathematical analysis were adopted to obtain a theoretical correlation equation of particle velocity versus solid characteristics :

$$u_s = 1.29 \times 10^{-3} p^{0.5} D^{0.488} L^{0.408} \rho_s^{-1.06} \rho_g^{0.56} d_s^{-0.896} \varepsilon^{6.55} w^{-0.559}$$

And this equation can be used to predict the tendency of particle velocity for different kind of solid precisely.

1. INTRODUCTION

With widespread application of technology of dense phase pneumatic conveying, its effect factors has been focused in recent research. It was found that the factors of effecting dense phase pneumatic conveying are many and also complex, mainly including performance of powder, feature of conveying pipeline, operated condition in conveying processing(such as pressure), etc(Mainwaring et al, 1987; Pan,1999). The effects of performance of powder on characteristic of dense phase pneumatic conveying which has greater effect on flow mode, characteristic parameter and pressure drop are critical（Pan, 1999). In this paper, the effects of characterization of powder on particle velocity are researched by testing powders on properties and dense phase pneumatic conveying experiments. Theoretical correlation equation of particle velocity versus powder performance is obtained.

2. INTRODUCTION OF EXPERIMENT

2.1 Experimental System

The pneumatic conveying system in this paper is a dense-phase pneumatic conveying and circulating experimental bench with long-distance pipeline whoes length is 203m and diameter is 80mm. The measuring instrument and data collected system consist of GP /DP(Differential Pressure) transmitters, Weight balance, Vortex Flowmeter and Paperless Recorder, as shown in Fig.1.

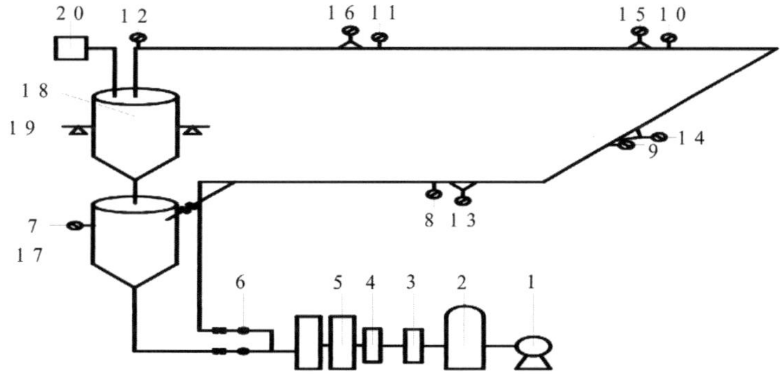

1. Air compressor 2. Gas container 3. Oil and water separator 4. Desiccator 5. Surge tank 6. Flowmeter 7—12. Gauge pressure transmitter 13—16. Differential pressure transmitter 17.Feeder 18. Collecting bin 19. Weight balance 20. Dust collector

Fig.1 Schematic diagram of experimental system

2.2 Experimental Material and Characterization

The fly ash, gypsum, cement and desulfurization gypsum are used for conveying materials, Whose performance such as granularity, density, shape, angle of repose, void fraction and moisture content are characterized, as shown in table.1.

Table.1 Result of material's characterization

Material	mean particle diameter (μm)	median particle diameter (μm)	particle density (kg/m^3)	packing density (kg/m^3)	shape	angle of repose (°)	void fraction (%)	percent moisture content (%)
fly ash	58.73	50.51	2078	771.5	spherical	36	66.8	0.68
gypsum	40.3	30.65	2703	910.9	horny	41	65.9	1.07
cement	47.57	27.24	2881	1001.2	cluster	39	63.2	0.7
desulfu-rization gypsum	13.46	5.751	2476	663.6	granular	44	66.2	0.86

It can be seen from photoes of SEM that shape of fly ash is nearly spherical whose degree of sphericity is maximal. So its fluidity is better than other materials.

2.3 Experimental Procedure

The whole experimental to be divided four phases are materials being put into the feeder, being fluidized by compressed air(as shown t_1 in Fig.2), conveying materials (as shown t_2 in Fig.2)and cleaning pipe(as shown t_3 in Fig.2). The whole process can be operated circularly.

Variable curve of data of each measurer in conveying process is shown in Fig.2. It can accurately reflect changed trends of pressure in the feeder, gas instantaneous mass flow and solid mass flow in the conveying process .

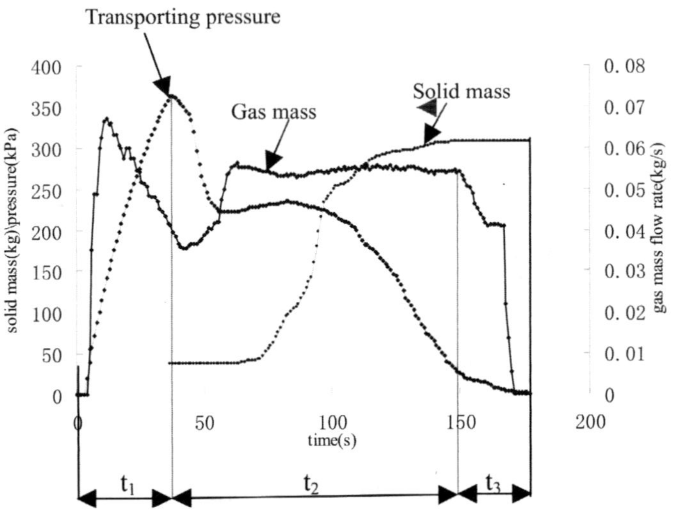

Fig.2 Variable curve of data of each measurer in conveying process

2.4 Experimental Scheme

Five test segments along horizontal pipeline were employed averagely to analyze the tendency of velocity, pressure and pressure drop. A differential pressure transmitter and a gauge pressure transmitter were assembled in the place of 52.2 m-53.4 m、90.5 m-91.7 m、126.5 m-127.7 m、176m-177.2 m.And two gauge pressure transmitter were separately assembled at the point of the initial and terminal of pipe. The layout of transmitters was shown in Fig.1.

In this experiment, operating condition is mainly controlled by changing the pressure of feeder. Feeder pressure was set from 0.26MPa to 0.42MPa and each 0.02MPa interval is as work condition. The experiments of 9 operating conditions were carried out in total and several repeated experiments were carried out in each operating condition.

3. EXPERIMENTAL RESULT

3.1 Average Particle Velocity

Particle velocity is one of the most important conveying characteristic parameter in dense phase pneumatic conveying. It was found that as a result of speed-variable moving of particle in the pipe(Peter et al, 2003), average particle velocity in whole pipeline is researched firstly. The average particle velocities is defined by the ratio of total length of pipe to total time of conveying. The relationship between average particle velocity of four materials and feeding pressure was shown in Fig.3.

From Fig.3, it can be seen that average particle velocity is low at the range of 3～8m/s. And the solid loading ratio of all materials is more than 25, up to maximum 47. It is shown that the conveying process is dense phase pneumatic conveying. Particle velocity increase gradually with pressure increasing. Because the driving power to particles of gas in pipe increases as the continuous increase of pressure. When feeding pressure is constant, particle velocity increase with reduction of density, and decrease with decreasing of void fraction. It is shown that density and void fraction affect average particle velocity obviously. However, about the effect of particle size, average particle velocity of fly ash with maximum particle size is higher than other three materiel, and average particle velocity of desulfurization gypsum with minimum particle size is higher than gypsum and cement. It was shown that particle diameter have not obvious effect on average particle velocity.

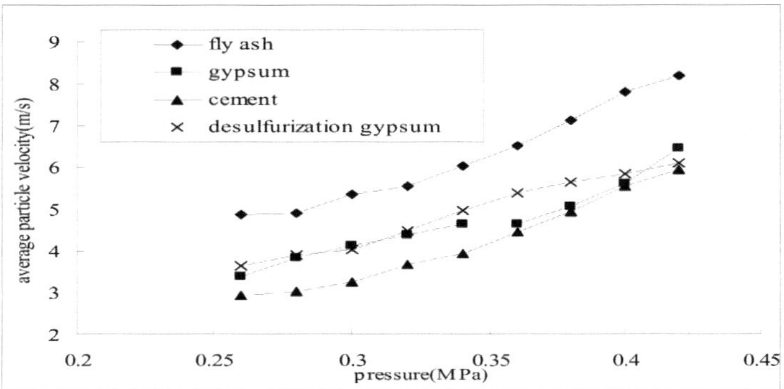

Fig.3 Relationship between average particle velocity and feeding pressure

3.2 Particle Velocity along Pipeline

About long distance pneumatic conveying, particle velocity is changed along pipeline. The particle velocity is changed with the distance of along pipeline changing which is analyzed by using gauge pressure transmitter.

All signals from each pressure transmitter were transmitted to MultiFR4000 recorder in the conveying process. The reading of each gauge pressure transmitter changing with time is shown in Fig.4. The six curves from left to right express the changes of static pressure of 7-12 in Fig.1.

Based on time interval of signal emergence of every pressure transmitter in different place(Fig.4), particle average velocities were achieved as follows:

$$u_{ij} = \frac{L_{ij}}{t_{ij}} \tag{1}$$

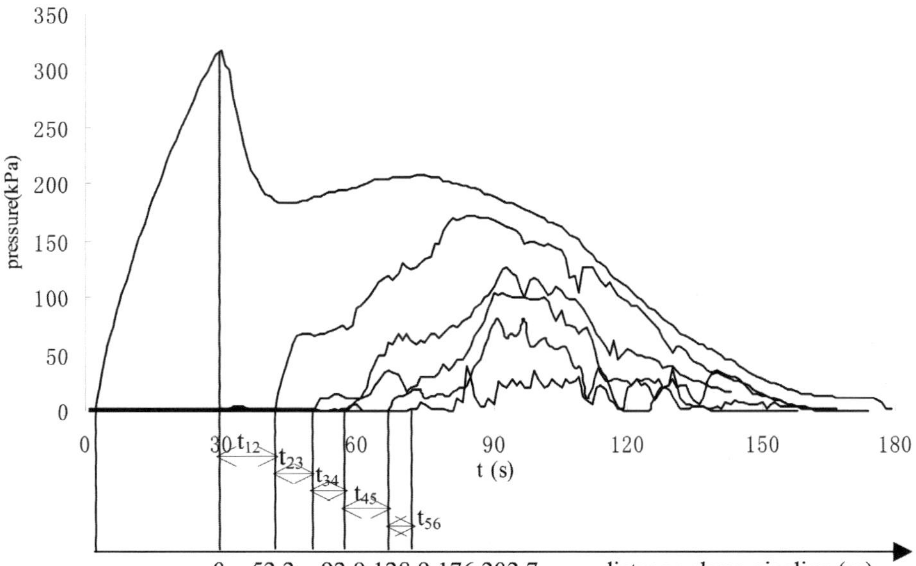

Fig.4 Variable curve of static pressures with the feed pressure 0.32MPa

Variable curve of four materiel particle velocity along pipeline with the feed pressure 0.34MPa was shown in Fig.5. It can be seen that particle velocity increase gradually along pipeline because of the continuous decrease of pressure and the expansion of gas. When feeding pressure is a constant, on the same point, particle velocity increase with the reduction of density.

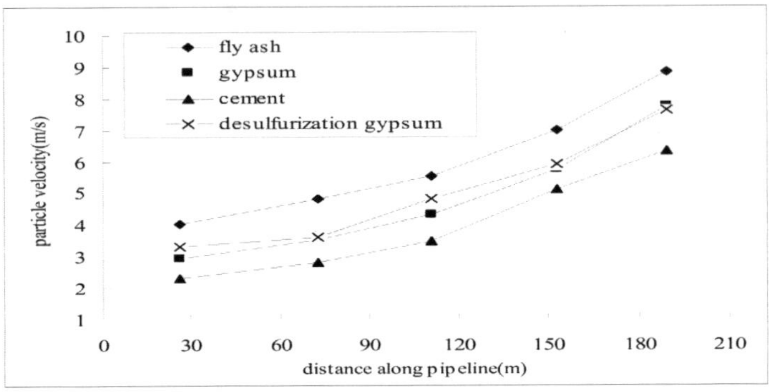

Fig.5 Variable curve of particle velocity along pipeline(0.34MPa)

4. CORRELATION BETWEEN CHARACTERIZATION OF POWDER AND PARTICLE VELOCITY

4.1 Deduction of Theoretical Correlation Equation

It was found that complexity of gas-solid flow makes particle velocity be difficult to calculate(Liu et al, 2006). And it should be related to physic characterization of powder(such as size, shape, density and void fraction of powders), system performance and operating conditions, physical properties and velocity of conveying gas (Gilbertson et al, 2003; Luis Sanchez et al, 2003). In this section, theoretical correlation equation of particle velocity versus powder performance is deduced base on the assumption.

As mentioned previous, influencing factors of particle velocity u_s are p, L, D, ρ_g, ρ_s, d_s, ε, σ, w, etc. Suppose the expression of u_s as follows:

$$u_s = f\left(p, L, D, \rho_s, \rho_g, d_s, \varepsilon, \tan\sigma, w\right) \tag{2}$$

The expression of power product can be expressed as follows:

$$u_s = K p^a L^b D^c \rho_s^d \rho_g^e d_s^f \varepsilon^g (\tan\sigma)^h w^i \tag{3}$$

Base on the dimensional analysis, it can be expressed as follows:

$$LT^{-1} = (ML^{-1}T^{-2})^a (L)^{b+c+f} (ML^{-3})^{d+e} \tag{4}$$

Equations set was obtained.

$$\left.\begin{array}{ll} M & 0=a+d+e \\ L & 1=a+b+c-3d-3e+f \\ T & -1=-2a \end{array}\right\} \tag{5}$$

Undetermined coefficients are b, c, e. Then

$$a=0.5,\ d=-0.5-e,\quad f=-b-c \tag{6}$$

Eq. (6) was substituted into Eq. (4).

$$u_s = Kp^{0.5}\left(\frac{L}{d_s}\right)^b\left(\frac{D}{d_s}\right)^c\frac{1}{\rho_s^{0.5}}\left(\frac{\rho_g}{\rho_s}\right)^e\varepsilon^g(\tan\sigma)^h w^i \qquad (7)$$

where $E=L/d_s$; $\psi=D/d_s$; $Y=\rho_g/\rho_s$

Eq. (7)can be changed as followed:

$$u_s = \frac{Kp^{0.5}}{\rho_s^{0.5}}E^b\psi^c Y^e\varepsilon^g(\tan\sigma)^h w^i \qquad (8)$$

Base on the mathematical analysis, substituting 144 groups of data into Eq.(8) and adopting linear regression analysis in seven unknown, regression equation was obtained as follows:

$$\ln u_s = -3.375+0.153\ln E+0.662\ln Y+1.563\ln\psi+8.55\ln\varepsilon-1.47\ln(\tan\sigma)-1.118\ln w \qquad (9)$$

Significance test of regression equation showed that $\ln(\tan\sigma)$ has no obvious effect on $\ln u_s$.So it is eliminated from equation and quadratic linear regression should be performed. Result was gained as follows:

$$\ln u_s = -6.653+0.408\ln E+0.56\ln Y+0.488\ln\psi+6.55\ln\varepsilon-0.559\ln w \qquad (10)$$

Significance test of previous equation was performed again.It was found that significant probability of each coefficient is less than 0.05. linear regression of the equation wasn't needed again. The multiple correlation coefficient of the equation R is 0.964 which means regressive result is good. Regression equation of particle velocity u_s was gained as follows:

$$u_s = 1.29\times10^{-3}p^{0.5}D^{0.488}L^{0.408}\rho_s^{-1.06}\rho_g^{0.56}d_s^{-0.896}\varepsilon^{6.55}w^{-0.559} \qquad (11)$$

From Eq.(11), it can be seen that particle velocity was affected by some parameters such as feeding pressure, pipe diameter, pipe length, particle diameter, density, void fraction, moisture content. Comparing each parameter of the theorical formula, it can not be directly compared, because each parameter has different units and orders of magnitude. In this paper, we get partial regression analysis of Eq.(11) and standardize the coefficients, making it easy to compare. From all of the standard partial regression coefficient, the standard partial regression coefficients of the density, size, void fraction, moisture content were obtained:

$b_i'(\rho_s)$=-0.763, $b_i'(d_s)$=-0.496, $b_i'(w)$=-0.235, $b_i'(\varepsilon)$=3.256

where $b_i'<0$ shows that independent variable increase a standard deviation unit. dependent variable will reduce b_i' standard deviation units. It introduced the increasing of density has greater effect than moisture content and particle size on particle velocity reducing and the increasing of void fraction has more obvious effect than other influencing factors on the increasing of particle velocity.

For this experimental system, particle velocity increase gradually with the increase of pipe diameter and feeding pressure, while system operated parameters are invariable,particle velocity increase with the reduction of density and the increasing of void fraction .That is consistent with experimental result.

4.2 Error Analysis

We compared theoretical value with experimental value. It was found that good accuracy is achieved with the relative error range -13.6%—12.38% between academic fitting equation and experiment data.The relationship of particle velocity between theoretical and experimental value was shown in Fig.6.It can be seen that theoretical result of particle velocity is consistent with trend of particle velocity along pipeline in experiment, and the theoretical arithmetic value is identical with experiment data. Therefore, it can be used to predict trends of particle velocity in the conveying process for different kinds of powder materials preferably.

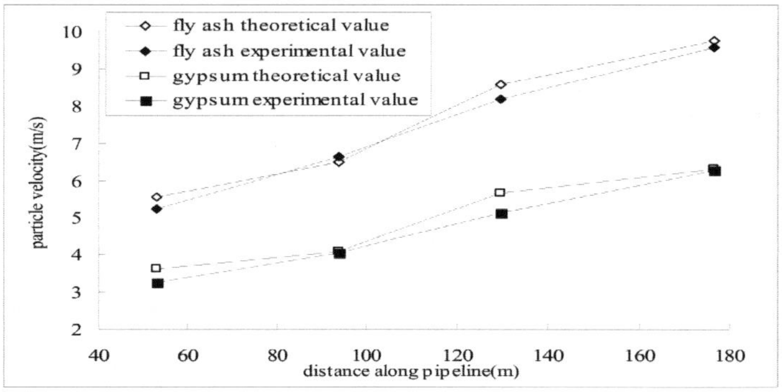

Fig.6 Relationship of particle velocity between theoretical and experimental value(0.36MPa)

5. CONCLUSIONS

1) Average particle velocity and particle velocity along pipeline can be measured by Gauge Pressure transmitters.
2) Average particle velocity increase with the increase of feeding pressure. When feeding pressure is a constant, average particle velocity increase with the reduction of density of materials, while particle velocity decreases with the reduction of void fraction.
3) Particle velocity increase gradually along pipeline conveying direction and feeding pressure. When feeding pressure is a constant, on the same point, particle velocity increase with the reduction of density, while high void fraction increase particle velocity.
4) Theoretical correlation equation of particle velocity versus powder characterization has been obtained with dimensional analysis and mathematical analysis as follows:

$$u_s = 1.29 \times 10^{-3} p^{0.5} D^{0.488} L^{0.408} \rho_s^{-1.06} \rho_g^{0.56} d_s^{-0.896} \varepsilon^{6.55} w^{-0.559}$$

It was shown that good accuracy is achieved by Error analysis. Therefore, it can be used to predict trends of particle velocity in the conveying process for different kinds of powder materials precisely.

ACKNOWLEDGEMENTS

This work was supported by the Natural Science Foundation of Shandong Province(grant no.Y2006F04) and Jinan Science & Technology Young Star Project(grant no.07103)

NAMENCLATURE

u_s	particle velocity	[m/s]
L_{ij}	distance between two gauge pressure transmitters	[m]
t_{ij}	the time interval of signal emergence of two gauge pressure transmitters	[s]

p	conveying pressure	[MPa]
L	length of pipeline	[m]
D	diameter of pipeline	[mm]
ρ_g	gas density	[kg/m^3]
ρ_s	particle density	[kg/m^3]
d_s	diameter of particle	[μm]
ε	void fraction	[%]
σ	angle of repose	[°]
w	moisture content	[%]
E	the ratio of pipe length and particle diameter	
Ψ	the ratio of pipe diameter and particle diameter	
Y	the ratio of gas density and particle density	

REFERENCES

Mainwaring, N. J. et al. (1987). "Permeability and Air Retention Characteristics of Bulk Solids Materials in Relation to Modes of Dense Phase Pneumatic Conveying," Bulk Solids Handling, 7(3), pp. 415-425.

Pan, R. (1999). "Material Properties and Flow Modes in Pneumatic Conveying," Powder Technology, 104, pp.157-163.

Peter, W. et al. (2003). "Minimum Transport Boundary for Horizontal dense Phase Pneumatic Conveying of Granular Materials," Powder Technology, 119, pp. 111-121.

Gilbertson, M. A. et al. (2003). "The Influence of Particle Size on the Flow of Fluidized Powders," Powder Technology, 131, pp. 197-205.

Luis Sanchez. et al. (2003). "Characterization of Bulk Solids to Assess Dense Phase Pneumatic Conveying," Powder Technology, 138, pp. 93-117.

Liu, Z. M. et al. (2006). "Particle Velocity and Pressure Drop in Long-distance Dense-phase Pneumatic Conveying of Fly Ash," The 5th International Symposium on Measurement Techniques for Multiphase Flows, Macau, China, Dec 10-13, pp.107-112

Journal of Physics: Conference Series **147** (2009) 012069 doi:10.1088/1742-6596/147/1/012069

Experimental Characterization of Solid Particle Transport by Slug Flow Using Particle Image Velocimetry

A Goharzadeh, P Rodgers

Department of Mechanical Engineering, The Petroleum Institute,
Abu Dhabi, United Arab Emirates

agoharzadeh@pi.ac.ae

Abstract. This paper presents an experimental study of gas-liquid slug flow on solid particle transport inside a horizontal pipe with two types of experiments conducted. The influence of slug length on solid particle transportation is characterized using high speed photography. Using combined Particle Image Velocimetry (PIV) with Refractive Index Matching (RIM) and fluorescent tracers (two-phase oil-air loop) the velocity distribution inside the slug body is measured. Combining these experimental analyses, an insight is provided into the physical mechanism of solid particle transportation due to slug flow. It was observed that the slug body significantly influences solid particle mobility. The physical mechanism of solid particle transportation was found to be discontinuous. The inactive region (in terms of solid particle transport) upstream of the slug nose was quantified as a function of gas-liquid composition and solid particle size. Measured velocity distributions showed a significant drop in velocity magnitude immediately upstream of the slug nose and therefore the critical velocity for solid particle lifting is reached further upstream.

1. INTRODUCTION

Many industrial processes, such as hydrocarbon production and processing, in the petroleum industry involve complex multiphase flows such as (oil/water, gas and sand mixtures). Understanding the mechanism of sand transport in multiphase flow lines has direct impact on estimation, design and detailed analysis of new generation of horizontal oil wells. For instance the increasing amount of sand in horizontal pipelines produces a stationary sand deposit which creates a pressure drop and affects the rate of production. The formation of a sand bed inside the pipeline during the shutdown process creates many engineering challenges particularly during the startup process (Takahashi *et al.*, 1989; Takahashi & Masuyama, 1991). The sand transport during the resuming operation can cause pipe blockage if the gas-liquid velocity becomes excessive (Takahashi & Masuyama, 1991). The frequently occurring gas-liquid flow situation appearing in pipeline is slug flow, classified as an intermittent flow of liquid and gas. The slug flow is defined as an intermittent phenomenon where a succession of long gas bubbles and liquid slugs appears in the pipeline, figure 1.

Note: *l* = Slug Nose Inactive Length (SNIL).

Fig. 1 Mechanistic representation of slug flow.

For near horizontal flow, the investigation of slug flow on sand transportation represents a key topic for design and improvement of production rate in oil and gas wells. The majority of studies published to-date on pipeline multiphase flows have focused on the dynamics of two-phases liquid-gas (Mandhane & Aziz, 1974; Fabre & Line, 1992; Manolis, 1995; Gopal & Jepson, 1998; Rosa, 2004, Woods *et al.,* 2006;) or liquid-solid transport (Doron & Barnea, 1995; Ould-Dris *et al.*, 1996; Matousek, 2001; Skudarnov & Lin, 2004; Ramadana *et al.*, 2005). Limited studies exist on the simultaneous transport of three phase flow containing solid particles, gas-liquid flow inside horizontal pipes (Turian *et al.* 1987; Angelsen *et al.*, 1989; Oudeman, 1993; Gillies *et al.*, 1997; Salama, 1998; Stevenson et *al.*, 2001; Stevenson & Thorpe, 2003; Oladele *et al.*, 2005; Danielson, 2007; Orell, 2007). Stevenson et *al.*, (2001) and Stevenson & Thorpe (2003) studied the dynamic of isolated particles in a multiphase flow line for a critical sand concentration $C < 50$ ppm. They predicted with good accuracy the sand particle velocity in intermittent slug flow by considering the slug flow as a hybrid between two steady flows: (i) stratified flow and (ii) simple hydraulic conveying. In addition they found that gas fraction does not influence directly the velocity of isolated sand particles. However their predictions are not appropriate for the present study which contains a high concentration of sand particles inside the pipeline. The influence of gas fraction for higher sand concentration has been studied by Oudeman (1993) for continuous feeding. Contrary to Stevenson *et al.*, (2001) and Stevenson & Thorpe (2003); Oudeman (1993) found that the gas fraction influences greatly the sand transport. However few studies have focused on the influence of slug flow on sand migration with initial stationary sand bed geometry. At low flow rates below the critical velocity U_c, it is assumed that the sand bed is stationary (with no sand particle motion) and occupies the bottom of the pipeline with a specific height. At the critical velocity U_c the sand particle is entrained from the stationary layer, transported along the pipe generating the bed load pattern.

It has been shown in previous published studies (Stevenson et *al.*, 2001; Stevenson & Thorpe, 2003) that particle transport in intermittent flow occurs primarily in "slug" phase rather than the film phase. The particle does not move immediately on the slug nose arrival. Instead the slug passes a distance (*l*) downstream before it entrains the first particle on the top of the bed. In this study the distance between the slug front and the first moving solid particles is referred to as the Slug Nose Inactive Length (SNIL), figure 1. The term inactive is related to static behavior of solid particle inside the SNIL. The importance of this distance (*l*) is significant for numerical simulation of particle transport. If this phenomenon is not taken into account, in simulation, prediction of particle transport will be over estimated. Previous studies (Stevenson et *al.*, 2001; Stevenson & Thorpe, 2003) have observed this phenomenon but no existing study has quantified this parameter in terms of flow rate conditions. Stevenson *et al.*, (2001) investigated the influence of the slug flow on particle transport in near horizontal pipeline. They assumed that the observed delay in pick-up of particles on arrival of the slug nose can be explained by considering the length required for the turbulence experienced in the slug nose to diffuse into the calmer region. They exposed an analogy based on a turbulent 2D jet flow where there is a region of turbulent, high-velocity fluid meeting a slower fluid. This analogy lead Stevenson & Thorpe (2003) to propose a range of 3 to 4 pipe diameters for the SNIL, without details about the flow condition. No further investigation has since been published on the influence of the flow conditions on the SNIL.

In order to understand the influence of slug flow on solid particle migration, it is essential to experimentally measure the velocity distribution inside the slug body and at bottom liquid layer. Previous experimental studies have measured local velocity distributions inside slug flow using Laser Doppler Velocimetry (Kvernvold *et al.*, 1984), Hot-Film Anemometry (Sharma *et al.*, 1998; Lewis *et al.*, 2002), Photomatic Dye Activation (Kawaji *et al.*, 1995), Nuclear Magnetic Resonance (Barberon & Leblond, 2001) and Digital Image Analysis (Gopal & Jepson, 1997). The local mean velocity profiles obtained from the various methods showed an axisymmetric behavior with highest velocity located at the upper region of the pipe. However, full 2D field velocity distribution is required to obtain fundamental insight into particle solid migration inside slug flow body. For vertical slug flow, a limited numbers of studies have presented 2D flow field measurements using techniques such as Particle Image Velocimetry (van Hout *et al.*, 2002; Nogueira *et al.*, 2003, 2006) and Advanced Electromagnetic Flowmetry (Kang *et al.*, 2004). Nogueira *et al.* (2003, 2006) combined the Particle Image Velocimetry (PIV) with Pulsed Shadowgraphy (PS) and obtained the 2D velocity distribution behind rising Taylor Bubble inside vertical pipe. The same method has been performed by Carpintero-Rogero *et al.*, (2006) for horizontal two phase flow. Experimental results obtained by Carpintero-Rogero *et al.*, (2006) were focused on the validity of the method for measuring 2D velocity field. They concluded that the simultaneous PIV (using fluorescent particle as markers) and PS is suitable for the investigation of stratified wavy and elongated bubble flow and conditionally acceptable for slug flows. They also observed that measured velocity distribution decreases in accuracy if the gas phase is highly mixed with the liquid phase, as only a part of tracer particles can be seen. Consequently measurement of the velocity distribution is challenging near the gas-liquid interface.

The present study investigates the influence of gas fraction in slug flow on preloaded solid particle bed geometry during startup process. Two experiments are presented: (i) Influence of slug length on solid particle transportation using high speed CCD camera (multiphase air-water-solid particle loop); (ii) Measurement of the velocity distribution inside slug body using combined Particle Image Velocimetry (PIV) with Refractive Index Matching (RIM) and fluorescent tracers (two-phase oil-air loop). Combining these experimental analyses, the results provide an insight into the physical mechanism of solid particle transportation due to slug flow.

2. EXPERIMENTAL SETUP AND FLOW VISUALIZATION SYSTEMS

Particle transport and flow field measurements are undertaken in three-phase and two-phase flow loops, respectively. Solid particle transportation measurements are performed using digital image analysis. Slug flow velocity measurements are obtained using Particle Image Velocimetry (PIV).

2.1 Three Phase Flow Loop (Air-Water-Glass Beads)

The overall design of the horizontal multiphase flow loop was specified according to the laboratory space available to house the test facility. The test facility was limited to 3m in width and 6m in length. To optimize the utilization of this space constraint, a U-shape test facility was designed, with the layout shown in figure 2. The total flow length of the loop is 14 m, with the inner pipe diameter equal to 25.4 mm. Piping is constructed from Plexiglas to permit flow visualization, with the exception of the curved section which was made from copper to permit fabrication. Water is circulated using a centrifugal pump, which can supply single phase flow rate of up to 120 lpm (Re = 100,000) to the working section, measured by a digital mass flow meter range, 0 to 10 lpm (GPI 09 Series). The water reservoir tanks have capacity of approximately 450 liters each. Air flow is obtained from a utilities compressed air line that supplies clean air, free of lubricants, and measured by a digital flow meter, 0 to 50 lpm (Cole-Parmer CP5990). Two phase air-water composition is controlled by separate flow valves on the air and water lines. Air flow was injected downstream of water flowmeter using a concentric annuli design. In order to remove potential surface curvature effects of the pipeline for flow visualization, the working section was submerged inside a transparent rectangular box filled with silicon oil (Dow Corning 556) having a refractive index of n = 1.46 which is close to Plexiglas n = 1.48. Three-phase flow measurements were realized at 24 °C.

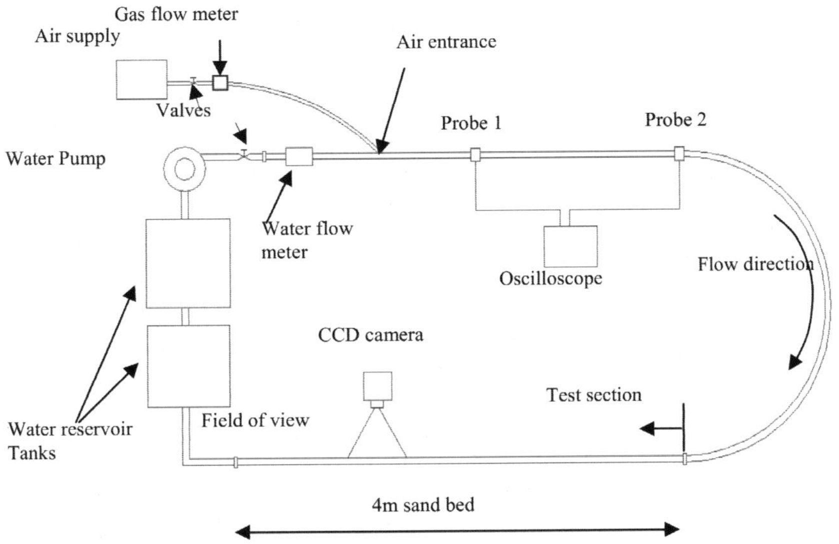

Fig. 2 Horizontal Multiphase Flow Loop (Air-Water-Solid Particle).

The gas-liquid flow structures were visualized in the vertical plane of the working section using high speed CCD camera (Flowsense M2/E 8bit). Mixture compositions were chosen to ensure that air-water flow was within the slug pattern zone (Mandhane & Aziz, 1974; Goharzadeh *et al.*, 2008). Various slug air-water flow rate mixture compositions of 3 to 11 lpm (air) and 3 to 5 lpm (water) over preloaded solid particle bed were visualized. As outlined by Bain & Bonnington (1970), sand transport pattern classification in terms of particle size, density, and shape have a considerable influence on the particle bed geometry. It was therefore decided to investigate the behavior of glass beads with a mean diameter of 0.4 - 0.8 mm and 2.00 mm for the above flow compositions. A uniform glass bead bed of h = 5 mm thickness was laid in the test section over a 4 m length, which represents approximately 160 pipe diameters.

2.2 Two Phase Flow Loop (Air-Oil)

The test facility (figure 3) consisted of a 1.5m horizontal transparent cylindrical working section having an inner diameter of 25 mm. Made from Plexiglas, the working section is connected to a centrifugal pump (Q_{max} =30 lpm) with a regulating valve and flowmeter. Air is injected through a 2 mm nozzle fitted to the top of the pipeline, downstream of the working section. To obtain steady laminar flow with the minimum of bubbles generated from air injection, a viscous silicon oil (Dow Corning 556) was selected as the working liquid. The kinematic viscosity of the silicon oil was measured to be $\nu = 18.6\ 10^{-6}$ m²/s at the working temperature of T = 34.7 °C. As for the previous test facility, surface curvature effects of the pipeline were eliminated using a transparent rectangular box fitted around the working section and filled with the same silicon oil as the working liquid.

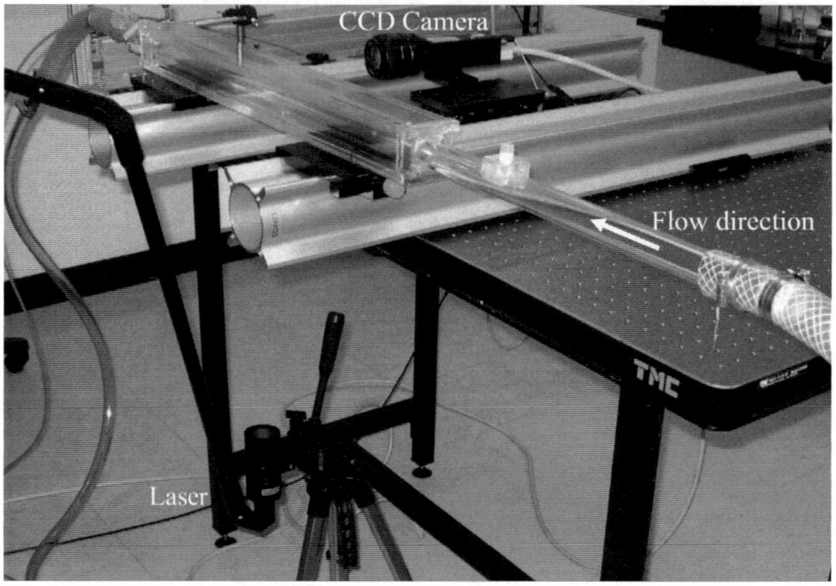

Fig. 3 Horizontal two phase flow (air-oil) experimental setup.

The Particle Image Velocimetry (PIV) was employed using fluorescent particles tracers (Rhodamine 6G) excited by a planar laser sheet from a pulsed YAG laser (NewWave Solo $\lambda = 532$ nm, P = 1.5 W) with a wavelength suitable for the absorption band of the dye ($\approx 500 - 600$ nm). The light is re-emitted with an emission maximum wavelength at approximately 650 nm. The planar laser sheet penetrated the central vertical plane of the working section, with the CCD camera (Flowsense M2/E 8bit) placed perpendicular to the laser sheet. To minimize the light scattering effect of traveling air bubbles at the top part of the pipeline the laser sheet was illuminated beneath the working section. In addition when recording visualization of two phase flow (air-oil), the CCD camera was fitted with a standard red interference filter to eliminate scattering light received from the two phase air-oil interface. A field of view of 85 × 25 mm was obtained using a focal length of $f = 28$ mm. Full-frame images of 1600 × 1186 pixels were acquired and transferred to a computer via a frame grabber.

Using the Flowmap System software provided by the Dantec Dynamics, the 2-D PIV image is divided into 16 × 16 pixels sub-regions with 50% overlap and average particle velocities is calculated using 100 images with the cross correlation method. The time interval between each image couple was 125 ms. The pulse separation time was adjusted to 0.1 ms.

Two phase velocity profiles in the slug body were obtained using a fluid mixture composition of $Q_{oil} = 6.7$ lpm, corresponding to a Reynolds number Re = 306, and $Q_g = 2.4$ lpm.

3. RESULTS

3.1 PIV Validation

To verify the upstream single phase flow conditions as fully developed laminar flow, as well as PIV velocity measurement accuracy, the obtained velocity profile was compared with an exact analytical solution, Munson *et al.* (2006):

$$u(r) = U_{max}\left[1 - \left(\frac{2r}{D}\right)^2\right] \tag{1}$$

where U_{max} is the maximum velocity at the center of the pipe, r and D are the radius and diameter of the pipe, respectively.

Journal of Physics: Conference Series **147** (2009) 012069 doi:10.1088/1742-6596/147/1/012069

For a Reynolds number (U_{mean} D/v) = 306, excellent agreement was obtained between the measured velocity profile and corresponding analytical prediction, as illustrated in figure 4. As an additional error analysis assessment, the pipe volumetric flow rate (Q) was measured using a flowmeter (GPI 09 Series). Based on the measured volumetric flow rate, mean velocity U_{mean} was calculated to be 0.224 m/s. For the corresponding PIV analysis, the measured maximum velocity is U_{max} = 0.416 m/s, which corresponds to an averaged pipe velocity of $U_{mean} = \frac{U_{max}}{2} = 0.208 m/s$. Thus the estimated error between the mean velocity measurements obtained using the flowmeter and PIV is approximately 7 %. Combined, both measurement error assessment approaches provide confidence in the accuracy of the PIV velocity measurement technique employed in this study.

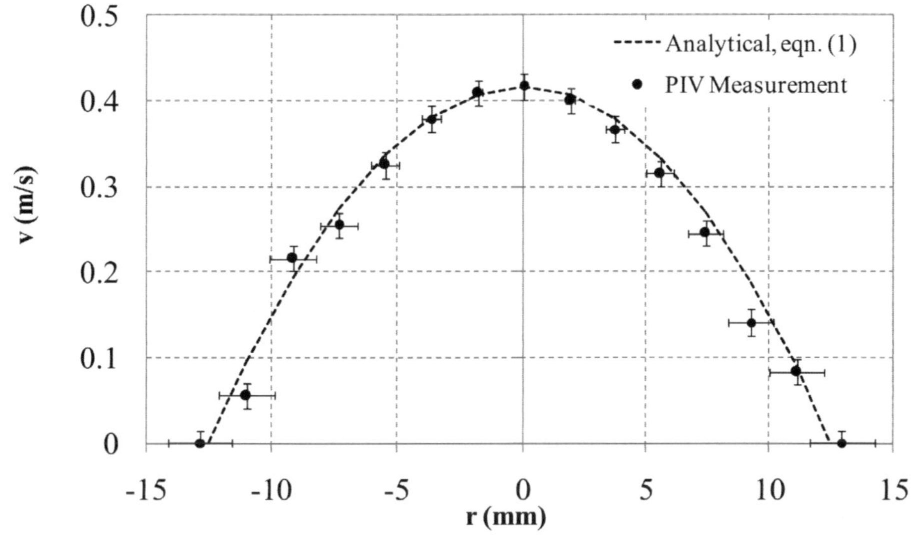

Fig. 4 Comparison of measured and predicted velocity profiles for fully developed single phase laminar flow (Re = 306).

3.2 Slug Nose Inactive Length

Measurement of SNIL (*l*) is illustrated in figure 5. The distinction between the inactive and active zones in terms of solid particle transport, upstream of the slug nose, is characterized by a disturbance at the liquid-porous interface.

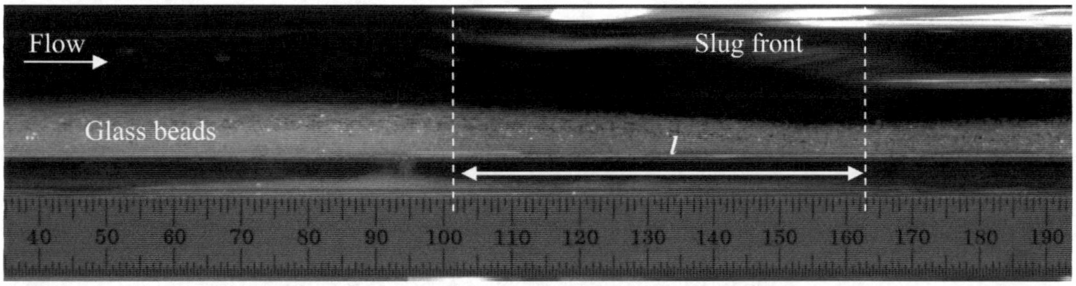

Note: Glass bead bed (0.4 - 0.8 mm diameter) 5.5 mm thick, slug flow Q_w= 4.0 lpm; Q_g= 7.0 lpm

Fig. 5 Measured Slug Nose Inactive Length.

The SNIL measurements given in figure 6 are average values for the number of slugs passing over the particle bed in the acquisition session. The SNIL results are presented in a non-dimensional ratio (l/D) with D the inner pipe diameter.

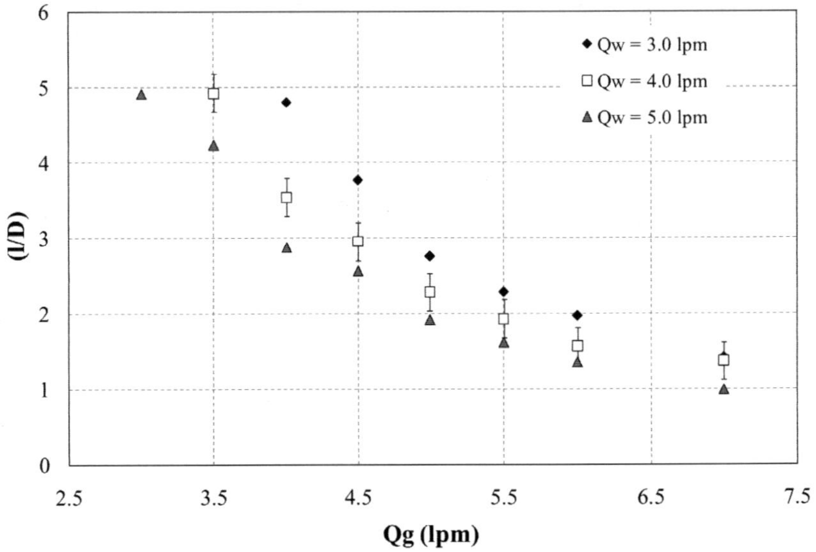

a) Glass beads, d = 0.4 - 0.8 mm.

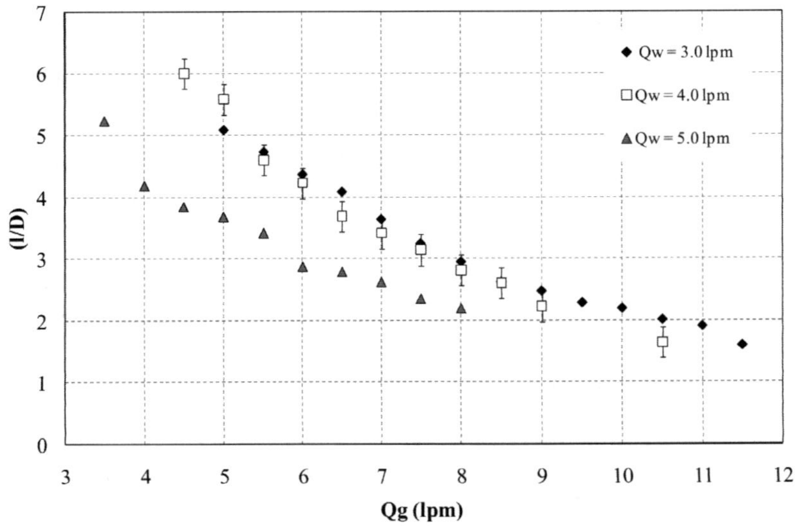

b) Glass beads, d = 2.0 mm.
Note: Particle bed average thickness = 5.5mm

Fig. 6 Measured slug nose inactive length (SNIL) ratio as a function of liquid gas composition and solid particle size.

The SNIL range measured is between 1 and 6 times the pipe diameter (figure 6). This is contrary to Stevenson *et al.* (2001) who postulated that SNIL can be generally simplified to be a factor 3 and 4 times the pipe diameter.

It was found that the SNIL depends significantly on the air-gas flow composition. For a given water flow rate, SNIL decreases with increasing gas flowrate. This trends also holds for a given gas flowrate with increasing liquid flowrate. In addition the influence of the particle size has been highlighted in figures 6 (a) and (b). For a fixed mixture composition, SNIL increase with particle diameter increase, indicating lower solid particle transportation rate for the 2 mm diameter glass beads particles.

3.3 Velocity Distribution inside Slug Body

In figure 7 a visualized slug flow sequence with corresponding velocity field measurements in the liquid region is presented. The velocity distribution in the slug nose is shown in figure 7(a) where a significant change on the velocity magnitude is observed along the stream-wise flow direction. Corresponding vertical velocity profiles of the horizontal component at four different downstream positions (x coordinate) are plotted in figure 8. The velocity magnitude in the liquid film is half that in the slug nose region (figure 8(a)). The same phenomenon appears in the slug front region (figure 7(b)) with greater magnitude change. The velocity profile is significantly affected when the air bubble passes above the liquid flow. The local velocity profiles, obtained at four different steam-wise positions (figure 8(b)), shows that the velocity magnitude increases by factor of five at the upper region of the pipe. However, in the same locations the velocity magnitude changes smaller at the lower region of the pipe. Upstream from the air bubble (figure 7(c-d)), the velocity magnitude increases and a laminar velocity profile is observed in this single phase flow region (figure 8(c)). It should be noted that the averaged velocity profile reaches its maximum magnitude in the region furthest upstream of the slug front (figure 7(d)). In terms of assessing the SNIL, the velocity distribution shows a drop in velocity magnitude immediately upstream of the slug nose and therefore the critical velocity for solid particle lifting is reached further upstream.

The slug flow is characterized by the intermittent flow of slugs, and therefore the observed velocity profile formation is cyclic. This observation is confirmed by comparing the velocity profiles in figure 8(a) at x = 80 mm with figure 8(c) which are comparable in magnitude. In addition this mechanism results in intermittent solid particle transportation in slug flow (Goharzadeh *et al.,* 2008).

(i) PIV Image.

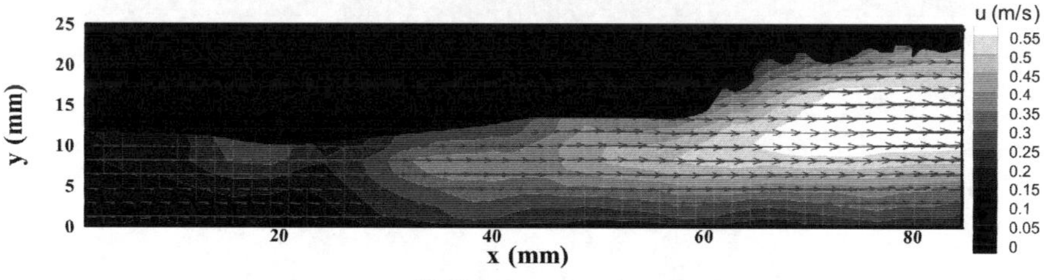

(ii) Velocity vector plot.

a) Bubble front and slug nose (t = 0 s).

(i) PIV Image.

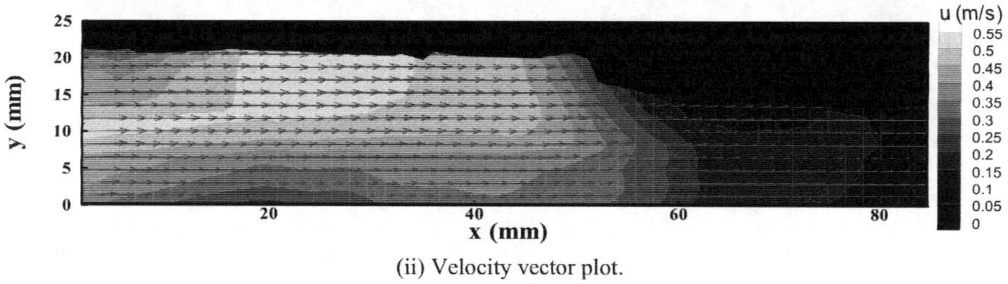

(ii) Velocity vector plot.

b) Slug front and bubble end (t = 0.125 s).

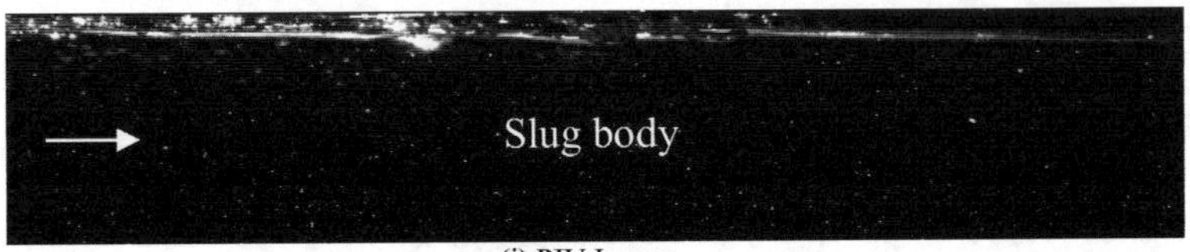

(i) PIV Image.

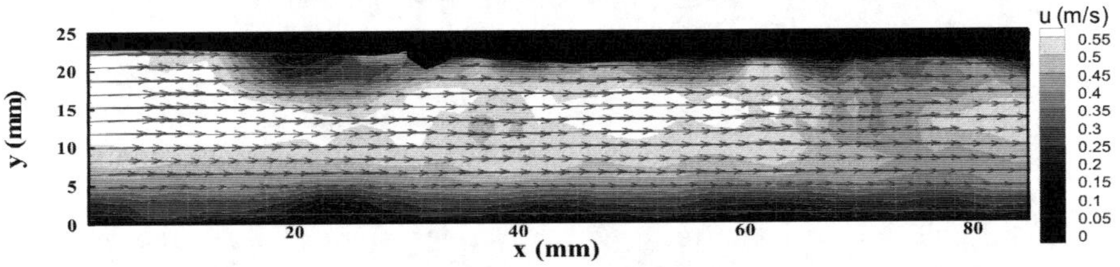

(ii) Velocity vector plot.

c) Slug body (t = 0.25 s).

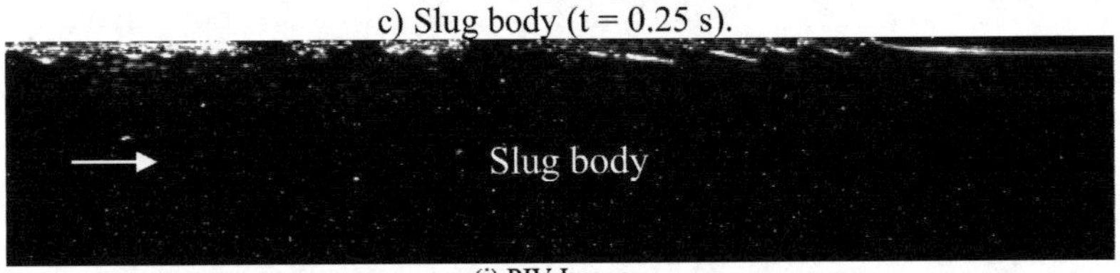

(i) PIV Image.

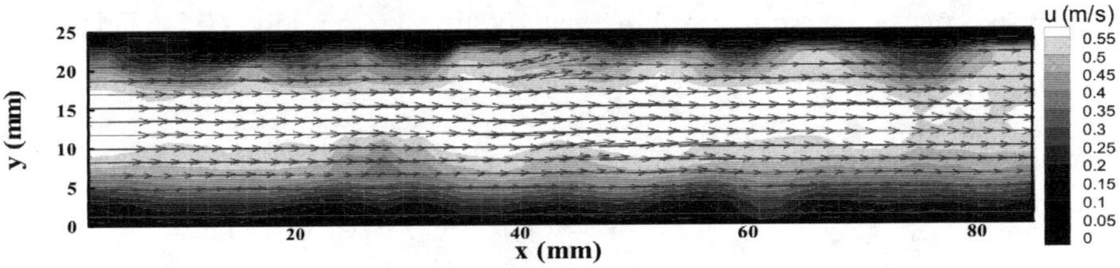

(ii) Velocity vector plot.

d) Slug body (t = 0.375 s).

Fig. 7 Measured slug flow velocity distribution (Q_{air} = 2.4 lpm, Q_{oil} = 6.7 lpm).

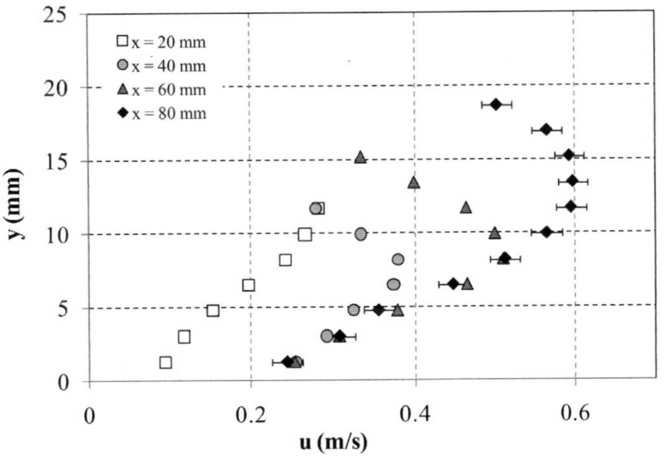

a) In the liquid film (x = 20 and 40 mm) and slug nose (x = 60 and 80 mm) (t=0s).

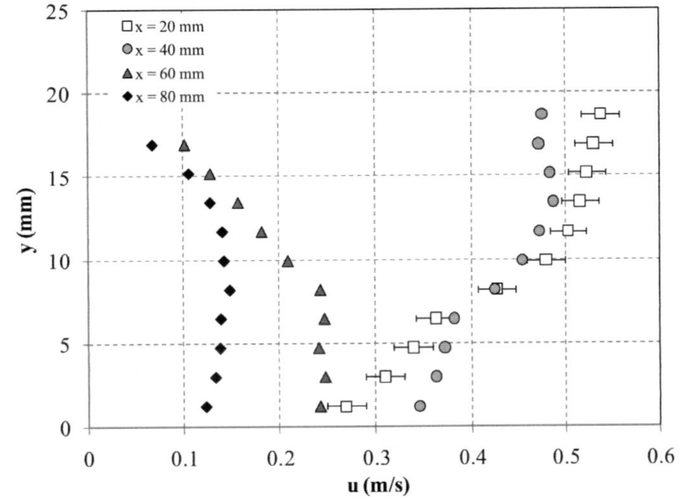

b) In slug front body (x = 20 and 40 mm) and liquid film (x = 60 and 80 mm) (t=0.125s).

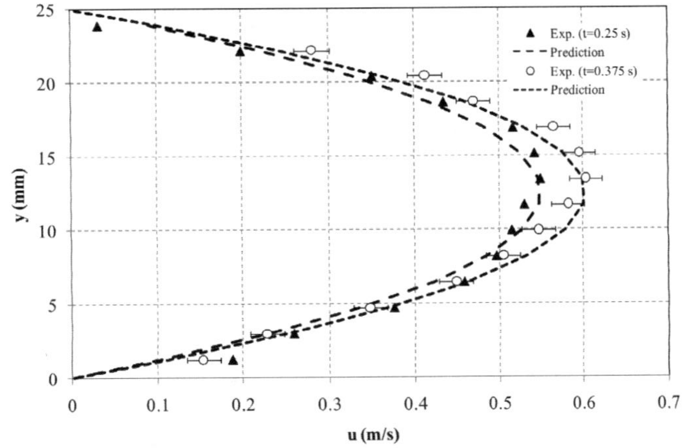

c) In the slug body for two different time (t = 0.25s and t = 0.375s).

Fig. 8 Measured vertical profile of horizontal velocity component in slug flow. (Q_g = 2.4 lpm, Q_{oil} = 6.7 lpm).

4. CONCLUSIONS

Particle transport and flow field measurements are undertaken in two-phase and three phase flow loops, respectively. Solid particle transportation measurements are performed using digital image analysis. Slug flow velocity measurements are obtained using Particle Image Velocimetry (PIV). Combining these two experimental analyses, an insight was provided into the physical mechanism of solid particle transportation due to slug flow. It was shown that the slug body significantly influences solid particle mobility. The physical mechanism of solid particle transportation was found to be discontinuous. The inactive region (in terms of solid particle transport) upstream of the slug nose was quantified as function of gas-liquid composition and solid particle size. Measured velocity distributions show a significant drop in velocity magnitude immediately upstream of the slug nose and therefore the critical velocity for solid particle lifting is reached further upstream.

ACKNOWLEDGEMENTS

This study was financially supported by ADNOC (Abu Dhabi National Oil Company). The authors are grateful to Mr. Chokri Touati and Mr. Shrinivas Bojanampati for their assistance in the measurements presented.

NOMENCLATURE

C	Sand concentration [ppm]
d	Sand particle diameter [m]
D	Pipe diameter [m]
f	Focal length of the camera [m]
h	Height of glass bead bed [m]
n	Refractive index n [-]
Re	Reynolds number [-]
Q_g	Gas flow rate [lpm]
Q_w	Water flow rate [lpm]
Q_{oil}	Oil flow rate [lpm]
t	Time [s]
U_{mean}	Averaged Velocity of liquid particles [m/s]
U_c	Critical velocity for lifting solid particles [m/s]
U_{max}	Maximum velocity of liquid particles [m/s]
x	Axial coordinate [m]
Greek letters	
λ	Wavelength [nm]
ν	Kinematic viscosity [m^2/s]

REFERENCES

Angelsen, S., Kvernvold, O., Lingelem, M., Olsen, S. (1989). "Long-distance transport of unprocessed HC sand settling in multiphase flow lines," Proceedings of the 4th international Conference on Multiphase Flow, Cannes, France, pp. 149-170.

Bain, A.G., and Bonnington S.T. (1970). The hydraulic transport of solids by pipeline, Pergamon Press.

Barberon, F., and Leblond, J. (2001). "Intermittent two-phase flow study by NMR," C. R. Acad. Sci. Paris, Chimie - Chemistry 4, pp. 853–856

Carpintero-Rogero, E., Kross, B., Sattelmayer, T. (2006). "Simultaneous HS-PIV and shadowgraph measurements of gas-liquid flows in a horizontal pipe," 13th Int. Symp. on Applications of Laser Techniques to Fluid Mechanics Lisbon, Portugal, 26-29 June.

Danielson, T.J. (2007). "Sand transport modeling in multiphase pipeline," Offshore Technology Conference.

Doron, P. and Barnea, D. (1995). "Flow pattern maps for solid-liquid flow in pipes," Int. J. Multiphase Flow, 22(2), pp. 273-83.

Fabre, J., and Line, A. (1992). "Modeling of two-phase slug flow," Annu. Rev. Fluid Mech. 24, pp. 21-46.

Gillies, R.G., McKibben, M.J., Shook, C.A. (1997). "Pipeline flow of gas, liquid and sand mixture at low velocities," The Journal of Canadian Petroleum Technology, 36(9), pp. 36-42.

Goharzadeh, A., Rodgers, P., Touati, C. (2008). "Influence of slug flow on hydraulic sand dune migration in horizontal pipelines," Proceedings of IMECE 2008 ASME International Mechanical Engineering Congress and Exposition October 31 - November 6, Boston, Massachusetts, USA, IMECE2008-67187.

Gopal, M. and Jepson, W.P. (1998). "The study of dynamic slug flow characteristics," Transactions of the ASME-Journal of Energy Resources Technology, 120, pp. 102-105.

Kang, D.K., Ahn, Y.C., Oh, B.D., Kim, M.H. (2004). "Advanced electromagnetic flowmetry for slug flow: numerical signal prediction and calibration," Int. J. Multiphase Flow, 30, pp. 585–614

Kawaji, M., Ali, M., Ciastek, A., Lorencez, C. (1995). "Study of liquid flow structure in horizontal concurrent gas–liquid slug flow," Proceedings of the ANS-THD 8, National Heat Transfer Conference, August 5–9, Portland, Oregon, pp. 79–88.

Kvernvold, O., Vindøy, V., Søntvedt, T., Saasen, A., Selmer-Olsen, S. (1984). "Velocity distribution in horizontal slug flow," Int. J. Multiphase Flow, 10, pp. 441–457.

Lewis, S., Fu, W.L., Kojasoy, G. (2002). "Internal flow structure description of slug flow-pattern in a horizontal pipe," Int. J. Heat Mass Transfer, 45, pp. 3897-3910.

Mandhane, J. M. and Aziz, K. (1974). "A flow pattern map for gas-liquid flow in horizontal pipes," Int. J. Multiphase Flow, 1(4), pp. 537-553.

Manolis, I.G. (1995). "High pressure gas – liquid slug flow," Ph.D. Thesis, University of London, London, UK.

Matousek, V. (2001). "Pressure drops and flow patterns in sand mixture pipes," Experimental and Thermal Fluid Science, 26, pp. 693-702.

Munson, B. R., Young, D. F., Okiishi, T. H. (2006). Fundamentals of fluid mechanics, John Wiley & Sons, New York.

Nogueira, S., Sousa, R.G., Pinto, A.M.F.R., Riethmuller, M.L., Campos, J.B.L.M. (2003). "Simultaneous PIV and pulsed shadow technique in slug flow: a solution for optical problems," Exp. Fluids 35, pp. 598-609.

Journal of Physics: Conference Series **147** (2009) 012069 doi:10.1088/1742-6596/147/1/012069

Nogueira, S., Riethmuller, M.L., Campos, J.B.L.M., Pinto, A.M.F.R. (2006). "Flow in the nose region and annular film around a Taylor bubble rising through vertical columns of stagnant and flowing Newtonian liquids," Chemical Engineering Science, 61, pp. 845–857.

Oladele, O.B., Kurt, M.R., Catalin, T. (2005). "Particle holdup profile in horizontal gas-liquid-solid multiphase flow pipeline," Chemical Engineering &Technology, 28(12), pp. 1546-1653.

Oudeman, P. (1993). "Sand transport and deposition in horizontal multiphase trunklines of subsea satellite developments," SPE Production & Facilities, 8(4), pp. 237-241.

Ould-Dris, A., Molodtsof, Y., Large, J.F. (1996). "A classification and design method for moving bed flow in pipes," Powder Technology, 87, pp. 49-57.

Orell, A. (2007). "The effect of gas injection on the hydraulic transport of slurries in horizontal pipes," Chemical Engineering Science, 62, pp. 6659–6676.

Ramadana, A., Skalleb, P., Saasenc, A. (2005). "Application of a three-layer modeling approach for solids transport in horizontal and inclined channels," Chemical Engineering Science, 60, pp. 2557–2570.

Rosa, E.S. (2004). "Slug flow structure in horizontal slug flow," Thermal Engineering, 3(2), pp. 151-160.

Salama, M.M. (1998). "Sand production management," Offshore Technology Conference, Annual Proceedings, Field Drilling and Development Systems, 4, pp. 743-751.

Sharma, S. Lewis, S., G. Kojasoy, G. (1998). "Local studies in horizontal gas–liquid slug flow," J. Nucl. Eng. Des. 184, pp. 305–318.

Skudarnov, P.V. and Lin, C.X. (2004). "Double-species slurry flow in a horizontal pipeline," Transactions of the ASME-Journal of Fluids Engineering, 126, pp. 125-132.

Stevenson, P., Thorpe, R.B., Kennedy, J.E., McDermott, C. (2001). "The transport of particles at low loading in near-horizontal pipes by intermittent flow," Chemical Engineering Science, 56, pp. 2149-2159.

Stevenson, P. and Thorpe, R.B. (2003). "Energy dissipation at the slug nose and the modeling of solids transport in intermittent flow," Canadian Journal of Chemical Engineering, 81(2), pp. 271-278.

Takahashi, H., Masuyama, T., Noda, K. (1989). "Unstable flow of a solid-liquid mixture in a horizontal pipe," Int. J. of Multiphase Flow, 15(5), pp. 831-841.

Takahashi, H. and Masuyama, T. (1991). "Stability criterion for the surface of the deposit bed of solid-liquid flows in horizontal pipes," Journal of Chemical Engineering of Japan, 24(3), pp. 319-325.

Turian, R.M., Hsu, F.L ., Ma, T.W. (1987). "Estimation of the critical velocity in pipeline flow of slurries," Powder Tech., 51, pp. 35–47.

van Hout, R.V., Gulitsky, A., Barnea, D., Shemer, L. (2002). "Experimental investigation of the velocity field induced by a Taylor bubble in stagnant water," Int. J. of Multiphase Flow 29, pp. 579–596.

Woods, B.D., Fan, Z., Hanratty, T.J. (2006). "Frequency and development of slugs in a horizontal pipe at large liquid flows," Int. J. of Multiphase Flow, 2(20), pp. 902-925.

Journal of Physics: Conference Series **147** (2009) 012070 doi:10.1088/1742-6596/147/1/012070

Free Fall Characteristics of Particle Clusters in a Vertical Pipe

K Nakashima [1], Y Johno [1] and T Shigematsu [1]

[1] Sasebo National College of Technology 1-1, Okishin-cho, Sasebo-shi, Nagasaki, 857-1193, JAPAN

nakash@sasebo.ac.jp

Abstract. When powder forms a weak interaction particle cluster in a gas-solid flow, the rate of fall of the cluster exceeds the terminal velocity of the individual particles (Slack, 1963; Marzocchella et al., 1991). However, the relationship between the unsteady characteristics of the free-fall of the particle cluster and the geometric condition of the experiment is not clear. We performed a simple experiment in which powder of a certain mass falls in a vertical pipe. When the powder falls in the vertical pipe, the distribution length of the powder expands, and the particle volume fraction is dense in the lower part, and is thin in the upper part. The fall velocity of the lower edge of the powder cluster and the flow rate of air generated by the powder fall were measured. We obtained the following results. The relative velocity of free-fall of the particle cluster has no relation to the individual particle diameters. The characteristic of a particle cluster exists unless the cluster has very high void fraction.

1. INTRODUCTION

In industry, we can often see an operation in which powder of a certain mass falls freely in a vertical pipe, for example, a powder packaging machine. A powder packaging machine is one which fills up small bags with a powdery material, such as food, medical supplies or a chemical industry product, individually with a certain constant mass. When the powdery material is supplied to a small bag, a vertical pipe is used for a supply chute in the powder packaging machine, as in Fig. 1. In such a case, powder of a certain mass falls in a vertical pipe.

Concerning a problem in this operation, if the particle diameter or the particle density is small, automatic control of the powder packaging machine is difficult, because particles are scattered and the distribution length of the powder is prolonged. In the real operation of the powder packaging machine, most of the powdery materials are A or B particles of Geldart's classification. And if the physical properties of particles approach the C group of Geldart's classification, the powder packaging machine causes a serious sanitary error in production activity because of the powder scattering. Therefore, a quantitative survey of the operation in which powder of a certain mass falls in a vertical pipe is important in the field of food machine industry for determining the most suitable design, development and use.

Concerning the flow characteristics of sedimentation of solid particles under gravity while still fluid, if the particles behave with uniform dispersion and uniform velocity, a method that corrects the drag force of a single particle based on Stoke's law by function of void fraction is generally used (Steinour, 1944; Richardson and Zaki, 1954; Richardson and Meikle, 1961; Famularo and Happel, 1965). And when the void fraction is comparatively high this method is effective, for example $\varepsilon > 0.6$ according to the study of Steinour (1944).

When particles fall with a low void fraction and non-uniform dispersion, the powder forms particle clusters (Slack, 1963). When the powder forms a particle cluster, it is thought that the particles and the fluid included in the cluster, stand still relatively, and move together. It is thought that the particle cluster has begun to attract attention from Slack's experiments (1963). In this experiment, the spherical clusters (25-90 [cm] diameter) made up of spherical particles belonging to A group of Geldart's classification were caused to fall through still air. Slack (1963) showed that the fall velocity of the cluster exceeded the terminal velocity of the individual particles in all cases of the experiment, and he expected that if powder forms a particle cluster the cluster can be stable with a bowl shape. On the other hand, Marzocchella et al. (1991) investigated the release of a cylindrical cluster of solid particles in still air inside a two-dimensional column observed by a high-speed video camera. Powders belonging to A group, B group and D group of Geldart's classification were used in this experiment. Marzocchella et al. (1991) expected that clusters consisting of A group particles fall almost like rigid bodies and that clusters can be considered as bodies impermeable to gas. Subsequently, particle clusters attracted attention in the field of fluidized beds. Some investigation was performed using a high-speed camera, and the shape and the scale structure of clusters formed in a fluidized bed are clarified by the following studies.

Rhodes et al. (1992) investigated the flow pattern of a cluster near the wall of a fluidized bed. Hatano et al. (1994) investigated the flow pattern of a cluster near the wall and central area of a fluidized bed by using a microscope visualization. Tsukada et al. (1997) investigated the three-dimensional structure of clusters formed in a circulating fluidized bed by using a scanning laser sheet technique. Lackermeier et al. (2001) investigated the internal flow structure of the circulating fluidized bed by using optical endoscopes. Three characteristic shapes of the particle cluster were observed: paraboloid heading downward, parabola in vertical cross-section heading upward and vertical film along the channel.

There are a few studies on the flow characteristics of the free-fall of the particle cluster. Marzocchella et al. (1991) evaluated the fall velocity of the particle cluster quantitatively in an experiment of circular clusters with two-dimensional flow. They showed that the nonsteady curve of the free-fall velocity of a particle cluster could be calculated based on a one-dimensional equation of motion on a single particle. In that calculation, the cluster diameter and the bulk density which were measured before the fall were substituted for the particle diameter and the true density respectively.

While not an experiment for the particle cluster, Ogata et al. (2001) investigated a powder jet exhaust positioned downward from a hopper. In this study, the velocity distribution of particles at the jet central axis was compared with the curve calculated based on the one-dimensional equation of motion of a single particle as in Marzocchella et al. (1991). The powder jet resembles the free-fall of the particle cluster at the points that the void fraction is low at the hopper exit and that the state of low void fraction is kept downward. Therefore, though Ogata et al. (2001) measured in a steady state, it seems that the curve of a single particle suggests free-fall characteristics of the particle clusters.

On the other hand, approximately realistic simulation of numerical analysis considering interplay between fluid and particles became possible by the contribution of the development of calculation technology. Zhang et al. (2008) calculated a two-dimensional fluidized bed by coupling the Discrete Element Method (DEM) and Direct Numerical Simulation (DNS). Tsuji et al. (2008) calculated a three-dimensional fluidized bed by coupling a hard sphere model and DNS. These studies explained visually the formation of the cluster and the process of the dispersion. The U-shaped cluster structure that was observed in experiments with a high-speed camera appeared in these calculation, too.

If limited to a vertical pipe flow, there are few studies on powder free-fall. Most studies of powder flow in a vertical pipe are analyzed for pneumatic transport or for a standpipe of the circulating fluidized bed. As an example of powder free-fall in a vertical pipe, there is Matsushita's experiment (2002, 2003) on transition of powder slugging in a field of the statistical physics of the pattern formation. In this experiment, it was found that there exists a critical in-flow rate for the formation of density waves, and this result may suggest the condition for generation of the particle cluster.

These conventional studies often pay attention to dynamic mechanisms, for example the mathematical characteristics of the phenomenon, the shape of the cluster or the process of formation and dispersion. However most of these studies are not useful in the design of a powder packaging machine. Therefore, in this study, we performed a simple experiment in which powder of a certain mass falls through still air in a vertical pipe, and measured the velocity of the lower edge of the powder cluster and the discharged air volume generated by falling powder by video camera, and investigated the characteristic distribution of the powder free-fall. The results in this paper are basic data which are useful for the design of powder packaging machines.

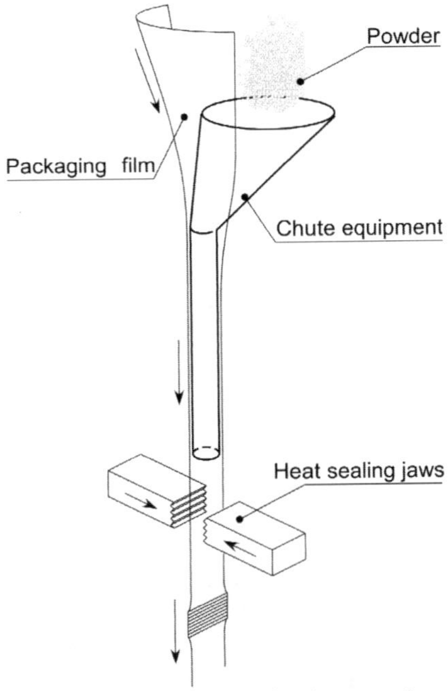

Fig. 1 Schematic view of a supply chute in the powder packaging machine

2. EXPERIMENTAL APPARATUS AND CONDITIONS

Fig. 2 shows a schematic of the experimental apparatus in this work. It is constituted from transparent acrylic pipe for the powder fall section and glass pipe for the volumetric flowmeter. The diameter and length of the pipe through which powder falls is ϕ 18 [mm] and 1000 [mm] respectively. The lower part of the acrylic pipe is connected to conductive materials which are grounded, so that the experiment was not affected by static electricity. The cross section of the device where powder starts to fall is shown in the upper left in Fig. 2. It is constituted from two acrylic pipes which are connected by a flange. PET resin film is inserted between the flange, and the powder is built up naturally on the film. When the film between the flange is pulled out the powder starts falling. The diameter of the volumetric flowmeter is ϕ 9.2 [mm]. In all experiments, the tube connected to the flowmeter is removed and the soap film is prepared in the upper part of the flowmeter. The flowmeter is set up so that the soap film moves from the upper to the lower, so that movement resistance by the weight of the soap film is reduced. The experiments are recorded on videotape (the frame interval is 1/30 [s]). The coordinate z is for vertical downward. The z-axis of the lower edge of the powder cluster, of the upper edge of the powder cluster and of the soap film of the flowmeter were measured.

Table 1 shows powder properties and experimental conditions. The mean particle diameter and the powder mass per experiment are d_p and M respectively. The powders are glass beads which were sieved (mesh sizes are 53 [μ m], 63 [μ m], 75 [μ m], 90 [μ m], 125 [μ m], 150 [μ m], 180 [μ m], 212 [μ m] and 250 [μ m]) and separated into a uniform size for each group. The mean particle diameters are the average of ten particles that were picked out from individual groups. An electron microscope was used for the measurement of the diameter. The conditions of the powder mass per experiment were 5 [g], 10 [g] and 15 [g]. Concerning these mass conditions, when the powder filled up a ϕ 18 [mm] pipe, the filling heights became 13 [mm], 26 [mm] and 39 [mm] respectively. The variety of powder had six different diameters in Table 1 all belonging to A or B group of Geldart's classification. The powder particles which have a diameter from 58.3 [μ m] to 84.0 [μ m] belong to A group of Geldart's classification, and the powder particles which have a diameter from 123.2 [μ m] to 220.4 [μ m] belong to B. A total of 18 different experiments were performed in all combinations for six kinds of d_p and three kinds of M.

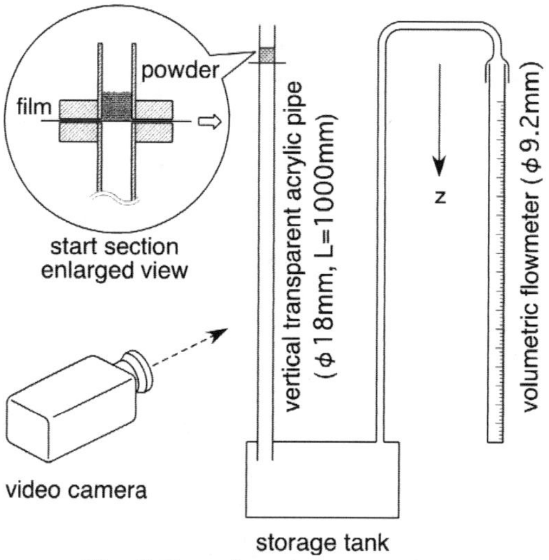

Fig. 2 Experimental apparatus.

Table 1. Experimental materials properties and experimental conditions.

	Particle characteristics
Material	Glass beads
ρ_p [g/cm^3]	2.35
d_p [μ m]	58.3, 69.4, 84.0, 123.2, 160.4, 220.4
M [g]	5, 10, 15

3. RESULTS AND DISCUSSION

3.1 Measurement of the height (movement distance)

Fig. 3 (a) shows the heights of the lower edge of powder cluster, of the upper edge of powder cluster and of the soap-film of the volumetric flowmeter as a function of elapsed time. The values plotted in Fig. 3 (a) are the averages of ten experiments. The experiment conditions of Fig. 3 (a) are d_p = 58.3 [μ m] and M = 15 [g]. Concerning experimental conditions, the influence that particles had on air was the largest. Consequently the discharged air volume from the vertical pipe was the largest influence in all experiments. The origin of the height of powder is the position of the lower edge of the powder cluster before fall start (the position of PET film), and the origin of the height of soap film is the position of the soap film before starting a fall. The explanatory notes of plotted signs are shown on the upper left in Fig. 3 (a). The height of the soap film in Fig. 3 (a) is converted by multiplying the measured value and the area ratio, because it is easy to compare the discharged air volume with the volume that the powder moved. The area ratio is calculated by dividing the cross section of the flowmeter by the cross section of the pipe. In Fig. 3 (a), it is found that the lower edge of powder cluster ● is increasing depending on the elapsed time, and that the increase rate of z is increasing at ranges from t = 0.0 [s] to t = 0.45 [s]. It appears that this tendency conforms to the general scheme of the free-fall of a single particle. At about t = 0.45 [s] the lower edge of the powder cluster is drained from the pipe. The upper edge of powder cluster ○ is increasing depending on elapsed time, and the increase rate of z is increasing at ranges from t = 0.0 [s] to t = 0.5 [s], as well. However it is found that the increase rate of z is decreasing after t = 0.5 [s]. When we see Fig. 3 (a) as z = constant, at the same z, the time difference between the lower edge ● and the upper edge ○ indicates the passage time of the powder at the position of z. In Fig.3 (a), it is found that the passage time of the powder gets longer depending on the increase of z in excess of about z = 800 [mm]. It appears that the particle velocity approaches the terminal velocity of individual particles, because the particle volume fraction is low (void fraction is high) near the upper edge. Therefore, the atmosphere that is just forming sedimentation of the single particles may exist near the upper edge of the powder cluster. On the other hand, it is found that the discharged air volume from the vertical pipe is very low in comparison with the volume of powder moved. It appears that there was less volume of discharged air so that most of the air passes outside or inside of the powder fall.

Fig. 3 (b) shows the heights of the lower edge of the powder cluster, of the upper edge of the powder cluster and of the soap film of the volumetric flowmeter as a function of elapsed time. The experiment conditions are d_p = 220.4 [μ m] and M = 15 [g]. The method of the data reduction is totally the same as in the case of Fig. 3 (a). In Fig. 3 (b), it is found that each distribution (●, ○, △) has the same tendency as in Fig.3 (a) at the ranges from t = 0.0 [s] to t = 0.6 [s]. However it is found that the slowdown of the upper edge of the powder cluster is small after t = 0.6 [s]. It appears that there are few scattered particles near the upper edge of the powder cluster in the condition of d_p = 220.4 [μ m].

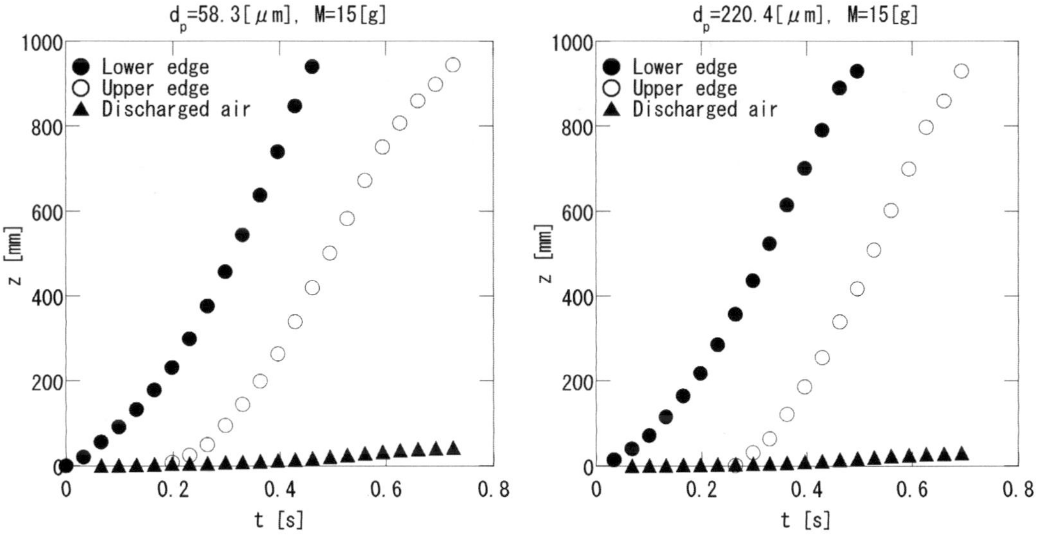

(a) d_p = 58.3 [μ m], M = 15 [g] (b) d_p = 220.4 [μ m], M = 15 [g]

Fig. 3 The heights of the lower edge of powder cluster, of the upper edge of powder cluster and of the soap film of the volumetric flowmeter as a function of elapsed time. The height of the soap film ▲ is converted by multiplying the measured value and the area ratio. The area ratio is calculated by dividing the cross section of the flowmeter by the cross section of the pipe.

3.2 Discharged air volume generated by the powder fall

Fig.4 shows the relationship between the air volume that was discharged from the pipe by the fall of powder and elapsed time. All six kinds of mean particle diameters (58.3 [μ m], 69.4 [μ m], 84.0 [μ m], 123.2 [μ m], 160.4 [μ m], 220.4 [μ m]) were plotted. The result of the experiment of M = 15

[g], and the influence that particles give to air is the largest. In Fig. 4, it is found that the increase rate of V (i.e. flow rate) is increasing in the ranges from t = 0.4 [s] to t = 0.6 [s] temporarily, and that the increase rate of V is decreasing after t = 0.6 [s], at all d_p. Further details of this will be described later. Almost all powder is discharged upon exit of the pipe (z = 1000 [mm]) at t = 0.8 [s], and the discharge of air by the fall stops. When the maximum value of V at t = 0.8 [s] is compared with the volume of the pipe (= 254.5 [cm^3]), it was 4.7 [%] at d_p = 58.3 [μ m] it appears that there was less volume of discharged air because most of the air (95.3 [%]) passes outside or inside of the powder cluster.

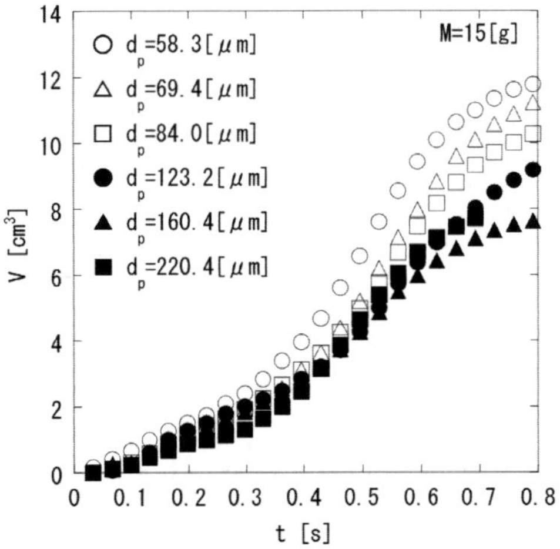

Fig. 4 Relationship between the discharged air volume generated by the powder fall and elapsed time.

3.3 Relative velocity of the particle cluster to the air flow

Fig. 5 (a)-(f) show the relationship between the relative velocity of the particle cluster to the air flow and the height of the z-axis. The powder velocity v_p was measured at the lower edge of the powder cluster. The air velocity v_f was calculated from the discharged air volume in Fig. 2. The bold line represents the calculated local free-fall velocity of a single particle as a sphere by Eq. (1) where the gravitational, buoyant, and drag forces are considered:

$$m_p \frac{dv_p}{dt} = \frac{m_p}{\rho_p}(\rho_p - \rho_f)g - \frac{1}{2} C_D \rho_f v_s^2 A \qquad (1)$$

A, C_D, g, m_p, v_s, ρ_p and ρ_f are the projected area of a particle, the drag coefficient, gravitational acceleration, the mass of one particle, local free-fall velocity of a single particle in unbounded fluid, the particle true density and the air density respectively. Morsi and Alexander's (1972) approximate equations were used for the C_D. The fine line represents the calculated local free-fall velocity of a single particle as a sphere by Eq. (1) with no drag force. It is plotted as the factual basis for the decision to judge measurement. It is evident that the measurement result is approximately correct, because of the measurement results in Fig. 5 (a)-(f) are distributed between the bold line and the fine line. It is also found that the results in Fig. 5 (a)-(f) have the same tendency. The distributions of v_p-v_f exceed the terminal velocity of the individual particles in all cases as well as those of Slack (1963) and Marzocchella et al. (1991). It is inferred that the powder is forming a particle cluster at the lower edge of the powder cluster. Moreover, the v_p-v_f increased from z = 0 [mm] to z = 600-800 [mm] according to the law of free-fall, and decreased after that. It is inferred that the decay of the particle cluster causes the slowdown.

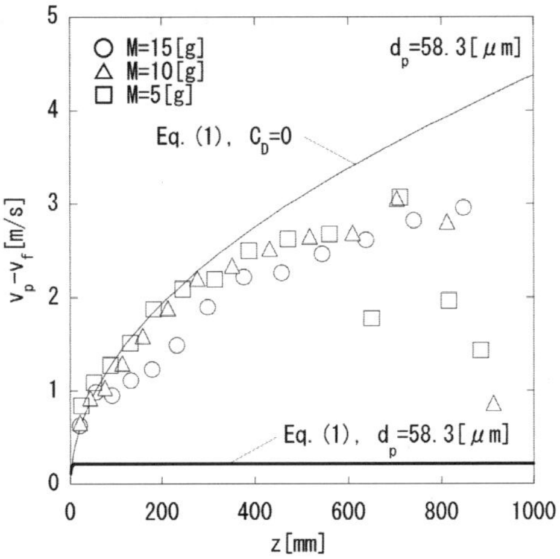

(a) d_p = 58.3 [μ m]

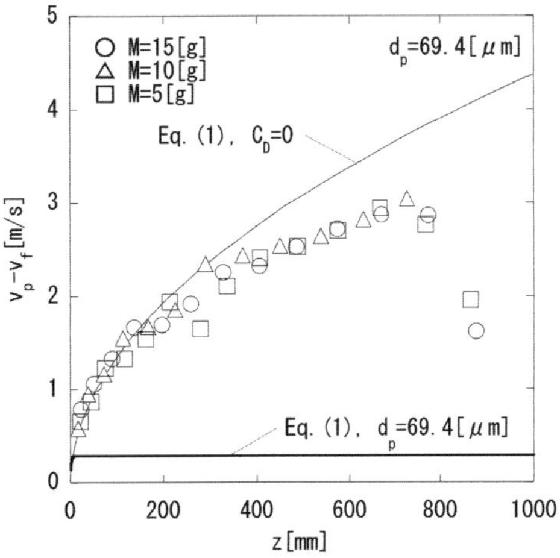

(b) d_p = 69.4 [μ m]

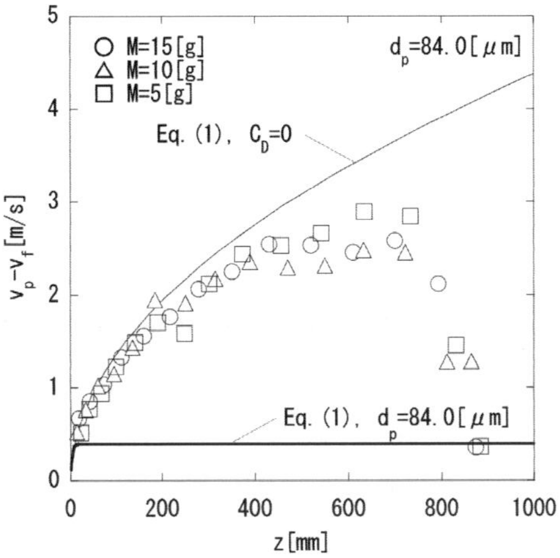

(c) d_p = 84.0 [μ m]

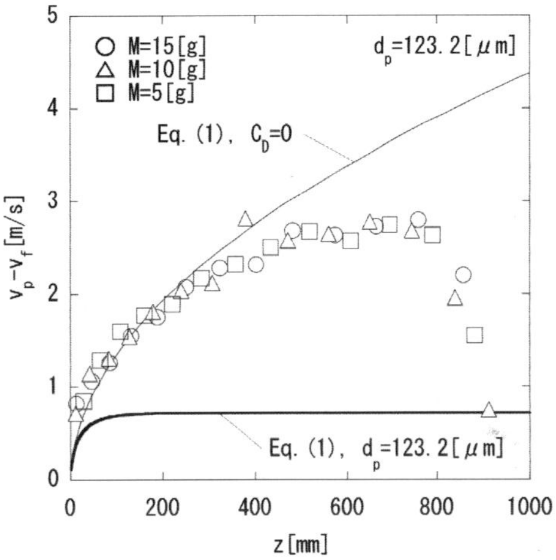

(d) d_p = 123.2 [μ m]

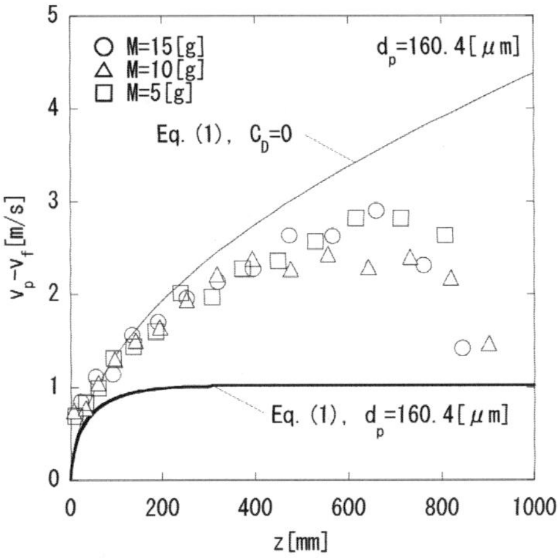

(e) d_p = 160.4 [μ m]

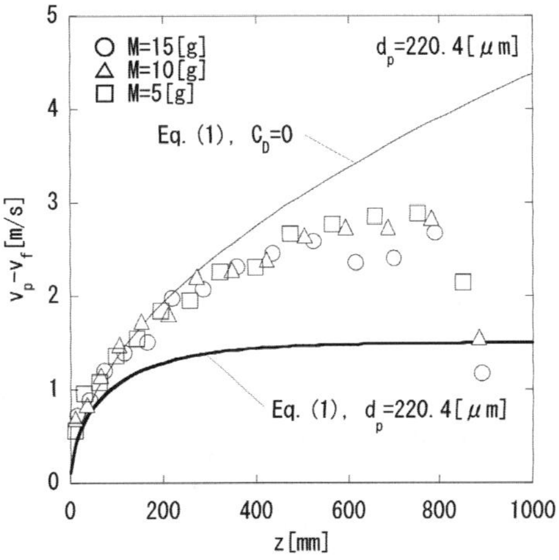

(f) d_p = 220.4 [μ m]

Fig. 5 Relationship between the relative velocity of the particle cluster to the air flow and z. These are plotted for all experiment conditions.

Fig. 6 is the relationship between the relative velocity of the particle cluster to the air flow and the height z-axis, and is a graph that showed only results for M = 15[g] in Fig.5 (a)-(f). The solid lines are the case which neglects aerodynamic drag, the case of d_p = 1.0 [mm] and the case of d_p = 0.5 [mm] respectively, which are the fall velocity of a single particle calculated with Eq. (1). The distributions for v_p-v_f are almost the same, however the particle diameters are different. And it closely resembles the distribution of the fall velocity for a single particle of d_p = 0.5 [mm]. When the powder forms the particle cluster, the fall velocity of the particle cluster may depend on the void fraction as well as sedimentation under uniform dispersion.

When we see Fig.5 (a)-(f) and Fig. 6 it is thought that the particle cluster decays near z = 600-800 [mm] because the relative velocity has maximum value. For calculating void fraction at the maximum point, if it is assumed that the particle cluster forms a sphere shape, the balance of force is given as Eq. (2):

$$\frac{\pi D_c^3}{6}(1-\varepsilon)(\rho_p-\rho_f)g-\frac{\pi D_c^2}{4}\frac{1}{2}C_D\rho_f(v_p-v_f)^2=0 \qquad (2)$$

Reynolds number based on the particle cluster is given as Eq. (3):

$$\mathrm{Re}_c=\frac{\rho_f D_c(v_p-v_f)}{\mu} \qquad (3)$$

The void fraction was about 99.4 [%], in our calculation (refer to APPENDIX 1) based on Eq. (2), Eq. (3) and Morsi and Alexander's (1972) approximate equation. It is expected that the particle cluster maintains its characteristic as a particle cluster until high void fraction comparatively, however this is a slightly rough calculation.

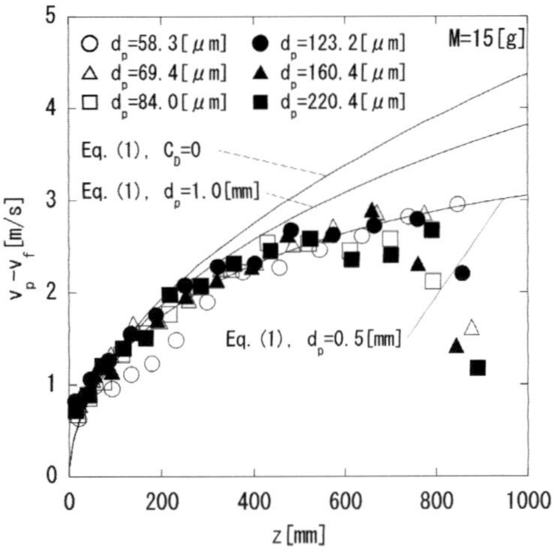

Fig. 6 Relationship between the relative velocity of the particle cluster to the air flow and z. This is gathered data from Fig.5 (a)-(f) as results of M = 15[g].

4. CONCLUSIONS

The experiment concerning how powder of a certain mass falls in a vertical pipe was performed, using glass beads whose mean diameters were 58.3 [μm], 69.4 [μm], 84.0 [μm], 123.2 [μm], 160.4 [μm], 220.4 [μm] respectively. The velocity of the lower edge of the powder cluster and the velocity of air flow generated by the powder were measured by video camera, and obtained the following results.

(1) The fall velocity of the particle cluster has no relation to the particle diameter in the range of 58.3 [μ m] < d_p < 220.4 [μ m], and approximates to the fall velocity of single particle of 0.5 [mm] in diameter.

(2) The free-fall velocity of the particle cluster has the characteristic of exceeding the terminal velocity of the individual particles (Slack, 1963; Marzocchella et al., 1991). It is expected that this characteristic exists unless the cluster has high void fraction.

(3) The discharged air volume generated by a powder fall from the pipe exit is low, and most of the air (95.3 [%]) passes to the outside or inside of the particle cluster.

ACKNOWLEDGEMENTS

This work was sponsored by Sanko machinery Co., Ltd (Japan). We would like to acknowledge their support and express our gratitude.

NOMENCLATURE

A	particle projected area
C_D	drag coefficient of a particle in unbounded fluid
d_p	particle diameter
D_c	cluster diameter
g	gravity acceleration
M	mass of powder falling at one time
m_p	mass of a particle
Re_c	Reynolds number based on a particle cluster
Re_p	Reynolds number based on a particle
t	elapsed time from powder fall start
V	discharged air volume from vertical pipe generated by powder fall
v_f	actual velocity of air
v_p	actual velocity of powder (measured at lower edge of powder cluster)
v_s	local free-fall velocity of a single particle in unbounded fluid
z	fall height from fall start point

Greek Letters

ε	void fraction
μ	viscosity of air
ρ_f	density of air
ρ_p	true density of particle

REFERENCES

Famularo, J. and Happel, J. (1965). "Sedimentation of dilute suspension in creeping motion," A. I. Ch. E. J., 11-6, pp. 981-988.

Hatano, H. et al. (1994). "Microscope visualization of solid particles in circulating fluidized beds," Powder Technol., 78, pp. 115-119.

Lackermeier, U. et al. (2001). "Visualization of flow structures inside a circulating fluidized bed by means of laser sheet and image processing," Powder Technol., 114, pp. 71-83.

Marzocchella, A. et al. (1991). "Break-up of cylindrical clusters of solid particles under gravity flow in a two-dimensional column," Powder Technol., 65, pp. 453-460.

Matsushita, M. et al. (2002). "Dynamics of Granular Flow through a Vertical Pipe under the Control of Inflow Rate," J. Phys. Soc. Jpn., 71, pp. 2859-2862.

Matsushita, M. et al. (2003). "Transition to Density Waves of Granular Particles Flowing through a Vertical Pipe," Prog. Theor. Phys. Suppl., 150, pp. 136-146.

Morsi, S. A. and Alexander, A. J. (1972). "An investigation of particle trajectories in two-phase flow systems," J. Fluid Mech., 55-2, pp. 193-208.

Ogata, K. et al. (2001). "Experimental investigation of a free falling powder jet and the air entrainment," Powder Technol., 115, pp. 90-95.

Rhodes, M. et al. (1992). "Particle motion at the wall of a circulating fluidized bed," Powder Technol., 70, pp. 207-214.

Richardson, J. F. and Zaki, W. N. (1954). "Sedimentation and fluidization Part 1," Trans. Inst. Chem. Eng., 32, pp. 35-53.

Richardson, J. F. and Meikle, R. A. (1961). "Sedimentation and fluidization Part 3: the sedimentation of uniform fine particles and of two-component mixtures of solids," Trans. Inst. Chem. Eng., 39, pp. 349-356.

Slack, G. W. (1963). "Sedimentation of compact clusters of uniform spheres," Nature, 200, pp. 466-467.

Slack, G. W. (1963). "Sedimentation of a large number of particles as a cluster in air," Nature, 200, p. 1306.

Steinour, H. H. (1944). "Rate of sedimentation, nonflocculated suspensions of uniform spheres," Ind. Eng. Chem., 36-7, pp. 618-624.

Steinour, H. H. (1944). "Rate of sedimentation, suspensions of uniform-size angular particles," Ind. Eng. Chem., 36-9, pp. 840-847.

Tsuji, T. et al. (2008). "Multi-scale structure of clustering particles," Powder Technol., 179, pp. 115-125.

Tsukada, M. et al. (1997). "Three-dimension imaging of particle clusters in dilute gas-solid suspension flow," Can. J. Chem. Eng., 75, pp. 466-470.

Zhang, M. H. et al. (2008). "A CFD-DEM study of the cluster behavior in riser and downer reactors," Powder Technol, 184, pp. 151-165.

APPENDIX 1

The void fraction calculation when the particle cluster decays

It is assumed that the particle cluster decays immediately before the particle cluster slows down (at a maximum point in Fig. 5 (a)-(f) or Fig.6).

Then, from Fig. 5 (a)-(f) or Fig.6

$$v_p - v_f = 2.6[m/s] \text{ (at } z = 700\ [mm])$$

In addition, it is assumed that D_c is nearly equal to the pipe diameter.

$$D_c = 18[mm]$$

The other conditions that are necessary for the calculation are as follows.

$$g = 9.80665[m/s^2]$$
$$\rho_f = 1.2465[kg/m^3]$$
$$\rho_p = 2346.518[kg/m^3]$$
$$\mu = 17.74\times10^{-6}[Pa\cdot s]$$

Therefore, Reynolds number based on the particle cluster is calculated by Eq. (3)

$$\mathrm{Re}_c = \frac{\rho_f D_c (v_p - v_f)}{\mu} = 3288$$

C_D is given by the following from Morsi and Alexander's (1972) approximate equations.

$$C_D = \frac{148.62}{\mathrm{Re}_p} - \frac{4.75\times10^4}{\mathrm{Re}_p^2} + 0.357 \quad (1000 < \mathrm{Re}_c < 5000)$$

Where, we substitute Reynolds number based on particle cluster for Reynolds number based on a single particle.

$$C_D = \frac{148.62}{\mathrm{Re}_c} - \frac{4.75\times10^4}{\mathrm{Re}_c^2} + 0.357 = 0.3978$$

When Eq. (2) is arranged

$$\varepsilon = 1 - \frac{3}{4}\frac{C_D \rho_f (v_p - v_f)^2}{D_c g (\rho_p - \rho_f)}$$

When all the conditions are substituted, we obtain ε.

$$\varepsilon = 0.994$$

Journal of Physics: Conference Series **147** (2009) 012071 doi:10.1088/1742-6596/147/1/012071

Shannon Entropy Increment Analysis on Dynamic Behavior of A Biomass Fluidized Bed

WQ Zhong, BS Jin, Y Zhang, XF Wang, MY Zhang, R Xiao
School of Energy & Environment, Southeast University, Nanjing 210096, P. R. China
wqzhong@seu.edu.cn

Abstract. Information theory based Shannon entropy increment analysis of the differential pressure fluctuations was successfully developed to recognize dynamic behavior of gas-solid flow and different flow patterns. Experiments were carried out in a biomass fluidized bed with cross-section of 100 mm×30 mm and height of 500 mm was carried out. Cylinder-shaped biomass particles with the size of 2.6 mm in diameter and 6 mm in length were used to as bed materials. Pressure fluctuations and flow patterns were obtained by a multi-channel differential pressure sampling system and a high-resolution digital CCD camera, respectively. It was found that both Shannon entropy and the present developed Shannon entropy increment analyses were pronounced with the flow patterns and transitions. Besies, Shannon entropy increment analyses, i.e. Shannon entropy increment and Shannon entropy increment rate, are very helpful to obtain the chaotic nature especially the change of chaotic nature with operating condition in gas-solid flow.

1. INTRODUCTION

Pressure fluctuations in multiphase flows contain much dynamic information. Many methods, e.g. statistical analysis (Fan et al., 1984; Bi et al., 1995; Xu et al., 2004), power spectrum analysis (Vaccaro et al., 1997), chaotic analysis (Van den Bleek et al., 1993; Marzocchella et al., 1997; Ellis et al., 2003), fractal analysis (Fan et al.,1993; Lowe et al., 1999)and wavelet analysis(Lu and Li, 1999; Ellis et al., 2003)have been proposed to describe the dynamic behavior of two-phase flow successfully.

Besides, as a measure of the information content or complexity of measurement series, Shannon entropy analysis has been active in the multiphase flows since last decade. Kang et al. (1999) used the Shannon entropy of pressure fluctuations to detect the flow pattern transitions in three-phase fluidized beds. Zhang and Shi (1999) introduced the calculation of Shannon entropy to study the flow characteristics of two-phase flow systems. Shi et al. (2000) used Shannon entropy to characterize the air-water two phase flows in vertical pipes. Cho et al. (1999) studied the heat transfer and temperature fluctuations between an immersed heater and the bed proper in the riser of a three-phase circulating fluidized bed. Shi et al. (2004) employed Shannon entropy to describe the two-phase flow density wave instability for 200MW nuclear heating reactor. Zhong and Zhang (2005) attempted to depict the dynamic behaviors and flow patterns of a rectangular spout-fluid bed at ambient conditions by Shannon entropy. Camesasca et al. (2006) computed the Shannon entropy in order to find a suitable chosen group/complex of particles for a given species at a given location. Besides, measurements of the mixing performance based on Shannon entropy were employed to quantify the mixing of two miscible fluids by Chang and Yang (2006). Liang et al. (2007) analyze the dense-phase pneumatic conveying of pulverized coal with variable moisture content at high pressure by Shannon entropy. These efforts showed that it is promising to grasp the dynamic behaviors of multiphase flows by Shannon entropy analysis of pressure fluctuations.

Journal of Physics: Conference Series **147** (2009) 012071 doi:10.1088/1742-6596/147/1/012071

It is known that entropy increment represents an event that transits from ordered to disorder in the second law of thermodynamics. Thus, we wonder whether the dynamic behavior especially the flow patterns and transitions could be described pronouncedly by Shannon entropy increment. The present work attempted to develop Shannon entropy increment of dynamic behavior of gas-solid flow and different flow patterns. It focused on studying dynamic behavior of gas-solid flow in the aspect of Shannon entropy increment.

2. THEORETICAL

2.1 Shannon Entropy

It is well known that the concept of entropy was firstly used in the second law of thermodynamics. As a state function, entropy reveals that the direction of any un-reversible spontaneous process in an isolated system is the direction of increased entropy. Then, the well-known physical scientist, Boltzmann, investigated the entropy in molecular movement theory and indicated that entropy was the reflection and the measurement of the chaos degree of molecular motion (Araujo et al., 2003). The founder of information theory, Shannon, introduced the concept of entropy into information theory, suggesting entropy is a measurement of the amount of information in a certain information source and the degree of indeterminacy in a certain system (Shannon, 1948), called Shannon entropy resulting in it enriching the meaning of entropy.

Shannon entropy provides a scientific method to the essential state of things (Otwinowski, 2006), which can be utilized to express the degree of uncertainty involved in predicting the output of a probabilistic event (Camesasca et al., 2006). That is to say, if one predicts the outcomes exactly before it happens, the probability will be a maximum value and, as a result, the Shannon entropy will be a minimum value. Besides, if one is absolutely able to predict about the outcomes of an event, the Shannon entropy will be zero.

The Shannon entropy of any differential pressure time series in a fluidized bed can be defined as

$$S=-\sum_{i=1}^{n}p(x_i)\lg[p(x_i)] \quad (1)$$

Where, n is the length of time series signal, $p(x_i)$ is the probability of every component in the signal, satisfying the constraint $\Sigma p(x_i)=1$. $p(x_i)$ can be estimated by joint probability density formula. The unit of S is decibels (dB). It can be seen that the more disorder in a system, the larger Shannon entropy, which implies more complex and stochastic chaos nature resulting in turbulent motion of the gas or particles, gas-solid intensive interactions, flow instability, and etc.

2.2 Shannon Entropy Increment and Shannon Entropy Increment Rate

In the second law of thermodynamics, thermodynamics entropy increment represents an event that transits from order to disorder. For Shannon entropy, the value of entropy expresses the degree of uncertainty in a certain system. Similar to the thermodynamics entropy increment, Shannon entropy increment should also represent an event that transits from order to disorder.

The Shannon entropy increment in a multiphase flow system when the operating condition changes from i to j is defined as

$$\Delta S_{ij}=S_i(x)-S_j(x) \quad (2)$$

Where $S_i(x)$ and $S_j(x)$ are the Shannon entropy of differential pressure time series at i operating condition and j operating condition, respectively.

$$S_i(x)=-\sum_{i=1}^{m}p(x_i)\lg[p(x_i)] \quad (3)$$

$$S_j(x) = -\sum_{i=1}^{n} p(x_i)\lg[p(x_i)] \quad (4)$$

in which, m and n are the lengths of sampled time series.

In order to study the speed of Shannon entropy increment when operating condition changes, we define the Shannon entropy increment rate here. The Shannon entropy increment rate for the operating condition transiting from i to j is defined as

$$\delta_{ij} = \frac{\Delta S_{ij}}{\Delta C_{ij}} = \frac{S_i(x) - S_j(x)}{C_i(x) - C_j(x)} \quad (5)$$

Where δ_{ij} is the Shannon entropy increment rate, and ΔC_{ij} is the operating condition increment. $C_i(x)$ and $C_j(x)$are the operating conditions i and j, respectively. The unit of Shannon entropy increment rate δ_{ij} is dB/ (unit of operating condition). If the unit of operating condition is m/s, the unit of Shannon entropy increment rate is dB/(m s^{-1}).

3. EXPERIMRENTAL

An experimental setup was established to verify the simulations. The schematic diagram of experimental setup is shown in Fig.1. This set-up consists of a fluidized bed column, a gas supply system and some sampling instruments. The column has a cross-section of 100 mmX30 mm and a height of 500 mm. It was made of 6mm thick Plexiglas. The orifices in the air distributor are 1 mm in diameter, and the total area of all orifices is 1.1% of the gas distributor.

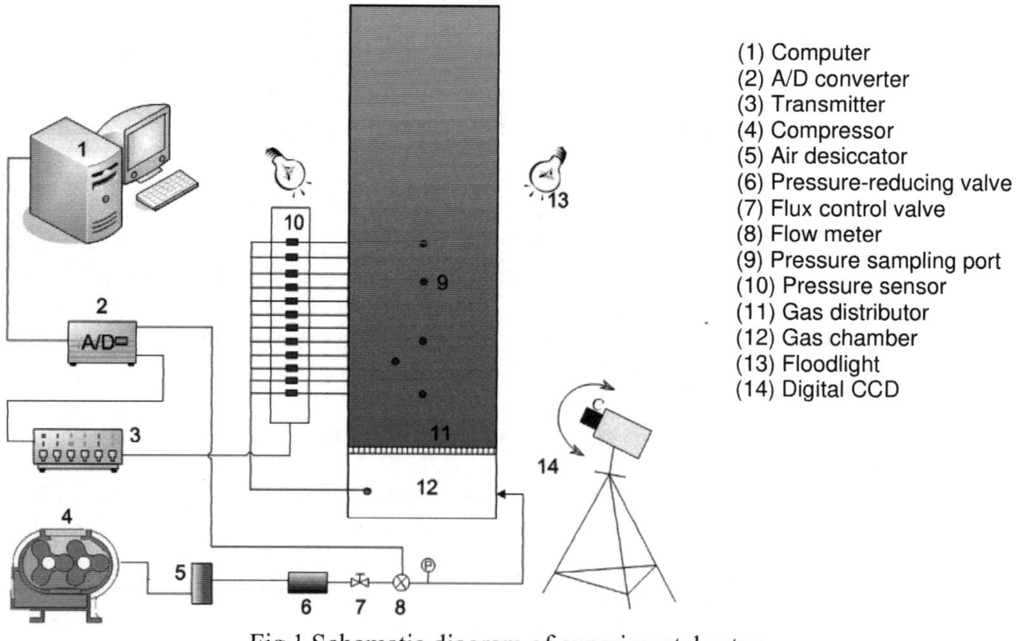

Fig.1 Schematic diagram of experimental setup

The fluidizing gas was supplied by a compressor supplied. A pressure-reducing valve was installed to avoid pressure oscillations and achieve a steady gas flow. Besides, an air desiccator was installed in the gas supplied pipe to avoid moisture in the gas. The gas flow rates were measured by flow meter. The fluidizing gas was sent into the gas chamber, and then entered the bed via the orifices in the gas distributor.

Pressure fluctuations in the bed were obtained by a multi-channel differential pressure signal sampling system. There were five pressure-measuring holes located on the back wall of the column, with heights of 30 mm, 70 mm, 110 mm, 160 mm and 260 mm above the bottom of the bed. Every differential pressure sensor has two ports, one was connected to the pressure-measuring hole in the column wall, and the other was connected to a fluidizing gas chamber. The differential pressures were measured and then converted into voltage signals by multi-channel differential pressure signal transmitter. The voltage signals were sent to a computer through an A/D converter. A digital camera (Nikon 5000) was employed to photograph the flow patterns through the transparent wall during the experiments. Two groups of 2000W floodlights were used to enhance the photo definition when photographing.

Branches were cut to the cylinder-shaped particle with the size of 2.6 mm in diameter (d_p) and 6 mm in length (L_p). They were used as bed materials. The real density (ρ_p) of the biomass particle is 850 kg/m^3, the initial packed bed height (H_0) is 80mm.

4. RESULTS AND DISCUSSION

4.1 Total Pressure Drops with Superficial Gas Velocities

Pressure fluctuations were obtained at various superficial gas velocities. The mean values of total pressure drops with increasing superficial gas velocities are plot in Fig.2. The total pressure drop sharply increases with increasing superficial gas velocity increasing to about 1.25 m/s, while the increasing of total pressure drop becomes slightly when the superficial gas velocity increases from 1.25 m/s to 2.25 m/s. For the superficial gas velocity increases over 2.25 m/s, the increasing of total pressure drop becomes sharply. The tendency of total pressure drop with increasing superficial gas velocity is similar to the previous report by Abdullah et al.(2003). According to the pressure drops with increasing superficial gas velocities, the minimum fluidizing velocity was determined to be about 1.24m/s.

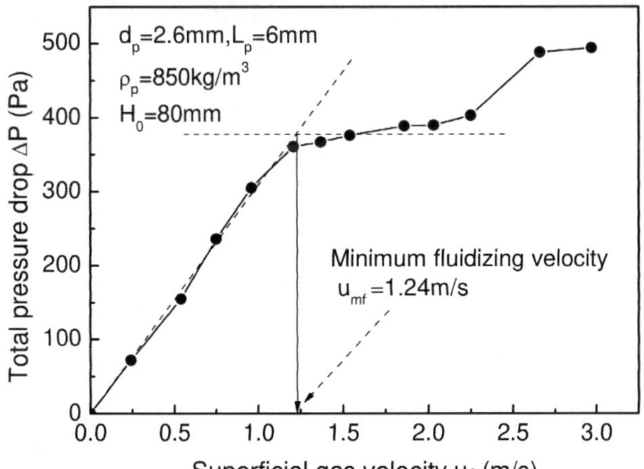

Fig.2 The total pressure drop with increasing superficial gas velocity.

4.2 Flow Patterns at Ddifferent Superficial Gas Velocities

The flow patterns through the transparent wall were photographed by a digital camera (Nikon 5000) during the experiments. Fig.3 shows the flow patterns at different superficial gas velocities. The flow patterns are distinct with increasing superficial gas velocities. According to the images of flow patterns and the total pressure drop, the gas-solid flows can be divided into three patterns. They are under-fluidization (Fig.3a-e), steady fluidization (Fig.3f-i) and turbulent fluidization (Fig.3j-l).

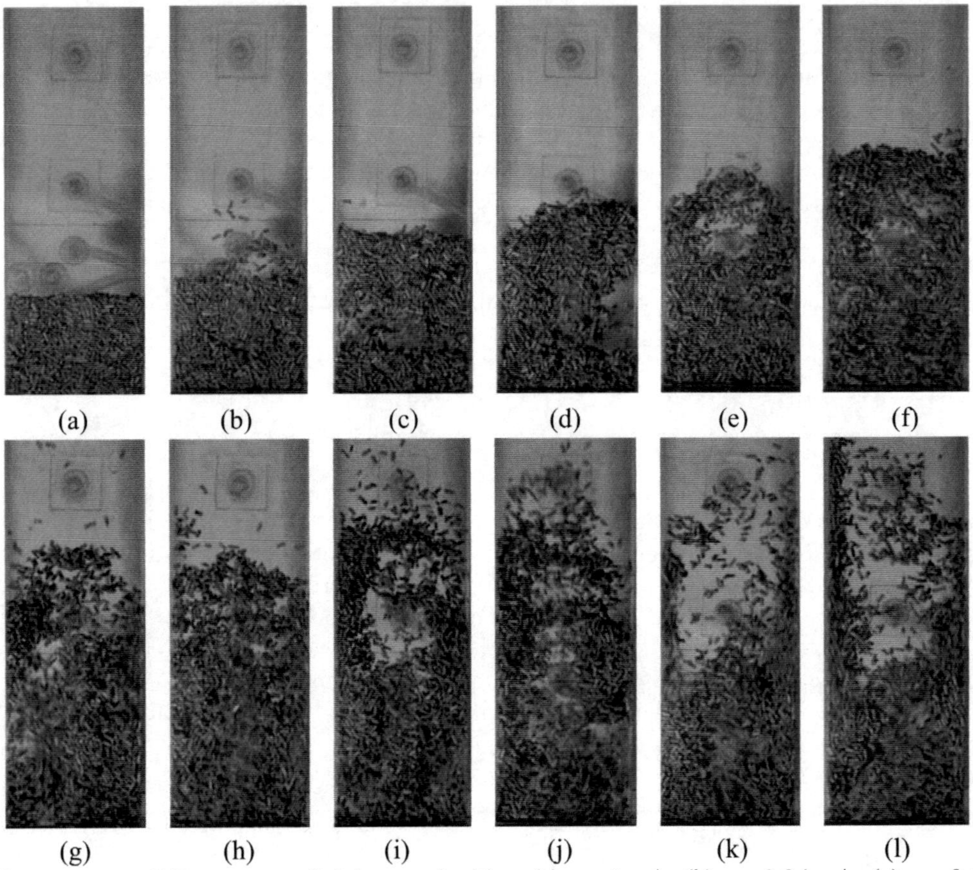

Fig.3 Flow patterns different superficial gas velocities: (a) u_f=0 m/s; (b) u_f=0.24 m/s; (c) u_f=0.54 m/s; (d) u_f=0.96 m/s; (e) u_f=1.21 m/s; (f) u_f=1.37 m/s; (g) u_f=1.54 m/s; (h) u_f=1.86 m/s; (i) u_f=2.03m/s; (j) u_f=2.25 m/s; (k) u_f=2.66 m/s; (l) u_f=2.97 m/s.

For the flow pattern of under-fluidization, the gas-solid flow comes through the flow patterns of fixed bed (Fig.3a, b), bubbling with slugging (Fig.3c, d) and bubbling (Fig.3e) with the increasing of superficial gas velocity. Unlike spherical particles, larger bubbles can be seen in the middle of the bed when fluidizing of non-spherical biomass when the superficial gas velocity is less than the minimum fluidizing velocity. At this time, slugging can be often observed due to the movement of large bubbles. When the superficial gas velocity is beyond the minimum fluidizing velocity, the bed is fluidized. Here, we defined this kind of flow pattern as steady fluidization. Because the bed is not well fluidized but steady according to the phenomena observed in the experiments. Agglomeration of particles can be clear seen in the bed.

In this case, the bed could not be well fluidized. The phenomena agree well with the previous indication that biomass particles are difficult to fluidize due to their peculiar shapes, sizes and densities by Cui and Grace(2007). Besides, large bubbles can be seen in this flow pattern but the bubble size is less and the bed is much steady than those at the superficial gas velocity less than the minimum fluidizing velocity. When the superficial gas velocity is larger than 2.25 m/s, the bed is turbulent. In this case, the particles are fluidized but the interaction of gas and particles are unsteadily turbulent. Thus, we define this kind of flow pattern as turbulent fluidization.

4.3 Shannon Entropy

The Shannon entropies at different superficial gas velocities for three times measurements and their mean values are plotted in Fig.4. It can be seen that the Shannon entropies increase with superficial gas velocities, implying the fact that the gas-solid interaction in the bed is more and more complex and stochastic when the superficial gas velocity increases, because Shannon entropy reflects the characteristic of dynamic behaviors (e.g. turbulent motion of the gas or particles, gas-solid intensive interactions and flow instability).

Besides, the Shannon entropies are found to be significant different for different flow patterns. In the flow pattern of under-fluidization, the Shannon entropies of pressure fluctuations sharply when the superficial gas velocity increases. While the increasing of Shannon entropy for the flow pattern of steady fluidization becomes slightly with increasing superficial gas velocity. When the flow pattern transits into turbulent fluidization, the increasing of Shannon entropy with increasing superficial gas velocity becomes sharply again. The results indicate that the changes of Shannon entropies of pressure fluctuations were pronounced with the transitions of flow patterns, which is similar to the previous findings (e.g. Kang et al.,1999; Zhong and Zhang, 2005).

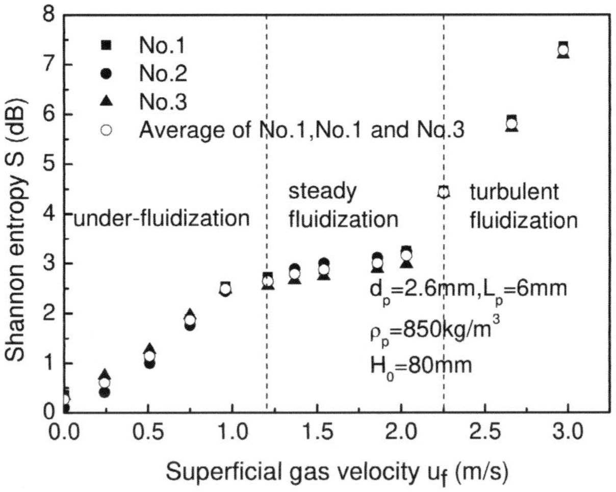

Fig.4 Shannon entropy derived from pressure fluctuations for three times measurements at various superficial gas velocities.

4.4 Shannon Entropy Increment

Fig.5 plots the Shannon entropy increments for three flow patterns. The Shannon entropy increments were calculated by equation (2). The Shannon entropy increment is found to be larger than zero in all the cases. This means that the chaotic nature occurs in these flow patterns, and a fluidized bed was a deterministic chaos system since the Shannon entropy increment was positive. The Shannon entropy increment of steady fluidization is the lowest, while the Shannon entropy increment of turbulent fluidization is the highest. The pressure fluctuations are much more random in the turbulent fluidization than under-fluidization and steady fluidization, and the system appears a chaotic nature. The gas-solid turbulent motions were clearly observed during the experiments, and sometimes the growth of surface disturbances could be seen. However in the case of steady fluidization, the particles are fluidized steady, the pressure fluctuations are less random. As a result, the Shannon entropy increment is lower.

It can be seen that Shannon entropy increments are also very distinct difference to those flow patterns. For the flow pattern of under-fluidization, the mean Shannon entropy increments is about 0.5 dB. While for steady fluidization and turbulent fluidization, the mean values are about 0.1 dB and 1.4 dB, respectively.

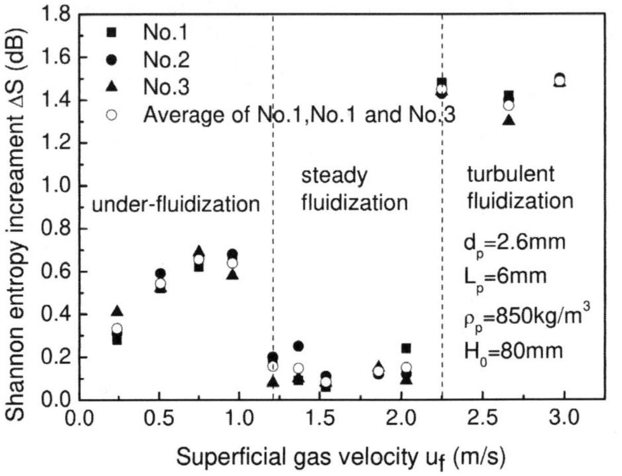

Fig.5 Shannon entropy increments with various superficial gas velocities.

4.5 Shannon Entropy Increment Rate

The changes of Shannon entropy with operating conditions can be described by Shannon entropy increment rate. The Shannon entropy increment rates for the flow patterns of under-fluidization, steady fluidization and turbulent fluidization with increasing superficial gas velocity are plotted in Fig.6.The Shannon entropy increment rate with the increasing of superficial gas velocity is found to be about 2.5 dB/ (m s^{-1}) from the linear fitting line at the flow pattern of under-fluidization. However, the increment rate of Shannon entropy drops at a low level for about 0.6 dB/ (m s^{-1}) in the flow pattern of steady fluidization. For turbulent fluidization, the increment rate of Shannon entropy is about 4.9 dB/ (m s^{-1}) when the superficial gas velocity increases.

The results indicated that turbulent fluidization appears largest insentive in the change of chaotic nature at the flow pattern of turbulent fluidization, but lest insentive in change of chaotic nature in the flow pattern of steady fluidization. As a result, Shannon entropy increment analyses, i.e. Shannon entropy increment and Shannon entropy increment rate, are distinct difference to the flow patterns, besides they are very helpful to obtain the chaotic nature especially the change of chaotic nature with operating condition in gas-solid flow.

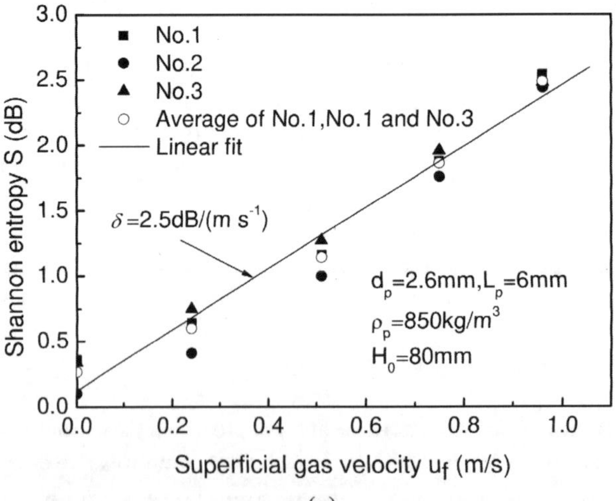

(a)

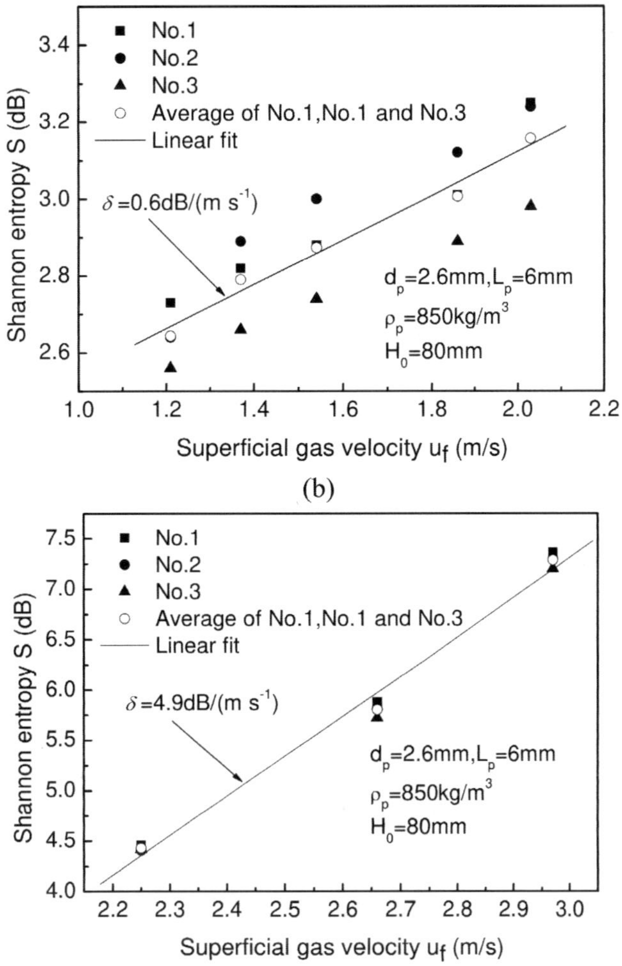

(c)

Fig.6 Shannon entropy increment rate with flow patterns: (a) under-fluidization, (b) steady fluidization, (c) turbulent fluidization.

5. CONCLUSIONS

As a kind of information, pressure fluctuations contain much dynamic information of multiphase flows. Thus, it is conceivable to grasp the dynamic behaviors of multiphase flows by information theory based Shannon entropy analysis. In the present work, information theory based Shannon entropy increment analysis of the differential pressure fluctuations was successfully developed to recognize dynamic behavior of gas-solid flow and different flow patterns. Experiments were carried out in a biomass fluidized bed with cross-section of 100 mm×30 mm and height of 500 mm was carried out. It was found that both Shannon entropy and the present developed Shannon entropy increment analyses were pronounced with the flow patterns and transitions. Besies, Shannon entropy increment analyses, i.e. Shannon entropy increment and Shannon entropy increment rate, are very helpful to obtain the chaotic nature especially the change of chaotic nature with operating condition in gas-solid flow.

ACKNOWLEDGEMENTS

Financial support from the National Natural Science Foundation of China (50706007) was sincerely acknowledged.

NOMENCLATURE

$C_i(x)$	operating condition i
d_p	diameter of particles, [mm]
D_i	diameter of spout nozzle, [mm]
D_t	diameter of bed column, [mm]
H_0	static bed height, [mm]
L_p	length of cylinder-shaped particle, [mm]
P	operating Bed pressure, [MPa]
$P(x_i)$	probability of certain component in pressure fluctuations
Q_s	spouting gas flow rate, [Nm^3/h]
Q_f	fluidizing gas flow rate, [Nm^3/h]
Q_{mf}	minimum fluidizing gas flow rate at , [Nm^3/h]
S	Shannon entropy, [dB]
n	length of pressure fluctuation time series
	Greek Letters
ΔC_{ij}	operating condition increment
ΔS_{ij}	Shannon entropy increment, [dB]
δ_{ij}	Shannon entropy increment rate, [dB/(operating condition)]
ρ_p	particle density, [kg/m^3]

REFERENCES

Abdullah, M.Z. et al. (2003). "Analysis of cold flow fluidization test results for various biomass fuels," Biomass and Bioenergy, 24, pp.487-494.

Araujo, D.B. et al. (2003). "Shannon entropy applied to the analysis of event-related MRI time series," NeuroImage, 20, pp.311-317.

Camesasca, M. et al. (2006)."Quantifying fluid mixing with the Shannon entropy," Macromolecular Theory and Simulations, 15, pp.595-607.

Chang, C. C. et al. (2006). "A particle tracking method for analyzing chaotic electroosmotic flow mixing in 3D microchannels with patterned charged surfaces," Journal of Micromechanics and Microengineering, 16, pp.1453-1462.

Cho, Y. J. et al. (1999). "Heat transfer and bubble properties in three-phase circulating fluidized beds," Chem.Eng.Sci., 75, pp.113-119.

Cui, H. et al. (1999). "Fluidization of biomass particles: A reviewof experimental multiphase flow aspects," Chem. Eng. Sci., 62, pp.45-55.

Ellis, N. et al. (2003). "Characteristics of dynamic behavior in gas-solid turbulent fluidized bed using chaos and wavelet analyses," Chem. Eng. J., 96, pp.105-116.

Fan, L.T. et al. (1984). "Analysis of pressure fluctuations in a gas-solid fluidized bed," AIChE Journal, 30, pp.346-349.

Fan, L.T. et al. (1993). "Fractal analysis of fluidized particle behavior in Liquid-solid fluidized bed," AIChE Journal, 39, pp.513-517.

Fuhrman, S. et al. (2000). "The application of shannon entropy in the identification of putative drug targets," BioSystems, 55, pp. 5-14.

Kang, Y. et al. (1999). "Particle flow behavior in three-phase fluidized beds," Korean Journal of Chemical Engineering, 16, pp.784-788.

Liang, C. et al. (2007). " Flow characteristics and Shannon entropy analysis of dense-phase pneumatic conveying of pulverized coal with variable moisture content at high pressure," Chem. Eng. & Tech., 30, pp.926-931.

Lu, X. et al. (1999). "Wavelet analysis of pressure fluctuation signals in a bubbling fluidized bed," Chem. Eng. J., 75, pp.113-119.

Marzocchella, A. et al. (1997). "Chaotic behavior of gas-solid flow in the riser of a laboratory –scale circulating fluidized bed," AIChE Journal, 43, pp. 1458-1468.

Otwinowski, H. (2006). "Maximum entropy method in comminution modeling," Granular Matter, 8, pp.239-249.

Shannon, C.E. (1948). "A mathematical theory of communication," Bell System Tech. J., 27, pp.379-423.

Shi, L. et al. (2000). "Shannon entropy characteristics of air-water two phases flow regimes in vertical pipes," Chinese Journal of Nuclear Power Engineering, 5, pp.411-415.

Shi, L. et al. (2004). "Shannon entropy characteristics of two-phase flow density wave instability experiments for 200MW nuclear heating reactor," Chinese Journal of Nuclear Power Engineering, 1, pp.18-21.

Vaccaro, S. et al. (1997). "A technique for measurement of the jet penetration height in fluidized bed by pressure signal analysis," Powder Technol., 92, pp.223-231.

Van den Bleek, M. et al. (1993). "Deterministic chaos: a new tool in fluidized bed design and operation," Chem. Eng. J., 53, pp.75-87.

Xu, J. et al. (2004). "Statistical and frequency analysis of pressure fluctuations in spouted beds," Powder Technol., 140, pp.141-154.

Zhang, Z.Y. et al. (1999). "Shannon entropy characteristics of two-phase flow systems," Journal of Applied Physics, 85, pp.7544-7551.

Zhong, W. et al. (2005). "Characterization of dynamic behavior of a spout-fluid bed with Shannon entropy analysis," Powder Technol., 159, pp.121-126.

Journal of Physics: Conference Series **147** (2009) 012072 doi:10.1088/1742-6596/147/1/012072

Numerical Simulation of Particle Settling and Cohesion in Liquid

Y Johno [1], **M Satomi** [2], **K Nakashima** [1], **T Shigematsu** [1], **B Ono** [1]
[1] SASEBO National College of Technology, 1-1 Okishin, Sasebo, Nagasaki, 857-1193 JAPAN
[2] Sony Semiconductor Kyushu Corporation, Kikuchigun, Kumamoto, JAPAN

yjohno@post.cc.sasebo.ac.jp

Abstract. In this study, the motions of particles and particle clusters in liquid were numerically simulated. The particles of two sizes (Dp=40μm and 20μm) settle while repeating cohesion and dispersion, and finally the sediment of particles are formed at the bottom of a hexahedron container which is filled up with pure water. The flow field was solved with the Navier-Stokes equations and the particle motions were solved with the Lagrangian-type motion equations, where the interaction between fluid and particles due to drag forces were taken into account. The collision among particles was calculated using Distinct Element Method (DEM), and the effects of cohesive forces by van der Waals force acting on particle contact points were taken into account. Numerical simulations were performed under conditions in still flow and in shear flow. It was found that the simulation results enable us to know the state of the particle settling and the particle condensation.

1. INTRODUCTION

Particulate operation has played the important role in chemical industry, electricity and electronic industry, medicine manufacture, and the product manufacture process of the information industry. In the design of fluid and particle process, solid liquid separation process and the manufacture process of solid dispersion system, it is required to grasp the state of particle motion and particle formation in fluid correctly.

For example, in the manufacturing process of a ceramic product, it is important to predict the state of the sediment after materials particles sediment in liquid. Although the materials particles of the ceramics of a sediment serve as the end products according to processes, such as compression, drying, dryness, fabrication, and sintering, the character of the end products is greatly influenced by the state of the sediment made first. The unevenness of void made when settling materials particles in liquid causes problems, such as an internal defect in the end products, and material strength unevenness. Therefore, in order to predict the quality of the end products, it is important to grasp the state of particle settling, particle condensation, and sediment of particles.

In this study, the numerical simulation of process in which the particles of two sizes sediment in the case of still flow and in the case of simple shear flow was performed. The particles of two sizes form a particle cluster, and they are distributed depending on the case. While the particles which sediment in liquid repeated condensation and distribution, process until it fabricates a sediment was pursued using the three-dimensional numerical simulation. In order to calculate the contact force of particles, Distinct Element Method (DEM) was used. The DEM is one of the trajectory models, is a simple numerical model by which interactions due to multi-body collisions can be calculated. The DEM has

Journal of Physics: Conference Series **147** (2009) 012072 doi:10.1088/1742-6596/147/1/012072

been applied to many kinds of multi-body complex problems (Kawaguchi et al., 1992; Yuu et al., 1995). The effects of cohesive forces by van der Waals force acting on particle contact points were taken into account. Condensation of particles, the state of distribution, and the state of a sediment were considered from the simulation result.

2. CALCULATION METHOD

In this calculation, the three-dimensional Navier-Stokes equation which is a governing equation of fluid was used for the calculation for fluid phase, and the Lagrangian-type equation of motion was used for the calculation of the particle motion. The motion of fluid and particles was calculated by being allied and solving the equation of motion of fluid and particles. The drag force produced with both relative velocity as a dynamic interaction of fluid and particles was taken into account. In order to calculate the contact force of particles, DEM was used, and in order to calculate the adhesion between particles, van der Waals force was used. Moreover, in order to calculate the settling velocity of a particle cluster, the sphere equivalent diameter of a particle cluster was used.

2.1 Calculation of fluid phase

The governing equations for the fluid phase are the three-dimensional Navier-Stokes equations for the incompressible fluid with particle source terms, and the equation of continuity. These equations are expressed with a non-dimension as follows in consideration of existence of particles (Anderson et al., 1967).

$$\frac{\partial \varepsilon \mathbf{u}}{\partial t} + \nabla \cdot \varepsilon \mathbf{u}\mathbf{u} = -\varepsilon \nabla p - \frac{\varepsilon}{\mathrm{Re}} \nabla \cdot \tau - S_p \tag{1}$$

$$\frac{\partial \varepsilon}{\partial t} + \nabla \cdot \varepsilon \mathbf{u} = 0 \tag{2}$$

τ in Eq. (1) is the viscous stress tensor expressed by Newton's law of viscosity. S_p in Eq. (1) is the particle source term, which indicate the mutual interaction between particles and fluid. S_p is resistance of Stokes produced with the relative velocity of particles and fluid, and motions of particles and fluid are linked through the interaction term S_p. The equation of S_p term is expressed with a non-dimension as follows.

$$S_p = \frac{3\pi\mu D_P D\, n}{U_o \rho}(\mathbf{u} - \mathbf{u}_p) \tag{3}$$

In actual numerical computation, by taking divergence of both sides of Eq. (1), the Poisson's equation about pressure P was obtained and the Poisson's equation was solved by the Relaxation method (Harlow et al., 1965; Rai et al., 1991). The fourth-order central difference scheme is used for the convection terms, and the second-order central difference scheme is used for other spatial derivative terms. The forward difference scheme is used for the time derivative terms. In addition, since it was difficult to calculate the flow between particles strictly, physical quantity, such as velocity, calculated the amount of partial averages, and calculated it in consideration of void fraction.

2.2 Calculation of particle motion

2.2.1 The equation for particle motion

The motion of the particles in the fluid is pursued with the Lagrangian type equation of motion. The l agrangian type equation for particle motion is described as the following equation (Umekage et al., 2001).

$$\frac{\pi}{6}\rho_p D_p^3 \frac{d\mathbf{u}_p}{dt} = \sum_j [F_j + D_j - F_{Cj}] + 3\pi\mu (D_p)_{deq}(\mathbf{u} - \mathbf{u}_p) + \frac{\pi}{6}\rho D_p^3 \frac{d\mathbf{u}}{dt} + \frac{\pi}{12}\rho D_p^3\left(\frac{d\mathbf{u}}{dt} - \frac{d\mathbf{u}_p}{dt}\right) + \frac{\pi}{6}(\rho - \rho_p)D_p^3 g \tag{4}$$

In Eq. (4), the left side is the acceleration term of particles, the 1st term of the right-hand side is interaction force between particles, the 2nd term of the right-hand side is a drag force based on Stokes's law of resistance, the 3rd term is the force by the pressure slope of the circumference of the particles produced in order that fluid may accelerate by particles, the 4th term is the force required to accelerate added mass and the 5th term is gravity and lift. The Eq. (4) is an equation of motion about the translation motion of particles, and is not taking rotational motion of particle into consideration in this calculation.

2.2.2 Calculation of contact force using Distinct Element Method

In the process of particle settling, Particles collide with many other particles repeatedly. Therefore, it is important to calculate particles collision. In this study, the particle-particle interactions due to multi-body contacts were simulated using Distinct Element Method (DEM) (Cundall et al., 1979). The DEM, which is one of the trajectory models, is a simple numerical model to calculate the multi-body contacts (Yuu et al., 2000). In DEM, the contact force of particles is calculated in all contact points. In order to estimate the contact force of particles, an elastic spring and viscous dashpot are introduced. The contact model of DEM is shown in Fig. 1. The increments of the normal and shear forces, ΔF_n and ΔF_s, modeled with the elastic spring are calculated as follows using the normal and shear relative displacement increments, Δn and Δs.

$$\Delta F_n = k_n \Delta n \quad , \quad \Delta F_s = k_s \Delta s \tag{5}$$

The normal stiffness k_n is estimated from the Hertz contact theory. The shear stiffness k_s is $k_s = (1/5)k_n$ in this calculation. The increments of the normal and shear damping forces, ΔD_n and ΔD_s, modeled with the dashpots are calculated as follows using the normal and shear relative velocity increments, $\Delta \dot{n}$ and $\Delta \dot{s}$.

$$\Delta D_n = -\eta_n \Delta \dot{n} \quad , \quad \Delta D_s = -\eta_s \Delta \dot{s} \tag{6}$$

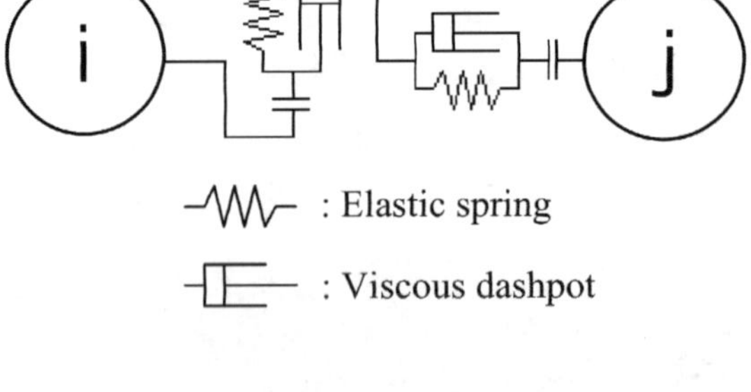

Fig. 1 The contact model of DEM

Journal of Physics: Conference Series **147** (2009) 012072 doi:10.1088/1742-6596/147/1/012072

The normal and tangential contact and damping forces at time t are

$$(F_n)_t = (F_n)_{t-1} + \Delta F_n \quad , \quad (F_s)_t = (F_s)_{t-1} + \Delta F_s \tag{7}$$

$$(D_n)_t = (D_n)_{t-1} + \Delta D_n \quad , \quad (D_s)_t = (D_s)_{t-1} + \Delta D_s \tag{8}$$

The summation of these contact force components in each direction gives the resultant force.
The normal damping coefficient η_n, and the shear damping coefficient η_s must satisfy the following relation, respectively.

$$\eta_n = 2\sqrt{m_p k_n} \quad , \quad \eta_s = 2\sqrt{m_p k_s} \tag{9}$$

2.2.3 Calculation of cohesive force using Van der Waals force

Van der Waals force was taken into consideration as the attraction and adhesion between particles. When the distance between the surfaces of particles approaches 0.1 μm or less, we calculate using the following equation.

$$F_C = \frac{-AD_{pi}D_{pj}}{12(D_{pi}+D_{pj})\,Z_p^2} \tag{10}$$

A in a Eq. (10) is a Hamaker constant. When the distance between the particle surfaces was 4Å or less, it was calculated by setting the value of distance to 4Å, and when particles contacted, it was calculated so that van der Waals force act as adhesion.

2.2.4 Calculation of drag force using the sphere equivalent diameter

The settling velocity of a particle cluster differs from the settling velocity of an isolated particle. In this calculation, when calculating the drag force between particle and fluid, we calculated using the sphere equivalent diameter of a particle cluster. The sphere equivalent diameter of a particle cluster is calculated from the volume of the composition particles of a particle cluster. The definition of the sphere equivalent diameter of the particles which consists of N particles is shown by the following equation.

$$(D_p)_{eq} = \left(\sum_{j=1}^{N}(D_{pj})^3\right)^{\frac{1}{3}} \tag{11}$$

In this calculation, the equation of motion is solved not for particle clusters but for individual particles. Therefore, calculation of a drag force must be performed to individual particles which constitute a particle cluster. The drag force of particles calculated using the value which carried out dignity attachment of the sphere equivalent diameter obtained by Eq. (11) with the diameter of the particles which constitute a particle cluster. The diameter for calculating the drag force of the particles which constitute a particle cluster is shown by the following equation.

$$(D_p)_{deq} = \frac{D_{pi}}{\sum_{j=1}^{N}(D_{pj})}(D_p)_{eq} \tag{12}$$

About the fluid velocity for calculating drag force, the value in the central point of each particle which calculated fluid velocity using proportional distribution was used.

3. CALCULATION CONDITIONS

The calculation conditions and the computational domain are shown in Table 1 and Figs. 2, respectively. The particles of two kinds of sizes (40 μm and 20 μm) were calculated in this calculation. The particles of the number of 5000 were calculated in total using the particles of the number of 2500, respectively. In this calculation, in order to calculate motions of particles and fluid, three-dimensional calculational domain was used. The size of the sedimentation container which is calculation space was made into horizontal length 1200 μm, depth length 480 μm, vertical length 1200 μm. When the particles of the number of 5000 were put in, the average void fraction in the container was set to 0.864. In this study, calculation in the flow state of two types, the case of still flow and the case of shear flow, was performed. In calculation of still flow, the no slip wall boundary condition was used for all walls. In calculation of shear flow, the periodic boundary condition was used for the wall of right-and-left both sides, and used the no slip wall boundary condition except it. Furthermore, in the case of shear flow, the flow velocity of X direction at the bottom was set to 0, and the flow velocity of X direction at the upper wall was given as fixed at the velocity of 6 times the terminal velocity of a large particle. And in the initial state, the flow velocity of the X direction in calculation space was set up so that the shear rate is fixed in the Z direction. Initial state of particles were formed to state of noncontact using a random number series. In this study, calculation was performed to use the physical properties of pure water for fluid and the physical properties of a glass bead for particles.

Table 1 Calculation conditions

Particle diameter D_p	Large	40.0 μm
	Small	20.0 μm
Number of particles	Large	2500
	Small	2500
Terminal velocity of a single particle	Large	1305 μm/s
	Small	330 μm/s
Particle density ρ_p		2.5×10^3 kg/m^3
Fluid density ρ		0.988×10^3 kg/m^3
Fluid viscosity μ		0.001 Pa·s
Averaged void fraction		0.864
Computational cell sizes $\Delta x\times\Delta y\times\Delta z$		120×120×120μm
Numbers of cells		10×4×10
Elastic coefficient E		1.0×10^5 N/m^2
Friction coefficient	Particle	0.25
	Wall	0.3
Poisson ratio ν		0.25
Time step Δt		1.0×10^{-6} sec
Height of sedimentation vessel D		1200 μm
Hamaker constant A		1.5×10^{-12} gcm^{2}s^{-2}
Shear rate of fluid in the initial state $\frac{du_x}{dz}$		6.53 s^{-1}

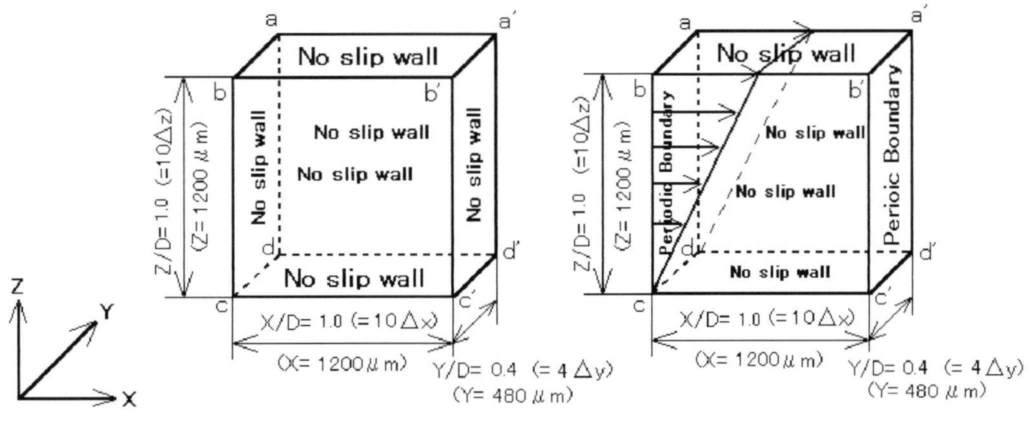

(a) Still flow (b) Shear flow

Figs. 2 Computational domains

4. RESULTS

4.1 Settling velocity of particles

Fig. 3 shows the settling velocity of a certain large particle with time. Comparison of the result calculated using the sphere equivalent diameter and the result calculated without using it is shown. Furthermore, Fig. 4 shows the number of the particles adhering to the large particle shown in Fig. 3 with time. From these results, it is thought that the settling velocity according to a condensed state is calculable by using the sphere equivalent diameter. The theoretical terminal velocity of a large particle calculated from Stokes equation is 0.131 cm/sec, and in the case of isorated particle (i.e., when it calculates without using the sphere equivalent diameter), it sediments at the velocity almost equal to it. For example, the 5th data from the left of Figs. 3 and 4 are considered. The particle cluster consists of the three large particles and the one small particle, and the value which calculated theoretical terminal velocity using the sphere equivalent diameter is 0.279 cm/sec. It is shown that the settling velocity of this particle cluster in Fig. 3 is almost equal to the theoretical terminal velocity.

Fig. 5 shows the fluid mean velocity of the Z direction in the time for 0.1 sec, 0.2 sec, and 0.3 sec. A horizontal axis is non-dimension height, a vertical axis is fluid mean velocity, and fluid mean velocity is the value which averaged in each section of the Z direction. The direction of fluid flow is opposite to direction of particle settling. It is in the hindered settling state and the realistic phenomenon was shown in this calculation. It is thought that the fluid velocity near the bottom is large by 0.1sec is the influence which particles deposited on the bottom.

4.2 The snapshots of instantaneous particle configurations

Figs. 6 (a) - (f) show calculated snapshots of instantaneous particle configurations in the case of still flow. Figures are sectional view around the depth direction (Y-axis) central part, a vertical axis expresses the Z-axis and the horizontal axis expresses the X-axis. In these results, the state of particle settling and the state of forming the sediment near the bottom are shown. Since the settling velocity of large particles is 4 times of the settling velocity of small particles, particles contact in settling process and it forms a particle cluster. Since the particle clusters have higher settling velocity than the particles which exist independently, they sediment more quickly than independent particles. It is thought that these results are obtained since the sphere equivalent diameter is used for calculation of the settling velocity of the particle clusters.

Journal of Physics: Conference Series **147** (2009) 012072 doi:10.1088/1742-6596/147/1/012072

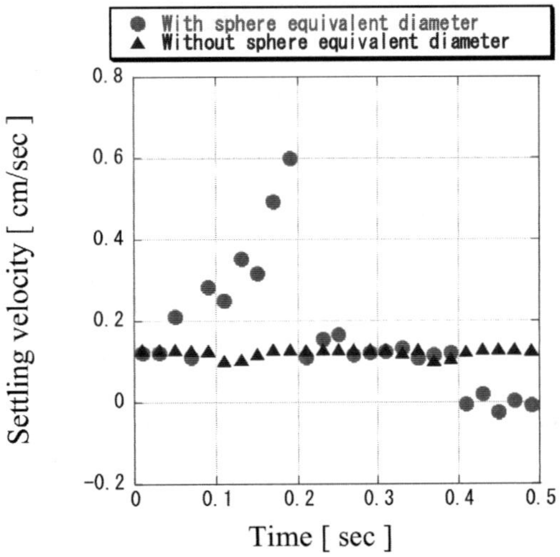

Fig. 3 Settling velocity of large particle in a particle cluster

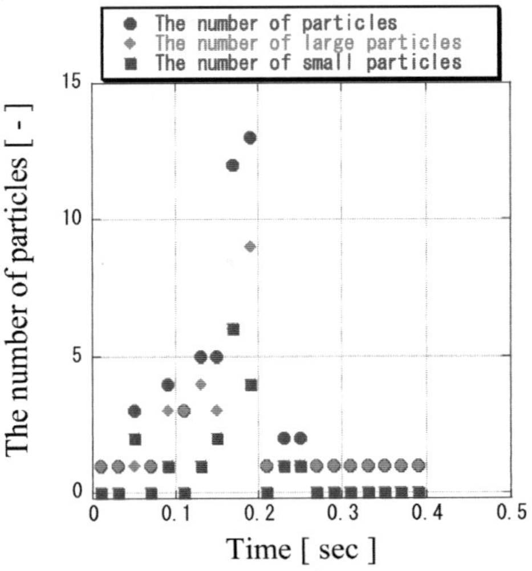

Fig. 4 The number of the particles in a particle cluster

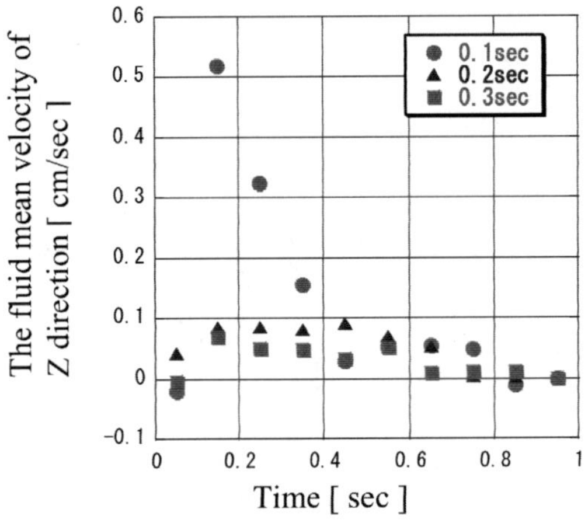

Fig. 5 The fluid mean velocity of Z direction

Figs. 7 (a) - (f) show calculated snapshots of instantaneous particle configurations in the case of simple shear flow. Particles sediment moving to the right from the left by shear flow. From the result shown in Figs. 7, In the case of shear flow, many particle clusters are formed rather than the case of still flow. Since settling velocity becomes high by condensation of particles, particle cluster arrives early at the bottom, and the number of floating particles in the case of shear flow has become less than that in the case of still flow. In the case of shear flow, it is thought that the motion of particles becomes large by shear flow, and the contact opportunity of particles increase.

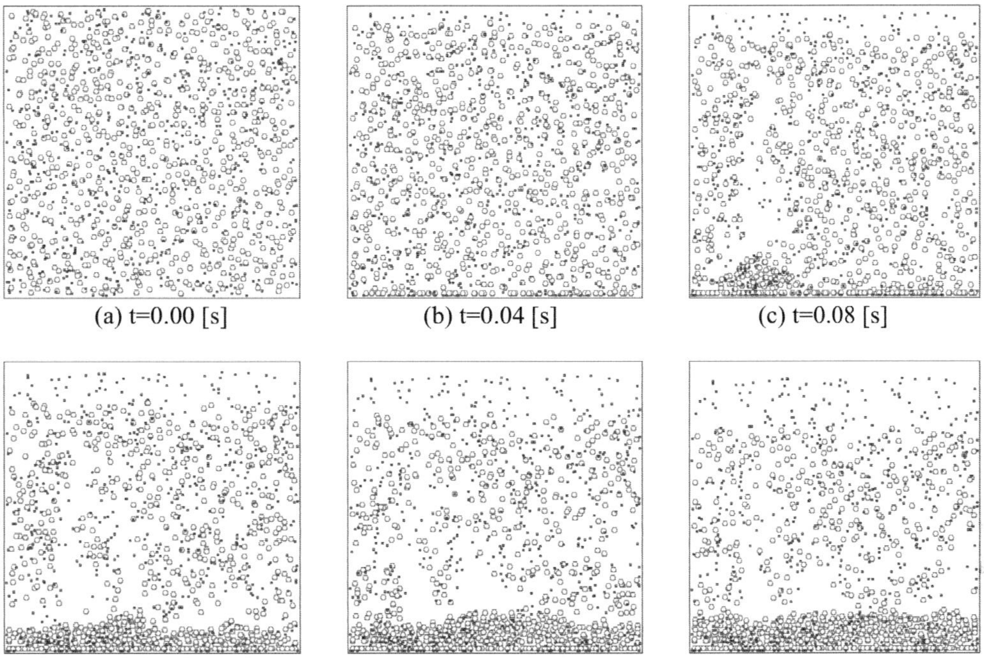

(a) t=0.00 [s] (b) t=0.04 [s] (c) t=0.08 [s]

(d) t=0.12 [s] (e) t=0.16 [s] (f) t=0.20 [s]

Figs. 6 Calculated instantaneous particle configurations in still flow

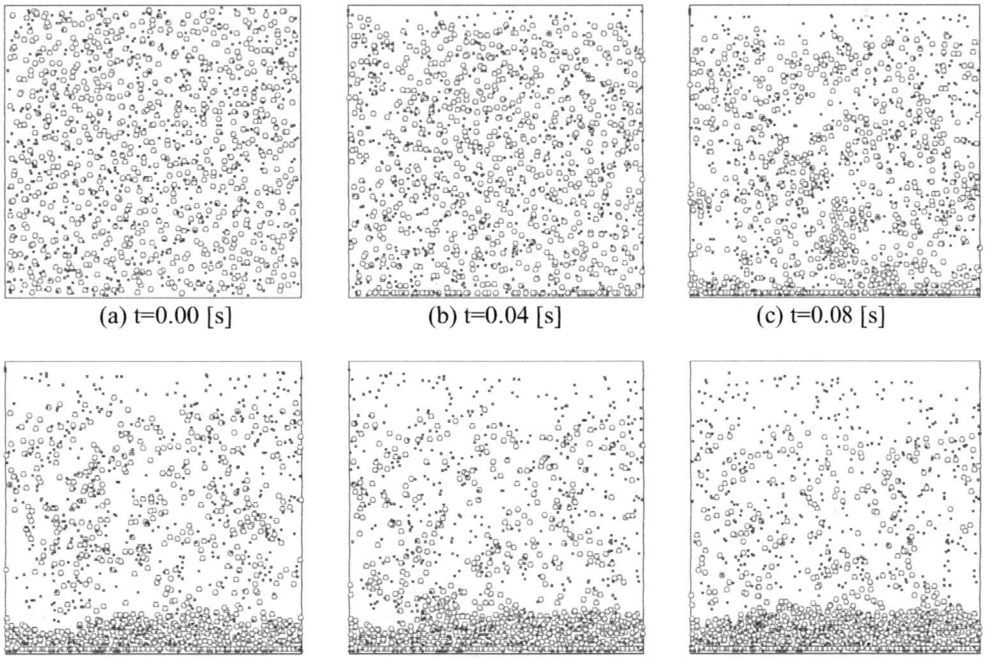

(a) t=0.00 [s] (b) t=0.04 [s] (c) t=0.08 [s]

(d) t=0.12 [s] (e) t=0.16 [s] (f) t=0.20 [s]

Figs. 7 Calculated instantaneous particle configurations in shear flow

4.3 The state of condensation and the fraction of particle clusters

Figs. 8(a) , (b) show the number of particle clusters formed by 2 to 5 particles with time. The number of the particle clusters formed by 2 to 5 particles is shown in the vertical axis, and the partic cluster formed from 6 or more particles is not shown here. The number of particle clusters that consist of 2 particles increases rapidly at once when the calculation is started. Afterwards, the number of particle clusters that consist of 3 particles increases, and, in addition, the number of particle clusters that consist of 4 particles increases afterwards. The number of particles of the particle cluster increases by uniting it to other isolated particles or particle clusters. Therefore, the number of the particle clusters which consist of many particles increases with time. Oppositely, the number of the particle clusters which consist of few particles decreases as the particle cluster which consists of many particles increases. These results show that particle clusters are formed early in the case of shear flow.

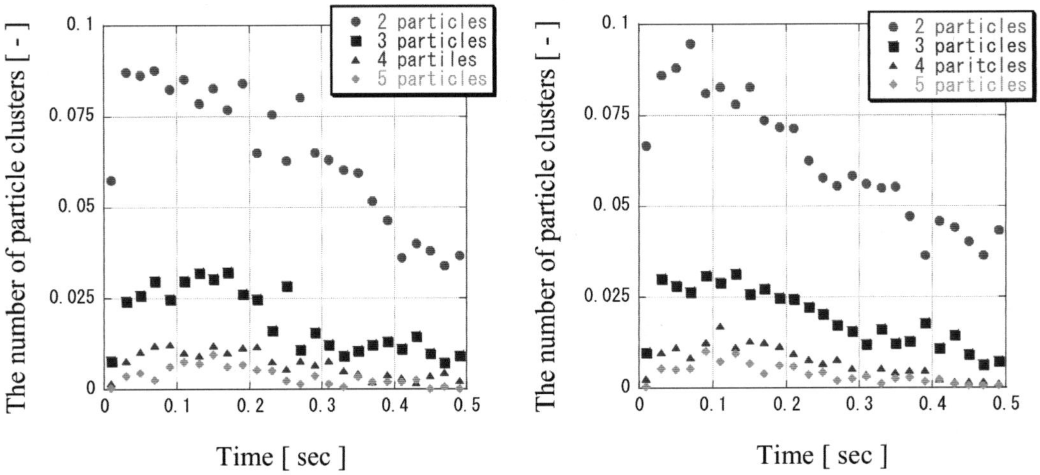

(a) still flow (b) shear flow

Figs. 8 The number of particle clusters formed by 2 to 5 particles

Fig. 9 shows the fraction of the particle clusters in all particles with time. The fraction of the particle clusters increases rapidly immediately after starting settling, and the slope is loose after passing over about 0.1 sec. The tendency of graphs is nearly equal in both. On the whole, the fraction of particle clusters in the case of shear flow is higher than that in the case of still flow.

Fig. 10 shows the fraction of particle clusters in the floating particles with time. Here, the number of the floating particles is sum total of the number of the floating particle clusters and the number of the floating isorated particles. The value in the case of shear flow is small in time after 0.15sec. It is thought that the particle cluster arrives at the bottom early since the settling velocity of particle clusters is large, and the number of the floating particle clusters decrease.

Fig. 11 shows the distribution of void fraction in the sediment made at the bottom. Table 2 shows the average value, max value, min value of the data in Fig. 11. It was obtained from the result after all particles sediment. The value averaged in the depth direction (Y direction) is shown. Under the conditions of this study, there was no difference clearly made into the average value of void fraction. Moreover, the void fraction became an almost fixed value in the X direction.

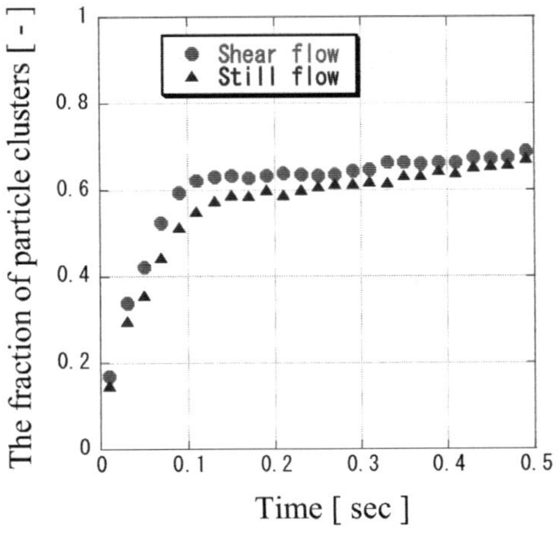

Fig. 9 Fraction of Particle clusters

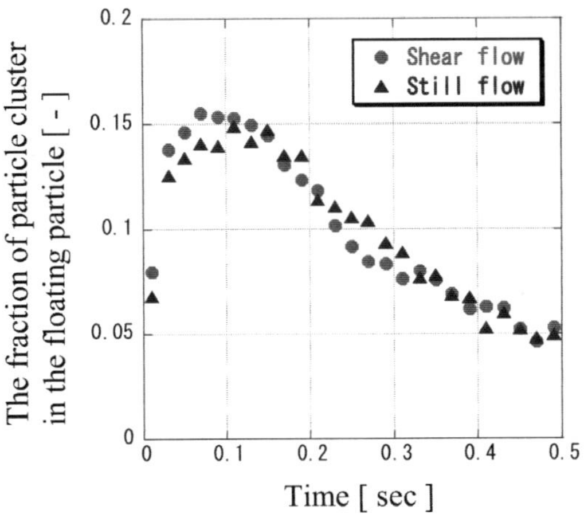

Fig. 10 Fraction of particle clusters in the floating particle

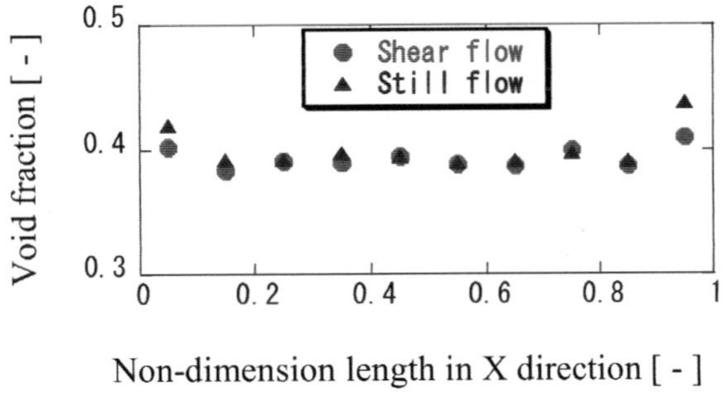

Fig. 11 Distribution of void fraction at the bottom

Table 2 Void fraction at the bottom

	Average	Max	Min
still flow	0.400	0.437	0.390
shear flow	0.393	0.409	0.383

5. CONCLUSION

The numerical simulation of particle settling process in the case of still flow and the case of shear flow was performed. In order to calculate the contact force of particles, DEM was used, and in order to calculate the adhesion between particles, van der Waals force was used. Moreover, in order to calculate the settling velocity of particle clusters, the sphere equivalent diameter was used. The following conclusion was obtained from the calculation results.

Simulation results well describe various phenomena, such as particle condensation, particle dispersion and the hindered settling state.

Journal of Physics: Conference Series **147** (2009) 012072 doi:10.1088/1742-6596/147/1/012072

The settling velocity according to the condensed state of particles was obtained by using a sphere equivalent diameter. The settling velocity of this simulation result is almost equal to the theoretical terminal velocity.

Comparison of the results in the case of still flow and the results in the case of shear flow, the fraction of particle clusters in the case of shear flow is higher than that in the case of still flow.

The difference of void fraction in a sediment did not appear clearly.

NOMENCLATURE

A	Hamaker constant	[J]
D	height of sedimentation vessel	[m]
D_j	damping force at contact point j	[N]
D_p	particle diameter	[m]
$(D_p)_{eq}$	sphere equivalent diameter	[m]
$(D_p)_{deq}$	diameter of dignity attachment of $(D_p)_{eq}$	[m]
F_{Cj}	cohesive force at contact point j	[N]
F_j	dispersion force at contact point j	[N]
g	gravitation acceleration	[m/s^2]
k	stiffness	[N/m]
m_p	particle mass	[kg]
N	particle number of cluster	[-]
n	number of particles per unit volume	[-]
$\Delta n, \Delta s$	increments of normal and shear relative displacement	[m]
$\Delta \dot{n}, \Delta \dot{s}$	increments of normal and shear relative velocities	[m]
P	pressure	[Pa]
Re	Reynolds number	[-]
Re_p	particle Reynolds number	[-]
Sp	interaction term due to fluid drag force	[-]
t	time	[sec]
U_0	terminal velocity of a large particle	[m/s]
$\mathbf{u}$	fluid velocity vector (=(u, v, w))	[m/s]
$\mathbf{u}_p$	particle velocity vector (=(u_p, v_p, w_p))	[m/s]
X, Y, Z	coordinates	[-]
$\Delta X, \Delta Y, \Delta Z$	computational cell sizes	[-]
Z_p	Distance between the particle surfaces	[m]
∇	non-dimensional nabla operator	[-]

Greek Letters

ε	void fraction	[-]
η	damping coefficient	[kg/s]
μ	fluid viscosity	[Pa·s]
ρ	fluid density	[kg/m^3]
ρ_p	particle density	[kg/m^3]
τ	molecular viscous stress tensor	[-]

Subscripts

i	particle i	[-]
j	particle j	[-]
n	normal direction	[-]
s	tangential direction	[-]

REFERENCES

Cundall, P. A. et al. (1979). "A Discrete Numerical Model for Granular Assemblies", Geotechnique.29, pp.47-65.

Harlow, F. H. et al. (1965). "Numerical Calculation of Time-dependent Viscous Incompressible Flow of Fluid with Free Surface", Phys. Fluids, 8, pp.2182-3289.

Kawaguchi, T. et al. (1992). "Numerical Simulation of Fluidized Bed using the Discrete Element Method (the Case of Spouting Bed)", Jpn. Soc. Mech. Eng., Ser. B, 58, pp.79-85.

Rai, M. M. et al. (1991). "Direct Simulations of Turbulent Flow Using Finite-difference Schemes", Comput. Phys., 96, pp.15-53.

T. B. Anderson, et al. (1967). "A fluid mechanical description of fluidized beds" and I&EC Fundamentals, 6, pp.527-539.

Umekage, T. et al. (2001). "Numerical Shimulation of Particle Sedimentation in Liquid and Experimental Verification", J. Soc. Powder Technol., Japan, 38, 3, pp.140-149.

Yuu, S. et al. (1995). "Three-dimensional numerical simulation of the motion of particles discharging from a rectangular hopper using distinct element method and comparison with experimental data (effect of time steps and material properties)", Advanced Powder Technol., 6, 4, pp.259-269.

Yuu, S. et al. (2000). "Numericcal Simulation of Air and Particle Motions in Bubbling Fluidized Bed of Small Particles", Powder Technology, 110, pp.158-168.

Journal of Physics: Conference Series **147** (2009) 012073 doi:10.1088/1742-6596/147/1/012073

Cultivation of the photosynthesis microorganism in a Taylor-Couette vortex flow with a small aspect ratio

H Kawai [1], S Yasui [1], H Takahashi [2], H Kikura [2], M Aritomi [2]

[1] Muroran Institute of Technology, 27-1 Mizumoto-cho, Muroran, 050-8585 JAPAN
[2] Tokyo Institute of Technology, 2-12-1 Ohokayama, Meguro, 152-8550 JAPAN

takahashi@2phase.nr.titech.ac.jp

Abstract. This study focuses on the dynamics of the Taylor-Couette Vortex Flow (TVF) in a photo-bioreactor in which CO_2 is changed to O_2 with high efficiency by the photosynthesis ability of micro algae. Stirring by means of a screw propeller is generally used for a simple agitation. However, the problem is that there exists a very high shearing flow region just near the propeller, which causes the destruction of the alga cell by the shearing force. In contrast, the TVF mixing is expected to reduce such a local and random shearing force because of their column of steady and orderly vortices. In this study, the relationship between the microorganism growth rate and the flow structures in dilute suspensions of a TVF is investigated and the flow characteristics are measured by using an ultrasonic velocity profiler with a small aspect ratio of 3.

1. INTRODUCTION

Bio-reactors are utilized in many applications, for example, food, medical and environmental areas, because it provides desired products without complex processes and produces less by-products (secondary-compounds) by use of biocatalysts; animals and plants cells, and microorganisms. Bioreactors require agitating the animal and plant cells, otherwise those cells may be precipiatated with coagulation, and nutrition doesn't spread uniformly. Generally sedimentation with coagulation is an obstacle to the culture multiplication. Therefore, an appropriate stirring method is expected without destructing the cells; and mixing them uniformly with low shearing force. We adopted a stirring method based on Taylor-Couette Vortex Flow (TVF), which enables uniform and mild mixing without causing local shearing flows seen in screw-type stirring. Besides, recently, TVF has been applied intensively to bioindustry and medical field (Wreley et al., 1999).

TVF can be obtained typically as a vortex flow produced by two concentric cylinders, with the inner one rotating. When the aspect ratio between their gap and height is infinite, TVF has been investigated as the chaos with the spectrum evolution of the disturbance. The practical usage accelerates its compaction, and the study of TVF flow with finite boundaries (end-walls) and small aspect ratio. In the case of TVF with a small aspect ratio, the influence of the Ekman boundary is significant, and its flow pattern becomes more complex even with comparatively low Reynolds number, resulting in the decreased original mixing ability.

Nakamura (1988) studied the flow patterns with a variety of aspect ratios (1-8). Benjamin (1978) studied the mutation of primary flow by changing the length of comparatively short annulus. Mullin (1982) investigated the evolution of primary flow and the transition from N-cell mode to N+2-cell mode by the flow visualization. However the dfficulty is in understanding the mode bifurcation by measuring the spatiotemporal internal flow completely. For this reason, direct measurement is necessary to clarify the internal velocity field structures.

The conventional methods, i.e. Laser Doppler Anemometry (LDA) and Particle Image Velocimetry (PIV) cannot be applied to opaque fluids, because their intransparency prevent the light incidence. Therefore, Ultrasonic Velocity Profiler (UVP) has been developed (Takeda , 1986; Kikura et al., 2004) for optically non-transparent liquid flows. Recently the velocity of a liquid-metal (opaque) flow, i.e. mercury was obtained by Takeda et al.(2002). Takada (1999) measured the spatiotemporal velocity field using an ultrasonic velocity profiler (UVP). The flow fields were analyzed by using two-dimensional Fourier transform and the orthogonal decomposition technique; the intensities of coherent structural modes were quantitatively obtained. With the respective Reynolds number, the intensities of various modes clearly showed that a transition behavior of the quasi-periodic state results from the wavy vortex mode, and the modulating waves were found to disappear suddenly. Kikura, et al. (1999) also used the UVP method and measured the dynamics of magnetic fluid (opaque) Taylor-Couette flows produced by two concentric cylinders with the inner one rotating. Hence, this UVP method for the flow fields has also been applicable to "the turbidity solution". And also, the method has a merit that it can measure the fluid flow characteristics without intruding the flows.

The objective of this research is to investigate the influence of flow fields on the microorganism (biocatalyst) growth in the bioreactor. Recently life supporting systems in closed space have been intensively reseached in the aerospace industries. In those areas, the reseaches on food supply and oxygen supply systems are of no small concern. We paid an attention to the oxgen supply system. Generally the oxgen supply sytem is operated mainly by the electric process or using O_2 gas bottle. On the other hand we adopted the photosynthesis of plant cells that produces O_2 from CO_2. The cells are expected to produce O_2 with high efficiency because of their additional abilities such as growing in the photosynthesis process. Many researchers have been studying photosynthetic bioreactors which use micro algae such as Chlorella and Spirulina. Mitsuhashi et al.(1995) examined the photosynthetic growth rate in a suspension medium of Spirulina at the temperature between 30℃ and 40℃, and reported that the oxygen production rate was increased with the shear stress until a plateau was reached at about 0.3 Pa. This study could be the first precise report on the relationship between an oxygen production rate and shearing flow of Spirulina.

In this paper, we describe the relationship between the microorganism "Spirulina" growth rate and the flow structures in dilute suspensions of a TVF with a small aspect ratio of 3, moreover, we explain flow characteristics measured by an ultrasonic velocity profiler.

2. EXPERIMENT

2.1 Experimental Apparatus and Method

In this study, *Spirulina Platensis* (IAM M-135) was chosen as a sample of plant cells (Fig.1). *Spirulina Platensis* is one of high efficiency photosynthetic algae that are fast growing. The shape of this microorganism is spiral, and the length is about 200-500μm. Spirulina has been also spotlighted from the viewpoint of food problems. In the United Nations World Food Cenference of 1974 it has been lauded as the best food for the future, and proposed by NASA (CELSS) as one of the primary food to be cultivated during long-term space missions.

We adjusted an initial Spirulina weight concentration, about 0.3g/liter in SOT medium, and then put its liquid into the TVF cylinders gap. SOT medium is alkaline aqueous solution containing minerals (Table.1). The schematic diagram of TVF between two concentric cylinders is shown in Fig.2. The apparatus consists of two concentric cylinders made of transparent acrylic material. The length of the cylinders is H = 75 mm, the outer radius of the inner cylinder is R_1 = 15 mm, and the inner radius of the outer cylinder is R_2 = 40 mm. They are positioned vertically and the space between the two cylinders is filled with medium contaning Spirulina. In other words, the aspect ratio Γ is $H/(R_2-R_1)=3$, the radius ratio η is $R_1/R_2=0.375$. The outer cylinder is fixed, and the inner cylinder is rotated with the stepping motor located on the upper wall.

The experimental cultivation conditions are shown in Table 2. Illumination condition is kept constant by the incubator fluorescent light controller (3.5klx). The temperature is also controlled at 303K by the incubator. The experimental vortex mode is 2 cell, and the growth phase condition is liner phase.

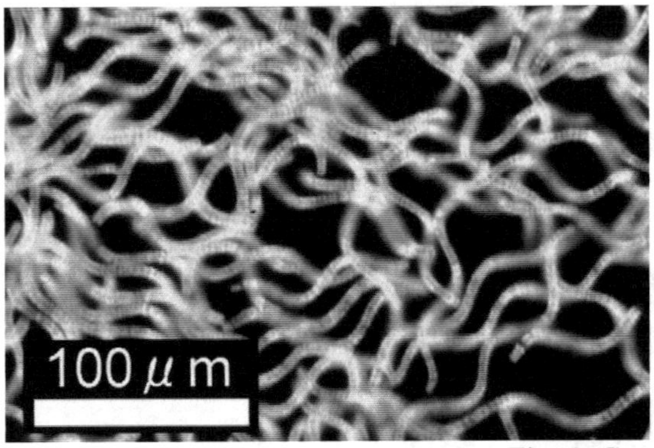

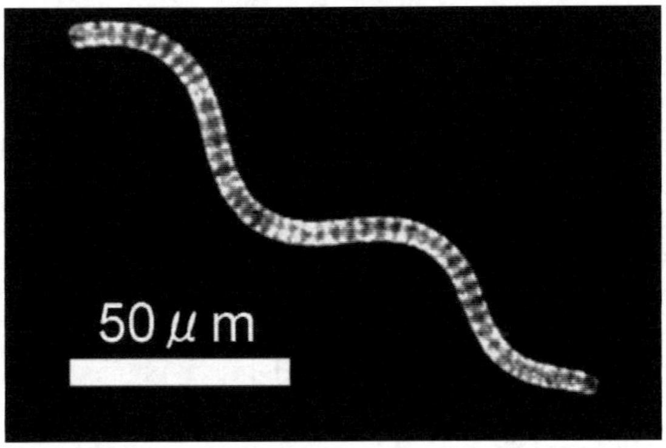

Fig. 1 *Spirulina platensis*

Table 1 SOT medium

$FeSO_4 \cdot 7H_2O$ solution (0.1g/50ml)
$Na_2 \cdot$ EDTA $\cdot 2H_2O$ solution (8g/500ml)
$MgSO_4 \cdot 7H_2O$ solution (20g/500ml)
$CaCl_2 \cdot 2H_2O$ solution (4g/500ml)
$NaNO_3$
$NaCl$
K_2SO_4
$NaHCO_3$
K_2HPO_4
H_3BO_3
$MnSO_4 \cdot 5H_2O$
$ZnSO_4 \cdot 7H_2O$
$CuSO_4 \cdot 5H_2O$
$Na_2MoO_4 \cdot 2H_2O$
Distilled water

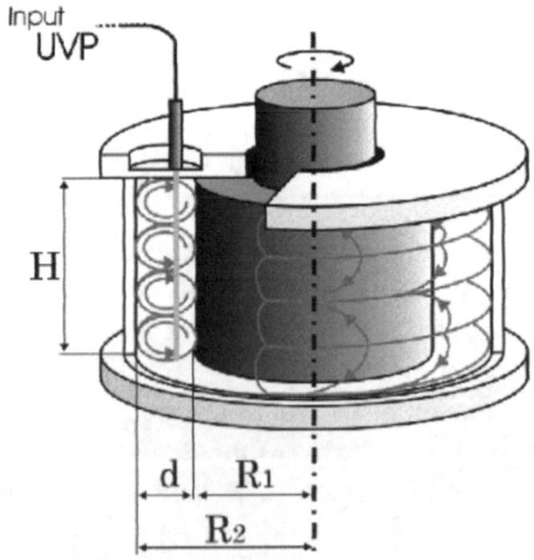

Fig. 2 UVP measurement of Taylor-Couette Vortex Flow (TVF)

Table 2 Experimental cultivation conditions

Microorganism breed	*Spirulina platensis* (IAM Culture collection No. M-135)
Medium	SOT
Initial concentration of Spirurina in SOT [g/liter]	0.3
Illumination [klx]	3.5
Temperature [K]	303
TVF mode	Two cell
Growth phase	Linear growth phase

The ultrasonic transducer is installed on top of the vessel to measure the axial velocity distribution, as shown in Fig.2. In the system with the outer cylinder fixed, and the small Reynolds number, the fluid in the annular gap moves as a plane in the perpendicular direction to the cylinder axis. Reynolds number was described by Eq.(1), where, Ω is the rotation rate of an inner cylinder and ν is the kinematic viscosity of a medium contaning Spirulina.

$$\mathrm{Re} = \frac{d \cdot U}{\nu} = \frac{(R_2 - R_1) \cdot \Omega R_1}{\nu} \quad (1)$$

The UVP monitor used in this study is the UVP-DUO model (Met Flow Inc.). The principle of the ultrasonic Doppler method is based on the echography for position detection, and the Doppler shift for velocity detection. The ultrasound transducer was operated with a basic frequency of 8MHz and a beam diameter of 3mm. The channel distance (the distance between two adjacent measument volumes) was 0.75mm. Liquid measurement requires the ultrasonic reflectors, because UVP is based on Doppler shift. In the case of Spirulina, by means of using them as the reflectors, we suppose it is possible to measure culture solution flows and the organism behaviors without additional reflectors, because the Spirulina size is larger than ultrasonic wavelength. However, when Spirulina is used as reflectors, it is feared that acoustic impedance is not strong enough since the organisms contain lots of aqueous fluid internally.

2.2 Multiplication Characteristics Evaluation

For evaluating the multiplication characteristics, the growth rate Q of the Spirulina was described by Eq.(2), where, C_0 and C_1 are the initial concentration and the 1-day later concentration of Spirulina respectively. The concentration is measured by absorption spectrum of 560nm (before this measurement, we made the calibration line between Spirulina concentration and absorption spectrum). The sampling unit of about 5 ml was taken from the test section after 1-day cultivation.

$$Q = \frac{C_1 - C_0}{C_0} \quad (2)$$

3. RESULTS AND DISCUSSION

3.1 Multiplication Experiment Results

Fig.3 shows the photographs of the cylinders side views under several Re flow conditions. Fig.4 shows the relation between the growth rate Q and Reynolds number Re. Q increases with the increase of Reynolds number. Shown in the photographs, after one day with no stirring, Spirulina coagulate with each other as connecting upper and bottom wall. In contrast, after 1-day TVF agitation, it was observed that the floating algae increase with Re increase. Therefore, we can guess that the growth rate Q increases with Re increase because the floating algae don't shade the light each other, in other words, they improve the optical conditions.

(a) Initial condition

(b) Without agitation (after 1-day)

(c)Re=500 (after 1-day)

(d) Re=1000 (after 1-day)

(e) Re=1600 (after 1-day)

(f) Re=7000 (after 1-day)

Fig. 3 The side view of TVF cylinders

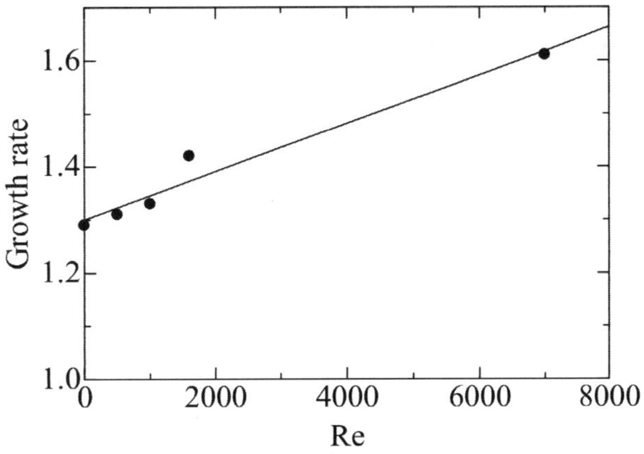

Fig. 4 Spirulina growth rate after 1-day incubation

3.2 Flow Characteristics of a Diluted TVF

3.2.1 Ultrasonic Velocity Profile Measurement using Spirulina Plaensis

Fig.5(a) shows the profiles (by UVP) of time averaged velocities in Spirulina medium. The horizontal axis is a distance from the upper wall, the vertical axis is velocity. Plots in the figure are time averaged values, and the vertical short lines are the standard deviations from them. For the comparison under the same conditions, the time averaged velocity results are shown in Fig.5(b), when 80μm nylon particles (Japan Laser Corp.) are used as ultrasonic reflectors in SOT medium without Spirulina.

In the case of TVF of 0.37 g/litter Spirulina medium, as well as the results of the medium fluid velocity with nylon particles reflector, the formation of two vortices was observed in the cylinder axial direction. As the results, in the flow field measurement in a bio-reactor, the applicability of Spirulina as an ultrasonic reflector is verified.

In the case of spirurina medium, the velocity profile differences were seen at the near plus and minus maximum peak. This phenomenon are most likely due to the several effects such as the Spirulina's weight, shape and population changing phenomena while incubation, and the effects of O_2 small bubbles produced by Spirulina, and the acoustic effects derived by the relative angle of the spiral-shaped Spirulina against the ultrasonic beam axis. However, further investigation will be required for the accurate understanding of this phenomenon.

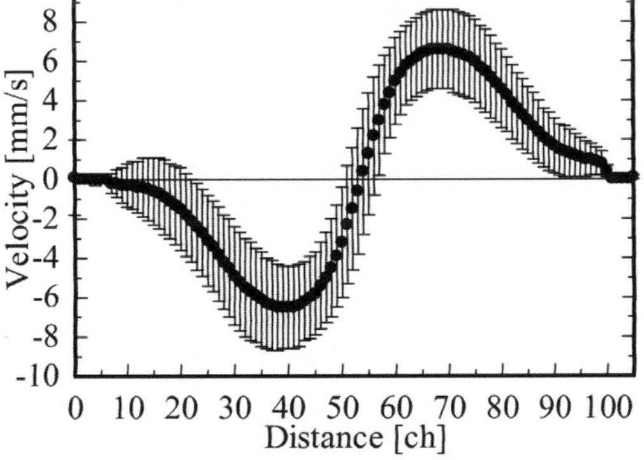

(a)*Spirulina platensis*

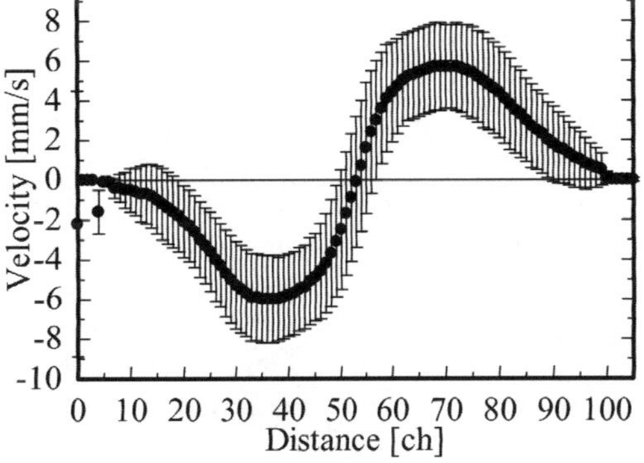

(b)Nylon particles

Fig. 5 Velocity profile difference between Spirulina plaensis and nylon particles reflector (Re=1600)

3.2.2 Time Averaged Velocity Profile Measurement Results of Spirulina Medium (under Multiplication Experiment Re Condition)

The flow patterns were measured using an Ultrasound Doppler Profiler. Fig.6 shows the time averaged velocity profiles (concentration of Spirurina in SOT is about 0.3 g/liter). The transducer was set up on the top at the radial direction 3mm position inward from the outer cylinder and the ultrasonic wave is emitted from the upper to the bottom region. From Fig.6 in each Re condition, it is well observed that the flow has two vortices, and the averaged velocity are larger as Reynolds number increase. Therefore, in the case of dilute suspensions of a TVF with a small aspect ratio of 3, we found that the agitation effect becomes larger as Reynolds number increase within the range of Re between 0 and 7000. As a result, we guess that these flow structures can influence the growth rate. However, further study is necessary to precisely understand the effects of flow structures on the growth rate.

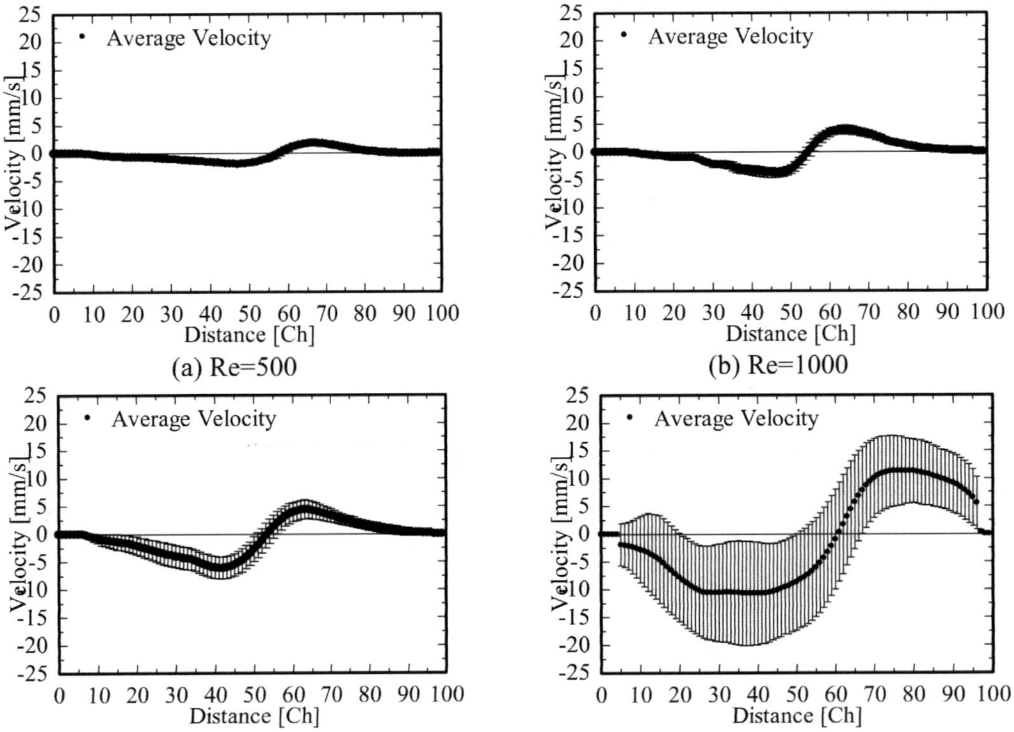

(a) Re=500 (b) Re=1000

(c) Re=1600 (d) Re=7000

Fig. 6 Time averaged velocity profiles of SOT medium containing *Spirurina platensis*

4. CONCLUSIONS

Our purpose is to investigate the effects of TVF on the bio-catalytic microoraganism growth --- in bio-reactors. In the case of the dilute suspensions of a TVF with a small aspect ratio of 3, we found that the agitation effects become larger as Reynolds number increase. As a result, it was found that the growth rate Q increases as Reynolds number increase within the range of Re = 0 to 7000.

For furher understanding of the flow structure effects on the microorganism growth, we will carry out the simultaneous 2 phase measurement (Spirulina:solid and medium:liquid) in the near future, by using 2 types of wavelength transducers.

NOMENCLATURE

H	length of cylinders	[m]
R_1	outer radius of inner cylinder	[m]

R_2	inner radius of outer cylinder	[m]
d	gap between inner and outer cylinder	[m]
Re	Raynolds number	[-]
C_0	initial Spirulina concentration in SOT	[g/liter]
C_1	Spirulina concentration in SOT after 1-day incubation	[g/liter]
Q	Spirulina growth rate	[-]
	Greek Letters	
Γ	aspect ratio	[-]
η	radius ratio	[-]
ν	kinematic viscosity of medium	[m^2/sec]
Ω	rotation rate of inner cylinder	[1/sec]

REFERENCES

Benjamin, T.B. (1978). "Bifurcation phenomena in steady flows of a viscous fluid. II. Experiments," Proc. R. Soc. London A, 359, pp.27-43.

Kikura, H. et al. (1999). "Velocity profile measurement of the Taylor vortex flow of a magnetic fluid using the ultrasonic Doppler method," Exp. Fluids, 26(3), pp. 1432-1114.

Kikura, H. et al. (2004). "Effect of measurement volume size on turbulent flow measurement using ultrasonic Doppler method," Exp. Fluids, 36(1), pp. 187-196.

Mitsuhashi, S. et al. (1995). "Effects of shear flow on photosynthesis in a dilute suspension of microalgae," Appl.Microbiol.Biotechnol., 42, pp. 744 -749.

Mullin, T. (1982) "Mutations of steady cellular flows in the Taylor experiment," J. Fluid. Mech., 121, pp. 207-218.

Nakamura, I. et al. (1988). "An Experiment on a Taylor Vortex Flow in a Gap with a Small Aspect Ratio (2nd Report, Instability of Taylor Vortex Flows)," Trans. JSME B, 54(505), pp. 2425-2432.

Takeda, Y. (1986). "Velocity profile measurement by ultrasound Doppler shift method" Int. J. Heat Fluid Flow, 7(4), pp. 313-318.

Takeda, Y. (1999). "Quasi-periodic state and transition to turbulence in a rotating Couette system," J. Fluid Mech., 389, pp.81-99.

Takeda, Y. et al. (2002). "Flow mapping of the mercury flow," Exp. Fluids, 32(2), pp. 161-169.

Wreley, S.T. et al. (1999). "Internal particle motion in a Taylor Couette rotating filter," Phys. Fluids, 11(2), pp. 325-333.

Journal of Physics: Conference Series **147** (2009) 012074 doi:10.1088/1742-6596/147/1/012074

Characteristics of movement of solid particle in snow-water mixtures

Mikio SASAKI [1]

[1] Hachinohe Institute of Technology
88-1 Ohbiraki, Myo, Hachinohe 031-8501, Japan
Tel: +81-0178-25-8074 Fax: +81-0178-25-0722

msasaki@hi-tech.ac.jp

Abstract. In the present study, experimental observations for the movement of solid particles in the solid-water mixture flow were carried out in the horizontal pipe with a high-speed camera. When flow is slow, the picture was taken at 1000 scenes per second, and when flow is fast, the flow observation catches the solid particle very clearly at 4000 scenes per second. From the flow observation of the solid-water multi-phase flow that the solid specific gravity is near that of the transportation fluid, the change of the solid particle position and the speed change of the solid particle were shown in the present study. Within the scope of the present study, the following conclusions were derived. The solid particle flows with moving up and down. The change of the speed is large. The speeds in the lower layer and upper layer are both great fluctuations because the change of the speed near the wall is large. On the other hand, the vicinity of the solid particle in the middle layer of the pipe fluctuates a little and a steady speed continues. The solid particle near the wall rotates greatly, however, the solid particle in the middle layer of the pipe shows a small rotation. The rotation means that the flow of the mixtures is not Bingham flow..

1. INTRODUCTION

In this experimental study, polystyrene particles are used. The density of solid is nearly equal to that of the carrier fluid, namely, water. Using the particles with the density near 1, we have got good observations in the high degree of exactness for the snow-water mixture flows. Solid particles with the specific gravity lower than 1 are chosen to imitate the pure snow masses. On the other hand, solid particles with the specific gravity denser than water are taken to observe the behavior of snow masses including sand or mud.

Heavy snowfall occurs in the north districts in Japan. To remove snow, snow drains are utilized by townspeople in the north cities. The snow drains are constructed along streets, and used as a hydraulic conveyer for removing snow from the urban area to a river. Recently, the number of cities utilizing the snow drain has increased greatly, and many more cities are planning to build new snow drains. However, there are many locations where such snow drains are not feasible due to lack of adequate slope to the road. If the end of the snow drain is lower than a river, snow conveyed by the drain is not transported by water flow from to a river. Nowadays, a snow drain system with pipeline in final transportation has been investigated to pump up the mixtures from the snow drain to a river when the end of the snow drain is lower than a river. Then, the water slurries of snow in pipe are of considerable importance in final transportation in the snow drain system.

The solid-liquid multi-phase flow is observed with the high-speed camera. The observation is to examine the rotation behavior characteristic of the solid particle. In this experiment, the pipeline was used, and the diameter of the pipe is 5 cm, the specific gravity of the solid particle is 1.06.

2. EXPERIMENTAL EQUIPMENT

The equipment in the present experiment is shown in Figure 1. The test section, which is 13 meters in long, is made of transparent acrylic pipe with the diameter, d, d=49.7 mm. A part of the pipe goes through a quadrilateral water jacket. Within the jacket, visual observations for the behavior of the slid particles can be made with a high-speed video camera.

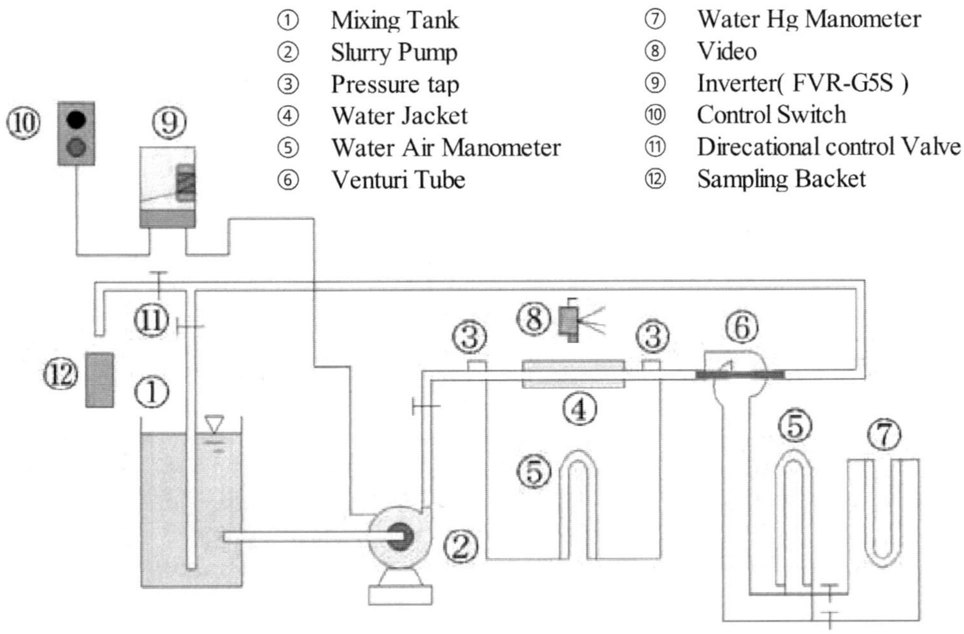

Figure 1. Experimental equipment.

Table 1. Experimental conditions.

Diameter of polystyrene particle, $d_{s\ (mm)}$	3.09
Specific gravity of solid particle, S (-)	1.04
Mean velocity, v (m/s)	0.5, 1.0, 1.5, 2.0 and 2.5
Reynolds number, Re (-)	From 25700 to 128500
Solid concentration, Cv (-)	0.025 and 0.05

The diameter and the specific gravity of the polystyrene particles are 3.09 mm and 1.04. In this present experiment, mean velocity and Reynolds number and the solid concentration by volume are in the ranges from 0.5 to 2.5 m/s, from 25700 to 128500, respectively. The solid concentrations, Cv are 0.025 and 0.05, where Cv is defined by Cv=Vs/Q, where Vs is the flow rate of solid particles and Q is the flow rate of the mixtures. Table 1 shows the experimental conditions in this study. In Table 1,

Journal of Physics: Conference Series **147** (2009) 012074 doi:10.1088/1742-6596/147/1/012074

Reynolds number, Re, is defined by Re = vd/ν_w, where v is the mean velocity and ν_w is the kinematic viscosity of water.

In general, it is better to set the solid pump at the beginning of the pipeline. When the slurry pump is set at the start station in the snow-water mixture flows, the pump crushes snow blocks into much smaller ones. Hence, the snow-water mixture flows become similar to polystyrene solid-water mixtures.

When the velocity is higher than 0.5 m/s, the snow blocks are flowing down individually without combining together. In the present experiment, there is no combine by meltdown of the snow blocks in the mixtures as long as the snow blocks are flowing through the pipe with a speed beyond 0.5 m/s. The results of observations in the present study are available for snow-water mixtures.

3. SOLID PARTICLE VELOCITY

Figures 2(a), 2(b) and 2(c) show an example of the observations with the high speed camera for the solid particle at solid concentration Cv = 2.5%. The solid particle moves quick in order of (a), (b), and (c), that is, the mean velocities are 0.6, 1.5 and 2.5 m/s in the figures. Figure 2(a) is showing that solid particles are flowing in the lower part of the pipe. The reason for the movement of a particle that is heavier than the fluid is that the particle flows in the lower layer of the pipeline because the shear stress working on the solid particle is small as the flow velocity becomes lower. Figure 2(b) shows the movement of solid particles that are heavier than the fluid. As shown Figure 2(b), the solid particle extends to the entire pipe section and flows when the mean flow velocity becomes 1.5 m/s. Figure 2(c) shows that solid particles that are heavier than the fluid move to the upper layer of the pipe and extend to the entire pipe section and are flowing down in the entire pipe without the distinctions of upper layer and lower layer as the average velocity becomes 2.0 m/s. This means that the shear stress working on the solid particle is growing enough, as the mean velocity becomes higher. The shear stress is caused by the speed difference between the solid particle and the fluid particle. As the mean flow velocity grows, this speed difference, that is, slipping velocity becomes high. Of course, there is a speed difference between the solid particle and the fluid particle in the flow of a low-speed region like case 2. The velocity difference is small also in the low-speed region though the slipping velocity exists. It doesn't make a large shear stress that the solid particle defies gravity to flow up. As shown in the figures, the solid particle extends to the entire pipe as flow velocity becomes higher.

Figure 2(a). Observations for solid particle in Case 1 , mean velocity, v = 0.6 m/s, solid concentration, Cv = 0.025. The flow is from the right to left.

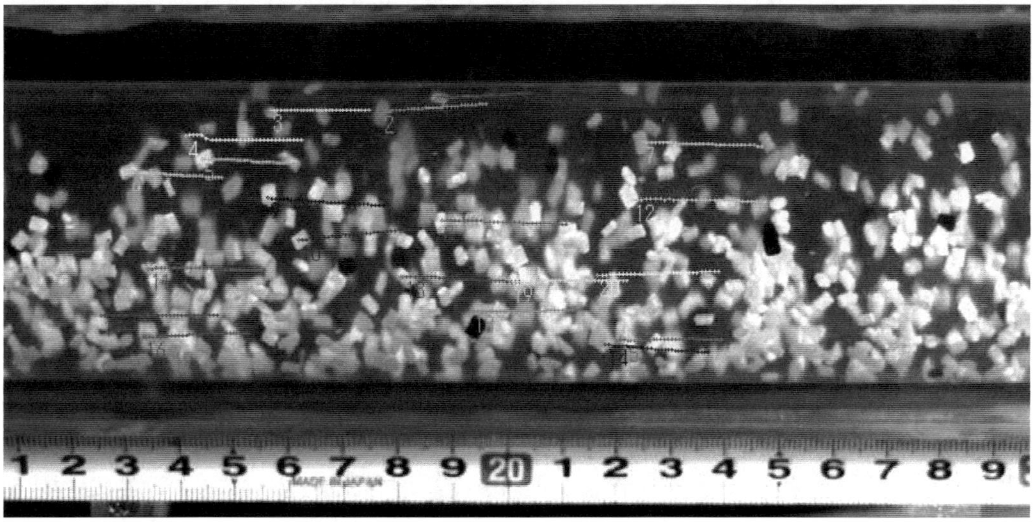

Figure 2(b). Solid particle observations in Case 7, mean velocity, v=1.5 m/s, solid concentration, Cv = 0.025. The flow is from the right to left.

Figure 2 (c). Solid particle observations in Case9, mean velocity, v = 2.5 m/s, solid concentration, Cv = 0.025. The flow is from the right to the left.

When the solid specific gravity is larger than the specific gravity of the transportation fluid, when flow velocity is small, the solid particle flows accumulating in the lower layer, the distortion of the speed distribution grows, and the energy loss grows. When the solid specific gravity becomes smaller than one, the solid particle accumulates in the upper layer in the low-speed region. As the mean velocity becomes large, the solid particle extends to the entire pipe section because the speed difference between the solid particle and the fluid particle grows, and the shearing force grows.

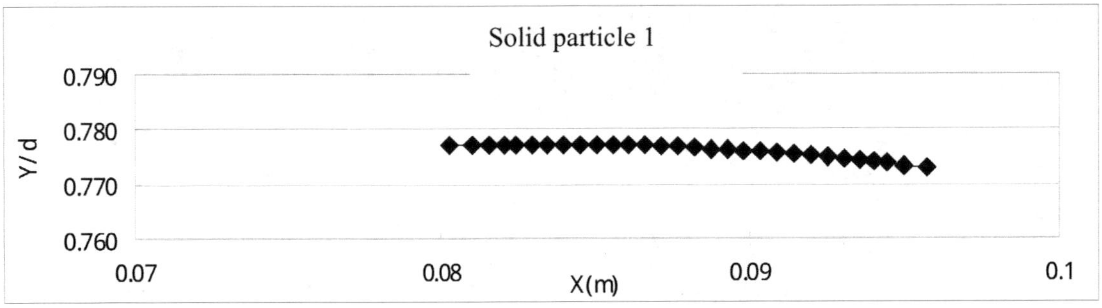

Figure 3(a) Movement of solid particle in upper layer along pipe direction

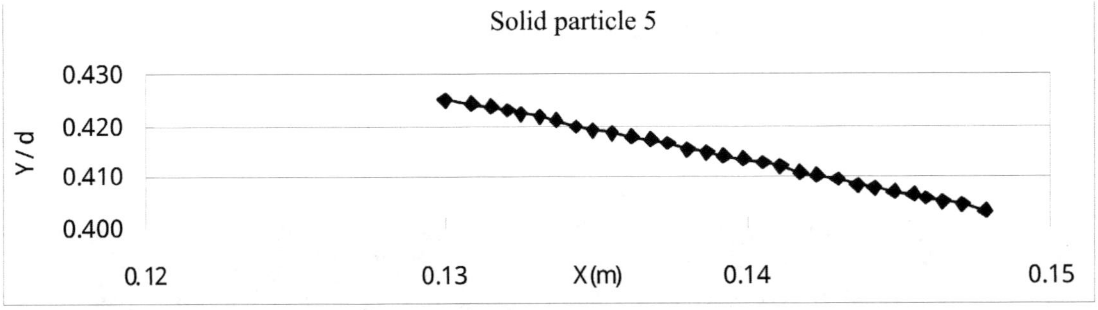

Figure 3(b) Movement of solid particle in middle layer along pipe direction

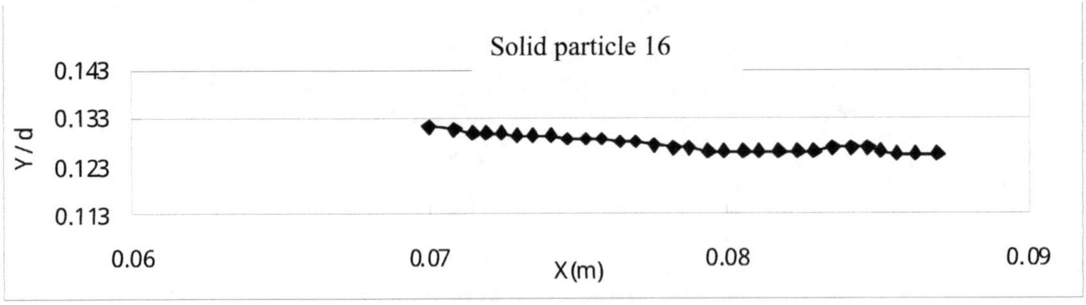

Figure 3(c) Movement of solid particle in lower layer along pipe direction

Figure 3. Movement of solid particle in the upper, middle and lower layers in Case 1, mean velocity, v = 0.6 m/s, solid concentration, Cv=0.025

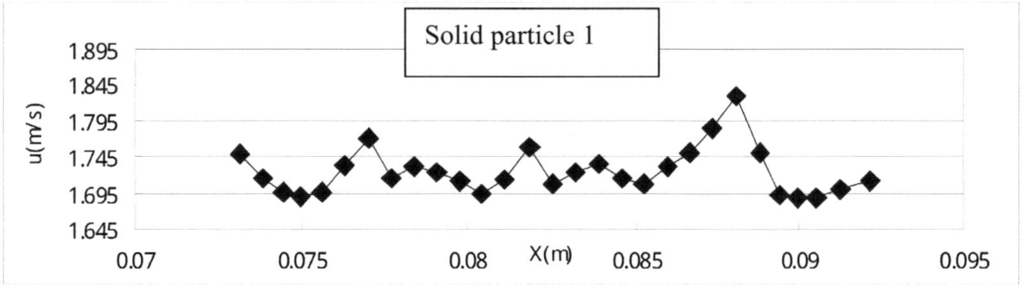

Figure 4(a). Velocity of solid particle in upper layer along pipe

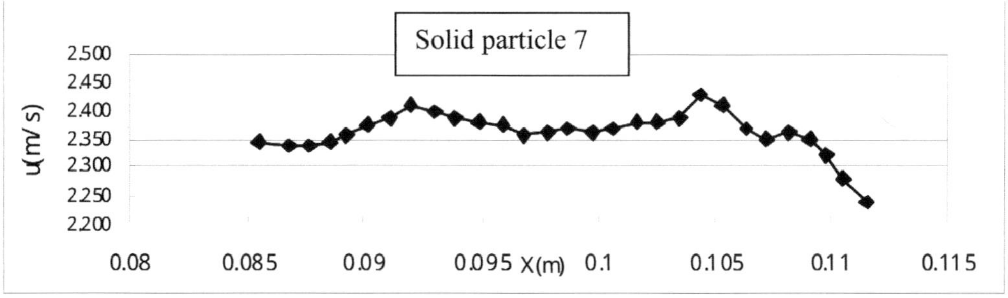

Figure 4(b). Velocity of solid particle in middle layer along pipe

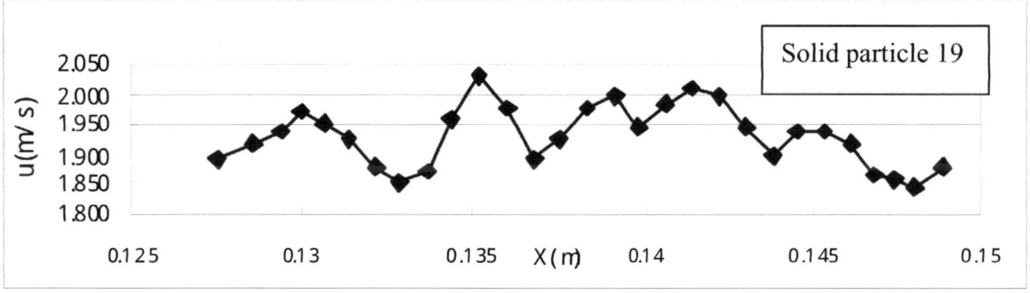

Figure 4(c). Velocity of solid particle in lower layer along pipe

Figure 4. Observation of solid particle velocity in Case 8, mean velocity, v = 2.0 m/s, solid concentration, Cv = 0.025

Figure 3 shows an example of the result of examining the vertical movement of the solid particle. The solid particle flows with repeating the vertical movement in the solid size scale that becomes 0.06

in the value of the vertical axis because the pipe diameter d is 50mm and the size of solid particle ds is 3mm. As shown figure 3(a), the solid particle moves from 0.778 to 0.772 in a perpendicular direction while it moves horizontally by 15mm. This becomes a perpendicular distance 0.3 mm. Therefore, the figure 3(a) shows that the solid particle flows down horizontally little moving in a perpendicular direction in the upper layer in the pipeline. On the other hand, the solid particle moves from 0.426 to 0.407 in a perpendicular direction in the middle layer while it moves horizontally by 15 mm as shown in Figure 3(b) and this distance is 0.5mm. The movement of a perpendicular direction of the solid particle in the middle layer is larger than the one of the solid particle in the upper layer. And, a perpendicular movement of the solid particle is small in the lower layer as shown in Figure 3(c) as well as the upper layer. The solid particle in the lower layer moves by the same degree of the distance as the upper layer. The difference of the movement of the perpendicular distance means the difference of the force to act on the solid particle, the force on the solid particle becomes small in the middle layer and the force is strong near the upper and lower layers. The nearer to the wall the solid particle goes, the stronger the force acting on the solid particle. This force is the shear stress caused from the speed difference between the solid particle and the transportation fluid.

Figure 4 shows an example of the result of observing the speed of the solid particle in the upper, middle and lower layers, where the horizontal axis of figure is a distance taken in the direction of the flow. As shown in the figure, the speed of the solid particle changes in the direction of the flow. As shown in Figure 4(a), the speed change is caused in 8 cm/s in the upper layer, however, as shown in Figure 4(b), the speed changes in a range of 5 cm/s in the middle layer. As well as the upper layer, the speed change is caused in a range of 15 cm/s. Thus, the speed change is growing in the upper and lower layers more than the middle layer. The difference of the width of the speed change between the middle layer and wall layer is due to the gradient of the velocity distribution in the direction of the pipe section. As shown in Figure 4, there is the cycle in the speed change. As the solid particle moves by 1 cm, the cycle is one in the middle layer and the cycle comes to two in the upper and lower layers.

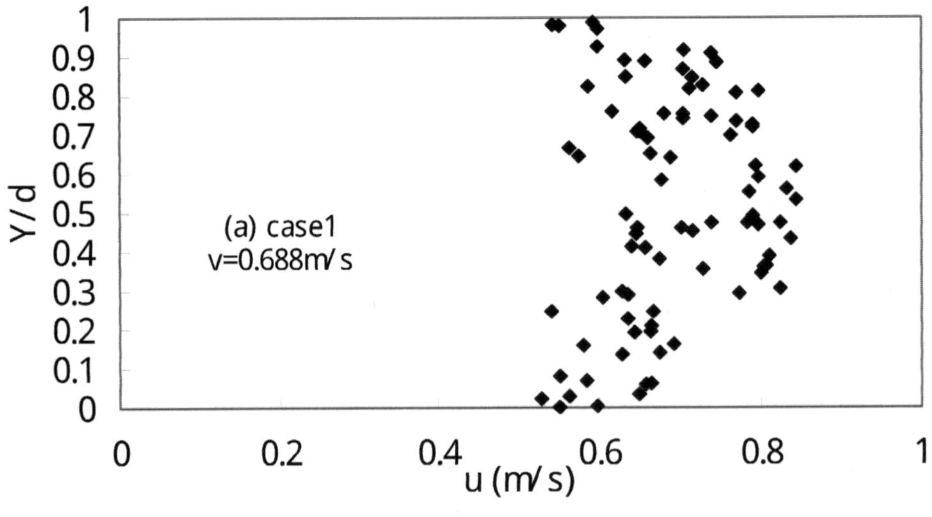

Figure 5. Velocity distributions

Figure 5 shows the velocity distribution in the lower mean velocity. As shown in the figure, the distortion of the velocity distribution causes a little because the solid concentration is small. In this case, the energy loss of the mixtures becomes small due to the small solid concentration. Because the solid particle flows with accumulating in the lower layer in a low-velocity region (See Figure 2(a)), the velocity of the solid particle is relatively low. On the other hand, the velocity of the solid particle in the upper layer has quickened comparatively because most upper layers are liquid layers. As shown in Figure 5, the velocity distribution doesn't become symmetry for the centre of the pipe, and the maximum value of flow velocity is caused within the range of 0.6-0.7 of the vertical axis in the figure.

4. ROTATION OF SOLID PARTICLE

Fig.6 shows the result of the rotational angle speed of the solid particle in the low mean velocity. In the figure, y is the distance from the pipe bed to the top of the pipe. The horizontal axis is a rotation angle of the solid particle per 10 cm. The rotation angle shows the rotation of the solid particle as the solid particle moves by 10 cm. It is observed by using the present high-speed camera. As shown in the figure, the solid particle near the wall rotates greatly compared with the solid particle in the middle layer. The solid particle's rotating shows that the flow of this solid-water mixture flow is not the Bingham flow but the Dilatant flow. Then, It will be necessary to show the velocity distribution of the solid-water mixture flow shown in Figure 6 by the Dilatant flow.

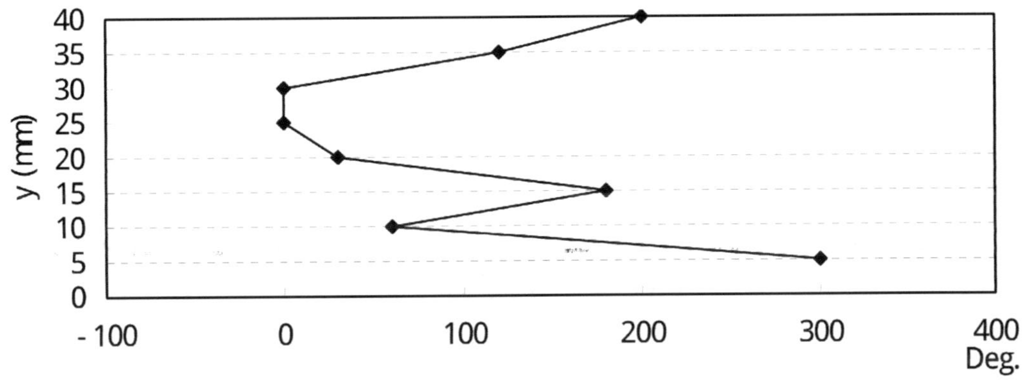

Figure 6 Rotation of solid particle per horizontal distance, 10 cm along the pipe in low-speed region

3. CONCLUSIONS

In the present study, experimental observations for the energy losses were carried out in the solid-water mixture flows. Within the scope of the present study, the following conclusions were derived:

(1) The solid particle flows with moving up and down. The movement of a perpendicular direction of the solid particle in the middle layer is larger than the one of the solid particle in the upper layer. The difference of the movement of the perpendicular distance is due to the difference of the force to act on the solid particle, the force on the solid particle becomes small in the middle layer and the force is strong near the upper and lower layers.
(2) The change of the speed is large. The speeds in the lower layer and upper layer are both great fluctuations because the change of the speed near the wall is large. On the other hand, the vicinity of the solid particle in the middle layer fluctuates a little and a steady speed continues.
(3) The solid particle near the wall rotates greatly, however, the solid particle in the middle layer shows a small rotation. The rotation means that the flow of the mixtures is not Bingham flow but Dilatant flow.

Journal of Physics: Conference Series **147** (2009) 012074 doi:10.1088/1742-6596/147/1/012074

NOMENCLATURE

d	=	diameter
Re	=	Reynolds number
v	=	velocity
vw	=	kinematic viscosity of water
ds	=	Diameter of polystyrene particle
S	=	Specific gravity of solid particle
Cv	=	Solid concentration
Fr	=	Froude number
f*	=	friction factor
f	=	wall friction coefficients
fw	=	wall friction coefficients
I	=	hydraulic gradient
m	=	proportion coefficient
A and B	=	coefficients which change depending on the specific gravity
S	=	specific gravity
φ	=	function

REFERENCES

(1) Kawashima T., Sasaki M. and Takahashi H. (1993) : Experimental study of snow-water mixture flows in horizontal pipes, Hydrotransport 12,pp.525-533.

(2) Mikio Sasaki, Toshio Kawashima and Hiroshi Takahashi (1992): Flow characteristics of snow-water mixture flows in horizontal pipes, Proceedings of 2nd International conference of Snow Engineering, pp.301-312.

(3) Mikio Sasaki, Toshio Kawashima and Hiroshi Takahashi (1995): slip velocities of solid-liquid mixture flows in horizontal pipe, Two-Phase Flow Modelling and Experimentation, pp.1035-1039.

(4) Mikio Sasaki and Takahiro Takeuchi (1999): Eddy viscosity of solid-liquid flow with nearly equal density in horizontal pipe, Hydrotransport 14, pp.359-370.

(5) Mikio Sasaki and Takahiro Takeuchi and Hiroshi Takahashi (2001): Coefficients of wall friction in snow-water mixture, The CD-ROM Proceedings of the ICMF-2001, the 4th International Conference on Multiphase, Session AM, No264, 2001.

Journal of Physics: Conference Series **147** (2009) 012075 doi:10.1088/1742-6596/147/1/012075

Early agglomeration monitor of coarse cohesive particles

Xi Wang, Changfu You, Ruichang Yang

Key laboratory for Thermal Science and Power Engineering of Ministry of Education, Department of Thermal Engineering, Tsinghua University, Beijing 100084, China

xiwang06@mails.tsinghua.edu.cn
(youcf@tsinghua.edu.cn, yangrc@mail.tsinghua.edu.cn)

Abstract. This paper established a three-dimensional visualization experimental system. Paraffin particles were chosen as a test of raw material to simulate the movement and bonded process of coarse cohesive particles. Experimental study on fluidization process of coarse cohesive particles has been done by high-speed camera. The study shows that: from the normal fluidization to shut-down, it can be divided into two phases. The former is a slower phase of the bond and the formation of core grew up phase; the second stage is the huge bonded core turn to bridge fast and shut-down happens with great velocity. Then, comparison of different methods has been done to deal with the experimental date of pressure, the results show that: Compared to the average pressure and pressure deviation method, S attractor method that is based on chaotic theory could monitor the agglomeration much earlier, reflect the degrees of bond with the specific numerical value and also demonstrate the velocity of agglomeration. Thus, it could be applied to the early monitor of agglomeration of coarse cohesive particles.

1. INTRODUCTION

Fluidization technology as a basic technology has penetrated into the national economy in many sectors. It has been widely used and has very good prospects for industry. According to the Gerldart's classification of particles, particles can be divided into four types by size and density of material, namely type A, B, C and D. Type D particle generally refers to the particles larger than 1 mm in diameter and is too coarse or spray used particle. With big bubbles, it is difficult to stabilize operations, so agglomeration and shut-down often happens in the actual industry production. The research on Type D particle is still in the trial of a simple description and does not receive relatively common laws. Furthermore, the research on coarse cohesive particles in high temperature is rather small and different scholars in the same issue come to the different conclusions (ZHOU Yong-min et al., 2005; Aihua Chen. Hsiaotao T.Bi., 2003). Although many scholars have done a lot of research on the influence of temperature to the behaviour of particle fluidization, many of them concentrated in the Type B particle and did more research on the impact of fluid flow by temperature, rarely considering the viscosity of particles(B.Formisani et al., 1998; GUO Qing jie et al., 2002). In view of the fluidized operation must be done against coarse cohesive particles in actual industrial production, it is of great theoretical and practical value to establish the general law of fluidization and agglomeration of coarse cohesive particles as well as the method for early monitor of agglomeration.

Journal of Physics: Conference Series **147** (2009) 012075 doi:10.1088/1742-6596/147/1/012075

2. Experimental System and the Process

2.1 Experimental System

The three-dimensional visualization experimental system of coarse cohesive particles is shown in Figure 1. It consists of three-dimensional fluidized bed body, measurement system and power system. The fluidized bed body is a 3-D cylinder made from plexiglas, whose diameter is 156 mm and height is the 1200 mm, in order to make visual observation to the flow of material and bonded process in bed. Bed material is supported in bed screen. A total of four measuring points, as shown in Figure 2, are: the measuring point 1 at the top of the exit in the centre of the bed, the measuring point 2 above the screen at 10 cm in the centre of the bed, the measuring points 3 above the screen at 1 cm and the measuring point 4 below the screen at 7 cm in the side wall of the bed. They are used to measure pressure information in order to better study the pressure fluctuation in different locations with different flow state and bonded status. To control the temperature of bed, there is also a measuring point below the screen to measure the temperature of fluid by thermocouples. Rotor flowmeter is used to flow measurement. To measure accurately, pressure sensor calibration has been done. The experimental system also includes blower, high-speed camera and electric heater.

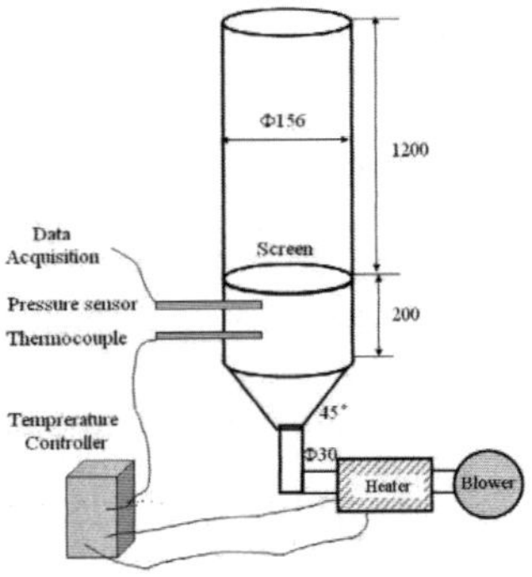

Fig. 1 Sketch of Experimental System

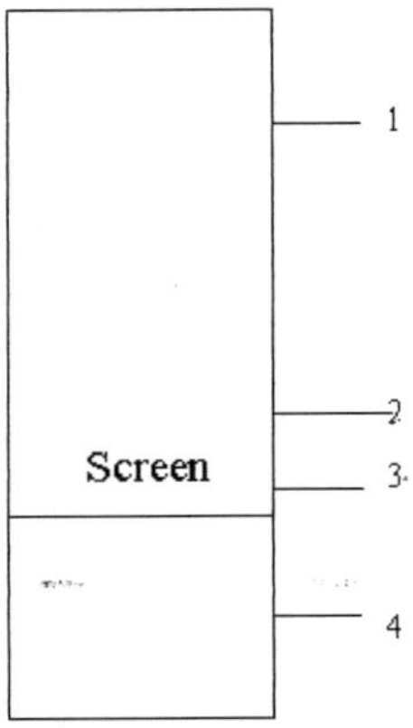

Fig. 2 The Layout of Measuring Points

2.2 The Raw Material used in Experiment

To be able to simulate the movement and bonded process of coarse cohesive particles, paraffin particles with the average size of 5 mm and the melting point of 58 °C were chosen as a test of raw material. Compared with cement, it can be softened at a lower temperature (nearly 40 °C) and has a good cohesive to meet the needs of experimental research purpose. With this raw material, visual observation of the bonded process of coarse cohesive particles in low temperature can be easily realized.

2.3 3-D Experimental Process

3-D experimental process as follows: First add to paraffin particles in bed. Then start blower and set a fixed speed with the rotor flowmeter. When the movement of the fluidized bed has stabilized, record the pressure information with pressure sensors. The computer sampling frequency of pressure signal is 500 Hz and the effective time of each sample is 60 s. Then, stop data collection and use vacuum cleaners to aspirate the paraffin particles from bed. Heat up the gas by electric heater without closing blower until the temperature is up to pre-set value. After the system is in stable, put paraffin particles into bed. At the same time, pressure sensors begin to record pressure information in bed, until shut-down happens.

Fig. 3 Flow State in Slugging

Figure 3 shows the flow state in slugging of coarse cohesive particles. It can be seen from the figure that there is only one big bubble throughout the bed cross-section, in the recycling course of generating, rising and rupturing in entire bed. Particles at the bottom of the bubble are almost static. It shows that the gas-solid contact with bad conditions in the flow state of coarse cohesive particles, causing the flow of deterioration easily. If the flow parameters are selected unreasonable, it will have a bonded slagging and make shut-down happen, as shown in Figure 4. Therefore, the flow parameters must select the appropriate scope to make sure the flow state is in normal fluidization.

Fig. 4 Bonded Block in Shut-down State

3. The General Law of Agglomeration of Coarse Cohesive Particles

When the temperature is up to pre-set value and the system is in stable, paraffin particles in room temperature are put into bed. In the early stages of the flow, paraffin particles are dispersed in bed and the flow state is in a normal fluidization. As the increase of the stay time in bed, heat exchange between the paraffin particles and the gas is becoming more and stronger, so that the temperature gradually increases and paraffin particles viscosity also increases. Paraffin particles in movement collide with each other constantly. When the cohesive force comes to a certain value with the increasing viscosity, some paraffin particles bond together and no longer separate, forming a bigger particle. While the gravity of them overcomes the other forces, they will be at the bottom of the bed and be static. They will continue growing up and have great influence in bonded process, so they are called bonded core. As more obvious side effect of the wall, agglomeration is very likely to happen in the area where the gas velocity is slow, such as the area near the wall. The bond often starts from the low-speed zone of side wall and extends to the centre of bed, showing layers. When the viscidity of paraffin particles reaches a certain value, particles suddenly bond together, land to the centre of bed and a large number of particles quickly adherence to the surface of bonded core, so the material in bed collapse to make shut-down happen. Figure 5 is the sketch of bonded process.

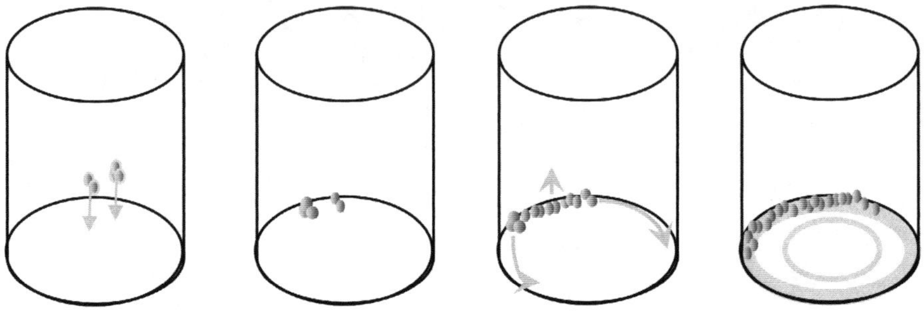

(a)Some Particle Bond (b)Particle Decline (c)Circumferential Development (d)Radial Development

Fig. 5 The Sketch of Bonded Process

4. The Analysis of Pressure Information

If it can predict in advance that the deteriorative trend of flow state, it would be possible to avoid the occurrence of the shut-down phenomenon by taking certain methods, such as: changing the bed temperature or increasing the apparent wind speed of gas to prevent further reunion. Therefore, it is very necessary to monitor the flow state of fluidized bed. Pressure fluctuation signals are often used to analyze and judge the flow state of bed. Many research on early monitor of agglomeration with different methods scholars have been conducted (Jaap C. Schouten., 1998; Kai. T et al., 1985). In this paper, research and comparison of these methods have been done by using them to deal with the experimental date of pressure, judging whether the method can be used for early monitor of agglomeration of coarse cohesive particles.

4.1 The Average Pressure Method

The average pressure method is often used in the first experimental study, so this paper also uses this method to analyze bed pressure information in different flow state primarily. Figure 6 shows the pressure information of different measuring points. As the constraints of experimental system, the pressure information of four measuring points comes from different experiments. Therefore, this paper only has a qualitative analysis to laws that exist in a large number of experiments. From the curve of the changes in Figure 6, it can be seen that the average pressure change of the measuring point 2 is on a declining trend and the average pressure changes of the measuring point 1 and point 4 are irregular with the different bonded state. Although the average pressure change of measuring point 3 shows a declining trend, shut-down has been formed before the average pressure decrease. It can not reflect the changes of bonded state in due course. Therefore, after choosing the right measuring point, it has applicability to the average pressure method to realize the early monitor of agglomeration of coarse cohesive particles.

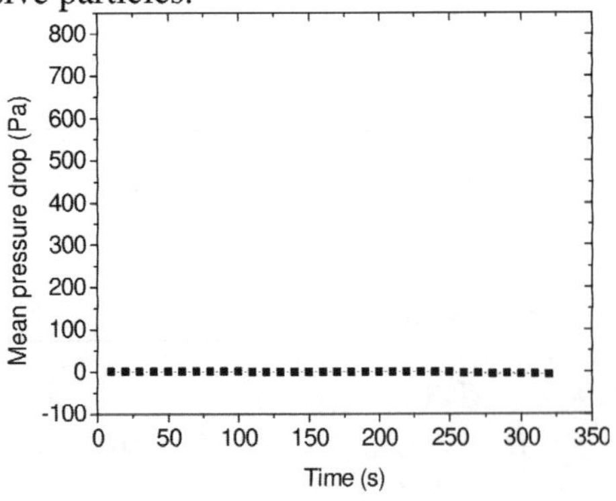

(a)Pressure Information of Point 1

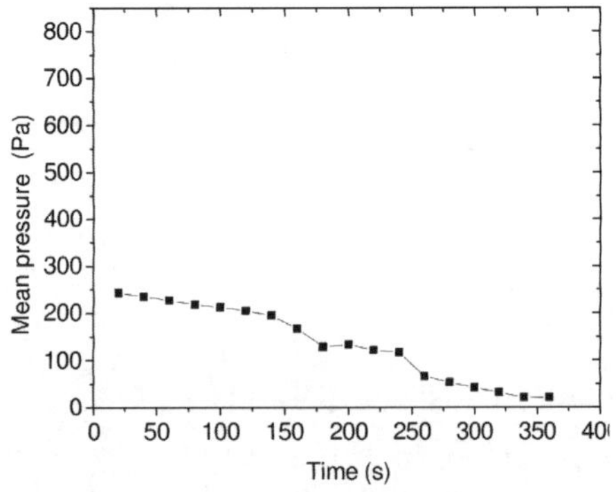

(b) Pressure Information of Point 2

Journal of Physics: Conference Series **147** (2009) 012075 doi:10.1088/1742-6596/147/1/012075

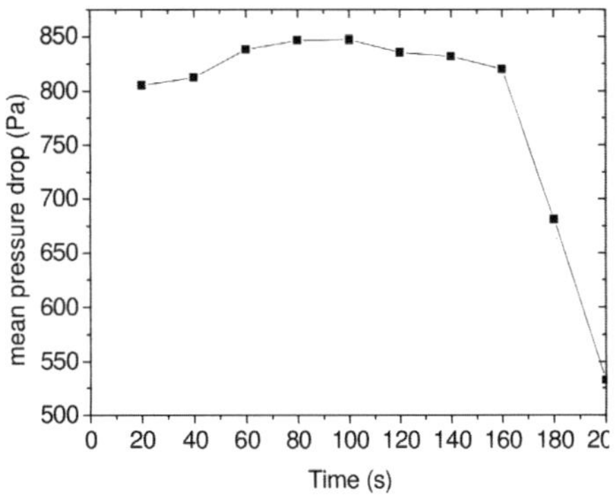

(c) Pressure Information of Point 3

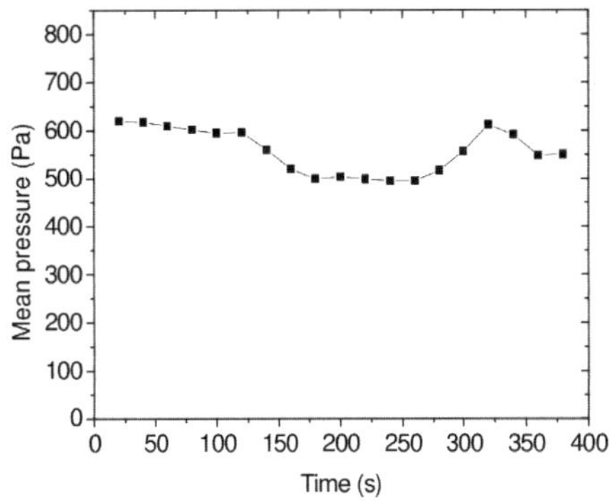

(d) Pressure Information of Point 4

Fig. 6 The Average Pressure of Different Points

From this figure, pressure information of different measuring points has different sensitivity to the flow and bonded state. Measuring point 2 near the top of the material is very sensitive to the change of bonded state. Its pressure information can accurately reflect the bonded state of paraffin particles in bed. This is mainly because the changes of pressure are influenced by the recycling course of the bubbles' generating, rising and rupturing in entire bed. A high degree of material will be reduced due to the bonded deterioration, so it is very sensitive to the change of pressure. At the same time, It has enough distance to weaken the impact of the pressure fluctuation for measuring point 1, so the pressure fluctuation is rather small. Measuring point 3 is in the bubbles' generating area, so the bubble is relatively small and moves quickly and the point is not sensitive to pressure changes. Measuring point 4 under the screen is almost not influenced by the bubbles, so the pressure fluctuation is small, either. Thus, the point which is very sensitive to the change of pressure is the point which is higher in the impact range of material. It can distinguish between different flow state and bonded state conveniently.

4.2 The Standard Deviation Method

In this paper, the standard deviation method is used to analyze the pressure information of different measuring points, and the results are shown as below. With the bond increasing, the changes of standard deviation of measuring point 1 and point 2 are in certain regularity, except for measuring point 3 and point 4. As mentioned above, it is because pressure information of different measuring points have different sensitivity to the flow and bonded state. After choosing the right measuring point, it can realize the early monitor of agglomeration of coarse cohesive particles, either.

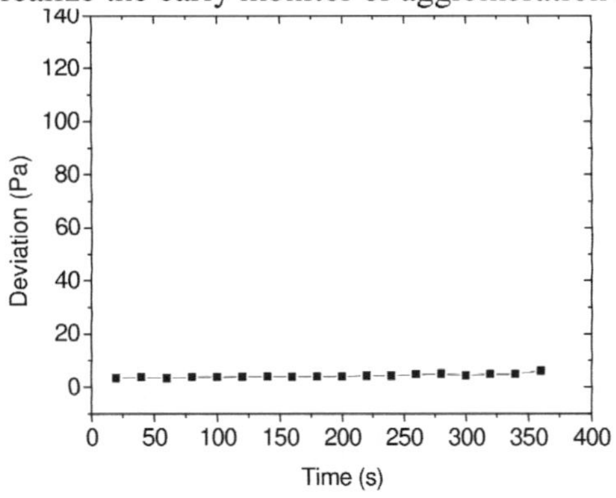

(a) Standard Deviation of Point 1

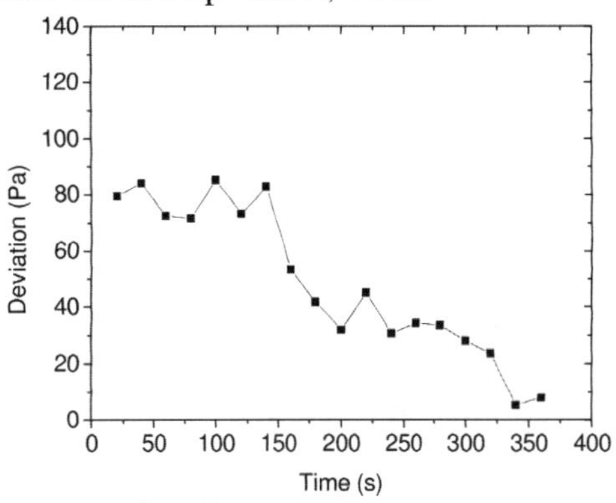

(b) Standard Deviation of Point 2

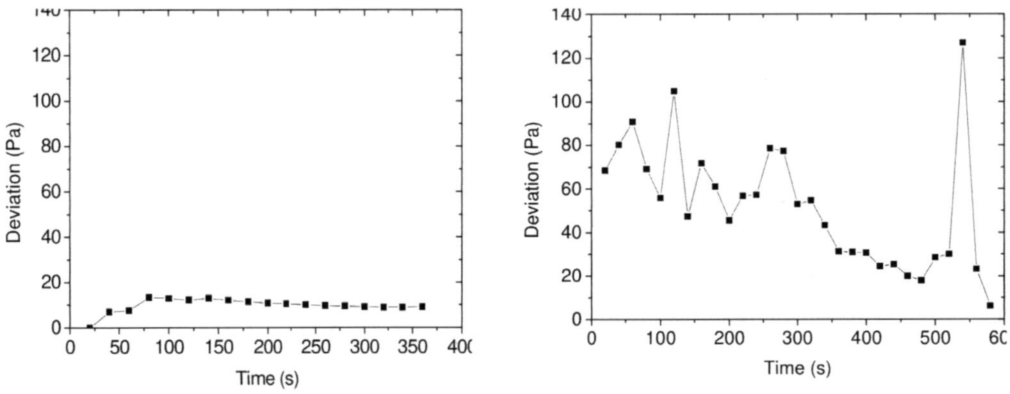

(c) Standard Deviation of Point 3 (d) Standard Deviation of Point 4

Fig. 7 The Standard Deviation of Different Points

4.3 The Wavelet Analysis Method

Wavelet analysis method is based on Fourier analysis. The basic idea is to express or approach signal with function family. This function family is composed of the parallel movement and flex of basic wavelet function $\Psi(t)$ with different scales. There are still some shortcomings with wavelet analysis, such as the wavelet is unsound theoretically. This paper attempts to use wavelet analysis, and the results are shown in Figure 8. The components of characteristics vector from 1 to 5 got by wavelet analysis are on a declining trend and the components of characteristics vector 6 and 7 got by wavelet analysis are irregular with the different bonded state. In addition, each of the changes in the law is not the same components of characteristics vector. The components of characteristics vector got by wavelet analysis cannot show a pattern of changes accompanying the progressive deterioration of the situation or ease, so it is very difficult for using wavelet analysis in early monitor of agglomeration of coarse cohesive particles.

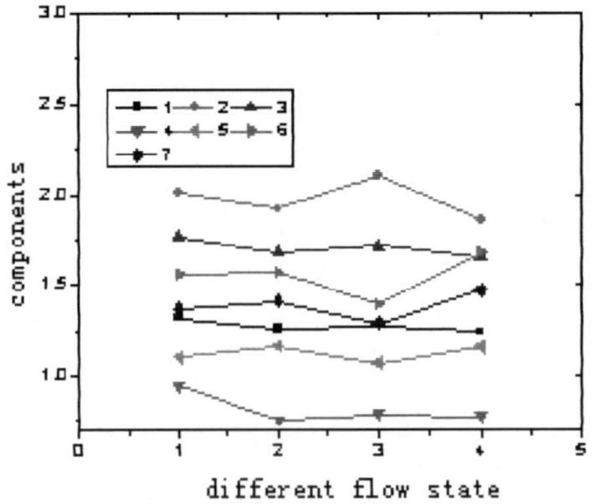

Fig. 8 Components with Different Flow State

4.4 S Attractor Method

4.4.1 Introduction of S Attractor Method

Many non-linear methods are based on a so-called state-space projection of a dynamical system, such as a fluidized bed. In general, the state of a fluidized bed at a certain time can be determined by projecting all variables governing the system into a multi-dimensional space; the collection of the successive states of the system during its evolution in time is called the attractor and can be considered as a characteristic measure for a dynamical system. It has been shown by Takens(Takens F., 1981) that the dynamic state of a system can be reconstructed from the time series of only one characteristic variable such as the local pressure in a fluidized bed. Using so-called time-delay coordinates it is possible to convert a pressure time series consisting of N values into a set of N – m delay vectors P with m elements. The subsequent delay vectors can be regarded as points in an m-dimensional state space yielding a reconstructed attractor, for which Takens has proven that it has the same dynamic characteristics as the true attractor, obtained from all variables governing the system. The attractor can therefore also be seen as a characteristic measure of the hydrodynamics of a fluidized bed.

S attractor method (van Ommen JR et al., 2000; Malte Bartels et al., 2008) is an attractor comparison method that is based on chaotic theory. It is a step beyond the short-term predictability in state space and has been applied to monitoring agglomeration effects in fluidized beds. The principle of this method consists of reconstructing and comparing attractors of the system. One first has to record a reference attractor representing a desired (e.g. well-fluidized) state of the system. Subsequently, the attractor of the current operating state is reconstructed and compared to the reference attractor. This comparison is based on a statistical test developed by Diks et al. (Diks C et al., 1996), which evaluates the dimensionless squared distance S between two attractors. The S-value is defined as

$$S = \frac{\hat{Q}}{\sqrt{V_c\left(\hat{Q}\right)}} \qquad (1)$$

with $\hat{Q}$ is the estimator of the squared distance between the two smoothed distributions of the two attractors and V_c is the variance of $\hat{Q}$.

For attractors being generated by the same mechanism, S has an expectation of 0 and a standard deviation of 1. An S-value larger than 3 indicates with at least 95% confidence that the two attractors have been generated by a different system. As the test is in fact one-sided, the actual confidence is even higher, 98 – 99%, as shown in numerical simulations by Diks et al(1996). The principle is shown in Figure 9.

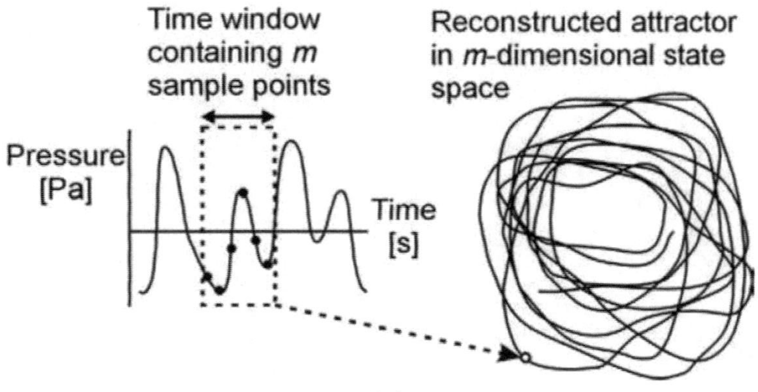

(a)

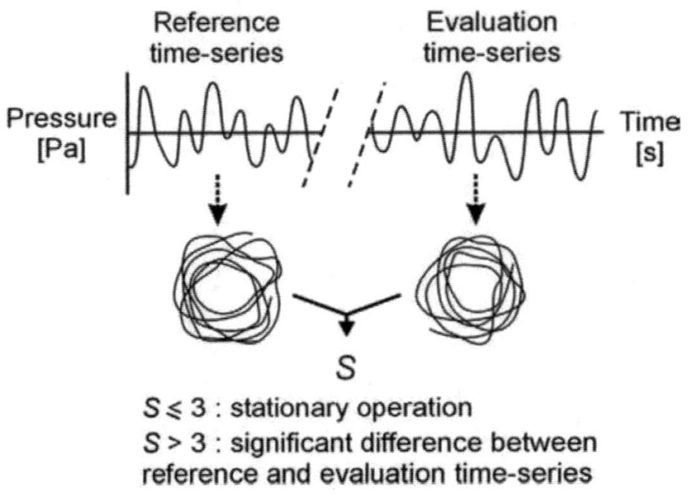

(b)

Fig. 9 The Principle of S Attractor Method

4.4.2 The Correlative Preferences

In the reconstruction phase space, there are four important parameters that is Delay Time τ 、Embedding Dimension m , Segment Length L and Length Scale d .These preferences are of great significance which has a direct impact on the quality of S attractor. Many selection methods have been studied by many scholars (Lin Jiayu et al., 1999; Huang E N., 1998; Liangyue Cao., 1997; van Ommen JR et al., 2000). In this paper, appropriate method has been chosen.

(1) Delay Time τ

Multiple autocorrelation method without excursion(Lin Jiayu et al., 1999) has been adopted. It has a strong theoretical basis, simple mathematical expression, and moderate complexity. Most importantly, the demand of its data length is not strong. The expressions is:

$$R_{xx} = \frac{1}{N}\sum_{i=1}^{N-\tau}(x_i - \bar{x})(x_{i+\tau} - \bar{x}) \tag{2}$$

Where N is the number of samples, $\bar{x}$ is the average value of sequence. When R_{xx} falls to the initial value of $1-1/e$ with the increasing of τ , τ is the delay time of the reconstruction phase space. This paper establishes a procedure to calculate this parameter. The result is 1 which is same to references (Huang E N., 1998). A pressure time-series in actual experiment is chosen to calculate the delay time, and the result is also 1.

(2) Embedding Dimension m

Cao method (Liangyue Cao., 1997) is used widely at present to calculate this parameter. The method overcomes the shortcomings of the other methods. It is shown as below. Suppose that there is a time series:

$$x_1, x_2, x_3, \ldots, x_N \tag{3}$$

The time-delay vectors can be reconstructed as follows:

$$y_i(d) = (x_i, x_{i+T}, \ldots, x_{i+(d-1)T}), i = 1,2,\ldots, N-(d-1)\tau \quad (4)$$

Where d is the embedding dimension and r is the time delay. Define

$$a(i,d) = \frac{\left\| y_i(d+1) - y_{n(i,d)}(d+1) \right\|}{\left\| y_i(d) - y_{n(i,d)}(d) \right\|}, i = 1,2,\ldots, N-d\tau \quad (5)$$

$$\left\| y_k(m) - y_i(m) \right\| = \max_{0 \le j \le m-1} \left| x_{k+j\tau} - x_{i+j\tau} \right| \quad (6)$$

$y_i(d+1)$ is the ith reconstructed vector with embedding dimension d + 1.

$$y_i(d+1) = (x_i, x_{i+\tau}, \ldots, x_{i+d\tau}), 1 \le n(i,d) \le N - d\tau \quad (7)$$

The $n(i,d)$ in the numerator of Eq. (5) is the same as that in the denominator. If $y_{n(i,d)}(d)$ equals $y_i(d)$, the second nearest neighbour is taken instead of it.

To get more accurate results, the value of m is determined by $E1(d)$ instead of $a(i,d)$.

$$E(d) = \frac{1}{N-d\tau} \sum_{i=1}^{N-d\tau} a(i,d) \quad (8)$$

$$E1(d) = E(d+1) / E(d) \quad (9)$$

It is found that $E1(d)$ stops changing when d is greater than some value d_0 if the time series comes from an attractor. Then $d_0 + 1$ is the minimum embedding dimension looking for.

This paper use 1000 points to compare the results with this reference. From the Figure 10, the difference of E1-value between them is rather small. That is to say, the procedure in this paper can calculate this value correctly. In this experiment, the m-value is 6.

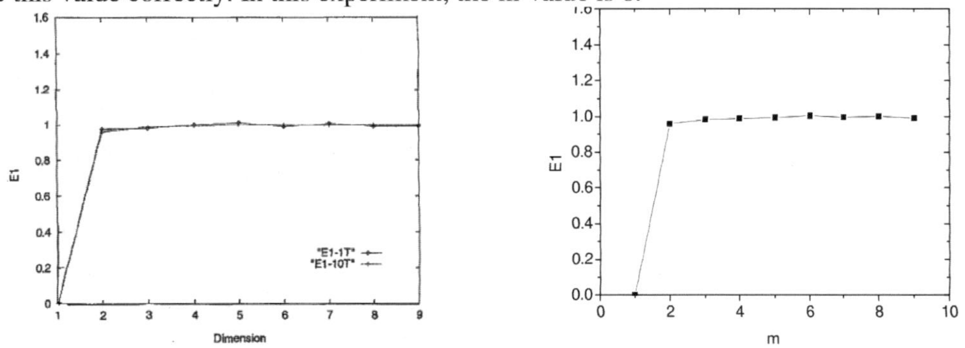

(a) The E1-value of Henon Function in Reference

(b) The E1-value of Henon Function in This Paper

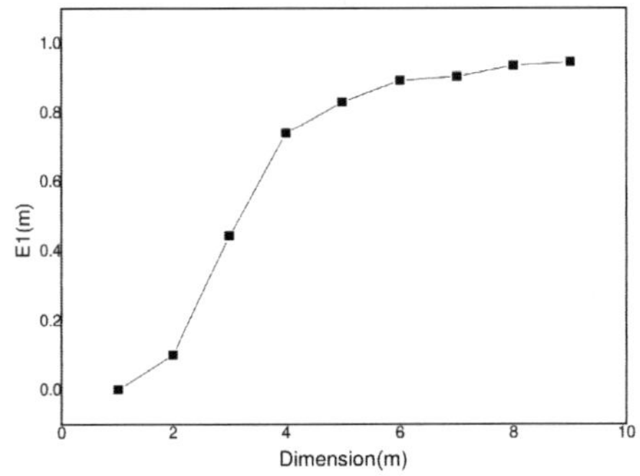

(c) The E1-value in actual experiment with a pressure time-series
Fig. 10 The E1-value in different situation

(3) Segment Length L

The selection of L-value is to ensure that the vectors in the reconstruction phase space are unrelated. It is better that a time series contains at least not less than eight fragments, and each fragment contains a number of signals in the cycle of the pressure information. The method in this reference (van Ommen JR et al., 2000) is chosen. When S begins to fall with the increasing L, L is the suitable value. According to this method, L-value of pressure information in this experiment is 80.

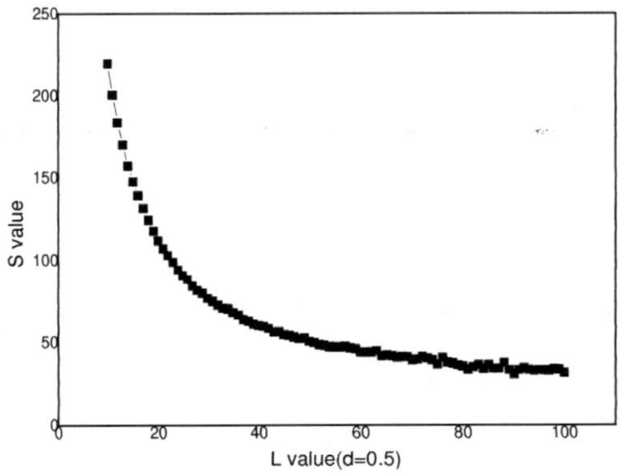

Fig. 10 The Selection of L

(4) Length Scale d

d is the length scale to compare the two vectors. Optimal value depends on the number of research object and decreases with the increasing data quantity. The method in this reference (van Ommen JR et al., 2000) is chosen, either. When S begins to rise with the increasing d, d is the suitable value, as shown in Figure 11. According to this method, d-value of pressure information in this experiment is 0.5.

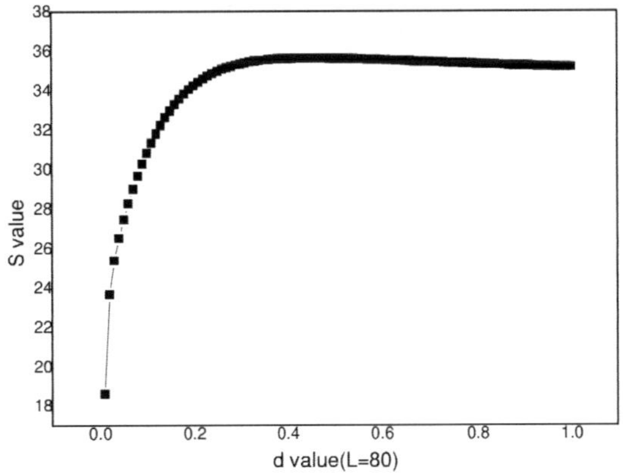

Fig. 11 The Selection of d

4.4.3 The Calculation Process

The process of the procedure is shown as follows:

First get a time series of pressure information from normal fluidization, setting as a reference sequence.

$$P_1, P_2, P_3, \ldots, P_N \tag{10}$$

In order to eliminate the impact of the apparent wind speed, standardization has been made to this sequence.

$$x_k = \frac{P_k - \overline{P}}{\sigma_P} \tag{11}$$

The vectors come from the construction of its space:

$$X_i = (x_{(i-1)m+1}, x_{(i-1)m+2}, \ldots, x_{im})', \quad N_x \text{ vectors Matrix A} \tag{12}$$

Similarly, get the Matrix B of the vectors to be detected. Matrix A and B form a new Matrix Z. Then calculate h function:

$$h(Z_{i,} Z_j) = e^{-\left|Z - Z_j\right|^2 /(4d^2)} \tag{13}$$

To eliminate the relevance of the vectors, get the average value of $h(Z_i, Z_j)$ in every L vectors of Matrix Z:

$$H_{pq} = \frac{1}{L^2}\sum_{i=1}^{L}\sum_{j=1}^{L} h(Z_{(p-1)L+i}, Z_{(q-1)L+j}) \tag{14}$$

Then, S Attractor can be calculated as follows:

$$\tilde{Q} = \frac{2}{N_1(N_1-1)}\sum\sum_{1\le p<q\le N_1} H_{pq} + \frac{2}{N_1(N_1-1)}\sum\sum_{1+N_1\le p<q\le N} H_{pq} - \frac{2}{N_1N_2}\sum_{P=1}^{N_1}\sum_{q=N_1+1}^{N} H_{pq} \tag{15}$$

$$V(\tilde{Q}) = \frac{4(N-1)(N-2)}{N_1(N_1-1)N_2(N_2-1)N(N-3)}\sum\sum_{1\le P<Q\le n} \psi_{pq}^{\ 2} \tag{16}$$

$$\psi_{pq} = HH_{pq} - g_p - g_q \tag{17}$$

$$g_p = \frac{1}{N-2}\sum_{\substack{q\\ p\ne q}} HH_{pq} \tag{18}$$

$$S = \frac{\tilde{Q}}{\sqrt{V_C(\tilde{Q})}} \tag{19}$$

To test the accuracy of this procedure, comparison has been done between this paper and this reference(C. Diks et al., 1996). Two time series come from Henon function of different coefficient.

$$X_{n+1} = 1 - aX_n^{\ 2} + bX_{n-1} \tag{20}$$

The first time series is:

$$a = 1.35, b = 0.31 \tag{21}$$

The second time series is:

$$a = 1.4, b = 0.3 \tag{22}$$

$L = 18, d = 0.0025, \tau = 1, m = 3$

The results show that the more differences between the two series are, the greater the S-value is, in line with the reference. When the coefficient of time series is the same, S-value is less than 3. So it is credible for using this procedure to calculate S-value.

4.4.4 The Impact of the Number of Points

As shown in Figure 12, the more the number of points is, the greater S-value is. The procedure can detect changes in the flow state more easily. If the conditions permit, it is better to use more points. Considering computer performance and sensitivity, it is enough to use 3000 points in this experiment.

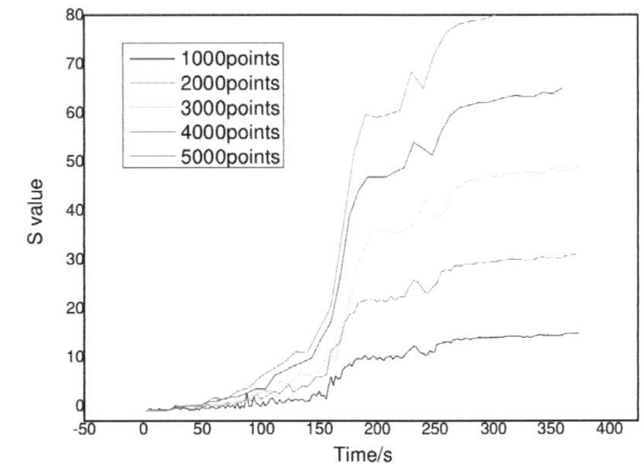

Fig. 12 The S-value in Different Calculation Points

4.4.5 The Comparison of Different Pressure Analysis Method

Figure 13(a) shows the change of S-value in the same experiment. According to experimental records, the bond begins gradually. The bond has a certain scale at 150 seconds. At 300 seconds the bond is already very serious. S-value has the corresponding changes which can be seen from Figure 13(a). S-curve is rising gradually, reflecting the particles begin to bond. S-curve rises very quickly from 150 seconds to 300 seconds, reflecting particles bond have a certain scale and are at a faster bonded velocity. After 300 seconds, bond is relatively stable and shut-down happens. Therefore, the use of S Attractor Method not only reflects the particles from the beginning when to bond, but also reflects the velocity and trend of agglomeration.

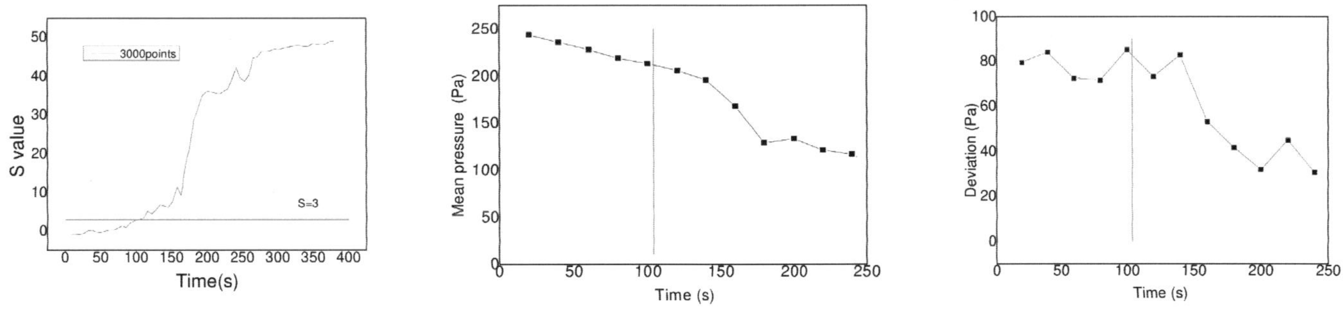

(a)S Attractor Method (b)Average Pressure Method (c) Standard Deviation Method

Fig. 13 Comparison of Different Methods

The results in Figure 13 come from the same experiment as mentioned above. The average pressure and the standard deviation method can show a pattern of changes when bond begins, but the results of the bond changes in the specific numerical value are difficult to determine. When S-curve shows a significant change, the trends in Figure 13(b) and (c) are not obvious, so it cannot quickly determine these changes whether indicate the occurrence and degrees of bond; Compared to the average pressure and pressure deviation method, S-value is about 3 at 104 seconds. It shows that the current state is different from normal fluidization, which is consistent with the experimental phenomenon. That is to say, S attractor method could monitor the agglomeration much earlier, reflect the degrees of bond with the specific numerical value and also demonstrate the velocity of agglomeration.

5. CONCLUSIONS

This paper established a three-dimensional visualization experimental system to monitor the bonded process of coarse cohesive particles and study the method of early monitor of agglomeration. The conclusions are shown as follows:

(1) From the normal fluidization to shut-down, it can be divided into two phases. The former is a slower phase of the bond and the formation of core grew up phase; the second stage is the huge bonded core turn to bridge fast and shut-down happens with great velocity. Through the research of general law of agglomeration, it lays the foundation to early monitor of agglomeration.

(2) Through the study on pressure information of different points, the point which is very sensitive to the change of pressure is the point which is higher in the impact range of material. It can distinguish between different flow state and bonded state conveniently. After choosing the right measuring point, it has applicability to the average pressure method to realize the early monitor of agglomeration of coarse cohesive particles.

(3) For the pressure information of different bonded degrees, the components of characteristics vector got by wavelet analysis cannot show a pattern of changes accompanying the progressive deterioration of the situation or ease, so it is very difficult for using wavelet analysis in early monitor of agglomeration of coarse cohesive particles; The average pressure and the standard deviation method can show a pattern of changes with all the different degrees of bond, but the result of the bonded changes in the specific numerical value is difficult to determine, so it cannot quickly determine these changes whether indicate the occurrence and degrees of bond.

(4) Compared to other methods, S attractor method that is based on chaotic theory could monitor the agglomeration much earlier, reflect the degrees of bond with the specific numerical value and also demonstrate the velocity of agglomeration. Thus, it could be applied to the early monitor of agglomeration of coarse cohesive particles more conveniently.

ACKNOWLEDGEMENTS

This project is funded by The National Natural Science Foundation Projects (NO. 50432040).

NOMENCLATURE

S the dimensionless squared distance S between two attractors

$\hat{Q}$ the estimator of the squared distance

V_c the variance of $\hat{Q}$

m Embedding Dimension

L Segment Length

d Length Scale

N the number of samples

$\bar{x}$ the average value of sequence

$y(d)$ the reconstructed vector

Greek Letters

τ Delay Time

Subscripts

i i-th coordinates

j j-th coordinates

k *k*-th coordinates

REFERENCES

Aihua Chen. Hsiaotao T.Bi. (2003). "Pressure fluctuations and transition from bubbling to turbulent fluidization," Powder Technology, 2003.133, pp. 237–246.

B.Formisani et al. (1998). "Analysis of the fluidization process of partical bodes at high temperature, " Chemical Engineering Science, 1998, 53(5), pp. 951-961.

C. Diks et al. (1996). "Detecting differences between delay vector distributions," Phys. Rev E. 1996, 53. pp. 2169-2176.

Diks C et al. (1996). "Detecting the differences between delay vector distributions," Phys Rev E 1996(53), pp. 2169.

GUO Qing jie et al. (2002). "Flow Characteristics and Defluidization Behavior with Agglomeration at High Temperature Fluidized Bed," Journal of Combustion Science and Technology, 2002(2).

Huang E N. (1998). "The empirical mode decomposition and Hilbert spectrum for nonlinear and non-stationary time series analysis [A]," Proc R Soc Lond A [C]. 1998. pp. 903-995.

Jaap C. Schouten. (1998). "Monitoring the Quality of Fluidization Using the Short-Term Predictability of Pressure Fluctuations, " Particle Technology and Fluidization, 1998, 44, pp 48-60.

Kai. T et al. (1985). "Behavior of Fluidized Beds of Small Particles at Elevated Temperatures, " J. Chem. Eng. 1985, 18(113).

Lin Jiayu et al. (1999). "Selection of Proper Time-Delay in Phase Space Reconstruction of Speech Signals," SIGNAL PROCESSING, 1999(03).

Liangyue Cao. (1997). "Practical method for determining the minimum embedding dimension of a scalar time series," Phys. Elsevier. 1997,110. pp. 43-53.

Malte Bartels et al. (2008). "Agglomeration in fluidized beds at high temperatures:Mechanisms, detection and prevention," Progress in Energy and Combustion Science 34 (2008), pp.633-666.

Takens F. (1981). "Detecting strange attractors in turbulence. Lecture notes in mathematics, " Dynamical systems and turbulence. Berlin, Germany: Springer; 1981. p. 366.

van Ommen JR et al. (2000). "Early warning of agglomeration in fluidized beds by attractor comparison. AIChE J 2000(46), pp. 2183- 2197.

ZHOU Yong-min et al. (2005). "Investigation on characteristics of fluidization of large particles, " Journal of Nanjing University of Technology, 27(2), pp. 34—38.

On-line Measurement of Pneumatic Conveying of Pulverized Coal in pipes

JIA Zhi-hai[1], FAN Xue-liang[1], LI Jun-feng[1], CAI Xiao-shu[1], LIU Ji-ze[2], CONG Xiao[2], JIA Jing-chen[2]

1.University of Shanghai for Science & Technology, Shanghai, 200093, China
2.Zouxian Power Plant, Huadian International Electric Power Co. Ltd. ,Zoucheng, 273522, China

jzhihai@163.com

Abstract. Running conditions of pulverized coal pipes and mills are of greatly effecting on the combustion and safely operation of utilities in coal-fired power plants. However, there are less effective ways to on-line measure or diagnose the pipes or mills. Based on light fluctuation method, the on-line measurement of size and concentration of pulverized coal were presented in this work. Moreover, the velocity of pulverized coal in pipes was measured by cross correlation method. A novel on-line monitoring system was designed with above measurement principles. The on-line monitoring system was applied to measure pulverized coal pipes and mills in-situ in a 1000MW ultra supercritical unit. The unit had 48 pulverized coal pipes and adopted three–layer opposed firing pattern. The experimental results showed the on-line monitoring system can simultaneously measure the size, concentration and velocity of pulverized coal in different coal pipes accurately and quickly. Using these parameters, the running condition of pulverized coal pipes and mills can be diagnosed. Therefore, this work provided a practical method and instrument for on-line monitoring technology of pulverized coal pipes and mills.

1. INTRODUCTION

Pneumatic conveying is based on the physical principle that air, under certain conditions, is able to convey heavy materials. The importance of efficient pneumatic conveying of particulate materials is widely accepted in many industries. Pulverized coal flowing in pipes is a typical phenomenon of pneumatic conveying. Detailed information on the flowing particles other than the mass flow rate, such as the particle size, concentration and the local velocity, is recognized as important to process optimization and control as well as to improvement in the understanding of multiphase flow fundamentals. A common requirement exists for the continuous on-line measurement of these parameters. Much research work has been carried out in the field of pulverized coal flow metering and a wide range of techniques. Many authors tried to use different methods to on-line measurement of pneumatic conveying of pulverized coal in pipes. A number of methods have been proposed for sizing, concentration and mass flow rate, including those based on capacitance method, electrostatic method, differential pressure, acoustic emission, imaging method and so on.

Capacitance transducers are the early technique to measure mass flow rate. In general, these transducers are designed to determine the capacitance arising from the average permittivity of the mixture of conveyed and conveying phases flowing within the sensing volume of the capacitor electrode. For example, Beck et al(1990) designed a capacitance sensor to monitor mass flow measurement in pneumatic conveyors with very low solids loading. Later, electrostatic method was adopted to measure the parameters of pulverized coal. Yan and Li et al(1996, 2000, 2001) tested the velocity of particles with cross correlating two signals derived from a pair of electrostatic sensors. They designed a electrostatic sensors to evaluate the mass flow measurement of pneumatically conveyed solids. Digital imaging is becoming increasingly attractive because of the availability of low-cost high performance imaging devices and computing hardware. An advantage of using the imaging approach is that non-intrusive real-time measurement of many individual particles can be measured concurrently and continuously. Some researchers had done much work in this field, such as Carter and Yan(2003, 2005), Li and Tomita(2000), Nishino and Hiroyuki(2000). They studied the mass flow measurement of particles, velocity and volumetric concentration of particles using the digital imaging sensors. Some measurement equipments based on these methods have had development to measure various flow parameters. However, few systems have achieved commercial success to date.

In recent years, optical techniques have become one of the major contenders for being on-line multiphase flow measurement methods, such as Black and McQuay's work(1996). One of their main advantages is that they are nearly free of interference by most environmental factors, such as temperature, humidity and electrostatics. On the other hand, they are highly sensitive to change in either the concentration or the size of the measured particles. Therefore, optical techniques have been recognized as more appropriate method to on-line measure particle parameters. Teipel(2002) used laser light diffraction spectrometry technique to determine particle size distribution. Cole et al(2001) used Mie scattering theory to measure hypersonic microparticle characteristics. Recently, Xu and Cai et al(2004,2005) studied the forward scattering theory and designed an advanced coal measurement sensor to measure the pulverized coal characteristic parameters..

However, gas-particle two-phase flow is an unsteady and complicated nonlinear dynamics system. A detailed understanding of the behavior of particles is important for design, optimization, and operation of the pneumatic conveying system. Most of the proposed methods are still on the stage of laboratory research and few of them are commercialized and practically used in industry. There are still less effective ways to on-line monitor or diagnose the pulverized coal pipes. It is clear that an instrumentation system must be developed that can measure these important parameters online.

This paper presents a new solution for the measurement of size, concentration and velocity of particulates within a gas-solid pneumatic conveying pipeline using laser light fluctuation and cross-correlation method. An advanced on-line pulverized coal monitoring system was devised and applied to measure pulverized coal pipe in-situ in a 1000MW ultra supercritical unit and measurement results was going to be discussed.

2. MEASUREMENT PRINCIPLES

2.1 Measurement of the size and concentration of pulverized coal

When a bouquet of parallel single wavelength laser light passed through the measured particles, due to the scattering and absorption of light, the transmitted light intensity will decay. According to Lambert-Beer's Law, the transmitted light intensity can be described as

$$\ln\left(\frac{I}{I_0}\right) = -\frac{\pi}{4} L D^2 N E(\lambda, m, D) \tag{1}$$

Where I and I_0 are the transmitted and incident light intensity respectively, L is the path that the measuring beam traverses through the sensed volume, D is the mean diameter of particles ; N is the concentration of particles per unit volume, $E(\lambda, m, D)$—the coefficient of decay light and related with wavelength of laser light λ, relative refraction index m, and D. The coefficient of decay, E, can be calculated according to Mie scattering theory.

If the cross-sector area of light beam is A, then the volume of measurement zone is equal to AL. The equ. (1) can be modified into

Journal of Physics: Conference Series **147** (2009) 012076 doi:10.1088/1742-6596/147/1/012076

$$\ln\left(\frac{I}{I_0}\right) = -\frac{\pi}{4A} nD^2 E(\lambda, m, D) \tag{2}$$

Where n is the number of particles through the measurement zone, $n = NAL$.

Further researches showed the beam diameter A had complex relation with the size and concentration of pulverized coal. When A diminished to a certain extent, the number of particles through the test zone was stochastic and varied with the measurement time. That resulted in the transmitted light intensity increasing. Because the transmitted light intensity I was caused by the number of the particles and size of pulverized coal, it contains the information of the concentration and size of pulverized coal. Therefore, through analyzing the transmitted light intensity, the size and concentration of pulverized coal can be acquired by inversion algorithm.

In most cases, the fluctuation of the transmitted light intensity can be mostly ascribed to the variation in the number of the sensed particles. If the particle/beam size ratio is much smaller than 1, the "boundary effect" can be neglected. Then, the mean transmitted light intensity $\overline{I}$ and its variance σ^2 can be calculated with equ. (3) and equ. (4).

$$\overline{I} = I_0 e^{-\tau} \tag{3}$$

$$\sigma^2 = \overline{(\Delta I)^2} = I_0^2 \Phi(\tau) / \overline{N} \tag{4}$$

Where τ is the absorption coefficient of particles, $\tau = -\ln(\overline{I} / I_0)$, $\overline{N}$ is the average number of the particles in the measurement zone, and $\overline{N} = \overline{n}L$, $\overline{n}$ is the average concentration of the particles, $\overline{n} = \tau / LS_1$, $\Phi(\tau)$ is a single-peaked Shifrin function.

$$\Phi(\tau) = 2\tau e^{-2\tau} \int_0^{\pi} x(\tau, \Psi) d\Psi \tag{5}$$

$$x(\tau, \Psi) = \sin(\Psi) \exp\left[\frac{\tau(\Psi - \sin(\Psi))}{\pi} - 1\right] \tag{6}$$

Hence, the projection area of the particle of pulverized coal is

$$S_1 = \frac{\overline{(\Delta I)^2}}{I_0^2}\left[A \frac{\tau}{\Phi(\tau)}\right] \tag{7}$$

or the average size of the particle is

$$D = \sqrt{4S_1 / \pi} \tag{8}$$

and mass concentration of pulverized coal is

$$W = \frac{\pi}{6} \rho \overline{n} D^3 \tag{9}$$

Where ρ is the density of pulverized coal.

2.2 Measurement of the velocity of pulverized coal

The cross-correlation method is a technique widely used for measurement of flow rate, which may also be applied to measure the velocity of the flowing particles: two measuring light beams are aligned at a short distance, l, along the flowline and the time lag between the upstream and downstream transmission signals, τ, is determined by extremum point estimation of the cross-correlation function, then the local flow velocity, v, is given as follows:

$$R_{12}(\tau) = \lim_{T \to \infty} \frac{1}{T} \int_0^T y_1(t - \tau) y_2(t) dt \tag{10}$$

The cross-correlation function $R_{12}(\tau)$ and transmit time τ can be calculated with FFT. Then the velocity of pulverized coal can be acquired with equ. (11).

$$v = l / \tau \tag{11}$$

3. PROBE STRUCTURE AND INSTALLATION

In this work, a novel on-line monitoring system based on the laser light fluctuation method and cross-correlation method has been developed for on-line measurement of pneumatically conveyed solid particles, as shown in Fig. 1. It is of great significance to consider how to enable enough flexibility and easy install/uninstall processes for the intrusive probe. The laser diode, the optical fibers, the photodetectors and the preamplifier circuit are all integrated into a rugged package at the probe tip, making the install/uninstall process quite convenient. In addition, the intrusive probe is made of ceramics to ensure excellent work reliability and resistance to abrasive erosion by flowing particles. This makes the probe appropriate for in-line measurement of hard solids, even with a high flow rate. To prevent contamination and blockage of the beam aperture by the flowing solids, a purging air flow is injected into the probe, which also helps with probe cooling. When used for continuous measurement of long duration, little maintenance work is required besides occasional self-calibration with no particle load.

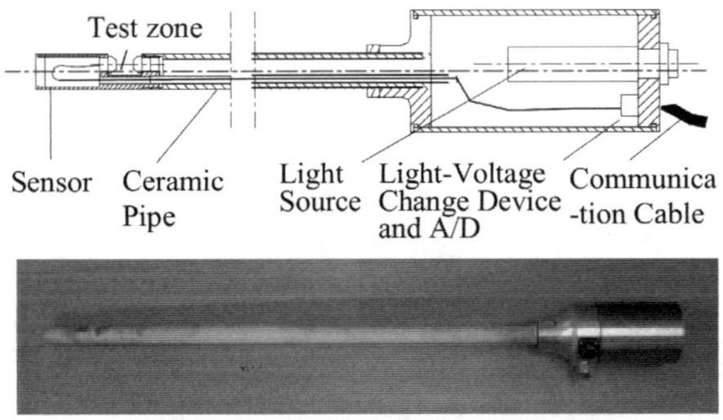

Fig.1 Structure of the measurement probe

Fig. 2 Measurement probes installed in pulverized coal pipes for a in-situ test

The instrument can measure simultaneously the concentration, size and velocity of pulverized coal. It works as follows: a light beam produced by a laser diode traverses through the sensed volume and undergoes attenuation due to absorption and scattering by the sensed particles; the transmitted light is received by an optical fiber and then falls onto a photodetector, giving rise to a series of voltage signals, which are pre-amplified and sent to the AD unit whose output is analyzed by a computer. A compact portable type or a stationary type of the instrument is available upon request. The former consists of an intrusive probe for user-defined sizes compatible with a certain application, a pre-amplifier/AD unit with a USB output port and an optional laptop. The instrument is capable of simultaneously driving 48 individual probes.

Experiments were carried out in a 1000MW ultra supercritical unit in Zouxian power plant, China. The boiler adopted three–layer opposed firing pattern. The supercritical boiler had six double charge-discharge coal ball mills and a mill provided pulverized coal to eight pipes. The total coal pipes are 48. The measurement probes installed in pluverized coal pipes in situ were shown in Fig. 2.

The running parameters of pulverized coal are as follow: diameter of pipe: Φ480×10mm; operating temperature: 70~100℃; primary air flow velocity: 18~25m/s; concentration of pulverized coal (coal/air): 0.45~0.61kg/kg; probes working environment: working temperature <100℃, working pressure <6kPa; surrounding working environment: temperature 0~40℃.

4. MEASUREMENT RESULTS AND DISCUSSION

The measurement probe was installed on the coal pipes at Zouxian coal-fired power plants for medium term in situ tests. Real-time, on-line monitoring of pulverized coal flow in transport pipes may provide an indication of how to retain the proper air/fuel balance at each burner as well as control and diagnosis of mills running. A real-time experiment results by the on-line monitoring system of pulverized coal were shown in Tab. 1.

Tab.1 A real-time measurement data for D and B layers

Pipes	**D** layer			**B** layer		
	C (kg/m^3)	S (μm)	V (m/s)	C (kg/m^3)	S (μm)	V (m/s)
1	0.67	19.25	20.00	0.65	19.60	15.62
2	0.66	13.62	18.23	0.57	20.10	21.79
3	0.61	29.90	27.82	0.65	39.18	20.00
4	0.65	57.87	23.00	0.68	20.14	19.84
5	0.59	49.83	20.19	0.65	36.08	20.79
6	0.66	10.97	15.47	0.64	15.94	19.23
7	0.63	40.56	16.78	0.62	39.28	15.91
8	0.65	14.50	18.52	0.57	25.94	15.07

Note: C-Concentration; S-Size; V-Velocity

The D and B layers are the lower opposite layers of the three layer burners. From the table 1, the measured parameters are in general accord with the outlet parameters of mills. That proved the monitoring system has the characteristic of high accuracy. In addition, the parameters of pulverized coal have the characteristic of fluctuation. For example, the concentration of pulverized coal from D layer varied from 0.59～0.67kg/m^3, however, 0.57～0.68kg/m^3 from B layer. These parameters provided valuable data to monitor the running condition of coal pipes and mills.

In order to further study the characteristics of pulverized coal, the long-time measurement was carried out for D1 and B1 coal pipes. The measurement results were shown in Fig. 3, Fig. 4 and Fig. 5. For the 1000MW ultra supercritical unit, the concentration of pulverized coal from D1 and B1 coal pipes from the same mill varied from 0.55～0.70kg/m^3. That proved that the primary air velocity had an effect on the concentration of pulverized coal. Because the primary air velocity was unsteady, that resulted in the concentration of pulverized coal varies, but the mean concentration was steady.

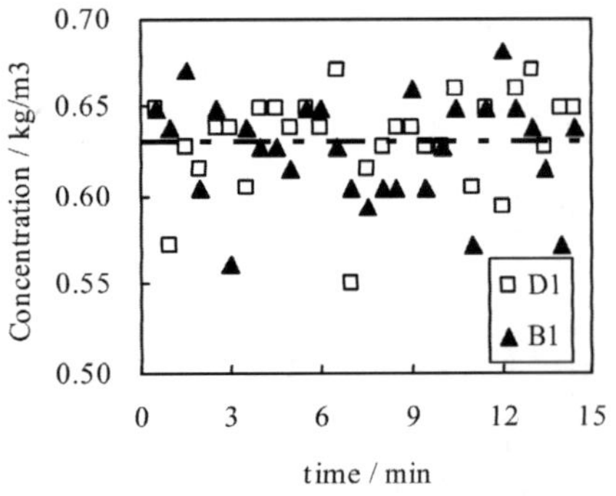

Fig.3 The concentration of pulverized coal

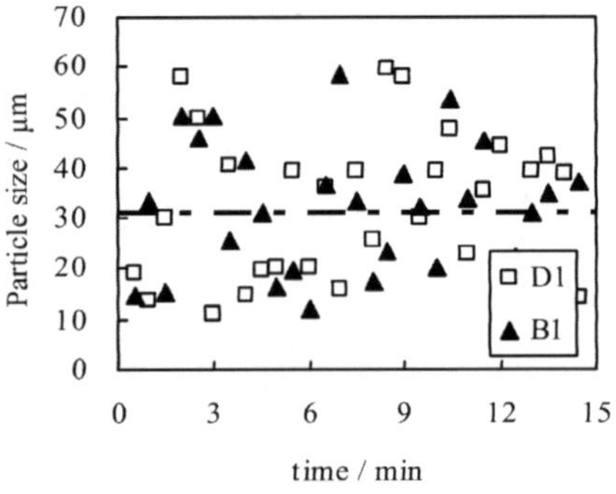

Fig.4 The size of pulverized coal

The size of pulverized coal has relation with safe and steady combustion. According to the Fig. 5, the size of particles from the mill varied from 10～60μm and the size was small in general. The sizes of D1 and B1 coal pipes were 31.25μm and 32.31μm, respectively. The pulverized coal was measured by off-line instrument, particle imaging analyzer, and the mean size was 34.65μm. The on-line monitoring system's error was less than 10% and satisfied the accuracy of measurement in situ. Moreover, the smaller the mill produced the size of pulverized coal, the higher the pulverized coal contained volatile content and calorific value. If the concentration of pulverized coal is higher than critical concentration (1.2~2.0kg/m^3), the explosion will be caused. According to the measurement results, the size of pulverized coal was in a safe level and assured the boiler safe running.

The mass flow rate is of great important to monitor the running operation of pipes and boiler. The flow velocity directly related to mass flow. The velocity of pulverized coal in D1 and B1 pipes was shown in Fig. 5. Because of turbulence effect, the flow velocity of particulates varied with air flow. Their mean velocities were less than primary air velocity. The difference of mean velocity between the D1 and B1 pipe was small and the mean velocity was 20.63m/s. Therefore, the boiler lay in balanced combustion.

Through testing the 1000MW ultra supercritical unit with the on-line monitoring system, the unit lay in a good running status.

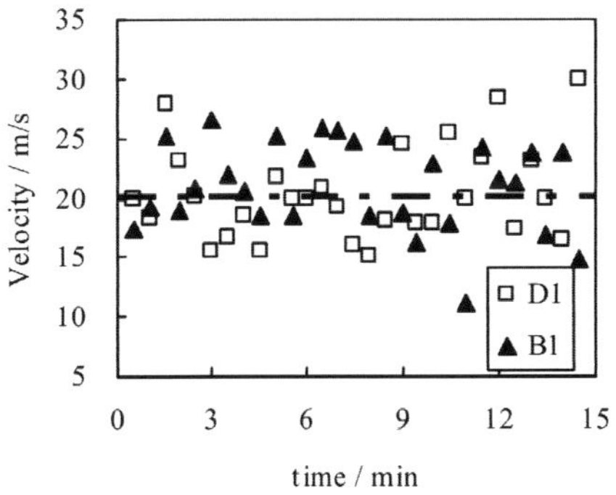

Fig.5 The velocity of pulverized coal

5. CONCLUSIONS

Through monitoring the running condition of pulverized coal pipes and mills of the 1000MW Ultra supercritical unit, the following conclusions can be made.

(1) The on-line monitoring system can measure accurately the concentration, size and velocity of pulverized coal. A novel on-line monitoring system of pulverized coal was provided to test in situ.

(2) It is different that the concentration, size and velocity of pulverized coal from the different pipes of the same mill. Using the fluctuation signals, the running status of mill can be evaluated.

(3) It is helpful for boiler by monitoring these parameters and helpful to know the running status of pulverized coal pipes.

(4) The on-line monitoring system has the advantage of high accuracy, simplicity, reliability, and low cost and it is convenient to install/uninstall in pipes.

ACKNOWLEDGEMENTS

The authors like to thank National Natural Science Foundation of China (No.50706029), 863 plan of china (2006AA03Z349) and Shanghai special fund for choosing and cultivating young scholars.

REFERENCES

Beck M S, Green R G, plaskowski A B, Scott A L.(1990). Capacitance measurement applied to a pneumatic conveyor with very low solids loading, 1(7): 561-564.

Black D L, McQuay M Q, Bonin M P. (1996). Laser-based techniques for particle-size measurement: A review of sizing methods and their industrial applications. Progress in Energy and Combustion Science, 22(3): 267-306.

Cai X S, Li J F, Ouyang X, Zhao Z J, Su M X. (2005). In-line measurement of pneumatically conveyed particles by a light transmission fluctuation method. Flow Measurement and Instrumentation, 16 (5): 315–320.

Carter R M, Yan Y. (2003). On-line particle sizing of pulverized and granular fuels using digital imaging techniques. Measurement Science and Technology, 14(7): 1099-1109.

Carter R M, Yan Y, Cameronb S D. (2005). On-line measurement of particle size distribution and mass flow rate of particles in a pneumatic suspension using combined imaging and electrostatic sensors. Flow Measurement and Instrumentation, 16(5): 309–314.

Cole J B, Jagadeesh G, Takayama K. (2001). Optical measurement of hypersonic microparticles using Mie scattering. Proceedings of SPIE-The international Society for Optical Engineering, Vol. 4183, pp552-555.

Li H, Tomita Y. (2000). Particle velocity and concentration characteristics in a horizontal dilute swirling flow pneumatic conveying. Powder Technology, 107(1-2): 144–152.

Ma J, Yan Y. (2000). Design and evaluation of electrostatic sensors for the mass flow measurement of pneumatically conveyed solids, Flow Measurement and Instrumentation, 11(3): 195–204.

Nishino K, Hiroyuki K, Kahoru T. (2000). Stereo imaging for simultaneous measurement of size and velocity of particles in dispersed two-phase flow. Measurement Science and Technology, 11(6): 633-645.

Teipel U. (2002). Problems in Characterizing Transparent Particles by Laser Light Diffraction Spectrometry. Chemical Engineering & Technology, 25(1): 13–21.

Xu F, Cai X S, Ren K F. (2004). Geometrical Optics Approximation of Forward Scattering by Coated Particle. Applied Optics, 43(9):1870－1879.

Yan Y. (1996). Mass flow measurement of bulk solids in pmeumatic pipelines[J]. Measurement Science and Technology, 7(12): 1687-1706.

Yan Y. (2001). Guide to the flow measurement of particulate solids in pipelines, Part I: Fundamentals and principles. Powder Handling and Processing, 13 (4): 343–352.

Journal of Physics: Conference Series **147** (2009) 012077 doi:10.1088/1742-6596/147/1/012077

Measurement of characteristics of solid flow in the cyclone separators with fiber optical probe

Shaohua Li, Yan Li, Jinjing Li, Shi Yang, Hairui Yang, Hai Zhang, Junfu Lu, Guangxi Yue

Key Laboratory for Thermal Science and Power Engineering of Ministry of Education, Department of Thermal Engineering, Tsinghua University, Beijing , 100084 CHINA

lishaohua00@mails.tsinghua.edu.cn, li-y-05@mails.tsinghua.edu.cn, lijinjing00@mails.tsinghua.edu.cn, yangshi07@mails.tsinghua.edu.cn, yhr@mail.tsinghua.edu.cn, haizhang@mail.tsinghua.edu.cn, Lvjf@mail.tsinghua.edu.cn, ygx-dte@mail.tsinghua.edu.cn

Abstract. In some applications, e.g. circulating fluidized beds (CFB), cyclones are usually operated at high solid loadings. Under high inlet solid concentration, most of the particles are collected at the wall and form a dense particle spiral band because of high separation efficiency. As a result, gas-solid reactions should occur mostly in the near-wall region. To understand the gas-solid reaction mechanism in the cyclone, an experimental study was conducted in a plexiglass CFB cold apparatus, with a riser of 0.2m I.D. and 5m high, and a standard Lapple cyclone. Fiber optical probe was used to measure the characteristics of solid flow in the cyclone, including particle velocity and volumetric solid concentration, especially in the near-wall region of the cyclone. Based on the experiment results, the combustion of carbon particles in the cyclone of a CFB boiler was estimated with group combustion theory. The calculated results show that combustion effectiveness factor η eff of near-wall particle cloud is smaller than 1/25, which means the combustion rate of a carbon particle in the near-wall region is greatly restricted by other particles in the cloud.

1. INTRODUCTION

Cyclones are widely used as gas-solid separators in the chemical engineering area because of their high separation efficiency, simple structure, good adaptability and low cost. Most of the studies are mainly focused on separation efficiency and pressure drop. A few of them were experimentally (Kessler and Leith, 1991; Patterson and Munz, 1996) or theoretically (Barth, 1956; Mothes and Löffler, 1988) conducted on the gas flow field in the cyclone.

However, gas-solid reactions in a cyclone are also important in many cases, such as when cyclones are used in coal-fired circulating fluidized bed (CFB) boilers (Yue et al., 2006) or the gasifier for fast pyrolysis of biomass (Lede, 2000). But the investigations on gas-solid reactions in a cyclone are very limited.

In CFB boilers and some other applications, cyclones are usually operated at high solid loadings, e g., 5 kg particles/kg-gas (Muschelknautz and Greif, 1997). Under such conditions, most of the particles are collected at the wall and form a dense particle spiral band because of high separation efficiency. As a result, gas-solid reactions should occur mostly in the near-wall region. The investigation of gas-solid reaction should be based on the knowledge of solid flow behavior in the cyclone. Therefore, well understanding the characteristics of gas-solid flow, including volumetric solid concentration and particle velocity distribution, in the near-wall region of the cyclone is of particular interest.

Nevertheless, the experimental studies on the volumetric solid concentration and velocity distribution of near-wall particles are relatively sparse. The available experimental results on the particle concentration and velocity distribution are mainly obtained in the center of the cyclone when the solid concentration is dilute: Zhou and Soo (1990) measured tangential velocity distribution of particles in the center of the cyclone where the solid concentration was dilute by using Laser Doppler Velocimeter (LDV). Mothes and Löffler (1985) measured the particle concentration distribution at the location relatively far away from the wall (r/R<0.8) by probe sampling, where r is the radial distance from the center and R is the cylindrical diameter of the cyclone. Recently, the motion of near-wall particles in the cyclone was experimentally studied.

Recently, Chan et al. (2008) used the technique of positron emission particle tracking (PEPT) to determine the particle motion in the cyclone. They found that the average tangential and axial velocity of near-wall particles over the length of the cylindrical body was between 1.5~2.0m/s and 0.9~1.3m/s respectively. However, no volumetric solid concentration of near-wall particles was measured. Muschelknautz and Röper (2008) have made an assumption that the solids velocity near the wall is about several meters per second and the layer thickness is of several millimeters for the CFB boiler condition. However, no experimental data were reported.

Fiber optical probe (FOP) has been used for the measurement of particle velocity and volumetric solid concentration by several authors (Johnsson and Johnsson, 2001; Tayebi et al., 1999; Werther, 1999). The fiber optical probes are simple and relatively inexpensive, yield high signal-to-noise ratios and create a minimum disturbance to the flow. They can measure from dilute to dense conditions for different particles in gas or liquid media. And they are more advantageous than some other techniques for the measurements of solid velocity and volume-fraction at the condition of dense gas-solid two phase flow (Tayebi et al., 1999; Zhang et al., 1998; Zhu et al., 2001). Therefore, fiber optical probes are used to measure the volumetric solid concentration and velocity distribution of near-wall particles in this study.

In this research, experiments were conducted to investigate the volumetric solid concentration and velocity distribution of near-wall particles. Based on the experimental results, the combustion of particles in the cyclone of a CFB boiler was estimated with group combustion theory.

2. EXPERIMENTAL

2.1 Experimental System

The cold CFB apparatus is shown in Fig. 1, in which the riser, cyclone, standpipe and loopseal form a circulating loop. Most of the apparatus was made of plexiglass for observation convenience. The riser is 5m in height and 0.2m in inner diameter. The geometry of the cyclone is illustrated in Fig. 2.

During the experiments, the fluidizing gas velocity was measured by a vortex flowmeter and controlled by adjusting the damper openness of the valves connected to the forced draft fan and the induced fan. The circulating rate, G_s (kg/s) was measured by weighting the particles accumulation on an electric stop valve installed at the middle of the downcomer in a given time. The pressure drops at different locations of the system were measured by pressure sensors. More details about the experimental system were described in previous studies (Li et al., 2007; Li et al., 2008).

As shown in Fig. 1, several measuring holes were opened at different locations on the cyclone surface, and 3 of them (point 1, 2 and 3) located along the movement direction of particles spiral band in the cylindrical section of the cyclone.

The system was operated at required fluidizing gas velocity and initial static bed height. When the system was under the condition of stable fluidizing gas velocity and stable static height of bed material in the downcomer, $G_s$was measured. Then the volumetric solid concentrations and particles velocities at different radial positions of a measuring point (if it is not specified, the measuring point is point 2 as shown in Fig. 1) in the cyclone were measured by fiber optical probes.

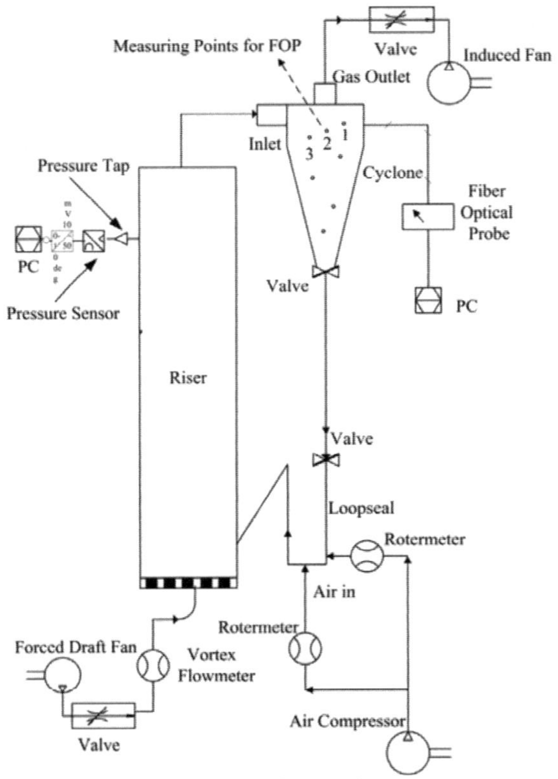

Figure 1. Experimental system

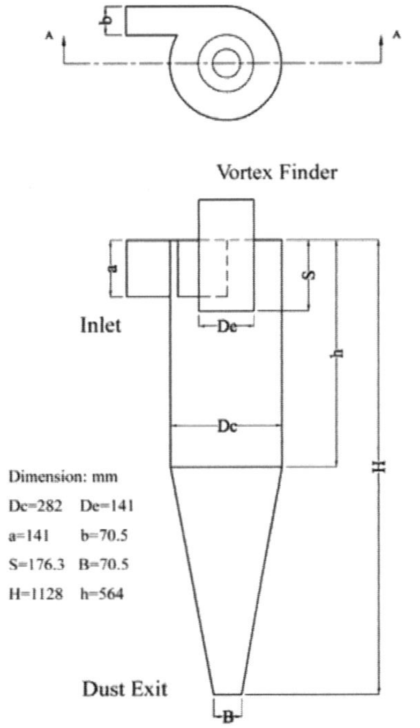

Figure 2. Cyclone geometry

The experimental variables in the present study, include gas velocity at the cyclone inlet V_{in} (m/s) and solid concentration at cyclone inlet $C_{s,in}$ (kg/m^3). Quartz sand was used as bed material. The values of the process parameters can be found in Tab. 1.

Table. 1 Process parameters

d_p (mm)	ρ_b (kg/m^3)	ρ_r (kg/m^3)	$C_{v,0}$	V_{in} (m/s)	$C_{s,In}$ (kg/m^3)
260	1550	2650	0.585	8.9,13.3 17.4,19.7,22.2 and 25.3	2.5~5.7

2.2 Fiber Optical Probe System

Fiber optical probe (FOP) was used to measure particle velocity and volumetric solid concentration near the wall in the cyclone. This technique was used in our and some others' previous studies (Johnsson and Johnsson, 2001; Li et al., 2008; Tayebi et al., 1999). With proper design and application setup, FOP can have high signal-to-noise ratios while with minor disturbance to the flow. Particularly, FOP is more advantageous than other techniques to measure particle velocity and volumetric solid concentration for a dense gas-solid flow (Tayebi et al., 1999; Zhang et al., 1998; Zhu et al., 2001).
The probes used in this study are based on the backscattering principle. The tips of the fiber optical velocity probe and volume-fraction probe both have an outer diameter of 3.8mm. The probes contain thousands of emitting and receiving quartz fibers, each of which has a diameter of 24μm. The velocity probe has two bundles of fibers, and the distance of the bundles, $\triangle x$, is 1.93mm. In each bundle, the fibers are arranged in an alternating array, corresponding to emitting and receiving layers of fibers. The volume-fraction probe has a square area of 2×2mm^2 at the center of the tip, in which the fibers are arranged.

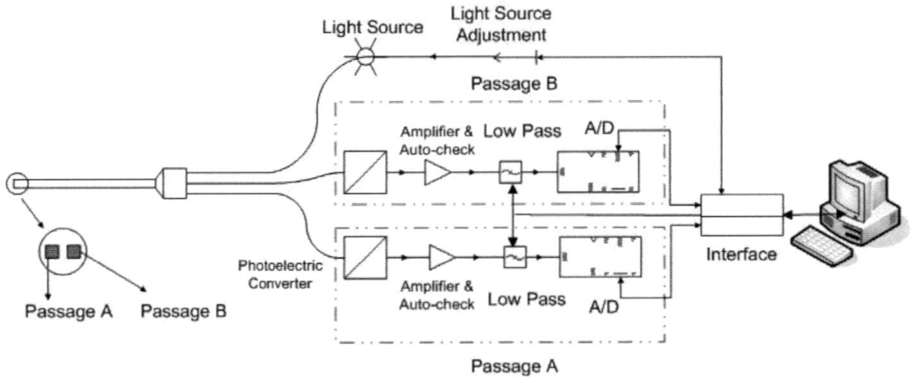

Figure 3. Velocity probe

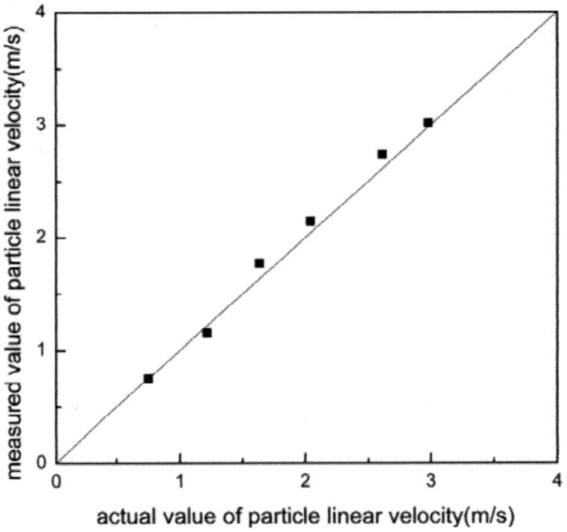

Figure 4. Comparison between measured values and actual values of u_s

The schematic view of the velocity probe can be seen in Fig.3. In the measurement, the probe was inserted into the cyclone with its tip perpendicular to the movement direction of the particles, illuminated a small volume of particles. The two passages of the velocity probe were arranged in the direction of the particle movement and Passage A was in the front, so that one particle passed in front of Passage A first, and after a time delay of $\triangle t$, the particle passed in front of Passage B. The reflected light signals received by each of the passage were converted into an electric signal by a photoelectric converter. And $\triangle t$ was calculated on the basis of cross-correlating the two electric signals with a real-time hardware processor. Then u_s can be calculated by Eq. (1). The detailed measurement principle of the fiber optical probes can be found in the literatures (Chang and Louge, 1992; Werther, 1999).

$$u_s = \triangle x / \triangle t \qquad (1)$$

Calibration of FOP is very important. The calibration procedure of the velocity probe is similar to that done by Zhu et al. (2001) and Tayebi et al. (1999). An average-size particle was glued at a radius of r (m) on a thin rotating disk, the surface of which was painted to be black. The disk was driven by a variable speed motor. And the rotational velocities, n (1/s), were measured by a tachometer. Thus, the actual linear velocity of the particle u_s (m/s) is the product of r and n. By changing the rotating speed of the disk, different particle linear velocities were obtained. The measured values of u_s with velocity probe and the actual values were compared as shown in Fig. 4. It can be seen that the measured values were in very good agreement with the actual values.

With a prior accurate calibration, the reflected light intensity received by the volume-fraction probe, i.e., the output voltage was correlated to the volumetric solid concentration, in the form of power function as Eq. (2) (Johnsson and Johnsson, 2001; Tayebi et al., 1999).

$$U - U_0 = aC_v^k \qquad (2)$$

Where, U is measured signal intensity (V) and U_0 is signal intensity at C_v=0. C_v is volumetric solid concentration. a and k are constants. The calibration was done in a fixed bed and a dilute region of a well-fluidized bed as the previous works (Chang and Louge, 1992; Lu et al., 2005). For our study, k=0.588 and a=6.55.

3. MEASUREMENT RESULTS AND DISCUSSION

In the experiments, under high inlet solid concentration, it was found that solid particles are mostly concentrated in the near-wall particle band, nearly free in the center, and C_v at different radial locations in the particle band changed with V_{in} and $C_{s,in}$. As shown in Fig. 5 and Fig. 6, at the same location, C_v increases with the increase of V_{in} when $C_{s,in}$ is constant, while C_v increases with $C_{s,in}$ when V_{in} is constant. The influence of V_{in} and $C_{s,in}$ on C_v is straightforward.

The thickness of the continuous particle band at the wall, δ (mm), is defined as the distance from the wall where C_v=0.01. Shown in Fig. 5 and 6, δ is in the range of 4-12mm in this experiment, validating the assumption of Muschelknautz and Röper (2008).

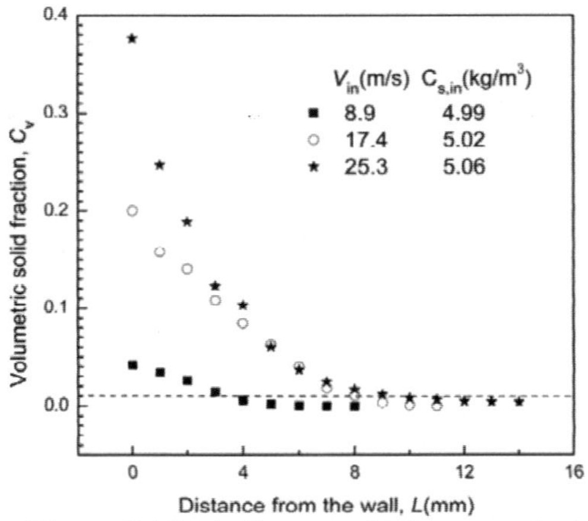

Figure 5. The influence of inlet gas velocity on volumetric solid concentration

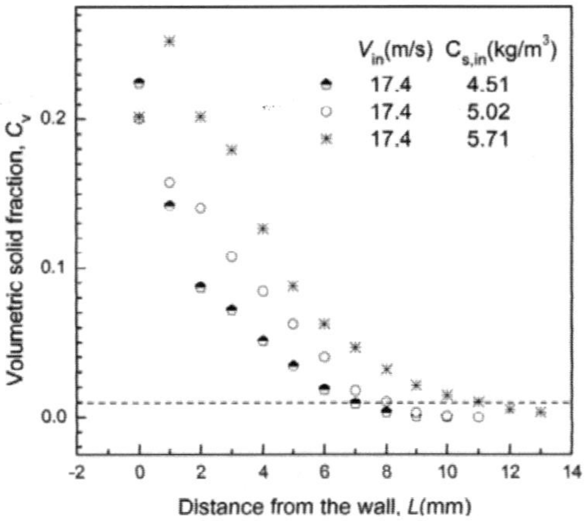

Figure 6. The influence of inlet solid concentration on volumetric solid concentration

It is worth to point out for the cyclone used in the CFB boiler, the width of the cyclone inlet is much larger than that in current experiments. Thus, for the same $C_{s,in}$, the thickness of the particle band could be much larger, even in one order.

The radial distributions of near-wall particle velocity u_s were measured under different V_{in} and $C_{s,in}$. Since the radial velocity of near-wall particles was very small, u_s was the resultant velocity of tangential velocity and axial velocity. Fig. 7 shows that radial distribution of particle velocity under different V_{in} and $C_{s,in}$ are similar: u_s first decreases then increases with the increase of L. And u_s is in the range of 0.5~2.5m/s. The range of u_s is similar to that of Chan et al. (2008). The results of this study confirm the assumption of Muschelknautz and Röper (2008).

According to the results of Chan et al. (2008), the near-wall particles' average tangential velocity over the length of the cyclone body was not obviously influenced by the change of V_{in} and $C_{s,in}$. Based on our previous study (Li et al., 2008), particle average residence time in the cyclone is nearly constant. Therefore, it is expected that the near-wall particles' average axial velocity over the length of the cyclone body is also inappreciably influenced by the change of V_{in} and $C_{s,in}$. In this study, the mass-based average velocity at point 2, $\overline{u}_s$, is in a small range of 1.5~2.5m/s. Thus, u_s=2m/s is proposed for the estimation of gas-solid reaction in the near-wall region of a cyclone. The detail influences of V_{in} and $C_{s,in}$ on u_s at different positions need to be investigated further.

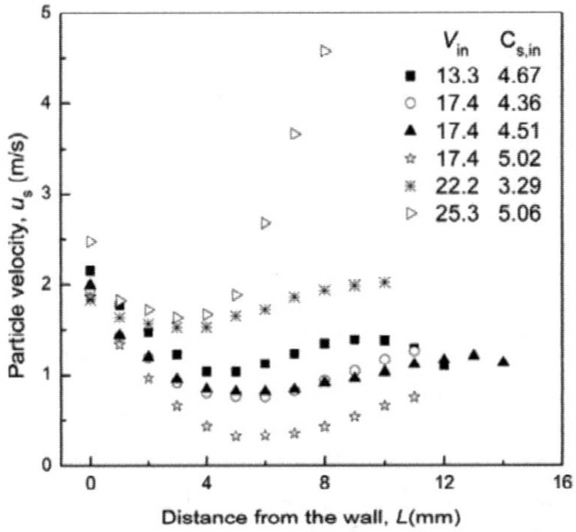

Figure 7. Particle velocities at different distances from the wall

The measured volumetric solid concentration $C_{v,i}$ and particle velocity $u_{s,i}$ can be validated through circulating rate G_s.

$$G_s = \sum_{i=1}^{k} l_t W \rho_r u_{s,i} C_{v,i} \qquad (3)$$

In Eq. (3), k is the total number of measuring intervals the probe moved along the radial direction; l_t is the distance interval the probe moved in each step, in our study l_t=1mm; W (m) is the width of the particle spiral band, which is about 0.2~0.3m in the present experiment. $u_{s,i}$ (m/s) and $C_{v,i}$ correspond to the particle velocity and volumetric solid concentration measured in the ith step.
For the conditions that V_{in} =17.4m/s, $C_{s,in}$ =4.51kg/m^3 and V_{in} =25.3m/s, $C_{s,in}$ =5.06kg/m^3, the calculated G_s are 0.73kg/s and 1.40kg/s respectively, while the measured values of G_s are 0.78kg/s and 1.27kg/s respectively. It can be seen the experimental results agree well with the theoretical calculations.

4. COMBUSTION OF PARTICLES IN A CYCLONE OF A CFB BOILER

Group Combustion theory (Annamalai et al., 1994) can be used to estimate the combustion of carbon particles in a cyclone. While the temperature is higher than 1000K, $2C+O_2=2CO$ is the primary reaction for carbon combustion (Law, 2006). Since the ambience temperature in the cyclone of a CFB boiler is about 1200K, *single film model* can be used to estimate carbon combustion (Annamalai et al., 1994). For *single film model*, the combustion of particle cloud is divided into sevral stages, which can be seen in Tab.2 : If the cloud is very dilute, the buring rate of each particle is the same as that of an isolated particle, this stage is called as *isolated particle combustion* (ISOC). With the increment of number density, the particle cloud undergoes the stages of *interactive combustion* and then *partial group combustion*. When the number density is very large, the oxygen mass fraction at the cloud surface approaches zero, and the entire cloud can be treated as a single large particle with the density the same as the cloud mass density. This stage is called *sheath combustion*.

In Tab. 2, group combustion number $G^{'}$, is defined as the ratio of transport rate between the particles in the cloud to transport rate between the cloud and ambience.

Table 2. Division of particle cloud combustion

$G^{'}$	Combustion stage of particle cloud
$G^{'} < 0.3$	*isolated particle combustion*
$0.3 < G^{'} < 9$	*interactive combustion*
$9 < G^{'} < 100$	*partial group combustion*
$G^{'} > 100$	*sheath combustion*

According to the experimental results, the particles are nearly free in the center of the cyclone, e.g. the particles number density is very small in the center. Consequently, in the cyclone of a CFB boiler, the combustion of carbon particles in the center region can be treated as *isolated particle combustion*.

For using group combustion theory to estimate the near-wall particles' combustion in the cyclone of a CFB boiler, the follow simplification and assumption should be made:

(1) In fact, bed material in a CFB boiler is a mixture of carbon particles and inertia particles, and the ratio of carbon particles, f, is as small as 2%~5%. f=3% is used for the estimation.

(2) Near-wall particles cloud can be simplified as a slab, whose character length is the thickness of the band, δ.

(3) It is assumed the near-wall particles cloud is monosized, which means the diameters of particles in the cyclone are the same.

(4) To simplify the estimation, it is assumed that the near-wall particles cloud is uniform, e.g. the volumetric solid concentration at different locations in the near-wall particle band equals to the average value of volumetric solid concentration, $1-\bar{\varepsilon}$. In fact, according to our experimental results, the distribution of volumetric solid concentration is not even.

(5) Near-wall particles cloud is isothermal.

The species conservation equation of oxygen can be written as Eq. (4). $S_{V,C}$, the surface area of particles per unit volume, can be calculated with Eq. (5). R_{ch} and R_d, which can be calculated with Eq. (6) and (7), are chemical resistance and diffusion resistance respectively. Because the diffusion resistance of oxygen to the surface of near-wall particle cloud can be omitted, only the diffusion resistance of oxygen in the particle cloud is considered.

For a slab particle cloud, Eq. (4) is converted to Eq. (8). Using $\xi = L/\delta$, $\Phi^{'} = Y_{O2}/(Y_{O2})_C$ to simplify Eq. (8), we can get a second-order linear differential equation Eq. (9). The expression of G' and G can be seen in Eq. (10) and (11). $\Phi^{'}$ in Eq. (12) is the solution of Eq. (9).

$$\nabla \cdot (\rho D \nabla Y_{O2}) = \dot{m}_{O2}''' = S_{V,C} Y_{O2} / (R_{ch} + R_d) \tag{4}$$

$$S_{V,C} = \pi d_p^2 M = \frac{6(1-\bar{\varepsilon})}{d_p} \tag{5}$$

$$R_{ch} = \frac{1}{k\rho} \tag{6}$$

$$R_d = \frac{f d_p}{\rho Sh_p D} \tag{7}$$

$$\frac{d}{dL}(\rho D dY_{O2} / dL) = S_{V,C} Y_{O2} / (R_{ch} + R_d) \qquad L < \delta \tag{8}$$

$$\frac{d^2\Phi'}{d\xi^2} = G'\Phi' \qquad \xi < 1 \tag{9}$$

$$G' = G/(1 + \frac{Sh_p D}{fkd_p}) \tag{10}$$

$$G = S_{V,C}\delta^2 Sh_p / (fd_p) \tag{11}$$

$$\Phi' = \cosh[(G')^{\frac{1}{2}}\xi] / \cosh[(G')^{\frac{1}{2}}] \tag{12}$$

$$\eta_{eff} = \int_0^\delta \frac{\dot{m}_{O2,r}}{\dot{m}_{O2,C}} dL = \int_0^\delta \frac{A4\pi M d_p^2 Y_{O2}}{A\delta 4\pi M d_p^2 Y_{O2,C}} dL = \frac{\tanh[(G')^{\frac{1}{2}}]}{(G')^{\frac{1}{2}}} \tag{13}$$

In Eq. (13), η_{eff}, combustion effectiveness factor, is the average ratio of reaction rate of particles in the cloud to that of particle at cloud surface. Therefore, provided that G' is calculated, combustion of near-wall particle cloud in the cyclone of a CFB boiler can be estimated.

Then, we use the experimental results and group combustion theory to estimate the near-wall particle combustion in a cyclone of a 135MWe CFB boiler. The dimensions of the cyclone can be found in Tab. 3.

Table. 3 Dimensions of the cyclone of a 135MWe CFB boiler (m)

D	h	H	a	b	S	De	B
8.35	10.45	23	6.7	2.7	6.4	3.67	3

Table. 4 Process parameters of the cyclone of a 135MWe CFB boiler

d_p (μm)	f	u_s (m/s)	D (m^2/s)	k (m/s)	T (K)	V_{in} (m/s)	G_s (kg/s)	W (m)
100	3%	2	2.15e-4	0.031	1173	25	340	8.1
η	ρ_p (kg/m^3)	Re_p	Sc_p	Sh_p	δ (m)	$\bar{\varepsilon}$	G'	η_{eff}
0.99	2650	1.31	0.71	2.61	0.1	0.922	6500	1.24%

$$Sh_p = 2 + 0.6 Re_p^{\frac{1}{2}} Sc_p^{\frac{1}{3}} \tag{14}$$

$$\delta = \beta \frac{0.008b}{0.0705} \tag{15}$$

$$\eta G_s = C_{s,in} V_{in} ab = W\delta u_s \rho_p (1 - \bar{\varepsilon}) \tag{16}$$

The process parameters of this cyclone are shown in Tab. 4. Since $0<Re_p<200$, Sh_p can be calculated with Eq. (14). Eq. (15) is used to calculate δ in the cyclone of the 135MWe CFB boiler. In Eq. (15), b and 0.0705 are the inlet widths of cyclones in the 135MWe boiler and experimental system respectively. 0.008m is the average value of δ in our experiment. β is correction coefficient and $0<\beta<1$. $\bar{\varepsilon}$ can be calculated with Eq. (16). In Eq. (16), W is assumed to be 8.1m.

The corresponding values of G' and η_{eff} under different value of δ and $\bar{\varepsilon}$ can be seen in Fig. 8. Since $G'>>100$ and η_{eff} is very small, the near-wall particles in the cyclone of the 135MWe CFB boiler are in *sheath combustion* stage. The results indicate that the agglomeration of particles at the near-wall region in the cyclone of a CFB boiler prevent oxygen diffusion into the particle cloud and greatly hinder the combustion of carbon in the near-wall particles.

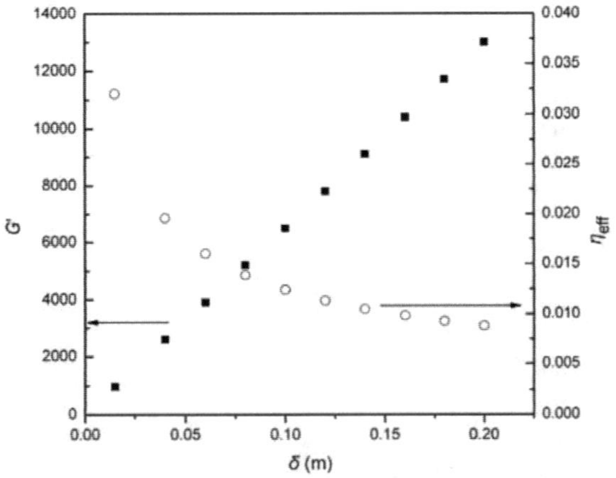

Figure 8. Values of G' and η_{eff} under different value of δ

5. CONCLUSION

In this study, the distributions of volumetric solid concentration C_v and particle velocity u_s in the near-wall region of a cyclone were experimentally investigated. Then the combustion of particles in the cyclone of a CFB boiler was estimated. Following conclusions are drawn:

The C_v is higher at the positions closer to the wall. It increases with the increment of gas velocity or solid concentration at cyclone inlet. The particle band in the cylindrical part of the cyclone is about 4-12mm thick under present experimental conditions.

The velocity of near-wall particles is in the range of 0.5~2.5m/s. Its average value over the cyclone body is insensitive to the operation parameters. u_s=2m/s is proposed to be used in the estimation of gas-solid reaction in the near-wall region.

Based on the experimental results, group combustion theory is used to estimate the combustion of near-wall particles in the cyclone of a 135MWe CFB boiler. It is found that the reaction rate of carbon and oxygen in the near-wall region is smaller than 1/25 of the reaction rate of the particle at cloud surface.

ACKNOWLEDGEMENTS

Financial supports of this work by Key Project of the National Eleventh-Five Year Research Program of China (2006BAA03B02) and National Science Fund Committee (50406002) are gratefully acknowledged.

NOMENCLATURE

a	coefficient in the calibration function (Eq. (1))	[-]
A	surface area	[m^2]

$C_{s,in}$	solids concentration at cyclone inlet	[kg/m^3]
C_V	volumetric solid concentration	[-]
$C_{V,0}$	volumetric solid concentration in a fixed bed	[-]
D	diffusion coefficient	[m^2/s]
d_p	mean diameter of particles	[μ m]
G	group combustion number	[-]
G_s	circulating rate	[kg/s]
k	chemical reaction rate constant	[m/s]
L	distance from the wall	[mm]
M	number density	[-]
$\dot{m}_{O2}^{'''}$	consumption rate of oxygen	[kg/m^3-s]
n	rotational velocity	[1/s]
r	radial distance of the particle to the center	[m]
R	cylindrical diameter of cyclone	[m]
$S_{V,C}$	surface area of particles per unit volume	[m^2/m^3]
T	ambience temperature in a cyclone	[K]
$u_{g,s}$	gas-solid slip velocity	[m/s]
u_s	particle velocity	[m/s]
$\overline{u}_s$	mass-based average velocity of near-wall particles	[m/s]
U	measured signal intensity	[V]
U_0	measured signal intensity at C_v =0	[V]
V_{in}	gas velocity at cyclone inlet	[m/s]
W	width of the particle layer	[m]
Y	gas phase mass fraction	[-]
Sh_p	particle Sherwood number	[-]

Greek Letters

δ	thickness of the continuous particle layer at the wall	[mm]
$\overline{\varepsilon}$	average value of voidage of near-wall particle cloud	[-]
η	separation efficiency	[-]
η_{eff}	combustion effectiveness factor	[-]
ρ	particle group density	[kg/m^3]
ρ_b	bulk density of particles	[kg/m^3]
ρ_r	real density of particles	[kg/m^3]

Subscripts

c	cloud surface
ch	chemical
d	diffusion
g	gas
in	cyclone inlet
p	particle
s	solid

REFERENCES

Annamalai, K., Ryan, W. and Dhanapalan, S. (1994) "Interactive processes in gasification and combustion-part III: coal/char particle arrays, streams and clouds" Prog. Energ. Combust., 20(6), pp.487-618.

Barth, W. (1956) "Berechnung und Auslegung von Zyklonabscheidern auf Grund neuerer Unterscuhungen" Brennst-warme-kraft., 8(1), pp.1-9.

Chan, C. W., Seville, J. P. K., Fan, X., Dewil, R. and Baeyens, J. (2008). "CFB Cyclones: Pressure Drop and particle motion viewed by positron emission particle tracking". 9th International Conference on Circulating Fluidized Beds Hamburg, Germany.

Chang, H. and Louge, M. (1992) "Fluid dynamic similarity of Circulating fluidized beds" Powder Technol., 70(3), pp.259-270.

Johnsson, H. and Johnsson, F. (2001) "Measurements of local solids volume-fraction in fluidized bed boilers" Powder Technol., 115(1), pp.13-26.

Kessler, M. and Leith, D. (1991) "Flow Measurement and Efficiency Modeling of Cyclones for Particle Collection" Aerosol. Sci. Tech., 15, pp.8-18.

Law, C. K. (2006). Combustion Physics. Cambridge University Press. New York.

Lede, J. (2000) "The cyclone: a multifunctional reactor for the fast pyrolysis of biomass" Ind. Eng. Chem. Res., 39(4), pp.893-903.

Li, S., Zhang, H., Yang, H., Yang, S., Lu, J. and Yue, G. (2007) "Determining Cyclone Particle Hold-up by the Pressure Drop for a CFB Boiler" Chem. Eng. Technol., 30(12), pp.1726-1731.

Li, S., Yang, S., Yang, H., Zhang, H., Liu, Q., Lu, J. and Yue, G. (2008) "Particle holdup and average residence time in the cyclone of a circulating fluidized bed boiler" Chem. Eng. Technol., 31(2), pp.224-230.

Li, S. H., Yang, H. R., Zhang, H., Lu, J. F. and Yue, G. X. (2008). "Hydrodynamics of two-phase flow and particle motion in the cyclone of a CFB boiler". 9th International Conference on Circulating Fluidized Beds Hamburg, Germany.

Li, S. H., Yang, S., Yang, H. R., Zhang, H., Liu, Q., Lu, J. F. and Yue, G. X. (2008) "Particle holdup and average residence time in the cyclone of a circulating fluidized bed boiler" Chem. Eng. Technol.(2).

Lu, X., Li, S., Du, L., Yao, J., Lin, W. and Li, H. (2005) "Flow structures in the downer circulating fluidized bed" Chem. Eng. J., 112, pp.23-31.

Mothes, H. and Löffler, F. (1985) "Motion and deposition of particle in cyclones" Ger. Chem. Eng, 8, pp.223-233.

Mothes, H. and Löffler, F. (1988) "Prediction of Particle Removal in Cyclone Separators" Int. Chem. Eng., 28(2), pp.231-240.

Muschelknautz, E. and Greif, V. (1997). "Cyclone and other gas-solid separators" in "Circulating Fluidized Beds". Grace, J. R., Avidan, A. A. and Knowlton, T. M., Blackie Academic & Professional, pp.181-213.

Muschelknautz, U. and Röper, B. (2008). "Online regulation of the solids circulation rate in a circulating fluidized bed combustor". 4th International VGB workshop "operating experience with fluidized bed firing systems", Hamburg, Germany,

Patterson, P. A. and Munz, R. J. (1996) "Gas and particle flow patterns in cyclones at room and eleavated temperatures" Can. J. Chem. Eng., 74, pp.213-221.

Tayebi, D., Svendsen, H. F., Grislingas, A., Mejdell, T. and Johannessen, K. (1999) "Dynamics of fluidized-bed reactors. Development and application of a new multi-fiber optical probe" Chem. Eng. Sci., 54(13-14), pp.2113-2122.

Werther, J. (1999) "Measurement techniques in fluidized beds" Powder Technol., 102(1), pp.15-36.

Yue, G. X., Yang, H. R. and Zhang, H. (2006). "Post Combustion in Circulating Fluidized Bed Boilers". 19TH International Fluidized Bed Combustion Conference, Vienna, Austria, May 21-24, 2006,

Zhang, H., Johnston, P. M., Zhu, J. X., Lasa, H. I. d. and Bergougnou, M. A. (1998) "A novel calibration procedure for a fiber optic solids concentration probe" Powder Technol., 100(2-3), pp.260-272.

Zhou, L. X. and Soo, S. L. (1990) "Gas-solid flow and collection of solids in a cyclone separator" Powder Technol., 63, pp.45-53.

Zhu, J. X., Li, G. Z., Qin, S. Z., Li, F. Y., Zhang, H. and Yang, Y. L. (2001) "Direct measurements of particle velocities in gas-solids suspension flow using a novel five-fiber optical probe" Powder Technol., 115(2), pp.184-192.

Journal of Physics: Conference Series **147** (2009) 012078 doi:10.1088/1742-6596/147/1/012078

Study on Flow Regimes of High-pressure and Dense-phase Pneumatic Conveying

Lu Peng, Chen Xiaoping, Liang Cai, Pu Wenhao, Zhou Yun, Xu Pan, Zhao Changsui*

School of Energy & Environment, Southeast University
Si Pailou 2, Nanjing 210096, China
seu_lupeng@yahoo.com.cn, xpchen@seu.edu.cn, cszhao@seu.edu.cn

Abstract. High-pressure and dense-phase pneumatic conveying of pulverized coal is a key technology in the field of large-scale entrained bed coal gasification. Flow regime plays an important role in two-phase flow because it affects not only flow behavior and safety operation, but also the reliability of practical processes. Few references and experiences in high-pressure and dense-phase conveying are available, especially for the flow regimes. And because of the high stickiness and electrostatic attraction of pulverized coal to the pipe wall, it is very difficult to make out the flow regimes in the conveying pipe by visualization method. Thus quartz powder was chosen as the conveyed material to study the flow regime. High-speed digital video camera was employed to photograph the flow patterns. Experiments were conducted on a pilot scale experimental setup at the pressure up to 3.6MPa. With the decrease in superficial gas velocity, three distinguishable flow regimes were observed: stratified flow, dune flow and plug flow. The characteristics of pressure traces acquired by high frequency response pressure transmitter and their EMD (Empirical Mode Decomposition) characteristics were correlated strongly with the flow regimes. Combining high-speed photography and pressure signal analysis together can make the recognition of flow patterns in the high-pressure and dense-phase pneumatic conveying system more accurate. The present work will lead to better understanding of the flow regime transition under high-pressure.

1. INTRODUCTION

High-pressure and dense-phase pneumatic conveying of pulverized coal is a vital technology in the field of large-scale entrained bed coal gasification. The pneumatic conveying has been successfully used in the chemical engineering, energy, metallurgy and other industrial processes with advantages, such as reliability, flexibility of layout, ease of automation, low maintenance, hygienic and environmental friendly. In recent years, numerous studies, for instance, experiment, signal analysis and numerical simulation, have been conducted on different pneumatic conveying systems. Many valuable achievements on pneumatic conveying have been obtained. Gong et al [1] investigated the flow characteristic of high solid loading pneumatic conveying under low pressure and obtained some fitted empirical equations. Wypych [2] explored the mechanism for the formation of the unstable zone experimentally and theoretically. Li [3] adopted photographic technology to measure the particle velocity and concentration profiles in the horizontal dilute-phase pneumatic conveying. Masona [4] developed a two-layer model to simulate dense phase pneumatic transport of fine powders and the model showed good quantitative agreement with experimentally determined pressure profiles for fully developed flows in straight horizontal pipes. Laouar [5,6]'s aims were to characterize the differential pressure in dense-phase pneumatic conveying line at a very low velocity and a general differential pressure law was obtained which proves to be independent of both flow regimes and pipe diameter.

However, the above achievements are mainly about the low pressure pneumatic conveying [1-10]. At present the large-scale coal gasification technology is gaining attention and developed, and dense-phase pneumatic conveying of pulverized coal under high pressure is one of its key technologies [11-12]. Few references and experiences in high-pressure and dense-phase conveying are available, especially for the flow regimes.

EMD is capable of extracting all the oscillatory modes present in signal. Each extracted mode is referred to as an Intrinsic Mode Function (IMF), which has a unique local characteristic. EMD has been widely applied in recent years in the fields of meteorology, ocean engineering, earthquake studies, etc. H.Ding et al. has used the method for characterization of gas-liquid two-phase flow and achieved satisfactory results [13]. This paper is devoted to flow regime visualization and pressure signal analysis of solid-gas flow in high-pressure and dense-phase pneumatic conveying.

2. EXPERIMENTAL APPARATUS

The pressurized experimental facility is shown schematically in Fig.1. High pressure nitrogen from the buffer tank is divided into pressurizing gas, fluidizing gas and supplement gas. The feeding hopper adopts the bottom-fluidization and top-discharge arrangement. Pulverized coal in the feeding hopper is fluidized by fluidizing gas and enters the conveying pipeline through the accelerating segment. Supplement gas is introduced at the outlet of the feeding hopper to enhance the conveying ability. Pressurizing gas is used to keep pressure constant in the feeding hopper. The pressure of the receiving hopper is controlled by the motor-drive control valve. Both the feeding hopper and the receiving hopper have capacity of 0.648m3. The conveying pipeline is stainless steel tube with an inside diameter of 10mm and a length of 53m. The gas volume rates are measured by the metal tube rotary flow meters, and solid mass in hopper is measured by the weight cells. Pressure and differential pressure are measured by the semiconductor pressure transducers with frequency response of 200Hz and precision of 0.3%. Signals of differential pressure, pressure, weight and gas volume flow rate are acquired by a multi-channel sampling system and sent to a computer through an A/D converter. Physical properties of conveyed materials are listed in Tab.1.

Journal of Physics: Conference Series **147** (2009) 012078 doi:10.1088/1742-6596/147/1/012078

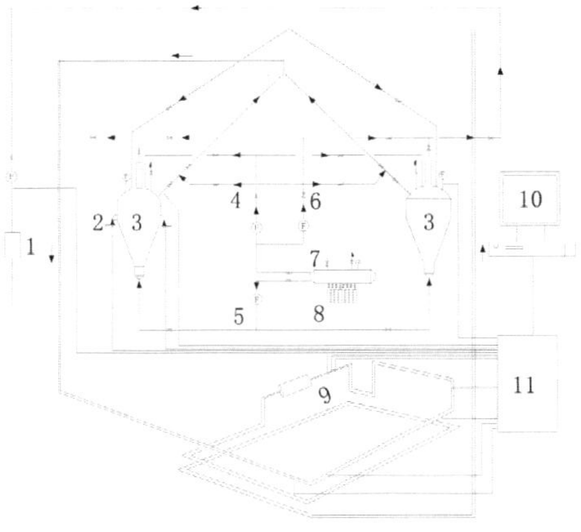

Fig.1 Schematic diagram of dense-phase pneumatic conveying of pulverized coal under high pressure
1-Motor-drived control valve 2-Weigh cell 3-Hopper 4- Pressurizing gas 5-Fluidizing gas 6-Supplemental gas 7-Buffer tank 8-Nitrogen cylinder 9-Visualization section 10-Computer 11-Sensors and A/D converter

Tab.1 Physical properties of pulverized coal

Material	Density/(kg.m^{-3})	Mean d_p/(μm)	Moisture/%
Quartz powder	2650	400	0
Inner Mongolia pulverized coal	1400	52	3.74

3. EMPIRICAL MODE DECOMPOSITION (EMD)

EMD is an adaptive technique that decomposes the original signal into a family of IMF components. Each component emphasizes the local embedded characteristics of the signal. An IMF must satisfy two constraints: (i) the number of extrema and the number of zero crossings must either equal each other or differ at most by one; and (ii) at any time, the mean value of the envelope defined by the local maxima and the envelope defined by the local minima is zero.

The first constraint is similar to the traditional narrow-band requirements for a stationary Gaussian process. The second constraint modifies the global requirement to a local one and is necessary so that the instantaneous frequency will not include unwanted fluctuations.

The decomposition of data series x(t) follows a number of logical steps[14,15]. These are described below.

1. Identify all of the local maxima and local minima of the signal x(t). Then use cubic spline interpolation to define the upper envelope xmax(t) and the lower envelope xmin(t) of the original data series. The mean m1(t) of the upper and lower envelopes is calculated as follows:

$$m_1(t) = \frac{x_{\max}(t) + x_{\min}(t)}{2}.$$

2. Subtracting m1(t) from the data series x(t) gives:

$$h_1(t) = x(t) - m_1(t).$$

3. In general, $h_1(t)$ is just an IMF candidate that is unlikely to satisfy all the requirements of IMF. Iterate the above procedure k times until the mean envelope is close to zero, so the first IMF component $I_1(t)$ that contains the highest frequency component of the signal can be designated as

$$h_{1(k-1)}(t) - m_{1k}(t) = h_{1k}(t), \qquad I_1(t) = h_{1k}(t).$$

The above procedure is a sifting process that has two effects: (i) to eliminate riding waves; and

(ii) to smooth uneven amplitude. In order to avoid obliterating the physically meaningful amplitude fluctuations, a stop criterion for the sifting process is defined [16].

$$SD = \sum_{t=0}^{T}\left[\frac{\left|(h_{1(k-1)}(t) - h_{1k}(t)\right|^2}{h_{1(k-1)}^2(t)}\right]$$

4. Separate $I_1(t)$ from the rest of the data series. The residue, $r_1(t)$, is treated as the new data series.

$$r_1(t) = x(t) - I_1(t)$$

Since the residue $r_1(t)$ still contains much information from lower frequency components, it is subjected to the same sifting process. Repeat the above procedure on the subsequent residual $r_j(t)$ until the range of the residue is below a predetermined value or the residue contains the lowest frequency component, and the result is

$$r_1(t) - I_2(t) = r_2(t), \ldots, r_{n-1}(t) - I_n(t) = r_n(t)$$

where $r_n(t)$ denotes the trend of the signal from which no more IMF can be extracted and I_n is the nth IMF. Each IMF contains lower frequency components than the one extracted just before. At the end of the EMD procedure, the signal $x(t)$ can be exactly reconstructed using a linear combination:

$$x(t) = \sum_{i=1}^{n} I_i(t) + r_n(t).$$

4. RESULTS AND DISCUSSION

4.1 Photos and pressure traces

Because of the high stickiness and electrostatic attraction of pulverized coal to the pipe wall, it is very difficult to make out the flow regimes in the conveying pipe through visualization method. So quartz powder with average diameter of 400μm and density of 2650kg/m3 was adopted as the substitute conveyed material to study the flow regime. High-speed digital video camera (125fps, 4096pics/run) was employed to photograph the flow patterns. Length of visualization section is 0.16m and interior diameter is 10mm.

Fig.2 depicts that flow regime transits from stratified flow to plug flow as superficial gas velocity diminishes under the basic condition (P=3.6MPa, △PT=0.8MPa). When superficial gas velocity is relatively high, a suspended phase and a settled layer of material are generally observed. With decrease in superficial gas velocity, dunes or clusters can be seen riding on a settled layer of material. At further lower gas velocity the material may flow as discreet or as a packed bed. The characteristics of pressure traces acquired by high frequency response pressure sensor were correlated closely with the flow regimes. When superficial gas velocity was relatively high, flow pattern was stratified flow and pressure signal fluctuated more quickly than that of the other flow regimes. As superficial gas velocity diminished, flow pattern transited to slug flow and pressure signal fluctuated like sine wave. Further decrease in superficial gas velocity yielded plug flow, meanwhile sine-like pressure signal fluctuated with further lower frequency and higher amplitude.

4.2 EMD characteristics of pressure fluctuation signals with quartz

4.2.1 IMF components

As presented above, pressure signal traces of different flow regimes show distinguishable characteristics. In order to investigate the characteristics of pressure more deeply, EMD analysis

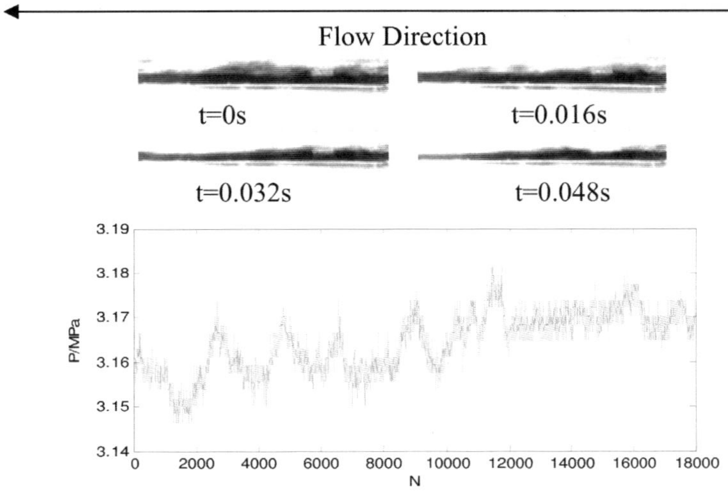

(a) Vg=9.4m/s, μ=101kg/m3, Stratified flow

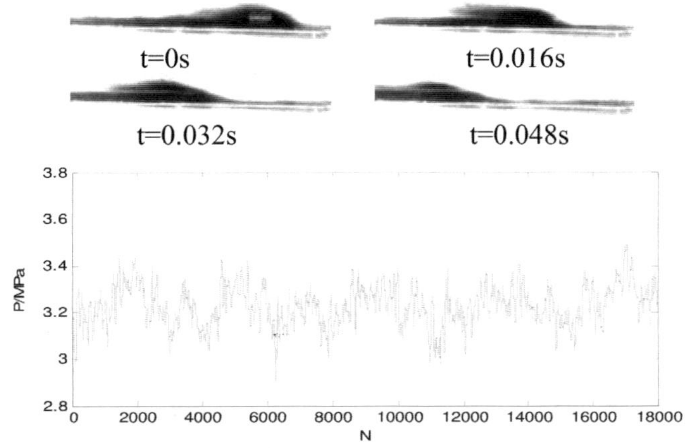

(b) Vg=5.6m/s, μ=146kg/m3, Dune speed 1.5~2m/s

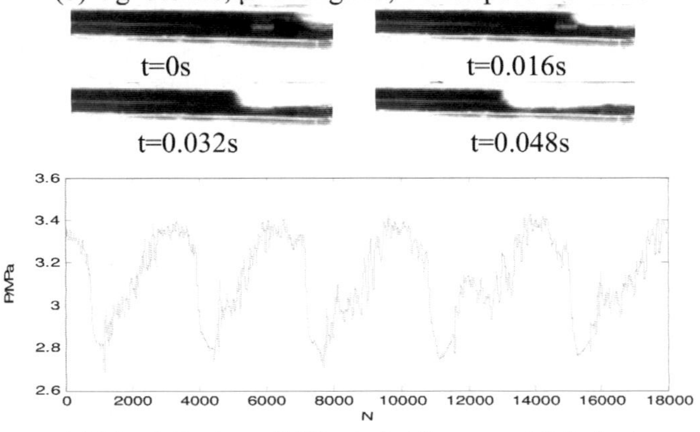

(c) Vg=1.7m/s, μ=348kg/m3, Plug speed 0.8~1m/s

Fig.2 Visualization photos of flow regimes and pressure traces at different superficial gas velocity

was employed on the pressure signals. Fig.3 shows side by side the IMF components of the pressure fluctuation signal. IMFs have different local frequencies. The EMD method picks out the highest frequency component present in the signal. Thus, each IMF component that contains a lower frequency than the preceding one extracted is presented in turn. The smaller scales represent the high-frequency information in the signal and the larger scales represent the low-frequency information in the signal.

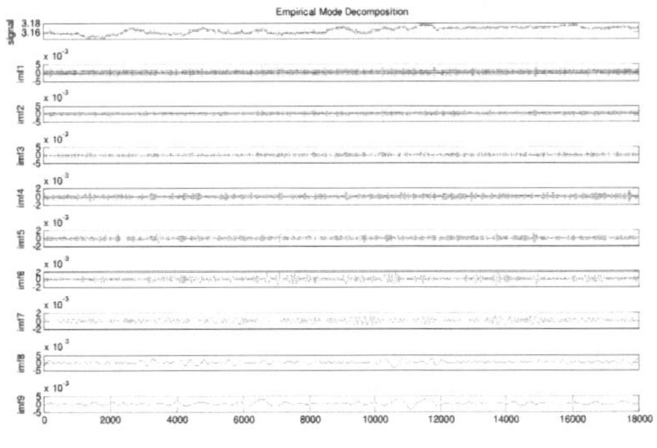

(a) IMF components (dilute stratified flow)

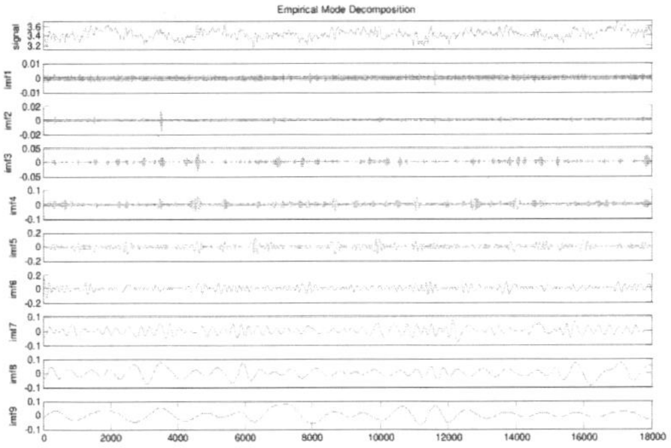

(b) IMF components (dune flow)

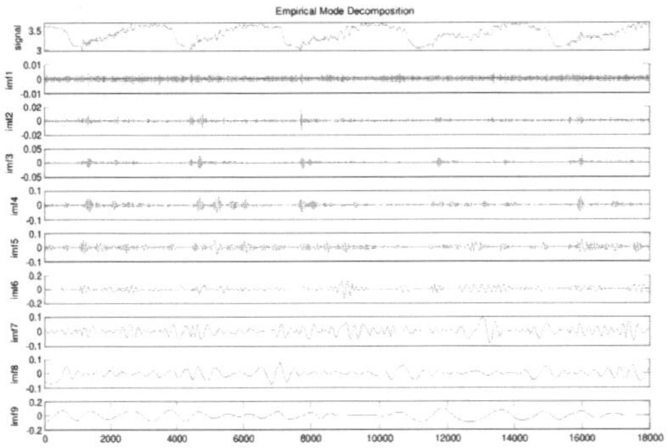

(c) IMF components (plug flow)

Fig.3 IMF components of pressure fluctuation signal of different flow pattern

4.2.2 Energy distribution of IMF

In this study, the energy distribution Hj of different frequency bands is defined as $H_j = \frac{E_j}{E}$ (j = 1, 2, 3), where E_j (measured by Standard Deviation) stands for the local energy of the different frequency bands (j=1 denotes the high-frequency band, j=2 the middle-frequency band and j=3 the low-frequency band) and E is the total energy(measured by Standard Deviation also). The energy distribution Hj is selected as the characteristics of the pressure fluctuation signal. In this study a number of logical steps were followed:

(1) Acquire the pressure fluctuation signal;

(2) Apply EMD to the signal;

(3) Calculate the total signal energy, each frequency band of local energy and the energy distribution;

(4) Determine the energy distribution of the signal.

As per the extensive experimental work undertaken and literature published, the IMF components are divided into three frequency bands in terms of their inherent frequencies for quantitative analysis of the energy distribution: high-frequency band H1 with IMF components of IMF1, IMF 2 and IMF 3, middle-frequency band H2 with IMF components of IMF 4, IMF 5 and IMF 6, and low-frequency band H3 with IMF components of IMF 7 and IMF 8. E1 represents the total energy of IMF 1, IMF 2 and IMF 3, E2 is the sum of the energy of IMF 4, IMF 5 and IMF 6, and E3 is the total energy of IMF7 and IMF8. Tab.2 summarizes the variations of energy characteristics Hj with flow pattern.

Tab. 2 Conveying condition and energy distribution with quartz

P (MPa)	ΔPT (MPa)	μ (kg/m3)	Vg (m/s)	Flow pattern	Energy characteristics Hj (%)		
					H1	H2	H3
3.6	2.8	101	9.4	Stratified flow	46.95	26.11	26.94
		146	5.6	Dune flow	4.79	55.23	39.98
		348	1.7	Plug flow	3.14	54.85	42.02

Tab.2 illustrates the variation of energy distribution in different frequency bands. With decreasing gas velocity, the flow pattern of solid-gas flow is continually changing and the energy characteristics Hj changes too. According to Table2, I1, I2 and I3 carry almost half of the embedded energy for the dilute stratified flow. In the case of dune flow, the energy diverts from the high-frequency band to the lower-frequency band. Consequently, the ratio of energy embedded in the high-frequency band decreases dramatically and the ratio of energy embedded in the middle and low-frequency band increases. As for plug flow, the ratio of energy embedded in the high-frequency band decreases further and the ratio of energy embedded in the low-frequency band increases slightly.

The above results imply that the energy characteristic Hj changes with the flow pattern and can therefore be regarded as an indicator for the identification of the flow pattern of high-pressure and dense-phase solid-gas flow. The result also demonstrates that EMD is suitable for analyzing the pressure fluctuation signal. However, it has to be pointed out that Hj changes from dune flow to plug flow not as evidently as from stratified flow to dune flow.

4.3 Energy distribution of IMF with pulverized coal

Inner Mongolia pulverized coal with mean diameter of 52μm was conveyed in the experimental facility. Conveying condition is shown in Tab.3. Because of the high stickiness and electrostatic attraction of pulverized coal to the pipe wall, it is very difficult to recognize the flow regimes in the conveying pipe through visualization method. It's also unclear and impossible to identify the flow pattern through the pressure signal trace alone. Through the preceding results, EMD is considered as an effective method to analyze the pressure fluctuation signal in order to identify the flow regime in the pipeline indirectly.

Similar varying tendency of the energy characteristic Hj with the deceasing gas velocity was found. With decreasing gas velocity, the energy characteristics Hj changes obviously. According to Tab.3, H1 carry more than 70% of the embedded energy at the relatively high Vg which is 11.1m/s. When Vg changes from 6.2m/s to 10.2m/s, the energy diverts from the high-frequency band to the lower-frequency band. Consequently, the ratio of energy embedded in the high-frequency band decreases to about 60% and the ratio of energy embedded in the middle and low-frequency band increases. Further decreasing Vg to 4.1m/s or below, the ratio of energy embedded in the high-frequency band decreases and the ratio of energy embedded in the low-frequency band increases significantly.

As a result, though impossible by visualization method, three flow patterns can be classified approximately according to the energy characteristics Hj when conveying pulverized coal.

Tab. 3 Conveying condition and energy distribution with pulverized coal

P (MPa)	ΔPT (MPa)	μ (kg/m3)	Vg (m/s)	Energy characteristics Hj (%)		
				H1	H2	H3
3.4	0.4	102	11.1	70.17	15.04	14.79
		126	10.2	56.28	11.52	32.20
		166	9	58.60	11.71	29.69
		213	7.9	61.85	13.00	25.15
		309	6.2	61.17	16.01	22.81
		531	4.1	28.60	14.51	56.89

5. CONCLUSIONS

(1) Three distinguishable flow regimes, dilute stratified flow, dune flow and plug flow were observed visually when conveying quartz powder under high-pressure and dense-phase pneumatic conveying.
(2) When conveying quartz powder, the characteristics of pressure traces acquired by high frequency response pressure transmitter were correlated closely with the flow regimes. When superficial gas velocity was relatively high, flow pattern was dilute stratified flow and pressure signal fluctuated more quickly than that of the other flow regimes. As superficial gas velocity diminished, flow pattern transited to dune flow and pressure signal fluctuated like sine wave. Further decrease in superficial gas velocity yielded plug flow, meanwhile sine-like pressure signal fluctuated with lower frequency and higher amplitude.
(3) In order to investigate the characteristics of pressure more deeply, EMD has been applied in the analysis of a pressure fluctuation signal. Results have demonstrated that the extracted energy characteristics reflect the energy shift with the variation of flow pattern. The energy characteristic Hj changes with the flow regimes and can therefore be regarded as an indicator for the identification of the flow regimes.
(4) It's unclear and almost impossible to identify the flow pattern through the pressure signal trace alone. Using EMD analysis on the pressure signal, the flow regime in the pipeline can be identified indirectly.

ACKNOWLEDGMENT

This work was financially supported by the National Key Basic Research Program of China (No. 2004CB217702)

NOMENCLATURE

Vg	superficial gas velocity	$[\mathrm{m}\cdot\mathrm{s}^{-1}]$
μ	solid-gas ratio	[kg·m-3]
P	ending pressure (pressure in the sending hopper)	[MPa]
ΔP_T	total conveying differential pressure	[MPa]

REFERENCES

X. Gong , X. Guo and et al. (2006). "High solids loading pneumatic conveying of pulverized coal," Journal of Chemical Industry and Engineering (china).,Vol.57, pp640-644.

P.W. Wypych and J. Yi(2003). "Minimum transport boundary for horizontal dense-phase pneumatic conveying of granular materials," Powder technology, Vol.129, pp111-121.

H. Li and Y Tomita(2000). "Particle velocity and concentration characteristics in a horizontal dilute swirling flow pneumatic conveying,"powder technology, Vol.107, pp144-152.

D J. Masona, A. Lev(2001). "A model for non-suspension gas-solids flow of fine powders in pipes," International Journal of Multiphase Flow, Vol.27, pp415-435.

S. Laouar, Y. Molodtsof (1998)."Experimental drop in dense phase pneumatic transport at very low velocity, "Powder Technology, Vol.95, pp165-173.

O. Molerus(1996). "Overview: pneumatic transport of solids," Powder technology, Vol.88, pp309-321.

Pavol Rajniak, Kumar Dhanasekharan, Csaba Sinka, et al(2008). "Modeling and measurement of granule attrition during pneumatic conveying in a laboratory scale system,"Powder Technology, 185(3), pp.202-210.

Santiago Laín, Martin Sommerfeld. (2008). "Euler/Lagrange computations of pneumatic conveying in a horizontal channel with different wall roughness," Powder Technology, 184(1), pp.76-88.

B.J. Azzopardi, K. Jackson, J.P. Robinson, et al. (2008). "Fluctuations in dense phase pneumatic conveying of pulverised coal measured using electrical capacitance tomography," Chemical Engineering Science, 63(9),pp2548-2558.

J.A. Woods, R.B. Thorpe, S.E. Johnson. (2008). "Horizontal pneumatic conveying from a fluidized bed," Chemical Engineering Science, 63(7), pp1741-1760.

X. Shen and Y. Xiong, (2005). "Experimental study on dense-phase pneumatic conveying of pulverized coal at high pressures," Proceedings of the CSEE, Vol.25, pp103-107.

Chen Xiaoping, Fan Chunlei, Liang Cai, et al. (2006). "Investigation on Characteristics of Pulverized Coal Dense-phase Pneumatic Conveying under High Pressure,"Korean Journal of Chemical Engineering, 24(3), pp499-502.

Hao Ding, Zhiyao Huang, Zhihuan Song, et al. (2007). "Hilbert-Huang transform based signal analysis for the characterization of gas-liquid two-phase flow,"Flow Measurement and Instrumentation 18, pp37-46.

Bouzid A, Ellouze N. (2004). "Empirica mode decompostion of voiced speech signal." In: First international symposium on control, communications and signal processing, pp.603-606.

Flandrin P, Rilling G, Goncalves P. (2004). "Empirical mode decompostion as a filter bank," IEEE Signal Prcocessing Letters, 11(2), pp112-114.

Huang NE, Shen Z, Long SR. (1999). "A new view of nonliner water waves: The Hilbert sprectrum,"Annual Review of Fluid Mechanis, 31, pp417-457.

Journal of Physics: Conference Series **147** (2009) 012079 doi:10.1088/1742-6596/147/1/012079

Experimental study on steam plume and temperature distribution for sonic steam jet

Xinzhuang Wu, Junjie Yan, Wenjun Li, Dongdong Pan, Ying Li

State Key Laboratory of Multiphase Flow in Power Engineering, Xi'an Jiaotong University, Xi'an, 710049, P.R.China.

yanjj@mail.xjtu.edu.cn

Abstract. The sonic steam jet in subcooled water was investigated experimentally over a wide range of steam mass flux and water temperature conditions. Four different steam plume shapes were observed in present test conditions, and the condensation form was mainly controlled by the steam mass flux and water temperature. Moreover, the unstable jet was observed on the condition of low steam mass flux and high water temperature. The transition criterion of unstable-stable jet was also given. The temperature fields in the steam plume and in the surrounding water were measured. Axial temperature distributions represented the four typical steam plumes, and the fluctuation of axial temperature confirmed the existence of expansion and compression waves. Additionally, the radial temperature distributions were independent of water temperature for small radial distance at nozzle exit, and further the axial location was apart from the nozzle exit, longer the radial distance affected by the momentum diffusion.

1. INTRODUCTION

The phenomenon of steam-water direct contact condensation has been investigated extensively, and used in engineering applications such as direct contact feed water heater, steam jet driven injector and nuclear reactor safety system, etc. When the steam was injected into the quiescent subcooled water, several condensation modes including chugging, bubbling and jet appeared according to the thermal hydraulic conditions (Chan et al., 1982; Liang et al., 1994).

The condensation of subsonic or sonic steam jet submerged in subcooled water has been studied by many researchers. The investigation about the jet mainly included condensation regime map, pressure and temperature distributions, and condensation heat transfer coefficient. At the high steam mass flux, the oscillatory cone jet and ellipsoidal jet were described (Chan et al., 1982). The three-dimensional condensation regime of steam injected into stagnant water was investigated (Petrovic et al., 2007), the nozzle diameter was also considered besides the steam mass flux and water temperature. The penetration length of sonic steam jet horizontally in quiescent subcooled water was experimentally and theoretically investigated (Kerney et al., 1972; Weimer et al., 1973), and the correlations to calculate the dimensionless penetration length was also given. The turbulent condensing vapor jets submerged in subcooled liquids was theoretically investigated (Chen et al., 1982), and they gave a model of this process for numerical simulation assuming an idealized plume shape and a homogeneous two phase flow. Giovanni et al. (1984) experimentally measured the pressure and temperature in the surrounding zone inside a steam jet condensing in a subcooled water pool and gave an adopted model to describe the phenomena. Chun et al. (1996) experimentally studied the direct contact condensation of steam injected into the subcooled water, two shapes of steam plume (conical and ellipsoidal) were observed,

Journal of Physics: Conference Series **147** (2009) 012079 doi:10.1088/1742-6596/147/1/012079

and found divergent plume existed at relatively small subcooling. Eden et al. (1998) investigated the centerline pressure and cavity shape of horizontal plane under-expanded vapor jets with low condensation potential, the pressure measurements showed the existence of periodic expansion/compression cells associated with under-expanded non-condensing gas jets. Kim et al. (2001) observed conical and ellipsoidal shapes of steam jet in the experiment, established the empirical correlations for dimensionless steam jet length, and the axial and radial temperature distributions in the steam jet were measured. The supersonic steam jet in subcooled water was studied (Wu et al., 2007), six different steam plume shapes were observed besides conical and elliptical shapes, and the correlations to predict expansion ratio, penetration length and average heat transfer coefficient were given.

Almost all the research on the steam jet was for subsonic and sonic, and two typical conical and elliptical shapes were reported. In present work, four different steam plume shapes were observed in sonic steam jet, and the unstable jet was discussed. The axial and radial temperature distributions in the steam plume and surrounding subcooled water were investigated at various test conditions. Moreover, the effects of steam mass flux and subcooled water temperature on condensation form and temperature distributions were discussed.

2. EXPERIMENTAL APPARATUS AND METHODS

The experimental system for investigating the steam jet is schematically presented in Fig. 1. The apparatus mainly consists of an electric steam generator, a surge tank, a water vessel of 3000mm×1000mm×1200mm and some valves. The electric steam generator with electric heaters of 330 kW supplies continuously the steam. The steam was injected into the subcooled water through a sonic nozzle which was fixed on the wall of water vessel by a flange. In present work, the upper part of the water vessel is open to the ambient and two observation windows with the same size were designed for observation and taking pictures. The test conditions are shown in Table 1.

The steam flow rate was measured by a vortex type steam flowmeter (accuracy 0.5%FS). The temperature in the jet exit region was measured by a mobile test block which was equipped with 21 K-type thermocouples (diameter 1.0mm, accuracy 0.5%FS), as shown in Fig. 2. Actually, the test block was cuspate and behind the thermocouple, which would not affect the upstream temperature fields. The water temperature in water vessel was measured by the same thermocouple. The inlet pressure in the nozzle was measured by high-temperature pressure transducer (MSI, accuracy 0.1%FS). A high-speed camera was used to take pictures of steam plume. All signals were processed by the data acquisition system consisting of PC and A/D converter.

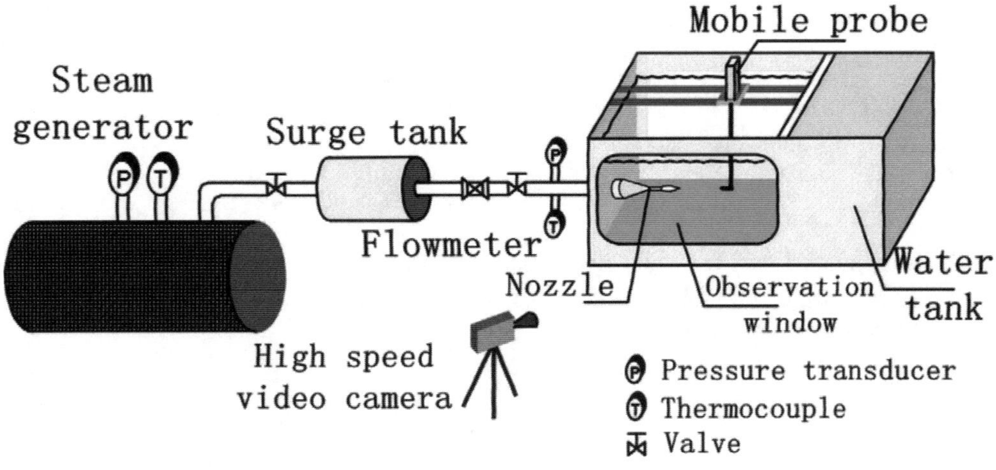

Fig. 1 Schematic diagram of experimental system

Tab 1. Test conditions in experiment

Parameters	Values
Steam mass flux at nozzle exit G_e, kg/m^{2}s	298-865
Water temperature T_f, K	293-343
Ambient pressure p_a, MPa	0.102
Exit diameter of nozzle d_e, mm	6.0

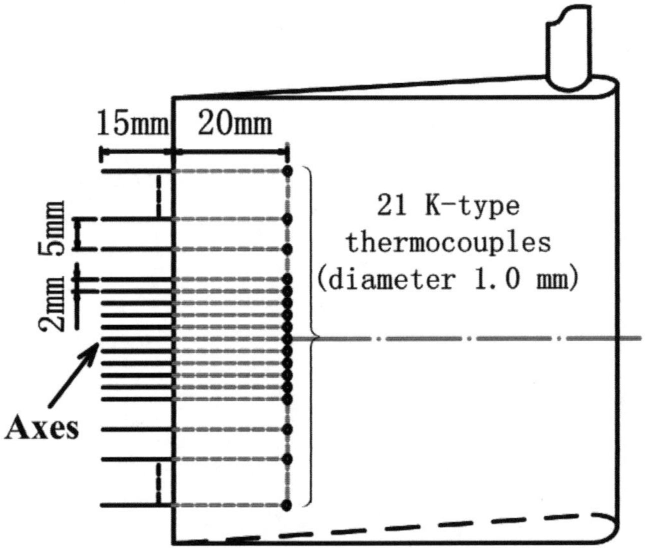

Fig. 2 Schematic diagram temperature test block

3. EXPERIMENTAL RESULTS AND DISCUSSIONS

3.1 Steam Plume Shape

In previous investigation on steam jet in subcooled water, the steam plume shape was observed mainly by high-speed camera, and the same method was used in present study. Almost only two typical steam plume shapes were reported for sonic jet by several researchers. The two steam plume shapes including conical and elliptical shapes, was also observed in present experiment, as shown in Fig. 3 (Shapes A and B). Additionally, another two steam plume shapes (Shapes C and D) were observed in sonic steam jet. The two shapes were reported in supersonic steam jet (Wu et al., 2007). Here, the four steam plumes (Shapes A, B, C and D in Fig. 3), were called constriction shape, expansion-constriction shape, double expansion-constriction shape, and double expansion-emanative shape, respectively.

According to the observation, the steam plume shape was influenced by steam mass flux and water temperature, actually. For low steam mass flux and low water temperature, the steam plume constricted directly. For the high steam mass flux and low water temperature, the steam plume expanded at the nozzle exit, then constricted due to the high condensation capability. With the increase of the water temperature, the steam plume expanded again, and the second expansion tended to emanative for the high water temperature (low condensation capability).

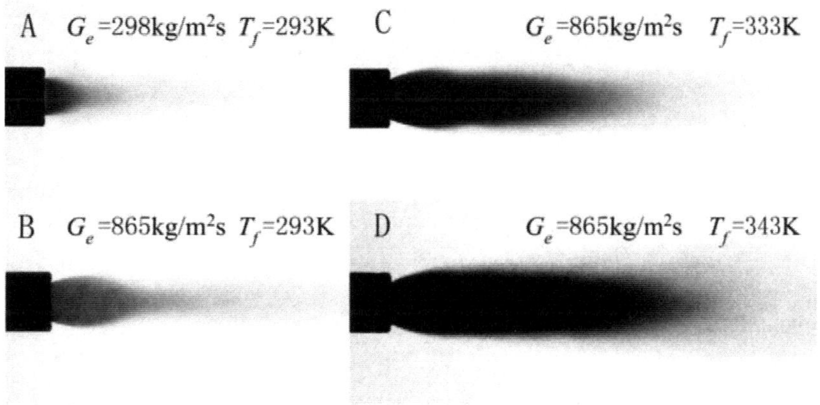

Fig. 3 Four typical steam plume shapes at various test conditions

Additionally, the jet tended to be unstable with the increase of the water temperature for low steam mass flux (298kg/m^{2}s), as shown in Fig. 4. It is obvious that the transition of unstable-stable jet is between 318K and 323K. When the water temperature was below 318K, the steam plume was obvious. But as the water temperature was above 323K, the steam plume was emanative and unstable.

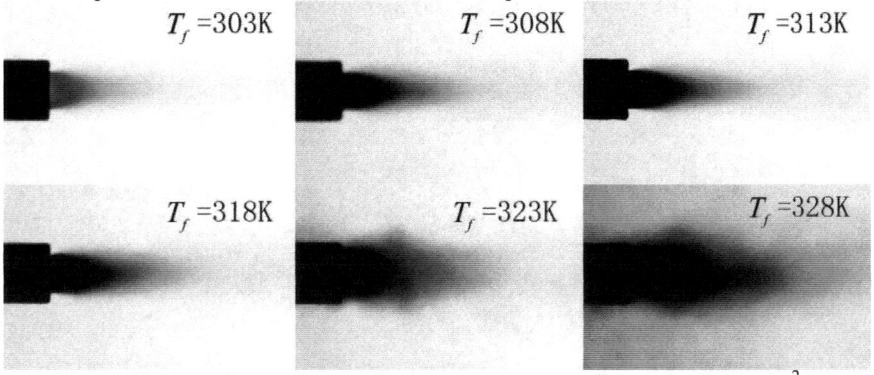

Fig. 4 Steam plume shapes for low steam mass flux (G_e=298kg/m^{2}s)

The criterion for bubbling-jetting transition was given (Liang et al. 1994). A criterion for the unstable-stable jet transition was developed based on the above method (Liang et al. 1994). The energy balance was used as follows:

$$\frac{\rho_s v_s A_e h_{fg}}{h_i A_i \Delta T_{sub}} \geq 1.0 \tag{1}$$

where h_{fg} is latent heat of vaporization, ΔT_{sub} is water subcooling. The area of the interface could be obtained by the photo, and the heat transfer coefficient could be expressed by interfacial transport due to turbulence intensity (Kim et al., 2004).

$$h_i = St(\frac{1}{2n})^{1/3}(\frac{\rho_f}{\rho_g})^{2/3} c_p G_e \tag{2}$$

where n is the ratio of the thermal boundary layer thickness to the eddy size, and it is recognized that the eddy size is about 1/5 of the thermal boundary layer thickness. G_e is the steam mass flux of nozzle exit. The Stanton number is the function of Nusselt number, Reynolds number and Prandtl number.

Journal of Physics: Conference Series **147** (2009) 012079 doi:10.1088/1742-6596/147/1/012079

Here, the Jacob number, which was adopted to describe the ratio of the specific energy absorption capabilities of the liquid and the energy density of the steam, was defined as:

$$\mathrm{Ja} = \frac{\rho_f c_p \Delta T_{sub}}{\rho_s h_{fg}} \tag{3}$$

According to the experimental data, the final form of the criterion for unstable-stable jet transition is:

$$0.35\left(\frac{\rho_f}{\rho_s}\right)^{1/3} \mathrm{Re}_f^{s\,1/4}\,\mathrm{Pr}^2\,\mathrm{Ja}^{-1.5} \geq 1.0 \tag{4}$$

Fig. 5 gives the calculated transition criterion of unstable-stable jet for various test conditions in present study. It was actual that the stability was directly proportional to the value of $0.35(\rho_f/\rho_s)^{1/3}\mathrm{Re}_f^{s\,1/4}\mathrm{Pr}^2\mathrm{Ja}^{-1.5}$. Although the steam mass flux was separated in other parameters, here, it was still used as the various test conditions, in order to compare the calculation with the experimental observation. As shown in Fig. 5, with the decrease of the water temperature and increase of the steam mass flux, the jet became more stable. When the water temperature was high and steam mass flux was small, the jet was unstable, which was in accordance with the observation of the experimental results, as shown in Figs. 3 and 4.

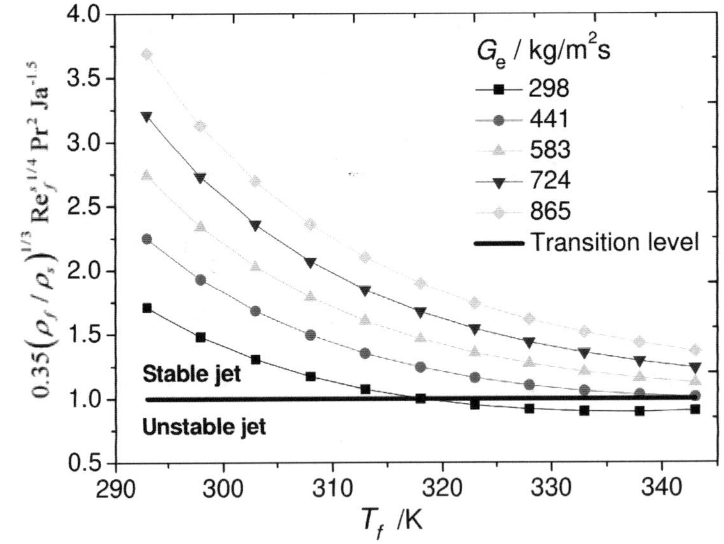

Fig. 5 The Unstable-stable jet transition criterion

3.2 Axial Temperature Distributions

The temperature distributions along the steam jet axis and in the surrounding water were measured for various test conditions. Fig. 6 shows the axial temperature distributions for two different steam mass flux and water temperature in the range of 293K to 343K. For low mass flux, the axial temperature distributions decreased to the ambient water temperature directly, which represented the trend of the constriction steam plume shape. For high mass flux, near the nozzle exit inside the steam jet, the axial temperature variation was independent of the water temperature. When the water temperature was low, the axial temperature decreased first then increased, after a peak, the temperature decreased again to the ambient water temperature. Such a temperature distribution represented the expansion-constriction steam plume shape. For high water temperature, after the first peak, the axial temperature tended to increase again, then decreased to the ambient water temperature due to condensation, which represented the double expansion-constriction steam plume shape. When the water temperature was above 343K, the second peak of axial temperature became smooth, which represented the double expansion-emanative steam plume shape.

In fact, the axial temperature distributions were affected by the expansion and compression waves for under-expanded jet (Kim et al., 2001; Wu et al., 2007). When the nozzle exit pressure is higher than surrounding water pressure, the expansion wave may occur at nozzle exit, which leaded to supersonic flow and steam flowing outward. When the supersonic flow was compressed by the ambient water, the compression wave occurred. The expansion and compression waves could reflect periodically for ideal condition. However, the reflection only occurred one or two times due to the condensation and viscosity. When the steam flow was expanded, the axial temperature would decrease, whereas the steam flow was compressed, the axial temperature would increase. Accordingly, the axial temperature distributions reflected the steam flowing characteristics. Fig. 7 shows the corresponding relation of axial temperature distributions and steam plume shapes. The peak and nadir of axial temperature were all in accord with the position of compression and expansion of steam plume, approximately.

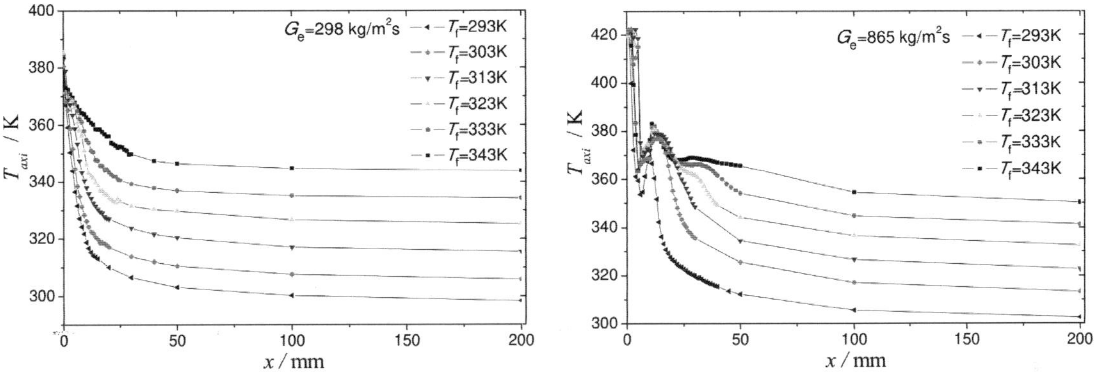

Fig. 6 Axial temperature distributions at various test conditions

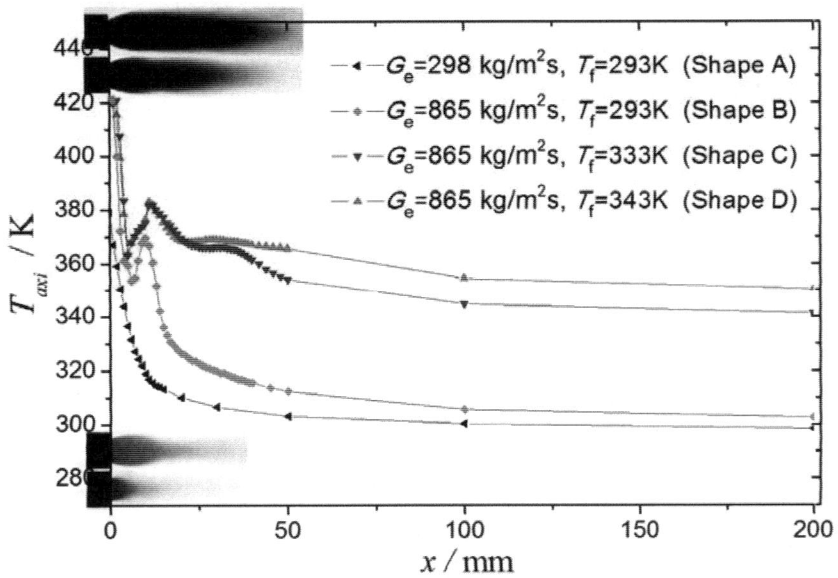

Fig. 7 Relations of axial temperature distributions and steam plumes

3.3 Radial Temperature Distributions

The temperature distributions in the radial direction have been obtained in the various test conditions. Radial temperature distributions at four different axial locations, for two steam mass flux and different water temperatures, are shown in Figs. 8 and 9. It can be seen that the radial temperatures were independent of water temperature for radial distance at axial location of 0, but for the far axial location, the radial temperatures were influenced by the water temperature significantly. Moreover, the influencing region was expansible with increasing the axial distance.

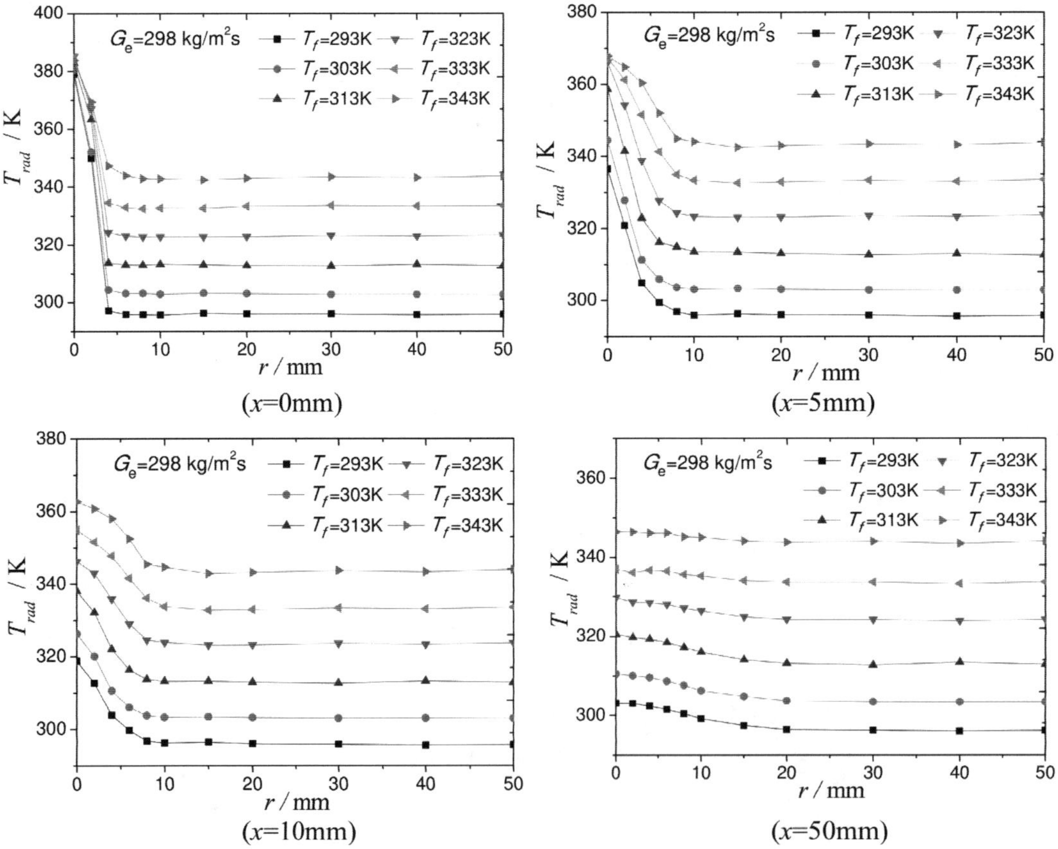

Fig. 8 Radial temperature distributions (G_e=298kg/m^{2}s)

In the case of the constriction steam plume shape, as shown in Fig. 8, the radial temperatures indicated the maximum values at the axial direction and then decrease toward the outer radial direction. But for the other cases of steam plume shapes, as shown in Fig. 9, the radial temperatures tended to increase first and then decreased toward the outer radial direction in this region, and this radial temperature characteristic was also reported for the case of elliptical steam plume shape (Kim et al., 2001). Based on the experimental data of expansion-constriction, double expansion-constriction and double expansion-emanative steam jet, a conclusion was drawn that the variation of the radial temperature was due to the expansion and compression waves of the under-expanded jet.

 doi:10.1088/1742-6596/147/1/012079

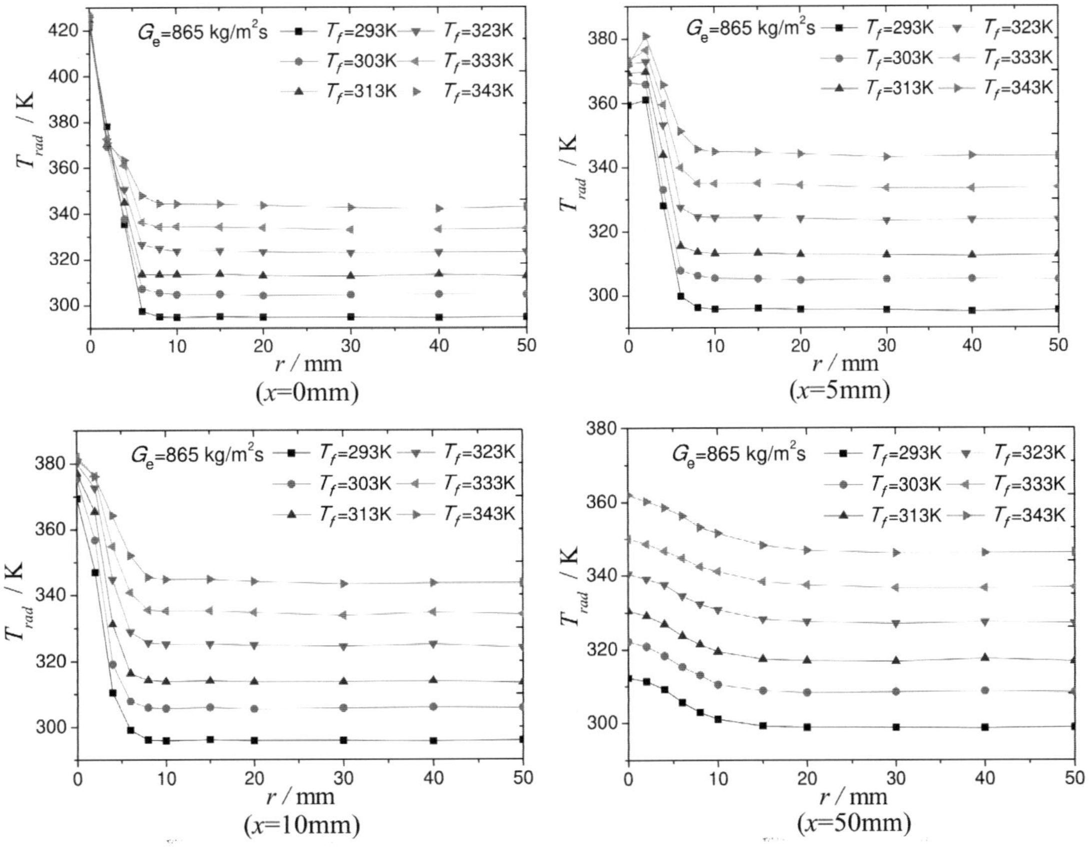

Fig. 9 Radial temperature distributions (G_e=865kg/m^{2}s)

4. CONCLUSIONS

In present work, experimental study on sonic steam jet in subcooled water for wide ranges of steam mass flux and water temperature was carried out. The new characteristics of steam plume and temperature fields were obtained. The main results could be summarized as follows:

(1) Four different steam plume shapes were observed for various test conditions for sonic nozzle. Besides the constriction and expansion-constriction shape reported by previous investigation, the double expansion-constriction and double expansion-emanative shapes were also observed in sonic steam jet. And the condensation form was mainly controlled by the steam mass flux and water temperature. Moreover, the transition criterion of unstable-stable jet was given based on the present experimental data.

(2) The temperature distributions in the steam plume and in the surrounding water were measured for different test conditions. The influencing area was affected by steam mass flux and water temperature. The temperature fluctuation at axial direction and radial direction were evident due to expansion and compression wave, and the axial temperature distributions also represented the four typical steam plumes. Additionally, the radial temperatures were independent of water temperature for small radial distance at nozzle exit, and further the axial location was apart form the nozzle exit, longer the radial distance affected by the momentum diffusion.

ACKNOWLEDGEMENTS

This work was supported by National Natural Science Foundation of China (No. 50676078) and

National High-Tech Research and Development Program of China (863 Program) (No. 2006AA05Z230).

NOMENCLATURE

A	heat transfer area	[m^2]
c_p	water specific heat	[J/(kgK)]
d	diameter of nozzle	[m]
G	steam mass flux	[kg/(m^{2}s)]
h	heat transfer coefficient	[W/(m^{2}K)]
h_{fg}	latent heat of vaporization	[J/(kg)]
n	ratio of δ to λ	
T	temperature	[K]
p	pressure	[MPa]
Ja	Jacob number	
Re_f^s	steam-liquid Reynolds number, $v_s d_e / \nu_f$	
Pr	Prandtl number, $\mu c_p/k$	
S_t	Stanton number	
v	velocity	[m/s]

Greek Letters

ρ	density	[kg/m^3]
δ	thermal boundary layer thickness	[m]
λ	eddy size	[m]
ΔT_{sub}	water subcooling	[K]

Subscripts

a	ambient
axi	axial direction
e	nozzle exit
i	interface
f	liquid
rad	radial direction
s	steam

REFERENCES

Chan, C. K. et al. (1982). "A Regime Map for Direct Contact Condensation," Int. J. Multiphase Flow, 8(1), pp. 11-20.

Chen, L. D. et al. (1982). "Condensation of Submerged Vapor Jets in Subcooled Liquids," Transactions of ASME, J. Heat Transfer, 104(11), pp. 774-780.

Chun, M. H. et al. (1996). "An Investigation of Direct Condensation of Steam Jet in Subcooled Water," Int. Comm. Heat Mass Transfer, 23(7), pp. 947-958.

Eden, T. J. et al. (1998). "The Centerline Pressure and Cavity Shape of Horizontal Plane Choked Vapor Jets with Low Condensation Potential," J. Heat Transfer, 120(11), pp. 999-1007.

Kerney, P. J. et al. (1972). "Penetration Characteristics of a Submerged Steam Jet," AICHE J., 18(5), pp. 548-553.

Kim, H.Y. et al. (2001). "Experimental Study on Stable Steam Condensation in a Quenching Tank," Int. J. Energy Res., 25, pp. 239-252.

Kim, Y. S. et al. (2004). "Investigation of the Steam-Water Direct Contact Condensation Heat Transfer Coefficients Using Interfacial Transport Model," Int. Comm. Heat Mass Transfer, 31(3), pp. 397-408.

Liang, K. S. et al. (1994). "Experimental and Analytical Study of Direct Contact Condensation of Steam in Water," Nucl. Eng. Des., 147, pp. 425-435.

Petrovic, A. et al. (2007). "Three-dimensional condensation regime diagram for direct contact condensation of steam injected into water," Int. J. Heat Mass Transfer, 50(9-10), pp. 1762-1770.

Weimer, J. C. et al. (1973). "Penetration of Vapor Jets Submerged in Subcooled Liquids," AICHE J., 19(3), pp. 552-558.

Wu, X. Z. et al. (2007). "Experimental study on the condensation of supersonic steam jet submerged in quiescent subcooled water: steam plume shape and heat transfer," Int. J. Multiphase Flow, 33(12), pp. 1296-1307.

Journal of Physics: Conference Series **147** (2009) 012080 doi:10.1088/1742-6596/147/1/012080

Study on Measurement Lower Limit of DOAS System Based on kalman Filter

Guanghua Tang[1], Chuanlong Xu[1], Litang Shao[1], Shimin Wang[1] , Li Tao[2]

[1] Southeast University, Si Pai Lou 2#, Nanjing, China

[2] Nanjing Rongchuan E&P, Equipment Co., Ltd., RM2202, RuiHua BD NO.315, South ZhongShan RD, Nanjing, China

tgh164@163.com

Abstract. Differential Optical Absorption Spectroscopy (DOAS) technique put forward by Platt in 1970s has been widely applied to the measurement of atmospheric trace gas. The key procedures in the retrieval algorithm are the separation of the absorption cross section into two parts and the solution of linear equations about concetration with the least-squares method. However, the above procedures can't be effectively applied to the monitoring of low concentration gases at short light path length. Here, an improved inversion algorithm based on kalman filter is applied to the optimal estimation of gas concentration instead of applying the least squares. Thus, it is not necessary to select specific wavelengths and subtraction of background spectroscopy. The concentration is treated as the state variable and is calculated with linear recursion. Research on experiment of quantitative measurement for SO_2 show that the accuracy of inversion is greatly improved with kalman filter at low ratio of signal to noise and the measurement lower limit is less than 28.6mg/m^3 at a light path length of 0.5 m. Zero drift of the DOAS system based on the improved algorithm is less than 2.9mg/m^3. It is also found that dynamic measurement range can be broadened by combining traditional DOAS algorithm with the improved algorithm by setting an appropriate threshold for the differential optical density.

1. INTRODUCTION

In the past decades, Differential Optical Absorption Spectroscopy(DOAS) has been widely used for monitoring of trace gases (Platt et al., 1979; Platt et al., 1988; Platt et al., 1984; Plane et al., 1992) in the stratosphere or troposphere. It was platt et al. who first (Platt et al., 1979) started to utilize the differential absorption for tropospheric gas composition measurement simultaneously without disturbing the chemical behaviors of substances to be investigated. The most straightforward atmospheric molecules to detect with DOAS are SO_2, NO_2 and O_3 (Plane et al., 1992; Axelsson et al., 1990; Platt et al., 1980) in the u.v./visible. In the tropospheric application, the light from a broad band xenon lamp is transmitted through the atmosphere for distances up to several kilometers. After the light is received, the measurement spectrum is matched to the spectral absorption of known molecules applying the least-squares fitting to derive gas concentrations. A new and interesting application (Mellqvist et al., 1996) of the DOAS technique is to be used for the in-situ detection of flue gases across the stack. However, it is very difficult to be used for the monitoring of low concentration flue gas or for the flue gas mixed with dust particles of high concentration at so short light path length with traditional DOAS technique. In the present article, an improved DOAS algorithm based on kalman filter was investigated and developed to greatly improve the accuracy and sensitivity at short light path length. Different from the traditional DOAS algorithm, the kalman filter applied linear recursion at all wavelengths in a certain wavelength region. The improved DOAS technique can be used to lower the detection limit. Also, if it is combined with traditional DOAS by setting an appropriate threshold for the differential optical density, then it can be used to enlarge the dynamic measurement range.

2. FUNDAMENTAL PRINCIPLE OF DOAS AND RETRIEVAL ALGORITHM

2.1 Fundamental Principle of DOAS

DOAS technique utilizes the configuration of differential absorption in the selected wavelength region to eliminate the interference of other gases and dust particles. The basic principle of DOAS is based on the Lambert-Beer law and the model can be expressed as follows:

$$I(\lambda,L) = I_0(\lambda,L)\exp[-L\bullet C\bullet\sigma(\lambda,T,P)] \tag{1}$$

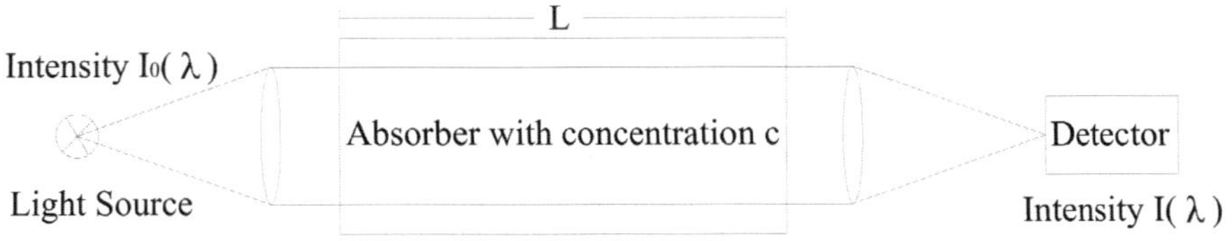

Fig. 1 Schematic of Lambert-Beer law

In the actual measurement, the pollutant gases to be measured usually coexist with other gases and dust particles. So the attenuation after the light is transmitted through the absorber includes the absorption of gas molecules and dust particles as well as the scattering of them.

2.2 Retrieval Algorithm

Though the Mie and Rayleigh scattering caused by dust particles and gases don't belong to absorption process in equation (1), diffused light derived from them can not be detected by the detector. So the Mie and Raileigh scattering can be equally viewed as the absorption process. In equation (1), the absorption cross section was separated into two parts which included the rapid part $\sigma_i^{'}$ varying rapidly with wavelength and the slow part σ_{ib} showing a general slope with wavelength. As a result, σ_i was expressed as:

$$\sigma_i = \sigma_i^{'} + \sigma_{ib} \tag{2}$$

If equation (2) is put in equation (1), then the calculation model can be written as equation (3):

$$\begin{aligned} I(\lambda,L) = I_0(\lambda,L)\exp[-(\sum_i c_i \bullet \sigma_{ib}(\lambda,T,P) + \varepsilon_R(\lambda) \\ + \varepsilon_M(\lambda)) \bullet L]\exp[-(\sum_i c_i \cdot \sigma_i^{'}(\lambda,T,P)) \bullet L] \end{aligned} \tag{3}$$

In equation (3), defining $I_0^{'}(\lambda,L) = I_0(\lambda,L)\exp[-(\sum_i c_i \bullet \sigma_{ib}(\lambda,T,P) + \varepsilon_R(\lambda) + \varepsilon_M(\lambda)) \bullet L]$, it denotes the absorption without any differential absorption and the slow features are derived by fitting a low-order polynomial(1st-6th order) (Mellqvist et al., 1996), I_p, to the recorded spectrum, then equation (3) is modified as:

$$D^{'}(\lambda) = \ln\frac{I_0^{'}(\lambda)}{I(\lambda)} = \ln\frac{I_p(\lambda)}{I(\lambda)} = \sum_i c_i \cdot \sigma_i^{'}(\lambda,T,P) \bullet L \tag{4}$$

After that, some featured points are chose to form the equations about concentrations:

$$\begin{cases} D'(\lambda_1) = \ln\dfrac{I_0^{'}(\lambda_1)}{I(\lambda_1)} = [C_1 \cdot \sigma_1^{'}(\lambda_1) + C_2 \cdot \sigma_2^{'}(\lambda_1) \cdot \quad \cdot \quad \cdot + C_n \cdot \sigma_n^{'}(\lambda_1)] \\ D'(\lambda_2) = \ln\dfrac{I_0^{'}(\lambda_2)}{I(\lambda_2)} = [C_1 \cdot \sigma_1^{'}(\lambda_2) + C_2 \cdot \sigma_2^{'}(\lambda_2) \cdot \quad \cdot \quad \cdot + C_n \cdot \sigma_n^{'}(\lambda_2)] \\ \quad \vdots \qquad\qquad\qquad\qquad \vdots \\ D'(\lambda_m) = \ln\dfrac{I_0^{'}(\lambda_m)}{I(\lambda_m)} = [C_1 \cdot \sigma_1^{'}(\lambda_m) + C_2 \cdot \sigma_2^{'}(\lambda_m) \cdot \quad \cdot \quad \cdot + C_n \cdot \sigma_n^{'}(\lambda_m)] \end{cases} \tag{5}$$

2.3 Retrieval Algorithm of DOAS Based on Kalman Filter

Here, the improved DOAS algorithm based on kalman filter was put forward. It can be applied to compensate for the deficiency of traditional DOAS at short light path length. Good results could also be achieved even under the strong background spectrum and severe overlapping of spectrum lines. Compared to the traditional DOAS algorithm, the DOAS based on kalman filter performed linear recursion i.e. the concentration was derived from the previous calculated results and current observed values at every wavelength in a continuous wavelength region. The kalman filter can be described with two equations which are state equation and measurement equation respectively introduced by Van Veen et al (Veen et al., 1990; Veen et al., 1991; Veen et al., 1993).

State equation:

$$C(k) = IC(k-1) \tag{6}$$

measurement equation:

$$x(k) = S^T(k)C(k) + n(k) \tag{7}$$

Where, the concentration is treated as the state variable denoted by $C(k)$ and it is a constant not varying with wavelength. I is the transfer matrix of state between $C(k)$ and $C(k-1)$. Measurement value $x(k)$ is a scalar denoting the differential optical density at wavelength λ_k. $S^T(k)$ is the transpose of the vector of sensitivity coefficients composed of differential absorption cross sections. $n(k)$ is the white measurement noise with a mean value of zero.

3. EXPERIMENTAL SYSTEM

A schematic of the setup is shown in Figure 2. In the system, the impulse and short arc xenon lamp was selected as the light source which can supply the continuous and broad band spectrum between 200 and 1100 nm. The resolution of the fiber spectrometer is 0.4 nm. The system is separated into two parts: the upper part in the dashed frame was used for obtaining the mixture of different component concentration and the lower part below it was the optical setup which was also the main body used for measurement.

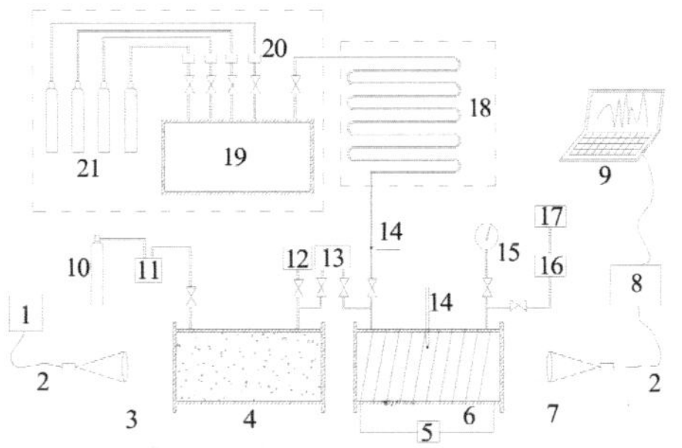

Fig. 2 Schematic of experimental system for measuring concentration of flue gas
1 xenon lamp; 2 optical fibre; 3、7 collimating lens; 4 measurement cell; 5 equipment of controlling temperature; 6 heating belt; 8 spectrometer; 9 pc; 10 high purity nitrogen; 11 aerosol generator; 12 dust collector; 13 vacuum pump; 14 K-type thermocouple; 15 pressure gauge; 16 analytical instrument for flue gas; 17 device for absorbing exhaust gas; 18 preheating pipe; 19 mixing pipe; 20 flowmeter; 21 standard gas;

Measurement of SO_2 was performed in this paper at room temperature and atmospheric pressure. The measurement cell shown in Figure 2 was cylindrical with a length of 0.5 m and a diameter of 0.06 m. Two fused quartz windows were attached to each side of the cell sealed with graphite. In order to conveniently add dust particles in the gases, a separate dust cell with the same length and diameter was put before the measurement cell in the light path. The light was transmitted through the dust cell first and then was transmitted through the measurement cell. The effect is the same as that of the coexistence of gases and dust particles in the measurement cell, which can be derived from equation (8).

$$
\begin{aligned}
I(\lambda,L) &= I_0(\lambda,L)\exp[-(\sum_i C_i \cdot \sigma_i(\lambda,T,P) + \varepsilon_R(\lambda) + \varepsilon_M(\lambda)) \cdot L]A(\lambda) \\
&= I_0(\lambda,L)\exp[-(\varepsilon_R(\lambda) + \varepsilon_M(\lambda)) \cdot L]A_1(\lambda) \cdot \exp[-\sum_i C_i \cdot \sigma_i(\lambda,T,P) \cdot L]A_2(\lambda) \qquad (8) \\
&= I_1(\lambda,L)\exp[-\sum_i C_i \cdot \sigma_i(\lambda,T,P) \cdot L]A_2(\lambda)
\end{aligned}
$$

In order to obtain the uniform mixture of different component before entering the measurement cell, standard gases and pure nitrogen gas were flowed into flowmeters according to volume ratio. Then the mixture was flowed into the pre-heating pipe with a length of 7 m and a diameter of 0.06 m. After that, the mixture was divided into two parts: one passed through the flue gas analyzer for monitoring the concentration and the other passed through the measurement cell, finally both entered the bottle for absorbing exhaust gases. The work flow of the measurement part is described as below: Firstly, the light emitted by xenon lamp entered the dust cell after it was collimated by lens 2, then the light entered the measurement cell, finally the light was focused on the optical fiber connected with spectrometer. Dust particles in the dust cell were supplied uniformly by the aerosol generator which utilized the density difference between dust particle and copper powder. They were fluidized by high pressure industrial nitrogen just like fluidized bed in a cylindrical container and were mixed according to mass ratio. The particle number concentration could be monitored by ELPI (electrical low-pressure impactor) as well as the mass concentration, shown in Figure 3 a and b respectively. The unit of l/h denotes the flux of industrial nitrogen and percentage denotes the mass ratio of dust particles to copper powders.

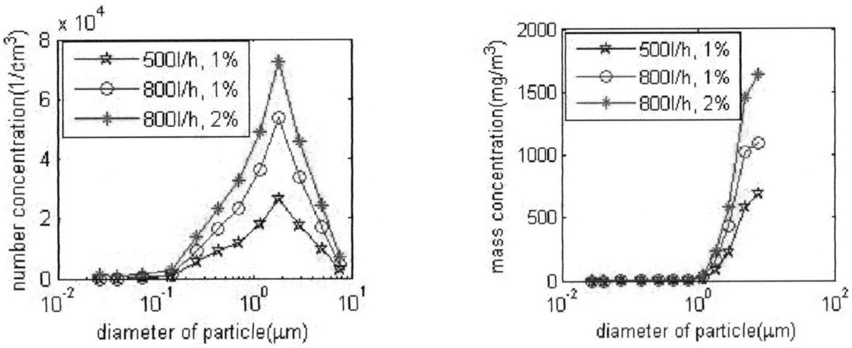

a. particle number concentration b. mass concentration

Fig. 3 Concentration distribution of particles

4. EXPERIMENTAL RESULTS AND ANALYSIS

The experiment was performed when gases were flowed through the measurement cell with different algorithms. The concentration inversion of SO_2 was performed between 290 and 310 nm. The desired gas concentration was achieved by diluting high concentration gas with pure nitrogen in the volume ratio, simultaneously, the concentration was monitored by a commercial analytical instrument for the flue gas. The traditional DOAS and the DOAS based on kalman filter were marked as $DOAS^1$ and $DOAS^2$ respectively. All the experiments were performed at the light path length of 0.5 m.

4.1 Comparison of SO_2 Concentration Measurement with Different Algorithms

Journal of Physics: Conference Series **147** (2009) 012080 doi:10.1088/1742-6596/147/1/012080

The relative errors based on different retrieval algorithms are shown in Figure 4. The solid line denotes the relative errors derived from $DOAS^1$ corresponding to the value listed in the left y-axis and the dashed line denotes the relative errors derived from $DOAS^2$ corresponding to the value listed in the right y-axis. The value at the bottom of Figure 4 denotes the concentration and the upper value is the maximum differential optical density corresponding to the concentration.

As we can see, the measurement error of the traditional DOAS algorithm varies from 0% to 228.6% for the concentration of SO_2 in the range of 20 to 12889 mg/m^3. When the concentration of SO_2 is less than 326 mg/m^3, the relative error of the traditional algorithm varies between 5.3% and 228.6%. Contrarily, on the same experimental conditions, the maximum relative error of the improved algorithm approximates to 30% and 10% respectively. Over the concentration of SO_2 from 326 to 7877 mg/m^3, the relative error is less than 8% for the traditional algorithm, and less than 19% for the improved algorithm. When the SO_2 concentration exceeds 7877 mg/m^3, the errors increase gradually for both the traditional algorithm and the improved algorithm and are more than 10%. This reason may be that the Lambert-Beer law is not suitable for the absorption measurement of high concentration gas. Therefore, it is necessary that the linearity correction to the law must be adopted when measuring the gas with relatively high concentration. It is noteworthy that a symbol $OD_m^{'}$ defined as the maximum value of the differential optical density is listed in Figure 4. Since the performances of different algorithms are greatly different when $OD_m^{'}$ changes, two values of $OD_m^{'}$, 0.0481 and 0.9272 corresponding to the SO_2 concentration of 326 mg/m^3 and 7877 mg/m^3 respectively, are selected as the threshold of $OD_m^{'}$. So it is suggested that the measurement of gas concentration should be performed in three different areas by judging $OD_m^{'}$. Consequently, If $OD_m^{'}$ is less than 0.0481, the improved algorithm is adopted and if $OD_m^{'}$ is between 0.0481 and 0.9272, the traditional algorithm is selected. When $OD_m^{'}$ is more than 0.9272, the traditional algorithm is superior to the improved algorithms. Furthermore, from the above analysis of experimental results, the dynamic measurement range of SO_2 concentration may be broaden by the combination of the traditional DOAS algorithm with the improved algorithm in DOAS system.

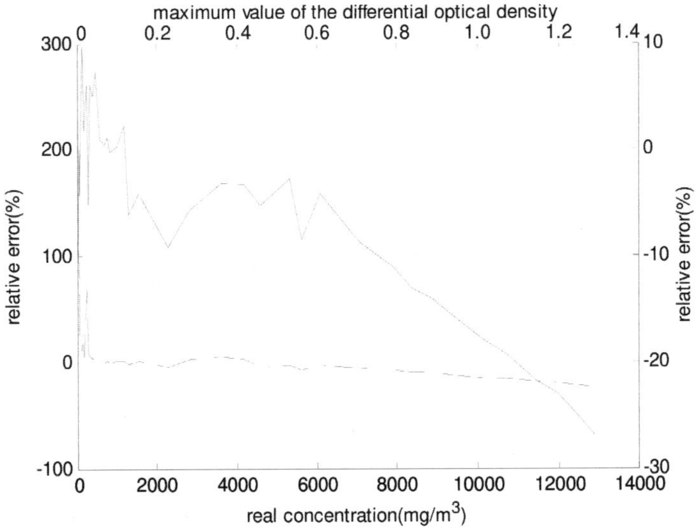

Fig. 4 Errors of measurement results based on different retrieval algorithms

4.2 Comparison of SO_2 Concentration Measurement Results with The Interference of NO

Table 1 shows the comparison of SO_2 concentration measurement results with the interference of standard 1304 mg/m^3 NO at a flux of 3l/min based on different algorithms of DOAS. Judging from the $OD_m^{'}$, the concentrations listed in Table 1 belong to the second area. It is proved once again that the calculation results from the traditional algorithm are generally superior to that from the improved algorithm in the defined second area. As can be seen, although the measurement of SO_2 is interfered by NO, the maximum relative error of the calculation results is less than 9% applying the traditional DOAS algorithm.

Table 1 Calculation results and errors based on three different algorithms

Real value (mg/m^3)	DOAS[1] (mg/m^3)	error (%)	DOAS[2] (mg/m^3)	error (%)	$OD_m^{'}$
1483	1604	8.1	1627	9.6	0.1987
1796	1842	2.5	1836	2.2	0.2218
2099	2214	5.4	2231	6.3	0.2722
2534	2637	4.1	2646	4.4	0.3185
3177	3366	5.9	3361	5.8	0.4078

3644	3724	2.2	3692	1.3	0.455
4081	4050	-0.8	3981	-2.5	0.4956
6438	5872	-8.8	5706	-11.4	0.7393

5. ZERO DRIFT CALIBRATION OF THE DOAS SYSTEM

The repetitive experiments of calibrating the zero drift for SO_2 were also carried out at the same light path length. The dust cell in the light path is full of high concentration dust particles which lead to the distribution of transmittance light intensity shown in Figure 5. It can be seen that after the light is transmitted through the dust cell and measurement cell, few light will enter the spectrometer. Table 2 lists the calibrating results of repetitive experiments. The zero drift from the traditional algorithm reaches about 57.2 mg/m^3, however, the value presented by the improved algorithm is less than 2.86 mg/m^3. So the improved algorithm can be used for the measurement of extremely low concentration gas including the calibration of zero drift.

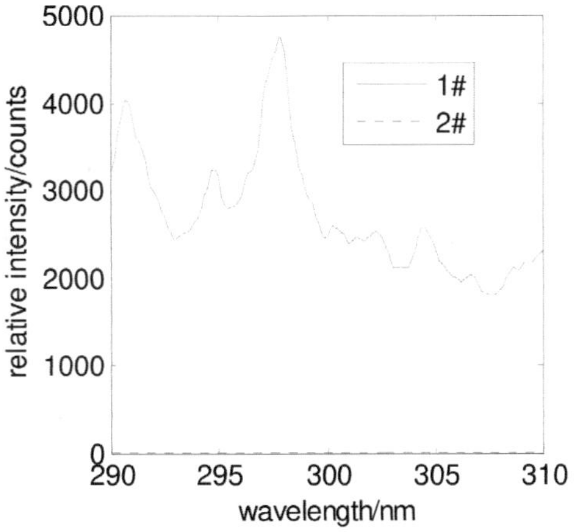

1#—original light intensity;

2#—transmitted light intensity with the existence of dust particles;

Fig. 5 Comparison of the original and transmitted light intensity

Table 2 zero drift with different algorithms for the concentration inversion of SO_2

Real value SO_2 (mg/m^3)	DOAS[1] SO_2 (mg/m^3)	DOAS[2] SO_2 (mg/m^3)
0	57.2	-0.09

0	40.0	0.09
0	60.1	1.14

6. CONCLUSIONS

The improved DOAS algorithm based on kalman filter was investigated to perform the accurate measurement of SO_2 from flue gases. Experimental results obtained on a bench-scale rig show that the improved algorithm have a detection limit less than 28.6 mg/m^3 at the light path length of 0.5 m even if there exists the interference of other background gases and dust particles. Also, the zero drift of the DOAS system based on improved algorithm is less than 2.86 mg/m^3. It is also found that the accurate measurement of SO_2 can be carried out by applying the improved DOAS algorithm according to the maximum value of differential optical density. Therefore, two values of 0.0481 and 0.9272 were put forward as the threshold to decide which DOAS algorithm is adopted. Thus the measurement of flue gases is performed in three different areas according to $OD_m^{'}$. If $OD_m^{'}$ is less than 0.0481, then the improved algorithm is adopted and if $OD_m^{'}$ is between 0.0481 and 0.9272, the traditional DOAS algorithm is adopted. When $OD_m^{'}$ exceeds 0.9272, both the traditional and improved algorithm bring significant relative errors. Fortunately, the traditional least-square retrieval algorithm enables the accurate measurement of relatively high concentration gas at a relatively short light path length. Therefore, it is advisable that the improved algorithm combined with the least-square retrieval algorithm of DOAS are used, which greatly lowers the detection limit and broaden the dynamic measurement range of flue gas concentrations by setting an appropriate threshold for the maximum value of differential optical density.

ACKNOWLEDGEMENTS

The authors would like to acknowledge the financial support of National Natural Science Foundation of China (Project No. 50576010).

NOMENCLATURE

$\sigma_i^{'}$ differential absorption cross section of gas molecule [cm^2/mole]

σ_{ib} a pseudo absorption cross section denoting the slow features of the molecular absorption [cm^2/mole]

$D^{'}(\lambda)$ the differential optical density at the wavelength of λ

$I_0^{'}(\lambda)$ the received light intensity excluding the differential absorption [counts]

$I_p(\lambda)$ a low-order polynomial denoting the slow features [counts]

m the total number of discrete wavelength;
n the species of gases to be measured.
λ wavelength [nm]
L light path length [m]
T temperature [K]
P pressure [pa]
i the index of gases to be measured

$I(\lambda, L)$ transmittance light intensity [counts]

$I_0(\lambda, L)$ original light intensity [counts]

$I_1(\lambda, L)$ light intensity received after light is transmitted through the dust cell [counts]

$A(\lambda)$ the attenuation caused by the system such as the lens and the spectrometer etc.

$A_1(\lambda)$ the attenuation after the light is transmitted through the dust cell

$A_2(\lambda)$ the attenuation caused by the measurement cell

C_i concentration of the ith gas [mg/m^3]

$\sigma_i(\lambda, T, P)$ absorption cross section depending on temperature and pressure [m^2/mole]

$\varepsilon_R(\lambda)$ coefficient of Rayleigh scattering [1/m]

$\varepsilon_M(\lambda)$ coefficient of Mie scattering [1/m]

REFERENCES

Axelsson, H. et al. (1990). "Differential optical absorption spectroscopy measurements of ozone in the 280-290 nm wavelength region," Appl. Spectrosc, 40, PP. 1654-1658.

MELLQVIST, J. et al. (1996). "DOAS for flue gas monitoring—I. Temperature effects in the U.V./Visible absorption spectra of NO, NO_2, SO_2 and NH_3," Journal of Quantitative Spectroscopy and Radiative Transfer, 56, PP. 187-208.

MELLQVIST, J. et al. (1996). "DOAS for flue gas monitoring—II. Deviations from the Beer Lambert law for the U.V./Visible absorption spectra of NO, NO_2, SO_2 and NH_3," Journal of Quantitative Spectroscopy and Radiative Transfer, 56, PP. 209-224.

MELLQVIST, J. et al. (1996). "DOAS for flue gas monitoring--III. In-situ monitoring of sulfur dioxide, nitrogen monoxide and ammonia," Journal of Quantitative Spectroscopy and Radiative Transfer, 56, PP. 225-240.

Platt, U. et al. (1979). "Simultaneous Measurement of Atmospheric CH_2O, O_3 and NO_2 by differential optical absorption," JOUNRNAL OF GEOPHYSICAL RESEARCH, 84C, PP. 6329-6334.

Platt, U. et al. (1979). "Determination of nitrous acid in the atmosphere by differential optical absorption," Geophysical Research Letters, 6, PP. 917-920.

Platt, U. et al. (1988). "New tropospheric OH measurements," Journal of Geophysical Research, 93D, PP. 5159-5166.

Platt, U. (1984). "Measurements of atmospheric trace gases by long path differential UV/visible absorption spectroscopy," Optical and Laser Remote Sensing, D. K. Killinger and A. Mooradien, eds(Springer, New York,), 39, PP. 95-105.

Plane, J. M. C et al. (1992). "Differential Optical Absorption Spectrometer for Measuring Atmospheric Trace Gases," Rev. Sci. Instrum, 63, PP. 1867-1876.

Platt, U. et al. (1980). "Direct measurements of atmospheric CH_2O, HNO_2, O_3 and SO_2 by differential absorption in the near UV," Geophys Res, 85, PP. 7453-7458.

Veen V. E. H. et al. (1990). "Kalman filtering of data from overlapping lines in inductively coupled plasma-atomic emission spectrometry," Spectrochimca Acta, 45B, PP. 313-328.

Veen, V. E. H. et al. (1990). "Some spectral interference studies using Kalman filtering in inductively coupled plasma-atomic emission spectrometry," Spectrochimca Acta, 45B, PP. 1109-1120.

Veen V. E. H et al. (1991). "Kalman Filtering for data reduction in inductively coupled plasma atomic emission spectrometry," Analytical Chemistry, 63, PP. 1441-1448.

Veen V. E. H et al. (1993). "Quantitative line selection with the Kalman filter approach for inductively coupled plasma atomic emission spectrometry," Spectrochimca Acta, 48B, PP. 1691-1701.

Journal of Physics: Conference Series **147** (2009) 012081 doi:10.1088/1742-6596/147/1/012081

Nonlinear CARS measurement of nitrogen vibrational and rotational temperatures behind hypervelocity strong shock wave

Takashi Osada[1], Youichi Endo[1], Chikara Kanazawa[2], Masanori Ota[3] and Kazuo Maeno[3]

Graduate Student[1], Undergraduate[2], Graduate School of Engineering[3], Chiba University
1-33 Yayoi, Inage, Chiba, #263-8522, JAPAN
maeno@faculty.chiba-u.jp

Abstract. The hypervelocity strong shock waves are generated, when the space vehicles reenter the atmosphere from space. Behind the shock wave radiative and non-equilibrium flow is generated in front of the surface of the space vehicle. Many studies have been reported to investigate the phenomena for the aerospace exploit and reentry. The research information and data on the high temperature flows have been available to the rational heatproof design of the space vehicles. Recent development of measurement techniques with laser systems and photo-electronics now enables us to investigate the hypervelocity phenomena with greatly advanced accuracy. In this research strong shock waves are generated in low-density gas to simulate the reentry range gas flow with a free-piston double-diaphragm shock tube, and CARS (Coherent Anti-stokes Raman Spectroscopy) measurement method is applied to the hypervelocity flows behind the shock waves, where spectral signals of high space/time resolution are acquired. The CARS system consists of YAG and dye lasers, a spectroscope, and a CCD camera system. We obtain the CARS signal spectrum data by this special time-resolving experiment, and the vibrational and rotational temperatures of N_2 are determined by fitting between the experimental spectroscopic profile data and theoretically estimated spectroscopic data.

1. INTRODUCTION

With the development in aerospace science and technology, the opportunities of space explanation and reentry to the atmosphere have been increasing. When the spacecrafts reenter the atmosphere, the hypervelocity strong shock waves are generated in front of the vehicles. Behind the strong shock waves, the high temperature and non-equilibrium radiating flow are generated. This high temperature and non-equilibrium flow field is a very important subject to design the heat shields of the spacecrafts.

In previous study we simulated this hypervelocity strong shock wave with a free-piston double diaphragm shock tube, and applied CARS system to the direct measurement of the vibrational/rotational temperatures of nitrogen molecule behind the radiating shock wave. This flow field can be regarded of a multiphase flow of vibrational and rotational exited gas molecules and ionaized plasma. In this study we obtain more CARS spectra data and measure the vibrational and rotational temperature of N_2 molecule in high temperature and radiating multiphase flow of plasma and air.

2. THEORETICAL CALCULATION OF CARS

The theory and experiment of CARS measurement have been developed extensively (Alan C.

Journal of Physics: Conference Series **147** (2009) 012081 doi:10.1088/1742-6596/147/1/012081

Eckbreth.,1984; R. J. Hall.,1979; P. R. R´egnier and P. E Taran., 1973), though they are restricted relatively slow gasdynamic phenomena from ms-orders as combustion experiments. According to the theory on CARS, incident YAG laser beams at frequencies ω_1 (pump beam) and ω_2 dye laser (Stokes beam) interact through the third order nonlinear electric susceptibility $\chi^{(3)}_{CARS}$ to produce a coherent radiation $I_3(\omega_3)$ at $\omega_3 = 2\omega_1 - \omega_2$ of the intensity written, as follows

$$I_3(\omega_3) = \frac{\omega_3^2}{n_1^2 n_2 n_3 c^4 \varepsilon_0^2} \left|\chi^{(3)}_{CARS}\right|^2 I_1^2(\omega_1) I_2(\omega_2) l^2 \left(\frac{\sin \frac{\Delta kl}{2}}{\frac{\Delta kl}{2}} \right)^2 \qquad (1)$$

$$\Delta k = 2k_1 - k_2 - k_3 \qquad (2)$$

where n_1 , n_2 and n_3 are the refractive indexes at ω_1, ω_2, ω_3, respectively; c is the velocity of light; l is interaction length; k_1, k_2 and k_3 are the wave vectors of the pump, Stokes and CARS beams, respectively; ε_0 is the permittivity of free space. The CARS signal is enhanced on the condition of phase matching, $\Delta k = 0$. The incident laser beams, i.e. two pump beams (k_1) and a Stokes beam (k_2), are aligned in order to satisfy vector relation, where $\chi^{(3)}_{CARS}$ is shown by

$$\chi^{(3)}_{CARS} = \sum_j K_j \frac{\Gamma_j}{2\Delta\omega_j - i\Gamma_j} + \chi_{nr} \qquad (3)$$

Here the j summation is over vibration-rotation transitions in the vicinity of $\omega_1 - \omega_2$; Γ_j is the Raman line width (FWHM); the non-resonant term χ_{nr} is a background contribution due to electrons and other remote resonances; and $\boldsymbol{K}_j$ is shown to be related to the Raman cross-section

$$K_j = \frac{2n_1 c^4}{n_2 \hbar \omega_2^4} N \Delta_j \left. \frac{d\sigma}{d\Omega} \right|_j \Gamma_j^{-1} \qquad (4)$$

where N is the number density of the Raman active molecule; Δ_j is the population difference between the upper and lower vibration-rotation states; and $(d\sigma/d\Omega)|_j$ is the cross-section for spontaneous Raman scattering.

On the assumption that the molecules have Boltzmann distributions based on the rotational (T_r) and vibrational (T_v) temperatures, Δ_j can be expressed as

$$\Delta_j = \frac{(2J+1)g_I}{Q_r Q_v} \left[\exp(\frac{-F_{v,J}hc}{kT_r}) \exp(\frac{-G_v hc}{kT_v}) - \exp(\frac{-F_{v+1,J}hc}{kT_r}) \exp(\frac{-G_{v+1}hc}{kT_v}) \right] \qquad (5)$$

where $F_{v,j}$ and G_v are the rotational and vibrational energy terms, respectively; g_I is the spin degeneracy, 6 for even-J rotational levels and 3 for odd-J levels in N_2; and Q_r and Q_v are the rotational and vibrational partition functions, respectively.

From equations (1), (3) and (4), CARS signal is approximately proportional to the square of number density of molecule. Therefore, it is very difficult to detect the CARS signal from low-pressure and strongly radiating gas of hypervelocity.

As written in the following section, we have been successful to detect the CARS signals fron hypervelocity air flow The measured CARS spectra are fitted with calculated CARS signals by treating T_v, T_r, G as free parameters to decide the temperatures.

Journal of Physics: Conference Series **147** (2009) 012081 doi:10.1088/1742-6596/147/1/012081

3. EXPERIMENTAL APPRATUS AND DIAGNOSTIC SYSTEM

A free-piston, double-diaphragm shock tube has been used to generate strong shock waves in low-density gas as shown in Fig.1. This shock tube consists of high-pressure chamber (driver gas is nitrogen), compression tube (helium gas is supplied) and a free piston, buffer tube (Supplied gas is also helium), low-pressure tube (test gas is air or nitrogen), and vacuum chamber. These tubes are divided off by a quick action valve, the first diaphragm (steel), and the second diaphragm (aluminum). The cross section of the low-pressure tube is 40 mm×40 mm square. The observation windows of the test section are mounted near a focal lens with some distance from the sidewall of the shock tube. This distance prevents high power laser from destroying the window. The sidewall of shock tube has two small halls along the optical path of laser beam. The shock velocity is measured by ion probes mounted on the sidewall of the test section.

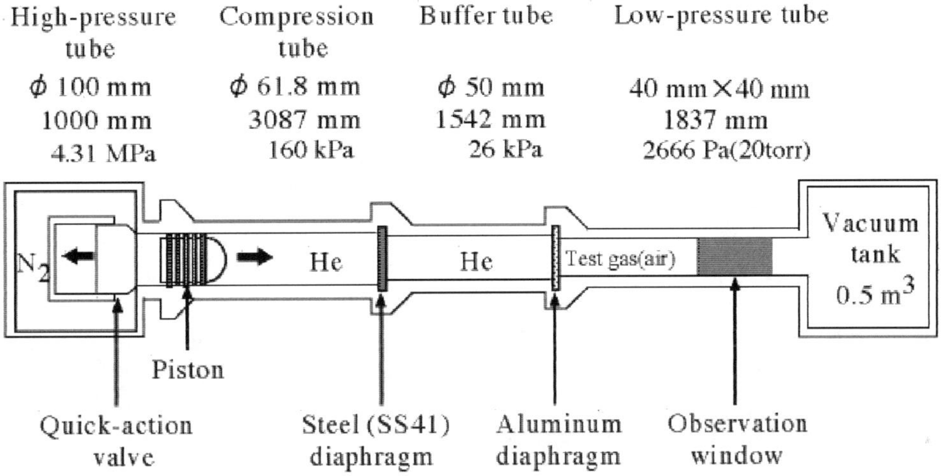

Fig. 1 Free-piston, double-diaphragm shock tube

Figure2 shows a layout of diagnostic system for the CARS measurement in shock waves (CARS system). CARS system consists of a second harmonics Nd: YAG laser (ω_1), a dye laser (ω_2), optical systems, a spectrograph, and an ICCD camera. The laser beam (ω_1) is divided in two by a beam splitter (BS). These beams are directed to the dye laser beam (ω_2) by a beam combiner (BC2). Then three laser beams are crossed and focused with the desirable angles in the shock tube observation section. The CARS spectra are detected by the spectroscopy ICCD camera. The entrance slit width is set to $100\,\mu m$ throughout our observation. The ICCD is mounted on the focal exit of the spectrograph.

Journal of Physics: Conference Series **147** (2009) 012081 doi:10.1088/1742-6596/147/1/012081

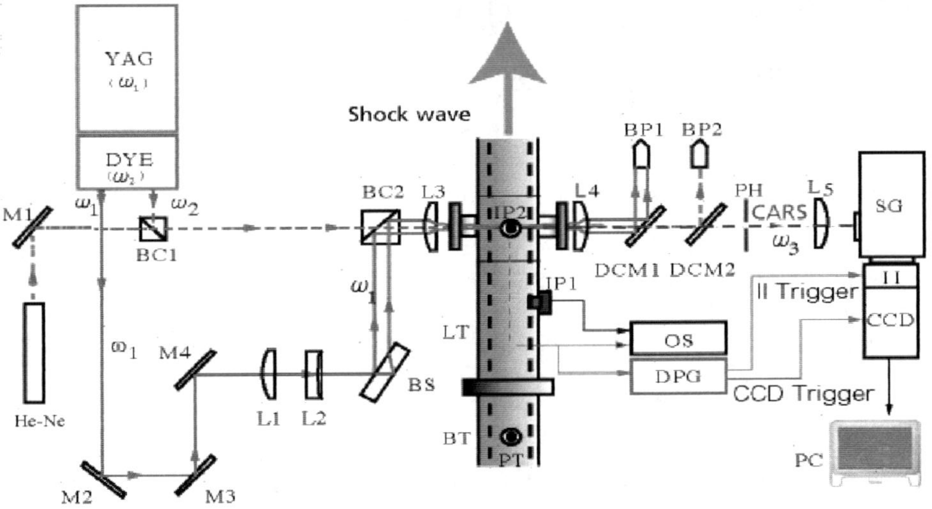

Fig.2 Layout of CARS measurement system

Code: HN : He-Ne laser, M : mirror, BC : beam combiner, L : lens, BS : beam splitter, DCM : diachronic mirrors, BP : beam pocket, PH : pinhole, IP : Ion Probe, PT : pressure transducer, DPG : delay pulse generator SG : spectrograph, OS : oscilloscope, LT : low-pressure tube, BT : buffer tube ST : shock tube.

4. RESULTS AND DISCUSSION

In this experimental paper we indicate the typical four CARS spectrum data of the multiphase flow behind strongly shocked air. The experimental and theoretical spectrum data are shown in Fig.3~5. The velocity, Mach number and measuring position of about 12mm behind shock front of each shock wave are listed on the Table1 below.

Table1 Details of typical measured shock waves

Fig	Velocity of shock wave [km/s]	Mach number	Measuring position behind shock front [mm]	Fitting error
3	4.27	12.56	12.10	5.07×10^{-3}
4	4.27	12.56	12.13	4.69×10^{-3}
5	4.19	12.32	11.90	1.18×10^{-2}
6	4.14	12.18	11.76	1.76×10^{-2}

The spectrum data in Figs.3 and 4 were measured under almost similar conditions. The calculated vibrational and rotational temperatures in Fig.3 are estimated 7000K, on the other hand, in Fig.4, the vibrational and rotational temperatures can be estimated 6000K. These results indicate that the N_2 molecules behind shock front are in strong radiation equilibrium but thermally non-equilibrium, and the vibrational and rotational temperatures behind shock front of 4.27km/s are expected between 6000K and 7000K.

Journal of Physics: Conference Series **147** (2009) 012081 doi:10.1088/1742-6596/147/1/012081

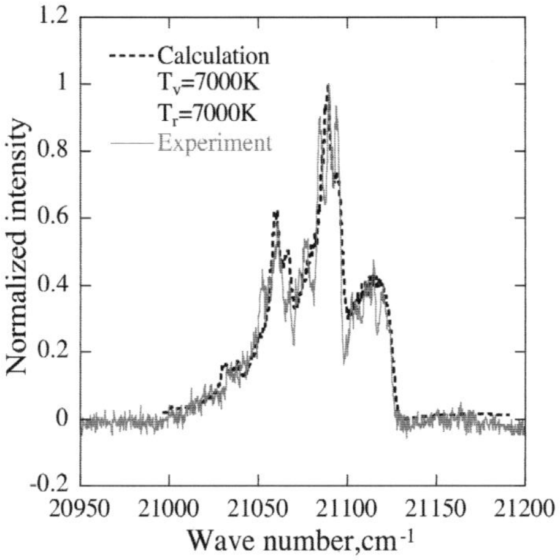

Fig.3 CARS spectra (12.10mm behind)

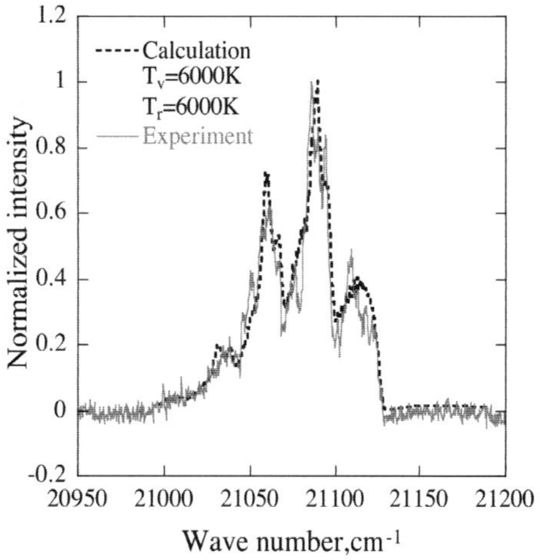

Fig.4 CARS spectra(12.13mm behind)

On the other hand, the spectrum data in Fig.5 and Fig.6 are different spectrum shapes from Fig.3 or Fig.4. The vibrational and rotational temperatures of Fig.5 are calculated 5500K and 7000K, respectively, while in Fig.6 they are calculated 3000K and 3000K, respectively. The variety of these values are typically detected at the 12mm behind the shock front. It seems that the results in fig.6 are influenced the cold driver helium gas may be mixed in the buffer tube. Through these results, activated high temperature non-equilibrium gas and low temperature gas which influenced from cold driver gas are mixed in this measuring position. In order to clarify these results around the 12mm behind shock front, we need to obtain the total image of shock waves simultaneously at the velocity about 4km/s.

The fitting error of Fig.4 is the smallest in the four results. Figure4 seems to be, therefore, more reliable than other data. There is a room for recalculate the spectral data to reduce the fitting error in Fig.5 and Fig.6.

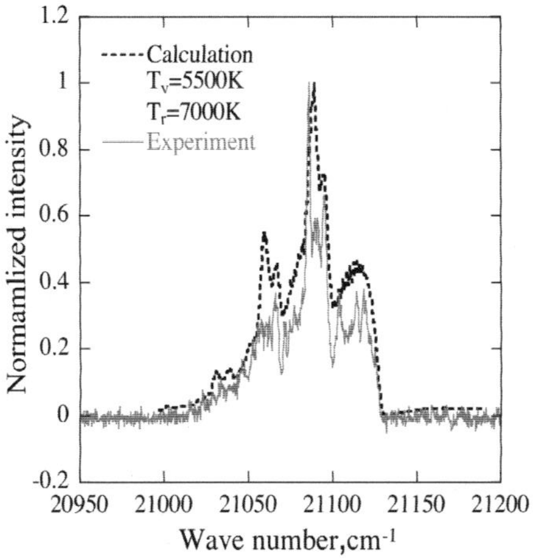

Fig.5 CARS spectra (11.90mm behind)

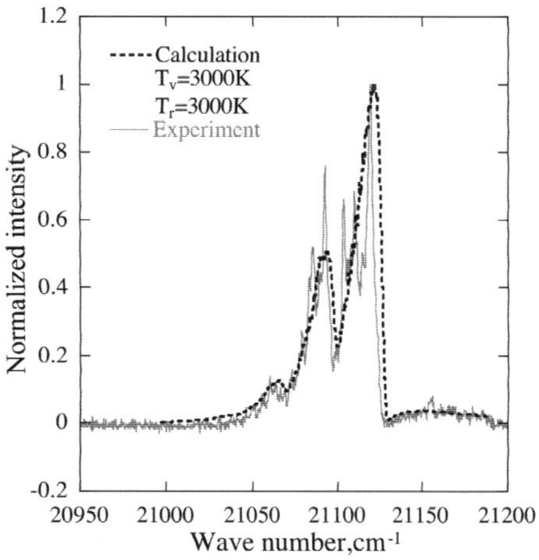

Fig.6 CARS spectra (11.76mm behind

5.CONCLUSION

In this research we have obtained the four cases of spectral data and the vibrational and rotational temperatures behind the almost fixed conditions of strong shock wave with CARS measurement system. These spectral results have shown some difference in spite of almost similar experimental conditions. From these results, the multiphase flow of cold driver gas, radiating plasma, and vibrationaly and rotationally excited gases exists around 12mm behind shock wave of the Mach number 12. We can determine the temperatures from direct CARS measurement, though, the fitting figures include a little gaps between experimental spectrum data and theoretical spectrum data. So, we have to try more precise fitting to clarify the reason why the calculated temperatures are steady and different from translational temperature.

More precise and larger numbers of experimental data shock waves under the variety of conditions near to the reentry surroundings need to be obtained.

To support these activities, it's necessary to take some image photographs of the shock wave and flow.

REFERENCES

Alan C. Eckbreth, Gregory M. Dobbs, John H. Stufflebeam, and Peter A. Tellex. "CARS temperature and species measurements in augmented jet engine exhausts". Applied Optics, 23(9):1328–1339, 1984.

P. Magre, G. Collin, O. Pin, J. M. Badie, G. Olalde and Cl´ement. "Temperature measurements by CARS and intrusive probe in an air-hydrogen supersonic combustion". International Journal of Heat and Mass Transfer, 44:4095–4105, 2001.

P. R. R´egnier and P. E Taran. "On the possibility of measureing gas concentrations by stimulated anti-Stokes scattering". Applied Physics Letters, 23(5):240–242, 1973.

R. J. Hall. "CARS spectra of combustion gases. Combustion and Flame", 35:47–60, 1979.

Journal of Physics: Conference Series **147** (2009) 012082 doi:10.1088/1742-6596/147/1/012082

Experimental study on boiling flow of liquid nitrogen in inclined tube—liquid slug length distribution and velocity

S. H. Wang, H. Zhang, J. Wang

Institute of Engineering Thermo-physics, Shanghai Jiaotong University
800 Dongchuan Road, Shanghai, 200240, China

wangj@sjtu.edu.cn

Abstract. An experimental study was carried out to understand the phenomena of the boiling flow of liquid nitrogen in an inclined tube with closed bottom by using a high speed motion analyzer. The experimental tube is 0.018 m ID and 1.0 m in length. The range of the inclination angle is 45-90° from the horizontal. The liquid slug length and velocity were studied. The mean liquid slug lengths increased first, and then decreased with decreasing θ, maximum at 60°, which showed the Taylor bubble was easier to coalescence from vertical to inclined, but coalescence lessening at 45°. The standard deviations of liquid slug lengths increase with increasing x/D at all inclination angles. The liquid slug propagation velocity increases first, and then decreases with decreasing θ, maximum at θ=60°, and is almost stable at 50D and 55D positions. These conclusions provide a basis for further study of the cryogenic two-phase slug flow.

1. INTRODUCTION

Gas–liquid slug flow is highly complex with an inherent unsteady behavior. It is characterized by long bullet-shaped bubbles separated by liquid slugs that may be aerated by small dispersed bubbles. Slug flow is found in many industrial applications one of which is transport and handling of cryogenic fluids.

With the development of current aerospace technology, cryogenic propellants are increasingly used in the missile industry. In cryogenic engineering, superheating always exists in conveyor and storage system of cryogenic liquid. So cryogenic two-phase flow is unavoidable . The propagation and storage of cryogenic liquids make many problems, such as stratification, geysering and rollover (Hands, 1988). Phenomena allied to geysering can cause high transient pressures and vapor flow rates, in some cases large enough to damage equipment. These bring new challenges on the application of the multi-phase flow theory in cryogenic engineering. Many researches were carried out to understand the liquid slug lengths which used normal atmospheric temperature liquid such as air-water and air- kerosene, and were carried out mainly for horizontal or slightly inclined slug flow and for vertical flow in developed slug flow (Nydal et al.1992; Brill et al.1981; Nicholson et al.1978; Andreussi et al. 1993; Cook et al. 2000; Griffith et al. 1961; Bernicot et al. 1989; van Hout et al. 1992; Costigan et al. 1997; van Hout et al. 2001; Mao et al.1989; Felizola et al. 1995; van Hout et al.2003). The liquid slug lengths distribution can be described by positively skewed distributions, such as the log-normal, the gamma, or the inverse Gaussian [Nydal et al.1992; Brill et al.1981; van Hout et al. 2001; van Hout et al.2003).

On the liquid slug lengths, the mean and maximum lengths were focused on study. For the horizontal tubes, the range of the mean liquid slug lengths varies from 10 to 100*D* (Nydal et al.1992; Brill et al.1981; Nicholson et al.1978; Andreussi et al. 1993). For the vertical tubes, the normalized mean liquid slug lengths vary between 10 and 20*D* with standard deviations between 30% and 50% (Griffith et al. 1961; Bernicot et al. 1989; van Hout et al. 1992; Costigan et al. 1997; van Hout et al. 2001). For the inclined tubes, the mean liquid slug length has a minimum of about 16*D* at 60° for air-kerosene slug flow (Felizola et al. 1995). R.van Hout, et al (2003) studied the evolution of the mean liquid slug lengths along the tubes at various inclination angles for air-water slug flow. The result shows the mean liquid slug lengths decrease with decreasing inclination angles.

Cryogenic vapor-liquid slug flow is seldom studied in inclined tube with closed bottom. Compared with normal atmospheric temperature liquid, cryogenic liquid has high compressibility, low density difference between vapor and liquid and low latent heat of vaporization. There are large differences on bubble motion in cryogenic two-phase flow and normal temperature two–phase flow.

The purpose of the present study is to investigate experimentally the distributions of the liquid slug length and the liquid slug velocity in inclined tube with closed bottom. The liquid nitrogen is used as working medium.

2. EXPERIMENTAL APPARATUS AND PROCESS

2.1 Experimental Set–up

Fig. 1 shows the schematic diagram of the experimental apparatus. The main body of the experiment, which is made of double layer Pyrex glass, includes a 0.4 m long stock tank with inner diameter 0.1 m, a 1.0 m long test sections with inner diameters, *D*, 0.018m. The main body of the experiment can be rotated around its axis and fixed at 40-90° inclination angles from the horizontal. The vacuum interlayer is 0.021 m, which is vacuumized by vacuum pump to serve as the thermal insulation to decrease the convection heat transfer. The degree of vacuum in vacuum interlayer is 6×10^{-2} Pa. The test part as a whole, only the upper end pipeline connected, so the heat leakage caused by the heat conduction is very small. The vacuum interlayer is vacuumized, so convection heat transfer is also very small. The main heat leakage on the pipeline is the radiation heat transfer.

The liquid nitrogen stored in a Dewar, which is heated by the electric heating rod, is supplied to the test section with the help of the high-pressure nitrogen gas in Dewar. The heating is controlled by a power Switch. The power switch is off when the liquid level of the upper tank is about 1.18 m, and is on when the liquid level of the upper tank is about 1.16 m.

2.2 High Speed Motion Analyzer

Fig. 2 is the schematic design of image processing system. The high speed motion analyzer (REDLAKE Motion-Pro®X3, 1280 ×1024 pixels resolution, 1000 frames/s with the full resolution) is employed in the experiment, together with a lens (AI NIKKOR 50/F1.2S).In the experiment, 512×512 pixels resolution is used with 1000 frames/s. The recorded images are transmitted to the computer for further analysis. Two photoflood lamps are used as light source, whose power is 1000W.

2.3 Experimental Condition

During the experiment, the range of inclination angles is 40–90°. The positions of 20*D*–55*D* from the bottom of tube are measured by using high speed motion analyzer.

2.4 Image Processing

The local propagation slug velocity of the bubble interface is calculated as the shift of the corresponding interface, divided by the time elapsed between the frames:

$$U_{Sinst} = \frac{x_2 - x_1}{n\Delta t} \quad (1)$$

Where, x_1 is the bubble's tail position of Frame n_1, x_2 is the bubble's tail position of Frame n_2, see Fig. 3, $n = n_2 - n_1$, $\Delta t = \frac{1}{f}$ (f is the Frame rate).

The liquid slug length is determined by multiplying the residence time of the liquid slug over the measured position by the liquid slug velocity.

3. RESULTS

3.1 Liquid Slug Video Images Along The Tube at Various Inclination Angles

Fig.4 presents the evolution of the liquid slug along the tube at various inclination angles. The left of the video images (Fig.4) are the bottom of the leading nitrogen Taylor bubbles, and the right are the nose of the trailing nitrogen Taylor bubbles. Fig.4 indicates that small bubbles of liquid slug regions are more and more near the tube upper with decreasing angles.

3.2 Liquid Slug Velocity

The statistics data of liquid slug propagation velocity are listed in Tab. 1. Tab. 1 shows that liquid slug propagation velocity $< U_{NLS} >$(m/s)increases first, and then decreases with decreasing θ, maximum at θ=60°, and is almost stable at 50D and 55D positions. It should be noted that the standard deviation of U_{NLS}(m/s) can approach about 0.10. This indicates small variability between individual slugs.

3.3 Liquid Slug Length Distributions

The histograms showing the distribution of liquid slug lengths are given in Fig.5. In general, the mean and the mode of the length distributions increase along the tube and the liquid slug length distributions are left-skewed in all cases.

The effect of the inclination angle on the measured liquid slug length distributions at different locations along the tube is shown in Fig. 5. The mean and the most mode liquid slug lengths increased first, and then decreased with decreasing θ, maximum at θ=60°, and the distributions become more inhomogeneous with decreasing θ.

3.4 Mean Liquid Slug Lengths and Standard Deviation

Fig. 6 present the evolution of the dimensionless mean liquid slug lengths as a function of inclination angle. In all cases, the values of L_{Smean} are 3.5-5.5D at x/D=20 and increase to 7.5–9.5D at x/D=50, x the axial pipe distance measured from the bottom of the tube.The evolution of the slug length along the tube is hardly affected by the inclination angle, while the mean liquid slug lengths increased first, and then decreased with decreasing θ, maximum at θ=60°, which shows Taylor bubble is easier to coalescence from vertical to inclined, but coalescence lessening at 45°. Similarly to L_{Smean}, the values of L_{Bmean} depend on the tube inclination as well, with the longest bubbles observed at 60°.

Fig. 7 presents the evolution of the dimensionless liquid slug lengths standard deviations as a function of inclination angle. Liquid slug lengths standard deviation increases with increasing x/D in all cases which shows liquid slug length distribution is inhomogeneous with increasing x/D.

3.5 The Log-normal Distributions of The Liquid Slug Length

Fig.5 shows the liquid slug length distributions are left-skewed in all cases. The log-normal shape is fitted to the measured distributions and is depicted in Fig.5 as a solid line. The probability density function of the log-normal distribution is

$$f(y)=\frac{1}{\sqrt{2\pi}\lambda}\left(\frac{L_S}{D}\right)^{-1}\exp\left[-\frac{1}{2}\left(\frac{\ln\left(\frac{L_S}{D}\right)-\xi}{\lambda}\right)^2\right] \quad (2)$$

where L_S is the values of the liquid slug lengths along the tube at various inclination angles, $y>0, \lambda>0$ .The parameters λ and ξ in Eq. (2) are given in Tab. 2 for all cases.

4. SUMMARY AND CONCLUSIONS

An experimental study of the evolution of continuous liquid nitrogen boiling slug flow along inclined tube with internal diameters 0.018 m is presented. The hydrodynamic and statistical parameters include liquid slug length distributions and instantaneous velocity of liquid slug. The measurements were carried out by high speed motion analyzer at different positions along the tube.

Through images analyzing, the small bubbles of liquid slug regions were more and more near the tube upper with decreasing angles.

Liquid slug propagation velocity increases first, and then decreases with decreasing θ, maximum at θ=60°, and is almost stable at 50*D* and 55*D* positions. The standard deviation of U_{NLS}(m/s) can approach about 0.10. This indicates small variability between individual slugs.

For all cases, measured length distributions were well described by the log–normal shape. The mean and the most mode liquid slug lengths increased first, and then decreased with decreasing θ, and the distributions became more inhomogeneous with decreasing θ, maximum at θ=60°.

In all cases, the values of L_{Smean} were 3.5–5.5*D* at *x/D*=20 and increased to 7.5–9.5*D* at *x/D*=50.The evolution of the slug length along the tube was hardly affected by the inclination angle, while the mean liquid slug lengths increased first, and then decreased with decreasing θ, maximum at θ=60°, which showed Taylor bubble was easier to coalescence from vertical to inclined, but coalescence lessening at 45°. Liquid slug lengths standard deviation increased with increasing *x/D* in all cases which showed liquid slug length distribution was more inhomogeneous with increasing *x/D*.

Tab. 1 Liquid slug velocity and standard deviation along the tube at various angles

x/D	parameters	90°	80°	70°	60°	45°	40°
20	$\langle U_{NLS}\rangle$	0.30	0.33	0.34	0.44	0.35	0.33
	S.D.	0.10	0.11	0.09	0.12	0.13	0.11
30	$\langle U_{NLS}\rangle$	0.33	0.36	0.40	0.46	0.36	0.35
	S.D.	0.09	0.14	0.11	0.13	0.10	0.13
40	$\langle U_{NLS}\rangle$	0.34	0.36	0.39	0.43	0.39	0.39
	S.D.	0.10	0.12	0.11	0.12	0.11	0.10
50	$\langle U_{NLS}\rangle$	0.38	0.37	0.38	0.39	0.34	0.30
	S.D.	0.13	0.09	0.11	0.10	0.12	0.08
55	$\langle U_{NLS}\rangle$	0.39	0.38	0.38	0.39	0.35	0.31
	S.D.	0.09	0.10	0.09	0.09	0.11	0.12

Tab. 2 Parameters a and b of log–normal fit

angles	parameters	20*D*	30*D*	40*D*	50*D*
90^0	λ	0.56	0.62	0.57	0.67
	ξ	1.34	1.38	1.66	1.77
80^0	λ	0.48	0.8	0.29	0.5
	ξ	1.42	1.64	1.58	1.76
70^0	λ	1.11	0.5	0.35	0.36
	ξ	1.51	1.55	1.57	2.06
60^0	λ	0.43	0.52	0.4	0.23
	ξ	1.77	1.74	1.9	1.76
45^0	λ	1.07	0.27	0.43	0.36
	ξ	1.29	1.51	1.79	1.76

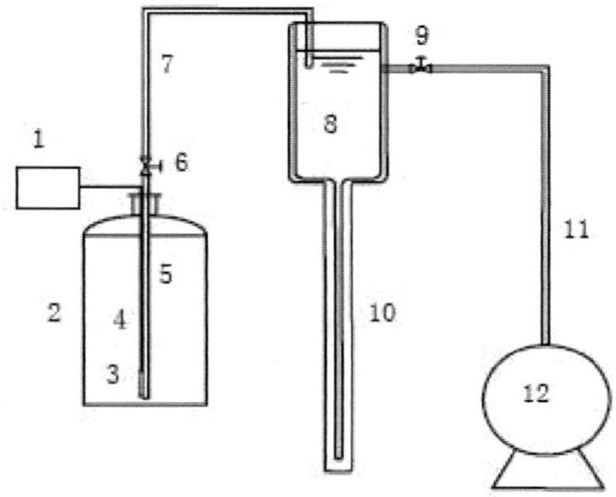

1. Power supply 2. Nitrogen dewar 3. Electric heating rod 4. Power cord 5. Liquid nitrogen delivery tube 6. Ball valve 7. Flexible Tube 8. Upper tank 9. Vacuum valve 10. Test section 11. Vacuum tube 12. Vacuum pump

Fig. 1 Experimental apparatus

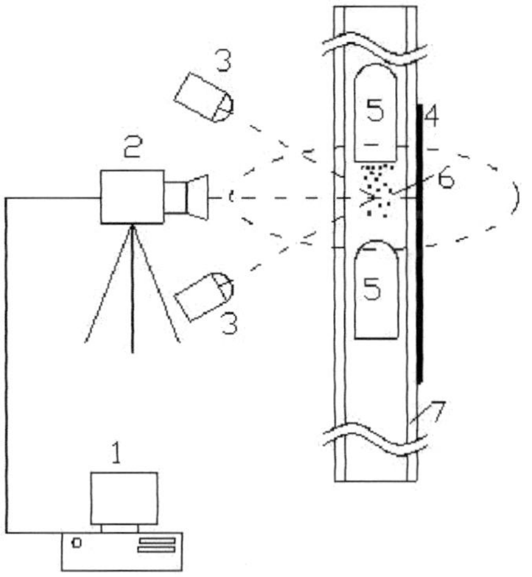

1. Computer 2. High speed motion analyzer 3. Light 4. Screen 5. Taylor bubble 6. Liquid slug 7. Vacuum interlayer

Fig. 2 The schematic design of Image processing system

Journal of Physics: Conference Series **147** (2009) 012082 doi:10.1088/1742-6596/147/1/012082

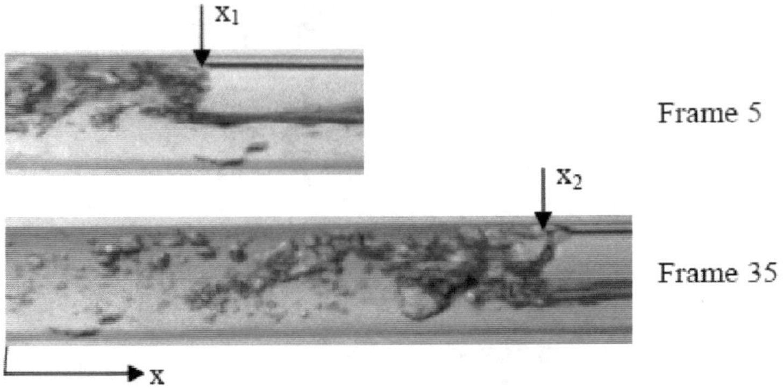

Fig. 3 Determination of the liquid slug length by image processing. Example for θ=45°, frame numbers 5 and 35. Frame rate =1000fps

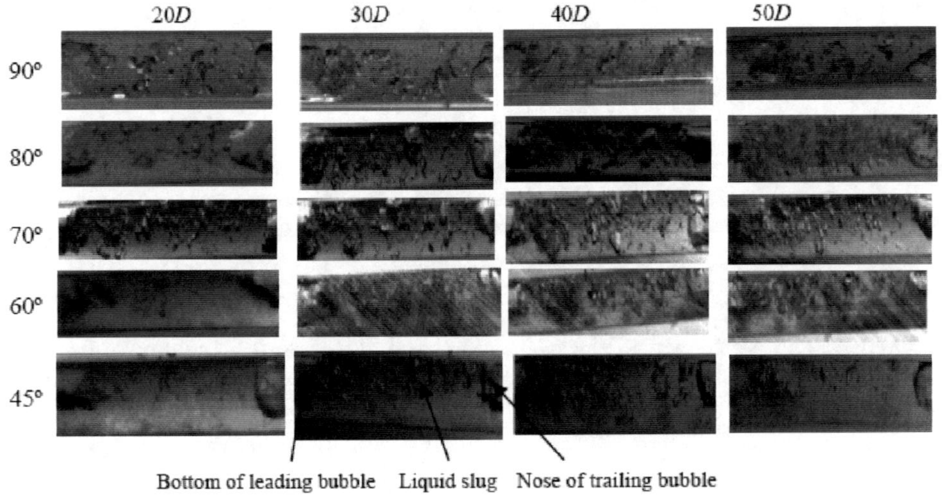

Fig. 4 Liquid slug images along the tube at various inclination angles

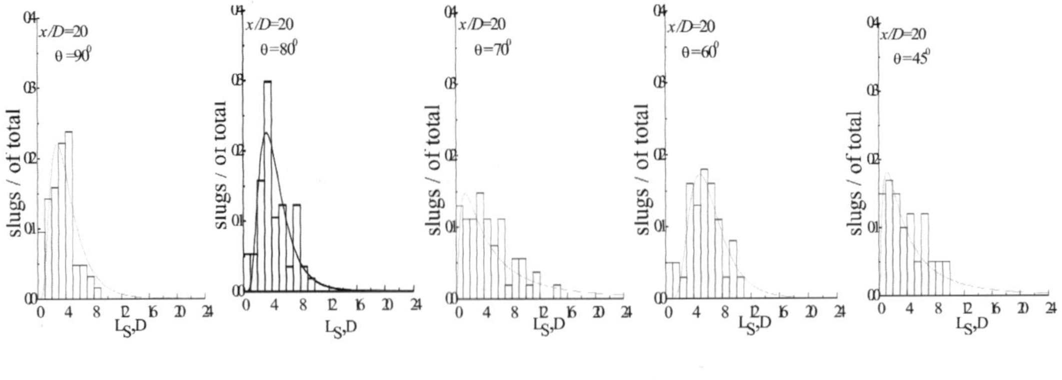

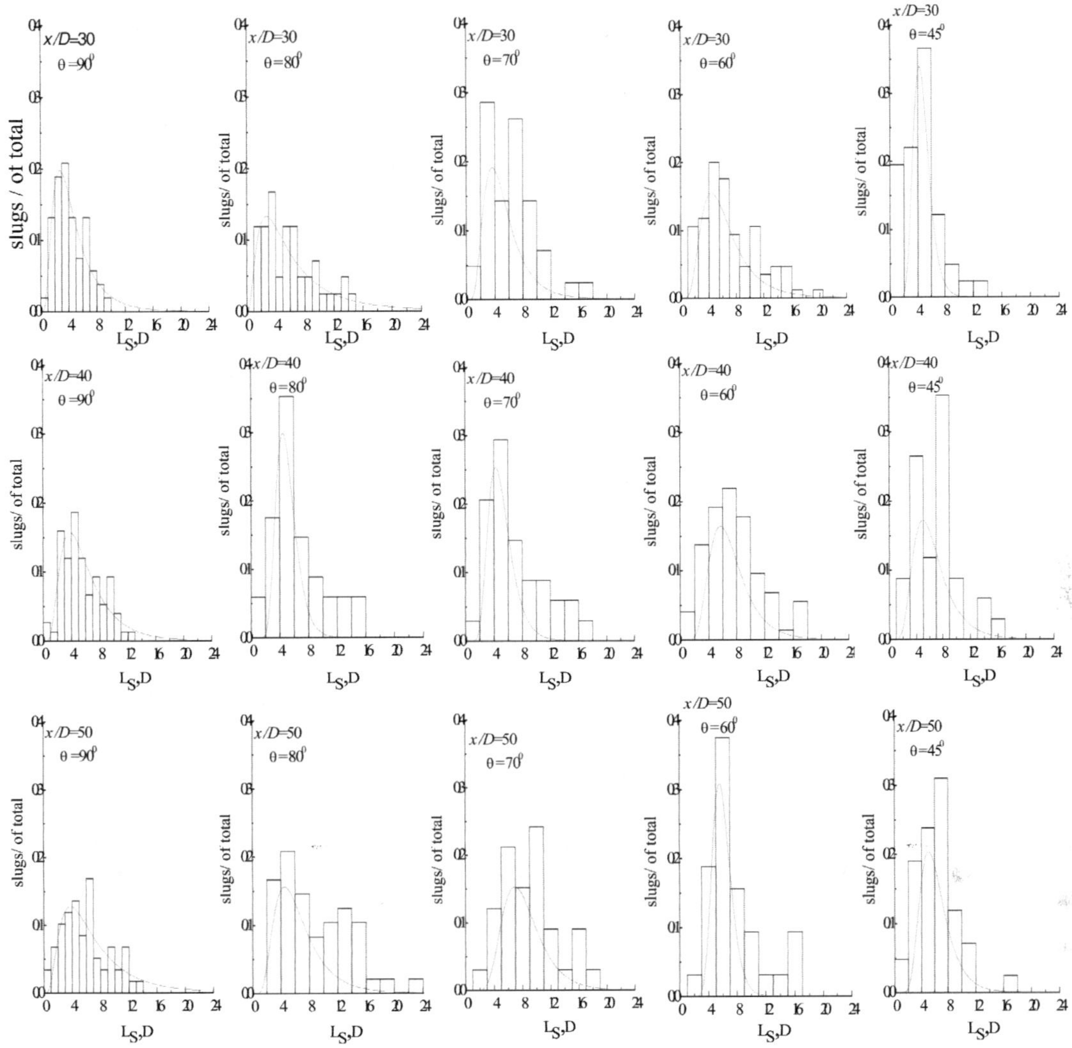

Fig. 5 Liquid slug length distribution along the tube at various inclination angles

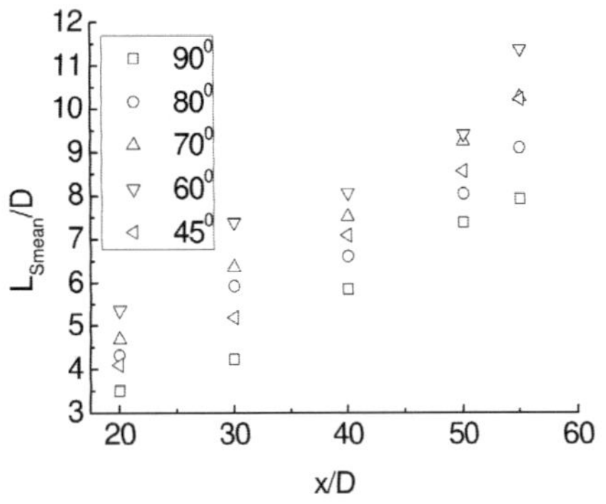

Fig. 6 Liquid slug mean length along the tube at various inclination angles

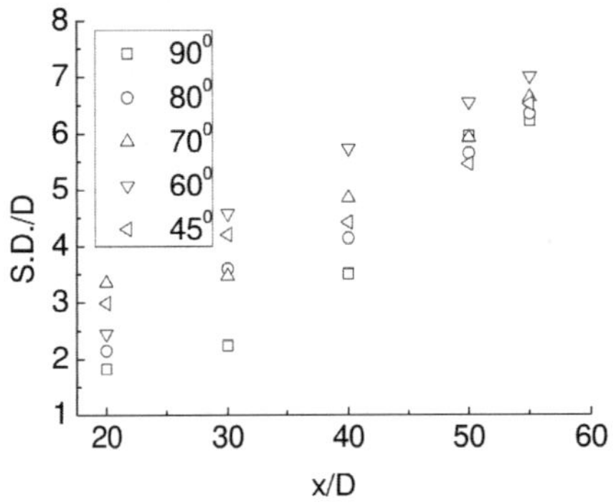

Fig. 7 Liquid slug lengths standard deviation along the tube at various inclination angles

ACKNOWLEDGEMENTS

This research is supported by National Natural Science Foundation of China (50476015) and National high-tech research development plan (863) Project of China (2006AA09Z333).

REFERENCES

Andreussi, P. et al. (1993). "Void distribution in slug flow," Int. J. Multiphase Flow, 19(5), pp.817–828.

Bernicot, M. et al. (1989). " Slug length distribution in two-phase transportation systems," '89 Proceedings of the fourth international conference on multiphase flow, Nice, France, BHRA,Cranfield, Beds ,pp.485–493.

Brill, J. P. et al.(1981). "Analysis of two–phase tests in large–diameter flow lines in prudhoe bay field," Society of Petroleum Engineers Journal, 271(6),pp.363–378.

Cook, M. et al. (2000). "Slug length prediction in near horizontal gas–liquid intermittent flow," Chemical Engineering Science, 55(11),pp.2009–2018.

Costigan, G. et al. (1997). "Slug flow regime identification from dynamic void fraction measurements in vertical air–water flows," Int. J. Multiphase Flow, 23(2), pp.263–282.

Felizola, H. et al. (1995). "A unified model for slug flow in upward inclined pipes," Journal of Energy Resources Technology, 117(1),pp.7–12.

Griffith, P. et al. (1961). "Two-phase flow," Journal of Heat Transfer, 83(3),pp. 307–320.

Hands,B.A.(1988)"Problems due to superheating of cryogenic liquids, " Cryogenics, 28(12),pp.823–829.

Nicholson, M. K. et al.(1978). " Intermittent two phase flow in horizontal pipes: Predictive models," The Canadian Journal of Chemical Engineering, 56(12),pp. 653–663.

Nydal, O. J. et al.(1992). "Statistical characterization of slug flow in horizontal pipes," Int. J. Multiphase Flow , 18(3),pp.439–453.

van Hout, R. et al. (1992). "Spatial distribution of void fraction within the liquid slug and some other related slug parameters," Int. J. Multiphase Flow, 18(6), pp.831–845.

van Hout, R. et al. (2001). "Evolution of statistical parameters of gas–liquid slug flow along vertical pipes," Int. J. Multiphase Flow, 27(9), pp.1579–1602.

van Hout, R. et al.(2003). "Evolution of hydrodynamic and statistical parameters of gas–liquid slug flow along inclined pipes," Int. J. Multiphase Flow , 29(1),pp.115–133.

Zai–Sha Mao et al.(1989). "An experimental study of gas-liquid slug flow," Experiments in Fluids,8(2),pp.169–182.

The 6th International Symposium on Measurement Techniques for Multiphase Flows IOP Publishing
Journal of Physics: Conference Series **147** (2009) 012083 doi:10.1088/1742-6596/147/1/012083

Error Transfer Analysis of Normalized Multichannel Measurement

X F Cheng[1], Z J Yang[1], T R Fu[2], L P Wang[1]

[1] University of Science and Technology of China, Hefei, Anhui, 230027, CHINA
[2] Tsinghua University, Beijing, CHINA

xfcheng@ustc.edu.cn

Abstract. In multichannel measurement, multiple signals affect the output result through the normalization process. It is difficult to analyze the error transfer, which represents the relation between the error of input and output, in this kind of measurement, especially in the normalization process. Thus, a new method of error transfer analysis is needed. Taking the primary spectrum pyrometry for example, this paper establishes a model for error transfer analysis in three-channel measurement by introducing the concept of solid angle. In the model, the error transfer is represented by the concept of error magnification, which is a dimensionless ratio of projection area of error sphere on the normalized plane to the cross section area of the error sphere. The error magnification can not be measured actually through experiment, but can be obtained from theoretical calculation. It is useful in the design of measurement system, especially the analysis of influence of different gain in different channel. The influence factors of the error magnification, including the absolute value and relation between them, are analyzed base on the model. The formula for analysis of two-channel and n-channel measurement is also derived from the three-channel model. Finally, some principles of operation and improvements in normalized multichannel measurement are proposed in order to reduce error.

1. INTRODUCTION

The measurement process can be generally expressed as

$$y = f(x) \tag{1}$$

Where, y is output of instrument, x is the quantity to be measured, f is the relation between them.

The error transfer means the relation between $\pm\Delta y$, the error of y produced in measurement, and $\pm\Delta x$, the error of x caused by $\pm\Delta y$.

$$x \pm \Delta x = f^{-1}(y \pm \Delta y) \tag{2}$$

Error transfer can be analyzed as

$$\Delta x = \frac{df^{-1}}{dy}\Delta y \tag{3}$$

However, there are many multichannel measurements used in practical application. For example, the multi-wavelength thermometry reviewed by Dai (2004), the multichannel detection of echo from underwater target (Wang et al., 2002), and two-wavelength fluorescent microarray scanning (Fu et al., 2007). Normalization is often necessary in multichannel measurement especially those base on color theory: the primary spectrum pyrometry (Cheng et al., 2004), color recognition (Li et al., 2004), shadow detection (Chen et al., 2006) and plants' diagnosis (Feng et al., 2007; Dong et al., 2008).

The relation between output signals and the quantity to be measured is

$$y_i = f_i(x)\,(i = 1, 2, \cdots, n) \tag{4}$$

The normalization is

$$Y_i = y_i \Big/ \sum_{i=1}^{n} y_i = f_i(x) \Big/ \sum_{i=1}^{n} f_i(x) \tag{5}$$

Finally, x can be calculated as

$$x = g(Y_1, Y_2, ..., Y_n) \tag{6}$$

Similarly, the error transfer can be expressed as

$$Y_i \pm \Delta Y_i = \frac{y_i \pm \Delta y_i}{\sum_{i=1}^{n} y_i \pm \Delta y_i} (i = 1, 2, ..., n) \tag{7}$$

$$x \pm \Delta x = g(Y_1 \pm \Delta Y_1, Y_2 \pm \Delta Y_2, ..., Y_n \pm \Delta Y_n) \tag{8}$$

In multichannel measurement, the error propagates to the normalized quantities first and then to the final result. It is difficult to analyze the error transfer in traditional way of derivation because the final error is affected by multi-quantities and nothing significant result can be obtained from the analysis.

Taking the primary spectrum pyrometry for example, a model for error transfer analysis of multichannel measurement will be proposed next, and some principles of design and operation will be obtained. In this model, error transfer is not the exact one mentioned foregoing, it is represented by the parameter error magnification. The error magnification can not be measured actually from experiment, it is used just in theoretical analysis of error transfer, and finding a better design of system.

2. ERROR TRANSFER MODEL OF NORMALIZATION

2.1 Case Analysis of the Primary Spectrum Thermometry

The primary spectrum pyrometry takes three-channel outputs, R, G and B, of CCD camera as the basis of calculation. The calculation process can be divided into to two steps.

(a) The normalization of (R, G, B)

$$r = \frac{R}{R+G+B}, g = \frac{G}{R+G+B}, b = \frac{B}{R+G+B} \tag{9}$$

(b) Searching for the isotherm where the point (r, g) is located on the normalized plane. The temperature of this isotherm is the quantity to be measured.

Among them, step (a) is the process of normalization and it is a general operation. Essentially, step (b) is the process of solving equations and its detail is different from other measurement method.

According to the independence and difference of the two steps, we can analyze the error transfer in step (a) first and then consider its influence on step (b).

The error transfer model of normalization in the primary spectrum pyrometry is a three-channel model. But its theory can be extended to two-channel or n-channel measurement.

2.2 The Three-Channel Model

Taking the three-channel outputs *R*, *G* and *B* as the coordinate axis, a 3D measurement space *R-G-B* can be constructed as shown in Fig.1.

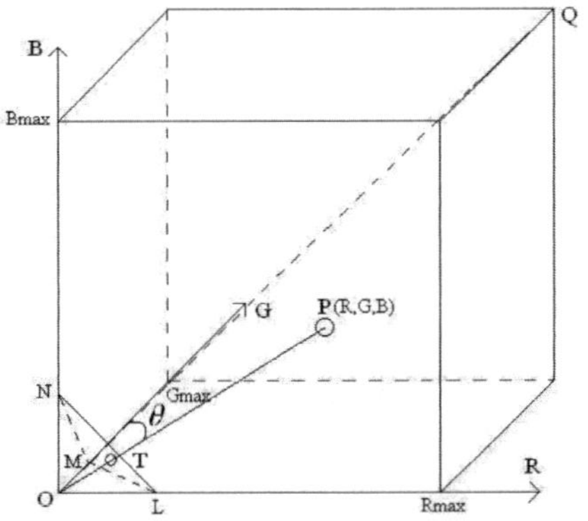

Fig. 1 3D measurement space *R-G-B*

In Fig.1, the cube represents the value range of every channel. The point P represents the true value of measurement. After normalization *r*, *g* and *b* are obtained. As the equation $r+g+b=1$ is satisfied, the point (r,g,b) must locate at the plane LMN(the normalized plane), where $\overline{OL}=\overline{OM}=\overline{ON}=1$.

If error exists, points of measured value will depart from the point P of true value and a cube centred at P appears. The cube represents the max range of error. But if the probability is considered, a sphere distribution which is the inscribed sphere of the cube is more reasonable.

After normalization the projection of the sphere on the normalized plane is an irregular area which can be regarded as infinitesimal element. In condition of constant error, the larger the projection area is, the bigger the transmission error is.

Now the analysis of error transfer has transformed into a solid geometry problem and it is necessary to introduce the concept of solid angle. Let ΔS_0 be the cross section area of the error sphere and ΔS_1 be the projection area, the solid angle subtended by ΔS_0 and ΔS_1 to P can be expressed as

$$\omega=\frac{\Delta S_0}{\overline{OP}^2}=\frac{\Delta S_1\cos\theta}{\overline{OT}^2}(0\le\theta\le\arccos\frac{\sqrt{3}}{3}) \tag{10}$$

Where θ is the angle between line OQ and OP. So the ratio of projection area to cross section area is

$$f=\frac{\Delta S_1}{\Delta S_0}=\frac{h^2/\cos^3\theta}{R^2+G^2+B^2} \tag{11}$$

Where, h is the distance between origin O and normalized plane LMN($h=\sqrt{3}/3$), and cosine is

$\cos\theta = h\big/\sqrt{r^2+g^2+b^2}$. So

$$f = \frac{r^2+g^2+b^2}{R^2+G^2+B^2}\frac{1}{\cos\theta} = \frac{\sqrt{3}\sqrt{R^2+G^2+B^2}}{(R+G+B)^3} \tag{12}$$

f is the ratio of two area and represents the error transfer in normalization. So it is defined as error magnification. The bigger the *f*, the larger the transmission error is.

2.3 The Two-Channel Model

There is something different for two-channel model as shown in Fig.2.

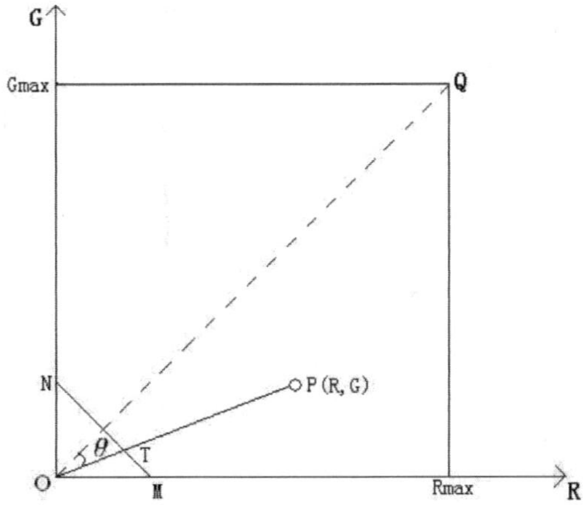

Fig. 2 2D measurement space R-G

After normalization, *r* and *g* are obtained from *R* and *G*.. As equation $r+g=1$ is satisfied, the point (r,g) must locate at line MN where $\overline{OM}=\overline{ON}=1$. If error exists, points of measured value will distribute in a circle centre at point P of true value. The projection of the circle on line MN is a line segment. So the error magnification can be defined as the ratio between length of the line segment and the diameter of the circle. Let Δd be the diameter and Δl be the length of line segment, the angle subtended by Δd and Δl to P can be expressed as

$$\varphi = \frac{\Delta d}{\overline{OP}} = \frac{\Delta l\cos\theta}{\overline{OT}}(0 \le \theta \le \frac{\pi}{4}) \tag{13}$$

So the error magnification is

$$f = \frac{\Delta l}{\Delta d} = \frac{h\big/\cos^2\theta}{\sqrt{R^2+G^2}} \tag{14}$$

Where, *h* is the distance between origin O and line MN($h=\sqrt{2}/2$), and cosine is $\cos\theta = h\big/\sqrt{r^2+g^2}$. So

$$f = \frac{\sqrt{r^2+g^2}}{\sqrt{R^2+G^2}}\frac{1}{\cos\theta} = \frac{\sqrt{2}\sqrt{R^2+G^2}}{(R+G)^2} \tag{15}$$

Obviously, the error magnification in two-channel measurement is similar to the three-channel one.

2.4 The N-Channel Model

From above analysis, it is obvious that the essence of error transfer is projection of dimension reduction. If the number of channel is greater than three, the definition of error magnification can not be derived from model because it is difficult to image the space more than three dimensions. But the definition can be obtained from some common relations.

The basic of the construction of Euclidean space is the distance (Lin, 2004) which can be expressed as

$$\rho = \sqrt{\sum_{i=1}^{n}(a_i - b_i)^2} \tag{16}$$

where a_i and b_i are the coordinates of two points in the space.

The definition of error magnification in 2D and 3D space includes three distance values, L the distance between origin and point of measured value, l the distance between origin and point of normalized value, h the minimum distance between origin and all points of normalized value. The relation between error magnification and the three distance values is inducted as

$$f_n = \left(\frac{l}{L}\right)^{n-1} \times \frac{l}{h} \tag{17}$$

where, $(l/L)^{n-1}$ represents the influence of distance, and l/h which is the reciprocal of cosine represents the influence of angle.

The minimum distance between origin and all points of normalized value can be derived from Cauchy inequality

$$\left(\sum_{i=1}^{n} Y_i^2\right)\left(\sum_{i=1}^{n} Z_i^2\right) \geq \left(\sum_{i=1}^{n} Y_i Z_i\right)^2 \tag{18}$$

Let Y_i be the normalized value and satisfy $\sum_{i=1}^{n} Y_i = 1$. When $Z_i = 1$, yield

$$\sqrt{\sum_{i=1}^{n} Y_i^2} \geq \frac{1}{\sqrt{n}} = h \tag{19}$$

Finally, the error magnification in n-channel measurement is expressed as

$$\begin{aligned} f_n &= \sqrt{n} \times \left(\sum_{i=1}^{n} Y_i^2\right)^{\frac{n}{2}} \Big/ \left(\sum_{i=1}^{n} y_i^2\right)^{\frac{n-1}{2}} \\ &= \sqrt{n}\sqrt{\sum_{i=1}^{n} y_i^2} \Big/ \left(\sum_{j=1}^{n} y_j\right)^n \end{aligned} \tag{20}$$

3. THE EFFECT OF DISTANCE AND ANGLE

f is the error magnification and represents the error transfer in normalization. Taking the three-channel for example, it is obvious from Eq.(11) that the distance decreases f while the angle increases it. When $\theta = 0$ is satisfied the relation between f and distance $\overline{OP}$ is shown in Fig.3. When $R^2 + G^2 + B^2 = 10^4$ the relation between f and θ is shown in Fig.4.

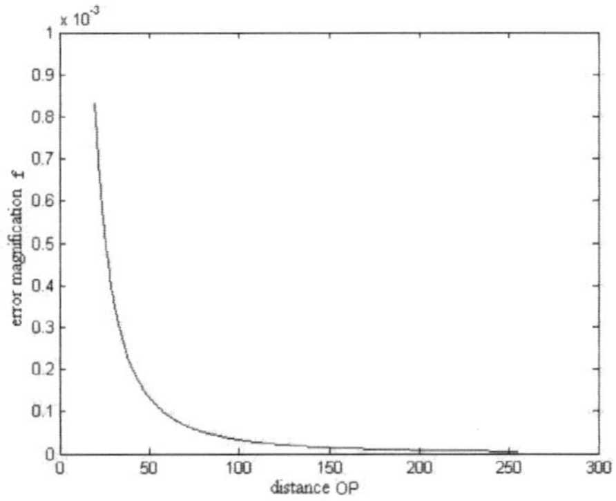

Fig. 3 Relation between f and $\overline{OP}$

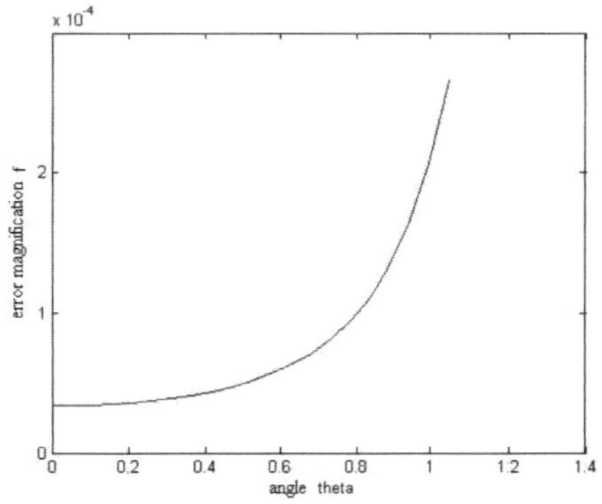

Fig. 4 Relation between f and θ

As the error magnification attenuates with increase of square distance, the measured value should be as large as possible but not saturated in measurement. For example, we should prolong the exposure time and increase the aperture while using CCD in multichannel measurement. As every output is increased in same rate, the operation will not change anything in subsequent solving process of equations even the process is unknown.

On the other hand, error magnification increases with increase of angle θ in the form of $1/\cos^3\theta$. This means that output from every channel had better keep approximate in measurement. As it is necessary to use the instrument function of separate signal gain in order to change the normalized value and the subsequent process will be affected, the complicated analysis will not be given in this paper.

In above analysis it is assumed that the absolute error keeps constant while the system is changed. Research of Xu (2004) and Li (2007) shows that common noise in CCD includes shot noise, dark current noise, transfer noise, reset noise, nonuniform response noise, clutter noise and quantization noise. All these noises are independent of output value except for the shot noise which is proportional to the square root of signal (Xu, 2004). So the assumption is reasonable.

Compared to the error analysis of single signal, the distance corresponds to the absolute value of signal and the error sphere corresponds to the absolute error. It is similar in two methods that if the absolute error keeps constant, the increase of absolute signal reduces the relative error. However, the relation between signals what is represented by angle θ should be considered in multichannel measurement. This is the main difference between two methods.

4. CONCLUSIONS

Taking the primary spectrum pyrometry for example, the analysis model of error transfer in normalized multichannel measurement is established and the effect of signals on error magnification is analyzed base on the model in this paper. The analysis points out that outputs should keep large and approximate to each other in order to reduce the error magnification.

Although the accurate relation between error and result can not be obtained, the effect of normalization on error is shown clearly. This method simplifies the analysis of error transfer and plays a guidance role in practical operation. The separate signal gain is useful in reducing error and its effect can also be analyzed base on this model.

ACKNOWLEDGEMENTS

The paper is supported by the National Natural Science Foundation of China(Grant No.50606033) and National High Technology Research and Development Program of China(2007AA04Z178).

NOMENCLATURE

R, G, B	output of red, green and blue channel of CCD
r, g, b	normalized signal of red, green and blue channel of CCD
ΔS_0	the cross section area of the error sphere
ΔS_1	the projection area
Δd	the diameter of error circle in two-channel measurement
Δl	the projection line segment of error circle in two-channel measurement
f	error magnification
h	the minimum distance between origin and all points of normalized value

Greek Letters

φ	the angle subtended by diameter of error circle in two-channel measurement
ω	solid angle subtended by cross section area of error sphere

REFERENCES

Chen, B. S. et al. (2006). "Normalized rgb Color Model Based Shadow Detection," J. Comput. App. 26(8), pp.1879-1881.

Cheng, X. F. et al. (2004). "Principle of Primary Spectrum Pyrometry," Sci. China(Series G), 34(6), pp.639-647

Dai, J. M. (2004). "Survey of Radiation Thermometry," Tech. Auto. App., 23(3), pp.1-7.

Dong, J. Y. et al. (2008). "Determination of Chlorophyll a and b Using Absorption Spectrum," Spectroscopy and Spectral Analysis, 28(1), pp.141-144.

Feng, J. et al. (2007). "Application of Multi-spectral Color Reproduction for Diagnosis of Plant Diseases and Inspect Pests," Image Tech., (6), pp.6-9.

Fu, D. X. et al. (2007). "Normalization of Two-wavelength Microarray Fluorescence Image," J. Optoelec. Laser, 18(6):721-724.

Li, G. J. et al. (2004). "Car-body Color Recognition Algorithm Based on Color Difference and Color Normalization," Comput. App. 24(9), pp.47-49.

Li, Y. F. (2007). "Noise Analyzing and Processing of TDI-CCD Image Sensor," Opt. Prec. Engi, 15(8), pp.1196-1202.

Lin, J. K. (2004). Fundamentals of Topology, Science Press, Beijing, China.

Wang, F. et al. (2002). "Research and Design of High Precision General Underwater Sound Receiver," Chinese J. Sci. Instru., 23(3), pp.249-252.

Xu, X. Z. (2004). "Analysis an Processing of CCD Noise," Infr. Laser Eng, 33(4), pp.343-346.

Journal of Physics: Conference Series **147** (2009) 012084 doi:10.1088/1742-6596/147/1/012084

Experimental Investigation on Soot Volume Fraction in an Ethylene Diffusion Flame by Emission Spectrometry without Optically-Thin Assumption

Jing Lu [1], Chun Lou [1], Huai-Chun Zhou *[1]

[1] Huazhong University of Science and Technology, 1037 Luoyu Road, Hongshan, Wuhan, Hubei, 430074, P. R. CHINA

hczhou@mail.hust.edu.cn

Abstract. An improved soot diagnostics technique based on tomographic reconstruction of flame emission spectra has been developed for an axisymmetric laminar diffusion flame without optically-thin assumption. Emission from the flame is scanned along the horizontal lateral axis of flame at several altitudes above the burner. At each measurement position, the local line-of-sight flame emission spectra is collected by a spectrometry over a spectral range of 700-1100 nm. Inversion of these data through one-dimensional tomography using a three-point Abel inversion yields radial distributions of the soot radiation from which soot volume fraction profiles are extracted. Traditionally, this procedure was applied only in the optically-thin flames for which the self-attenuation term of emitted radiation is ignored. However, this self-attenuation term is considered by iterative calculations in the paper. The configuration of the investigation flame is similar to the flames reported in the literatures. The results by the conventional method is found to be almost twice to the results by the improved method. This discrepancy revealed that the optically-thin assumption for the flames will cause serious errors in the conventional method, and the improved method could overcome this deficiency effectively.

1. INTRODUCTION

The understanding of soot formation mechanisms and knowledge of soot distribution within combustion systems is a key factor in many areas of combustion research. Flame radiation, heat transfer properties, particulate emission from combustion systems, and combustion efficiency all depend upon soot formation. As a consequence, optical diagnostics are recommended for measurement of radiant emission from hot soot due to their non-intrusiveness (Kennedy, 1997). They are a key source of information on the soot concentration field in many combustion applications.

The most established method is based on light extinction, which is a path-integrated or line-of-sight technique that has been widely used to measure soot volume fractions in flames, and frequently associated with scattering measurement for the simultaneous determination of soot particle size and number density (Santoro et al., 1983; Dasch et al., 1991; Gulder et al., 1993; Choi et al., 1995; Greenberg et al., 1997; Konsur et al., 1999). The technique provides field-integrated values of the soot volume fractions and needs an inversion method to reconstruct its spatial distribution in non-uniform fields, e.g., the Abel inversion method, for axisymmetric configurations (Vest, 1979), whose pitfall and advantages have been discussed in detail by Snelling et al. (1999).

Other nonintrusive methods can also be used. Laser-induced incandescence (LII) of soot provides spatially and temporally resolved measurements of soot distribution (Queay et al., 1994; Shaddix et al., 1996; Mitrovic et al., 1998). It requires a calibration procedure based upon extinction measurements or gravimetric sampling for quantitative data of soot volume fraction distribution. Laser-induced fluorescence (LIF) is based upon measurements of the fluorescence of C2 molecules from vaporized soot (Bengstsson, 1991; Westblom, 1991; Bengtsson et al., 1995; Bengtsson, 1996).

Spectral flame emission measurements have also been employed to obtain soot concentration distribution. Hall et al. (1990) used single wavelength emission/transmission tomography to two-dimensional maps of soot concentration in laminar diffusion flame(Snelling et al,2002). Two-color emission pyrometry is widely used in practical situations, especially for studies in internal combustion engines. (Sivathanu et al., 1991; Sivathanu et al., 1992; Klassen et al., 1990; Choi et al., 1994; Sivathanu, 1990; Xu et al., 1998). However, it only has provided data on spatially averaged soot concentration and its accuracy is significantly limited by the uncertainty in the emissivity model, which is based on knowledge of soot optical properties (i.e., refractive index). De Iuliis et.al. used a multi-wavelength diagnostic method in which multiwavelength soot emission intensities were measured by a low resolution spectrograph, and optical tomography (Able inversion) was applied to reconstruct soot volume fraction distribution in an axisymmeric flame(Snelling et al,2002). Ayrancı et. al.(2008) also performed FTIR emission spectrometry to measure line-of-sight intensity spectra emitted by a co-annular ethylene/air diffusion flame and infer soot volume profiles from tomographic reconstruction. Nevertheless, when the tomographic technique was applied, they did not consider flame absorption of radiation and neglected the self-absorption term in the path integral for line-of-sight emission intensity.

In this paper, the soot emission concentration measurement method is improved by introducing a correction procedure for flame absorption of radiation, focuses on the measurement for line-of-sight flame emission spectrometry within 700-1100nm range on an axisymmetric ethylene/air diffusion flame. The paper presents descriptions of the experimental methodology in section 2 and a comparison of conventional soot volume fraction reconstruction method to our modulated method with the self-absorption term correction procedure in section 3. Finally, the accuracy of the modulated method is evaluated through comparisons of the results to previously published data for similar flames by Santoro et al. who used combined light extinction and scattering to measure the soot distribution (Santoro et al., 1983)

2. EXPERIMENTAL METHODOLOGY

2.1 Experimental Apparatus

The experimental setup is shown schematically in Fig.1. A laboratory grade axisymmetric ethylene/air diffusion flame was used to perform emission spectrometry measurements. Radiation emission from soot in flame passes through an optical fiber and is directed to the entrance slit of a spectrometer. The both sides of fiber are designed elaborately, at the side near the flame the head of fiber is covered by a cylindrical cap with a small hole of 0.25mm diameter in its center to pass the emission from flame and at the side near the spectrometer the end of fiber is attached to a splint which assembled on front of the entrance slit of the spectrometer. A cycloidal hole in the center of splint is designed by the size of the end of fiber to allow the radiation from the fiber passing and isolate the radiation from circumstance. Transverse resolution is defined by the 0.25mm pinhole and the spatial angle is 2~3°. This design is very similar with Conventional Kurlbaum which can give fine longitudinal resolution. So did our design. The spectrometer slit is oriented vertically with a height of 1.5mm and a width of 1.5mm for all emission could be directed to spectrometer. Emission from the flame scanned along its horizontal lateral axis at several altitudes above the burner is collected over a spectral range from 700 to 1100nm. At each position, the spectrometer processes emitted radiation to yield local line-of-sight flame emission spectra and is calibrated with a blackbody furnace. The blackbody furnace is calibrated against a standard photoelectric pyrometer and the uncertainty in the spectral radiance temperature is ±5K.This calibration allows the spectra to be scaled to units of absolute spectral intensity.

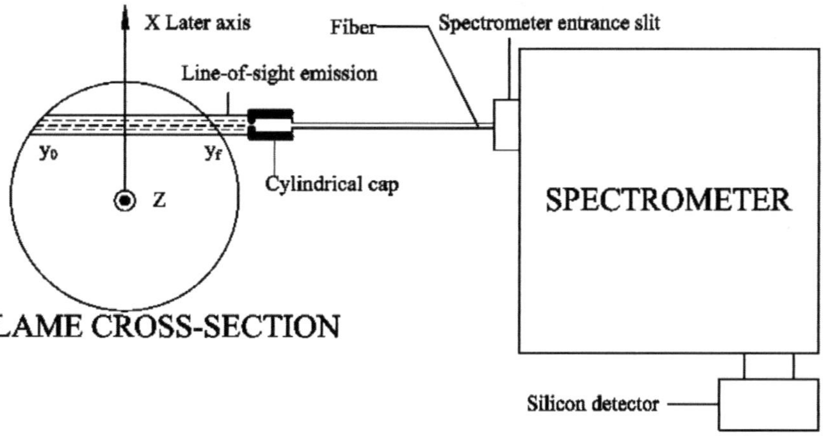

Fig. 1 Schematic diagram of the experimental setup

The burner for the laboratory grade axisymmetric laminar diffusion flame pictured in Fig2 is of similar design to that previously reported by the National Research Council of Canada and others. It consists of a 10.7-mm-inner-diameter fuel tube, centred in a 88.6-mm-diameter air nozzle. The air passes through packed beds of glass beads and honeycomb to smooth the flow and prevent flame instabilities. It is different from other similar experiments (Hall et al., 1990; Deluliis et al., 1998; Ayranci et al., 2008), in the present paper, the burner is fixed on the platform and the head of fiber with a cylindrical cap is attached to a motorized translation stage (MT-310, BOIF, China) that enable

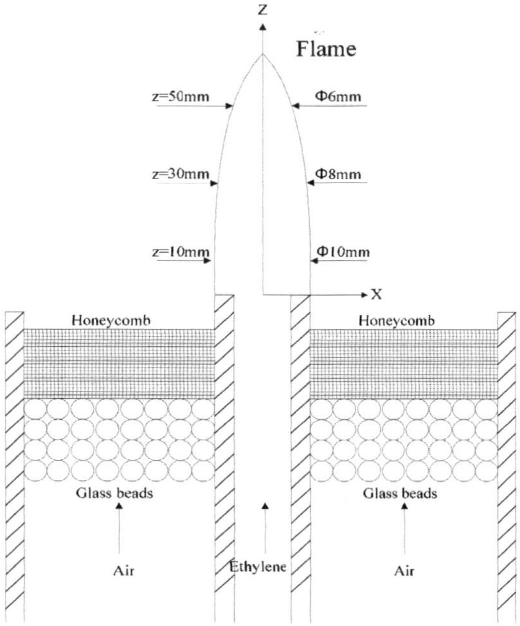

Fig 2. Schematic cross-section of the burner

spatial exploration of the flame along vertical and horizontal directions. In other experiments, the burner is situated on micrometric positioning stages. In order to eliminate external air circulations, the burner is protected by a flexible steel mesh. An appropriate viewing port in the mesh provides optical access.

Journal of Physics: Conference Series **147** (2009) 012084 doi:10.1088/1742-6596/147/1/012084

A blackbody furnace with 800-1800℃ temperature range was used for calibration of emission spectra by replacing the burner with the blackbody. Calibration of the blackbody for operating temperatures was checked by a pyrometer with certified calibration.

The spectrometer used in this study is a 300mm focal length monochromator (Spectra-Pro-2300i, Acton, US). It features an astigmatism-corrected optical system, triple indexable gratings and triple grating turret. The SpectraPro-2300i includes a direct digital grating scan mechanism with full wavelength scanning capabilities. It can be configured for various spectral ranges from 185nm to far infrared with available gratings and accessories. The configuration for the range under consideration in this study is a Silicon based photodiode detector for use in the wavelength range of 400 to 1100nm.[Model:SI-440].

2.2 Experiment Conditions

The experiment parameters of the measurements are summarized in Table 1. The C2H4 fuel flow rate is set to 129 ml/min (21℃,1 atm), and the air coflow is 189 l/min, producing an overventilated flame 6.5 cm in height, which is studied by many researchers (Santoro et al., 1983; DeIuliis et al., 1998; Koylu et al., 1997). A larger airflow significantly reduced a relatively slow pulsing global instability that is characteristic of buoyant nonpremixed flames, but also reduced the visible flame height. This, in turn, truncated the spatial resolution of the experiments. For the reason that the soot emission spectrum which is well know to display continuum characteristics that can be adequately captured by low resolution measurements and the flame flickering would induce the possible dynamic changes for the recorded soot emission intensity during the long recoding times required for higher resolutions, the spectrometer was set to a low spectral resolution of 2nm.

2.3 Calibration Methodology

The spectra recoded by the spectrometer needs to be calibrated so that the effects of instruments can be eliminated and measured instrument units can be related to physical units of intensity. Calibration is carried out by measuring spectra from a blackbody furnace in the present study. A instrument response function which is deduced from the theoretical evaluation of reference measurement enables relate the spectrum recorded in instrument units to emission intensity spectrum in physical units. Ayrancı et. al. (1994) described the detailed procedure and emission spectrum calibration presented in this study are based on the same principle as them.

Table 1 Experiment conditions

Spectrometer			
Spectral range	700-1100nm	Spectral resolution	2nm
Number of scans	200	Total scan time	60s
Flame properties			
Fuel	Ethylene,C_2H_4	Air flow rate	189l/min
Purity	$\geq$ 99.99%	Luminous flame height	65mm
Fuel flow rate	129ml/min		
Flame scanning parameters			
Horizontal spatial resolution			0.25±0.1mm
Vertical spatial increment			20 ±0.5mm

3. THEORY

3.1 Conventional Soot Property Reconstruction Method

The reconstruction of internal flame temperatures from external, path-integrated intensity measurements has long been a problem of interest. In the spectral intervals where combustion gases can be assumed to transparent, flame soot emits, absorbs and scatters radiation continuously along the spectra. This isolated radiative activity of soot enables its characterization from emission spectroscopy at certain wavelength intervals. The generalized path integral for line-of-sight soot emission intensity along a chord y , crossing the flame at fixed lateral position, x, as shown in Fig1 is given in Eq. (1) (Hall et al., 1990; Ayranci et al., 2008).

$$I_\lambda(x) = \int_0^{y_f}\left[\kappa_\lambda^{(a)}(y)\cdot I_{b,\lambda}[T(y)]\right]\cdot\exp\left[-\int_y^{y_f}\kappa_\lambda^{(e)}(y')dy'\right]dy \qquad (1)$$

where $\kappa_\lambda^{(a)}$ and $\kappa_\lambda^{(e)}$ are the local absorption and extinction coefficients, as follows from Rayleigh-Debye-Gans/polydisperse-fractal aggregate(RDG/PFA) theory of particle light scattering and extinction (Koylu et al., 1994a; Koylu et al., 1994b; Koylu et al., 1997), for primary particles in the Rayleigh range($\pi d/\lambda<0.3$) in the present study, scattering will be negligible and $\kappa_\lambda^{(e)} = \kappa_\lambda^{(a)} = \kappa_\lambda$. $I_{b,\lambda}(T)$ is the Planck function

$$I_{b,\lambda}(T) = (2\pi hc_0^{\ 2}/\lambda^5)[\exp(hc_0/\lambda kT)-1]^{-1} \approx (2\pi hc_0^{\ 2}/\lambda^5)[\exp(hc_0/\lambda kT)] \qquad (2)$$

where the Wien's approximation is a valid representation for Planck's distribution at present temperature and spectral range.

The term before the exponential, $\kappa_\lambda^{(a)}(y)\cdot I_{b,\lambda}[T(y)]$, models the local radiation emission rate, and the exponential term, $-\int_y^{y_f}\kappa_\lambda^{(e)}(y')dy'$, models the attenuation experienced by this emitted radiation as it propagates to the detector side of the flame along the chord y. In the conventional soot property reconstruction method, the flame which is studied usually is assumed to be optically thin flame, thus the attenuation of emitted radiation is ignored:

$$\int_y^{y_f}\kappa_\lambda^{(e)}(y')dy' \approx 0 \qquad (3)$$

Assuming the medium is axisymmetric, the line-of-sight soot emission intensity at a fixed lateral position is modeled as

$$I_\lambda(x) \approx \int_0^{y_f}\left[\kappa_\lambda^{(a)}(y)\cdot I_{b,\lambda}[T(y)]\right] = \int_0^{y_0} B(y)dy \qquad (4)$$

In the Rayleigh limit, the absorption coefficient of soot and its dependence on wavelength are

$$\kappa_\lambda = 6\pi E(m) f_v/\lambda \qquad (5)$$

where f_v is the local volume fraction of soot and $E(m)\{\equiv \mathrm{Im}[(m^2-1)/(m^2+2)]\}$ is a function of wavelength dependent soot complex index of refraction for which the correct value is much debated (Koylu et al., 1994). If assuming E(m) is constant in the present spectral range and spatial domain, then substituting Eqs (2) and (5) in Eq(4) yields

$$B_\lambda(y) = (12\pi hc_0^2/\lambda^4) f_v(y) E(m)\exp\{hc_0/[\lambda kT(y)]\} \qquad (6)$$

Taking natural logarithm of both sides and rearranging gives following linear function in the wavenumber

$$N_\lambda(y) = \ln[B_\lambda(y)\lambda^4] = \{hc_0/[kT(y)]\}\cdot(1/\lambda) + \ln[12\pi hc_0^2 f_v(y)E(m)] \qquad (7)$$

Spectral emission source term B_λ is reconstructed from measured emission intensities in lateral domain by employing 1-D tomography and soot properties at each radial location is retrieval by performing linear regression to $\ln[B_\lambda(y)\lambda^4] vs.(1/\lambda)$ plot. Volume fraction is obtained from the intercept for a gray refractive index, m.

3.2 Modulated Soot Property Reconstruction Method

Since soot is both a good emitter and absorber of radiation, a portion of the light emission radiation along the measurement chord is reabsorbed by soot particles. The percentage of emission attenuation is dependent on the position of the emitter along the chord through the flame(Snelling et al,2002). In the development of the equations for tomographic inversion, the use of tomography in soot property measurements is based on the assumption that the percentage of emission re-absorption by soot is vary low, which is called optically thin as its optical thickness is vary small. When the optical thickness beyond a critical value(almost 2 for Fourier transform reconstructive algorithm (Hall et al., 1990), errors started to appear. Strictly speaking, this means that emission measurements are not line integrals of a local property field as is required for tomographic reconstruction, because each emitter does not experience the same percentage attenuation, the line-of-sight measurements become technically unsuitable for tomographic inversion using the Abel inversion. Therefore, when the medium is not optically thin, corrections for the self-attenuation term in Eq.1should be introduced. To these flames investigated in many papers (Hall et al., 1990; DeIuliis et al., 1998; Aranci et al., 2004), although their configuration is not the same, all of them are assumed to be optically thin flames and the self-attenuation term is ignored even if the relative large-size flame of which the fuel flow rate is set to the smoke point and the top soot volume fraction is up to 14 ppm (Aranci et al., 2004). As the reconstructive algorithms(Fourier transform or three-point Abel inversion algorithm, onion-peeling method and so on) is sensitive to the optical thickness, the optically thin flames assumption is not suitable in many experiment conditions and the self-attenuation term can be introduced by iterative calculation. The detailed procedure included the following 3 steps:

1) Assuming the self-attenuation term is ignored, Eq.3 and Eq. 4 is established and the soot local volume fraction are obtained by the conventional reconstruction method.

2) As particle light scattering is negligible in this paper, the value of self-attenuation term can be calculated out using the soot local volume fraction gained in Step 1, that is $\exp\left[-\int_{y}^{y_f}\kappa_{\lambda}^{(e)}(y')dy'\right]=A$.

3)Introducing the value of self-attenuation term A to the local property field $I_{\lambda}(x)$, thus Eq.1 changed to be $I_{\lambda}(x)=\int_{0}^{y_f}\left[\kappa_{\lambda}^{(a)}(y)\cdot I_{b,\lambda}[T(y)]\right]dy\cdot A$, since A is constant, the soot local volume fraction in Eq.(1) can be solved by conventional method. Then the new result will be taken back to Step 2 and a new iterative procedure from Step 2 to 3 will begin until it end up when the condition of convergence is satisfied.

The condition of convergence in the present study is set to be less 0.001ppm for the soot local volume fraction between two adjacent iterative results.

For the one-dimensional tomography reconstructive algorithms performed on the data , there existed 4 approaches at least (three-point Abel inversion method (Dasch, 1992), Fouries transform(Hall et al., 1990), onion-peeling and filtered backprojection methods (Dasch, 1992)). Three-point Abel inversion algorithm is selected to deconvolve soot emission spectrometry projections as it was previously found to be more accurate and efficient to onion-peeling and filtered backprojection methods and less complicated data reduction procedures to Fourier transform .

4. RESULTS AND DISCUSSION

4.1 Blackbody Furnace and Flame Emission Measurements

Blackbody furnace measurements were carried out at blackbody temperatures from 1000 ℃ to 1500 ℃ of 20 ℃ interval which are expectedly close to flame temperature. Fig 3 depicts blackbody furnace spectra measured at 1200℃ and theoretical blackbody intensities obtained from Plank function. An iron covering attached to the window of blackbody furnace is made in particular for calibration to ensure background instrument emission spectra are measured to be null.

Calibrated flame emission intensity spectra of 800 nm wavelength collected at 3 vertical positions above the burner are displayed in Fig. 4. Evolution of lateral profiles with vertical position is displayed in Fig. 5 at 800nm wavelength which is adequate to represent the results within 700-1100nm range as the spectral variations are monotonic in this range. The intensity profiles confirm that the flame is acceptably axisymmetric.

4.2 Soot Volume Fraction Reconstruction without Self-Attenuation Term

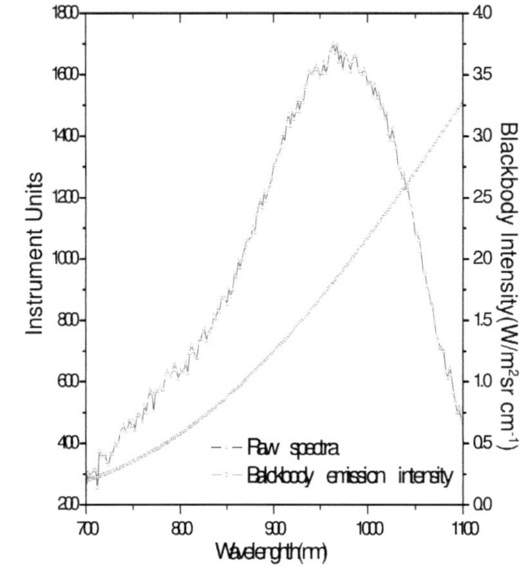

Fig. 3 Raw blackbody spectra and theoretical intensity

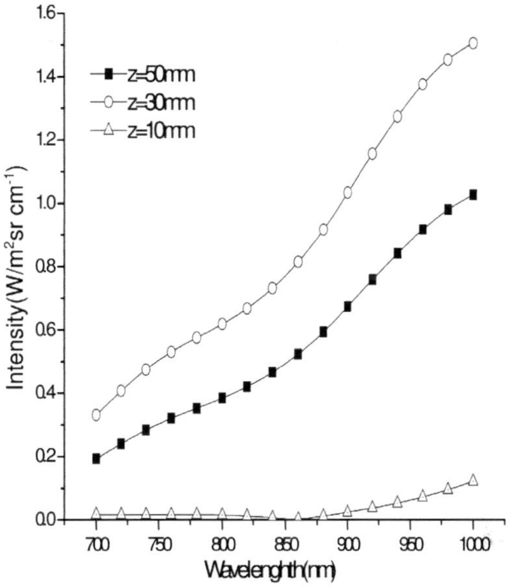

Fig.4 Calibrated flame emission intensity spectra of 800nm wavelength at position x=0mm for three heights

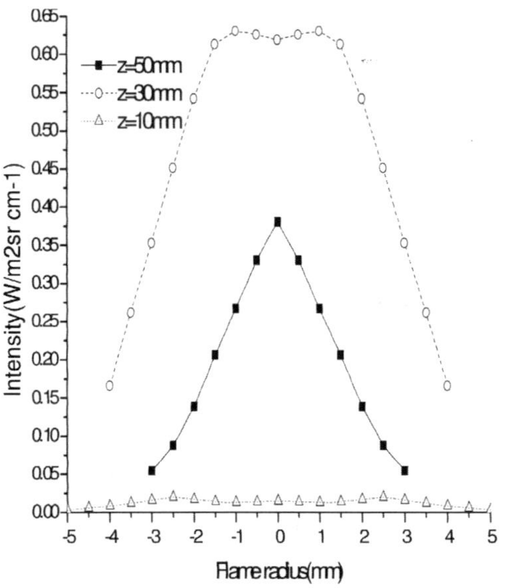

Fig. 5 Flame emission intensity profiles with vertical position at 800nm wavelength

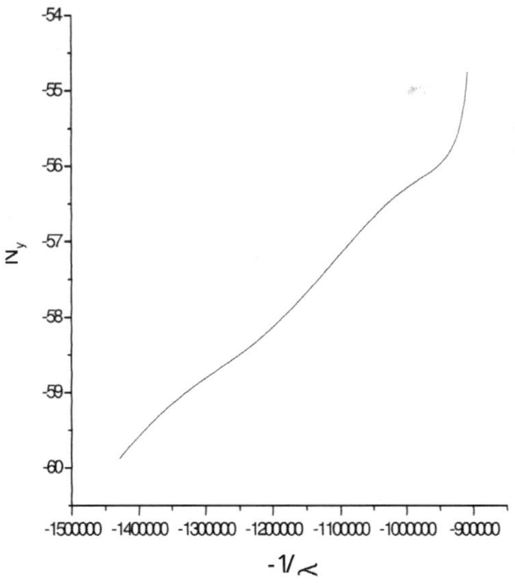

Fig. 6 N_λ versus wavenumber (-1/ λ)plot at x=0mm, z=30mm above the burner

The regression to N_λ versus wavenumber (-1/ λ)plot at x=0mm, z=30mm above the burner is displayed in Fig6. N_λ is linear relationship with (-1/ λ) within 700~1000nm spectral range indicates the soot refractive index function E(m) is constant in the present study. Consequently, the assumption that E(m) is independent of wavelength is valid and a magnitude of 0.26(n=1.57 and k=0.56 (Dalzell, 1959)) is acceptable in this paper.

Journal of Physics: Conference Series **147** (2009) 012084 doi:10.1088/1742-6596/147/1/012084

Soot volume fraction profiles inferred by optically thin flame assumption are plotted in Fig7. The results are compared to LE(light extinction) measurements. C.P. Arana et al. (Aranci et al., 2004) used full-field light extinction to measure the soot distribution in various ethylene/air partially premixed flames (including nonpremixed flame),the structure of their burner is similar with ours and the size of investigated flame corresponds to our flame. they used three-point Abel inversion method to calculate soot volume fraction based on a constant E(m) value assumption, specially obtained two results using two different E(m) values. We chose the results of nonpremixed flame with the same E(m) value to compare.. Since there is some discrepancy in the size of investigated flame between our and the earlier researchers, the results of lateral profiles at vertical position of the same H/F(height/flame size)ratio is chosen to compare.

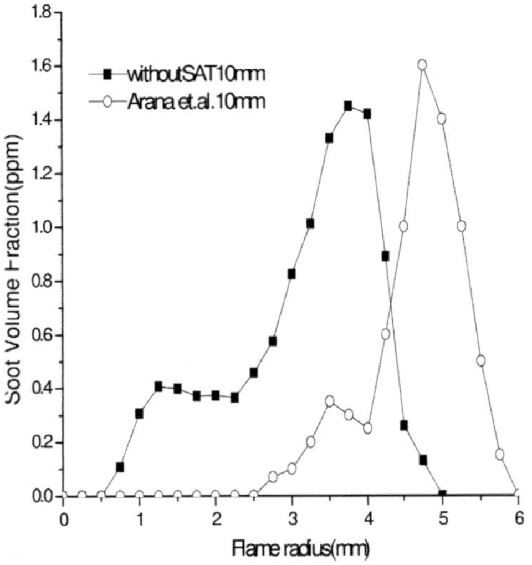

(a)

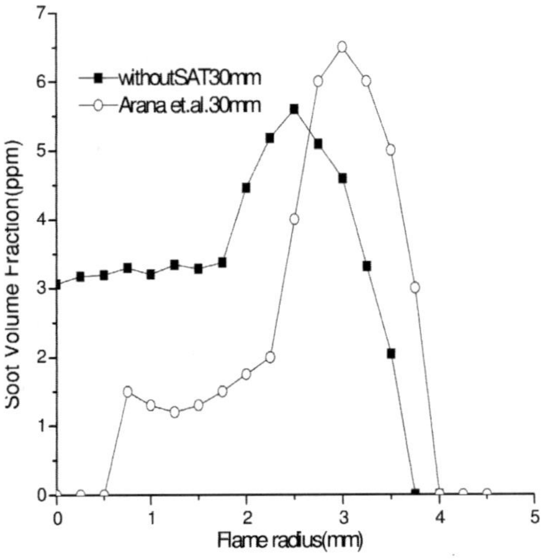

(b)

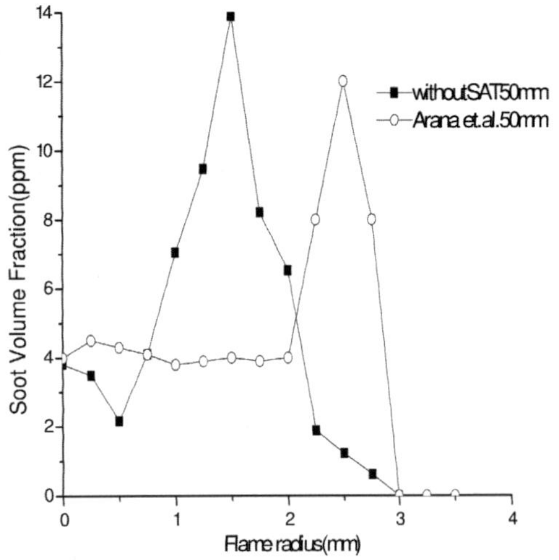

(c)

Table 2 Comparison of the max soot volume fraction of our and Arana et.al's data at three height above the burner

Heights	Max Volume Fraction	
	Without SAT(ppm)	Arana et.al (ppm)
Z=10mm	1.4	1.6
Z=30mm	8.2	9
Z=50mm	13.9	12

Fig. 7.Comparison of soot volume fraction distributions for a nonpremixed flame inferred by optically thin flame assumption with those of Arana et al.[33] at various heights above the burner(HAB).(a) through (c) ,correspond respectively to HAB=10,30 and 50mm

At z=10mm above the burner the inspection of soot volume fraction profile along the lateral axis indicates that flame contains no soot at the center but within an annular region at 2-4mm radius. This profiles are nearly flat at flame center and increase towards the flame edge. Our results at the flame center were found to be in line with observations in the literature and the variation range of volume fraction are in good agreement with previously published data, only the small discrepancy lies in the location of peak temperature and volume fraction since the radial dimension of flame is not same between our and earlier studies. At higher locations of the flame, soot concentration increases but is still low around the flame axis. At z=30mm height, the maximum soot concentration is in the outer layer at 3mm radius. this trend changed gradually with height, at z=50mm above the fuel outlet, the centerline of flame is the region of maximum soot concentration. Along the axis a narrowing of the visible flame is observed, which is well described by the sequence of the soot profiles. The comparison of the max soot volume fraction of our and Arana et.al's data at three height above the burner are shown in Table.2.

4.3 The Variation of the Value of Self-Attenuation Term

Fig. 8 shows the average value of self-attenuation term for the three heights at 800nm with attenuation correction applied in the first and second interation procedures. All values along the lateral axis at three height of flame before interation are less than 1.At z=10mm above the fuel tube, the self-attenuation term value is close to 1 due to the low soot volume fraction existed there, furthermore it changes a little after two interation procedures. Neverthless, at z=30 and 50mm, the self-attenuation term value has a significant change between before and after interation procedures since there are higher soot volume fraction relatively. The observation of Fig. 8 obviously reveals that the self-attenuation term value is close to 1 after two interation procedures, which means that the error of the assumption of optically thin flame could decrease to a minimum value after our correction procedures.

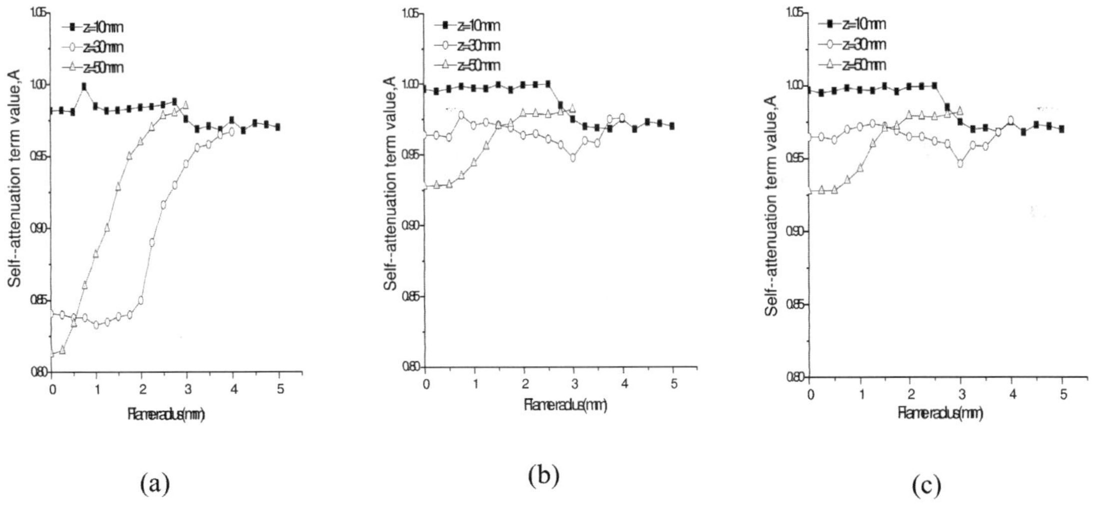

Fig. 8 Comparison of the average value of self-attenuation term with one time of iteration (b) and two times of iterations (c) and without iteration (a) for the three heights at 800 nm.

4.4 Soot Volume Fraction Reconstruction with Self-Attenuation Term

In Fig. 9, the soot volume fraction with self-attenuation correction applied after two iteration procedures are shown. The profile is compared to previously published data by R.J.Santoro et.al , they used combined light extinction and scattering to measure the soot distribution in nonpremixed ethylene coflow flames, which is not related to optically thin and a E(m) constant value assumption and has been extensively discussed in the literatures. Considering the different size of investigated flames, the results are in good agreement with earlier data. For example (Santoro et al., 1983), the maximum value for the soot volume fraction in the annular soot layer at z=30mm height (H/F≈ 0.50)is

7ppm, whereas the present value was 5.5ppm, the error of LE measurement is ±0.5-1ppm ,These two values are in best agreement within errors. The radial profile at z=30mm above the burner outlet reveals that the agreement of the variation trend and range along the lateral axis between our results and those by Santoro et.al (1983) is very good, the serious discrepancy is the data at z=50mm height, the region of maximum soot concentration is flame center for our result correspond to flame edge for data by Santoro et.al (1983).

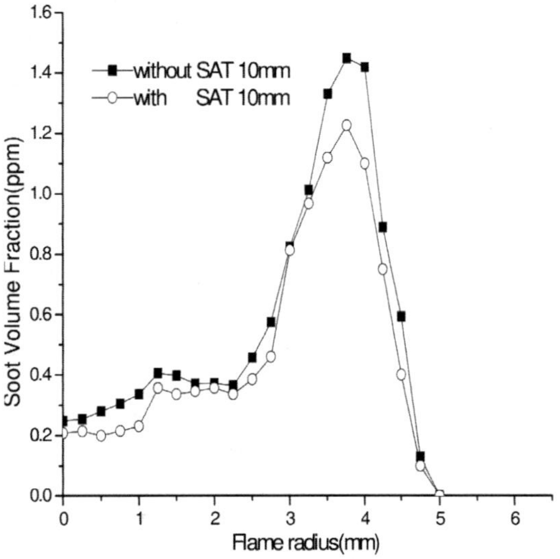

(a)

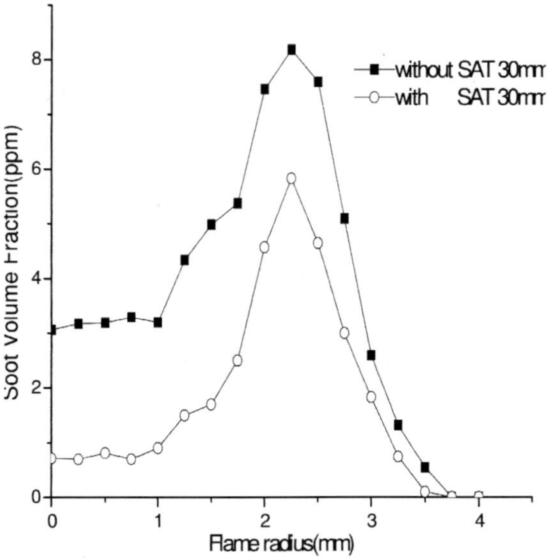

(b)

Table 3 Comparison of the max soot volume fraction of our and R.J.Santoro et.al's data at three heights above the burner

Height	Max Volume Fraction	
	With SAT (ppm)	R.J.Santoro et.al(ppm)
Z=10mm	1.2	1
Z=30mm	5.5	7
Z=50mm	7.4	9

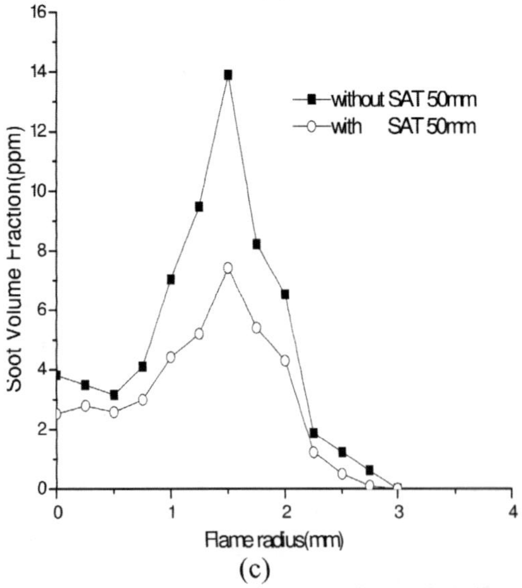

(c)

Fig. 9.Comparison of soot volume fraction distributions for a nonpremixed flame with and without self-attenation term and those of R.J.Santoro et.al.[2] at various heights above the burner(HAB).(a) through (c) ,correspond respectively to HAB=10,30 and 50mm

This is attributed to the different height of flame, which influence the results of the upper half of flame significantly. Table 2 summarizes the comparison of the max soot volume fraction of our and data by Santoro et.al (1983) at three heights above the burner. A comparison of soot volume fraction reconstruction with and without attenuation correction indicates that considering the self-attenuation term could reduce the value of soot volume fraction, these improve the measurement precision specially for the flames with higher soot loading.

5. CONCLUSIONS

An improved soot diagnostics technique based on tomographic reconstruction of flame emission spectra has been developed for nonintrusive characterization of soot volume fraction fields within an optically thick, axisymmetric laminar diffusion flame. Emission from the flame is scanned along the horizontal lateral axis of flame at several altitudes above the burner. At each measurement position, the local line-of-sight flame emission spectra is collected by the spectrometry over a spectral range of 700-1100nm. Inversion of these data through one-dimensional tomography using a three-point Abel inversion yields radial distributions of the soot radiation from which soot volume fraction profiles are extracted. With the conventional method, these procedures is applied only in the optically thin flames for which the self-attenuation term of emitted radiation is ignored, however, with our improved method, this self-attenuation term is considered by iterative calculations in the flames with no assumption of optically thin. The results by the conventional method are found to be in agreement with literatures but almost twice to the results by the improved method. The comparison of soot volume fraction reconstruction with and without attenuation correction indicates that considering the self-attenuation term could reduce the value of soot volume fraction, these improve the measurement precision specially for the flames with higher soot loading.

ACKNOWLEDGEMENTS

The present research was partially supported by the National Natrual Science Foundation of China (No. 50636010). Dr. Fengshan Liu from Combustion Technology Group, Institute for Chemical Process & Environmental Technology, National Research Council, Canada, helped author's group to carry out combustion simulation and experiment work, is gratefully appreciated.

NOMENCLATURE

I	radiation intensity	[W/m²/sr]
l	distance	[m]
d	radiation intensity	[W/m²/sr]
f_v	soot volume fraction	[ppm]
m	real part of the refractive index	[/]
n	imaginary of the refractive index	[/]
T	temperature	[K]

Greek Letters

κ	absorption coefficiency	[m^{-1}]
λ	wavelength	[nm]

Subscripts

b	black body

REFERENCES

Ayrancı I et.al (2008) "Near-infrared emission spectrometry measurements for nonintrusive soot diagnostics in flames", J. Quantitative Spectroscopy& Radiative Transfer pp.349-361

Arana.C.P. et.al (2004) "Field measurements of soot volume fractions in laminar partially premixed coflow ethylene/air flames", Combust. Flame 138, pp.362-372.

Bengtsson ,P.E (1991) "Carbon atom fluorescence and C2 emission detected in fuel-ric-h flames using a UV laser", Combust. Sci. Technol. 77 pp.307-318.

Bengtsson P.E .et.al (1995) "Soot-visualization strategies using laser techniques Laser-induced fluorescence in C2 from laser-vaporized soot and laser-induced soot incandescence", Appl. Phys. B 60 pp.51-59

Bengtsson P.E . (1996) "Simultaneous Two-Dimensional Visualization of Soot and OH in Flames Using Laser-Induced Fluorescence", Appl. Spectrosc.50pp.1182-1186.

Choi, M. Y. et.al, (1995) "Comparisons of the soot volume fraction using gravimetric and light extinction techniques", Combust. Flame 102 pp. 161-169

Choi, M.Y, et.al (1994) "Simultaneous optical measurement of soot volume fraction and temperature in premixed flames", Combust. Flame 99 pp.174 -186

Dasch, C.J et.al (1991) "Planar imaging of soot formation in turbulent ethylene diffusion flames: Fluctuations and integral scales" ,Combust.Flame 85,pp.389-402

DeIuliis S. et.al (1998) "Determination of the soot volume fraction in an ethylene diffusion flame by multiwavelength analysis of soot radiation", Combust. Flame 115 pp.253 –261

Dasch,C.J. (1992) "One-dimensional tomography: a comparison of able, Onion-Peeling, and Filtered Backprojection methods" Appl. Opt., 31, pp.1146-1152.

Dalzell, W.H. (1959) "Optical constants of soot and their application to heat flux calculation", J. Heat Transfer, 91, pp.100-104.

Greenberg ,P.S.et.al. (1997) "Soot volume fraction imaging", Appl. Opt. 36, p. 5514.-5522

Gulder,O.L.et.al (1993) "Influence of nitrogen dilution and flame temperature on soot formation in diffusion flames", Combust. Flame 92 pp. 115-124.

Hall, R.J et.al, (1990) "Sooting flame thermometry using emission/absorption tomography", Appl.Opt.29 pp.4590-4598.

Kennedy, I.M. (1997) "Models of soot formation and oxidation", Prog. Energy Combust.Sci.17 pp..95-132

Klassen,M et.al (1992) "Simultaneous emission absorption measurements in toluene-fueled pool flames: Mean and RMS Properties", Combust. Flame 90 pp.34-44.

Konsur,B.et.al, (1999) "Fuel preheat effects on soot-field structure in laminar gas jet diffusion flames burning in 0-g and 1-g", Combust. Flame 116 pp.334-347

Köylü, Ü.Ö et.al. (1994) "Optical properties of soot in buoyant laminar diffusion flames", J. Heat Transfer 116 pp.971-979.

Köylü.,Ü.Ö. (1997)"Quantitative analysis of in situ optical diagnostics for inferring particle/aggregate parameters in flames:implications for soot surface growth and total emissivity",Combust.Flame,109, pp.488-500

Köylü, Ü.Ö.et.al(1994) "Optical properties of overfire soot in buoyant turbulent diffusion flames at long residence times", J. Heat Transfer 116 pp.152-159.

Mitrovic, A.et.al, (1998) "Soot formation characteristics of laminar partially premixed flames", Combust. Flame 115pp.437-442.

Quay, B et.al. (1994) "Spatially resolved measurements of soot volume fraction using laser-induced incandescence", Combust, Flame 97 pp.384-392

Snelling, D.R et.al, (1999) "Two-Dimensional Imaging of Soot Volume Fraction in Laminar Diffusion Flames", Appl. Opt.38 pp.2478-2485.

Snelling, D.R et.al,(2002). "Spectrally Resolved Measurement of Flame Radiation to Determine Soot Temperature and Concentration" AIAA Journal 9,pp 1789-1795

Shaddix,C.R.et.al, (1996) "Laser-induced incandescence measurements of soot production in steady and flickering methane, propane, and ethylene diffusion flames", Combust. Flame 105 pp.418-452.

Santoro,R.J. et.al (1983), "Soot particle measurements in diffusion flames", Combust. Flame 51 pp.203-218

Smyth, K.C.et.al., (1996), "Elusive history of m=1.57-0.56i for the refractive index of soot", Combust. Flame.107 pp.314-320

Sivathanu, Y.R. et.al, (1992) "Transient Structure and Radiation Properties of Strongly Radiating Buoyant Flames" J. Heat Transfer 114 pp.659-665

Sivathanu, Y.R. et.al (1991) "Simultaneous Multiline Emission Absorption Measurements in Optically Thick Turbulent Flames" , Combust. Sci. Technol. 80 pp.1-21

Sivathanu,Y.R. (1990) "Temperature / soot volume fraction correlations in the fuel-rich region of buoyant turbulent diffusion flames", Combust. Flame 81 pp. 150-165.

Vest, C.M. (1979) "Holographic Interferometry", Wiley, New York.

Westblom , U. (1991) "Carbon atom fluorescence and C2 emission detected in fuel-rich flames using a UV laser", Appl. Phys. B 52.pp.371-375

Xu, F. et.al (1998) "Soot Formation in Laminar Premixed Methane/Oxygen Flames at Atmospheric Pressure" , Combust. Flame 115 pp.195-209.

High Concentration Effects of Transmission Fluctuation Spectrometry with Temporal Correlation

Bin Yu, Jianqi Shen, Yamin Xu, Huarui Wang

University of Shanghai for Science and Technology, 516 Jungong Rd, Shanghai 200093, CHINA

yubin7909@163.com, shenjq@online.sh.cn

Abstract. Transmission fluctuation spectrometry (TFS) with temporal or spatial correlation is a recently-developed method for real-time, online/inline particle analysis in two-phase flows, whereby the particle size distribution (PSD) and particle concentration can be measured simultaneously. On the basis of a layer model and the assumption of geometrical ray propagation, an analytical expression of the transmission fluctuation spectrum can be obtained. It is found that the measurement matches the theory very well as long as the particle concentration is low enough. However, as the particle concentration increases, the measured transmission fluctuation spectrum deviates gradually from the theory, which is a result from the high concentration effects. The high concentration effects include those from particle overlapping and the monolayer structure. In this presentation, the high concentration effects are studied with numerical simulation. As a result, the empirical expressions for the correction of the high concentration effects are obtained.

1. INTRODUCTION

Real-time, online/inline particle size analysis plays an important role in many areas of industry. For instance, the size of pulverized coal used in the power plant is an important parameter that affects both combustion efficiency and atmospheric emissions of the plant. Particle size is particularly important in quality control applications and is necessary for control of critical aspects of a process. Till now, the development of the reliable technique for real-time, online/inline particle sizing remains a challenge.

The theory of transmission fluctuation spectrometry (TFS) with correlation has recently been developed as a new method of particle size analysis, which can be used for real-time, online/inline applications. The theory is based on the layer model and the assumption of the geometrical ray propagation. An analytical expression of the transmission fluctuation correlation spectrum can be obtained. Two approaches to the TFS have been developed. One is the spatial correlation (transmission fluctuation spectrometry with spatial correlation, TFS-SC), in which two parallel narrow beams with a variable separation between each other are employed (Shen, et al., 2005a). However, the mechanical operation on the beam separation becomes challenging, especially for the measurements on small particles. The beams and the detectors should be able to move synchronously and fast enough in a micrometer scale. Furthermore, the beams should be consistent with each other both in the beam diameters and in propagating directions, which is very difficult to realize. So in real measurements, the temporal correlation is used with only one narrow beam (Shen et al., 2005b; Guo et al., 2005). This method is called as transmission fluctuation spectrometry with temporal correlation or auto-correlation (shortly TFS-TC or TFS-AC) (Yu et al., 2007a).

Journal of Physics: Conference Series **147** (2009) 012085 doi:10.1088/1742-6596/147/1/012085

It is found that the measurement matches the theory very well as long as the particle concentration is low enough. However, if the particle concentration is high, the measurement deviates from the theory. So in this presentation we concentrate on the high concentration effects. Similarly to what we have done before, effects such as multiple scattering and dependent scattering can be excluded so that only the steric interactions between particles are accounted for (Shen et al., 2004, 2005c; Riebel et al., 2004). Steric interactions between particles are classified into those between particles within the same monolayer (i.e. the monolayer structure) and those between particles from different monolayers (known as particle overlapping), which cause different effects on the transmission fluctuation correlation spectrum. Theoretical and experimental studies show that, if the beam diameter is infinitely small compared to the particle size, the effect from particle overlapping can be excluded so that the transmission fluctuation spectrum is affected by the monolayer structure only (Shen et al., 2005a, b; Yu et al., 2007b). However, while the beam diameter is not very small compared to the particle size, effects from particle overlapping are much stronger than those from the monolayer structure (Guo et al., 2005; Yu et al., 2007a). Due to the mathematical difficulties, the high concentration is studied with simulation. As the result, the empirical expression is obtained for the correction of the high concentration effects on the TFS.

2. PRINCIPLE OF THE MEASUREMENTS

In the TFS-TC, the fluctuating transmission signal $T(t)$ of a narrow beam is recorded in a very long period of sampling time t_s. The expectancy of transmission product (ETP) is then calculated with a variable time difference (or the correlation time) τ:

$$\text{ETP} = \lim_{t_s \to \infty} \frac{1}{t_s} \int_0^{t_s} T(t) T(t+\tau) dt \tag{1}$$

By varying the correlation time, the ETP is obtained as a spectrum, which can be further solved to extract the information on particle size distribution and particle concentration simultaneously.

When the particle concentration is not high, the ETP of a 3-dimensional particle system can be expressed as (Shen et al., 2005a):

$$\begin{aligned} \text{ETP} &= e\{T(t)T(t+\tau)\} \approx \prod_{i=1}^{N} e\{T_{ML_i}(t) T_{ML_i}(t+\tau)\} \\ &= e\{T_{ML}(t) T_{ML}(t+\tau)\}^N = \langle T(t)T(t+\tau)\rangle = (\text{ETP}_{ML})^N \end{aligned} \tag{2}$$

which means that the particle system can be taken as a pile of several monolayers. Therefore, we may start from a monolayer containing the mono-dispersed spherical particles. After a tedious deduction, the analytical solution of the ETP through a monolayer is obtained (Shen et al., 2007):

$$\text{ETP}_{ML} = 1 - \beta[2 - \chi(\Delta, \Lambda)] + \beta^2[1 + \varepsilon(\Delta, \Lambda, \beta)] \tag{3}$$

Here, β is the monolayer density, which is related to the volume particle concentration C_V (i.e. $\beta = PC_V$, $P \geq 1.5$ is the parameter of the suspension structure). The transition function $\chi(\Delta, \Lambda)$ and $\varepsilon(\Delta, \Lambda, \beta)$ are defined as:

$$\chi(\Delta, \Lambda) = \int_0^{+\infty} J_0(2u\Delta) \cdot \left[\frac{2J_1(\Lambda \cdot u)}{\Lambda \cdot u}\right]^2 \cdot \frac{2J_1^2(u)}{u} du \tag{4}$$

$$\varepsilon(\Delta, \Lambda, \beta) = \int_0^{+\infty} J_0(2u\Delta) \cdot \left[\frac{2J_1(\Lambda \cdot u)}{\Lambda \cdot u}\right]^2 \cdot \frac{2J_1^2(u)}{u} \cdot F_{MS} \cdot du \tag{5}$$

Here, $\Delta = v \cdot \tau / x$ (v is the flow velocity) is called as the dimensionless correlation time, $\Lambda = D/x$

(D is the beam diameter) is the dimensionless beam diameter (or beam-to-particle diameter ratio), and x is the particle diameter. The factor $\left[2J_1(\Lambda,u)/(\Lambda\cdot u)\right]^2$ originates from spatial averaging over the cross section of the circular uniform beam, $2J_1^2(u)/u$ describes the property of the spherical particle, and $J_0(2u\Delta)$ is the effect of temporal correlation. The factor F_{MS} describes the monolayer structure, which is determined by the particle concentration and the particle interaction potential (Shen et al., 2003).

When the particle concentration is not high, the logarithm of the ETP of a 3-dimensional mono-dispersed particle system can be expressed as:

$$\ln\text{ETP} = -\frac{1.5}{x}\Delta Z\cdot C_V\left[2-\chi(\Delta,\Lambda)\right] \tag{6}$$

Here, ΔZ is the beam path in the measuring zone (i.e. the thickness of the particle system in the direction of ray propagation).

Eq.(6) can be extended to a poly-dispersed particle system, expressed as

$$\ln\text{ETP}(\tau_j) = -\sum_i \frac{1.5}{\overline{x}_i} C_V(\overline{x}_i)\left\{2-\chi\left[\Delta_{i,j},\Lambda_i(Z,\overline{x}_i)\right]\right\}\Delta Z \tag{7}$$

τ_j ($j=1,2,\cdots,N$) is the variable time difference for signal correlation. $\overline{x}_i$ is the mean particle diameter in the ith fraction of particle size ($i=1,2,\cdots,M$) and $C_V(\overline{x}_i)$ is the corresponding volume concentration. $\Delta_{i,j}=v\tau_j/x_i$ is the dimensionless correlation time.

Eq.(7) shows that the logarithm of the ETP is linear with respect to the particle concentration and the effects from the different particle size fractions superimpose linearly to form the logarithm of ETP. Therefore, particle size distribution and particle concentration can be retrieved by deconvoluting from the measurement of transmission fluctuations.

3. EFFECTS FROM THE MONOLAYER STRUCTURE

3.1 A Brief Discussion on the Monolayer Structure

The monolayer structure is determined by the particle concentration and the interaction potentials between particles. It is difficult to give an analytical expression exactly even for the simplest case of a system containing hard spherical particles, which corresponds to an interaction potential of infinity for interpenetrating particles and an interaction potential of zero for non-contacting particles. Only approximate solutions can be obtained. The monolayer structure is, in the theory of granularity of a monolayer, statistically characterized by means of the 2-dimensional particle pair correlation function (PPCF), $g(r,\beta)$. Provided that the arrangement of spherical particles in the monolayer is isotropic and macroscopically homogeneous, the particle pair correlation function may be defined in terms of the radial distribution function:

$$g(r,\beta) = \frac{dN(r)}{C_N\cdot 2\pi r\cdot dr} \tag{8}$$

where $dN(r)$ is the number of particles found in an annular zone of width dr at a distance r from a central particle, $C_N = 4\beta/\pi x^2$ is the particle number concentration of the monolayer.

The PPCF can be numerically calculated with the Percus-Yevick approximation (Ripoll & Tejero, 1995; Shen et al., 2004). As shown in Fig.1, the PPCF varies as the monolayer density increases. In the limit of low particle concentration, the PPCF is

$$g(r,0) = g(r,\beta)\big|_{\beta=0} = \begin{cases} 0 & \text{for } r/x \le 1 \\ 1 & \text{else} \end{cases} \tag{9}$$

which means that an inter-penetration between hard particles in the single monolayer is prohibited and the particles can arrange randomly for distances of $r > x$.

As the particle concentration increases, an increasingly oscillatory deviation from $g(r,0)=1$ occurs in proximity to $r = x$. Both the amplitude and the reach of the oscillation grow with increasing concentration, indicating the gradual transition from disorder to low-range order and further to far-range order in the monolayer structure.

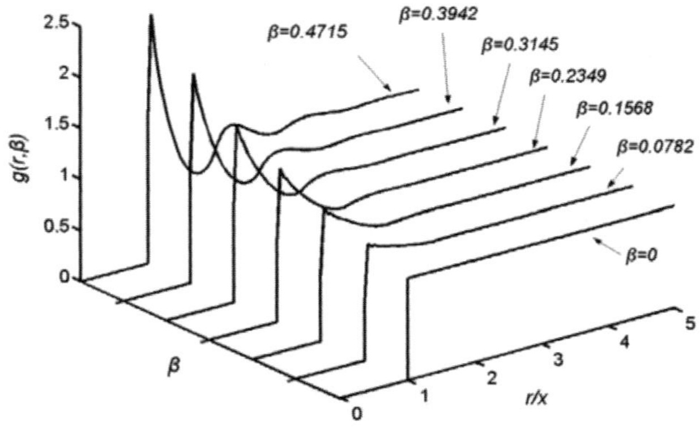

Fig.1 The particle pair correlation function at different concentrations, $g(r,\beta)$.

3.2 Simulations on the Monolayer Structure

Numerical simulation on the monolayer is performed with monodisperse spherical particles contained in a monolayer of square shape. The extension of the monolayer is perpendicular to the radiation and is very large compared with the particle size so that the deviation from statistics could be somewhat tiny. The outline of the simulation was detailedly introduced by Shen et al., 2003.

A new transition function of ETP_{ML} is introduced, $\Xi_{ML} = \ln\{ETP_{ML}\}/\ln\{ET_{ML}\}$, whereby $ET_{ML} = 1-\beta$ is the average transmission. Accordingly, the step height of the transition function is defined as:

$$\Xi_{ML,SH} = \Xi_{ML}\big|_{\Delta\to\infty} - \Xi_{ML}\big|_{\Delta=0} = 2 - \Xi_{ML}\big|_{\Delta=0} \tag{10}$$

The relationship between Ξ_{ML} and $\chi(\Delta,\Lambda)$ is found to be $\lim_{\beta\to 0}\Xi_{ML} = 2-\chi$ in the limit of low particle concentration.

As an example, Fig.2 shows the theoretical and simulated results in which the beam-to-particle diameter ratio is $\Lambda = 1.0$. It can be found that, as the monolayer density increases, the step height of the transition function decreases and the transition function shifts to the range of smaller dimensionless correlation time gradually. The peak of the transition function Ξ_{ML} appears close to $\Delta = 1$ with a slender fluctuant. Comparing Fig.2a with Fig.2b, it can be found that they show the similar change on the step height of the transition function and the dimensionless correlation time response. No more than, the change of the theoretical results is more obvious than that of the simulation results. The difference between the theoretical results and the simulated ones might be caused by both the approximation used in theoretical calculation of the PPCF and the deviation from statistics in the simulation due to the limited dimensions of the monolayer.

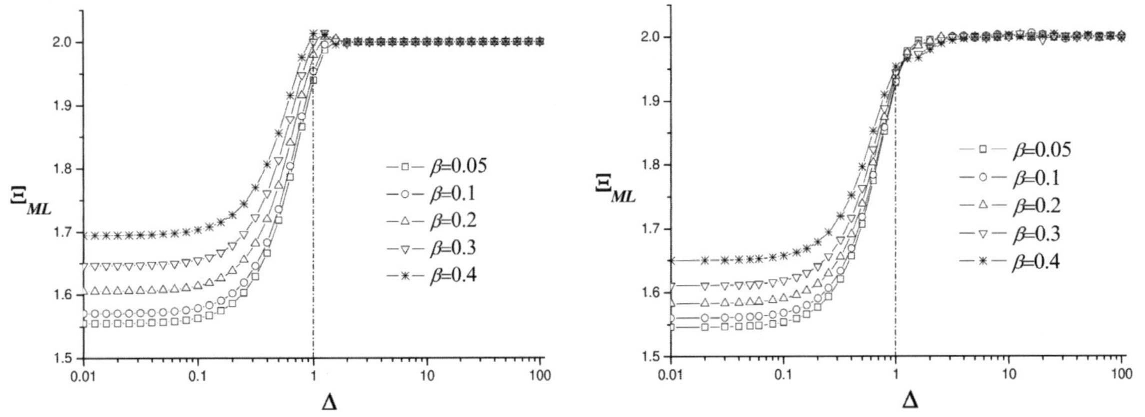

(a) The theoretical results. (b) The simulated results.

Fig.2 Transition functions of the ETP through a monolayer irradiated by a finite narrow beam.

4. EFFECTS FROM PARTICLE OVERLAPPING

4.1 A Brief Discussion on Particle Overlapping

Shielding effects can be traced back to near-field phenomena. Considering the near field diffraction, the radiation incident upon the second monolayer is strongly dependent on particle arrangement in the first monolayer. Even at a very low concentration, when the particle arrangement in each monolayer is random, there are statistically a certain part of the particles in the second layer located in the shadows produced by those particles in the first layer. This interaction might be called as particle intersection or particle overlapping.

4.2 Simulations of Particle Overlapping

The high concentration effects due to particle overlapping can be quantified by the extinction of the irradiation *E*, which is approximately a product of the monolayer density β and the number of layers N (Yu et al., 2007a).

$$E = -\ln T = -\ln\left(1-\beta\right)^{N} \approx \beta \cdot N \tag{11}$$

From this point of view, an increase of particle concentration is approximately equivalent to an increase of the layer number. However, the increase of particle concentration additionally leads to increasing effects from the monolayer structure. Therefore, in order to avoid any disturbance from the monolayer structure in the simulations, the particle concentration is assumed to be very low. Alternatively, this may be simulated either by assuming a large number of layers, each layer containing very few particles only, or by assuming a single, dense monolayer containing non-interacting particles, whereby particles are allowed to interpenetrate with each other. The dense monolayer containing *N* non-interacting particles is equivalent to a 3-dimensional particle suspension containing *N* layers, each layer containing one particle only and hence the monolayer density is

$$\beta = \frac{\pi x^2}{4 \cdot l_1 l_2} \tag{12}$$

As long as the dimensions of the monolayer, l_1 and l_2, are large enough, the monolayer density tends to zero so that the average transmission is expected to be:

$$e\left(T\right) = \left(1-\beta\right)^{N} \rightarrow e^{-\beta \cdot N} \tag{13}$$

A detailed description of the simulation on particle overlapping was given by Yu et al., 2007a so that only simulated results are presented here. As is given in Fig.3a, when the beam-to-particle diameter ratio is $\Lambda = 0$, effects from particle overlapping are invisible. However, if the beam-to-particle diameter ratio equals 1, the step height of the transition function increases and the transition function shifts to the range of larger dimensionless correlation time gradually as the extinction increases. Similar effects are found for different values of Λ, which means the effects from particle overlapping have the same tendency and are dependent on the beam-to-particle diameter ratio Λ.

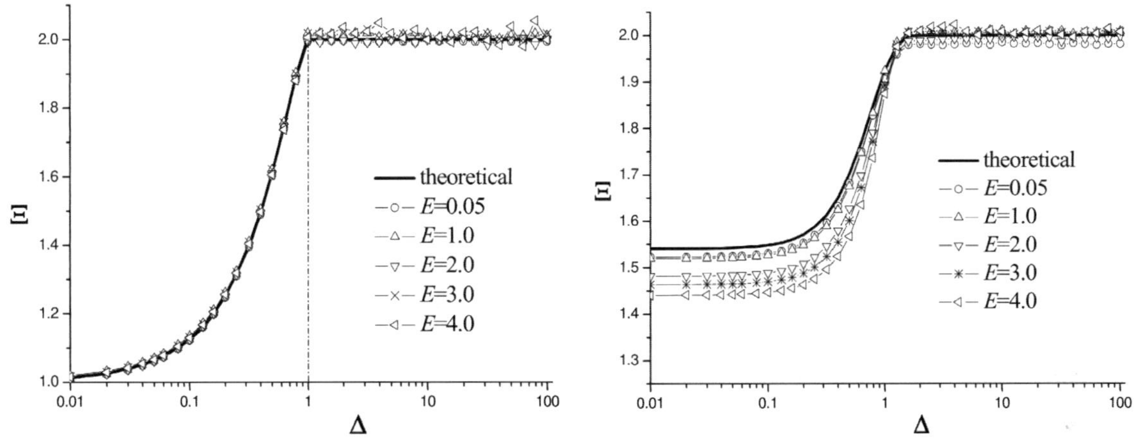

(a) $\Lambda = 0$ (b) $\Lambda = 1.0$

Fig.3 Simulated results of particle overlapping.

5. THE HIGH CONCENTRATION EFFECTS

Usually, the particle concentration is not low so that particle overlapping and the monolayer structure exist simultaneously, leading the high concentration effects to be more complicated. Fig.4a shows the dependence of the transition functions on the extinction, in which the number of layers is $N = 40$ and the beam-to-particle diameter ratio is $\Lambda = 1.0$. It can be found that, as the extinction increases, the step height of the transition function increases and the transition function shifts to the range of larger dimensionless correlation time gradually. And the simulated curves are below the theoretical curve in the range of the dimensionless correlation time $\Delta < 1$. This means that the transmission fluctuation spectrum is mainly affected by particle overlapping. As a comparison, the transition functions for the number of layers $N = 5$ and the beam-to-particle diameter ratio $\Lambda = 1.0$ are simulated, as shown in Fig.4b. It can be found that, as the extinction increases, the step height of the transition function decreases and the transition function shifts to the range of smaller dimensionless correlation time gradually. All the simulated curves are over the theoretical curve in the range of the dimensionless correlation time $\Delta < 1$. This is because that, in the former case, the number of layers is large and the monolayer density (or the particle concentration) is not very high so that effects from the monolayer structure are weaker than those from particle overlapping. However, in the latter case, the layer number is small and the particle concentration is high so that effects from monolayer structure are much stronger than those from particle overlapping.

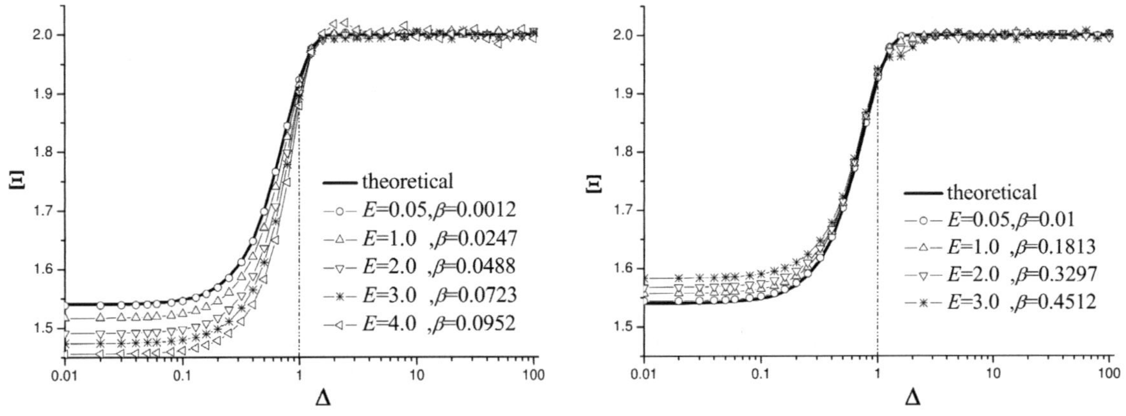

(*a*) $N = 40$ (*b*) $N = 5$

Fig.4 Dependence of the transition functions on the extinction. ($\Lambda = 1.0$)

6. EMPIRICAL CORRECTION FROM MEASUREMENTS

6.1 Correction of the High Concentration Effects on the Spectrum Response

In order to see the high concentration effects on the response of the transmission fluctuation spectrum, the transition function is introduced into a normalized transition function, separating out the change of the step height. The normalized transition function $\Xi_N(\Delta)$ is defined as:

$$\Xi_N(\Delta) = \frac{\Xi(\Delta) - \Xi(\Delta \to 0)}{\Xi(\Delta \to \infty) - \Xi(\Delta \to 0)} + 1 \tag{14}$$

When $0 < \Lambda \leq 4$, the high concentration effects on the normalized transition function can be empirically expressed as the deviation from theoretical normalized transition function in low concentration limit, defined as:

$$\Delta\Xi_N(\Delta) = \left[\Xi_N(\Delta) - \Xi_N(\Delta, C_V \to 0) \cdot e^{-(\Lambda+1)\cdot\Delta}\right] \cdot e^{-\left[(4-\Lambda)+\frac{1}{\Lambda+1}\right]\cdot\Delta} \tag{15}$$

Here, $\Xi_N(\Delta, C_V \to 0)$ is the theoretical normalized transition function in low concentration limit.

In order to take a deeper look, the deviation of the normalized transition function is factorized into a product of the Module M and the normalized deviation $\zeta(\Delta)$. M is the average of the maximum of the deviation of the normalized transition function $\Delta\Xi_N(\Delta)$:

$$\Delta\Xi_N(\Delta) = M \cdot \zeta(\Delta) \tag{16}$$

The normalized deviation $\zeta(\Delta)$ depends on the dimensionless beam diameter, which can be described with a translation of the dimensionless correlation time of:

$$\Delta' = \Delta \cdot \left[\frac{1 + 0.285 \cdot (\Lambda + 0.45) \cdot e^{-0.51\cdot\Lambda\cdot\Delta}}{1 + 0.546 \cdot (\Lambda + 0.1)}\right] \tag{17}$$

and the distribution of the normalized deviation might be described with an empirical expression:

$$\zeta(\Delta') = \left(\frac{e}{a}\right)^a \cdot \xi^a \cdot e^{-\xi}, \quad \text{here } \xi = \left(\frac{\Delta'}{b}\right)^c \tag{18}$$

The parameters are $a = 1.31843$, $b = 0.162711$ and $c = 0.860572$.

Fig.5 shows the dependence of M on Λ, with different values of E and N. As shown in Fig.5a, when N keeps unchanged (e.g. $N = 40$) and E is 0.05, 2.0 and 4.0 respectively, M increases along with the increase of Λ. These three curves are approximately consistent with each other. Similarly, when E is unchanged (e.g. $E = 0.05$) and N is 5, 10, 40 respectively, M increases along with the increase of Λ (see Fig.5b). Therefore, M is related to the beam-to-particle diameter ratio Λ only and almost independent of E and N. This is quite different from what we obtained before in TFS with signal averaging or filtering (Xu et al., 2006). The approximate expression is obtained

$$M = M(\Lambda) = 0.028\Lambda^2 + 0.061\Lambda + 0.087 \qquad (19)$$

From the simulation above, we may conclude that effects from particle overlapping are closely related to the extinction E, the layer number N and the beam-to-particle diameter ratio Λ.

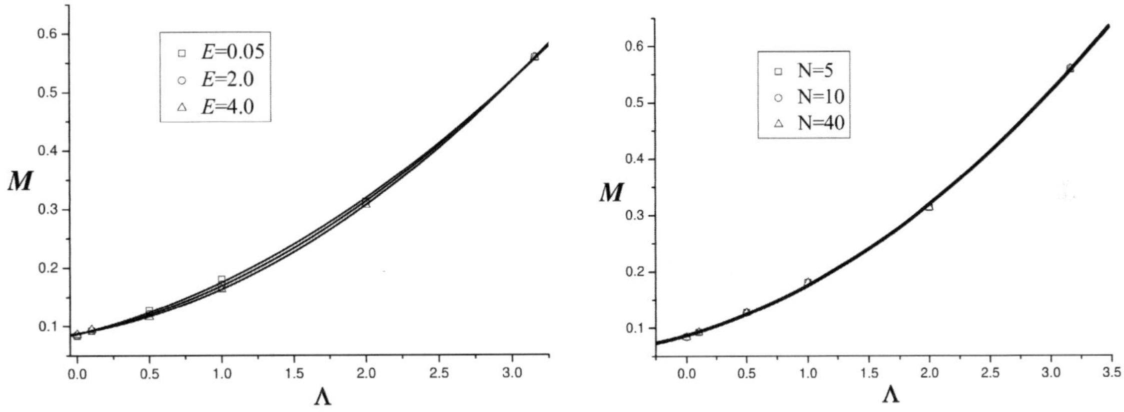

(a) $N = 40$
The curves correspond to E= 0.05, 2.0, 4.0

(b) $E = 0.05$
The curves correspond to N=5, 10, 40

Fig.5 The module of the normalized deviation

6.2 Correction of the High Concentration Effects on the Step Height

In order to correct the high concentration effects on the step height of the transition function, another function K is introduced, which is defined as $K = \Xi(\Delta = 0)$. The empirical expression is found from the simulation:

$$K = K(E) = f \cdot E^2 + g \cdot E + h \qquad (20)$$

Here, f, g, h are found to depend on the beam-to-particle diameter ratio Λ:

$$\begin{aligned} f &= -0.002\Lambda \\ g &= 0.012\Lambda^2 - 0.044\Lambda \qquad (21) \\ h &= -0.115\Lambda^2 + 0.650\Lambda + 1 \end{aligned}$$

One special case is that the particle is much larger than the beam diameter, i.e. the beam-to-particle diameter ratio is $\Lambda = 0$. In this case, no matter the particle system is thin or thick, the function is found to be $K|_{\Lambda=0} = 1$, which was proved by both the simulations of $N = 5, 10, 20, 30, 40$ and the theory (Shen et al., 2005a, b; Yu et al.,2007b). Since the beam-to-particle diameter ratio is equal to 0, the step height of the transition function keeps unchanged. This implies that, if the beam diameter is infinitely small compared to the particle size, effects from particle overlapping are so small that the transmission fluctuation spectrum is affected by the monolayer structure only. It should be pointed out that another special case occurs when the number of layers is very small, wherein the high concentration effects arise mainly from the monolayer structure. As it was given in the simulation (see Fig.2), in this case the step height decreases when the particle concentration increases. The variation of the step height caused by the monolayer structure depends closely on the beam-to-particle diameter ratio. However, this situation is not very important because in real measurements the particle system generally includes a lot of layers.

Fig.6 shows the simulated effects on the step height from particle overlapping (i.e. effects from the monolayer structure are excluded, please refer to sec.4.2), wherein the values calculated with the empirical expression are plotted in lines and the simulated results are given with dots. The match between the empirical expression and the simulation is satisfactory.

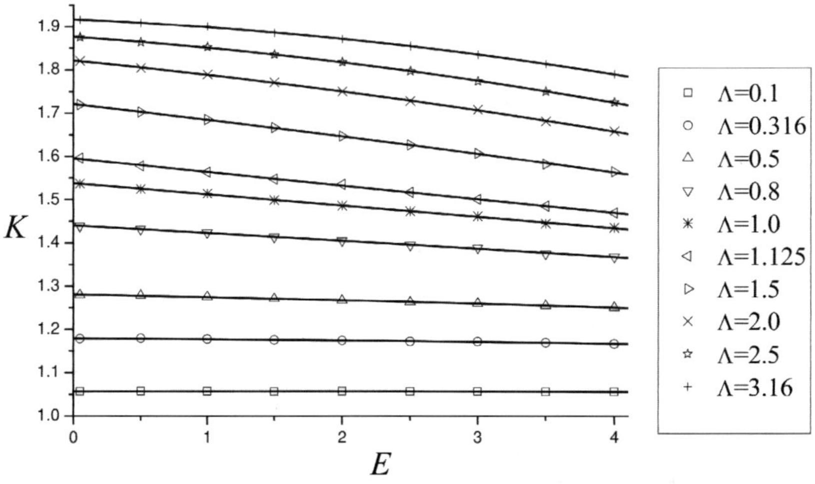

Fig.6 Dependence of K on the extinct on particle overlapping.

For particle systems which contain both the monolayer structure and particle overlapping, simulations are given in Figs.7 & 8. All the values calculated with the empirical expression are plotted in lines and the simulated results are given with dots. In Fig.7, the number of layers is $N = 20$ and the beam-to-particle diameter ratio is 0.1, 0.5, 1.0, 2.0 and 3.16 respectively.

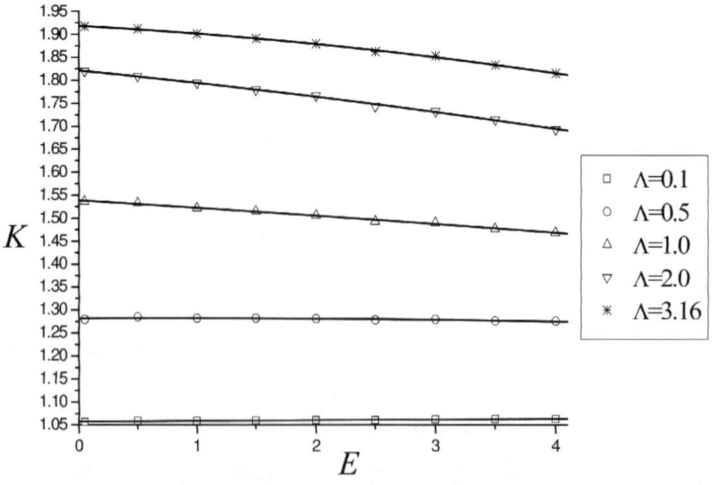

Fig.7 Dependence of K on the extinct E ($N = 20$).

In Fig.8, the beam-to-particle diameter ratio is $\Lambda = 1$ and the number of layers is $N = 5, 10, 20, 30, 40$ respectively. It can be found that all these curves except for $N = 5$ decreases along with the increase of the extinction. This is because that, when $N = 5$, the effects from the monolayer structure are stronger than those from particle overlapping. In all the cases plotted in Figs. 6-8, the empirical expression for the step height can match the simulated results very well.

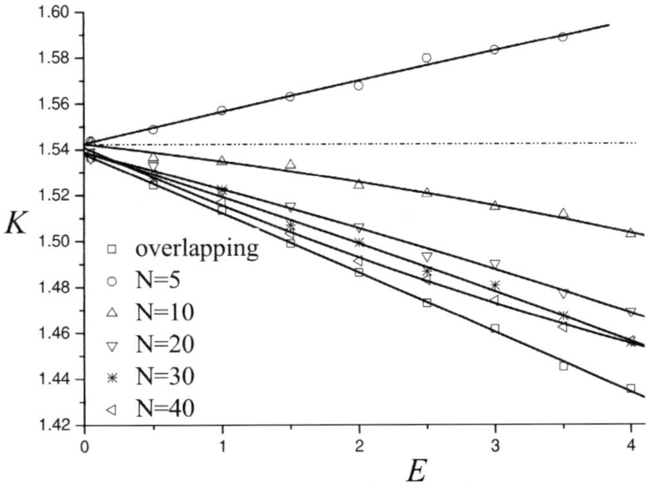

Fig.8 Dependence of K on the extinct E ($\Lambda = 1$).

7. CONCLUSIONS

In this presentation, the high concentration effects on the transmission fluctuation spectrum with temporal correlation are studied. It is found that, in usual cases, these effects arise from both particle overlapping and the monolayer structure. The increase of particle concentration enhances the effects from both the monolayer structure and particle overlapping, while the increase of the layer number enhances the effects from particle overlapping only. The monolayer structure shifts the transmission fluctuation spectrum to the range of lower correlation time and reduces the step height of the transition function. On the contrary, particle overlapping shifts the transmission fluctuation spectrum to the range of higher correlation time and increases the step height of the transition function. Therefore, in real measurements, there is a balance between the high concentration effects from the monolayer structure and those from particle overlapping. Usually, the particle concentration is not too high and the particle system is thick enough to contain a lot of layers. Thus, effects from particle overlapping are much stronger than those from the monolayer structure.

The simulated results show that the high concentration effects are closely related to the beam-to-particle diameter ratio, the number of layers and the particle concentration (or the extinction). Analyzing the simulated results, the empirical expressions are ackieved which indicate both the variation of the temporal response of spectrum and the variation of the step height of the transition function. The empirical expressions may be hopefully used to correct the high concentration effects in the inverse problem to extract particle size distribution and particle concentration.

ACKNOWLEDGEMENTS

The authors thank for the financial support from the National Natural Science Foundation of China (NSFC 50876069) and the Innovation Fund Project for Graduate Student of Shanghai.

NOMENCLATURE

C_N	particle number concentration	
C_V	particle volume concentration	
D	beam diameter	[m]
$e(\ )$	expectancy of	

Journal of Physics: Conference Series **147** (2009) 012085 doi:10.1088/1742-6596/147/1/012085

E	extinction	
ET	the average transmission	
ETP	the expectancy of transmission product, $e(T_1T_2)$	
F_{MS}	factor of the monolayer structure	
$g(r,\beta)$	particle pair correlation function	
J_0, J_1	Bessel function	
K	the function with parameters of *f*, *g* and *h*	
l_1, l_2	length of the edges of the monolayer in simulations	[m]
M	module of the deviation (with parameters of k_1, k_2 and k_3)	
N	number of monolayers	
P	monolayer structural parameter $P \geq 1.5$	
PSD	the particle size distribution	
T	transmission	
$T(t)$	fluctuation transmission signal	
TFS	transmission fluctuation spectrometry	
TFS-AC	transmission fluctuation spectrometry with auto-correlation	
TFS-SC	transmission fluctuation spectrometry with spatial correlation	
TFS-TC	transmission fluctuation spectrometry with temporal correlation	
t_s	sampling time	[s]
u	integration variable	
v	particle flow velocity	[m/s]
x	particle diameter	[m]
$\overline{x}_i$	the mean particle diameter in the ith fraction of particle size	[m]

Greek Letters

β	monolayer density	
Δ	dimensionless correlation time, $\Delta = v \cdot \tau / x$	
ΔZ	the beam path in the measuring zone	[m]
$\Delta\Xi, \Delta\Xi_N$	deviation of the transition function, of the normalized transition function	
$\varepsilon(\ )$	transition function of ETP	
$\zeta(\Delta)$	distribution of the normalized deviation (with parameters of a, b and c)	
Λ	dimensionless spatial averaging parameter / beam-to-particle diameter ratio $\Lambda = D/x$	
Ξ, Ξ_N	transition function of ETP, normalized transition function	
Ξ_{SH}	step height of the transition function, defined as the difference between the higher level and the lower level of the transition function, i.e. $\Xi_{SH} = 2 - \Xi(\Delta \to 0)$	
τ	correlation time	
$\chi(\ )$	transition function	

Subscripts

i, j	variables of the number

ML	monolayer
N	normalized
SH	step height, defined as the difference between the higher level and the lower level of the transition function
V	volume

REFERENCES

Guo, X. et al. (2005). " Experimental Study on Particle Size Distribution and Concentration Using Transmission Fluctuation Spectrometry with the Autocorrelation Technique", Part. Part. Syst. Charact., 22, pp. 161-171.

Riebel, U. et al. (2004). " Transmission Fluctuation Spectrometry in Concentrated Suspensions. Part 2: Effects of Particle Overlapping", Part. Part. Syst. Charact., 21, pp.440-454.

Ripoll, M. S. & Tejero, C. F. (1995). "Approximate analytical expression for the direct correlation function of hard discs within the Percus-Yevick equation", Molecular Physics 85, pp.423-428.

Shen, J. et al. (2003). " Particle Size Analysis by Transmission Fluctuation Spectrometry: Fundamentals and Case Studies", Dissertation, Cuvillier Verlag Göttingen.

Shen, J. et al. (2004). " Transmission Fluctuation Spectrometry in Concentrated Suspensions. Part 1: Effects of the Monolayer Structure", Part. Part. Syst. Charact., 21, pp. 429-439.

Shen, J. et al. (2005a). " Transmission Fluctuation Spectrometry with Spatial Correlation", Part. Part. Syst. Charact., 22, pp. 24-37.

Shen, J.et al. (2005b). " Measurements on Particle Size Distribution and Concentration by Transmission Fluctuation Spectrometry with Temporal Correlation", Opt. Lett., 30, pp.2098-2100.

Shen, J. et al. (2005c). " Transmission Fluctuation Spectrometry in Concentrated Suspensions. Part 3: Measurements", Part. Part. Syst. Charact., 22, pp.14-23.

Shen, J. et al. (2007). " Signal Processing in Transmission Fluctuation Spectrometry with Band-Pass Filters", America Institute of Physics Conference Proceeding, 914, pp. 226-231.

Xu, Y. et al. (2006). " Simulation on High Concentration Effects of Transmission Fluctuation Spectrometry", Chinese Journal of Lasers, 33, pp. 253-260.

Yu, Bin et al. (2007a). " Simulations on Particle Overlapping of Transmission Fluctuation Spectrometry with Temporal Correlation", America Institute of Physics Conference Proceeding, 914, pp. 341-349.

Yu, Bin et al. (2007b). " Simulations on High Concentration Effects of Transmission Fluctuation Spectrometry with Temporal Correlation with Infinitesimally Thin Beam", ACTA OPTICA SINICA, 27, pp. 1309-1315. (in Chinese)

Journal of Physics: Conference Series **147** (2009) 012086 doi:10.1088/1742-6596/147/1/012086

Recent Achievements in Measurements of Soot Volume Fraction and Temperatures in a Coflow, Diffuse Ethylene-air Flame by Visible Image Processing

Huai-Chun Zhou *[1], Chun Lou [1], Jing Lu [1]

[1] Huazhong University of Science and Technology, 1037 Luoyu Road, Hongshan, Wuhan, Hebei, 430074, P. R. CHINA

hczhou@mail.hust.edu.cn

Abstract. In this review paper, the recent achievements in measurements of soot volume fraction and temperatures in a coflow, diffuse Ethylene-air flame by visible image processing are briefly outlined. For the inverse analysis of the radiative properties and temperatures, different methods show different features. The least-squares method, a regularization method and a linear programming method are all suitable for this problem, and a linear programming method can give more reasonable results. The red, green and blue flame images, which can be captured by some colour CCD camera, can be taken approximately as monochromatic images, and can be used to reconstruct temperature and soot volume fraction. But more ideal is the true monochromatic images filtered by filters at certain wavelengths. Finally, the optically-thin assumption, which is adopted widely, will cause large errors, about 100 K for temperature and 50% for soot volume fraction, as the absorption of the flame medium is neglected.

1. INTRODUCTION

Soot emission from hydrocarbon flames plays an important role in relation to both heat transfer by radiation and air pollution (Haroun et al., 2000). Knowledge about temperature and soot distribution in a flame is essential for gaining fundamental insight into the combustion process. Numerical simulation, capable of predicting the field variables, still requires extensive validation using experimental data. The experimental approach can be divided into two categories (Bhattacharjee, 2000): intrusive probe methods and non-intrusive optical or acoustic methods. The former methods suffer from a relatively large time response in addition to disturbing the combustion reaction. Optical methods utilize emission, absorption, or refraction of light by the burnt gases, or laser interaction such as coherent anti-Stokes Raman scattering (CARS) (Robert et al., 1997) or laser-induced fluorescence (LIF) (Bengtsson, 1996), while the acoustic method exploits the transit time of sound (Passechnik et al., 1996).

Lasor Induced Incandescence (LII) is a newly developing method to measure the soot volume fraction and its size distributions in flames by more and more researchers and institutes (Schulz et al., 2006). Schulz et al. (2006) presents controversial topics and areas of disagreement identified in the comparison of different models and measurement approaches. Despite extensive application of LII to many practical diagnostics problems and development of commercial instruments for routine implementation of LII, these comparisons demonstrate significant shortcomings in the fundamental understanding of LII and, hence, in the interpretation of measurement results (Schulz et al., 2006).

Most of these methods, however, require radiation sources, which restricts the measurement environment (Bhattacharjee, 2000). They are also subject to interference from background radiation.

Two-color emission pyrometry (di Stasio et al., 1994; Iuliis et al., 1998; Panagiotpu et al., 1996) is widely used in practical situations, especially for studies in internal combustion engines. It readily allows the estimation of averaged values of soot volume fraction and temperature. The reliability of the data depends on the calibration procedure and knowledge of the index of refraction. But this method is mostly used in strictly controlled flames with uniform flames.

For non-uniform flames, Liu et al (2003) presented a multi-wavelength inversion method to reconstruct the temperature distribution in non-axisymmetric turbulent unconfined sooting flame by the multi-wavelength measured data of outgoing emission and transmission radiation intensities. Since the additional laser light should be adopted in this method, it is undesirable for practical applications, such as large-scale furnaces, diesel engines.

An improved Tikhonov regularization method with radiation energy images detected from the boundary of the furnace to reconstruct the temperature distribution, absorptivity of wall surface and absorption coefficient of medium in a 2-D furnace system with a gray medium has been simulated (Zhou et al., 2000; Zhou et al., 2002; Zhou et al., 2003). The combustion monitoring technologies based on flame image processing are now showing more and more attractive potential for on-line monitoring of combustion in large-scale furnaces. Recently, much work has been made in this filed by the authors (Zhou, 2005; Zhou et al., 2005; Lou et al., 2007), including two papers presented in the recent International Symposiums on Combustion (Zhou et al., 2005; Lou et al., 2007). The issues related to the online visualization of 2-D/3-D temperature distributions in large-scale furnaces by flame image processing, including radiative image formation models and reconstruction methods of temperature distributions, the experiments of visualization of 2-D/3-D temperature distributions in coal-fired boiler furnaces in several power plants, and the application of the radiative energy information got from the monitoring system into the combustion and load control of boiler-turbine unit, was reviewed in detail (Zhou et al., 2007).

In this paper, the theories for the flame image processing and the reconstruction of the temperature and soot volume fraction distributions in a coflow, diffuse Ethylene-air flame will be briefly introduced. Then, experiments carried out to get the radiative properties, temperatures and soot volume fraction of the flame will be outlined. Finally, some concluding remarkes will be given.

2. THEORIES AND SIMULATION

For the reconstruction of temperature and soot volume fraction distributions inside a non-uniform sooting flame, good selection of inverse algorithms is a key step to get desirable results. Different inverse algorithms and their features in reconstruction of temperatures and soot volume fraction have be studied by simulations (Ai et al., 2005; Zhou et al., 2007). This is a fundamental work for analysis of experimental images and data.

2.1 Simulation on Simultaneous Reconstruction

An absorbing, emitting, non-scattering, axisymmetric sooting flame was studied (Ai et al., 2005). As shown in Fig. 1, the radiation intensities received by CCD camera were simulated for the inverse analysis. The radiation intensity of flame received by the jth line of sight can be expressed as:

$$I_\lambda(j)=\int_{l_0(j)}^{l_1(j)}\kappa_\lambda(l)I_{b\lambda}(l)\exp\left[-\int_{l}^{l_1(j)}\kappa_\lambda(l')\mathrm{d}l'\right]\mathrm{d}l \qquad (1)$$

where κ_λ is the spectral absorption coefficient, given by Rayleigh limit of the Mie theory:

$$\kappa_\lambda(l)=\frac{36\pi F(\lambda)}{\lambda}f_v(l) \qquad (2)$$

where $F(\lambda)$ is a function of the real and imaginary parts of the refractive index:

$$F(\lambda)=\frac{1}{6}\mathrm{Im}\left|\frac{m^2-1}{m^2+2}\right|=\frac{nk}{\left(n^2-k^2+2\right)^2+4n^2k^2} \qquad (3)$$

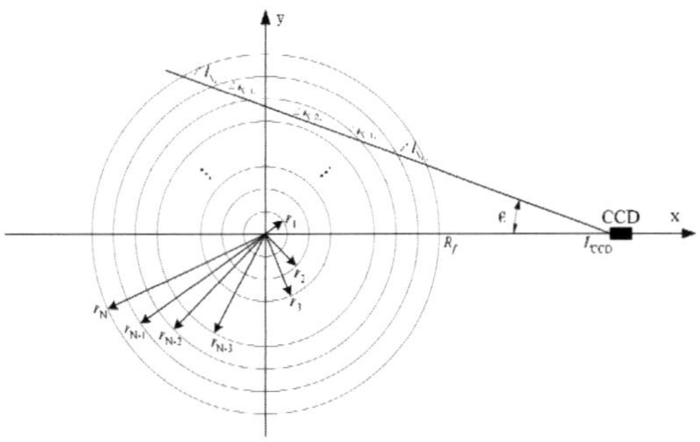

Fig. 1 Geometry in a cross section of the axisymmetric flame

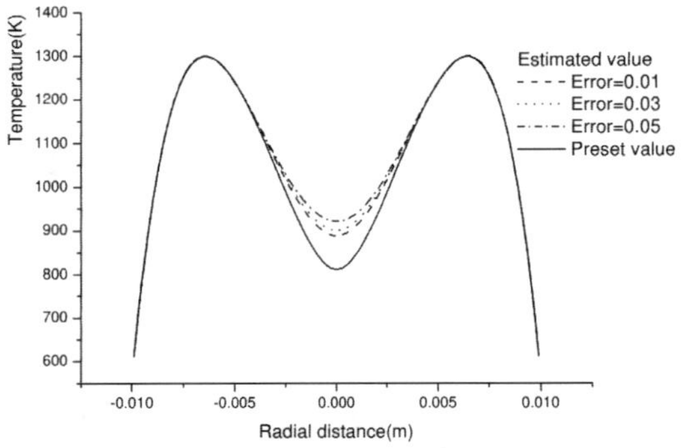

Fig. 2 The temperatures estimaed from noised measurement of intensity compared with the preset temperatures.

$I_{b\lambda}$ is the spectral radiation intensity of blackbody, given by Planck's formula:

$$I_{b\lambda}(l)=\frac{C_1\lambda^{-5}}{\pi\left(e^{C_2/(\lambda T(l))}-1\right)} \quad (4)$$

where C_1 and C_2 are blackbody radiation constants, λ is the wavelength.

According to the radiation energy image received by CCD, we could obtain three monochromatic intensity at three wavelengths, R(red), G(green), B(Blue). When the profiles of soot temperature and volume fraction were known, they could be estimated by Eq. (1). Using a Newton-type iteration algorithm and the least-squares method, simultaneous estimation of temperature and soot volume fraction in the flame was conducted. A factor is introduced to balance the magnitude difference between the radiation intensities received by different region of CCD. The effects of different temperature and soot concentration distributions and measurement errors were investigated. The result shown in Fig. 2 shows that the simultaneous estimation method have enough accuracy with various distributions and different measurement error. As shown in Fig. 2, the large deviation in the flame center is caused by the non-invasive detect technique which only provides accumulation intensity data in the measured lines. If the low temperature zone locates inside the center of the flame, which emits low radiation, it is hard to detect the temperature from the accumulation radiation measured from outside.

2.2 Inverse Methods for Optically-thin Flame Medium

In most studies of flame emission tomography, the diffuse Ethylene-air flame is taken as optically-thin medium (Ayranc , 2008). In this case, the intensity in Eq. (1) is simplified as

$$I_\lambda(j)\approx\int_{l_0(j)}^{l_1(j)}\kappa_\lambda(l)I_{b\lambda}(l)\mathrm{d}l=\int_{l_0(j)}^{l_1(j)}H_\lambda(l)\mathrm{d}l \quad (5)$$

After being discretized, the equation above is transformed into a matrix equation as (Zhou et al., 2007)

$$\mathbf{AH}=\mathbf{I} \quad (6)$$

Basing on the monochromatic radiation intensity images, a least-squares method, a regularization method (Zhou et al., 2002) and a linear programming method (Zhou et al., 2000b) were separately used to solve the inverse problem of radiative transfer equation for the axisymmetric optically-thin flame (Zhou et al., 2007). In this study, the central wavelength of the filter is 690 λm, the mass flow rates of Ethylene and air are 100 (ml/min), and 112 (l/min), respectively. The diameter of the flame is about 1 cm, and its height is about 2.3 cm. Three typical cross sections at different heights of the flame are measured, which locate in the bottom, the center and the top parts of the flame. The results are shown in Fig. 3. From the results it can be seen that all the three methods can reveal the main features of the flame.

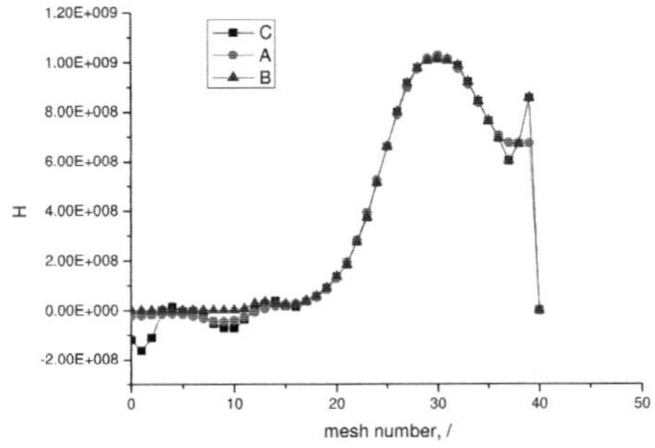

(a)

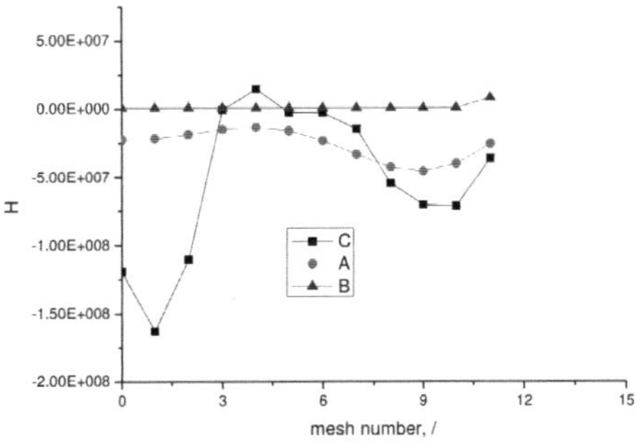

(b)

Fig. 3 Reconstruction of ***H*** profiles by the three methods (a) and enlarged display in the local area of the flame center (b). C: least-squares method; A: regularization method; B: linear programming method.

From Fig. 3b it is also very obvious that significant errors occur at the center part of the flame, where the emmittance is relatively low in this case. Among these three methods, the linear programming method both having the characteristics of lubricity and restriction and is the best appropriate method compared with the other two methods for the region of axisymmetric flame where has a relatively low temperature core. For the least square method, the intensity will be negative at some points, which is not reasonable. For the regularization method, the smoothness is better than that by the least square method, but the negative data still occur.

3. EXPERIMENTS AND ANALYSIS

Generally, there are two main methods to conduct measurements of temperature and soot volume fraction inside a flame using visible radiation image processing, one using directly the colour images (Ai et al., 2006), and the other using monochromatic radiation images. For the first method, the colour images are always taken as approximately monochromatic images with representative wavelengths for Red (R), Green (G) and Blue (B) images. In this case, the bands of the image sensors (generally with CCD cameras) for the different monochromatic images are not so narrow in order to get strictly monochromatic images, and errors will be included.

3.1 Measurements by Colour Flame Image Processing

A new emission CT method was used to estimate temperatures and soot volume fractions simultaneously in laminar ethylene flames from the monochromatic radiation intensity images captured by a colour flame image detector (Ai et al., 2006). Fig. 4 shows the four flame images captured in different conditions. For the first flame image which was captured at the smallest mass flow rate of ethylene, the intensity images from red and green monochromatic images, and the temperature and the soot volume fraction images are all shown in Fig. 5. It can be seen that the zone with the greater soot concentration lies inside the higher flame temperature zones in the flame, both inside the flame front and outside the flame axis. In addition, with the fuel flow rate increasing, the amount of soot in the flame increases, thus increasing radiation losses. This, in turn, causes a lower flame temperature.

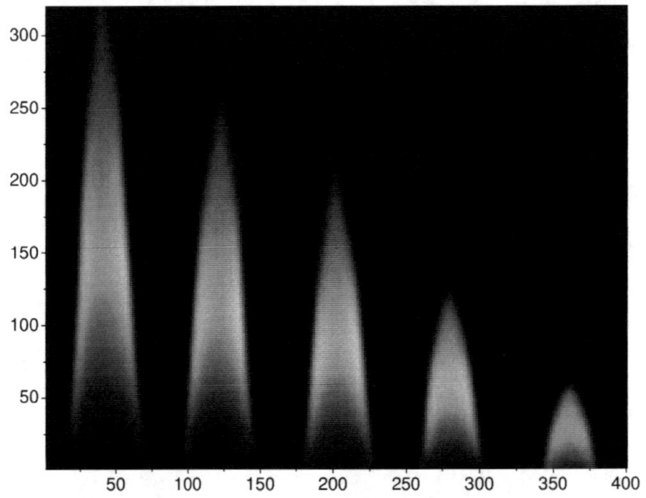

Fig. 4 Flame images under five different cases

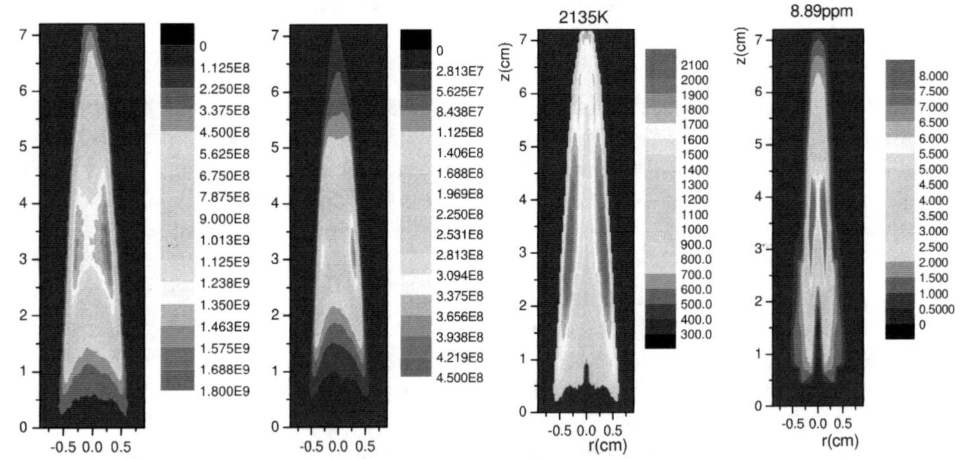

Fig. 5 The intensity image (W/m3•sr) from red image (a) and green image (b), the temperature (c) and the soot volume fraction (d) images.

3.2 Measurements by Monochromatic Flame Image Processing

As radiative flame images from soot in the laminar diffusion flame were captured by the Samsung SCC-833P CCD camera, a red interferometric filter (10 nm FWHM, 20% transmissivity) was placed between the lens and CCD sensor in order to capture monochromatic radiative images (Lou et al., 2008). In the experiment, two kinds of filter with different wavelength of 600 nm and 618 nm were used. The camera had a 1/3" interline-transfer CCD sensor (type S-HAD) with a resolution of 752(H) × 582(V) pixel elements. The electronic shutter of the camera is set as 1/5000 s^{-1}. Dedicated application software developed by the authors was used to process monochromatic flame images and to calculate the distributions of soot temperature and volume fraction.

Three different cases of the diffusion flame were generated with air and fuel (ethylene) flow rates of 94 l/min and 64 ml/min (case 1), 142 l/min and 97 ml/min (case 2), and 189 l/min and 129 ml/min (case 3), respectively.

Fig. 6 shows the monochromatic flame images captured by the CCD camera in the three cases at the wavelengths of 600 nm (a) and 618 nm (b), respectively. The corresponding monochromatic radiative intensity images of 600 nm and 618 nm calculated by image processing are given in Fig. 7. The flame temperature images and emissivity images calculated by image processing in the three cases are given in Fig. 8.

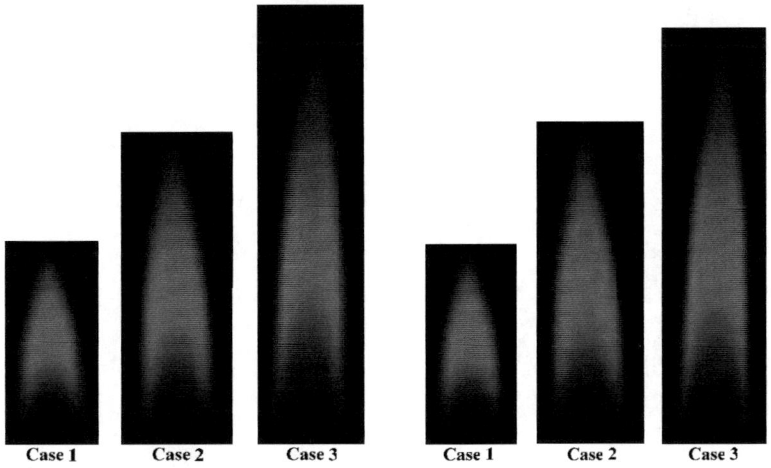

(a) (b)

Fig. 6 Representative flame images at the wavelength of 600 nm (a) and 618 nm (b), respectively.

The distributions of soot temperature and volume fraction in three cases have been reconstructed simultaneously, and the results are shown in Fig. 9. It should be mentioned that, at the bottom of the flame (5 mm above the burner), there is no radiation received from center region of flame and it is hard for the reconstruction of soot temperature. The radiation imaging model will be improved in the near future. The maximum soot volume fraction shown in Fig. 7(b) is nearly 5 ppm, which is similar to that given in (Iuliis, S. D. et al (1998)). The maximum temperature in the flame is about 2000 K, which has been reported in literature. From Fig. 9, it can also be seen that, with the increase of the fuel flow rate the soot volume fraction in the flame also increases. In addition, as the fuel mole fraction increases, the location of peak soot shifts from the center of the flame to the edge of the flame. This movement of location of the peak soot is like that mentioned in (M.D. Smooke, M. D. et al (2005).). Besides that, the experimental results indicated that much more soot lies inside the flame zone with higher temperatures, which was also measured in (Snelling, 2002).

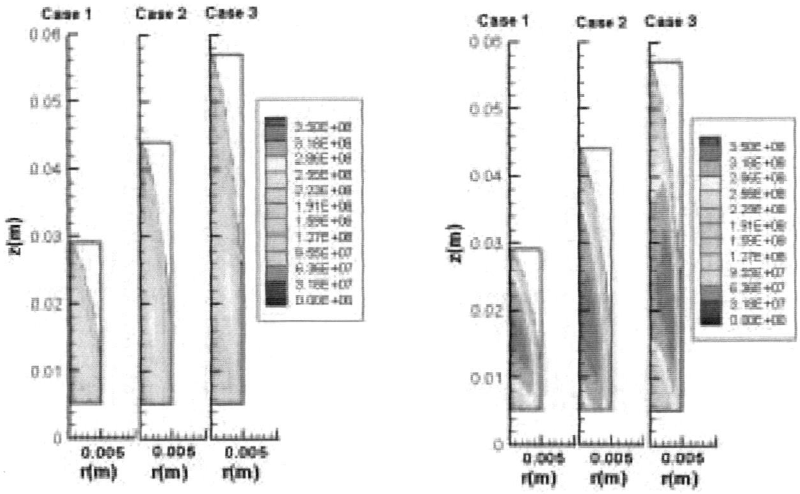

(a) (b)

Fig. 7 Monochromatic radiative intensity images (W/m^2/sr) in the three cases at the wavelength of 600 num (a) and 618 nm (b), respectively.

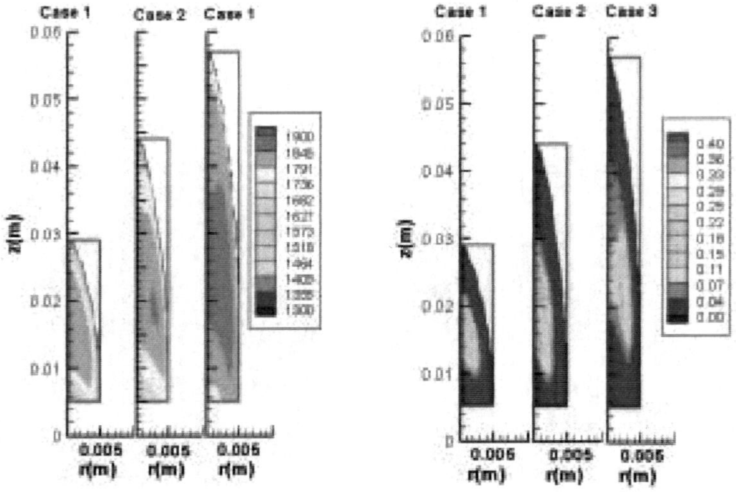

(a) (b)

Fig. 8 Flame temperature images (K) (a) and emissivity images (b) in the three cases.

3.3 Effects of Absorption of Medium on Reconstruction

With the reconstruction results for temperature and soot volume fraction with the optically-thin assumption of the medium as a initial guess, the absorption of the medium on the reconstruction can be considered in the inverse analysis procedure. Without optically-thin assumption,

$$I_\lambda(j)=\int_{l_0(j)}^{l_1(j)} H_\lambda(l)\cdot\exp[-\int_{l}^{l_1(j)}\kappa_\lambda(l^{'})\mathrm{d}l^{'}]\mathrm{d}l \qquad (7)$$

Compared with Eq. (5), the term $\exp[-\int_{l}^{l_1(j)}\kappa_\lambda(l^{'})\mathrm{d}l^{'}]$ in Eq. (7) can be taken as a compensation factor term, which can be updated from the initial guess of soot volume fraction. Then, Eq. (7) can be solved by the same method, and an updated results for the temperature and the soot volume fraction can be got. After three times of iteration, the convergence can be achieved in calculation.

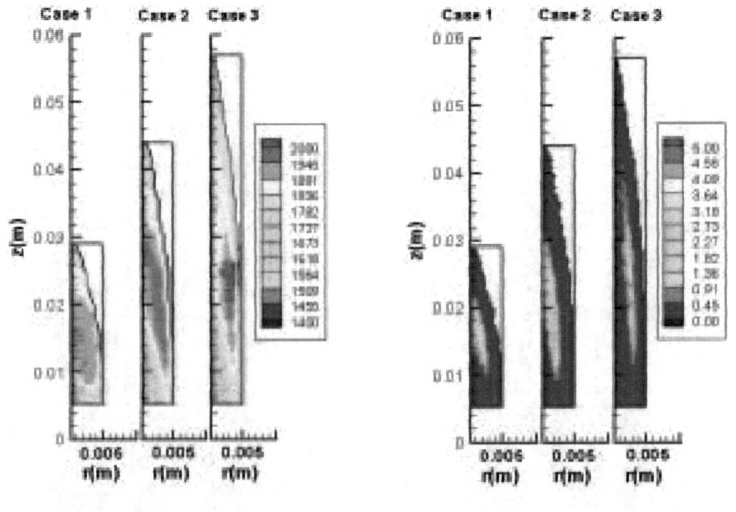

(a) (b)

Fig. 9 The calculated distributions of soot temperature (K) (a) and volume fraction (ppm) (b) in the three cases, respectively.

Journal of Physics: Conference Series **147** (2009) 012086 doi:10.1088/1742-6596/147/1/012086

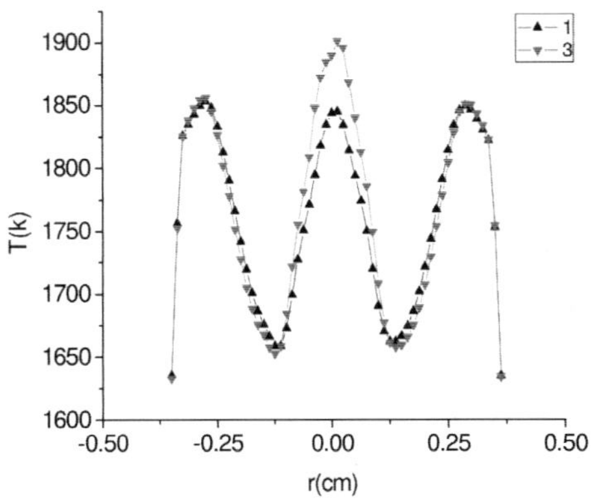

(a)

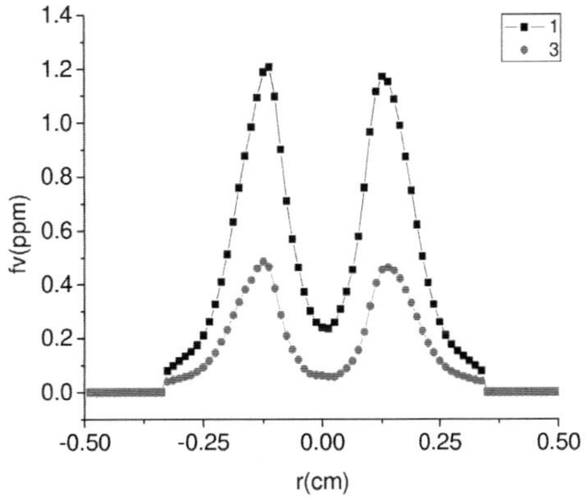

(b)

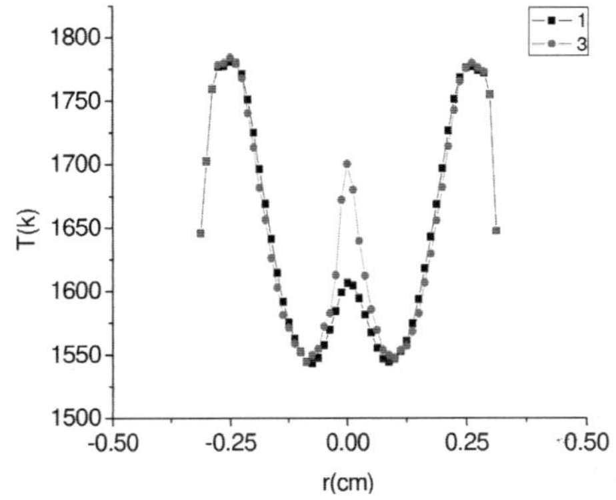

(c)

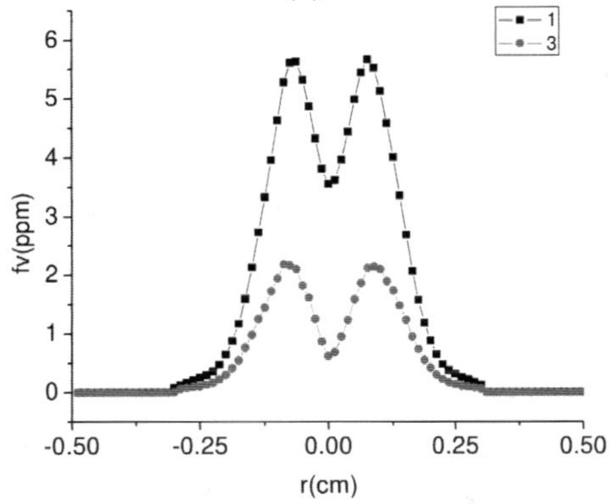

(d)

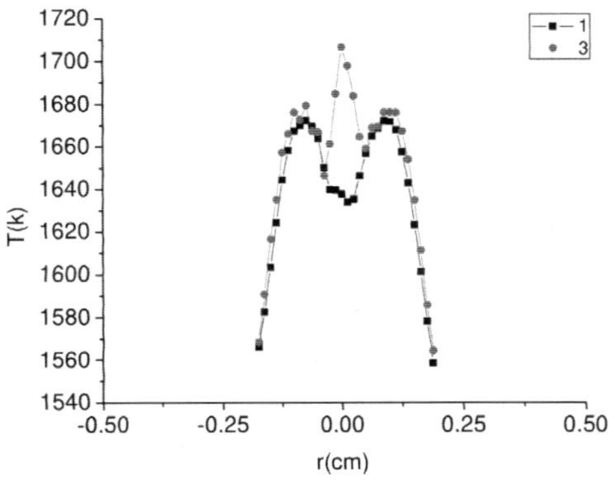

(e)

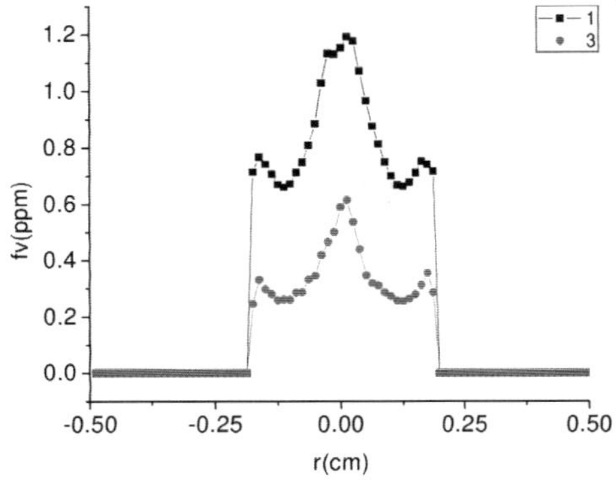

(f)

Fig. 10 Temperatures (a, c, e) and soot volume fraction (b, d, f) in three cross sections at the top (a, b), middle (c, d) and bottom (e, f) parts of the flame. 1: the initial guess with optically-thin assumption; 3: converged results three times of iteration without optically-thin assumption.

From Fig. 10 it can be seen that the temperatures close to the edges of the flame will keep unchanged even if the optically-thin assumption is given up. But the temperatures at the center of the flame change significantly, and the peak temperature will increase about 100 K. Consequently, the peak concentration of the soot decreases about 1 ppm, which should be taken into account.

4. CONCLUSIONS

In this review paper, the recent achievements in measurements of soot volume fraction and temperatures in a coflow, diffuse Ethylene-air flame by visible image processing are briefly outlined. For the inverse analysis of the radiative properties and temperatures, different methods show different features. The least-squares method, a regularization method and a linear programming method are all suitable for this problem, and a linear programming method can give more reasonable results. The red, green and blue flame images, which can be captured by some colour CCD camera, can be taken approximately as monochromatic images, and can be used to reconstruct temperature and soot volume fraction. But more ideal is the true monochromatic images filtered by filters at certain wavelengths. Finally, the optically-thin assumption, which is adopted widely, will cause large errors, about 100 K for temperature and 50% for soot volume fraction, as the absorption of the flame medium is neglected.

ACKNOWLEDGEMENTS

The study has been supported by the National Natural Science Foundation of China (No. 50636010, 50721005) and the Programme of Introducing Talents of Discipline to Universities ("111" project No.B06019).

NOMENCLATURE

I	radiation intensity	[W/m^2/sr]
l	distance	[m]
d	radiation intensity	[W/m^2/sr]
f_v	soot volume fraction	[ppm]
m	real part of the refractive index	[/]
n	imaginary of the refractive index	[/]
T	temperature	[K]
	Greek Letters	
κ	absorption coefficiency	[m^{-1}]
λ	wavelength	[nm]
	Subscripts	
b	black body	

REFERENCES

Ai, Y. H. et al (2005). "Simulation on Simultaneous Estimation of Nonuniform Temperature and Soot Volume Fraction Distributions in Axisymmetric Sooting Flames," Journal of Quantitative Spectroscopy and Radiative Transfer, 91(1), pp.11-26.

Ai, Y. H. et al (2006). "Simultaneous Measurement of Distributions of Temperature and Soot Volume Fraction in Laminar Ethylene Flames by Emission CT," Journal of Engineering Thermophysics, 27(4), pp.717-719.

Ayranc , I. et.al (2008). "Near-infrared emission spectrometry measurements for nonintrusive soot diagnostics in flames ".J. Quantitative Spectroscopy& Radiative Transfer pp.349-361

Bhattacharjee, S. (2000). "Approximate Two-color Pyrometry," Journal of Heat Transfer, 122, pp.15-20.

Bengtsson, P. E. (1996). "Simultaneous Two-Dimensional Visualization of Soot and OH in Flames Using Laser-Induced Fluorescence," Applied Spectroscopy, 50(9), pp.1182-1186.

Haroun, A. K. et al (2000). "Investigation of Soot Formation and Temperature Field in Laminar Diffusion Flames of LPG-air Mixture," Energy Conversion & Management, 14, pp. 1897-1916.

Iuliis, S. D. et al (1998). "Estimation of the Soot Volume Fraction in an Ethylene Diffusion Flame by Multiwavelength Analysis of Soot Radiation," Combustion and Flame, 115, pp.253-261.

Liu, L. H. et al (2003). "Reconstruction of Time-averaged Temperature of Non-axisymmetric Turbulent Unconfined Sooting Flame by Inverse Radiation Analysis," Journal of Quantitative Spectroscopy & Radiative Transfer, 78, pp.139-149.

Lou, C. et al (2007). "Measurements of the Flame Emissivity and Radiative Properties of Particulate Medium in Pulverized-coal-fired Boiler Furnaces by Image Processing of Visible Radiation," Proceedings of the Combustion Institute, 31(2), pp.2771–2778.

Lou, C. et al (2008). "Distributions of Temperature and Soot Volume Fraction in an Ethylene/Air Diffusion Flame Measured by Image Processing of Visible Radiation," 32nd International Symposium on Combustion, Work-in-Progress Session, Toronto, Canada, Auguest 3-8.

Panagiotpu, T. et al (1996). "Measurements of Particle Flame Temperatures Using Three-color Optical Pyrometry," Combustion and Flame, 104, pp.272-287.

Passechnik, V. I. et al (1996). "Physical Basis and Perspectives of Acoustothermography," Ultrasonics, 34, pp.511-512.

Robert, D. et al (1997). "Nitrogen and Hydrogen CARS Temperature Measurements in a Hydrogen/Air Flame Using a Near-Adiabatic Flat-Flame Burner," Combustion and Flame, 109, pp. 323-331.

Schulz, C. et al (2006). "Laser-induced Incandescence: Recent Trends and Current Questions," Appl. Phys. B, 83, pp.333–354.

Smooke, M. D. et al (2005). "Soot Formation in Laminar Diffusion Flames," Combust. Flame, 143, pp.613-628.

Snelling, D. R. et al (2002). "Spectrally Resolved Measurement of Flame Radiation to Determine Soot Temperature and Concentration," AIAA Journal, 40, pp.1789-1795.

Stasio, S. D. et al (1994). "Influence of the Soot Property Uncertainties in Temperature and Volume-fraction measurements by two-colour pyrometry, Meas. Sci. Technol., 5, pp.1453-1465.

Zhou, H. C. (2005), Measurement for Visualization of Flames inside Furnaces: Fundamental and Technology (in Chinese), China Science Press, Beijing, China.

Zhou, H. C. et al (2000a). “Visualization of Three-dimensional Temperature Distributions in a Large-scale Furnace via Regularized Reconstruction from Radiative Energy Images: Numerical Studies,” Journal of Quantitative Spectroscopy and Radiative Transfer, 72, pp.361-383.

Zhou, H. C. et al., (2000b). “Reconstruction of Temperature Distributions in a 2-D Absorbing-emitting System from Radiant eEnergy Images,” JSME International Journal , Series B, 43(1), pp.104-109.

Zhou, H. C. et al (2002). “An Inverse Radiative Transfer Problem of Simultaneously Estimating Profiles of Temperature and Radiative Parameters from Boundary Intensity and Temperature Measurements,” Journal of Quantitative Spectroscopy & Radiative Transfer, 74, pp.605-620.

Zhou, H. C. et al (2003). “Simultaneous Reconstruction of Temperature Distribution, Absorptivity of Wall Surface and Absorption Coefficient of Medium in a 2-D Furnace System,” International Journal of Heat and Mass Transfer, 46(14), pp.2645-2653.

Zhou, H. C. et al (2005). “Experimental Investigations on Visualization of Three-dimensional Temperature Distributions in a Large-scale Pulverized-coal-fired Boiler Furnace,” Proceedings of the Combustion Institute, 30(1), pp.1699-1706.

Zhou, H. C. et al (2007a). “On-line Visualization of Temperature Distributions by Flame Image Processing in Furnaces for Combustion Diagnosis and Control,” Invited Keynote Lecture, 6th Asia-Pacific Conference on Combustion, Nagoya Congress Center, Nagoya, Japan, May 20-23.

Zhou, H. C. et al (2007b). “Comparison of Three Algorithms for Solving the Inverse Problem of Axi-symmetric Radiative Transfers,” J . Huazhong Univ. of Sci. & Tech. (Nature Science Edition), 35(11), pp.121-123, 132.

AUTHOR INDEX

AUTHOR INDEX

AUTHOR INDEX

Institute of Physics
Dirac House, Temple Back
Bristol BS1 6BE UK